The Permian Timescale

Geological Society books refereeing procedures

The Society makes every effort to ensure that the scientific and production quality of its books matches that of its journals. Since 1997, all book proposals have been refereed by specialist reviewers as well as by the Society's Books Editorial Committee. If the referees identify weaknesses in the proposal, these must be addressed before the proposal is accepted.

Once the book is accepted, the Society Book Editors ensure that the volume editors follow strict guidelines on refereeing and quality control. We insist that individual papers can only be accepted after satisfactory review by two independent referees. The questions on the review forms are similar to those for *Journal of the Geological Society*. The referees' forms and comments must be available to the Society's Book Editors on request.

Although many of the books result from meetings, the editors are expected to commission papers that were not presented at the meeting to ensure that the book provides a balanced coverage of the subject. Being accepted for presentation at the meeting does not guarantee inclusion in the book.

More information about submitting a proposal and producing a book for the Society can be found on its website: www.geolsoc.org.uk.

It is recommended that reference to all or part of this book should be made in one of the following ways:

Lucas, S. G. & Shen, S. Z. (eds) 2018. *The Permian Timescale*. Geological Society, London, Special Publications, **450**.

Vachard, D. 2016. Permian smaller foraminifers: taxonomy, biostratigraphy and biogeography. *In*: Lucas, S. G. & Shen, S. Z. (eds) *The Permian Timescale*. Geological Society, London, Special Publications, **450**, 205–252. First published online December 8, 2016, https://doi.org/10.1144/SP450.1

GEOLOGICAL SOCIETY SPECIAL PUBLICATION NO. 450

The Permian Timescale

EDITED BY

S. G. LUCAS
New Mexico Museum of Natural History, USA

and

S. Z. SHEN
Nanjing Institute of Geology and Palaeontology, China

2018
Published by
The Geological Society
London

THE GEOLOGICAL SOCIETY

The Geological Society of London (GSL) was founded in 1807. It is the oldest national geological society in the world and the largest in Europe. It was incorporated under Royal Charter in 1825 and is Registered Charity 210161.

The Society is the UK national learned and professional society for geology with a worldwide Fellowship (FGS) of over 10 000. The Society has the power to confer Chartered status on suitably qualified Fellows, and about 2000 of the Fellowship carry the title (CGeol). Chartered Geologists may also obtain the equivalent European title, European Geologist (EurGeol). One fifth of the Society's fellowship resides outside the UK. To find out more about the Society, log on to www.geolsoc.org.uk.

The Geological Society Publishing House (Bath, UK) produces the Society's international journals and books, and acts as European distributor for selected publications of the American Association of Petroleum Geologists (AAPG), the Indonesian Petroleum Association (IPA), the Geological Society of America (GSA), the Society for Sedimentary Geology (SEPM) and the Geologists' Association (GA). Joint marketing agreements ensure that GSL Fellows may purchase these societies' publications at a discount. The Society's online bookshop (accessible from www.geolsoc.org.uk) offers secure book purchasing with your credit or debit card.

To find out about joining the Society and benefiting from substantial discounts on publications of GSL and other societies worldwide, consult www.geolsoc.org.uk, or contact the Fellowship Department at: The Geological Society, Burlington House, Piccadilly, London W1J 0BG: Tel. +44 (0)20 7434 9944; Fax +44 (0)20 7439 8975; E-mail: enquiries@geolsoc.org.uk.

For information about the Society's meetings, consult *Events* on www.geolsoc.org.uk. To find out more about the Society's Corporate Affiliates Scheme, write to enquiries@geolsoc.org.uk.

Published by The Geological Society from:
The Geological Society Publishing House, Unit 7, Brassmill Enterprise Centre, Brassmill Lane, Bath BA1 3JN, UK

The Lyell Collection: www.lyellcollection.org
Online bookshop: www.geolsoc.org.uk/bookshop
Orders: Tel. +44 (0)1225 445046, Fax +44 (0)1225 442836

British Library Cataloguing in Publication Data

A catalogue record for this book is available from the British Library.
ISBN 978-1-78620-282-6
ISSN 0305-8719

Distributors
For details of international agents and distributors see:
www.geolsoc.org.uk/agentsdistributors

Typeset by Nova Techset Private Limited, Bengaluru & Chennai, India
Printed and bound by CPI Group (UK) Ltd, Croydon CR0 4YY

Contents

The Permian timescale: an introduction

SPENCER G. LUCAS[1]* & SHU-ZHONG SHEN[2]

[1]*New Mexico Museum of Natural History and Science, 1801 Mountain Road NW, Albuquerque, NM 87104-1375, USA*

[2]*State Key Laboratory of Palaeobiology and Stratigraphy, Nanjing Institute of Geology and Palaeontology, 39 East Beijing Road, Nanjing, Jiangsu 210008, China*

**Correspondence: spencer.lucas@state.nm.us*

Abstract: The Permian timescale has developed over about two centuries of research to the current chronostratigraphic scale advocated by the Subcommission on Permian Stratigraphy of three series and nine stages: Cisuralian (lower Permian) – Asselian, Sakmarian, Artinskian, Kungurian; Guadalupian (middle Permian) – Roadian, Wordian, Capitanian; and Lopingian (upper Permian) – Wuchiapingian and Changhsingian. The boundaries of the Permian System are defined by global stratotype sections and points (GSSPs) and the numerical ages of those boundaries appear to be determined with a precision better than 1‰. Nevertheless, much work remains to be done to refine the Permian timescale. Precise numerical age control within the Permian is very uneven and a global polarity timescale for the Permian is far from established. Chronostratigraphic definitions of three of the nine Permian stages remain unfinished and various issues of marine biostratigraphy are still unresolved. In the non-marine Permian realm, much progress has been made in correlation, especially using palynomorphs, megafossil plants, conchostracans and both the footprints and bones of tetrapods (amphibians and reptiles), but many problems of correlation remain, especially the cross-correlation of non-marine and marine chronologies. The further development of a Permian chronostratigraphic scale faces various problems, including those of stability and priority of nomenclature and concepts, disagreements over changing taxonomy, ammonoid v. fusulinid v. conodont biostratigraphy, differences in the perceived significance of biotic events for chronostratigraphic classification and correlation problems between provinces. Future research on the Permian timescale should focus on GSSP selection for the remaining undefined stage bases, the definition and characterization of substages, and further development and integration of the Permian chronostratigraphic scale with radioisotopic, magnetostratigraphic and chemostratigraphic tools for calibration and correlation.

The Subcommission on Permian Stratigraphy (SPS), part of the International Union of Geological Sciences International Commission on Stratigraphy, currently advocates a Permian chronostratigraphic scale of three series and nine stages (Fig. 1). The boundaries of the Permian System and six of its nine stages are defined by global stratotype sections and points (GSSPs). The numerical ages of the system boundaries appear to be determined with a precision better than 1‰, but precise numerical age control within the Permian is generally sparse and uneven. A global polarity timescale for the Permian is being developed, but is not complete. Chronostratigraphic definitions of most of the 13 substages used by some workers to subdivide the Permian stages remain unfinished. For the non-marine Permian strata, correlations based on palynomorphs, conchostracans and tetrapods (amphibians and reptiles) have been proposed, but many problems of correlation remain, especially the cross-correlation of Permian non-marine and marine chronologies.

This volume reviews the state of the art of the Permian timescale and this introductory chapter provides an overview of the book. It also presents the current Permian timescale of the SPS.

Permian chronostratigraphy

Lucas & Shen (2016) review the nearly two century long development of the Permian chronostratigraphic scale, which is now a hierarchy of three series and nine stages (Fig. 1). In 1841, Murchison coined the term Permian for strata in the Russian Urals. Recognition of the Permian outside Russia and Central Europe soon followed, but it took about a century for the Permian to be accepted globally as a distinct geological system.

The work of the SPS began in the 1970s and resulted in the current recognition of nine Permian stages in three series: Cisuralian (lower Permian) – Asselian, Sakmarian, Artinskian, Kungurian;

From: Lucas, S. G. & Shen, S. Z. (eds) 2018. *The Permian Timescale*. Geological Society, London, Special Publications, **450**, 1–19.
First published online November 23, 2017, updated November 28, 2017, https://doi.org/10.1144/SP450.15

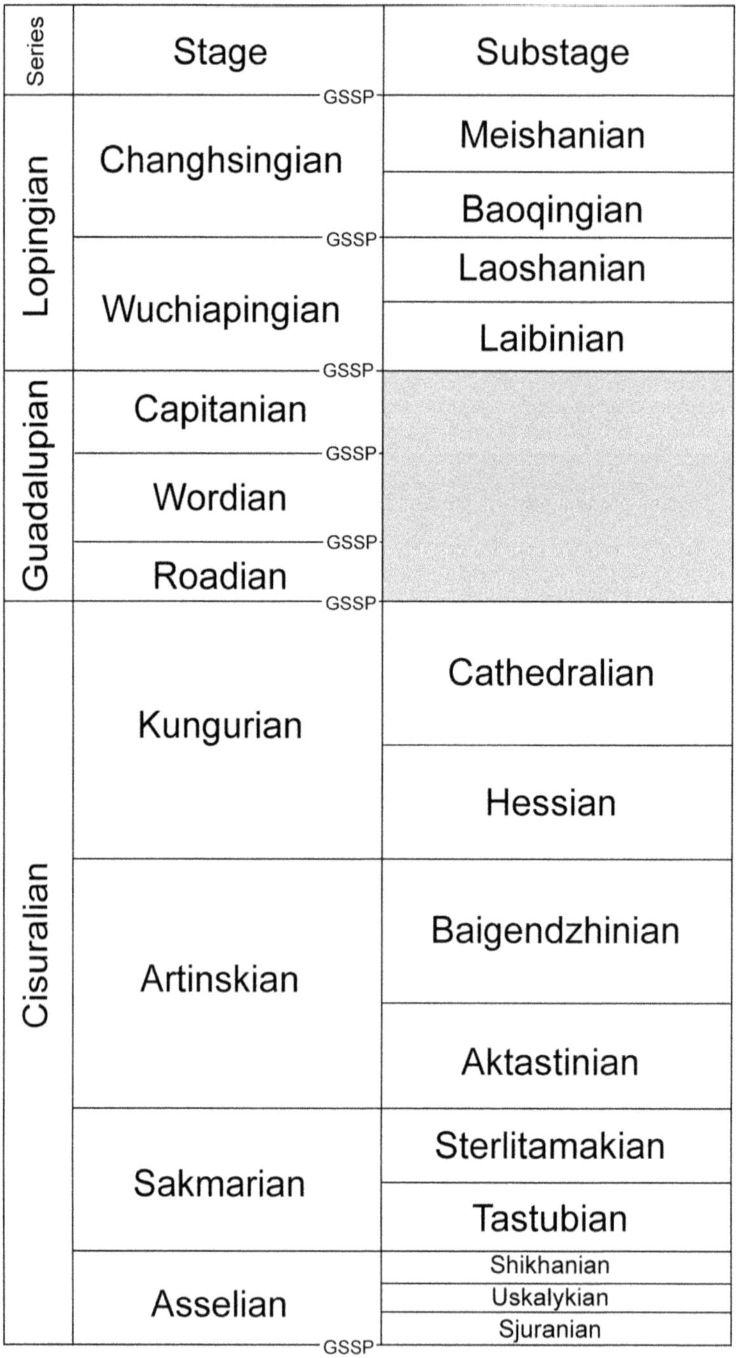

Series	Stage	Substage
	GSSP	
Lopingian	Changhsingian	Meishanian
		Baoqingian
	GSSP	
	Wuchiapingian	Laoshanian
		Laibinian
	GSSP	
Guadalupian	Capitanian	
	GSSP	
	Wordian	
	GSSP	
	Roadian	
	GSSP	
Cisuralian	Kungurian	Cathedralian
		Hessian
	Artinskian	Baigendzhinian
		Aktastinian
	Sakmarian	Sterlitamakian
		Tastubian
	Asselian	Shikhanian
		Uskalykian
		Sjuranian
	GSSP	

Fig. 1. Permian chronostratigraphic scale showing the ratified GSSPs of the stage bases.

Guadalupian (middle Permian) – Roadian, Wordian, Capitanian; and Lopingian (upper Permian) – Wuchiapingian and Changhsingian. The 1990s saw the rise of Permian conodont biostratigraphy so that all Permian GSSPs now use conodont evolutionary events as the primary signals for correlation. Most of the bases of the Permian stages have been defined by GSSPs (Figs 1 & 2):

(1) In 1998, the base of the Asselian (=base of the Permian) was defined by the GSSP at Aidaralash Creek in western Kazakstan (Davydov *et al.* 1998). The primary criterion (signal) for correlation of the GSSP is the first appearance datum (FAD) of the conodont *Streptognathodus isolatus* in the *S. wabaunensis* chronomorphocline. At Aidaralash Creek, the lowest occurrence (LO) of *S. isolatus* is *c.* 6 m below the secondary signal, which is the LO of the fusulinid *Sphaeroschwagerina fusiformis*, and it is *c.* 27 m below the traditional Asselian base determined by ammonoid biostratigraphy (Bogoslovskaya *et al.* 1995).
(2) There is no official GSSP to define the base of the Sakmarian, one of three Permian GSSPs that remain to be agreed. The most recent proposal is the Usolka section in southern Russia, where the primary signal for correlation is the FAD of the conodont *Mesogondolella monstra* in the hypothesized evolutionary lineage *M. uralensis–M. monstra–M. manifesta* (Chernykh *et al.* 2016).
(3) There is no ratified GSSP for the base of the Artinskian. The currently proposed GSSP for the base of the Artinskian Stage is the Dal'ny Tulkas section in southern Russia, with its primary signal for correlation the FAD of the conodont *Sweetognathus* aff. *S. whitei* in the hypothesized chronomorphocline *Sw.* aff. *Sw. merrilli–Sw. binodosus–Sw. anceps–Sw.* aff. *Sw. whitei* (Chuvashov *et al.* 2013).
(4) There is no agreed GSSP for the base of the Kungurian. Henderson *et al.* (2012*a*) proposed the Rockland section near Wells, Nevada, USA, as a GSSP for the base of the Kungurian, where the primary signal for correlation is the FAD of the conodont *Neostreptognathodus pnevi* in a lineage from *Neostreptognathodus pequopensis* to *N. pnevi*. Chernykh *et al.* (2012) advocated the Mechetlino section in Russia as the GSSP candidate, with the same conodont event as its primary signal. They recently proposed moving the GSSP candidate section to the nearby Mechetlino Quarry section, which has better rock quality for conodonts and chemostratigraphy.
(5) The base of the Roadian Stage is defined by its GSSP in Stratotype Canyon, Guadalupe Mountains National Park, Texas, USA. The primary signal for correlation is the FAD of the conodont *Jinogondolella nankingensis*, hypothesized to have descended from its ancestors among *Mesogondolella idahoensis lamberti* (Glenister *et al.* 1999; Mei & Henderson 2002; Henderson *et al.* 2012*b*), but the precise first occurrence of the serrated *Jinogondolella* needs to be investigated further.
(6) The base of the Wordian is now defined by its GSSP at Gateway near Guadalupe Pass in the Guadalupe Mountains National Park.

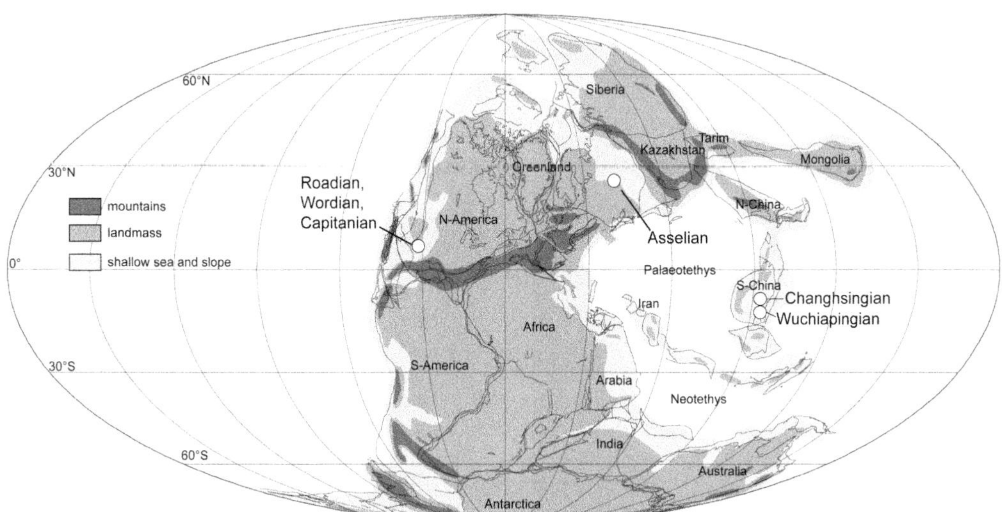

Fig. 2. Permian world map showing the locations of ratified GSSPs of Permian stage bases after Lucas *et al.* (2006).

Its primary signal for correlation is the FAD of the conodont *Jinogondolella aserrata* in a hypothesized lineage as the descendant of *J. nankingensis* (Glenister *et al.* 1999; Mei & Henderson 2002; Henderson *et al.* 2012*b*). However, this definition has not been confirmed by recent studies and needs to be studied further.

(7) The base of the Capitanian is defined by its GSSP at Nipple Hill in the Guadalupe Mountains National Park. Its primary signal for correlation is the FAD of the conodont *Jinogondolella postserrata* within the hypothesized lineage from *J. nankingensis* to *J. aserrata* to *J. postserrata*. However, very rare specimens (*c.* 1%) among abundant *J. aserrata* can be recovered at the top of Nipple Hill and further collecting is not possible.

(8) The Wuchiapingian base (=base of the Lopingian) is defined by the GSSP in the Penglaitan section in southern China. Its primary signal for correlation is the FAD of the conodont *Clarkina postbitteri postbitteri* within a hypothesized evolutionary lineage from *C. postbitteri hongshuiensis* to *C. dukouensis* (Jin *et al.* 2001, 2006*a*; Henderson *et al.* 2012*b*).

(9) The base of the Changhsingian is defined at the GSSP at Meishan Section D, southeastern China, where the primary signal is the FAD of the conodont *Clarkina wangi* within the hypothesized *C. longicuspidata–C. wangi* lineage (Mei *et al.* 2004; Jin *et al.* 2006*b*; Henderson *et al.* 2012*b*).

(10) The base of the Induan (base of the Triassic = end of the Permian) is defined by the FAD of the conodont *Hindeodus parvus* at the Meishan Section D in southern China (Yin 1996; Yin *et al.* 1996, 2001).

Thus ratified GSSPs define the boundaries of six of the nine Permian stages recognized by the SPS and also define the boundaries of the three Permian Series and of the Permian System. The bases of most of the Permian substages (Fig. 1) lack formal definition. They provide a more refined subdivision of Permian time than the stages and should be the focus of future chronostratigraphic research.

Radioisotopic ages

A precise and detailed numerical timescale does not yet exist for the Permian. This is partly due to the relatively low level and sporadic distribution of late early–middle Permian volcanism recorded in fossiliferous rocks. This has resulted in a dearth of datable volcanic ash beds, in contrast with some of the other geological systems (such as the Cretaceous), which have a much more extensive record of volcanism. Nevertheless, some important advances have been made in the last two decades. The Lopingian and the Permian–Triassic boundary have been best dated with high-precision chemical abrasion isotope dilution thermal ionization mass spectrometry (Shen *et al.* 2011; Burgess *et al.* 2014). A series of high-precision U–Pb dates were also obtained from many volcanic ash beds in the Asselian, Sakmarian and Artinskian in the southern Urals (Schmitz & Davydov 2012). A high-precision U–Pb age for the base of the Guadalupian in South China has been published (Wu *et al.* 2017).

Ramezani & Bowring (2017) provide a concise and up to date review of the numerical calibration of the Permian timescale. This review demonstrates that the age of the base and top of the Permian are well constrained at 298.92 ± 0.19 and 251.90 ± 0.10 Ma, respectively. The ages of the bases of the Sakmarian and Artinskian stages are also determined with some precision at 293.52 ± 0.17 and 290.1 ± 0.2 Ma, respectively. There are extensive data to calibrate the Lopingian stage boundaries. However, numerical age control from the Artinskian through Capitanian remains sparse, although the newly published age by Wu *et al.* (2017) calibrates the base of the Guadalupian with some precision at 272.95 ± 0.11 Ma.

Ramezani & Bowring (2017) also provide a new age to calibrate a statigraphic level *c.* 23 m below the base of the Capitanian at 265.2 ± 0.3 Ma, which is within the error range of the older published age, but much more precisely constrained. More numerical age data are clearly needed to precisely calibrate the Artinskian–Capitanian interval and this will bring the next great advance in numerical calibration of the Permian timescale.

Magnetostratigraphy

The global polarity timescale for rocks of Late Jurassic, Cretaceous and Cenozoic age provides a valuable tool for evaluating and refining correlations that are based primarily on radioisotopic ages or biostratigraphy. Permian magnetostratigraphy has long been thought to consist of the Kiaman Reversed Polarity Superchron, which lasted from the Pennsylvanian through to the early part of the middle Permian (about 50 myr), followed by the Illawara Mixed Polarity Superchron of middle–late Permian age. However, there is no agreement on a geomagnetic polarity timescale for the Permian, although a composite geomagnetic polarity timescale is now becoming available based on successions correlated to each other from marine and non-marine sections in North America, Europe and Asia.

Hounslow & Balabanov (2016) review the current status of Permian magnetostratigraphy. They note that within the reverse polarity Kiaman Superchron, long considered to have no normal polarity magnetochrons, there appear to be three normal magnetochrons during the early Permian: during the early Asselian, late Artinskian and mid-Kungurian. According to **Hounslow & Balabanov (2016)**, the mixed polarity Illawara Superchron begins in the early Wordian at about 267.08 ± 0.35 Ma. The Wordian to Capitanian interval has a strong bias to normal polarity, but the basal Wuchiapingian marks the beginning of a significant interval dominated by reverse polarity. The late Wuchiapingian and Changhsingian have roughly equal durations of normal and reverse magnetochrons. No significant gap in magnetostratigraphic data exists in the Permian geomagnetic polarity record.

Steiner (2001, 2006) identified the beginning of the Illawara Superchron as late Wordian and its identification as early Wordian by **Hounslow & Balabanov (2016)** appears to be based on a miscorrelation. Thus the Illawara Superchron begins in the shallow marine to shelfal Grayburg Formation of New Mexico, USA, which was deposited landwards of the famous Capitanian reef complex. All agree that the Grayburg Formation is correlative to some part of the basinal Cherry Canyon Formation, which indicates it is of Guadalupian age. However, more important is the fact that the Grayburg Formation directly overlies the San Andres Formation, which is Roadian–Wordian in age in its upper part (e.g. Kerans *et al.* 1993). Thus an age of the Grayburg of early Wordian is unlikely and the late Wordian age assigned to it, and thus to the base of the Illawara Superchron, is correct (Steiner 2001, 2006).

Hounslow & Balabanov (2016) calibrate the early Cisuralian magnetochrons to a succession of fusulinid zones and they calibrate the later Cisuralian and Guadalupian magnetochrons to fusulinid and conodont biostratigraphy. The Lopingian magnetochrons are calibrated to conodont zonations. **Hounslow & Balabanov (2016)** also provide numerical age calibration of the magnetochrons based on more than 15 U–Pb numerical ages. The numerically dated control points are most numerous in the Gzhelian–Asselian and Changhsingian intervals, giving 95% confidence intervals to the chron ages. In their summary of Permian magnetostratigraphy, **Hounslow & Balabanov (2016)** identify three or four very short normal polarity chrons in the Kiaman interval, followed by five normal multichrons in the Guadalupian and three normal multichrons in the Lopingian.

Carbon isotope stratigraphy

The use of carbon isotopes in stratigraphic correlation has grown dramatically during the last decade, but this volume lacks a review of Permian carbon isotope stratigraphy. Therefore we present a brief review here. First, we note that isotope curves that plot the composition of carbon or changes in the ratio of δ^{13}C to δ^{12}C have the potential to provide a means of correlation essentially independent of other methods. However, like magnetostratigraphy, this record needs calibration to a datum or to datums, either biostratigraphic or radioisotopic.

Saltzman & Thomas (2012, fig. 11.5) published a compiled carbon isotope record for the Permian based primarily on records from the USA (Nevada) and China (Saltzman 2003). That compilation shows δ^{13}C fluctuating around 2‰, with major negative excursions at the Asselian–Sakmarian, Artinskian–Kungurian, Wordian–Capitanian, Capitanian–Wuchiapingian and Changhsingian–Induan (Permo-Triassic) boundaries. However, as Buggisch *et al.* (2015) observed, the values of the Saltzman and Thomas compilation are lower than the mean values of other δ^{13}C records.

Henderson *et al.* (2012*b*, fig. 24.9) presented a generalized δ^{13}C curve for the Permian based primarily on data published by Shen *et al.* (2010) and Buggisch *et al.* (2011). This curve shows a general trend of decreasing values through the Permian, with notable positive excursions at about the Sakmarian–Artinskian, Kungurian–Roadian, Roadian–Wordian, Capitanian–Wuchiapingian and Permo-Triassic boundaries.

The most recent compilation, by Buggisch *et al.* (2015, fig. 9), does not show an overall negative trend (Fig. 3). Buggisch *et al.* (2015) reviewed data from sections in South China, the Moscow basin, the Southern Alps and Kansas, USA. Their review shows generally high values throughout the Permian, punctuated by sharp negative excursions of several per mil. It also shows substantial gaps of what Buggisch *et al.* (2015) consider to be reliable carbon isotope data during the entire middle Permian and parts of the early Permian.

Carbon isotope excursions during the Asselian–Sakmarian have been attributed to the change in oceanic carbon isotope composition during the final phase of the late Palaeozoic ice ages (Isbell *et al.* 2003). As reviewed by Buggisch *et al.* (2011), the waxing and waning of continental ice sheets is known to coincide with positive or negative shifts in δ^{13}C, although there is no simple causal connection between ice volume and δ^{13}C values. Two negative shifts in C isotope values near the Asselian–Sakmarian boundary correspond to the major collapse of Gondwana ice sheets (Zeng *et al.* 2012).

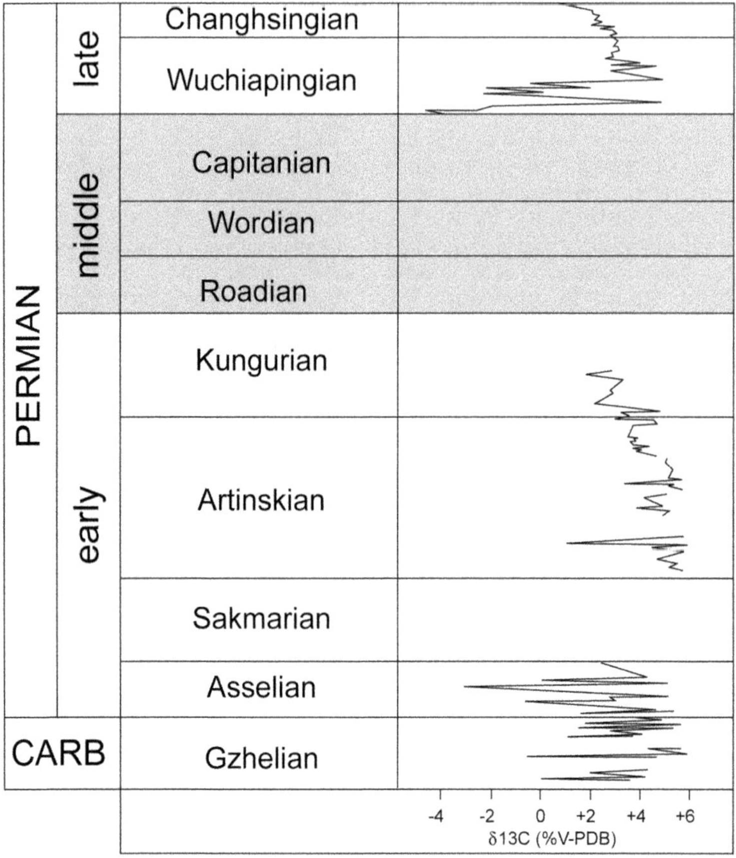

Fig. 3. Permian carbon isotope curve after Buggisch *et al.* (2015).

Analyses of carbon isotope records across the Artinskian–Kungurian boundary have produced strikingly different results. Thus some sections (USA Midcontinent; Naqing, China) show a negative shift across the Artinskian–Kungurian boundary, whereas others (Guadalupe Mountains, Southern Alps, Urals) show a positive shift (Buggisch *et al.* 2015, fig. 9). The compilation of Henderson *et al.* (2012*b*) shows a negative shift, but there is no clear global pattern (Buggisch *et al.* 2015).

The end-Guadalupian extinction has long been viewed as coinciding with (causing) a major negative shift in carbon isotopes (e.g. Baud *et al.* 1989; Bond *et al.* 2010). Thus it has been argued that the excursion is a global signal of large fluctuations right in the extinction, which has been dated as mid-Capitanian by some researchers (Wignall *et al.* 2009; Bond *et al.* 2010). The carbon isotope excursion has been linked to the release of methane by volcanism. This predates an interval of unusually heavy $\delta^{13}C$ isotope values that has been called the Kamura event, hypothesized to reflect long-term cooling due to extensive high marine productivity drawing down atmospheric CO_2 values (Isozaki *et al.* 2007*a*, *b*). The inference here is a catastrophic release of methane to produce a major carbon excursion after the extinction. If correct, this is a global event best seen in sections in southern China. However, whether or not there was a single mass extinction has been questioned (Clapham *et al.* 2009) and carbon isotope excursions near the Guadalupian–Lopingian boundary are highly variable in both pattern and magnitude (Shen *et al.* 2013).

Korte & Kozur (2010) provided a comprehensive review of the carbon isotope record across the Permo-Triassic boundary (PTB). They compiled at least 40 globally widespread records of the PTB carbon isotope record in both marine and non-marine sections. There is a marked negative excursion at this boundary of a 4–7‰ decrease over about half a million years. Korte & Kozur (2010) argued that this trend was gradual, beginning about 500 000 years before the PTB, but was interrupted by a short-term positive event at the marine extinction event, which then decreased to a first minimum just below the PTB, followed by a one- or two-peaked minimum. The positive spike at the main extinction boundary and the decrease beginning before the main extinction obviate the extinction as the cause of the negative carbon trend. Thus Korte & Kozur (2010) argued that the negative carbon excursion had multiple causes, including the effects of Siberian Trap volcanism and the possibility of anoxic waters reaching shallow sea-levels. However, a recent study has shown that significant diagenesis has also affected carbon isotope values around the PTB (Li & Jones 2017), so the analysis of Korte & Kozur (2010) may need to be re-evaluated.

Thus the $\delta^{13}C$ records from Permian strata in different locations contain evident contradictions that are, in part, underlain by regional differences. Buggisch *et al.* (2015) identified the following factors as responsible: (1) stagnant basin conditions; (2) restricted circulation between shallow water carbonate platforms, epicontinental seas and open oceans; (3) local up- and downwelling of waters; and (4) freshwater input in coastal areas. Thus they suggested that contradictory (opposite) trends in $\delta^{13}C$ in the Midcontinent USA and the Russian Urals, and in the Guadalupe Mountains of Texas and the Naqing section in southern China, during the late Artinskian–Kungurian could be due to closure of the Ural gateway, which may have caused the upwelling of cold, nutrient-rich deep water and a change to colder water environments.

Buggisch *et al.* (2015) concluded that, at the current state of knowledge, Permian carbon isotope records are not particularly useful in stratigraphic correlation (Fig. 3). This is in part due to a general lack of study of multiple records for many Permian time intervals. They state that the only definitive global negative trend during the Permian is at the PTB. More carbon isotope stratigraphy is clearly needed from more sections to establish a robust carbon isotope curve for all of Permian time.

Strontium isotope stratigraphy

Korte & Ullmann (2016) review the strontium isotope stratigraphy of the Permian. They note that the secular evolution of the $^{87}Sr/^{86}Sr$ ratios in Permian seawater record information about global tectonic processes, palaeoclimate and palaeoenvironments, including the early Permian deglaciation, the amalgamation of Pangaea and the PTB mass extinctions (also see Dudás *et al.* 2017). The marine $^{87}Sr/^{86}Sr$ curve can also be used for robust correlations when other bio-, litho- and/or chemostratigraphic markers are inadequate. The accuracy of marine $^{87}Sr/^{86}Sr$ reconstructions, however, depends on high-quality age control of the reference data and on sample preservation, both of which generally decrease in quality for older time intervals.

According to **Korte & Ullmann (2016)**, Permian seawater $^{87}Sr/^{86}Sr$ values show a decrease from *c.* 0.7080 in the earliest Permian (Asselian) to *c.* 0.7069 in the latest Guadalupian (Capitanian), followed by a steepening increase from the latest Guadalupian towards the PTB (*c.* 0.7071–0.7072) and into the Early Triassic. Various higher order changes in the slope of the Permian $^{87}Sr/^{86}Sr$ curve are indicated, but cannot currently be verified due to lack of sampling and disagreement over published $^{87}Sr/^{86}Sr$ records. Thus, like carbon isotope stratigraphy, the strontium isotope stratigraphy of the Permian is in an early phase of development.

Marine biostratigraphy

The distribution of fossils in marine Permian strata has provided the primary basis for construction of the Permian chronostratigraphic scale. The most important taxa in this regard are conodonts, fusulinids, non-fusulinid forams and ammonoids. Brachiopods, rugose corals and radiolarians also provide some biostratigraphy of Permian marine strata. These groups are reviewed by papers in this volume and we also briefly discuss the biostratigraphic value of Permian marine bivalves and gastropods.

Conodonts

Conodonts are microscopic tooth-like structures composed of calcium phosphate and are abundant and widespread in Permian marine strata. Although the biological source of conodonts was long unknown, they are now clearly associated with chordates. In the 1990s, conodonts became the preferred tool for defining Permian chronostratigraphic boundaries. Thus the six defined GSSPs of the Permian stages have conodont events as their primary correlation signals and the three stages awaiting GSSP definition will also use conodont events as their primary correlation signals.

Henderson (2016) reviews the state of the art of Permian conodont biostratigraphy. He summarizes 40 Permian conodont biozones and correlates an additional 35 regional zones. As Henderson notes, the lower Permian is largely zoned by partial range–lineage–interval zones of species of *Streptognathodus*, *Sweetognathus*, *Neostreptognathodus* and *Mesogondolella*. The middle and upper Permian are zoned on the basis of partial range–lineage–interval zones of species of *Jinogondolella* and *Clarkina*, respectively. The ranges of key taxa to develop Permian conodont zonations are depicted by **Henderson (2016)** in relationship to the geochronological ages of Permian stages.

Thus a total of 75 biozones based on the ranges of conodonts are summarized by **Henderson (2016)**. Forty of these are regarded as standard international biozones and the other 35 are referred to as correlative regional biozones. However, due to several factors, only a few of the biozones have a truly global character. These factors include regional glaciation in the earliest Permian, major lowstands of sea-level in the Kungurian and Capitanian–Wuchiapingian boundary interval, significant levels of provincialism, especially from the early Kungurian to the early Changhsingian, and the high degree of morphological plasticity displayed by key taxa. The last factor underlies widely different opinions on conodont taxonomy, which affect the correlation potential of Permian conodonts. As **Henderson (2016)** concludes, Permian conodonts, regardless of the taxonomic approach, exhibit low to moderate diversity and two major lineages provide most of the biostratigraphic control: the sweetognathid and gondolellid lineages. Future progress in Permian conodont biostratigraphy will come from more taxonomic refinement of these lineages to delineate better evolutionary events by which to define zonation.

Radiolarians

Radiolarians are marine zooplankton that secrete a skeleton of opaline silica. In the modern oceans they form massive skeletal accumulations (radiolarian oozes) on the seafloor in deep waters (up to 4000 m deep). Their Permian fossils are typically found in deep marine deposits associated with chert horizons.

Zhang *et al.* (2017) review the Permian radiolarian record and its application to biostratigraphy. They recognize and review 17 Permian radiolarian zones. Among these, seven zones are recognized in the uppermost Carboniferous to lower Permian (in ascending order): *Pseudoalbaillella bulbosa* Assemblage Zone, *Pseudoalbaillella uforma–Pseudoalbaillella elegans* Abundance Zone, *Pseudoalbaillella lomentaria–Pseudoalbaillella sakmarianensis* Assemblage Zone, *Pseudoalbaillella rhombothoracata* Interval Zone, *Albaillella xiaodongensis* Assemblage Zone, *Albaillella sinuate* Abundance Zone and *Pseudoalbaillella ishigai* Abundance Zone. Five zones are recognized in the middle Permian (ascending order): *Pseudoalbaillella globosa* Interval Zone, *Follicucullus monacanthus* Interval Zone, *Follicucullus porrectus* Interval Zone, *Follicucullus scholasticus* Interval Zone and *Follicucullus charveti* Interval Zone. Five zones belong to the uppermost middle Permian to upper Permian (ascending order): *Albaillella cavitata* Interval Zone, *Albaillella levis* Interval Zone, *Albaillella excelsa* Interval Zone, *Albaillella triangularis* Interval Zone and *Albaillella yaoi* Abundance Zone. **Zhang *et al.* (2017)** discuss these Permian radiolarian biozones and their correlations with the conondont zones and other chronostratigraphic schemes. They conclude that more continuous successions are needed to produce a more refined and robust radiolarian biostratigraphy of the Permian.

Rugose corals

Wang *et al.* (2017) review the Permian record of rugose corals. They recognize two coral realms, Tethyan and Cordilleran–Arctic–Uralian. The Tethyan realm is characterized by the families Kepingpphyllidae and Waagenophyllidae during the Cisuralian, Waagenophyllidae during the Guadalupian and the subfamily Waagenophyllinae during the Lopingian. By contrast, the Cordilleran–Arctic–Uralian realm is characterized by the families Durhaminidae and Kleopatrinidae during the Cisuralian and the almost total disappearance of colonial and dissepimented solitary rugose corals from the Guadalupian to the Lopingian.

According to **Wang *et al.* (2017)**, the geographical barrier resulting from the amalgamation of Pangaea controlled the development of these coral realms. Changes in the composition and diversity of Permian rugose corals suggest that an evolutionary turnover event might have occurred at the end of the Sakmarian, probably due to a global drop in sea-level. It was characterized by the change from mixed Pennsylvanian and Permian coral faunas to typical Permian coral faunas. In addition, three coral extinction events are evident (end-Kungurian, end-Guadalupian and end-Permian), which **Wang *et al.* (2017)** argue were triggered (respectively) by the northwards movement of Pangea, Emeishan volcanism and subsequent global regression, and global warming induced by the Siberian Trap eruptions. The Permian record of rugose corals thus documents important evolutionary and palaeobiogeographical events, but does not provide a robust biostratigraphy.

Ammonoids

Ammonoids were long the workhorses of Permian marine biostratigraphy, so that much of the Permian timescale was early built on ammonoid biostratigraphy. **Leonova (2016)** reviews the application of ammonoids to Permian chronostratigraphy.

Leonova (2016) notes that almost all stages of the Permian System were originally based on ammonoids. Thus, traditionally, ammonoid ‘zones’ of the lower and middle Permian were used as the equivalents of stages (Glenister 1981), whereas for the upper Permian these zonal subdivisions have been much more detailed. Two 'zones' of the scale of Böse (1917) have been used by stratigraphers for many decades: the *Perrinites* and *Waagenoceras* zones, corresponding to the Artinskian and Wordian stages, respectively. Miller (1938), working in North America, introduced two more zones so that the resulting zonal scheme was *Properrinites* and *Perrinites* for the lower Permian and *Waagenoceras* and *Timorites* for the upper Permian.

Ruzhentsev (1955) recognized seven successive ammonoid assemblages corresponding to the Asselian, Sakmarian, Aktastinian, Baigendzhinian, Sicilian (or Wordian), Capitanian and Dzhulfian stages. On the basis of these assemblages, Glenister & Furnish (1961) recognized six stages: Asselian, Sakmarian, Artinskian (with Aktastinian and Baigendzhinian substages), Wordian, Capitanian and Dzhulfian. Furnish (1966) introduced the Roadian Stage for the uppermost beds of the lower Permian. In addition, he subdivided the upper Permian series into three stages: Guadalupian, Chidruan and Dzhulfian. Furnish (1973) later proposed a more refined Permian chronostratigraphy based on ammonoids and Ruzhentsev (1976) proposed a less refined version of the global scale.

The ammonoid-based definition of the Carboniferous–Permian boundary is based on the first appearance of four new families – Paragastrioceratidae (*Svetlanoceras*), Metalegoceratidae (*Juresanites*), Popanoceratidae (*Protopopanoceras*) and Perrinitidae (*Subperrinites*, *Properrinites*) – and seven new genera that appeared in previously existing families – *Vanartinskia, Mescalites, Kargalites, Cardiella, Martoceras, Prostacheoceras* and *Tabantalites* (Bogoslovskaya *et al.* 1995). The lower Permian stages were established in the South Urals, beginning with the Artinskian ammonoids discovered in the 1800s (Murchison *et al.* 1845). The Artinskian was the first Permian stage to be recognized (Karpinsky 1874, 1889; Krotow 1885). Ruzhentsev (1938, 1951, 1954) recognized three diverse, temporally successive ammonoid assemblages that became the bases of the Asselian, Sakmarian and Artinskian stages.

Thus, on a global scale, all nine stages of the Permian are relatively completely characterized by ammonoids, which are successfully used for biostratigraphy and interregional correlations. Nevertheless, ammonoid localities in Permian sections are relatively rare and sections that document changes in species and genera, characterizing long intervals of geological time, are extremely rare. This diminishes the use of ammonoids as primary signals in the GSSP definition of Permian stages, although their value as secondary signals for GSSP correlation remains undiminished.

Non-fusulinid forams

Non-fusulinid forams are abundant in many Permian marine strata and some taxa have very broad distributions in shallow marine carbonate and evaporate facies. This has led to the use of non-fusulinid forams in Permian biostratigraphy, especially in Europe, North America and Asia.

Vachard (2016) reviews the biostratigraphy of Permian smaller forams belonging to four classes: Fusulinata, Miliolata, Nodosariata and Textulariata. Biostratigraphic markers of these classes are mainly found in the orders and superfamilies Lasiodiscoidea, Bradyinoidea and Globivalvulinoidea (Fusulinata), Cornuspirida (Miliolata) and in the class Nodosariata. The class Textulariata is too little known during the Permian to be of biostratigraphic significance, although the appearance of the order Verneulinida is probably an important bioevent. The main genera among the lasiodiscids include *Mesolasiodiscus*, *Lasiodiscus*, *Lasiotrochus*, *Asselodiscus*, *Pseudovidalina* and *Xingshandiscus*. Among the bradyinoids, the main genera are *Bradyina* and *Postendothyra*; and, among the globivalvulinoids, *Globivalvulina*, *Septoglobivalvulina*, *Labioglobivalvulina*, *Paraglobivalvulina*, *Sengoerina*, *Dagmarita*, *Danielita*, *Louisettita*, *Paradagmarita*, *Paradagmaritopsis* and *Paremiratella*. The biostratigraphically significant genera of miliolates are *Rectogordius*, *Okimuraites*, *Neodiscus*, *Multidiscus*, *Hemigordiopsis*, *Lysites*, *Shanita* and *Glomomidiellopsis,* and the genera of tubiphytids and ellesmerellids, which may be specialized miliolate and cyanobacterium consortia. The Nodosariata markers belong to *Nodosinelloides*, *Tezaquina*, *Polarisella*, *Geinitzina*, *Pachyphloia* and *Rectoglandulina*, the first true *Nodosaria*, *Langella*, *Pseudolangella*, *Calvezina*, *Cryptoseptida*, *Cylindrocolaniella*, *Colaniella*, *Frondina* and *Ichthyofrondina*, but their lineages are too poorly understood to allow their accurate use in Permian biostratigraphy. **Vachard (2016)** thus concludes that broad correlations using these genera allow the recognition of 12 intervals of Permian time, many roughly equivalent to a stage.

Fusulinids

The study of Permian fusulinids began in the early to middle 1800s and the first attempt at a Permian fusulinid biostratigraphy was published by Schellwien (1898) based on his studies in the Carnic Alps of southern Europe (Douglass 1977). Detailed Permian biostratigraphy based on fusulinids has been developed in the twentieth century, primarily in Russia,

China, the USA, Japan and Tajikistan. It has long been recognized that Permian fusulinids were provincialized and this has hampered global correlations using fusulinids.

Zhang & Wang (2017) review Permian fusulinid biostratigraphy. They note that the rapid evolution of Permian fusulinids has rendered them important taxa in correlating Permian strata globally. Thus **Zhang & Wang (2017)** draw attention to key evolutionary events among fusulinids that define biostratigraphic boundaries. The FADs of *Pseudoschwagerina* or *Sphaeroschwagerina* mark the base of the Asselian stage. These schwagerinids and pseudoschwagerinids gave rise to *Paraschwagerina, Zellia, Darvasites* and *Robustoschwagerina* during the Sakmarian. During the Yakhtashian, *Levenella* and *Pamirina* originated from the Tethyan region and formed the basis of the neoschwagerinids. The first occurrences of *Brevaxina*, advanced *Misellina*, *Neoschwagerina* and *Yabeina/Lepidolina* define the bases of the Bolorian, Kubergandian, Murgabian and Midian stages, respectively. Both neoschwagerinids and schwagerinids became extinct at the end of the Midian stage. The remaining fusulinids of the late Permian are small-sized *Codonofusiella* and *Reichelina* and a few new genera such as *Palaeofusulina* and *Gallowayinella*. *Palaeofusulina*, in particular the advanced species *P. sinensis,* is characteristic of the Changhsingian. According to **Zhang & Wang (2017)**, these are the main evolutionary milestones of fusulinids that provide the basis of fusulinid biostratigraphy across Tethys.

By contrast, there is a different evolutionary history of fusulinids in the North American cratonal region. The main differences between the Tethyan region and the North American craton are seen in the fusulinid faunal compositions from the Leonardian Stage to the Capitanian Stage. The fusulinids of North America are characterized during this interval by the dominance of schwagerinids such as *Parafusulina* and *Polydiexodina.* However, *Parafusulina* plays a minor part from the Kubergandian to Midian in the Tethyan region. Therefore the direct and precise correlation of middle Permian fusulinid biostratigraphy between the Tethyan region and the North American craton region requires further detailed investigation. Thus, as has long been known, Permian fusulinids are powerful biostratigraphic tools, but this biostratigraphy is challenged by their Permian provinciality.

Brachiopods

Brachiopoda is a phylum of marine animals with two valves known from more than 12 000 fossil species in more than 5000 genera. Brachiopods were common shelly benthos during the Permian, mostly as seafloor filter-feeders. They suffered major losses as one of the animal groups hardest hit by the end-Permian extinctions.

In his review of Permian brachiopod biostratigraphy, **Shen (2016)** stresses the difficulty in establishing a Permian biochronological scheme for global correlation based on brachiopods due to provincialism and endemism. However, numerous new brachiopod assemblages have been described during the last 40 years, making it possible to improve and update the brachiopod biostratigraphy in different regions. The Permian biogeography of brachiopods is characterized by three distinctive realms (Boreal, Palaeoequatorial and Gondwanan) and two transitional zones developed between these realms. The brachiopods from the two transitional zones show a mixture of the cold water brachiopods of the two anti-tropical realms and the warm water brachiopods of the Palaeoequatorial realm. Therefore they provide a bridge to facilitate correlation between the different realms.

Shen (2016) gives a brief overview of Permian brachiopod successions in the five major palaeobiogeographical realms. Based on a global database of Permian brachiopods, a characteristic brachiopod assemblage represented by the genera *Bandoproductus*, *Punctocyrtella* and *Cimmeriella* in the lower Cisuralian (probably upper Asselian to lower Artinskian) of Gondwana, the peri-Gondwanan region and the Cimmerian blocks was widely distributed, indicative of the acme of the late Palaeozoic ice ages.

As a result of the amelioration of the palaeoclimate during the late Kungurian and/or the northwards drift of the Cimmerian blocks, brachiopods show a distinct shift from their cold water affinity to mixed or warm water affinity late in the Cisuralian. By contrast, brachiopods in the northern transitional zone are warm water faunas associated with fusulinids in the lower Cisuralian. The Guadalupian brachiopods of the northern transitional zone were clearly mixed, with boreal and palaeoequatorial affinities indicated by the presence of some antitropical or boreal genera (e.g. *Gypospirifer, Kaninospirifer*, *Rhombospirifer, Spiriferella*, *Yakovlevia)* associated with numerous typical Cathaysian elements, suggesting that the northern transitional zone drifted to a more northerly location during the Guadalupian.

The onset of the end-Permian mass extinctions in the latest Changhsingian is exhibited globally by the occurrence of a dwarf brachiopod assemblage characterized by small sizes and thin shells. Typical representatives are *Paracrurithyris*, *Fusichonetes, Paryphella, Spinomarnigifera, Martinia* and numerous lingulids. In addition, some brachiopods with bipolar/bitemperate distributions may be useful for correlation between the Boreal and Gondwanan realms.

Other marine biostratigraphy

Non-cephalopod molluscs, the bivalves and gastropods, were, like the brachiopods, common denizens of Permian seafloors. However, they have only been used in a limited fashion in Permian biostratigraphy.

The most abundant Permian bivalves were epifaunal pteriomorphs; much less common were semi-infaunal to infaunal suspension feeders such as Myophoriidae, Astartidae, Permophoridae, Megadesmidae and Pholadomyidae (e.g. Nakazawa & Runnegar 1973; Miller 1988, 1989). Bivalves became increasingly important members of marine communities through Permian time and the late Permian was a time of increased generic richness, ecological diversity and cosmopolitanism (Clapham & Bottjer 2007; Mondal & Harries 2016). However, Permian biostratigraphy based on marine bivalves suffers from the problems that beset the brachiopods, especially provincialization and facies control of distributions, so marine bivalves are little used in Permian correlations.

Permian marine gastropods reached their peak diversity during the Guadalupian, but this may reflect a bias due to the extremely prolific assemblages from the Guadalupian of West Texas (Batten 1973). As an example of the limited use of marine gastropods in Permian biostratigraphy, we note that Knight (1940, p. 1128) stated that the euomphaline gastropod genus *Omphalotrochus* ‘makes its first appearance throughout the world at the base of what appear to be the equivalents of the Wolfcamp series in the United States’. However, this includes records of *Omphalotrochus* in the Gzhelian *Uddenites* shale of the Gaptank Formation in West Texas and Gzhelian records in Russia (e.g. Yochelson 1954, 1956; Mazaev 1994). So, although most records of *Omphalotrochus* appear to be lower Permian, the FAD of the genus is below the long accepted base of the Permian and not obviously synchronous globally.

Non-marine biostratigraphy

Non-marine Permian biostratigraphy has also been developed, based primarily on palynomorphs, megafossil plants, conchostracans, tetrapod (amphibian and reptile) footprints and tetrapod body fossils. Some other non-marine fossil groups (charophytes, ostracods, freshwater bivalves, insects and fishes) have been used for Permian biostratigraphy and are also briefly discussed here.

Palynomorphs

Spores and pollen are the microscopic reproductive structures of vascular plants. Their organic walls resist pressure, desiccation and microbial decomposition, so they are often well preserved in sedimentary rocks and Permian strata are no exception. Because of their abundance (one plant may produce thousands of palynomorphs), durability and easy dispersal (often by wind), palynomorphs are found in both non-marine and marine strata. They provide an important method for the cross-correlation of non-marine and marine strata based on shared palynomorph taxa. However, most palynomorphs are only dispersed within a few kilometres or less of the plant that produces them, and any provincialization of the palaeoflora hinders their use in broad-scale correlation. Plants are also environmentally sensitive, so palaeoenvironmental and facies restrictions of extinct plants affect the distribution of their palynomorphs.

As **Stephenson (2016)** notes, during the Permian the palaeoflora was provincialized into three or more provinces. Thus correlations based on fossil plants and the palynoflora have proved to be difficult between portions of Permian Pangaea that were in relatively close proximity, but in different palaeo-provinces, such as Western Europe and eastern North America. Permian palynostratigraphic schemes have been mainly used to correlate coal- and hydrocarbon-bearing rocks within and between basins, sometimes at high levels of biostratigraphic resolution. Up to now, their main limitations have been the lack of correlation with schemes outside the basins, coalfields and hydrocarbon fields where they occur, and a lack of correlation with the Permian Standard Global Chronostratigraphic Scale (SGCS). This is partly because phytogeographical provinciality from the Guadalupian through Lopingian makes correlation between regional palynostratigraphic schemes difficult. However, local high-resolution palynostratigraphic schemes for different regions are now being linked to each other, either by assemblage level quantitative taxonomic comparison or by the use of single, well-characterized palynological taxa that occur across Permian phytogeographical provinces. Such taxa include *Scutasporites*, *Vittatina*, *Weylandites*, *Lueckisporites virkkiae*, *Otynisporites eotriassicus* and *Converrucosisporites confluens*. These palynological correlations are facilitated and supplemented with radioisotopic dating, magnetostratigraphy, independent faunal correlations and strontium isotope stratigraphy. Thus the future of Permian palynostratigraphy looks promising.

Megafossil plants

Megafossil plants have been part of Permian chronostratigraphy and biostratigraphy back to the beginning of the concept of the Permian, due, in part, to the identification of a Permian flora distinct from that of the Carboniferous or the Triassic

(Murchison 1841; see also Murchison *et al.* 1845). This is a flora of late Palaeozoic aspect, dominated by primitive conifers, peltasperms, true ferns, sphenopsids and cordaites that co-occur with the last lycopods. However, the Permian floras were very provincialized due to the complex topography and steep climatic gradients that characterized Permian Pangaea. This hinders the development of a global biostratigraphy based on Permian plants.

Cleal (2016) reviews the use of megafossil plants in Permian biostratigraphy and correlations. He notes that separate biostratigraphic schemes have been developed for Permian macrofloras in the five main phytochoria (palaeokingdoms) of Permian Pangaea, reflecting the general lack of overlap in taxonomic composition between the phytochoria. Two biozones are normally recognized in Europe; in North America three zones, in Cathaysia three or four zones, in Gondwana four zones and in Angara five zones. The stratigraphic resolution is thus far less than for palynology and up to an order of magnitude coarser than the macrofloral biozones of the Pennsylvanian. This is probably due (at least in part) to the lack of rigour in the way that the Permian macrofloral zones have been defined. Nevertheless, the existing zones do provide evidence of the overarching trajectory of change in vegetation through the Permian Period, as it responded at all palaeolatitudes to a combination of climate change, large-scale volcanic eruptions and tectonically driven landscape changes. Thus, as is generally true of fossil plants, Permian plant fossils remain excellent indicators of palaeoclimates and palaeoenvironments, but are not robust biostratigraphic tools.

Conchostracans

Conchostracans are bivalved crustaceans that have lived in freshwater lakes and ponds over the last few hundred million years. Their minute, drought-resistant eggs can be dispersed by the wind and this guaranteed a broad geographical range to some conchostracan taxa across much of Permian Pangaea. Their habitats ranged from perennial lakes of the Carboniferous and early Permian to seasonal playa lakes and temporary ponds and puddles of the late early Permian through the Triassic, when they could form mass death assemblages. This, together with relatively high speciation rates, make them ideal guide fossils, especially in otherwise non-fossiliferous wet and dry red beds.

Schneider & Scholze (2016) present a review of Permian conchostracan biostratigraphy in Western Europe. A preliminary conchostracan zonation is proposed based on material and data collected from surface outcrops and well cores in Central Europe since the 1980s. This is based on assemblage zones consisting of two or three species instead of species-range zones with one or two forms only. The average temporal resolution of the proposed zones during the interval from the late Bashkirian (Westphalian A) to the Early Triassic (Induan) is approximately 5 myr, but it deteriorates to 15 myr during the late early to middle Permian because of a discontinuous fossil record and sampling biases. Isotopically dated occurrences of conchostracan zone species or co-occurrences of conchostracans, insect zone species and marine index fossils, such as conodonts and fusulinids, are used for correlations with the SGCS. The conchostracan zones are cross-correlated to the SGCS based on selected radioisotopic ages and the co-occurrence of some conchostracan index taxa with marine index fossils. Nevertheless, much work remains to improve Permian conchostracan biostratigraphy, particularly to resolve their generally oversplit taxonomy.

Tetrapod footprints

Fossil footprints of Permian tetrapods, which have been studied since the early 1800s, are common in some Permian non-marine strata and had broad palaeogeographical distributions. Some Permian non-marine strata that lack or nearly lack a tetrapod bone record have an extensive footprint record. Therefore various workers have used Permian tetrapod footprints in biostratigraphy.

Voigt & Lucas (2016) review the use of tetrapod footprints in Permian biostratigraphy. They argue that several characteristic Permian footprint assemblages and ichnotaxa have restricted stratigraphic ranges and thus represent distinct time intervals that they term footprint biochrons (after Lucas 2007). Based on the temporal distribution of the 13 best known Permian tetrapod ichnotaxa, three footprint biochrons are recognized: (1) *Dromopus* biochron – latest Carboniferous (roughly Gzhelian) to late early Permian (roughly Artinskian), encompassing ichnoassemblages dominated by tracks of temnospondyls, reptiliomorphs, pelycosaurs and early diapsids; (2) *Erpetopus* biochron – late early Permian (roughly Kungurian) to late middle Permian (roughly Capitanian), including ichnoassemblages dominated by tracks of non-diapsid eureptiles; and (3) *Paradoxichnium* biochron – late Permian (Wuchiapingian and Changhsingian), encompassing ichnoassemblages dominated by tracks of medium- and large-sized parareptiles, non-diapsid eureptiles and early saurians.

This is a conservative ichnostratigraphic scheme for Permian tetrapod tracks and should be refined to almost stage-level resolution by future comprehensive analysis, especially of Permian captorhinomorph and therapsid footprints. Other major tasks to improve Permian tetrapod footprint ichnostratigraphy include augmenting our knowledge of middle Permian tetrapod footprints and clarification of the palaeoenvironmental factors that control the distribution of tetrapod footprints in space and time.

Tetrapod footprints thus have some use for Permian biostratigraphy and biochronology. However, compared with the tetrapod body fossil record, which has been used to discriminate 11 biochrons, the three footprint-based biochrons provide much less temporal resolution. Nevertheless, in non-marine Permian strata where body fossils are rare, tetrapod footprints will remain important for biostratigraphy and biochronology.

Tetrapods

Permian tetrapod (amphibian and reptile) fossils have long been used in non-marine biostratigraphy, with a tradition extending back to at least the 1870s. Lucas (1998) advocated developing a global Permian timescale based on tetrapod evolution and Lucas (2006) presented a comprehensive global Permian tetrapod biochronology that divided the Permian into 11 time intervals (land vertebrate faunachrons) based on tetrapod evolution. **Lucas (2017)** presents the current status of the Permian tetrapod-based timescale.

The most extensive Permian tetrapod (amphibian and reptile) fossil records from the western USA (New Mexico and Texas) and South Africa define 11 land vertebrate faunachrons (in ascending order): Coyotean, Seymouran, Mitchellcreekian, Redtankian, Littlecrotonian, Kapteinskraalian, Gamkan, Hoedemakeran, Steilkransian, Platbergian and Lootsbergian. These faunachrons provide a biochronological framework with which to assign ages to and correlate Permian tetrapod fossil assemblages. Intercalated marine strata, radioisotopic ages and magnetostratigraphy correlate the Permian land vertebrate faunachrons to the SGCS with varying degrees of precision. Such correlations identify the following significant events in Permian tetrapod evolution: a Coyotean chronofaunal event (persistence of the Coyotean chronofauna for about 10 myr), a series of extinctions, the Redtankian events (Mitchellcreekian–Littlecrotonian), Olson's gap (the global absence of tetrapod fossils during the late Littlecrotonian), a therapsid event (sudden appearance of therapsid-dominated assemblages during the Kapteinskraalian), the dinocephalian extinction event (the largest Permian mass extinction of tetrapods at about the end of the Gamkan) and a latest Permian extinction event (late Platbergian to Lootsbergian).

Problems of incompleteness, endemism and taxonomy, and the relative lack of non-biochronological age control, continue to hinder refinement and correlation of a Permian timescale based on tetrapod biochronology. Nevertheless, the global Permian timescale based on tetrapod biochronology is a robust tool for both global and regional age assignment and correlation. Advances in Permian tetrapod biochronology will come from new fossil discoveries, more detailed biostratigraphy and additional alpha taxonomic studies based on sound evolutionary taxonomic principles.

Other non-marine biostratigraphy

Some other non-marine Permian fossils have been used in biostratigraphy, including charophytes, ostracods, bivalves and fish. None of these groups has provided what can be considered a robust global or even provincial biostratigraphy, but all have some potential to aid in Permian correlations.

Charophytes are the calcified egg cases (gyrogonites) of characeous algae and have been documented from some Permian lacustrine deposits (e.g. Feist *et al.* 2005). However, too little is known of the Permian charophyte record to allow its use in biostratigraphy. The only substantial record is from China (e.g. Wang & Wang 1986) and the biozonation based on it (four Permian assemblages: Feist *et al.* 2005, table 4) needs to be tested with data from other regions.

Much of the variation in gyrogonite morphology is ecophenotypic, so it is more a function of environmental variation than a consistent evolutionary signal (e.g. Lucas & Johnson 2016). Thus we suspect that the long-ranging charophyte genera now known from the Permian (e.g. *Stellatochara*) will not segregate into temporally successive species useful in biostratigraphy.

Non-marine ostracods are common in various lacustrine deposits of the Permian, ranging from the black shales and limestones of perennial lakes to claystones and micritic limestones of temporary ponds and pools. They can even be common in strata deposited by brackish waters or environments of higher salinity. Mass death assemblages in shales and limestones, sometimes rock-forming, may be linked to ecological factors that prevent the co-occurrence of other inhabitants of the same or similar guild, as well as the occurrence of ostracod-feeding predators.

Nevertheless, the use of non-marine ostracods in Permian biostratigraphy is hampered by two factors. First, freshwater ostracods have a simple shell morphology and the state of preservation (lack of preserved muscle scars, deformation up to complete flattening during sediment compaction) often precludes precise identification. The second problem is their oversplit alpha taxonomy. Nevertheless, in modern semi-arid and arid environments in Africa

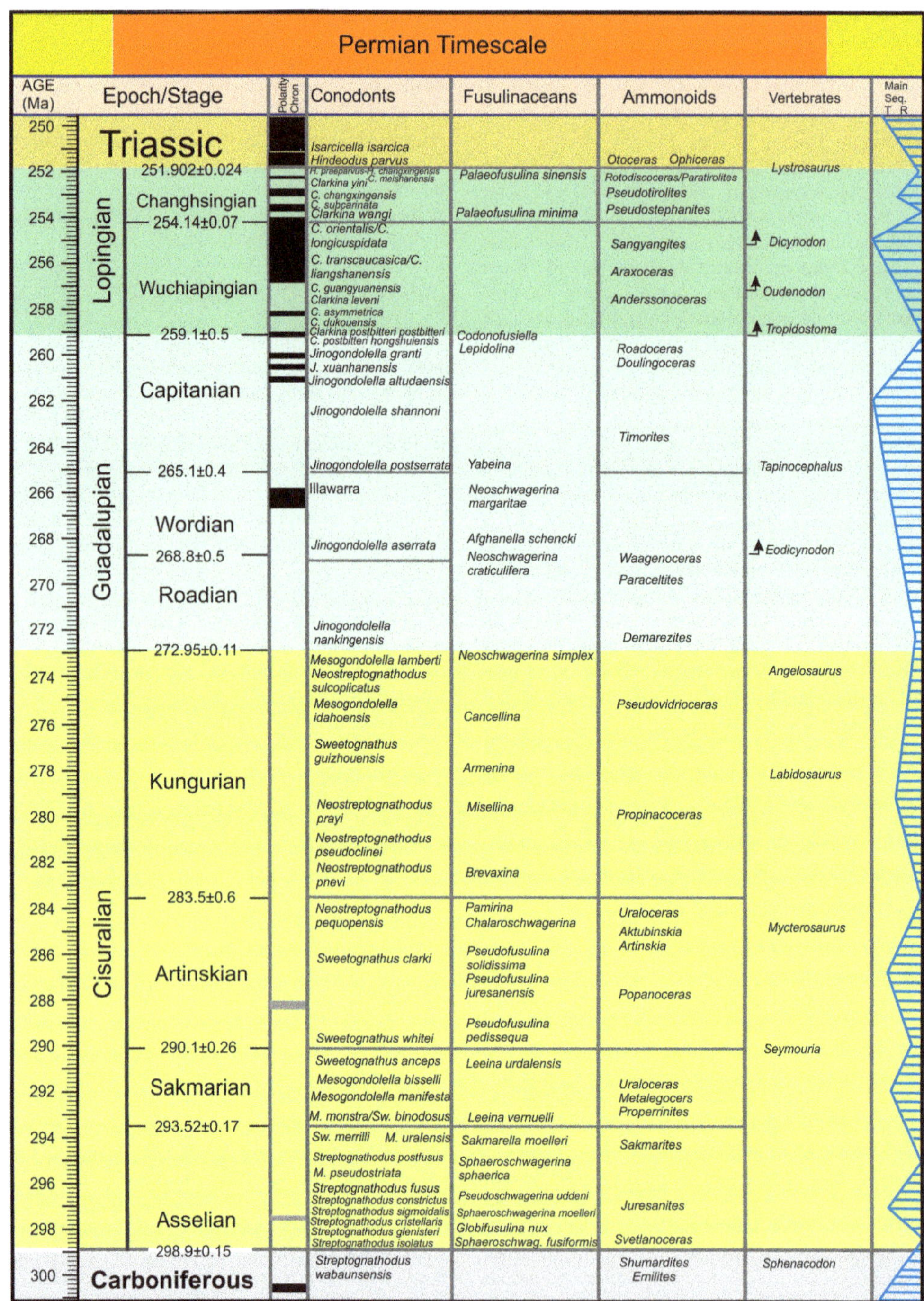

Fig. 4. Current Permian timescale of the Subcommission on Permian Stratigraphy.

and Arabia, the minute eggs of freshwater ostracods are drought resistant. They can thus be easily distributed over hundreds of kilometres by air currents and this may have happened during the Permian. Therefore Permian non-marine ostracods could have promise for biostratigraphy.

Non-marine bivalves, including the anthracosiids, palaeomutelids and some myalinids (brackish water), had a worldwide distribution during the Permian. Some biostratigraphic correlations have been based on these bivalves (e.g. Eagar 1984), but their alpha taxonomy seems extremely oversplit, as most variation is ecophenotypic, not interspecific, in origin. It seems unlikely that the stratigraphic ranges of all non-marine Permian bivalves are well established. Thus, for example, Lucas & Rinehart (2005) documented *Palaeanodonta* in the lower Permian of North America, whereas the genus is otherwise known from the middle or late Permian of Antarctica, South Africa, Kenya, Russia, Myanmar and Siberia, among other places. This substantial range extension suggests to us that the true distributions in time and space of Permian freshwater bivalves are not well known. This and the taxonomic problems should make us cautious in using non-marine bivalves for Permian biostratigraphy.

Scudder (1879, 1885) first attempted to use insect wings for Permian biostratigraphy, recognizing the common occurrence of genera and species of blattid insects (cockroaches) in North America and Europe. Later, Durden (1969, 1984) proposed blattid zonations of the Pennsylvanian and Permian, but his correlations were problematic because of inadequate taxonomy (Lucas *et al.* 2013). Schneider (1983) published a revised classification of Pennsylvanian and Permian blattids and from this came the first proposal of spiloblattinid zones and later of archimylacrid/spiloblattinid/conchostracan zones for the Early Pennsylvanian (Westphalian A) through to the late early Permian (Artinskian) (Schneider 1982; Schneider & Werneburg 1993, 2006, 2012). This biostratigraphy has been cross-correlated to parts of the SGCS where the cockroach fossils co-occur or are intercalated with marine strata that yield fusulinids and/or conodonts (e.g. Lucas *et al.* 2013).

Freshwater fish have never provided a robust biostratigraphy in non marine strata. This is because of the limitations of these fish and their fossils to specific lithofacies and locations, so that their record is heavily affected by facies control and endemism. Permian xenacanth shark teeth have been applied to regional correlations between some European basins, but their wider use is limited because the migration of fish is restricted to river systems connecting the basins (e.g. Schneider 1996; Schneider *et al.* 2000). Freshwater fish do provide valuable information on basin interconnections and drainage systems (Schneider & Zajic 1994) as well as about ecological changes and events (Boy & Schindler 2000). Thus freshwater fish biozonations, such as that of Zajíc (2014), are actually local ecostratigraphy, not robust biostratigraphy.

A Permian timescale

The Permian timescale presented here (Fig. 4) is that of the SPS, well reflected by many of the papers in this volume. Issues in the further development of a Permian chronostratigraphic scale include those of stability and priority of nomenclature and concepts, disagreements over changing taxonomy, ammonoid v. fusulinid v. conodont biostratigraphy, differences in the perceived significance of biotic events for chronostratigraphic classification and correlation problems between provinces. Further development of the Permian chronostratigraphic scale should focus on GSSP selection for the remaining, undefined stage bases, the definition and characterization of substages, and further integration of the Permian chronostratigraphic scale with radioisotopic, magnetostratigraphic and chemostratigraphic tools for calibration and correlation.

We thank all the contributors to this volume for their perspicacity and patience. We also thank the reviewers for their reviews of this manuscript. Shuzhong Shen's work is supported by the Strategic Priority Research Program (B) of the Chinese Academy of Sciences (XDB18000000) and National Natural Science Foundation of China (grant no. 41290260).

Correction notice: The catchline of the original version was incorrect.

References

Batten, R.L. 1973. The vicissitudes of the gastropods during the interval of Guadalupian-Ladinian time. *In*: Logan, A. & Hills, L.V. (eds) *The Permian and Triassic Systems and Their Mutual Boundary*. Canadian Society of Petroleum Geologists, Memoirs, **2**, 596–607.

Baud, A., Margaritz, M. & Holser, W.T. 1989. Permian–Triassic of the Tethys; carbon isotope studies. *Geologische Rundschau*, **78**, 649–677.

Bogoslovskaya, M.F., Leonova, T.B. & Shkolin, A.A. 1995. The Carboniferous-Permian boundary and ammonoids from the Aidaralash section, southern Urals. *Journal of Paleontology*, **69**, 288–301.

Bond, D.P.G., Hilton, J., Wignall, P.B., Ali, J.R., Stevens, L.G., Sun, Y. & Lai, X. 2010. The middle Permian (Capitanian) mass extinction on land and in the oceans. *Earth-Science Reviews*, **102**, 100–116.

Böse, E. 1917. The Permo-Carboniferous ammonoids of the Glass Mountains, west Texas and their stratigraphical significance. *Texas University Bulletin*, **1762**, 1–241.

Boy, J.A. & Schindler, T. 2000. Ökostratigraphische Bioevents im Grenzbereich Stephanium/Autunium (höchstes Karbon) des Saar-Nahe-Beckens (SW-Deutschland) und benachbarter Gebiete. *Neues Jahrbuch für Geologie und Paläontologie Abhandlungen*, **216**, 89–152.

Buggisch, W., Wang, X.D., Alekseev, A.S. & Joachimski, M. 2011. Carboniferous-Permian carbon isotope stratigraphy of successions from China (Yangtze platform), USA (Kansas) and Russia (Moscow basin and Urals). *Palaeogeography, Paleoclimatology, Paleoecology*, **301**, 18–38.

Buggisch, W., Krainer, K., Schaffhauser, M., Joachimski, M. & Korte, K. 2015. Late Carboniferous to late Permian carbon isotope stratigraphy: a new record from post-Variscan carbonates from the Southern Alps (Austria and Italy). *Palaeogeography, Paleoclimatology, Paleoecology*, **433**, 174–190.

Burgess, S.D., Bowring, S.A. & Shen, S.Z. 2014. High-precision timeline for Earth's most severe extinction. *Proceedings of the National Academy of Sciences*, **111**, 3316–3321.

Chernykh, V.V., Chuvashov, B.I., Davydov, V.I. & Schmitz, M.D. 2012. Mechetlino section: a candidate for the Global Stratotype [Section] and Point (GSSP) of the Kungurian Stage (Cisuralian, lower Permian). *Permophiles*, **56**, 21–34.

Chernykh, V.V., Chuvashov, B.I., Shen, S.Z. & Henderson, C.M. 2016. Proposal for the Global Stratotype Section and Point (GSSP) for the base-Sakmarian Stage (lower Permian). *Permophiles*, **63**, 4–18.

Chuvashov, B.I., Chernykh, V.V., Shen, S.Z. & Henderson, C.M. 2013. Proposal for the Global Stratotype Section and Point (GSSP) for the base-Artinskian Stage (lower Permian). *Permophiles*, **58**, 26–34.

Clapham, M.E. & Bottjer, D.J. 2007. Permian marine paleoecology and its implications for large-scale decoupling of brachiopod and bivalve abundance and diversity during the Lopingian. *Palaeogeography, Palaeoclimatology, Palaeoecology*, **249**, 283–301.

Clapham, M.E., Shen, S.Z. & Bottjer, D.J. 2009. The double mass extinction revisted: reassessing the severity, selectivity, and causes of the end-Guadalupian biotic crisis (late Permian). *Paleobiology*, **35**, 32–50.

Cleal, C.J. 2016. A global review of Permian macrofloral biostratigraphical schemes. *In*: Lucas, S.G. & Shen, S.Z. (eds) *The Permian Timescale*. Geological Society, London, Special Publications, **450**. First published online December 8, 2016, https://doi.org/10.1144/SP450.4

Davydov, V.I., Glenister, B.F., Spinosa, C., Ritter, S.M., Chernykh, V.V., Wardlaw, B.R. & Snyder, W.S. 1998. Proposal of Aidaralash as Global Stratotype Section and Point (GSSP) for base of the Permian System. *Episodes*, **21**, 11–17.

Douglass, R.C. 1977. The development of fusulinid biostratigraphy. *In*: Kauffman, E.G. & Hazel, J.E. (eds) *Concepts and Methods of Biostratigraphy*. Dowden, Hutchinson, and Ross, Stroudsburg, 463–481.

Dudás, F.Ö., Yuan, D.X., Shen, S.Z. & Bowring, S.A. 2017. A conodont-based revision of the $^{87}Sr/^{86}Sr$ seawater curve across the Permian–Triassic boundary. *Palaeogeography, Palaeoclimatology, Palaeoecology*, **470**, 40–53.

Durden, C.J. 1969. Pennsylvanian correlation using blattoid insects. *Canadian Journal of Earth Sciences*, **6**, 1159–1177.

Durden, C.J. 1984. North American provincial insect ages for the continental last half of the Carboniferous and first half of the Permian. *In*: *Compte Rendus 9. Congrès International de Stratigraphie et de Géologie du Carbonifère, Nanking*. Vol. **2**, 606–612.

Eagar, R.M.C. 1984. Late Carboniferous-early Permian non-marine bivalve faunas of northern Europe and eastern North America. *In*: *Compte Rendu Neuvième Congrès International de Stratigraphie et de Géologie de Carbonifère, Washington and Champaign-Urbana*. Vol. **2**, 559–576.

Feist, M., Grambast-Fessard, N. et al. 2005. *Charophyta: Treatise on Invertebrate Paleontology. Part B. Protoctista 1*. Vol. 1. Geological Society of America/University of Kansas, Boulder/Lawrence.

Furnish, W.M. 1966. Ammonoids of the upper Permian *Cyclolobus* Zone. *Neues Jahrbuch für Mineralogie, Geologie und Paläontologie Abhandlungen*, **125**, 265–296.

Furnish, W.M. 1973. Permian stage names. *In*: Logan, A. & Hills, L.V. (eds) *The Permian and Triassic Systems and Their Mutual Boundary*. Canadian Society of Petroleum Geologists, Memoirs, **2**, 522–548.

Glenister, B.F. 1981. Permian ammonoid 'zones'. *In*: House, M.R. (ed.) *The Ammonoidea*. Academic Press, New York, 386–396.

Glenister, B.F. & Furnish, W.M. 1961. The Permian ammonoids of Australia. *Journal of Paleontology*, **35**, 673–736.

Glenister, B.F., Wardlaw, B.R. et al. 1999. Proposal of Guadalupian and component Roadian, Wordian and Capitanian stages as international standards for the middle Permian series. *Permophiles*, **34**, 3–11.

Henderson, C.M. 2016. Permian conodont biostratigraphy. *In*: Lucas, S.G. & Shen, S.Z. (eds) *The Permian Timescale*. Geological Society, London, Special Publications, **450**. First published online December 14, 2016, https://doi.org/10.1144/SP450.9

Henderson, C.M., Wardlaw, B.R., Davydov, V.I., Schmitz, M.D., Schiappa, T.A., Tierney, K.E. & Shen, S.Z. 2012*a*. Proposal for base-Kungurian GSSP. *Permophiles*, **56**, 8–21.

Henderson, C.M., Davydov, V.I. & Wardlaw, B.R. 2012*b*. The Permian Period. *In*: Gradstein, F.M., Ogg, J.G., Schmitz, M.D. & Ogg, G.M. (eds) *The Geologic Time Scale 2012*. Vol. **2**. Elsevier, Amsterdam, 653–679.

Hounslow, M. & Balabanov, Y.P. 2016. A geomagnetic polarity timescale for the Permian, calibrated to stage boundaries. *In*: Lucas, S.G. & Shen, S.Z. (eds) *The Permian Timescale*. Geological Society, London, Special Publications, **450**. First published online December 8, 2016, https://doi.org/10.1144/SP450.8

Isbell, J.L., Lenaker, P.A., Askin, R.A., Miller, M.F. & Babcock, L.E. 2003. Reevaluation of the timing and extent of late Paleozoic glaciation in Gondwana: Role of the Transantarctic Mountains. *Geology*, **31**, 977–980.

Isozaki, Y., Kawahata, H. & Minoshima, K. 2007*a*. The Capitanian (Permian) Kamura cooling event; the beginning of the Paleozoic–Mesozoic transition. *Palaeoworld*, **16**, 16–30.

Isozaki, Y., Kawahata, H. & Ota, A. 2007*b*. A unique carbon isotope record across the Guadalupian-Lopingian (middle-upper Permian) boundary in mid-oceanic paleo-atoll carbonates: The high productivity 'Kamura event' and its collapse in Panthalassa. *Global and Planetary Change*, **55**, 21–38.

JIN, Y.G., HENDERSON, C.M., WARDLAW, B.R., GLENISTER, B.F., MEI, S.L., SHEN, S.Z. & WANG, X.D. 2001. Proposal for the Global Stratotype Section and Point (GSSP) for the Guadalupian-Lopingian boundary. *Permophiles*, **39**, 32–42.

JIN, Y.G., SHEN, S.Z. *ET AL.* 2006*a*. The Global Stratotype Section and Point (GSSP) for the boundary between the Capitanian and Wuchiapingian stages (Permian). *Episodes*, **29**, 253–262.

JIN, Y.G., WANG, Y., HENDERSON, C.M., WARDLAW, B.R., SHEN, S.Z. & CAO, C.Q. 2006*b*. The Global Stratotype Section and Point (GSSP) for the base of the Changhsingian Stage (upper Permian). *Episodes*, **29**, 175–182.

KARPINSKY, A.P. 1874. Geologische Untersuchungen im Gouvernment Orenburg. *Verhandlung Mineralogische Gesellschaft St Petersbourg*, **9**, 212–330.

KARPINSKY, A.P. 1889. Über die Ammoneen der Artinsk-Stufe und einige mit denselben verwandte carbonische Formen. *Memoires de l'Academic Imperiale des Sciences de St Petersbourg*, **37**, 1–104.

KERANS, C., FITCHEN, W.M., GARDNER, M.H. & WARDLAW, B.R. 1993. A contribution to the evolving stratigraphic framework of middle Permian strata of the Delaware basin, Texas and New Mexico. *New Mexico Geological Society Guidebook*, **44**, 175–184.

KNIGHT, J.B. 1940. Are the '*Omphalotrochus* beds' of the U.S.S.R. Permian. *American Association of Petroleum Geologists Bulletin*, **24**, 1128–1133.

KORTE, C. & KOZUR, H.W. 2010. Carbon-isotope stratigraphy across the Permian–Triassic boundary: a review. *Journal of Asian Earth Sciences*, **39**, 215–235.

KORTE, C. & ULLMANN, C.V. 2016. Permian strontium isotope stratigraphy. *In*: LUCAS, S.G. & SHEN, S.Z. (eds) *The Permian Timescale*. Geological Society, London, Special Publications, **450**. First published online December 12, 2016, https://doi.org/10.1144/SP450.5

KROTOW, P. 1885. Artinskian Stage: geological-paleontological monograph on the Artinskian sands. *Kazan Universitet Obshchestvo Estestvoispytatele Trudy*, **13**, 1–314 [in Russian].

LEONOVA, T.B. 2016. Permian ammonoid biostratigraphy. *In*: LUCAS, S.G. & SHEN, S.Z. (eds) *The Permian Timescale*. Geological Society, London, Special Publications, **450**. First published online December 8, 2016, https://doi.org/10.1144/SP450.7

LI, R. & JONES, B. 2017. Diagenetic overprint on negative $\delta^{13}C$ excursions across the Permian/Triassic boundary: a case study from Meishan section, China. *Palaeogeography, Palaeoclimatology, Palaeoecology*, **468**, 18–33.

LUCAS, S.G. 1998. Toward a tetrapod biochronology of the Permian. *New Mexico Museum of Natural History and Science Bulletin*, **12**, 71–91.

LUCAS, S.G. 2006. Global Permian tetrapod biostratigraphy and biochronology. *In*: LUCAS, S.G., CASSINIS, G. & SCHNEIDER, J.W. (eds) *Non-marine Permian Biostratigraphy and Biochronology*. Geological Society, London, Special Publications, **265**, 65–93, https://doi.org/10.1144/GSL.SP.2006.265.01.04

LUCAS, S.G. 2007. Tetrapod footprint biostratigraphy and biochronology. *Ichnos*, **14**, 5–38.

LUCAS, S.G. 2017. Permian tetrapod biochronology, correlation, evolutionary events. *In*: LUCAS, S.G. & SHEN, S.Z. (eds) *The Permian Timescale*. Geological Society, London, Special Publications, **450**. First published online May 15, 2017, https://doi.org/10.1144/SP450.12

LUCAS, S.G. & JOHNSON, G.D. 2016. *Palaeochara* from the lower Permian of Oklahoma, USA. *New Mexico Museum of Natural History and Science Bulletin*, **74**, 141–143.

LUCAS, S.G. & RINEHART, L.F. 2005. Nonmarine bivalves from the lower Permian (Wolfcampian) of the Chama basin, New Mexico. *In*: LUCAS, S.G., ZEIGLER, K.E., LUETH, V.W. & OWEN, D.E. (eds) *Geology of the Chama Basin*. New Mexico Geological Society, Fall Field Conference Guidebook, **56**, 283–287.

LUCAS, S.G. & SHEN, S.Z. 2016. The Permian chronostratigraphic scale: history, status, prospectus. *In*: LUCAS, S.G. & SHEN, S.Z. (eds) *The Permian Timescale*. Geological Society, London, Special Publications, **450**. First published online December 9, 2016, https://doi.org/10.1144/SP450.3

LUCAS, S.G., SCHNEIDER, J.W. & CASSINIS, G. 2006. Non-marine Permian biostratigraphy and biochronology: an introduction. *In*: LUCAS, S.G., CASSINIS, G. & SCHNEIDER, J.W. (eds) *Non-marine Permian Biostratigraphy and Biochronology*. Geological Society, London, Special Publications, **265**, 1–14, https://doi.org/10.1144/GSL.SP.2006.265.01.01

LUCAS, S.G., BARRICK, J.E., KRAINER, K. & SCHNEIDER, J.W. 2013. The Carboniferous-Permian boundary at Carrizo Arroyo, central New Mexico, USA. *Stratigraphy*, **10**, 153–170.

MAZAEV, A.V. 1994. Middle and Upper Carboniferous gastropods from the central part of the Russian plate: Part 1, Euomphalacea. *Ruthenica*, **4**, 143–168.

MEI, S.L. & HENDERSON, C.M. 2002. Conodont definition of the Kungurian (Cisuralian) and Roadian (Guadalupian) boundary. *In*: HILLS, L.V., HENDERSON, C.M. & BAMBER, E.W. (eds) *Carboniferous and Permian of the World*. Canadian Society of Petroleum Geologists, Memoirs, **19**, 529–551.

MEI, S.L., HENDERSON, C.M. & CAO, C.Q. 2004. Conodont sample-population approach to defining the base of the Changhsingian Stage, Lopingian Series, upper Permian. *In*: BEAUDOIN, A. & HEAD, M. (eds) *The Palynology and Micropaleontology of Boundaries*. Geological Society, London, Special Publications, **230**, 105–121, https://doi.org/10.1144/GSL.SP.2004.230.01.06

MILLER, A.I. 1988. Spatio-temporal transitions in Paleozoic Bivalvia: an analysis of North American fossil assemblages. *Historical Biology*, **1**, 251–273.

MILLER, A.I. 1989. Spatio-temporal transitions in Paleozoic Bivalvia: a field comparison of Upper Ordovician and upper Paleozoic bivalve-dominated fossil assemblages. *Historical Biology*, **2**, 227–260.

MILLER, A.K. 1938. Comparison of Permian ammonoid zones of Soviet Russia with those of North America. *American Association of Petroleum Geologists Bulletin*, **22**, 1014–1019.

MONDAL, S. & HARRIES, P.J. 2016. Phanerozoic trends in ecospace utilization: the bivalve perspective. *Earth-Science Reviews*, **152**, 106–118.

MURCHISON, R.I. 1841. First sketch of some of the principal results of a second geological survey of Russia. *The Philosophical Magazine*, **19**, 417–422.

MURCHISON, R.I., DE VERNEUIL, E. & KEYSERLING, A. VON. 1845. *The Geology of Russia in Europe and the Ural Mountains*. Vol. **1**. *Geology*. John Murray, London.

NAKAZAWA, K. & RUNNEGAR, B. 1973. The Permian–Triassic boundary: a crisis for bivalves. *In*: LOGAN, A. & HILLS, L.V. (eds) *The Permian and Triassic Systems and their Mutual Boundary. Canadian Society of Petroleum Geologists, Memoirs*, **2**, 608–621.

RAMEZANI, J. & BOWRING, S.A. 2017. Advances in numerical calibration of the Permian timescale based on radioisotopic geochronology. *In*: LUCAS, S.G. & SHEN, S.Z. (eds) *The Permian Timescale*. Geological Society, London, Special Publications, **450**. First published online November 1, 2017, https://doi.org/10.1144/SP450.17

RUZHENTSEV, V.E. 1938. Ammonoids of the Sakmarian stage and their stratigraphic significance. *Problemy Paleontologii*, **4**, 187–285.

RUZHENTSEV, V.E. 1951. Lower Permian ammonoids of the South Urals. I. Ammonoids of the Sakmarian Stage. *Akademiya Nauk SSSR Paleontologicheska Instituta Trudy*, **33**, 1–188 [in Russian].

RUZHENTSEV, V.E. 1954. Asselian Stage of the Permian System. *Doklady Akademii Nauk SSSR*, **99**, 1079–1082 [in Russian].

RUZHENTSEV, V.E. 1955. On the family Cyclolobidae Zittel. *Doklady Akademiya Nauk SSSR*, **103**, 701–703 [in Russian].

RUZHENTSEV, V.E. 1976. Late Permian ammonoids in the Far East. *Paleontological Journal*, **1976**, 36–50.

SALTZMAN, M.R. 2003. The late Paleozoic ice age: oceanic gateway or pCO_2? *Geology*, **31**, 151–154.

SALTZMAN, M.R. & THOMAS, E. 2012. Carbon isotope stratigraphy. *In*: GRADSTEIN, F.M., OGG, J.G., SCHMITZ, M.D. & OGG, G.M. (eds) *The Geologic Time Scale 2012*. Vol. **1**. Elsevier, Amsterdam, 207–232.

SCHELLWIEN, E. 1898. Die Fauna des karnischen Fusulinenkalks. II: Foraminifera. *Palaeontographica*, **44**, 237–282.

SCHMITZ, M.D. & DAVYDOV, V.I. 2012. Quantitative radiometric and biostratigraphic calibration of the Pennsylvanian–early Permian (Cisuralian) time scale and pan-Euramerican chronostratigraphic correlation. *Geological Society of America Bulletin*, **124**, 549–577.

SCHNEIDER, J. 1982. Entwurf einer Zonengliederung für das euramerische Permokarbon mittels der Spiloblattinidae (Blattodea, Insecta). *Freiberger Forschungshefte*, **375**, 27–47.

SCHNEIDER, J. 1983. Die Blattodea (Insecta) des Paläozoikum, Teil 1: Systematik, Ökologie und Biostratigraphie. *Freiberger Forschungshefte*, **382**, 106–145.

SCHNEIDER, J.W. 1996. Xenacanth teeth – a key for taxonomy and biostratigraphy. *Modern Geology*, **20**, 321–340.

SCHNEIDER, J.W. & SCHOLZE, F. 2016. Late Pennsylvanian–Early Triassic conchostracan biostratigraphy: a preliminary approach. *In*: LUCAS, S.G. & SHEN, S.Z. (eds) *The Permian Timescale*. Geological Society, London, Special Publications, **450**. First published online December 8, 2016, https://doi.org/10.1144/SP450.6

SCHNEIDER, J. & WERNEBURG, R. 1993. Neue Spiloblattinidae (Insecta, Blattodea) aus dem Oberkarbon und Unterperm von Mitteleuropa sowie die Biostratigraphie des Rotliegend. *Veröffentlichungen Naturhistorisches Museum Schleusingen*, **7/8**, 31–52.

SCHNEIDER, J.W. & WERNEBURG, R. 2006. Insect biostratigraphy of the Euramerican continental Late Pennsylvanian and early Permian. *In*: LUCAS, S.G., CASSINIS, G. & SCHNEIDER, J.W. (eds) *Non-marine Permian Biostratigraphy and Biochronology*. Geological Society, London, Special Publications, **265**, 325–336, https://doi.org/10.1144/GSL.SP.2006.265.01.15

SCHNEIDER, J.W. & WERNEBURG, R. 2012. Biostratigraphie des Rotliegend mit Insekten und Amphibien. *Schriftenreihe der Deutschen Gesellschaft fur Geowissenschaften*, **61**, 110–142.

SCHNEIDER, J.W. & ZAJIC, J. 1994. Xenacanthodier (Pisces, Chondrichthyes) des mitteleuropäischen Oberkarbon und Perm - Revision der Originale zu Goldfuss 1847, Beyrich 1848, Kner 1867 und Fritsch 1879–90. *Freiberger Forschungshefte*, **C452**, 101–151.

SCHNEIDER, J.W., HAMPE, O. & SOLER-GIJÓN, R. 2000. The Late Carboniferous and Permian aquatic vertebrate zonation in southern Spain and German basins. *Courier Forschungsinstitut Senckenberg*, **223**, 543–561.

SCUDDER, S.H. 1879. Paleozoic cockroaches: a complete revision of the species of both world, with an essay toward their classification. *Memoirs of the Boston Society of Natural History*, **3**, 23–134.

SCUDDER, S.H. 1885. New genera and species of fossil cockroaches, from the older American rocks. *Proceedings of the Academy of Natural Sciences Philadelphia*, **1885**, 34–39.

SHEN, S.Z. 2016. Global Permian brachiopod biostratigraphy: an overview. *In*: LUCAS, S.G. & SHEN, S.Z. (eds) *The Permian Timescale*. Geological Society, London, Special Publications, **450**. First published online December 9, 2016, https://doi.org/10.1144/SP450.11

SHEN, S.Z., HENDERSON, C.M. ET AL. 2010. High-resolution Lopingian (late Permian) timescale of South China. *Geological Journal*, **45**, 122–134.

SHEN, S.Z., CROWLEY, J.L. ET AL. 2011. Calibrating the end-Permian mass extinction. *Science*, **334**, 1367–1372.

SHEN, S.Z., CAO, C.Q. ET AL. 2013. High-resolution $\delta^{13}C_{carb}$ chemostratigraphy from latest Guadalupian through earliest Triassic in South China and Iran. *Earth and Planetary Science Letters*, **375**, 156–165.

STEINER, M.B. 2001. Magnetostratigraphic correlation and dating of west Texas and New Mexico late Permian strata. *New Mexico Geological Society Guidebook*, **52**, 59–68.

STEINER, M.B. 2006. The magnetic polarity timescale across the Permian–Triassic boundary. *In*: LUCAS, S.G., CASSINIS, G. & SCHNEIDER, J.W. (eds) *Non-marine Permian Biostratigraphy and Biochronology*. Geological Society, London, Special Publications, **265**, 15–38, https://doi.org/10.1144/GSL.SP.2006.265.01.02

STEPHENSON, M.H. 2016. Permian palynostratigraphy: a global overview. *In*: LUCAS, S.G. & SHEN, S.Z. (eds) *The Permian Timescale*. Geological Society, London,

Special Publications, **450**. First published online December 8, 2016, https://doi.org/10.1144/SP450.2

Vachard, D. 2016. Permian smaller foraminifers: taxonomy, biostratigraphy, biogeography. *In*: Lucas, S.G. & Shen, S.Z. (eds) *The Permian Timescale*. Geological Society, London, Special Publications, **450**. First published online December 8, 2016, https://doi.org/10.1144/SP450.1

Voigt, S. & Lucas, S.G. 2016. Outline of a Permian tetrapod footprint ichnostratigraphy. *In*: Lucas, S.G. & Shen, S.Z. (eds) *The Permian Timescale*. Geological Society, London, Special Publications, **450**. First published online December 9, 2016, https://doi.org/10.1144/SP450.10

Wang, X.D., Yao, L. & Lin, W. 2017. Permian rugose corals of the world. *In*: Lucas, S.G. & Shen, S.Z. (eds) *The Permian Timescale*. Geological Society, London, Special Publications, **450**. First published online June 8, 2017, https://doi.org/10.1144/SP450.13

Wang, Z. & Wang, R.N. 1986. Permian charophytes from the southeastern area of the North China platform. *Acta Micropaleontologica Sinica*, **3**, 273–278.

Wignall, P.B., Sun, Y.D. *et al.* 2009. Volcanism, mass extinction, and carbon isotope fluctuations in the middle Permian of China. *Science*, **324**, 1179–1182.

Wu, Q., Ramezani, J. *et al.* 2017. Calibrating the Guadalupian Series (middle Permian) of South China. *Palaeogeography, Palaeoclimatology, Palaeoecology*, **466**, 361–372.

Yin, H.F. (ed.) 1996. *The Palaeozoic–Mesozoic Boundary: Candidates of the Global Stratotype Section and Point of the Permian–Triassic Boundary*. University of Geoscience Press, Wuhan.

Yin, H.F., Sweet, W.C. *et al.* 1996. Recommendation of the Meishan section as global stratotype section and point for basal boundary of Triassic System. *Newsletters on Stratigraphy*, **34**, 81–108.

Yin, H.F., Zhang, K.X., Tong, J.N., Yang, Z.Y. & Wu, S.B. 2001. The global stratotype section and point (GSSP) of the Permian–Triassic boundary. *Episodes*, **24**, 102–114.

Yochelson, E.L. 1954. Some problems concerning the distribution of the late Paleozoic gastropod *Omphalotrochus*. *Science*, **120**, 233–234.

Yochelson, E.L. 1956. Permian Gastropoda of the southwestern United States. 2. Bellerophontacea and Patellacea. *Bulletin of the American Museum of Natural History*, **119**, 205–294.

Zajíc, J. 2014. Permian fauna of the Krkonoše piedmont basin (Bohemian massif, central Europe). *Acta Musei Nationalis Pragae Serie B Historia Naturalis*, **70**, 131–142.

Zeng, J., Cao, C.Q., Davydov, V.I. & Shen, S.Z. 2012. Carbon isotope chemostratigraphy and implications of palaeoclimatic changes during the Cisuralian (Early Permian) in the southern Urals, Russia. *Gondwana Research*, **21**, 601–610.

Zhang, L., Feng, Q.L. & He, W.H. 2017. Permian radiolarian biostratigraphy. *In*: Lucas, S.G. & Shen, S.Z. (eds) *The Permian Timescale*. Geological Society, London, Special Publications, **450**. First published online October 30, 2017, https://doi.org/10.1144/SP450.16

Zhang, Y.C. & Wang, Y. 2017. Permian fusuline biostratigraphy. *In*: Lucas, S.G. & Shen, S.Z. (eds) *The Permian Timescale*. Geological Society, London, Special Publications, **450**. First published online June 8, 2017, https://doi.org/10.1144/SP450.14

The Permian chronostratigraphic scale: history, status and prospectus

SPENCER G. LUCAS[1]* & SHU-ZHONG SHEN[2]

[1]*New Mexico Museum of Natural History and Science, 1801 Mountain Road NW, Albuquerque, NM 87104-1375, USA*

[2]*State Key Laboratory of Palaeobiology and Stratigraphy, Nanjing Institute of Geology and Palaeontology, 39 East Beijing Road, Nanjing, Jiangsu 210008 China*

**Correspondence: spencer.lucas@state.nm.us*

Abstract: In 1841, Murchison coined the term Permian for strata in the Russian Urals. Recognition of the Permian outside of Russia and central Europe soon followed, but it took about a century for the Permian to be accepted globally as a distinct geological system. The work of the Subcommission on Permian Stratigraphy began in the 1970s and resulted in current recognition of nine Permian stages in three series: the Cisuralian (lower Permian) – Asselian, Sakmarian, Artinskian and Kungurian; the Guadalupian (middle Permian) – Roadian, Wordian and Capitanian; and the Lopingian (upper Permian) – Wuchiapingian and Changhsingian. The 1990s saw the rise of Permian conodont biostratigraphy, so that all Permian Global Stratigraphic Sections and Points (GSSPs) use conodont evolutionary events as the primary signal for correlation.

Issues in the development of a Permian chronostratigraphic scale include those of stability and priority of nomenclature and concepts, disagreements over changing taxonomy, ammonoid v. fusulinid v. conodont biostratigraphy, differences in the perceived significance of biotic events for chronostratigraphic classification, and correlation problems between provinces. Further development of the Permian chronostratigraphic scale should focus on GSSP selection for the remaining, undefined stage bases, definition and characterization of substages, and further integration of the Permian chronostratigraphic scale with radioisotopic, magnetostratigraphic and chemostratigraphic tools for calibration and correlation.

The Palaeozoic Era ended with the approximately 47 myr-long Permian Period, a major juncture in Earth history when the vast Pangean supercontinent continued its assembly (Fig. 1), and the global biota faced its greatest diversity crisis, the end-Permian mass extinction, the most extensive biotic decimation of the Phanerozoic. The temporal ordering of geological and biotic events during Permian time is thus critical to the interpretation of some unique and pivotal events in Earth history.

This temporal ordering is mostly based on the Permian chronostratigraphic scale – a relative scale of series and stages that has been developed and refined for nearly two centuries. Here, we offer a brief history of that scale (see Zittel 1901; Sherlock 1947; Furnish 1973; Waterhouse 1976 for more in-depth treatments of some aspects of this history). Our principal goal in presenting this history of the Permian chronostratigraphic scale is to explain the origin and development of the currently used Permian chronostratigraphic terminology. We also present some analysis and evaluation of this process, and advance a prospectus for further refinement of the current Permian chronostratigraphic scale.

Abbreviations

In this article, GSSP is Global Stratotype Section and Point, ICS is the International Commission on Stratigraphy of the IUGS and SPS is the ICS Sub commission on Permian Stratigraphy. The standard global chronostratigraphic scale (SGCS) is the series and stages used as the global framework for ordering geological time.

We distinguish biostratigraphic datums from biochronological events. Biostratigraphic datums are the lowest occurrence (LO) and highest occurrence (HO) of a fossil in a stratigraphic section. Biochronological events are the first appearance datum (FAD) and last appearance datum (LAD) of a taxon, its evolutionary origination and extinction, respectively. For chronostratigraphic definition, it is hoped that the LO and the FAD of a taxon coincide, at least if it is the primary signal for correlation of a

From: Lucas, S. G. & Shen, S. Z. (eds) 2018. *The Permian Timescale*. Geological Society, London, Special Publications, **450**, 21–50.
First published online December 9, 2016, https://doi.org/10.1144/SP450.3

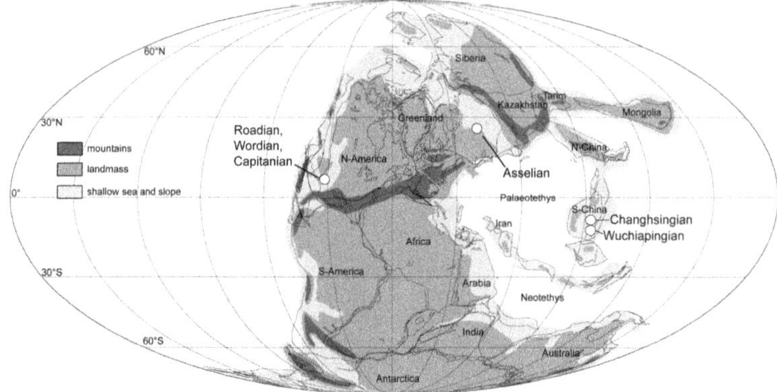

Fig. 1. Map of Permian Pangea showing locations of ratified GSSPs of Permian stages. Map modified from Lucas *et al.* (2006).

GSSP, although, given the problems of sampling and facies, it is highly unlikely that this will ever be the case.

History

Recognition of a distinctive interval in Earth history (originally identified as a distinct succession of stratified rocks) that corresponds to the current concept of Permian began in Germany more than 200 years ago. The history of the development of the Permian chronostratigraphic scale can be divided into five distinct phases:

- The initial studies of the Permian strata of Germany during the 1700s.
- In 1841, Murchison coins the term 'Permian' to refer to a succession of rocks in the Ural Mountains region of European Russia.
- The century-long process that ended with recognition of the Permian as a separate system of the geological timescale.
- Development of diverse Permian stage nomenclatures and alternative Permian chronostratigraphic scales between about 1874 and 1976.
- Beginning in 1975, the work of the SPS to develop the current Permian SGCS.

Germany 1700s

Permian rocks were some of the first rocks studied stratigraphically, in the late 1700s by the first German stratigraphic geologists Johann Gottlieb Lehman (1719–67) and Georg Christian Füchsel (1722–73). Thus, continental European geologists long united the German Rotliegend and Zechstein into one 'group' or 'system'. These were a portion of the 'Flötz-Gebirge' (Flötz = seam; Gebirge = rock sequence) rocks of the eighteenth century German miners and geologists, an economically important stratigraphic interval (Flöz means lode or seam). Indeed, they were a part of Abraham Werner's (1749–1817) 'Flötzformationen' of the 1780s because they included important sources of copper mined from the famed copper slates (Kupferschiefer of the Zechstein: Zeche means mine; Stein means rock; together they referred to the rocks on which the mine buildings were constructed). The underlying rocks without metal ores were originally termed the 'Rothe Todte Liegende' by Mylius (1720), literally the 'red dead underlayer', the ore-free Permian beds of the old German mining terminology. The term was later shortened to Rotliegend. Thus, inclusion of the Rotliegend and Zechstein in a single system found long precedence in German geological research (Zittel 1901) and, without doubt, facilitated later acceptance of the Permian System, which united them.

Murchison's Permian

All students of the Permian know that legendary British geologist Roderick Murchison (1792–1871) (Fig. 2) named the Permian as a result of fieldwork he undertook in Russia. This fieldwork, in 1840 and 1841, is well documented by Murchison's own publications (especially Murchison *et al.* 1845), his principal biography (Geike 1875) and Collie & Diener (2004), who published an edited and annotated version of Murchison's previously unpublished narrative description of his work in Russia.

To summarize, while in Paris in 1840, Murchison became aware of an extensive, flat-lying (little deformed) and fossiliferous Palaeozoic section in the Baltic region of European Russia. That summer he went to Russia with French palaeontologist Édouard de Verneuil (1805–73), and they were

Fig. 2. Roderick Impey Murchison, who coined the term Permian in 1841.

accompanied by Russian government official Baron Alexander von Meyendorff (1798–1865) and Russian geologist Count Alexander von Keyserling (1815–91). Their goal was to examine the older Palaeozoic strata and to confirm their succession, and, in particular, to establish further the validity of the Devonian System. The 1840 excursion visited the shores of the White Sea, then travelled SW along the Dvina River, and finally proceeded to the west and south via the Volga River to Moscow.

The same team returned to Russia the next summer, for 5 months in 1841, but this time with the financial support of the Russian Czar, ostensibly to evaluate coal resources. They travelled from Moscow east across the Russian platform via Perm to the Ural Mountains, then south along the Urals, SW to the Sea of Azov and north back to Moscow. The second trip fulfilled Murchison's original purpose, which was to recognize Silurian, Devonian and Carboniferous rocks in Russia based on their fossil content.

An incidental byproduct of the second trip was the naming of the Permian, first proposed in a letter Murchison wrote in the autumn of 1841 to the Russian government official Johann Gotthelf Fischer von Waldheim in St Petersburg. Fischer von Waldheim (1841) published this letter in Russian (it was originally written in French), and identified Murchison as its author. Later, in 1841, Murchison published an article (essentially an English translation of the letter) in the *Philosophical Magazine* establishing the Permian Period for

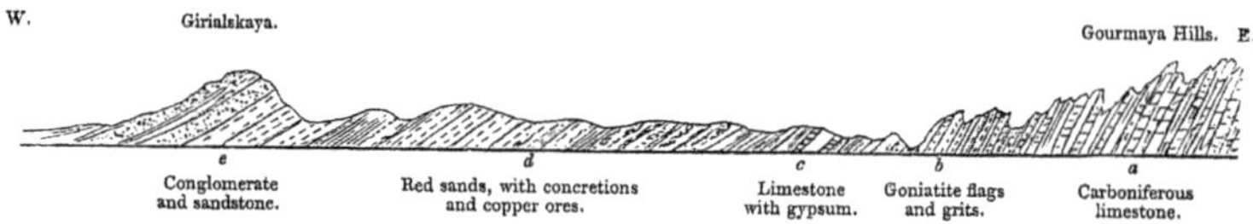

Fig. 3. The Murchison *et al.* (1845, p. 132) section of Permian strata on the Samarka River in Russia. The Carboniferous limestone on the far right is overlain by 'goniatite flags and grits' considered correlative to the 'grits of Artinsk' and of Carboniferous age. The overlying strata, beginning with the 'limestone with gypsum', were deemed Permian by Murchison *et al.* (1845).

a succession of marls, limestones, sandstones and conglomerates on the western flank of the Urals. He thus wrote:

> The Carboniferous System is surmounted, to the east of the Volga, by a vast series of beds of marls, schists, limestones, sandstones and conglomerates, to which I propose to give the name of 'Permian System', because, although this series represents as a whole, the lower new red sandstone (Rothe-todte liegende [sic]) and the Magnesian Limestone or Zechstein, yet it cannot be classified exactly . . . with either of the German or British subdivisions of this age. . . . To this 'Permian System' we refer the chief deposits of gypsum of Arzamas, of Kazan, and of the rivers Piana, Kama and Oufa, and of the environs of Orenbourg; we also place in it the saline sources of Solikamsk and Sergiefsk, and the rock salt of Iletsk and other localities in the government of Orenbourg, as well as all the copper mines and the large accumulations of plants and petrified wood, of which you have given a list in the 'Bulletin' of your Society (*anno* 1840).
>
> (Murchison 1841, p. 419)

Murchison *et al.* (1845, p. 7*) went on to affirm the significance of their 'establishing under the name of 'Permian' a copious series of deposits which form the true termination of the long Palaeozoic periods' (Fig. 3).

According to Murchison, the Permian System in Russia overlies Carboniferous rocks (including the 'grits of Artinsk') that he correlated to the British Millstone Grit. However, Murchison judged fossil fishes and amphibians from the Russian Permian to be similar to those of the German Zechstein, which supported correlation of the Russian Permian to the British Magnesian Limestone. Murchison also considered the Permian fossil plants to be intermediate between those of Carboniferous and Triassic ages. He thus equated part of the Permian to the British 'lower New Red Sandstone', which supported correlation to the German Rotliegend.

It has always been clear that Murchison's type Permian is not the entirety of the Permian of most later usage. Thus, the strata in Russia that Murchison identified as Permian are now assigned to the regional Kungurian, Ufimian, Kazanian and Tatarian stages, and thus encompass the uppermost lower Permian (Cisuralian), middle Permian (Guadalupian), upper Permian (Lopingian) and even a part of the lowermost Triassic in current usage (Fig. 4). Murchison regarded as Carboniferous the underlying strata of what are now considered the majority of the lower Permian (Cisuralian) Series. This means that the original base of the Permian *sensu* Murchison was much higher than the current base of the Permian.

Extension of that base downwards took place in two ways. First, inclusion of the central European Rotliegend in the Permian, a mis-correlation first advocated by Murchison, immediately brought strata older than the 'type' Permian into the system. Second, subsequent studies of ammonoids by Russian palaeontologist Alexander Karpinsky

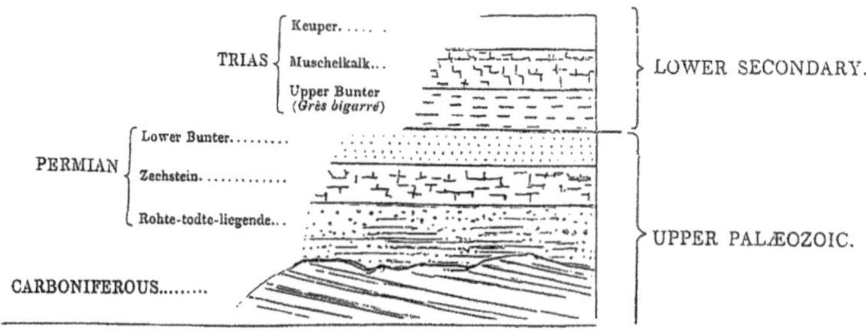

Fig. 4. Woodcut diagram showing the inferred European equivalents of the Russian type Permian (from Murchison *et al.* 1845, p. 204). Note that the 'Rothe-todte-liegende' is actually older than the Russian type Permian, and that most of the 'Lower Bunter' is Triassic.

(1847–1936) (see especially the monograph of Karpinsky 1889) included the Russian Artinskian strata (the 'grits of Artinsk', considered Carboniferous by Murchison) and its much later recognized, older subdivisions, the Asselian and Sakmarian, in the Permian (see below).

Global Permian System

The Permian of Murchison (1841) soon gained wide use. Thus, in 1858, Permian rocks and fossils were recognized in North America (Foster 1989). Blanford *et al.* (1856) identified them in India, Bain (1856) in South Africa and von Richthofen (1888) in China. Indeed, the idea of 'Gondwánaland' published by Austrian geologist Eduard Suess (1831–1914) in his classic book *Das Antlitz der Erde* (1885) was based in part on his recognition of Permian rocks in Australia, India and Africa.

Despite the acceptance of Permian by many, alternative names for the system were proposed throughout the 1800s. Prior to Murchison's 1841 name Permian, Belgian geologist Jean Baptiste Julien d'Omalius d'Halloy (1783–1875), in his 1834 book *Elemente der Geologie,* had proposed the name 'Terrain Penéen' to refer to the Rotliegend plus Zechstein strata. According to d'Omalius d'Halloy, the term 'penéen' referred to the poor fossil record of these rocks. In 1859, French-American geologist and palaeontologist Marcou (1824–98) proposed the term Dyas ('two parts') as a supposedly more appropriate term than Permian (Marcou 1862; Murchison 1862).

For those unable or unwilling to separate the Carboniferous and Permian, German geologist and palaeontologist Wilhelm Waagen (1841–1900) combined them into one system that he named Anthracolithic, because of the 'intimate connection between the two systems' (Waagen 1891, p. 294). In 1896, American stratigrapher Charles Rollin Keyes (1864–1942) proposed the term Oklahoman as a North American term to replace Permian (Keyes 1896). Of these alternative terms, only Dyas achieved limited use – Penéen, Anthracolithic and Oklahoman were quickly forgotten.

Accepting the Permian as a separate system seems to have been most difficult among North American stratigraphers, who generally combined it with the Carboniferous until about the time of World War II. Thus, the United States Geological Survey long recognized a Carboniferous System divided into three series – Mississippian, Pennsylvanian and Permian (Wilmarth 1925); and some American workers used Anthracolithic well into the 1900s (e.g. Prosser 1910; Wheeler 1934). As an aside, and perhaps most unusual, was British stratigrapher R.L. Sherlock (1928, 1947), who advocated combining most of the Permian and most of the Triassic into one system he called Epiric, but this gained no followers.

Acceptance of the Permian as a separate system seems to have been universal by World War II. In the USA, during the late 1930s, the American Association of Petroleum Geologists formed a committee devoted to Permian stratigraphy. That committee (Adams *et al.* 1939; Tomlinson *et al.* 1940) recognized Permian as a separate system/period and proposed a North American standard for Permian time that remains the basis for the provincial series/stages still in use (see below). In 1941, the United States Geological Survey finally recognized the Permian as a separate system (Cohee 1960). With that late recognition by American geologists, the Permian came into universal use as a distinct system/period of the geological timescale.

Permian stages

Permian stage terminology began with Karpinsky (1874), who used the term Artinskian to refer to the ammonoid-rich clastic succession of strata immediately below Murchison's type Permian (the 'grits of Artinsk' of Murchison *et al.* 1845). Almost all other Permian stage terminology was developed by about 1976.

This development took place primarily in three separate countries, at least in terms of stage names that have attained global usage. Thus, Russian regional stages of the type Permian – the Kungurian, Ufimian, Kazanian and Tatarian – were coined. The early Permian stages Asselian and Sakmarian were mostly developed by the Soviet expert on late Palaeozoic ammonoids, V.E. Ruzhentsev (1899–1978) in the 1930s–50s (see below). Chinese Permian stage names began with Huang (1932). The North American series terms (often treated as stages) were introduced by Adams *et al.* (1939); and Tethyan stage names were originally based on work in Tajikistan by Miklucho-Maklay (1958).

Likharev (1959), Glenister & Furnish (1961), Furnish (1973) and Waterhouse (1976, 1978) provided useful reviews of Permian stage terminology. They presented three Permian chronostratigraphic scales similar to each other but different in detail (Fig. 5). It is fair to say that these scales reflect the state of Permian chronostratigraphy when the SPS began its work.

We review the development of both global and regional Permian stage terminology below. Important points are:

- Almost all of the Permian stages developed originally as concepts based on either ammonoid or fusulinid biostratigraphy.
- Many of the Permian regional stages are routinely used within the regions for which they were created.

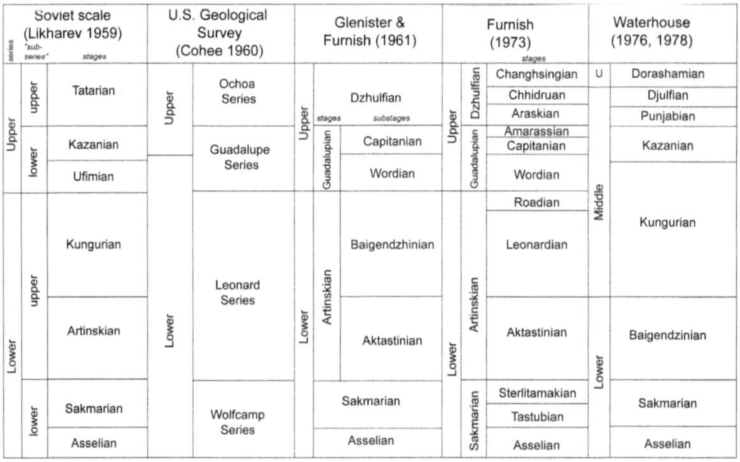

Soviet scale (Likharev 1959) series	"sub-series"	stages	U.S. Geological Survey (Cohee 1960)		Glenister & Furnish (1961)	stages	substages	Furnish (1973)		stages	Waterhouse (1976, 1978)	
Upper	upper	Tatarian	Upper	Ochoa Series	Upper		Dzhulfian	Upper	Dzhulfian	Changhsingian	U	Dorashamian
										Chhidruan		Djulfian
										Araskian		Punjabian
	lower	Kazanian		Guadalupe Series		Guadalupian	Capitanian		Guadalupian	Amarassian		Kazanian
										Capitanian		
		Ufimian					Wordian			Wordian		
Lower	upper	Kungurian	Lower	Leonard Series	Lower	Artinskian	Baigendzhinian	Lower	Artinskian	Roadian	Middle	Kungurian
										Leonardian		
		Artinskian					Aktastinian			Aktastinian	Lower	Baigendzinian
	lower	Sakmarian		Wolfcamp Series		Sakmarian			Sakmarian	Sterlitamakian		Sakmarian
										Tastubian		
		Asselian				Asselian				Asselian		Asselian

Fig. 5. Some of the Permian chronostratigraphic scales available when the SPS began its work in the 1970s. Note that this is not intended to be a precise correlation chart of the different scales. Only parts of the five scales are approximately correlative to each other.

- The surfeit of Permian stage concepts/names provided ample material from which to select a set of global stages with generally excellent boundary stratotype sections.

The SPS timescale

The SPS was authorized in 1972 by the ICS at its Montreal meeting, and had its inaugural meeting in 1975 at the 8th International Carboniferous Congress in Moscow (Glenister & Nassichuk 1998). Its newsletter, soon to be named *Permophiles* by Yugan Jin, first appeared in 1978. The first vice-chairs of the SPS were S.V. Meyen (1936–87), W.W. Nassichuk and D.L. Stepanov. In 1980, Brian Glenister (1928–2012) was added to the list of vice-chairs. At the time of the first *Permophiles*, in 1978, the SPS had 23 members. It was in that very first *Permophiles* that Meyen wrote, 'presently it is absolutely impossible to suggest a unified Permian timescale, even tentative'. Much has happened since that pronouncement (Fig. 6).

Discussion of the number of Permian series began quickly, with Grant (1979) and Waterhouse (1979) who supported a three-series Permian, and by Movshovich (1980) who opposed it, favouring the priority of a two-series Permian. In *Permophiles* 4, Glenister published a 'call to action' that urged improved communication between Permian specialists, further recruiting of SPS members and the development of projects, especially those devoted to defining stage stratotypes. Indeed, by *Permophiles* 5, it was clear that Glenister's call had been heeded, as various working groups had been created, including one headed by Grant to search for Middle Permian stratotypes in the American Southwest. Furthermore, in 1983, Nassichuk reported in *Permophiles* 7 on the activities of Carboniferous–Permian and Permian–Triassic boundary working groups, as well as advances in the search for Middle Permian stratotypes in Texas (Cys 1983 recommended early a Guadalupian stratotype in the Guadalupe Mountains) and Upper Permian stratotypes in China. In *Permophiles* 8 (1984), Nassichuk even anticipated agreement on all Permian chronostratigraphic boundary stratotypes by the time of the International Geological Congress in 1988.

In 1985, new SPS officers were Shen Jinzhang (Chair), J.M. Dickins (1923–2005) (Vice-chair) and Jin Yugan (1937–2006) (Secretary). There were 16 voting ('titular') members and 66 corresponding members of the SPS. In *Permophiles* 12 (1987), the newsletter changed from a 'news-and-views' publication to a more content-rich format, when it first published correlation charts. In the next issue, the first Permian stage stratotype proposal was published, to define the top of the Permian (base of the Triassic) at Selong in Tibet (Wang *et al.* 1988) (Fig. 6). Indeed, *Permophiles* 13 contained many lengthy research articles. It also announced the creation of SPS working groups on Permian magnetostratigraphy and on the continental Permian.

New officers of the SPS in 1989 were Jin Yugan (Chair), Vice-chairs Boris Chuvashov and Dickins, and Secretary John Utting. Yancey & Yang (1989) proposed that the type section of the Road Canyon Formation in Texas serves as the unit stratotype of the Roadian Stage. In 1991, Glenister published the first (brief) advocacy of using Guadalupian for the Middle Permian series, with three component

2013 new GSSPs base Sakmarian, Artinskian proposed

2012 new GSSP base Kungurian proposed

2007 new GSSP base Sakmarian proposed

2005 SPS Permian timescale

2004 new GSSP base Artinskian proposed

2003 GSSP base Changxingian proposed

2002 new GSSPs bases Sakmarian, Artinskian, Kungurian proposed

2001 ICS approves Guadalupian Series

2000 GSSPs bases of Asselian, Roadian, Wordian, Capitanian approved

1999 GSSP base Sakmarian proposed

1999 GSSPs bases of Roadian, Wordian and Capitanian proposed

1995 GSSP base of Asselian proposed

1994 First SPS chronostratigraphic scale

1993 Base of Lopingian proposed in China or Iran

1992 First conodont biostratigraphy article appears in *Permophiles*

1991 Guadalupian Series with three stages proposed

1988 First boundary stratotype proposed (base of Triassic)

1983 Guadalupian stratotype in Guadalupe Mountains proposed

1978 First issue of Permophiles

1975 SPS inaugural meeting

1972 SPS authorized by ICS

Fig. 6. Some milestones in the history of the SPS.

stages – the Roadian, Wordian and Capitanian (Glenister 1991).

In 1992, new officers of the SPS were Jin (Chair), Vice-chairs Chuvashov and Claude Spinosa, and Utting (Secretary). In that year, the first article on Permian conodont biostratigraphy to appear in *Permophiles* was published in issue 21.

It is fair to say that conodont biostratigraphy revolutionized the development of Permian chronostratigraphy during the 1990s (see below). Prior to that time, all Permian chronostratigraphic concepts were rooted in ammonoid and/or fusulinid biostratigraphy. However, the SPS in the 1990s began to define chronostratigraphic boundaries that had the FAD of conodont species or subspecies as primary signals for correlation in perceived evolutionary lineages, and the pages of *Permophiles* reflected that.

The next year saw Glenister (1993*a*) again publishing a call to action, urging resolution of Permian boundary stratotypes. That year, the primary working groups of the SPS were dedicated to finding boundary stratotypes for all the Permian chronostratigraphic units. Thus, Jin *et al.* (1993) advocated defining a Lopingian base in China or Iran. Glenister (1993*b*) published a letter from the United States National Park Service guaranteeing unfettered access to the Guadalupian stage stratotypes in the Guadalupe Mountains National Park of West Texas.

In 1994, *Permophiles* 24 reported on an informal questionnaire that indicated broad support (but some vocal opposition) to the Guadalupian, and Ross & Ross (1994) published one of the last Permian (or Carboniferous) stages to be named, the Bursumian (it had been used earlier by Ross & Ross 1987 without definition). Chernykh & Ritter (1994) also published the first analysis of the conodont record across the Carboniferous–Permian boundary at the Aidaralash section in western Kazakstan. This section became the GSSP for the base of the Asselian (Permian), and the recognition of the evolution of nodose streptognathodids, first articulated by Chernykh & Ritter (1994), ultimately became the primary signal for correlation of the GSSP.

Jin *et al.* (1994*a*, *c*) presented the first complete Permian chronostratigraphic scale for the Permian based on the work of the SPS (Fig. 7). They divided the Permian into four series (Uralian, Leonardian (or Chihsian), Guadalupian and Lopingian). The Leonardian was divided into the Hessian (Ross & Ross 1987) and Cathedralian (Ross 1986) stages, and the Lopingian into the Wuchiapingian and Changhsingian stages. The new scale was soon critiqued at length, particularly with regard to: its middle series, the Guadalupian, as not reflecting well the major Permian events in the evolution of ammonoids and fusulinids (Bogoslovskaya & Leonova 1994); because its compound character (it utilized Russian, American and Chinese standards) insured problems of correlation between regions (Leven 1994); the supposition that the Tethyan scale is superior to the American scale for the Middle Permian (Davydov 1994); problems with the Guadalupian correlation of an American standard to Tethys (Kotlyar 1995); and the assertion that the Guadalupian reef and associated strata are not a good facies for chronostratigraphic definition (Taraz 1995). Shevelev *et al.* (1996) represented well the viewpoint that the Russian chronostratigraphic scale long in use was sufficient for the Permian; they even argued that continental stratotypes are superior to marine. Archbold & Dickins (1997) also argued for the priority of retaining a two-series Permian (also see Dickins 1994).

Davydov *et al.* (1995) published a proposal to place the GSSP of the Asselian base (=base of Permian System) at Aidaralash Creek in western Kazakstan, the first complete GSSP proposal of the SPS (Fig. 6). They noted that an informal vote of SPS working group members approved this, with 21 in favour, zero opposed and two abstentions. A later formal vote of the titular members of the SPS approved the proposal with 15 for and one abstention.

In June 1996, new SPS officers took over: Bruce Wardlaw (1947–2016) (Chair), Ernst Leven and Clint Foster (Vice-chairs), and Spinosa continued as Secretary. Later in 1996, *Permophiles* 29 published diverse letters to then ICS Chairman Jürgen Remane (1934–2004) and SPS Chairman

Harland *et al.* (1982)		Harland *et al.* (1990)			Jin *et al.* (1994*c*)		Jin *et al.* (1997)	
Upper	Tatarian	Zechstein	Longtanian	Changhsingian	Lopingian	Changhsingian	Lopingian	Changhsingian
				Longtanian		Dzhulfian (Wuchiapingian)		Wuchiapingian
	Kazanian		Guadalupian	Wordian	Guadalupian	Capitanian	Guadalupian	Capitanian
	Ufimian			Ufimian		Wordian		Wordian
						Roadian		Roadian
Lower	Kungurian	Rotliegendes		Kungurian	Leonardian	Cathedralian	Cisuralian	Kungurian
						Hessian		
	Artinskian			Artinskian	Uralian	Artinskian		Artinskian
	Sakmarian			Sakmarian		Sakmarian		Sakmarian
	Asselian			Asselian		Asselian		Asselian

Fig. 7. Various chronostratigraphic scales advanced during the first two decades of the work of the SPS, including that of Jin *et al.* (1997), which is the chronostratigraphic scale in use today.

Wardlaw, further critiquing the proposed Permian chronostratigraphic scale. The letters focused on the priority of all Russian types, argued against the disproportionate length of the three proposed Permian Series (stressing the short Lopingian) and raised various issues of correlation. It could be said that these letters represented the last, unified opposition to a three-series Permian. Both Remane and Wardlaw, as well as Glenister, Jin and German magnetostratigrapher Manfred Menning, answered these criticisms, arguing that the proposed chronostratigraphic scale (Fig. 7) was the best among the alternatives and deserved to be further defined. However, objections to abandoning the Russian Upper Permian stages and a two-series Permian continued, particularly from Russian workers (e.g. Lazarev 1999).

Much of the success of the efforts of the SPS was fuelled by *Permophiles*, which, by the 1990s, had become a venue for publication of new data and analyses in journal-style articles, and of lengthy discussions by proponents of various issues regarding Permian chronostratigraphy. Glenister *et al.* (1999) first published a full proposal of the GSSPs of the base Roadian, Wordian and Capitanian, and thereby the base of the Guadalupian (Fig. 6). In the same year, Wardlaw *et al.* (1999) proposed a GSSP for the base of the Sakmarian at the Kondurovsky section in Russia.

In 1999, the SPS could only identify five reliable numerical ages by which to calibrate the Permian SGCS (see *Permophiles* 35, p. 2). That year, approval of the Guadalupian was by a small margin of 62% of the working group members (60% is the minimum necessary), whereas the GSSP for the base of the Triassic at Meishan section D in southern China received an 87% positive vote. In 2001, the ICS approved the Guadalupian and its component stages as defined by the proposed GSSPs.

Charles Henderson became SPS Secretary in 2000, and that year the 16 voting members approved the GSSPs for the bases of the Asselian, Roadian, Wordian and Capitanian, as well as the Guadalupian as the middle Permian series (Fig. 6). Lengthy (and, in part, acrimonious) discussion of differences of opinion on conodont taxonomy in *Permophiles* 40 and 41 focused on the conodont signal for correlation of the basal Lopingian GSSP (e.g. Henderson 2000; Wang 2000, 2001; Jin *et al.* 2001). Chuvashov *et al.* (2002*a*) proposed a GSSP for the Sakmarian base at the Kondurovsky section in Russia, with its primary signal for correlation being the FAD of the conodont *Sweetognathus merrilli* Kozur. The Kondurovsky section was also advanced as a possible GSSP for the base of the Artinskian, and the GSSP for the base of the Kungurian was suggested to be near Mechetlino in Russia (Chuvashov *et al.* 2002*b*). Jin *et al.* (2003) published a first proposal for the GSSP of the base of the Changhsingian (Fig. 6).

In 2004, new SPS officers were Charles Henderson (Chair), Vladimir Davydov (Vice-chair) and Shuzhong Shen (Secretary). In 2004, the Dal'ny Tulkas section in Russia was first mentioned as a possible GSSP for the base of the Artinskian. Henderson (2005, fig. 1; also see Wardlaw *et al.* 2004) presented the then agreed on international Permian timescale of the SPS (Fig. 8).

Permophiles 50 in 2007 celebrated 30 years of work by the SPS. In 2009, in *Permophiles* 54, Charles Henderson noted problems with the Kondurovsky section (e.g. reworked conodonts and an inability to confirm the LO of *Sweetognathus merrilli* in the section, as well as the fact that the LO of that taxon is detectably diachronous) that supported moving the proposed base Sakmarian GSSP to the Usolka section. He also noted that there were problems with the Mechetlino section as a proposed GSSP for the Kungurian base, including the rarity of conodonts in the proposed FAD sample, no datable ashes, reworked zircons, heavily weathered rocks and the difficulty in establishing a chemostratigraphy. Henderson suggested looking at sections in the USA (Idaho, Nevada) as possible GSSP candidates for the Kungurian base. Dal'ny Tulkas, however, remained the prime candidate section for the GSSP of the Artinskian base (see also Henderson 2010).

Thus, Henderson *et al.* (2012*b*) proposed the Rockland section near Wells, Nevada, USA, as a GSSP for the base of the Kungurian, with its primary signal for correlation being the FAD of the conodont *Neostreptognathodus pnevi*. They suggested Mechetlino as a reference section, but Chernykh *et al.* (2012) advocated Mechetlino as the GSSP candidate.

Permophiles 56 announced new SPS officers – Shen (Chair), Joerg W. Schneider (Vice-chair) and Lucia Angiolini (Secretary). It also published a new SPS timescale with much more detail and more extensive numerical calibration than previous Permian timescales. In the new SPS Chair's message, Shen urged completion of the three remaining Permian GSSPs at the top of a list of priorities.

Lucas (2013) critiqued the Aidaralash GSSP for the base of the Permian as having a primary signal for correlation (FAD of the conodont *Streptognathodus isolatus*) that has problematic taxonomy and phylogeny, and is demonstrably diachronous. Davydov (2013) replied, defending the Aidaralash GSSP. Chernykh *et al.* (2013) proposed a GSSP for the base of the Sakmarian, and Chuvashov *et al.* (2013) proposed a GSSP for the base of the Artinksian. Diverse discussions followed in *Permophiles* 59–60 (2014).

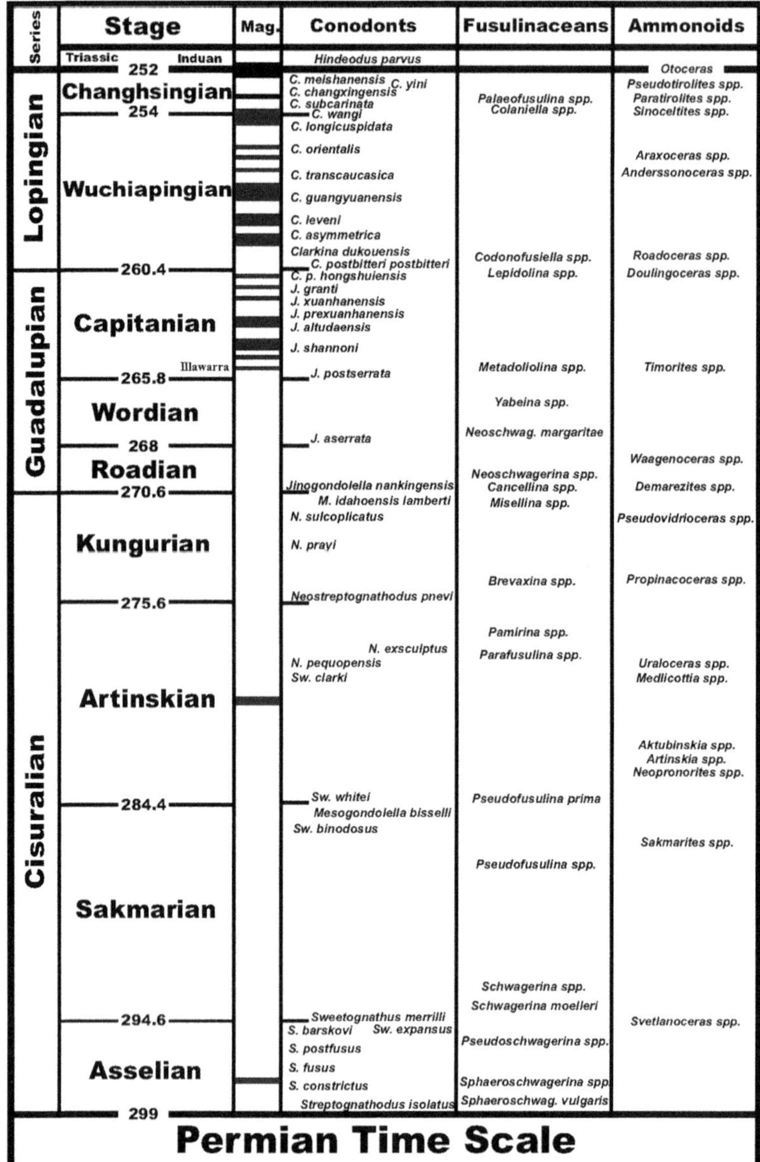

Fig. 8. The Henderson (2005) SPS Permian timescale.

At present, the SPS has 17 voting members, five honorary members and 92 corresponding members. The current SPS chronostratigraphic scale (Fig. 9) is nearly finished, recognizing three series and nine stages, six of which have ratified GSSPs for their basal boundaries. Much of the ongoing work of the SPS is focused on resolving the GSSPs for the bases of the Sakmarian, Artinskian and Kungurian stages (Fig. 6).

Subdivisions of the Permian

Introduction

As is clear from the historical review above, division of the Permian into two series – the lower and the upper, corresponding in some sense to the European Rotliegend and Zechstein – had a tremendous amount of precedence to as far back as the

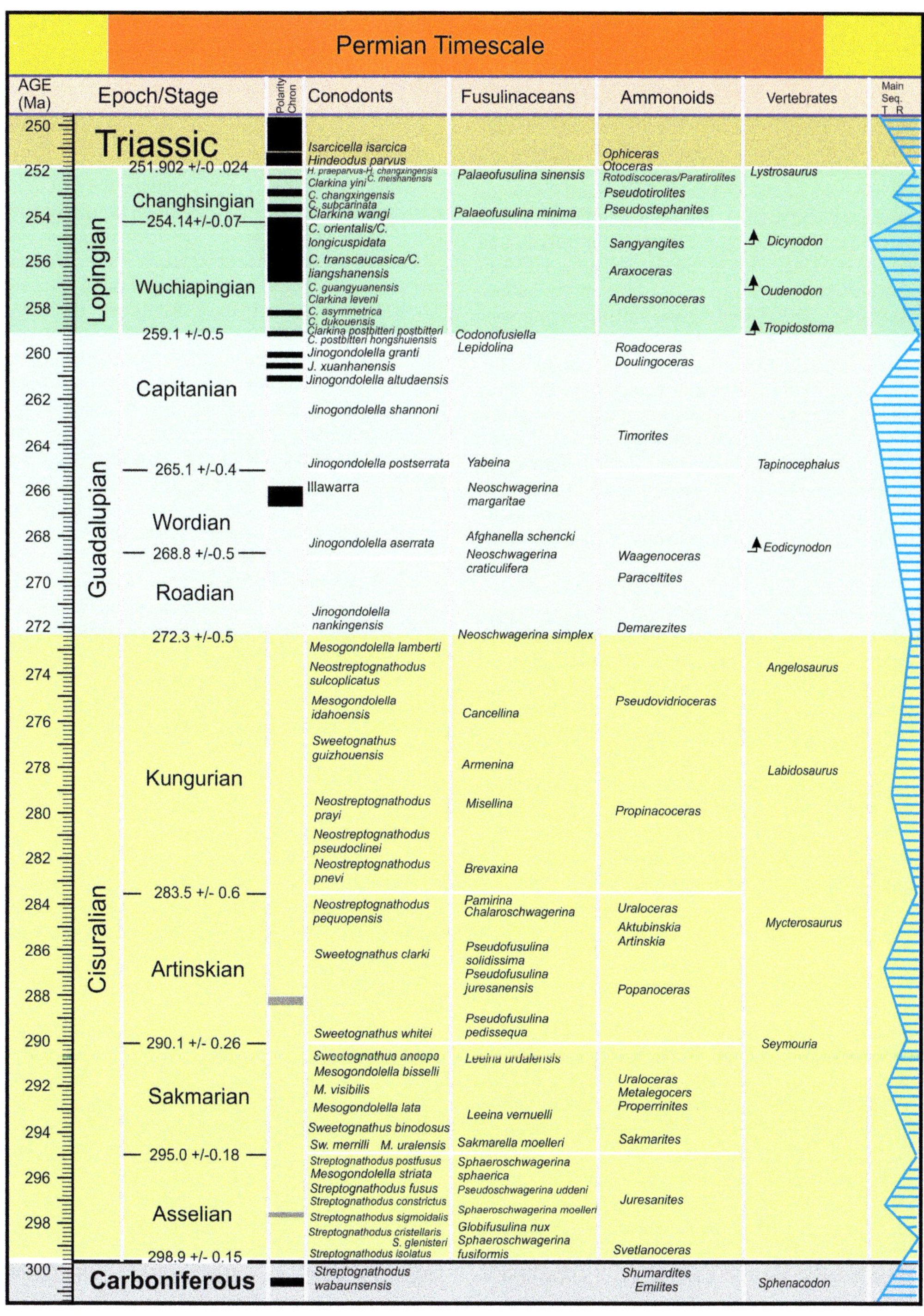

Fig. 9. Current SPS Permian timescale.

1700s. The term 'Dyas', which was coined by Marcou (1859), captured that concept, and the Permian was viewed as two series until the 1990s. Indeed, Harland *et al.* (1990, fig. 3.7) used the terms Rotliegendes and Zechstein to denote the two Permian series (Fig. 7) (also see Harland *et al.* 1982).

Occasionally, a three-fold division of the Permian had been proposed, but with a limited following. In an early expression, Munier-Chalmas & de Lapparent (1893) divided the European (primarily non-marine) Permian into Autunian, Saxonian and Thuringian (the former two = Rotliegend; the latter = Zechstein). These terms found favour among some palaeobotanists working in Europe (see below). Much later, Waterhouse (1978) suggested a pre-Kungurian Early Permian, a Kungurian–Dzhulfian Middle Permian and a Late Permian equivalent to the Griesbachian, which is now regarded as Triassic. This met with no followers. In North America, four series were identified by Adams *et al.* (1939) – the Wolfcamp, Leonard, Guadalupe and Ochoa – but the first two of these were primarily used as stages by subsequent workers.

In the 1990s, it became clear to many students of the Permian timescale that dividing the Permian into three series would better represent biotic events during the period. Thus, they advocated and ultimately agreed on a lower, middle and upper Permian series – the Cisuralian, Guadalupian and Lopingian (Fig. 7). Major changes in fusulinid and ammonoid evolution, in particular, mark the boundaries of the Guadalupian, as does the Middle Permian marine transgression and global highstand so evident in Tethys and western North America. Furthermore, the Guadalupian ends with a major marine regression and a substantial extinction.

Cisuralian Series

Waterhouse (1982) introduced the term 'Cisuralian Series' to encompass the Asselian, Sakmarian and Artinskian stages. Jin *et al.* (1997) added the Kungurian to the Cisuralian. Cisuralian is now the official term for the lower Permian series (Figs 7–9).

De Lapparent (1900) had introduced the term 'Uralian' to refer to the '*Schwagerina* horizon' in the Urals, but incorrectly correlated it to the Stephanian–Gzhelian (Upper Carboniferous). Ruzhentsev (1936) later identified the '*Schwagerina* horizon' as the Sakmarian Stage, and Gerasimov (1937) used the term Uralian for the pre-Kungurian Permian. Leven (2003) revived Uralian as a series term to encompass the Asselian and Sakmarian. But, as Henderson *et al.* (2012*a*) noted, Uralian has generally been abandoned as a little used term of varied, inconsistent usage. Cisuralian is the accepted lower Permian series term to encompass the Asselian, Sakmarian, Artinskian and Kungurian (Fig. 9).

Asselian Stage

Karpinsky (1874) first used the term 'Artinskian' to refer to much of what became the lower Permian. About half a century later, Ruzhentsev recognized two separate stages – the Asselian and Sakmarian – in the lower part of the original Artinskian.

Thus, Ruzhentsev (1937) used the term Asselian horizon, later (Ruzhentsev 1950) recommended use of an Asselian substage of the Sakmarian and then advocated an independent Asselian Stage (Ruzhentsev 1951, 1954). The stratotype section of Ruzhentsev's (1937) Asselian (Assel River near Orenburg) is a section rich in ammonoids, and Ruzhentsev defined the base of the Asselian (his Permian base) by the LOs of four ammonoid families: the Metalegoceratidae, Paragastrioceratidae, Perrinitidae and Popanoceratidae.

In 1998, the base of the Asselian (=base of the Permian) was defined by the GSSP located at Aidaralash Creek in western Kazakstan (Davydov *et al.* 1998). The primary criterion (signal) for correlation of the GSSP is the FAD of the conodont *Streptognathodus isolatus* in the *S. wabaunensis* chronomorphocline. At Aidaralash Creek, the LO of *S. isolatus* is about 6 m below the secondary signal, which is the LO of the fusulinid *Sphaeroschwagerina fusiformis*, and it is about 27 m below the traditional Asselian base determined by ammonoid biostratigraphy (Bogoslovskaya *et al.* 1995).

Sakmarian Stage

Within his Artinskian, Karpinsky (1874) recognized two stratigraphic intervals with ammonoids, a lower interval at the Sakmara River and an upper interval at the Ufa River. Furthermore, Karpinsky (1874, p. 268) had referred to the basal carbonates of the Artinskian as the Sakmarian limestone, naming it for exposures along the Sakmara River. This Sakmarian limestone also contains numerous fusulinids (inflated schwagerinids) termed early on as *Schwagerina* (later *Pseudoschwagerina*): hence, the limestone was sometimes referred to as the *Schwagerina* or *Pseudoschwagerina* horizon or zone. Karpinsky did not, however, intend to recognize a Sakmarian Stage distinct from the Artinskian (Karpinsky 1891).

Ruzhentsev (1936) restricted the Artinskian to the strata above what he called the *Pseudoschwagerina* zone and below the Kungurian base. The stratotype of the Sakmarian was the Kondurovka section, and its traditional base was the FO of the fusulinid '*Schwagerina*' (or '*Pseudofusulina*') *moelleri*. Indeed, Ruzhentsev proposed the Sakmarian for his zone of *Pseudoschwagerina* and only later (Ruzhentsev 1950, 1955) equated it to the *Properrinites* ammonoid zone.

There is no defined GSSP to define the base of the Sakmarian, one of three Permian GSSPs that remain to be agreed on. The most recent proposal is the Usolka section in southern Russia, with its primary signal for correlation being the FAD of the conodont *Mesogondolella uralensis* in the hypothesized evolutionary lineage *M. pseudostriata–M. arcuata–M. uralensis–M. monstra* (Chernykh *et al.* 2013, fig. 5).

Sakmarian has long been divided into two substages – the Tastubian and Sterlitamakian – which are sometimes treated as stages (e.g. Furnish 1973). Rauser-Chernousova (1940) introduced Tastubian (for Tastuba Mountain on the Ufa Plateau in Bashkiria) for essentially the zone of *'Schwagerina' ('Pseudofusulina') moelleri*, and to mark the LOs of the ammonoids *Metalegoceras* and *Uraloceras*. Rauser-Chernousova (1938) had earlier introduced Sterlitamakian, and Ruzhentsev (1951, 1952, 1955) treated it as a substage of the Sakmarian. In the southern Urals, the base of the Sterlitamakian is marked by the LOs of the fusulinid *Pseudofusulina uralense*, the conodont *Sweetognathus primus* and the ammonoid *Sakmarites inflatus*.

Artinskian Stage

Murchison *et al.* (1845) had referred to the 'grits of Artinsk' as the youngest strata beneath the type Permian. Karpinsky (1874) used the term 'Artinskian' to refer to these strata, a thick succession of clastic-dominated rocks, some intervals rich in ammonoids, exposed on the western flank of the Urals. The name is from the town of Arti, a settlement on the Ufa River about 200 km west of Yekaterinaberg. More specifically, Karpinsky's Artinskian was developed primarily in sandstone quarries on Kashkabash Mountain near Arti. In the 1800s, the Artinskian was generally regarded as Carboniferous (see Krotow 1885). Indeed, Murchison *et al.* (1845) had regarded the thick succession that Karpinsky called Artinskian as being of Carboniferous age, correlating it to the British Millstone Grit.

Years later, Karpinsky (1889) was uncertain about the position of the Artinskian, frequently referring to it as 'Permo Carboniferous', but concluding that 'this boundary [Carboniferous–Permian boundary] probably most correctly must be placed beneath the Artinskian beds' (Karpinsky 1889, p. 101: our translation from the German). He based this conclusion largely on correlation of the Artinskian strata to ammonoid-bearing units elsewhere that were already regarded as Permian. Thus, Karpinsky viewed the base of the Artinskian as a turning point in ammonoid evolution, and also thought that it was marked by the extinction of the trilobites.

There is no ratified GSSP for the base of the Artinskian. The currently proposed GSSP for the base of the Artinskian Stage is the Dal'ny Tulkas section in southern Russia, with its primary signal for correlation being the FAD of the conodont *Sweetognathus* aff. *Sw. whitei* in the hypothesized chronomorphocline *Sw.* aff. *Sw. merrilli–Sw. binodosus–Sw. ancep–Sw.* aff. *Sw. whitei* (Chuvashov *et al.* 2013).

The Artinskian has long been divided into two substages (Ruzhentsev 1956). The lowest, the Aktastinian, was introduced by Ruzhentsev (1934) as the Aktastin horizon and later (Ruzhentsev 1955) treated as a substage. It is named after the Aktasty River in western Kazakstan.

Ruzhentsev (1956) introduced the Baigendzhinian substage for the Baigendzhin region on the border of the Aktyubinsk (now Aktobe) and Orenburg districts of the southern Urals. Studies of Baigendzhinian ammonoids go back to de Verneuil (1845) and Karpinsky (1889), and the LOs of ammonoids such as *Pseudoschistoceras*, *Paraprionites*, *Sicanites*, *Atsabites*, *Paraceltites* and *Neocrimites* mark the beginning of the Baigendzhinian. Waterhouse (1976, 1978) used Baigendzhinian (consistently misspelt as 'Baigendzinian': Fig. 5) as a stage instead of Artinskian to refer to what he called the 'classic' Artinskian (i.e. the post-Sakmarian Artinskian of Karpinsky), abandoning the Artinskian as a facies term of varied usage.

Kungurian Stage

Kungurian joins Artinskian as one of the few Permian stages named before 1900. Stuckenberg (1890) named it after the Russian town of Kungur near Perm, and a stratotype section was later identified along the nearby Sylva River. The traditional Kungurian included two horizons: the Filipovian (lower) and Irenian (upper). However, because of the lack of biostratigraphically useful fossils in these horizons, Chuvashov (1994) proposed to move the base of the Kungurian down to include what had traditionally been the upper horizon of the Artinskian – the Sarana. The Sarana horizon includes conodonts (particularly *Neostreptognathodus pnevi*) by which the Kungurian base could be defined. An important point for North American workers is that moving the Kungurian base downwards brought that base to essentially the base of the Leonardian, so the Kungurian and Leonardian became correlative.

There is no agreed GSSP for the base of the Kungurian. As noted earlier, Henderson *et al.* (2012*b*) proposed the Rockland section near Wells, Nevada, USA, as a GSSP for the base of the Kungurian, with its primary signal for correlation being the FAD of the conodont *Neostreptognathodus pnevi* in a mesogondolellid lineage from *Sweetognathus* to *Neostreptognathodus*. Chernykh *et al.* (2012) advocated the Mechetlino section in Russia as the GSSP candidate, with the same conodont event as its primary signal, and they recently moved it to the Mechetlino Quarry section, which has a better rock quality for conodonts and chemostratigraphy. The new section is only about 500 m from the Mechetlino section and is only about 20 m thick, so it, too, may be a problematic GSSP candidate.

Guadalupian Series

Girty (1902, p. 368) coined the term 'Guadalupian period' as 'a regional name which shall be employed in a force similar to Mississippian and Pennsylvanian' and subsequently (Girty 1909) documented its macrofossil assemblages. Adams *et al.* (1939) used Guadalupian as a series, but some other workers (e.g. Schenck *et al.* 1941; Glenister & Furnish 1961) used it as a stage. Glenister *et al.* (1992) reviewed use of the term 'Guadalupian' and formally proposed it as a series to include the Roadian, Wordian and Capitanian stages. This proposal was ratified by ICS in early 2001 (Glenister *et al.* 1999; Henderson *et al.* 2012*a*) (Fig. 6).

The Guadalupian was the time of the most extensive marine transgression and the warmest climates of the Permian. Among ammonoids, the cyclolobids and ceratatidans first appeared during the Guadalupian. The advanced fusulinids (Verbeekinidae and Pseudodoliniacea) diversified, and the Neoschwagerinida, Polydexodinidae and Tangchienidae appeared. The goniatitids and most of the fusulinids disappeared at or by the end of the Guadalupian at a substantial mass extinction event. However, note that Zhou *et al.* (1996; also see Furnish *et al.* 2009) plotted the stratigraphic ranges of all Permian ammonoids and concluded that there was no extinction peak among ammonoids at the end of the Guadalupian.

Roadian Stage

In the Glass Mountains of West Texas, USA, the 'first limestone member' of the Word Formation became the Road Canyon Member (Cooper & Grant 1964). It provided the basis for the Roadian Stage of Furnish (1973), who drew attention to its distinctive ammonoid assemblage, which includes diverse adrianitids, abundant *Eumedlicottia burckhardti* and *Perrinites hilli*. The base of the Roadian Stage is defined by its GSSP in Stratotype Canyon, Guadalupe Mountains National Park, Texas, USA, with its primary signal for correlation being the FAD of the conodont *Jinogondolella nankingensis*, hypothesized to have descended from its ancestors among *Mesogondolella idahoensis* (Glenister *et al.* 1999; Mei & Henderson 2002; Henderson *et al.* 2012*a*).

Wordian Stage

Furnish (1973) first used Wordian as a stage to refer to the distinctive ammonoid fauna of the Word Formation in the Glass Mountains of West Texas, USA (also see Böse 1917). Notably, this was the base of the Guadalupian Series of Furnish (1973), marked by the appearance of the widespread ammonoid *Waagenoceras* (Fig. 5). The base of the Wordian is now defined by its GSSP at Guadalupe Pass in the Guadalupe Mountains National Park, with its primary signal for correlation being the FAD of the conodont *Jinogondolella aserrata* in a hypothesized lineage as the descendant of *J. nankingensis* (Glenister *et al.* 1999; Mei & Henderson 2002; Henderson *et al.* 2012*a*).

Capitanian Stage

The Capitan limestone of Richardson (1904) was named after the famous El Capitan promontory, a mountain of reef carbonate at the southern end of the Guadalupe Mountains escarpment in West Texas, USA (Fig. 10). Miller & Furnish (1940) first used Capitanian as a biostratigraphic concept to refer to the *Timorites* ammonoid zone, followed by Ruzhentsev (1955) as the 'Capitanian complex' and by Glenister & Furnish (1961), who used it as a substage (Fig. 5). Furnish (1973) recognized the Capitanian as a stage based largely on ammonoid biostratigraphy. The base of the Capitanian is defined by its GSSP in the Guadalupe Mountains National Park, with its primary signal for correlation being the FAD of the conodont *Jinogondolella postserrata* within the hypothesized lineage from *J. nankingensis* to *J. aserrata* to *J. postserrata*.

Lopingian Series

Grabau (1923) used the term 'Loping horizon' for a limestone stratigraphically beneath the Changhsing Limestone in southern China. Huang (1932) used the Loping as a series to refer to all Chinese Permian strata above the Maokou Formation. The Lopingian of Huang (1932) is thus the oldest name for an Upper Permian series based on a relatively complete marine succession, in preference to Ochoan (Adams *et al.* 1939), Djulfian (also spelt Dzhulfian) (Schenck *et al.* 1941) and Yichangian or Transcaucasian (Waterhouse 1982). Lopingian is now the official upper Permian series name and encompasses the Wuchiapingian and Changhsingian stages (Fig. 9). The Lopingian base (=base of the Wuchiapingian) is defined by the GSSP in the Penglaitan section in southern China, with its primary

Fig. 10. The first published image of the famous Guadalupian section at El Capitan and Guadalupe Peak, now in the Guadalupe Mountains National Park of West Texas, USA (from Bartlett 1854).

signal for correlation being the FAD of the conodont *Clarkina postbitteri postbitteri* within a hypothesized evolutionary lineage from *C. postbitteri hongshuiensis* to *C. dukouensis* (Jin *et al.* 2001, 2006*a*; Henderson *et al.* 2012*a*).

The Lopingian is bracketed by two severe biotic crises: the end-Guadalupian (or pre-Lopingian) and end-Permian extinctions (Jin 1993; Jin *et al.* 1994*a*, *b*, *c*). The base of the Lopingian is also a major global regression of sea level that marks the boundary between the middle and upper Absaroka megasequences. Global correlation of the Lopingian is precise, detailed and is numerically calibrated by numerous datable ash beds in sections in southern China (Shen *et al.* 2010).

Wuchiapingian Stage

Some have attributed the name Wuchiapingian Stage to Kanmera & Nakazawa (1973), but they credit Sheng (1962) for proposing Wuchiapingian as a stage. It was named after the Wuchiaping Limestone in Hanzhong, Shaanxi Province in central China, classically considered the zone of the fusulinid *Codonofusiella* (Sheng 1963). The Longtanian (Yang *et al.* 1986) is a near synonym of Wuchiapingian named after the Longtan coal-bearing series near Nanjing City, with a type section at Tianbaoshan consisting of continental strata probably in part correlative with the Maokouan (Jin & Shang 2000).

The base of the Wuchiapingian is defined by the GSSP in the Penglaitan section in southern China that also defines the base of the Lopingian (see above). Substages of the Wuchiapingian are the Laibinian and Laoshanian, divided by the FAD of the conodont *Clarkina leveni* (Jin *et al.* 1994*a*, *c*, 1998).

Changhsingian Stage

Grabau (1923) named the Changhsing Limestone as a lithostratigraphic unit. (Note that the Pinyin transliteration is Changxing, whereas the older Wade-Giles transliteration is Changhsing.) Huang (1932) referred to it as a 'standard unit' of the Chinese Permian, and Furnish & Glenister (1970) introduced the Changhsingian Stage to identify the youngest Permian ammonoid faunas, as represented in the Changhsing Limestone of southern China (Fig. 5). Zhao *et al.* (1981) formally proposed Changhsingian as the last Permian stage. The base of the Changhsingian is defined at the GSSP at Meishan Section D, southern China, with its primary signal being the FAD of the conodont *Clarkina wangi* within the hypothesized *C. longicuspidata–C. wangi* lineage (Mei *et al.* 2004; Jin *et al.* 2006*b*; Henderson *et al.* 2012*a*).

Substages of the Changhsingian are the Baoqingian and Meishanian. They are divided by the FAD of the conodont *Clarkina changxingensis* (Jin & Shang 2000).

Base of the Triassic

The base of the Induan (base of the Triassic = end of the Permian) is defined by the FAD of the conodont *Hindeodus parvus* at the Meishan Section D in SE China (Yin *et al.* 1996, 2001; see also Yin 1996).

Regional Permian chronostratigraphic scales

Current stratigraphic practice seeks to recognize a single global stage for each interval of geological time, and each series and system base corresponds to the base of a stage (this is the SGCS). Furthermore, the definition of stages is now based on the GSSP concept and the practise of integrated stratigraphy that applies multiple datasets to the definition of chronostratigraphic units (e.g. Salvador 1994; Remane *et al.* 1996; Walsh *et al.* 2004). However, the provinciality of fossil taxa compounded by limitations of facies distributions (rarely is any taxon or facies global in extent) have hindered universal recognition and the use of a single chronostratigraphic terminology. This is why regional Permian stages, which are specific to palaeoprovinces and/or facies, are of great utility in regional correlations. Thus, there is great value in regional stages, which Cope (1996) has aptly called the 'secondary standard' in stratigraphy.

The Permian has a variety of secondary standards (Fig. 11), and these provided a rich source for the nine stages ultimately chosen to define the SGCS for the system. Here, we review the regional

SGCS		Russia	Tajikistan	China	Eur.	USA
Triassic		Vetlugan	Dorashamian	Induan	Bunt-ss	
Lopingian	Changhsingian		Dorashamian	Wuchiapingian: Meishanian, Baogomgian	Thuringian	Ochoan
Lopingian	Wuchiapingian	Vyatkian	Dzhulfian	Wuchiapingian: Laoshanian, Laibinian	Thuringian	Ochoan
Guadalupian	Capitanian	Tatarian: Severodvinian, Urzhumian	Midian	Yangsingian / Maokouan: Lengwuan	Saxonian	Capitanian
Guadalupian	Wordian	Kazanian: Povolzhian, Sokian	Midian / Murgabian	Yangsingian / Maokouan: Kuhfengian	Saxonian	Wordian
Guadalupian	Roadian	Ufimian: Sheshmian	Murgabian	Yangsingian / Maokouan: Kuhfengian	Saxonian	Roadian
Cisuralian	Kungurian	Ufimian: Solikamian; Kungurian: Irenian, Filippovian, Saranian	Kubergandian, Bolorian	Yangsingian / Chihsian: Xiangboan, Luodianian	Saxonian	Leonardian: Cathedralian
Cisuralian	Artinskian	Artinskian: Sarginian, Irginian, Burtsevian	Yakhtashian	Longlinian	Saxonian / Autunian	Leonardian: Hessian; Wolfcampian: Lenoxian
Cisuralian	Sakmarian	Sakmarian: Sterlitamakian, Tastubian	Sakmarian	Chuanshanian: Zisongian	Autunian	Wolfcampian: Nealian
Cisuralian	Asselian	Asselian: Shikhanian, Uskalykian, Sjuranian	Asselian	Chuanshanian: Zisongian	Autunian	Wolfcampian: Newwellian

Fig. 11. The most widely used Permian regional chronostratigraphic scales (after Henderson *et al.* 2012*a*). Note that this is not intended to be a precise correlation chart of the different scales to each other.

scales that remain in regular use, and note that their regional utility will guarantee their continued use (also see Menning *et al.* 2006).

Russian (East European) regional scale

The Russian Early Permian stages – the Asselian, Sakmarian, Artinskian and Kungurian – are used in the SGCS and have been reviewed earlier in this paper. The younger regional stages – the Ufimian, Kazanian and Tatarian – were long in use as the global stages of the Upper Permian Series (Fig. 5). These stages were the bulk of Murchison's type Permian, but they are based on terrestrial, marginal marine and evaporitic facies that lack age-diagnostic fossils useful for global correlation. The Ufimian, Kazanian and Tatarian remain in use in Russia as part of what is usually referred to as the 'East European scale' (e.g. Esaulova *et al.* 1998).

Nechaev (1915) named both the Ufimian and Kazanian stages, and later monographed their palaeontology (Nechaev 1921). Ufimian was named after Ufa, and Nechaev regarded it as the 'sub-Zechstein' red beds. Kazanian was named after the town of Kazan on the Volga River. Nechaev placed the base of the Kazanian at the base of the '*Spirifer* Series' (=*Licharewa* Beds).

The Ufimian has proven to be a source of much disagreement. It is divided into two horizons, Solikamsk overlain by Sheshma. But, some workers put the Solikamsk into the upper Kungurian (e.g. Kotlyar 1977; Klets *et al.* 2001), and others include the Shesma in the Kazanian (e.g. Chuvashov *et al.* 2002*b*; Kotlyar *et al.* 2004). As reviewed by Lozovsky *et al.* (2006), the Russian Interdepartmental Stratigraphic Committee abandoned the Ufimian, absorbing the Solikamsk horizon into the Kungurian, and placing the base of the Middle Permian at the base of the Sheshma horizon, which was absorbed into the Kazanian. Lozovsky *et al.* (2006), nevertheless, argued for the retention of the Ufimian as a distinctive chronostratigraphic unit. They noted that the Committee's dismemberment of the Ufimian was largely a decision to allow the Russian regional stages to make a stage boundary (Kungurian–Kazanian) correspond to the Cisuralian–Guadalupian series boundary, instead of having the series fall within a stage (Ufimian; see the discussion of the Bursumian below). Nevertheless, there seems to be little disagreement on how to correlate the Ufimian horizons – the stage straddles the Cisuralian–Guadalupian series boundary (Fig. 11).

Kazanian was long correlated to the Wordian, but recent analyses suggest it also encompasses at least part of Roadian time (e.g. Leven & Bogoslovskaya 2006; Leonova 2007). In the Russian section, Kazanian encompasses two horizons: the Sokian and overlying Povolzhian. Nevertheless, the Kazanian stratotypical strata may represent very little time. One cyclostratigraphic analysis suggests that it is equivalent to as little as 6 kyr (Nourgaliev & Nourgalieva 1999).

Nikitin (1887) named the Tatarian (after the Tatar people) to refer to most of the red-bed succession at the top of Murchison's type Permian. He distinguished it by an extensive non-marine fossil assemblage of plants, bivalves, ostracods, conchostracans, fishes and tetrapods. The original Tatarian also included some non-marine strata now regarded as Triassic – the Vetluga horizon. It now encompasses three horizons (ascending), the Urzhumian, Severodivinian and Vyatkian, mostly non-marine red beds with an important fossil record of terrestrial plants, invertebrates and verebrates (e.g. Esaulova *et al.* 1998).

Each of the Russian regional stages is divided into substages. Those shown here (Fig. 11) are for the Russian platform; different substages are used in other parts of Russia (e.g. Kotlyar 2000). In general, these Russian substages correspond to horizons or svitas, 'suites', which are somewhat analogous to rock formations, but are posited to have isochronous boundaries. Some of these substages have already being used as stages or as subdivisions of the SGCS stages, such as the Tastubian and Sterlitamakian substages of the Sakmarian or the Baigendzhinian substage of the Artinskian. The most recent revision to the Russian regional scale, initiated by Grunt *et al.* (1999), proposed a Biarmian Series to encompass the Ufimian and Kazanian, and raised Tatarian to series rank and its constituent horizons to stages (Kotlyar & Pronina-Nestell 2005).

Chinese regional scale

Regional Permian chronostratigraphy in China (Fig. 11) began with the work of Huang (1932). Three series are recognized: the Chuanshanian, Yangsingian and Lopingian (e.g. Yang *et al.* 1986; Sheng & Jin 1994; Jin & Shang 2000), but the Chuanshanian was assigned to the Upper Carboniferous in China during the 1960–80s (e.g. Sheng 1962). Stage-level subdivisions are fusulinid based. Lopingian, and its stages Wuchiapingian and Changhsingian, are part of the SGCS (see above).

The Chuanshanian Series (Huang 1932) encompasses the Zisongian (Zhang *et al.* 1988) and Longlinian (Huang *et al.* 1982) stages. These stages are primarily delimited by the LO of the fusulinid *Pseudoschwagerina* (base Zisongian), the LO of the fusulinid *Pamirina davasica* (base of Longlinian) and the LO of the fusulinid *Misellina* (top of Longlinian).

The Yangsingian Series of Huang (1932) encompasses two subseries – the Chihsian and Maokouan – that were long considered stages, but were elevated to series level by Jin & Shang (2000) because of their perceived long duration. Chihsian includes the Luodianian Stage of Sheng & Jin (1994), equivalent to the stratigraphic range of the *Brevaxina dyhrenfurthi* Zone and two *Misellina* zones (Jin *et al.* 2001); and the Xiangboan Stage (Fan *et al.* 1990), which is equivalent to the *Cancellina* and *Neoschwagerina simplex* fusulinid zones.

The Maokouan includes the Kuhfengian and Lengwuan stages, defined by conodonts (Jin *et al.* 1994*a*). Thus, the base of the Kuhfengian is the LO of *Jinogondolella nankingensis*, so its base is the same as the base of the Roadian on the SGCS. The Lengwuan base is the LO of *J. postserrata*, so its base is the same as the base of the Capitanian on the SGCS. Therefore, the Maokouan is the equivalent of the Guadalupian (Fig. 11).

Tethyan regional scale

Fusulinacean biozonation in the Pamirs of Tajikistan has long provided a standard for Tethyan shallow-marine correlations (Fig. 12). The first scheme, that of Miklucho-Maklay (1958), divided the Tajik Permian into (in ascending order) the Darvasian, Murgabian and Pamirian stages. Leven (1963) added the Kubergandian, and in 1975 named the Chihsian (based on Chinese strata). Leven (1979) replaced the Chihsian with the Bolorian, and Leven (1980) added the Yaktashian. All of these stages have stratotypes in Tajikistan, either in the Darvas (a district of south-central Tajikistan) or the Pamir Mountains (e.g. Leven 1967, 1975, 1993; Grunt & Dmitrieva 1973).

These stages are fusulinid defined, and thus have been treated as fusulinid stages for the Eurasian Tethys province (e.g. Leven 2003). For the Cisuralian, the Asselian and Sakmarian are applied in Tajikistan, followed by the Yaktashian, named after Mount Yak Tash. The Bolorian of Leven (1979) begins with the FAD of the fusulinid *Misellina*. At its simplest, the Murgabian is the *Neoschwagerina* genozone, whereas the Midian is the *Yabeina–Lepidolina* genozone.

The Kubergandian, Murgabian and Midian are approximately equivalent to the Guadalupian (Figs 11 & 12). In fusulinid evolution, the Kubergandian–Midian is the time of the diversification of the Neoschwagerinida. The Midian saw the diversification of Schubertellida, the last diversification of Schwagerinida and the appearance of Ozawainellida.

Most recently, Leven (2003) used four series in the Tethyan chronstratigraphy: Uralian (Asselian + Sakmarian), Darvasian (Yaktashian + Bolorian), Yangsingian (Kubergandian, Murgabian + Midian) and Lopingian (Dzhulfian + Dorashamian). Kotlyar & Pronina (1995) reviewed the history of naming and refining the Tethyan Permian scale (Fig. 12), noting the various modifications and problems with the ranges of fusulinid zones. Correlation of the Tethyan scale to the SGCS continues to be worked on (Angiolini & Henderson 2013; Anglioni & Vachard 2015), as this is a valuable regional scale in Permian Tethyan correlations from southern Europe to south China.

Western European regional scale

As mentioned earlier, use of Autunian, Saxonian and Thuringian has gained some traction with European workers, especially palaeobotanists. However, these terms are neither a suitable chronostratigraphic division of the Permian System nor useful for correlation with non-European regions or with the SGCS (Schneider 2001). Despite this, Broutin *et al.* (1999), for example, indicated that significant 'Autunian' floras (characterized by the presence of

<table>
<tr><th>SGCS series</th><th>Miklucho-Maklay (1958)</th><th>Leven (1963)</th><th colspan="2">Leven (1967, 1975)</th><th>Grunt & Dmitrieva (1973)</th><th>Leven (1979, 1980)</th><th>Leven (1993)</th><th>Leven (2003)</th><th>series</th></tr>
<tr><td rowspan="3">Lopingian</td><td rowspan="3">Pamirian</td><td rowspan="5">Pamirian</td><td rowspan="5">Pamirian</td><td rowspan="3">Dzhulfian</td><td rowspan="2">Dzhulfian</td><td rowspan="2">Dzhulfian</td><td rowspan="4">Dzhulfian</td><td>Dorashamian</td><td rowspan="3">Lopingian</td></tr>
<tr><td rowspan="2">Dzhulfian</td></tr>
<tr><td rowspan="5">Murgabian</td><td rowspan="3">Midian</td></tr>
<tr><td rowspan="9">Guadalupian</td><td rowspan="4">Murgabian</td><td rowspan="2">Capitanian</td><td rowspan="3">Midian</td><td rowspan="8">Yangsinigian</td></tr>
<tr><td rowspan="2">Midian</td></tr>
<tr><td rowspan="4">Murgabian</td><td colspan="2" rowspan="4">Murgabian</td><td rowspan="4">Murgabian</td></tr>
<tr><td rowspan="3">Murgabian</td><td rowspan="3">Murgabian</td></tr>
<tr><td rowspan="6">Darvasian</td><td rowspan="6">Darvasian</td></tr>
<tr></tr>
<tr><td rowspan="3">Kubergandian</td><td colspan="2">Kubergandian</td><td rowspan="2">Kubergandian</td><td rowspan="2">Kubergandian</td><td rowspan="2">Kubergandian</td></tr>
<tr><td colspan="2" rowspan="3">Chihsian</td></tr>
<tr><td rowspan="2">Bolorian</td><td rowspan="2">Bolorian</td><td>Bolorian</td><td rowspan="2">Darvasian</td></tr>
<tr><td>Cisuralian</td><td>Artinskian</td><td>Yaktashian</td></tr>
</table>

Fig. 12. Evolution of Tethyan regional Middle–Upper Permian chronostratigraphy (modified from Kotlyar & Pronina 1995).

Autunia, *Rhachiphyllum*, *Lodevia*, *Arnhardtia* and *Gracilopteris*) are widespread and well known in many continental basins of Europe, and should be regarded as characteristic of the continental Carboniferous–Permian (latest Ghzelian–early Sakmarian) transition (Fig. 11). Therefore, even though the temporal boundaries of the Autunian are not defined in the typical Autun Basin, it may have some biostratigraphic value in palaeobotanical biostratigraphy. Nevertheless, the problematic nature of the Autunian is underscored by DiMichele *et al.* (2013), who recently judged a North American flora Autunian, although it is directly associated with conodonts and fusulinids of early late Pennsylvanian (Kasimovian–Missourian) age (Lucas *et al.* 2011).

'Saxonian' has a long history of vague and inconsistent usage, so it should be abandoned (e.g. Kozur 1986, 1993; Cassinis *et al.* 1992, 1995; Lucas *et al.* 2006). In the type area it seems to be equivalent to more than six stages, from Sakmarian to basal Wuchiapingian (Fig. 11). Thus, 'Saxonian' is too extensive temporally to be of chronostratigraphic value.

'Thuringian' has up to now been used in different countries of Europe to refer to both marine and continental strata, which often bear various palaeontological elements that generally pertain to the Upper Permian. The type area of this unit, which represents an equivalent of the German Zechstein, corresponds to the Wuchiapingian and early Changhsingian (Kozur 1988, 1993) (Fig. 11). However, the Thuringian has been subject to different stratigraphic interpretations, some equating it to the Russian Kazanian and Ufimian, so it should be abandoned.

The Dyas has been reintroduced in Germany, but not as a chronostratigraphic term. Instead, it is a lithostratigraphic supergroup that encompasses the Rotliegend and Zechstein (Schneider 2000; Menning *et al.* 2006).

North American regional scale

As noted earlier, Adams *et al.* (1939) introduced a set of North American standard series for Permian time (in ascending order): the Wolfcamp, Leonard, Guadalupe and Ochoa series (Fig. 11). However, the Wolfcamp Formation in the Glass Mountains of West Texas provided a most problematic basis for the Wolfcampian, as this unit has a basal unconformity and another substantial unconformity separating its two currently recognized divisions, the Neal Ranch and overlying Lenox Hill formations (e.g. Ross & Ross 2003). Most American fusulinid biostratigraphers have long divided the Wolfcampian into three substages, early, middle and late – also termed Newwellian (or Bursumian), Nealian and Lenoxian (e.g. Thompson 1954; Dunbar *et al.* 1960; Wilde 1990, 2002, 2006). However, the Newwellian is not part of the type Wolfcampian, but instead is at most represented by the fusulinid assemblage from the upper grey limestone member of the Gaptank Formation, beneath the sub-Wolfcampian unconformity. Thus, there are two concepts of the Wolfcampian, the original more restricted sense of Adams *et al.* (1939) and a fusulinid-based concept that includes an interval older than the base of the type Wolfcampian.

The Leonard Series of Adams *et al.* (1939) was based on the Leonard Formation in the Glass Mountains. This unit now comprises the Hess Formation, the overlying Cathedral Mountain Formation and lateral equivalents. These stratigraphic intervals have provided the basis for fusulinid-based substages, the Hessian and Cathedralian (Ross 1986; Ross & Ross 1987), also sometimes used as stages. Thus, Furnish (1973) used Leonardian as a restricted stage, dividing the Leonard Series into the Aktastinian (=lower Leonard), Leonardian (=upper Leonard) and Roadian (=Road Canyon Formation) stages (Fig. 5). Ross (1986, p. 548) introduced the Cathedralian Stage to replace the restricted Leonardian Stage of Furnish (1973), and Ross & Ross (1987) proposed the Hessian Stage to replace the Aktastinian.

The Guadalupian and its three North American stages (Roadian, Wordian and Capitanian) were discussed above, as they provide the international standard for most of Middle Permian time. Only the topmost part of the Guadalupian, equivalent to the *Jinogondolella granti* and *Clarkina postbitteri hongshuiensis* in southern China, is missing or represented by the evaporite deposits in the basal part of the Castile Formation in West Texas, USA.

The Ochoa Series of Adams *et al.* (1939) was based on a very thick (up to 1700 m), but evaporite-dominated, section in West Texas–SE New Mexico (in ascending order: the Castile, Salado, Rustler and Dewey Lake/Quartermaster formations). Ochoan strata yield very few fossils of biostratigraphic value, so Lucas & Anderson (1994, 1997) advocated recognizing an Ochoa Group as a lithostratigraphic unit, rather than as a chronostratigraphic unit.

The Schenck *et al.* (1941) critique of the Permian chrononostratigraphy of Adams *et al.* (1939) and Tomlinson *et al.* (1940) is often overlooked. They objected to equating the Wolfcampian base, and unconformity, with the Permian base, stating that this 'is not a progressive move and is not likely to be a stable placement' (p. 2197). They thus noted in the Texas section that *Schwagerina* s. str. appears before *Pseudoschwagerina* s. str., so that placing the base at the LO of *Pseudoschwagerina* in the Texas section equated it to that base in Russia, which to them was the Sakmarian base. Schenck *et al.* (1941) also recommended that Guadalupian be a stage, not a series, and they objected to using the evaporite-dominated section of the Ochoan as a chronostratigraphic standard. They thus proposed a three-series Permian in which the Upper Permian encompasses the 'Penjabian' (Punjabian) of de Lapparent (1893) and the Djulfian, a term they introduced to refer to the '*Prototoceras* beds' of Armenia. Furthermore, Schenck *et al.* (1941) objected to using lithostratigraphic formation names (Wolfcamp, Leonard) as the names of chronostratigraphic units, although this was and remains common practice.

In North America, the Carboniferous (Pennsylvanian)–Permian boundary has long corresponded to the base of the Wolfcampian Stage (Virgilian–Wolfcampian boundary). The establishment in western Kazakstan of the GSSP for the Carboniferous–Permian boundary (Davydov *et al.* 1998) forced the position to change in the North American regional scale. Correlation of this new boundary to the North American fusulinacean zonation indicated that the base of the Permian is close to the LO of the inflated schwagerinids (*Paraschwagerina* or *Pseudoschwagerina*), which is a level within the Wolfcampian (e.g. Baars *et al.* 1994*a*; Wahlman 1998; Wilde 2002, 2006).

Thus, the SPS Carboniferous–Permian boundary corresponds to the lower–middle Wolfcampian boundary of earlier usage. In the SGCS, each system base corresponds to the base of a stage, so to many American workers the secondary standard provided by the North American regional stages should also have the Carboniferous (Pennsylvanian)–Permian System boundary correspond to the base of a stage, although this requires some modification or redefinition of the regional stages. Baars *et al.* (1992, 1994*a*, *b*), working in Kansas, proposed to solve this problem by redefining the highest part of the Pennsylvanian Virgilian Stage to encompass strata previously included in the lower Wolfcampian. A second solution, advocated by Ross & Ross (1994), was to recognize a new, uppermost Carboniferous Bursumian Stage equivalent to the lower Wolfcampian of earlier usage.

Lucas & Wilde (2000; see also Davydov 2001; Lucas *et al.* 2001) critiqued the Bursumian Stage as lacking an adequate stratotype and representing only one fusulinid zone. This was culminated by Wilde (2002, 2006), who rejected Bursumian as a stage and called the same fusulinid concept the Newwellian Substage of the Wolfcampian.

Other regional scales

There are a variety of other Permian regional stage terms, too many to review here, and most are of very local significance or synonymous with other terms more widely used. These include:

- The regional Permian chronostratigraphy used only in New Zealand (e.g. Waterhouse 1967; Carter 1974).
- The Araksian Stage (for strata with diverse araxoceratid ammonoids at Araks near Dzhulfa in Transcausia) and Amarassian Stage (ammonoid fauna from western Timor, Indonesia) introduced by Furnish (1973).
- The Chidruan (also spelt Chihdruan) Stage for strata in the Pakistani Salt Range with a characteristic ammonoid fauna (Gerth 1950; Furnish 1966, 1973), as well as the overlying Punjabian ('Penjabian') of de Lapparent (1893; also see Stepanov 1973).
- Stages/series used for the Kansas (USA) Permian, the Gearyan, Cimarron and Custer (O'Connor 1963; O'Connor *et al.* 1968), which have been long abandoned (Rasscoe & Baars 1972; Baars 1990).
- The Svalbardian Stage, which united the Kungurian and Ufimian (Stepanov 1957, 1973), or the Paykhoyan Stage for marine equivalents of the Ufimian (Ustritsky 1971).
- The Dorashamian Stage (Rostovtsev & Azaryan 1973) for the uppermost Permian strata in Transcaucasia, still used in the Tethyan chronostratigraphic scale of Leven (2003) (Fig. 12).

Discussion

This review of the historical development of the Permian chronostratigraphic scale reveals some issues of stratigraphic philosophy and procedure that have been central to developing a Permian timescale, and will continue to be factors in its future refinement. Here, we briefly comment on these issues.

Stability and priority

There are no strict rules of priority in chronostratigraphic nomenclature, as there are in zoological nomenclature. However, most stratigraphers do attempt to employ the oldest term proposed for a time interval, redefining those terms as needed in the light of current data.

Dividing the Permian into two series had tremendous priority extending back to the beginning of stratigraphic research in the 1700s (see earlier in this paper). Therefore, as discussed above, there was much opposition based on priority to a three-series Permian. However, it is clear that a three-series Permian better represents the main biotic and abiotic events of Permian history than does a two-series Permian, which really was little more than an artefact of the stratigraphic architecture of European Permian strata. Furthermore, many forget that a goal of timescale research must be to discriminate shorter intervals of time, to refine the timescale. Therefore, *a priori*, a three-series Permian is an advance over a two-series Permian. Indeed, a four-series Permian, as suggested by several workers (e.g. Leven 1992, 2003; Jin *et al.* 1994*c*), may be a goal of future chronostratigraphic subdivision of the Permian.

It seems that after about a quarter of a century of study and debate by the SPS, the nine Permian stage names they endorse merit continued use. None of these stages is perfect – all have their defects. Furthermore, we do not believe that current definitions of their bases should be immutable – they, too, will be subject to further scientific testing and, if deemed advisable, redefinition. However, in the interest of stability, the nine Permian stage names are long used and widely published terms that should continue to be used. Permian chronostratigraphic research now needs to focus on subdividing the nine stages into substages and defining their boundaries (Fig. 13), as these will provide an even more detailed basis for the subdivision of Permian time than do the stages.

Series	Stage	Substage
Lopingian	Changhsingian	Meishanian
		Baoqingian
	Wuchiapingian	Laoshanian
		Laibinian
Guadalupian	Capitanian	
	Wordian	
	Roadian	
Cisuralian	Kungurian	Cathedralian
		Hessian
	Artinskian	Baigendzhinian
		Aktastinian
	Sakmarian	Sterlitamakian
		Tastubian
	Asselian	Shikhanian
		Uskalykian
		Sjuranian

Fig. 13. The Permian chronostratigraphic scale showing a possible set of Permian substages. Note that no substages of the Guadalupian have been proposed.

Taxonomy

Arguments over biostratigraphy commonly reduce to arguments over taxonomy. This is because biostratigraphic correlation relies on what Lucas (2010) called the 'index taxon hypothesis' – taxon A is the same age wherever its fossils are present. This hypothesis is usually tested by comparing the distribution of taxon A to that of other taxa, or by using another method of correlation (e.g. radioisotopic dates or magnetostratigraphy) to try to demonstrate diachroneity in the distribution of taxon A. However, palaeontological taxonomists sometimes decide that taxon A is no longer a single taxon, so they split it into two or more taxa, or they synonymize a taxon with another taxon with a longer strati graphic range, thus reducing (or eliminating) its utility as an index taxon.

These kinds of taxonomic changes destabilize existing biostratigraphy and exasperate many geologists who are forced to contend with resulting changes in the chronostratigraphy (see Shaw 1969 for an excellent expression of this). However, taxonomic changes will always be with us, as new data, techniques and epistemologies are employed in palaeontological taxonomy.

To our mind, in Permian biostratigraphy this favours the ammonoids and fusulinids, whose long history of taxonomic study has tested many taxonomic concepts. Permian conodonts, in contrast, have a much younger and less tested taxonomy. We predict that further testing of their taxonomy will produce destabilizing changes that will modify existing Permian conodont biostratigraphy, and this will impact conodont-based chronostratigraphic definitions.

Ammonoid v. fusulinid v. conodont biostratigraphy

The Permian chronostratigraphic scale was originally built on ammonoid and fusulinid biostratigraphy (see the historical reviews of Furnish 1973; Douglass 1977). Some brachiopod experts have developed Permian brachiopod biostratigraphy and advocated its use in chronostratigraphic definitions (e.g. Grabau 1931; Cooper & Grant 1964; Waterhouse 1976), but Permian brachiopods have long been regarded as too provincialized to serve in defining global chronostratigraphy. In the 1990s, a movement to define Permian stage boundaries with conodonts began, and, at present, six stage boundaries (bases of the Asselian, Roadian, Wordian, Capitanian, Wuchiapingian and Changhsingian) are defined by the FADs of conodont taxa, and it is likely that the bases of the other Permian stages will also be defined by conodont datums.

For the Permian ammonoids and fusulinids, many problems of taxonomy, facies restrictions and provinciality have long been known and are well discussed in a diverse and often fractious literature. No such problems apparently beset the Permian conodont record in the early 1990s – too little was known (for some of the earliest efforts, see Gunnell 1933; Clark 1972; Behnken 1975). Thus, conodonts appeared as a 'deus ex machina' by which to define Permian chronostratigraphic boundaries.

Nevertheless, we see various problems with conodont-based Permian biostratigraphy, including: (1) as already noted, the relative youth of Permian conodont taxonomy, which remains unstable for many taxa, in particular owing to the different taxonomic approaches employed by different experts, which are based either on individual form characters or on sample populations (see a detailed discussion by Mei *et al.* 2004; Shen & Mei 2010); (2) reworking of conodonts, which is not always recognized and rarely addressed (Macke & Nichols 2007); (3) problems of facies restrictions, diachroneity and provinciality, which do affect Permian conodont distributions (Clark 1984; Mei *et al.* 1999; Mei & Henderson 2001); and (4) the invisibility of conodonts on outcrop, so that they cannot be used in the field to determine the ages of strata.

Against this, we recognize that the ubiquity of conodonts in some strata, like other microfossils, presents an advantage over macrofossils, which have much more limited records. However, the ammonoid/fusulinid-based Permian timescale is underpinned by nearly two centuries of collecting and taxonomic work on Permian ammonoids and fusulinids. It provides macrofossil-based age assignments that can be used in the field where ammonoids and fusulinids occur, and suffers from few reworking issues. Moreover, the species-level evolution through time of ammonoids and fusulinids is commonly obvious from the morphology of their shells, which record their entire life history. Instead, conodont evolution is necessarily interpreted from morphological change in a particular element among the tooth-like elements that constitute their only commonly fossilized record. In addition, Permian conodont chronostratigraphy has been based on a model of conodont species evolution by gradual anagenesis, one of several possible mechanisms by which species may originate. Perhaps that is how all Permian conodont evolution took place, but alternative models (e.g. see the 'rectangular' phylogenies of Boardman *et al.* 2009) merit further evaluation.

To a large extent, Permian conodont-based GSSPs were an answer to longstanding disagreements over taxonomy and correlation among ammonoid and fusulinid specialists. They were also an important part of developing an integrated chronostratigraphic scale. As a relatively newly studied taxonomic group, Permian conodonts did not have the perceived 'excess baggage' of ammonoids – a long history of taxonomic changes and disagreements, known provinciality, and demonstrably diachronous distributions of some taxa. Nevertheless, we are certain that further studies of Permian conodonts will reveal that they, too, have all of the 'excess baggage' of the ammonoids and fusulinids, and are not inherently superior biostratigraphic tools with which to define Permian chronostratigraphy.

Biotic events

Typically, the beginnings of stages have been defined by biotic events. GSSP definitions of stage bases now focus on one biotic event, marked by the FAD of a single taxon, in order to identify unambiguously a single point in time at a single location. However, many of the Permian stages were originally conceived of as turning points in ammonoid or fusulinid evolution – typically the appearance of new ammonoid/fusulinids families and/or the beginnings of new evolutionary diversifications of ammonoids/fusulinids. Arguments about the definition and recognition of Permian stages thus often were arguments over the timing and magnitude of a biotic event – is it significant enough to mark the beginning of a new stage?

Stage recognition also relies heavily on correlateability, both of the stage boundaries and of the stage itself. Currently, there are few problems assigning marine strata to the Permian SGCS stages based on biostratigraphic criteria. More difficult are substage correlations and, as stated earlier, these need to be the focus of further refinement of the Permian chronostratigraphic scale.

Provinciality

The Permian chronostratigraphic scale is built on fossils collected from marine strata deposited primarily around the periphery of the Pangean supercontinent (Fig. 1). Many of these rocks were deposited in the palaeoequatorial Tethyan region, extending from the European Alps to southern China (Fig. 1). This has created a problem of provinciality among some of the fossil taxa used for stage definition and characterization. Thus, key ammonoid and fusulinid taxa known in Tethys are not all known in western North America (or vice versa), and there is suspected diachroneity of some taxa across the provinces. This provinciality was, as noted earlier, one of the deficits in using ammonoids and fusulinids in Permian chronostratigraphy, and helped to advance the apparently more cosmopolitan conodonts into a pivotal role in chronostratigraphic definitions. However, provinciality of conodonts was also present during the Permian (Mei & Henderson 2001).

Thus, provinciality has created some problems for Permian timescale construction because each province may have a different evolutionary lineage, as it almost always does anywhere in the fossil record. The Permian is one of the periods with strong provinciality (e.g. Mei & Henderson 2001; Shen *et al.* 2009, 2013*b*) compared to some other time intervals, such as the Late Devonian and Mississippian. So, the correlation between different realms is always a substantial problem (e.g. Gondwanan, Boreal and Tethyan realms) and will sustain the regional Permian chronostratigraphic scales.

Radioisotopic ages, magnetostratigraphy and chemostratigraphy

Nearly two centuries of Permian biostratigraphic research have produced remarkably detailed and reasonably precise correlations within a nearly complete, stage-level chronostratigraphic framework. Nevertheless, the biostratigraphy has its limitations, mostly due to facies restrictions of fossils, diachroneity often related to provinciality, variable evolutionary turnover rates and continuous developing (improving?) taxonomy. The use of other chronological tools to calibrate and correlate Permian rocks is thus critical to refinement of a Permian timescale.

Fortunately, the Permian was a time of diverse magmatism, especially in Eurasia. This is leading to a plethora of radioisotopic ages, many from ash beds that can be directly correlated to marine biostratigraphy (e.g. Ramezani *et al.* 2007; Shen *et al.* 2010; Schmitz & Davydov 2012). However, in contrast, much of the Permian was a time of relatively little apparent activity of the magnetic pole – the Kiaman superchron of the Pennsylvanian–Middle Permian.

The field became active again during the Wordian (*c.* 266 Ma), the Illawara event or reversal, and this was the beginning of the Permian–Triassic mixed superchron of frequent polarity changes during the Late Permian and Triassic (Embleton *et al.* 1996; Jin *et al.* 1999; Vozarova & Tunyi 2003; Isozaki 2009). The Illawara event also provides a valuable tie point for the correlation of marine and non-marine Permian strata (Lucas 2004, 2006; Lucas *et al.* 2006).

The last decade has also seen great advances in chemostratigraphy of the Permian, particularly of carbon and strontium isotopes (e.g. Isozaki *et al.* 2007; Wignall *et al.* 2009; Chen *et al.* 2011; McArthur *et al.* 2012; Zeng *et al.* 2012; Kani *et al.* 2013; Liu *et al.* 2013; Shen *et al.* 2013*a*). These data provide relatively new tools for the correlation and timescale subdivision that are currently being studied and utilized.

Prospectus

The two-century history of the development of the Permian chronostratigraphic scale has produced a scheme of three series and nine stages (Fig. 13). Many of the contentious issues surrounding this chronostratigraphic scale have been resolved, although not always for the 'best' reasons nor always to the satisfaction of all workers. Despite this, we believe that the three series and nine stages of the Permian currently recognized do a good job of honouring priority and stability in Permian chronostratigraphic nomenclature. The formal definition of the stage boundaries is nearly complete. The frontier of Permian chronostratigraphy is in the definition and characterization of substages (Fig. 13) – this is the way forward.

We dedicate this review to the memory of Brian Glenister, Jin Yugan and Bruce Wardlaw, three knights errant of the Permian timescale who did much to build the current Permian chronostratigraphy. SZS's work is supported by NSFC (41420104003 and 41290260). Joerg Schneider provided a detailed and helpful review of the manuscript.

References

ADAMS, J.E., CHENEY, M.G. ET AL. 1939. Standard Permian section of North America. *American Association of Petroleum Geologists Bulletin*, **23**, 1673–1681.

ANGIOLINI, L. & HENDERSON, C.M. 2013. The once and future quest: looking for mid-Permian correlation between the Tethyan and international time scales. *Permophiles*, **57**, 19–20.

ANGLIONI, L. & VACHARD, D. 2015. The Permian sedimentary successions of the Pamir Mountains, Tajikistan. *Permophiles*, **62**, 15–18.

ARCHBOLD, N.W. & DICKINS, J.M. 1997. Comments on subdivisions of the Permian and a standard world scale. *Permophiles*, **30**, 4–5.

BAARS, D.L. 1990. Permian chronostratigraphy in Kansas. *Geology*, **18**, 687–690.

BAARS, D.L., MAPLES, C.G., RITTER, S.M. & ROSS, C.A. 1992. Redefinition of the Pennsylvanian–Permian boundary in Kansas, mid-continent USA. *International Geology Review*, **34**, 1021–1025.

BAARS, D.L., RITTER, S.M., MAPLES, C.G. & ROSS, C.A. 1994*a*. Redefinition of the Upper Pennsylvanian Virgilian Series in Kansas. *Kansas Geological Survey Bulletin*, **230**, 11–16.

BAARS, D.L., ROSS, C.A., RITTER, S.M. & MAPLES, C.G. 1994*b*. Proposed repositioning of the Pennsylvanian–Permian boundary in Kansas. *Kansas Geological Survey Bulletin*, **230**, 5–10.

BAIN, A.G. 1856. On the geology of southern Africa. *Transactions of the Geological Society of London 2nd Series*, **7**, 175–192.

BARTLETT, J.R. 1854. *Personal Narrative of Explorations and Incidents in Texas, New Mexico, California, Sonora, and Chihuahua, Connected with the United States and Mexican Boundary Commission, during the years 1850, 1851, 1852, and 1853*. D. Appleton & Company, New York.

BEHNKEN, F.H. 1975. Conodonts as Permian biostratigraphic indices. *In*: CYS, J.M. & TOOMEY, D.F. (eds) *Permian Exploration, Boundaries, and Stratigraphy*. West Texas Geological Society and Permian Basin Section SEPM, Publications, **75-65**, 84–90.

BLANFORD, W.T., BLANFORD, H.F. & THEOBALD, W. 1856. On the geological structure and relations of the Talcheer Coal Field, in the District of Cuttock. *Memoirs of the Geological Survey of India*, **1**, 33–89.

BOARDMAN, D.R., WARDLAW, B.R. & NESTELL, M.K. 2009. Stratigraphy and conodont biostratigraphy of the uppermost Carboniferous and Lower Permian from the North American Midcontinent. *Kansas Geological Survey Bulletin*, **255**, 1–42.

BOGOSLOVSKAYA, M.F. & LEONOVA, T.B. 1994. Comments on the proposed operational scheme of Permian chronostratigraphy by M. F. Bogosolvskaya and T. B. Leonova. *Permophiles*, **25**, 15–17.

BOGOSLOVSKAYA, M.F., LEONOVA, T.B. & SHKOLIN, A.A. 1995. The Carboniferous–Permian boundary and ammonoids from the Aidaralash section, southern Urals. *Journal of Paleontology*, **69**, 288–301.

BÖSE, E. 1917. *The Permo-Carboniferous Ammonoids of the Glass Mountains, West Texas and Their Stratigraphical Significance*. Texas University Bulletin, **1762**.

BROUTIN, J., CHATEAUNEUF, J.J., GALTIER, J. & RONCHI, A. 1999. L'Autunian d'Autun reste-t-il une référence pour les dépôts continentaux du Permien inférieur d'Europe? Apport des données paléobotaniques. *Géologie de la France*, **2**, 17–31.

CARTER, R.M. 1974. A New Zealand case-study of the need for local time-scales. *Lethaia*, **7**, 181–202.

CASSINIS, G., TOUTIN-MORIN, N. & VIRGILI, C. 1992. Permian and Triassic events in the continental domains of Mediterranean Europe. *In*: SWEET, W.C., YANG, Z.Y., DICKINS, J.M. & YIN, H.F. (eds) *Permo-Triassic Events in the Eastern Tethys*. Cambridge University Press, Cambridge, 60–77.

CASSINIS, G., TOUTIN-MORIN, N. & VIRGILI, C. 1995. A general outline of the Permian Continental Basins in Southwestern Europe. *In*: SCHOLLE, P.A., PERYT, T.M. & ULMER-SCHOLLE, D. (eds) *The Permian of Northern Pangea, Volume 2, Sedimentary Basins and Economic Resources*. Springer, Berlin, 37–157.

CHEN, B., JOACHIMSKI, M.M., SUN, Y.D., SHEN, S.Z. & LAI, X.L. 2011. Carbon and conodont apatite oxygen isotope records of Guadalupian–Lopingian boundary sections: climatic or sea-level signal? *Palaeogeography, Palaeoclimatology, Palaeoecology*, **311**, 145–153.

CHERNYKH, V.M. & RITTER, S.M. 1994. Preliminary biostratigraphic assessment of conodonts from the proposed Carboniferous–Permian boundary stratotype, Aidaralash Creek, northern Kazakhstan. *Permophiles*, **25**, 4–6.

CHERNYKH, V.V., CHUVASHOV, B.I., DAVYDOV, V.I. & SCHMITZ, M.D. 2012. Mechetlino Section: a candidate for the Global Stratotype [Section] and Point (GSSP) of the Kungurian Stage (Cisuralian, Lower Permian). *Permophiles*, **56**, 21–34.

CHERNYKH, V.V., CHUVASHOV, B.I., SHEN, S.Z. & HENDERSON, C.M. 2013. Proposal for the Global Stratotype Section and Point (GSSP) for the base-Sakmarian Stage (Lower Permian). *Permophiles*, **58**, 16–26.

CHUVASHOV, B.I. 1994. Progress report of Permian Stratotypes Working Group. *Permophiles*, **25**, 7–8.

CHUVASHOV, B.I., CHERNYKH, V.V. ET AL. 2002*a*. Proposal for the base of the Sakmarian Stage: GSSP in the Kondurovsky section, southern Urals, Russia. *Permophiles*, **41**, 4–13.

CHUVASHOV, B.I., CHERNYKH, V.V. ET AL. 2002*b*. Progress report on the base of the Artinskian and base of the Kungurian by the Cisuralian Working Group. *Permophiles*, **41**, 13–16.

CHUVASHOV, B.I., CHERNYKH, V.V., SHEN, S.Z. & HENDERSON, C.M. 2013. Proposal for the Global Stratotype Section and Point (GSSP) for the base-Artinskian Stage (Lower Permian). *Permophiles*, **58**, 26–34.

CLARK, D.L. 1972. Early Permian crisis and its bearing on Permo-Triassic taxonomy. *Geologica et Palaeontologica*, **1**, 147–158.

CLARK, D.L. (ed.). 1984. *Conodont Biofacies and Provincialism*. Geological Society of America, Special Papers, **196**.

COHEE, G.V. 1960. Series subdivisions of the Permian. *American Association of Petroleum Geologists Bulletin*, **44**, 1578–1579.

COLLIE, M. & DIENER, J. (eds) 2004. *Murchison's Wanderings in Russia*. Halstan & Co. Ltd, Amersham.

COOPER, G.A. & GRANT, R.E. 1964. New Permian stratigraphic units in Glass Mountains, West Texas. *American Association of Petroleum Geologists Bulletin*, **48**, 1581–1588.

COPE, J.C.W. 1996. The role of the secondary standard in stratigraphy. *Geological Magazine*, **133**, 107–110.

CYS, J.M. 1983. Potential Middle Permian (Guadalupian) stratotype in West Texas. *Permophiles*, **7**, 5.

DAVYDOV, V.I. 1994. Comments by V. I. Davydov on the proposed operational scheme of Permian chronostratigraphy. *Permophiles*, **25**, 19–21.

DAVYDOV, V.I. 2001. The terminal stage of the Carboniferous: Orenburgian v. Bursumian. *Newsletter on Carboniferous Stratigraphy*, **19**, 58–64.

DAVYDOV, V.I. 2013. The GSSP at Aidaralash is solid and has no alternative. *Permophiles*, **58**, 13–15.

DAVYDOV, V.I., GLENISTER, B.F., SPINOSA, C., RITTER, S.M., CHERNYKH, V.V., WARDLAW, B.R. & SNYDER, W.S. 1995. Proposal of Aidaralash as GSSP for base of the Permian System. *Permophiles*, **26**, 1–9.

DAVYDOV, V.I., GLENISTER, B.F., SPINOSA, C., RITTER, S.M., CHERNYKH, V.V., WARDLAW, B.R. & SNYDER, W.S. 1998. Proposal of Aidaralash as Global Stratotype Section and Point (GSSP) for base of the Permian System. *Episodes*, **21**, 11–17.

DICKINS, J.M. 1994. Major divisions of the Permian and 'Middle Permian'. *Permophiles*, **24**, 60–61.

DIMICHELE, W.A., WAGNER, R.H., BASHFORTH, A.R. & ÁLVAREZ-VÁZQUEZ, C. 2013. An update on the flora of the Kinney Quarry of central New Mexico (Upper Pennsylvanian), its preservation and environmental significance. *New Mexico Museum of Natural History and Science Bulletin*, **59**, 289–325.

DOUGLASS, R.C. 1977. The development of fusulinid biostratigraphy. *In*: KAUFFMAN, E.G. & HAZEL, J.E. (eds) *Concepts and Methods of Biostratigraphy*. Dowden, Hutchinson, and Ross, Inc., Stroudsburg, PA, 463–481.

DUNBAR, C.O., BAKER, A.A. *ET AL*. 1960. Correlation of the Permian formations of North America. *Geological Society of America Bulletin*, **71**, 1763–1806.

EMBLETON, B.J.J., MCELHINNY, M.W., MA, X.H., ZHANG, Z.K. & LI, Z.X. 1996. Permo-Triassic magnetostratigraphy in China: the type section near Taiyuan, Shanxi Province, North China. *Geophysical Journal International*, **126**, 382–388.

ESAULOVA, N.K., LOZOVSKY, V.R. & ROZANOV, A.Yu. (eds) 1998. *Stratotypes and Reference Sections of the Upper Permian in the Regions of the Volga and Kama Rivers*. GEOS, Moscow.

FAN, J.S., QI, J.W., ZHOU, T.M. & ZHANG, X.L. 1990. *Permian Reef in Longli, Guangxi*. Geological Publishing House, Beijing.

FISCHER VON WALDHEIM, G. 1841. Observations géologiques sur la Russie. Lettre addressée à Son Excellence Monsieur G. Fischer de Waldheim, Conseiller d'état etc. de R. I. Murchison. *Bulletin of the Moscow Society of Naturalists*, **14**, 901–909.

FOSTER, M. 1989. The Permian controversy of 1858: an affair of the heart. *Proceedings of the American Philosophical Society*, **133**, 370–390.

FURNISH, W.M. 1966. Ammonoids of the Upper Permian Cyclolobus Zone. *Neues Jahrbuch für Mineralogie, Geologie und Paläontologie Abhandlungen*, **125**, 265–296.

FURNISH, W.M. 1973. Permian stage names. *In*: LOGAN, A. & HILLS, L.V. (eds) *The Permian and Triassic Systems and Their Mutual Boundary*. Society of Petroleum Geology of Canada, Memoirs, **2**, 522–548.

FURNISH, W.M. & GLENISTER, B.F. 1970. Permian ammonoid *Cyclolobus* from the Salt Range, West Pakistan. *In*: KUMMEL, B. & TEICHERT, C. (eds) *Stratigraphic Problems: Permian and Triassic of West Pakistan*. University Press of Kansas, Lawrence, KS, 153–175.

FURNISH, W.M., GLENISTER, B.F., KULLMAN, J. & ZHOU, Z. 2009. *Treatise on Invertebrate Paleontology Part L Mollusca 4 Revised Volume 2: Carboniferous and Permian Ammonoidea (Goniatitida and Prolecaitida)*. University Press of Kansas, Lawrence, KS.

GEIKE, A. 1875. *Life of Sir Roderick I. Murchison Based on His Journals and Letters, Volume 1*. John Murray, London.

GERASIMOV, L.P. 1937. The Uralian Series of the Permian system. *Scientific Reports of Kazan University Geology*, **97**, 1–68 [in Russian].

GERTH, H. 1950. Die Ammonoideen des Perm von Timor und ihre Bedeutung für die stratigraphische Gliederung der Permformation. *Neues Jahrbuch für Mineralogie, Geologie und Paläontologie*, **91**, 233–320.

GIRTY, G.A. 1902. The Upper Permian in western Texas. *American Journal of Science*, **14**, 363–368.

GIRTY, G.H. 1909. *The Guadalupian Fauna*. United States Geological Survey, Professional Papers, **58**.

GLENISTER, B.F. 1991. Proposal of Guadalupian as international standard for the Middle Permian series. *Permophiles*, **18**, 10–11.

GLENISTER, B.F. 1993*a*. A call to action. *Permophiles*, **22**, 2–5.

GLENISTER, B.F. 1993*b*. Stratotype of Guadalupian Series. *Permophiles*, **23**, 17–20.

GLENISTER, B.F. & FURNISH, W.M. 1961. The Permian ammonoids of Australia. *Journal of Paleontology*, **35**, 673–736.

GLENISTER, B.F. & NASSICHUK, W.W. 1998. Preface to compilation of Permophiles. *Permophiles*, **31**, 8–10.

GLENISTER, B.F., BOYD, D.W. *ET AL*. 1992. The Guadalupian: proposed international standard for a Middle Permian series. *International Geology Review*, **34**, 857–888.

GLENISTER, B.F., WARDLAW, B.R. *ET AL*. 1999. Proposal of Guadalupian and component Roadian, Wordian and Capitanian stages as international standards for the Middle Permian series. *Permophiles*, **34**, 3–11.

GRABAU, A.W. 1923. *Stratigraphy of China, Part I: Paleozoic and Older*. China Geological Survey, Peking, China.

GRABAU, A.W. 1931. *The Permian of Mongolia*. Natural History of Central Asia, **4**. American Museum of Natural History, New York.

GRANT, R.E. 1979. [Comments on a Permian subcommission]. *Permophiles*, **1**, 6–7.

GRUNT, T. & DMITRIEVA, V.Y. 1973. Permian brachiopods of the Pamirs. *Transactions of the Paleontological Institute*, **136**, 1–211 [in Russian].

GRUNT, T., MOLOSTOVSKY, E. *ET AL*. 1999. Alternative proposal of international standard references for the middle and late Permian series. *Permophiles*, **35**, 25–26.

GUNNELL, F.H. 1933. Conodonts and fish remains from the Cherokee, Kansas City, and Wabaunsee groups of Missouri and Kansas. *Journal of Paleontology*, **7**, 261–297.

HARLAND, W.B., COX, A.V., LLEWELLYN, P.G., PICKTON, C.A.G., SMITH, A.G. & WALTERS, R. 1982. *A Geologic Time Scale*. Cambridge University Press, Cambridge.

HARLAND, W.B., ARMSTRONG, R.L., COX, A.V., CRAIG, L.E., SMITH, A.G. & SMITH, D.G. 1990. *A Geologic Time Scale*. Cambridge University Press, Cambridge.

HENDERSON, C.M. 2000. Reply to 'a discussion on the definition for the base of the Lopingian'. *Permophiles*, **37**, 21–22.

HENDERSON, C.M. 2005. International correlation of the marine Permian timescale. *Permophiles*, **46**, 6–9.

HENDERSON, C.M. 2010. Update on base-Artinskian GSSP. *Permophiles*, **55**, 15–17.

HENDERSON, C.M., DAVYDOV, V.I. & WARDLAW, B.R. 2012*a*. The Permian Period. *In*: GRADSTEIN, F.M., OGG, J.G., SCHMITZ, M.D. & OGG, G.M. (eds) *The Geologic Time Scale 2012, Volume 2*. Elsevier, Amsterdam, 653–679.

HENDERSON, C.M., WARDLAW, B.R., DAVYDOV, V.I., SCHMITZ, M.D., SCHIAPPA, T.A., TIERNEY, K.E. & SHEN, S.Z. 2012*b*. Proposal for base-Kungurian GSSP. *Permophiles*, **56**, 8–21.

HUANG, T.K. 1932. *The Permian Formations of Southern China*. Geology Memoir of the Geological Survey of China Series A, **10**.

HUANG, Z.X., SHI, Y. & WEI, M.C. 1982. A new stratigraphic unit of the Permian-Longlinian Stage. *Journal of the Chengdu College of Geology*, **4**, 63–73.

ISOZAKI, Y. 2009. Illawarra Reversal: the fingerprint of a superplume that triggered Pangean breakup and the end-Guadalupian (Permian) mass extinction. *Gondwana Research*, **15**, 421–432.

ISOZAKI, Y., KAWAHATA, H. & MINOSHIMA, K. 2007. The Capitanian (Permian) Kamura cooling event: the beginning of the Paleozoic–Mesozoic transition. *Palaeoworld*, **16**, 16–30.

JIN, Y.G. 1993. Pre-Lopingian benthos crisis. *Comptes Rendus XII ICCP, Buenos Aires*, **2**, 269–278.

JIN, Y.G. & SHANG, Q.H. 2000. The Permian of China and its interregional correlation. *In*: YIN, H.F., DICKINS, J.M., SHI, G.R. & TONG, J.N. (eds) *Permian–Triassic Evolution of Tethys and Western Circum-Pacific*. Elsevier, Amsterdam, 71–98.

JIN, Y.G., MEI, S.L. & ZHU, Z.L. 1993. The potential stratigraphic levels for Guadalupian/Lopingian boundary. *Permophiles*, **23**, 17–20.

JIN, Y.G., GLENISTER, B.F., KOTLYAR, G.V. & SHENG, J.Z. 1994*a*. An operational scheme of Permian chronostratigraphy. *Palaeoworld*, **4**, 1–13.

JIN, Y.G., MEI, S.L. & ZHU, Z.L. 1994*b*. The Makouan–Lopingian boundary sequences in south China. *Palaeoworld*, **4**, 119–132.

JIN, Y.G., SHENG, J.G. ET AL. 1994*c*. Revised operational scheme of Permian chronostratigraphy. *Permophiles*, **25**, 12–15.

JIN, Y.G., WARDLAW, B.R., GLENISTER, B.F. & KOTLYAR, C.V. 1997. Permian chronostratigraphic subdivisions. *Episodes*, **20**, 6–10.

JIN, Y.G., MEI, S.L., WANG, W., WANG, X.D., SHEN, S.Z., SHANG, Q.H. & CHEN, Z.Q. 1998. On the Lopingian series of the Permian System. *Palaeoworld*, **9**, 1–18.

JIN, Y.G., SHANG, Q.H., WANG, X.D., WANG, Y. & SHENG, J.Z. 1999. Chronostratigraphic subdivision and correlation of the Permian in China. *Acta Geologica Sinica (English Edition)*, **73**, 127–138.

JIN, Y.G., HENDERSON, C.M., WARDLAW, B.R., GLENISTER, B.F., MEI, S.L., SHEN, S.Z. & WANG, X.D. 2001. Proposal for the Global Stratotype Section and Point (GSSP) for the Guadalupian–Lopingian boundary. *Permophiles*, **39**, 32–42.

JIN, Y.G., SAHNG, Q.H. & WANG, X.D. 2003. Permian stratigraphy of China. *In*: ZHANG, W.T., CHEN, P.J. & PALMER, A.R. (eds) *Biostratigraphy of China*. Science Press, Beijing, 331–378.

JIN, Y.G., SHEN, S.Z. ET AL. 2006*a*. The Global Stratotype Section and Point (GSSP) for the boundary between the Capitanian and Wuchiapingian stages (Permian). *Episodes*, **29**, 253–262.

JIN, Y.G., WANG, Y., HENDERSON, C.M., WARDLAW, B.R., SHEN, S.Z. & CAO, C.Q. 2006*b*. The Global Stratotype Section and Point (GSSP) for the base of the Changshingian Stage (Upper Permian). *Episodes*, **29**, 175–182.

KANI, T., HISANABE, C. & ISOZAKI, Y. 2013. The Capitanian (Permian) minimum of $^{87}Sr/^{86}Sr$ ratio in the mid-Panthalassan paleo-atoll carbonates and its demise by the deglaciation and continental doming. *Gondwana Research*, **24**, 212–221.

KANMERA, K. & NAKAZAWA, K. 1973. Permian–Triassic relationships and faunal changes in the eastern Tethys. *In*: LOGAN, A. & HILLS, L.V. (eds) *The Permian and Triassic Systems and Their Mutual Boundary*. Society of Petroleum Geology of Canada, Memoirs, **2**, 100–119.

KARPINSKY, A.P. 1874. Geologische Untersuchungen im Gouvernment Orenburg. *Verhandlung Mineralogische Gesellschaft St Petersbourg*, **9**, 212–330.

KARPINSKY, A.P. 1889. Über die Ammoneen der Artinsk-Stufe und einige mit denselben verwandte carbonische Formen. *Memoires de l'Academic Imperiale des Sciences de St-Petersbourg*, **37**, 1–104.

KARPINSKY, A.P. 1891. On ammonoids of the Artinskian Stage and some similar Carboniferous forms. *Proceedings of the St Petersburg Mineralogical Society*, **2**, 1–192 [in Russian].

KEYES, C.R. 1896. Serial nomenclature of the Carboniferous. *The American Geologist*, **18**, 22–28.

KLETS, A.G., BUDNIKOV, I.B., KUTYGIN, R.V. & GRINENKO, V.S. 2001. The reference section of the Lower–Upper Permian boundary beds in the Verkhoyansk region and its correlation. *Stratigraphy and Geological Correlation*, **9**, 247–262.

KOTLYAR, G.V. 1977. *Stratigraphic Dictionary of the USSR, Carboniferous, Permian*. Nedra, Leningrad [in Russian].

KOTLYAR, G.V. 1995. Comments on the proposed operational scheme of Permian chronostratigraphy. *Permophiles*, **26**, 23–25.

KOTLYAR, G.V. 2000. The Permian of Russia and CIS and its interregional correlation. *In*: YIN, H.F., DICKINS, J.M., SHI, G.R. & TONG, J.N. (eds) *Permian–Triassic Evolution of Tethys and Western Circum-Pacific*. Elsevier, Amsterdam, 17–35.

KOTLYAR, G.V. & PRONINA, G.P. 1995. Murgabian and Midian stages of the Tethyan realm. *Permophiles*, **27**, 23–26.

KOTLYAR, G.V. & PRONINA-NESTELL, G.P. 2005. Report on the committee on the Permian System of Russia. *Permophiles*, **46**, 9–13.

KOTLYAR, G.V., KOSSOVSKAYA, O.L., SHISHLOV, S.B., ZHRAVLEV, A.R. & PUKHONTO, S.K. 2004. Boundary between Permian Series in diverse sedimentary facies of north European Russia: constraints of event stratigraphy. *Stratigraphy and Geological Correlation*, **12**, 460–484.

KOZUR, H. 1986. Boundaries, subdivision and correlation of the marine and continental Permian. *In*: *S.G.I. and IGCP Project No. 203, Field Conference on Permian and Permian–Triassic Boundary in the South-Alpine Segment of the Western Tethys, and Additional Regional Reports*, 4–12 July 1986, Societa Geologica Italiana, Brescia, Italy, *Abstracts*, 29–31.

KOZUR, H. 1988. The age of the central European Rotliegendes. *Zeitschrift für Geologische Wissenschaften*, **16**, 907–915.

KOZUR, H. 1993. Application of the 'parastratigraphic' and international scale for the continental Permian of Europe. *In*: HEIDTKE, U. (compiler) *New Research on Permo-Carboniferous Faunas*. Pollichia-Buch, Bad Dürkheim, **29**, 9–22.

KROTOW, P. 1885. *Artinskian Stage: Geological–Paleontological Monograph on the Artinskian Sands*. Kazan Universitet Obshchestvo Estestvoispytatele Trudy, **13** [in Russian].

DE LAPPARENT, A. 1893. *Traité de géologie: Avec gravure dans le texte*. Savy, Paris.

DE LAPPARENT, A. 1900. *Traité de géologie, 4e edition refondue et considerablement augmentee*. Savy, Paris.

LAZAREV, S.S. 1999. How to save the traditional nomenclature of Upper Permian stages. *Permophiles*, **33**, 10–11.

LEONOVA, T.B. 2007. Correlation of the Kazanian of the Volga–Urals with the Roadian of the global Permian scale. *Palaeoworld*, **16**, 246–253.

LEVEN, E.YA. 1963. On the phylogeny of higher fusulinids and the distribution of Upper Permian outcrops of the Tethys. *Voprosy Mikropaleontologiy*, **7**, 57–70 [in Russian].

LEVEN, E.YA. 1967. *Stratigraphy and Fusulinids of Permian Deposits of the Pamirs*. Transactions of the Geological Institute of the Academy of Science of the U.S.S.R., **167** [in Russian].

LEVEN, E.YA. 1975. Stage scale of the Permian outcrops of Tethys. *Byull MOIP*, **50**, 5–21 [in Russian].

LEVEN, E.YA. 1979. Bolorian stage of the Permian: basis, characteristics, correlation. *Akademia Nauk SSSR Izvestiya Geologicheskaya Seriya*, **1979**, 53–65 [in Russian].

LEVEN, E.YA. 1980. Yaktashian stage of the Permian: basis, characteristics, correlation. *Akademia Nauk SSSR Izvestiya Geologicheskaya Seriya*, **1980**, 50–60 [in Russian].

LEVEN, E.YA. 1992. The division of the Permian System at a series level. *Permophiles*, **21**, 8–10.

LEVEN, E.YA. 1993. Early Permian fusulinids from the central Pamirs. *Rivista Italiana di Paleontologia e Stratigrafia*, **104**, 3–42.

LEVEN, E.YA. 1994. Comments by E. Y. Leven on the proposed operational scheme of Permian chronostratigraphy. *Permophiles*, **25**, 17–19.

LEVEN, E.YA. 2003. The Permian stratigraphy and fusulinids of Tethys. *Rivista Italiana di Paleontologia e Stratigrafia*, **101**, 267–280.

LEVEN, E.YA. & BOGOSLOVSKAYA, M.F. 2006. The Roadian Stage of the Permian and problems of its global correlation. *Stratigraphy and Geological Correlation*, **14**, 161–173.

LIKHAREV, B.K. 1959. The boundaries and principal subdivisions of the Permian System. *Soviet Geology*, **6**, 13–30 [in Russian].

LIU, X.C., WANG, W. ET AL. 2013. Late Guadalupian to Lopingian (Permian) carbon and strontium isotopic chemostratigraphy in the Abadeh section, central Iran. *Gondwana Research*, **24**, 222–232.

LOZOVSKY, V.R., MINIKH, M.G., GRUNT, T.A., KUKHTINOV, D.A., PONOMARENKO, A.G. & SUKACEVA, I.D. 2006. The Ufimian Stage of the East European scale: status, validity, and correlation potential. *Stratigraphy and Geological Correlation*, **17**, 602–614.

LUCAS, S.G. 2004. A global hiatus in the Middle Permian tetrapod fossil record. *Stratigraphy*, **1**, 47–64.

LUCAS, S.G. 2006. Global Permian tetrapod biostratigraphy and biochronology. *In*: LUCAS, S.G., CASSINIS, G. & SCHNEIDER, J.W. (eds) *Non-Marine Permian Biostratigraphy and Biochronology*. Geological Society, London, Special Publications, **265**, 65–93, https://doi.org/10.1144/GSL.SP.2006.265.01.04

LUCAS, S.G. 2010. The Triassic chronostratigraphic scale: history and status. *In*: LUCAS, S.G. (ed.) *The Triassic Timescale*. Geological Society, London, Special Publications, **334**, 17–39, https://doi.org/10.1144/SP334.2

LUCAS, S.G. 2013. We need a new GSSP for the base of the Permian. *Permophiles*, **58**, 8–12.

LUCAS, S.G. & ANDERSON, O.J. 1994. Ochoan (Late Permian) stratigraphy and chronology, southeastern New Mexico and West Texas. *New Mexico Bureau of Mines and Mineral Resources Bulletin*, **50**, 29–36.

LUCAS, S.G. & ANDERSON, O.J. 1997. Ochoa as a lithostratigraphic unit, not a chronostratigraphic unit of the Permian. *West Texas Geological Society Bulletin*, **36**, 5–10.

LUCAS, S.G. & WILDE, G.L. 2000. The Bursum Formation stratotype, Upper Carboniferous of New Mexico, and the Bursumian Stage. *Permophiles*, **36**, 7–10.

LUCAS, S.G., KUES, B.S. & KRAINER, K. 2001. The Bursumian Stage. *Permophiles*, **39**, 23–28.

LUCAS, S.G., SCHNEIDER, J.W. & CASSINIS, G. 2006. Non-marine Permian biostratigraphy and biochronology: an introduction. *In*: LUCAS, S.G., CASSINIS, G. & SCHNEIDER, J.W. (eds) *Non-Marine Permian Biostratigraphy and Biochronology*. Geological Society, London, Special Publications, **265**, 1–14, https://doi.org/10.1144/GSL.SP.2006.265.01.01

LUCAS, S.G., ALLEN, B.D. ET AL. 2011. Precise age and biostratigraphic significance of the Kinney Brick Quarry Lagerstätte, Pennsylvanian of New Mexico, USA. *Stratigraphy*, **8**, 7–27.

MACKE, D.L. & NICHOLS, K.M. 2007. Conodonts as evolving heavy-mineral grains. *New Mexico Museum of Natural History and Science Bulletin*, **41**, 262–267.

MARCOU, J. 1859. *Dyas und Trias*. Archives de la Bibliothèque Universite, Genêve.

MARCOU, J. 1862. Observations on the terms 'Pénéen,' 'Permian,' and 'Dyas'. *Proceedings of the Boston Society of Natural History*, **9**, 33–336.

MCARTHUR, J.M., HOWARTH, R.J. & SHIELDS, G.A. 2012. Strotium isotope stratigraphy. *In*: GRADSTEIN, F.M., OGG, J.G., SCHMITZ, M.D. & OGG, G.M. (eds) *The Geological Time Scale 2012, Volume 2*. Elsevier, Amsterdam, 127–144.

MEI, S.L. & HENDERSON, C.M. 2001. Evolution of Permian conodont provincialism and its significance in global correlation and paleoclimate implication. *Palaeogeography, Palaeoclimatology, Palaeoecology*, **170**, 237–260.

MEI, S.L. & HENDERSON, C.M. 2002. Conodont definition of the Kungurian (Cisuralian) and Roadian (Guadalupian) boundary. *In*: HILLS, L.V., HENDERSON, C.M. & BAMBER, E.W. (eds) *Carboniferous and Permian of the World*. Canadian Society of Petroleum Geologists, Memoirs, **19**, 529–551.

MEI, S.L., HENDERSON, C.M. & JIN, Y.G. 1999. Permian conodont provincialism, zonation and global correlation. *Permophiles*, **35**, 9–16.

MEI, S.L., HENDERSON, C.M. & CAO, C.Q. 2004. Conodont sample-population approach to defining the base of the Changshingian Stage, Lopingian Series, Upper Permian. *In*: BEAUDOIN, A. & HEAD, M. (eds)

The Palynology and Micropaleontology of Boundaries. Geological Society, London, Special Publications, **230**, 105–121, https://doi.org/10.1144/GSL.SP.2004.230.01.06

MENNING, M., ALESEEV, A.S. *ET AL*. 2006. Global time scale and regional stratigraphic reference scales of Central and West Europe, East Europe, Tethys, South China, and North America as used in the Devonian–Carboniferous–Permian Correlation Chart 2003 (DCP 2003). *Palaeogeography, Palaeoclimatology, Palaeoecology*, **240**, 318–372.

MIKLUCHO-MAKLAY, A.D. 1958. On the delineation of stages for the marine Permian outcrops of the southern region of the USSR. *Doklady Akademiya Nauk SSSR*, **120**, 175–178 [in Russian].

MILLER, A.K. & FURNISH, W.M. 1940. *Permian Ammonoids of the Guadalupe Mountain Region and Adjacent Areas*. Geological Society of America, Special Papers, **26**, 1–238.

MOVSHOVICH, E.V. 1980. Is the three-fold subdivision of the Permian really necessary? *Permophiles*, **4**, 44–45.

MUNIER-CHALMAS, E. & DE LAPPARENT, A. 1893. Note sur la nomenclature des terrains sédimentaires. *Bulletin de la Societe Géologique de France*, **21**, 438–493.

MURCHISON, R.I. 1841. First sketch of some of the principal results of a second geological survey of Russia. *The Philosophical Magazine*, **19**, 417–422.

MURCHISON, R.I. 1862. On the inapplicability of the new term 'Dyas' to the 'Permian' group of rocks, as proposed by Dr. Geinitz. *The Geologist*, **5**, 4–10.

MURCHISON, R.I., DE VERNEUIL, E. & VON KEYSERLING, A. 1845. *The Geology of Russia in Europe and the Ural Mountains, Volume 1, Geology*. John Murray, London.

MYLIUS, G.F. 1720. Fernere Nachricht zum Eisslebischen Bergwerck gehörig. *In*: MYLIUS, G.F. (ed.) *Memorabilia Saxoniae Subterraneaea, i. e. des unterirdischen Sachsens seltsame Wunder der Natur*. G. Weidmann, Leipzig, 9–16.

NECHAEV, A.V. 1915. Kazanian and Ufimian stages of the Permian System. *Geological Vestnik*, **1**, 4–6 [in Russian].

NECHAEV, A.V. 1921. *Geology of Russia. Upper Permian Sediments*. Geologicheskaya Komitet, Petrograd.

NIKITIN, S. 1887. Observations geologiques efectuees le long de la voire ferree Samara-Oufa. *Zechstein et Tatarien. Izvestiya Geologicheskiy Komitet*, **6**, 245.

NOURGALIEV, D.K. & NOURGALIEVA, N.G. 1999. Astronomical calibration of the east Russian plate Upper Permian sedimentary cycles: preliminary data about duration of the Kazanian Stage. *Permophiles*, **34**, 15–19.

O'CONNOR, H.G. 1963. Changes in Kansas stratigraphic nomenclature. *American Association of Petroleum Geologist Bulletin*, **47**, 1873–1877.

O'CONNOR, H.G., ZELLER, D.E., BOYNE, C.K., JEWETT, J.M. & SWINEFORD, A. 1968. Permian System. *Kansas Geological Survey Bulletin*, **189**, 43–53.

PROSSER, C.S. 1910. The Anthracolithic or upper Paleozoic rocks of Kansas and related regions. *Journal of Geology*, **18**, 125–161.

RAMEZANI, J., SCHMITZ, M.D., DAVYDOV, V.I., BOWRING, S.A., SNYDER, W.S. & NORTHRUP, C.J. 2007. High-precision U–Pb zircon age constraints on the Carboniferous–Permian boundary in the southern Urals stratotype. *Earth and Planetary Science Letters*, **256**, 244–257.

RASSCOE, B. & BAARS, D.L. 1972. Permian system. *In*: MALLORY, W.W. (ed.) *Geological Atlas of the Rocky Mountain Region*. Rocky Mountain Association of Geologists, Denver, CO, 143–165.

RAUSER-CHERNOUSOVA, D. 1938. The upper Paleozoic Foraminifera of the Samara Bend and the Trans-Volga Region. *Trudy Geologicheskaya Instituta Akademia Nauk SSSR*, **7**, 69–167 [in Russian].

RAUSER-CHERNOUSOVA, D.M. 1940. Stratigraphy of the Upper Carboniferous and Artinskian Stage of the west slopes of the Urals and data on fusulinacean faunas. *Trudy Geologicheskaya Instituta Akademia Nauk SSSR*, **2**, 37–101 [in Russian].

REMANE, J., BASSET, M.G., COWIE, J.W., GOHRANDT, K.H., LANE, H.R., MICHELSEN, O. & NAIWEN, W. 1996. Revised guidelines for the establishment of global chronostratigraphic standards by the International Commission of Stratigraphy (ICS). *Episodes*, **19**, 77–81.

RICHARDSON, G.B. 1904. Reconnaissance in Trans-Pecos north of the Texas and Pacific Railway. *University of Texas Mineral Survey Bulletin*, **9**, 1–117.

VON RICHTHOFEN, A. 1888. *China, Ergebnisse eigener Reisen und darauf begründeter Studien, Bd 2*. D. Reimer, Berlin.

ROSS, C.A. 1986. Paleozoic evolution of southern margin of Permian basin. *Geological Society of America Bulletin*, **97**, 536–554.

ROSS, C.A. & ROSS, J.P. 1987. *Biostratigraphic Zonation of Late Paleozoic Depositional Sequences*. Cushman Foundation for Foraminiferal Research, Special Publications, **24**, 151–168.

ROSS, C.A. & ROSS, J.P. 1994. The need for a Bursumian Stage, uppermost Carboniferous, North America. *Permophiles*, **24**, 3–6.

ROSS, C.A. & ROSS, J.P. 2003. Sequence evolution and sequence extinction: fusulinid biostratigraphy and species-level recognition of depositional sequences, Lower Permian, Glass Mountains, West Texas, U.S.A. *In*: OLSON, H.C. & LECKIE, R.M. (eds) *Micropaleontologic Proxies for Sea-Level Change and Stratigraphic Discontinuities*. SEPM (Society for Sedimentary Geology), Special Publications, **75**, 317–359.

ROSTOVTSEV, K.O. & AZARYAN, N.R. 1973. The Permian–Triassic boundary in Transcaucasia. *In*: LOGAN, A. & HILLS, L.V. (eds) *The Permian and Triassic Systems and Their Mutual Boundary*. Society of Petroleum Geology of Canada, Memoirs, **2**, 89–99.

RUZHENTSEV, V.E. 1934. New data on the stratigraphy of the Artinskian Stage on the west slope of the Urals. *Oil Economy*, **6**, 16–23 [in Russian].

RUZHENTSEV, V.E. 1936. New data on the stratigraphy of the Carboniferous and Lower Permian of the Orenburg and Aktyubinsk District. *Problemy Sovyetskoi Geologi*, **6**, 476–506 [in Russian].

RUZHENTSEV, V.E. 1937. Stratigraphic outline of the Upper Carboniferous and Lower Permian deposits in the Orenburg district (U.S.S.R.). *Bulletin de la Societe Naturelle Moscou Series Geologique*, **15**, 187–214.

RUZHENTSEV, V.E. 1950. Type section and biostratigraphy of the Sakmarian Stage. *Doklady Akademiya Nauk SSSR*, **71**, 1101–1104 [in Russian].

RUZHENTSEV, V.E. 1951. *Lower Permian Ammonoids of the Southern Urals. I. Ammonoids of the Sakmarian Stage.* Akademiya Nauk SSSR Paleontologicheska Instituta Trudy, **33** [in Russian].

RUZHENTSEV, V.E. 1952. *Biostratigraphy of the Sakmar Stage in the Aktyubin Region of the Kazakhstan SSR.* Akademiya Nauk SSSR Paleontologicheska Instituta, **42** [in Russian].

RUZHENTSEV, V.E. 1954. Asselian Stage of the Permian System. *Doklady Akademii Nauk SSSR*, **99**, 1079–1082 [in Russian].

RUZHENTSEV, V.E. 1955. On the family Cyclolobidae Zittel. *Doklady Akademiya Nauk SSSR*, **103**, 701–703 [in Russian].

RUZHENTSEV, V.E. 1956. *Lower Permian Ammonites of the Southern Urals. II. Ammonites of the Artinskian Stage.* Akademiya Nauk SSSR Paleontologicheskaya Instituta Trudy, **60** [in Russian].

SALVADOR, A. (ed.) 1994. *International Stratigraphic Guide.* 2nd edn. Geological Society of America, Boulder, CO.

SCHENCK, H.G., CHILDS, T.S., JR. ET AL. 1941. Stratigraphic nomenclature. *American Association of Petroleum Geologists Bulletin*, **25**, 2195–2202.

SCHMITZ, M.D. & DAVYDOV, V.I. 2012. Quantitative radiometric and bio-stratigraphic calibration of the Pennsylvanian–Early Permian (Cisuralian) time scale and pan-Euramerican chronostratigraphic correlation. *Geological Society of America Bulletin*, **124**, 549–577.

SCHNEIDER, J.W. 2000. Geinitz and the Dyas. *Schriften des Staatlichen Museums für Mineralogie und Geologie zu Dresden*, **11**, 107–108.

SCHNEIDER, J.W. 2001. Rotliegend stratigraphy – principles and problems. *Beiträge zur Geologie von Thuringen Neue Folge*, **8**, 7–42.

SHAW, A.B. 1969. Adam and Eve, paleontology and the non-objective arts. *Journal of Paleontology*, **43**, 1085–1098.

SHEN, S.Z. & MEI, S.L. 2010. Lopingian (Late Permian) high-resolution conodont biostratigraphy in Iran with comparison to South China zonation. *Geological Journal*, **45**, 135–161.

SHEN, S.Z., XIE, J.F., ZHANG, H. & SHI, G.R. 2009. Roadian–Wordian (Guadalupian, Middle Permian) global palaeobiogeography of brachiopods. *Global and Planetary Change*, **65**, 166–181.

SHEN, S.Z., HENDERSON, C.M. ET AL. 2010. High-resolution Lopingian (Late Permian) timescale of South China. *Geological Journal*, **45**, 122–134.

SHEN, S.Z., CAO, C.Q. ET AL. 2013*a*. High-resolution $\delta^{13}C_{carb}$ chemostratigraphy from latest Guadalupian through earliest Triassic in South China and Iran. *Earth Planetary Science Letters*, **375**, 156–165.

SHEN, S.Z., ZHANG, H., SHI, G.R., LI, W.Z., XIE, J.F., MU, L. & FAN, J.X. 2013*b*. Early Permian (Cisuralian) global brachiopod palaeobiogeography. *Gondwana Research*, **24**, 104–124.

SHENG, J.Z. 1962. *Permian Stratigraphy of China. A Compilation of Scientific Reports of the National Congress on Stratigraphy.* Science Press, Beijing.

SHENG, J.Z. 1963. Permian fusulinids of Kwangsi, Kueichouw and Szechuan. *Palaeontologia Sinica New Series*, **10**, 1–119 [in Chinese], 129–147 [in English].

SHENG, J.Z. & JIN, Y.G. 1994. Correlation of Permian deposits in China. *Palaeoworld*, **4**, 14–113.

SHERLOCK, R.L. 1928. A correlation of the Permo-Triassic rocks. Part II. *Proceedings of the Geologists Association*, **39**, 49–95.

SHERLOCK, R.L. 1947. *The Permo-Triassic Formations: A World Review.* Hutchinson's Scientific and Technical Publications, London.

SHEVELEV, A.I., KHALAMBADZHA, V.G., BUROV, B.V., ESAULOVA, N.K. & GUSEV, A.K. 1996. On the problems of the international Permian stratigraphic scale. *Permophiles*, **27**, 38–39.

STEPANOV, D.L. 1957. A new stage of the Permian System in the Arctic. *Messenger Leningrad State University Series Geology–Geography*, **24**, 20–24.

STEPANOV, D.L. 1973. The Permian System in the USSR. *In*: LOGAN, A. & HILLS, L.V. (eds) *The Permian and Triassic Systems and their mutual boundary.* Society of Petroleum Geology of Canada, Memoirs, **2**, 120–136.

STUCKENBERG, A.A. 1890. *Geological Map of Russia: Sheet 13, Geological Studies in the Northwestern Part of Sheet 138.* Trudy Geologicheskovo Komiteta, **4** [in Russian].

TARAZ, H. 1995. Comments by H. Taraz on the proposed operational scheme of Permian chronostratigraphy. *Permophiles*, **26**, 25–26.

THOMPSON, M.L. 1954. *American Wolfcampian Fusulinids.* University of Kansas Paleontological Contributions, **5**.

TOMLINSON, C.W., MOORE, R.C., DOTT, R.H., CHENEY, M.G. & ADAMS, J.E. 1940. Classification of Permian rocks. *American Association of Petroleum Geologists Bulletin*, **24**, 337–358.

USTRITSKY, V.I. 1971. *Upper Paleozoic Biostratigraphy of the Arctic.* Trudy Arktik Nauchno-issled Instituta, **164** [in Russian].

DE VERNEUIL, E. 1845. *Geologie de la Russie d'Europe et des Montagnes de l'Oural (by Murchison, de Verneuil, and Keyserling). Volume 2, Paleontologie by de Verneuil.* John Murray, Paris.

VOZAROVA, A. & TUNYI, I. 2003. Evidence of the Illawarra reversal in the Peiraiian sequence of the Hronic Nappe (Western Carpathians, Slovakia). *Geologica Carpathica*, **54**, 229–236.

WAAGEN, W. 1891. Salt Range fossils, geological results. *Palaeontologica Indica Series 13*, **4**, 89–242.

WAHLMAN, G.P. 1998. Fusulinid biostratigraphy of the new Pennsylvanian–Permian boundary in the southwest and midcontinent U.S.A. *Geological Society of America Abstracts with Programs*, **30**, 34.

WALSH, S.L., GRADSTEIN, F.M. & OGG, J.G. 2004. History, philosophy, and application of the Global Stratotype Section and Point. *Lethaia*, **37**, 201–218.

WANG, C.Y. 2000. A discussion on the definition for the base of the Lopingian Series. *Permophiles*, **37**, 19–21.

WANG, C.Y. 2001. Re-discussion of the base of the Lopingian Series. *Permophiles*, **38**, 27–30.

WANG, Y., CHEN, C.Z. & RUI, L. 1988. A potential global stratotype of the Permian–Triassic boundary. *Permophiles*, **13**, 3–7.

Wardlaw, B.R., Leve, E.Ya., Davydov, V.I., Schiappa, T.A. & Snyder, W.S. 1999. The base of the Sakmarian Stage: call for discussion (possible GSSP in the Kondurovsky section, southern Urals, Russia). *Permophiles*, **34**, 19–26.

Wardlaw, B.R., Davydov, V. & Gradstein, F.M. 2004. The Permian Period. *In*: Gradstein, F.M., Ogg, J.G. & Smith, A.G. (eds) *A Geologic Time Scale 2004*. Cambridge University Press, Cambridge, 249–270.

Waterhouse, J.B. 1967. Proposal of series and stages for the Permian of New Zealand. *Transactions of the Royal Society of New Zealand Geology*, **5**, 161–176.

Waterhouse, J.B. 1976. *World Correlations for Permian Marine Faunas*. University of Queensland Papers Department of Geology, **7**.

Waterhouse, J.B. 1978. Chronostratigraphy for the world Permian. *In*: Cohee, G.V., Glaessner, M.F. & Hedberg, H.D. (eds) *Contributions to the Geologic Time Scale*. American Association of Petroleum Geologists, Studies in Geology, **6**, 299–322.

Waterhouse, J.B. 1979. Permian subdivisions and correlations. *Permophiles*, **2**, 15–17.

Waterhouse, J.B. 1982. An early Djulfian (Permian) brachiopod faunule from Upper Shyok Valley, Karakorum Range, and the implications for dating of allied faunas from Iran and Pakistan. *Contributions to Himalayan Geology*, **2**, 188–233.

Wheeler, H.F. 1934. The Carboniferous–Permian dilemma. *Journal of Geology*, **42**, 62–70.

Wignall, P.B., Sun, Y.D. *et al.* 2009. Volcanism, mass extinction, and carbon isotope fluctuations in the Middle Permian of China. *Science*, **324**, 1179–1182.

Wilde, G.L. 1990. Practical fusulinid zonation – the species concept; with Permian Basin emphasis. *West Texas Geological Society Bulletin*, **29**, 5–33.

Wilde, G.L. 2002. The Newwellian substage: rejection of the Bursumian Stage. *Permophiles*, **41**, 53–62.

Wilde, G.L. 2006. Pennsylvanian–Permian fusulinaceans of the Big Hatchet Mountains, New Mexico. *New Mexico Museum of Natural History and Science Bulletin*, **38**, 1–321.

Wilmarth, M.G. 1925. *The Geologic Time Classification of the United States Geological Survey Compared with Other Classifications*. United States Geological Survey Bulletin, **769**.

Yancey, T.E. & Yang, Z. 1989. Permian stratotype. *Permophiles*, **15**, 22–24.

Yang, Z.Y., Cheng, Y.Q. & Wang, H.Z. 1986. *The Geology of China*. Oxford Monographs on Geology & Geophysics, **3**.

Yin, H.F. (ed.). 1996. *The Palaeozoic–Mesozoic boundary: Candidates of the Global Stratotype Section and Point of the Permian–Triassic Boundary*. University of Geoscience Press, Wuhan, China.

Yin, H.F., Sweet, W.C. *et al.* 1996. Recommendation of the Meishan section as global stratotype section and point for basal boundary of Triassic System. *Newsletters on Stratigraphy*, **34**, 81–108.

Yin, H.F., Zhang, K.X., Tong, J.N., Yang, Z.Y. & Wu, S.B. 2001. The global stratotype section and point (GSSP) of the Permian–Triassic boundary. *Episodes*, **24**, 102–114.

Zeng, J., Cao, C.Q., Davydov, V.I. & Shen, S.Z. 2012. Carbon isotope chemostratigraphy and implications of palaeoclimatic changes during the Cisuralian (Early Permian) in the southern Urals, Russia. *Gondwana Research*, **21**, 601–610.

Zhang, Z.H., Wang, Z.H. & Li, C.Q. 1988. *A Suggestion for Classification of Permian in South Guizhou*. The People's Publishing House of Guizhou, Guiyang.

Zhao, J.K., Sheng, J.Z., Yao, Z.Q., Liang, X.L., Chen, C.Z., Rui, L. & Liao, Z.T. 1981. The Changhsingian and the Permian–Triassic boundary in South China. *Bulletin of the Nanjing Institute of Geology and Palaeontology*, **2**, 1–112.

Zhou, Z.R., Glenister, B.F., Furnish, W.M. & Spinosa, C. 1996. Multi-episodal extinction and ecological differentiation of Permian ammonoids. *Permophiles*, **29**, 52–62.

Zittel, K.A. 1901. *History of Geology and Palaeontology to the End of the Nineteenth Century*. Walter Scott, London.

Advances in numerical calibration of the Permian timescale based on radioisotopic geochronology

JAHANDAR RAMEZANI* & SAMUEL A. BOWRING

Earth, Atmospheric and Planetary Sciences, Massachusetts Institute of Technology, Cambridge, MA 02139, USA

**Correspondence: ramezani@mit.edu*

Abstract: Radioisotopic age determinations targeted at key stratigraphic successions worldwide continue to refine the geological timescale with increasing precision and accuracy and to unravel the tempo of global geological, palaeoclimatic and palaeobiotic processes that have shaped our planet. The last decade has witnessed significant progress in the calibration of the Permian Period through integrated stratigraphic, palaeontological and high-precision geochronological investigations. These studies have largely focused on the Cisuralian and Lopingian stages, particularly the end-Permian mass extinction, whereas much of the Guadalupian and its associated events remain inadequately calibrated. A compilation of the high-precision U–Pb geochronology generated in the past ten years yields ages of 298.92 ± 0.19 Ma for the onset of the Permian, 293.52 ± 0.17 Ma for the base-Sakmarian, 290.10 ± 0.14 Ma for the base-Artinskian, 272.95 ± 0.11 Ma for the Cisuralian–Guadalupian boundary, 265.22 ± 0.34 Ma for the base-Capitanian, 254.14 ± 0.12 Ma for the base-Changhsingian and 251.90 ± 0.10 Ma for the Permian–Triassic boundary. Extension of modern astrochronological methods to the Palaeozoic Era presents new opportunities for broader stratigraphic correlations and enhanced calibration of the Permian timescale.

Supplementary material: Table S1 (U–Pb data) is available at https://doi.org/10.6084/m9.figshare.c.3917425

The Permian is the last and shortest period of the Palaeozoic and spans the geological time between *c.* 298.9 and *c.* 251.9 Ma (Cohen *et al.* 2013, updated). Former divisions of the Permian into 'Early', 'Middle' and 'Late' are typified by three series (epochs) – the Cisuralian, Guadalupian and Lopingian in modern timescales – following the convention of Jin *et al.* (1997). These are, in turn, subdivided into four, three and two stages (ages), respectively (Fig. 1). The marine biostratigraphic schemes that underpin the Permian subdivisions have significantly improved in the last three decades through more methodical investigations into the global distributions and stratigraphic ranges of fossil ammonoid cephalopods, fusulinid forams and, more recently, conodonts. At present, all nine stages of the Permian are recognized based on the first appearance datum (FAD) of conodont species, of which six have a formally ratified global boundary stratotype section and point (GSSP).

During the Permian, the Earth's major land masses had converged to form the supercontinent Pangea. The early Permian also marks a global transition from an icehouse to a greenhouse climate during which vast domains of the Gondwanan subcontinent emerged from the late Palaeozoic ice age. The Permian Period was concomitant with the first appearance of modern trees transitioning from the swamp-dwelling plants of the Carboniferous. The diversification of early amniotes also occurred in the Permian, giving rise to therapsids that dominated the Permian continents, and the first archosaurs that would lead to the evolution of dinosaurs in the Triassic.

The Permian is particularly known for its spectacular mass extinctions and associated reorganizations of marine and terrestrial ecosystems that occurred near the end of the period and led to the establishment of new faunal and floral assemblages. Understanding the patterns, timing and mechanisms of extinction and recovery, as well as the tempo of observed perturbations in the global ocean–atmosphere system, require methodical integrations of sedimentological, palaeontological and geochemical records, often from distant successions. This will not be possible without precise and accurate age models based on state of the art radioisotopic geochronology.

Significant technical improvements in both U–Pb and $^{40}Ar/^{39}Ar$ radioisotopic analyses, the two chronometer schemes most commonly used in calibration of geological time, have come about in the past decade or so. Despite major advances in U–Pb analysis by *in situ* (i.e. microbeam) techniques, such as laser ablation inductively coupled plasma or secondary ion mass spectrometry, isotope

From: Lucas, S. G. & Shen, S. Z. (eds) 2018. *The Permian Timescale*. Geological Society, London, Special Publications, **450**, 51–60.
First published online November 1, 2017, https://doi.org/10.1144/SP450.17

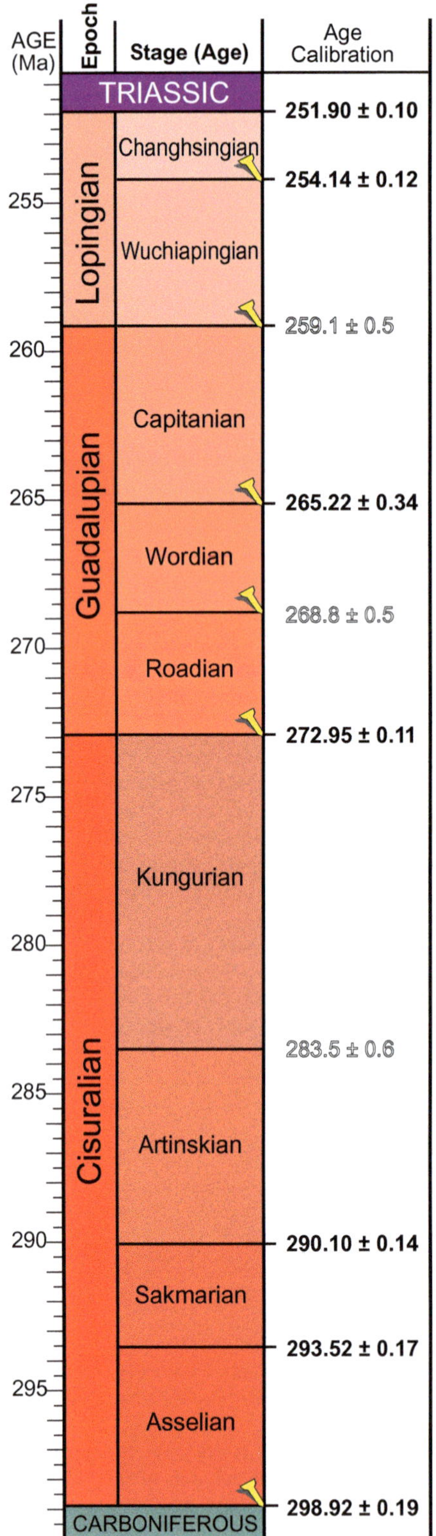

Fig. 1. The Permian timescale, including the numerical calibration of its stage boundaries (right-hand column), based on the present compilation of the geochronological results. Numbers in bold are boundary ages based on the direct radioisotopic calibration discussed in this paper; see text for sources of data. Golden spikes indicate boundaries with formally ratified global stratotype sections and points.

dilution thermal ionization mass spectrometry (ID-TIMS) has remained the method of choice for high-precision U–Pb geochronology applied to the stratigraphic record. The development of single-zircon U–Pb analysis by ID-TIMS and the advent of thermal annealing and chemical leaching pre-treatment method (the chemical abrasion or CA-TIMS of Mattinson 2005) as an effective remedy for radiation-induced Pb loss in zircon have greatly improved the accuracy of the ID-TIMS geochronology (see Bowring *et al.* 2006). The quest for improved precision and accuracy in timescale calibration crystallized into the EARTHTIME initiative and its community-wide drive to eliminate inter-laboratory and inter-chronometer biases. Among the achievements credited to EARTHTIME are the development of standard algorithms and software for the reduction U–Pb data (Bowring *et al.* 2011; McLean *et al.* 2011), re-evaluation of the isotopic composition of natural uranium (Hiess *et al.* 2012) and the production of calibrated U–Pb tracer solutions for ID-TIMS analyses (Condon *et al.* 2015; McLean *et al.* 2015). Because of the steady nature of these developments, the literature today contains multiple vintages of reported geochronological data that are sometimes difficult to reconcile in terms of quality and accuracy and, in some cases, require re-analysis in accordance with new protocols.

This paper reviews the set of modern geochronological data that underpins the present temporal calibration of the Permian Period and is most relevant to the numerical sequencing of its geological and palaeobiotic events. Emphasis has been put on reported geochronology that directly constrains the epoch and stage boundaries or provides tie points for the global correlation of Permian events. We also present new U–Pb geochronology from a previously dated ash bed from the base of the Capitanian Stage in its type locality in western Texas to validate the legacy age. All age uncertainties stated herein are 2σ unless noted otherwise.

Cisuralian Epoch

Carboniferous–Permian boundary

The onset of the Permian coincides with the base of the Asselian Stage (basal Cisuralian) and has been traditionally defined by the appearance of the characteristic ammonoid families, as well as the earliest inflated fusulinid forms attributed to 'Schwagerina' (see Davydov *et al.* 1995). The latter is of practical significance in global correlation of the base-Asselian. Unlike the geographically restricted and largely endemic ammonoid taxa, however, the Carboniferous–Permian conodont zonation has been broadly recognized and thus serves as a more reliable boundary definition. The FAD of the conodont *Streptognathodus isolatus* formally defines the Carboniferous–Permian boundary at its GSSP at Aidalarash Creek in northeastern Kazakhstan (Davydov *et al.* 1998).

Considerable geochronological data on the Cisuralian sections of the southern Urals of Russia and Kazakhstan have been accumulated over the past decade. Ramezani *et al.* (2007) reported high-precision U–Pb geochronology by the ID-TIMS method from the Usolka section in Russia, the basinal correlative of the GSSP *c.* 500 km to the south, which constrained the biostratigraphically defined Carboniferous–Permian boundary (Fig. 1). Four ash beds that straddle the boundary at Usolka produced weighted mean $^{206}Pb/^{238}U$ zircon dates that ranged from 299.22 $\pm$ 0.14 to 298.05 $\pm$ 0.44 Ma (including tracer calibration errors) and yielded a boundary age of 298.92 $\pm$ 0.19 Ma (recalculated after Ramezani *et al.* 2007). The single-zircon analyses of that study used a combination of mechanical and chemical abrasion techniques and predated the EARTHTIME U–Pb tracers and analytical protocols. Re-analysis of the dated samples using the latter recommendations can potentially improve the precision and accuracy of the boundary age.

Cisuralian calibration

Schmitz & Davydov (2012) expanded the lower Cisuralian chronostratigraphy in the southern Urals with additional U–Pb geochronology by the CA-ID-TIMS method using the new EARTHTIME analytical protocols and in the context of a composite biostratigraphic section for the Urals. Their results include three supplementary Permian ash beds from the Usolka reference section, as well as another three from the Dal'ny Tulkas road-cut section with weighted mean $^{206}Pb/^{238}U$ zircon dates (296.69 $\pm$ 0.12 Ma to 288.21 $\pm$ 0.06 Ma) that constrain the lower Cisuralian stage boundaries. Accordingly, the Asselian–Sakmarian stage boundary was calibrated at 295.0 ($\pm$0.3) Ma and the Sakmarian–Artinskian boundary at 290.1 ($\pm$0.2) Ma (Henderson *et al.* 2012*a*; Schmitz & Davydov 2012).

The calibration of the base of the Sakmarian Stage by Schmitz & Davydov (2012) was based on the FAD of the conodonts *Sweetognathus merrilli* and/or *Mesogondolella uralensis* at the Usolka section, which were bracketed by two dated ash beds. A more detailed analysis of the conodont distribution throughout the Usolka section as part of a new GSSP proposal (Chernykh *et al.* 2016) identified a new level marked by the FAD of the conodont *Mesogondolella monstra*, $<$4 m higher in the stratigraphy, as a more suitable definition for the base of the Sakmarian (Henderson 2016). Extrapolation to the new level using the same dated ash beds yields a revised age of 293.52 $\pm$ 0.17 Ma (Fig. 1) for the Asselian–Sakmarian boundary (Chernykh *et al.* 2016).

Age calibration of the upper Cisuralian has been hampered by a general absence of dated ash beds from the candidate GSSPs in the southern Urals and northeastern Nevada, where the FAD of conodont *Neostreptognathodus pnevi* defines the potential GSSP level. Alternatively, a presumed global, unidirectional, shift in the strontium isotopic composition of Cisuralian–Guadalupian seawater, as recorded in conodont elements, has been considered as a potential chronostratigraphic proxy for this interval (Henderson *et al.* 2012*b*). Henderson *et al.* (2012*b*) presented an age-calibrated, lower Cisuralian, strontium isotope curve that was extrapolated up-section and, when correlated with the biostratigraphically defined base-Kungurian at the proposed GSSP, yielded an apparent boundary age of 283.5 $\pm$ 0.5 Ma. This indirect constraint presently serves as the best age estimate for the Artinskian–Kungurian boundary (Fig. 1).

Guadalupian Epoch

Cisuralian–Guadalupian boundary

Despite the abundance of geochronological studies, a large gap in radioisotopic data available for timescale calibration has existed throughout the late Cisuralian and early Guadalupian, spanning nearly 23 myr, close to half the duration of the Permian. As a consequence, temporal constraints on the bases of the Kungurian, Roadian (the Cisuralian–Guadalupian boundary) and Wordian stages have been estimated only by means of broad extrapolations (see Henderson *et al.* 2012*a*). A recent U–Pb geochronological study on the Cisuralian–Guadalupian stage boundary strata in eastern China by Wu *et al.* (2017) has added a key data point to this age gap. They provided a U–Pb zircon date by the CA-ID-TIMS method from an ash bed that marks the boundary between strata containing the conodonts *Sweetognathus subsymmetricus* below and *Jinogondolella nankingensis* above, in the Pingdingshan East Section near Chaohu City, Anhui Province. The reported weighted mean $^{206}Pb/^{238}U$ date of 272.95 $\pm$ 0.11 Ma from this ash bed therefore provides a direct age constraint on the Cisuralian–Guadalupian boundary in South China.

At present, complexities in conodont taxonomy and lineages, as well as uncertainties in the FAD of *J. nankingensis*, prevent a robust correlation between the exact base of the Guadalupian Series in China and the Roadian GSSP in the Guadalupe Mountains National Park, Texas, USA. Pending more rigorous studies, however, the reported date of 272.95 ± 0.11 Ma by Wu *et al.* (2017) currently serves as the best direct radioisotopic age constraint on the global Cisuralian–Guadalupian boundary (Fig. 1). Accordingly, the duration of the Guadalupian Epoch is estimated to be 13.85 ± 0.52 myr, given a Guadalupian–Lopingian boundary age of 259.1 ± 0.5 Ma.

Guadalupian calibrations

The base of the Wordian Stage as defined by the FAD of conodont *Jinogondolella aserrata* at its GSSP at Guadalupe Pass, Guadalupe Mountains National Park in Texas (Glenister *et al.* 1999), remains uncalibrated. The only existing direct radioisotopic age constraint for the base of the Capitanian Stage has come from an ash bed from the lower Bell Canyon Formation in the Guadalupe Mountains National Park in Texas. The ash bed is located stratigraphically 20 m below the *Jinogondolella postserrata* conodont zone, which defines the base of the Capitanian Stage at its GSSP in Nipple Hill (Glenister *et al.* 1999). Bowring *et al.* (1998) reported six zircon U–Pb analyses by the ID-TIMS method from the Nipple Hill ash bed that produced a weighted mean $^{206}Pb/^{238}U$ date of 265.3 ± 0.2 Ma and provided a maximum age estimate for the Capitanian Stage at its base. The latter analyses were based on multi-grain fractions of zircon that were pre-treated by mechanical abrasion; a technique that, in general, is of questionable accuracy compared with the modern CA-ID-TIMS analytical procedures because of the possibility of unresolved Pb loss.

To verify and reinforce the age constraint for the Capitanian GSSP, we analysed additional zircon grains from the same Nipple Hill ash bed sample of Bowring *et al.* (1998) by the CA-ID-TIMS technique at the MIT Isotope Laboratory (Supplementary Table S1, available online). The U–Pb analytical procedures are described in detail in Ramezani *et al.* (2011) and are essentially the same as those used by Burgess *et al.* (2014) in calibrating the age of the Permian–Triassic boundary. Seven analysed single zircons with no outliers produced a weighted mean $^{206}Pb/^{238}U$ date of 265.46 ± 0.27 Ma (2σ analytical uncertainty only), or ± 0.33 Ma if tracer calibration errors are considered, with a MSWD of 1.0 (Fig. 2). The new U–Pb date is consistent within uncertainty with the published date of Bowring *et al.* (1998) and does not necessitate any major revision to the Wordian–Capitanian boundary age, although additional age constraints from above the stage boundary would improve the age model.

Emeishan volcanism and mid-Permian (end-Guadalupian) extinction

Massive eruptions of basaltic lava and emplacement of associated intrusive rocks occurred over an

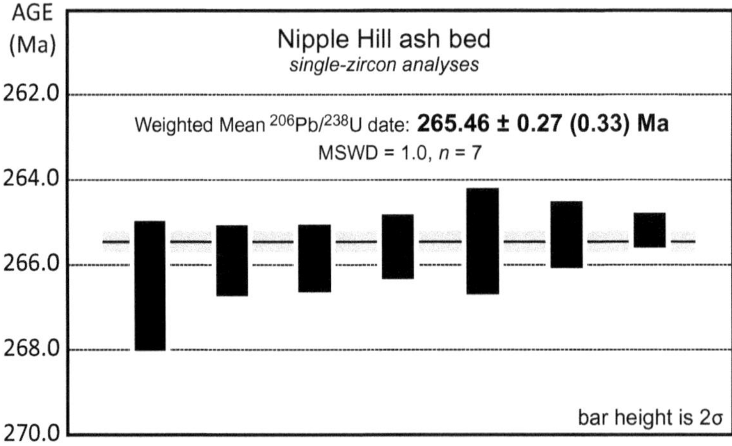

Fig. 2. Date distribution plot of new U–Pb analyses of the Nipple Hill ash bed, originally reported in Bowring *et al.* (1998). Vertical bars represent 2σ analytical uncertainties of individual zircon analyses. The horizontal line and shaded band signify the weighted mean date and its 2σ internal error. The uncertainty in parentheses includes the U–Pb tracer calibration errors; see Table S1 (Supplementary Material, available online) for supporting U–Pb data.

area of *c.* 250 000 km^2 in SW China near the end of the Guadalupian Epoch, which is known as the Emeishan large igneous province (LIP) (e.g. Chung & Jahn 1995; Ali *et al.* 2005). It is believed that the LIP magmatism may have triggered an extinction of mid-Permian age among marine fauna, and possibly land fauna and flora (Courtillot *et al.* 1999; Wignall 2001; Bond *et al.* 2010*a*; Day *et al.* 2015), predating the end-Permian mass extinction. The mid-Permian extinction was locally prevalent among forams, brachiopods, ammonoids, large bivalves, corals and plants, and coincided with a major excursion in seawater carbon isotope composition, together with evidence of global cooling at low latitudes and a drop in sea-level (Stanley & Yang 1994; Jin *et al.* 1995; Isozaki 2009; Isozaki & Aljinovic 2009; Bond et al. 2010*a*).

High-precision radioisotopic dates are generally difficult to obtain from mafic lavas associated with LIP volcanism and geochronological studies on the Emeishan LIP have instead focused on the intrusive dykes and sills that are considered to be contemporaneous feeders to the flood basalts. Initial U–Pb geochronology by the sensitive high-resolution microprobe (SHRIMP) technique from a gabbro belonging to a layered intrusion emplaced within the Emeishan basalts (Zhou *et al.* 2002) produced an average $^{206}Pb/^{238}U$ date of 259 ± 3 Ma (1σ). This age indicated that the LIP magmatism occurred near the Guadalupian–Lopingian boundary, confirming a temporal link between the latter and what has been known as the end-Guadalupian mass extinction (Stanley & Yang 1994; Courtillot *et al.* 1999). Subsequent U–Pb geochronology by SHRIMP and laser ablation inductively coupled mass spectrometry techniques yielded zircon dates from the Emeishan plutonic rocks that ranged from 266 ± 5 to 251 ± 6 Ma, i.e. longer than the entire duration of the Lopingian (Zhou *et al.* 2002, 2008; Guo *et al.* 2004; He *et al.* 2007; Zhong *et al.* 2007, 2009, 2011; Xu *et al.* 2008). These results appear to be in accord with earlier $^{40}Ar/^{39}Ar$ geochronology, which had speculated that the Emeishan magmatism took place in two stages: a precursory stage at *c.* 258–256 Ma and the main stage at *c.* 252.8–251.2 Ma (Lo *et al.* 2002) and thus a temporal link to the end-Permian extinction. Nevertheless, U–Pb geochronology by the CA-ID-TIMS method from a range of Emeishan plutons and mafic dykes displayed a much narrower range of emplacement ages (weighted mean $^{206}Pb/^{238}U$ dates) from 259.6 ± 0.5 Ma to >257 Ma (Shellnutt *et al.* 2012), suggesting that the main pulse of magmatism did not last longer than *c.* 2 myr. Further temporal constraints on the termination of the Emeishan volcanism has been provided by a CA-ID-TIMS weighted mean $^{206}Pb/^{238}U$ date of 259.1 ± 0.5 Ma from a felsic ignimbrite at the uppermost part of its lava succession (Zhong *et al.* 2014). In summary, high-precision U–Pb geochronology appears to confine the Emeishan LIP magmatism to less than 1 myr, with its extrusive activity ending *c.* 7.2 myr prior to the end-Permian extinction.

Some workers have argued that the earlier of the two late Permian extinctions preceded the Lopingian and occurred during the uppermost Capitanian Stage of the Guadalupian (Wignall *et al.* 2009; Bond *et al.* 2010*b*; Sun *et al.* 2010). This is largely based on fossil evidence (conodonts) from SW China that points to the onset of Emeishan eruptions, as well as the collapse of carbonate platforms linked to plume activity, in the mid-Capitanian (Sun *et al.* 2010). Specifically, LIP eruption began in the mid-Capitanian *Jinogondolella altudaensis* conodont zone and continued through the *Jinogondolella xuanhanensis* Zone, the fifth and third biozones below the base of Lopingian, respectively (Bond *et al.* 2010*a*; Sun *et al.* 2010). Although some microfossil losses have been reported from the same biostratigraphic level in South China (Shen & Shi 2009; Wignall *et al.* 2009), the overall temporal correlation of shallow marine biotic crises, plant extinction and carbon isotope excursion to Emeishan eruptions has been difficult to establish. This is, in part, due to the geographical distance between outcrops with well-resolved biostratigraphy (e.g. base-Lopingian GSSP at Penglaitan) and those of the magmatic units suitable for geochronology.

Lopingian Epoch

Guadalupian–Lopingian boundary

The base of the Lopingian Series (Wuchiapingian Stage) is well defined by the FAD of the conodont *Clarkina postbitteri postbitteri* at its GSSP in the Penglaitan section near Laibin, Guangxi Province, South China (Jin *et al.* 2006). Zhong *et al.* (2013) conducted U–Pb analyses by laser ablation inductively coupled mass spectrometry on zircon from the interbedded Guadalupian–Lopingian boundary claystones at Penglaitan and concluded that they were largely clastic in origin and unsuitable for age calibration. Therefore no direct calibration of the boundary exists at this time. The only indirect radioisotopic age constraints on the Guadalupian–Lopingian boundary are from the presumed coeval Emeishan LIP rocks. The current extrapolated age for the Guadalupian–Lopingian stage boundary is 259.1 ± 0.5 Ma (Cohen *et al.* 2013, updated), giving the Lopingian a total duration of 7.2 ± 0.5 myr (Fig. 1).

Lopingian calibrations

Extensive U–Pb geochronology by the CA-ID-TIMS method on 29 ash beds from nine Lopingian successions throughout China by Shen *et al.* (2011) has provided a wealth of age data for the calibration of the Lopingian timescale, as well as the Permian–Triassic boundary. Although these data were produced using the chemical abrasion technique adapted from Mattinson (2005), they predate the EARTHTIME U–Pb tracers and recommended analytical protocols. Therefore slight age biases relative to the more recent U–Pb geochronologicalal data (<0.08%) are not unexpected. Two stratigraphically well-constrained weighted mean $^{206}Pb/^{238}U$ dates of 254.31 ± 0.07 and 253.60 ± 0.08 Ma from the Shangsi section in central China and their biostratigraphic correlation to the Meishan section (Shen *et al.* 2011) place the Wuchiapingian–Changhsingian stage boundary at 254.14 ± 0.12 Ma (Fig. 1). This boundary is defined in China by the FAD of conodont *Clarkina wangi* at its GSSP in Meishan (Jin *et al.* 2006).

Permian–Triassic boundary, end-Permian mass extinction and the Siberian Traps volcanism

The onset of the Triassic has been defined by the FAD of the conodont *Hindeodus parvus* at its GSSP in Meishan, Zhejiang Province, eastern China (see Mei *et al.* 1998). The end of the Permian Period is marked by a severe mass extinction, when an estimated 90% of the Permian marine species disappeared in what is known as the greatest ecological catastrophe of the Phanerozoic (Erwin 2015). In the past two decades, several major U–Pb geochronological studies with emphasis on the Chinese Permian–Triassic sedimentary records, and the Meishan GSSP in particular, have attempted to constrain the timing and abruptness of the end-Permian extinction (Bowring *et al.* 1998; Mundil *et al.* 2004; Shen *et al.* 2011; Burgess *et al.* 2014). These studies have been concomitant with new developments in radioisotopic geochronology, most notably the increasing precision and accuracy of the U–Pb analytical techniques.

The latest calibration of the end-Permian mass extinction at Meishan using modern U–Pb analytical methods and protocols has placed the Permian–Triassic boundary at 251.902 ± 0.024 Ma (±0.104 Ma including U–Pb tracer calibration errors; Fig. 1), whereas the onset of mass extinction is determined to be at 251.941 ± 0.037 Ma, with an estimated duration for the extinction interval of 61 ± 48 kyr (Burgess *et al.* 2014). Recent U–Pb geochronology from more extended platform Permian–Triassic sections in South China have yielded boundary ages generally consistent with that in the deeper slope section at Meishan. These include a boundary age of 251.985 ± 0.097 Ma from Taiping (Lehrmann *et al.* 2015). Newly reported U–Pb boundary ages of 251.938 ± 0.029 Ma from Dongpan and 251.982 ± 0.031 Ma from Penglaitan (Baresel *et al.* 2017) were, however, focused mainly on lithostratigraphic boundaries and not based on a high-resolution biostratigraphy of diagnostic fossils.

The large-scale eruption of continental flood basalts at the end of the Permian, which formed the Siberian Traps, has long invoked notions of a causal connection to the end-Permian ecological collapse and biotic crisis (see Svensen *et al.* 2009; Black *et al.* 2012). The detailed U–Pb geochronology of Burgess & Bowring (2015) showed that explosive volcanism associated with the Siberian Traps started on or after 255.6 Ma and two-thirds of the total lava volume erupted prior to, and coeval with, the documented timing of the end-Permian mass extinction. In addition, the massive emplacement of mafic sills into supracrustal rocks began concomitant with the extinction and continued for at least 500 myr into the Early Triassic. These have reinforced the role of LIP magmatism as both a viable trigger for the end-Permian mass extinction and a suppression mechanism for post-extinction recovery. Full-scale ecological recovery from the mass extinction took nearly 4.6 myr (Lehrmann *et al.* 2006, 2015).

Other stratigraphic aspects of the Permian system

Permian magnetostratigraphy

The Permian is characterized by a prominent period of stable, reverse, geomagnetic polarity known as the Kiaman Reverse Superchron, which started from the mid-Carboniferous and spans nearly 50 myr into the Guadalupian (Irving & Parry 1963). It was interrupted prior to the Lopingian Stage by the Illawarra reversal, a magnetostratigraphic turnover that marks the onset of the late Permian and Triassic mixed polarity time. The Illawarra reversal serves as an important tie point for global correlation of the Guadalupian Stage. Unfortunately, rocks within the Guadalupian type locality in Texas do not exhibit a well-preserved magnetism. However, the shelf/back-reef facies of the Guadalupe Mountains in southeastern New Mexico records a polarity reversal (Peterson & Nairn 1971), which, when projected onto the GSSP section, suggests a reversal to normal polarity in the uppermost Wordian and near the base of the Capitanian (Menning 2000). The U–Pb age calibration of Bowring *et al.* (1998), supported by additional data presented herein from the Nipple Hill GSSP, therefore, places the Illawarra reversal at *c.* 265.3 Ma. A possible causal relationship between the Illawarra reversal, Emeishan LIP volcanism, sea-level drop and global environmental change at the end of Guadalupian has been proposed by some workers (Isozaki 2009).

Permian tetrapod biostratigraphy

Although the earliest amniotes are known from the mid-Carboniferous, major diversification of tetrapods occurred in Permian time. The record of Permian and Early Triassic land vertebrates is best preserved in the intracratonic sedimentary rocks of the Karoo Supergroup of South Africa. The Karoo tetrapods were dominated by therapsids, 'mammal-like reptiles' that roamed the Permian landscape in abundance and form the basis of the Permian (Guadalupian–Lopingian) land vertebrate biostratigraphy. Radioisotopic geochronology with good biostratigraphic control from the Karoo Basin has been reported from the fluvial–deltaic complex of the Beaufort Group of Guadalupian to Early Triassic age (Coney *et al.* 2007; Rubidge *et al.* 2013; Gastaldo *et al.* 2015). These studies have provided direct temporal calibration for four of the six late Permian vertebrate biozones of global significance, namely the *Pristerognathus*, *Tropidostoma*, *Cistecephalus* and *Dicynodon* assemblage zones, spanning the Capitanian to the late Lopingian. The Permian–Triassic boundary has been biostratigraphically correlated to a faunal turnover at the end of the *Dicynodon* Assemblage Zone (Ward *et al.* 2005).

The most recent analysis of the Karoo tetrapod fauna, along with U–Pb geochronology, reported by Day *et al.* (2015) correlated the Guadalupian–Lopingian boundary with the *Pristerognathus* Assemblage Zone and identified a significant extinction event among dinocephalian therapsids at *c.* 260 Ma. The close temporal relationship of the latter with the Emeishan LIP magmatism has raised the feasibility of concurrent marine and terrestrial mass extinctions of mid-Permian (end-Guadalupian) age (Day *et al.* 2015). By contrast, the coincidence of the presumed end-Permian tetrapod turnover with marine ecosystem collapse and floral extinction remains contested (e.g. Gastaldo *et al.* 2015).

Permian cyclostratigraphy and astrochronology

Astronomical timescales based on cyclostratigraphic records of orbitally forced climate that are tuned to astronomical solutions are widely used in sequencing of the Cenozoic time. Their application to pre-Cenozoic records, however, is hampered by ambiguous cyclostratigraphic interpretations, as well as uncertainties in astronomical models due to the chaotic behaviour of the solar system in deep time. Therefore, it is imperative for the astronomical timescales to be substantiated by independent radioisotopic geochronology at high precision if they are to be extended further back to Mesozoic and Palaeozoic times (e.g. Wang *et al.* 2016).

Cyclic sedimentary successions (cyclothems) are widespread throughout the upper Carboniferous–Permian deposits of Europe, South America, South China and the North American Midcontinent (e.g. Isbell *et al.* 2003; Heckel *et al.* 2007; Franco *et al.* 2011; Wang *et al.* 2013). Models of orbital control on cyclothem deposition, along with high-resolution age calibration of conodont biostratigraphy, allow the construction of cyclothem-based chronostratigraphic frameworks that can be extended to the lower Cisuralian successions, even where direct radioisotopic geochronology is not available (Schmitz & Davydov 2012). This has been essential to understanding the Cisuralian transition from an icehouse to greenhouse climate associated with the termination of the late Palaeozoic ice age.

Wu *et al.* (2013) applied cyclostratigraphic methods to the Lopingian successions of Meishan and Shangsi in eastern and south-central China and constructed an astronomical timescale by empirically extending the Earth's orbital parameters back into the Permian. They used U–Pb geochronology established at these localities by Shen *et al.* (2011) to test the Milankovitch forcing of the cyclic magnetic properties of sedimentary rocks and to anchor the cyclicity in absolute time. The resulting age-calibrated astronomical timescale based on the Shangsi section yields a total duration of 7.793 myr for the Lopingian Epoch (Wu *et al.* 2013). This duration is only slightly longer than the estimated 7.2 $\pm$ 0.5 myr based on the extrapolated age of the Guadalupian–Lopingian boundary.

This paper benefited from helpful discussions with Shuzhong Shen and his research group. Reviews by Charles Henderson and Spencer Lucas improved the quality of the paper.

References

Ali, J.R., Thompson, G.M., Zhou, M.F. & Song, X.Y. 2005. Emeishan large igneous province, SW China. *Lithos*, **79**, 475–489, https://doi.org/10.1016/j.lithos.2004.09.013

Baresel, B., Bucher, H., Brosse, M., Cordey, F., Guodun, K. & Schaltegger, U. 2017. Precise age for the Permian–Triassic boundary in South China from high precision U–Pb geochronology and Bayesian age-depth modelling. *Solid Earth*, **8**, 361–378, https://doi.org/10.5194/se-2016-145

Black, B.A., Elkins-Tanton, L.T., Rowe, M.C. & Peate, I.U. 2012. Magnitude and consequences of volatile release from the Siberian Traps. *Earth and Planetary Science Letters*, **317**, 363–373, https://doi.org/10.1016/j.epsl.2011.12.001

Bond, D.P.G., Hilton, J., Wignall, P.B., Ali, J.R., Stevens, L.G., Sun, Y.D. & Lai, X.L. 2010*a*. The Middle Permian (Capitanian) mass extinction on land and in the oceans. *Earth-Science Reviews*, **102**, 100–116, https://doi.org/10.1016/j.earscirev.2010.07.004

Bond, D.P.G., Wignall, P.B. *et al.* 2010*b*. The mid-Capitanian (Middle Permian) mass extinction and carbon isotope record of South China. *Palaeogeography, Palaeoclimatology, Palaeoecology*, **292**, 282–294, https://doi.org/10.1016/j.palaeo.2010.03.056

Bowring, J.F., McLean, N.M. & Bowring, S.A. 2011. Engineering cyber infrastructure for U–Pb geochronology: Tripoli and U–Pb_Redux. *Geochemistry, Geophysics, Geosystems*, **12**, Q0AA19.

Bowring, S.A., Erwin, D.H., Jin, Y.G., Martin, M.W., Davidek, K. & Wang, W. 1998. U/Pb zircon geochronology and tempo of the end-Permian mass extinction. *Science*, **280**, 1039–1045.

Bowring, S.A., Schoene, B., Crowley, J.L., Ramezani, J. & Condon, D.J. 2006. High-precision U–Pb zircon geochronology and the stratigraphic record: progress and promise. *In*: Olszewski, T.D. (ed.) *Geochronology: Emerging Opportunities*. Paleontological Society, Papers, **12**, 25–45.

Burgess, S.D., Bowring, S. & Shen, S.Z. 2014. High-precision timeline for Earth's most severe extinction. *Proceedings of the National Academy of Sciences of the United States of America*, **111**, 3316–3321, https://doi.org/10.1073/pnas.1317692111

Burgess, S.D. & Bowring, S.A. 2015. High-precision geochronology confirms voluminous magmatism before, during, and after Earth's most severe extinction. *Science Advances*, **1**, https://doi.org/10.1126/sciadv.1500470

Chernykh, V.V., Chuvashov, B.I., Shen, S.Z. & Henderson, C. 2016. Proposal for the Global Stratotype Section and Point (GSSP) for the base-Sakmarian Stage (Lower Permian). *Permophiles*, **63**, 4–17.

Chung, S.L. & Jahn, B.M. 1995. Plume-lithosphere interaction in generation of the Emeishan flood basalts at the Permian–Triassic boundary. *Geology*, **23**, 889–892, https://doi.org/10.1130/0091-7613(1995)0232.3.Co;2

Cohen, K.M., Finney, S.C., Gibbard, P.L. & Fan, J.X. 2013 (updated). The ICS International Chronostratigraphic Chart. *Episodes*, **36**, 199–204.

Condon, D.J., Schoene, B., McLean, N.M., Bowring, S.A. & Parrish, R.R. 2015. Metrology and traceability of U–Pb isotope dilution geochronology (EARTHTIME Tracer Calibration Part I). *Geochimica et Cosmochimica Acta*, **164**, 464–480.

Coney, L., Reimold, W.U. *et al.* 2007. The continental Permian-Triassic boundary in the Karoo Basin, South Africa: no evidence for meteorite impact. *Meteoritics & Planetary Science*, **42**, A32–A32.

Courtillot, V., Jaupart, C., Manighetti, I., Tapponnier, P. & Besse, J. 1999. On causal links between flood basalts and continental breakup. *Earth and Planetary Science Letters*, **166**, 177–195, https://doi.org/10.1016/S0012-821x(98)00282-9

Davydov, V.I., Glenister, B.F., Spinosa, C., Ritter, S.M., Chernykh, V.V., Wardlaw, B.R. & Snyder, W.S. 1995. Proposal of Aidaralash as GSSP for the base of the Permian System. *Permophiles*, **26**, 1–9.

Davydov, V.I., Glenister, B.F., Spinosa, C., Ritter, S.M., Chernykh, V.V., Wardlaw, B.R. & Snyder, W.S. 1998. Proposal of Aidaralash as Global Stratotype Section and Point (GSSP) for base of the Permian System. *Episodes*, **21**, 11–18.

Day, M.O., Ramezani, J., Bowring, S.A., Sadler, P.M., Erwin, D.H., Abdala, F. & Rubidge, B.S. 2015. When and how did the terrestrial mid-Permian mass extinction occur? Evidence from the tetrapod record of the Karoo Basin, South Africa. *Proceedings of the Royal Society B, Biological Sciences*, **282**, https://doi.org/10.1098/Rspb.2015.0834

Erwin, D.H. 2015. *Extinction: How Life on Earth Nearly Ended 250 million Years Ago*. Princeton University Press, Princeton, NJ.

Franco, D.R., Hinnov, L.A. & Ernesto, M. 2011. Spectral analysis and modeling of microcyclostratigraphy in late Paleozoic glaciogenic rhythmites, Paraná Basin, Brazil. *Geochemistry, Geophysics, Geosystems*, **12**, https://doi.org/10.1029/2011GC003602

Gastaldo, R.A., Kamo, S.L., Neveling, J., Geissman, J.W., Bamford, M. & Looy, C.V. 2015. Is the vertebrate-defined Permian–Triassic boundary in the Karoo Basin, South Africa, the terrestrial expression of the end-Permian marine event? *Geology*, **43**, 939–942, https://doi.org/10.1130/G37040.1

Glenister, B.F., Wardlaw, B.R. *et al.* 1999. Proposal of Guadalupian and component Roadian, Wordian and Capitanian Stages as international standards for the Middle Permian series. *Permophiles*, **34**, 3–11.

Guo, F., Fan, W.M., Wang, Y.J. & Li, C.W. 2004. When did the Emeishan mantle plume activity start? Geochronological and geochemical evidence from ultramafic-mafic dikes in southwestern China. *International Geology Review*, **46**, 226–234, https://doi.org/10.2747/0020-6814.46.3.226

He, B., Xu, Y.G., Huang, X.L., Luo, Z.Y., Shi, Y.R., Yang, Q.J. & Yu, S.Y. 2007. Age and duration of the Emeishan flood volcanism, SW China: geochemistry and SHRIMP zircon U–Pb dating of silicic ignimbrites, post-volcanic Xuanwei Formation and clay tuff at the Chaotian section. *Earth and Planetary Science Letters*, **255**, 306–323, https://doi.org/10.1016/j.epsl.2006.12.021

Heckel, P.H., Alekseev, A.S. *et al.* 2007. Cyclothem ['digital'] correlation and biostratigraphy across the global Moscovian-Kasimovian-Gzhelian stage boundary interval (Middle-Upper Pennsylvanian) in North America and Eastern Europe. *Geology*, **35**, 607–610, https://doi.org/10.1130/G23564a.1

Henderson, C.M., Davydov, V.I. & Wardlaw, B.R. 2012*a*. The Permian period. *In*: Gradstein, F.M., Ogg, J.G., Schmitz, M.D. & Ogg, G.M. (eds) *The Geological Time Scale 2012*. Vol. **2**. Elsevier, Amsterdam, 653–680.

Henderson, C.M., Wardlaw, B.R., Davydov, V.I., Schmitz, M.D., Schiappa, T.A., Tierney, K.E. & Shen, S.Z. 2012*b*. Proposal for base-Kungurian GSSP. *Permophiles*, **56**, 8–21.

Henderson, C.M. 2016. Permian conodont biostratigraphy. In: Lucas, S.G. & Shen, S.Z. (eds) *The Permian Timescale*. Geological Society, London, Special Publications, **450**. First published online December 14, 2016, https://doi.org/10.1144/SP450.9

Hiess, J., Condon, D.J., McLean, N. & Noble, S.R. 2012. $^{238}U/^{235}U$ systematics in terrestrial uranium-bearing minerals. *Science*, **335**, 1610–1614, https://doi.org/10.1126/science.1215507

Irving, E. & Parry, L.G. 1963. The magnetism of some Permian rocks from New-South-Wales. *Geophysical Journal of the Royal Astronomical Society*, **7**, 395–411, https://doi.org/10.1111/j.1365-246X.1963.tb07084.x

Isbell, J.L., Miller, M.F., Wolfe, K.L. & Lenaker, P.A. 2003. Timing of late Paleozoic glaciation in Gondwana: was glaciation responsible for the development of Northern Hemisphere cyclothems? *In*: Chan, M.A. & Archer, A.W. (eds) *Extreme Depositional Environments: Mega End Members in Geologic Time*. Geological Society of America, Boulder, CO.

Isozaki, Y. 2009. Illawarra Reversal: the fingerprint of a superplume that triggered Pangean breakup and the end-Guadalupian (Permian) mass extinction. *Gondwana Research*, **15**, 421–432, https://doi.org/10.1016/j.gr.2008.12.007

Isozaki, Y. & Aljinovic, D. 2009. End-Guadalupian extinction of the Permian gigantic bivalve Alatoconchidae: end of gigantism in tropical seas by cooling. *Palaeogeography, Palaeoclimatology, Palaeoecology*, **284**, 11–21, https://doi.org/10.1016/j.palaeo.2009.08.022

Jin, Y.G., Zhang, J. & Shang, Q.-H. 1995. Pre-Lopingian catastrophic event of marine faunas. *Acta Palaeontologica Sinica*, **34**, 410–427.

Jin, Y.G., Wardlaw, B.R., Glenister, B.F. & Kotlyar, G.V. 1997. Permian chronostratigraphic subdivisions. *Episodes*, **20**, 10–15.

Jin, Y.G., Wang, Y., Henderson, C., Wardlaw, B.R., Shen, S.Z. & Cao, C.Q. 2006. The Global Boundary Stratotype Section and Point (GSSP) for the base of Changhsingian Stage (Upper Permian). *Episodes*, **29**, 175–182.

Lehrmann, D.J., Ramezani, J. et al. 2006. Timing of recovery from the end-Permian extinction; geochronologic and biostratigraphic constraints from south China. *Geology*, **34**, 1053–1056.

Lehrmann, D.J., Stepchinski, L. et al. 2015. An integrated biostratigraphy (conodonts and foraminifers) and chronostratigraphy (paleomagnetic reversals, magnetic susceptibility, elemental chemistry, carbon isotopes and geochronology) for the Permian-Upper Triassic strata of Guandao section, Nanpanjiang Basin, south China. *Journal of Asian Earth Sciences*, **108**, 117–135, https://doi.org/10.1016/j.jseaes.2015.04.030

Lo, C.H., Chung, S.L., Lee, T.Y. & Wu, G.Y. 2002. Age of the Emeishan flood magmatism and relations to Permian–Triassic boundary events. *Earth and Planetary Science Letters*, **198**, 449–458, https://doi.org/10.1016/S0012-821x(02)00535-6

Mattinson, J.M. 2005. Zircon U/Pb chemical abrasion (CA-TIMS) method; combined annealing and multi-step partial dissolution analysis for improved precision and accuracy of zircon ages. *Chemical Geology*, **220**, 47–66.

McLean, N.M., Bowring, J.F. & Bowring, S.A. 2011. An algorithm for U–Pb isotope dilution data reduction and uncertainty propagation. *Geochemistry, Geophysics, Geosystems*, **12**, Q0AA18.

McLean, N.M., Condon, D.J., Schoene, B. & Bowring, S.A. 2015. Evaluating uncertainties in the calibration of isotopic reference materials and multi-element isotopic tracers (EARTHTIME Tracer Calibration Part II). *Geochimica et Cosmochimica Acta*, **164**, 481–501.

Mei, S., Zhang, K. & Wardlaw, B.R. 1998. A refined succession of Changhsingian and Griesbachian neogondolellid conodonts from the Meishan section, candidate of the global stratotype section and point of the Permian–Triassic boundary. *Palaeogeography, Palaeoclimatology, Palaeoecology*, **143**, 213–226, https://doi.org/10.1016/S0031-0182(98)00112-6

Menning, M. 2000. Magnetostratigraphic results from the Middle Permian type section, Guadalupe Mountains, West Texas. *Permophiles*, **37**, 16.

Mundil, R., Ludwig, K.R., Metcalfe, I. & Renne, P.R. 2004. Age and timing of the Permian mass extinctions: U/Pb dating of closed-system zircons. *Science*, **305**, 1760–1763.

Peterson, D.N. & Nairn, A.E.M. 1971. Palaeomagnetism of Permian Redbeds from South-Western United-States. *Geophysical Journal of the Royal Astronomical Society*, **23**, 191–205, https://doi.org/10.1111/j.1365-246X.1971.tb01812.x

Ramezani, J., Schmitz, M.D., Davydov, V.I., Bowring, S.A., Snyder, W.S. & Northrup, C.J. 2007. High-precision U–Pb zircon age constraints on the Carboniferous–Permian boundary in the Southern Urals stratotype. *Earth and Planetary Science Letters*, **256**, 244–257.

Ramezani, J., Hoke, G.D. et al. 2011. High-precision U–Pb zircon geochronology of the Late Triassic Chinle Formation, Petrified Forest National Park (Arizona, USA): temporal constraints on the early evolution of dinosaurs. *Geological Society of America Bulletin*, **123**, 2142–2159, https://doi.org/10.1130/b30433.1

Rubidge, B.S., Erwin, D.H., Ramezani, J., Bowring, S.A. & de Klerk, W.J. 2013. High-precision temporal calibration of Late Permian vertebrate biostratigraphy: U–Pb zircon constraints from the Karoo Supergroup, South Africa. *Geology*, **41**, 363–366, https://doi.org/10.1130/g33622.1

Schmitz, M.D. & Davydov, V.I. 2012. Quantitative radiometric and biostratigraphic calibration of the Pennsylvanian–Early Permian (Cisuralian) time scale and pan-Euramerican chronostratigraphic correlation. *Geological Society of America Bulletin*, **124**, 549–577, https://doi.org/10.1130/b30385.1

Shellnutt, J.G., Denyszyn, S.W. & Mundil, R. 2012. Precise age determination of mafic and felsic intrusive rocks from the Permian Emeishan large igneous province (SW China). *Gondwana Research*, **22**, 118–126, https://doi.org/10.1016/j.gr.2011.10.009

SHEN, S.Z. & SHI, G.R. 2009. Latest Guadalupian brachiopods from the Guadalupian/Lopingian boundary GSSP section at Penglaitan in Laibin, Guangxi, South China and implications for the timing of the pre-Lopingian crisis. *Palaeoworld*, **18**, 152–161.

SHEN, S.-Z., CROWLEY, J.L. *ET AL.* 2011. Calibrating the End-Permian mass extinction. *Science*, **334**, 1367–1372, https://doi.org/10.1126/science.1213454

STANLEY, S.M. & YANG, X. 1994. A double mass extinction at the end of the Paleozoic Era. *Science*, **266**, 1340–1344, https://doi.org/10.1126/science.266.5189.1340

SUN, Y.D., LAI, X.L. *ET AL.* 2010. Dating the onset and nature of the Middle Permian Emeishan large igneous province eruptions in SW China using conodont biostratigraphy and its bearing on mantle plume uplift models. *Lithos*, **119**, 20–33, https://doi.org/10.1016/j.lithos.2010.05.012

SVENSEN, H., PLANKE, S., POLOZOV, A.G., SCHMIDBAUER, N., CORFU, F., PODLADCHIKOV, Y.Y. & JAMTVEIT, B. 2009. Siberian gas venting and the end-Permian environmental crisis. *Earth and Planetary Science Letters*, **277**, 490–500, https://doi.org/10.1016/j.epsl.2008.11.015

WANG, T., RAMEZANI, J., WANG, C., WU, H., HE, H. & BOWRING, S.A. 2016. High-precision U–Pb geochronologic constraints on the Late Cretaceous terrestrial cyclostratigraphy and geomagnetic polarity from the Songliao Basin, Northeast China. *Earth and Planetary Science Letters*, **446**, 37–44, https://doi.org/10.1016/j.epsl.2016.04.007

WANG, X., QIE, W. *ET AL.* 2013. Carboniferous and lower Permian sedimentological cycles and biotic events of South China. *In*: GĄSIEWICZ, A. & SŁOWAKIEWICZ, M. (eds) *Palaeozoic Climate Cycles: Their Evolutionary and Sedimentological Impact*. Geological Society, London, Special Publications, **376**, 33–46, https://doi.org/10.1144/sp376.11

WARD, P.D., BOTHA, J. *ET AL.* 2005. Abrupt and gradual extinction among Late Permian land vertebrates in the Karoo Basin, South Africa. *Science*, **307**, 709–714, https://doi.org/10.1126/science.1107068

WIGNALL, P.B. 2001. Large igneous provinces and mass extinctions. *Earth-Science Reviews*, **53**, https://doi.org/10.1016/S0012-8252(00)00037-4

WIGNALL, P.B., SUN, Y.D. *ET AL.* 2009. Volcanism, mass extinction, and carbon isotope fluctuations in the Middle Permian of China. *Science*, **324**, 1179–1182, https://doi.org/10.1126/science.1171956

WU, H.C., ZHANG, S.H., HINNOV, L.A., JIANG, G.Q., FENG, Q.L., LI, H.Y. & YANG, T.S. 2013. Time-calibrated Milankovitch cycles for the late Permian. *Nature Communications*, **4, 2452**, https://doi.org/10.1038/ncomms3452

WU, Q., RAMEZANI, J. *ET AL.* 2017. Calibrating the Guadalupian Series (Middle Permian) of South China. *Palaeogeography, Palaeoclimatology, Palaeoecology*, **466**, 361–372, https://doi.org/10.1016/j.palaeo.2016.11.011

XU, Y.G., LUO, Z.Y., HUANG, X.L., HE, B., XIAO, L., XIE, L.W. & SHI, Y.R. 2008. Zircon U–Pb and Hf isotope constraints on crustal melting associated with the Emeishan mantle plume. *Geochimica et Cosmochimica Acta*, **72**, 3084–3104, https://doi.org/10.1016/j.gca.2008.04.019

ZHONG, H., ZHU, W.G., CHU, Z.Y., HE, D.F. & SONG, X.Y. 2007. Shrimp U–Pb zircon geochronology, geochemistry, and Nd-Sr isotopic study of contrasting granites in the Emeishan large igneous province, SW China. *Chemical Geology*, **236**, 112–133, https://doi.org/10.1016/j.chemgeo.2006.09.004

ZHONG, H., ZHU, W.G., HU, R.Z., XIE, L.W., HE, D.F., LIU, F. & CHU, Z.Y. 2009. Zircon U–Pb age and Sr-Nd-Hf isotope geochemistry of the Panzhihua A-type syenitic intrusion in the Emeishan large igneous province, southwest China and implications for growth of juvenile crust. *Lithos*, **110**, 109–128, https://doi.org/10.1016/j.lithos.2008.12.006

ZHONG, H., CAMPBELL, I.H., ZHU, W.G., ALLEN, C.M., HU, R.Z., XIE, L.W. & HE, D.F. 2011. Timing and source constraints on the relationship between mafic and felsic intrusions in the Emeishan large igneous province. *Geochimica et Cosmochimica Acta*, **75**, 1374–1395, https://doi.org/10.1016/j.gca.2010.12.016

ZHONG, Y.T., HE, B. & XU, Y.G. 2013. Mineralogy and geochemistry of claystones from the Guadalupian–Lopingian boundary at Penglaitan, South China: insights into the pre-Lopingian geological events. *Journal of Asian Earth Sciences*, **62**, 438–462, https://doi.org/10.1016/j.jseaes.2012.10.028

ZHONG, Y.T., HE, B., MUNDIL, R. & XU, Y.G. 2014. CA-TIMS zircon U–Pb dating of felsic ignimbrite from the Binchuan section: implications for the termination age of Emeishan large igneous province. *Lithos*, **204**, 14–19, https://doi.org/10.1016/j.lithos.2014.03.005

ZHOU, M.F., MALPAS, J. *ET AL.* 2002. A temporal link between the Emeishan large igneous province (SW China) and the end-Guadalupian mass extinction. *Earth and Planetary Science Letters*, **196**, 113–122.

ZHOU, M.F., ARNDT, N.T., MALPAS, J., WANG, C.Y. & KENNEDY, A.K. 2008. Two magma series and associated ore deposit types in the Permian Emeishan large igneous province, SW China. *Lithos*, **103**, 352–368, https://doi.org/10.1016/j.lithos.2007.10.006

A geomagnetic polarity timescale for the Permian, calibrated to stage boundaries

MARK W. HOUNSLOW[1]* & YURI P. BALABANOV[2]

[1]*Palaeomagnetism, Environmental Magnetism, Lancaster Environment Centre, Lancaster University, Bailrigg, Lancaster LA1 4YW, UK*

[2]*Institute of Geology and Petroleum Technologies, Kazan Federal University, Kremlyovskaya ul. 18, Kazan 420008, Republic of Tatarstan, Russian Federation*

**Correspondence: m.hounslow@lancaster.ac.uk*

Abstract: The reverse polarity Kiaman Superchron has strong evidence for at least three, or probably four, normal magnetochrons during the early Permian. Normal magnetochrons are during the early Asselian (base CI1r.1n at 297.94 ± 0.33 Ma), late Artinskian (CI2n at 281.24 ± 2.3 Ma), mid-Kungurian (CI3n at 275.86 ± 2.0 Ma) and Roadian (CI3r.an at 269.54 ± 1.6 Ma). The mixed-polarity Illawarra Superchron begins in the early Wordian at 266.66 ± 0.76 Ma. The Wordian–Capitanian interval is biased to normal polarity, but the basal Wuchiapingian begins the beginning of a significant reverse polarity magnetochron LP0r, with an overlying mixed-polarity interval through the later Lopingian. No significant magnetostratigraphic data gaps exist in the Permian geomagnetic polarity record. The early Cisuralian magnetochrons are calibrated to a succession of fusulinid zones, the later Cisuralian and Guadalupian to a conodont and fusulinid biostratigraphy, and Lopingian magnetochrons to conodont zonations. Age calibration of the magnetochrons is obtained through a Bayesian approach using 35 radiometric dates, and 95% confidence intervals on the ages and chron durations are obtained. The dating control points are most numerous in the Gzhelian–Asselian, Wordian and Changhsingian intervals. This significant advance should provide a framework for better correlation and dating of the marine and non-marine Permian.

Igneous and sedimentary rocks record the Earth's magnetic field at the time of their formation, via their small content of mostly Fe-oxides. This is recorded as a remanent magnetization, which needs to be stable with time, to resist potential later remagnetization events and be subsequently extracted using palaeomagnetic measurements. The first study of remanent magnetization of Permian rocks was by Mercanton (1926) using volcanic rocks from near Kiama on the New South Wales coast in Australia. This followed earlier studies on much younger rock by Brunhes (1906), in which the remanent magnetization directions recorded had orientations similar to the modern geomagnetic field, which is now defined as having a *normal* polarity. However, some volcanic rocks recorded a remanent magnetization direction opposite to the modern field (*reverse* polarity), which Matuyama (1929) suggested recorded a reversal in the main (i.e. dipole) component of the Earth's magnetic field (see discussion of these early developments in Jacobs 1963). Mercanton (1926) was the first to identify remanent magnetization in Permian rocks with a reverse polarity, but science did not recognize the significance of these Australian volcanics until the re-study of the New South Wales coastal sections by Irving & Parry (1963). The reverse polarity of other early Permian volcanics were studied earlier by Creer *et al.* (1955), and red-bed sediments by Doell (1955), Graham (1955) and Khramov (1958).

The pioneer in our understanding of using changes in the polarity of the Earth's magnetic field for correlation and dating was A.N. Khramov, who, in Khramov (1958), outlined a rudimentary polarity stratigraphy from late Permian and early Triassic sections in the Vyatka River region of the Moscow Basin, with details of this work later appearing in Khramov (1963*b*). Khramov (1958) discussed issues of data quality and cross-validation by exploring the concepts of utilizing data from multiple sections, with minimum sampling requirements to define intervals (magnetozones) of single polarity, concepts which are now embodied in the magnetostratigraphic quality criteria proposed by Opdyke & Channell (1996).

Irving & Parry (1963) later defined a polarity stratigraphy from the late Carboniferous through to the Permian, and into the Triassic, using Permian palaeopole-type palaeomagnetic data coming from sedimentary, volcanic and igneous units from most of the major continents. They proposed using the name Kiaman (from Mercanton's early work near

From: LUCAS, S. G. & SHEN, S. Z. (eds) 2018. *The Permian Timescale*. Geological Society, London, Special Publications, **450**, 61–103.
First published online December 8, 2016, https://doi.org/10.1144/SP450.8

Kiama) for the predominantly reverse polarity interval from the late Carboniferous until the mid-Permian. Later, Irving (1971) suggested substituting the more cumbersome 'late Palaeozoic Reversed Interval' for the Kiaman interval, based on limiting proliferation of new names for geomagnetic chron intervals (see a longer discussion in Klootwijk *et al.* 1994). This work will refer to this long duration polarity interval as the Kiaman Superchron (as opposed to the more cumbersome 'Permo-Carboniferous quiet interval' or superchron of Irving & Pullaiah 1976). The start of the reverse and normal polarity interval following the end of the Kiaman Superchron in the mid-Permian is what Irving & Parry (1963) referred to as the 'Illawarra reversal': a confusing term as the 'reversal' by definition is the base of the first major overlying normal magnetozone, which they hypothesized occurred in approximately 100 m of unsampled strata. We like others (e.g. Klootwijk *et al.* 1994) refer to this interval beginning in the mid-Permian as the 'Illawarra Superchron' (hyperchron in Russian literature: Molostovsky *et al.* 1976), which is composed of normal and reverse polarity intervals that extend into the Triassic (Hounslow & Muttoni 2010). Although, perhaps, from a historical precedent, a better term for this interval might be the 'Volga–Kama Superchron', as the best type area and first identification of the Illawarra Superchron was in these Russian river basins. In Australian sections, the first normal polarity in the Narrabeen Shale, originally defining the upper boundary of the 'Illawarra reversal' of Irving & Parry (1963), was studied by Embleton & McDonnell (1980) in the Kiama area and shown to be Triassic in age. Later studies of the units equivalent to the Illawarra Coal Measures, however, do appear to show both reverse and normal polarity in the Illawarra Superchron (Klootwijk *et al.* 1994).

Development of a Permian geomagnetic polarity timescale

There have been several previous attempts at construction of a Permian polarity stratigraphy, such as Khramov (1963*a*, *b*, 1967), McElhinny & Burek (1971), Irving & Pullaiah (1976), Molostovsky *et al.* (1976), Klootwijk *et al.* (1994), Opdyke (1995), Jin *et al.* (2000) and Molostovsky (2005). The latest comprehensive attempt for the mid- and late Permian is that of Steiner (2006), with Shen *et al.* (2010), Henderson *et al.* (2012) and Hounslow (2016) attempting integration with geochronology to produce a geomagnetic polarity timescale (GPTS). The 2012 Permian polarity timescale (Henderson *et al.* 2012) uses data from only a small number of key sections, plus several of the pre-1996 composites.

Over the half century since the first Permian magnetostratigraphy, palaeomagnetic methods that extract the original remanent magnetization (i.e. characteristic remanence) of the geomagnetic field have improved. There has been an increasing focus on improving the sensitivity of magnetometers (Kirschvink *et al.* 2015), the magnetic cleaning techniques (i.e. demagnetization) and the rate of specimen measurements (Kirschvink *et al.* 2008). Measurements on Permian sediments in the 1960s–80s often focused on red-bed successions, since these provided both large remanent magnetization intensity and stable magnetizations, but often lacked detailed biochronology. This evolved during the 1990s to examination of carbonate and non-red clastic rocks, with weaker characteristic remanences, but often much better biochronology. These improvements need to be borne in mind when considering Permian magnetostratigraphic data; it is not that early datasets are necessarily more unreliable than recent data, it is that they need to be considered in this wider improvement in palaeomagnetic techniques and associated chronology.

In this work, we primarily utilize the original magnetostratigraphic or palaeomagnetic datasets rather than rely on previously constructed composites. Some of the section magnetozones boundaries have been modified from the original publications, to maintain a consistent data style. The associated biochronology and correlations have been supplemented by additional available biostratigraphic data since the original publication. Finally, a GPTS for the Permian in constructed using radiometric dates where available, starting from the section composting procedures in Hounslow (2016).

A magnetochron labelling scheme

Naming conventions for pre-late Jurassic magnetochrons have not been standardized, with Permian conventions based on either stage–abbreviation–number labels (Ogg *et al.* 2008; Ogg 2012) or labelling individual magnetochrons (Creer *et al.* 1971; Davydov *et al.* 1992; Steiner 2006). The mid- to late Permian Russian labelling system is, perhaps, the most widely used (Molostovsky 1996), but is not easily adaptable to the early Permian or to areas outside Russian sections, since correlations are somewhat debatable. Like the Triassic (Hounslow & Muttoni 2010), the stability in the stage-boundary dating of Permian magnetochrons has not solidified sufficiently at this time, so it is not always crystal clear to what stage, every magnetochron belongs. Hence, applying the stage–abbreviation–number labels of Ogg *et al.* (2008) could require major future changes, whereas stability with respect to Series is more stable. For ease of description, the

Table 1. *Studies showing reliable normal polarity data in the early Permian and latest Carboniferous*

Location/age	Lithology, lithostratigraphy	N_S [N_{MZ}]	D_m/FT/S	h_{MZ} (m)	Reference sources
Arizona, USA/mid-Kungurian	Clastic red-beds, Schnebly Hill Formation	30 [1?]	0/0/PP	?	Graham (1955)
Spitsbergen, Norway/Kungurian	Cherts, spiculitic shales, Kapp Starostin Formation	4 [1]	1/0/MS	*c.* 12	Nawrocki (1999); Nawrocki & Grabowski (2000)
Ellesmere Island, Canada/early–middle Kungurian	Basaltic lavas, Esayoo Volcanics	5 [1]	1/0/PP	*c.* 10–30	Wynne *et al.* (1983); LePage *et al.* (2003); Morris (2013)
Prince Edward Island, Canada/Late Artkinskian	Red beds, Pictou Group, Orby Head Formation	9 [2]	2/0/PP	?	Symons (1990); Ziegler *et al.* (2002)
Oklahoma, USA/mid-Artkinskian	Red Sandstone, Garber Sandstone	7 [1]	2/0/PP	?	Peterson & Nairn (1971); Giles *et al.* (2013)
Spitsbergen, Norway/late Artkinskian	Cherts, spiculitic shales, Kapp Starostin Formation	3 [1]	1/0/MS	*c.* 18	Nawrocki (1999); Nawrocki & Grabowski (2000)
Paganzo Basin, Argentina/Artinskian	Red beds, La Colina Formation	8 [1]	2/0/MS	*c.* 15	Valencio *et al.* (1977); Valencio (1980); Irving & Monger (1987); Césari & Gutiérrez (2000); Césari *et al.* (2011);
British Columbia/late Sakmarian–early Artinskian	Tuffs, Asitka Group	15 [1?]	2/0/PP	?	MacIntyre *et al.* (2001)
West Virginia, USA/Asselian	Dunkard Group, Washington Formation	2 [1]	2/0/PP	?	Helsley (1965); Gose & Helsley (1972); Schneider *et al.* (2013)
Karachatyr, Tajikistan/Asselian	Marine limestones and clastics	≥2? [1]	2/F+/MS	<200	Davydov & Khramov (1991)
Aidaralash, Kazakhstan/early Asselian	Marine limestones and clastics	2 [1]	2/0/MS	10	Khramov & Davydov (1984, 1993)
Saar–Nahe Basin, Germany/300–290 Ma (Sakmarian–Asselian)	Nohfelden and Donnersberg rhyolites	11 [1?]	1/0/PP	?	Berthold *et al.* (1975); Schmidberger & Hegner (1999)
Thuringia, Germany/Gzhelian?	Grey, coal-bearing sandstones, Manebach Formation	5 [?]	2/0/MS	?	Menning (1987); Menning *et al.* (1988)
Aidaralash, Urals/late Gzhelian	Marine limestones and clastics	2 [1?]	2/0/MS	<15	Khramov & Davydov (1984, 1993)
Nikolskyi, Urals/late Gzhelian	Marine limestones and clastics	2?[2]	2/0/MS	<~20	Khramov & Davydov (1984, 1993)
Spitsbergen, Norway/late Gzhelian	Limestones, dolomites Tyrrellfjellet Member	2 [1]	1/0/MS	*c.* 8	Nawrocki (1999); Nawrocki & Grabowski (2000)
Fergana, Tajikistan/Gzhelian	Marine limestones and clastics	*c.* 6 [3]	2/F+/MS	*c.* 3 to <300	Davydov & Khramov (1991)
Donets Basin, Suhoj-Jaz, Ukraine/late Gzhelian	Red beds/Kartamysh Suite	17 [1]	0/F+/PP	<100	Khramov (1963*a*); Khramov & Davydov (1984); Davydov & Leven (2003); Iosifidi *et al.* (2010)

N_s, Number of specimens with normal polarity; N_{MZ}, number of normal magnetozones; Dm/FT/S, demagnetization method/fold test/study type; $D_m = 1$, if full demagnetization applied to all samples, with principle component or great circle extraction; $D_m = 2$, pilot demagnetizations of simple magnetization behaviour, with stable point averaging or single step; $D_m = 0$, no demagnetization; F+, fold test positive (or demonstrate pre-folding magnetization); F− fold test negative; F = 0, no fold test; S = PP or MS for palaeopole or magnetostratigraphic study, h_{MZ}, normal magnetozone thickness; ?, unknown; respectively.

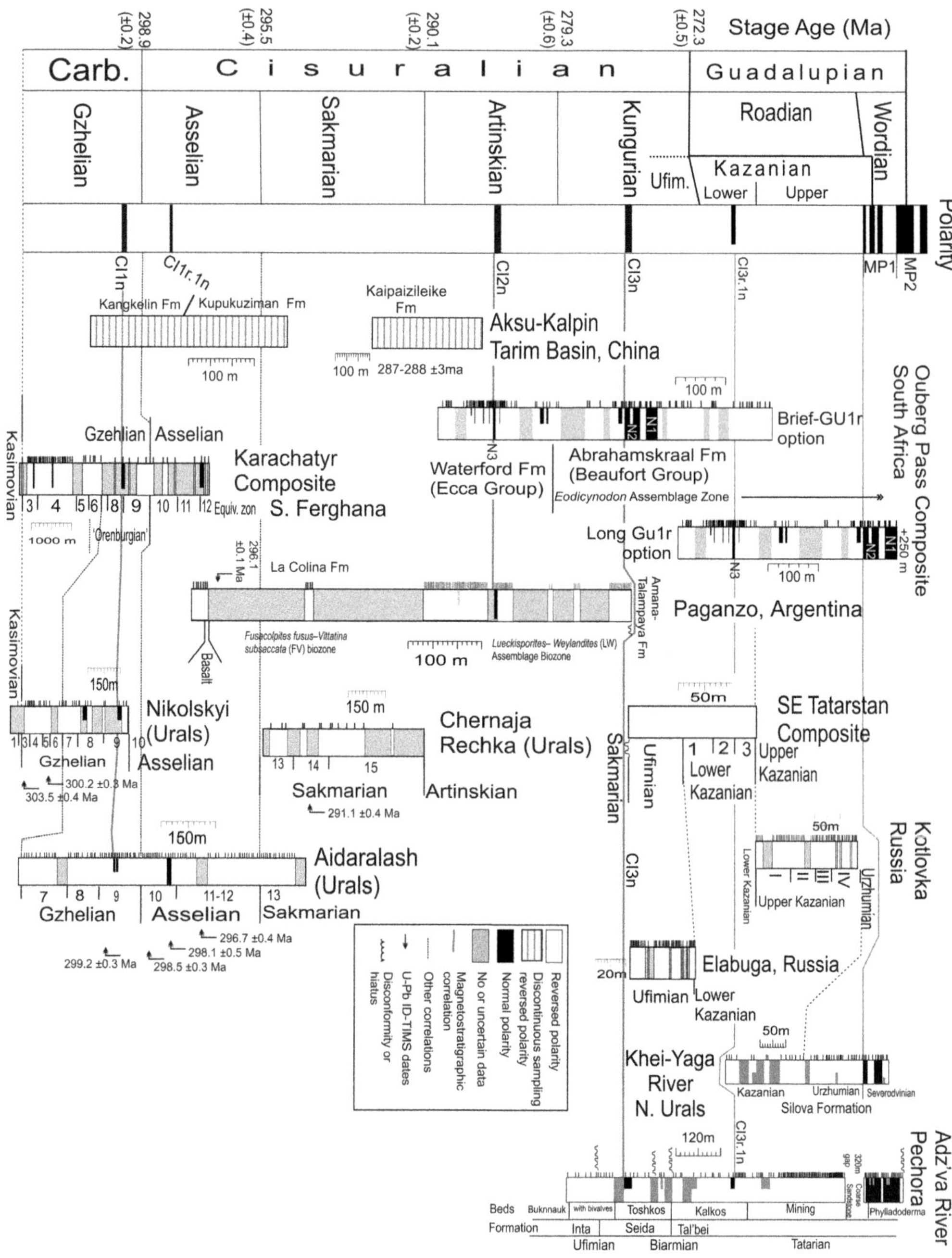

Stage Age (Ma)
272.3 (±0.5)
279.3 (±0.6)
290.1 (±0.2)
295.5 (±0.4)
298.9 (±0.2)
Carb.
Cisuralian
Guadalupian
Gzhelian
Asselian
Sakmarian
Artinskian
Kungurian
Ufim.
Kazanian
Lower
Upper
Roadian
Wordian
Polarity
Cl1n
Cl1r.1n
Cl2n
Cl3n
Cl3r.1n
MP1
MP2
Kangkelin Fm / Kupukuziman Fm
Kaipaizileike Fm
Aksu-Kalpin Tarim Basin, China
100 m
287-288 ±3ma
Ouberg Pass Composite South Africa
Brief-GU1r option
Abrahamskraal Fm (Beaufort Group)
Waterford Fm (Ecca Group)
Eodicynodon Assemblage Zone
Long Gu1r option
+250 m
Karachatyr Composite S. Ferghana
Equiv. zon
'Orenburgian'
1000 m
Kasimovian
La Colina Fm
296.1 ±0.1 Ma
Basalt
Fusacolpites fusus–Vittatina subsaccata (FV) biozone
Lueckisporites–Weylandites (LW) Assemblage Biozone
Amana-Talampaya Fm
Paganzo, Argentina
Nikolskyi (Urals)
150m
300.2 ±0.8 Ma
303.5 ±0.4 Ma
Chernaja Rechka (Urals)
150 m
291.1 ±0.4 Ma
SE Tatarstan Composite
50m
Ufimian
Lower Kazanian
Upper Kazanian
Kotlovka Russia
Lower Kazanian
Upper Kazanian
Urzhumian
Aidaralash (Urals)
296.7 ±0.4 Ma
298.1 ±0.5 Ma
298.5 ±0.3 Ma
299.2 ±0.3 Ma
Reversed polarity
Discontinuous sampling reversed polarity
Normal polarity
No or uncertain data
Magnetostratigraphic correlation
Other correlations
U-Pb ID-TIMS dates
Disconformity or hiatus
Elabuga, Russia
20m
Khei-Yaga River N. Urals
Kazanian
Urzhumian
Severodvinian
Silova Formation
Adz'va River Pechora
120m
320m gap
Coarse Sandstone
Beds
Buknnauk
with bivalves
Toshkos
Kalkos
Mining
Phylladoderma
Formation
Inta
Seida
Tal'bei
Ufimian
Biarmian
Tatarian

Permian magnetozones have been formally numbered in couplets (i.e. a normal with overlying reverse) for each of the Permian Series: from CI1 to CI3 (Cisuralian), from GU1 to GU3n (Guadalupian); and from LP0r to LP3 (Lopingian shown as LP, so as to not confuse with the Lower Ordovician, LO: Hounslow 2016). The basal Triassic magnetochron labelling is after Hounslow & Muttoni (2010). Chrons are grouped according to polarity dominance in the section data, except in the Cisuralian (see Murphy & Salvador 1999 for chronostratigraphic definition of magnetochrons and their sub-divisions). Sub-magnetochron labelling is applied (i.e. n.1r or r.1n) to less dominant chrons or those with less supporting data, but seen in multiple sections. Tentative sub-chrons are labelled .ar and .an if the subchron is considered to possess insufficiently strong evidence from multiple sections. This hierarchical labelling gives a clue to the strength of evidence available, and allows easier relabelling in later studies. The chron numbering is in the opposite direction (i.e. younger magnetochrons given larger number) to the Cenozoic and late Mesozoic chron labelling (Ogg 2012), which starts from 0 Ma. This follows the procedure suggested by Kent & Olsen (1999), but widely adopted in other Mesozoic and Palaeozoic studies since the studies of Khramov (1967) and McElhinny & Burek (1971).

The early Permian and the Kiaman Superchron

The early Permian is characterized by the Kiaman Superchron, the interval of predominantly reverse polarity, well known from studies in the 1960s and 1970s (Irving & Parry 1963; Irving & Pullaiah 1976). The main issue for defining the nature of the GPTS for the early Permian is therefore the age and duration of any normal polarity magnetozones in the Kiaman Superchron. There have been a great many (in excess of 400) palaeomagnetic studies of the early Permian, primarily focusing on palaeopole type studies (i.e. defining tectonic motions, etc.). These have shown that if there are normal polarity magnetozones in the early Permian, they are likely to be short in duration (Irving & Pullaiah 1976; Opdyke 1995). Sampling density and stratigraphic dating issues with palaeopoles-type studies often mean that stratigraphic relationships between samples may be poorly defined, ages poorly defined, sampled horizons may be few and widely spread out through a large stratigraphic range, so they cannot be used to build a reliable polarity stratigraphy (but can indicate polarity bias). However, sampled sites with normal polarity from such studies do give strong evidence for the presence of a limited number of normal polarity magnetozones in the Kiaman Superchron (Table 1). In spite of the very large number of early Permian palaeomagnetic palaeopole-type studies, there is a much small number of conventional magnetostratigraphic studies in this interval that have used closely spaced stratigraphic sampling.

In spite of an often perceived lack of normal magnetozones in the Kiaman, expressed in polarity composites like Opdyke (1995), there are sufficient datasets that show a consistent pattern of normal magnetozones in the early Permian, which are reasonably well dated (Table 1). These data suggest at least three, probably four, normal magnetozones in the early Permian, during the early Asselian (CI1r.1n), late Artinskian (CI2n), mid-Kungurian (CI3n) and mid-Roadian (C3r.1n). As a result of the occasional difficulty in distinguishing CI1r.1n from a normal magnetozone in the underlying (Carboniferous) late Gzhelian strata (CI1n), we discuss the data relating to CI1n and CI1r.1n together. We take the late Gzhelian CI1n normal magnetochron as the start of the labelled Permian chrons, since the CI1n–CI1r.1n interval straddles the Carboniferous–Permian boundary.

Gzhelian and Asselian magnetochrons CI1n and CI1r.1n

The study of Khramov (1963*a*) was the first to identify a likely normal polarity magnetozone in the Kiaman Superchron (here called CI1n), from

Fig. 1. Lower Permian magnetic polarity data from Russia, Asia, South America and Africa. Ticks on the columns are sample positions. Data sources for magnetostratigraphy and supporting stratigraphic details: Tarim Basin from Sharps *et al.* (1989), Wang & Yang (1993), Li *et al.* (2011) and Xu *et al.* (2014). South Ferghana composite from Davydov & Khramov (1991), with equivalent numbered foraminifera zones from the Urals successions, mapped using their figure 5 (see Table 2). Ouberg Pass, South Africa from Lanci *et al.* (2013) and Modie & Le Hérissé (2009). Nikolskyi, Chernaja Rechka and Aidaralash from Khramov & Davydov (1984, 1993), Davydov *et al.* (1998) and Davydov & Leven (2003), with numbered foraminifera zones as indicated in Table 2. Paganzo Basin magnetostratigraphy and radiometric date from Valencio *et al.* (1977), Césari & Gutiérrez (2000) and Césari *et al.* (2011). SE Tatarstan composite, Kotlovka and Elabuga from Burov *et al.* (1998), Silantiev *et al.* (2015*c*). Khei-yaga River section from Iosifidi & Khramov (2009). Adz'va River section data from Balabanov (1998), with additional stratigraphy from Rasnitsyn *et al.* (2005), Lozovsky *et al.* (2009) and Kotlyar (2015). Stage base ages are those of Henderson *et al.* (2012).

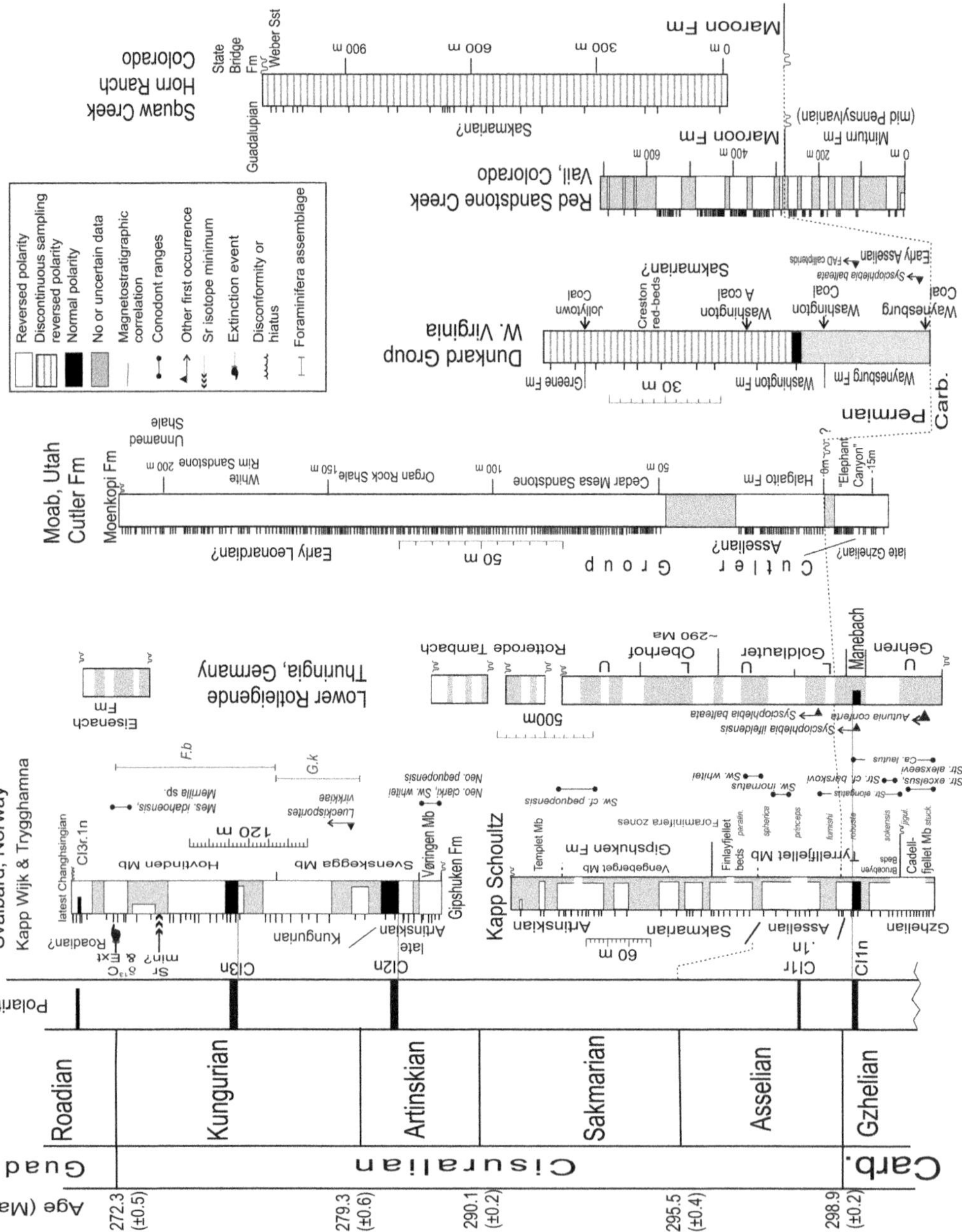
Age (Ma)
272.3 (±0.5)
279.3 (±0.6)
290.1 (±0.2)
295.5 (±0.4)
298.9 (±0.2)
Guad.
Cisuralian
Carb.
Roadian
Kungurian
Artinskian
Sakmarian
Asselian
Gzhelian
Polarity
Svalbard, Norway
Kapp Wijk & Trygghamna
Kapp Schoultz
Lower Rotleigende
Thuringia, Germany
Moab, Utah
Cutler Fm
Dunkard Group
W. Virginia
Red Sandstone Creek
Vail, Colorado
Squaw Creek
Horn Ranch
Colorado
Reversed polarity
Discontinuous sampling reversed polarity
Normal polarity
No or uncertain data
Magnetostratigraphic correlation
Conodont ranges
Other first occurrence
Sr isotope minimum
Extinction event
Disconformity or hiatus
Foraminifera assemblage

the Donets Basin, located in the Kartamysh Suite (Kartamyshskaya Formation) in the upper Gzhelian between limestones Q4 and Q8 (Davydov & Leven 2003) (Fig. 1). In spite of it being established with many specimens (Table 1), it was only located in the Suhoj-Jaz section, with the specimens not subject to conventional modern demagnetization techniques. Fusulinids found in marine analogues of the Kartamyshskaya Formation in the Predonets Trough suggest correlation of limestones Q1–Q6 with the late Gzhelian *Ultradaixina bosbytauensis–Schwagerina robusta* fusulinid zone and limestones Q7–Q12 with the early Asselian *Sphaeroschwagerina vulgaris–Sch. fusiformis* fusulinid zone (Davydov *et al.* 1992). However, the palaeopole-type study of Iosifidi *et al.* (2010), which sampled this same formation and the same section, failed to find evidence of normal polarity. However, this may relate to the wide sample spacing used, indicating that the equivalent of CI1n found by Khramov (1963*a*) is brief in duration, as suggested by other studies.

The base Permian Global Boundary Stratotype Section and Point (GSSP) section at Aidaralash contains a tentative normal magnetozone that is restricted to the *U. bosbytauensis–Sch. robusta* fusulinid zone, directly below the Carboniferous–Permian boundary (Fig. 1). This normal magnetozone, which was named the 'Kartamyshian' by Davydov & Khramov (1991), has also been detected in the Nikolsky section of the Southern Urals, the Belaya River section of the northern Caucasus and the Ivano-Darievka section of the Donets Basin (Khramov 1963*a*; Khramov & Davydov 1984; Davydov *et al.* 1998; Davydov & Leven 2003). A study with widely spaced samples, from three overlapping sections (Dzhingilsaj, Uchbulak and Dastarsaj) in Ferghana (Uzbekistan: Davydov & Khramov 1991), identified four normal polarity intervals (all based on single samples, multiple specimens) in the Gzhelian–Asselian, dated by a fusulinid zonation (Fig. 1). The data from the oldest section (Dzhingilsaj) being the best defined, with the closest spaced sampling in the Gzhelian parts of these sections. Like the Suhoj-Jaz, Nikolsky and Aidaralash sections, the southern Ferghana Uchbulak section contains a tentative normal magnetozone approximately within age-equivalent foraminifera zones, indicating substantive evidence for CI1n.

Higher in the Aidaralash section, a normal magnetozone, CI1r.1n (defined by two sample level), occurs in the early Asselian *Sph. vulgaris–Sch. fusiformis* Zone (equivalent to the *Sph. aktjubensis–Sph. fusiformis* Zone of Schmitz & Davydov 2012). The youngest tentative normal polarity magnetozone in the southern Ferghana, Dastarsaj, section is in the *Sph. sphaerica–Sch. firma* Zone, equivalent to the late Asselian *Sph. gigas* Zone of Schmitz & Davydov (2012). Hence, it is not clear whether this is the same magnetozone as at the Aidaralash section, in spite of Davydov & Leven (2003) 'moving' the Dastarsaj section normal magnetozone into the early Asselian. The interval containing the equivalent *Sp. vulgaris–Sc. fusiformis* Zone in the Dastarsaj section does not have close sample spacing, so it is possible that the equivalent of CI1r.1n was unsampled.

Nawrocki & Grabowski (2000) collected some 300 samples, supporting a detailed magnetostratigraphy through the early Permian in Svalbard (Fig. 2). Three short normal polarity intervals occur, one within the base of the Tyrrellfjellet Member, one in the lower parts of the Svenskegga Member and a third in the base of the Hovtinden Member (Fig. 2) (data of Nawrocki & Grabowski 2000, but using the lithostratigraphy in Hounslow & Nawrocki 2008). The normal magnetozone in the Tyrrellfjellet Member is just below the *Palaeoaplysina* build-ups in the upper part of the Brucebyen Beds. At levels below the top of the Brucebyen Beds, there are a succession of Gzhelian fusulinid zones, with the boundary between the *Zigarella furnishi* and the *Sch. robusta* zones marking the

Fig. 2. Lower Permian magnetic polarity data from Europe and North America. Ticks on the columns are sample positions. Data sources used are: Svalbard (Norway), magnetostratigraphy from Nawrocki & Grabowski (2000) and Hounslow & Nawrocki (2008), with additional stratigraphic details from Nakrem *et al.* (1992), Nilsson & Davydov (1997), Bond *et al.* (2015) and Ehrenberg *et al.* (2010). Lower Rotliegend (Germany), magnetostratigraphy from Menning (1987) and Menning *et al.* (1988), and additional stratigraphy from Schneider *et al.* (2013). Moab (Utah, USA), magnetostratigraphy from Gose & Helsley (1972), with additional age constraints from Condon (1997), Soreghan *et al.* (2002) and Lucas (2006). Dunkard Group (West Virginia, USA), magnetostratigraphy from Helsley (1965) and Gose & Helsley (1972), with additional stratigraphy from DiMichele *et al.* (2013) and Lucas (2013). Red Sandstone Creek (Colorado, USA), magnetostratigraphy from Miller & Opdyke (1985), with additional stratigraphy from Johnson *et al.* (1990). Squaw Creek (Colorado, USA), magnetostratigraphy from Miller & Opdyke (1985). Foraminifera zone names on the Svalbard column modified by Davydov *et al.* (2001), from Nilsson & Davydov (1997): stuck., *Rauserites stuckenbergi*; jigul., *Jugulites jigulensis*; sokensis, *Daixina sokensis*; robusta, *Schwagerina robusta*; furnishi, *Zigarella furnishi*; princeps, *Sch. princeps*; spherical, *Sch. sphaerica*; paralin, *Eoparafusulina paralinearis*. G.k, *Gerkeina komiensis*; F.d, *Frondicularia bajcurica* foraminifera assemblage zones. Stage base ages from Henderson *et al.* (2012).

probable Gzhelian–Asselian boundary (Nilsson & Davydov 1997; Davydov *et al.* 2001). In the underlying Cadellfjellet Member, the conodont *Streptognathodus alekseevi* also indicates a Gzhelian age (Nakrem *et al.* 1992) (Fig. 2). However, in contrast, the conodont *Str. barskovi* (Fig. 2) is normally considered to be indicative of the mid-Asselian in the Urals (Nakrem *et al.* 1992). The overlying part of the Tyrrelfjlellet Member has two further Asselian foraminifera zones (*Sch. princeps* and *Sch. sphaerica*), with the uppermost *Eoparafusulina paralinearis* assigned to the Sakmarian by Nilsson & Davydov (1997). This suggests the normal magnetozone in the Tyrellfjellet Member probably represents the equivalent of the late Gzhelian normal magnetozone CI1n (Fig. 2). If the Asselian magnetozone CI1r.1n is present, it is rather too brief to have been detected by the approximately 5–10 m spaced samples of Nawrocki & Grabowski (2000). There is a notable disparity between the foraminifera-based ages in the upper part of the Tyrrellfjellet Member and the presence of *Sweetognathus* sp., which usually suggests an Artinskian age (Nakrem *et al.* 1992), although there are taxonomic issues with *Sw. inornatus* (Mei et al. 2002).

Several studies have examined the reversal stratigraphy through the Lower Rotliegend, which should include the Gzhelian–Asselian interval (Fig. 2). Menning *et al.* (1988) summarized and synthesized these studies, which appear to show a tentative normal polarity interval in the mid-parts of the Manebach Formation in non-red-mudstone samples from the 'Hinteres schulzental' locality, isolated with alternating field (AF) demagnetization (Menning 1987; Menning *et al.* 1988). Representatives of the insect zone *Sysciophlebia ilfeldensis* occur as fragments in the Manebach Formation, suggesting that the formation spans the Gzhelian–Asselian boundary (Schneider *et al.* 2013). Therefore, it is not totally clear whether this normal polarity magnetozone represents CI1n or CI1r.1n, although it is most likely to be equivalent to CI1n (see below).

The CI1r.1n magnetozone may have been detected in the Nohfelden and Donnersberg rhyolites in the Saar–Nahe Basin (Berthold *et al.* 1975). More recent dating of associated extrusives and intrusives associated with these volcanic centres, using Rb–Sr, K–Ar and $^{40}Ar/^{39}Ar$ radiometric ages from rhyolites, yields ages of 300–290 Ma (Schmidberger & Hegner 1999), suggesting an Asselian age.

The coal-bearing Dunkard Group in West Virginia, USA, was reconnaissance sampled by Helsley (1965), with his lowest sample level approximately 8 m above the Washington Coal, with data displaying tentative normal polarity using undemagnetized samples (Fig. 2). Gose & Helsley (1972) subsequently demagnetized these normal polarity samples and found two of the three samples to be stable to demagnetization, which indicates the good likelihood of a normal magnetozone. The highest resolution biochronology data for these units appears to be spiloblattinid insects, with *Sysciophlebia balteata* occurring in the earliest part of the Dunkard Group (Schneider *et al.* 2013), suggesting that the entire Dunkard Group is early Permian. This probably places the Dunkard normal magnetozone in the Asselian, equivalent to CI1r.1n. The parts sampled by Helsley (1965), which did not include the youngest Dunkard Group, probably extend into the Sakmarian (DiMichele *et al.* 2013; Lucas 2013). The occurrence of *S. ilfeldensis* in the German Manebach Formation of the Lower Rotliegend places the Dunkard Group normal magnetozone as probably younger than that in the Manebach Formation (Fig. 2).

The study of Diehl & Shive (1979), on the Ingleside Formation in northern Colorado (in Owl Canyon), tried to locate normal polarity intervals in the early Permian by collecting samples through this formation at an average spacing of 0.28 m. In the original study, the Ingleside Formation was assigned to the early Permian: however, the fusulinid *Triticites ventricosus* in the base of the formation (Hoyt & Chronic 1961) suggests a Virgilian age (late Gzhelian), according to Gomez-Espinosa *et al.* (2008) and Wahlman & West (2010). The formation's younger age is not clear, as it is overlain unconformably by the Owl Canyon Formation of early Permian age, although the formation presumably covers the Carboniferous–Permian boundary interval into the Sakmarian (Sweet *et al.* 2015). However, Diehl & Shive (1979) failed to find normal polarity samples in the complete 70 m of the formation, which should have covered magnetochron interval CI1n–CI1r.1n.

Sakmarian–Artinskian

The Sakmarian is consistently reverse polarity in all studies. The earliest study to detect the equivalent of normal magnetochron CI2n in the Artinskian was the palaeopoles-type study of Peterson & Nairn (1971) on the Garber Formation of Oklahoma, USA, who performed thermal demagnetization up to 600°C to isolate normal polarity in seven specimens (Table 1). According to Giles *et al.* (2013), the Garber Formation is mid-Artinskian in age, based on regional correlation of the laterally equivalent Hennessey Shale. A younger age straddling the Artinskian–Kungurian boundary was suggest by May *et al.* (2011), based on vertebrate (dissorophids) ranges. Other palaeopole-type studies in red beds of Artinskian age with normal polarity intervals are from the Pictou Group of Prince Edwards Island, Canada (Symons 1990). The Pictou Group data were from megasequence IV (Orby Head Formation: Ziegler *et al.* 2002) with nine specimens from three blocks, demagnetized to 650°C, showing apparently two normal polarity intervals. One of these is from near the base of the formation, but with most of the normal polarity data from two sites near the top of the formation. Plant fossil data suggest a late Artinskian age for the Orby Head Formation (Ziegler *et al.* 2002). Considering the uncertainty in age assignment for the Orby Head Formation, it is possible that the lower normal polarity level is CI2n and the upper one is CI3n.

Irving & Monger (1987) found normal polarity samples in their palaeopole-type study of the volcanic units of the Asitka Group (British Columbia). Modern demagnetization techniques were employed, and normal polarity was found in multiple specimens (Table 1). The Asitka Group is dated by overlying limestones containing Sakmarian and Artinskian conodonts (MacIntyre *et al.* 2001), but fusulinids suggest a late Sakmarian–early Artinskian age (Ross & Monger 1978). This suggests that the magnetochron detected in the Asitka Group is probably CI2n.

Palaeopole and magnetostratigraphic studies of Valencio *et al.* (1977), Sinito *et al.* (1979) and Valencio (1980) measured a predominantly reverse polarity stratigraphy through the La Colina Formation from the Paganzo Basin in Argentina. Based on palynological and radiometric dating, their data probably range in age from the Asselian to the Artinskian (Césari & Gutiérrez 2000; Césari *et al.* 2011). Valencio *et al.* (1977) detected a single normal polarity interval, which is correlated to Artinskian CI2n. Normal polarity samples below this level were detected by Sinito *et al.* (1979), but are less reliably located stratigraphically and appear to have less reliable palaeomagnetic data. In the same area, normal polarity samples measured by Thompson (1972) were from the overlying Amana Formation, which is now assigned to the Triassic (Césari *et al.* 2011).

The magnetostratigraphic data from Spitsbergen of Nawrocki & Grabowski (2000), through the upper part of the Tyrrellfjellet Member into the Gipshuken Member, shows only reverse polarity (Fig. 2). The Tyrrellfjellet Member contains the conodont *Sweetognathus inornatus*, indicating a Sakmarian–Artinskian age, whereas the rich fusulinid assemblages suggest age ranges from the Asselian to the Sakmarian (Nakrem *et al.* 1992). The more restricted range of conodont and foraminifera faunas from the Gipshuken Formation suggests a probable age range into the Artinskian. A regional hiatus is widely concluded at the base of the overlying Kapp Starostin Formation (base Vøringen Member) (Blomeier *et al.* 2011), but the age gap is below the resolution of biostratigraphy. Nawrocki & Grabowski (2000) found normal polarity in three specimens from the lower part of the Svenskegga Member (above the Vøringen Member): two at Kapp Wijk (30 m from the base of the Kapp Starostin Formation) (Fig. 2) and one at Trygghamna, which probably represents the equivalent normal magnetozone. The normal magnetozone in the lower parts of the Svenskegga Member is CI2n (Table 1). The Vøringen Member contains a diverse marine fauna, with conodonts including *Sweetognathus whitei* and *S. clarki*, indicating a probable late Artinskian age (Nakrem *et al.* 1992; Nakrem 1994). A Sr-isotope value of 0.70746 from the Vøringen Member also suggests an Artinskian age (Ehrenberg *et al.* 2010). The overlying middle and upper parts of the Svenskegga Member contain foraminifera assigned to the *Gerkeina komiensis* assemblage zone (Sosipatrova 1967; Nakrem *et al.* 1992), correlated to the Iren Horizon in the Uralian successions, where it is assigned a mid-Kungurian age (Lozovsky *et al.* 2009). This suggests the Artinskian–Kungurian boundary occurs in the lower–middle parts of the Svenskegga Member (Fig. 2).

Kungurian to Roadian

Graham (1955) was the first to identify a normal polarity interval in the Kungurian. His palaeopole-type study (using undemagnetized specimens) identified both reverse and normal polarity in samples, from the Supai Group in the Oak Creek and Carrizo Creek sections in Arizona, USA. Although precise details of stratigraphic levels sampled are not clear, both these locations have good sections through the upper part of the 'Supai' (Corduroy, Big A Butte members, Esplanade Sandstone, Hermit Formation: Winters 1962; Blakey & Middleton 1987), which probably locate Graham's samples in the uppermost Supai Group and overlying Schnebly Hill Formation, using the modern lithostratigraphy (Blakey 1990). Conodonts within the Fort Apache Member of the Schnebly Hill Formation date Graham's data to the mid-Leonardian (Blakey 1990; Eagar & Peirce 1993), which is early–mid-Kungurian (Henderson *et al.* 2012). The study of Graham (1955) has not been re-evaluated using modern palaeomagnetic techniques.

Wynne *et al.* (1983) performed a palaeopole-type study of the Esayoo Volcanic Formation on Ellesmere Island, northern Canada, in which they initially assumed an Artinskian age, but has been re-dated as Kungurian (Table 1). The Esayoo Volcanic Formation is sandwiched between the Great Bear Cape and the overlying Sabine Bay formations (Morris 2013), although no data on stratigraphic position in the lava succession are described by

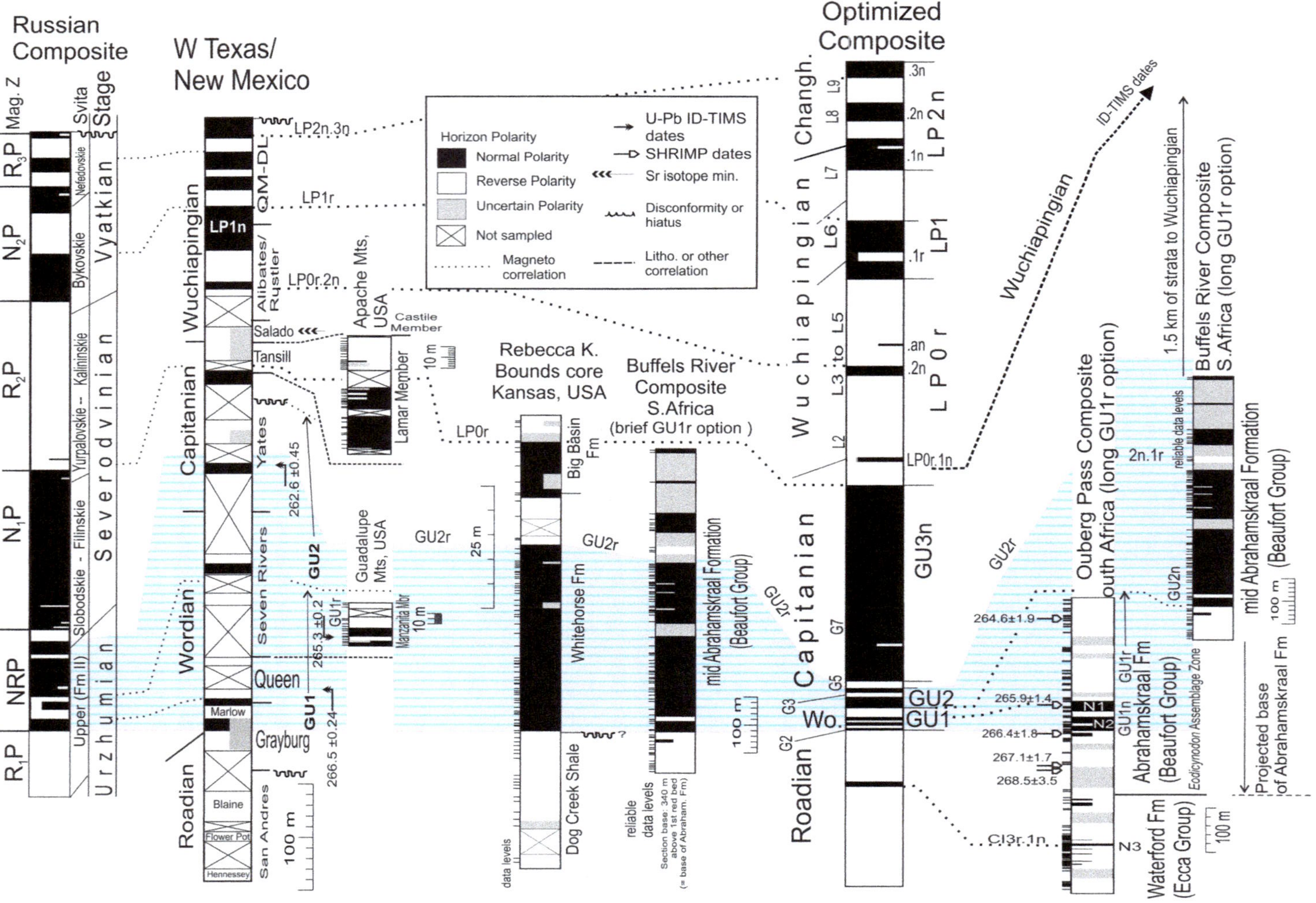

Russian Composite
Mag. Z
Svita
Stage
Urzhumian
Severodvinian
Vyatkian
Upper (Fm II)
Slobodskie - Filinskie
Yurpalovskie -- Kalininskie
Bykovskie
Nefedovskie
W Texas/ New Mexico
Roadian
Wordian
Capitanian
Wuchiapingian
San Andres
Blaine
Flower Pot
Hennessey
Grayburg
Marlow
Queen
Seven Rivers
Yates
Tansill
Salado
Alibates/ Rustler
QM-DL
LP1n
100 m
266.5 ±0.24
265.3 ±0.2
262.6 ±0.45
GU1
GU2
GU1r
LP2n.3n
LP1r
LP0r.2n
Guadalupe Mts, USA
Manzanita Mbr
10 m
Apache Mts, USA
Lamar Member
Castile Member
GU2r
Rebecca K. Bounds core Kansas, USA
Dog Creek Shale
Whitehorse Fm
Big Basin Fm
data levels
25 m
LP0r
Buffels River Composite S.Africa (brief GU1r option)
mid Abrahamskraal Formation (Beaufort Group)
reliable data levels
Section base: 340 m above 1st red bed (= base of Abraham. Fm)
Horizon Polarity
Normal Polarity
Reverse Polarity
Uncertain Polarity
Not sampled
Magneto correlation
U-Pb ID-TIMS dates
SHRIMP dates
Sr isotope min.
Disconformity or hiatus
Litho. or other correlation
Optimized Composite
Roadian
Wo.
Capitanian
Wuchiapingian
Changh.
GU1
GU2
GU3n
LP0r
LP0r.1n
LP1
LP2n
Wuchiapingian
ID-TIMS dates
Ouberg Pass Composite South Africa (long GU1r option)
Waterford Fm (Ecca Group)
Abrahamskraal Fm (Beaufort Group)
Eodicynodon Assemblage Zone
Projected base of Abrahamskraal Fm
264.6±1.9
265.9±1.4
266.4±1.8
267.1±1.7
268.5±3.5
C13r.1n
Buffels River Composite S.Africa (long GU1r option)
mid Abrahamskraal Formation (Beaufort Group)
1.5 km of strata to Wuchiapingian

Wynne *et al.* (1983). LePage *et al.* (2003) suggested the sediments overlying the Esayoo Volcanics are mid- to late Kungurian, based on plant megafossils, and the youngest part of the Sabine Bay Formation is late Kungurian, based on conodonts such as *Mesogondolella idahoensis* (Henderson & Mei 2000). The youngest part of the Great Bear Cape Formation, underlying the Esayoo Volcanic Formation, is earliest Kungurian in age (Mei *et al.* 2002), suggesting the normal polarity interval is early–mid-Kungurian.

In the Spitsbergen magnetostratigraphic data of Nawrocki & Grabowski (2000), the best quality data showing normal polarity in these Permian successions are in cherts from the base of the Hovtinden Member at Trygghamna (Hounslow & Nawrocki 2008). Brachiopod and bryozoans faunas from the youngest parts of the Kapp Starostin Formation (i.e. Hovtinden Member) suggest equivalence with Ufimian and Kazanian faunas from Greenland and Novaya Zemlya, suggesting possible late Kungurian–Roadian ages (Stemmerik 1988; Nakrem *et al.* 1992). Foraminiferal and coral assemblages suggest Kungurian–Ufimian ages when compared to the Urals successions (Nakrem *et al.* 1992; Chwieduk 2007). A conodont fauna of *Mesogondolella idahoensis* and *Merrillina* sp. from the upper most part of the Kapp Starostin (Nakrem *et al.* 1992) suggests latest Kungurian–early Roadian. Since reverse polarity dominates to the topmost part of the Kapp Starostin Formation (Fig. 2) (Hounslow *et al.* 2008), without major intervals of normal polarity, it suggests, like the faunal data, that most of the Wordian, Capitanian and late Lopingian (and their normal polarity intervals) are missing on Spitsbergen. This suggests the Hovtinden Member normal magnetozone is probably Kungurian in age. However, approximately 70 m above this normal magnetozone, an interpreted late Capitanian low in Sr-isotope data has been detected (Ehrenberg *et al.* 2010; Bond *et al.* 2015) that contradicts the faunal and magnetostratigraphic data. This occurs prior to a brachiopod extinction and negative excursion in $\delta^{13}C_{org}$ in the Kapp Starostin Formation, which occur approximately 45 m below the top of the formation (Bond *et al.* 2015). A partial reconciliation of the magnetostratigraphic and Sr-isotope is provided if the normal polarity intervals in the Wordian and Capitanian are missing, so the reverse polarity in the early Wuchiapingian (to preserve lows in the Sr-isotope and early Wuchiapingian $\delta^{13}C$ excursion) sits on Roadian or late Kungurian strata in the upper part of the Hovtinden Member. However, this option remains incompatible with the key conodont data, and there is no evidence of a major mid-Kapp Starostin hiatus in the Barents Sea (Ehrenberg *et al.* 2001). The Spitsbergen brachiopod extinction is quite dramatic and, using the age model proposed here, likely corresponds instead with a latest Kungurian brachiopod and bivalve extinction event seen in NE Asia (Biakov 2012).

The detailed magnetostratigraphy from the Adz'va River section in the Pechora Basin through the Tal'bei and upper-most Inta formations (Balabanov 1988) shows the Illawarra Superchron in the *Phylladoderma* beds, underlain by predominantly reverse polarity down into the Inta Formation (Fig. 1). The biostratigraphic ages of these units in the Pechora Basin has been much debated (Rasnitsyn *et al.* 2005; Lozovsky *et al.* 2009; Kotlyar 2015). Based on floral, fish and insect remains, the Inta Formation is probably placed in the Ufimian (late Kungurian?). This suggests that the tentative normal polarity interval in the lower part of the Seida Formation may be the equivalent of CI3n (Fig. 1). There is tentative (a single sample) evidence for a normal polarity magnetozone in the Tal'bei Formation (CI3r.1r) that may equate with a tentative normal polarity level in the Trygghamna section from Spitsbergen (Figs 1 & 2), although other extensive data through the Russian Ufimian–Kazanian sections show no substantiated evidence of normal polarity (Burov *et al.* 1998).

The age of CI3n is, perhaps, best constrained in the Esayoo Volcanic Formation, by the overlying and underlying sedimentary units with conodont ages, along with its relationship to the magnetostratigraphy from Spitsbergen sections, which suggests that the age of CI3n is mid-Kungurian.

Fig. 3. Magnetic polarity data from North American and South African sections through the mid and late Permian. West Texas/New Mexico data are a composite from several studies discussed in Steiner (2006), with the partly unpublished Guadalupian data from the back-reef facies of the Guadalupe Mountains. The Guadalupe basinal facies (Apache Mountains and Guadalupe Mountains sections) from Burov *et al.* (2002). Additional stratigraphic details from Lambert *et al.* (2007), Olszewski & Erwin (2009) and Rush & Kerans (2010). Radiometric dates from Bowring *et al.* (1998) and Nicklen (2011), related to the lithostratigraphy via the sequence correlation of Rush & Kerans (2010). Rebecca K Bounds core magnetostratigraphy from Soreghan *et al.* (2015), and additional details from Sawin *et al.* (2008). Buffels River composite (South Africa) from Tohver *et al.* (2015), in which the grey (uncertain) intervals represent sampled intervals that yielded no polarity data. Individual section height scales on each section. The two options for correlation of the Abrahamskraal Formation data are discussed in the text.

Other normal polarity intervals in the early Permian?

The palaeopole-type study of Rakotosolofu *et al.* (1999) found normal polarity in the Lower Sakamena and Lower Sakoa formations from Madagascar, originally allocated to the Permian. However, the basal tillites sampled from the Lower Sakoa Group are probably early Pennsylvanian in age (Wescott & Diggens 1998) and those from the lower Sakamena Formation are from the late Permian (Illawarra Superchron) according to palynological dating (Wescott & Diggens 1998).

Halvorsen *et al.* (1989) published work on dual-polarity magnetizations from the Karkonosze Granite, SW Poland, which was originally dated to the 305–281 Ma interval, but now has a more precise chronology (Kryza *et al.* 2014), with a main intrusion age of 311 $\pm$ 3 Ma placing it in the Moscovian.

Creer *et al.* (1971) reported normal polarity in 31 Permian andesitic and basaltic specimens from the San Rafael area of Argentina. These are now assigned to the Cerro Carrizalito Formation of the upper part of the Choiyoi volcanics (Rocha-Campos *et al.* 2011), and dated using SHRIMP U–Pb zircon ages to the mid-Guadalupian and younger (265 $\pm$ 2.9–252 $\pm$ 2.7 Ma), not so different from the K–Ar age (263 $\pm$ 5 Ma) determined by Creer *et al.* (1971). These indicate that these normal polarity data are from the Illawarra Superchron.

There have been several other reported normal polarity sample sets in the Permian (e.g. Klootwijk *et al.* 1983; Pruner 1992; Vozárová & Túnyi 2003; Geuna & Escosteguy 2004). These share the characteristics of having very poor age control and a very wide spacing of stratigraphic sampling, in palaeopole-type studies, so it is impossible to evaluate their usefulness for construction of a magnetostratigraphy.

McMahon & Strangway (1968) identified normal polarity samples in the red-bed Maroon Formation in Colorado, but with inadequate AF demagnetization. These were in the lower parts of the Maroon Formation and underlying (Pennsylvanian) Minturn Formation (Fig. 2). The youngest age of the Maroon Formation is constrained by the overlying State Bridge Formation, which contains Guadalupian fossils (Johnson *et al.* 1990). The youngest detrital zircons from the Maroon Formation suggest an age no older than Wolfcampian (Soreghan *et al.* 2015). However, large age uncertainties from the zircon populations and similar mean ages (*c.* 293.1 $\pm$ 4.5 Ma), from the top and bottom of the formation, do not help in constraining its age duration, but rather suggest that it is restricted to a Sakmarian age. The Maroon Formation sits unconformably on the mid-Pennsylvanian Minturn Formation, so the Gzhelian–Asselian boundary interval may be missing. A later, approximately 1 m spaced, magnetostratigraphic sampling of the Maroon Formation by Deon (1974) found that 99.2% of the samples were of reverse polarity, with only three specimens of interpreted normal polarity (but not in adjacent strata). Miller & Opdyke (1985) purposefully re-sampled the Red Sandstone Creek section used by McMahon & Strangway (1968) to try to locate the tentative normal polarity intervals, but found no normal polarity samples. These data may indicate, like the zircon populations, that the Maroon Formation occupies the reverse polarity late Asselian–early Artinskian interval (Fig. 2).

North American early Permian studies

Red-bed and limestone-bearing Permian strata in the American SW in Utah, Colorado and Wyoming have a distinct absence of early Permian normal polarity intervals, in spite of several detailed studies and apparently appropriate ages of strata. We critically examine these data, since they clearly have a bearing on the reliability of these studies, as it opens the question of the reliability of the normal polarity intervals in the Cisuralian, seen in studies outside the American SW. These American studies have critically influenced the conventional hypothesis of the reverse-only character of the Permian part of the Kiaman Superchron.

In the marine sandstone and limestone beds in the Casper Formation of Wyoming (at Horse Creek), Diehl & Shive (1981) sampled 190 m in total of the 220 m of this formation at 0.33 m spacing and found only reverse polarity. The age of the Casper Formation sampled is Desmoinesian (late Moscovian) to Wolfcampian (Sakmarian?), based on brachiopod, fusulinids and conodonts. Red-bed units of the Cutler Group at Moab in Utah were also extensively sampled (Fig. 2) at a close stratigraphic spacing by Gose & Helsley (1972), but again no normal polarity samples were found through a Wolfcampian (possibly Virgilian: Condon 1997; Soreghan *et al.* 2002; Scott 2013) to Leonardian interval (i.e. Gzhelian–Kungurian). Based on vertebrate data, Scott (2013) has suggested that the Carboniferous–Permian boundary is in the lower 10 m of the Halgaito Formation in SE Utah (Fig. 2). Vertebrates from the Organ Rock Shale indicate a Seymouran land vertebrate zone (Lucas 2006), implying that the section sampled by Gose & Helsley (1972) at Moab may extend into the Kungurian. However, this does not seem to be borne out by the detailed sampling that shows only reverse polarity (Fig. 2), which implies that the section may end before the Artinskian. Reasons for the absence of normal polarity in the Cutler Group are unclear, possibly due to unsampled intervals in the Halgaito Formation, an unsuspected hiatus and a shorter age range than anticipated that does not extend into the Artinskian–Kungurian.

Further east in northern Colorado, a magnetostratigraphic study of the red beds of the Ingleside Formation (Diehl & Shive 1979) specifically tried to find the normal polarity intervals in the Gzhelian–Asselian interval, but failed. The same study-targeting issue applied by Miller & Opdyke (1985) to the Maroon Formation in Colorado. Steiner (1988) also sampled extensively the lower and central portions of the Laborcita Formation (Gzhelian–early Asselian: Krainer *et al.* 2003), and about a third of the overlying Abo Formation (Asselian–late Sakmarian) in New Mexico, but found only reverse polarity.

The reasons for the inability of these studies to detect the brief early Permian normal polarity intervals in North American sections that are seen in other areas are not clear, but there may be several possibilities:

- The stratigraphic complexity and often poor-dating resolution in the red beds may mean that the Carboniferous–Permian boundary interval, containing the latest Gzhelian–early Asselian, may be missing (although this does not apply to the Laborcita Formation: Krainer *et al.* 2003). Likewise, in some cases, the red-bed units may not extend up to the CI2n magnetochron, as usually implied by the low-resolution biochronology from these strata.
- Issues with diagenetically delayed magnetizations (Turner 1979; Kruiver *et al.* 2003; Van der Voo & Torsvik 2012) or late Kiaman remagnetizations (e.g. Magnus & Opdyke 1991) may be more common in these units than currently realized. In the front ranges of the Rocky Mountains, Kiaman-age remagnetizations, carried by hematite, appear to be widespread and associated with modest burial, connected with deformation of the ancestral Rocky Mountains (Geissman & Harlan 2002). It is not clear whether this situation in Colorado applies also to the Permian in the Paradox Basin in Utah or to the Casper Formation of Wyoming. However, there have been suggestions that a late Permian–Triassic remagnetization may be affecting some datasets from the North American Craton (Steiner 1988; Pan & Symons 1993).

Guadalupian

Age of the start of the Illawarra Superchron

The chronostratigraphic age of the end of the Kiaman Superchron is in the early Wordian. The first normal polarity magnetochron of the Illawarra Superchron appears to be shown in the middle and upper parts of the back-reef Grayburg Formation (and overlying Queen Formation) in the Guadalupe Mountains in western Texas (Fig. 3) (Steiner 2006). The Grayburg Formation is inferred to be early Wordian in age, based on its lateral relationship to conodont- and fusulinid-dated units. This is based on the basinal to back-reef stratigraphic correlations of Barnaby & Ward (2007), Lambert *et al.* (2007), Olszewski & Erwin (2009) and Rush & Kerans (2010). Nicklen (2011) has suggested the Queen and Grayburg formations correlate to the basinal South Wells Member (of the Cherry Canyon Formation), which has an associated U–Pb isotope-dilution thermal ionization mass spectrometry (ID-TIMS) date (using EARTHTIME standards) of 266.5 $\pm$ 0.24 Ma, potentially directly dating the start of the Illawarra Superchron (Fig. 3). Alternatively, Olszewski & Erwin (2009) correlated the South Wells Member to a level higher than the Queen Formation. Normal and reverse polarity intervals in the Manzanita Member of the Cherry Canyon Formation in the Guadalupe Mountains (Burov *et al.* 2002) derive from the late Wordian (Olszewski & Erwin 2009), probably corresponding to the GU2 magnetochron. Nicklen (2011) suggested that the zircon U–Pb date of 265.3 $\pm$ 0.2 Ma of Bowring *et al.* (1998) provides a date for the bentonites in the Manzanita Member.

The end of the Kiaman Superchron is also shown in the Kyushu sections in Japan (Fig. 4), occurring in the *Neoschwagerina craticulifera* fusulinid assemblage zone (Kirschvink *et al.* 2015). *N. craticulifera* has its first appearance in the late Roadian (Henderson *et al.* 2012), but Kasuya *et al.* (2012) correlated the *N. craticulifera Zone* in these Japanese sections to the early Wordian.

The end of the Kiaman Superchron is very well defined in numerous sections from Russia (Fig. 5), in the upper Urzhumian Stage within the Biarmian Series (Molostovsky 1996; Burov *et al.* 1998; Molostovsky *et al.* 1998). The base of the underlying Kazanian Stage and the Biarmian Series is marked by the first occurrence of the Roadian conodont *Kamagnathus khalimbadzhae*, and this is further emphasized by an assemblage of ammonoids, slightly above the base of the Kazanian, which dates it to the Roadian (Silantiev *et al.* 2015*b*). The regional stages Urzhumian, Severodvinian and Vytakian are demarcated by the first occurrence of non-marine ostracod species in continuous phylogenetic lineages (Tverdokhlebov *et al.* 2005; Silantiev *et al.* 2015*b*). These series are also subdivided by detailed freshwater bivalve, tetrapod and fish biozonations (Tverdokhlebov *et al.* 2005; Silantiev *et al.* 2015*b*). As such, the Biarmian and Tatarian series have a very detailed internal biozonation, but

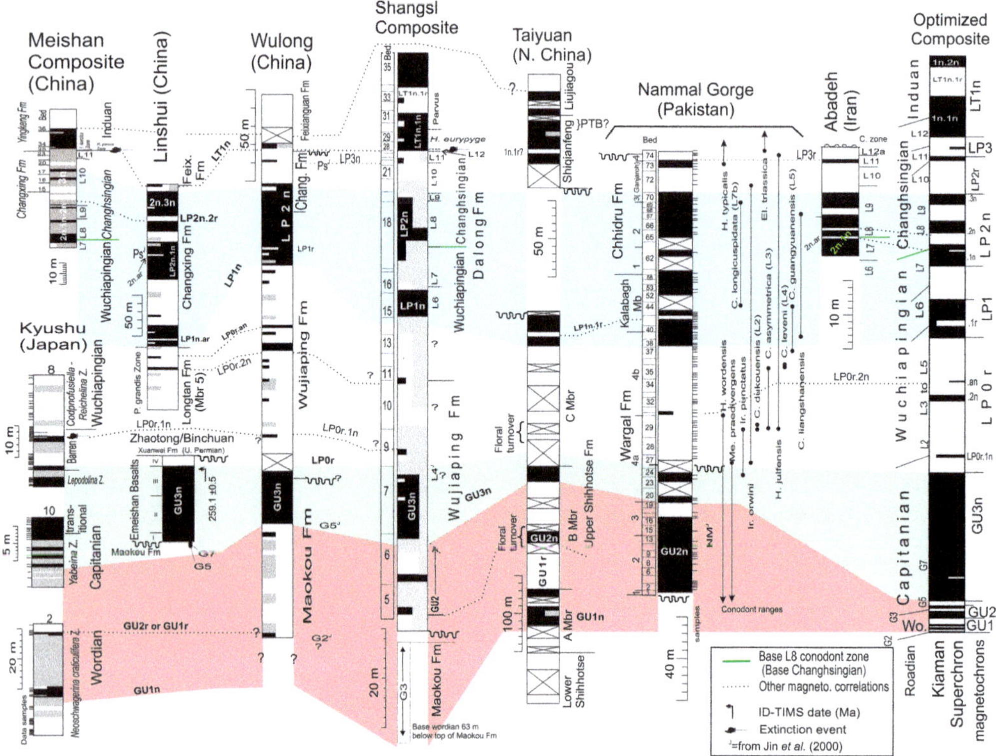

Fig. 4. Summary of the middle and late Permian magnetostratigraphy from marine and non-marine sections (modified from Hounslow 2016). The Meishan magnetic polarity composite from Figure 8. The Kyushu section magnetostratigraphy and fusulinid zones are from Kirschvink *et al.* (2015). The Linshui section magnetostratigraphy is from Heller *et al.* (1995), other stratigraphy modified from that originally published (see the text for details). Emeishan Basalt magnetostratigraphy from Ali *et al.* (2002), Zheng *et al.* (2010), Liu *et al.* (2012) and Zhong *et al.* (2014), and associated stratigraphy from He *et al.* (2007) and Sun *et al.* (2010). The biostratigraphy of the Wulong section (Jin *et al.* 2000) is inadequately documented, but the magnetostratigraphy (Chen *et al.* 1994; Heller *et al.* 1995) appears to range into the lower Capitanian. The Shangsi magnetostratigraphy composite is from Figure 6. The Taiyuan (a non-marine section) magnetostratigraphy is from Embleton *et al.* (1996), with additional stratigraphic details from Menning & Jin (1998) and Stevens *et al.* (2011). The Nammal Gorge section magnetostratigraphy is from Haag & Heller (1991), with conodont ranges projected from nearby sections based on Wardlaw & Pogue (1995), Wardlaw & Mei (1999) and Waterhouse (2010). The Abedah section magnetostratigraphy is from Gallet *et al.* (2000) and Szurlies (2013), and its associated conodont biostratigraphy is from Shen & Mei (2010). Fusulinids: Ps, *Palaeofusulina* spp.; Nm, *Neoschwagerina margaritae* (Jin *et al.* 2000). Conodont zones: G2, *J. asserata* (base Wordian); G3, *J. postserrata* (base Capitianian); G5, *J. altudaensis* (mid-Capitanian); G7, *J. xuanhanensis* (upper Capitanian). L1–L12 are the standard Lopingian conodont zones from Shen *et al.* (2010).

wider correlation to the international stages is reliant on Eurasian-wide correlation of these non-marine faunas (Kotlyar 2015). Multiple sections, borehole cores and studies through the Kazanian (Silantiev *et al.* 2015*c*) and lower Urzhumian have failed to substantiate any normal polarity intervals below the Russian normal and reversed polarity (NRP) mixed-polarity magnetozone (Fig. 5), so the top of the Kiaman Superchron is very clearly expressed (Burov *et al.* 1998). However, the Russian regional stages have long been problematic to correlate in detail to marine sections with conodont and fusulinid zonations, but the Wordian is widely inferred to correlate approximately to the Urzhumian (Lozovsky *et al.* 2009; Henderson *et al.* 2012; Kotlyar 2015). Although not commonly discussed, it is clear that, at least locally, there are a number of hiatus or unconformities in the Tatarian successions (of unknown duration), such as the Urzhumian erosion contact on the Kazanian and, locally, the Vyatkian on the Severodvinian (Tverdokhlebov *et al.* 2005). Integration of sequence stratigraphic concepts in these successions with the magnetostratigraphy needs to evolve in this

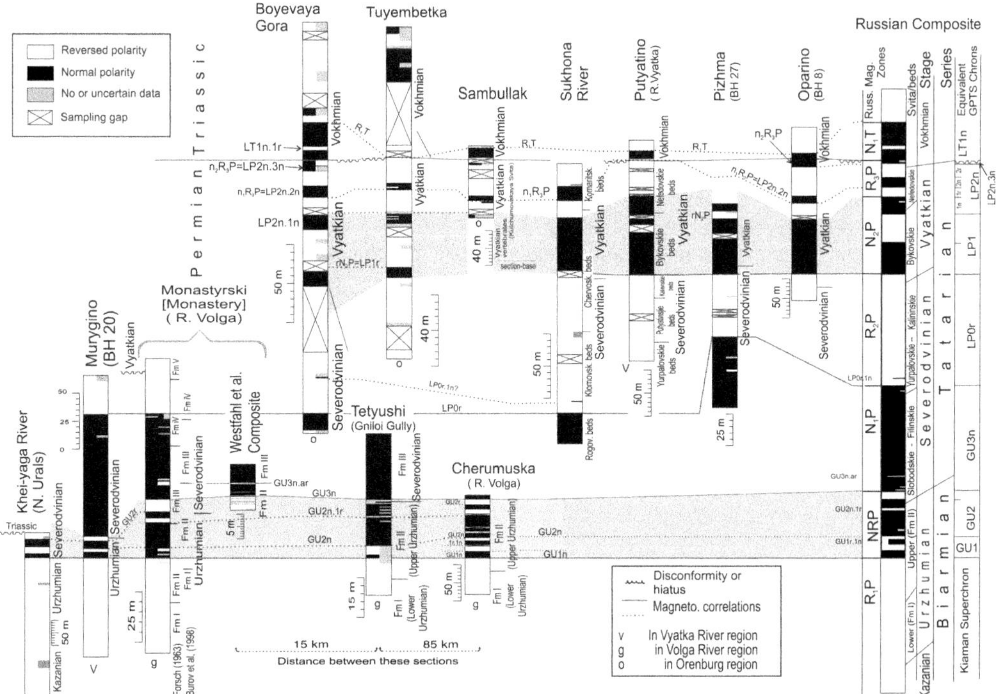

Fig. 5. Summary of the mid and late Permian magnetostratigraphy from the Russian East European Basin successions west of the Urals (modified from Hounslow 2016). Boyevaya Gora, Tuyembetka and Sambullak sections are near Orenburg, and are from Taylor *et al.* (2009). Murygino, Tetyushi, Cheremushka, Putyatino, Pizhma, Oparino sections are near the Kama, Volga and Vyatka rivers (SW Tataria, Kazan region: Silantiev *et al.* 2015*a*, *b*), approximately 700 km NE of the Orenburg area, and are from Burov *et al.* (1998). Monastyrski (Volga River) section (Mouraviev *et al.* 2015) from the Kazan region based on Gialanella *et al.* (1997), Burov *et al.* (1998), Westfahl *et al.* (2005) and Balabanov (2014). The Sukhona River section from NE Russian, approximately 600 km north of Kazan, is from Khramov *et al.* (2006). The Khei-Yaga River section is as in Figure 2. Each section has a thickness scale in metres. Composite magnetochrons are also labelled with the Russian naming convention (Molostovsky 1996; Molostovsky *et al.* 1998) and the Russian regional stratigraphy (Kotlyar & Pronina-Nestell 2005). BH, borehole number.

respect in order to better understand issues of missing strata.

In the Monastyrski Ravine (Monastery Ravine, type section of the basal Severodvinian) section (Fig. 5), the base of the Illawarra Superchron corresponds to the *Paleodarwinula tuba–P. arida–P. torensis* ostracod Zone (Mouraviev *et al.* 2015; Kotlyar 2015). The better biostratigraphic dating of the end of the Kiaman from sections in Texas and Japan suggest that the base of the NRP magnetozone (in the late Urzhumian) is slightly older than commonly inferred (e.g. Golubev 2015), and should equate to the earliest Wordian or latest Roadian.

Other more poorly dated, non-marine, sections probably also display the end of the Kiaman Superchron, such as the Taiyuan section in China, within the lower member of the Upper Shihhotse Formation (Embleton *et al.* 1996; Stevens *et al.* 2011). This occurs between two floral extinction events. The earlier one is in the Lower Shihhotse Formation (inferred to be not shown on Fig. 4), and lower one of two later extinction events in the middle and upper members of the Upper Shihhote Formation and inferred to be late Guadalupian (Stevens *et al.* 2011) (Fig. 4).

The start of the Illawarra Superchron is present in European red-bed successions in the German Upper Rotliegend, Parchim Formation (Langereis *et al.* 2010), and in southern England in the Exeter Group (Hounslow *et al.* 2016). The biostratigraphic age dating of these units is low resolution, largely based on tetrapods (Rotleigend only), footprints and, occasionally, long-ranging palynomophs, such as *Lueckisporites virkkiae* (Edwards *et al.* 1997; Słowakiewicz *et al.* 2009). Generally, the end of the Kiaman provides a higher-resolution dating tool in these successions. The base of the Illawarra Superchron has also probably been detected in

Kansas (USA) in the Rebecca K Bounds core (Soreghan *et al.* 2015), in a succession that lacks independent evidence of age, but whose age is approximately constrained by subsurface regional relationships (Sawin *et al.* 2008).

In the type region of the Illawarra Superchron in Australia, magnetic polarity details and ages are less clear. The base of the Illawarra Superchron is thought to be within the Mulbring Siltstone in the Hunter Valley region of New South Wales (Idnurum *et al.* 1996; Foster & Archibold 2001). The Mulbring Siltstone correlates to the Broughton and underlying Berry formations of the southern Sydney Basin in SE Australia around the Kiama area, this is because of the *Echinalosia wassi* brachiopod range zone and palynological zones in these two areas (Campbell & Conaghan 2001; Cottrell *et al.* 2008). The lateral equivalent to the Broughton Formation (Campbell & Conaghan 2001) is the lower part the Gerringong Volcanics (Blowhole, Bumbo, Dapto and Cambewarra flows), which is reverse polarity and widely considered to be within the end of the Kiaman Superchron (Irving & Parry 1963; Cottrell *et al.* 2008). Irving & Parry (1963) also found reverse polarity in the youngest (Berkeley) flow of the Gerringong Volcanics. This suggests the base of the Illawarra Superchron may be within the laterally equivalent and overlying Pheasants Nest Formation (of the Illawarra Coal Measures: Campbell & Conaghan 2001; Metcalfe *et al.* 2014) in the southern Sydney Basin. Foster & Archibold (2001) inferred that the brachiopod faunas of the Broughton Formation have a similarity to latest Ufimian–Kazanian brachiopod assemblages. However, the Mulbring Siltstone has U–Pb SHRIMP ages of approximately 264 $\pm$ 2.2 Ma (Retallack *et al.* 2011), and the laterally equivalent uppermost part of the Broughton Formation has a U–Pb ID-TIMS date of 263.5 $\pm$ 0.31 Ma (Metcalfe *et al.* 2014), suggesting an early Capitanian age in the timescale of Henderson *et al.* (2012) and that proposed here. These inconsistencies probably indicate that the Sydney Basin brachiopod faunas are of little use for international correlation (as suggested by Metcalfe *et al.* 2014), and the new radiometric dates suggest that the reverse polarity Gerringong Volcanics may not be within the Kiaman Superchron, but instead correlate to GU2r.

Guadalupian data from marine sections

The Nammal Gorge section (Haag & Heller 1991) is a key marine section for the mid-Permian magnetostratigraphy as it has an associated conodont biostratigraphy, but in its original publication had very little supporting biostratigraphic detail (Fig. 4). However, based on nearby sections (Saidu Wali, Kotla Lodhian, Zalucj Nala, Chihidru Nala and Kathwai), conodont ranges (Fig. 4) can be related to the magnetostratigraphic data in the Nammal Gorge section (Wardlaw & Pogue 1995, Wardlaw & Mei 1999). These conodont ranges are correlated onto the magnetostratigraphy, using the lithostratigraphy and bed numbers from published sedimentary logs (Baud *et al.* 1995; Waterhouse 2010). A hiatus in the Nammal Gorge section is present in the late Capitanian (i.e. missing conodont zones) between the Lakriki and the Sakesar members of the Wargal Formation (Mei & Henderson 2002; Mertmann 2003; Waterhouse 2010). This hiatus separates dominantly normal polarity below from reverse polarity in the upper part of the Wargal Formation (Fig. 4). Hence, the oldest normal polarity interval in the original published data (Haag & Heller 1991) is probably magnetozone GU2n in the late Wordian–earliest Capitanian. No magnetostratigraphy was measured from the underlying Amb Formation, which possesses an array of conodonts indicating a Wordian age (Wardlaw & Mei 1999). The early Wordian–Capitanian fusulinid *Neoschwagerina margaritae* is found in unit 2 of the Wargal Formation (Jin *et al.* 2000; Waterhouse 2010) (Fig. 4).

The Shangsi section magnetostratigraphy is key to using radiometric dating for age calibration of the Lopingian, and the section probably extends down into the Capitanian. Unfortunately, the three studies of the Permian magnetostratigraphy (Heller *et al.* 1988; Steiner *et al.* 1989; Glen *et al.* 2009) in this section, display differences in the interpretations of the polarity (Fig. 6). A composite magnetostratigraphy was constructed using the agreement between these, based on the sampling positions. The study of Glen *et al.* (2009) has many sampling levels in the Wujiaping Formation that failed to yield polarity information, whereas the study of Steiner *et al.* (1989) yielded a relatively simple polarity pattern through this formation. The age of the lower part of the Wujiaping Formation is not clear from the faunal data owing to a barren interval (Sun *et al.* 2008). U–Pb ID-TIMS radiometric dates (260.4 $\pm$ 0.8 and 259.1 $\pm$ 0.9 Ma), which appear to be from reworked material from the Emeishan volcanics (Zhong *et al.* 2014), suggest a maximum age, but are consistent with the normal polarity interval in bed 7 being of late Capitanian age. The underlying Maokou Formation at Shangsi contains the late Roadian through to the Wordian and on to earliest Capitanian, with a major hiatus at the base of the Wujiaping Formation (Sun *et al.* 2008). The cyclostratigraphy at Shangsi suggests large changes in sedimentation rates (Fig. 6). The biostratigraphy of the Maokou Formation in the Wulong section (Fig. 4) is based on unattributed conodont data in Jin *et al.* (2000).

Details of the Guadalupian and Wordian magnetostratigraphy are generally poorly defined from

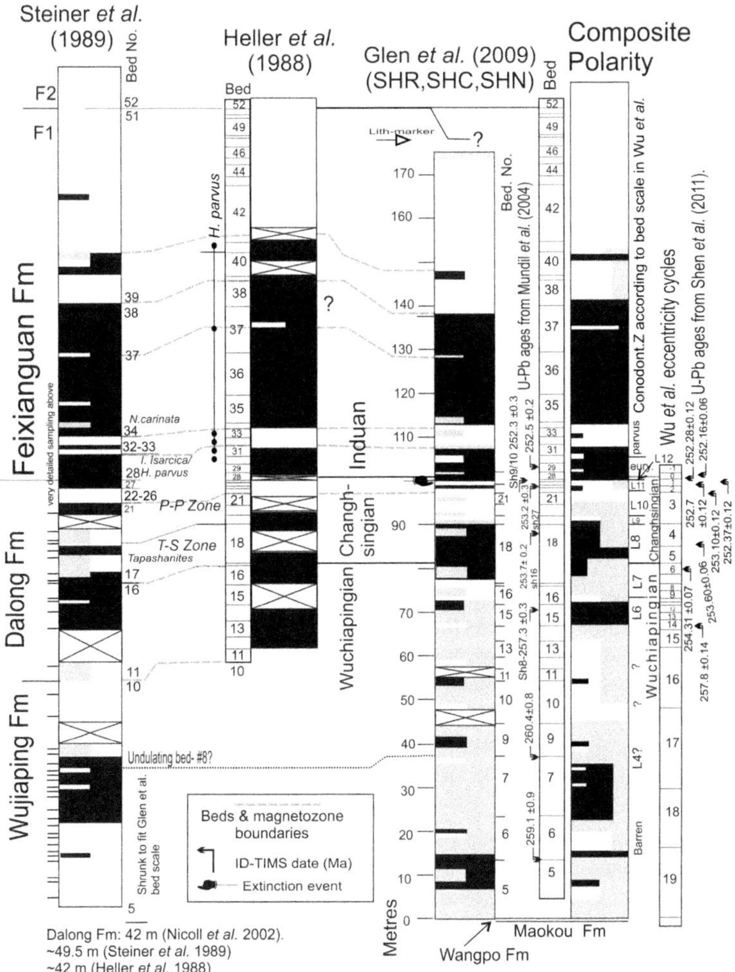

Fig. 6. Summary of magnetic polarity data for the Shangsi section. The composite magnetostratigraphy on the right is derived from three magnetostratigraphic studies of Heller *et al.* (1988), Steiner *et al.* (1989) and Glen *et al.* (2009). Radiometric dates from the section are from Mundil *et al.* (2004) and Shen *et al.* (2010). Biostratigraphy is from Lai *et al.* (1996), Jin *et al.* (2000) and Sun *et al.* (2008). There are inconsistencies in the thickness of units between the three studies, but generally the datasets can be related using the bed number stratigraphy. Many of the uncertain (grey) intervals from the study of Glen *et al.* (2009) represent sample levels that yielded no polarity information. The cyclostratigraphy and conodont zonal boundaries are from Wu *et al.* (2013). Key as in Figure 4. Ammonoid zones: T-S, *Tapashanites–Shevyrevites* Assemblage Zone; P-P, *Pseudotirolites–Pleuronodoceras* Assemblage Zone.

marine sections alone, but show both polarities in the early Wordian and early Capitanian (Figs 4 & 3). The upper Capitanian is normal polarity dominated as the chron GU3n (the 'Capitan-N' chron of Steiner 2006). This is shown by the data from the Wulong section (Heller *et al.* 1995), the Emeishan Basalts (Zheng *et al.* 2010; Liu *et al.* 2012) and data from the *Yabiena–Lepodolina* fusulinid zones from Kyushu in Japan (Fig. 4). The Ebian County magnetostratigraphy through the Emeishan Basalts and overlying units (Ali *et al.* 2002), together with the radiometric dates, suggest a middle–late Capitanian age for the Emeishan Basalts (He *et al.* 2007; Zheng *et al.* 2010; Liu *et al.* 2012). This is supported by the middle–late Capitanian age suggested by the conodonts *J. altudaensis* (conodont zone G5) and *J. xuanhanensis* (zone G7) from the few metres of the Maokou Formation that underlie the Emeishan Basalts (Sun *et al.* 2010). The predominantly normal polarity

Emeishan Basalts continue into an overlying reverse polarity magnetozone (Ali *et al.* 2002), which is inferred to be the latest Capitanian base to LP0r (Fig. 4).

Guadalupian data from non-marine sections

Magnetic polarity data from Russian sections through the Urzhumian and early parts of the Severodvinian provide detail through the earliest part of the Illawarra Superchron, suggesting that the Wordian–Capitanian interval has a bias towards normal polarity (Fig. 5). The Russian NRP mixed-polarity magnetozone appears to show two major reverse polarity intervals, the upper one of which is subdivided by a normal polarity sub-magnetozone. In these sections, the structure of the earliest normal magnetozones in the Illawarra Superchron are best represented by the thick Cheremushka section (Silantiev *et al.* 2015*a*), which is the parastratotype of the Urzhumian. Similar polarity structure is shown in other Russian sections, such as Tetyushi, Monastyrski and Murygino (Gialanella *et al.* 1997; Burov *et al.* 1998; Balabanov 2014), which allows a subdivision into two major normal magnetochrons (GU1n and GU2n), most clearly seen in the Murygino core and the Khei-yaga River section (Fig. 5). However, the NRP polarity interval has problems with partial normal overprints, making magnetozones in the NRP zone difficult to define (Westfahl *et al.* 2005; Silantiev *et al.* 2015*a*). However, normal polarity intervals detected in the many sections in the upper Urzhumian suggests a Permian geomagnetic signature rather than a later overprint.

Like the marine-section data, and the Russian sections, the dominance of normal polarity through the later parts of the Guadalupian (i.e. GU3n) are well displayed in other non-marine sections, such as the Whitehorse Formation in Kansas (Fig. 3) and the Havel Subgroup, and the Exeter Mudstone and Sandstone Formation in the Rotliegend equivalent in Europe (Fig. 7).

Options for the magnetostratigraphy of the Wordian

A key problem in comparing marine and non-marine sections in the earliest part (i.e. Wordian) of the Illawarra Superchron is that there are two likely magnetic polarity models for this interval: a 'long-GU1r' option and a 'brief-GU1r' option.

Long-GU1r option. In sections such as at Wulong, Taiyuan and those in western Texas (Figs 4 & 3), thicker intervals of reverse polarity are displayed, compared to the associated normal magnetozones in the GU1–GU2 interval. Sections through the Abrahamskraal Formation in the lowermost Beaufort Group (South Africa) have similar characteristics. Crucially, the South African sections have SHRIMP U–Pb dates that overlap the ID-TIMS radiometric dates from the Guadalupian type area, allowing fuller integration of the geochronology and magnetostratigraphy. This is the option used here in the Permian GPTS; Hounslow (2016), however, uses the 'brief GU1r option'.

Brief-GU1r option. This is exemplified by the Russian Urzhumian data (Fig. 5), where there is a dominance of normal polarity in the earliest parts of the Illawarra (GU1–GU2 interval), and the reverse polarity magnetozones appear generally briefer than the normal magnetozones (e.g. Russian composite: Fig. 5). The Wargal Formation, Whitehorse Formation and the SW English coast data share similar characteristics (Figs 3, 4 & 7).

Lanci *et al.* (2013) measured a magnetostratigraphy through the Waterford Formation (Ecca Group) and the overlying lower parts of the Abrahamskraal Formation (Beafort Group), and interpreted these data as evidence of the base of the Illawarra Superchron because of three normal polarity magnetozones (N1–N3: Figs 1 & 3). They interpreted the N3 magnetozone (identified in two separate sections) as the start of the Illawarra Superchron. Normal polarity dominates the overlying argillaceous mid-parts of the Abrahamskraal Formation in the Buffels River area (Tohver *et al.* 2015) (Fig. 3). Tohver *et al.* (2015) estimated the base of the Abrahamskraal Formation to be some 340 m below their lowest sampled levels, suggesting that the youngest polarity data in the Ouberg Pass study of Lanci *et al.* (2013) is approximately equivalent with the oldest strata sampled by Tohver *et al.* (2015) at Buffels River (Fig. 3). A correlation more likely than that proposed by Lanci *et al.* (2013) is that magnetozone interval N2–N1 is the equivalent of GU1n, marking the base of the Illawarra Superchron (Fig. 1), and indicating one reverse subzone (GU1n.1r) in GU1n. This 'long-GU1r' option suggests magnetozone N3 is the magnetochron CI3r.1n (Figs 1 & 3). In the same general area as the study of Tohver *et al.* (2015), Jirah & Rubidge (2014) measured the total stratigraphic thickness of the Abrahamskraal Formation as 2565 m, suggesting that the uppermost sampled levels of Tohver *et al.* (2015) at Buffel River are approximately 920 m above the base of the Abrahamskraal Formation. These upper samples are therefore approximately at the upper range of the *Eodicynodon* Assemblage Zone (Jirah & Rubidge 2014). The 'long-GU1r' option is supported by the similarity in U–Pb SHRIMP dates of 266.4 $\pm$ 1.8 Ma (Lanci *et al.* 2013) from near the base of N2 (GU1n) and from the ID-TIMS date 266.5 $\pm$ 0.24 Ma near the base of the Wordian in the Texas/New Mexico sections (Bowring *et al.*

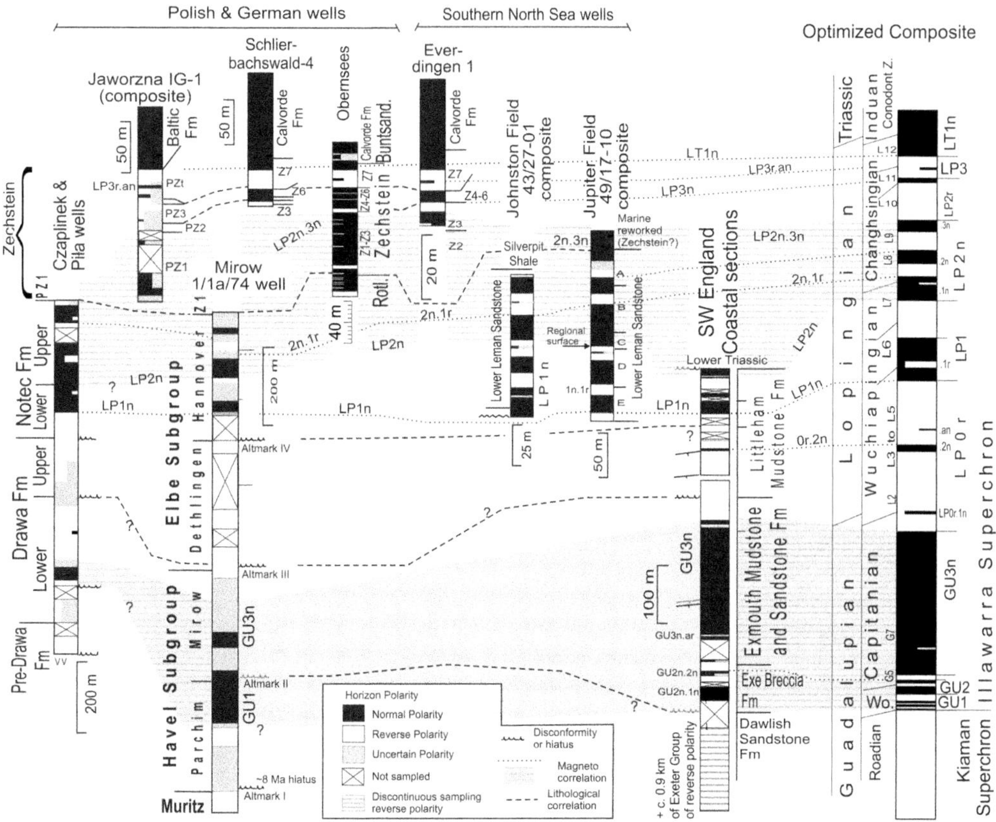

Fig. 7. Magnetostratigraphic data from the Upper Rotleigend–Zechstein equivalent Permian-age sections in Europe. Czaplinek, Piła and Jaworzna IG-1 well magnetostratigraphy composite derived from Nawrocki (1997), with additional stratigraphic details from Słowakiewicz *et al.* (2009). Mirow well 1/1a/74 is from Menning *et al.* (1988) and Langereis *et al.* (2010). Obernsees well composite polarity reinterpretation is from Szurlies (2013). The Schlierbachswald-4 and Everdingen 1 wells are from Szurlies *et al.* (2003) and Szurlies (2013). The Southern North Sea data for the Leman Sandstone Formation are from Turner *et al.* (1999) and Lawton & Robertson (2003). The SW England coast section data are from Hounslow *et al.* (2016).

1998) (Fig. 3). The youngest U–Pb SHRIMP date in the Ouberg Pass section of 264.4 ± 1.9 Ma indicates a level in GU1r. Zircon ID-TIMS dates from approximately 1.5 km higher in the Beaufort Group than for the Buffels River magnetostratigraphy (Fig. 3) suggests that the Capitanian–Wuchiapingian boundary (at *c.* 260 Ma) approximates the boundary between the tetrapod *Tropidostoma* and *Pristerognathus* assemblage zones (Rubidge *et al.* 2013). Using this date and the 'long-GU1r' option suggest that the bulk of this additional 1.5 km of strata is predicted to be normal polarity, corresponding to most of GU3n (Fig. 3).

Palynological zonations of the underlying Ecca Group generally support the 'long-GU1r' option, suggesting that the youngest parts may be late Cisuralian or, possibly, Roadian in age (Modie & Le Hérissé 2009). This is largely based on correlation of the Ecca Group assemblage zones (in the upper half of the Ecca Group) to the *Lueckisporites virkkiae* Interval Zone of the Parana Basin, where the base of the interval zone in Argentina is dated (using SHRIMP U–Pb on zircons) to 278.4 Ma (Modie & Le Hérissé 2009), placing its base in the Kungurian. This is a similar position to the first occurrence of *L. virkkiae* in the Svalbard sections (Fig. 2).

Tetrapod fauna of the Beaufort Group *Eodicynodon* Assemblage Zone forms the key components of the Kapteinskraalian land vertebrate faunachron (LVF) of Lucas (2006). The fauna of this LVF is most similar to the Ocher and part of the Mezen tetrapod assemblages from Russia (Lucas 2006), which occur within the Shesmian (upper interval of the Ufimian) to Kazanian to the mid-Urzhumian (Golubev 2015). In Russian sections, this interval is reverse polarity only (Figs 1 & 5), whereas the assumed equivalent *Eodicynodon* Assemblage interval is associated with both polarities. Hence, the 'long-GU1r' option indicates diachroneity of the Kapteinskraalian LVF, with the Russian faunas being the oldest representatives of this LVF.

The alternative 'brief-GU1r' option places the start of the Illawarra Superchron approximately 400 m above in the mid-parts of the Abrahamskraal Formation (Buffels River section), at the base of the interval of normal polarity dominance (Fig. 3). This option suggests that the Ouberg Pass section N2–N3 magnetozones represent the Kungurian magnetochron CI3n, and magnetozone N3 is possibly CI2n (or a tentative magnetozone between CI2n and CI3n: Fig. 1). This 'brief-GU1r' option is compatible with the normal polarity dominance in the mid-parts of the Abrahamskraal Formation in the Buffels River area (Fig. 3). However, it requires the overlying 1.5 km of strata to the Wuchiapingian boundary in the Beaufort Group to be largely normal polarity, corresponding to the younger part of GU3n. This option makes the correlations between the Russian and South African expression of the Kapteinskraalian LVF more consistent in terms of the reverse polarity dominance in the inferred late Ufimian–mid-Urzhumian age for the *Eodicynodon* Assemblage Zone. However, it does push the base of the *Eodicynodon* Assemblage Zone into the Kungurian, potentially as early as the Kungurian–Artinskian boundary, which is counter to current thinking that suggests tetrapod assemblages yielding 'bona fide therapsids' are mid-Permian (Lucas 2006). The two older Littlecrotian and Redtankian LVFs (Lucas 2006) have little independent age control. The older LVF, the Redtankian, has equivalent tetrapod fauna from the Garber Formation (in which CI2n has been inferred: Table 1), suggesting that the Waterford Formation magnetozone N3 is a good deal younger than late Artinskian. Supporting evidence for the 'brief-GU1r' option is the reassessment of the detrital zircon SHRIMP ages (due to suspected lead loss) from the top of the Ecca Group (Tohver *et al.* 2015), which suggest ages as old as 275 Ma (i.e. Kungurian) for deposition of the upperparts of the Ecca Group.

Broadly, the 'long-GU1r' option implies that polarity dominance over the GU1 magnetochron is poorly represented by the Russian Urzhumian dataset (Fig. 5), supporting suspected normal polarity overprints in this dataset. It also implies a large diachroneity of the Kapteinskraalian LVF. The crucial supporting data are the age overlap between the radiometric dates from the Abrahamskraal Formation and those from the Guadalupian type area (Fig. 3).

The 'brief-GU1r' option relies on the large wealth of data from the Russian sections through the Urzhumian, and requires that the U–Pb SHRIMP ages from the Abrahamskraal Formation are too young, probably impacted by lead loss (e.g. Tohver *et al.* 2015). It also indicates the Kapteinskraalian LVF extends into the Kungurian, counter to the hypotheses of vertebrate workers.

Lopingian and the Permian–Triassic boundary

The key part of the Lopingian magnetochron pattern is the reverse-polarity-dominated early Wuchiapingian (i.e. LP0r), a key feature clearly seen in many marine and non-marine datasets. This reverse polarity interval and its transition from GU3n is seen by the relatively thick LP0r, overlying a relatively thick GU3n in many sections. The LP0r is followed through the late Wuchiapingian and Changhsingian by a pattern of reverse and normal magnetozones with similar relative thickness (Figs 4 & 5).

Biochronological control of the magnetic polarity changes across the Guadalupian–Lopingian boundary is probably best defined in the Kyushu sections (Kirschvink *et al.* 2015), where an extinction level and change to the Wuchiapingian fusulinid *Codonofusiella–Reichelina* Zone is seen (Fig. 4). The extinction level appears to be located in an approximately 2–3 m-thick normal polarity interval (LP0r.1n), within an interval of predominant reverse polarity. There are tentative brief normal polarity intervals in other sections (e.g. Wulong, Shangsi and Sukhona River) at around this level following GU3n (Figs 4 & 5), but none of them have a better biochronology. Magnetostratigraphic studies of the Laibin section (and the Wuchiapingian GSSP section) by M. Menning and S. Shen have only recovered remagnetizations (Jin *et al.* 2006*a*).

The normal magnetochron LP0r.2n is clearly shown in the Wulong and Linshui section in China, and tentatively in the Shangsi and Nammal Gorge sections (Fig. 4). This magnetozone is the 'P3' normal chron of Steiner (2006). It occurs within the range of the conodont *Clarkina asymmetrica* (L3 standard conodont zone) in the Nammal Gorge section, placing it in the early Wuchiapingian. The age of unit 5 of the Longtan Formation at the base of the Linshui section is based on regional correlations of brachiopod assemblages, suggesting a late Wuchiapingian age (Chen *et al.* 2005). This age is supported by the presence of the conodont *C. liangshanensis* (equivalent to conodont zones L6–L7: Shen *et al.* 2010) in the basal beds of the Longtan Formation, approximately 300 m below the measured magnetostratigraphy (Shuzhong Shen pers comm. 2010). Equivalents to magnetochron LP0r.2n also occur in the Rustler

Formation in New Mexico and the Littleham Mudstone Formation in England (Figs 3 & 7).

The base of magnetochron LP1n is a clear stratigraphic marker in many Lopingian marine sections, following the LP0r chron (Fig. 4). In non-marine sections in Russia and Europe, this is a very clear boundary to an overlying interval with several major normal polarity intervals (Figs 5 & 7). The base of LP1n is within the range of the Wuchiapingian conodont *C. guangyuanensis* (L5 standard conodont zone) at Nammal Gorge, with LP1n extending to near the top of the late Wuchiapingian *C. transcaucasica* Zone (conodont zone L6) in the Shangsi section (Fig. 4).

The interval LP1n–base LP2r shows a pattern of polarity changes that tend to be dominated by normal polarity in marine sections, yet include regular reverse polarity intervals. This interval is the 'Chang-N' chron of Steiner (2006). The Linshui section (which has a high accumulation rate) displays this interval particularly well, whereas the Wulong, Shangsi and Nammal Gorge sections do not display the intervening reverse magnetozones clearly (Fig. 4). In New Mexico, the Quartermaster and Dewey Lake formations clearly show a pattern of three major reverse magnetozones (Fig. 3), as in the Linshui section. The upper boundary of the LP2n.3n magnetochron is within the Changhsingian *C. subcarinata* Zone (L9) of the Abedah section, but probably within the *C. changxingensis* Zone (L10) in the Shangsi section.

The three studies on the Changhsingian and Induan GSSPs at Meishan (Fig. 8) show a poor degree of similarity in the magnetic polarity through the section (Li & Wang 1989; Liu *et al.* 1999; Meng *et al.* 2000). An additional summary in Yin *et al.* (2001) shows some additional details, although the source data are not published. The low degree of consistency between the magnetostratigraphic data does suggest a normal polarity interval (the LP2n.2n–LP2n.3n interval?) in the *C. wangi*–*C. subcarinata* zones (L8–L9) and mixed polarity in the *C. changxingensis* Zone: possibly corresponding to the LP2r–LP3r interval (Fig. 8).

Lopingian non-marine sections

Magnetic polarity data from marine sections display more detail in magnetozones through the Lopingian than the Russian non-marine sections (Figs 4 & 5). The simplest interpretation of this is the absence of most of the late Changhsingian, often inferred in Russian sections (Lozovsky 1998; Tverdokhlebov *et al.* 2005; Lozovsky *et al.* 2014). The oldest units of the Vetlugian (i.e. Vokhmian, considered to be early Triassic) have a transitional latest Changhsingian flora and reverse polarity (i.e. upper part of LP3r), clearly resting on an eroded surface of the late Vyatkian (Lozovsky *et al.* 2001). The Permian–Triassic boundary is therefore clearly within the basal-most Vokhmian.

In Russian Tatarian sections (Fig. 5), the uppermost normal polarity parts of magnetozone R_3P (i.e. n_1R_3P and n_2R_3P) are missing from some sections, but are clearly present at the Oparino and Boyevaya Gora sections and other sections shown in Burov *et al.* (1998). This probably reflects the variable erosion at the base of the Vokhmian. Both in the marine and non-marine sections, the three reverse magnetochrons in the magnetochron interval LP1–LP2 vary greatly in thickness (Figs 4 & 5). Some of this variation in the Russian sections may be due to channel bodies, which can give variable accumulation rates, together with likely local hiatus, features that are being investigated in more detail (Arefiev *et al.* 2015).

In Europe, magnetostratigraphic studies in the Upper Rotliegend of the southern Permian Basin (well Mirow 1/1a/74: Menning *et al.* 1988; Langereis *et al.* 2010) and wells in Poland (Nawrocki 1997) clearly show the reverse polarity LP0r. Above this is a mixed-polarity interval, which includes the Zechstein (Fig. 7). The incomplete sampling of the normal and reverse magnetozones in the Notec and Hannover formations are more fully represented by studies from the laterally equivalent Lower Leman Sandstone from the Johnston and Jupiter gas fields in the Southern North Sea (Turner *et al.* 1999; Lawton & Robertson 2003). In the Southern Permian Basin, these European-wide correlations are strongly constrained by the overlying Zechstein, the base of which is usually inferred to be an isochronous lithostratigraphic marker. In the southern German Obernsees core, normal polarity dominates the Z1–Z3 interval (Szurlies 2013), with a briefer reverse polarity magnetozone near the base of the Z1 interval that may correlate to the uppermost tentative reverse seen in the Polish Czaplinkek, Pila and Jaworzna IG-1 well (Fig. 7). Like the Everdingen-1 and Schlierbachswald-4 wells, the Z4–Z6 interval is dominated by normal polarity in the Obernsees core (Szurlies 2013).

Correlations in Figure 7 imply that the base of the Zechstein (basal Z1 cycle) occurs in the oldest parts of magnetochron LP2n.3n in the mid-Changhsingian. The equivalent of LP2n.3n seems to be exceptionally thick in the Zechstein successions (*c.* Z1–Z3 interval), which may be explained by the rapid infilling of the Zechstein Basin upon initial flooding. Additional support for the Changhsingian age of the Zechstein comes from Sr-isotope data, which indicate a short duration for the Zechstein of approximately 2 myr, and an age range in the interval 255–251.5 Ma, placing it firmly in the Changhsingian (Denison & Peryt 2009). Attempts at direct dating of the Kupferschiefer (the base

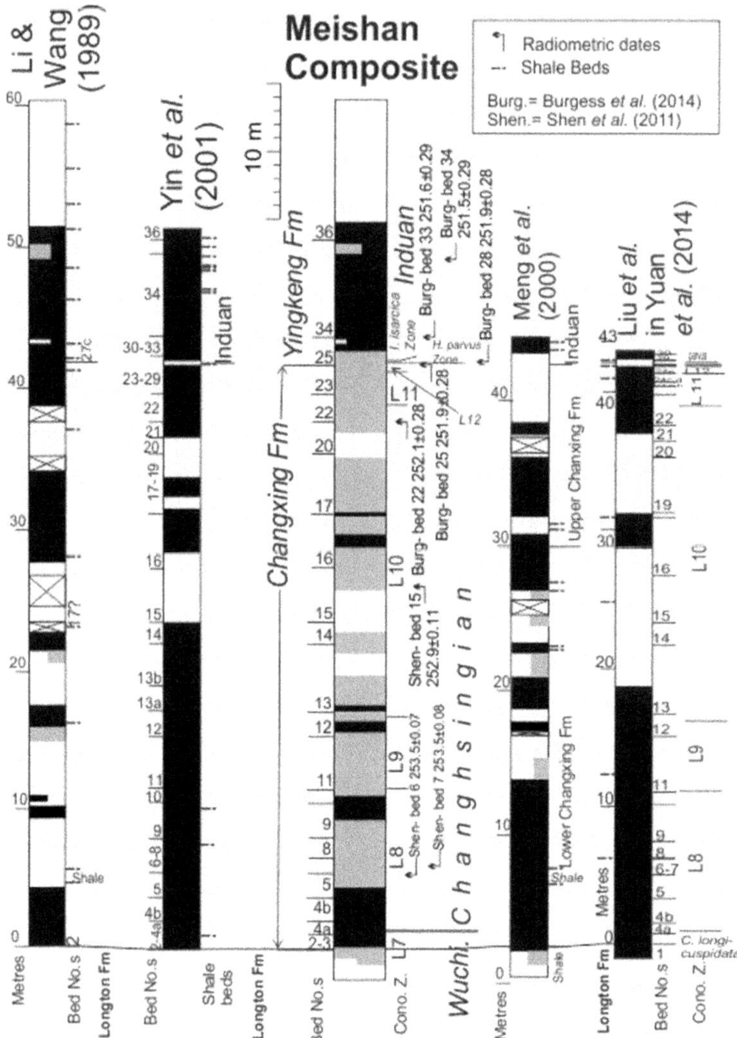

Fig. 8. Summary of the magnetic polarity data for the Meishan section. The composite magnetic polarity is derived from the three published studies of the Meishan section from Li & Wang (1989), Liu *et al.* (1999) in Yuan *et al.* (2014) and Meng *et al.* (2000). Associated radiometric dates and biostratigraphy are from Jin *et al.* (2006*b*), Mundil *et al.* (2010), Shen *et al.* (2010) and Burgess *et al.* (2014). The data for the polarity composite shown in Yin *et al.* (2001) have never been published. There is some ambiguity about how to relate these datasets as thicknesses vary and bed numbers are not shown in Li & Wang (1989) and Meng *et al.* (2000). Data relationships were attempted using the shale beds in the section logs.

of the Zechstein Z1 cycle) have failed to yield consistent results, with Re–Os ages giving wide 95% confidence intervals (Pašava *et al.* 2010). The Changhsingian age conflicts with conventional age interpretation of the basal Zechstein, which is usually assigned to the early Wuchiapingian (Szurlies 2013). This is primarily based on the conodonts *Merrillina divergens* and *Mesogondolella britannica* from the Kupferschiefer and Zechsteinkalk of the Z1 Formation (Swift 1986; Korte *et al.* 2005; Legler *et al.* 2005; Słowakiewicz *et al.* 2009; Szurlies 2013), as, according to Kozur in Szurlies (2013), *Mer. divergens* occurs in the range interval of *Clarkina leveni* (conodont L4 standard zone) in Iran. However, *Mer. divergens* is found in the uppermost Alibashi Formation in the Changhsingian *C. yini–C. zhangi* Zone in Iran (Kozur 2007), and from Wordian, Capitanian and late Cisuralian strata (Swift 1986; Nakrem *et al.* 1991). Therefore, Zechstein conodont faunas do not provide a precise biochronology – owing to differences between cold- and warm-water faunas, they only provide an approximate Lopingian age (Henderson & Mei 2000).

The Permian–Triassic boundary

The late Changhsingian transition towards the Permian–Triassic boundary has been well documented in terms of magnetic polarity in both marine (Gallet *et al.* 2000; Glen *et al.* 2009; Li *et al.* 2016) and non-marine successions (Glen *et al.* 2009; Hounslow & Muttoni 2010; Szurlies 2013), where a reverse-polarity-dominated interval (LP2r–LP3r) occupies the late Changhsingian. This occupies the *C. yini* (L11) and *C. meishanensis* (L12) conodont zones (and parts of the *C. changxingensis* in some sections), prior to the main extinction event in the latest Changhsingian. In spite of the well-studied nature of this interval, the conodont zonal boundaries are not consistently located with respect to the polarity boundaries, perhaps indicating placement issues with the conodont standard zones. In this interval, the normal magnetozone LP3n is the 'P5' chron of Steiner (2006), and is clearly seen in several marine and non-marine sections (Figs 4 & 7).

In the Induan GSSP at Meishan (Fig. 8), the exact relationship between the polarity stratigraphy and the first occurrence of *Hindeodus parvus* is not clear, but the Shangsi and Abedah sections indicate that the inferred base of the Induan is consistently in the lower part of the LT1n.1n magnetochron (Glen *et al.* 2009; Hounslow & Muttoni 2010; Szurlies 2013). The Shangsi section probably provides the most precise placement of the Permian–Triassic boundary interval with respect to the magnetostratigraphy (Fig. 9). At Shangsi, the base of LT1n is near the base of the *C. meishanensis* conodont zone, within 0.5 m of the extinction event bed (Glen *et al.* 2009). A variety of chemical abrasion ID-TIMS (CA-ID-TIMS) U–Pb radiometric dates indicate approximately 252.3 Ma for the age of the base of LT1n, in the latest Changhsingian (Fig. 9). At Shangsi, the precisely correlated base of the Induan (base of the *H. eurypyge* Zone: Shen *et al.* 2011) is based on constrained optimization (CONOP) correlation and the occurrence of *H. changxingensis*

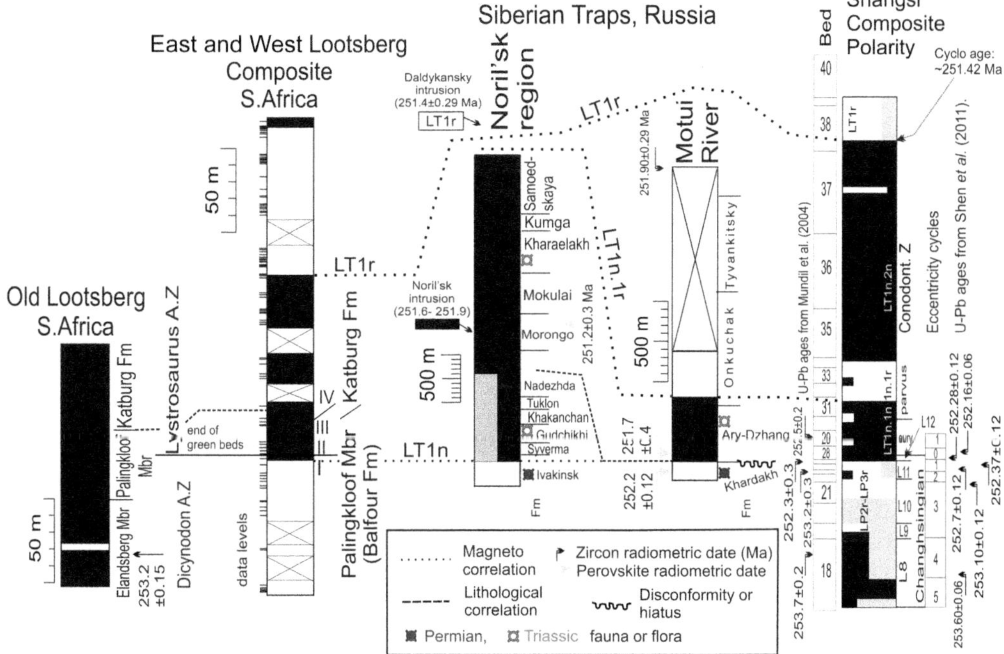

Fig. 9. Geomagnetic polarity datasets for non-marine sections that span the Changhsingian–Induan boundary compared to the data from the Shangsi section (which shows the clearest relationship between the magnetostratigraphy and a precise bio- and geochronology). Section data from Old Lootsberg and East and West Lootsberg are from Gastaldo *et al.* (2015) and Ward *et al.* (2005), respectively: these are drawn using the same vertical scale. Polarity data for the Siberian traps using the composites in Fetisova *et al.* (2014), supported by magnetic and geochronological data in Gurevitch *et al.* (2004) and Burgess & Bowring (2015). The Shangsi section data are from Figure 6. Cyclostratigraphic age on the base of LT1r is from Li *et al.* (2016).

rather than *H. parvus*, the first occurrence of which is younger in the section (Metcalfe *et al.* 2007).

In Russian Platform sections, there is dispute about the continuity of the successions across the Permian–Triassic boundary, with some preferring a lack of hiatus (Sennikov & Golubev 2006; Krassilov & Karasev 2009; Taylor *et al.* 2009; Newell *et al.* 2010) but others suggesting hiatus (Lozovsky 1998; Tverdokhlebov *et al.* 2005); much depends on the stratigraphic resolution of the dating tools. However, it is clear in the magnetostratigraphy from the Russian sections that there are insufficient magnetozones following LP1n (N_2P in Russian magnetozones: Fig. 5) to accommodate the entire Lopingian, indicating a major hiatus at the base of the Vokhmian or, locally, in the Vyatkian. The basal Vokhmian typically shows an increase in magnetite abundance, expressed as increases in magnetic susceptibility and remanence intensity (Burov *et al.* 1998; Lozovsky *et al.* 2014), which appears to be associated with an enhanced volcanic ash contribution (Burov 2004). In some other sections, where magnetozone n_2R_3P is not seen, the late Permian magnetozones are variably removed by erosion at the base of the Vokhmian, indicating that Russian magnetozone n_2R_3P is the equivalent of LP2n.3n (Fig. 5). However, in the Yug River Basin, the transition of LP3r into LT1n (or, perhaps, LT1n.1r into LT1n.2n) and the transition into the Triassic may be preserved in the Nedubrovo Member. This member has plant and spore remains typical of the Tatarian and the Zechstein, as well as megaspores *Otynisporites eotriassicus* and *O. tuberculatus* typical of the earliest Triassic (Burov 2004; Lozovsky *et al.* 2014; Arefiev *et al.* 2015).

In sections (the East and West Lootsberg Pass and Komandodriftdam sections) from the Karoo Basin (South Africa), the turnover in vertebrate assemblages is seen just below the Balfour Formation–Katburg Formation boundary (Fig. 9). This change is inferred to represent the Permian–Triassic boundary because of the association between the vertebrate biochronology, the expected magnetostratigraphy (Fig. 9) and negative $^{13}C_{org}$ isotopic excursions (De Kock & Kirschvink 2004; Ward *et al.* 2005). However, magnetostratigraphy and U–Pb ID-TIMS dating from the nearby Old Lootsberg Pass (Gastaldo *et al.* 2015) suggest that these supposed boundary successions are older: probably Changhsingian in age, at around 253.2 ± 0.15 Ma (Fig. 9). This may relate to difficulties in defining the Permian–Triassic boundary based on tetrapods alone (Lucas 2006). However, there are serious disagreements about the polarity in the upper part of the Balfour Formation, which either indicates problems with local hiatus (Gastaldo *et al.* 2015) or issues in the palaeomagnetic data from Old Lootsberg Pass in distinguishing the present-day overprints from the normal polarity Permian directions, which are similar to modern field directions (De Kock & Kirschvink 2004). It is not clear how these magnetic polarity datasets relate to each other, but there is not sufficiently strong evidence to invalidate the original interpretations of Ward *et al.* (2005).

There have been many magnetostratigraphic studies on the Siberian Traps (Gurevitch *et al.* 2004; Fetisova *et al.* 2014) and several attempts at a synthesis (Steiner 2006; Fetisova *et al.* 2014; Burgess & Bowring 2015). The successions indicate a simple pattern of magnetic polarity changes dominated by normal polarity in the Noril'sk region, but with reverse magnetozones in the Kotui River region and at the base of the successions in the Ivaninsky and Khardakh formations (Fig. 9). Inadequately described fossil spores, pollen and brachiopod remains constrain the succession into an older Permian and younger Triassic set of units (Fetisova *et al.* 2014). Based on the combination of biostratigraphic data, radiometric dating evidence and palaeomagnetic data, Fetisova *et al.* (2014) suggested that the oldest units (i.e. the Ivakinsk (at Noril'sk) and Khardakh formations) are late Permian. The overlying, predominantly normal polarity basalts at Noril'sk are likely to correspond to LT1n.1n (Fig. 9). The Syverma–Nadezhda suites of the Noril'sk succession record the transitional geomagnetic field behaviour across the boundary of the LP3r and LT1n.1n magnetochrons (Gurevitch *et al.* 2004), implying that these units have a rapid (6–20 m ka^{-1}) accumulation rate. This transitional field interval is not shown in the Motui River sections, suggesting that there may be a hiatus (or poorly sampled interval) at the base of the Ary-Dzhang Formation (Kamo *et al.* 2003; Fetisova *et al.* 2014). Radiometric data have consistently indicated the brief duration of the Siberian traps, which are constrained by dates from perovskite of 252.2 ± 0.2 Ma from the Khardakh basal flows (Kamo *et al.* 2003) to 251.4 ± 0.29 Ma for the Daldykansky intrusion, which cuts the lava flows in the Noril'sk region. Burgess & Bowring (2015) argued that the lava eruptions were approximately 0.8 myr in duration, with some two-thirds of the volume erupted in the 0.3 myr prior to the end-Permian extinction. The Permian–Triassic boundary is therefore within the middle–upper parts of the flood basalt succession at Noril'sk (Burgess & Bowring 2015).

A calibrated Permian geomagnetic polarity timescale

To generate a Permian geomagnetic polarity pattern in a million year (Ma) scale, we first utilize the section compositing method proposed by Hounslow (2016). This first produces a magnetic polarity composite using numerical optimization, in a composite scaled to relative height (Fig. 10b, e). This is, in effect, a numerical version of the hand-drawn composites produced by syntheses such as Opdyke (1995), Steiner (2006) and Hounslow & Muttoni (2010). The optimized composite utilizes the proxy for time embedded in the relative height of magnetozones in the data from the source sections, and so smoothes the between-section sedimentation rate changes by averaging magnetozone boundary positions across sections (Fig. 10b, e). This requires simple choices about relative sedimentation rates in the sections.

Secondly, the resulting optimized composite is scaled to Ma, using appropriate radiometric data (i.e. an age model is applied to the optimized scale), from which an age estimate of the magnetochron bases is determined (Table 2; Figs 10c & 11). To construct the age model, we use the Bayesian-based approach of Haslett & Parnell (2008) and Parnell *et al.* (2008), as implemented in the Bchron functions in R (Chambers 1998). This constructs an age model

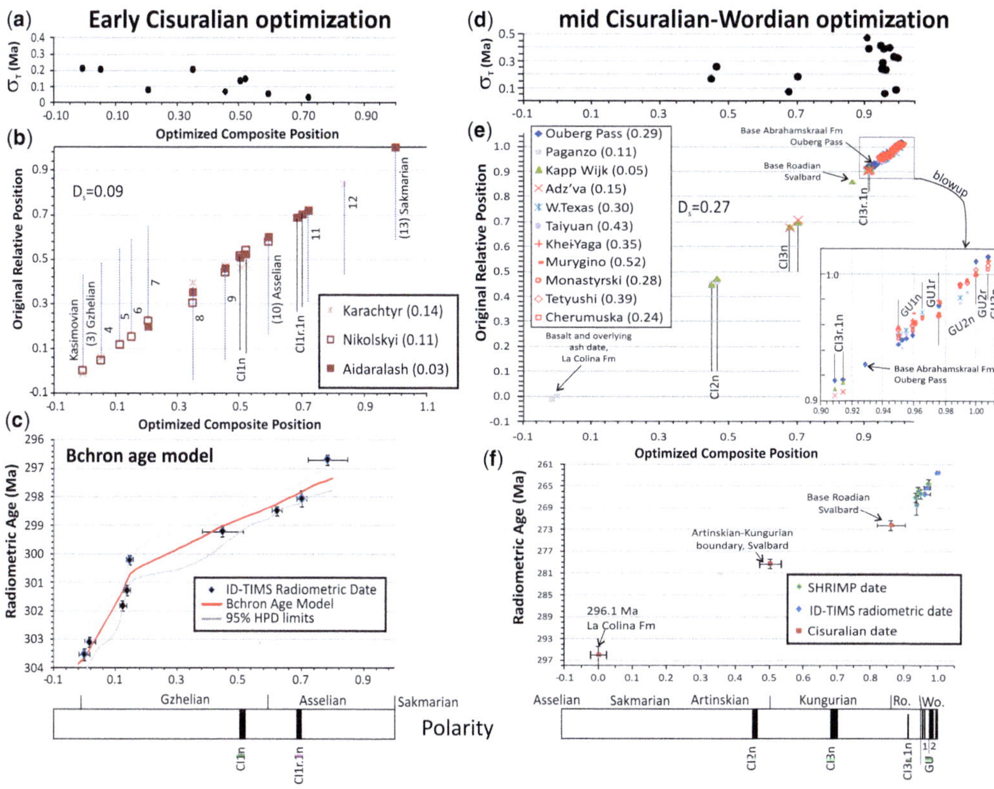

Fig. 10. Optimized composites (**a** & **b**) and age model (**c**) for the Gzhelian–Asselian and likewise for the Artkinskian–Wordian interval (**d, e** & **f**). Optimized composites based on methodology in Hounslow (2016). (a) & (d) show the standard deviation (σ_T) for the levels used in the optimized scaling procedure (scaled to Ma, using the final age model). This is a measure of the correlated level misfit. The correlated levels are shown in (b) & (e). No σ_T values for a corresponding level shown in (b) & (e) indicate that the level was not used to constrain the optimized model, but simply scaled with the section. (b) & (e) are the original section data shown on the *y*-axis (in a relative height scale), along with the final composite position of the levels on the *x*-axis. Scatter in the *y*-axis relates to the degree of between-section misfit shown in the overlying panel as σ_T. Numbers in brackets next to section names are the D_j values of Hounslow (2016), which express the misfit of the section data to the optimized composite (i.e. the Karachtyr data has a mean residual of 14% per average 'chron width' for the optimized model). D_s is the average of the D_j values across all sections. (c) The Bchron age model for the Carboniferous–Permian boundary, showing the scaling of the optimized position to Ma, using the radiometric dates (magnetochrons in the scale of the optimized composite shown at the bottom). (**f**) The radiometric dates used to scale the optimized composite scale. In (c) & (f) error bars on the *y*- and *x*-axis are, respectively, the radiometric (σ_R) and stratigraphic (e_s) uncertainty values in Table 2.

Table 2. *Permian radiometric dates used*

1 Code, age (Ma)	2 $\pm 2\sigma_R$ (Ma)	3 $\pm e_s$	4 Location [estimated position]	5 Biostratigraphy, stratigraphy {position in zone}	6 P_{out}	7 References
SH03, 260.74	0.9	10% of GU3n	36.3 m above base of Wujiaping Formation, Shangsi, bed 8	Base Lopingian {base of LP0r}	0.12	Mundil *et al.* (2004); Schmitz (2012); Zhong *et al.* (2014)
JW1, 259.1	0.5+	10% of GU3n	Emeishan Basalts, Zhaotong	*c.* 100 m below the top of unit III {95% into GU3n}	0.45	Zhong *et al.* (2014)
GM-20, 262.58	0.45	100% of GU2r	20 m above Rader Limestone (Patterson Hills)	Within *Polydiexodina* fusulinid Zone (i.e. *c. J. postserrata* Zone) {95% into GU2n}	0.03	Nicklen (2011)
OPA483, 264.6*	1.9	10% of GU1r	484 m above base of Abrahamskraal Formation, Ouberg Pass. South Africa	Within the mid-*Eodicynodon* assemblage {75.4% into GU2r}	0.01	Lanci *et al.* (2013)
OPA292, 265.9*	1.4	5% of GU1n	195 m above base of Abrahamskraal Formation, Ouberg Pass. South Africa	Within base *Eodicynodon* assemblage {86.7% into GU1n}	0.01	Lanci *et al.* (2013)
NH, 265.35	0.5	100% of GU1r.1n	Nipple Hill, Guadalupian Mountains	37.2 m below base Capitanian, 2 m above top of the Hegler Member {base GU1r.1n}	0.04	Bowring *et al.* (1998); Nicklen (2011, fig. 1.8)
OPA230, 266.4*	1.8	5% of GU1n	132 m above base of Abrahamskraal Formation, Ouberg Pass. South Africa	Within base *Eodicynodon* assemblage {97.4% into CI3r.2r}	0.01	Lanci *et al.* (2013)
GM-29, 266.50	0.24	100% of GU1r	Below South Wells Limestone ('Monolith Canyon')	Within *J. asserrata* Zone {10% into MP1r}	0.17	Nicklen (2011)
OPA160, 267.1*	1.7	10% of CI3r.2r	62 m above base of Abrahamskraal Formation, Ouberg Pass. South Africa	Within base *Eodicynodon* assemblage {69.1% into CI3r.2r}	0.00	Lanci *et al.* (2013)
OPA151, 268.5*	3.5	10% of CI3r.2r	52 m above base of Abrahamskraal Formation, Ouberg Pass. South Africa	Within base *Eodicynodon* assemblage {65.0% into CI3r.2r}	0.01	Lanci *et al.* (2013)

PPAsh-1 296.09	0.35	300% of CI2n	La Colina Formation, Pagenzo Basin, Argentina [tens of metres above basalt flow/sill]	Pagenzo Group, *Fusacolpites fusus–Vitattina subsaccata* Interval Biozone	0.01	Gulbranson *et al.* (2010); Césari *et al.* (2011)
01DES212, 296.69	0.37	400% of CI1r.1n	Usolka section, Russia	Mid-Asselian {54% into Zone 11}	0.01	Schmitz & Davydov (2012)
01DES202, 298.05	0.54	100% of CI1r.1n	Usolka section, Russia	Early Asselian {83% into Zone 10}	0.01	Ramezani *et al.* (2007)
01DES194, 298.49	0.34	100% of CI1r.1n	Usolka section, Russia	Earliest Asselian {21% into Zone 10}	0.01	Ramezani *et al.* (2007)
01DES144, 299.22	0.34	400% of CI1n	Usolka section, Russia	Latest Gzhelian {63% into zones 8 & 9}	0.01	Ramezani *et al.* (2007)
97USO-23.3, 300.22	0.35	30% of Zone 5	Usolka section, Russia	Mid-Gzhelian {83% into Zone 5}	0.01	Schmitz & Davydov (2012)
01DES121, 301.29	0.36	30% of Zone 5	Usolka section, Russia	Mid-Gzhelian {61% into Zone 5}	0.01	Schmitz & Davydov (2012)
01DES112, 301.82	0.36	30% of Zone 5	Usolka section, Russia	Mid-Gzhelian {26% into Zone 5}	0.01	Schmitz & Davydov (2012)
01DES63, 303.10	0.36	30% of Zone 3	Usolka section, Russia	Basal Gzhelian {40% into Zone 3}	0.01	Schmitz & Davydov (2012)
97USO-2.7, 303.54	0.39	30% of Zone 3	Usolka section, Russia	Basal Gzhelian {10% into Zone 3}	0.01	Schmitz & Davydov (2012)

Column 1: Analysis code and date (in Ma). **Column 2:** $\pm 2\sigma_R$, two-sigma error on age. **Column 3**: $\pm e_s$, estimated stratigraphic error in placing the date onto the magnetostratigraphy in units of magnetochron or foraminifera zone widths. **Column 4:** section name, location. **Column 5:** Stratigraphic age or location, {..} = correlated position of date from base of chron, zone or interval. **Column 6:** P_{out}, probability (0–1.0), the date is an outlier (from Bchron); larger values suggest that the probability of being at outlier is more likely. For those dates not displayed here, but in supplementary table 2 in Hounslow (2016), all have $P_{out} < 0.2$, except for those at 253.47, 251.1 and 252.85 Ma giving P_{out} values of 0.998, 0.992 and 0.207, respectively. **Column 7:** source reference for the radiometric and age information. Foraminifera zone numbers in columns 3 and 5, based on Khramov & Davydov (1993), Davydov & Leven (2003) and Schmitz & Davydov (2012): 2, *Rauserites quasiarcticus*; 3, *Daixina fragilis*; 4, *D. crispa*; 5, *D. ruzhenzevi*; 6 & 7, *D. sokensis*; 8 & 9, *Ultradaixina bosbytauensis*; 10, *Sphaeroschwagerina aktjubensis–Sp. fusiformis*; 11, *Schwagerina nux–Pseudoschwagerina robusta*; 12, *Sp. Gigas*; 13, *S. moelleri*; 14, *S. verneulli*; 15, *Ps. pilicatissima–Ps. urdalensis*. Nicklen (2011) used hand-picked acicular, clear zircons, annealed at 900°C for 48 h then chemically abraded and spiked with EARTHTIME tracer solution. GM-20 has 100 crystals picked, with the weighted mean using two multi-crystal and two single-crystal analyses combined. GM-29 had 100 crystals separated, which produced a weighted mean using eight concordant single crystals.

*Monto Carlo simulation of best-fit SHRIMP ages and associated uncertainties.

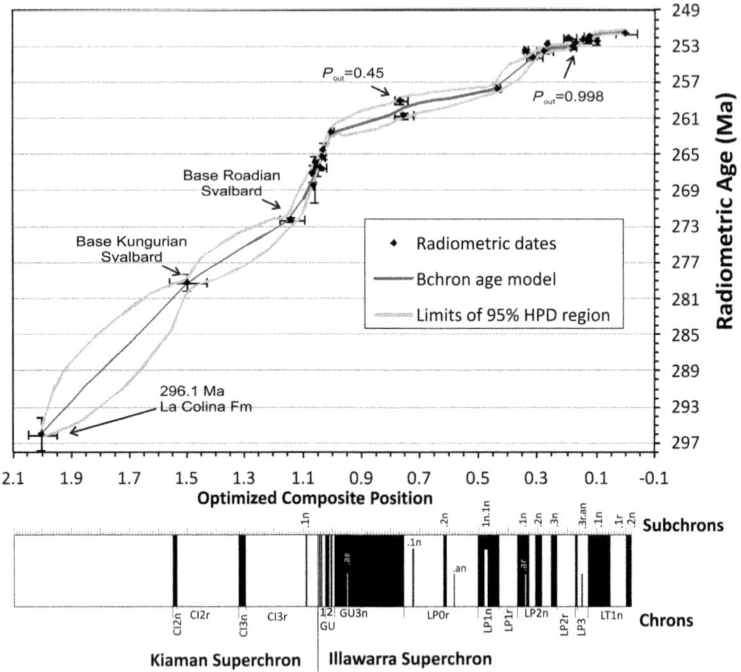

Fig. 11. Bchron age model for the Kungurian–earliest Triassic. The optimized composite position scale is that of Figure 10f, ranging from the radiometric date at 296.1 Ma to GU3n, joined to that from GU3n to LT1n.2n from Hounslow (2016).

based on piecemeal linear segments constructed by simulating the sedimentation process by small increments that are random in both duration and sedimentation rate. The method handles radiometric date uncertainties (as normally distributed values) and uses the procedure of Christen & Perez (2009) to deal with radiometric date outliers, which flags the dates with a probability of being an outlier (P_{out} in Table 2). Uncertainties in placing the radiometric date onto the optimized polarity composite are handled as a defined range ('sample depth' range in Parnell *et al.* 2008) in the composite scale, in which the date occurs ($\pm e_s$: Table 2), and treated as coming from a uniform distribution. Stratigraphic (e_s: Hounslow 2016) and radiometric uncertainties (σ_R) on the dates are listed in Table 2, and in the supplementary table 2 of Hounslow (2016).

Confidence intervals on the magnetochron ages are obtained from the Monto Carlo simulations used in Bchron, using the limits of the 95% highest posterior density region (HPD) from the age model (Haslett & Parnell 2008) (Fig. 10c). However, confidence intervals derived from Bchron may be overly pessimistic in intervals without age control points (Blaauw & Christen 2011). In the age models, the measure of uncertainty (i.e. σ_T: Hounslow 2016) (Table 3) in the position of the magnetochrons in the optimized composite scale is also included (Fig. 10a, d) as the 'uniform range' ($d_{max} - d_{min}$ of Parnell *et al.* 2008) corresponding to $\pm\sigma_T$. The method therefore takes account of all the major uncertainties in the GPTS. The Permian optimized composite (Table 3) is scaled to age in two segments because no sections span the CI1r.1n–CI2n interval.

Gzhelian–Asselian age scaling

The Gzhelian–Asselian magnetozone optimized composite uses data from the Karachatyr, Nikolskyi and Aidaralash sections. These can be tied together as they have a well-defined fusulinid zonation that is also utilized in the scaling (Fig. 10b). Linear rate scaling for the sections (Hounslow 2016) was used in the optimized composite. The Kapp Schoultz section from Svalbard was not used, as the relationships between the biostratigraphy and the magnetostratigraphy are not sufficiently well defined to accurately identify positions of either biozones or stage boundaries with respect to the polarity or to the biozones in the Uralian sections. The ID-TIMS U–Pb radiometric ages from the Usolka section were used (Table 2), which are directly related to the Urals foraminifera zones via the conodont

Table 3. *Permian chron base ages and durations*

Chron	Age (Ma)	Chron duration (myr)	C_{95} (Ma)	σ_T (ka)	$\%D_{95}$	Chron	Age (Ma)	Chron duration (myr)	C_{95} (Ma)	σ_T (ka)	$\%D_{95}$
LT1n.2n	251.444		0.28	–		GU3n.an	262.129	2.297	0.89	–	9.6
LT1n.1r	251.634	0.190	0.23	–	19.1	GU3n.ar	262.160	0.031	0.87	–	31.4
LT1n.1n	252.242	0.608	0.23	40	15.1	GU3n	262.592	0.432	0.57	38	15.7
LP3r.ar	252.54	0.298	0.17	–	16.8	GU2r	262.740	0.148	0.55	78	20.5
LP3r.an	252.571	0.031	0.17	–	31.4	GU2n.2n	263.134	0.394	0.84	73	15.9
LP3r	252.668	0.097	0.19	28	23.2	GU2n.1r	263.446	0.312	0.90	280	16.6
LP3n	252.796	0.128	0.23	166	21.4	GU2n.1n	264.375	0.929	0.95	346	12.5
LP2r	253.196	0.400	0.34	99	15.9	GU1r	265.746	1.371	0.69	394	10.4
LP2n.3n	253.242	0.046	0.37	20	28.7	GU1n.3n	266.274	0.528	0.73	94	15.4
LP2n.2r	253.802	0.560	0.43	164	15.3	GU1n.2r	266.374	0.100	0.76	251	23.0
LP2n.2n	254.194	0.392	0.40	229	15.9	GU1n.2n	266.496	0.122	0.70	110	21.7
LP2n.1r	254.637	0.443	0.66	400	15.6	GU1n.1r	266.566	0.070	0.70	77	25.6
LP2n.an	254.876	0.239	0.88	–	17.8	GU1n	266.659	0.093	0.76	220	23.5
LP2n.ar	255.106	0.230	0.99	–	18.0	CI3r.2r	269.240	2.581	1.59	254	9.6
LP2n.1n	255.922	0.816	1.12	286	13.5	CI3r.1n	269.542	0.302	1.61	355	16.7
LP1r	257.584	1.662	0.75	424	9.9	CI3r.1r	275.386	5.844	2.02	132	9.6
LP1n.2n	258.002	0.418	0.58	177	15.8	CI3n	275.862	0.476	1.99	45	15.4
LP1n.1r	258.072	0.070	0.58	111	25.6	CI2r	280.736	4.874	1.97	210	9.6
LP1n	258.214	0.142	0.66	115	20.8	CI2n	281.242	0.506	2.26	184	15.5
LP0r.ar	258.683	0.469	0.86	–	15.8	CI1r.2r	297.835	16.593	0.34	–	9.6
LP0r.an	258.731	0.048	0.89	–	28.4	CI1r.1n	297.938	0.103	0.33	–	22.8
LP0r.3r	258.842	0.111	0.94	21	22.3	CI1r.1r	298.694	0.733	0.37	123	14.4
LP0r.2n	258.922	0.080	0.96	127	24.6	CI1n	298.774	0.081	0.37	140	24.5
LP0r.2r	259.316	0.394	1.08	–	15.9						
LP0r.1n	259.396	0.080	1.10	–	24.6						
LP0r.1r	259.832	0.436	1.13	32	15.7						

C_{95}, 95% highest posterior density intervals on the age of the chron, estimated using Bchron in two age segments (shown in Figs 10c & 11). σ_T, standard deviation of the chron position in the sections for the chron (from the optimization method), scaled by the duration of the optimized chron. σ_T is a measure of the uncertainty in defining the chron position in the optimized GPTS. $\%D_{95}$ is the 95% confidence interval on the duration (expressed as the percentage of the chron duration: Fig. 13b). The age models define the base of the stages at the following: Gzhelian, 303.79 Ma; Asselian, 298.41 Ma; Sakmarian, 295.5 Ma; Artinskian, 290.1 Ma; Kungurian, 279.3 Ma; Roadian, 272.13 Ma; Wordian, *c.* 266.7 Ma; Capitanian, *c.* 263.5 Ma; Wuchiapingian, 259.7 Ma; Changhsingian, 255.4 Ma; Induan, 252.1 ± 0.23 Ma, from which the relative position of the chrons in the stages can be determined. The age of some tentative subchrons designated were estimated using relative locations within the main chrons at the Monastyrski (for GU3n.ar), Wulong (for LP0r.1n), Linshui (for LP0r.an) and Everdingen (for LP3r.an) sections. The differing Wordian–Changhsingian age model and method of Hounslow (2016) give slightly different age and uncertainty values compared to the data here.

ranges in the Usolka section, and the conodont–foraminiferal biozonal correlations in Schmitz & Davydov (2012). The optimized composite scale was then related to the radiometric ages using Bchron (Fig. 10c). None of the radiometric dates were flagged as potential outliers (Table 2). The 95% HPD regions from Bchron show bowing and pinching related to the distribution of age control points, expressing the greater uncertainty between the more widely spaced dates (Fig. 10c), which is also expressed in the chron age uncertainty (C_{95}, Table 3).

Kungurian–earliest Induan age scaling

A Kungurian–Capitanian optimized magnetozone composite (CI2n–GU3n) was constructed using the Paganzo, Ouberg Pass (long-GU1r option), Adz'va (Fig. 1), Kapp Wijk/Trygghamna (Fig. 2), western Texas (Fig. 3) and the Taiyuan sections (Fig. 4), along with the Russian Khei-yaga, Muygino, Monastrski, Tetyushi and Cherumuska sections (Figs 5 & 10c). The optimization tends to 'compress' the composite scale in the CI3r.1n–GU2n interval owing to a higher number of data points and magnetozones. Scale compression was controlled by expressing the minimized value (where E_{tot} = sum of the squared mismatch per magnetozone boundary; Hounslow 2016), as E_{tot} divided by the median chron duration. Linear rate scaling (Hounslow 2016) was used for all but the Monastyrki, Tetyushi and Cherumuska sections, for which transgressive rate functions were used. Transgressive rate functions account for the apparently condensed GU1n (Fig. 5). Overall, the optimized component produces a poorer model (large D_s, where D_s = mean measure of the misfit of all the section data) than the Gzhelian–Asselian model owing to the widely varying relative durations of the magnetozones in the Wordian, which is shown as larger σ_T and D_j values (Fig. 10d,e where D_j = mean mismatch between the optimized composite and the section magnetozone boundaries).

The Cisuralian part of this range is sparse in radiometric dates. One ID-TIMS date from an ash in the base of the La Colina Formation was used (Gulbranson *et al.* 2010) (Table 2), together with the Kungurian–Roadian boundary age from Henderson *et al.* (2012), which was inferred to coincide with the brachiopod extinction and $\delta^{13}C$ excursion in the Hovtinden Member on Spitsbergen (Figs 2 & 10f). The Artinskian–Kungurian boundary has an array of dates (Henderson *et al.* 2012), but cannot be clearly related to the polarity in any section. To constrain the Cisuralian, the Artinksian–Kungurian boundary age from Henderson *et al.* (2012) was used for the base of the Kungurian in the mid-part of the Svenkegga Member at Kapp Wijk/Tryghamna (Fig. 2). In the Wordian–Capitanian, zircon SHRIMP dates from the Abrahamskaal Formation are supplemented by additional radiometric dates from the Texas sections of Bowring *et al.* (1998) and Nicklen (2011). These radiometric dates have been placed onto the magnetostratigraphy (Table 2; Figs 3 & 11) using the magnetic polarity data of Burov *et al.* (2002) and stratigraphic relationships discussed by Nicklen (2011).

The late Capitanian–earliest Triassic optimized composite (GU3n–LT1n.2n) is that derived by Hounslow (2016). This uses the magnetozone data from the Khei-yaga, Murygino, Monastyrki, Boyevaya Gora, Tuyembetka, Sambullak, Tetyushi, Cheremushka, Sukhona, Pizhma, Oparino, western Texas, Linsui, Wulong, Shangsi, Taiyuan, Nammal Gorge and Abadeh sections to construct the optimized composite. This optimized composite is joined to that from the CI2n–GU3n interval at the base of GU3n (Fig. 11). This compound optimized composite is then scaled to age with Bchron using 28 dates (the 11 upper ones in Table 2), plus the 17 ID-TIMS listed in the supplementary table 2 of Hounslow (2016). Bchron identified two probable outliers in the age model, at 252.1 and 253.47 Ma (P_{out} of 0.992 and 0.998, respectively: Fig. 11).

Two intervals giving possibly unrealistic chron age estimates are the LP0r–LP1r and LP2n.2r–LP2n.3n intervals, as the Bchron age scaling does not accord with the relative thickness of section chrons in these two intervals. The former interval is strongly influence by the late Wuchiapingian date of 257.79 Ma from the Shangsi section (Figs 6 & 11) that gives a probably too brief LP0r chron. This date may be incorrectly located with respect to the polarity stratigraphy. Attempts at correcting the LP2n.2r–LP2n.3n interval by excluding the possible outlier at 253.47 Ma failed to produce much improvement as the age model using Bchron had already accounted for its outlier status.

Chron and stage ages, and the relationship to biozones

The earliest Permian age model (Figs 10c & 12) gives an age for the base of fusulinid zone 10 (correlated base Asselian) of 298.41 ± 0.36 Ma, similar to the 298.9 ± 0.15 Ma proposed by Schmitz & Davydov (2012). The age differences probably relate to assumptions of conformity of the fusulinid and conodont zonal boundaries (Schmitz & Davydov 2012), the different means of scale compositing (faunal range-top and bottom scaling in CONOP), and the method of scaling the composite to age. The derived dates of the magnetochrons (Table 3) are broadly what would be expected based on the biozonal stage–age relationships proposed by Henderson *et al.* (2012). This is not surprising considering that we largely use the same sets of controlling radiometric dates, and we have pinned the base Roadian and base Kungurian to that inferred by Henderson *et al.* (2012). However, our age control through the Wordian is considerably improved over the 2012 timescale, and we estimate the base Wordian to be at approximately 266.7 Ma and the base Capitanian at approximately 263.5 Ma, significantly displaced from the 2012 timescale by approximately 2 Ma (Table 3). The base of the Lopingian stages and the Induan are similar to those inferred in the 2012 timescale, as there are many radiometric dates in this interval. Like the Asselian, the small differences likely to relate to the different methods used.

The relationships between the stage biozones and the magnetochrons have a variable amount of precision through the Permian. In the earliest part of the Cisuralian, the relationships between CI1r.1n and the Urals foraminifera biozones is fairly well defined (Fig. 12), but becomes much less precise for CI2n and CI3n, where relationships to conodonts zones seem to hold the best future promise for refinement (Figs 1 & 2). For the mid-Permian, there is a slightly more refined biozone–magnetochron relationship. The Lopingian has a comparatively well-defined conodont biozone–magnetochron relationship (Figs 4 & 12).

Chron duration uncertainties

Apparent magnetozone durations (and zonal intervals in millions of years (myr)) in the sections can be 'back-calculated' from the relationship between optimized composite chron duration and age. If the duration (in myr) of a magnetochron (or chron interval) is C_m in the GPTS, and the pseudo-height in the composite of this interval is

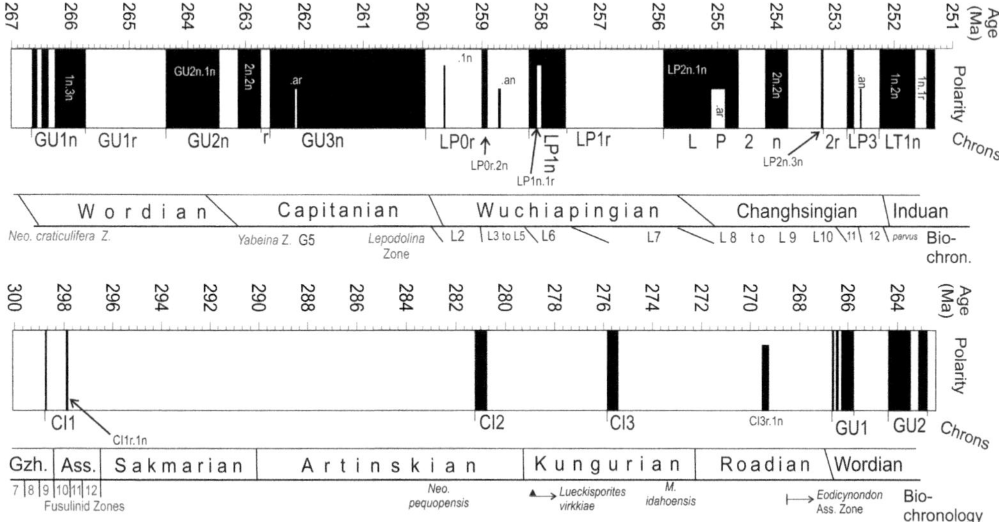

Fig. 12. Summary Permian geomagnetic polarity timescale. Chron scale in Ma derived from Bchron models in Figures 10c and 11. Numbered fusulinid zones in the earliest Permian are those of Figure 1 and as detailed in Table 2. Standard conodont zones L2–L12 are from Shen *et al.* (2010), derived from data in Figure 4. Selected other key biochronology are from Figures 2 and 4. Estimated radiometric ages of stages are indicated in Table 3.

Y_m, then the apparent duration (C_S in myr) of the equivalent magnetozone (or zonal interval) in the section can be estimated by: $C_s = C_m(Y_s/Y_m)$, where Y_s is the pseudo-height of the magnetozone (or interval) in the section in the units of the optimized composite. Linear scaling is appropriate, as the segment age models are approximately linear at a timescale comparable to the magnetochron intervals used. This gives a cloud of points (Fig. 13a) that expresses the apparent age duration of chrons in the sections (i.e. C_s), visually showing the scatter in the original data, which for each chron is also expressed by σ_T (in Fig. 10a,d).

Uncertainties on the chron durations can also be determined by the 95% HPD intervals derived directly from the differences in the simulated age determinations for each chron ('events' in Bchron: Parnell *et al.* 2008) (Fig. 13a). However, these Bchron 95% HPD estimates express more a prediction interval than a confidence interval on the 'mean' age model as they only consider the simulated data from a single magnetochron duration (rather than the whole age model: Dybowski & Roberts 2001). This can be seen in that the HPD bands largely encompass the cloud of points from the section estimates (Fig. 13a).

One estimate of the confidence intervals (D_{95}) on the durations (i.e. on C_m) can be determined from a conventional regression of C_s v. C_m (a 'section estimate': Fig. 13a). This approach is conceptually similar to that used by Agterberg (2004) for estimating confidence intervals on stage ages since the estimated magnetozone duration in the section (C_s) is an independent estimate compared to that 'average' derived from the optimized chron scale. Statistically, it is preferable to utilize a log–log regression for this 'section estimate' as durations are typically exponentially distributed (Lowrie & Kent 2004). For shorter chron durations, the percentage uncertainty increases (Fig. 13b) because there is a proportionally larger impact of changes in deposition rates in sedimentary systems (Sadler & Strauss 1990; Talling & Burbank 1993) and sampling density (i.e. fluctuations in the sedimentation rate and palaeomagnetic sampling density section introduces additional variance in the section's chrons duration). However, for longer chrons, the log–log regression produces unrealistically large confidence intervals because of the spreading of the confidence bands at the tails (Fig. 13b), and a linear $C_s - C_m$ relationship is probably more appropriate (shown as a linear model in Fig. 13b). Uncertainty on longer chrons is more impacted by the uncertainty in the age model from the radiometric dates (σ_R in Table 2) and their uncertainty of position with respect to the magnetochrons (i.e. e_s in Table 2). However, neither of these 'section estimates' (i.e. log or linear models in Fig. 13b) takes account of uncertainty in the age model.

Agterberg (2004) proposed that an estimate of D_{95} can be obtained from the confidence interval on a regression of calculated radiometric age v. actual radiometric age derived from the age model. A 'sample point distribution' correction factor

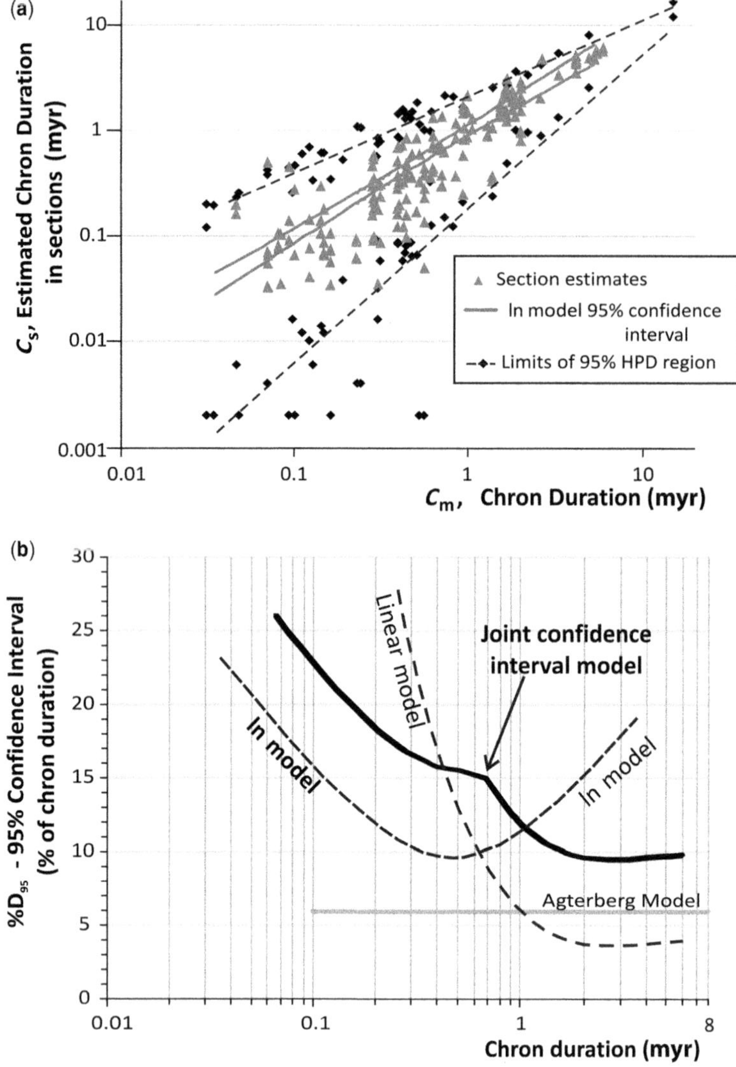

Fig. 13. Confidence interval data for chron durations. (**a**) Estimated magnetozone durations (and zone intervals) from each of the sections (triangles) used in the optimized composites (*y*-axis) v. the duration of the equivalent chron. Data for 173 magnetozones and zone intervals are shown. The 95% confidence intervals on the linear regression relationship using the ln model (solid line). The 95% HPD limits, from Bchron, for each of the Permian chrons are shown (as diamonds), along with lines (dashed) expressing this variation with duration. (**b**) Estimates of the 95% confidence intervals using uncertainty in the age model (grey line), using the approach of Agterberg (2004) and symmetrical 95% confidence intervals using the section magnetozone data (dashed lines) shown in (a). The final 95% confidence interval model (black line) adds the Agterberg estimates to the linear model when the chron duration is longer than 0.7 myr and ln model is added to the Agterberg estimate when durations are shorter than 0.5 myr (i.e. 0.5–0.7 myr is linearly interpolated). Regression and confidence intervals used the linear model routines in R, version 3.2.4 (Becker *et al.* 1988).

should also be applied to correct for the Ma range of the age model (Agterberg 2004). We determined this 'Agterberg estimate' using the data for the Wordian–early Triassic (i.e. data in Fig. 11, from 269 to 251 Ma) interval, as the larger number of dates in this interval probably best expresses the uncertainty in the age model. This estimate gives values for $\%D_{95}$ similar to the linear model 'section estimates' at chron durations of more than 1 myr (Fig. 13b). The final confidence intervals for the durations is a joint model (Fig. 13b; $\%D_{95}$ in Table 3) that adds the 'Agterberg estimate' (for the age model uncertainty) to the log and linear model 'section estimates' for chron durations (Fig. 12b). This gives a balanced estimate that includes both the uncertainty from the optimized polarity and uncertainty in the age model.

Conclusions

A robust geomagnetic polarity timescale is constructed through the Permian, with no major intervals of missing polarity data (Fig. 12). The statistical compositing method of Hounslow (2016) allows the construction of a numerical magnetochron composite using data from many sections. This composite is calibrated against radiometric dates using Bayesian principles applied in the program Bchron using two segments, one for the Carboniferous–Permian boundary and a Kungurian–earliest Triassic interval. The Artinskian, Kungurian and Roadian interval are the least well constrained in terms of controlling radiometric dates, so two previous estimates of stage boundary age are utilized for this interval. Estimates of the 95% confidence intervals on the chron-base ages and chron durations are derived.

In spite of a long-held belief, by many, that the early Permian contains no substantiated normal polarity intervals, there is good evidence that the Cisuralian contains at least two, probably four, brief normal magnetochrons, and a further normal in the latest Carboniferous (latest Gzhelian). The Asselian magnetochrons CI1r.1n (base at 297.94 ± 0.33 Ma) and CI3r.1n (base at 269.54 ± 0.70 Ma) are least well validated of these, whereas CI2n (base at 281.24 ± 2.3 Ma) and CI3n (base at 275.86 ± 2.0 Ma) in the Artinskian and Kungurian are rather better identified in more studies. The age calibration of these clearly shows that these magnetochrons are brief (*c.* 81–506 ka) in duration, which has added to their difficulty in detection in the dominantly reverse polarity Kiaman Superchron. The presence of these magnetochrons holds promise as high-resolution time markers in the Cisuralian.

The start of the mixed-polarity Illawarra Superchron is at 266.7 ± 0.76 Ma in the early Wordian, long known to be a major chronostratigraphic marker in the mid-Permian. The European Russian upper Tatarian (Vyatkian) magnetostratigraphic data appear incomplete in comparison to the better dated marine successions, indicating that a part of the Changhsingian is missing from the European Russian sections. Magnetostratigraphic data from the European Upper Rotliegend and Zechstein clearly indicate the presence of the Guadalupian and Lopingian in these non-marine basins. However, magnetostratigraphic correlations suggest that the Zechstein represents a much shorter age interval than conventionally inferred, occupying only the middle–late Changhsingian. The magnetic polarity with respect to the many high-resolution stratigraphic studies across the Permian–Triassic boundary is well defined, and the best expressions of the linked polarity to faunal changes are found in the Shangsi section in China. Radiometric and magnetostratigraphic data suggest that the voluminous Siberian Traps were erupted rapidly, starting in the latest Permian magnetochron LP3r, into and through the earliest Triassic normal chron LT1n.

Key uncertainties and future refinements of the Permian GPTS needed are:

- Sub-magnetochrons CI1r.1n (early Asselian) and chron CI3r.1n (early Roadian) are the least well defined of the Permian chrons in the Kiaman Superchron, and need further work to consolidate the understanding of these. Permian sections on Svalbard or in the Urals may hold the best promise for better calibration of these against biostratigraphy. High-resolution studies of North American sections through the Laborcita Formation may aid investigations of chrons at the Carboniferous–Permian boundary.
- Other Permian chrons in the Kiaman Superchron, CI2n and CI3n, are magnetically well defined, but not well calibrated to biostratigraphy or radiometric dates. The Arctic Permian sections seem to hold the best promise for a better intercalibration of magnetochrons and biochronology.
- Details of the polarity through the Wordian, which includes GU1 and GU2, have two alternative scenarios: first, a long-GU1r model (the one preferred here), based on fragmentary marine and non-marine sections in the Beaufort Group from South African; and, secondly, a brief-GU1r model with more normal-polarity dominance, largely defined by the datasets from European Russian sections. The later scenario depends on the reliability of normal magnetozones in the Russian NRP mixed-polarity magnetozone and how to correlate marine and non-marine magnetostratigraphies in the Wordian. Detailed magnetostratigraphic data from the back-reef facies in the type region of the Guadalupian would help in this uncertainty.
- The interval LP1n–LP2n.3n (late Wuchiapingian–mid-Changhsingian) is normal polarity dominated in many sections, but the relative duration of reverse polarity magnetozones in this interval vary greatly between sections, particularly for LP1r. Better integration of regional sedimentological, cyclostratigraphic, radiometric and magnetostratigraphic studies in both marine

and non-marine would help to refine an improved polarity timescale through this interval.
- Radiometric date control points for age scaling of the GPTS appear to severely distort the relationship between the apparent relative duration of chrons in the sections in the early Wuchiapingian and mid-Changhsingian. Acquisition of more dates and/or the reassessment of either the radiometric dates or their position with respect to the composite magnetostratigraphy are needed to unravel the apparent conflict.

Vassil Karloukovski assisted with Russian translations. Andrew Parnell provided useful guidance on using Bchron. The reviewers Jim Ogg and Spencer Lucas provided much constructive comment, improving the text.

References

AGTERBERG, F.P. 2004. Statistical procedures. *In*: GRADSTEIN, F., OGG, J. & SMITH, A. (eds) *A Geologic Time Scale*. Cambridge University Press, Cambridge, 269–273.

ALI, J.R., THOMPSON, G.M., SONG, X. & WANG, Y. 2002. Emeishan basalts (SW China) and the end-Gudalupian crisis: magnetobiostratigraphic constraints. *Journal of the Geological Society, London*, **159**, 21–29, https://doi.org/10.1144/0016-764901086

AREFIEV, V., GOLUBEV, V.K. ET AL. 2015. Type and reference sections of the Permian–Triassic continental sequences of the East European Platform: main isotope, magnetic, and biotic events. Paper presented at the XVIII International Congress on Carboniferous and Permian, 11–15 August 2015, Kazan, Russia. Palaeontological Institute of the Russian Academy of Sciences, Moscow.

BALABANOV, YU.P. 1988. Paleomagnetic rock sequence and the magnetic properties of the Permian coal-bearing rocks and Triassic basalts in the Adz'va River area. *In*: MURAV'EV, I.S. (ed.) *The Permian System: Stratigraphy and the History of the Organic World*. Kazan University Press, Kazan, 126–134 [in Russian].

BALABANOV, YU.P. 1998. Paleomagnetic characterization of the Permian–Triassic boundary deposits in the northwestern Timan-Pechora plate, in Permian–Triassic boundary in continental series of Eastern Europe. *In*: LOZOVSKY, V.R. & ESAULOVA, N.K. (eds) *Upper Permian Stratotypes of the Volga Region*. GEOS, Moscow, 148–162 [in Russian].

BALABANOV, YU.P. 2014. Paleomagnetic characterization of the Middle and Upper Permian deposits based on the results from the key section in the Monastery Ravine. *In*: NURGALIEV, D.K., SILANTIEV, V.V. & URAZAEVA, M.N. (eds) *Carboniferous and Permian Earth Systems, Stratigraphic Events, Biotic Evolution, Sedimentary Basins and Resources. Kazan Golovkinsky Stratigraphic Meeting, Kazan*. Federal University, Kazan, 14–17.

BARNABY, R.J. & WARD, W.B. 2007. Outcrop analog for mixed siliciclastic–carbonate ramp reservoirs – stratigraphic hierarchy, facies architecture, and geologic heterogeneity: Grayburg Formation, Permian basin, USA. *Journal of Sedimentary Research*, **77**, 34–58.

BAUD, A., ATUDOREI, V. & SHARP, Z. 1995. The Upper Permian of the Salt Range area revisited: new stable isotope data. *Permophiles*, **27**, 39–41.

BECKER, R.A., CHAMBERS, J.M. & WILKS, A.R. 1988. *The New S Language*. Chapman & Hall, New York.

BERTHOLD, G., NAIRN, A.E.M. & NEGENDANK, J.F.W. 1975. A palaeomagnetic investigation of some of the igneous rocks of the Saar–Nahe Basin. *Neues Jahrbuch für Geologie und Paläontologie, Monatshefte*, **3**, 134–150.

BIAKOV, A.S. 2012. Permian biospheric events in North-east Asia. *Stratigraphy and Geological Correlation*, **20**, 199–210.

BLAAUW, M. & CHRISTEN, J.A. 2011. Flexible paleoclimate age–depth models using an autoregressive gamma process. *Bayesian Analysis*, **6**, 457–474.

BLAKEY, R.C. 1990. Stratigraphy and geologic history of Pennsylvanian and Permian rocks, Mogollon Rim region, central Arizona and vicinity. *Geological Society of America Bulletin*, **102**, 1189–1217.

BLAKEY, R.C. & MIDDLETON, L.T. 1987. Late Paleozoic depositional systems, Sedona-Jerome area, central Arizona. *In*: DAVIES, G.H. & VANDENDOLDER, E.M. (eds) *Geologic Diversity of Arizona and Its Margins: Excursions to Choice Areas*. Arizona Bureau of Geology & Mineral Technology, Special Papers, **5**, 143–157.

BLOMEIER, D., DUSTIRA, A., FORKE, H. & SCHEIBNER, C. 2011. Environmental change in the Early Permian of NE Svalbard: from a warm-water carbonate platform (Gipshuken Formation) to a temperate, mixed siliciclastic-carbonate ramp (Kapp Starostin Formation). *Facies*, **57**, 493–523.

BOND, D.P., WIGNALL, P.B. ET AL. 2015. An abrupt extinction in the Middle Permian (Capitanian) of the Boreal Realm (Spitsbergen) and its link to anoxia and acidification. *Geological Society of America Bulletin*, **127**, 1411–1421.

BOWRING, S.A., ERWIN, D.H., JIN, Y.G., MARTIN, M.W., DAVIDEK, K. & WANG, W. 1998. U–Pb zircon geochronology and Tempo of the End-Permian Mass extinction. *Science*, **280**, 1039–1045.

BRUNHES, B. 1906. Recherches sur la direction d'aimantation des roches volcaniques. *Journal de Physique Théorique et Appliquée*, **5**, 705–724.

BURGESS, S.D. & BOWRING, S.A. 2015. High-precision geochronology confirms voluminous magmatism before, during, and after Earth's most severe extinction. *Science Advances*, **1**, e1500470.

BURGESS, S.D., BOWRING, S. & SHEN, S.Z. 2014. High-precision timeline for Earth's most severe extinction. *Proceedings of the National Academy of Sciences of the United States of America*, **111**, 3316–3321.

BUROV, B.V. 2004. Boundary between the Permian and Triassic rocks in the Moscow Syneclise reconstructed from the rock sequences exposed in the Kichmenga River basin. *Russian Journal of Earth Sciences*, **7**, 1–8.

BUROV, B.V., ZHARKOV, I.Y., NURGALIEV, D.K., BALABONOV, Y.P., BORISOV, A.S. & YASONOV, P.G. 1998. Magnetostratigrphic characteristivs of Upper Permian sections in the Volga and the Kama areas.

In: ESAULOVA, N.K., LOZOVSKY, V.R. & ROZANOV, A.Y. (eds) *Stratotypes and Reference Sections of the Upper Permian in the Region of the Volga and Kama Rivers*. GEOS, Moscow, 236–270.

BUROV, B.V., ESAULOVA, N.K., ZHARKOV, I.Ya., YASONOV, P.G. & NURGALIEV, D.K. 2002. Tentative palaeomagnetic data on the Permian Lamar and Manzanita members of the upper part of the Guadalupian Series, Guadalupe and Apache Mountains (Texas, USA) and their comparison with the east European magnetostratigraphic scale. *International Journal of Georesources*, **6**, 24–28.

CAMPBELL, L.M. & CONAGHAN, P.J. 2001. Flow-field and palaeogeographic reconstruction of volcanic activity in the Permian Gerringong Volcanic Complex, southern Sydney Basin, Australia. *Australian Journal of Earth Sciences*, **48**, 357–375.

CÉSARI, S.N. & GUTIÉRREZ, P.R. 2000. Palynostratigraphy of upper Paleozoic sequences in central-western Argentina. *Palynology*, **24**, 113–146.

CÉSARI, S.N., LIMARINO, C.O. & GULBRANSON, E.L. 2011. An Upper Paleozoic bio-chronostratigraphic scheme for the western margin of Gondwana. *Earth-Science Reviews*, **106**, 149–160.

CHAMBERS, J.M. 1998. *Programming with Data*. Springer, New York.

CHEN, H.-H., SUN, S. & LI, J.-L. 1994. Permo-Triassic magnetostratigraphy in Wulong area, Sichuan, China. *Science in China (Series B)*, **37**, 203–212.

CHEN, Z.-Q., CAMPI, M.J., SHI, G.R. & KAIHO, K. 2005. Post extinction brachiopod faunas from the Late Permian Wuchiapingian coal series of South China. *Acta Palaeontologica Polonica*, **50**, 343–363.

CHRISTEN, J.A. & PEREZ, S. 2009. A new robust statistical model for radiocarbon data. *Radiocarbon*, **51**, 1047–1059.

CHWIEDUK, E. 2007. Middle Permian rugose corals from the Kapp Starostin Formation, South Spitsbergen (Treskelen Peninsula). *Acta Geologica Polonica*, **57**, 281–304.

CONDON, S.M. 1997. *Geology of the Pennsylvanian and Permian Cutler Group and Permian Kaibab Limestone in the Paradox Basin, Southeastern Utah and Southwestern Colorado*. United States Geological Survey Bulletin, **2000**.

COTTRELL, R.D., TARDUNO, J.A. & ROBERTS, J. 2008. The Kiaman Reversed Polarity Superchron at Kiama: toward a field strength estimate based on single silicate crystals. *Physics of the Earth and Planetary Interiors*, **169**, 49–58.

CREER, K.M., IRVING, E. & RUNCORN, S.K. 1955. The direction of the geomagnetic field in remote epochs in Great Britain. *Journal of Geomagnetism and Geoelectricity*, **6**, 163–168.

CREER, K.M., MITCHELL, J.G. & VALENCIO, D.A. 1971. Evidence for a normal geomagnetic field polarity event at 263±5 my BP within the Late Palaeozoic reversed interval. *Nature*, **233**, 87–89.

DAVYDOV, V.I. & KHRAMOV, A.N. 1991. Paleomagnetism of Upper Carboniferous and Lower Permian in the Karachatyr region (southern Ferghana) and the problems of correlation of the Kiama hyperzone. *In*: KHRAMOV, A.N. (ed.) *Paleomagnetizm i paleogeodinamika territorii SSSR (Palaeomagnetism and Palaeogeodynamics of the Territory of USSR)*. Transactions of the VNIGRI, 45–53 [in Russian].

DAVYDOV, V.I. & LEVEN, E.J. 2003. Correlation of Upper Carboniferous (Pennsylvanian) and Lower Permian (Cisuralian) marine deposits of the Peri-Tethys. *Palaeogeography, Palaeoclimatology, Palaeoecology*, **196**, 39–57.

DAVYDOV, V.I., BARSKOV, I.S., BOGOSLOVSKAYA, M.F., LEVEN, E.Y., POPOV, A.V., AKHMETSHINA, L.Z. & KOZITSKAYA, R.I. 1992. The Carboniferous–Permian boundary in the former USSR and its correlation. *International Geology Review*, **34**, 889–906.

DAVYDOV, V.I., GLENISTER, B.F., SPINOSA, C., RITTER, S.M., CHEMYKH, V.V., WARDLAW, B.R. & SNYDER, W.S. 1998. Proposal of Aidaralash as global stratotype section and point (GSSP) for base of the Permian System. *Episodes*, **21**, 11–18.

DAVYDOV, V.I., NILSSON, I. & STEMMERIK, L. 2001. Fusulinid zonation of the Upper Carboniferous Kap Jungersen and Foldedal formations, southern Amdrup Land, eastern North Greenland. *Bulletin of the Geological Society of Denmark*, **48**, 31–77.

DE KOCK, M. & KIRSCHVINK, J. 2004. Paleomagnetic constraints on the Permian–Triassic boundary in terrestrial strata of the Karoo Supergroup, SouthAfrica: implications for causes of the end-Permian extinction event. *Gondwana Research*, **7**, 175–183.

DENISON, R.E. & PERYT, T.M. 2009. Strontium isotopes in the Zechstein (Upper Permian) anhydrites of Poland: evidence of varied meteoric contributions to marine brines. *Geological Quarterly*, **53**, 159–166.

DEON, C.F. 1974. *A paleomagnetic investigation of the Permo-Carboniferous Maroon and upper Permian–Lower Triassic State Bridge formations in north central Colorado*. PhD thesis, University of Texas, Dallas.

DIEHL, J.F. & SHIVE, P.N. 1979. Palaeomagnetic studies of the Early Permian Ingelside Formation of northern Colorado. *Geophysical Journal International*, **56**, 271–282.

DIEHL, J.F. & SHIVE, P.N. 1981. Paleomagnetic results from the Late Carboniferous/Early Permian Casper Formation: implications for northern Appalachian tectonics. *Earth and Planetary Science Letters*, **54**, 281–292.

DIMICHELE, W.A., KERP, H., SIRMONS, R., FEDORKO, N., SKEMA, V., BLAKE, B.M. & CECIL, C.B. 2013. Callipterid peltasperms of the Dunkard Group, Central Appalachian Basin. *International Journal of Coal Geology*, **119**, 56–78.

DOELL, R.R. 1955. Palaeomagnetic study of rocks from the Grand Canyon of the Colorado River. *Nature*, **176**, 1167.

DYBOWSKI, R. & ROBERTS, S.J. 2001. Confidence intervals and prediction intervals for feed-forward neural networks. *In*: DYBOWSKI, R. & GANT, V. (eds) *Clinical Applications of Artificial Neural Networks*. Cambridge University Press, Cambridge, 298–326.

EAGAR, R.M.C. & PEIRCE, H.W. 1993. A nonmarine pelecypod assemblage in the Pennsylvanian of Arizona and its correlation with a horizon in Pennsylvania. *Journal of Paleontology*, **67**, 61–70.

EDWARDS, R.A., WARRINGTON, G., SCRIVENER, R.C., JONES, N.S., HASLAM, H.W. & AULT, L. 1997. The Exeter Group, south Devon, England: a contribution to the early post-Variscan stratigraphy of northwest Europe. *Geological Magazine*, **134**, 177–197.

EHRENBERG, S.N., PICKARD, N.A.H., HENRIKSEN, L.B., SVÅNÅ, T.A., GUTTERIDGE, P. & MCDONALD, D. 2001. A depositional and sequence stratigraphic model for cold-water, spiculitic strata based on the Kapp Starostin Formation (Permian) of Spitsbergen and equivalent deposits from the Barents Sea. *American Association of Petroleum Geologists Bulletin*, **85**, 2061–2087.

EHRENBERG, S.N., MCARTHUR, J.M. & THIRLWALL, M.F. 2010. Strontium isotope dating of spiculitic Permian strata from Spitsbergen outcrops and Barents sea well-cores. *Journal of Petroleum Geology*, **33**, 247–254.

EMBLETON, B.J.J. & MCDONNELL, K.L. 1980. Magnetostratigraphy in the Sydney Basin, southeastern Australia. *Journal of Geomagnetism and Geoelectricity*, **32**, (Suppl. 3), SIII1–SIII10.

EMBLETON, B.J.J., MCELHINNY, M.W., ZHANG, Z. & LI, Z.X. 1996. Permo-Triassic magnetostratigraphy in China: the type section near Taiyuan, Shanxi Province, North China. *Geophysical Journal International*, **126**, 382–388.

FETISOVA, A.M., VESELOVSKII, R.V., LATYSHEV, A.V., RAD'KO, V.A. & PAVLOV, V.E. 2014. Magnetic stratigraphy of the Permian–Triassic traps in the Kotui River valley (Siberian Platform): new paleomagnetic data. *Stratigraphy and Geological Correlation*, **22**, 377–390.

FOSTER, C.B. & ARCHIBOLD, N.W. 2001. Chronologic anchor points for the Permian Early Triassic of the eastern Australian basins. *In*: WEISS, R.H. (ed.) *Contributions to the Geology and Palaeontology of Gondwana in Honour of Helmut Wopfner*. University of Cologne Geological Institute, Cologne, 175–197.

GALLET, Y., KRYSTYN, L., BESSE, J., SAIDI, A. & RICOU, L.-E. 2000. New constraints on the upper Permian and Lower Triassic geomagnetic polarity timescale from the Abadeh section (central Iran). *Journal of Geophysical Research*, **105**, 2805–2815.

GASTALDO, R.A., KAMO, S.L., NEVELING, J., GEISSMAN, J.W., BAMFORD, M. & LOOY, C.V. 2015. Is the vertebrate-defined Permian–Triassic boundary in the Karoo Basin, South Africa, the terrestrial expression of the end-Permian marine event? *Geology*, **43**, 939–942.

GEISSMAN, J.W. & HARLAN, S.S. 2002. Late Paleozoic remagnetization of Precambrian crystalline rocks along the Precambrian/Carboniferous nonconformity, Rocky Mountains: a relationship among deformation, remagnetization, and fluid migration. *Earth and Planetary Science Letters*, **203**, 905–924.

GEUNA, S.E. & ESCOSTEGUY, L.D. 2004. Palaeomagnetism of the Upper Carboniferous–Lower Permian transition from Paganzo basin, Argentina. *Geophysical Journal International*, **157**, 1071–1089.

GIALANELLA, P.R., HELLER, F. *ET AL*. 1997. Late Permian magnetostratigraphy on the eastern Russian platform. *Geologie en Mijnbouw*, **76**, 145–154.

GILES, J.M., SOREGHAN, M.J., BENISON, K.C., SOREGHAN, G.S. & HASIOTIS, S.T. 2013. Lakes, loess, and paleosols in the Permian Wellington Formation of Oklahoma, U.S.A.: implications for paleoclimate and paleogeography of the Midcontinent. *Journal of Sedimentary Research*, **83**, 825–846.

GLEN, J.M., NOMADE, S., LYONS, J.L., METCALFE, I., MUNDIL, R. & RENNE, P.R. 2009. Magnetostratigraphic correlations of Permian–Triassic marine and terrestrial sediments from western China. *Journal of Asian Earth Sciences*, **36**, 521–540.

GOLUBEV, V.K. 2015. Dinocephalian stage in the history of the Permian tetrapod fauna of eastern Europe. *Paleontological Journal*, **49**, 1346–1352.

GOMEZ-ESPINOSA, C., VACHARD, D., BUITRÓN-SÁNCHEZ, B., ALMAZÁN-VAZQUEZ, E. & MENDOZA-MADERA, C. 2008. Pennsylvanian fusulinids and calcareous algae from Sonora (Northwestern Mexico), and their biostratigraphic and palaeobiogeographic implications. *Comptes Rendus Palevol*, **7**, 259–268.

GOSE, W.A. & HELSLEY, C.E. 1972. Paleomagnetic and rock-magnetic studies of the Permian Cutler and Elephant Canyon Formations in Utah. *Journal of Geophysical Research*, **77**, 1534–1548.

GRAHAM, J.W. 1955. Evidence of polar shift since Triassic time. *Journal of Geophysical Research*, **60**, 329–348.

GULBRANSON, E.L., MONTAÑEZ, I.P., SCHMITZ, M.D., LIMARINO, C.O., ISBELL, J.L., MARENSSI, S.A. & CROWLEY, J.L. 2010. High-precision U–Pb calibration of Carboniferous glaciation and climate history, Paganzo Group, NW Argentina. *Geological Society of America Bulletin*, **122**, 1480–1498.

GUREVITCH, E.L., HEUNEMANN, C., RAD'KO, V., WESTPHAL, M., BACHTADSE, V., POZZI, J.P. & FEINBERG, H. 2004. Palaeomagnetism and magnetostratigraphy of the Permian–Triassic northwest central Siberian Trap Basalts. *Tectonophysics*, **379**, 211–226.

HAAG, M. & HELLER, F. 1991. Late Permian to Early Triassic magnetostratigraphy. *Earth and Planetary Science Letters*, **107**, 42–54.

HALVORSEN, E., LEWANDOWSKI, M. & JELEŃSKA, M. 1989. Palaeomagnetism of the Upper Carboniferous Strzegom and Karkonosze Granites and the Kudowa Granitoid from the Sudet Mountains, Poland. *Physics of the Earth and Planetary Interiors*, **55**, 54–64.

HASLETT, J. & PARNELL, A. 2008. A simple monotone process with application to radiocarbon-dated depth chronologies. *Royal Statistical Society Journal, Series C*, **57**, 399–418.

HE, B., XU, Y.-G., HUANG, X.-L., LUO, Z.-Y., SHI, Y.-R., YANG, Q.-J. & YU, S.-Y. 2007. Age and duration of the Emeishan flood volcanism, SW China: geochemistry and SHRIMP zircon U–Pb dating of silicic ignimbrites, post-volcanic Xuanwei Formation and clay tuff at the Chaotian section. *Earth and Planetary Science Letters*, **255**, 306–323.

HELLER, F., LOWRIE, W., HUANMEI, L. & JUNDA, W. 1988. Magnetostratigraphy of the Permo-Triassic boundary section at Shangsi (Guangyuan, Sichuan Province, China). *Earth and Planetary Science Letters*, **88**, 348–356.

HELLER, F., CHEN, H., DOBSON, J. & HAAG, M. 1995. Permian–Triassic magnetostratigraphy – new results from South China. *Earth and Planetary Science Letters*, **89**, 281–295.

HELSLEY, C.E. 1965. Paleomagnetic results from the lower Permian Dunkard series of West Virginia. *Journal of Geophysical Research*, **70**, 413–424.

Henderson, C.M. & Mei, S. 2000. Preliminary cool water Permian conodont zonation in North Pangea: a review. *Permophiles*, **36**, 16–23.

Henderson, C.M., Davydov, V.I. & Wardlaw, B.R. 2012. The Permian period. *In*: Gradstein, F.M., Ogg, J.G., Schmitz, M. & Ogg, G. (eds) *The Geologic Time Scale, Volume 2*. Elsevier, Amsterdam, 653–679.

Hounslow, M.W. 2016. Geomagnetic reversal rates following Palaeozoic superchrons have a fast re-start mechanism. *Nature Communications*, **7**, 12507, https://doi.org/10.1038/ncomms12507

Hounslow, M.W. & Muttoni, G. 2010. The geomagnetic polarity timescale for the Triassic: linkage to stage boundary definitions. *In*: Lucas, S.G. (ed.) *The Triassic Timescale*. Geological Society, London, Special Publications, **334**, 61–102, https://doi.org/10.1144/SP334.4

Hounslow, M.W. & Nawrocki, J. 2008. Palaeomagnetism and magnetostratigraphy of the Permian and Triassic of Spitsbergen: a review of progress and challenges. *Polar Research*, **27**, 502–522.

Hounslow, M.W., Peters, C., Mørk, A., Weitschat, W. & Vigran, J.O. 2008. Biomagnetostratigraphy of the Vikinghøgda Formation, Svalbard (Arctic Norway), and the geomagnetic polarity timescale for the Lower Triassic. *Geological Society of America Bulletin*, **120**, 1305–1325.

Hounslow, M.W., McIntosh, G., Edwards, R.A., Laming, D.J.C. & Karloukovski, V. 2016. End of the Kiaman Superchron in the Permian of SW England: Magnetostratigraphy of the Aylesbeare Mudstone and Exeter groups. *Journal of the Geological Society, London*, first published online July 22, 2016, https://doi.org/10.1144/jgs2015-141

Hoyt, J.H. & Chronic, B.J. 1961. Wolfcampian fusulinids from the Ingleside Formation, Owl Canyon, Colorado. *Journal of Paleontology*, **35**, 1089.

Idnurum, M., Klootwijk, C., Théveniaut, H. & Trench, A. 1996. Magnetostratigraphy. *In*: Young, G.C. & Laurie, J.R. (eds) *The Australian Phanaerozoic Timescale*. Oxford University Press, Oxford, 23–51.

Iosifidi, A.G. & Khramov, A.N. 2009. Magnetostratigraphy of Upper Permian Sediments in the Southwestern Slope of Pai-Khoi (Khei-Yaga River Section): evidence for the global Permian–Triassic Crisis. *Izvestiya, Physics of the Solid Earth*, **45**, 3–13.

Iosifidi, A.G., Mac Niocaill, C., Khramov, A.N., Dekkers, M.J. & Popov, V.V. 2010. Palaeogeographic implications of differential inclination shallowing in permo-carboniferous sediments from the donets basin, Ukraine. *Tectonophysics*, **490**, 229–240.

Irving, E. 1971. Nomenclature in magnetic stratigraphy. *Geophysical Journal of the Royal Astronomical Society*, **24**, 529–531.

Irving, E. & Monger, J.W.H. 1987. Preliminary paleomagnetic results from the Permian Asitka Group, British Columbia. *Canadian Journal of Earth Sciences*, **24**, 1490–1497.

Irving, E. & Parry, L.G. 1963. The magnetism of some Permian rocks from New South Wales. *Geophysical Journal International*, **7**, 395–411.

Irving, E. & Pullaiah, G. 1976. Reversals of the geomagnetic field, magnetostratigraphy, and relative magnitude of paleosecular variation in the Phanerozoic. *Earth-Science Reviews*, **12**, 35–64.

Jacobs, J.A. 1963. *The Earth's Core and Geomagnetism, Volume 1*. Pergamon, New York.

Jin, Y.G., Shang, Q.H. & Cao, C.Q. 2000. Late Permian magnetostratigraphy and its global correlation. *Chinese Science Bulletin*, **45**, 698–704.

Jin, Y.G., Shen, S. et al. 2006*a*. The Global Stratotype Section and Point (GSSP) for the boundary between the Capitanian and Wuchiapingian stage (Permian). *Episodes*, **29**, 253.

Jin, Y.G., Wang, Y., Henderson, C., Wardlaw, B.R., Shen, S. & Cao, C. 2006*b*. The global boundary stratotype section and point (GSSP) for the base of the Changhsingian Stage (Upper Permian). *Episodes*, **29**, 175–182.

Jirah, S. & Rubidge, B. 2014. Refined stratigraphy of the Middle Permian Abrahamskraal Formation (Beaufort Group) in the southern Karoo Basin. *Journal of African Earth Sciences*, **100**, 121–135.

Johnson, S.Y., Schenk, C.J., Anders, D.L. & Tuttle, M.L. 1990. Sedimentology and petroleum occurrence, Schoolhouse Member, Maroon Formation (Lower Permian): Northwestern Colorado. *American Association of Petroleum Geologists Bulletin*, **74**, 135–150.

Kamo, S.L., Czamanske, G.K., Amelin, Y., Fedorenko, V.A., Davis, D.W. & Trofimov, V.R. 2003. Rapid eruption of Siberian flood-volcanic rocks and evidence for coincidence with the Permian–Triassic boundary and mass extinction at 251 Ma. *Earth and Planetary Science Letters*, **214**, 75–91.

Kasuya, A., Isozaki, Y. & Igo, H. 2012. Constraining paleo-latitude of a biogeographic boundary in mid-Panthalassa: Fusuline province shift on the Late Guadalupian (Permian) migrating seamount. *Gondwana Research*, **21**, 611–623.

Kent, D.V. & Olsen, P.E. 1999. Astronomically tuned geomagnetic polarity timescale for the Late Triassic. *Journal of Geophysical Research: Solid Earth*, **104**, 12,831–12,841.

Khramov, A.N. 1958. *Paleomagnetic Correlation of Sedimentary Rocks*. Gostoptehizdat, Leningrad [in Russian].

Khramov, A.N. 1963*a*. Palaeomagnetic investigations of the Upper Palaeozoic and the Triassic of the western part of the Donbass Basin. *In*: Khramov, A.N. (ed.) *Palaeomagnetism of the Palaeozoic. Transactions of the VNIGRI*, **204**, 97–117 [in Russian].

Khramov, A.N. 1963*b*. Palaeomagnetic investigations of Upper Permian and Lower Triassic sections on the northern and eastern Russian Platform. *In*: Khramov, A.N. (ed.) *Palaeomagnetism of the Palaeozoic. Transactions of the VNIGRI*, **204**, 145–174 [in Russian].

Khramov,A.N. 1967. Magnetic field of the Earth in the Late Palaeozoic. *Proceedings of the Academy of Sciences of the USSR, Series, Physics of the Earth*, **1**, 86–108 [in Russian].

Khramov, A.N. & Davydov, V.I. 1984. Paleomagnetism of Upper Carboniferous and Lower Permian in the south of USSR and the problems of structure of the Kiama Hyperzone. *In*: Khramov, A.N. (ed.) *Palaeomagnetic Methods in Stratigraphy*. Transactions of VNIGRI, 55–73 [in Russian].

Khramov, A.N. & Davydov, V.I. 1993. Results of palaeomagnetic investigations. Permian system: guides to geological excursions in the Uralian type localities. *In*: Chuvashov, B.I., Chernykh, V.V., Kipnin, V.J., Molin, V.A., Ozhgibesov, V.P. & Sofronitsky, P.A. (eds) *Permian System: Guides to Geological Excursions in the Uralian Type Localities*. Jointly published by Uralian Branch, Russian Academy of Sciences, Ekaterinburg, Russia and ESRI. Occasional Publications ESRI, New Series, **10**. Earth Sciences and Resources Institute (ESRI), University of South Carolina, Columbia, SC, 34–42.

Khramov, A.N., Komissarova, R.A., Iosifidi, A.G., Popov, V.V. & Bazhenov, M.L. 2006. Upper Tatarian magnetostratigraphy of the Sukhona River sequence: a re-study. *In*: Troyan, V.N., Semenov, V.S. & Kubyshkina, M.V. (eds) *Problems of the Geocosmos. 6th International Conference*. St Petersburg State University, St Petersburg, 317–321.

Kirschvink, J.L., Kopp, R.E., Raub, T.D., Baumgartner, C.T. & Holt, J.W. 2008. Rapid, precise, and high-sensitivity acquisition of paleomagnetic and rock-magnetic data: development of a low-noise automatic sample changing system for superconducting rock magnetometers. *Geochemistry, Geophysics, Geosystems*, **9**, 1–18.

Kirschvink, J.L., Isozaki, Y. *et al.* 2015. Challenging the sensitivity limits of Paleomagnetism: magnetostratigraphy of weakly magnetized Guadalupian–Lopingian (Permian) Limestone from Kyushu, Japan. *Palaeogeography, Palaeoclimatology, Palaeoecology*, **418**, 75–89.

Klootwijk, C.T., Shah, S.K., Gergan, J., Sharma, M.L., Tirkey, B. & Gupta, B.K. 1983. A palaeomagnetic reconnaissance of Kashmir, northwestern Himalaya, India. *Earth and Planetary Science Letters*, **63**, 305–324.

Klootwijk, C.T., Idnurm, M., Theveniaut, H. & Trench, A. 1994. *Phanerozoic Magnetostratigraphy: A Contribution to the Timescales Project*. Australian Geological Survey Organisation, Record, **1994/45**.

Korte, C., Jasper, T., Kozur, H.W. & Veizer, J. 2005. $\delta^{18}O$ and $\delta^{13}C$ of Permian brachiopods: a record of seawater evolution and continental glaciation. *Palaeogeography, Palaeoclimatology, Palaeoecology*, **224**, 333–351.

Kotlyar, G.V. 2015. Permian sections of southern Primorye: a link in correlation of stage units in the standard and general stratigraphic scales. *Russian Journal of Pacific Geology*, **9**, 254–273.

Kotlyar, G.V. & Pronina-Nestell, G.P. 2005. Report of the committee on the Permian system of Russian. *Permophiles*, **46**, 9–13.

Kozur, H.W. 2007. Biostratigraphy and event stratigraphy in Iran around the Permian–Triassic Boundary (PTB): implications for the causes of the PTB biotic crisis. *Global and Planetary Change*, **55**, 155–176.

Krainer, K., Vachard, D. & Lucas, S. 2003. Microfacies and microfossil assemblages (smaller foraminifers, algae, pseudoalgae) of the Hueco Group and Laborcita Formation (upper Pennsylvanian–lower Permian) south-central New Mexico, USA. *Rivista Italiana di Paleontologia e Stratigrafia*, **109**, 3–36.

Krassilov, V. & Karasev, E. 2009. Paleofloristic evidence of climate change near and beyond the Permian–Triassic boundary. *Palaeogeography, Palaeoclimatology, Paleoecology*, **284**, 326–336.

Kruiver, P.P., Langereis, C.G., Dekkers, M.J. & Krijgsman, W. 2003. Rock-magnetic properties of multicomponent natural remanent magnetization in alluvial red beds (NE Spain). *Geophysical Journal International*, **153**, 317–332.

Kryza, R., Pin, C., Oberc-Dziedzic, T., Crowley, Q.G. & Larionov, A. 2014. Deciphering the geochronology of a large granitoid pluton (Karkonosze Granite, SW Poland): an assessment of U–Pb zircon SIMS and Rb–Sr whole-rock dates relative to U–Pb zircon CA-ID-TIMS. *International Geology Review*, **56**, 756–782.

Lai, X., Yang, F., Hallam, A. & Wignall, P.B. 1996. The Shangsi section candidate of the global stratotype section and point of the Permian–Triassic boundary. *In*: Yin, H.F. (ed.) *The Paleozoic–Mesozoic Boundary*. China University of Geosciences Press, Beijing, 113–124.

Lambert, L.L., Wardlaw, B.R. & Henderson, C.H. 2007. *Mesogondolella* and *Jinogondolella* (Conodonta): multielement definition of the taxa that bracket the basal Guadalupian (Middle Permian Series) GSSP. *Palaeoworld*, **16**, 208–221.

Lanci, L., Tohver, E., Wilson, A. & Flint, S. 2013. Upper Permian magnetic stratigraphy of the lower Beaufort group, Karoo basin. *Earth and Planetary Science Letters*, **375**, 123–134.

Langereis, C.G., Krijgsman, W., Muttoni, G. & Menning, M. 2010. Magnetostratigraphy–concepts, definitions, and applications. *Newsletters on Stratigraphy*, **43**, 207–233.

Lawton, D.E. & Robertson, P.P. 2003. The Johnston Gas Field, Blocks 43/26a, 43/27a, UK Southern North Sea. *In*: Gluyas, J. & Hichens, H.M. (eds) *United Kingdom Oil and Gas Fields, Commemorative Millennium Volume*. Geological Society, London, Memoirs, **20**, 749–759, https://doi.org/10.1144/GSL.MEM.2003.020.01.62

Legler, B., Gebhardt, U. & Schneider, J.W. 2005. Late Permian non marine to marine transitional profiles in central southern Permian Basin, northern Germany. *International Journal of Earth Sciences*, **94**, 851–862.

LePage, B.A., Beauchamp, B., Pfefferkorn, H.W. & Utting, J. 2003. Late Early Permian plant fossils from the Canadian High Arctic: a rare paleoenvironmental/climatic window in northwest Pangea. *Palaeogeography, Palaeoclimatology, Palaeoecology*, **191**, 345–372.

Li, H. & Wang, J. 1989. Magnetostratigraphy of the Permo-Triassic boundary section of Meishan of Changxing, Zhejang. *Science in China*, **8**, 652–658.

Li, M., Ogg, J.G., Zhang, Y., Huang, C., Hinnov, L., Chen, Z.-Q. & Zou, Z. 2016. Astronomical tuning of the end-Permian extinction and the Early Triassic Epoch of South China and Germany. *Earth and Planetary Science Letters*, **441**, 10–25.

Li, Z., Chen, H., Song, B., Li, Y., Yang, S. & Yu, X. 2011. Temporal evolution of the Permian large igneous province in Tarim Basin in northwestern China. *Journal of Asian Earth Sciences*, **42**, 917–927.

LIU, C., PAN, Y. & ZHU, R. 2012. New paleomagnetic investigations of the Emeishan basalts in NE Yunnan, southwestern China: constraints on eruption history. *Journal of Asian Earth Sciences*, **52**, 88–97.

LIU, Y.Y., ZHU, Y.M. & TIAN, W.H. 1999. New magnetostratigraphic results from the Meishan section, Changxing County, Zheijiang, China. *Earth Science Journal of China University of Geosciences*, **24**, 151–154.

LOWRIE, W. & KENT, D.V. 2004. Geomagnetic polarity timescales and reversal frequency regimes. *In*: CHANNELL, J.E.T., KENT, D.V., LOWRIE, W. & MEERT, J. (eds) *Timescales of the Palaeomagnetic Field*. America Geophysical Union, Geophysical Monograph Series, **145**, 117–129.

LOZOVSKY, V.R. 1998. Chapter 7: The Permian–Triassic boundary. *In*: ESAULOVA, N.K., LOZOVSKY, V.R. & ROZANOV, A.Y. (eds) *Stratotypes and Reference Sections of the Upper Permian in the Region of the Volga and Kama Rivers*. GEOS, Moscow, 271–281.

LOZOVSKY, V.R., KRASSILOV, V.A., AFONIN, S.A., BUROV, B.V. & YAROSHENKO, O.P. 2001. Transitional Permian–Triassic deposits in European Russia, and nonmarine correlations. *Natura Bresciana, Annuario del Museo Civico di Storia Naturale di Brescia*, **25**, 301–310.

LOZOVSKY, V.R., MINIKH, M.G., GRUNT, T.A. & KUKHTINOV, D.A., PONOMARENKO, A.G. & SUKACHEVA, I.D. 2009. The Ufimian Stage of the East European scale: status, validity, and correlation potential. *Stratigraphy and Geological Correlation*, **17**, 602–614.

LOZOVSKY, V.R., BALABANOV, YU.P., PONOMARENKO, A.G., NOVOKOV, I.V., BUSLOVICH, A.L., MORKOVIN, B.I. & YAROSHENKO, O.P. 2014. Stratigraphy, palaeomagnetism and petromagnetism of the Lower Triassic in the Moscow Syneclise. 1, Yug River Basin. *Bulletin of the Moscow Society of Naturalists. Geological Series*, **89**, 61–72 [in Russian].

LUCAS, S.G. 2006. Global Permian tetrapod biostratigraphy and biochronology. *In*: LUCAS, S.G., CASSINIS, G. & SCHNEIDER, J.W. (eds) *Non-Marine Permian Biostratigraphy and Biochronology*. Geological Society, London, Special Publications, **265**, 65–93, https://doi.org/10.1144/GSL.SP.2006.265.01.04

LUCAS, S.G. 2013. Vertebrate biostratigraphy and biochronology of the upper Paleozoic Dunkard Group, Pennsylvania West Virginia–Ohio, USA. *International Journal of Coal Geology*, **119**, 79–87.

MACINTYRE, D.G., VILLENEUVE, M.E. & SCHIARIZZA, P. 2001. Timing and tectonic setting of Stikine Terrane magmatism, Babine-Takla lakes area, central British Columbia. *Canadian Journal of Earth Sciences*, **38**, 579–601.

MAGNUS, G. & OPDYKE, N.D. 1991. A paleomagnetic investigation of the Minturn Formation, Colorado: a study in establishing the timing of remanence acquisition. *Tectonophysics*, **187**, 181–189.

MATUYAMA, M. 1929. On the direction of magnetisation of basalt in Japan, Tyosen and Manchuria. *Proceedings of the Imperial Academy*, **5**, 203–205.

MAY, W., HUTTENLOCKER, A.K., PARDO, J.D., BENCA, J. & SMALL, B.J. 2011. New Upper Pennsylvanian armored dissorophid records (Temnospondyli, Dissorophoidea) from the US midcontinent and the stratigraphic distributions of dissorophids. *Journal of Vertebrate Paleontology*, **31**, 907–912.

MCELHINNY, M.W. & BUREK, P.J. 1971. Mesozoic palaeomagnetic stratigraphy. *Nature*, **232**, 98–102.

MCMAHON, B.E. & STRANGWAY, D.W. 1968. Investigation of Kiaman Magnetic Divison in Colorado Redbeds. *Geophysical Journal International*, **15**, 265–285.

MEI, S. & HENDERSON, C.M. 2002. Comments on some Permian conodont faunas reported from SE Asia and adjacent areas and their global correlation. *Journal of Asian Earth Sciences*, **20**, 599–608.

MEI, S., HENDERSON, C.M. & WARDLAW, B.R. 2002. Evolution and distribution of the conodonts Sweetognathus and Iranognathus and related genera during the Permian, and their implications for climate change. *Palaeogeography, Palaeoclimatology, Palaeoecology*, **180**, 57–91.

MENG, X., HU, C., WANG, W. & LIU, H. 2000. Magnetostratigraphic Study of Meishan Permian–Triassic Section, Changxing, Zhejiang Province, China. *Journal of China University of Geosciences*, **11**, 361–365.

MENNING, M. 1987. Magnetostratigraphy. *In*: LÜTZNER, H. (ed.) *Sedimentary and Volcanic Rotliegend of the Saale Depression: Excursion Guidebook*. Symposimum on Rotliegend in Central Europe. Academy of Science GDR, Potsdam, 92–96.

MENNING, M. & JIN, Y. 1998. Comment on 'Permo-Triassic magnetostratigraphy in China: the type section near Taiyuan, Shanxi Province, North China' by B.J.J. Embleton, M.W. McElhinny, X. Ma, Z. Zhang & Z.X. Li. *Geophysical Journal International*, **133**, 213–216.

MENNING, M., KATZUNG, G. & LUTZNER, H. 1988. Magnetostratigraphic investigations in the Rotliegend (300–252 Ma) of central-Europe. *Zeitschrift fur Geologische Wissenschaften*, **16**, 1045–1063.

MERCANTON, P.L. 1926. Inversion de l'inclinaison magnètique terrestre aux ages geologiques. *Terrestrial Magnetism and Atmospheric Electricity*, **31**, 187.

MERTMANN, D. 2003. Evolution of the marine Permian carbonate platform in the Salt Range (Pakistan). *Palaeogeography, Palaeoclimatology, Palaeoecology*, **191**, 373–384.

METCALFE, I., NICOLL, R.S. & WARDLAW, B.R. 2007. Conodont index fossil *Hindeodus changxingensis* Wang fingers greatest mass extinction event. *Palaeoworld*, **16**, 202–207.

METCALFE, I., CROWLEY, J.L., NICOLL, R.S. & SCHMITZ, M. 2014. High-precision U–Pb CA-TIMS calibration of Middle Permian to Lower Triassic sequences, mass extinction and extreme climate-change in eastern Australian Gondwana. *Gondwana Research*, **28**, 61–81.

MILLER, J.D. & OPDYKE, N.D. 1985. Magnetostratigraphy of the Red Sandstone Creek section – Vail, Colorado. *Geophysical Research Letters*, **12**, 133–136.

MODIE, B.N. & LE HÉRISSÉ, A. 2009. Late Palaeozoic palynomorph assemblages from the Karoo Supergroup and their potential for biostratigraphic correlation, Kalahari Karoo Basin, Botswana. *Bulletin of Geosciences*, **84**, 337–358.

MOLOSTOVSKY, E.A. 1996. Some aspects of magnetostratigraphic correlation. *Stratigraphy and Geological Correlation*, **4**, 231–237.

MOLOSTOVSKY, E.A. 2005. Magnetostratigrphic correlation of Upper Permian Marine and Continental Formations. *Stratigraphy and Geological Correlation*, **13**, 49–58.

MOLOSTOVSKY, E.A., PEVZNER, M.A., PECHERSKY, D.M., RODIONOV, V.P. & KHRAMOV, A.N. 1976. Magnetostratigraphic timescale of the Phanerozoic and the geomagnetic field reversal regime. *In*: *Geomagnitnye issledovania*. Nauka, Moscow, **17**, 45–52 [in Russian].

MOLOSTOVSKY, E.A., MOLOSTOVSKAYA, I.I. & MINIKH, M.G. 1998. Stratigraphic correlations of the Upper Permian and Triassic beds from the Volga-Ural and Cis-Caspian. *In*: CRASQUIN-SOLEAU, S. & BARRIER, É. (eds) *Peri-Tethys Memoir 3: Stratigraphy and Evolution of Peri-Tethyan Platforms*. Mémoires du Muséum National d'histoire Naturelle, **177**, 35–44.

MORRIS, N.J. 2013. *Stratigraphy and geochemistry of Lower Permian volcanics in the Sverdrup Basin, Northwest Ellesmere Island, Nunavut*. Msc thesis, University of Calgary, Calgary, Canada.

MOURAVIEV, F.A., AREFIEV, M.P. *ET AL*. 2015. Monastery Ravine section: stratotype of the Urzhumian and limitotype of the Severodvinian Stage. *In*: NURGALIEV, D.K., SILANTIEV, V.V. & NIKOLAEV, S.V. (eds) *Type and Reference Sections of the Middle and Upper Permian of the Volga and Kama River Regions*. A Field Guidebook of XVIII International Congress on Carboniferous and Permian. Kazan University Press, Kazan, 120–141.

MUNDIL, R., LUDWIG, K.R., METCALFE, I. & RENNE, P.R. 2004. Age and timing of the Permian mass extinctions: U/Pb dating of closed-system zircons. *Science*, **305**, 1760–1763.

MUNDIL, R., PÁLFY, J., RENNE, P.R. & BRACK, P. 2010. The Triassic timescale: new constraints and a review of geochronological data. *In*: LUCAS, S.G. (ed.) *The Triassic Timescale*. Geological Society, London, Special Publications, **334**, 41–60, https://doi.org/10.1144/SP334.3

MURPHY, M.A. & SALVADOR, A. 1999. Special International Stratigraphic Guide – an abridged version. *Episodes*, **22**, 255–271.

NAKREM, H.A. 1994. Bryozoans from the Lower Permian Vøringen Member (Kapp Starostin Formation), Spitsbergen, Svalbard. *Norsk Polarinstitutt Skrifter*, **196**, 5–93.

NAKREM, H.A., RASMUSSEN, J.A. & SWIFT, A. 1991. Late Carboniferous to earliest Triassic conodonts of Svalbard. *Geonytt*, **1**, 38–39.

NAKREM, H.A., NILSSON, I. & MANGERUD, G. 1992. Permian biostratigraphy of Svalbard (Arctic Norway) – a review. *International Geology Review*, **34**, 933–959.

NAWROCKI, J. 1997. Permian to early Triassic magnetostratigraphy from the Central European Basin in Poland: implications on regional and worldwide correlations. *Earth and Planetary Science Letters*, **152**, 37–58.

NAWROCKI, J. 1999. Paleomagnetism of Permian through Early Triassic sequences in central Spitsbergen: implications for paleogeography. *Earth and Planetary Science Letters*, **169**, 59–70.

NAWROCKI, J. & GRABOWSKI, J. 2000. Palaeomagnetism of Permian through Early Triassic sequences in central Spitsbergen: contribution to magnetostratigraphy. *Geological Quarterly*, **44**, 109–118.

NEWELL,A.J., SENNIKOV, A.G., BENTON, M.J., MOLOSTOVSKAYA, I.A., GOLUBEV, V.K., MINIKH, A.V. & MINIKH, M.G. 2010. Disruption of playa–lacustrine depositional systems at the Permo-Triassic boundary: evidence from Vyazniki and Gorokhovets on the Russian Platform. *Journal of the Geological Society, London*, **167**, 695–716, https://doi.org/10.1144/0016-76492009-103

NICKLEN, B.L. 2011. *Establishing a Tephrochronologic Framework for the Middle Permian (Guadalupian) Type Area and Adjacent Portions of the Delaware Basin and Northwestern Shelf, West Texas and Southeastern New Mexico, USA*. PhD thesis, University of Cincinnati, Cincinnati, OH.

NILSSON, I. & DAVYDOV, V.I. 1997. Fusulinid biostratigraphy in Upper Carboniferous (Gzhelian) and Lower Permian (Asselian–Sakmarian) successions of Spitsbergen. Arctic Norway. *Permophiles*, **30**, 18–24.

OGG, J.G. 2012. Magnetostratigraphy. *In*: GRADSTEIN, F.M., OGG, J.G., SCHMITZ, M.D. & OGG, G.M. (eds) *The Geologic Time Scale 2012*. Elsevier, Amsterdam, 85–114.

OGG, J.G., OGG, G. & GRADSTEIN, F.M. 2008. *The Concise Geologic Time Scale*. Cambridge University Press, Cambridge.

OLSZEWSKI, T.D. & ERWIN, D.H. 2009. Change and stability in Permian brachiopod communities from western Texas. *Palaios*, **24**, 27–40.

OPDYKE, N.D. & CHANNELL, J.E. 1996. *Magnetic Stratigraphy*. Academic Press, New York.

OPDYKE, N.D. 1995. Permo-Carboniferous magnetostratigraphy. *In*: BERGGREN, W.A., KENT, D.V., AUBRY, M.P. & HARDENBOL, J. (eds) *Geochronology, Time Scales and Global Stratigraphic Correlation*. Society for Sedimentary Geology (SEPM), Special Publications, **54**, 41–50.

PAN, H. & SYMONS, D.T.A. 1993. The Pictou red beds' Pennsylvanian pole: could Phanaerozoic rocks in the interior United States be remagnetized? *Journal of Geophysical Research*, **98**, 6227–6235.

PARNELL, A.C., HASLETT, J., ALLEN, J.R.M., BUCK, C.E. & HUNTLEY, B. 2008. A flexible approach to assessing synchroneity of past events using Bayesian reconstructions of sedimentation history. *Quaternary Science Reviews*, **27**, 1872–1885.

PAŠAVA, J., OSZCZEPALSKI, S. & DU, A. 2010. Re–Os age of non-mineralized black shale from the Kupferschiefer, Poland, and implications for metal enrichment. *Mineralium Deposita*, **45**, 189–199.

PETERSON, D.N. & NAIRN, A.E.M. 1971. Palaeomagnetism of Permian redbeds from the south-western United States. *Geophysical Journal International*, **23**, 191–205.

PRUNER, P. 1992. Palaeomagnetism and palaeogeography of Mongolia from the Carboniferous to the Cretaceous – final report. *Physics of the Earth and Planetary Interiors*, **70**, 169–177.

RAKOTOSOLOFU, N.A., TORSVIK, T.H., ASHWAL, L.D., EIDE, E.A. & DE WIT, M.J. 1999. The Karoo Supergroup revisited and Madagascar-Africa fits. *Journal of African Earth Sciences*, **29**, 135–151.

RAMEZANI, J., SCHMITZ, M.D., DAVYDOV, V.I., BOWRING, S.A., SNYDER, W.S. & NORTHRUP, C.J. 2007.

High-precision U–Pb zircon age constraints on the Carboniferous–Permian boundary in the southern Urals stratotype. *Earth and Planetary Science Letters*, **256**, 244–257.

Rasnitsyn, A.P., Sukacheva, I.D. & Aristov, D.S. 2005. Permian insects of the Vorkuta Group in the Pechora Basin, and their stratigraphic implications. *Paleontological Journal*, **39**, 404–416 [translated from Paleontologicheskii Zhurnal, **4**, 63–75, 2005].

Retallack, G.J., Sheldon, N.D. *et al.* 2011. Multiple Early Triassic greenhouse crises impeded recovery from Late Permian mass extinction. *Palaeogeography, Palaeoclimatology, Palaeoecology*, **308**, 233–251.

Rocha-Campos, A.C., Basei, M.A. *et al.* 2011. 30 million years of Permian volcanism recorded in the Choiyoi igneous province (W Argentina) and their source for younger ash fall deposits in the Paraná Basin: SHRIMP U–Pb zircon geochronology evidence. *Gondwana Research*, **19**, 509–523.

Ross, C.A. & Monger, J.W.H. 1978. Carboniferous and Permian fusulinaceans from the Omineca Mountains, British Columbia. Contributions to Canadian Paleontology. *Geological Survey of Canada Bulletin*, **267**, 43–63.

Rubidge, B.S., Erwin, D.H., Ramezani, J., Bowring, S.A. & De Klerk, W.J. 2013. High-precision temporal calibration of Late Permian vertebrate biostratigraphy: U–Pb zircon constraints from the Karoo Supergroup, South Africa. *Geology*, **41**, 363–366.

Rush, J. & Kerans, C. 2010. Stratigraphic response across a structurally dynamic shelf: the latest Guadalupian composite sequence at Walnut Canyon, New Mexico, USA. *Journal of Sedimentary Research*, **80**, 808–828.

Sadler, P.M. & Strauss, D.J. 1990. Estimation of completeness of stratigraphical sections using empirical data and theoretical models. *Journal of the Geological Society, London*, **147**, 471–485, https://doi.org/10.1144/gsjgs.147.3.0471

Sawin, R.S., Franseen, E.K., West, R.R., Ludvigson, G.A. & Watney, W.L. 2008. Clarification and changes in Permian stratigraphic nomenclature in Kansas. *Current Research in Earth Sciences: Kansas Geological Survey Bulletin*, **254**, 1–4, https://www.kgs.ku.edu/Current/2008/Sawin/index.html

Schmidberger, S.S. & Hegner, E. 1999. Geochemistry and isotope systematics of calc-alkaline volcanic rocks from the Saar-Nahe basin (SW Germany) – implications for Late-Variscan orogenic development. *Contributions to Mineralogy and Petrology*, **135**, 373–385.

Schmitz, M.D. 2012. Radiogenic isotope geochronology. *In*: Gradstein, F.M., Ogg, J.G., Schmitz, M.D. & Ogg, G.M. (eds) *The Geologic Time Scale 2012*. Elsevier, Amsterdam, 115–126.

Schmitz, M.D. & Davydov, V.I. 2012. Quantitative radiometric and biostratigraphic calibration of the Pennsylvanian–Early Permian (Cisuralian) time scale and pan-Euramerican chronostratigraphic correlation. *Geological Society of America Bulletin*, **124**, 549–577.

Schneider, J.W., Lucas, S.G. & Barrick, J.E. 2013. The Early Permian age of the Dunkard Group, Appalachian basin, USA, based on spiloblattinid insect biostratigraphy. *International Journal of Coal Geology*, **119**, 88–92.

Scott, K.M. 2013. Carboniferous–Permian boundary in the Halgaito Formation, Cutler Group, Valley of the Gods and surrounding area, southeastern Utah. *In*: Lucas, S.G., DiMichele, W.A., Barrick, J.E., Schneider, J.W. & Spielmann, J.A. (eds) *The Carboniferous–Permian Transition. New Mexico Museum of Natural History and Science Bulletin*, **60**, 398–409.

Sennikov, A.G. & Golubev, V.K. 2006. Vyazniki Biotic Assemblage of the terminal Permian. *Paleontological Journal*, **40**, S475–S481.

Sharps, R., McWilliams, M. *et al.* 1989. Lower Permian paleomagnetism of the Tarim block, northwestern China. *Earth and Planetary Science Letters*, **92**, 275–291.

Shen, S.-Z. & Mei, S.H. 2010. Lopingian (Late Permian) high-resolution conodont biostratigraphy in Iran with comparison to South China zonation. *Geological Journal*, **45**, 135–161.

Shen, S.-Z., Henderson, C.M. *et al.* 2010. High resolution Lopingian (late Permian) timescale of South China. *Geological Journal*, **45**, 122–134.

Shen, S.Z., Crowley, J.L. *et al.* 2011. Calibrating the end-Permian mass extinction. *Science*, **334**, 1367–1372.

Silantiev, VV.V., Arefiev, M.P. *et al.* 2015*a*. Cheremushka Section: parastratotype of the Urzhumian Stage. *In*: Nurgaliev, D.K., Silantiev, V.V. & Nikolaev, S.V. (eds) *Type and Reference Sections of the Middle and Upper Permian of the Volga and Kama River Regions*. A Field Guidebook of XVIII International Congress on Carboniferous and Permian. Kazan University Press, Kazan, 70–119.

Silantiev, V.V., Kotlyar, G.V., Zorina, S.O., Golubev, V.K. & Liberman, V.B. 2015*b*. The geological setting and Permian stratigraphy of the Volga and Kama river regions. *In*: Nurgaliev, D.K., Silantiev, V.V. & Nikolaev, S.V. (eds) *Type and Reference Sections of the Middle and Upper Permian of the Volga and Kama River Regions*. A Field Guidebook of XVIII International Congress on Carboniferous and Permian. Kazan University Press, Kazan, 10–23.

Silantiev, V.V., Nurgalieva, N.G. *et al.* 2015*c*. Elabuga section, Ufimian/Kazanian boundary. *In*: Nurgaliev, D.K., Silantiev, V.V. & Nikolaev, S.V. (eds) *Type and Reference Sections of the Middle and Upper Permian of the Volga and Kama River Regions*. A Field Guidebook of XVIII International Congress on Carboniferous and Permian. Kazan University Press, Kazan, 144–153.

Sinito, A.M., Valencio, D.A. & Vilas, J.F. 1979. Palaeomagnetism of a sequence of Upper Palaeozoic–Lower Mesozoic red beds from Argentina. *Geophysical Journal International*, **58**, 237–247.

Słowakiewicz, M., Kiersnowski, H. & Wagner, R. 2009. Correlation of the Middle and Upper Permian marine and terrestrial sedimentary sequences in Polish, German, and USA Western Interior Basins with reference to global time markers. *Palaeoworld*, **18**, 193–211.

Soreghan, G.L., Elmore, R.D. & Lewchuk, M.T. 2002. Sedimentologic-magnetic record of western Pangean climate in upper Paleozoic loessite (lower Cutler beds, Utah). *Geological Society of America Bulletin*, **114**, 1019–1035.

Soreghan, G.S., Benison, K.C., Foster, T.M., Zambito, J. & Soreghan, M.J. 2015. The paleoclimatic and geochronologic utility of coring red beds and evaporites: a case study from the RKB core (Permian, Kansas, USA). *International Journal of Earth Sciences*, **104**, 1589–1603.

Sosipatrova, G.P. 1967. Upper Paleozoic foraminifera of Spitsbergen. *In*: Sokolov, V.N. (ed.) *Stratigraphy of Spitsbergen*. Institut Geologii Arktiki, Leningrad, 125–163 [in Russian].

Steiner, M.B. 1988. Paleomagnetism of the Late Pennsylvanian and Permian: a test of the rotation of the Colorado Plateau. *Journal of Geophysical Research: Solid Earth*, **93**, 2201–2215.

Steiner, M.B. 2006. The magnetic polarity time scale across the Permian–Triassic boundary. *In*: Lucas, S.G., Cassinis, G. & Schneider, J.W. (eds) *Non-Marine Permian Biostratigraphy and Biochronology*. Geological Society, London, Special Publications, **265**, 15–38, https://doi.org/10.1144/GSL.SP.2006.265.01.02

Steiner, M.B., Ogg, J., Zhang, Z. & Sun, S. 1989. The Late Permian/early Triassic magnetic polarity time scale and plate motions of south China. *Journal of Geophysical Research*, **94**, 7343–7363.

Stemmerik, L. 1988. Discussion. Brachiopod zonation and age of the Permian Kapp Starostin Formation (Central Spitsbergen). *Polar Research*, **6**, 179–180.

Stevens, L.G., Hilton, J., Bond, D.P.G., Glasspool, I.J. & Jardine, P.E. 2011. Radiation and extinction patterns in Permian floras from North China as indicators for environmental and climate change. *Journal of the Geological Society, London*, **168**, 607–619, https://doi.org/10.1144/0016-76492010-042

Sun, Y., Lai, X., Jiang, H., Luo, G., Sun, S., Yan, C. & Wignall, P.B. 2008. Guadalupian (Middle Permian) conodont Faunas at Shangsi section, Northeast Sichuan Province. *Journal of China University of Geosciences*, **19**, 451–460.

Sun, Y., Lai, X. et al. 2010. Dating the onset and nature of the Middle Permian Emeishan large igneous province eruptions in SW China using conodont biostratigraphy and its bearing on mantle plume uplift models. *Lithos*, **119**, 20–33.

Sweet, D.E., Carsrud, C.R. & Watters, A.J. 2015. Proposing an entirely Pennsylvanian age for the Fountain Formation through new lithostratigraphic correlation along the Front Range. *The Mountain Geologist*, **52**, 43–70.

Swift, A. 1986. The conodont *Merrillina divergens* (Bender & Stoppel) from the Upper Permian of England. *In*: Harwood, G.M. & Smith, D.B. (eds) *The English Zechstein and Related Topics*. Geological Society, London, Special Publications, **22**, 55–62, https://doi.org/10.1144/GSL.SP.1986.022.01.05

Symons, D.T.A. 1990. Early Permian pole: evidence from the Pictou red beds, Prince Edward Island, Canada. *Geology*, **18**, 234–237.

Szurlies, M. 2013. Late Permian (Zechstein) magnetostratigraphy in western and central Europe. *In*: Gąsiewicz, A. & Słowakiewicz, M. (eds) *Palaeozoic Climate Cycles: Their Evolutionary and Sedimentological Impact*. Geological Society, London, Special Publications, **376**, 73–85, https://doi.org/10.1144/SP376.7

Szurlies, M., Bachmann, G.H., Menning, M., Nowaczyk, N.R. & Käding, K.-C. 2003. Magnetostratigraphy and high resolution lithostratigraphy of the Permian–Triassic boundary interval in Central Germany. *Earth and Planetary Science Letters*, **212**, 263–278.

Talling, P. & Burbank, D. 1993. Assessment of uncertainties in magnetostratigraphic dating of strata. *In*: Aissaoui, D.M., McNeill, D.F. & Hurley, N.F. (eds) *Application of Paleomagnetism to Sedimentary Geology*. Society for Sedimentary Geology (SEPM), Tulsa, OK, 59–70.

Taylor, G.K., Tucker, C. et al. 2009. Magnetostratigraphy of Permian/Triassic boundary sequences in the Cis-Urals, Russia: no evidence for a major temporal hiatus. *Earth and Planetary Science Letters*, **281**, 36–47.

Thompson, R. 1972. Palaeomagnetic results from the Paganzo Basin of north-west Argentina. *Earth and Planetary Science Letters*, **15**, 145–156.

Tohver, E., Lanci, L., Wilson, A., Hansma, J. & Flint, S. 2015. Magnetostratigraphic constraints on the age of the lower Beaufort Group, western Karoo basin, South Africa, and a critical analysis of existing U–Pb geochronological data. *Geochemistry, Geophysics, Geosystems*, **16**, 3649–3665.

Turner, P. 1979. The palaeomagnetic evolution of continental red beds. *Geological Magazine*, **116**, 289–301.

Turner, P., Chandler, P., Ellis, D., Leveille, G.P. & Heywood, M.L. 1999. Remanance acquisition and magnetostratigraphy of the Leman Sandstone Formation: Jupiter Fields, southern North Sea. *In*: Tarling, D.H. & Turner, P. (eds) *Palaeomagnetism and Diagenesis in Sediments*. Geological Society, London, Special Publications, **151**, 109–124, https://doi.org/10.1144/GSL.SP.1999.151.01.11

Tverdokhlebov, V.P., Tverdokhlebova, G.I., Minikh, A.V., Surkov, M.V. & Benton, M.J. 2005. Upper Permian vertebrates and their sedimentological context in the South Urals, Russia. *Earth-Science Reviews*, **69**, 27–77.

Valencio, D.A. 1980. Reversals and excursions of the geomagnetic field as defined by palaeomagnetic data from Upper Palaeozoic-Lower Mesozoic sediments and igneous rocks from Argentina. *Journal of Geomagnetism and Geoelectricity*, **32**, (Suppl. 3), SIII137–SIII142.

Valencio, D.A., Vilas, J.F. & Mendía, J.E. 1977. Palaeomagnetism of a sequence of red beds of the Middle and Upper Sections of Pagnazo Group (Argentina) and the correlation of Upper Palaeozoic-Lower Mesozoic rocks. *Geophysical Journal International*, **51**, 59–74.

Van Der Voo, R. & Torsvik, T.H. 2012. The history of remagnetization of sedimentary rocks: deceptions, developments and discoveries. *In*: Elmore, R.D., Muxworthy, A.R., Aldana, M.M. & Mena, M. (eds) *Remagnetization and Chemical Alteration of Sedimentary Rocks*. Geological Society, London,

Special Publications, **371**, 23–53, https://doi.org/10.1144/SP371.2

Vozárová, A. & Túnyi, I. 2003. Evidence of the Illawarra Reversal in the Permian sequence of the Hornic nappe (Western Carpathians, Slovakia). *Geologica Carpathica*, **54**, 229–236.

Wahlman, G.P. & West, R.R. 2010. Fusulinids from the Howe Limestone Member (Red Eagle Limestone, Council Grove Group) in northeastern Kansas and their significance to the North American Carboniferous (Pennsylvanian)–Permian boundary. *Current Research in Earth Sciences: Kansas Geological Survey Bulletin*, **258**, 1–13, https://www.kgs.ku.edu/Current/2010/Wahlman/index.html

Wang, C. & Yang, S. 1993. Brachiopod fauna around the Carboniferous–Permian boundary from the Balikelike Formation in Keping, Xinjiang. *Journal of Changchun University of Earth Sciences*, **23**, 1–9.

Ward, P.D., Botha, J. *et al.* 2005. Abrupt and gradual extinction among Late Permian land vertebrates in the Karoo Basin, South Africa. *Science*, **307**, 709–714.

Wardlaw, B.R. & Mei, S. 1999. Refined conodont biostratigraphy of the Permian and lowest Triassic of the Salt and Khizor Ranges, Pakistan. *In*: Yin, H. & Tong, J. (eds) *Proceedings of the International Conference on Pangea and the Palaeozoic-Mesozoic Transition*. China University of Geosciences Press, Wuhan, 154–156.

Wardlaw, B.R. & Pogue, K.R. 1995. The Permian of Pakistan. *In*: Scholle, P.A., Peryt, T.M. & Ulmer-Scholle, D.S. (eds) *The Permian of Northern Pangea, Volume 2: Sedimentary Basins and Economic Resources*. Springer, Berlin, 215–224.

Waterhouse, J.B. 2010. Lopingian (Late Permian) stratigraphy of the Salt Range, Pakistan and Himalayan region. *Geological Journal*, **45**, 264–284.

Wescott, W.A. & Diggens, J.N. 1998. Depositional history and stratigraphical evolution of the Sakamena Group (Middle Karoo Supergroup) in the southern Morondava Basin, Madagascar. *Journal of African Earth Sciences*, **27**, 461–479.

Westfahl, M., Surkis, Y.F., Gurevich, E.L. & Khramov, A.N. 2005. Kiama–Illawarra geomagnetic reversal recorded in the Tatarian Stratotype (the Kazan region). *Izvestiya, Physics of the Solid Earth*, **41**, 634–653.

Winters, S.S. 1962. Lithology and stratigraphy of the Supai Formation, Fort Apache Indian Reservation, Arizona. *In*: Weber, R.H. & Peirce, H.W. (eds) *Mogollon Rim Region (East-Central Arizona)*. New Mexico Geological Society 13th Annual Fall Field Conference Guidebook. New Mexico Geological Society, Socorro, NM, 87–88.

Wu, H., Zhang, S., Hinnov, L.A., Jiang, G., Feng, Q., Li, H. & Yang, T. 2013. Time-calibrated Milankovitch cycles for the late Permian. *Nature Communications*, **4**, 2452, https://doi.org/10.1038/ncomms3452

Wynne, P.J., Irving, E. & Osadetz, K. 1983. Paleomagnetism of the Esayoo Formation (Permian) of northern Ellesmere Island: possible clue to the solution of the Nares Strait dilemma. *Tectonophysics*, **100**, 241–256.

Xu, Y.-G., Wei, X., Luo, Z.-Y., Liu, H.-Q. & Cao, J. 2014. The Early Permian Tarim Large Igneous Province: main characteristics and a plume incubation model. *Lithos*, **204**, 20–35.

Yin, H., Zhang, K., Tong, J., Yang, Z. & Wu, S. 2001. The global stratotype section and point (GSSP) of the Permian–Triassic boundary. *Episodes*, **24**, 102–114.

Yuan, D.X., Shen, S.Z., Henderson, C.M., Chen, J., Zhang, H. & Feng, H.Z. 2014. Revised conodont-based integrated high-resolution timescale for the Changhsingian Stage and end-Permian extinction interval at the Meishan sections, South China. *Lithos*, **204**, 220–245.

Zheng, L., Yang, Z., Tong, Y. & Yuan, W. 2010. Magnetostratigraphic constraints on two-stage eruptions of the Emeishan continental flood basalts. *Geochemistry, Geophysics, Geosystems*, **11**, Q12014, https://doi.org/10.1029/2010GC003267

Zhong, Y.T., He, B., Mundil, R. & Xu, Y.G. 2014. CA-TIMS zircon U–Pb dating of felsic ignimbrite from the Binchuan section: implications for the termination age of Emeishan large igneous province. *Lithos*, **204**, 14–19.

Ziegler, A.M., Rees, P.M. & Naugolnykh, S.V. 2002. The Early Permian floras of Prince Edward Island, Canada: differentiating global from local effects of climate change. *Canadian Journal of Earth Sciences*, **39**, 223–238.

Permian strontium isotope stratigraphy

CHRISTOPH KORTE[1]* & CLEMENS V. ULLMANN[2]

[1]*Department of Geosciences and Natural Resource Management, University of Copenhagen, Øster Voldgade 10, 1350 Copenhagen-K, Denmark*

[2]*Camborne School of Mines and Environment and Sustainability Institute, University of Exeter, Penryn Campus, Treliever Road, Penryn, Cornwall TR10 9FE, UK*

**Correspondence: korte@ign.ku.dk*

Abstract: The secular evolution of the Permian seawater $^{87}Sr/^{86}Sr$ ratios carries information about global tectonic processes, palaeoclimate and palaeoenvironments, such as occurred during the Early Permian deglaciation, the formation of Pangaea and the Permian–Triassic (P–Tr) mass extinction. Besides this application for discovering geological aspects of Earth history, the marine $^{87}Sr/^{86}Sr$ curve can also be used for robust correlations when other bio-, litho- and/or chemostratigraphic markers are inadequate. The accuracy of marine $^{87}Sr/^{86}Sr$ reconstructions, however, depends on high-quality age control of the reference data, and on sample preservation, both of which generally deteriorate with the age of the studied interval. The first-order Permian seawater $^{87}Sr/^{86}Sr$ trend shows a monotonous decline from approximately 0.7080 in the earliest Permian (Asselian) to approximately 0.7069 in the latest Guadalupian (Capitanian), followed by a steepening increase from the latest Guadalupian towards the P–Tr boundary (*c.* 0.7071–0.7072) and into the Early Triassic. Various higher-order changes in slope of the Permian $^{87}Sr/^{86}Sr$ curve are indicated, but cannot currently be verified owing to a lack of sample coverage and significant disagreement of published $^{87}Sr/^{86}Sr$ records.

Supplementary material: Numbers, information, data and references of the samples discussed are available at https://doi.org/10.6084/m9.figshare.c.3589460

Seawater $^{87}Sr/^{86}Sr$ ratios have varied throughout Earth's history (Peterman *et al.* 1970; Veizer & Compston 1974), and are mostly defined by the relative influence of unradiogenic Sr derived from Earth's mantle (mantle sources) and radiogenic Sr of the continental crust (riverine input) (Palmer & Edmond 1989; Taylor & Lasaga 1999). Owing to radioactive decay of ^{87}Rb to ^{87}Sr, both have increased from an initial ratio of 0.69897 ± 0.00003 at Earth's accretion (Hans *et al.* 2013) to current approximate values of 0.704 and 0.713 (Pearce *et al.* 2015), respectively (Fig. 1d). Secular fluctuations of the marine $^{87}Sr/^{86}Sr$ ratio, incorporated into authigenic marine sediments and hard parts of fossils, can be used as chemostratigraphic markers for global correlations (e.g. Steuber 2001, 2003; Denison *et al.* 2003; Burla *et al.* 2009; Ehrenberg *et al.* 2010; see also McArthur *et al.* 2012). This application makes use of the observations that seawater has a globally uniform $^{87}Sr/^{86}Sr$ ratio, and shows secular drifts of measureable magnitude. Globally uniform $^{87}Sr/^{86}Sr$ ratios in biominerals of marine origin (e.g. Brand *et al.* 2003) are observed in most fully marine settings due to mixing and the long residence time of Sr in the oceans of >1000 kyr (e.g. Li 1982; Elderfield 1986; Veizer 1989; McArthur 1994; Pearce *et al.* 2015). Deviations from the marine ratios in marginal and estuarine settings can occur, however, where salinities are significantly reduced (Fig. 1c, Bryant *et al.* 1995; Sharma *et al.* 2007), and such deviations have been inferred, for example, for the Middle Jurassic of Poland (Wierzbowski *et al.* 2012).

The first detailed Phanerozoic strontium isotope curve, revealing distinct fluctuations, was generated by Burke *et al.* (1982) using mostly whole-rock carbonate as an $^{87}Sr/^{86}Sr$ archive. This record was later successively refined for specific time intervals (e.g. DePaolo & Ingram 1985; Palmer & Elderfield 1985; Hodell *et al.* 1989, 1991). For the Carboniferous and Permian, additional work on the marine $^{87}Sr/^{86}Sr$ trends was carried out by Popp *et al.* (1986), Brookins (1988), Kramm & Wedepohl (1991) and Nishioka *et al.* (1991), and later re-evaluated by Denison *et al.* (1994) and Denison & Koepnick (1995). McArthur (1994) emphasized that excellent sample preservation is crucial for high-resolution Sr isotope stratigraphy. Subsequent work on various intervals of the Phanerozoic has consequently increasingly used hard parts of organisms precipitating low-Mg calcite (LMC) (e.g. brachiopods, belemnites, oysters) or bioapatite

From: Lucas, S. G. & Shen, S. Z. (eds) 2018. *The Permian Timescale*. Geological Society, London, Special Publications, **450**, 105–118.
First published online December 12, 2016, https://doi.org/10.1144/SP450.5

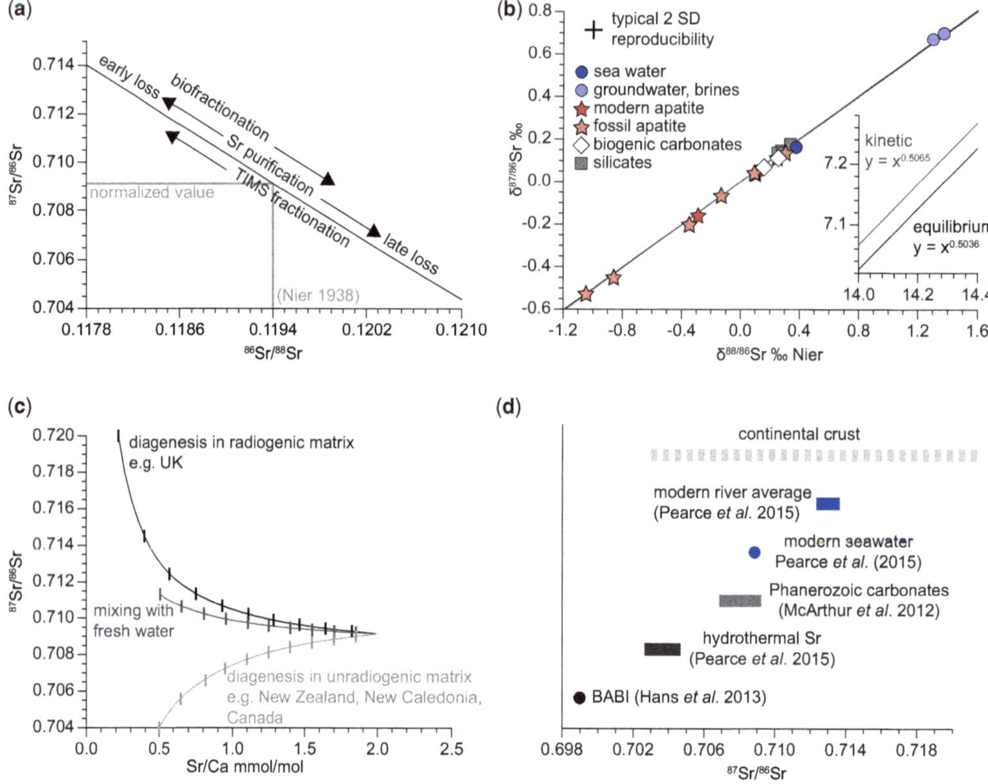

Fig. 1. Sr isotope systematics. (**a**) Coupled fractionation of the $^{87}Sr/^{86}Sr$ and $^{86}Sr/^{88}Sr$ ratio in various processes. Biofractionation generally enriches the biomineral in the light isotopes, leading to a lower $^{87}Sr/^{86}Sr$ but higher $^{86}Sr/^{88}Sr$ ratio. During ion-exchange chemistry in the laboratory, loss of strontium (e.g. due to too small a collection window or residual Sr on the ion-exchange resin) can lead to enrichment in the light and the heavy Sr isotopes. During thermal ionization mass spectrometry, the light Sr isotopes evaporate first from the filament and the sample successively evolves to heavier Sr isotopic composition. All of these effects can be corrected by applying a fractionation law and fixing the $^{86}Sr/^{88}Sr$ ratio of Sr at the accepted value of 0.1194 (Nier 1938). (**b**) $\delta^{87/86}Sr$ and $\delta^{88/86}Sr$ values of natural samples measured by Neymark *et al.* (2014) together with potential law equations describing equilibrium and kinetic Sr isotope fractionation computed from formulae given in Young & Galy (2004). Significant isotope fractionation is observed especially for fossil apatite and brines, but all values can be corrected using either the equilibrium or kinetic fractionation law without introduction of significant errors. Kinetic and equilibrium fractionation curves become measurably different only at large isotopic fractionation (inset). (**c**) Deviations from hypothetical, arbitrary marine biogenic calcite signal due to diagenesis or seawater–freshwater mixing. Depicted curves are calculated using arbitrary values of $^{87}Sr/^{86}Sr$ and Sr/Ca of end members and Sr distribution coefficients, which are partially specific to site, biomineral, calcite precipitation rate and geological interval. (**d**) $^{87}Sr/^{86}Sr$ ratio of selected Sr reservoirs on Earth.

(e.g. conodonts) that were checked for sample preservation (e.g. Jurassic: Jones *et al.* 1994; Triassic: Martin & Macdougall 1995; Korte *et al.* 2003, 2004; Permian: Martin & Macdougall 1995; Korte *et al.* 2004, 2006; Carboniferous: Bruckschen *et al.* 1999; Devonian: Diener *et al.* 1996; Ebneth *et al.* 1997; Ordovician: Shields *et al.* 2003). Such macrofossil remains are thought to be comparatively resistant to diagenesis (Veizer 1989; Blake *et al.* 1997; Veizer *et al.* 1999; Zazzo *et al.* 2004), representing therefore the most robust available archives for past seawater $^{87}Sr/^{86}Sr$ (e.g. Veizer *et al.* 1997, 1999; Jenkyns *et al.* 2002).

Here, we review existing Permian $^{87}Sr/^{86}Sr$ data constraining the evolution of the marine $^{87}Sr/^{86}Sr$ ratio throughout this period. We discuss possible pitfalls of diagenetic alteration or local environmental influences and evaluate the data, in particular with regard to the potential stratigraphic resolution. Finally, we briefly discuss the impact of global tectonic events and environmental change on the Permian seawater $^{87}Sr/^{86}Sr$.

Theoretical background

Physical processes during mass spectrometry, strontium loss during sample preparation in the laboratory and the biomineralization process all affect the isotopic composition of Sr (Fig. 1a). In order to correct for these fractionation effects, traditionally all measured $^{87}Sr/^{86}Sr$ ratios have been recalculated using the synchronously measured $^{86}Sr/^{88}Sr$ ratio (Fig. 1a) (Elderfield 1986). This latter ratio is assumed to be 0.1194, as measured on 99.9% pure Eimer & Amend Sr metal (Nier 1938), and later adopted and recommended for general use by the IUGS Subcommission on Geochronology (Steiger & Jäger 1977). When this ratio is assumed to be constant, all fractionation-related deviations of $^{87}Sr/^{86}Sr$ in a sample from coeval seawater can be removed as long as the correct fractionation law is employed for calculations.

Figure 1b shows measured pairs of $\delta^{88/86}Sr$ and $\delta^{87/86}Sr$ values of natural samples (Neymark *et al.* 2014) analysed using a double-spike protocol so that the Nier correction for instrumental fractionation could be avoided. Also shown are fractionation curves for kinetic and equilibrium isotope fractionation of strontium calculated from equations given in Young & Galy (2004), which yield virtually the same results in the range of observed natural variations of strontium isotope ratios. It is thus evident that biological fractionation of the $^{87}Sr/^{86}Sr$ ratio can be significant (especially in apatite) but can conveniently be approximated using either a kinetic or an equilibrium fractionation law for samples with $^{87}Sr/^{86}Sr$ close to the accepted reference (NIST SRM 987, $^{87}Sr/^{86}Sr = 0.710248$: McArthur *et al.* 2001; see also Neymark *et al.* 2014). In order to generate a measurable bias through the assumption of a wrong fractionation law (at currently common levels of precision), biological fractionation effects would need to be an order of magnitude larger than those observed (see inset of Fig. 1b).

Disequilibrium effects of biomineralization that can severely complicate interpretation of other isotopic systems in biominerals (e.g. Wefer & Berger 1991 for O and C isotopes) therefore play no role in their $^{87}Sr/^{86}Sr$ ratio – an invaluable advantage.

Uncertainties in numerical age assignment and correlation

The problem of assigning precise and accurate relative and numerical ages to globally distributed samples has a direct impact on the fidelity with which the marine $^{87}Sr/^{86}Sr$ reference curve can be constrained. This problem is particularly relevant in the Permian, for which a significant lack of internal age constraint has been noted (Henderson *et al.* 2012). An accurate relative sample sequence can be ensured when sourcing samples from drill cores (Morante 1996), continuous successions at one locality (Sedlacek *et al.* 2014) or small geographical areas (Denison *et al.* 1994). Correlation problems of local biostratigraphic and lithostratigraphic schemes to other regions, however, are highly likely to lead to erroneous age interpretations (Denison *et al.* 1994; Martin & Macdougall 1995; Henderson *et al.* 2012). Such wrong/distorted age assignments are particularly problematic where the $^{87}Sr/^{86}Sr$ curve changes rapidly (Burke *et al.* 1982). High-fidelity correlation in the Permian is hampered by low faunal diversity (coarse biostratigraphic schemes) and low sea level, generating isolated basins with endemic faunas, in particular in the Late Permian (Martin & Macdougall 1995). Analysing only materials that are firmly tied by biostratigraphic schemes (e.g. conodonts: Korte *et al.* 2003, 2006) is currently the best way of ensuring acceptable age resolution and correlation between regions. For example, conodont biozonations of the Permian give an average age resolution of approximately 1.3 myr (Korte *et al.* 2006; Henderson *et al.* 2012). Such biozonal schemes have been integrated into a Permian Composite Standard (Henderson *et al.* 2012), which is an attempt at an integrated stratigraphic framework in the absence of cyclostratigraphic and high-resolution chemostratigraphic constraints. Biases of approximately 1 myr in age assignments for samples entering the Permian database are currently acceptable for most stages, because limited analytical precision of the available data and rate of change of the marine $^{87}Sr/^{86}Sr$ curve do not allow the generation of a higher-resolution curve. Future research, however, should focus on improving both the fidelity of $^{87}Sr/^{86}Sr$ data and the age constraints of samples that are used for $^{87}Sr/^{86}Sr$ reference curves.

Diagenesis

Besides poor age control and contributions of non-marine strontium to ambient water (e.g. in marginal-marine environments and estuaries), diagenetically induced alterations of the $^{87}Sr/^{86}Sr$ ratios in fossil materials constitute the largest obstacle for generating a well-defined marine $^{87}Sr/^{86}Sr$ record. Even minor alteration of LMC and bioapatite can have a significant influence on the strontium isotopes because $^{87}Sr/^{86}Sr$ ratios are measured to a great level of precision (usually $\pm$ 0.00002), while secular drifts of the marine $^{87}Sr/^{86}Sr$ curve are comparatively slow, averaging 0.000026 per myr for the last 500 myr (McArthur *et al.* 2012).

It is commonly considered that diagenesis leads to lower Sr concentrations in LMC (Brand & Veizer

1980) and it is often implicitly assumed that alteration leads to more radiogenic (i.e. higher $^{87}Sr/^{86}Sr$) $^{87}Sr/^{86}Sr$ ratios (e.g. Veizer & Compston 1974; Denison *et al.* 1994; Shields *et al.* 2003). While these trends usually hold, they are not universal. It has been observed that calcite cements can be rich in Sr (Ullmann *et al.* 2015) and that diagenesis can lead to ^{87}Sr depletions (=lower $^{87}Sr/^{86}Sr$ ratios) in altered calcite (Fig. 1c) (Burke *et al.* 1982; Brand 1991; Steuber & Schlüter 2012; Ullmann *et al.* 2013, 2014). Two schematic trends for diagenetic alteration of $^{87}Sr/^{86}Sr$ and Sr/Ca ratios are shown in Figure 1c together with countries in which these relative trends are observed. In order to reconstruct original seawater Sr concentrations and isotope compositions, it is necessary to take into account the specific locality and lithology patterns of diagenesis (Ullmann & Korte 2015).

Sample preservation of conodonts is assessed using the Conodont Alteration Index (CAI: Martin & Macdougall 1995; Veizer *et al.* 1997; Korte *et al.* 2003), a measure of the discoloration of conodont elements through thermal maturation of organic matter in the bioapatite (Martin & Macdougall 1995). Increases in CAI have been found to co-vary with significant changes of $^{87}Sr/^{86}Sr$ in conodonts (e.g. Veizer *et al.* 1997), suggesting that this alteration proxy has a good potential for excluding data of altered specimens from interpretation. Some doubts about the general suitability of conodonts for reconstructing marine $^{87}Sr/^{86}Sr$ ratios, however, have been voiced (Korte *et al.* 2003; McArthur *et al.* 2012). These doubts are fuelled by unresolvable offsets from coeval records generated on other fossil materials that might be related to early Sr exchange with the sediment matrix (Veizer *et al.* 1999).

Data accuracy, comparability and confidence

Strontium isotope stratigraphy relies on accurate high-precision data. Adequate sample control, laboratory processes and analytical routines are therefore of utmost importance for retrieving meaningful numerical ages from marine carbonates.

Since the 1980s, almost all laboratories use NIST SRM 987 (strontium carbonate) as the international standard for data control and to ensure interlaboratory comparability of $^{87}Sr/^{86}Sr$ data. For the purpose of comparing results from different studies, we here adjust published ratios to $^{87}Sr/^{86}Sr_{NIST\ SRM\ 987}$ = 0.710248 (McArthur *et al.* 2001, 2012). This adjustment ensures that biases arising during the analytical process can be corrected, and data should, in theory, be accurate and comparable. It is, however, common practice to measure aliquots of pure NIST SRM 987 that have not undergone any previous laboratory processing, so that absolute values and precision of this material only relate to processes after sample dissolution and Sr purification (i.e. mass spectrometry). Contamination of samples during processing cannot be excluded when using this protocol and some of the observed disagreements between coeval $^{87}Sr/^{86}Sr$ records (see interlaboratory biases of McArthur *et al.* 2001, 2012) might arise from such contamination.

It is therefore encouraged to control the performance of sample processing, Sr purification and spectrometric analysis using an additional international reference material with matrix comparable to the sample matrix (e.g. JLs-1 for carbonates: Rasmussen *et al.* 2016). It is further encouraged to analyse element concentrations from a split of each sample solution for preservation control (Ullmann *et al.* 2013, 2014; Rasmussen *et al.* 2016).

Permian $^{87}Sr/^{86}Sr$ trend

The published marine $^{87}Sr/^{86}Sr$ data for the Permian are plotted in Figure 2, and show a first-order long-term decreasing trend throughout the Early Permian until the late Middle Permian (Capitanian), followed by a steepening increase across the P–Tr boundary that continues throughout the Early Triassic. These overall trends have been known since the early work of, for example, Veizer & Compston (1974), Burke *et al.* (1982), Popp *et al.* (1986), Brookins (1988), Kramm & Wedepohl (1991), Nishioka *et al.* (1991) and Denison *et al.* (1994). The results of these pioneering studies were confirmed by data generated from conodonts (e.g. Martin & Macdougall 1995), and macrofossil calcite (brachiopods) screened for diagenesis and with improved biostratigraphic control (e.g. Korte *et al.* 2006). While the Early Permian Asselian–Sakmarian $^{87}Sr/^{86}Sr$ trend is constrained only by data from three northern hemisphere regions, the late Early (Kungurian)–Late Permian interval is covered by data from a multitude of the southern hemisphere and northern hemisphere mid- to low-palaeolatitude successions (Fig. 3).

The Asselian–Capitanian $^{87}Sr/^{86}Sr$ decrease

A decreasing $^{87}Sr/^{86}Sr$ trend that continues until the Capitanian commences in the Moscovian (Middle Pennsylvanian, *c.* 310 Ma: Denison *et al.* 1994; Bruckschen *et al.* 1999; McArthur *et al.* 2012). This trend starts from a ratio of approximately 0.7083, and reaches approximately 0.7080–0.7082 at the Carboniferous–Permian boundary (298.9 Ma) (Burke *et al.* 1982; Nishioka *et al.* 1991; Denison *et al.* 1994; Bruckschen *et al.* 1999; Korte

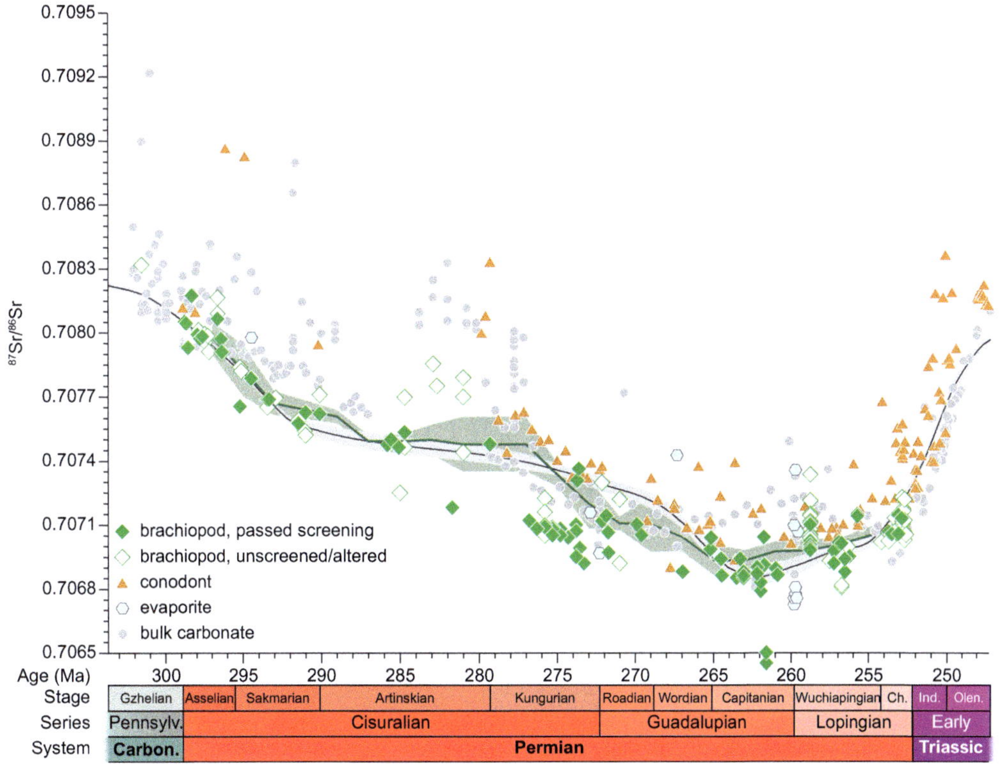

Fig. 2. Available $^{87}Sr/^{86}Sr$ data for which independent age constraints allow for robust age assignment. Grey circles, whole-rock carbonate; blue hexagons, evaporates; brown triangles, conodont elements; open diamonds, unscreened or poorly preserved brachiopods; green diamonds, brachiopod calcite that passed screening for diagenesis. Values taken from Veizer & Compston (1974), Denison *et al.* (1994), Martin & Macdougall (1995), Morante (1996), Korte *et al.* (2003, 2004, 2006), Tierney (2010) and Sedlacek *et al.* (2014). All data are adjusted to NIST SRM 987 = 0.710248 (McArthur *et al.* 2001). All data are recalculated to the Geologic Time Scale 2012 (Gradstein *et al.* 2012). Black curve with a grey 95% confidence envelope: Sr look-up table Version 5 (McArthur *et al.* 2001, 2012); green curve with a light green 95% confidence envelope: Running average of biostratigraphically well-defined (conodont biozonation) and well-preserved (diagenetically screened) brachiopod data from Korte *et al.* (2006) with 2 myr steps and a 5 myr window. Carbon., Carboniferous; Pennsylv, Pennsylvanian; Ch., Changhsingian; Ind., Induan; Olen., Olenekian.

et al. 2006; Tierney 2010). Canadian (Sverdrup Basin) and Russian (Moscow Basin) brachiopods of the Late Carboniferous (Bruckschen *et al.* 1999), as well as whole-rock carbonate data from the Carboniferous–Permian transition in the USA (Tierney 2010), suggest a more radiogenic value of 0.7082. Lower values of approximately 0.7080 for this boundary are supported by whole-rock carbonate data of the Akiyoshi Limestone (Nishioka *et al.* 1991), and well-preserved earliest Permian brachiopods from the USA and Russia (Korte *et al.* 2006).

The $^{87}Sr/^{86}Sr$ decrease continues throughout the Early Permian and most of Middle Permian (Fig. 2), representing the largest sustained downwards trend of the marine $^{87}Sr/^{86}Sr$ curve in the Phanerozoic (Veizer *et al.* 1999; McArthur *et al.* 2012) and this is well documented globally (Burke *et al.* 1982; Nishioka *et al.* 1991; Denison *et al.* 1994; Morante 1996; Korte *et al.* 2006; Tierney 2010). The slope of this downwards trend is not constant through time: relatively steep intervals of the Asselian–Sakmarian and Wordian–Capitanian are separated by a more gentle decrease during the Artinskian–Roadian (Fig. 2) (Korte *et al.* 2006; McArthur *et al.* 2012). Whole-rock carbonate data from the USA (Denison *et al.* 1994) even indicate a temporary increase in seawater $^{87}Sr/^{86}Sr$ in the Artinskian. The more radiogenic whole-rock data of Denison *et al.* (1994) may be explained by elevated Sakmarian–Roadian $^{87}Sr/^{86}Sr$ ratios of shark teeth from the USA, which have been inferred to be related to basinal restriction and a significant admixture of terrestrial radiogenic Sr (Fischer *et al.* 2014). Most

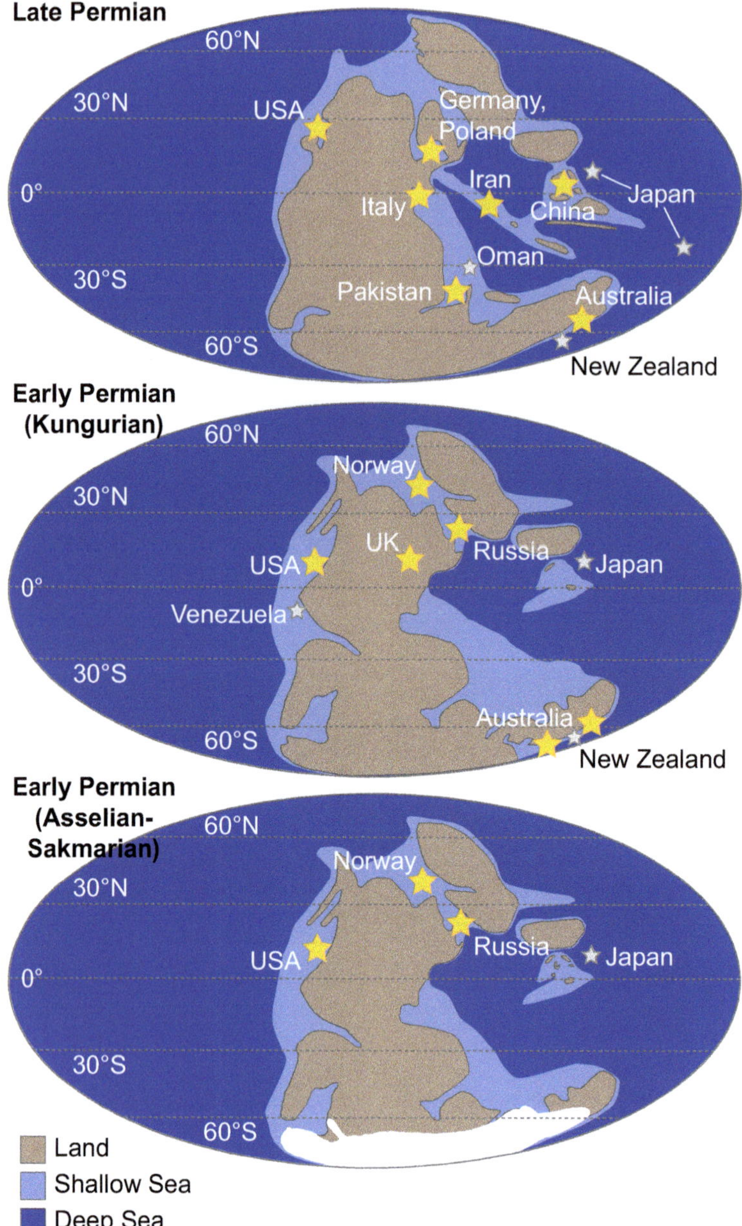

Fig. 3. Palaeogeographical maps with sample localities for the Asselian–Sakmarian (*c.* 299–290 Ma), Kungurian (*c.* 284–272 Ma) and Wuchiapingian–Changhsingian (*c.* 260–252 Ma) interval. Maps modified from Metcalfe (2011). Yellow stars depict regions studied to constrain the marine $^{87}Sr/^{86}Sr$ curve for the respective intervals. Grey stars denote studied regions for which $^{87}Sr/^{86}Sr$ analyses were used to either estimate depositional ages or where age constraints are too weak to precisely define sample age.

Kungurian–Capitanian $^{87}Sr/^{86}Sr$ datasets show comparatively large data variability, but are in accord with each other regardless of the sample material analysed. Australian brachiopod data that reach extremely low values of approximately 0.7065 in the Capitanian (Morante 1996) are, however, consistently more unradiogenic than time-equivalent values from elsewhere (Fig. 2).

The Early–Middle Permian downwards trend of the marine $^{87}Sr/^{86}Sr$ curve has been used to improve age constraints in multiple localities: for example, Japan (Miura *et al.* 2004), Russia (Nurgalieva *et al.* 2007), Spitsbergen (Ehrenberg *et al.* 2010), Oman (Stephenson *et al.* 2012) and Venezuela (Laya *et al.* 2013).

The Capitanian $^{87}Sr/^{86}Sr$ minimum

The Pennsylvanian–Guadalupian (*c.* 313–262 Ma: McArthur *et al.* 2012) (Fig. 2) trend of decreasing $^{87}Sr/^{86}Sr$ stops in the Capitanian (late Guadalupian) at ratios of approximately 0.7068–0.7069 (Burke *et al.* 1982; Denison *et al.* 1994; Morante 1996; Korte *et al.* 2006; Kani *et al.* 2008, 2013; Wignall *et al.* 2009; Shen *et al.* 2010; Tierney 2010; Liu *et al.* 2013). These low values are the lowest recorded for the entire Palaeozoic and very close to the Phanerozoic minimum at the Middle–Late Jurassic transition (Jones *et al.* 1994; McArthur *et al.* 2012), and therefore an important chemostratigraphic marker. The Capitanian minimum has consequently been used as a stratigraphic marker: for example, in carbonate successions now exposed in China (Huang *et al.* 2008) and Japan (Kani *et al.* 2008, 2013).

The Wuchiapingian–Changhsingian increasing $^{87}Sr/^{86}Sr$ trend

The $^{87}Sr/^{86}Sr$ ratios for the Late Permian and the P–Tr boundary were obtained from different archives: magnesites of the Eastern Alps (Frimmel & Niedermayr 1991); evaporites (Kampschulte *et al.* 1998; Denison & Peryt 2009) for the Zechstein of the Germanic Basin; gypsum and anhydrite for several North Alpine localities (Spötl & Pak 1996), and bulk carbonate for the Gartnerkofel core (Austria) (Kralik 1991); and at the GSSP site at Meishan in China (Kaiho *et al.* 2001; Cao *et al.* 2009). Records of $^{87}Sr/^{86}Sr$ across the P–Tr boundary, obtained solely from conodonts, were published for different sections of the Salt Range (Pakistan) (Martin & Macdougall 1995), for Abadeh (Iran) and Sosio (Italy) (Korte *et al.* 2003, 2004), and for Meishan (Twitchett 2007). Late Permian $^{87}Sr/^{86}Sr$ ratios of well-preserved brachiopods have been reported by Gruszczynski *et al.* (1992) and Stemmerik *et al.* (2001) for high-latitude localities from Greenland and Spitsbergen, and by Korte *et al.* (2006) for the low-latitude successions (Meishan and other Chinese localities, and Jolfa, Iran). Throughout the Wuchiapingian and Changhsingian (Lopingian = Late Permian), $^{87}Sr/^{86}Sr$ ratios show a moderate increase that steepens in the latest Permian to its Phanerozoic maximum rate of increase around the P–Tr boundary (Fig. 2) (e.g. Martin & Macdougall 1995; Korte *et al.* 2003, 2004; McArthur *et al.* 2012). At the P–Tr boundary, the marine $^{87}Sr/^{86}Sr$ curve reaches a value of approximately 0.7071–0.7072 (e.g. Korte *et al.* 2003, 2004, 2006; Sedlacek *et al.* 2014).

In conodont records, the Late Permian–Early Triassic increasing trend has been observed to be interrupted by a phase of relatively little change in seawater $^{87}Sr/^{86}Sr$ across the P–Tr boundary (Twitchett 2007; Korte *et al.* 2010) (Fig. 4). This interruption of the long-term trend has been suggested to last from the early Dorashamian (Changhsingian) up to the higher Induan *I. isarcica* Zone (Fig. 4). In bulk carbonate records from Meishan, however, a general positive $^{87}Sr/^{86}Sr$ trend is visible in the same interval (Kaiho *et al.* 2001), but not in acetic acid leachates from carbonates of the same area (Cao *et al.* 2009). Dubious data related to diagenesis are evident in whole-rock datasets: for example, contradictory values just below the event horizon (Meishan bed 21: *c.* 0.708 (Kaiho *et al.* 2001) v. *c.* 0.7072 (Cao *et al.* 2009)). The samples of Cao *et al.* (2009) originate from a core rather than from the deeply weathered outcrops at Meishan. The core-derived $^{87}Sr/^{86}Sr$ values are similar to those of the conodonts from Iran (Korte *et al.* 2004) and may therefore represent the primary past seawater signal more closely. A transient $^{87}Sr/^{86}Sr$ positive bulge of approximately 0.0008 in the Meishan-1 core throughout the lower half of the Griesbachian (early Induan stage), however, also indicates some preservation issues within this record.

Up to the early Middle Triassic, $^{87}Sr/^{86}Sr$ values continue to rise (Martin & Macdougall 1995; Korte *et al.* 2003; McArthur *et al.* 2012), reaching a maximum value of approximately 0.7080 – similar to the value at the Carboniferous–Permian boundary.

Age resolution of the Permian $^{87}Sr/^{86}Sr$ curve

The marine $^{87}Sr/^{86}Sr$ curve as, for example, approximated by Howarth & McArthur (1997) and McArthur *et al.* (2001, 2012) permits the assignment of numerical ages to samples that cannot otherwise be constrained precisely in time owing to a lack of suitable alternative methods. The main factors affecting this potential for deriving stratigraphic information are: (I) the accuracy of the curve; (II) the precision of the curve; (III) its rate of change; (IV) the precision of laboratory analyses; and (V) the fidelity of the $^{87}Sr/^{86}Sr$ data derived from the materials, the age of which is to be evaluated.

(I) & (II) Biases in the marine $^{87}Sr/^{86}Sr$ curve of some magnitude have been present and will persist unless the evolution of seawater can be

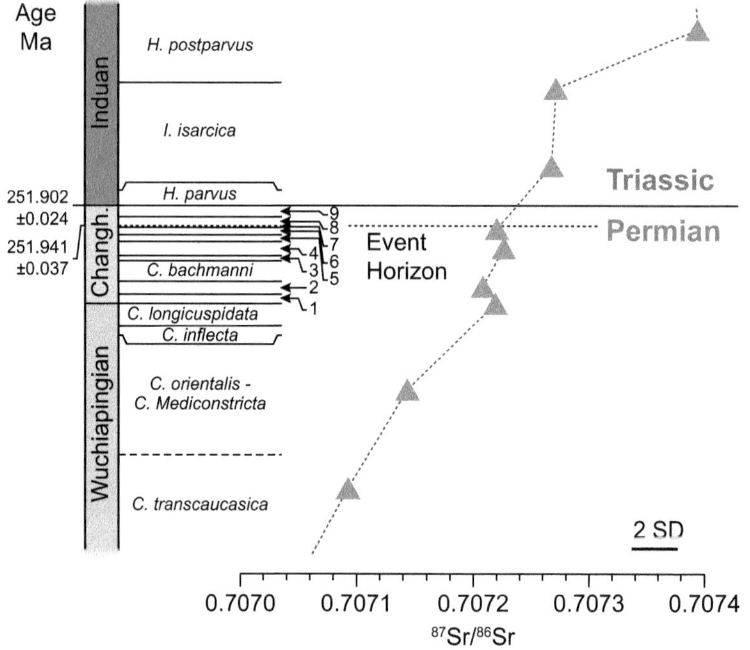

Fig. 4. Wuchiapingian–Induan $^{87}Sr/^{86}Sr$ record from conodonts at Abadeh, Iran (modified from Korte *et al.* 2010). 1, *C. hambastensis* Zone; 2, *C. subcarinata* Zone; 3, *C. nodosa* Zone; 4, *C. changxingensis–C. deflecta* Zone; 5, *C. zhangi* Zone; 6, *C. iranica* Zone; 7, *C. hauschkei* Zone; 8, *C. meishanensis–H. praeparvus* Zone; 9, *M. ultima–S. ?mostleri* Zone; Changh., Changhsingian; EH, event horizon. Ages are from Burgess *et al.* (2014).

constrained more tightly through a renewed effort in covering the entire Phanerozoic with multiple high-quality, high-resolution $^{87}Sr/^{86}Sr$ datasets. For the Permian period, the Artinskian–Wordian interval is currently especially underconstrained and has yielded conflicting proxy data (Fig. 2). The precision of the often-employed strontium look-up table (Howarth & McArthur 1997; McArthur *et al.* 2001) is only given by the uncertainty of the LOWESS fit to the data that were chosen for its construction. This precision should therefore not be taken as meaningful for either confidence in the curve or for results obtained from measuring $^{87}Sr/^{86}Sr$ ratios of unknown materials (see also McArthur *et al.* 2001, 2012).

(III) & (IV) The faster the seawater $^{87}Sr/^{86}Sr$ ratio changed through time, the higher the theoretical resolution that can be expected, as long as this change is unidirectional, as seen, for example, since the late Eocene (see also McArthur *et al.* 2012 for an overview). Ignoring artefacts of diagenesis and biases in the $^{87}Sr/^{86}Sr$ curve, the age resolution of a datum is given as the ratio of analytical uncertainty with the rate of change of the marine $^{87}Sr/^{86}Sr$ curve (0.000035 per myr on average in the Permian). Except for the steepening Changhsingian $^{87}Sr/^{86}Sr$ increase leading up to the fastest observed rate of change in the Phanerozoic in the Early Triassic, the marine $^{87}Sr/^{86}Sr$ ratio appears to have changed throughout the Permian at a rate that is quite typical for the Phanerozoic (McArthur *et al.* 2012). From this rate of change, an average age resolution of 0.6 myr for a typical uncertainty value of 0.000020 can be expected for the whole period (Table 1). The Artinskian fares worst, with a rate of change of approximately 0.000011 per myr, whereas the Changhsingian is characterized by a very rapid rate of change of approximately 0.00011 per myr, leading to the best nominal resolution.

(V) The above calculations yield encouragingly high nominal temporal precision – especially when new, high-precision laboratory protocols are implemented. Such numbers, however, are unrealistic: besides the high likelihood that the best estimate of the marine $^{87}Sr/^{86}Sr$ curve is wrong in places (the curve can only be as good as the data taken to constrain it), the uncertainty of fossil $^{87}Sr/^{86}Sr$ ratios is never defined by analytical precision alone. A crude test for this effect is illustrated in Table 1. Here the median deviation of a datum from the $^{87}Sr/^{86}Sr$ reference

Table 1. *Theoretical and practical age resolution of the Permian $^{87}Sr/^{86}Sr$ curve*

Stage	Average rate of change per myr	Nominal time resolution (Ma)	Median deviation (Ma)	*n*
Changhsingian	0.000109	0.2	0.6	5
Wuchiapingian	0.000029	0.7	2.2	14
Capitanian	0.000032	0.6	2.1	17
Wordian	0.000063	0.3	0.8	3
Roadian	0.000025	0.8	4.8	6
Kungurian	0.000019	1.1	8.7	17
Artinskian	0.000011	1.9	2.3	7
Sakmarian	0.000056	0.4	0.1	6
Asselian	0.000063	0.3	0.3	10
Permian	0.000035	0.6	2.3	85

The nominal age resolution is computed assuming an analytical uncertainty of ± 0.00002. Median deviations of published data and number of published data are from screened brachiopods.

curve (Sr look-up table Version 5: McArthur *et al.* 2001, 2012) for brachiopod samples that have passed screening for diagenesis is given for each Permian stage and the whole period. The resulting deviations are only comparable to the nominal age resolution of the reference curve where it has been constructed from the same samples used here to test its resolution. Otherwise, median age disagreements up to 8 times as large (almost 9 myr for the Kungurian) and 2.3 myr (4 times worse than nominal resolution) for the whole Permian are observed. To counteract this inherent noise in the fossil data, it is therefore not enough to measure single $^{87}Sr/^{86}Sr$ data for age determination. Instead, multiple analyses and site-specific studies of magnitude and direction of isotopic change through post-depositional processes are paramount for robust age estimates.

Geological interpretation of the Permian $^{87}Sr/^{86}Sr$ curve

Contrasting conclusions have been drawn about the $^{87}Sr/^{86}Sr$ evolution of the Permian due to the use of different carbonate archives. Bulk carbonate samples are readily available, for example, from the best-studied P–Tr boundary sequences, and allow determination of the $^{87}Sr/^{86}Sr$ ratio at superior stratigraphic resolution. Their susceptibility for diagenetic alteration, however, often makes it hard to assess data quality objectively. Confidence in $^{87}Sr/^{86}Sr$ ratios of fossil materials can be bolstered by testing preservation through various geochemical and optical techniques. Phosphatic conodonts and low-Mg calcitic brachiopods have therefore played a major role in constraining the Permian $^{87}Sr/^{86}Sr$ curve. Their occurrence, however, is bound to certain depositional environments and, especially during biotic crises, the quality and abundance of fossils deteriorates dramatically. For example, brachiopods can only found on very rare occasions in earliest Triassic strata. Erroneous inferences about past seawater $^{87}Sr/^{86}Sr$ can therefore primarily arise from unconstrained amounts of alteration in bulk rock records, and from data sparsity and minor diagenetic effects in macrofossil records.

The geological interpretation of the marine $^{87}Sr/^{86}Sr$ curve faces the additional problem of an underconstrained system, where flux magnitudes and isotopic ratios of strontium sources to the ocean and the net effect of geological processes on these parameters are not well known. It is, furthermore, doubtful that the marine $^{87}Sr/^{86}Sr$ ratios have ever reached a dynamic equilibrium – a prerequisite for interpreting absolute changes in the $^{87}Sr/^{86}Sr$ curve. Evaluation of the modern situation rather suggests that the Sr cycle is out of equilibrium (e.g. Pearce *et al.* 2015) and it may be instructive to also evaluate relevant forcings leading to changes in the slope of the $^{87}Sr/^{86}Sr$ curve. Explaining the marine $^{87}Sr/^{86}Sr$ in a tectonic or palaeoenvironmental framework therefore most often remains conjecture, but can constitute one of multiple lines of evidence for reconstructing models of past Earth systems.

The Early–Middle Permian trend

Few studies have addressed the geological significance of the Early–Middle Permian decrease in seawater $^{87}Sr/^{86}Sr$. Initially, this decrease was related to enhanced igneous activity (Denison *et al.* 1994). A later interpretation put forward a combination of the Permian–Carboniferous glaciation, Early–Middle Permian aridity and enhanced seafloor spreading related to the opening of the Neotethys (Korte *et al.* 2006). In particular, the change from ‘worldwide’ generally more humid conditions in the Asselian and early Sakmarian – suggested by widespread coal seams in Gondwana and the wet climate in the Euramerian province – to progressively more arid conditions (e.g. 'White Band' of South Africa: Kozur 1984) predominantly during the Artinskian and Kungurian have been pointed out (Korte *et al.* 2006). The associated decrease in chemical weathering and probable reduction in radiogenic riverine Sr fluxes can explain well the evolution of the strontium isotope curve.

The Capitanian $^{87}Sr/^{86}Sr$ minimum

The $^{87}Sr/^{86}Sr$ minimum of the Capitanian has received ample attention in the literature (e.g. Huang *et al.* 2008; Kani *et al.* 2008, 2013; Isozaki 2009; Wignall *et al.* 2009). The Capitanian turning point has been related to the initiation of the Pangaea break-up (Kani *et al.* 2008), a mantle plume (Isozaki 2009), cessation of basaltic volcanism, the humidification of Pangaea (Korte *et al.* 2006; Wignall *et al.* 2009), or deglaciation and continental doming (Kani *et al.* 2013).

Late Permian–Early Triassic trend

The increase of marine $^{87}Sr/^{86}Sr$ ratios commencing in the Capitanian has received attention because it reaches a steepness apparently not matched at any other time in the Phanerozoic (McArthur *et al.* 2012). It has been pointed out, however, that the exact slope is subject to uncertainties in the absolute interval of time recorded by the Late Permian–Early Triassic strata and might, therefore, be overestimated (Denison & Koepnick 1995; McArthur *et al.* 2012; see also Korte *et al.* 2003). The $^{87}Sr/^{86}Sr$ increase has been explained by changes in the Sr cycle, namely a combination of increasing riverine Sr flux, an increasing riverine $^{87}Sr/^{86}Sr$ ratio and a reduction in hydrothermal Sr flux commencing in the Wordian–Capitanian (Martin & Macdougall 1995; see also Morante 1996). The latest Permian destruction of dense land vegetation and associated enhanced continental erosion that lasted until the Anisian has been put forward as an additional influence (Korte *et al.* 2003; Huang *et al.* 2008).

Permian–Triassic boundary

An intermittent retardation of the generally increasing Late Permian–Early Triassic $^{87}Sr/^{86}Sr$ trend at the P–Tr boundary has been suggested on the grounds of Permian brachiopod (Korte *et al.* 2006) and conodont data (Korte *et al.* 2010) (Fig. 4). Whilst its presence is still a matter of debate, a relatively short-lived (a few 100 kyrs) modulation of the marine $^{87}Sr/^{86}Sr$ curve would have potentially far-reaching implications for the latest Permian Earth system behaviour. Such retardation requires temporary changes in the balance of Sr fluxes or in the $^{87}Sr/^{86}Sr$ ratio of riverine runoff. This change in balance could be related to either: (1) a relative decrease in the riverine Sr flux, by a reduced weathering and hydrological cycle or increased mid-ocean ridge activity; or (2) a lowering of the $^{87}Sr/^{86}Sr$ ratio of continental runoff.

A reduction in the continental Sr flux could be related to global sea-level change, altering the contribution of Sr from continental weathering. Global sea-level rise, however, starts no earlier than the late Changhsingian *C. hauschkei* Zone, distinctly later than the anticipated change in slope of the seawater $^{87}Sr/^{86}Sr$ curve. A change in slope of the Sr-isotope curve related to a decrease in the river flux to the oceans is also at odds with sedimentological and geochemical evidence (Twitchett 2007). Geochemical records rather suggest an increase in the river discharge in several regions. The Germanic Basin (Kozur 1998*a*, *b*), Russia (Newell *et al.* 1999), South Africa (Ward *et al.* 2000) and eastern Australia (Michaelsen 2002) all show this effect, but detailed correlation of facies changes in the latter three with the Germanic Basin is not feasible. Twitchett (2007) therefore suggested instead that a change in river water $^{87}Sr/^{86}Sr$ ratios towards less radiogenic Sr may have been generated by enhanced weathering of carbonates or evaporites due to catchment extension. Also, large amounts of strontium from basaltic rocks of the Siberian Traps would have been supplied to the oceans by rapid weathering (Holser & Magaritz 1987; Grard *et al.* 2005). In addition, volcanic aerosols causing acid rain may have accelerated weathering rates (Kozur 1998*a*, *b*; Krassilov & Karasev 2009; Korte & Kozur 2010; Korte *et al.* 2010; see also Visscher *et al.* 2004; Sephton *et al.* 2005; Wignall 2007; Kraus *et al.* 2013; Schobben *et al.* 2014; Sedlacek *et al.* 2014).

These processes, forcing the global continental runoff to less radiogenic $^{87}Sr/^{86}Sr$ ratios, may match temporally and might, therefore, be the cause of the interruption in the seawater $^{87}Sr/^{86}Sr$ increase at the P–Tr boundary (Korte *et al.* 2010). Age data for the Siberian trap volcanism, however, are currently not available at high precision and fidelity (Burgess *et al.* 2014), and it has been suggested that the hypothetical transient interruption of the seawater $^{87}Sr/^{86}Sr$ increase precedes the onset of volcanism (Korte *et al.* 2010). A direct connection between volcanic forcing and the $^{87}Sr/^{86}Sr$ response can therefore only be made ambiguously. In addition, ε_{Nd} values in this interval in records for Pakistan and the USA (Martin & Macdougall 1995) indicate an intensified Nd contribution of old continental crust to the local seawater, which predicts an enhanced flux of radiogenic Sr.

Conclusions

The Permian marine $^{87}Sr/^{86}Sr$ curve is trough-shaped, falling from the Moscovian Stage to the Capitanian Stage and subsequently rising to the Middle Triassic Anisian Stage. The falling limb constitutes the strongest sustained decrease in marine $^{87}Sr/^{86}Sr$ ratios of the Phanerozoic and reaches values close the Phanerozoic minimum. The following increase accelerates in the Late Permian, reaching a rate of increase around the P–Tr boundary that is unique in the Phanerozoic.

The $^{87}Sr/^{86}Sr$ ratios of marine fossils are an important chemostratigraphic tool for the correlation of Permian strata, and a precision of approximately $\pm$ 2 myr can be expected from single analyses for this period.

Geological interpretation of marine $^{87}Sr/^{86}Sr$ ratios through the Permian interval is still problematic owing to partially contradictory published $^{87}Sr/^{86}Sr$ records and the sparsity of available data. Further refinement of the Permian $^{87}Sr/^{86}Sr$ trend is necessary before less ambiguous interpretations of its significance are possible.

We acknowledge the editorial work of Spencer G. Lucas, and comments by Hubert Wierzbowski and Stephen Ruppel that helped to significantly improve the quality of the manuscript. CVU acknowledges funding from the Leopoldina - German National Academy of Sciences (grant No. LPDS 2014-08).

References

Blake, R.E., O'Neil, J.R. & Garcia, G. 1997. Oxygen isotope systematics of biologically mediated reactions of phosphate; I, Microbial degradation of organophosphorus compounds. *Geochimica et Cosmochimica Acta*, **61**, 4411–4422.

Brand, U. 1991. Strontium isotope diagenesis of biogenic aragonite and low-Mg calcite. *Geochimica et Cosmochimica Acta*, **55**, 505–513.

Brand, U. & Veizer, J. 1980. Chemical diagenesis of a multicomponent carbonate system – 1: trace elements. *Journal of Sedimentary Petrology*, **50**, 1219–1236.

Brand, U., Logan, A., Hiller, N. & Richardson, J. 2003. Geochemistry of modern brachiopods: applications and implications for oceanography and paleoceanography. *Chemical Geology*, **198**, 305–334.

Brookins, D.G. 1988. Seawater $^{87}Sr/^{86}Sr$ for the Late Permian Delaware Basin evaporates (New Mexico, U.S.A.). *Chemical Geology*, **69**, 209–214.

Bruckschen, P., Oesmann, S. & Veizer, J. 1999. Isotope stratigraphy of the European Carboniferous: proxy signals for ocean chemistry, climate and tectonics. *Chemical Geology*, **161**, 127–163.

Bryant, J.D., Jones, D.S. & Mueller, P.A. 1995. Influence of freshwater flux on $^{87}Sr/^{86}Sr$ chronostratigraphy in marginal marine environments and dating of vertebrate and invertebrate faunas. *Journal of Paleontology*, **69**, 1–6.

Burgess, S.D., Bowring, S. & Shen, S.-Z. 2014. High-precision timeline for Earth's most severe extinction. *Proceedings of the National Academy of Sciences of the United States of America*, **111**, 3316–3321.

Burke, W.H., Denison, R.E., Hetherington, E.A., Koepnick, R.B., Nelson, H.F. & Otto, J.B. 1982. Variation of seawater $^{87}Sr/^{86}Sr$ throughout Phanerozoic time. *Geology*, **10**, 516–519.

Burla, S., Oberli, F., Heimhofer, U., Wiechert, U. & Weissert, H. 2009. Improved time control on Cretaceous coastal deposits: new results from Sr isotope measurements using laser ablation. *Terra Nova*, **21**, 401–409.

Cao, C.-Q., Love, G.D., Hays, L.E., Wang, W., Shen, S.-Z. & Summons, R.E. 2009. Biogeochemical evidence for euxinic oceans and ecological disturbance presaging the end-Permian mass extinction event. *Earth and Planetary Science Letters*, **281**, 188–201.

Denison, R.E. & Koepnick, R.B. 1995. Variations in $^{87}Sr/^{86}Sr$ of Permian seawater: an overview. *In*: Scholle, P.A., Peryt, T.M. & Ulmer-Scholle, D.S. (eds) *The Permian of Northern Pangea*. Springer, Berlin, 124–132.

Denison, R.E. & Peryt, T.M. 2009. Strontium isotopes in the Zechstein (Upper Permian) anhydrites of Poland: evidence of varied meteoric contributions to marine brines. *Geological Quarterly*, **53**, 159–166.

Denison, R.E., Koepnick, R.B., Burke, W.H., Hetherington, E.A. & Fletcher, A. 1994. Construction of the Mississippian, Pennsylvanian and Permian seawater $^{87}Sr/^{86}Sr$ curve. *Chemical Geology*, **112**, 146–167.

Denison, R.E., Miller, N.R., Scott, R.W. & Reaser, D.F. 2003. Strontium isotope stratigraphy of the Comanchean Series in north Texas and southern Oklahoma. *Geological Society of America Bulletin*, **115**, 669–682.

DePaolo, D.J. & Ingram, B.L. 1985. High-resolution stratigraphy with strontium isotopes. *Science*, **227**, 938–941.

Diener, A., Ebneth, S., Veizer, J. & Buhl, D. 1996. Strontium isotope stratigraphy of the middle Devonian: brachiopods and conodonts. *Geochimica et Cosmochimia Acta*, **60**, 639–652.

Ebneth, S., Diener, A., Buhl, D. & Veizer, J. 1997. Strontium isotope systematics of conodonts: Middle Devonian, Eifel Mountains, Germany. *Palaeogeography, Palaeoclimatology, Palaeoecology*, **132**, 79–96.

Ehrenberg, S.N., McArthur, J.M. & Thirwall, M.F. 2010. Strontium isotope dating of spiculitic Permian strata from Spitsbergen outcrops and Barents Sea well-cores. *Journal of Petroleum Geology*, **33**, 247–254.

Elderfield, H. 1986. Strontium isotope stratigraphy. *Palaeogeography, Palaeoclimatology, Palaeoecology*, **57**, 71–90.

Fischer, J., Schneider, J.W. *et al.* 2014. Stable and radiogenic isotope analyses on shark teeth from the Early to the Middle Permian (Sakmarian–Roadian) of the southwestern USA. *Historical Biology*, **26**, 710–727.

Frimmel, H.E.E. & Niedermayr, G. 1991. Strontium isotopes in magnesites from Permian and Triassic strata, Eastern Alps. *Applied Geochemistry*, **6**, 89–96.

GRADSTEIN, F.M., OGG, G. & SCHMITZ, M. 2012. *The Geologic Time Scale 2012, Volumes 1 and 2*. Elsevier, Amsterdam.

GRARD, A., FRANÇOIS, L.M., DESSERT, C., DUPRÉ, B. & GODDERIS, Y. 2005. Basaltic volcanism and mass extinction at the Permo-Triassic boundary: environmental impact and modelling of the global carbon cycle. *Earth and Planetary Science Letters*, **234**, 207–221.

GRUSZCZYNSKI, M., HOFFMAN, A., MAŁKOWSKI, K. & VEIZER, J. 1992. Seawater strontium isotopic perturbations at the Permian-Triassic boundary, West Spitsbergen, and its implications for the interpretation of strontium isotopic data. *Geology*, **20**, 779–782.

HANS, U., KLEINE, T. & BOURDON, B. 2013. Rb–Sr chronology of volatile depletion in differentiated protoplanets: BABI, ADOR and ALL revisited. *Earth and Planetary Science Letters*, **374**, 204–214.

HENDERSON, C.M., DAVYDOV, V.I. & WARDLAW, B.R. 2012. The Permian period. *In*: GRADSTEIN, F.M., OGG, G. & SCHMITZ, M. (eds) *The Geologic Time Scale 2012, Volume 2*. Elsevier, Amsterdam, 653–679.

HODELL, D.A., MUELLER, P.A., MCKENZIE, J.A. & MEAD, G.A. 1989. Strontium isotope stratigraphy and geochemistry of the late Neogene ocean. *Earth and Planetary Science Letters*, **92**, 165–178.

HODELL, D.A., MUELLER, P.A. & GARRIDO, J.R. 1991. Variations in the strontium isotopic composition of seawater during the Neogene. *Geology*, **19**, 24–27.

HOLSER, W.T. & MAGARITZ, M. 1987. Events near the Permian–Triassic boundary. *Modern Geology*, **11**, 155–180.

HOWARTH, R.J. & MCARTHUR, J.M. 1997. Statistics for strontium isotope stratigraphy: a robust LOWESS fit to the marine Sr-isotope curve for 0 to 206 Ma, with look-up table for derivation of numeric age. *Journal of Geology*, **105**, 441–456.

HUANG, S.-J., QING, H.-R., HUANG, P.-P., HU, Z.-W., WANG, Q.-D., ZOU, M.-L. & LIU, H.-N. 2008. Evolution of strontium isotopic composition of seawater from Late Permian to Early Triassic based on study of marine carbonates, Zhongliang Mountain, Chongqing, China. *Science in China Series D: Earth Sciences*, **51**, 528–539.

ISOZAKI, Y. 2009. Illawarra Reversal: the fingerprint of a superplume that triggered Pangean breakup and the end-Guadalupian (Permian) mass exctinction. *Gondwana Research*, **15**, 421–432.

JENKYNS, H.C., JONES, C.E., GRÖCKE, D.R., HESSELBO, S.P. & PARKINSON, D.N. 2002. Chemostratigraphy of the Jurassic System: applications, limitations and implications of palaeoceanography. *Journal of the Geological Society, London*, **159**, 351–378, https://doi.org/10.1144/0016-764901-130

JONES, C.E., JENKYNS, H.C., COE, A.L. & HESSELBO, S.P. 1994. Strontium isotopic variations in Jurassic and Cretaceous seawater. *Geochimica et Cosmochimia Acta*, **58**, 3061–3074.

KAIHO, K., KAJIWARA, Y. *ET AL*. 2001. End-Permian catastrophe by a bolide impact: evidence of a gigantic release of sulfur from the mantle. *Geology*, **29**, 815–818.

KAMPSCHULTE, A., BUHL, D. & STRAUSS, H. 1998. The sulfur and strontium isotopic compositions of Permian evaporates from the Zechstein basin, northern Germany. *Geologische Rundschau*, **87**, 192–199.

KANI, T., FUKUI, M., ISOZAKI, Y. & NOHDA, S. 2008. The Paleozoic minimum of $^{87}Sr/^{86}Sr$ ratio in the Capitanian (Permian) mid-oceanic carbonates: a critical turning point in the Late Paleozoic. *Journal of Asian Earth Sciences*, **32**, 22–33.

KANI, T., HISANABE, C. & ISOZAKI, Y. 2013. The Capitanian (Permian) minimum of $^{87}Sr/^{86}Sr$ ratio in the mid-Panthalassan paleo-atoll carbonates and its demise by deglatiation and continental doming. *Gondwana Research*, **24**, 212–221.

KORTE, C. & KOZUR, H.W. 2010. Carbon-isotope stratigraphy across the Permian–Triassic boundary: a review. *Journal of Asian Earth Sciences*, **39**, 215–235.

KORTE, C., KOZUR, H.W., BRUCKSCHEN, P. & VEIZER, J. 2003. Strontium isotope evolution of Late Permian and Triassic seawater. *Geochimica et Cosmochimica Acta*, **67**, 47–62.

KORTE, C., KOZUR, H.W., JOACHIMSKI, M.M., STRAUSS, H., VEIZER, J. & SCHARK, L. 2004. Carbon, sulfur, oxygen and strontium isotope records, organic geochemistry and biostratigraphy across the Permian/Triassic boundary in Abadeh, Iran. *International Journal of Earth Sciences*, **93**, 565–581.

KORTE, C., JASPER, T., KOZUR, H.W. & VEIZER, J. 2006. $^{87}Sr/^{86}Sr$ record of Permian seawater. *Palaeogeography, Palaeoclimatology, Palaeoecology*, **240**, 89–107.

KORTE, C., PANDE, P., KALIA, P., KOZUR, H.W., JOACHIMSKI, M.M. & OBERHÄNSLI, H. 2010. Massive volcanism at the Permian-Triassic boundary and its impact on the isotopic composition of the ocean and atmosphere. *Journal of Asian Earth Sciences*, **37**, 293–311.

KOZUR, H. 1984. Perm. *In*: TRÖGER, K.-A. (ed.) *Abriß der Historischen Geologie*. Akademie, Berlin, 270–307.

KOZUR, H.W. 1998*a*. Some aspects of the Permian–Triassic boundary (PTB) and of the possible causes for the biotic crisis around this boundary. *Palaeogeography, Palaeoclimatology, Palaeoecology*, **143**, 227–272.

KOZUR, H.W. 1998*b*. Problems for evaluations of the scenario of the Permian–Triassic boundary biotic crisis and of its causes. *Geologia Croatica*, **51**, 135–162.

KRALIK, M. 1991. The Permian–Triassic of the Gartnerkofel-1 core (Carnic Alps, Austria): strontium isotopes and carbonate chemistry. *Abhandlungen der Geologischen Bundesanstalt in Wien*, **45**, 169–174.

KRAMM, U. & WEDEPOHL, K.H. 1991. The isotopic composition of strontium and sulfur in seawater of Late Permian (Zechstein) age. *Chemical Geology*, **90**, 253–262.

KRASSILOV, V. & KARASEV, E. 2009. Paleofloristic evidence of climate change near and beyond the Permian-Triassic boundary. *Palaeogeography, Palaeoclimatology, Palaeoecology*, **284**, 326–336.

KRAUS, S.H., BRANDNER, R., HEUBECK, C., KOZUR, H.W., STRUCK, U. & KORTE, C. 2013. Carbon isotope signatures of latest Permian marine successions of the Southern Alps suggest a continental runoff pulse enriched in land plant material. *Fossil Record*, **16**, 97–109.

LAYA, J.C., TUCKER, M.E., GRÖCKE, D.R. & PEREZ-HUERTA, A. 2013. Carbon, oxygen and strontium isotopic composition of low-latitude Permian carbonates

(Venezuelan Andes): climate proxies of tropical Pangea. *In*: GĄSIEWICZ, A. & SŁOWAKIEWICZ, M. (eds) *Palaeozoic Climate Cycles: Their Evolutionary and Sedimentological Impact*. Geological Society, London, Special Publications, **376**, 367–385, https://doi.org/10.1144/SP376.10

LI, Y.-H. 1982. A brief discussion on the mean oceanic residence time of elements. *Geochimica et Cosmochimica Acta*, **46**, 2671–2675.

LIU, X.-c., WANG, W. *ET AL.* 2013. Late Guadalupian to Lopingian (Permian) carbon and strontium isotopic chemostratigraphy in the Abadeh section, central Iran. *Gondwana Research*, **24**, 222–232.

MARTIN, E.E. & MACDOUGALL, J.D. 1995. Sr and Nd isotopes at the Permian/Triassic boundary: a record of climate change. *Chemical Geology*, **125**, 73–99.

MCARTHUR, J.M. 1994. Recent trends in strontium isotope stratigraphy. *Terra Nova*, **6**, 331–358.

MCARTHUR, J.M., HOWARTH, R.J. & BAILEY, T.R. 2001. Strontium isotope stratigraphy: LOWESS version 3: best fit to the marine Sr-isotope curve for 0–509 Ma and accompanying look-up table for deriving numerical age. *Journal of Geology*, **109**, 155–170.

MCARTHUR, J.M., HOWARTH, R.J. & SHIELDS, G.A. 2012. Strontium isotope stratigraphy. *In*: GRADSTEIN, F.M., OGG, J.G., SCHMITZ, M. & OGG, G. (eds) *The Geologic Time Scale, Volume 1*. Elsevier, Amsterdam, 127–144.

METCALFE, I. 2011. Tectonic framework and Phanerozoic evolution of Sundaland. *Gondwana Research*, **19**, 3–21.

MICHAELSEN, P. 2002. Mass extinction of peat-formation plants and the effect on fluvial styles across the Permian–Triassic boundary, northern Bowen Basin, Australia. *Palaeogeography, Palaeoclimatology, Palaeoecology*, **179**, 173–188.

MIURA, N., ASAHARA, Y. & KAWABE, I. 2004. Rare earth element and Sr isotopic study of the Middle Permian limestone–dolostone sequence in Kuzuu area, central Japan: seawater tetrad effect and Sr isotopic signatures of seamount-type carbonate rocks. *Journal of Earth and Planetary Sciences, Nagoya University*, **51**, 11–35.

MORANTE, R. 1996. Permian and Early Triassic isotopic records of carbon and strontium in Australia and a scenario of events about the Permian-Triassic boundary. *Historical Biology*, **11**, 289–310.

NEWELL, A.J., TVERDOKHLEBOV, V.P. & BENTON, M.J. 1999. Interplay of tectonics and climate on a transverse fluvial system, Upper Permian, Southern Uralian Foreland Basin, Russia. *Sedimentary Geology*, **127**, 11–29.

NEYMARK, L.A., PREMO, W.R., MEL'NIKOV, N.N. & EMSBO, P. 2014. Precise determination of δ^{88}Sr in rocks, minerals, and waters by double-spike TIMS: a powerful tool in the study of geological, hydrological and biological processes. *Journal of Analytical Atomic Spectrometry*, **29**, 65–75.

NIER, A.O. 1938. The isotopic constitution of strontium, barium, bismuth, thallium and mercury. *Physical Review*, **54**, 275–278.

NISHIOKA, S., ARAKAWA, Y. & KOBAYASHI, Y. 1991. Strontium isotope profile of Carboniferous–Permian Akiyoshi Limestone in southwest Japan. *Geochemical Journal*, **25**, 137–146.

NURGALIEVA, N.G., PONOMARCHUK, V.A. & NURGALIEV, D.K. 2007. Strontium isotope stratigraphy: possible applications for age estimation and global correlation of Late Permian carbonates of the Pechishchi type section, Volga River. *Russian Journal of Earth Sciences*, **9**, ES1002.

PALMER, M.R. & EDMOND, J.M. 1989. The strontium isotopic budget of the modern ocean. *Earth and Planetary Science Letters*, **92**, 11–26.

PALMER, M.R. & ELDERFIELD, H. 1985. Sr isotope composition of sea water over the past 75 Myr. *Nature*, **314**, 526–528.

PEARCE, C.R., PARKINSON, I.J., GAILLARDET, J., CHARLIER, B.L.A., MOKADEM, F. & BURTON, K.W. 2015. Reassessing the stable ($\delta^{88/86}$Sr) and radiogenic (^{87}Sr/^{86}Sr) strontium isotopic composition of marine inputs. *Geochimica et Cosmochimica Acta*, **157**, 125–146.

PETERMAN, Z.E., HEDGE, C.E. & TOURTELOT, H.A. 1970. Isotopic composition of strontium in sea water throughout Phanerozoic time. *Geochimica et Cosmochimica Acta*, **34**, 105–120.

POPP, B.N., ANDERSON, T.F. & SANDBERG, P.A. 1986. Brachiopods as indicators of original isotopic compositions in some Paleozoic limestones. *Geological Society of America Bulletin*, **97**, 1262–1269.

RASMUSSEN, C.M.Ø., ULLMANN, C.V. *ET AL.* 2016. Onset of main Phanerozoic marine radiation sparked by emerging Mid Ordovician icehouse. *Scientific Reports*, **6**, 18884.

SCHOBBEN, M., JOACHIMSKI, M.M., KORN, D., LEDA, L. & KORTE, C. 2014. Palaeotethys seawater temperature rise and an intensified hydrological cycle following the end-Permian mass extinction. *Gondwana Research*, **26**, 675–683.

SEDLACEK, A.R.C., SALTZMAN, M.R., ALGEO, T.J., HORACEK, M., BRANDNER, R., FOLAND, K. & DENNISTON, R.F. 2014. ^{87}Sr/^{86}Sr stratigraphy from the Early Triassic of Zal, Iran: linking temperature to weathering rates and the tempo of ecosystem recovery. *Geology*, **42**, 779–782.

SEPHTON, M.A., LOOY, C.V., BRINKHUIS, H., WIGNALL, P.B., DE LEEUW, J.W. & VISSCHER, H. 2005. Catastrophic soil erosion during the end-Permian biotic crisis. *Geology*, **33**, 941–944.

SHARMA, M., BALAKRISHNA, K., HOFMANN, A.W. & SHANKAR, R. 2007. The transport of Osmium and Strontium isotopes through a tropical estuary. *Geochimica et Cosmochimica Acta*, **71**, 4856–4867.

SHEN, Shu-zhong, HENDERSON, C.M. *ET AL.* 2010. High-resolution Lopingian (Late Permian) timescale of South China. *Geological Journal*, **45**, 122–134.

SHIELDS, G.A., CARDEN, G.A.F., VEIZER, J., MEIDLA, T., RONG, J.-Y. & LI, R.-Y. 2003. Sr, C, and O isotope geochemistry of Ordovician brachiopods: a major isotopic event around the Middle-Late Ordovician transition. *Geochimica et Cosmochimica Acta*, **67**, 2005–2025.

SPÖTL, C. & PAK, E. 1996. A strontium and sulfur isotopic study of Permo-Triassic evaporates in the Northern Calcareous Alps, Austria. *Chemical Geology*, **131**, 219–234.

STEIGER, R.H. & JÄGER, E. 1977. Subcommission on geochronology: convention on the use of decay constants in geo- and cosmochronology. *Earth and Planetary Science Letters*, **36**, 359–362.

Stemmerik, L., Bendix-Almgreen, S.E. & Piasecki, S. 2001. The Permian-Triassic boundary in central East Greenland: past and present views. *Bulletin of the Geological Society of Denmark*, **48**, 159–167.

Stephenson, M.H., Angiolini, L., Leng, M.J. & Darbyshire, F. 2012. Geochemistry, and carbon, oxygen and strontium isotope composition of brachiopods from the Khuff Formation of Oman and Saudi Arabia. *GeoArabia*, **17**, 61–76.

Steuber, T. 2001. Strontium isotope stratigraphy of Turonian–Campanian Gosau-type rudist formations in the Northern Calcareous and Central Alps (Austria and Germany). *Cretaceous Research*, **22**, 429–441.

Steuber, T. 2003. Strontium isotope stratigraphy of Cretaceous hippuritid rudist bivalves: rates of morphological change and heterochronic evolution. *Palaeogeography, Palaeoclimatology, Palaeoecology*, **200**, 221–243.

Steuber, T. & Schlüter, M. 2012. Strontium-isotope stratigraphy of Upper Cretaceous rudist bivalves: Biozones, evolutionary patterns and sea-level change calibrated to numerical ages. *Earth-Science Reviews*, **114**, 42–60.

Taylor, A.S. & Lasaga, A.C. 1999. The role of basalt weathering in the Sr isotope budget of the oceans. *Chemical Geology*, **161**, 199–214.

Tierney, K.E. 2010. *Carbon and strontium isotope stratigraphy of the Permian from Nevada and China: Implications from an icehouse to greenhouse transition*. PhD thesis, Ohio State University.

Twitchett, R.J. 2007. Climate change across the Permian/Triassic boundary. *In*: Williams, M., Haywood, A.M., Gregory, F.J. & Schmidt, D.N. (eds) *Deep-Time Perspective on Climate Change: Marrying the Signal from Computer Models and Biological Proxies*. The Micropalaeontological Society, Special Publications. Geological Society, London, 191–200.

Ullmann, C.V. & Korte, C. 2015. Diagenetic alteration in low-Mg calcite from macrofossils: a review. *Geological Quarterly*, **59**, 3–20.

Ullmann, C.V., Campbell, H.J., Frei, R., Hesselbo, S.P., Pogge von Strandmann, P.A.E. & Korte, C. 2013. Partial diagenetic overprint of Late Jurassic belemnites from New Zealand: implications for the preservation potential of δ^7Li values in calcite fossils. *Geochimica et Cosmochimica Acta*, **120**, 80–96.

Ullmann, C.V., Campbell, H.J., Frei, R. & Korte, C. 2014. Geochemical signatures in Late Triassic brachiopods from New Caledonia. *New Zealand Journal of Geology and Geophysics*, **57**, 420–431.

Ullmann, C.V., Frei, R., Korte, C. & Hesselbo, S.P. 2015. Chemical and isotopic architecture of the belemnite rostrum. *Geochimica et Cosmochimia Acta*, **159**, 231–243.

Veizer, J. 1989. Strontium isotopes in seawater through time. *Annual Reviews in Earth and Planetary Sciences*, **17**, 141–167.

Veizer, J. & Compston, W. 1974. ^{87}Sr/^{86}Sr composition of seawater during the Phanerozoic. *Geochimica et Cosmochimica Acta*, **38**, 1461–1484.

Veizer, J., Buhl, D. et al. 1997. Strontium isotope stratigraphy: potential resolution and event correlation. *Palaeogeography, Palaeoclimatology, Palaeoecology*, **132**, 65–77.

Veizer, J., Ala, D. et al. 1999. ^{87}Sr/^{86}Sr, δ^{13}C and δ^{18}O evolution of Phanerozoic seawater. *Chemical Geology*, **161**, 59–88.

Visscher, H., Looy, C.V., Collinson, M.J., Brinkhuis, H., Konijnenburg-van Cittert, J.H.A., Kürschner, W.M. & Sephton, M.A. 2004. Environmental mutagenesis during the end-Permian ecological crisis. *Proceedings of the National Academy of Sciences of the United States of America*, **101**, 12952–12956.

Ward, P.D., Montgomery, D.R. & Smith, R. 2000. Altered river morphology in South Africa related to the Permian–Triassic Extinction. *Science*, **289**, 1740–1743.

Wefer, G. & Berger, W.H. 1991. Isotope paleontology: growth and composition of extant calcareous species. *Marine Geology*, **100**, 207–248.

Wierzbowski, H., Anczkiewicz, R., Bazarnik, J. & Pawlak, J. 2012. Strontium isotope variations in Middle Jurassic (Late Bajocian–Callovian) seawater: implications for Earth's tectonic activity and marine environments. *Chemical Geology*, **334**, 171–181.

Wignall, P.B. 2007. The end-Permian mass extinction – how bad did it get? *Geobiology*, **5**, 303–309.

Wignall, P.B., Védrine, S., Bond, D.P.G., Wang, W., Lai, X.-L., Ali, J.R. & Jiang, H.-S. 2009. Facies analysis and sea-level change at the Guadalupian–Lopingian Global Stratotype (Laibin, South China), and its bearing on the end-Guadalupian mass extinction. *Journal of the Geological Society, London*, **166**, 655– 666, https://doi.org/10.1144/0016-76492008-118

Young, E.D. & Galy, A. 2004. The isotope geochemistry and cosmochemistry of magnesium. *Reviews in Mineralogy and Geochemistry*, **55**, 197–230.

Zazzo, A., Lécuyer, C. & Mariotti, A. 2004. Experimentally-controlled carbon and oxygen isotope exchange between bioapatites and water under inorganic and microbially mediated conditions. *Geochimica et Cosmochimica Acta*, **68**, 1–12.

Permian conodont biostratigraphy

CHARLES M. HENDERSON

Department of Geoscience, University of Calgary, Calgary, Alberta, Canada T2N 1N4
cmhender@ucalgary.ca

Abstract: Forty Permian conodont biozones are summarized and 35 additional regional zones are correlated. The Lower Permian is largely zoned on the basis of partial range lineage interval zones of species of *Streptognathodus*, *Sweetognathus*, *Neostreptognathodus* and *Mesogondolella*. The Middle and Upper Permian are zoned, respectively, on the basis of partial range lineage interval zones of species of *Jinogondolella* and *Clarkina*. The ranges of all key taxa to develop this zonation are depicted in relation to the geochronological ages of Permian stages. The lowest Permian succession has been astronomically tuned with 400 kyr cyclothems linked to interpolated ages in the Uralian succession.

This paper reviews the current state of conodont biostratigraphy for the entire Permian System. The intention is to provide a biozonation summary, and to discuss some of the problems and successes of correlation on a global basis. The biozonation (Fig. 1) provides a template to test correlation using other important fossil groups, such as brachiopods, fusulinaceans and ammonoids, which will be a difficult task given the differing controls on their distribution (see Henderson & Mei 2003). This has not been done here, since the goal of this paper was to provide an independent zonation tied to key sections and geochronological ages. The zonal scheme can also serve as a template to test correlation using other stratigraphic tools, including sequence stratigraphy, astronomical tuning, isotope stratigraphy, magnetostratigraphy and geochronological age data.

The utility of Permian conodonts for biostratigraphy was established by a number of early studies (Youngquist *et al.* 1951; Clark & Behnken 1971; Behnken 1975; Perlmutter 1975; Kozur 1978; Movshovich *et al.* 1979; Igo 1981; Wang & Wang 1981) and later summarized by Kozur *et al.* (1995). Most of these studies involved sections with poorly documented location data and only a few illustrations of the microfossils. Today, with GPS and digital scanning electron microscopy (SEM) capabilities, locations can be accurately duplicated for testing and large numbers of specimens can be illustrated.

The temporal ranges of conodont species has proven to be vital for the definition of many Palaeozoic Global Stratotype Sections and Points (GSSPs) and this is exemplified in the Permian in which all nine stages are, or will soon, be recognized using the first appearance datum (FAD) of conodont species as the primary signal. Correlation is not a matter of matching up local first occurrences (FO), but rather a complex task utilizing all available means (Henderson 2006), including detailed taxonomic assessments. It is therefore important to review how Permian conodont species have been defined and how evolutionary relationships have been recognized, as well as recommending potential steps forwards to improve methodology.

In fact, so much effort has been expended on establishing the GSSPs that the zonation between points has been somewhat neglected, which has added to the challenge of constructing the biozonation scheme for the entire Permian (Fig. 1). The scheme includes 40 standard international biozones and 35 additional regional biozones. Many biozones are well established and, in some cases, linked to a geochronological framework, whereas others remain poorly constrained and difficult to correlate. The determined effort of the International Commission on Stratigraphy (ICS) and the editors of the *Geologic Time Scale* volumes (Gradstein *et al.* 2012) to complete stage-level definitions has led to significant advancements in the field of stratigraphy. It is important that this GSSP process be completed so that investigations can focus on the entire history of each system, such as the Permian (Jin *et al.* 1994), and not just on boundaries. In the following discussion, some gaps will be identified where further collections are particularly needed.

Conodont species definition

One of the advantages of using conodonts for biostratigraphy is that most samples from marine successions will yield at least some specimens. The disadvantage is that recovery is unknown until substantial processing of the rock has been completed. Processing involves the dissolution of carbonate rock in a 10% solution of acetic or formic acid, or disaggregation of shale (physical or boiled in Quaternary O detergent) followed by sieving (retaining at least all sand-sized >75 μm). The residue is often boiled in bleach to remove any organic fibres

From: Lucas, S. G. & Shen, S. Z. (eds) 2018. *The Permian Timescale*. Geological Society, London, Special Publications, **450**, 119–142.
First published online December 14, 2016, https://doi.org/10.1144/SP450.9

Series	Stage	Base age	Ma	International Zonation	Correlative Regional Zonation
	Triassic / Induan		250	*Hindeodus parvus*; 3. *H. prae.-H.chang.*; 2. *C. mei.-C. zhe.*; 1. *C. yini*	*C. taylorae*
Lopingian	Changhsingian	251.902	252	1 2 3	*C. hauschkei*
Lopingian	Changhsingian			*C. changxingensis*	*M. sheni*
Lopingian	Wuchiapingian	254.14	254	*C. wangi* — *C. subcarinata*	*M. rosenkrantzi*
Lopingian	Wuchiapingian			*C. longicuspidata*; *I. tarazi*	
Lopingian	Wuchiapingian		256	*Clarkina orientalis*	
Lopingian	Wuchiapingian			*C. transcaucasica*	
Lopingian	Wuchiapingian		258	*C. guangyuanensis*; *C. leveni*	
Lopingian	Wuchiapingian			*C. postbitteri*, *C. dukouensis*, *C. asymmetrica*; *I. spp.*	
Guadalupian	Capitanian	259.1		*C. hongshuiensis*	*M. bitteri*
Guadalupian	Capitanian		260	*J. granti*; *Sw. feng.*	*M. omanensis*
Guadalupian	Capitanian			*J. xuanhanensis*; *Sw. hanzhongensis*	
Guadalupian	Capitanian		262	*J. prexuanhanensis*	
Guadalupian	Capitanian			*J. altudaensis*	
Guadalupian	Capitanian		264	*J. shannoni*	
Guadalupian	Capitanian			*J. postserrata*	
Guadalupian	Wordian	265.1	266		*M. wilcoxi*
Guadalupian	Wordian		268	*Jinogondolella aserrata*	*M. phosphoriensis - M. prolongata*
Guadalupian	Roadian	268.8	270	*Jinogondolella nankingensis*; *Sw. subsymmetricus*	*N. newelli*; *M. gracilis*; *M. siciliensis*
Cisuralian	Roadian		272		
Cisuralian	Kungurian	272.3		*N. sulcoplicatus*; *Mesogondolella lamberti*	*M. idahoensis*
Cisuralian	Kungurian		274	*N. prayi*	
Cisuralian	Kungurian		276		
Cisuralian	Kungurian		278	*N. clinei*	*M. zsuzsannae*
Cisuralian	Kungurian		280	*Sw. guizhouensis*	*Uraloceras cochleatus*
Cisuralian	Kungurian		282	*N. pnevi*	*M. glenisteri*
Cisuralian	Kungurian				*M. intermedia - M. asiatica*
Cisuralian	Artinskian	283.5	284	*Neostreptognathodus pequopensis*	*M. gugioensis - M. intermedia*
Cisuralian	Artinskian		286	*Sw. clarki*	*M. laevigata*
Cisuralian	Artinskian		288	*Sw. "whitei"*; *Sw. behnkeni*	*Mesogondolella bisselli*
Cisuralian	Sakmarian	290.1	290	*Sw. anceps*	
Cisuralian	Sakmarian		292	??	??
Cisuralian	Sakmarian			*Sweetognathus binodosus*	*M. longifoliosa - M obliquimarginatus - M. manifesta - M. monstra*; *Sw. inornatus*
Cisuralian	Sakmarian		294	*Sw. 'merrilli'*; *S. florensis*	*M. uralensis*; *Sw. bucaramangus*; *Sw. whitei*
Cisuralian	Asselian	295.0		*Streptognathodus postfusus - S. barskovi*	*M. striata - M. simulata - M. camilla - M. arcuata - M. pseudostriata*; *Sw. merrilli*
Cisuralian	Asselian		296	*Streptognathodus fusus*	*M. belladontae - M. dentiseparata*
Cisuralian	Asselian			*Streptognathodus constrictus*	
Cisuralian	Asselian		298	*S. isolatus*, *S. glenisteri*, *S. cristellaris*, *S. sigmoidalis*	*Sw. expansus*
	/ Gzhelian Carboniferous	298.9	300	*Streptognathodus wabaunsensis*	

or hydrocarbons and then sieved again. The clean residue is dried and then separated in a heavy liquid at a specific gravity of 2.85 (tetrabromoethane or sodium polytungstate). The conodonts are hand-picked from the heavy fraction using a wet fine brush or single bristle with a static charge (elements have a specific gravity of 2.9–3.0). Selected specimens are then photographed using SEM.

Many samples yield numerous specimens that typically exhibit the growth series because conodont elements grow during the lifetime of an individual conodont animal. The conodont animal (see Briggs *et al.* 1983) had numerous shaped elements constituting an apparatus: for example, ozarkodinids, which include the majority of Permian biostratigraphically useful genera, had 15 elements. Many samples are over-represented in terms of the platform elements, suggesting some degree of sorting on the seafloor prior to burial. Other samples show the full complement of elements in the anticipated numbers. Typically, the platform elements or P1 elements (sometimes called the Pa element in older literature) are described at the species level because these elements have the greatest range of morphological variability. They are typically arranged in bilateral symmetry and are described as left and right (or sinistral and dextral elements). The other 13 elements (six pairs and one symmetrical element) collectively can be useful in discriminating families and sometimes genera, and therefore could be used to discriminate potential homeomorphic development of the platform elements.

In a given sample, if conodonts are present, there may be hundreds or even thousands of specimens, typically ranging from 0 to 1000 per kg. The platform elements will have variable sizes and significant variation in ornament, making it difficult to determine the degree of variation necessary to discriminate species. It is fair to say that many species determined by experts may prove difficult to discriminate by non-experts. This may, in part, be related to descriptions that incompletely represent the morphological variability at a given site or fail to encompass the full range of geographical variability. Some species have been recognized as rare morphotypes or even as juvenile or gerontic forms in a sample and, unfortunately, these may be more easily discriminated by the non-expert. However, such 'species' will have long ranges, diminishing their value for biostratigraphy.

The fact that conodonts come in large numbers, exhibit growth series and display significant morphological variability demands a population approach. Mei *et al.* (2004) described the sample population concept that has proven especially important in discriminating Late Permian species and should be employed at all levels. The basic philosophy of this approach is to initially consider that a sample has a single species, and then to arrange the specimens into a growth series focusing primarily on the most consistent and stable morphological characteristics. In so doing, more than one growth series or species may become apparent. This approach mentions, but would not normally give importance to, rare unusual forms, especially if these are seen in many samples in a sequence. Typically, the juvenile forms of species within a given genus will appear very similar. Gerontic forms often have unusual changes to the posterior carina or overall platform thickening. Anyone familiar with Permian conodont biostratigraphy will know of the many ongoing debates on taxonomy and correlation. There are too many to discuss in this summary paper. Species have been discriminated by 'lumpers' and 'splitters', and some employ 'form' taxonomy, but the sample population concept is the most consistent method for species discrimination; it has been most successfully used for Lopingian species, but rarely used so far for Cisuralian species. The following represents my paradigm of how these Permian taxa are identified and correlated. The foundation for all of this work, regardless of the methods employed, will always be taxonomy.

Conodont evolution recognition

The reason why conodonts are especially valuable for biostratigraphy is that multiple closely spaced samples will yield numerous specimens in which morphological variation can be seen to change. If the same pattern is recognized at additional sites, it is increasingly safe to assume that the changes are a result of evolution and not merely ecophenotypic. In the case of GSSP definition (e.g. Jin *et al.* 2006*a*, *b*), the interval in question will have been continuously sampled, with samples typically 10 cm thick, depending on depositional rates and bed thickness. It is true that even such narrowly sampled intervals still time-average hundreds of

Fig. 1. Schematic framework for Permian conodont biozones. There are 40 zones described as international in scope, although only some can be traced globally. Thirty-five additional correlative zones are of regional scope because of endemism or provincialism. *S*, *Streptognathodus*; *Sw*, *Sweetognathus*; *M*, *Mesogondolella*; *N*, *Neostreptognathodus*; *J*, *Jinogondolella*; *C*, *Clarkina*; *H*, *Hindeodus*; *I*, *Iranognathus*. Zones *C.mei.-C.zhe.*, *Clarkina meishanensis–C. zhejiangensis*; *H. prae.- H. chang.*, *Hindeodus praeparvus–H. changxingensis*; *Sw. feng.*, *Sw. fengshanensis*; I.spp., several species of *Iranognathus* (see Fig. 3).

generations, and intermediate forms can be recognized within the speciation interval. An estimate of speciation rate can be made for the Changhsingian at Meishan, where high-resolution age dates are available (Jin *et al.* 2006*b*); 10 cm in the *Clarkina longicuspidata–C. subcarinata* interval represents, on average, about 6 kyr. Transitional forms between *Clarkina longicuspidata* and *C. wangi* are difficult to distinguish over a 50 cm-thick interval. Assuming selection is acting on the population and increasing the proportion of certain traits over that 50 cm, it is possible to calculate 30 kyr and possibly 15 000 generations in the speciation process. In these cases, evolution can essentially be 'seen' to occur, and various statistical techniques, like unitary associations, are not necessary and have been rarely employed by Permian conodont workers. Such statistical techniques are particularly useful for taxa that show less continuous, more disparate occurrences in the fossil record (Angiolini & Bucher 1999). The techniques may also be valuable for conodonts, especially in areas where occurrences are more disparate or where conodonts are less common, outside their optimal geographical ranges. For example, in the highly punctuated record in western Canada, only 16 zones are recognized spanning the full Permian (Zubin-Stathopoulos *et al.* 2013). In my view, statistical techniques, including unitary associations and CONOP9 (Sadler 2004), are valuable as a means of testing correlations determined by traditional methods, rather than as a replacement for the determination of biozones (see the later discussion of the Permian–Triassic boundary interval). CONOP9 is also a great tool for evaluating fossil ranges and integrating geochronological ages, and will provide a framework for estimating evolutionary rates or pointing out intervals in which we may have been conservative in our efforts to describe species. Davydov (in Henderson *et al.* 2012*a*) provided a table that includes a composite of geochronological ages and CONOP9-extended first occurrences of species: the ranges depicted in the figures in this paper honour these values. One problem with the age model for the Permian is that datable ashes are far more common in the lower and uppermost parts, and rare in the middle, meaning that extrapolations may be less accurate in the Middle Permian.

The Permian is a difficult system to zone because, for much of its 47 myr, provincialism (especially from the Kungurian to mid-Changhsingian) affected the geographical distribution of taxa (Mei & Henderson 2001). In such cases, geographical clines have been employed to link geographical provinces (Henderson & Mei 2007).

Furthermore, the majority of recent research has focused on GSSP definitions, which has limited the global testing of zonal schemes between those boundaries. The utility of species to define biozones at stage boundaries has been tested in different areas, but intra-stage biozones have rarely been tested. Some correlation issues will be discussed in the following zone-by-zone descriptions.

Lower Permian zones and key species

The Lower Permian (Cisuralian Series) conodont zonation is derived from a variety of published and some unpublished theses. Two important issues that focus research in this interval are the timing of Gondwana glaciation termination and the extent of the Kungurian sea-level lowstand.

Asselian–Lower Sakmarian

The Asselian and Lower Sakmarian biozonation is based on species of *Streptognathodus* (Chernykh *et al.* 1997; Boardman *et al.* 2009). Perhaps the best sources for this zonation are two books published in Russian (Chernykh 2005, 2006). Chernykh defined seven zones in the Asselian (duration of 3.9 myr; *c.* 0.56 myr/zone) on the basis of the ranges of *Streptognathodus* species including, in ascending order, *S. isolatus*, *S. glenisteri*, *S. cristellaris*, *S. sigmoidalis*, *S. constrictus*, *S. fusus* and *S. postfusus*. Many of these zones would be defined as lineage interval partial range zones as they are defined by the interval between respective FADs of these taxa and which are interpreted to have an evolutionary relationship. The most important source for this zonation is the Usolka section in the Southern Urals. These taxa are well illustrated in numerous plates by Chernykh (2005, 2006); in many respects, you can view each plate as a population, even though several species may be discriminated within a defined biozone and, as such, are very useful for comparison to other regions. I have provided brief descriptions below for key species because many are described only in Russian. Chernykh (2005, 2006) discriminated many additional species, but most have not been tested outside their type areas. The genus displays cyclic morphology through the latest Carboniferous and Early Permian. Mid-Gzhelian species from the *S. vitali* to *S. bellus* zones (Chernykh 2005, pls 5 & 7) lack accessory lobes on the platform. These accessory lobes develop quickly in the latest Gzhelian *S. wabaunsensis* Zone and continue through the *S. sigmoidalis* Zone, but some species still lack these lobes. This ornamentation is subsequently not present in the *S. constrictus–S. postfusus* zones, but it is not clear whether the latter lineages developed by losing the accessory lobes or from species lacking the lobes. Small accessory lobes are, again, present during the Early Sakmarian at the time that the genus became extinct (Fig. 2). Such cyclic morphology is a characteristic of other conodont lineages (e.g. *Clarkina* in the Late Permian) and seems to track relative sea-level changes associated with third-order sequences.

The genus *Streptognathodus* becomes extinct during the early Sakmarian and is ecologically replaced by *Sweetognathus*. There have been suggestions that *Streptognathodus* ranges into the early Artinskian in the mid-continent USA (Boardman *et al.* 2009), but this is really a function of incorrect correlation and species definition. Boardman *et al.* (2009) provided numerous plates of these taxa, thus providing an excellent source for comparison with the Uralian succession, except they did not reference the two books by Chernykh (2005, 2006), nor any of his new species. This means that some taxa will probably be synonymized, which will clarify some correlations. Some probable synonymies are mentioned in the zonal descriptions below, but a detailed analysis is beyond the scope of this paper. It is interesting that Chernykh (2006) does refer to some taxa named by Wardlaw, Boardman and Nestell in Boardman *et al.* (2009), suggesting that the manuscript may have been distributed at least 3 years prior to it being published.

The earliest occurrences of *Sweetognathus* are from Lower Asselian strata in midcontinent USA and Bolivia (Henderson 2014*a*, *b*), where they considerably overlap the range of *Streptognathodus* species. Later, during the latest Asselian (Fig. 2), *Sweetognathus* migrated towards many regions and evolved into a long lineage of species (Fig. 3). High-resolution ages are available for much of the Early Permian (Henderson *et al.* 2012*a*; Schmitz & Davydov 2012), which provides an opportunity to resolve these regional differences.

The most recognizable lithostratigraphic characteristics of Asselian–early Sakmarian strata are the cyclothems that can be used to tune the U–Pb ages with the biostratigraphy. These cyclothems abruptly terminate in the mid-Sakmarian as a result of the end of major Gondwana glaciation – above this level, broad third-order sequences dominate the relative sea-level record (Henderson in Chernykh *et al.* 2013). This provides an unparalleled sequence biostratigraphic signature, but the ramifications of this signature have not readily been accepted (Chernykh & Chuvashov 2014). This is a global signature that provides a valuable datum for correlation: for example, this datum correlates with the Chuanshan (or Maping)–Chihsia contact in South China (Zhu & Zhang 1994; Shen *et al.* 2007), although the defined formational contact is likely to be diachronous. This datum correlates with the timing of the end of widespread ice sheets (Fielding *et al.* 2008). These cyclothems or fourth-order sequences (see Boardman *et al.* 2009) are related to Milankovitch 400 kyr-long eccentricity cycles. Schmitz & Davydov (2012) tuned the entire 'Asselian–early Artinskian' interval in the mid-continent USA, and indicated that it should include 22 cycles given the 8.8 myr duration of the interval: however, there were only 10 cycles. They suggested that most of the missing cycles were represented within numerous palaeosols. In the classic biostratigraphic dilemma, tuning to 22 cycles results in a record that lines up all *Sweetognathus whitei* occurrences, but results in diachronous records for species of *Streptognathodus*. The interpretation that the duration of palaeosol development is included within each 400 kyr cyclothem is equally likely. Ten cycles fits within the 3.9 + myr of the Asselian and earliest Sakmarian, and this forms the basis for a new astronomical tuning of the mid-continent USA biostratigraphic succession with the Uralian geochronological age interpolations (Fig. 2). Tuning to 10 cycles results in the alignment of *Streptognathodus* species, as well as the timing of *Streptognathodus* extinction, and demonstrates that there are two forms referred to *Sweetognathus whitei*, one of which is 4–5 myr younger and present only in post-cyclothem successions.

Latest Carboniferous (latest Gzhelian) Streptognathodus wabaunensis *Zone*

Definition. Partial range lineage zone defined by the range of *S. wabaunsensis* below the first occurrence of *S. isolatus*.

Remarks. A chronomorphocline from *S. wabaunsensis* to *S. isolatus* defines the Carboniferous–Permian boundary (C–PB). This was defined at the Aidaralash section in northern Kazakhstan (Chernykh & Ritter 1997), which was later voted as the GSSP for the base of the Permian (Davydov *et al.* 1998).

Additional species. *Streptognathodus acuminatus*, *S. bellus*, *S. bonus*, *S. insignitus*, *S. nodulinearis*, *S. rectangulatus* and *S. longus* occur in addition to the named species at Usolka (Chernykh 2006).

1. Streptognathodus isolatus Zone

Age. Earliest Permian (earliest Asselian).

Definition. Partial range lineage zone defined by the range of *S. isolatus* below the first occurrence of *S. glenisteri*. *Streptognathodus isolatus* ranges upwards through the *S. sigmoidalis* Zone. The species was named by Chernykh *et al.* (1997) for material at Aidaralash.

Remarks. The FAD of *S. isolatus* within the chronomorphocline of *S. wabaunsensis* (accessory lobe with numerous nodes that sit adjacent to the inner platform parapet or central carina) to *S. isolatus* (accessory lobe with numerous nodes that are isolated from the inner platform parapet margin by a distinct furrow) defines base-Asselian GSSP and the C-PB. This zone is recognized in the mid-continent

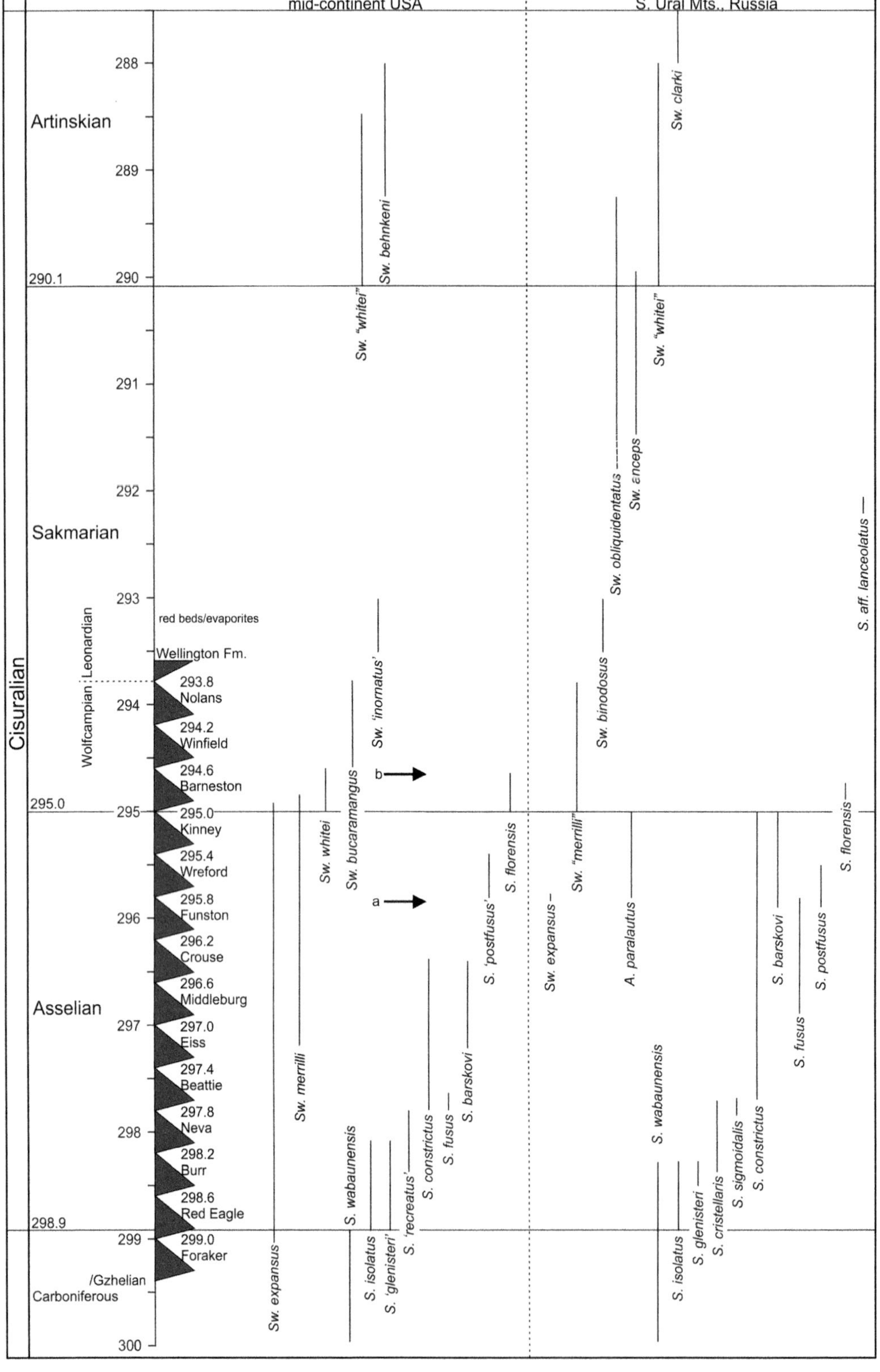
mid-continent USA
S. Ural Mts., Russia
Cisuralian
Artinskian
Sakmarian
Asselian
Carboniferous
/Gzhelian
Wolfcampian
Leonardian
290.1
295.0
298.9
288
289
290
291
292
293
294
295
296
297
298
299
300
red beds/evaporites
Wellington Fm.
293.8
Nolans
294.2
Winfield
294.6
Barneston
295.0
Kinney
295.4
Wreford
295.8
Funston
296.2
Crouse
296.6
Middleburg
297.0
Eiss
297.4
Beattie
297.8
Neva
298.2
Burr
298.6
Red Eagle
299.0
Foraker
Sw. "whitei"
Sw. behnkeni
Sw. 'inornatus'
Sw. bucaramangus
Sw. whitei
Sw. merrilli
Sw. expansus
S. wabaunensis
S. isolatus
S. 'glenisteri'
S. 'recreatus'
S. constrictus
S. fusus
S. barskovi
S. 'postfusus'
S. florensis
a
b
Sw. clarki
Sw. "whitei"
Sw. obliquidentatus
Sw. enceps
Sw. binodosus
S. aff. lanceolatus
Sw. expansus
Sw. "merrilli"
A. paralautus
S. wabaunensis
S. isolatus
S. glenisteri
S. cristellaris
S. sigmoidalis
S. constrictus
S. barskovi
S. fusus
S. postfusus
S. florensis

within the Bennett Shale of the Red Eagle cyclothem. The base of the Permian has been interpolated to 298.9 ± 0.15 Ma (Schmitz & Davydov 2012).

Chernykh (2005, 2006) defines a series of early Asselian zones with numerous named species that share a common characteristic; almost all taxa in these zones have an accessory lobe that consists of numerous nodes formed as a cluster or a linear row, or reduced to a single node. In practice, it is not clear that this complete lineage has been fully recognized outside of the Southern Urals. The zone and the base-Permian GSSP has been criticized (Lucas 2013; also see comments and replies in the same issue of *Permophiles*), suggesting that the zone is diachronous with respect to inflated fusulinaceans and that the species has been poorly documented outside the type region. Finally, Lucas (2013) argues that this is a GSSP in which definition preceded correlation. It is more likely that the inflated fusulinaceans, which include multiple genera, are diachronous – at the very least, the taxonomy of this group needs to be improved. *Streptognathodus isolatus* is recognized in Kansas and in the Urals, so at least some correlation preceded definition. This is the only way that the process can operate effectively. Technically, correlation precedes definition, but, practically, a concerted effort by many palaeontologists to correlate a boundary is unlikely before it is actually defined. Now that we have a boundary, I agree more work is definitely needed to improve correlatability. The group of lower Asselian zones (*S. isolatus*, *S. glenisteri*, *S. cristellaris* and *S. sigmoidalis*) could be recognized as a single zonal interval based on the characteristic morphology shared by these taxa.

Additional species. *Streptognathodus acuminatus, S. bipartitus, S. bonus, S. distortum, S. nodulinearis, S. russoflangulatus, S. invaginatus, S. semiglomus, S. wabaunsensis* and *S. longus* occur in addition to the named species at Usolka (Chernykh 2006). *Sweetognathus expansus* is present in the mid-continent USA (Boardman *et al.* 2009).

2. Streptognathodus glenisteri *Zone*

Age. Early Asselian.

Definition. Partial range lineage zone defined by the range of *S. glenisteri* below the first occurrence of *S. cristellaris*. *Streptognathodus glenisteri* ranges upwards through the *S. cristellaris* Zone. The species was named by Chernykh & Ritter (1997) after material at Aidaralash, but at this locality *S. glenisteri*, *S. cristellaris* and *S. sigmoidalis* share the same first occurrence level.

Remarks. The named species is characterized by outwardly deflected anterior parapets and a nearly occluded anterior median trough (Chernykh & Ritter 1997). Specimens assigned to *S. minacutus* in the mid-continent USA by Boardman *et al.* (2009) should probably be synonymized with *S. glenisteri* (Fig. 2).

Additional species. Streptognathodus acuminatus, S. bipartitus, S. deflexus, S. distortum, S. grandis, S. isolatus, S. paraisolatus and *S. longus* occur in addition to the named species at Usolka (Chernykh 2006). *Sweetognathus expansus* is present in the mid-continent USA (Boardman *et al.* 2009).

3. Streptognathodus cristellaris *Zone*

Age. Early Asselian.

Definition. Partial range lineage zone defined by the range of *S. cristellaris* below the first occurrence of *S. sigmoidalis* in the Usolka section. However, at Aidaralash, these two species have the same first occurrence (Chernykh & Ritter 1997), suggesting that appearances may be diachronous.

Remarks. The named species is characterized by a deep, symmetrically located trough and inner accessory lobe, and sometimes an outer lobe (Chernykh & Reshetkova 1987). *Streptognathodus recreatus* is a characteristic species of this zone, but apparently neither it nor the named zone species are present in Kansas. However, specimens assigned to *S. nevaensis* in the mid-continent (Kansas) USA by Boardman *et al.* (2009) are herein considered a junior synonym of *S. recreatus*, which has the same inner lobe reduced to one or two denticles (or lacking denticles in some forms) and a high inner adcarinal parapet.

Additional species. Streptognathodus glenisteri, S. isolatus, S. recreatus, S. plenus, S. semiglomus, S. russoflangulatus and *S. tumeous* occur in addition to the named species at Usolka (Chernykh 2006). *Sweetognathus expansus* is present in the mid-continent USA (Boardman *et al.* 2009).

Fig. 2. Detailed stratigraphic ranges for the Asselian–Early Artinskian in Kansas, USA based on Boardman *et al.* (2009) and the Usolka section in the Southern Urals based on Chernykh (2005, 2006). The Kansas material is astronomically tuned to the 404 kyr-long eccentricity cycle (simplified to 400 kyr) using cyclothems from Boardman *et al.* (2009) and follow a methodology from Schmitz & Davydov (2012). Each cyclothem has a rapid transgressive component and longer regressive components, with the maximum flooding surface indicated by the point. The Usolka material comes from a section with numerous dated ash beds with interpolated ages between (Schmitz & Davydov 2012). Arrows: (**a**) time of interpreted migration of the *Sweetognathus* lineage to the Urals; and (**b**) significant reduction in abundance of *Streptognathodus* near extinction level.

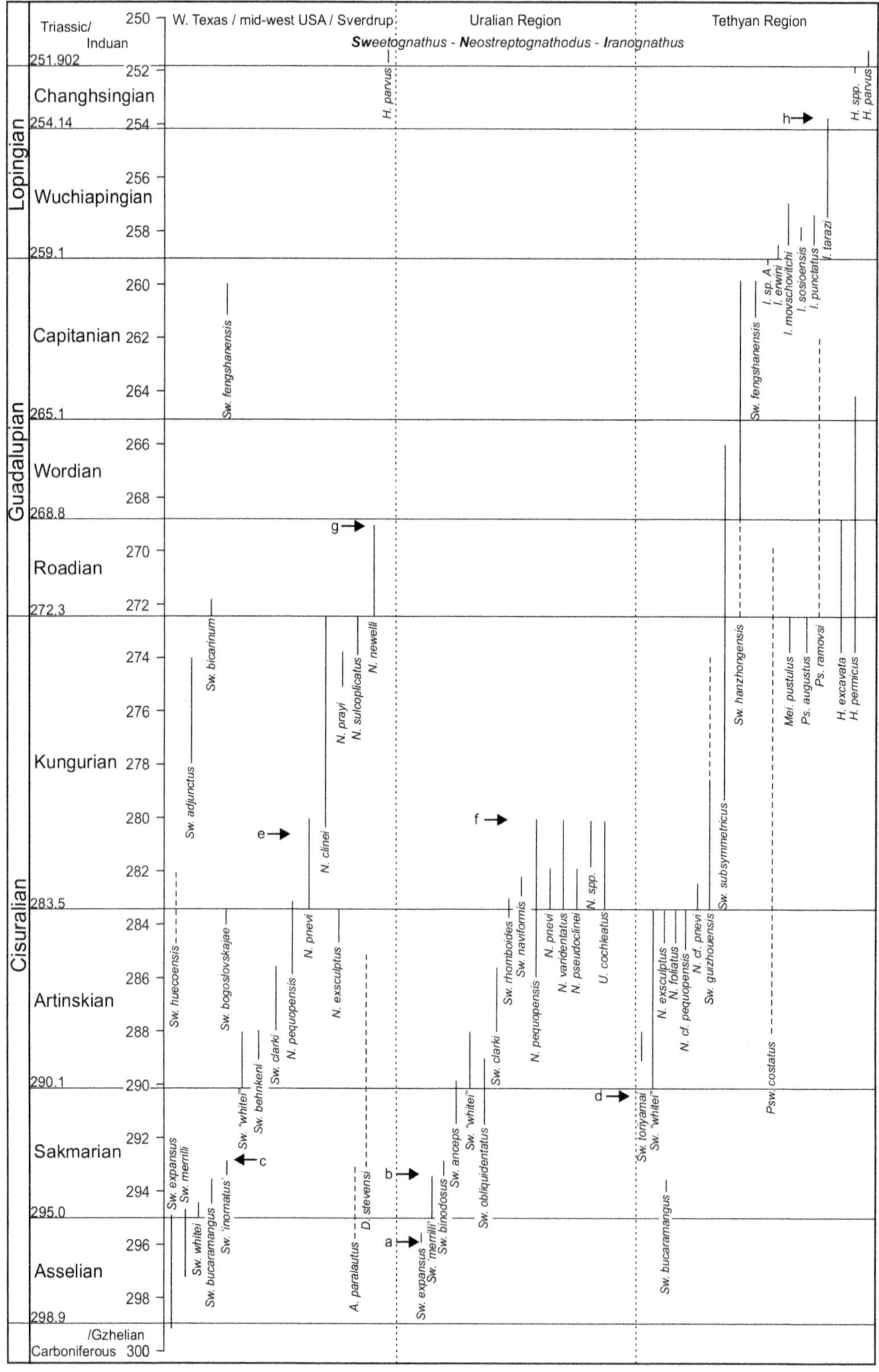
W. Texas / mid-west USA / Sverdrup
Uralian Region
Tethyan Region
Sweetognathus - Neostreptognathodus - Iranognathus
Triassic/ Induan
251.902
Changhsingian
254.14
Lopingian
Wuchiapingian
259.1
Capitanian
265.1
Guadalupian
Wordian
268.8
Roadian
272.3
Kungurian
283.5
Cisuralian
Artinskian
290.1
Sakmarian
295.0
Asselian
298.9
/Gzhelian
Carboniferous
250
252
254
256
258
260
262
264
266
268
270
272
274
276
278
280
282
284
286
288
290
292
294
296
298
300
H. parvus
H. spp.
h
I. tarazi
I. punctatus
I. sosioensis
I. movschovitchi
I. erwini
I. sp. A
Sw. fengshanensis
g
Sw. bicarinum
Sw. adjunctus
N. prayi
N. sulcoplicatus
N. newelli
N. clinei
e
f
Sw. hanzhongensis
Mei. pustulus
Ps. augustus
Ps. ramovsi
H. excavata
H. permicus
Sw. subsymmetricus
Sw. huecoensis
Sw. bogoslovskajae
N. pnevi
N. exsculptus
Sw. clarki
N. pequopensis
Sw. behnkeni
Sw. rhomboides
Sw. naviformis
N. varidentatus
N. pseudoclinei
N. spp.
U. cochleatus
N. foliatus
N. cf. pequopensis
N. cf. pnevi
Sw. guizhouensis
Psw. costatus
Sw. "whitei"
d
Sw. toriyamai
c
Sw. expansus
Sw. merrilli
Sw. 'inornatus'
Sw. bucaramangus
Sw. whitei
D. stevensi
A. paralautus
b
a
Sw. 'merrilli'
Sw. binodosus
Sw. anceps
Sw. obliquidentatus

4. Streptognathodus sigmoidalis *Zone*

Age. Early Asselian.

Definition. Partial range lineage zone defined by the range of *S. sigmoidalis* below the first occurrence of *S. constrictus* at Usolka. *Streptognathodus sigmoidalis* appears to have a restricted range (Chernykh & Ritter 1997; Chernykh 2006).

Remarks. The named species is characterized by a platform with an inner accessory lobe reduced to a single node or lacking nodes, as well as its sigmoidal shape and particularly deep, inner adcarinal trough (Chernykh & Ritter 1997).

Additional species. Streptognathodus cristellaris, S. isolatus, S. costalis, S. latus and *S. profundus* occur in addition to the named species at Usolka (Chernykh 2006).

5. Streptognathodus constrictus *Zone*

Age. Middle Asselian.

Definition. Partial range lineage zone defined by the range of *S. constrictus* below the first occurrence of *S. fusus* at Usolka. This zone also coincides with the first occurrences of two species of *Mesogondolella*, including *M. belladontae* and *M. dentiseparata* (Chernykh 2006), which would be more commonly found in offshore environments compared to *Streptognathodus*. This interval could be referred to an assemblage zone as *S. constrictus–M. belladontae*. *Mesogondolella* was not recovered in the mid-continent USA sections according to Boardman *et al.* (2009).

Remarks. The named species has a distinctly asymmetrical P1 element pair (Chernykh & Ritter 1997). The sinistral element is narrow, elongate and abruptly constricted at the carinal posterior termination. The dextral element is broader and parapets slope toward a central sulcus. Both elements have anterior parapets that are convex with relatively deep adcarinal troughs. *Mesogondolella* has not typically been used to define zones in the Asselian because of its relative rarity in other sections, but, with their recognition, some of these species are now being found in other regions. *Mesogondolella*, like *Streptognathodus*, displays a cyclic morphology temporally in terms of the carinal denticulation. Species in the *S. constrictus* Zone, like *M. dentiseparata*, have discrete denticles, whereas species in the Upper Asselian, like *M. pseudostriata, M. obliquimarginata*, and *M. camilla* have increasingly closely spaced to fused denticles. Sakmarian forms like *M. monstra* again have discrete denticles.

Additional species. Streptognathodus cristellaris, S. longissimus, S. mizensi, S. postsigmoidalis, S. ex gr. *barskovi, M. adentata, M. belladontae* and *M. dentiseparata* occur in addition to the named species at Usolka (Chernykh 2006). At Aidaralash, *S. constrictus* and *S. barskovi* co-occur (Chernykh & Ritter 1997). *Sweetognathus expansus* is present in the mid-continent USA (Boardman *et al.* 2009).

6. Streptognathodus fusus *Zone*

Age. Middle–upper Asselian.

Definition. Partial range lineage zone defined by the range of *S. fusus* below the first occurrence of *S. postfusus* at Usolka.

Remarks. The named species has a distinctly asymmetrical P1 element pair with the sinistral element narrow, elongate and somewhat constricted at the carinal posterior termination. The dextral element is broader, with wide parapets that are flat and bear numerous closely spaced transverse ridges that are separated by a very distinct central sulcus. Both elements have anterior parapets that are high, flared and weakly convex with relatively deep adcarinal troughs. This species and *S. postfusus* far outnumber all other taxa in samples from the type locality of *Sweetognathus whitei* in Wyoming (Rhodes 1963).

Additional species. Streptognathodus adversus, S. constrictus, S. mizensi, S. verus, M. belladontae and *M. dentiseparata* occur in addition to the named species at Usolka (Chernykh 2006). *Streptognathodus verus* occurs at the top of the Cottonwood Limestone Member of the Beattie Limestone in Kansas (Boardman *et al.* 2009, pl.20, figs 6 & 7; not *S. fusus* and *S. constrictus*; see Chernykh 2006; pl. IX, figs 12–21 for comparison). *Sweetognathus merrilli* is present in the mid-continent USA (Boardman *et al.* 2009).

Fig. 3. Ranges of conodont species of the sweetognathid lineage including *Sweetognathus* (*Sw.*), *Neostreptognathodus* (*N.*), *Iranognathus* (*I.*), *Pseudosweetognathus* (*Psw.*), and *Meiognathus* (*Mei.*). Additional taxon ranges include *Adetognathus paralautus* (for the Sverdrup Basin), *Diplognathodus stevensi* (see Ritter 1986), *Pseudohindeodus* (*Ps.*) and *Hindeodus* (*H.*). *H.* spp. = multiple species of *Hindeodus*; *N.* spp. = multiple endemic species of *Neostreptognathodus* (Chernykh 2006). Dashed ranges indicate that the species is rare, occurs rarely or has tentative extensions. Arrows: (**a**) time of interpreted migration of the *Sweetognathus* lineage to the Urals; (**b**) time of migration of Sweetognathid lineage into Sverdrup Basin; (**c**) extinction (Sverdrup) or near-extinction (Urals) of Carboniferous holdovers *Adetognathus* and *Streptognathodus*; (**d**) migration of the Sweetognathid lineage into South China; (**e**) local apparent extinction of *Neostreptognathodus* in the Sverdrup Basin owing to ocean cooling; (**f**) local apparent extinction of *Neostreptognathodus* in the Urals owing to restricted environments; (**g**) final extinction of *Neostreptognathodus* in the Phosphoria Basin; (**h**) final extinction of *Iranognathus* and Sweetognathid lineages.

7. Streptognathodus postfusus–S. barskovi *Zone*

Age. Upper Asselian.

Definition. Assemblage zone defined by the ranges of the two named taxa at Usolka.

Remarks. The named species have distinctly asymmetrical P1 element pairs. *Streptognathodus barskovi* is similar to *S. fusus*, having similarly wide parapets on the dextral element, but the parapets slope slightly in *S. barskovi* towards a weakly defined sulcus, and some of the numerous transverse ridges may touch in the sulcus. This zone and the immediately overlying zone are critical in the correlation between the mid-continent USA and the Uralian succession in Russia. Chernykh (2006) did not cite the paper, but notes two species named by Wardlaw, Boardman and Merrill in Boardman *et al.* (2009), including *S. postelongatus* and *S. florensis* that span the Asselian–Sakmarian boundary in the study by Chernykh (2006) and the Sakmarian–Artinskian boundary in the study by Boardman *et al.* (2009). There is no evidence of an unconformity at Usolka (see the age curve in fig. 9 of Schmitz & Davydov 2012) that could explain such a discrepancy. The type locality for *Sweetognathus whitei* in the Tensleep Sandstone of Wyoming yields abundant specimens of *S. fusus* and *S. postfusus* (Rhodes 1963) that he referred to as *S. elongatus*.

Additional species. Streptognathodus anaequalis, S. fusus, S. lanceolatus, S. latus, S. constrictus, S. postconstrictus and *S. postelongatus* occur in addition to the named species at Usolka (Chernykh 2006). In addition, *M. belladontae* occurs at the base of the zone, *M. dentiseparata* through the lower half, and *M. simulata, M. striata, M. pseudostriata, M. arcuata* and *M. camilla* appear throughout the upper half of the zone, with *M. uralensis* near the very top (depending on the ultimate zone definition). In addition, *M. foliosa* and *M. paralacerta* occur in the Kondurovsky section (Chernykh 2006). At Usolka, the first occurrence of *Sweetognathus expansus* coincides with the upper part of this zone and represents the migration of this genus to the region (Fig. 2, arrow a). The significant number of *Mesogondolella* species suggests that this interval may represent a maximum flooding interval at Usolka. However, there is also a single occurrence of the shallow-water taxon *Adetognathus paralautus* (Chernykh 2006) similar to an occurrence in the Canadian Arctic (Mei *et al.* 2002), although my previous correlation was a full stage too young (Henderson 1988).

8. Sweetognathus 'merrilli' *Zone*

Age. Early Sakmarian.

Definition. This zone is marked at the base by the sudden appearance of the named taxon at 54.3 m at Usolka. The top is marked by the appearance of the related form *Sw. binodosus* at 55.4 m at Usolka (Chernykh 2006). The occurrence of *Streptognathodus florensis* could be used to correlate the base of this zone.

Remarks. The correlation of this zone is highly problematic (Henderson 2014*a*, *b*; Chernykh 2014) for at least three reasons: first, the local first occurrence is likely to be a migration event; second, this species is very rare, reducing confidence in the appearance point; and, third, the range of the named taxon is apparently highly diachronous. This zone has been proposed to define the base Sakmarian (Chuvashov *et al.* 2003*a*, 2004). The taxonomic details of these taxa, *Sw. whitei* and *Sw. 'whitei'* (quotation marks on species indicate that the species will be revised and possibly renamed), are being prepared separately by the author. Other taxa like *Mesogondolella uralensis* or *M. monstra* are now under consideration for the GSSP definition of the base-Sakmarian (Henderson *et al.* 2012*a*; Chernykh *et al.* 2013). The *Sweetognathus* lineage appears to have evolved much earlier in the mid-continent USA, as well as in Bolivia (Suárez-Riglos *et al.* 1987). *Sweetognathus merrilli* first occurs in the Eiss Limestone Member of the Bader Limestone of Kansas associated with species of the *Streptognathodus fusus* Zone (Boardman *et al.* 2009, pl. 18 includes *S. fusus* fig. 5; *S. barskovi* figs 1, 3 & 4; *S.* cf. *constrictus* figs 6 & 7). Kozur (1975) named *Sw. merrilli* on the basis of illustrations in Merrill (1973) from the Eiss Limestone Member of Kansas. Astronomical tuning using cyclothems would suggest an age of 297.2 Ma (Fig. 2) for this species appearance in Kansas compared to 294.8 Ma at Usolka (Schmitz & Davydov 2012).

Additional species. Streptognathodus florensis, Mesogondolella camilla and *M. uralensis* occur in addition to the named species at Usolka (Chernykh 2006). *Mesogondolella gutta* and *M. lacerta* occur at this level in the Kondurovsky section (Chernykh 2006). The type *Sweetognathus whitei* occurs at this time in the mid-continent USA and at its type locality in Wyoming.

Mid-Sakmarian–Late Kungurian

This interval has largely been zoned on species of *Sweetognathus* and the related genus *Neostreptognathodus* – they represent the ecological replacement of the Late Carboniferous–earliest Permian *Streptognathodus* and, as such, typically occupy relatively shallow carbonate platform environments. This ecological replacement occurs during the meltdown of Gondwanan glaciations, and the taxa have a wide distribution associated with the accompanying transgression. Like many shallow-water taxa, these species exhibit a high degree of morphological plasticity. Such phenotypic plasticity does have a role in evolution as a diversifying agent (West-Eberhard

1989), but a number of correlation issues have stemmed from attempts to resolve this morphological variability at the species level. Increasingly, species of *Mesogondolella* have been utilized for the zonation of more open-marine outer-platform to slope environments. Some occurrences of *Mesogondolella* will need to be re-evaluated; *M. luouensis* from the Luodian section (Wang 1994) is very similar to *M. manifesta*, but the former is associated with *Sweetognathus 'whitei'* and the latter with *Sw. binodosus*.

9. Sweetognathus binodosus *Zone*

Age. Mid-Sakmarian.

Definition. Partial range lineage zone defined by the range of *Sw. binodosus* at Usolka below the first occurrence of *Sw. anceps* at Dalny Tulkas (Chernykh 2006).

Remarks. The named species P1 element has a carina consisting of a single row of low ovoid to narrow dumb-bell-shaped nodes with pustulose ornament. Ritter (1986) named a species *Sw. inornatus* in which the holotype is essentially identical to *Sw. binodosus*, but the paratypes are quite variable and long ranging, causing many workers to reject the taxon. This species determination should be resolved in a more taxonomic-based paper, but it is likely that the two species are synonymous with priority to *Sw. inornatus*. At Burbank Hills in western Utah, the *Sw. inornatus* holotype occurs 222 m above the base of the Riepetown Formation, which is in a stratigraphic position above *Sw. bucaramangus* and *Sw. whitei. Streptognathodus* was increasingly rare in the early Sakmarian and finally became extinct during the *Sw. binodosus* Zone. Another Carboniferous holdover genus, *Adetognathus*, also became extinct during this zone (Beauchamp & Henderson 1994; Mei *et al.* 2002). At the same time, the first *Sweetognathus* migrated into the Canadian Arctic (referred to as *Sw. inornatus* in Henderson 1988; Beauchamp & Henderson 1994; and *Sw.* aff. *merrilli* in Mei *et al.* 2002). Specimens of *Sweetognathus bucaramangus* (formerly *Rabeignathus*) in the mid-continent USA (Ritter 1986, 1987) may correlate with the base of this zone. This means that the co-occurrences of *S. whitei* and *S. bucaramangus* in the Elm Creek Limestone in the Midland Basin, western Texas, in the Winfield Limestone of the Chase Group in Kansas, and in the Dingjiazhai Formation of west Yunnan, SW China are early Sakmarian and not early Artinskian as previously reported (Boardman *et al.* 2009; Holterhoff *et al.* 2013; Ueno *et al.* 2002).

Additional species. Streptognathodus postelongatus and *S.* aff. *lanceatus* occur in the lower portion of the zone, and *M. obliquimarginata*, *M. monstra*, *M. longifoliosa*, *M. manifesta* and *M. striata* occur throughout in addition to the named species at Usolka (Chernykh 2006).

10. Sweetognathus anceps *Zone*

Age. Late Sakmarian.

Definition. Partial range lineage zone defined by the range of *Sw. anceps* below the first occurrence of *S. 'whitei'* at Dalny Tulkas (Chernykh 2006).

Remarks. The named species P1 element has a carina consisting of between six and 10 bi-lobed transverse ridges with distinct pustulose ornament. A pustulose ridge joins the midpoint of some transverse ridges or is completely absent. There seems to be a gap in the chronomorphocline from *Sw. binodosus* to *Sw. anceps* (Figs 1 & 2) that is presumably related to the switch of sections from Usolka to Dalny Tulkas. The origin of *Sweetognathus obliquidentatus* is uncertain. This is a *Neostreptognathodus* homeomorph with very irregular pustulose ornament which may be derived from the older *Sweetognathus* lineage that is characterized by such ornament, suggesting that it appears in the Uralian succession by migration.

Additional species. Sweetognathus obliquidentatus and *Mesogondolella bisselli* occur in addition to the named species at Dalny Tulkas (Chernykh 2006).

11. Sweetognathus 'whitei' *Zone*

Age. Early Artinskian.

Definition. Partial range lineage zone defined by the range of *Sw. 'whitei'* below the first occurrence of *Sw. clarki* at Dalny Tulkas (Chernykh 2006).

Remarks. The named species is similar to *Sw. anceps*, except that the pustulose ridge joins the midpoint of all transverse ridges. This zone has a very wide distribution, including South China, which is the first occurrence by migration of the genus in the region, and the Canadian Arctic (Mei *et al.* 2002). The global distribution indicates that it represents an excellent marker for GSSP definition for the base-Artinskian (Chuvashov *et al.* 2003*b*, 2013), except for the taxonomic issue. This lineage has very regular pustulose micro-ornament that is restricted to the transverse ridges. In my own samples from Dalny Tulkus, at the FAD of *Sw. 'whitei'*, the named species is present, but recovered specimens are dominated by *Sw. obliquidentatus*.

Additional species. Sweetognathus anceps, *Sw. obliquidentatus*, *Sw. gravis* and *Mesogondolella bisselli* occur in addition to the named species at Dalny Tulkas (Chernykh 2006). *Sweetognathus behnkeni* occurs in the western USA (Ritter 1986, 1987), in Bolivia (Suárez-Riglos *et al.* 1987) and the Harper Ranch Group in central British Columbian (Orchard & Forster 1988) associated with the upper part of this interval. The holotype of *Sweetognathus toriyamai* (Igo 1981) from Japan is similar to *Sw. behnkeni*.

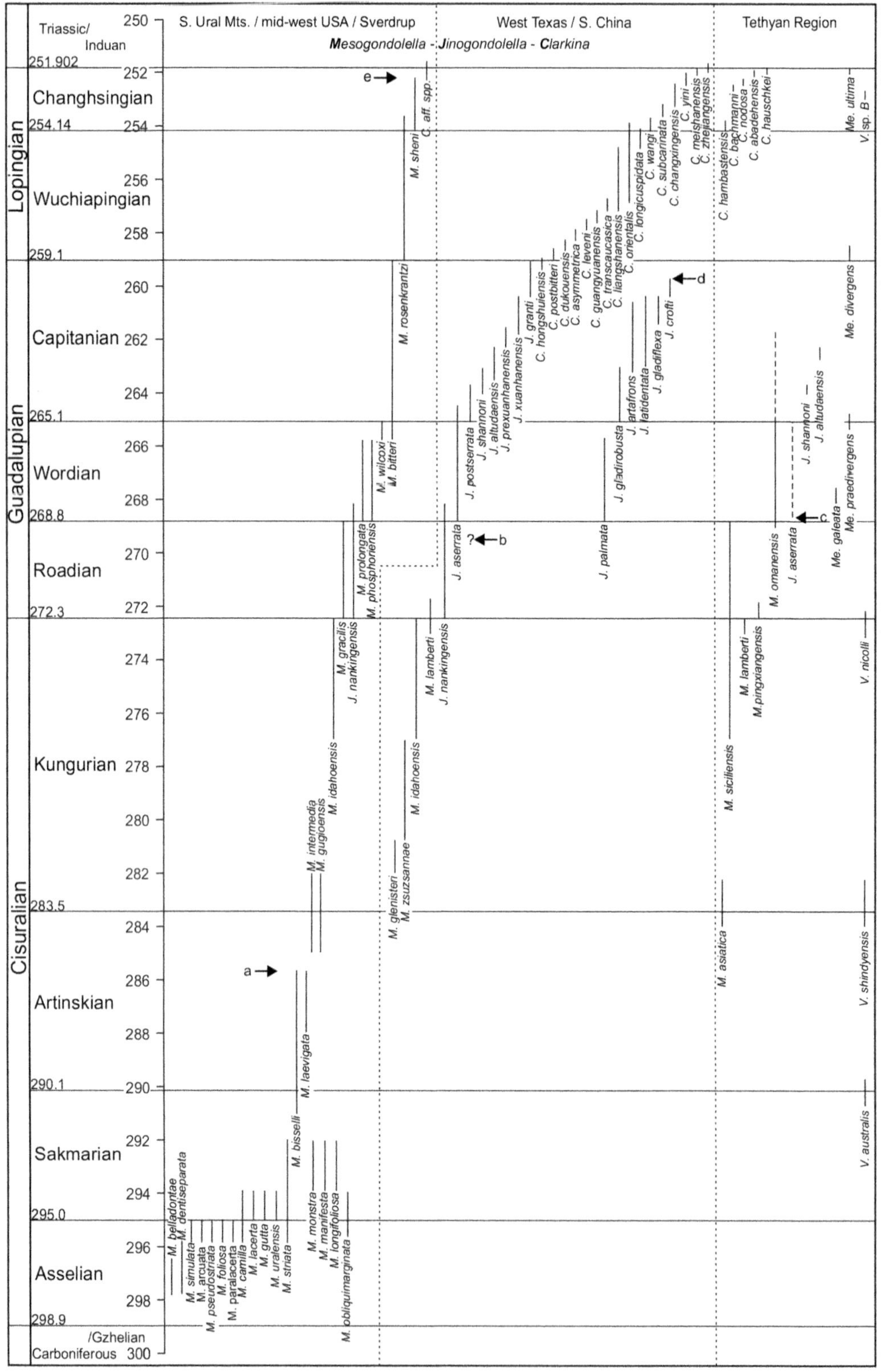
Triassic/ Induan
251.902
Lopingian
Changhsingian
254.14
Wuchiapingian
259.1
Guadalupian
Capitanian
265.1
Wordian
268.8
Roadian
272.3
Cisuralian
Kungurian
283.5
Artinskian
290.1
Sakmarian
295.0
Asselian
298.9
/Gzhelian Carboniferous
250
252
254
256
258
260
262
264
266
268
270
272
274
276
278
280
282
284
286
288
290
292
294
296
298
300
S. Ural Mts. / mid-west USA / Sverdrup
West Texas / S. China
Tethyan Region
Mesogondolella - Jinogondolella - Clarkina
a
b
c
d
e
?
M. belladontae
M. dentiseparata
M. simulata
M. arcuata
M. pseudostriata
M. foliosa
M. paralacerta
M. camilla
M. lacerta
M. gutta
M. uralensis
M. striata
M. bisselli
M. laevigata
M. monstra
M. manifesta
M. longifoliosa
M. obliquimarginata
M. intermedia
M. gujioensis
M. idahoensis
M. gracilis
J. nankingensis
M. prolongata
M. phosphoriensis
M. wilcoxi
M. bitteri
M. rosenkrantzi
M. sheni
C. aff. spp.
M. glenisteri
M. zsuzsannae
M. lamberti
J. aserrata
J. postserrata
J. shannoni
J. altudaensis
J. prexuanhanensis
J. xuanhanensis
J. granti
C. hongshuiensis
C. postbitteri
C. dukouensis
C. asymmetrica
C. leveni
C. guangyuanensis
C. transcaucasica
C. liangshanensis
C. orientalis
C. longicuspidata
C. wangi
C. subcarinata
C. changxingensis
C. yini
C. meishanensis
C. zhejiangensis
J. palmata
J. gladirobusta
J. artafrons
J. latidentata
J. gladiflexa
J. crofti
M. asiatica
M. siciliensis
M. pingxiangensis
M. omanensis
C. hambastensis
C. bachmanni
C. nodosa
C. abadehensis
C. hauschkei
Me. galeata
Me. praedivergens
Me. divergens
Me. ultima
V. sp. B
V. nicolli
V. shindyensis
V. australis

12. Sweetognathus clarki *Zone*

Age. Mid-Artinskian.

Definition. Partial range lineage zone defined by the range of *Sw. clarki* below the first occurrence of *Neostreptognathodus pequopensis* at Dalny Tulkas (Chernykh 2006).

Remarks. The named species is transitional between *Sweetognathus 'whitei'* and *Neostreptognathodus pequopensis* as the transverse ridges are beginning to split apart at their midpoint – typically, the anteriormost two or three ridges. This zone has a very wide distribution, including in the Canadian Arctic. Mei *et al.* (2002) synonymized several taxa including *N. ruzhencevi*, *N. transitus* and *N. tschuvaschovi* with *Sw. clarki*, which was reasonable given the high morphological plasticity exhibited by this group. However, given the duration of this zone (also other Artinskian zones) and the high morphological plasticity, this interval should be resampled to further test species discrimination using the sample population concept.

Additional species. Sweetognathus 'whitei', *Sw. binodosus* and *Mesogondolella laevigata* occur in addition to the named species at Dalny Tulkas (Chernykh 2006). This is the highest occurrence of *Mesogondolella* in the Uralian sections (Fig. 4).

13. Neostreptognathodus pequopensis *Zone*

Age. Late Artinskian.

Definition. Partial range lineage zone defined by the range of *N. pequopensis* below the first occurrence of *N. pnevi* (pronounced pe-nyo-vee) at the Mechetlino section (Chernykh 2006).

Remarks. Although the base of this zone is not clearly recognized in the Uralian region, it has been documented in western Canada (Henderson & McGugan 1986), in the Canadian Arctic (Henderson 1988) and in the USA (Behnken 1975).

Additional species. Sweetognathus aff. *whitei*, *Stepanovites* cf. *festiva* and an informal new species of *Sweetognathus* occur in addition to the named species at Mechetlino (Chernykh 2006). *Neostreptognathodus exsculptus*, *Mesogondolella intermedia*, *M. gugioensis* and *M. asiatica* correlate with this zone in Japan, and may also cross into the earliest Kungurian (Igo 1981); *M. intermedia* and *M. gugioensis* have the same range in the Sverdrup Basin (Henderson 1988; Beauchamp & Henderson 1994). Katvala & Henderson (2002) showed that *N. exsculptus* can be used as a palaeolatitudinal indicator by comparing SE British Columbia and Vancouver Island (Wrangellia Terrane) occurrences.

14. Neostreptognathodus pnevi *Zone*

Age. Early Kungurian.

Definition. Partial range lineage zone defined by the range of *N. pnevi* below the first occurrence of *N. clinei* in the Mechetlino section (Chernykh 2006).

Remarks. The chronomorphocline from *N. pequopensis* to *N. pnevi* is recognized in many regions, including the Canadian Arctic (Mei *et al.* 2002), and, as such, marks an excellent GSSP position for the base-Kungurian. At the moment, there are two competing locations – at Rockland in Nevada, USA (Henderson *et al.* 2012*b*) and at Mechetlino in the Urals of Russia – but there is consensus on the FAD of *N. pnevi* as the signal (Henderson *et al.* 2012*b*). Some Kungurian successions seem to be very thick, suggesting increased rates of tectonic subsidence in regions like Nevada and South China. This is also a time of increased restriction in the Uralian seaway and, as such, a considerable number of endemic taxa evolved in that region. This transition is also illustrated in Mytton *et al.* (1983) in southern Idaho. *Neostreptognathodus* undergoes an apparent extinction in the Canadian Arctic near the top of this zone (Fig. 3, arrow e). The genus continues through most of the Roadian in the mid-continent USA before its true extinction (Fig. 3, arrow g).

Additional species. Neostreptognathodus pequopensis, *N. kamajensis* (endemic), *N. pseudoclinei*, *N. varidentatus* (endemic) and *Uralognathus cochleatus* (endemic) occur in addition to the named species at Mechetlino (Chernykh 2006). Smooth gondolellids like *Mesogondolella glenisteri* and *M. zsuzsannae* seem to correlate with this interval and the overlying *N. clinei* Zone in the Glass Mountains of Texas (Wardlaw 2001). *Sweetognathus guizhouensis* corresponds to this interval in South China (Mei *et al.* 2002). In the peri-Gondwana province, *Vjalovognathus shindyensis* may correlate with this interval (Nicoll & Metcalfe 1998).

Fig. 4. Ranges of conodont species of the gondolellid lineage including *Mesogondolella* (*M.*), *Jinogondolella* (*J.*), and *Clarkina* (*C.*). Additional taxon ranges include *Vjalovognathus* (*V.*), *Merrillina* (Me.), and *C.* aff. spp. = species of *Clarkina* with affinity to latest Changhsingian tethyan species (Algeo *et al.* 2012). Dashed ranges indicate species is rare, occurs rarely, or tentative extensions. Arrows: (**a**) apparent local extinction of *Mesogondolella* in the Urals as a result of shallowing; (**b**) possible migration of *Jinogondolella nankingensis* into sections in China, indicating that the local FO is not equal to the FAD; (**c**) migration of rare *Jinogondolella* into the western Tethys (Oman); (**d**) local apparent extinction of the gondolellid lineage owing to the restriction of the basin (i.e. deposition of the Castille evaporites); (**e**) migration of *Clarkina* spp. into the Sverdrup Basin, probably as a result of global warming associated with the end-Permian extinction event interval.

15. Neostreptognathodus clinei *Zone*

Age. Mid-Kungurian.

Definition. Taxon range zone of *Neostreptognathodus clinei*, but without a defined top at Mechetlino (Chernykh 2006).

Remarks. Endemic taxa are common in the base of this zone, during which the Uralian seaway becomes fully restricted and haline, and continental deposits begin (Fig. 3, arrow f).

Additional species. Neostreptognathodus pequopensis, *N.* aff. *pequopensis*, *N. fastigatus* (endemic), *N. varidentatus* (endemic), several informal and endemic species of *Neostreptognathodus*, and *Uraloceras cochleatus* (endemic) occur in addition to the named species at Mechetlino (Chernykh 2006). Smooth gondolellids like *Mesogondolella zsuzsannae* and *M. idahoensis* seem to correlate with this interval in the Glass Mountains of Texas (Wardlaw 2001). The oldest forms of *Sweetognathus subsymmetricus* correspond to this interval in South China (Mei *et al.* 2002).

16. Neostreptognathodus prayi–N. sulcoplicatus *Zone*

Age. Mid–late Kungurian.

Definition. Assemblage zone defined essentially by the ranges of the two named taxa in the Phosphoria Basin (Behnken *et al.* 1986; Wardlaw & Collinson 1986). The ranges of these taxa are not well constrained, otherwise they may define separate distinct biozones.

Remarks. With the Uralian seaway becoming fully restricted with haline and continental deposits, it was necessary to establish a zonation elsewhere: in this case, mid-continent USA and West Texas.

Additional species. M. idahoensis, *Sweetina festiva*, and *N. clinei* occur in addition to the named species in the Phosphoria Basin (Behnken *et al.* 1986; Wardlaw & Collinson 1986). *Sweetognathus subsymmetricus* ranges through this interval in South China (Mei *et al.* 2002).

17. Mesogondolella lamberti *Zone*

Age. Latest Kungurian.

Definition. Partial range interval zone defined by the range of *M. lamberti* below the first occurrence of *Jinogondolella nankingensis*.

Remarks. This zone at least partially overlaps with the *N. prayi–N. sulcoplicatus* biozone. It is best established in the Delaware Basin of western Texas (Lambert *et al.* 2000). The named taxon was originally described as a subspecies (Mei & Henderson 2002*a*), *M. idahoensis lamberti*, but Lambert *et al.* (2007) raised it to species status.

Additional species. Occurs with transitional forms from *M. lamberti* to *J. serrata*. Correlative taxa may include *Meiognathus pustulus*, *Hindeodus excavatus*, *H. permicus* and *Pseudohindeodus augustus* in Japan (Shen *et al.* 2012), *Sweetognathus subsymmetricus* in China (Fig. 3), and *Vjalovognathus nicolli* in the peri-Gondwana province (see *V.* sp.nov. *A* in Nicoll & Metcalfe 1998; Yuan *et al.* 2016).

Middle Permian zones and key species

The Middle Permian has mostly been zoned on the evolution of serrated (Clark & Behnken 1979) *Jinogondolella* species. *Sweetognathus* may be used in shallow, warm-water platform environments, but exact age ranges are as yet poorly constrained and homeomorphy must be considered (Wang *et al.* 1987). The type region for the Guadalupian is located in the Delaware Basin of west Texas, and *The Guadalupian Symposium* publication, which includes conodont papers by Wardlaw (2000) and Lambert *et al.* (2000), represents a valuable resource. A number of papers show that the zonation established in this region is comparable to the Middle Permian succession in South China (Mei *et al.* 1994*a*, *b*, *c*, 1998*a*). It is also clear, when looking at plates in these publications, that the high degree of morphological plasticity makes consistent species discrimination difficult, resulting in taxonomic differences between workers (Lambert *et al.* 2010). In some cases, independent isotopic tests would be useful to fully establish the correlation between west Texas and South China. Deeper-water successions with cherty sequences yield fewer conodonts, but can be integrated with radiolarians for correlation (Ning *et al.* 2010; Nishikane *et al.* 2011). Furthermore, the zonation does not extend into the cool-water provinces, except for brief intervals, as shown, for example, by a geographical cline (Henderson & Mei 2007) where serrated forms are found in the Sverdrup Basin (basal Assistance Formation) and the NE China Jilin Province (Wang *et al.* 2000) immediately above a major flooding surface marking the base of the Guadalupian. Only rare specimens of long-ranging *Mesogondolella* species occur above this datum (Fig. 4) in the north cool-water province. In the peri-Gondwana cool-water province, rare *Mesogondolella* and endemic *Vjalovognathus* (Nicoll & Metcalfe 1998; Yuan *et al.* 2016) are recognized (Fig. 4). *Jinogondolella* is also somewhat rare in the western Tethys of Oman (Henderson & Mei 2003; Kozur & Wardlaw 2010), but does occur sporadically to provide a link with the standard zonation. The Neotethys Ocean opens during this interval and conodonts play a role in determining the timing (Baud *et al.* 2012).

18. Jinogondolella nankingensis *Zone*

Age. Roadian.

Definition. Partial range lineage zone defined by the range of *J. nankingensis* below the first occurrence of *J. aserrata*.

Remarks. The base-Roadian (base-Guadalupian Series) GSSP (Henderson *et al.* 2012*a*) is recognized by the FAD of *Jinogondolella nankingensis* in the Delaware Basin. The chronomorphocline from *Mesogondolella lamberti* to *Jinogondolella nankingensis* involves the development of strong serration that affects the lateral margin of the platform and a change in the S3 element morphology by adding a bifid process (Lambert *et al.* 2007).

Additional species. *Mesogondolella gracilis* and *Neostreptognathodus newelli* in the Phosphoria Basin (Wardlaw & Collinson 1984); *Sweetognathus bicarinum*, *Hindeodus excavatus* and *H. permicus* in western Texas (Wardlaw 2000), and *M. siciliensis* and *Sw. subsymmetricus* in China and western Tethys (Gullo & Kozur 1992; Henderson & Mei 2003). The genus *Neostreptognathodus* becomes extinct within this zone. The history of *Jinogondolella nankingensis* is interesting. It was discovered in China (Ching 1960) as *Gondolella nankingensis*, but the occurrence was unknown outside China when *Gondolella serrata* was named in the USA (Clark & Ethington 1962). Priority was later given to *G. nankingensis*, which has also been referred to as a species of *Neogondolella*, *Mesogondolella* and, finally, *Jinogondolella*. However, it is very likely that the FO in China is not equivalent to the FAD in western Texas. Equating the FO and FAD has resulted in significant debates (Henderson & Mei 2003), particularly with respect to the correlation of Tethyan stages (Mei & Henderson 2002*b*; Angiolini & Henderson 2013; Angiolini *et al.* 2015). The most likely source of resolution will come from geochronology and chemostratigraphy. Henderson & Mei (2003) suggested several potential correlations, but I now speculate that the FO of *Jinogondolella nankingensis* (Fig. 4, arrow b) at Luodian (bed 42) is actually closer to the latter part of its range, and shortly before the evolution of *J. aserrata*. Furthermore, the first serrated forms in the western Tethys (Oman) belong to *J. aserrata* (Fig. 4, arrow c) suggesting migration of this serrated lineage into the greater Tethys (including the Kuhfeng Formation near Nanjing) between 269.5 and 268.5 Ma.

19. Jinogondolella aserrata *Zone*

Age. Wordian.

Definition. Partial range lineage zone defined by the range of *J. aserrata* below the first occurrence of *J. postserrata*.

Remarks. The base-Wordian GSSP (Henderson *et al.* 2012*a*) is recognized by the FAD of *J. aserrata* in the Delaware Basin.

Additional species. *Mesogondolella phosphoriensis*, *M. prolongata* and *H. permicus* in the Phosphoria Basin (Behnken *et al.* 1986; Wardlaw & Collinson 1986), *M. omanensis* in Oman (Kozur & Wardlaw 2010), *Jinogondolella palmata* in west Texas (Nestell & *Wardlaw* 2002), and *Sweetognathus subsymmetricus* and *Sw. hanzhongensis* in China (Mei *et al.* 2002; Henderson & Mei 2003) and Tunisia (Angiolini *et al.* 2008). Wardlaw & Nestell (2010*b*) named a new species, *Jinogondolella errata*, and described an apparatus with three P elements (P1, P2 and P3) that is unknown in any other comparable species. It is possible that the animal was dimorphic with two different P2 elements paired with the P1. The species is synonymous with *J. aserrata*.

20. Jinogondolella postserrata *Zone*

Age. Earliest Capitanian.

Definition. Partial range lineage zone defined by the range of *J. postserrata* below the first occurrence of *J. shannoni* at various sections in South China (Mei *et al.* 1994*c*) and originally in the Delaware Basin of west Texas (Pinery to Lamar Limestone members of the Bell Canyon Formation).

Remarks. The FAD of *J. postserrata* in the Delaware Basin defines the base-Capitanian GSSP (Henderson *et al.* 2012*a*).

Additional species. *Mesogondolella omanensis*, *M. bitteri*, *Jinogondolella gladirobusta* in West Texas (Nestell *et al.* 2006) and *Sweetognathus hanzhongensis*.

21. Jinogondolella shannoni *Zone*

Age. Early Capitanian.

Definition. Partial range lineage zone defined by the range of *J. shannoni* below the first occurrence of *J. altudaensis* at various sections in South China (Mei *et al.* 1994*c*) and originally in the Delaware Basin of west Texas (Altuda Formation).

Remarks. Specimens assigned to this species are recognized at Wadi Wasit in Oman.

Additional species. *Mesogondolella omanensis*, *M. bitteri* and *Sweetognathus hanzhongensis*.

22. Jinogondolella altudaensis *Zone*

Age. Mid-Capitanian.

Definition. Partial range lineage zone defined by the range of *J. altudaensis* below the first occurrence of *J. prexuanhanensis* at various sections in South China (Mei *et al.* 1994*c*) and originally in the Delaware Basin of west Texas (near the top of the Altuda Formation).

Remarks. Interpretations of the named taxon vary, and it is in need of refinement. It has a strict definition and short range according to Mei & Henderson (2001), but Lambert *et al.* (2002, 2010) indicate that it has a long range to the top of the Guadalupian and represents the ancestor of *Clarkina hongshuiensis*. The origin of the genus *Clarkina* was also discussed by Wardlaw & Mei (1998*b*). Alternative scenarios are discussed below.

Additional species. *Mesogondolella omanensis*, *M. bitteri* and *Sweetognathus hanzhongensis*.

23. Jinogondolella prexuanhanensis *Zone*

Age. Mid–late Capitanian.

Definition. Partial range lineage zone defined by the range of *J. prexuanhanensis* below the first occurrence of *J. xuanhanensis* at various sections in South China (Mei *et al.* 1994*c*).

Remarks. Emeishan volcanism began during the *J. altudaensis* Zone, and a significant Middle Permian extinction occurs around the *J. prexuanhanensis* Zone (Wignall *et al.* 2009).

Additional species. Mesogondolella omanensis, M. bitteri and *Sweetognathus hanzhongensis.*

24. Jinogondolella xuanhanensis *Zone*

Age. Late Capitanian.

Definition. Partial range lineage zone defined by the range of *J. xuanhanensis* below the first occurrence of *J. granti* at various sections in South China (Mei *et al.* 1994*c*).

Remarks. This zone is recognized in various sections in South China (Mei *et al.* 1994*c*).

Additional species. Mesogondolella omanensis, M. bitteri, Sweetognathus hanzhongensis and *Sw. fengshanensis.* Lambert *et al.* (2002, 2010) and Wardlaw & Nestell (2010*a*) described a number of new *Jinogondolella* taxa (*J. artafrons, J. gladiflexa* and *J. latidentata*) corresponding to this zone down to the *J. shannoni* Zone in the Lamar Limestone of west Texas. It has not yet been demonstrated whether any of these taxa can also be recognized in South China – they may be endemic.

25. Jinogondolella granti *Zone*

Age. Late Capitanian.

Definition. Partial range lineage zone defined by the range of *J. granti* below the first occurrence of *C. hongshuiensis* at the Penglaitan section in South China (Mei *et al.* 1994*c*; Henderson *et al.* 2002).

Remarks. This species was named and the correlation first discussed in Mei *et al.* (1994*c*). It only occurs in a lowstand carbonate unit called the Laibin Limestone at the Penglaitan section. It may also occur immediately below the Castille Formation in western Texas just as the Delaware Basin was becoming restricted during a sea-level lowstand (Lambert *et al.* 2002), making it necessary to establish a zonation elsewhere: in this case, South China. West Texas specimens attributed to *J. granti* and *C. hongshuiensis* identified by Lambert *et al.* (2002) differ from type material in South China, indicating that they are either geographical variants or rare morphotypes of different species.

Additional species. The first occurrence of a species of *Iranognathus* appears at this level (Mei *et al.* 2002) and *Jinogondolella crofti* occurs in west Texas (Wardlaw & Mei 1998*a*, *b*).

26. Clarkina hongshuiensis *Zone*

Age. Latest Capitanian.

Definition. Partial range lineage zone defined by the range of *C. hongshuiensis* below the first occurrence of *C. postbitteri* at Penglaitan in South China (Henderson *et al.* 2002).

Remarks. Henderson *et al.* (2002) suggested that the named species evolved by sympatric isolation during the late Guadalupian lowstand from *Jinogondolella granti* by narrowing the anterior platform and loss of serration. It is also possible that this species evolved from *Mesogondolella omanensis* (Kozur & Wardlaw 2010) during a migration event from western Tethys by narrowing the anterior platform, but the apparatus will need to be investigated. Both *Clarkina* and *Jinogondolella* have a bifid S3 element, which is not present in, at least, Early Permian *Mesogondolella* (Henderson & Mei 2007). Henderson *et al.* (2002) named this population as a subspecies of *C. postbitteri* and Lambert *et al.* (2010) elevated it to species status. The point generated much discussion (Henderson & Mei 2002; Henderson 2003); it is fortunate that an arbitrary point within the anagenetic evolution of *C. postbitteri postbitteri* from *C. postbitteri hongshuiensis* was chosen as the primary signal to recognize the GSSP. In many respects, choosing this point near the top of the lowstand Laibin Limestone is a natural boundary interval, as more proximal sections that are unconformable will have Guadalupian below and Lopingian above the unconformity.

Additional species. Jinogondolella granti and *Iranognathus* sp. *A* at Penglaitan, South China.

Upper Permian zones and key species

The Upper Permian has generally been zoned on the basis of gondolellid (Kozur 1990) species of *Clarkina*, but, in shallow, warm-water environments, species of *Iranognathus*, which evolved from *Sweetognathus*, have been utilized. These taxa are all typical of the Equatorial Warm-Water Province (EWWP) and are not found in peri-Gondwana or in north cool-water provinces where long-ranging species of *Mesogondolella* continued to evolve (Wardlaw & Collinson 1979); a zonation is emerging, but it is poorly constrained as yet (Beauchamp *et al.* 2009; Algeo *et al.* 2012). Species of *Clarkina* have a much wider distribution in the latest Permian following significant warming events associated with the extinction. A high-resolution Lopingian timescale was constructed by Shen *et al.* (2010), and a discussion of Lopingian conodont zones in Iran, as well as correlation problems, was provided by Shen & Mei (2010). Kozur (2004, 2005) named several new taxa in the Iranian succession and suggested that this indicated gaps in the Meishan succession where the species were absent. However, this is more likely attributed to a level of endemism or geographical variation. Shen *et al.* (2013) provided a high-resolution correlation between Iran and South China using biostratigraphy and chemostratigraphy that indicates there are no gaps in the South China section, as suggested by Kozur. Yuan *et al.* (2014) provided a high-resolution timescale and numerous plates depicting Changhsingian *Clarkina* species populations; this is the type of work needed for all Permian species, including especially Cisuralian species of *Streptognathodus*.

27. Clarkina postbitteri *Zone*

Age. Earliest Wuchiapingian.

Definition. Partial range lineage zone defined by the range of *C. postbitteri* below the first occurrence of *C. dukouensis* at Penglaitan in South China, as well as the Dukou and Nanjiang sections (Mei *et al.* 1994*b*). The base-Wuchiapingian (base-Lopingian Series) GSSP at Penglaitan (Jin *et al.* 2006*a*) is recognized by the FAD of *C. postbitteri postbitteri*.

Remarks. This biozone appears to be missing in Iran (Shen & Mei 2010). *Clarkina postbitteri* has widely spaced discrete carinal denticles. Subsequent species bear denticles that become increasingly fused (e.g. *C. transcaucasica*). This cyclicity continues upwards with *C. longicuspidata* discrete, then *C. wangi* is fused and *C. zhejiangensis* discrete. Typically, the more discrete denticulation forms are recovered from transgressive deposits.

Additional species. Mesogondolella rosenkrantzi in the north cool-water province, and *Iranognathus erwini* in tropical settings.

28. Clarkina dukouensis *Zone*

Age. Early Wuchiapingian.

Definition. Partial range lineage zone defined by the range of *C. dukouensis* below the first occurrence of *C. asymmetrica* at Penglaitan in South China (Mei *et al.* 1998*a*) and in the Dukou and Nanjiang sections (Mei *et al.* 1994*b*).

Remarks. *Clarkina postbitteri* and *C. dukouensis* show only very minor differences of posterior denticle spacing and are, therefore, difficult to distinguish.

Additional species. Mesogondolella rosenkrantzi in the north cool-water province, and *Iranognathus erwini* and *I. movschovitschi* in tropical settings.

29. Clarkina asymmetrica *Zone*

Age. Early Wuchiapingian.

Definition. Partial range lineage zone defined by the range of *C. asymmetrica* below the first occurrence of *C. leveni* at Penglaitan in South China (Mei *et al.* 1998*a*) and in the Dukou and Nanjiang sections (Mei *et al.* 1994*b*).

Remarks. Present also in the Kuh-e-Ali Bashi, NW Iran (Shen & Mei 2010).

Additional species. Mesogondolella rosenkrantzi, Iranognathus punctatus and *I. sosioensis* occur in various regions at this time.

30. Clarkina leveni *Zone*

Age. Early Wuchiapingian.

Definition. Partial range lineage zone defined by the range of *C. leveni* below the first occurrence of *C. guangyuanensis* at Penglaitan in South China (Mei *et al.* 1998*a*, *b*) and in the Dukou and Nanjiang sections (Mei *et al.* 1994*b*).

Remarks. The zone was first recognized in the Transcaucasus by Kozur (1975).

Additional species. Mesogondolella rosenkrantzi and *Iranognathus punctatus* occur in various regions at this time.

31. Clarkina guangyuanensis *Zone*

Age. Early–mid-Wuchiapingian.

Definition. Partial range lineage zone defined by the range of *C. guangyuanensis* below the first occurrence of *C. transcaucasica* at Penglaitan in South China (Mei *et al.* 1998*a*, *b*) and in the Dukou and Nanjiang sections (Mei *et al.* 1994*b*).

Remarks. Present also in the Kuh-e-Ali Bashi, NW Iran (Shen & Mei 2010).

Additional species. Mesogondolella rosenkrantzi and *Iranognathus punctatus* occur in various regions at this time.

32. Clarkina transcaucasica *Zone*

Age. Mid-Wuchiapingian.

Definition. Partial range lineage zone defined by the range of *C. transcaucasica* below the first occurrence of *C. orientalis* at Penglaitan in South China (Mei *et al.* 1998*a*, *b*) and in the Dukou and Nanjiang sections (Mei *et al.* 1994*b*).

Remarks. The last species of *Iranognathus, I. tarazi*, ranges from this zone up to the lower *Clarkina subcarinata* Zone (Mei *et al.* 2002), where it apparently became extinct.

Additional species. Clarkina liangshanensis is common in this zone (Shen *et al.* 2010). *Mesogondolella rosenkrantzi* and *Iranognathus tarazi* occur in various regions at this time.

33. Clarkina orientalis *Zone*

Age. Mid–late Wuchiapingian.

Definition. Partial range interval zone defined by the range of *C. orientalis* below the first occurrence of *C. longicuspidata* at Meishan in South China (Mei *et al.* 1994*b*; Jin *et al.* 2006*b*).

Remarks. This is a relatively long-ranging zone in which the named taxon is very abundant and other species are very rare, as if the acme was controlled by a specialized environment or migration. The named species occurs as high as the *Clarkina wangi* Zone in the lower Changhsingian. This species is homeomorphic with *C. abadehensis* of the latest Permian, which caused a long debate (Sweet & Mei 1999*a*, *b*) that was resolved by Henderson *et al.* (2008). The species was described in more detail by Shen (2007).

Additional species. *Mesogondolella rosenkrantzi*, *Clarkina liangshanensis* and *Iranognathus tarazi* occur in various regions at this time.

34. Clarkina longicuspidata *Zone*

Age. Late Wuchiapingian.

Definition. Partial range interval zone defined by the range of *C. longicuspidata* below the first occurrence of *C. wangi* in the Meishan and Nanjiang sections in South China (Mei *et al.* 1994*b*; Mei *et al.* 2004; Jin *et al.* 2006*b*).

Remarks. It is uncertain whether this zone can be recognized in Iran (Shen & Mei 2010).

Additional species. *Mesogondolella rosenkrantzi* and *Iranognathus tarazi* occur in various regions at this time.

35. Clarkina wangi *Zone*

Age. Earliest Changhsingian.

Definition. Partial range lineage zone defined by the range of *C. wangi* (FAD at bed 4a-2) below the first occurrence of *C. subcarinata* at Meishan in South China (Mei *et al.* 2004; Yuan *et al.* 2014).

Remarks. The base-Changhsingian GSSP at Meishan (Jin *et al.* 2006*b*; Wang *et al.* 2006) is recognized by the FAD of *C. wangi*. Mei *et al.* (1998*b*) utilized a mosaic approach to taxonomy for the Changhsingian species and showed at least three parallel lineages separated by posterior platform morphology. This approach evolved into the population concept of Mei *et al.* (2004) that emphasized the carina as a uniting characteristic. Both are legitimate approaches, and should be tested by looking at geographical variation and how it impacts evolutionary interpretations. Apparently, only the pointed posterior form named *C. hambastensis* by Kozur (2005) occurs in Iran, in contrast to three morphotypes in South China.

Additional species. *Mesogondolella rosenkrantzi*, *M. sheni*, *Clarkina orientalis*, *C. hambastensis*, transitional forms of *C. longicuspidata*, and *Iranognathus tarazi* occur in various regions at this time.

36. Clarkina subcarinata *Zone*

Age. Early Changhsingian.

Definition. Partial range lineage zone defined by the range of *C. subcarinata* (FAD at bed 11-2b) below the first occurrence of *C. changxingensis* at Meishan in South China (Mei *et al.* 1998*a*, *b*; Yuan *et al.* 2014).

Remarks. *Iranognathus* and the Sweetognathid lineage become extinct at this level (Fig. 3, arrow h).

Additional species. *Mesogondolella sheni* and transitional forms of *Clarkina wangi* occur in various regions at this time.

37. Clarkina changxingensis *Zone*

Age. Mid-Changhsingian.

Definition. Partial range lineage zone defined by the range of *C. changxingensis* (FAD at bed 12-8b) below the first occurrence of *C. yini* at Meishan in South China (Wang & Wang in Zhao *et al.* 1981; Mei *et al.* 1998*a*, *b*; Yuan *et al.* 2014).

Remarks. This is the longest duration zone in the Changhsingian, representing about half of the stage duration. Evolutionary rates seem to be higher during the early Changhsingian sea-level rise and during the end-Permian crisis. Yuan *et al.* (2014) recognized three evolutionary stages in the zone, with the lower dominated by forms with a wide platform, the middle by posteriorly pointed forms with a gap anterior of the cusp, and the upper with forms transitional to *C. yini* and *C. meishanensis*.

Additional species. *Mesogondolella sheni*, transitional forms of *C. subcarinata*, and *C. bachmanni* occur near the top of the zone.

38. Clarkina yini *Zone*

Age. Late Changhsingian.

Definition. Partial range lineage zone defined by the range of *C. yini* (FAD at bed 22-10-2) below the first occurrence of *C. meishanensis* at Meishan in South China (Mei *et al.* 1998*a*, *b*; Mei & Henderson 2001; Yuan *et al.* 2014).

Remarks. The end-Permian extinction crisis begins during this zone interval (Wang *et al.* 2013).

Additional species. *Mesogondolella sheni*, transitional forms of *C. changxingensis* and *C. nodosa* near the base of the zone. Brosse *et al.* (2016) completed a unitary association study in which *Clarkina* (not *Neogondolella*) *changxingensis* extends into the Early Triassic above *Hindeodus parvus* and yet it does not occur higher than the base of the *C. yini* Zone, showing that statistical approaches cannot improve limited taxonomic assessment.

39. Clarkina meishanenis–C. zhejiangensis *Zone*

Age. Latest Changhsingian.

Definition. Assemblage zone defined essentially by the ranges of the two named taxa at Meishan (Yuan *et al.* 2014), but *C. meishanensis* appears in bed 24e and *C. zhejiangensis* begins in bed 26.

Remarks. Yuan *et al.* (2014) separated these into two zones of very short duration, but more work is warranted in sections that are less condensed than at Meishan to determine whether they truly have distinctive ranges.

Additional species. *Clarkina abadehensis* in Iran. In western Canada (Schoepfer *et al.* 2012) and the Canadian Arctic (Algeo *et al.* 2012), there is a sudden influx of new gondolellids that have been compared to late Changhsingian species of *Clarkina* (*C.* aff. spp. in Fig. 4), presumably because of warming events that ultimately led to the end-Permian extinction.

40. Hindeodus praeparvus–H. changxingensis *Zone*

Age. Latest Changhsingian.

Definition. Assemblage zone based on the ranges of the two named taxa below the first occurence of *H. parvus* (Jiang *et al.* 2007; Zhang *et al.* 1996, 2007).

Remarks. In many sections there is a dramatic change in biofacies and the gondolellid lineage is replaced temporarily by species of *Hindeodus* (Chen *et al.* 2009).

Additional species. *Hindeodus latidentatus*, *Clarkina zhejiangensis*, *C. hauschkei* and *Merrillina ultima* occur in various regions at this time.

Earliest Triassic (earliest Induan) Hindeodus parvus *Zone*

Definition. Partial range interval zone defined by the range of *H. parvus* below the first occurrence of *Isarcicella staechei* (Yin *et al.* 2001).

Remarks. The GSSP for the base-Induan and base-Triassic in the Meishan section D (Yin *et al.* 2001) is recognized by the FAD of *Hindeodus parvus* (bed 27c). The top of the Permian (i.e. base of the Triassic) has been interpolated at 251.902 ± 0.024 Ma (Shen *et al.* 2011; Burgess & Bowring 2015), making the Permian almost exactly 47 myr in duration.

Additional species. *Hindeodus changxingensis* and *Clarkina zhejiangensis* range across the boundary.

Summary

A total of 75 biozones based on the ranges of conodonts are summarized in this paper. Forty of these are regarded as standard international biozones and 35 others are referred to as correlative regional biozones. A number of factors, however, preclude only a few of the biozones from having truly global character. These factors include regional glaciation in the earliest Permian, major lowstands of sea level in the Kungurian and Capitanian–Wuchiapingian boundary interval, significant levels of provincialism, especially from the early Kungurian to the early Changhsingian, and the high degree of morphological plasticity displayed by key taxa. The latter aspect has led to widely different opinions on taxonomy, which affects the correlation potential. In general, Permian conodonts, regardless of taxonomic approach, exhibit low to moderate diversity throughout the Permian, and two major lineages provide most of the biostratigraphic control. The sweetognathid lineage (Fig. 3) and gondolellid lineage (Fig. 4), however, provide a remarkable evolutionary record.

The most significant difference with this biozonation summary from any previous scheme is that it is fully tied to the current geochronological scale as much as possible. Significant revision of the correlation of the Asselian–Artinskian interval has resulted from astronomical tuning of cyclothemic successions in the mid-continent USA to well constrained time curves for the Uralian succession. Numerous geochronological dates in the Lopingian provide a high-resolution scheme for that interval, but there is still a significant gap in ages for the Late Artinskian–Late Capitanian. A CONOP9 correlation scheme in Henderson *et al.* (2012*a*) does provide some resolution for the ages of biozones, but clearly more needs to be done. Apparent evolutionary rates vary significantly (see Figs 3 & 4), which may be a function of environmental stability (long ranges in the Kungurian–Wordian) or simply an artefact of study intensity (shorter ranges in the earliest and latest Permian). This could be tested by more intensive study of Artinskian–Wordian strata. There is also an apparent gap in the mid-Sakmarian of at least 1 myr where the transition from *Sweetognathus binodosus* to *Sw. anceps* has not been demonstrated.

Methods, like unitary associations and constrained optimization, should be employed to test all of these biozones and increase the accuracy of correlations. However, these techniques are dependent on accurate taxonomy, do not replace traditional biozonal methods, nor do they reduce the importance of continuing to get out into the field to find new sections, new ash beds and new conodonts.

I would like to acknowledge the research support provided over the years from the NSERC Discovery Grant program in Canada, as well as from the University of Calgary. I also thank Spencer Lucas and Shuzhong Shen for inviting my paper, for the reviews of this manuscript and for editing this special publication.

References

Algeo, T.J., Henderson, C.M. *et al.* 2012. Evidence for a diachronous Late Permian marine crisis from the Canadian Arctic region. *Geological Society of America Bulletin*, **124**, 1424–1448.

Angiolini, L. & Bucher, H. 1999. Taxonomy and quantitative biochronology of Guadalupian brachiopods from the Khuff Formation, southeastern Oman. *Geobios*, **32**, 665–699.

Angiolini, L. & Henderson, C.M. 2013. The once and future quest: looking for mid-Permian correlation between the Tethyan and the International Time Scales. *Permophiles*, **57**, 19–20.

ANGIOLINI, L., CHAOUACHI, C. *ET AL.* 2008. New fossil findings and discovery of conodonts in the Guadalupian of Jebel Tebaga de Medenine: biostratigraphic implications. *Permophiles*, **51**, 10–21.

ANGIOLINI, L., ZANCHI, A. *ET AL.* 2015. From rift to drift in South Pamir (Tajikistan): Permian evolution of a Cimmerian terrane. *Journal of Asian Earth Sciences*, **102**, 146–169.

BAUD, A., RICHOZ, S., BEAUCHAMP, B., CORDEY, F., GRASBY, S., HENDERSON, C.M. & KRYSTYN, L. 2012. The Buday'ah Formation, Sultanate of Oman: a Middle Permian to Early Triassic oceanic record of the Neotethys and the late Induan microsphere bloom. *Journal of Asian Earth Sciences*, **43**, 130–144.

BEAUCHAMP, B. & HENDERSON, C.M. 1994. The Lower Permian Raanes, Great Bear Cape and Trappers Cove formations, Sverdrup Basin, Canadian Arctic: stratigraphy and conodont zonation. *Bulletin of Canadian Petroleum Geology*, **42**, 562–597.

BEAUCHAMP, B., HENDERSON, C.M., GRASBY, S.E., GATES, L., BEATTY, T., UTTING, J. & JAMES, N.P. 2009. Late Permian sedimentation in the Sverdrup Basin, Canadian Arctic: the Lindström and Black Stripe formations. *Bulletin of Canadian Society of Petroleum Geology*, **57**, 167–191.

BEHNKEN, F.H. 1975. Leonardian and Guadalupian (Permian) conodont biostratigraphy in western and southwestern United States. *Journal of Paleontology*, **49**, 284–315.

BEHNKEN, F.H., WARDLAW, B.R. & STOUT, L.N. 1986. Conodont biostratigraphy of the Permian Meade Peak Phosphatic Shale Member, Phosphoria Formation, southeastern Idaho. *Contributions to Geology, University of Wyoming*, **24**, 169–190.

BOARDMAN, D.R., WARDLAW, B.R. & NESTELL, M.K. 2009. Stratigraphy and Conodont Biostratigraphy of Uppermost Carboniferous and Lower Permian from North American Midcontinent. *Kansas Geological Survey Bulletin*, **25**, 42.

BRIGGS, D.E.G., CLARKSON, E.N.K. & ALDRIDGE, R.J. 1983. The conodont animal. *Lethaia*, **16**, 1–14.

BROSSE, M., BUCHER, H. & GOUDEMAND, N. 2016. Quantitative biochronology of the Permian–Triassic boundary in South China based on conodont unitary associations. *Earth Science Reviews*, **155**, 153–171, https://doi.org/10.1016/j.earscirev.2016.02.003

BURGESS, S.D. & BOWRING, S.A. 2015. High-precision geochronology confirms voluminous magmatism before, during, and after Earth's most severe extinction. *Science Advances*, **1**, 1–14.

CHEN, J., BEATTY, T., HENDERSON, C.M. & ROWE, H. 2009. Permian–Triassic boundary at the Dawen Section, Great Bank of Guizhou, Guizhou Province. *Journal of Asian Earth Sciences*, **36**, 442–458.

CHERNYKH, V.V. 2005. *Zonal Methods and Biostratigraphy – Zonal Scheme for the Lower Permian of the Urals According to Conodonts*. Institute of Geology and Geochemistry, Urals Branch of the Russian Academy of Science, Ekaterinburg [in Russian].

CHERNYKH, V.V. 2006. *Lower Permian Conodonts of the Urals*. Institute of Geology and Geochemistry, Urals Branch of the Russian Academy of Science, Ekaterinburg [in Russian].

CHERNYKH, V.V. 2014. Remarks on Henderson's comments in Permophiles, 59. *Permophiles*, **60**, 5–7.

CHERNYKH, V.V. & CHUVASHOV, B.I. 2014. Response and Comments: Uralian stratotypes of the stage boundaries of the Lower Series of the Permian System. *Permophiles*, **59**, 8–13.

CHERNYKH, V.V. & RESHETKOVA, N.P. 1987. *Biostratigraphy and Conodonts from the Frontier Deposits of the Carboniferous and Permian in the Western Slopes of Southern and Central Urals*. Akademia Nauk SSSR, Uralian Science Centre, Sverdlovsk.

CHERNYKH, V.V. & RITTER, S.M. 1997. *Streptognathodus* (Conodonta) succession at the proposed Carboniferous–Permian boundary stratotype section, Aidaralash Creek, northern Kazakhstan. *Journal of Paleontology*, **71**, 459–474.

CHERNYKH, V.V., RITTER, S.M. & WARDLAW, B.R. 1997. *Streptognathodus isolatus* new species (Conodonta): proposed index for the Carboniferous–Permian boundary. *Journal of Paleontology*, **71**, 162–164.

CHERNYKH, V.V., CHUVASHOV, B.I., SHEN, S. & HENDERSON, C.M. 2013. Proposal for the Global Stratotype Section and Point (GSSP) for the base-Sakmarian Stage (Lower Permian). *Permophiles*, **58**, 16–26.

CHING, Y. 1960. Conodonts from the Kufeng Formation of Lungtan, Nanking. *Acta Paleontologica Sinica*, **8**, 236–257.

CHUVASHOV, B., CHERNYKH, V. *ET AL.* 2003*a*. Proposal for the base of the Sakmarian Stage: GSSP in the Kondurovsky Section, southern Urals, Russia. *Permophiles*, **41**, 4–13.

CHUVASHOV, B., CHERNYKH, V. *ET AL.* 2003*b*. Progress report on the base of the Artinskian and base of the Kungurian by the Cisuralian Working Group. *Permophiles*, **41**, 13–16.

CHUVASHOV, B.I., CHERNYKH, V.V. & AMON, E.O. 2004. Sakmarian stage of Cis-Uralian division of Permian system: biostratigraphy and correlative potential. *Permophiles*, **44**, 10–11.

CHUVASHOV, B.I., CHERNYKH, V.V., SHEN, S. & HENDERSON, C.M. 2013. Proposal for the Global Stratotype Section and Point (GSSP) for the base-Artinskian Stage (Lower Permian). *Permophiles*, **58**, 26–34.

CLARK, D.L. & BEHNKEN, F.H. 1971. Conodonts and biostratigraphy of the Permian. *In*: STWEET, W.C. & BERGSTROM, S.M. (eds) *Symposium on Conodont Biostratigraphy*. Geological Society of America, Memoirs, **127**, 415–439.

CLARK, D.L. & BEHNKEN, F.H. 1979. Evolution and taxonomy of the North American Upper Permian *Neogondolella serrata* complex. *Journal of Paleontology*, **53**, 263–275.

CLARK, D.L. & ETHINGTON, R.L. 1962. Survey of Permian conodonts in western North America. *Brigham Young University Geology Studies*, **9**, 102–114.

DAVYDOV, V.I., GLENISTER, B.F., SPINOSA, C., RITTER, S.M., CHERNYKH, V.V., WARDLAW, B.R. & SNYDER, W.S. 1998. Proposal of Aidaralash as Global Stratotype Section and Point (GSSP) for base of the Permian System. *Episodes*, **21**, 11–17.

FIELDING, C.R., FRANK, T.D. & ISBELL, J.L. 2008. The late Paleozoic ice age – a review of current understanding and synthesis of global climate patterns. *In*: FIELDING, C.R., FRANK, T.D. & ISBELL, J.L. (eds) *Resolving the late Paleozoic Ice Age in Time and Space*. Geological Society of America, Special Papers, **441**, 343–354.

GRADSTEIN, F.M., OGG, J.G., SCHMITZ, M. & OGG, G. (eds) 2012. *Geologic Time Scale 2012, Volumes 1 and 2*. Elsevier, Amsterdam.

GULLO, M. & KOZUR, H.W. 1992. Conodonts from the pelagic deep-water Permian of central western Sicily (Italy). *Neues Jahrbuch fuer Geologie und Palaeontologie*, **184**, 203–234.

HENDERSON, C.M. 1988. *Conodont paleontology and biostratigraphy of the Upper Carboniferous to Lower Permian Canyon Fiord, Belcher Channel, Nansen, an unnamed, and Van Hauen formations, Canadian Arctic Archipelago*. PhD thesis, University of Calgary.

HENDERSON, C.M. 2003. Reply to Kozur 'Definition of the Lopingian Base with the FAD of *Clarkina postbitteri postbitteri*'. *Permophiles*, **41**, 52–53.

HENDERSON, C.M. 2006. Beware of your FO and be aware of the FAD. *Permophiles*, **47**, 8–9.

HENDERSON, C.M. 2014*a*. Response and comments: the GSSP process and the GSSPs proposed for base-Sakmarian and base-Artinskian stages. *Permophiles*, **59**, 13–17.

HENDERSON, C.M. 2014*b*. Remarks on Chernykh's comments (in Permophiles 60). *Permophiles*, **60**, 7–8.

HENDERSON, C.M. & MCGUGAN, A. 1986. Permian conodont biostratigraphy of the Ishbel Group, southwestern Alberta and southeastern British Columbia. *Contributions to Geology, University of Wyoming*, **24**, 219–235.

HENDERSON, C.M. & MEI, S. 2002. Reply to Kozur and Wang's 'Comments to the base of the Lopingian Series'. *Permophiles*, **40**, 30–31.

HENDERSON, C.M. & MEI, S. 2003. Stratigraphic v. environmental significance of Permian serrated conodonts around the Cisuralian–Guadalupian boundary: new evidence from Oman. *Palaeogeography, Palaeoclimatology, Palaeoecology*, **191**, 301–328.

HENDERSON, C.M. & MEI, S.L. 2007. Geographical clines in Permian and lower Triassic gondolellids and its role in taxonomy. *Palaeoworld*, **16**, 90–201.

HENDERSON, C.M., MEI, S. & WARDLAW, B.R. 2002. Conodonts at the Guadalupian–Lopingian boundary at Penglaitan, South China. *In*: HILLS, L.V., HENDERSON, C.M. & BAMBER, E.W. (eds) *Carboniferous and Permian of the World*. Canadian Society of Petroleum Geologists, Memoirs, **19**, 725–735.

HENDERSON, C.M., MEI, S.L., SHEN, S.Z. & WARDLAW, B.R. 2008. Resolution of the reported Upper Permian conodont occurrences from northwestern Iran. *Permophiles*, **51**, 2–9.

HENDERSON, C.M., DAVYDOV, V.I. & WARDLAW, B.R. 2012*a*. The Permian Period. *In*: GRADSTEIN, F.M., OGG, J.G., SCHMITZ, M. & OGG, G. (eds) *The Geologic Time Scale 2012*. Elsevier, Amsterdam, 653–679.

HENDERSON, C.M., WARDLAW, B.R., DAVYDOV, V.I., SCHMITZ, M.D., SCHIAPPA, T.A., TIERNEY, K.E. & SHEN, S.-Z. 2012*b*. Proposal for the base-Kungurian GSSP. *Permophiles*, **56**, 8–21.

HOLTERHOFF, P.F., WALSH, T.R. & BARRICK, J.E. 2013. Artinskian (Early Permian) conodonts from the Elm Creek Limestone, a heterozoan carbonate sequence on the eastern shelf of the Midland Basin, West Texas, USA. *In*: LUCAS, S.G., NELSON, W.J. *ET AL*. *The Carboniferous–Permian Transition*. New Mexico Museum of Natural History and Science Bulletin, **60**, 109–119.

IGO, H. 1981. *Permian Conodont Biostratigraphy of Japan*. Palaeontological Society of Japan Special Papers, **24**.

JIANG, H.S., LAI, X.L., LUO, G.M., ALDRIDGE, R.A., ZHANG, K.X. & WIGNALL, P.B. 2007. Restudy of conodont zonation and evolution across the P/T boundary at Meishan section, Changxing, Zhejiang, China, environmental and biotic changes during the Paleozoic–Mesozoic transition. *Global and Planetary Change*, **55**, 39–55.

JIN, Y.G., GLENISTER, B.F., KOTLYAR, G.V. & SHENG, J.Z. 1994. An operational scheme of Permian chronostratigraphy. *Palaeoworld*, **4**, 1–13.

JIN, Y.G., SHEN, S.Z. *ET AL*. 2006*a*. The Global Stratotype Section and Point (GSSP) for the boundary between the Capitanian and Wuchiapingian stage (Permian). *Episodes*, **29**, 253–262.

JIN, Y.G., WANG, Y., HENDERSON, C.M., WARDLAW, B.R., SHEN, S.Z. & CAO, C.Q. 2006*b*. The Global Boundary Stratotype Section and Point (GSSP) for the base of Changhsingian Stage (Upper Permian). *Episodes*, **29**, 175–182.

KATVALA, E.C. & HENDERSON, C.M. 2002. Conodont sequence biostratigraphy and paleogeography of the Pennsylvanian–Permian Mount Mark Formation, southern Vancouver Island. *In*: HILLS, L.V., HENDERSON, C.M. & BAMBER, E.W. (eds) *Carboniferous and Permian of the World*. Canadian Society of Petroleum Geologists Memoirs, **19**, 461–478.

KOZUR, H.W. 1975. Beitrage zur conodontenfauna des Perm. *Geologisch Palaeontologische Mitteilungen Innsbruck*, **5**, 1–44.

KOZUR, H.W. 1978. Beitraege zur Stratigraphie des Perms; Teil II, Die Conodontenchronologie des Perms. *Freiberger Forschungshefte, Reihe C: Geowissenschaften, Mineralogie Geochemie*, **334**, 85–161.

KOZUR, H.W. 1990. The taxonomy of the gondolellid conodonts in the Permian and Triassic. *Courier Forschungsinstitut Institut Senckenberg*, **117**, 409–469.

KOZUR, H.W. 2004. Pelagic uppermost Permian and the Permian–Triassic boundary conodonts of Iran. Part 1: taxonomy. *Hallesches Jahrbuch Für Geowissenschaften, Reihe B: Geologie, Paläontologie, Mineralogie*, **18**, 39–68.

KOZUR, H.W. 2005. Pelagic uppermost Permian and the Permian–Triassic boundary conodonts of Iran, Part II: investigated sections and evaluation of the conodont faunas. *Hallesches Jahrbuch Für Geowissenschaften, Reihe B: Geologie, Paläontologie, Mineralogie*, **19**, 49–86.

KOZUR, H. & WARDLAW, B.R. 2010. The Guadalupian conodont fauna of Rustaq and Wadi Wasit, Oman and a West Texas connection. *Micropaleontology*, **56**, 213–231.

KOZUR, H., BRANDNER, R., RESCH, W. & MOSTLER, H. 1995. Permian conodont zonation and its importance for the Permian stratigraphic standard scale. *Geologisch Palaeontologische Mitteilungen Innsbruck*, **20**, 165–205.

LAMBERT, L.L., LEHRMANN, D.J. & HARRIS, M.T. 2000. Correlation of the Road Canyon and Cutoff formations, West Texas, and its relevance to establishing and International Middle Permian (Guadalupian) Series. *In*: WARDLAW, B.R., GRANT, R.E. & ROHR, D.M. (eds) *The Guadalupian Symposium*. Smithsonian Contributions to the Earth Sciences, **22**, 37–87.

LAMBERT, L.L., WARDLAW, B.R., NESTELL, M.K. & NESTELL, G.P. 2002. Latest Guadalupian (Middle Permian) conodonts and foraminifers from West Texas. *Micropaleontology*, **48**, 343–364.

LAMBERT, L.L., WARDLAW, B.R. & HENDERSON, C.M. 2007. *Mesogondolella* and *Jinogondolella* (Conodonta): multi-element definition of the taxa that bracket the basal Guadalupian (Middle Permian Series) GSSP. *Palaeoworld*, **16**, 208–221.

LAMBERT, L.L., BELL, G.L., FRONIMOS, J.A., WARDLAW, B.R. & YISA, M.O. 2010. Conodont biostratigraphy of a more complete Reef Trail Member section near the type section, latest Guadalupian Series type region. *Micropaleontology*, **56**, 233–253.

LUCAS, S.G. 2013. We need a new GSSP for the base of the Permian. *Permophiles*, **58**, 8–13.

MEI, S.L. & HENDERSON, C.M. 2001. Evolution of Permian conodont provincialism and its significance in global correlation and paleoclimate implication. *Palaeogeography, Palaeoclimatology, Palaeoecology*, **170**, 237–260.

MEI, S.L. & HENDERSON, C.M. 2002*a*. Conodont definition of the Kungurian (Cisuralian) and Roadian (Guadalupian) Boundary. *In*: HILLS, L.V., HENDERSON, C.M. & BAMBER, E.W. (eds) *Carboniferous and Permian of the World*. Canadian Society of Petroleum Geologists, Memoirs, **19**, 529–551.

MEI, S.L. & HENDERSON, C.M. 2002*b*. Comments on some Permian conodont faunas reported from Southeast Asia and adjacent areas and their global correlation. *Journal of Asian Earth Sciences*, **20**, 599–608.

MEI, S.L., JIN, Y.G. & WARDLAW, B.R. 1994*a*. Succession of conodont zones from the Permian 'Kuhfeng' Formation, Xuanhan, Sichuan and its implications in global correlation. *Acta Palaeontologica Sinica*, **33**, 1–23.

MEI, S.L., JIN, Y.G. & WARDLAW, B.R. 1994*b*. Succession of Wuchiapingian conodonts from northeastern Sichuan and its worldwide correlation. *Acta Micropalaeontologica Sinica*, **11**, 121–139.

MEI, S.L., JIN, Y.G. & WARDLAW, B.R. 1994*c*. Zonation of conodonts from the Maokouan–Wuchiapingian boundary strata, South China. *In*: JIN, Y.G., UTTING, J. & WARDLAW, B.R. (eds) *Permian Stratigraphy, Environments and Resources. Palaeoworld*, **4**, 225–233.

MEI, S.L., JIN, Y.G. & WARDLAW, B.R. 1998*a*. Conodont succession of the Guadalupian–Lopingian boundary strata in Laibin of Guangxi, China and West Texas, USA. *In*: JIN, Y.G., WARDLAW, B.R. & WANG, Y. (eds) *Permian Stratigraphy, Environments and Resources. Palaeoworld*, **9**, 53–76.

MEI, S.L., ZHANG, K.X. & WARDLAW, B.R. 1998*b*. A refined succession of Changhsingian and Griesbachian neogondolellid conodonts from the Meishan section, candidate of the global stratotype section and point of the Permian–Triassic boundary. *Palaeogeography, Palaeoclimatology, Palaeoecology*, **143**, 213–226.

MEI, S.L., HENDERSON, C.M. & WARDLAW, B.R. 2002. Evolution and distribution of the conodont *Sweetognathus* and *Iranognathus* and related genera during the Permian, and their implications for climate changes. *Palaeogeography, Palaeoclimatology, Palaeoecology*, **180**, 57–91.

MEI, S.L., HENDERSON, C.M. & CAO, C.Q. 2004. Conodont sample-population approach to defining the base of the Changhsingian Stage, Lopingian Series, Upper Permian. *In*: BEAUDOIN, A.B. & HEAD, M.J. (eds) *The Palynology and Micropalaeontology of Boundaries*. Geological Society, London, Special Publications, **230**, 105–121, https://doi.org/10.1144/GSL.SP.2004.230.01.06

MERRILL, G.K. 1973. Pennsylvanian nonplatform conodont genera, 1 – *Spathognathodus*. *Journal of Paleontology*, **47**, 289–314.

MOVSHOVICH, E.V., KOZUR, H., PAVLOV, A.M., PNEV, V.P., POLOZOVA, A.N., CHUVASHOV, B.N. & BOGOSLOVSKAYA, M.R. 1979. Complexes of conodonts from the Lower Permian of the pre-Urals and problems of correlation of Lower Permian deposits. *In*: PAPULAR, G.N. & PUCHKOV, V.N. (eds) *Conodonts of the Urals and their Stratigraphic Significance*. Trudy Institute of Geology and Geochemistry, Urals Science Centre, Akademii Nauk SSSR, **145**, 94–133.

MYTTON, J.W., MORGAN, W.A. & WARDLAW, B.R. 1983. Stratigraphic relations of Permian units, Cassia Mountains, Idaho. *In*: MILLER, D.M., TODD, V.R. & HOWARD, K.A. (eds) *Tectonic and Stratigraphic Studies in the Eastern Great Basin*. Geological Society of America, Memoirs, **157**, 281–303.

NESTELL, M.K. & WARDLAW, B.R. 2002. *Jinogondolella palmata*, a new gondolellid conodont species from the Bell Canyon Formation, Middle Permian, Apache Mountains, West Texas. *Micropaleontology*, **56**, 185–194.

NESTELL, M.K., NESTELL, G.P., WARDLAW, B.R. & SWEATT, M.J. 2006. Integrated biostratigraphy of foraminifers, radiolarians and conodonts in shallow and deep water Middle Permian (Capitanian) deposits of the 'Radar Slide', Guadalupe Mountains, West Texas. *Stratigraphy*, **3**, 161–194.

NICOLL, R.S. & METCALFE, I. 1998. Early and Middle Permian conodonts from the Canning and southern Carnarvon basins, Western Australia: their implications for regional biogeography and palaeoclimatology. *Proceedings of the Royal Society of Victoria*, **110**, 419–461.

NING, Z., HENDERSON, C.M. & WENCHEN, X. 2010. Conodonts and radiolarian through the Cisuralian–Guadalupian boundary from the Pingxiang and Dachongling sections, Guangxi region, South China. *Alcheringia*, **34**, 35–160.

NISHIKANE, Y., KAIHO, K., TAKAHASHI, S., HENDERSON, C.M., SUZUKI, N. & KANNO, M. 2011. The Guadalupian–Lopingian boundary (Permian) in a pelagic sequence from Panthalassa recognized by integrated conodont and radiolarian biostratigraphy. *Marine Micropaleontology*, **78**, 84–95.

ORCHARD, M.J. & FORSTER, P.J.L. 1988. *Permian Conodont Biostratigraphy of the Harper Ranch Beds, Near Kamloops, Southern-Central British Columbia*. Geological Survey of Canada Paper, **88**.

PERLMUTTER, B. 1975. Conodonts from the uppermost Wabaunsee Group (Pennsylvanian) and the Admire and Council Grove groups (Permian) in Kansas. *Geologica et Palaeontologica*, **9**, 95–115.

RHODES, F.H.T. 1963. Conodonts from the topmost Tensleep sandstone of the Eastern Big Horn Mountains, Wyoming. *Journal of Paleontology*, **37**, 401–408.

RITTER, S.M. 1986. Taxonomic revision and Phylogeny of post-early crisis *bisselli–whitei* Zone conodonts with comments on Late Paleozoic diversity. *Geologica et Palaeontologica*, **20**, 139–165.

RITTER, S.M. 1987. Biofacies-based refinement of Early Permian conodont biostratigraphy, in central and western USA. *In*: AUSTIN, R.L. (ed.) *Conodonts: Investigative Techniques and Applications*. British Micropalaeontological Society, London & Ellis Horwood, Chichester, 382–403.

SADLER, P.M. 2004. Quantitative biostratigraphy – achieving finer resolution in global correlation. *Annual Review of Earth and Planetary Sciences*, **32**, 187–213.

SCHMITZ, M.D. & DAVYDOV, V.I. 2012. Quantitative radiometric and biostratigraphic calibration of the Pennsylvanian–Early Permian (Cisuralian) time scale and pan-Euramerican chronostratigraphic correlation. *Geological Society of America Bulletin*, **124**, 549–577.

SCHOEPFER, S., HENDERSON, C.M., GARRISON, G. & WARD, P. 2012. Cessation of a productive coastal upwelling system in the Panthalassic Ocean at the Permian–Triassic Boundary. *Palaeogeography, Palaeoclimatology, Palaeoecology*, **313–314**, 181–188.

SHEN, S.Z. 2007. The conodont species *Clarkina orientalis* (Barskov & Koroleva 1970) and its spatial and temporal distribution. *Permophiles*, **50**, 25–37.

SHEN, S.Z. & MEI, S.L. 2010. Lopingian (Late Permian) high-resolution conodont biostratigraphy in Iran with comparison to South China zonation. *Geological Journal*, **45**, 135–161.

SHEN, S.Z., WANG, Y., HENDERSON, C.M., CAO, C.Q. & WANG, W. 2007. Biostratigraphy and lithofacies of the Permian System in the Laibin–Heshan area of Guangxi, South China. *Palaeoworld*, **16**, 120–139.

SHEN, S.Z., HENDERSON, C.M. ET AL. 2010. High-resolution Lopingian (Late Permian) timescale of South China. *Geological Journal*, **45**, 122–134.

SHEN, S.Z., CROWLEY, J.L. ET AL. 2011. Calibrating the end-Permian mass extinction. *Science*, **334**, 1367–1372.

SHEN, S.Z., YUAN, D.X., HENDERSON, C.M., TAZAWA, J. & ZHANG, Y.C. 2012. Implications of Kungurian (Early Permian) conodonts from Hatahoko, Japan, for correlation between the Tethyan and International Timescales. *Micropaleontology*, **58**, 505–522.

SHEN, S.Z., CAO, C.Q. ET AL. 2013. High resolution Del13Ccarb chemostratigraphy from latest Guadalupian through earliest Triassic in South China and Iran. *Earth and Planetary Science Letters*, **375**, 156–165.

SUÁREZ-RIGLOS, M., HÜNICKEN, M.A. & MERINO, D. 1987. Conodont biostratigraphy of the Upper Carboniferous–Lower Permian rocks of Bolivia. *In*: AUSTIN, R.L. (ed.) *Conodonts: Investigative Techniques and Applications*. British Micropalaeontological Society, London & Ellis Horwood, Chichester, 317–325.

SWEET, W.C. & MEI, S.L. 1999*a*. Conodont succession of Permian Lopingian and basal Triassic in northwest Iran. *In*: YIN, H.F. & TONG, J.N. (eds) *Proceedings of the International Conference on Pangea and the Paleozoic–Mesozoic Transition*. China University of Geosciences Press, Wuhan, 43–47.

SWEET, W.C. & MEI, S.L. 1999*b*. The Permian Lopingian and basal Triassic sequence in Northwest Iran. *Permophiles*, **33**, 14–18.

UENO, K., MIZUNO, Y., WANG, X.-D. & MEI, S.-L. 2002. Artinskian conodonts from the Dingjiazhai Formation of the Baoshan Block, West Yunnan, southwest China. *Journal of Paleontology*, **76**, 741–750.

WANG, C.Y. & WANG, Z.H. 1981. Permian conodont biostratigraphy of China. *In*: TEICHERT, C., LIU, L. & CHEN, P.J. (eds) *Paleontology in China, 1979*. Geological Society of America, Special Papers, **187**, 227–236.

WANG, C.Y., RITTER, S.M. & CLARK, D.L. 1987. The *Sweetognathus* complex in the Permian of China: implications for evolution and homeomorphy. *Journal of Paleontology*, **61**, 1047–1057.

WANG, C.Y., ZHEN, C., PEN, Y. & WANG, G. 2000. A conodont fauna of Permian northern temperate zone from the Fanjiatun Formation at Lijiayao, Jilin. *Acta Micropalaeontologica Sinica*, **17**, 430–432.

WANG, Y., SHEN, S., CAO, C., WANG, W., HENDERSON, C.M. & JIN, Y.G. 2006. The Wuchiapingian–Changhsingian boundary (Upper Permian) at Meishan of Changxing County, South China. *Journal of Asian Earth Sciences*, **26**, 575–583.

WANG, Y., SADLER, P.M., SHEN, S.Z., ERWIN, D.H., ZHANG, Y.C., WANG, X.D. & HENDERSON, C.M. 2013. Quantifying the process and abruptness of the end-Permian mass extinction. *Paleobiology*, **40**, 113–129.

WANG, Z.H. 1994. Early Permian conodonts from the Nashui Section, Luodian of Guizhou. *Palaeoworld*, **4**, 203–224.

WARDLAW, B.R. 2000. Guadalupian conodont biostratigraphy of the Glass and Del Norte Mountains. *In*: WARDLAW, B.R., GRANT, R.E. & ROHR, D.M. (eds) *The Guadalupian Symposium*. Smithsonian Contributions to the Earth Sciences, **22**, 37–87.

WARDLAW, B.R. 2001. Smooth gondolellids from the Kungurian and Guadalupian of the western US. *Permophiles*, **38**, 22–24.

WARDLAW, B.R. & COLLINSON, J.W. 1979. Youngest Permian conodont faunas from the Great Basin and Rocky Mountain regions. *Brigham Young University Geology Studies*, **26**, 151–163.

WARDLAW, B.R. & COLLINSON, J.W. 1984. Conodont paleoecology of the Permian Phosphoria Formation and related rocks of Wyoming and adjacent areas. In: CLARK, D.L. (ed.) *Conodont Biofacies and Provincialism*. Geological Society of America, Special Papers, **196**, 263–281.

WARDLAW, B.R. & COLLINSON, J.W. 1986. Paleontology and deposition of the Phosphoria Formation. *Contributions to Geology, University of Wyoming*, **24**, 107–142.

WARDLAW, B.R. & MEI, S.L. 1998*a*. *Clarkina* (Conodont) zonation for the Upper Permian of China. *Permophiles*, **31**, 3–4.

WARDLAW, B.R. & MEI, S.L. 1998*b*. A discussion of the early reported species of *Clarkina* (Permian Conodonta) and the possible origin of the genus. *In*: JIN, Y.G., WARDLAW, B.R. & WANG, Y. (eds) *Permian Stratigraphy, Environments and Resources*. *Palaeoworld*, **9**, 33–52.

Wardlaw, B.R. & Nestell, M.K. 2010*a*. Latest Middle Permian conodonts from the Apache Mountains, West Texas. *Micropaleontology*, **56**, 149–184.

Wardlaw, B.R. & Nestell, M.K. 2010*b*. Three *Jinogondolella* apparatuses from a single bed of the Bell Canyon Formation in the Apache Mountains, West Texas. *Micropaleontology*, **56**, 195–212.

West-Eberhard, M.J. 1989. Phenotypic plasticity and the origins of diversity. *Annual Review of Ecology and Systematics*, **20**, 249–278.

Wignall, P.B., Sun, Y.D. et al. 2009. Volcanism, mass extinction, and carbon isotope fluctuations in the Middle Permian of China. *Science*, **324**, 1179–1182.

Yin, H.F., Zhang, K.X., Tong, J.N., Yang, Z.Y. & Wu, S.B. 2001. The Global Stratotype Section and Point (GSSP) of the Permian–Triassic boundary. *Episodes*, **24**, 275–275.

Youngquist, W., Hawley, R.W. & Miller, A.K. 1951. Phosphoria conodonts from southeastern Idaho. *Journal of Paleontology*, **25**, 356–364.

Yuan, D.X., Shen, S.Z., Henderson, C.M., Chen, J., Zhang, H. & Feng, H.Z. 2014. Revised conodont-based integrated high-resolution timescale for the Changhsingian Stage and end-Permian extinction interval at the Meishan sections, South China. *Lithos*, **204**, 220–245.

Yuan, D., Zhang, Y. et al. 2016. Early Permian conodonts from the Xainza area, central Lhasa Block, Tibet, and their palaeobiogeographical and palaeoclimatic implications. *Journal of Systematic Palaeontology*, **14**, 365–383.

Zhang, K.X., Tong, J.N., Yin, H.F. & Wu, S.B. 1996. Sequence stratigraphy of the Permian–Triassic boundary section of Changxing, Zhejiang. *Acta Geologica Sinica*, **70**, 270–283.

Zhang, K.X., Tong, J.N. et al. 2007. Early Triassic conodont–palynological biostratigraphy of the Meishan D Section in Changxing, Zhejiang Province, South China. *Palaeogeography, Palaeoclimatology, Palaeoecology*, **252**, 4–23.

Zhu, Z.L. & Zhang, L.X. 1994. On the Chihsian Successions in South China. *Palaeoworld*, **4**, 114–137.

Zubin-Stathopoulos, K.D., Beauchamp, B., Davydov, V.I. & Henderson, C.M. 2013. Variability of Pennsylvanian–Permian carbonate associations and implications for NW Pangea Palaeogeography, east-central British Columbia, Canada. *In*: Gąsiewicz, A. & Słowakiewicz, M. (eds) *Palaeozoic Climate Cycles: Their Evolutionary and Sedimentological Impact*. Geological Society, London, Special Publications, **376**, 47–72, https://doi.org/10.1144/SP376.1

Permian radiolarian biostratigraphy

LEI ZHANG[1]*, QINGLAI FENG[1,2] & WEIHONG HE[1,3]

[1]*School of Earth Sciences, China University of Geosciences, 388 Lumo Road, Wuhan 430074, China*

[2]*State Key Laboratory of Geological Processes and Mineral Resources, China University of Geosciences, 388 Lumo Road, Wuhan 430074, China*

[3]*State Key Laboratory of Biogeology and Environmental Geology, China University of Geosciences, 388 Lumo Road, Wuhan 430074, China*

**Correspondence: zhangleirad@cug.edu.cn*

Abstract: The Permian Period is one of the most varied times, not only in oceanic environments, but also for the evolution of the radiolarian assemblages. Permian radiolarians have been extensively studied, and great progress has been achieved recently in their biostratigraphy. Seventeen Permian radiolarian zones are summarized here. Among these, seven zones are assigned to the latest Carboniferous–early Permian, namely (in ascending order): the *Pseudoalbaillella bulbosa* Assemblage Zone, the *Pseudoalbaillella u-forma–Pseudoalbaillella elegans* Assemblage Zone, the *Pseudoalbaillella lomentaria–Pseudoalbaillella sakmarensis* Assemblage Zone, the *Pseudoalbaillella rhombothoracata* Interval Zone, the *Albaillella xiaodongensis* Assemblage Zone, the *Albaillella sinuata* Abundance Zone and the *Pseudoalbaillella ishigai* Abundance Zone. Five zones are attributed to the middle Permian, including the *Pseudoalbaillella globosa* Interval Zone, the *Follicucullus monacanthus* Interval Zone, the *Follicucullus porrectus* Interval Zone, the *Follicucullus scholasticus* Interval Zone and the *Follicucullus charveti* Interval Zone. Five zones belong to the middle Permian–late Permian, including the *Albaillella cavitata* Interval Zone, the *Albaillella levis* Interval Zone, the *Albaillella excelsa* Interval Zone, the *Albaillella triangularis* Interval Zone and the *Albaillella yaoi* Abundance Zone. These Permian radiolarian biozones and their correlations with conondont zones or other chronostratigraphic schemes are discussed. The phylogenetic model of radiolarians through the Permian Period is also revised.

The Permian Period is one of the most varied time intervals in Earth history in terms of events. Several episodes of chert events happened worldwide, with diversified radiolarian assemblages documented in the last decades (Murchey & Jones 1992; Beauchamp & Baud 2002). In addition, different scales of chert gaps, along with distinct biocrises of radiolarians, were discovered in several intervals in the Permian (Isozaki 1994, 2009; Racki 1999, 2003; De Wever *et al.* 2006; Knoll *et al.* 2007; Feng & Algeo 2014). Conodonts have been used for decades as index fossils for international bio-chronostratigraphic correlations of Permian. However, in general, they are rare or absent in Permian deep-water chert sequences, whereas radiolarians are usually common (e.g. Feng & Algeo 2014). The radiolarian biozones were discovered in the chert strata of open-marine facies, slope facies and platform-basinal facies; some occur in tectonic or sedimentary mélanges and ophiolitic sequences throughout the world (Sashida & Tonishi 1985; Catalano *et al.* 1989, 1991; Miyamoto & Tanimoto 1993; Feng & Liu 2002; Sashida & Salyapongse 2002; Saesaengseerung *et al.* 2007; Wang & Yang 2011). Therefore, they are crucial in determining the timing of plate collisions and usually serve as the sole evidence for dating ophiolites within collision belts (Sashida & Tonishi 1985; Catalano *et al.* 1989; Wang *et al.* 1994, 2006; Wu *et al.* 1994; Sashida & Salyapongse 2002; Saesaengseerung *et al.* 2007; Wang & Yang 2007, 2011; Feng & Algeo 2014).

The preliminary radiolarian zonations for the Permian were first introduced based on material from North America (Holdsworth & Jones 1980). Two assemblages have been proposed from the early–middle Permian: the *Pseudoalbaillella* and *Follicucullus* assemblages. The two assemblages were also reported in the Sasayama area in SW Japan (Ishiga & Imoto 1980). Furthermore, several sub-assemblages have been recognized, including (in ascending order): the *P. u-forma–P. elegans*, the *P. lomentaria–P. longicornis* and the *P.* sp. A–*P. rhombothoracata* sub-assemblages (Ishiga & Imoto 1980). Together with the discovery of new species of *Albaillella*, *Neoalbaillella* and *Pseudoalbaillella*, more assemblages or sub-assemblages

From: Lucas, S. G. & Shen, S. Z. (eds) 2018. *The Permian Timescale*. Geological Society, London, Special Publications, **450**, 143–163.
First published online October 30, 2017, https://doi.org/10.1144/SP450.16

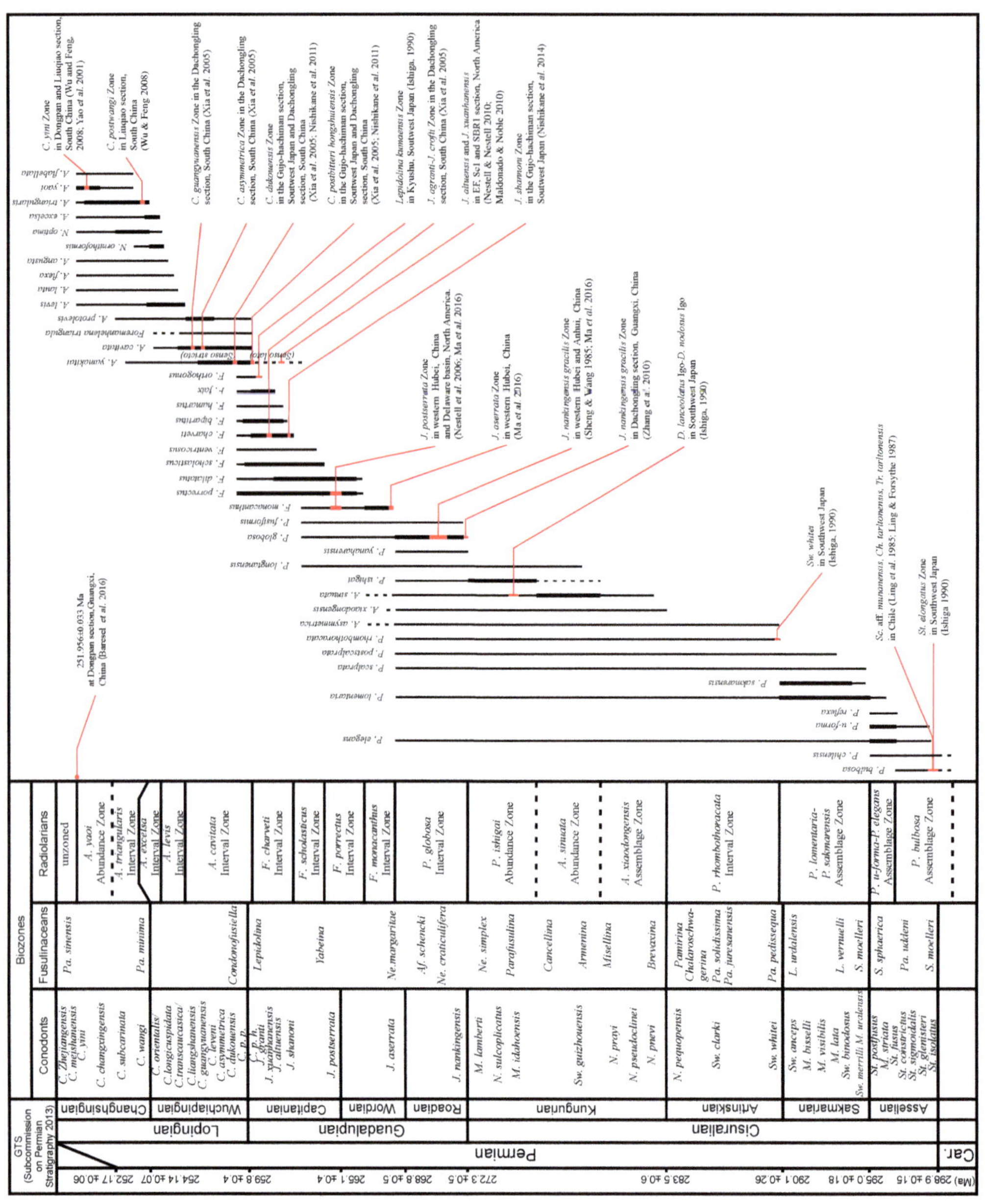

were recognized, namely the *Neoalbaillella* assemblage, the *Neoalbaillella–A. triangularis* sub-assemblage and the *N. ornithoformis* sub-assemblage in the late Permian (Ishiga *et al.* 1982*b*). Basically, the Permian albaillellids have received biostratigraphic importance because they display the most distinct shapes and short stratigraphic ranges (Wang *et al.* 2006; Wu & Feng 2008; Wang & Yang 2011). Among these, *Pseudoalbaillella* mainly represented the dominant species for early Permian biostratigraphy and *Follicucullus* includes the most important zone fossils for the middle Permian, whereas *Albaillella* and *Neoalbaillella* were found abundantly in late Permian strata, although some species, such as *A. xiaodongensis*, *A. sinuata* and *A. asymmetrica*, are also relatively common during the early and middle Permian.

Four assemblage zones were discriminated initially within the early Permian by Ishiga *et al.* (1982*a*), the *P. u-forma–P. elegans*, the *P. lomentaria*, the *P. rhombothoracata* and the *A. sinuata* assemblage zones. The *P. longtanensis* Assemblage Zone and the *A. xiaodongensis* Interval Zone, as well as some abundance zones, were later proposed based on material from SW Japan and South China (Ishiga 1990; Wang & Yang 2011). In addition, the *P. u-forma–P. elegans* Assemblage Zone was further divided into the *P. u-forma* m. I Zone and the *P. u-forma* m. II Zone (Ishiga 1986, 1990). In the South Urals and adjacent areas, numerous early Permian radiolarian zones have been recommended (Afanasieva 2000).

Regarding middle Permian biostratigraphy, four assemblage zones were known formerly: *P. globosa*, *F. monacanthus*, *F. scholasticus* and *F. bipartitus–F. charveti* assemblage zones (Ishiga *et al.* 1982*b*; Ishiga 1984, 1986, 1990). Later, a number of radiolarian interval zones were established for the middle Permian, based on the successive first appearance bioevents of some distinct albaillellid species, namely (in ascending order): *P. globosa*, *P. fusiformis*, *F. monacanthus*, *F. porrectus*, *F. scholasticus* and *F. charveti* (Sun & Xia 2006; Wang *et al.* 2012; Zhang *et al.* 2014).

Previously, two assemblage zones were widely recognized for the late Permian, namely the *N. ornithoformis* and *N. optima* zones (Ishiga *et al.* 1982*a*, *b*; Kuwahara *et al.* 1998). However, due to the relatively low diversity and evolutionary rate of the *Neoalbaillella* species, four *Albaillella* abundance Zones, including the *A. levis*, *A. flexa*, *A. excelsa* and *A. triangularis* zones, were established based on material from SW Japan (Kuwahara *et al.* 1998). Subsequently, six radiolarian assemblage zones were suggested based on data from the late Permian of South China, including the *F. bipartitus–F. charveti–F. orthogonus* Assemblage Zone, the *Foremanhelena triangula* Abundance Zone, the *A. protolevis* Assemblage Zone, the *A. levis–A. excelsa* Assemblage Zone, the *N. ornithoformis* Assemblage Zone and the *N. optima* Assemblage Zone (Wang *et al.* 2006). In addition, various representations of *Albaillella* zones have been reported from China, SE Asia and the USA (e.g. Blome & Reed 1992; Sashida *et al.* 2000*a*, *b*; Xia *et al.* 2004; Feng *et al.* 2009).

Regardless, the number of Permian radiolarian zones gradually increased from the original nine assemblage zones to at least 17 zones in about 30 years of development. In this study, the Permian radiolarian biozones and their correlations with the conondont zones or other chronostratigraphic schemes are summarized (Fig. 1) and the phylogenetic model of radiolarians through the Permian Period is also revised (Fig. 2).

Radiolarian zones

Pseudoalbaillella bulbosa Assemblage Zone

Definition. This zone was defined by Ishiga (1982) in SW Japan. The base of the zone is defined by the first occurrence (FO) of the zonal index fossil *P. bulbosa* without *P. nodosa* in the same stratigraphic section, and the top is marked by the FO of *P. u-forma* (=*P. u-forma* m. I according to Ling & Forsythe 1987) and *P. elegans* (Fig. 1).

Type section. Locations 1 and 2 in Maruyama, Sasayama Town, Hyogo Prefecture, SW Japan for the base, and the Tianche section in Bancheng Town, Guangxi, South China for the definition of the top of this zone.

Remarks. The lower part of the *P. bulbosa* Assemblage Zone can be correlated with the *P. bulbosa* Assemblage Zone in SW Japan (Ishiga 1986, 1990). Its upper part can be correlated with the *P. u-forma* m. I Assemblage Zone in South China (Yao *et al.* 2004; Shimakawa & Yao 2006), SW Japan (Ishiga 1986, 1990) and the Malay Peninsula (Spiller 1996) (Fig. 3). This zone can also be correlated with the *P. bulbosa* Assemblage in

Fig. 1. Chart showing Permian radiolarian zonations, the ranges of the zonal and common species, and the biozone correlations. The Permian Timescale 2013 (Shen *et al.* 2013) was followed in this study. Thick vertical bars are abundant intervals; dotted vertical lines are rare range intervals (the ranges of fossils are mainly based on Ling & Forsythe 1987; Kuwahara 1999; Jin *et al.* 2007; Feng *et al.* 2009; Wu *et al.* 2010; Wang & Yang 2011; Wang *et al.* 2012; Zhang *et al.* 2014).

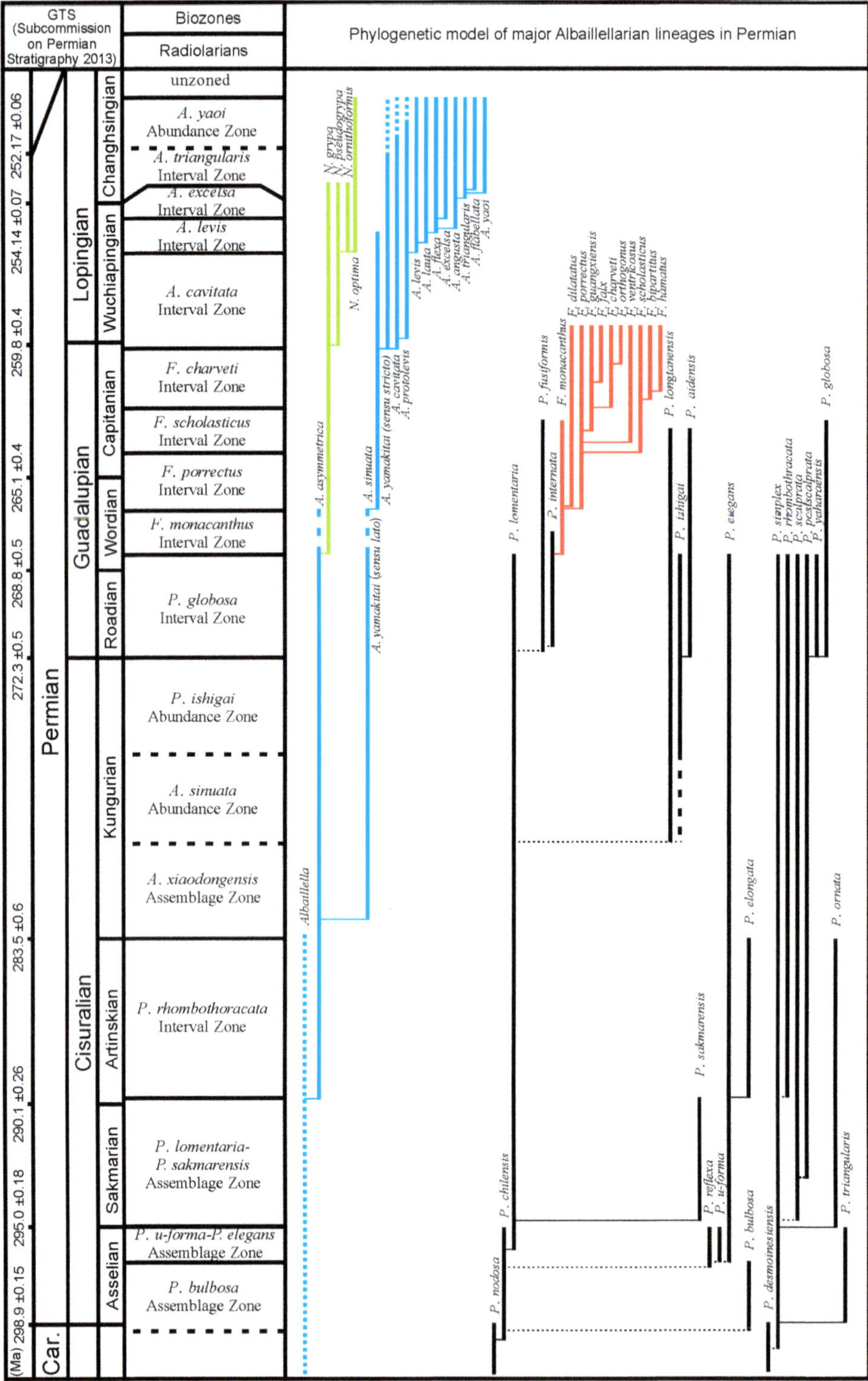

Fig. 2. The phylogenetic model of *Albaillella*, *Pseudoalbaillella*, *Follicucullus* and *Neoalbaillella* lineages in the Permian (modified from Kuwahara 1999 and Zhang *et al.* 2014).

GTS (Subcommission on Permian stratigraphy, 2016)			This study	Ishiga, 1986, 1990	Shimakawa and Yao, 2006	Wang and Yang, 2007, 2011
Permian	Guadalupian	Roadian				
Permian	Cisuralian	Kungurian	*P. ishigai* Abundance-Zone	*P. longtanensis* Zone	*P. longtanensis* Zone	*P. ishigai* Abundance-Zone
Permian	Cisuralian	Kungurian	*A. sinuata* Abundance-Zone	*A. sinuata* Zone	*A. sinuata* Zone	*A. sinuata* Abundance-Zone
Permian	Cisuralian	Kungurian	*A. xiaodongensis* Assemblage-Zone			*A. xiaodongensis* Interval-Zone
Permian	Cisuralian	Artinskian	*P. rhombothoracata* Interval-Zone	*P. scalprata* m. *rhombothoracata* Zone	*P. scalprata* m. *rhombothoracata* Zone	*P. rhombothoracata* Interval-Zone
Permian	Cisuralian	Sakmarian	*P. lomentaria*-*P. sakmarensis* Assemblage-Zone	*P. lomentaria* Zone	*P. lomentaria* Zone	*P. lomentaria*-*P. sakmarensis* Assemblage-Zone
Permian	Cisuralian	Asselian	*P. u-forma*-*P. elegans* Assemblage-Zone	*P. u-forma* m. II Zone	*P. u-forma* m. II Zone	*P. u-forma*-*P. elegans* Assemblage-Zone
Permian	Cisuralian	Asselian	*P. bulbosa* Assemblage-Zone	*P. u-forma* m. I Zone	*P. u-forma* m. I Zone	*P. bulbosa* Assemblage-Zone
Permian	Cisuralian	Asselian		*P. bulbosa* Zone		
Carb.	Late	Gzhelian				

Fig. 3. Correlation of the radiolarian biozones for the lower Permian.

Chile (Ling & Forsythe 1987) and Thailand (Sashida & Salyapongse 2002; Saesaengseerung *et al.* 2009). Besides the index species, *P. simplex*, *P. triangularis*, *P. chilensis* and *P. elegans* are also very common in this zone.

Distribution. South China (Zhang *et al.* 2002; Shimakawa & Yao 2006; Wang *et al.* 2006, 2012; Wang & Yang 2007, 2011), SW Japan (Ishiga 1986, 1990; Hori 2005), Thailand (Sashida & Nakornsri 1997), North America (Murchey 1991), Central British Columbia (Orchard *et al.* 2001), Chile (Ling & Forsythe 1987) and the South Urals in Russia (Nazarov & Ormiston 1985; Nazarov 1988).

Age. Probably late Gzhelian (latest Carboniferous) to early Asselian (earliest Permian).

Pseudoalbaillella u-forma–Pseudoalbaillella elegans Assemblage Zone

Definition. This zone was defined by Wang *et al.* (1994) in South China. The base of the zone is defined by the FO of the zonal species *P. u-forma* and *P. elegans*, and the top is marked by the FO of *P. sakmarensis* and abundant *P. lomentaria.*

Type section. Tianche section in Bancheng Town, Guangxi, South China.

Remarks. This zone can be correlated with the *P. u-forma* m. II (=*P. reflexa* according to Ling & Forsythe 1987) Assemblage Zone in South China (Shimakawa & Yao 2006) and SW Japan (Ishiga 1986, 1990) (Fig. 3). The main associated radiolarian species include *P. reflexa*, *P. triangularis* and *P. chilensis.*

Distribution. SE Guangxi, western Yunnan and southern Qinghai in China (Wang *et al.* 1994, 2006, 2012; Zhang *et al.* 2000; Wang & Yang 2007, 2011; Xie *et al.* 2011), SW Japan (Ishiga 1986, 1990; Hori 2005), the west coast of America (Holdsworth & Jones 1980; Murchey & Jones 1992), Canada (Harms & Murchey 1992), the Malay Peninsula (Spiller 1996, 2002; Jasin & Harun 2011), Thailand (Sashida *et al.* 1993; Sashida & Nakornsri 1997;

Saesaengseerung *et al.* 2008, 2009), Chile (Ling & Forsythe 1987), and Far East Russia (Rudenko & Panasenko 1990).

Age. Late Asselian (early Cisuralian).

Pseudoalbaillella lomentaria–Pseudoalbaillella sakmarensis Assemblage Zone

Definition. This zone was established by Wang *et al.* (1994) in South China. The base of the zone is defined by the abundance of *P. lomentaria* and the FO of *P. sakmarensis*, and the top is marked by the FO of *P. rhombothoracata* (Fig. 1).

Type section. Tianche section in Bancheng Town, Guangxi, South China.

Remarks. This zone can be correlated with the *P. lomentaria* Range Zone in South China (Shimakawa & Yao 2006) and SW Japan (Ishiga 1986, 1990) (Fig. 3). The common species of this zone are *P. scalprata* and *P. postscalprata.*

Distribution. South Guangxi (Wang *et al.* 1994, 2006, 2012; Shimakawa & Yao 2006; Wang & Yang 2007, 2011), western Yunnan (Feng & Ye 1996; Feng & Liu 2002; Xie *et al.* 2011), Tibet (Zhu *et al.* 2006) and Qinghai in China (Li & Bian 1993; Zhang *et al.* 2000), SW Japan (Ishiga 1986, 1990; Kuwahara 1992; Nakae 2001), Thailand (Saesaengseerung *et al.* 2009), the Malay Peninsula (Spiller 2002; Jasin & Harun 2011; Dzulkafli *et al.* 2012; Jasin *et al.* 2013), Oregon in the USA (Blome & Reed 1992), Canada (Harms & Murchey 1992), the Urals and Far East Russia (Kozur 1981; Nazarov & Ormiston 1985; Nazarov 1988; Rudenko & Panasenko 1990, 1997; Suzuki *et al.* 2005).

Age. Sakmarian (early middle Cisuralian).

Pseudoalbaillella rhombothoracata Interval Zone

Definition. Previously, this zone was established as an assemblage zone by Wang *et al.* (1994) in South China. Here, it is regarded as an interval zone. The base of the zone is defined by the FO of *P. rhombothoracata* and the top is marked by the FO of *A. xiaodongensis.*

Type section. Shijia reservoir section in Bancheng Town, Guangxi, South China.

Remarks. This zone can be correlated with the *P. scalprata* m. *rhombothoracata* Assemblage Zone in SW Japan (Ishiga 1986, 1990; Yamanaka 2001; Hori 2004) and the lower–middle part of the *P. scalprata* m. *rhombothoracata* Assemblage Zone in South China (Shimakawa & Yao 2006) (Fig. 3). The main associated radiolarians are *P. scalprata*, *P. longicornis*, *P. inflata* and *P. postscalprata.*

Distribution. SE Guangxi, SW Yunnan, Tibet, Inner Mongolia and South Qinghai in China (Wang *et al.* 1994, 2006, 2012; Bian *et al.* 1997; Sha 1998; Zhang *et al.* 2000; Wang & Yang 2007, 2011; Xie *et al.* 2011), SW Japan (Ishiga 1986, 1990; Yamanaka 2001; Hori 2004), the Malay Peninsula (Spiller & Metcalfe 1995; Jasin 1997; Jasin & Ali 1997; Spiller 2002; Jasin *et al.* 2005, 2013; Jasin & Harun 2007, 2011), Thailand (Sashida *et al.* 1993), central Oregon in the USA (Blome & Reed 1992), and Far East Russia (Rudenko & Panasenko 1997).

Age. Artinskian (middle late Cisuralian).

Albaillella xiaodongensis Assemblage Zone

Definition. This zone was established by Wang *et al.* (1994) in South China. The base of the zone is defined by the FO of *A. xiaodongensis* and the top is marked by the abundance of the *A. sinuata* (Figs 1 & 2).

Type section. Shiti reservoir section in Bancheng Town, Guangxi, South China.

Remarks. This zone can be correlated with the upper part of the *P. scalprata* m. *rhombothoracata* Assemblage Zone in South China (Shimakawa & Yao 2006) and the *A. xiaodongensis* Assemblage or Interval Zone in Guangxi, South China (Wang *et al.* 1994; Wang & Yang 2007, 2011). It can also be correlated with the lower part of the *A. sinuata* Zone in SW Japan (Ishiga 1986, 1990) (Fig. 3). The following radiolarian species are also developed during this zone: *P. scalprata*, *P. longicornis*, *P. fusiformis*, *Latentifistula texana*, *Nazarovella scalae* and *A. sinuata.*

Distribution. SE Guangxi (Wang *et al.* 1994, 2006, 2012; Shimakawa & Yao 2006; Wang & Yang 2007, 2011), western Yunnan (Ishiga 1986, 1990; Feng & Ye 1996) and Qinghai in China (Zhao *et al.* 2016), SW Japan (Ishiga 1986, 1990), the Malay Peninsula (Jasin & Harun 2011), Europe (Kozur 1993), and Far East Russia (Suzuki *et al.* 2005).

Age. Early Kungurian (late Cisuralian).

Albaillella sinuata Abundance Zone

Definition. This zone was established by Wang & Yang (2011) in South China. The base of the zone is defined by the abundant occurrence of the zonal species *A. sinuata* and the top is marked by the abundance of *P. ishigai.*

Type section. Shiti reservoir section in Bancheng Town, Guangxi, South China.

Remarks. This zone can be correlated with the *A. sinuata* Assemblage Zone in South China (Wang *et al.* 1994; Shimakawa & Yao 2006) and the middle part of *A. sinuata* Range Zone in SW Japan (Ishiga 1986, 1990; Hori 2004; Kurihara & Kametaka 2008) (Figs 1 & 3). The main associated species include *P. scalprata*, *P. longicornis*, *A. xiaodongensis* and *A. asymmetrica*.

Distribution. South China (Feng 1992; Kuwahara *et al.* 1997; Wang & Yang 2007, 2011), SW Japan (Ishiga 1986, 1990; Nakae 2001), the Malay Peninsula (Spiller & Metcalfe 1995; Spiller 2002), North America (Blome & Reed 1992; Murchey & Jones 1992) and Far East Russia (Suzuki *et al.* 2005).

Age. Middle Kungurian (late Cisuralian).

Pseudoalbaillella ishigai Abundance Zone

Definition. This zone is established by Wang & Yang (2011) in South China. The base of the zone is defined by the abundant presence of the species *P. ishigai* and the top is marked by the FO of *P. globosa* (Figs 1, 2 & 4).

Type section. Shiti reservoir section in Bancheng Town, Guangxi, South China.

Remarks. The zone is correlated with the *P. longtanensis* Assemblage Zone in South China (Wang *et al.* 1994), SW Japan (Ishiga 1986, 1990; Shimakawa & Yao 2006) and the Malay Peninsula (Spiller & Metcalfe 1995; Spiller 2002) (Fig. 3). This zone is mainly associated with species *P. scalprata*, *P. longicornis* and *A. inflata.*

Distribution. Guangxi, Qiangtang and Tibet in China (Wang *et al.* 1994; Zhu *et al.* 2006; Wang & Yang 2007, 2011), SW Japan (Ishiga 1986, 1990), and Malaysia (Spiller 2002).

Age. Late Kungurian (late Cisuralian).

Pseudoalbaillella globosa Interval Zone

Definition. This zone is established by Sun & Xia (2006). The base of the zone is defined by the FO of *P. globosa* and the top is marked by the FO of *F. monacanthus* (Figs 1, 2 & 4).

Type section. Dachongling section in Xiaodong Town, Guangxi, South China.

Remarks. The zone is widely distributed worldwide. It can be correlated with the *P. globosa* Assemblage Zone in many areas, such as South China (Xia *et al.* 2005; Wang *et al.* 2006, 2012; Wang & Yang 2007, 2011; Ma *et al.* 2016; Shi *et al.* 2016) and SW Japan (Ishiga 1986, 1990; Nakae 2001), etc. (Fig. 5).

This zone is associated mainly with the following radiolarian species: *P. fusiformis*, *P. yanaharensis*, *P. longtanensis*, *P. internata*, *P. scalprata*, *P. longicornis*, *P. simplex*, *P. lanceolata*, *P. lomentaria*, *A. inflata* and *Hegleria mammilla.*

Distribution. South and SW China (Wang *et al.* 1994, 2006, 2012; Wang 1995; Yao & Kuwahara 1999; Sun & Xia 2006; Wang & Yang 2007, 2011; Xie *et al.* 2011; Ito *et al.* 2013; Ma *et al.* 2016; Shi *et al.* 2016), SW Japan (Ishiga 1986, 1990; Nakae 2001), North America (Blome & Reed 1992; Murchey & Jones 1992), North Island of New Zealand (Murchey & Jones 1992), Malaysia (Spiller 2002), and Far East Russia (Rudenko & Panasenko 1990, 1997; Suzuki *et al.* 2005).

Age. Roadian (early Guadalupian).

Follicucullus monacanthus Interval Zone

Definition. Previously, this zone was established by Wang & Yang (2011). The base of this zone is marked by the FO of *F. monacanthus* and the top is marked by the FO of *F. scholasticus*. This zone was redefined by Zhang *et al.* (2014), and the renewed definition of this zone is as follows: the base of the zone is defined by the FO of *F. monacanthus* and the top is marked by the FO of *F. porrectus* (Figs 1, 2 & 4). Thus, the span of *F. monacanthus* Zone in this study was shorter than the previous one established by Wang & Yang (2011).

Type section. Gujingling section in Xiaodong Town, Guangxi, China.

Remarks. This zone is correlated with the lower–middle part of the previous *F. monacanthus* Zone in South China (Xia *et al.* 2005; Wang & Yang 2007, 2011; Ma *et al.* 2016) and SW Japan (Ishiga 1986, 1990). The basal boundary of this zone could be slightly above the Roadian–Wordian boundary according to Yao *et al.* (2015). Thus, the bottom of the *F. monacanthus* Zone is slightly higher than all previous *F. monacanthus* zones. *F. monacanthus*, *P. fusiformis*, *P. globosa*, *P.* cf. *longicornis* and *P. longtanensis* are common species in this zone.

Distribution. Hubei, Anhui, Jiangsu, Guizhou, Yunnan and Guangxi in China (Wang *et al.* 1994, 2006; Kuwahara *et al.* 1997, 2007; Yao & Kuwahara 2004; Yao *et al.* 2004; Sun & Xia 2006; Kametaka *et al.* 2009; Xie *et al.* 2011; Zhang *et al.* 2014;

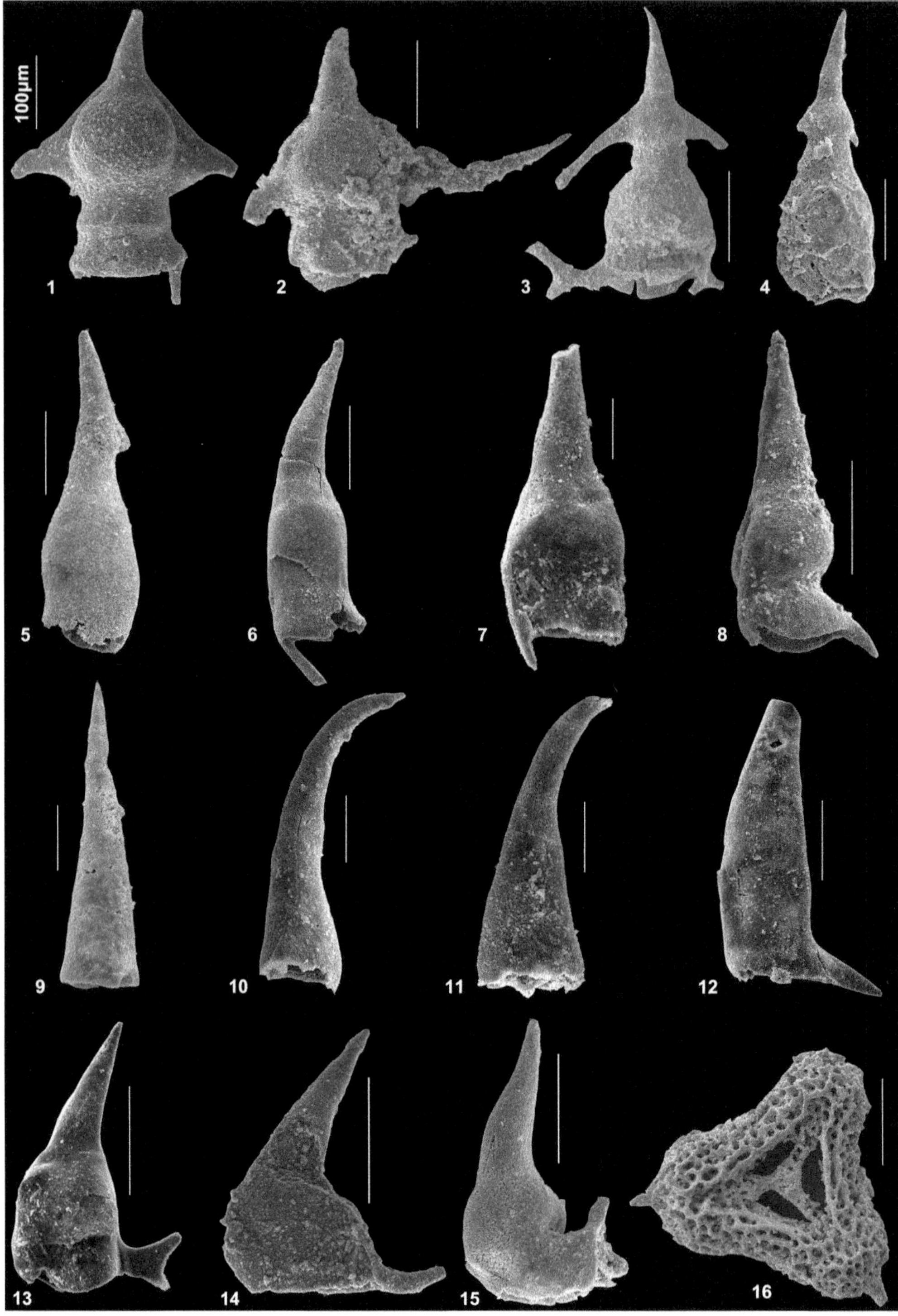

Fig. 4. The SEM photographs of representative species from the middle–late Permian radiolarian zones. All the specimens are from the Xiaodong area, Guangxi, South China. The localities of the sections mentioned below refer to Zhang *et al.* (2014). All the scale bars are 100 μm. **1**, *Pseudoalbaillella globosa* Ishiga and Imoto. GJL1-3/003,

GTS (Subcommission on Permian Stratigraphy 2016)			This study	Ishiga (1986, 1990)	Xia *et al.* (2005)	Wang & Yang (2007, 2011)	Ma *et al.* (2016)
Permian	Lopingian	Wuchiapingian	*A. cavitata* Interval Zone				
Permian	Guadalupian	Capitanian	*F. charveti* Interval Zone *F. scholasticus* Interval Zone	*F. scholasticus* m. II Zone: *F. charveti*-*F. bipartitus* Zone *F. scholasticus* m. I Zone	*F. falx*-*Foremanhelena triangula* Zone *F. charveti*-*F. bipartitus* Zone *F. dilatatus* Zone	*F. scholasticus*–*F. ventricosus* Assemblage Zone	*F. scholasticus* Interval Zone
Permian	Guadalupian	Wordian	*F. porrectus* Interval Zone *F. monacanthus* Interval Zone	*F. monacanthus* Zone	*F. scholasticus* Zone *F. monacanthus* Zone	*F. monacanthus* Interval Zone	*F. monacanthus* Interval Zone
Permian	Guadalupian	Roadian	*P. globosa* Interval Zone	*P. globosa* Zone	*P. fusiformis* Zone *P. globosa* Zone	*P. globosa* Assemblage Zone	*P. globosa* Assemblage Zone
Permian	Cisuralian	Kungurian					

Fig. 5. Correlation of the radiolarian biozones for the middle Permian.

Ma *et al.* 2016; Shi *et al.* 2016), SW Japan (Ishiga 1986, 1990; Kawai & Takeuchi 2001), North Thailand (Sashida *et al.* 1993; Sashida & Salyapongse 2002; Wonganan & Caridroit 2007), Far East Russia (Suzuki *et al.* 2005), and the Grindstone Terrane of central Oregon, USA (Blome & Reed 1992).

Age. Early Wordian (middle Guadalupian).

Follicucullus porrectus Interval Zone

Definition. The *F. porrectus* Interval Zone was defined by Zhang *et al.* (2014). The base of the zone is defined by the FO of *F. porrectus* and the top is marked by the FO of *F. scholasticus* (Figs 1, 2 & 4).

Type section. Gujingling section in Xiaodong Town, Guangxi, South China.

Remarks. This zone is correlated with the upper part of the previous *F. monacanthus* Zone in SW Japan and South China (Ishiga 1986, 1990; Wang & Yang 2007, 2011; Ma *et al.* 2016), and with the *F. scholasticus* Zone in South China (Xia *et al.* 2005) (Fig. 5). The co-occurring species are *F. porrectus*, *F. dilatatus*, *F. monacanthus*, *P.*

Fig. 4. (*Continued*) Gujingling section (pl. 1, fig. 1 in Zhang *et al.* 2014). **2**, *Pseudoalbaillella yaharaensis* Nishimura and Ishiga, GJL11-5/015, Gujingling section (pl. 1, fig. 3 in Zhang *et al.* 2014). **3**, *Pseudoalbaillella fusiformis* Holdsworth and Jones, GJL1-3/525, Gujingling section. **4**, *Pseudoalbaillella internata* Wang, GJL1-3/1001, Gujingling section. **5**, *Follicucullus monacanthus* Ishiga and Imoto, GJL11-5/003, Gujingling section (pl. 1, fig. 10 in Zhang *et al.* 2014). **6**, *Follicucullus porrectus* Rudenko, GY1-1/027, Guoyuan section (pl. 1, fig. 17 in Zhang *et al.* 2014). **7**, *Follicucullus dilatatus* Rudenko, YTL1-1/017, Yutouling section (pl. 1, fig. 13 in Zhang *et al.* 2014). **8**, *Follicucullus ventricosus* Ormiston and Babcock, GY1-1/018, Yutouling section (pl. 1, fig. 23 in Zhang *et al.* 2014). **9**, *Follicucullus scholasticus* Ormiston and Babcock, GJL11-5/012, Gujingling section (pl. 2, fig. 1 in Zhang *et al.* 2014). **10**, *Follicucullus bipartitus* Caridroit & De Wever, SPL6-8/002, Sanpaoling section (pl. 2, fig. 5 in Zhang *et al.* 2014). **11**, *Follicucullus hamatus* Caridroit & De Wever, SPL6-11/009, Sanpaoling section (pl. 2, fig. 7 in Zhang *et al.* 2014). **12**, *Follicucullus guangxiensis* Wang, SPL5-1/015, Sanpaoling section (pl. 2, fig. 10 in Zhang *et al.* 2014). **13**, *Follicucullus falx* Caridroit & De Wever, sample SPL6-24/036, Sanpaoling section (pl. 2, fig. 11 in Zhang *et al.* 2014). **14**, *Follicucullus charveti* Caridroit & De Wever, SPL6-24/030, Sanpaoling section (pl. 2, fig. 16 in Zhang *et al.* 2014). **15**, *Follicucullus orthogonus* Caridroit & De Wever, YTL1-1/035, Yutoulong section (pl. 2, fig. 20 in Zhang *et al.* 2014). **16**, *Foremanhelena triangula* De Wever and Caridroit, YTL1/032, Yutouling section.

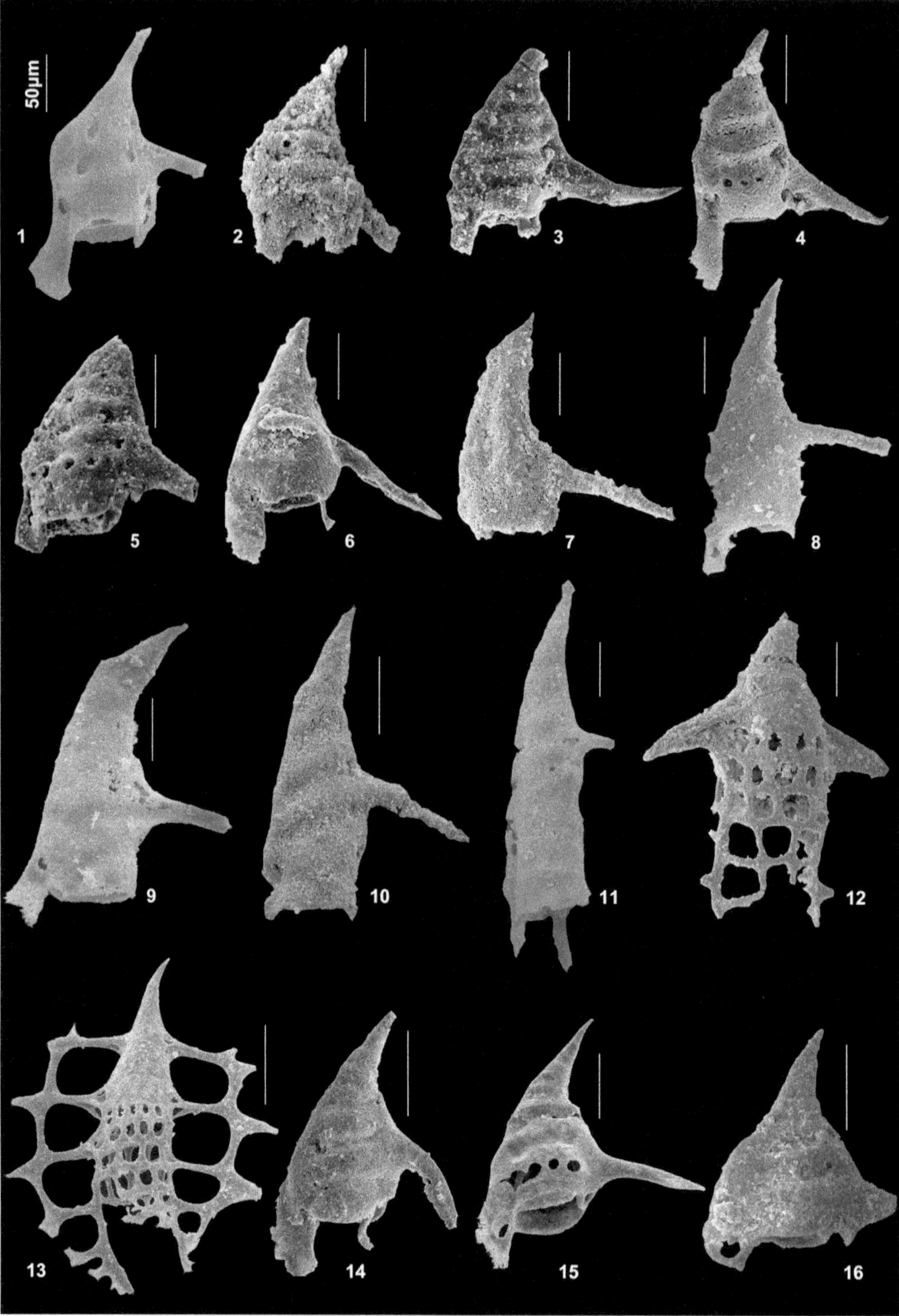

Fig. 6. SEM photographs of representative species from the late Permian radiolarian zones. All the specimens are from the Xiadong, Fusui and Nandan areas, Guangxi, South China. The localities of the sections mentioned below refer to Wu *et al.* (2010) and Zhang *et al.* (2014). All the scale bars are 50 μm. **1**, *Albaillella cavitata* Kuwahara.

fusiformis, *P. globosa*, *P. longicornis* and *P. longtanensis*. The *F. porrectus* Zone is equivalent to the *P. scalprata* m. *rhombothoracata* Zone described in Malaysia (Jasin 1997).

Distribution. Guangxi in China (Zhang *et al.* 2014), Central Japan (Kawai & Takeuchi 2001), North Thailand (Sashida & Salyapongse 2002; Wonganan & Caridroit 2007), Malaysia (Jasin 1997) and East Russia (Suzuki *et al.* 2005).

Age. Middle Wordian–early Capitanian (middle Guadalupian).

Follicucullus scholasticus Interval Zone

Definition. The *F. scholasticus* Zone was first established as the assemblage zone from the Capitanian of SW Japan by Ishiga *et al.* (1982*a*). It was changed into the interval zone by Zhang *et al.* (2014). The base is defined by the FO of *F. scholasticus* and the top by the FO of *F. charveti* (Zhang *et al.* 2014) (Figs 1, 2 & 4).

Type section. Gujingling section in Xiaodong Town, Guangxi, South China.

Remarks. The zone is correlated with the lower part of the *F. scholasticus–F. ventricosus* Assemblage Zone (Wang & Yang 2007, 2011), the *F. dilatatus* Zone and the lower part of the *F. charveti–F. bipartitus* Zone (Xia *et al.* 2005) in South China, and with the *F. scholasticus* m. I Zone and the lower part of the *F. scholasticus* m. II Zone in SW Japan (Ishiga 1986, 1990). It can also be correlated with the *F. sholasticus* Interval Zone in western Hubei, China (Ma *et al.* 2016). *F. porrectus* is particularly abundant, while the species *F. scholasticus*, *F. dilatatus*, *F. monacanthus*, *F. guangxiensis* and *F. ventricosus* are common.

Distribution. This biozone is widespread in South China (Wang & Li 1994; Wang *et al.* 1994, 2006; Wu *et al.* 1994; Kuwahara *et al.* 1997, 2007; Yao *et al.* 2004; Yao & Kuwahara 2004; Sun & Xia 2006; Wang & Yang 2007, 2011; Zhang *et al.* 2014; Ma *et al.* 2016), SW Japan (Ishiga 1986, 1990; Kuwahara & Yao 2001; Yamanaka 2001; Yao *et al.* 2001), central Oregon and Texas in the USA (Blome & Reed 1992; Nestell & Nestell 2010; Maldonado & Noble 2010; Noble & Jin 2010), and Far East Russia (Suzuki *et al.* 2005).

Age. Early–middle Capitanian (middle late Guadalupian).

Follicucullus charveti Interval Zone

Definition. The *F. charveti* Interval Zone was defined by Zhang *et al.* (2014). The base is defined by the FO of *F. charveti* and the top by the FO of *A. cavitata* (Figs 1, 2, 4 & 6).

Type section. Sanpaoling section in Xiaodong Town, Guangxi, China.

Remarks. This zone corresponds to the *F. falx–Foremanhelena triangula* Zone (Xia *et al.* 2005) and the upper part of the *F. scholasticus–F. ventricosus* Assemblage Zone in South China (Wang & Yang 2007, 2011), with the *F. charveti–F. bipartitus* Assemblage Zone and the upper part of the *F. scholasticus* m. II Zone in SW Japan (Ishiga 1986, 1990) (Fig. 5), and the lower part of the *F. charveti–A.* sp. F (=*A. yamakitai* in Kuwahara *et al.* 1998) Assemblage Zone (Kuwahara *et al.* 1998, 2003, 2007; Kuwahara & Yao 2001; Yao *et al.* 2001, 2004; Yao & Kuwahara 2004). Species *F. charveti*, *F. bipartitus*, *F. scholasticus*, *F. ventricosus*, *F. guangxiensis* and *F. orthogonus* are common in this zone.

Distribution. South China (Wang & Li 1994; Wang *et al.* 1994, 2006; Kuwahara *et al.* 1997, 2003; Yao & Kuwahara 2004; Yao *et al.* 2004; Xia *et al.* 2005; Sun & Xia 2006; Zhang *et al.* 2014), SW Japan (Ishiga 1986, 1990; Kuwahara & Yao 2001; Yao *et al.* 2001), East and Far East Russia (Suzuki *et al.* 2005).

Age. Late Capitanian (latest Guadalupian).

Fig. 6. (*Continued*) LWP5–9/181, Longwangpo section. **2**, *Albaillella yamakitai* (*sensu stricto*) Kuwahara, LWP1–11/006, Longwangpo section. 3, *Albaillella yamakitai* (*sensu lato*), SPL6-31(1)/001, Sanpaoling section. **4**, *Albaillella yamakitai* (*sensu lato*), XQL2-12/001, Xiaqianling section. **5**, *Albaillella yamakitai* (*sensu lato*), SPL6-31/001, Sanpaoling section. **6**, *Albaillella protolevis* Kuwahara, YTL6-9(3)/017, Yutouling section. **7**, *Albaillella levis* Ishiga, Kito and Imoto, LWP10–4/002, Longwangpo section. **8**, *Albaillella lauta* Kuwahara, LWP10–16/001, Longwangpo section. **9**, *Albaillella flexa* Kuwahara, LWP10–13/003, Longwangpo section. **10**, *Albaillella angusta* Kuwahara, LWP(2)10–18/007, Longwangpo section. **11**, *Albaillella excelsa* Ishiga, Kito and Imoto, LWP10–18/003, Longwangpo section. **12**, *Neoalbaillella ornithoformis* Takemura and Nakaseko, LWP(2)10–14/009, Longwangpo section. **13**, *Neoalbaillella optima* Ishiga, Kito and Imoto, BL-11/002, Balang section. **14**, *Albaillella triangularis* Ishiga, Kito and Imoto, LQ-22-4/023, Liuqiao section (Fig. 3.4 in Wu & Feng 2008). **15**, *Albaillella yaoi* Kuwahara, XC-4/001, Xichang section. **16**, *Albaillella flabellata* Jin and Feng, BL-3-2/001-01, Balang section.

GTS (Subcommission on Permian stratigraphy, 2016)			This study	Yao et al., 2001	Xia et al., 2004	Wang and Yang, 2007, 2011	Feng et al., 2009
Permian	Lopingian	Changhsingian	unzoned	*N. optima* Assemblage-Zone	*A. simplex* Lineage-Zone	*N. optima* Assemblage-Zone	unzoned
			A. yaoi Abundance-Zone		*A. degradans* Lineage-Zone		*A. yaoi* Range-Zone
					A. yaoi Lineage-Zone	*N. ornithoformis* Assemblage-Zone	
						A. levis-A. excelsa Abundance-Zone	
			A. triangularis Interval-Zone	*N. ornithoformis* Assemblage-Zone	*A. triangularis* Lineage-Zone		*A. triangularis* Interval-Zone
					A. angusta-A. flexa Lineage-Zone	*A. protolevis* Interval-Zone	*A. excelsa* Interval-Zone
		Wuchiapingian	*A. excelsa* Interval-Zone	*F. charveti-A. yamakitai* Assemblage-Zone	*N. optima-A. lauta* Lineage-Zone	*Foremanhelena triangula* Abundance-Zone	*A. levis* Interval-Zone
			A. levis Interval-Zone		*A. levis* Lineage-Zone		
			A. cavitata Interval-Zone		*A. protolevis* Lineage-Zone	*F. bipartitus-F. charveti-F. orthogonous* Assemblage-Zone	*A. cavitata* Interval-Zone
				F. scholasticus-F. ventricosus Assemblage-Zone	*A. postyamakitai* Lineage-Zone		
	Guadalupian	Capitanian					

Fig. 7. Correlation of the radiolarian biozones for the upper Permian.

Albaillella cavitata Interval Zone

Definition. This zone was established for Lopingian assemblages of South China by Feng *et al.* (2009). Its base is defined by the FO of *A. cavitata* and its top by the FO of *A. levis* (Figs 1, 2 & 6).

Type section. The Gujo-hachiman section in SW Japan.

Remarks. The FO of *A. cavitata* and *A. yamakitai* were both recognized as the basal Wuchiapingian (the base of the Lopingian) in both South China and SW Japan (Xia *et al.* 2004; Nishikane *et al.* 2011; Zhang *et al.* 2014). However, the species *A. yamakitai* (*sensu lato*) was also reported from the *Jinogondolella altudaensis* conodont zone of the uppermost Capitanian in the Delaware Basin (Maldonado & Noble 2010; Nestell & Nestell 2010; Noble & Jin 2010). In consideration of the disputes over both the stratigraphic range and the intraspecific variation of this species (Nestell & Nestell 2010), we employed the *A. cavitata* Interval Zone for the first biozone of the Lopingian. This zone can be correlated with the *F. bipartitus–F. charveti–F. orthogonus* Assemblage Zone and the lower part of *Foremanhelena triangula* Abundance Zone in South China (Wang & Yang 2007, 2011), the *F. scholasticus–F. ventricosus* Assemblage Zone and the lower part of the *F. charveti–A. yamakitai* Assemblage Zone (Yao *et al.* 2001), and the *A. postyamakitai* and *A. protolevis* lineage zones in SW Japan (Xia *et al.* 2004) (Fig. 7). This zone is characterized by abundant *A. cavitata*, *A. yamakitai*, *F. orthogonus* and *F. charveti*. Other common taxa in this zone are *F. bipartitus*, *A. protolevis*, *F. scholasticus* and *F. ventricosus*.

Distribution. It is widely recognized in South China (Xia *et al.* 2005; Sun & Xia 2006; Wang *et al.* 2006; Wang & Yang 2007, 2011; Feng *et al.* 2009), SW Japan (Kuwahara *et al.* 1998; Kuwahara 1999, Yao *et al.* 2001; Xia *et al.* 2004), Thailand (Saesaengseerung *et al.* 2007), Malaysia (Jasin 1997), Palawan Philippines (Marquez *et al.* 2006), Anatolia in northern Turkey (Göncüoğlu *et al.* 2004) and the USA (Blome & Reed 1995).

Age. Latest Capitanian–early Wuchiapingian (latest Guadalupian–early Lopingian).

Albaillella levis Interval Zone

Definition. The *A. levis* Abundance Zone was first established in SW Japan by Yao *et al.* (2001) and was correlated to the lower Changhsingian. It was later revised into a lineage zone by Xia *et al.* (2005). Feng *et al.* (2009) defined it as an interval zone. Its base is defined by the FO of *A. levis* and the top corresponds to the FO of the *A. excels* (Figs 1, 2 & 6).

Type section. The Gujo-hachiman section in SW Japan.

Remarks. The *A. levis* Interval Zone can be correlated with the middle part of the *Formanhelena triangula* Abundance Zone in South China (Wang & Yang 2007, 2011). It can also be correlated with the middle part of the *F. charveti–A. yamakitai* Assemblage Zone (Yao *et al.* 2001) and an *A. levis* Lineage Zone (Xia *et al.* 2004) in SW Japan (Fig. 7). Besides the index species, *A. lauta*, *A. flexa*, *A. angusta*, *N. optima* and *N. ornithoformis* are also very common in this zone.

Distribution. This zone is distributed in South China (Sun & Xia 2003, 2006; Xia *et al.* 2005; Wang *et al.* 2006; Wang & Yang 2007, 2011; Feng *et al.* 2009), SW Japan (Kuwahara *et al.* 1998; Yao *et al.* 2001; Xia *et al.* 2004), Thailand (Saesaengseerung *et al.* 2007) and the Malay Peninsula (Jasin 1997).

Age. Middle–late Wuchiapingian (early middle Lopingian).

Albaillella excelsa Interval Zone

Definition. This zone was erected as an abundance zone from the Lopingian of SW Japan by Yao *et al.* (2001). Feng *et al.* (2009) revised it as the interval zone. The base of this zone is defined by the FO of *A. excelsa* and the top by the FO of the *A. triangularis* (Figs 1, 2 & 6).

Type section. The Gujo-hachiman section in SW Japan.

Remarks. The *A. excelsa* Interval Zone in this study can be correlated with the upper part of the *Formanhelena triangula* Abundance Zone in South China (Wang & Yang 2007, 2011). It can also be correlated with the upper part of the *F. charveti–A. yamakitai* Assemblage Zone (Yao *et al.* 2001) and *N. optima–A. lauta* Lineage Zone (Xia *et al.* 2004) in SW Japan (Fig. 7). *A. excelsa*, *A. lauta*, *A. flexa*, *A. angusta* and *N. ornithoformis* are the dominant species throughout the zone.

Distribution. South China (Sun & Xia 2003, 2006; Wang *et al.* 2006; Wang & Yang 2007, 2011; Feng *et al.* 2009), SW Japan (Kuwahara *et al.* 1998; Yao *et al.* 2001; Xia *et al.* 2004), Thailand (Saesaengseerung *et al.* 2007), the Malay Peninsula (Sashida *et al.* 1995; Jasin *et al.* 2005; Jasin & Harun 2011) and Turkey (Göncüoğlu *et al.* 2004).

Age. Latest Wuchiapingian (middle Lopingian).

Albaillella triangularis Interval Zone

Definition. This zone was defined in Lopingian strata of SW Japan by Yao *et al.* (2001). It was originally defined as an abundance zone and later revised as a lineage zone (Xia *et al.* 2004). Feng *et al.* (2009) modified it as an interval zone. The base of this zone is marked by the FO of *A. triangularis* and the top by the FO of *A. yaoi* (Figs 1, 2 & 6).

Type section. The Gujo-hachiman section in SW Japan.

Remarks. The *A. triangularis* Interval Zone can be correlated with the *A. protolevis* Interval Zone and the lower part of the *A. levis–A. excelsa* Abundance Zone in South China (Wang & Yang 2007, 2011). This zone can be also correlated with the uppermost part of the *F. charveti–A. yamakitai* Assemblage Zone and the *N. ornithoformis* Assemblage Zone (Yao *et al.* 2001), and the *A. angusta–A. flexa* and the *A. triangularis* Lineage Zones in SW Japan. *A. excelsa*, *A. lauta*, *A. flexa*, *N. ornithoformis* and *N. optima* are dominant species of the zone (Xia *et al.* 2004).

Distribution. This zone is widely identified in South China (Yao & Kuwahara 2000; Sun & Xia 2003, 2006; Xia *et al.* 2005; Wang *et al.* 2006; Zhu *et al.* 2006; Wang & Yang 2007, 2011; Feng *et al.* 2009), SW Japan (Kuwahara *et al.* 1998; Yao *et al.* 2001; Xia *et al.* 2004), Thailand (Sashida *et al.* 2000*a*, *b*; Sashida & Salyapongse 2002) and the Malay Peninsula (Sashida *et al.* 1995; Jasin *et al.* 2005; Jasin & Harun 2011).

Age. Early Changhsingian (late Lopingian).

Albaillella yaoi Abundance Zone

Definition. The *A. yaoi* Zone was established in Lopingian strata of SW Japan by Xia *et al.* (2004) as a lineage zone. Feng *et al.* (2009) defined it as a range zone. The *A. yaoi* Abundance Zone was established in this study. Its base is defined by the abundant occurance of the species *A. yaoi*, and the top by the disappearance of this species.

Type section. The Gujo-hachiman section in SW Japan.

Remarks. The *A. yaoi* Abundance Zone can be correlated with the upper part of the *A. levis–A. excelsa* Abundance Zone, the *N. ornithoformis* Assemblage Zone, the lower to middle part of the *N. optima* Assemblage Zone (Wang & Yang 2007, 2011) and the *A. yaoi* Range Zone (Feng *et al.* 2009) in South China. It also can be correlated with the lower–middle part of the *N. optima* Zone (Yao *et al.* 2001), the *A. yaoi* Lineage Zone and the *A. degradans* Lineage Zone (Xia *et al.* 2004) in SW Japan. *A. yaoi yaoi*, *A. yaoi longa*, *A. excelsa*, *A. triangularis* and *N. optima* are the common species/subspecies throughout this zone.

Distribution. This zone is distributed in South China (Wang *et al.* 2006; Wang & Yang 2007, 2011; Feng *et al.* 2009) and SW Japan (Kuwahara *et al.* 1998; Kuwahara 1999; Xia *et al.* 2004).

Age. Changhsingian (late Lopingian).

Correlations with conodont chronostratigraphy and other chronostratigraphic schemes

Although 17 Permian radiolarian biozones have been established, the resolution of the radiolarian zones is still low compared to the 39 Permian conodont zones (Feng *et al.* 2009). The precise age constraints for most radiolarian zones mentioned above remain unavailable. Revision and refinement of these biozones are still needed because the correlation framework between the radiolarian and international standard conodont zonations or other chronostratigraphic schemes remains incomplete (Feng *et al.* 2009). Only a few papers on key stratigraphic intervals, such as the Carboniferous–Permian boundary (C-PB), the Guadalupian–Lopingian boundary (G-LB) and the Permo-Triassic boundary (P-TB), have been published that integrate the radiolarian and conodont biozonation schemes (e.g. Ling *et al.* 1985; Ling & Forsythe 1987; Ishiga 1990; Xia *et al.* 2004; Wu & Feng 2008; Nestell & Nestell 2010; Nishikane *et al.* 2011). Moreover, some of the conodont identifications are equivocal because they are based on poorly preserved or juvenile specimens (Nestell & Nestell 2010). The relatively reliable evidence on the correlations between the radiolarian zones, the conodont zones and other fossil zones are summarized as follows.

Pseudoalbaillella bulbosa Assemblage Zone

This zone is considered to mark the beginning of the Permian Period in South China (Wang & Yang 2011). However, the zonal species is associated with the fusulinids *Schwagerina* aff. *munaniensis*, *Chalaroschwagerina tarltonensis* and *Triticites tarltonensis* in the same section in Chile (Ling & Forsythe 1987). In SW Japan, the latest Carboniferous conodont species *Idiognathoides* sp. and early Asselian species *Streptognathodus elongatus* co-occur with the radiolarian zone species (Ishiga & Suzuki 1988). In the South Urals of Russia, the species *P. bulbosa* (= *Haplodiacanthus circinatus*) has also been found from the latest Carboniferous to the earliest Permian (Nazarov & Ormiston 1985; Nazarov 1988). Thus, the range of this zone is considered as late Gzhelian to early Asselian in this study.

Pseudoalbaillella u-forma–Pseudoalbaillella elegans Assemblage Zone

The range of *P. elegans* corresponds to the upper *Sphaeroschwagerina fusiformis* Zone or the upper *Streptognathodus elongatus* Zone in SW Japan, denoting that its geological age is late Asselian (Ishiga 1990; Zhang *et al.* 2000).

Pseudoalbaillella rhombothoracata Interval Zone

This zone is equal to the middle and upper *Chalaroschwagerina vulgaris* Zone and to the *Neogondolella bisselli–Sweetognathus whitei* Zone. Therefore, this zone corresponds to the early Artinskian (Ishiga 1990; Zhang *et al.* 2000).

Pseudoalbaillella globosa Interval Zone

This zone can be correlated with the *J. nanjingensis gracilis* Zone in Hubei, Anhui and Guangxi in China (Sheng & Wang 1985; Zhang *et al.* 2010; Ma *et al.* 2016). Therefore, this zone can be correlated with the Roadian.

Follicucullus monacanthus Interval Zone

The *F. monacanthus* Interval Zone is correlated with the *J. asserata* and the lower part of the *J. postserrata* conodont zones, as recorded in western Hubei, China (Kuwahara *et al.* 2008; Ma *et al.* 2016), and in the Delaware Basin, North America (Nestell *et al.* 2006). Therefore, this zone can be correlated to the Wordian and the earliest Capitanian.

Follicucullus scholasticus Interval Zone

The *F. scholasticus* Interval Zone was correlated with the fusulinid *Lepidolina kumaensis* Zone in the Mino Belt, SW Japan (Ishiga 1982). The *L. kumaensis* Zone is dated as late Capitanian

(Leven 1996; Kotylar 2008; Kasuya *et al.* 2012). Thus, the *F. scholasticus* Zone is assigned to the late Capitanian.

Follicucullus charveti Interval Zone

This zone can be correlated with the *J. shanoni* Zone in the Gujo-hachiman section, SW Japan (Nishikane *et al.* 2014) and the *J. agarant–J. crofti* Zone in the Dachongling section, South China (Xia *et al.* 2005). It is also equivalent to the *L. kumaensis* Zone in the Kyushu area, SW Japan (Ishiga 1990). Consequently, the *F. charveti* Interval Zone is assigned to the late Capitanian.

Albaillella cavitata Interval Zone

In South China and SW Japan, this zone is found to be correlated with the uppermost Guadalupian *Clarkina postbitteri hongshuiensis* Zone, and the Lopingian *C. postbitteri postbitteri*, *C. dukouensis*, *C. asymmetrica* and *C. guangyuanensis* Zones in ascending order (Xia *et al.* 2005; Sun & Xia 2006; Nishikane *et al.* 2011). Thus, this interval zone should not only be the last radiolarian zone of the Capitanian, but also the first radiolarian zone of the Wuchiapingian (Nishikane *et al.* 2011).

Albaillella triangularis Interval Zone

This zone can be correlated with the *C. postwangi* Zone in the Liuqiao section, South China (Wu & Feng 2008). Thus, it belongs to the early Changhsingian.

Albaillella yaoi Abundance Zone

A few specimens of *Neogondolella yini* have been found in SW Japan (Yao *et al.* 2001). This zone can be correlated with the middle–upper part of the *Neogondolella yini* Zone in the Dongpan and Liuqiao sections in Guangxi, China (Wu & Feng 2008). The top boundary of this zone is slightly below the P-TB because the Order Albaillellarian became extinct before the end of the Permian based on our current understanding (Feng *et al.* 2007; Feng & Algeo 2014). Their exact extinction time may vary slightly in different regions and with depth. Specific to the Dongpan section, a U–Pb age of 251.956 ± 0.033 Ma was obtained at the Albaillellarian extinction horizon (Baresel *et al.* 2016).

Discussion

Although studies of Permian radiolarian biostratigraphy have developed rapidly during the last decades, higher-resolution radiolarian zones are needed based on their evolutionary lineages (e.g. Caridroit & De Wever 1986; Kuwahara 1999; Xia *et al.* 2004; Wang & Yang 2011; Wang *et al.* 2012; Guex 2016; Guex *et al.* 2014; Zhang *et al.* 2014). Moreover, there are still many disputes on the definition and correlation between some radiolarian zones as well as their stratigraphic ranges, as demonstrated by the following examples:

Pseudoalbaillella disappears in the latest Capitanian, while *Follicucullus* is generally considered to have become extinct slightly above the G-LB in South China (Zhang *et al.* 2014). However, in some other works, *Follicucullus* and *Pseudoalbaillella* have been reported in the late Permian *N. optima* Zone of the Changhsingian (Ishiga *et al.* 1982*b*, *c*; Kuwahara & Yao 1998; Wang *et al.* 2006), and some *Albaillella*, *Follicucullus* and *Pseudoalbaillella* species have even been reported from Triassic or Jurassic strata (Bragin 1991; Sugiyama 1997; Takemura *et al.* 1998, 2002; Ishida & Murata 2006; Sano *et al.* 2010, 2012; Kamata *et al.* 2014). Among those examples, *Follicucullus* is the most frequent genus present in the Mesozoic strata. Sometimes they occur alone in the Triassic strata or together with other typical Triassic radiolarians: for example, *F. scholasticus* in the Hat Yai area, southern peninsular Thailand (Kamata *et al.* 2014) and *F. porrectus* in Far East Russia (Bragin 1991). In most cases, they coexist with other typical Permian radiolarians, such as *Entactinia*, *Entactinosphaera*, *Copicyntra*, *Copicyntroides*, *Hegleria*, *Cauletella*, *Quadriremis*, *Ishiguam*, *Latentifistula*, *Pseudotormentus*, *Stigmosphaerostylus* and *Archaeospongoprunum* (Takemura *et al.* 1998, 2002; Sano *et al.* 2010). Some *Follicucullus* might have been misinterpreted for slightly eroded fish teeth (e.g. pl. 3, figs 1 & 2 in Takemura *et al.* 2007). The second most common taxon is *Pseudoalbaillella*, including *P. longtanensis*, *P.* aff. *longicornis*, *P.* spp. (Ishida & Murata 2006), *P.* sp. cf. *globosa* (Sugiyama 1992) and *P. fusiformis* (Takemura *et al.* 1998). Only a few examples of *Albaillella* were found in Early Triassic strata: the species *A. aotearoa* in the Oruatemanu Formation of New Zealand (Takemura *et al.* 2007), *A. triangularis* in central Japan (Sugiyama 1992), *A.* spp. and *A.*? sp. in the Mino Terrane, central Japan (Sano *et al.* 2010). Some are considered to be 'Lazarus taxa' that survived in 'unknown refugees' (De Wever *et al.* 2006), and a few have been confirmed as reworked radiolarians that originated from the sediments on the flanks of submarine highs, such as seamounts or related to global-scale tectonic movements and erosion history, or from Permian exotic rocks and clastic grains (Ando *et al.* 1991; Sugiyama 1992; Kamata *et al.* 2000; Ishida & Murata 2006). Undoubtedly, further examinations are still needed for those reports (De Wever *et al.* 2006).

A. yamakitai (Kuwahara 1999) is a key species in the evolution of *Albaillella* as well as in its zonation at the Guadalupian-Lopingian transitions. Previously, the first occurrences of *A. yamakitai* were thought to mark the G-LB (Ishiga 1990; Kuwahara *et al.* 1998). Later, this species was moved below the G-LB after more detailed comparisons with conodont zones in South China and SW Japan (Xia *et al.* 2004; Nishikane *et al.* 2011). The species *A. yamakitai* (*sensu lato*) in Xia *et al.* (2005) was found in the late Guadalupian *Jinogodolella altudaensis* conodont zone in the Lamar Limestone and in the Oregon area (Maldonado & Noble 2010; Nestell & Nestell 2010; Noble & Jin 2010). Consequently, it cannot be a marker for the G-LB (Xia *et al.* 2005; Nishikane *et al.* 2011) (see the discussion in Nestell & Nestell 2010). The range of the *F. charveti–A. yamakitai* Zone or *A. yamakitai* Zone could be extended down to the middle Permian in this sense. Thus, the definition and the boundary of the intraspecific variation of *A. yamakitai* need to be examined and reconsidered before being applyed to biostratigraphy.

Earlier, the biostratigraphic ranges of the *Neoalbaillella* species in the upper Permian were reversed because they were found in overturned beds in SW Japan (e.g. Ishiga *et al.* 1982*b*; Ishiga 1984, 1986, 1990). The reversed ranges, in turn, led to reversed biostratigraphic arrangements as well as phylogenetic lineages (Ishiga 1986, 1990; Kuwahara 1997*a*, *b*). Upper Permian radiolarian biostratigraphy has been re-examined in the bedded chert sections of the Gujo-hachiman and Neo areas in the Mino Belt (Kuwahara *et al.* 1998). The *N. optima* Zone and *N. ornithoformis* Zone were treated upside down (Sashida 1997; Kuwahara *et al.* 1998), and the phylogenetic tree of late Permian *Albaillella* was redrawn (Kuwahara 1999). Nevertheless, some of these sections in SW Japan and South China are mostly composed of chert blocks, which are generally sliced into pieces by minor faults (Kuwahara *et al.* 1998; Nishikane *et al.* 2011). Small-scale hiatuses can be detected through the distributions of some radiolarians and conodonts (Kuwahara *et al.* 1998; Nishikane *et al.* 2011). Therefore, more continuous successions for both radiolarian biostratigraphy and their correlations with conodont zones or other chronostratigraphic schemes are needed for further investigation.

Conclusion

In this study, we summarize the Permian radiolarian zones across the world. Among them, seven zones are assigned to the early Permian, namely: the *Pseudoalbaillella bulbosa* Assemblage Zone, the *Pseudoalbaillella u-forma–Pseudoalbaillella elegans* Assemblage Zone, the *Pseudoalbaillella lomentaria–Pseudoalbaillella sakmarensis* Assemblage Zone, the *Pseudoalbaillella rhombothoracata* Interval Zone, the *Albaillella xiaodongensis* Assemblage Zone, the *Albaillella sinuata* Abundance Zone and the *Pseudoalbaillella ishigai* Abundance Zone, in ascending order. Five zones are attributed to the middle Permian: the *Pseudoalbaillella globosa* Interval Zone, the *Follicucullus monacanthus* Interval Zone, the *Follicucullus porrectus* Interval Zone, the *Follicucullus scholasticus* Interval Zone and the *Follicucullus charveti* Interval Zone. Five zones belong to the middle-late Permian: the *Albaillella cavitata* Interval Zone, the *Albaillella levis* Interval Zone, the *Albaillella excelsa* Interval Zone, the *Albaillella triangularis* Interval Zone and the *Albaillella yaoi* Abundance Zone. The *Pseudoalbaillella rhombothoracata* Interval Zone and *Albaillella yaoi* Abundance Zone were erected in this study. These Permian radiolarian biozones and their correlations with conodont zones and other chronostratigraphic schemes are discussed, and their phylogenetic models through the Permian Period are reviewed and updated.

This work was supported by NSFC (41402001), Ministry of Education of China (20110145130001), and State Key Laboratory of Geological Processes and Mineral Resources, China University of Geosciences in Wuhan (MSFGPMR201502). The authors thank Dr Kuwahara Kiyoko, Dr Ito Tsuyoshi, Chang Shan and Maliha Z. Khan for their constructive suggestions during writing, and Dr Marie B. Forel and Dr Spencer G. Lucas for editing the English.

References

Afanasieva, M.S. 2000. *Atlas of Paleozoic Radiolaria from the Russian Platform.* Scientific World, Moscow.

Ando, H., Tsukamoto, H. & Saito, M. 1991. Permian radiolarians in the Mt. Kinkazan area, Gifu city, central Japan. *Bulletin of the Mizunami Fossil Museum*, **18**, 101–106.

Baresel, B., Bucher, H., Brosse, M., Cordey, F., Kuang, G.D. & Schaltegger, U. 2016. Precise age for the Permian–Triassic boundary in South China from high precision U–Pb geochronology and Bayesian age–depth modelling. *Solid Earth Discussions*, **8**, https://doi.org/10.5194/se-2016-145

Beauchamp, B. & Baud, A. 2002. Growth and demise of Permian biogenic chert along northwest Pangea: evidence for end-Permian collapse of thermohaline circulation. *Palaeogeography, Palaeoclimatology, Palaeoecology*, **184**, 37–63.

Bian, Q.T., Zheng, X.S., Li, H.S. & Sha, J.G. 1997. Age and tectonic setting of ophiolite in the Hoh Xil region, Qinghai Province. *Geological Review*, **434**, 348.

Blome, C.D. & Reed, K.M. 1992. Permian and Early? Triassic radiolarian faunas from the Grindstone terrane, central Oregon. *Journal of Paleontology*, **66**, 351–383.

BLOME, C.D. & REED, K.M. 1995. Radiolarian biostratigraphy of the Quinn River Formation, Black Rock terrane, north-central Nevada: Correlations with eastern Klamath terrane geology. *Micropaleontology*, **41**, 49–68.

BRAGIN, N.Y. 1991. Radiolaria and Lower Mesozoic units of the USSR east regions. *Transaction of the Academy of Sciences of the USSR*, **469**, 1–125.

CARIDROIT, M. & DE WEVER, P. 1986. Some Late Permian radiolarians from pelitic rocks of the Tatsuno formation Hyogo Prefecture, southwest Japan. *Marine Micropaleontology*, **11**, 55–90.

CATALANO, R., DI STEFANO, P. & KOZUR, H. 1989. Lower Permian Albaillellacea (Radiolaria) from Sicily and their stratigraphic and paleogeographic significance. *Rendiconto dell'Accademia delle Scienze Fisiche e Matematiche serie*, **4**, 1–24.

CATALANO, R., DI STEFANO, P. & KOZUR, H. 1991. Permian circumpacific deep-water faunas from the western Tethys (Sicily, Italy) – New evidences for the position of the Permian Tethys. *Palaeogeography, Palaeoclimatology, Palaeoecology*, **87**, 75–108.

DE WEVER, P., O'DOGHERTY, L. & & GORICAN, S. 2006. The plankton turnover at the Permo-Triassic boundary, emphasis on radiolarians. *Swiss Journal of Geosciences*, **99**, 49–62.

DZULKAFLI, M.A., JASIN, B. & LEMAN, M.S. 2012. Radiolaria berusia Perm Awal Sakmarian dari singkapan baru di Pos Blau, Ulu Kelantan dan kepentingannya. *Bulletin of the Geological Society of Malaysia*, **58**, 67–73.

FENG, Q.L. 1992. Permian and Triassic Radiolarian biostratigraphy in South and Southwest China. *Journal of China University of Geosciences*, **3**, 51–62.

FENG, Q.L. & ALGEO, T.J. 2014. Evolution of oceanic redox conditions during the Permo-Triassic transition: Evidence from deepwater radiolarian facies. *Earth-Science Reviews*, **137**, 34–51.

FENG, Q.L. & LIU, B.L. 2002. Early Permian radiolarians from Babu ophiolitic mélange in southeastern Yunnan. *Earth Science – Journal of China University of Geosciences*, **27**, 1–3.

FENG, Q.L. & YE, M. 1996. Radiolarian stratigraphy of Devonian through Middle Triassic in southwestern Yunnan. *In*: FENG, N.Q. & FENG, Q.L. (eds) *Devonian to Triassic Tethys in Western Yunnan, China*. China University of Geosciences Press, Wuhan.

FENG, Q.L., HE, W.H., GU, S.Z., MENG, Y.Y., JIN, Y.X. & ZHANG, F. 2007. Radiolarian evolution during the latest Permian in South China. *Global and Planetary Change*, **55**, 177–192.

FENG, Q.L., WU, J., ZHANG, L., ZHANG, N., GU, S.Z. & HE, W.H. 2009. Progress in Lopingian Radiolarian Biostratigraphy of South China. *Acta Palaeontologica Sinica*, **48**, 465–473.

GÖNCÜOĞLU, M.C., KUWAHARA, K., TEKIN, U.K. & TURHAN, N. 2004. Upper Permian Changxingian radiolarian cherts within the clastic successions of the 'Karakaya Complex' in NW Anatolia. *Turkish Journal of Earth Sciences*, **13**, 201–213.

GUEX, J. 2016. *Retrograde Evolution during Major Extinction Crises*. Springer International, Cham, Switzerland.

GUEX, J., CARIDROIT, M., KUWAHARA, K. & O'DOGHERTY, L. 2014. Retrograde evolution of *Albaillella* during the Permian–Triassic crisis. *Revue de Micropaléontologie*, **57**, 39–43.

HARMS, T.A. & MURCHEY, B.L. 1992. Setting and occurrence of Late Paleozoic radiolarians in the Sylvester allochthon, part of a proto-Pacific ocean floor terrane in the Canadian Cordillera. *Palaeogeography, Palaeoclimatology, Palaeoecology*, **96**, 127–139.

HOLDSWORTH, B.K. & JONES, D.L. 1980. *A Provisional Radiolaria Biostratigraphy, Late Devonian through Late Permian*. United States Geological Survey, Open-File Report, **80-876**.

HORI, N. 2004. Permian radiolarians from chert of the Chichibu Belt in the Toyohashi district, ichi Prefecture, Southwest Japan. *Bulletin of the Geological Survey of Japan*, **55**, 287–301.

HORI, N. 2005. Paleozoic and Mesozoic radiolarians from the Chichibu Belt in the Iragomisaki istrict, Atsumi Peninsula, Aichi Prefecture, Southwest Japan. *Bulletin of the Geological Survey of Japan*, **56**, 37–83.

ISHIDA, N. & MURATA, M. 2006. Mixed radiolarian assemblages from the Middle Jurassic chert and siliceous mudstone in the Southern Chichibu terrane, southeastern part of the Kanto Mountains. *Journal of the Geological Society of Japan*, **112**, 197–209.

ISHIGA, H. 1982. Late Carboniferous and Early Permian radiolarians from the Tamba belt, southwest Japan. *Earth Science (Chikyu Kagaku)*, **36**, 333–339.

ISHIGA, H. 1984. Follicucullus (Permian Radiolaria) from Maizuru Group in Maizuru Belt, Southwest Japan. *Earth Science (Chikyu Kagaku)*, **38**, 427–434.

ISHIGA, H. 1986. Late Carboniferous and Permian radiolarian biostra-tigraphy of southwest Japan. *Journal of Geoscience, Osaka City University*, **29**, 89–100.

ISHIGA, H. 1990. Paleozoic radiolarians. *In*: ICHIKAWA, K., MIZUTANI, S., HARA, I., HADA, S. & YAO, A. (eds) *Pre-Cretaceous Terranes of Japan*. Nippon Insatsu Shuppan, Osaka, 285–293.

ISHIGA, H. & IMOTO, N. 1980. Some Permian radiolarians in the Tamba District, Southwest Japan. *Earth Science; Journal of the Association for the Geological Collaboration in Japan*, **34**, 333–345.

ISHIGA, H. & SUZUKI, S. 1988. Paleozoic radiolarian assemblages from the Shimomidani Formation in Akiyoshi Terrane, southwest Japan. *Journal of the Geological Society of Japan*, **94**, 493–499.

ISHIGA, H., KITO, N. & IMOTO, N. 1982*a*. Permian radiolarian biostratigraphy. *News of Osaka Micropaleontologists*, 17–26.

ISHIGA, H., KITO, N. & IMOTO, N. 1982*b*. Late Permian radiolarian assemblages in the Tamba district and an adjacent area, southwest Japan. *Earth Science (Chikyu Kagaku)*, **36**, 10–22.

ISHIGA, H., KITO, T. & IMOTO, N. 1982*c*. Middle Permian radiolarian assemblages in the Tamba District and an adjacent area, southwest Japan. *Earth Science (Chikyu Kagaku)*, **36**, 272–281.

ISOZAKI, Y. 1994. Superanoxia across the Permo-Triassic boundary: record in accreted deep-sea pelagic chert in Japan. *In*: EMBRY, A.F., BEAUCHAMP, B. & GLASS, D.J. (eds) *Pangea: Global Environments and Resources*. Canadian Society of Petroleum Geologists, Memoirs, **17**, 805–812.

Isozaki, Y. 2009. Integrated 'plume winter' scenario for the double-phased extinction during the Paleozoic–Mesozoic transition: the G-LB and P-TB events from a Panthalassan perspective. *Journal of Asian Earth Sciences*, **36**, 459–480.

Ito, T., Zhang, L., Feng, Q.L. & Matsuoka, A. 2013. Guadalupian Middle Permian radiolarian and sponge spicule faunas from the Bancheng Formation of the Qinzhou Allochthon, South China. *Journal of Earth Science*, **24**, 145–156.

Jasin, B. 1997. Permo-Triassic radiolarian from the Semanggol Formation, northwest Peninsular Malaysia. *Journal of Asian Earth Sciences*, **15**, 43–53.

Jasin, B. & Ali, C.A. 1997. Lower Permian Radiolaria from the Pos Blau area, Ulu Kelantan, Malaysia. *Journal of Asian Earth Sciences*, **15**, 327–337.

Jasin, B. & Harun, Z. 2007. Stratigraphy and sedimentology of the chert unit of the Semanggol Formation. *Geological Society of Malaysia Bulletin*, **53**, 103–109.

Jasin, B. & Harun, Z. 2011. Radiolarian biostratigraphy of Peninsular Malaysia: an update. *Bulletin of the Geological Society of Malaysia*, **57**, 27–38.

Jasin, B., Harun, Z., Said, U. & Saad, S. 2005. Permian radiolarian biostratigraphy of the Semanggol Formation, south Kedah, Peninsular Malaysia. *Geological Society of Malaysia Bulletin*, **51**, 19–30.

Jasin, B., Bashardin, A. & Harun, Z. 2013. Middle Permian Radiolarians from the siliceous mudstone block near Pos Blau, Ulu Kelantan and their significance. *Bulletin of the Geological Society of Malaysia*, **59**, 33–38.

Jin, Y.X., Feng, Q.L., Meng, Y.Y., He, W.H. & Gu, S.Z. 2007. Albaillellidae (Radiolaria) from the latest Permian in southern Guangxi, China. *Journal of Paleontology*, **81**, 9–18.

Kamata, Y., Hisada, K. & Lee, Y.I. 2000. Late Jurassic radiolarians from pebbles of Lower Cretaceous conglomerates of the Hayang Group, southeastern Korea. *Geosciences Journal*, **4**, 165–174.

Kamata, Y., Shirouzu, A. *et al.* 2014. Late Permian and Early to Middle Triassic radiolarians from the Hat Yai area, southern Peninsular Thailand: implications for the tectonic setting of the eastern margin of the Sibumasu Continental Block and closure timing of the Paleo-Tethys. *Marine Micropaleontology*, **110**, 8–24.

Kametaka, M., Nagai, H., Zhu, S. & Takebe, M. 2009. Middle Permian Radiolarians from Anmenkou, Chaohu, Northeastern Yangtze Platform, China. *Island Arc*, **18**, 108–125.

Kasuya, A., Isozaki, Y. & Igo, H. 2012. Constraining paleo-latitude of a biogeographic boundary in mid-Panthalassa: Fusuline province shift on the Late Guadalupian Permian migrating seamount. *Gondwana Research*, **21**, 611–623.

Kawai, M. & Takeuchi, M. 2001. Permian radiolarians from the Omi area in the Hida-gaien Tectonic Zone, central Japan. *News of Osaka Micropaleontologists, Special Volume*, **12**, 23–32.

Knoll, A.H., Bambach, R.K., Payne, J.L., Pruss, S. & Fischer, W.W. 2007. Paleophysiology and end-Permian mass extinction. *Earth and Planetary Science Letters*, **256**, 295–313.

Kotylar, G.V. 2008. Permskaya Sistema. *In*: Zhamoida, A.I. & Petrov, O.V. (eds) *Sostoyanie Izuchennosti Stratigrafii Dokembriya I Fanerozoya Rossii.* Zapadochi Dal'neyshikh Issledovaniy, VSEGEI, **38**, 69–76.

Kozur, H. 1981. Albaillellidea Radiolaria aus dem Unterperm des Vorurals. *Geologisch–Paläontologische Mitteilungen Innsbruck*, **10**, 263–274.

Kozur, H. 1993. Upper Permian radiolarians from the Sosio Valley area, western Sicily Italy and from the uppermost Lamar Limestone of West Texas. *Jahrbuch der Geologischen Bundesanstalt Wien*, **136**, 99–123.

Kurihara, T. & Kametaka, M. 2008. Radiolaria-dated Lower Permian clastic-rock sequence in the Fukuji area of the Hida-gaien terrane, central Japan, and its inter-terrane correlation across Southwest Japan. *Island Arc*, **174**, 531–545.

Kuwahara, K. 1992. Late Carboniferous to Early Permian radiolarian assemblages from Miyogawa area, Mie Prefecture, Japan. *New of Osaka Micropaleontologists, Special Volume*, **8**, 1–7.

Kuwahara, K. 1997*a*. Evolutionary patterns of Late Permian *Albaillella* Radiolaria as seen in bedded chert sections in the Mino Belt, Japan. *Marine Micropaleontology*, **30**, 65–78.

Kuwahara, K. 1997*b*. Upper Permian radiolarian biostratigraphy: abundance zones of *Albaillella. News of Osaka Micropaleontologists, Special Volume*, **10**, 55–75.

Kuwahara, K. 1999. Middle-Late Permian radiolarian assemblages from China and Japan. *In*: Yao, A., Ezaki, Y., Hao, W.C. & Wang, X.P. (eds) *Biotic and Geological Development of the Paleo-Tethys in China.* Peking University Press, Beijing, 43–54.

Kuwahara, K. & Yao, A. 1998. Diversity of Late Permian radiolarian assemblages. *News of Osaka Micropaleontologists, Special Volume*, **11**, 33–46.

Kuwahara, K. & Yao, A. 2001. Late Permian radiolarian faunal change in bedded chert of the Mino Belt, Japan. *News of Osaka Micropaleontology, Special Volume*, **12**, 33–49.

Kuwahara, K., Yao, A. & An, T.X. 1997. Paleozoic and Mesozoic complexes in the Yunnan area, China (part 1): Preliminary report of Middle–Late Permian radiolarian assemblages. *Journal of Geosciences, Osaka City University*, **40**, 37–49.

Kuwahara, K., Yao, A. & Yamakita, S. 1998. Reexamination of Upper Permian radiolarian biostratigraphy. *Earth Science (Chikyu Kagaku)*, **52**, 391–404.

Kuwahara, K., Yao, A., Ezaki, Y., Liu, J.B., Hao, W.C. & Kuang, G.D. 2003. Occurrence of Late Permian radiolarians from the chituao section, Laibin, Guangxi, China. *Journal of Geosciences, Osaka City University*, **46**, 13–23.

Kuwahara, K., Yao, A. & Yao, J.X. 2007. Middle Permian radiolarian biostratigraphy on the Gufeng Formation in the Songzi-Wufeng area Hubei Province China. *Journal of Geosciences, Osaka City University*, **50**, 55–66.

Kuwahara, K., Yao, A. & Yao, J.X., Feng, S.N., Ji, Z.S. & Yao, H.Z. 2008. Findings of Middle Permian radiolarian and conodont fossils from the Gufeng Formation of the Zigui area, Hubei Province, China. *Journal of Geosciences, Osaka City University*, **51**, 9–19.

LEVEN, E.Y. 1996. The Midian Stage of the Permian and its boundaries. *Stratigraphy and Geological Correlation*, **4**, 540–551.

LI, H.S. & BIAN, Q.T. 1993. Upper Paleozoic radiolarian of the Xijin Ulan-Gangqiqu Ophiolite Complex, Kekexili. *Geoscience, Journal of Graduate School, China University of Geosciences*, **7**, 411–420.

LING, H.Y. & FORSYTHE, R.D. 1987. Late Paleozoic pseudoalbaillellid radiolarians from southernmost Chile and their geological significance. *In*: MCKENZIE, G.D. (ed.) *Gondwana Six: Structure, Tectonics, and Geophysics*. American Geophysical Union, Geophysical Monograph Series, **40**, 253–260.

LING, H.Y., FORSYTHE, R.D. & DOUGLASS, R.C. 1985. Late Paleozoic microfaunas from southernmost Chile and their relation to Gondwanaland forearc development. *Geology*, **13**, 357–360.

MA, Q.F., FENG, Q.L., CARIDROIT, M., DANELIAN, T. & ZHANG, N. 2016. Integrated radiolarian and conodont biostratigraphy of the Middle Permian Gufeng Formation South China. *Comptes Rendus Palevol*, **15**, 453–459.

MALDONADO, A.L. & NOBLE, P.J. 2010. Radiolarians from the upper Guadalupian Middle Permian reef trail member of the Bell Canyon Formation, West Texas and their biostratigraphic implications. *Micropaleontology*, **56**, 69–115.

MARQUEZ, E.J., AITCHISON, J.C. & ZAMORAS, L.R. 2006. Upper Permian to Middle Jurassic radiolarian assemblages of Busuanga and surrounding islands, Palawan, Philippines. *Eclogae Geologicae Helvetiae*, **99**, 101–125.

MIYAMOTO, T. & TANIMOTO, Y. 1993. Late Permian olistostrome Kamoshishigawa Formation in the Chichibu Belt of South Kyushu, southwest Japan. *News of Osaka Micropaleontologists, Special Volume*, **9**, 19–33.

MURCHEY, B.L. 1991. Age and depositional setting of siliceous sediments in the upper Paleozoic Havallah sequence near Battle Mountain, Nevada; Implications for the paleogeography and structural evolution of the western margin of North America. *In*: HARWOOD, D.S. & MILLER, M.M. (eds) *Paleozoic and Early Mesozoic Paleogeographic Relations; Sierra Nevada, Klamath Mountains, and Related Terranes*. Geological Society of America, Special Papers, **255**, 137–156.

MURCHEY, B.L. & JONES, D.L. 1992. A mid-Permian chert event: widespread deposition of biogenic siliceous sediments in coastal, island arc and oceanic basins. *Palaeogeography, Palaeoclimatology, Palaeoecology*, **96**, 161–174.

NAKAE, S. 2001. Permian radiolarians from cherts of the Tamba Terrane in the Nishizu district, Fukui, Southwest Japan. *Bulletin of the Geological Survey of Japan*, **52**, 245–252.

NAZAROV, B.B. 1988. Paleozoic radiolaria. *Practical Manual of Microfauna of the USSR*, **2**, 1–232.

NAZAROV, B.B. & ORMISTON, A.R. 1985. Radiolaria from the Late Paleozoic of the Southern Urals, USSR and West Texas, USA. *Micropaleontology*, **31**, 1–54.

NESTELL, G.P. & NESTELL, M.K. 2010. Late Capitanian latest Guadalupian, Middle Permian radiolarians from the Apache Mountains, West Texas. *Micropaleontology*, **56**, 7–68.

NESTELL, M.K., NESTELL, G.P., WARDLAW, B.R. & SWEATT, M.J. 2006. Integrated biostratigraphy of foraminifers, radiolarians and conodonts in shallow and deep water Middle Permian (Capitanian) deposits of the 'Rader slide', Guadalupe Mountains, West Texas. *Stratigraphy*, **3**, 161–194.

NISHIKANE, Y., KAIHO, K., TAKAHASHI, S., HENDERSON, C.M., SUZUKI, N. & KANNO, M. 2011. The Guadalupian–Lopingian boundary Permian in a pelagic sequence from Panthalassa recognized by integrated conodont and radiolarian biostratigraphy. *Marine Micropaleontology*, **78**, 84–95.

NISHIKANE, Y., KAIHO, K., HENDERSON, C.M., TAKAHASHI, S. & SUZUKI, N. 2014. Guadalupian–Lopingian conodont and carbon isotope stratigraphies of a deep chert sequence in Japan. *Palaeogeography, Palaeoclimatology, Palaeoecology*, **403**, 16–29.

NOBLE, P.J. & JIN, Y.X. 2010. Radiolarians from the Lamar Limestone, Guadalupe Mountains, West Texas. *Micropaleontology*, **56**, 117–147.

ORCHARD, M.J., CORDEY, F. *ET AL*. 2001. Biostratigraphic and biogeographic constraints on the Carboniferous to Jurassic Cache Creek Terrane in central British Columbia. *Canadian Journal of Earth Sciences*, **38**, 551–578.

RACKI, G. 1999. Silica-secreting biota and mass extinctions: survival patterns and processes. *Palaeogeography, Palaeoclimatology, Palaeoecology*, **154**, 107–132.

RACKI, G. 2003. End-Permian mass extinction: oceanographic consequences of double catastrophic volcanism. *Lethaia*, **36**, 171–173.

RUDENKO, V.S. & PANASENKO, E.S. 1990. Permian Albaillellaria Radiolaria of the Pantovian sequence in Primorye. New data on Paleozoic and Mesozoic biostratigraphy of the south Far East. *In*: ZAKHAROV, Y.D., BELYAEVA, G.V. & NIKITINA, A.P. (eds) *New Data on Paleozoic and Mesozoic Biostratigraphy of the South Far East*. Far-Eastern Branch of the USSR Academy of Sciences, Vladivostok, 117–209.

RUDENKO, V.S. & PANASENKO, E.S. 1997. Biostratigraphy of Permian deposits of Sikhote-Alin based on radiolarians. *In*: BAUD, A., POPOVA, I., DICKINS, J.M., LUCAS, S. & ZAKHAROV, Y. (eds) *Late Paleozoic and Early Mesozoic Circum Pacific Events. Biostratigraphy, Tectonics and Ore Deposits of Primorye Far East Russia*. Mémoires de Géologie (Lausanne), **30**, 73–79.

SAESAENGSEERUNG, D., SASHIDA, K. & SARDSUD, A. 2007. Devonian to Triassic radiolarian faunas from northern and northeastern Thailand. *In*: TANTIWANIT, W. (ed.) *Proceedings of the International Conference on Geology of Thailand GEOTHAI'07: Towards Sustainable Development and Sufficiency Economy*. Department of Mineral Resources, Bangkok, 54–71.

SAESAENGSEERUNG, D., SASHIDA, K., SARDSUD, A. & SALYAPONGSE, S. 2008. Paleozoic and Mesozoic radiolarian faunas in Thailand. *In*: CHOOWONG, M. & THITIMAKORN, T. (eds) *Proceedings of the International Symposium on Geoscience Resources and Environments of Asian Terranes GREAT 2008 (4th IGCP and 5th APSEG)*. Chulalongkorn University, Chulalongkorn, 186–187.

SAESAENGSEERUNG, D., AGEMATSU, S., SASHIDA, K. & SARDSUD, A. 2009. Discovery of Lower Permian radiolarian and conodont faunas from the bedded chert of the Chanthaburi area along the Sra Kaeo suture zone, eastern Thailand. *Paleontological Research*, **13**, 119–138.

Sano, H., Kuwahara, K., Yao, A. & Agematsu, S. 2010. Panthalassan seamount-associated Permian–Triassic boundary siliceous rocks, Mino terrane, central Japan. *Paleontological Research*, **14**, 293–314.

Sano, H., Wada, T. & Naraoka, H. 2012. Late Permian to Early Triassic environmental changes in the Panthalassic Ocean: Record from the seamount-associated deep-marine siliceous rocks, central Japan. *Palaeogeography, Palaeoclimatology, Palaeoecology*, **363**, 1–10.

Sashida, K. 1997. Radiolaria biostratigraphy near the Permo-Triassic boundary. *Circular Society Science Form*, **12**, 30.

Sashida, K. & Nakornsri, N. 1997. Lower Permian radiolarian faunas from the Khanu Chert Formation distributed in the Sukhothai area, northern central Thailand. *In*: Kishimoto, I., Kishimoto, H. & Takahashi, T. (eds) *Proceedings of the International Conference on Stratigraphy and Tectonic Evolution of Southeast Asia and the South Pacific*. Department of Mineral Resources Thailand, Bangkok, 101–108.

Sashida, K. & Salyapongse, S. 2002. Permian radiolarian faunas from Thailand and their paleogeographic significance. *Journal of Asian Earth Sciences*, **20**, 691–701.

Sashida, K. & Tonishi, K. 1985. *Permian Radiolarians from the Kanto Mountains, Central Japan – Some Upper Permian Spumellaria from Itsukaichi, Western Part of Tokyo Prefecture*. Science Reports of the Institute of Geoscience, University of Tsukuba, 6.

Sashida, K., Igo, H., Hisafa, K.I., Nakornsri, N. & Ampornmaha, A. 1993. Occurrence of Paleozoic and Early Mesozoic radiolaria in Thailand preliminary report. *Journal of Southeast Asian Earth Sciences*, **8**, 97–108.

Sashida, K., Adachi, S., Igo, H., Koike, T. & Amnan, I.B. 1995. Middle and Late Permian radiolarians from the Semanggol Formation, Northwest Peninsular Malaysia. *Transactions and Proceedings of the Paleontological Society of Japan. New Series*, **177**, 43–58.

Sashida, K., Igo, H., Adachi, S., Ueno, K., Kajiwara, Y., Nakornsri, N. & Sardsud, A. 2000*a*. Late Permian to Middle Triassic radiolarian faunas from northern Thailand. *Journal of Paleontology*, **74**, 789–811.

Sashida, K., Salyapongse, S. & Nakornsri, N. 2000*b*. Latest Permian radiolarian fauna from Klaeng, eastern Thailand. *Micropaleontology*, **46**, 245–263.

Sha, J.G. 1998. Characteristics of stratigraphy and paleontology of Hoh Xil, Qinghai: geographic significance. *Acta Palaeontologica Sinica*, **37**, 85–96.

Shen, S.Z., Schneider, J.W., Angiolini, L. & Henderson, C.M. 2013. The international Permian timescale: March 2013 update. The Carboniferous–Permian transition. *New Mexico Museum of Natural History and Science, Bulletin*, **60**, 411–416.

Sheng, J.Z. & Wang, Y.J. 1985. Fossil radiolaria from Kuhfeng Formation at Lontan, Nanjing. *Acta Micropalaeontologica Sinica*, **242**, 174–180.

Shi, L., Feng, Q.L., Shen, J., Ito, T. & Chen, Z.Q. 2016. Proliferation of shallow-water radiolarians coinciding with enhanced oceanic productivity in reducing conditions during the Middle Permian, South China: evidence from the Gufeng Formation of western Hubei Province. *Palaeogeography, Palaeoclimatology, Palaeoecology*, **444**, 1–14.

Shimakawa, M. & Yao, A. 2006. Lower–Middle Permian radiolarian biostratigraphy in the Qinzhou area, South China. *Journal of Geoscience, Osaka City University*, **49**, 31–47.

Spiller, F.C. 1996. Late Paleozoic radiolarians from the Bentong-Raub suture zone, Peninsular Malaysia. *Island Arc*, **5**, 91–103.

Spiller, F.C. 2002. Radiolarian biostratigraphy of Peninsular Malaysia and implications for regional palaeotectonics and palaeogeography. *Palaeontographica Abteilung A – Palaozoologie–Stratigraphie*, **266**, 1–91.

Spiller, F.C. & Metcalfe, I. 1995. Late Palaeozoic radiolarians from the Bentong-Raub suture zone, and the Semanggol Formation of Peninsular Malaysia – initial results. *Journal of Southeast Asian Earth Sciences*, **11**, 217–224.

Sugiyama, K. 1992. Lower and Middle Triassic radiolarians from Mt. Kinkazan, Gifu Prefecture, central Japan. *Transactions and Proceedings Palaeontological Society of Japan, New series*, **167**, 1180–1223.

Sugiyama, K. 1997. Triassic and Lower Jurassic radiolarian biostratigraphy in the siliceous claystone and bedded chert units of the southeastern Mino Terrane, Central Japan. *Transactions and Proceedings of the Palaeontological Society of Japan, New Series*, **24**, 1773–1783.

Sun, D.Y. & Xia, W.C. 2003. Characteristics of Albaillellarian, radiolarian fauna from Guadalupian to Lopingian Series in Permian, South China. *Journal of China, University of Geosciences*, **14**, 314–320.

Sun, D.Y. & Xia, W.C. 2006. Identification of the Guadalupian–Lopingian boundary in the Permian in a bedded chert sequence, South China. *Palaeogeography, Palaeoclimatology, Palaeoecology*, **236**, 272–289.

Suzuki, N., Kojima, S. *et al.* 2005. Permian radiolarian faunas from chert in the Khabarovsk Complex, Far East Russia, and the age of each lithologic unit of the Khabarovsk Complex. *Journal of Paleontology*, **79**, 687–701.

Takemura, A., Aita, Y. *et al.* 1998. Preliminary report on the lithostratigraphy of the Arrow Rocks, and geologic age of the northern part of the Waipapa Terrane, New Zealand. *News of Osaka Micropaleontologists, Special Volume*, **11**, 47–57.

Takemura, A., Aita, Y. *et al.* 2002. Triassic radiolarians from the ocean-floor sequence of the Waipapa Terrane at Arrow Rocks, Northland, New Zealand. *New Zealand Journal of Geology and Geophysics*, **45**, 289–296.

Takemura, A., Sakai, M., Sakamoto, S., Aono, R., Takemura, S. & Yamakita, S. 2007. Earliest Triassic radiolarians from the ARH and ARF sections on Arrow Rocks, Waipapa terrane, Northland, New Zealand. *In*: Sporli, K.B., Takemura, A. & Hori, R.S. (eds) *The Oceanic Permian/Triassic Boundary Sequence at Arrow Rocks (Oruatemanu) Nothland, New Zealand*. GNS Science Monographs, **24**, 97–107.

WANG, R.J. 1995. Radiolarian fauna from Gufeng Formation (Lower Permian) in Hushan area of Nanjing, Jiangsu Province. *Scientia Geologica Sinica*, **30**, 139–146.

WANG, Y.J. & LI, J.X. 1994. Discovery of the *Follicucullus bipartitus–F.charveti* radiolarian assemblage zone and its geological significance. *Acta Micropalaeontologica Sinica*, **11**, 201–212.

WANG, Y.J. & YANG, Q. 2007. Carboniferous–Permian radiolarian biozones of China and their palaeobiogeographic implication. *Acta Micropalaeontologica Sinica*, **24**, 337–345 [in Chinese with English Abstract].

WANG, Y.J. & YANG, Q. 2011. Biostratigraphy, phylogeny and paleobiogeography of Carboniferous–Permian radiolarians in South China. *Palaeoworld*, **20**, 134–145.

WANG, Y.J., CHENG, Y.N. & YANG, Q. 1994. Biostratigraphy and systematics of Permian radiolarians in China. *Palaeoworld*, **4**, 172–202.

WANG, Y.J., YANG, Q., CHENG, Y.N. & LI, J.X. 2006. Lopingian Upper Permian radiolarian biostratigraphy of South China. *Palaeoworld*, **15**, 31–53.

WANG, Y.J., LUO, H. & YANG, Q. 2012. *Late Paleozoic Radiolarians in the Qinfang Area, Southeast Guangxi.* China University Science Technical Press, Hefei.

WONGANAN, N. & CARIDROIT, M. 2007. Middle to Upper Permian radiolarian faunas from chert blocks in Pai area, northwestern Thailand. *In*: BAUMGARTNER, P.O., AITCHISON, J.C., DE WEVER, P. & JACKETT, S.J. (eds) *Radiolaria*. Birkhäuser, Basel, 133–139.

WU, H.R., XIAN, X.Y. & KUANG, G.D. 1994. Late Paleozoic radiolarian assemblages of southern Guangxi and its geological significance. *Acta Geologica Sinica*, **29**, 339–345.

WU, J. & FENG, Q.L. 2008. Late Changhsingian radiolarian biostratigraphy from Guangxi, South China and its correlation to conodonts. *Science in China Series D: Earth Sciences*, **51**, 1601–1610.

WU, J., FENG, Q.L., GUI, B.W. & LIU, G.C. 2010. Some new radiolarian species and genus from Upper Permian in Guangxi Province, South China. *Journal of Paleontology*, **84**, 879–894.

XIA, W.C., ZHANG, N., WANG, G.Q. & KAKUWA, Y. 2004. Pelagic radiolarian and conodont biozonation in the Permo-Triassic boundary interval and correlation to the Meishan GSSP. *Micropaleontology*, **50**, 27–44.

XIA, W.C., ZHANG, N., KAKUWA, Y. & ZHANG, L.L. 2005. Radiolarian and conodont biozonation in the pelagic Guadalupian–Lopingian boundary interval at Dachongling, Guangxi, South China, and mid-upper Permian global correlation. *Stratigraphy*, **2**, 217–238.

XIE, L., YANG, W.Q., LIU, G.C. & FENG, Q.L. 2011. Late Paleozoic radiolaria from the Upper Triassic sedimentary mélange in Shangrila, Southwest China and its geological significance. *Palaeoworld*, **20**, 203–217.

YAMANAKA, M. 2001. Permian radiolarian biostratigraphy and radiolarian morphological change in bedded chert sequence of the Tamba Terrane in Sasayama area. *News of Osaka Micropaleontologists, Special Volume*, **12**, 13–22.

YAO, A. & KUWAHARA, K. 1999. Middle–late Permian radiolarians from the Guangyuan-Shangsi area, Sichuan province, China. *Journal of Geosciences, Osaka City University*, **42**, 69–83.

YAO, A. & KUWAHARA, K. 2000. Permian and Triassic radiolarians from the southern Guizhou Province, China. *Journal of Geosciences, Osaka City University*, **43**, 1–20.

YAO, A. & KUWAHARA, K. 2004. Radiolarian fossils from the Permian–Triassic of China. *News of Osaka Micropaleontologists, Special Volume*, **13**, 29–45.

YAO, A., KUWAHARA, K. & EZAKI, Y. 2004. Permian radiolarians from the Qinfang Terrane, south China, and its geological significance. *Journal of Geosciences, Osaka City University*, **47**, 71–83.

YAO, J.X., YAO, A. & KUWAHARA, K. 2001. Upper Permian biostratigraphic correlation between conodont and radiolarian zones in the Tamba-Mino Terrane, Southwest Japan. *Journal of Geosciences, Osaka City University*, **44**, 97–119.

YAO, X., ZHOU, Y.Q. & HINNOV, L.A. 2015. Astronomical forcing of a Middle Permian chert sequence in Chaohu, South China. *Earth and Planetary Science Letters*, **422**, 206–221.

ZHANG, K.X., HUANG, J.C., YIN, H.F., WANG, G.C., WANG, Y.B., FENG, Q.L. &, TIAN, J. 2000. Application of radiolarians and other fossils in non-Smith strata. *Science in China Series D: Earth Sciences*, **43**, 364–374.

ZHANG, L., ITO, T., FENG, Q.L., CARIDROIT, M. & DANELIAN, T. 2014. Phylogenetic model of *Follicucullus* lineages (Albaillellaria, Radiolaria) based on high-resolution biostratigraphy of the Permian Bancheng Formation, Guangxi, South China. *Journal of Micropalaeontology*, **33**, 179–192, https://doi.org/10.1144/jmpaleo2014-012

ZHANG, N., XIA, W.C. & SHAO, J. 2002. Radiolarian successional sequences and rare earth element variations in Late Paleozoic chert sequences of South China: An integrated approach for study of the evolution of Paleo-ocean basins. *Geomicrobiology Journal*, **19**, 439–460.

ZHANG, N., HENDERSON, C.M., XIA, W.C., WANG, G.Q. & SHANG, H.J. 2010. Conodonts and radiolarians through the Cisuralian–Guadalupian boundary from the Pingxiang and Dachongling sections, Guangxi region, South China. *Alcheringa*, **34**, 135–160.

ZHAO, Z.G., HUANG, X., ZHANG, X.H., YANG, B., CHEN, Z.Q. & XIONG, F.H. 2016. Permian radiolarians from the A'nyemaqen mélange zone in the Huashixia area of Madoi County, Qinghai Province, Western China, and their implications on regional tectonism. *Journal of Earth Science*, **27**, 623–630.

ZHU, T.X., ZHANG, Y.Q., DONG, H., WANG, Y.J., YU, Y.S. & FENG, X.T. 2006. Discovery of the Late Devonian and Late Permian radiolarian cherts in tectonic mélanges in the Cêdo Caka area, Shuanghu, northern Tibet, China. *Geological Bulletin of China*, **25**, 1413–1418.

Permian rugose corals of the world

XIANGDONG WANG[1]*, LE YAO[1,2] & WEI LIN[1]

[1]*Key Laboratory of Economic Stratigraphy and Palaeogeography, Nanjing Institute of Geology and Palaeontology, Chinese Academy of Sciences, Nanjing 210008, China*

[2]*University of Chinese Academy of Sciences, Beijing 100049, China*

**Correspondence: xdwang@nigpas.ac.cn*

Abstract: Permian rugose corals underwent evolutionary episodes of assemblage changeover, biogeographical separation and extinction, which are closely related to geological events during this time. Two coral realms were recognized, the Tethyan Realm and the Cordilleran–Arctic–Uralian Realm. These are characterized by the families Kepingophyllidae and Waagenophyllidae during the Cisuralian, Waagenophyllidae in the Guadalupian and the subfamily Waagenophyllinae in the Lopingian, and the families Durhaminidae and Kleopatrinidae during the Cisuralian and major disappearance of colonial and dissepimented solitary rugose corals from the Guadalupian to the Lopingian, respectively. The development of these coral realms is controlled by the geographical barrier resulting from the Pangaea formation. According to the changes in the composition and diversity of the Permian rugose corals, a changeover event might have occurred at the end-Sakmarian and is characterized by the mixed Pennsylvanian and Permian faunas to typical Permian faunas, probably related to a global regression. In addition, three extinction events are present at the end-Kungurian, the end-Guadalupian and the end-Permian, which are respectively triggered by the northward movement of Pangaea, the Emeishan volcanic eruptions and subsequent global regression, and the global climate warming induced by the Siberian Traps eruption.

The Permian is a critical interval when multiple global biological and geological events happened (e.g. Wang *et al.* 2006*a*; Shen *et al.* 2016), including the end-Sakmarian assemblage change (Erwin 1995; Wang *et al.* 2006*a*), the end-Guadalupian and end-Permian mass extinction events (Sepkoski 1996; Stanley 2007), palaeoclimatological changes (e.g. glacial and interglacial) (Fielding *et al.* 2008*a*, *b*; Isbell *et al.* 2012) with sea-level fluctuations (Ross & Ross 1987; Haq & Schutter 2008; Rygel *et al.* 2008) and carbon and oxygen isotopic variations (Chen *et al.* 2011, 2013; Grossman 2012; Saltzman & Thomas 2012), the Emeishan and Siberian Traps eruptions (Ali *et al.* 2005; Svensen *et al.* 2009) and Pangaea continent formation and dispersal accompanied by Neotethys Ocean opening (e.g. Scotese & McKerrow 1990; Domeier & Torsvik 2014). These events played important roles in the evolution of Permian rugose corals, in the aspects of changes in their abundance, diversity, composition and palaeobiogeographical distribution (e.g. Fedorowski 1997; Fedorowski *et al.* 1999; Wang & Sugiyama 2002; Wang *et al.* 2006*a*, *b*, 2013; Kossovaya 2009).

Rugose corals were abundant, diversified and globally distributed fossils during the Permian (e.g. Fedorowski 1997; Fedorowski *et al.* 1999; Wang & Sugiyama 2002; Wang *et al.* 2006*a*, *b*, 2013; Chwieduk 2007, 2009, 2013; Stevens 2008*a*, *b*, 2009, 2012; Kossovaya 2009). A comprehensive review of global Permian rugose corals was conducted by Fedorowski (1997). Two coral palaeobiogeographical provinces have been distinguished: the Tethyan Realm and the Cordilleran–Arctic–Uralian (CAU) Realm, which were separated by the Pangaea continent (Fedorowski 1997). However, the reviewed data are relatively scattered and/or incomplete under a poor age constraint, leading to insufficient knowledge about the coral fauna composition and comparisons of them between the Tethyan and CAU realms. In addition, some contributions have also been carried out to illustrate the evolutionary pattern of the Permian rugose corals and their relationships to the geological events (e.g. Fedorowski *et al.* 1999; Wang & Sugiyama 2002; Wang *et al.* 2006*a*, *b*). However, these studies were focused on regional scope, resulting in still unclear knowledge about global rugose coral evolution and the controlling factors during this time.

After Fedorowski (1997), more additional data on Permian rugose corals have been published in many areas, including South China (Wang & Sugiyama 2000, 2001; Wang *et al.* 2006*a*, 2010; Wang & Wang 2007), Inner Mongolia and NE China (Wang *et al.* 2006*b*), Japan (Yamagiwa *et al.* 1999; Wang *et al.* 2006*b*), Peri-Gondwana (Cimmerian continent) and Gondwana (Wang & Sugiyama 2002; Wang *et al.* 2003, 2010, 2013; Wang & Wang 2007; Aung *et al.* 2013), Timor (Sorauf 2004), the Ural Mountains and their adjacent areas (Kossovaya

From: Lucas, S. G. & Shen, S. Z. (eds) 2018. *The Permian Timescale*. Geological Society, London, Special Publications, **450**, 165–184.
First published online June 8, 2017, https://doi.org/10.1144/SP450.13

2007, 2009), Svalbard Archipelago (Fedorowski & Bamber 2001; Chwieduk 2007, 2009, 2013), Greenland (Fedorowski & Bamber 2001; Fedorowski *et al.* 2007), North America (Fedorowski *et al.* 2007; Stevens 2008*a*, *b*, 2009, 2012) and Central and South America (Fedorowski *et al.* 2007). Combined with the data from Fedorowski (1997), these relatively new data could provide a further basis for investigating the composition, distribution and evolution of the rugose corals in the Tethyan and CAU realms during the Permian.

In this paper, based on the published Permian coral data mentioned above, the diversity, abundance, composition and palaeogeographical distribution of the rugose corals were systematically and quantitatively reviewed in different time intervals from different localities, in three categories: colonial rugose coral; dissepimented solitary rugose coral; and non-dissepimented solitary rugose coral (Figs 1 & 2). However, the main aim of this study was to (1) distinguish global rugose coral realms during the Permian; (2) describe the composition and palaeogeographical distribution of the coral realms in the Permian; and (3) unravel the evolutionary pattern of the Permian rugose corals, and their relations to the geological events during this time.

Rugose coral composition and distribution

South China and Indochina

During the Permian, the South China and Indochina blocks were very close to the palaeoequator (Fig. 2), and had similar coral composition during this time (Fedorowski 1997). The Permian rugose corals of the Indochina Block have been inadequately studied, and the Permian strata are not continuous (Fedorowski 1997). However, the rugose coral faunas have been documented in detail under high-resolution biostratigraphic dating in the South China Block (e.g. Wang *et al.* 2006*a*). Hence, in this part, the composition of the rugose corals from South China is used to represent the characteristics of these areas, which are of typical Tethyan affinity (Wang *et al.* 2006*a*).

Cisuralian rugose corals

During the Cisuralian, the rugose corals have a high diversity in South China, comprising 25 families, 127 genera and 707 species. According to the development of rugose corals, the Cisuralian could moreover be divided into three intervals, which are the Asselian and Sakmarian stages, the Artinskian Stage and the Kungurian Stage (Fig. 1).

Asselian–Sakmarian

In South China, the Asselian–Sakmarian interval has a high diversity of rugose corals with 21 families, 88 genera and 311 species, including four new families and 44 new genera (Wang *et al.* 2006*a*). The Kepingophyllidae is one of the most important families with a very high diversity, within which are included all colonial genera, *Anfractophyllum, Antheria, Eokepingophyllum, Kepingophyllum, Nephelophyllum, Parawentzellophyllum* and *Szechuanophyllum* (Fig. 1). The other most important family is Waagenophyllidae, subdivided into two subfamilies, Waagenophyllinae and Wentzellophyllinae, based on the occurrence of tertiary septa. The Waagenophyllinae has no tertiary septa and comprises the dissepimented solitary rugose genera *Pavastehphyllum* and *Thomasiphyllum* and the non-dissepimented solitary rugose genus *Pseudocarniaphyllum*, as well as the colonial rugose genera *Akagophyllum* and *Yokoyamaella* (Fig. 1). The Wentzellophyllinae possesses tertiary septa and is an entirely Permian subfamily, including the dissepimented solitary rugose genus *Iranophyllum* and the colonial rugose genera *Lonsdaleiastraea, Polythecalis, Wentzelellites* and *Wentzellophyllum* (Wang *et al.* 2006*a*; Fig. 1). In addition, the families Aulophyllidae, Bothrophyllidae, Cyathopsidae and Petalaxidae continue from the uppermost Carboniferous with relatively high diversity in the Asselian–Sakmarian interval (Wang *et al.* 2006*a*). Notably, the Bothrophyllidae reached its highest diversity but became extinct at the end of this interval. Non-dissepimented solitary rugose corals are abundant and diversified in this interval, and the common genera are *Allotropiophyllum, Amplexocarinia, Asserculinia, Calophyllum, Timorphyllum, Verbeekiella* and *Zaphrentites* (Wang *et al.* 2006*a*; Fig. 1).

Artinskian

Compared with the Asselian–Sakmarian interval, the Artinskian stage has a relatively low diversity with 11 families, 47 genera and 149 species (Wang *et al.* 2006*a*). Typical Pennsylvanian taxa, such as the Bothrophyllidae, Cyathopsidae, Lithostrotionidae and Petalaxidae, disappeared in the Artinskian stage (Fig. 1). During the Artinskian stage, the Kepingophyllidae is also the dominant family, occupying about 37% of the total species. The most common genera of the Kepingophyllidae are *Anfractophyllum, Antheria, Kepingophyllum, Neoszechuanophyllum, Parawentzellophyllum* and *Szechuanophyllum* (Fig. 1). The second most common family is the Wentzellophyllinae, which contains 25% of the total species. The dominant genera in the Wentzellophyllinae are the colonial

Age			Tethyan Realm: South China and Indochina	Tethyan Realm: Inner Mongolia-NE China and Japan	Tethyan Realm: Peri-Gondwana and Gondwana	Tethyan Realm: Timor	Cordilleran-Arctic-Uralian Realm: Ural mountains and adjacent areas	Cordilleran-Arctic-Uralian Realm: Svalbard Archipelago	Cordilleran-Arctic-Uralian Realm: Greenland	Cordilleran-Arctic-Uralian Realm: North America	Cordilleran-Arctic-Uralian Realm: Central and South America
Permian	Lopingian	Changhsingian	CRC: Ipciphyllum, Liangshanophyllum, *Waagenophyllum, Paracaninia. DSRC: Zhenganophyllum. NDSRC: *Calophyllum, Lophocarinophyllum, Pentaphyllum (Plerophyllum), *Ufimia (Tachylasma), Asserculinia, Endothecium, Tetrelasma.		CRC: Huayunophyllum, Ipciphyllum, Liangshanophyllum, *Waagenophyllum, Gyanyimaphyllum. NDSRC: *Calophyllum, Lophocarinophyllum, Lophophyllidium, Pentaphyllum, *Ufimia, Amplexocarinia, Epiphanophyllum, Lytvolasma, Paracaninia, *Tachylasma, Timorcarinophyllum, Wannerophyllum.					CRC: *Waagenophyllum.	
		Wuchiapingian	CRC: **Huayunophyllum**, **Liangshanophyllum**, ***Waagenophyllum**, Ipciphyllum, Praewentzelella. NDSRC: Lophocarinophyllum, Lophophyllidium, Pentaphyllum, *Ufimia, Allotropiophyllum.						NDSRC: Allotropiochisma, *Calophyllum, Leonardophyllum, Soshkineophyllum, *Tachylasma, *Ufimia.		
	Guadalupian	Capitanian	CRC: Ipciphyllum, **Liangshanophyllum**, Multimurinus, ***Waagenophyllum**, **Parawentzelella**, Proipciphyllum. DSRC: Iranophyllum, **Paracaninia**. NDSRC: Asserculinia, *Ufimia, Allotropiophyllum.	CRC: Ipciphyllum, Lonsdaleiastraea, Wentzelella, Liangshanophyllum, Paraipciphyllum, Praewentzelella, Szechuanophyllum, *Waagenophyllum, Wentzelloides, Huangia, Huangyunophyllum, Maoriphyllum, Miyagiella, Tanbaella, Yatsengia, Yokoyamaella. DSRC: Iranophyllum, Geyerophyllum. NDSRC: Asserculinia, Duplophyllum, *Lophophyllidium, *Ufimia, *Calophyllum, Pentaphyllum, *Sochkineophyllum, Allotropiophyllum, Lophocarinophyllum, Paracaninia, Timorphyllum, Verbeekiella, Zaphrentis, Amygdalophylloides, Gerthia, Huangophyllum, Rotiphyllum.	CRC: Ipciphyllum, Lonsdaleiastraea, Wentzelella, Multimurinus, Paraipciphyllum, Praewentzelella, Wentzelellites, Wentzelloides, Wentzellophyllum, Akagophyllum, Monothecalis, Polythecalis, Pseudohuangia, Protolonsdaleiastraea, Prowentzelellites. DSRC: Iranophyllum, Pavastehphyllum, Sakamotosawanella, Thomasiphyllum, Zhenganophyllum. NDSRC: Asserculinia, Duplophyllum, *Lophophyllidium, *Ufimia, Amplexocarinia, *Calophyllum, Pentaphyllum, *Euryphyllum, Lophocarinophyllum, Paracaninia, Cyathaxonia, Cyathocarinia, *Lytvolasma, Neozaphrentis, Praetachylasma, Timorcarinophyllum.	CRC: Ipciphyllum, Lonsdaleiastraea. NDSRC: *Amplexocarinia, *Asserculinia, *Calophyllum, Duplophyllum, *Lophophyllidium, Pentaphyllum, *Ufimia, *Euryphyllum, *Sochkineophyllum, Timorphyllum, Verbeekiella, Basleophyllum, Duplocarinia, Endothecium, *Lytvolasma, *Paralleynia, Pleramplexus, Prosmilia, Zaphrentis, Wannerophyllum, Allophyllum, Endamplexus, *Falsiamplexus, Pentamplexus, Productiophyllum, Spineria, Tachylasma.		NDSRC: Allotropiochisma, *Calophyllum, *Euryphyllum, *Soshkineophyllum, *Ufimia, Fedorowskites.		CRC: Miyagiella, *Waagenophyllum. NDSRC: Allotropiochisma, *Calophyllum, *Euryphyllum, *Soshkineophyllum, *Ufimia, Assimulia, Leonardophyllum, *Lophophyllidium, Lophotichium, *Lytvolasma, Paraduplophyllum.	
		Wordian	CRC: Ipciphyllum, Wentzelella, Multimurinus, Szechuanophyllum, Wentzelellites, Wentzellophyllum, Anfractophyllum, Pseudopolythecalis. DSRC: Iranophyllum, Laophyllum. NDSRC: Asserculinia, Duplophyllum, *Lophophyllidium, *Ufimia, Amplexocarinia, *Sochkineophyllum, Allotropiophyllum, Paralleynia, Prosmilia, Bradyphyllum, Neozaphrentis, Zaphrentites.								
		Roadian									
	Cisuralian	Kungurian	CRC: Akagophyllum, Ipciphyllum, **Polythecalis**, Szechuanophyllum, **Wentzellophyllum**, Anfractophyllum, Kepingophyllum, Nephelophyllum, Pseudohuangia. DSRC: Pavastehphyllum, Thomasiphyllum. NDSRC: *Amplexocarinia, *Lophophyllidium, Pentaphyllum, *Ufimia, Duplophyllum, Lophocarinophyllum, Allotropiophyllum.	CRC: Akagophyllum, Lonsdaleiastraea, Polythecalis, Szechuanophyllum, Wentzellophyllum, Chusenophyllum, Multimurinus, Wentzelella, Wentzellites, Yatsengia, Yokoyamaella, Zhurihephyllum. NDSRC: *Amplexocarinia, *Lophophyllidium, Pentaphyllum, *Ufimia, *Calophyllum, *Euryphyllum, Lophocarinophyllum, Timorphyllum, Verbeekiella, Cyathocarinia, Endothecium, Metriophyllum, Rotiphyllum, *Bradyphyllum, Huangophyllum, ?Petraia, Zaphrentites.	NDSRC: *Amplexocarinia, *Lophophyllidium, Pentaphyllum, *Ufimia, *Calophyllum, Duplophyllum, *Euryphyllum, Lophocarinophyllum, Timorphyllum, Verbeekiella, Cyathacarinia, *Lytvolasma, Metriophyllum, *Paralleynia, Pleramplexus, Rotiphyllum, Amplexus, Cravenia, *Cyathaxonia, Paracaninia.		CRC: Pararachnastraea, Protowentzelella, Tschussovskenia, *Durhamina, Fomichevella, Heintzella, Kleopatrina, Parahentschioides, Pseudocystophora, Iskutella, Lytvophyllum, Permastraea, Protolonsdaleia. NDSRC: *Lophophyllidium, *Lytvolasma, Actinophrentis, *Calophyllum, Lophotichium, *Paralleynia, Pseudowannerophyllum, *Amplexocaninia, *Asserculinia, *Cyathaxonia, Cyathocaninia, Lopholasma, Hexalasma, *Soshkineophyllum, *Ufimia.	CRC: Pararachnastraea, Protowentzelella, Tschussovskenia, Fomichevella, Heintzella, Kleopatrina, Pseudocystophora, Stylastraea, Protolonsdaleiastraea. DSRC: *Bothrophyllum, Caninophyllum, *Timania, Amygdalophylloides, Arctophyllum, Barentsburgia, Gronfjordphyllum, Gshelia, Hornsundia, Linnephyllum, Pseudotimania, Svalbardphyllum, Yakovleviella. NDSRC: *Lytvolasma, Krusenella.		CRC: Pararachnastraea, Protowentzelella, Tschussovskenia, *Durhamina, Fomichevella, Heintzella, Kleopatrina, Parahentschioides, Pseudocystophora, Stylastraea, Iskutela, Lytvophyllum, Permastraea, *Petalaxis, Protolonsdaleiastraea, Cordillerastraea, Cystolonsdaleia, Darwasophyllum, Langenheimia, Sandolasma, Shastalasma, Wendoverella, Wilsonastraea. DSRC: *Bothrophyllum, Caninophyllum, *Timania, Heterocaninia, Siedlekia. NDSRC: *Lophophyllidium, *Lytvolasma, Actinophrentis, *Calophyllum, Lophotichium, *Paraleynia, Pseudowannerophyllum, Allotropiochisma, Assimulia, *Bradyphyllum, *Euryphyllum, *Falsiamplexus, Monophyllum, *Paraduplophyllum.	CRC: Pararachnastraea, Protowentzelella, Tschussovskenia, *Durhamina, Parahentschioides, Stylastraea, *Petalaxis. DSRC: *?Caninia. NDSRC: *Lophophyllidium.
		Artinskian	CRC: **Lonsdaleiastraea**, **Wentzellophyllum**, **Anfractophyllum**, **Antheria**, **Neoszechuanophyllum**, **Kepingophyllum**, **Parawentzellophyllum**, **Szechuanophyllum**, **Wentzelellites**. DSRC: Iranophyllum. NDSRC: Duplophyllum, Pentaphyllum, Allotropiophyllum, Lophamplexus, *Paraduplophyllum.	CRC: Akagophyllum, Wentzelophyllum, Yokoyamaella, Carinthiaphyllum, *Durhamina, Heritschioides, Huangia, Ivanovia, Yatsengia. DSRC: Amandophyllum, *Caninia, Iranophyllum, *Timania, Geyerophyllum, Laophyllum, Paracarruthersella, Sakamonosawanella, Sestrophyllum. NDSRC: *Calophyllum, Duplophyllum, *Lophophyllidium, Pentaphyllum, *Ufimia, Rotiphyllum, *Sochkineophyllum, Amplexizaphrentis, Barrandeophyllum, Cyathocarinia, Empodesma, Gerthia, Lophocarinophyllum.	DSRC: ?Amandophyllum, Caninia, ?Amygdalophylloides. NDSRC: Calophyllum, Duplophyllum, *Amplexocarinia, *Lophophyllidium, *Ufimia, Verbeekiella, Basleophyllum, Duplocarinia, *Euryphyllum, Rotiphyllum, Wannerophyllum, Amplexus, *Bradyphyllum, ?Claviphyllum, *Cyathaxonia, Meniscophyllum, Paracaninia, Pseudobradyphyllum, Pseudoverbeekiella.						
		Sakmarian	CRC: Akagophyllum, Lonsdaleiastraea, Wentzellophyllum, Yokoyamaella, **Anfractophyllum**, **Antheria**, **Eokepingophyllum**, **Nephelophyllum**, **Kepingophyllum**, **Parawentzellophyllum**, *Petalaxis, Polythecalis, **Szechuanophyllum**, Wentzelellites. DSRC: Iranophyllum, *Timania, *Bothrophyllum, Pavastehphyllum, Pseudocarniaphyllum, Thomasiphyllum. NDSRC: *Calophyllum, *Amplexocarinia, Verbeekiella, *Asserculinia, Timorphyllum, Allotropiophyllum, Zaphrentites.								
		Asselian									

Fig. 1. Stratigraphic distribution of the rugose coral faunas in the Tethyan and Cordilleran–Arctic–Uralian realms during the Permian. The orange, blue and green colored genera occurred in the four, three and two regions respectively among the Tethyan and Cordilleran–Arctic–Uralian realms. The asterisk represents the genus in both the Tethyan and Cordilleran–Arctic–Uralian realms. The genus in bold indicates that its abundance is very high. CRC, Colonial rugose coral; DSRC, dissepimented solitary rugose coral; NDSRC, non-dissepimented solitary rugose coral.

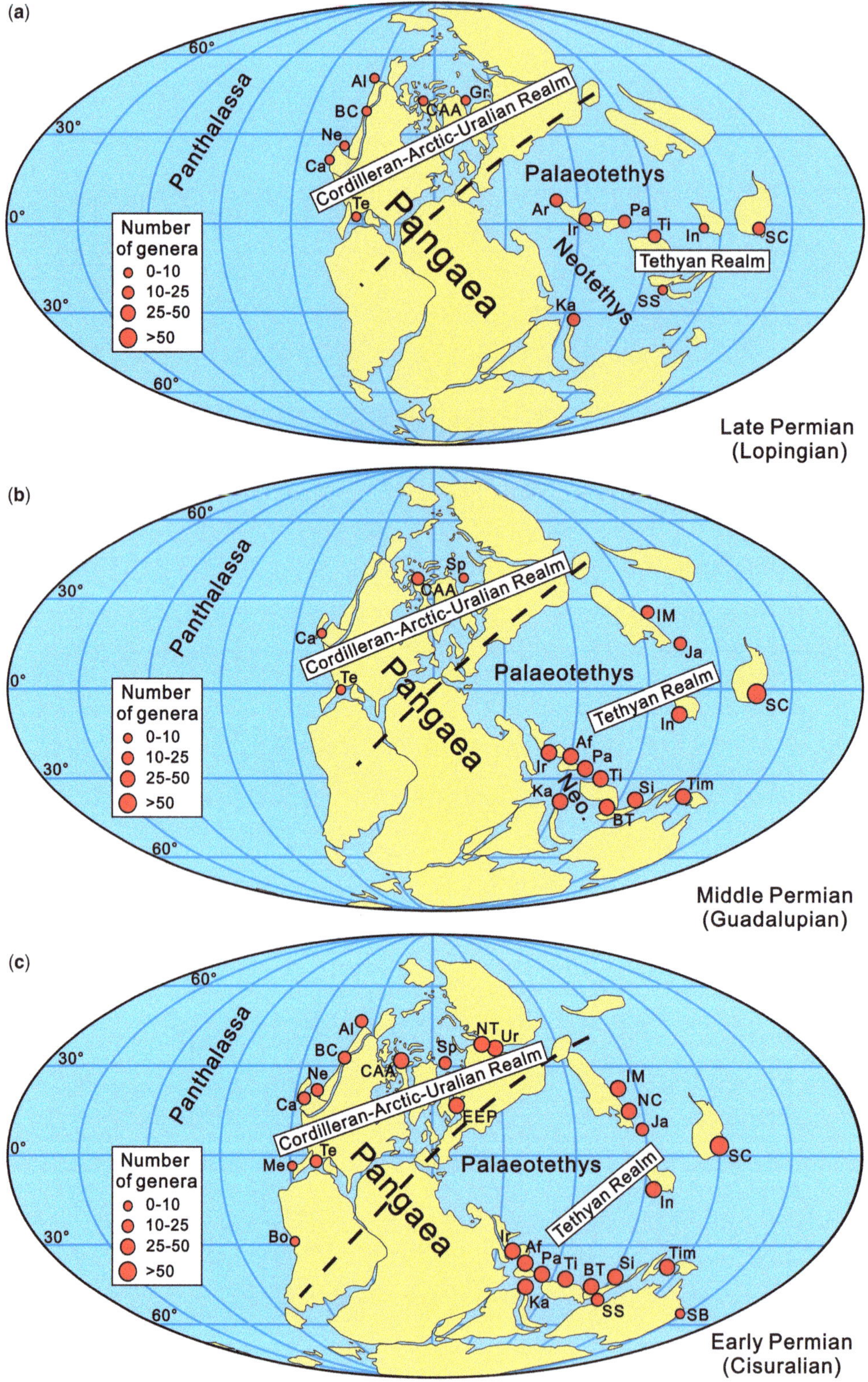
(a)
60°
30°
0°
30°
60°
Panthalassa
Al
BC
Ne
Ca
Te
CAA
Gr
Cordilleran-Arctic-Uralian Realm
Pangaea
Palaeotethys
Ar
Ir
Pa
Ti
In
SC
Tethyan Realm
Neotethys
SS
Ka
Number of genera
0-10
10-25
25-50
>50
Late Permian
(Lopingian)
(b)
60°
30°
0°
30°
60°
Panthalassa
Sp
CAA
Ca
Te
Cordilleran-Arctic-Uralian Realm
Pangaea
Palaeotethys
IM
Ja
SC
Tethyan Realm
In
Af
Pa
Ir
Ti
Ka
Neo.
Si
Tim
BT
Number of genera
0-10
10-25
25-50
>50
Middle Permian
(Guadalupian)
(c)
60°
30°
0°
30°
60°
Panthalassa
Al
BC
Ne
Ca
CAA
Sp
NT
Ur
Cordilleran-Arctic-Uralian Realm
EEP
Me
Te
Bo
Pangaea
Palaeotethys
IM
NC
Ja
SC
Tethyan Realm
In
Ir
Af
Pa
Ti
BT
Si
Tim
Ka
SS
SB
Number of genera
0-10
10-25
25-50
>50
Early Permian
(Cisuralian)

rugose genera *Lonsdaleiastraea*, *Wentzelellites* and *Wentzellophyllum*, and the dissepimented solitary rugose genus *Iranophyllum* (Fig. 1). This stage also consists of abundant endemic genera in southern Shanxi Province of northwestern South China, including the colonial rugose genera *Neoszechuanophyllum*, *Sanlichongophyllum*, *Shimenyaphyllum*, *Xikouphyllum* and *Yangshanophyllum*, and the dissepimented solitary rugose genus *Neogangamophyllum*. Non-dissepimented solitary rugose corals are also abundant and diversified during the Artinskian, among which the most common genera are *Allotropiophyllum*, *Duplophyllum*, *Lophamplexus*, *Paraduplophyllum* and *Pentaphyllum* (*Plerophyllum*) (Wang *et al.* 2006*a*; Fig. 1).

Kungurian

Similar to the Artinskian stage, this stage also contains a high diversity of rugose corals with 19 families, 50 genera and 271 species (Wang *et al.* 2006*a*; Fig. 1). During the Kungurian, the Wentzellophyllinae is highly diversified, including 49% of the total species. Dominant genera are the colonial rugose genera *Wentzellophyllum* and *Polythecalis* (Fig. 1). Another flourishing subfamily is the Waagenophyllinae, characterized by the colonial rugose genera *Akagophyllum*, *Ipciphyllum* and *Pseudohuangia*, and the dissepimented solitary rugose genera *Pavastehphyllum* and *Thomasiphyllum* (Fig. 1). The Kepingophyllidae lost its dominant position during this time, but it still comprise such common genera as *Anfractophyllum*, *Kepingophyllum*, *Nephelophyllum* and *Szechuanophyllum* (Fig. 1). Non-dissepimented solitary rugose corals are also important in this interval, including the flourishing genera *Allotropiophyllum*, *Amplexocarinia*, *Duplophyllum*, *Lophocarinophyllum*, *Lophophyllidium*, *Pentaphyllum* and *Ufimia* (*Tachylasma*) (Wang *et al.* 2006*a*; Fig. 1).

Guadalupian rugose corals

Compared with the Cisuralian, the diversity of the Guadalupian rugose corals obviously decreased in South China. They are composed of 18 families, 54 genera and 218 species. In terms of the development of rugose corals, the Guadalupian could be further divided into two intervals: the Roadian and Wordian stages and the Capitanian Stage (Fig. 1).

Roadian–Wordian

In total 13 families, 40 genera and 132 species were identified in this interval, with 19 genera becoming extinct (Wang *et al.* 2006*a*). The most important taxon is the Waagenophyllinae, including the highly diversified colonial rugose genus *Ipciphyllum* (Fig. 1). The Wentzellophyllinae still contains diversified genera, such as *Iranophyllum*, *Laophyllum*, *Multimurinus*, *Pseudopolythecalis*, *Wentzelella*, *Wentzelellites* and *Wentzellophyllum* (Fig. 1). The diversity of Kepingophyllidae greatly declined during this time, with only two remaining genera, *Anfractophyllum* and *Szechuanophyllum* (Fig. 1). The Kepingophyllidae became extinct at the end of this interval (Wang *et al.* 2006*a*). Non-dissepimented solitary rugose genera are highly diversified from the Roadian to Wordian, accompanied by the most common genus *Allotropiophyllum*, and the other common genera *Amplexocarinia*, *Asserculinia*, *Bradyphyllum*, *Duplophyllum*, *Lophophyllidium*, *Neozaphrentis*, *Paralleynia*, *Prosmilia*, *Sochkineophyllum*, *Ufimia* and *Zaphrentites* (Wang *et al.* 2006*a*; Fig. 1).

Capitanian

This stage has a similar rugose coral diversity to that of the Roadian–Wordian interval with 15 families, 30 genera and 104 species (Wang *et al.* 2006*a*). The number of extinct taxa (seven families and 17 genera) increased, in contrast to a low number of originating taxa (two families and seven genera). The latter consisted entirely of endemic genera, such as the dissepimented solitary rugose genus *Endamplexus* and non-dissepimented rugose genus *Endothecium* belonging to the Endamplexidae and Endotheciidae, respectively. During the Capitanian, the most diversified taxon is the Waagenophyllinae, comprising the common colonial rugose genera *Ipciphyllum*, *Parawentzelella*, *Proipciphyllum*, *Liangshanophyllum* and *Waagenophyllum* (Fig. 1). However, the Wentzellophyllinae dramatically decreased in diversity, with only the dissepimented

Fig. 2. Global distribution of rugose corals and subdivision of their palaeobiogeographical realms during the (**a**) late Permian (Lopingian), (**b**) middle Permian (Guadalupian) and (**c**) early Permian (Cisuralian). The base maps were revised from Domeier & Torsvik (2014). Af, Afghanistan; Al, Alaska; Ar, Armenia; BC, British Columbia; Bo, Bolivia; BT, Baoshan–Tengchong; Ca, California; CAA, Canadian Arctic Archipelago; EEP, Eastern Europe Platform; Gr, Greenland; IM, Inner Mongolia; In, Indochina; Ir, Iran; Ka, Karakoram; Ja, Japan; Me, Mexico; NC, northeast China; Ne, Nevada; Neo, Neotethys Ocean; NT, Northern Timan; Pa, Pamir; SB, Sydney Basin; SC, South China; Si, Sibumasu; Sp, Spitsbergen; SS, Shan States (Myanmar); Te, Texas; Ti, Tibet; Tim, Timor; Ur, Ural Mountains.

solitary rugose genus *Iranophyllum* and colonial rugose genus *Multimurinus* (Fig. 1). The solitary cyathopsid genus *Paracaninia* was recovered and highly diversified with 12 species during this time. Non-dissepimented solitary rugose corals are relatively abundant, and the major genera include *Allotropiophyllum*, *Asserculinia* and *Ufimia*. Common families, such as the Antiphyllidae, Cyathaxoniidae, Laccophyllidae and Timorphyllidae, became extinct at the end of this interval (Wang *et al.* 2006*a*; Fig. 1).

Lopingian rugose corals

The diversity of rugose corals dramatically declined from the Guadalupian to Lopingian in South China, containing nine families, 23 genera and 142 species. Based on the development of rugose corals, the Lopingian could be classified into two intervals: the Wuchiapingian Stage and the Changhsingian Stage (Fig. 1).

Wuchiapingian

The Wuchiapingian rugose corals have a low diversity, consisting of only eight families, 17 genera and 94 species (Wang *et al.* 2006*a*). The dominant taxon is the Waagenophyllinae, representing 49% of the total species and including the abundant colonial rugose genera *Huayunophyllum*, *Ipciphyllum*, *Liangshanophyllum* and *Waagenophyllum* (Fig. 1). The Wentzellophyllinae has only one survivor of the colonial rugose genus *Praewentzelella*, and became totally extinct at the end of this interval (Fig. 1). Non-dissepimented solitary rugose corals are relatively diversified, comprising the common genera *Allotropiophyllum*, *Lophocarinophyllum*, *Lophophyllidium*, *Pentaphyllum* and *Ufimia* (Wang *et al.* 2006*a*; Fig. 1).

Changhsingian

Similar to the Wuchiapingian, the diversity of the Changhsingian rugose corals is also very low, and only 62 species belonging to 14 genera in seven families occurred in this interval (Wang *et al.* 2006*a*). The Waagenophyllinae family is represented by the colonial rugose genera *Ipciphyllum*, *Liangshanophyllum* and *Waagenophyllum*, and the dissepimented solitary rugose genus *Zhenganophyllum* (Fig. 1). The Cyathopsidae family, represented by the non-dissepimented solitary rugose genus *Paracaninia*, was recovered at Anshun, Guizhou Province. Relatively high diversity is shown by non-dissepimented genera, including *Asserculinia*, *Calophyllum*, *Endothecium*, *Lophocarinophyllum*, *Lophophyllidium*, *Pentaphyllum*, *Tetralasma* and *Ufimia*. All rugose corals became extinct at the end of this interval (Wang *et al.* 2006*a*; Fig. 1).

Inner Mongolia–NE China and Japan

Inner Mongolia–NE China are regarded as transitional palaeobiogeographical regions based on the occurrence of antitropical brachiopods (Shi *et al.* 1995) and the fusulinid *Monodiexodina* (Ueno 2003), while Japan is regarded as part of the Panthalassa Ocean (Wang *et al.* 2006*b*). However, the coral faunas are typically of Tethyan affinity and widely developed from the Cisuralian to Guadalupian (Fedorowski 1997; Wang *et al.* 2006*b*).

Cisuralian rugose corals

During the Cisuralian, the rugose corals have a high diversity in NE China and Japan, which can be divided into two intervals in terms of their development: the Asselian–Artinskian Stage and the Kungurian Stage (Figs 1 & 2c).

Asselian–Artinskian

From the Asselian to Artinskian, the colonial and solitary rugose corals were widely distributed in Inner Mongolia and NE China, and consist of common Cathaysian faunas such as the colonial rugose genera *Akagophyllum*, *Carinthiaphyllum* and *Ivanovia*, the dissepimented solitary rugose genera *Amandophyllum*, *Paracarruthersella* and *Timania*, and the non-dissepimented solitary rugose genera *Amplexizaphrentis*, *Barrandeophyllum*, *Calophyllum*, *Cyathocarinia*, *Empodesma*, *Lophophyllidium*, *Rotiphyllum* and *Ufimia* (Guo 1983; Wang *et al.* 2006*b*; Fig. 1).

In Japan, the Asselian–Artinskian rugose corals were diversified and mainly occurred in the Akiyoshi, Mino and South Kitakami terranes. They mainly comprise the colonial rugose genera *Akagophyllum*, *Durhamina*, *Heritschioides*, *Huangia*, *Wentzelophyllum*, *Yatsengia* and *Yokoyamaella*, the dissepimented solitary rugose genera *Amandophyllum*, *Caninia*, *Geyerophyllum*, *Iranophyllum*, *Laophyllum*, *Sakamonosawanella* and *Sestrophyllum* (Sakaguchi & Yamagiwa 1958; Yokoyama 1960; Yamagiwa 1961; Minato & Kato 1965*a*; Minato & Rowett 1968; Kato 1972; Igo 1988; Igo *et al.* 2000; Igo & Adachi 2001), and the non-dissepimented solitary rugose genera *Duplophyllum*, *Gerthia*, *Lophocarinophyllum*, *Pentaphyllum* and *Sochkineophyllum* (Kamei 1957; Igo 1959; Yamada & Yamano 1980; Igo & Adachi 2000; Fig. 1).

Kungurian

In Kungurian time, both colonial and non-dissepimented solitary rugose corals flourished in NE China (Fig. 2c). Colonial rugose corals include the genera *Chusenophyllum*, *Multimurinus*, *Polythecalis*, *Szechuanophyllum*, *Wentzellites*, *Wentzellophyllum*, *Yatsengia* and *Zhurihephyllum* (Guo, 1980*a*, *b*; Yu *et al.* 1981; Xia & Ding 1983; Zhao & Yang 1985; Ding 1987; Fig. 1). In total 16 non-dissepimented solitary rugose coral genera have been documented, which are *Amplexocarinia*, *Bradyphyllum*, *Calophyllum*, *Cyathocarinia*, *Endothecium*, *Euryphyllum*, *Huangophyllum*, *Lophocarinophyllum*, *Lophophyllidium*, *Metriophyllum*, *Pentaphyllum*, *Rotiphyllum*, *Timorphyllum*, *Ufimia*, *Verbeekiella* and *Zaphrentites* (Guo 1980*a*, 1991; Yu *et al.* 1981; Fig. 1).

Compared with NE China, the Kungurian rugose corals in Japan are characterized by low diversity, containing the colonial rugose genera *Akagophyllum*, *Lonsdaleiastraea*, *Wentzelella*, *Yatsengia* and *Yokoyamaella*, and dissepimented solitary rugose genus ?*Petraia* (Minato 1955; Igo 1959; Yamagiwa 1961; Choi & Fujita 1970; Fig. 1).

Guadalupian rugose corals

During the Guadalupian, the diversity of rugose corals is also high and widely distributed in Inner Mongolia and NE China (Fig. 1). The colonial rugose corals have a relatively low diversity and are dominated by *Waagenophyllum*, which occurred in eastern Inner Mongolia (Guo, 1980*a*, *b*) and NE China (Yu *et al.* 1981; Figs 1 & 2b). The other colonial rugose coral genera include *Liangshanophyllum*, *Wentzelella* and *Yatsengia* (Fig. 2b), while the non-dissepimented solitary rugose corals are highly diversified in central Inner Mongolia, comprising the genera *Amygdalophylloides*, *Calophyllum*, *Duplophyllum*, *Gerthia*, *Huangophyllum*, *Lophocarinophyllum*, *Lophophyllidium*, *Paracaninia*, *Pentaphyllum*, *Timorphyllum*, *Ufimia* and *Verbeekiella* (Ding 1983; Ding *et al.* 1985; Fig. 1). The dissepimented solitary rugose genus *Pseudowaagenophyllum*, and the non-dissepimented solitary rugose genera *Calophyllum*, *Carinoverbeekiella*, *Damuqiphyllum*, *Diphycarinophyllum* and *Lophotichium* occurred in an upper horizon, which may be Wuchiapingian in age as the fusulinid *Codonofusiella asiatica* is recorded in these strata (Ding *et al.* 1985). However, the age of those rugose corals should be further verified.

In Japan, similar to Inner Mongolia–NE China, the Guadalupian rugose corals are also prosperous. The colonial rugose genera are highly diversified, including the genera *Huangia*, *Huayunophyllum*, *Ipciphyllum*, *Lonsdaleiastraea*, *Maoriphyllum*, *Miyagiella*, *Paraipciphyllum*, *Praewentzelella*, *Szechuanophyllum*, *Tanbaella*, *Waagenophyllum*, *Wentzelella*, *Wentzelloides*, *Yatsengia* and *Yokoyamaella*, while the diversity of the dissepimented solitary rugose genera is low and only comprises *Geyerophyllum* and *Iranophyllum* (Yabe & Minato 1945; Minato & Kato 1965*b*, 1980; Sakaguchi & Yamagiwa 1973; Yamagiwa & Suzuki 1976; Minato *et al.* 1979; Yamagiwa & Tsuda 1980; Yamagiwa & Yamano 1990; Igo *et al.* 2001; Fig. 1). Non-dissepimented solitary rugose corals consist of the genera *Allotropiophyllum*, *Asserculinia*, *Lophocarinophyllum*, *Lophophyllidium*, *Rotiphyllum*, *Sochkineophyllum*, *Ufimia*, *Verbeekiella* and *Zaphrentis* (Minato 1955; Igo 1959; Yanagisawa 1967; Fig. 1).

Peri-Gondwana and Gondwana

The Peri-Gondwana region, which equates to the term 'Cimmerian continent', consists of several allochthonous blocks (or terranes) located between Gondwana and Eurasia during the Permian (Sengor 1979, 1984; Fig. 2), which have a similar Late Palaeozoic evolutionary history. They were separated from the Gondwana continent during the early Permian, drifted northward in the Palaeotethys Ocean during the middle and late Permian, and amalgamated with southern Eurasia during the Triassic (Metcalfe 1991, 1999; Archbold & Shi 1996; Grunt & Shi 1997; Shi & Archbold 1998; Fig. 2). According to the definition by Sengor (1984), the Peri-Gondwana region (Cimmerian continent) includes the Mediterranean and Ghaznian (SW Asian) Cimmerides, and the Chinese and Indochinese (Southeast Asian) Cimmerides. In this study, the Permian rugose corals from the Peri-Gondwana are documented in Iran, Afghanistan, Pakistan, Pamir, Tibet, Southwestern China (Baoshan and Tengchong) and Myanmar (Fig. 2). However, the Gondwana region only contains the Sydney Basin, Australia, where few rugose corals were found (Wang *et al.* 2013; Fig. 2).

Cisuralian rugose corals

During the Cisuralian, the rugose corals from Peri-Gondwana have a high diversity (Fig. 1). They were described by two intervals of the Asselian–Artinskian stages and the Kungurian Stage, and classified in terms of their coral fauna development (Fig. 1).

Asselian–Artinskian

During the Asselian–Artinskian, the coral faunas are characterized by non-dissepimented solitary rugose corals, with comparatively rare dissepimented solitary rugose corals (Wang & Sugiyama 2002; Fig. 1). The non-dissepimented solitary corals, also named the *Cyathaxonia* faunas, widely occurred in Peri-Gondwana terranes such as Sibumasu, Baoshan, Tengchong, Central Tibet, South

Afghanistan, Pamir, Karakoram (Pakistan) and Iran. Typical genera are ?*Amandophyllum*, *Amplexocarinia*, *Amplexus*, ?*Amygdalophylloides*, *Basleophyllum*, *Bradyphyllum*, *Calophyllum*, *Caninia*, ?*Claviphyllum*, *Cyathaxonia*, *Duplocarinia*, *Duplophyllum*, *Euryphyllum*, *Lophophyllidium*, *Meniscophyllum*, *Paracaninia*, *Pseudobradyphyllum*, *Pseudoverbeekiella*, *Rotiphyllum*, *Ufimia*, *Verbeekiella* and *Wannerophyllum* (Waterhouse 1982; Yang & Fan 1982; Zhao & Wu 1986; Fan 1988; Flügel 1989; Yang & Nie 1990; Nie *et al.* 1993; Fedorowski 1997; Ilina 1997; Leven 1997; Wang *et al.* 1999; Fig. 1).

Kungurian

Consistent with the Asselian–Artinskian, the Kungurian rugose corals are also dominated by non-dissepimented rugose corals, comprising the genera *Amplexocarinia*, *Euryphyllum*, *Lophophyllidium*, *Lytvolasma*, *Paracaninia*, *Pleramplexus* and *Ufimia* from Peninsular Thailand (Fontaine *et al.* 1994), the Baoshan Block (Wang *et al.* 2001) and the southern Shan States of Myanmar (Smith 1941; Fig. 1). The most abundant non-dissepimented solitary rugose corals also occurred in the Lhasa Block, containing the genera *Amplexocarinia*, *Amplexus*, *Calophyllum*, *Cravenia*, *Cyathacarinia*, *Cyathaxonia*, *Duplophyllum*, *Lophocarinophyllum*, *Lophophyllidium*, *Lytvolasma*, *Metriophyllum*, *Paralleynia*, *Pentaphyllum*, *Rotiphyllum*, *Timorphyllum*, *Ufimia* and *Verbeekiella* (Wang & Liu 1982; Wu *et al.* 1982; Lin 1983; Fan 1988; Zhao 1991; Fig. 1). In addition, the Kungurian rugose corals were also present in the Wandrawandian Siltstone from the southern Sydney Basin, characterized by one non-dissepimented solitary rugose genus, *Euryphyllum* (Shi *et al.* 2010; Wang *et al.* 2013).

Guadalupian rugose corals

Different from the Cisuralian, very abundant colonial and large dissepimented solitary rugose corals (Cathaysian taxa) appeared in the Peri-Gondwana continent during the Guadalupian. They all belong to the typical middle Permian family Waagenophyllidae, including such common colonial rugose genera as *Akagophyllum*, *Ipciphyllum*, *Lonsdaleiastraea*, *Monothecalis*, *Multimurinus*, *Paraipciphyllum*, *Polythecalis*, *Praewentzelella*, *Protolonsdaleiastraea*, *Prowentzelellites*, *Pseudohuangia*, *Wentzelella*, *Wentzelellites*, *Wentzelloides* and *Wentzellophyllum*; dissepimented solitary rugose genera *Iranophyllum*, *Pavastehphyllum*, *Sakamotosawanella*, *Thomasiphyllum* and *Zhenganophyllum* (Douglas 1936; Hudson 1958; Fontaine *et al.* 1979; He & Weng 1982; Wu *et al.* 1982; Lin 1983; Zhao & Wu 1986; Fontaine & Suteethorn 1988; He 1990; Ezaki 1991; Zhao 1991; Flügel 1994; Fontaine & Jungyusuk 1995; Lin *et al.* 1995; Weidlich & Flügel 1995; Ueno *et al.* 1996; Fig. 1). *Thomasiphyllum* and *Wentzellophyllum* are the typical Cimmerian elements and developed from western Cimmeria to eastern Cimmeria during this time. Some endemic genera also occurred in this area, which are the colonial rugose genera *Ophryphyllum*, *Tibetophyllum* and *Xizangophyllum*, and the dissepimented solitary rugose genera *Sarcinophyllum* and *Axopinnophyllum* (He & Weng 1982; Wu *et al.* 1982; Lin 1983; Zhao & Wu 1986; Zhao 1991; Fig. 1). The non-dissepimented solitary rugose corals are diversified and contain the genera *Amplexocarinia*, *Asserculinia*, *Calophyllum*, *Cyathocarinia*, *Cyathaxonia*, *Duplophyllum*, *Euryphyllum*, *Lophocarinophyllum*, *Lophophyllidium*, *Lytvolasma*, *Neozaphrentis*, *Paracaninia*, *Pentaphyllum*, *Praetachylasma*, *Timorcarinophyllum* and *Ufimia* from Tibet, Iran, Pamir and Pakistan (Karakoram) (He & Weng 1982; He 1990; Ezaki 1991; Fedorowski 1997; Ilina 1997; Fig. 1).

Lopingian rugose corals

During the Lopingian, the rugose corals are dominated by Waagenophyllidae, comprising the genera *Liangshanophyllum*, *Huayunophyllum* and *Waagenophyllum* (Fig. 1). *Waagenophyllum* was discovered from the Plateau Limestone of the southern Shan States, indicating a possible Wuchiapingian age (Smith 1941). *Huayunophyllum* occurs in central and western Tibet, implying possible Lopingian strata in these areas (Wu *et al.* 1982; He 1990). In addition, exotic limestone blocks occurred in the Yarlung–Tsangpo suture zone, Tibet, accompanied by limestones probably deposited on a seamount in the Palaeotethys Ocean between the Gondwana and Cathaysian continents during the Lopingian (Shen *et al.* 2010; Zhang *et al.* 2013). These limestones contain very abundant colonial rugose corals such as the genera *Gyanyimaphyllum*, *Ipciphyllum*, *Liangshanophyllum* and *Waagenophyllum* in reefal facies (Fig. 1). Non-dissepimented solitary rugose corals also flourished, including *Amplexocarinia*, *Calophyllum*, *Epiphanophyllum*, *Lophocarinophyllum*, *Lophophyllidium*, *Lytvolasma*, *Paracaninia*, *Pentaphyllum*, *Timorcarinophyllum*, *Ufimia* and *Wannerophyllum* (Wu 1975; Flügel 1989; Fedorowski 1997; Ilina 1997; Fig. 1).

Timor

During the Permian, very abundant rugose corals were present in Timor Island, including 109 rugose species (Sorauf 2004; Figs 1 & 2). The rugose coral faunas are mainly composed of small, non-dissepimented solitary types, which are named

Cyathaxonia-fauna (Hill 1948; Kullman 1966). They consist of the following genera: *Allophyllum*, *Amplexocarinia*, *Asserculinia*, *Basleophyllum*, *Calophyllum*, *Duplocarinia*, *Duplophyllum*, *Endamplexus*, *Endothecium*, *Euryphyllum*, *Falsiamplexus*, *Lophophyllidium*, *Lytvolasma*, *Paralleynia*, *Pentamplexus*, *Pentaphyllum*, *Pleramplexus*, *Productiophyllum*, *Prosmilia*, *Sochkineophyllum*, *Spineria*, *Timorphyllum*, *Ufimia*, *Verbeekiella*, *Wannerophyllum* and *Zaphrentis* (Fig. 1). These non-dissepimented solitary rugose corals were collected from different localities of Timor and possibly from the Sakmarian to Capitanian strata, although lacking precise biostratigraphic control. Basleo, the southern part of western Timor, is the locality yielding the most abundant corals. Only two colonial rugose coral genera, *Ipciphyllum* and *Lonsdaleiastraea*, have been found from the Permian strata in Timor (Minato & Kato 1965*b*; Sorauf 2004; Fig. 1).

Ural Mountains and adjacent areas

Only Cisuralian rugose corals developed in these areas, including abundant and diversified colonial and non-dissepimented solitary rugose corals (Kossovaya 1996, 2007, 2009; Figs 1 & 2). A comprehensive review of the Cisuralian rugose corals has been conducted in the Urals, north Timan and the eastern part of Eastern European Platform (Kossovaya 1996). Based largely on Kossovaya (1996) with inputs from her two later papers concerning non-dissepimented solitary rugose corals (Kossovaya 2007, 2009), a list of generic names is summarized as follows: the colonial rugose genera *Durhamina*, *Fomichevella*, *Heintzella*, *Iskutella*, *Kleopatrina*, *Lytvophyllum*, *Paraheritschioides*, *Pararachnastraea*, *Permastraea*, *Protolonsdaleia*, *Protowentzelella*, *Pseudocystophora* and *Tchussovskenia*, and the non-dissepimented solitary rugose genera *Actinophrentis*, *Amplexocaninia*, *Asserculinia*, *Calophyllum*, *Cyathaxonia*, *Cyathocaninia*, *Hexalasma*, *Lopholasma*, *Lophophyllidium*, *Lophotichium*, *Lytvolasma*, *Paralleynia*, *Pseudowannerophyllum*, *Soshkineophyllum* and *Ufimia* (Fig. 1). The generic names of the colonial rugose corals presented here follow the taxonomical revision by Fedorowski *et al.* (2007), and some disputable definitions of genera such as *Ferganophyllum* and *Heritschioides*, were omitted in the above list.

Svalbard Archipelago

The Permian rugose corals were extensively exposed and studied in Spitsbergen (formerly known as Vestspitsbergen) of the Svalbard Archipelago, which is the largest island of the Svalbard Archipelago (e.g. Fedorowski 1997; Chwieduk 2013; Fig. 2).

Cisuralian rugose coral

During the Cisuralian, abundant rugose corals occurred in the Tyrrelfjellet Member (upper Wordiekammen Formation) of central Spitsbergen and the Trekelodden Formation of southern Spitsbergen (Fedorowski 1965, 1967, 1982*a*, 1997; Birkenmajer & Fedorowski 1980; Somerville 1997; Chwieduk 2009, 2013). They are dominated by colonial and dissepimented solitary rugose corals (Fig. 1). The colonial rugose corals contain the genera *Fomichevella*, *Heintzella*, *Kleopatrina*, *Pararachnastraea*, *Protolonsdaleiastraea*, *Protowentzelella*, *Pseudocystophora*, *Stylastraea* and *Tschussovskenia* (Fig. 1). The dissepimented solitary rugose corals comprise *Amygdalophylloides*, *Arctophyllum*, *Bothrophyllum*, *Barentsburgia*, *Caninophyllum*, *Gronfjordphyllum*, *Gshelia*, *Hornsundia*, *Linnephyllum*, *Pseudotimania*, *Svalbardphyllum*, *Timania* and *Yakovleviella* (Fig. 1). Only two non-dissepimented solitary rugose coral genera developed in the Cisuralian, which are *Krusenella* and *Lytvolasma* (Fig. 1).

Guadalupian rugose corals

A few non-dissepimented solitary rugose corals were reported from the Kapp Starostin Formation (KSF; Ezaki 1997; Chwieduk 2007, 2013), including the genera *Allotropiochisma*, *Calophyllum*, *Euryphyllum*, *Fedorowskites*, *Soshkineophyllum* and *Ufimia* (Fig. 1). In this paper, the age of this coral fauna is believed to be Guadalupian for the following two reasons: (1) the KSF coral genera described by Chwieduk (2007, 2013) from southern Spitsbergen were all considered as Wordian and/or Capitanian; and (2) Ezaki (1997) also documented several non-dissepimented solitary rugose genera in the KSF at its type locality in Festningen (with *Euryphyllum* occurring in the lower part). Although in this area the lower part of the KSF is usually dated as Kungurian (Chwieduk 2007, 2013), the precise age of the limestone beds bearing rugose corals has never been determined.

Greenland

Only a few Permian rugose corals were described from the Foldvik Creek Formation of eastern Greenland (Flügel 1973; Fedorowski 1982*b*; Fig. 2). They are all non-dissepimented solitary rugose corals, containing the genera *Allotropiochisma*, *Calophyllum*, *Leonardophyllum*, *Soshkineophyllum* and *Ufimia* (Fedorowski 1982*b*; Fedorowski & Bamber 2001; Fig. 1). Fedorowski & Bamber (2001) suggested that the coral fauna occurred from the

Wordian to early Capitanian (middle to late Guadalupian). However, the conodont data from a more recent study dated the fauna as Wuchiapingian (early Lopingian) (Mei *et al.* 2002).

North America

In North America, the Permian rugose corals were documented from numerous localities in the Sverdrup Basin and the western part of the North American continent, including the Canadian Arctic Archipelago, British Columbia, Alaska, California, Nevada and Texas (e.g. Wilson 1982, 1985, 1991, 1994; Nelson & Nelson 1985; Wu *et al.* 1985; Fedorowski 1987; Stevens *et al.* 1987; Stevens & Rycerski 1989; Wilson & Langenheim, Jr 1993; Fedorowski & Bamber 2001; Fedorowski *et al.* 2007; Stevens 2008*a*, *b*, 2009, 2010; Fig. 2).

Cisuralian rugose corals

The rugose coral fauna has its highest diversity in the Cisuralian, especially the colonial and non-dissepimented solitary rugose corals. The colonial rugose corals comprise 23 genera, which are *Cordilerastraea*, *Cystolonsdaleia*, *Darwasophyllum*, *Durhamina*, *Fomichevella*, *Heintzella*, *Iskutella*, *Kleopatrina*, *Langenheimia*, *Lytvophyllum*, *Pararachnastraea*, *Paraheritschioides*, *Petalaxis*, *Permastraea*, *Protolonsdaleiastraea*, *Protowentzelella*, *Pseudocystophora*, *Sandolasma*, *Shastalasma*, *Stylastraea*, *Tschussovskenia*, *Wendoverella* and *Wilsonastraea* (Fig. 1). The non-dissepimented solitary rugose corals contain 16 genera: *Actinophrentis*, *Allotropiochisma*, *Assimulia*, *Bradyphyllum*, *Calophyllum*, *Euryphyllum*, *Falsiamplexus*, *Lophophyllidium*, *Lophotichium*, *Lytvolasma*, *Monophyllum*, *Paraduplophyllum*, *Paralleynia*, *Pseudowannerophyllum* (Fig. 1), and three endemic genera *Diffingia*, *Plerodiffia* and *Turgidiffia* with restricted distributions in Texas. Only five dissepimented solitary rugose genera were recorded in North America, which are *Bothrophyllum*, *Caninophyllum*, *Heterocaninia*, *Siedlekia* and *Timania* (Fig. 1). The majority of these genera were found in the Canadian Arctic Archipelago, except for a few specimens of *Heterocaninia* from the Bilk Creek Terrane, Nevada (Stevens 2009).

Guadalupian rugose corals

The diversity of rugose corals greatly decreased in the Guadalupian, with merely 11 described genera (Fig. 1). Only two fasciculate genera/subgenera have been described, which are *Miyagiella* and *Waagenophyllum*, probably migrated from Palaeotethys (Fig. 1). The non-dissepimented solitary rugose corals consist of six genera *Allotropiochisma*, *Calophyllum*, *Euryphyllum*, *Lytvolasma*, *Soshkineophyllum* and *Ufimia* (Fig. 1). All of these genera were found in the Canadian Arctic Archipelago, two of them in Texas and one in California. In addition, several other non-dissepimented solitary rugose genera, *Assimulia*, *Leonardophyllum*, *Lophophyllidium*, *Lophotichium* and *Paraduplophyllum*, were found in the Roadian from Texas and its adjacent areas (Fedorowski 1997; Fig. 1). This coral fauna also continued to the Wordian and Capitanian, but with reduced diversity. To date, no dissepimented solitary rugose corals have been reported from North America (Fig. 1).

Lopingian corals

Only one colonial rugose genus, *Waagenophyllum*, was documented from the Lopingian strata in North America (Fig. 1). It is also the only colonial rugose genus recorded through the whole CAU Realm during this time interval (Fedorowski *et al.* 1999).

Central and South America

The Permian rugose corals were coarsely studied in Central and South America, mainly from the compiled data of this region (Wilson 1990), as well as some taxonomical revisions (e.g. Fedorowski *et al.* 2007). A total of nine rugose genera have been reported from the Cisuralian in Mexico and Bolivia, which are the colonial rugose genera *Durhamina*, *Pararachnastraea*, *Paraheritschioides*, *Petalaxis*, *Protowentzelella*, *Stylastraea*, and *Tschussovskenia*, the dissepimented solitary rugose genus ?*Caninia* and the non-dissepimented solitary rugose genus *Lophophyllidium* (Fig. 1).

Rugose coral palaeobiogeography

Classification

A comprehensive study of the global Permian rugose coral palaeobiogeography has been conducted by Fedorowski (1997), who classified two coral realms: the Tethyan Realm and the CAU Realm separated by the Pangaea continent (Fedorowski 1986, 1997; Fig. 2). In this study, consistent with Fedorowski (1997), two large and distinct coral realms have been identified based on a review of the composition and distribution of the global Permian rugose corals (Fig. 2). In addition, numerous smaller epicratonic basins could form obvious biogeographical regions within these realms (Sorauf 2004). Here, some smaller coral realms were also recognized within the Tethyan Realm and the CAU Realm, respectively (Fig. 2).

Tethyan Realm

The territory of the Tethyan Realm is very large, including South China and Indochina (Vietnam, Thailand and Timor), Inner Mongolia–NE China and Japan, Gondwana (Australia) and Peri-Gondwana (Iran, Afghanistan, Pakisan, Pamir, Tibet, Southwestern China and Myanmar; Fig. 2). These areas contain Tethyan affinity of rugose corals during the Permian. Except for Indochina, the Permian rugose corals have been thoroughly studied in the Tethyan realm under high-resolution biostratigraphy, especially for South China (Fig. 1). In general, four coral fauna intervals could be classified for the Tethyan Realm in the Permian times, which are the Asselian–Artinskian Stage, Kungurian Stage, Guadalupian Series and Lopingian Series (Fig. 1).

From the Asselian to Artinskian, the rugose corals have a high diversity, characterized by the colonial rugose genera *Akagophyllum*, *Wentzellophyllum* and *Yokoyamaella*, the dissepimented solitary rugose genera *Iranophyllum*, and the non-dissepimented solitary rugose genera *Amplexocarinia*, *Calophyllum*, *Duplophyllum*, *Pentaphyllum* and *Verbeekiella* (Fig. 1). During the Kungurian, the rugose corals are dominated by colonial and non-dissepimented solitary rugose corals, comprising the colonial rugose genera *Akagophyllum*, *Polythecalis*, *Szechuanophyllum* and *Wentzellophyllum*, and the non-dissepimented solitary rugose genera *Amplexocarinia*, *Lophocarinophyllum*, *Lophophyllidium*, *Pentaphyllum* and *Ufimia* (Fig. 1). In Guadalupian time, a high diversity of rugose corals occurred, which are mainly composed of the colonial rugose genera *Ipciphyllum*, *Liangshanophyllum*, *Lonsdaleiastraea*, *Praewentzelella*, *Szechuanophyllum*, *Waagenophyllum*, *Wentzelella* and *Wentzelloides*, the dissepimented solitary rugose genus *Iranophyllum*, and the non-dissepimented solitary rugose genera *Allotropiophyllum*, *Asserculinia*, *Paracaninia*, *Pentaphyllum*, *Lophophyllidium* and *Ufimia* (Fig. 1). Some of the above-mentioned genera also occur in Timor from the Cisuralian to Guadalupian, but they could not be dated to a more precise age, as for the other areas in the Tethyan Realm (Fig. 1). The diversity of rugose corals dramatically declined during the Lopingian, characterized by the colonial rugose genera *Huayunophyllum*, *Liangshanophyllum* and *Waagenophyllum*, and the non-dissepimented solitary rugose genera *Calophyllum*, *Lophocarinophyllum*, *Lophophyllidium*, *Pentaphyllum* and *Ufimia* (Fig. 1).

Cordilleran–Arctic–Uralian Realm

This coral realm was widely distributed on the western and northern shelves of the Pangaean continent, including the Svalbard Archipelago, Greenland, the Ural Mountains and adjacent area, North America (Canadian Arctic Archipelago, British Columbia, Alaska, California, Texas, etc.) and some regions of South America (Venezuela, Bolivia, Peru etc.) (Fedorowski 1997; Fig. 2). These areas contain CAU affinity of rugose corals in Permian time. Compared with the Tethyan Realm, the Permian rugose corals have been insufficiently studied under relatively low-resolution biostratigraphy (Fig. 1). Three coral fauna intervals could be distinguished for the CAU Realm, which are the Cisuralian, Guadalupian and Lopingian series (Fig. 1).

During the Cisuralian, a high diversity was present in the rugose corals, which are characterized by the colonial rugose genera *Durhamina*, *Fomichevella*, *Heintzella*, *Kleopatrina*, *Paraheritschioides*, *Pararachnastraea*, *Protowentzelella*, *Pseudocystophora*, *Stylastraea* and *Tchussovskenia*, the dissepimented solitary rugose genera *Bothrophyllum*, *Caninophyllum* and *Timania*, and the non-dissepimented solitary rugose genera *Actinophrentis*, *Calophyllum*, *Lophophyllidium*, *Lophotichium*, *Lytvolasma*, *Paralleynia* and *Pseudowannerophyllum* (Fig. 1). In Guadalupian time, the rugose coral diversity decreased dramatically in colonial and dissepimented solitary rugose corals (Fig. 1). However, the non-dissepimented rugose corals were still diversified in the Svalbard Archipelago and North America during this time, including the common genera *Allotropiochisma*, *Calophyllum*, *Euryphyllum*, *Soshkineophyllum* and *Ufimia* (Fig. 1). During the Lopingian, the rugose corals almost became extinct in the CAU Realm. The only known records were a non-dissepimented solitary coral assemblage in Greenland and a few surviving taxa of the colonial rugose genus *Waagenophyllum* in North America (Fig. 1). Hence, the characterized genera could not be concluded in this interval.

Other coral subrealms

Although the Tethyan Realm and the CAU Realm generally contain similar compositions of rugose coral faunas, some small coral realms (subrealms) could be also distinguished among the above two large realms in different times, based on their different coral compositions (Fig. 1).

Obviously different composition of rugose corals was present between the Peri-Gondwana (Cimmerian continent) and the other areas of the Tethyan Realm (e.g. South China, Inner Mongolia–NE China and Japan) during the Cisuralian (Fig. 1). Wang & Sugiyama (2002) classified two palaeobiogeographical provinces: (1) the Cathaysian Province, characterized by the occurrences of abundant colonial rugose and some dissepimented solitary rugose genera belonging to the families

Kepingophyllidae and Waagenophyllidae, which were widely distributed in South China; and (2) while the Cimmerian Province was represented by the *Cyathaxonia* (non-dissepimented solitary rugose coral) faunas, without the colonial and dissepimented solitary rugose genera mentioned above in South China (Figs 1 & 2c). This hypothesis was also supported by the other palaeogeographical conclusions based on other fossil groups, such as brachiopods (Shi *et al.* 1996), and spores and pollen (Yang 1999).

In the CAU Realm, based on the distribution of Guadalupian non-dissepimented solitary corals, Fedorowski & Bamber (2001) also recognized two palaeobiogeographical provinces: the Diffingiina Province and the *Calophyllum* Province. The Diffingiina Province is located in Texas and its adjacent areas, with only the occurrences of the suborder Diffingiina yielding a diversified and partly isolated (in species level) non-dissepimented solitary rugose coral fauna (Fedorowski 1997; Fedorowski & Bamber 2001; Fig. 1). The *Calophyllum* Province includes not only the Arctic region (Alaska of North America, Sverdrup Basin of Arctic Canada, Svalbard Archipelago and East Greenland), but also the Central European Basin and the Eastern European Platform (Fig. 2). This province is characterized by the occurrences of the non-dissepimented solitary rugose genus *Calophyllum* (Fig. 1). In Arctic regions, the Guadalupian rugose coral faunas were relatively diversified consisting of several genera, while in the last two regions only a few species of *Calophyllum* and a few specimens of the non-dissepimented solitary rugose genera *Tachylasma* and *Paracaninia* were recorded (e.g. Fedorowski 1997; Fedorowski *et al.* 1999; Fedorowski & Bamber 2001; Kossovaya 2009).

Comparison

Although the degree of studied precision is different between the Tethyan Realm and the CAU Realm, three intervals could still be compared in terms of the coral composition, which are the Cisuralian, Guadalupian and Lopingian series (Fig. 1). During the Cisuralian, the Tethyan Realm and the CAU Realm comprise distinctly different composition in colonial and dissepimented solitary rugose corals (Fig. 1). The colonial rugose corals in the Tethyan Realm are mainly composed of the families Wentzellophyllinae (e.g. *Iranophyllum, Lonsdaleiastraea, Polythecalis, Wentzellophyllum*), Waagenophyllinae (e.g. *Akagophyllum, Yokoyamaella*) and Kepingophyllidae (e.g. *Szechuanophyllum*) (Fig. 1), while in the CAU Realm, the colonial genera mainly belonged to the families Durhaminidae (e.g. *Durhamina, Pararachnastraea, Protowentzelella*), Kleopatrinidae (e.g. *Fomichevella, Heintzella, Kleopatrina, Paraheritschioides*) and Lithostrotionidae (e.g. *Pseudocystophora, Tchussovskenia*) (Fig. 1). The Tethyan and CAU realms contain their unique non-dissepimented solitary rugose genera of *Duplophyllum, Lophocarinophyllum, Pentaphyllum* and *Verbeekiella* in the Tethyan Realm and *Actinophrentis, Lophotichium* and *Pseudowannerophyllum* in the CAU Realm, although they had some of the same non-dissepimented solitary rugose genera, such as *Calophyllum* and *Lophophyllidium* (Fig. 1).

During the Guadalupian, the Tethyan Realm is obviously different from the CAU Realm, due to the presence of abundant and diversified colonial rugose corals and some dissepimented solitary rugose corals, which are characterized by the Waagenophyllinae (e.g. *Ipciphyllum, Liangshanophyllum, Waagenophyllum*) and Wentzellophyllinae (e.g. *Iranophyllum, Lonsdaleiastraea, Wentzelella*) (Fig. 1). They almost disappeared in the CAU Realm, where only two colonial rugose genera, *Miyagiella* and *Waagenophyllum*, occurred in North America (Fig. 1). The Tethyan and CAU realms all have a high diversity in the non-dissepimented solitary rugose corals, with the common genera *Soshkineophyllum* and *Ufimia* developed in both realms. Similar to the composition of non-dissepimented solitary rugose faunas in Cisuralian time, the Tethyan Realm and CAU Realm also contain their own non-dissepimented solitary rugose genera, such as *Amplexocarinia, Asseculinia, Duplophyllum* and *Pentaphyllum* in the Tethyan Realm, and *Allotropiochisma* in the CAU Realm (Fig. 1).

In Lopingian time, the compositions of rugose corals between the Tethyan Realm and CAU Realm are difficult to compare, due to the scarcity of rugose corals in the CAU Realm (Fig. 1). The Tethyan Realm is characterized by the occurrences of the common colonial rugose genera *Huayunophyllum, Liangshanophyllum* and *Waagenophyllum*, and the non-dissepimented solitary rugose genera *Calophyllum, Lophocarinophyllum, Lophophyllidium, Pentaphyllum* and *Ufimia* during this time (Fig. 1). In the CAU Realm, the colonial rugose genus *Waagenophyllum* and non-dissepimented solitary rugose genera *Calophyllum* and *Ufimia* are the only representatives in North America and Greenland during the Lopingian (Figs 1 & 2a).

Controlling factors

Tectonic movement and oceanic circulation played important roles in regulating marine fauna provincialism (Qiao & Shen 2014). The Gondwana continent joined with the Laurussia continent in the late Mississippian, and the two continents continued their collision during the Pennsylvanian and early

Permian, which gradually formed Pangaea (Scotese & McKerrow 1990; Shen *et al.* 2016; Fig. 2). Then, Pangaea underwent a dispersal process during the middle to late Permian, mainly including the northward movement of the Cimmerian continent (Scotese & McKerrow 1990; Domeier & Torsvik 2014; Shen *et al.* 2016; Fig. 2a, b). The collision of the Gondwana and Laurussia continents could have led to the closure of the Rheic Seaway and therefore formed a geographical barrier for various marine faunas between the west and east of Pangaea, such as the brachiopods (Qiao & Shen 2014), corals (Aretz 2010) and ammonoids (Korn *et al.* 2012). In addition, the closure of the Rheic Seaway could also deflect the palaeoequatorial east–west oceanic current to the north and south along the western coast of Palaeotethys, and lead to the current from north and south to the palaeoequator along the eastern coast of the Panthalassa Ocean (Qiao & Shen 2014). These current changes would enhance faunal similarity within the Palaeotethys and eastern Panthalassa oceans (Qiao & Shen 2014).

In this study, based on the composition and distribution of Permian rugose corals, two distinct coral realms have been recognized: the Tethyan Realm and the CAU Realm, which were located in the Palaeotethys Ocean and eastern Panthalassa Ocean, respectively (Fig. 2). The formation of these two coral realms may have been triggered by the Gondwana and Laurussia continents joining together, resulting in a geographical barrier and isolated current circulation between eastern and western Pangaea (Qiao & Shen 2014; Fig. 2). This hypothesis is also supported by the occurrence of coral subrealms within the Tethyan and CAU realms, which are also related to the movement and collision of blocks (Fedorowski & Bamber 2001; Wang & Sugiyama 2002; Fig. 2). In the Tethyan Realm, the coral composition of the Cimmerian Province is obviously different from that of the Cathaysian Province during the early Permian. Then, accompanied by the movement of the Cimmerian continent to the north and approaching to the South China Block, the coral composition of the Cimmerian continent shows a gradual increase in the rugose coral affinity with that of South China, implying that the dissimilarities of the coral composition between the Cimmerian and Cathaysian provinces were caused by their different palaeogeographical locations (Scotese & McKerrow 1990; Domeier & Torsvik 2014; Fig. 2c). In addition, the differences in the composition of rugose corals between the Diffingiina Province and the *Calophyllum* Province among the CAU Realm were also suggested to result from the configuration of the intervening coastline of western Pangaea (Fedorowski & Bamber 2001). However, some genera sometimes occurred in both provinces, indicating the existence of communications between the two provinces (Fedorowski & Bamber 2001). However, the co-occurrences of the same genera between the Tethyan and CAU realms may be explained by the migration of coral faunas across the Panthalassa Ocean using mid-ocean islands as dispersal points or transport on moving ocean plates, or by a combination of these two methods (Fedorowski *et al.* 1999; Fig. 1).

Rugose coral evolution and related environmental events

The diversity pattern of Permian rugose corals has been comprehensively conducted in South China, unravelling processes of assemblage changes and extinction events (Wang *et al.* 2006*a*). Mainly based on the publications of global and regional data on Permian rugose corals (e.g. Fedorowski 1997; Fedorowski *et al.* 1999; Wang & Sugiyama 2002; Wang *et al.* 2006*a*, *b*, 2013; Chwieduk 2007, 2009, 2013; Stevens 2008*a*, *b*, 2009, 2012; Kossovaya 2009), the abundance, diversity, composition and distribution of rugose corals have been reviewed (Figs 1 & 2), which provide profound insights into the changing pattern of global Permian rugose corals (Fig. 3). In addition, prominent global geological events in the Permian are displayed (Shen *et al.* 2016), including the changes in palaeoclimate (glacial and interglacial) (e.g. Fielding *et al.* 2008*a*, *b*; Chen *et al.* 2013), sea-level fluctuations (e.g. Ross & Ross 1987; Haq & Schutter 2008; Rygel *et al.* 2008), volcanic eruptions (e.g. Ali *et al.* 2005; Svensen *et al.* 2009) and tectonic movements (e.g. Scotese & McKerrow 1990; Domeier & Torsvik 2014). These global and/or regional events had a severe influence on the distribution and evolution of rugose corals during this time (Wang & Sugiyama 2002; Wang *et al.* 2006*a*, *b*).

During the Cisuralian, the colonial, dissepimented and non dissepimented solitary rugose corals have a high diversity and are globally distributed (Figs 1, 2c & 3). An end-Sakmarian change event of rugose coral assemblages has been documented in South China (Wang *et al.* 2006*a*). This study finds that this event may also be present in both the Tethyan Realm and the CAU Realm (Figs 1 & 3). The Tethyan Realm is characterized by mixed Pennsylvanian (e.g. Bothrophyllidae and Petalaxidae) and Permian (Kepingophyllidae and Waagenophyllidae) taxa during the Asselian–Sakmarian, and typical Permian taxa (Kepingophyllidae and Waagenophyllidae) from the Artinskian to Kungurian (Fig. 3). The typical Permian rugose coral family Kepingophyllidae and subfamily Waagenophyllinae originated during the

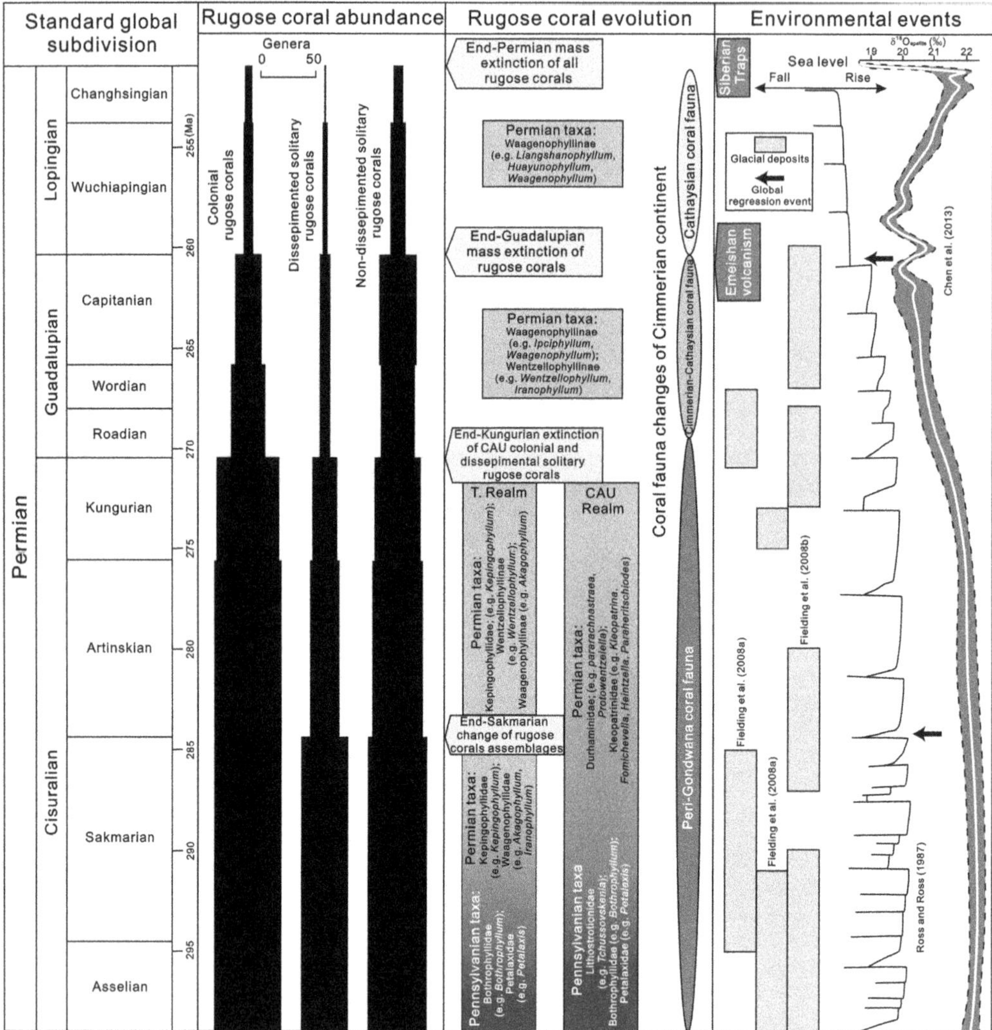

Fig. 3. Diversity and faunal changes of global rugose corals, and their relations to the geological events during the Permian. T., Tethyan.

Pennsylvanian, while the subfamily Wentzellophyllinae is first present in the earliest Permian (Wang *et al.* 2006*a*). In the CAU Realm, the rugose faunas are represented by the Pennsylvanian taxa (e.g. Bothrophyllidae, Lithostrotionidae and Petalaxidae) and the Permian taxa (Durhaminidae and Kleopatrinidae) in the early Permian (Fig. 3). However, these coral faunas are difficult to further distinguish, due to the low-resolution biostratigraphic dating in the present study (Fig. 1). The coral assemblage change at the end-Sakmarian may be related to a major, worldwide regression event (Wang *et al.* 2006*a*; Fig. 3). In addition, the early Permian *Cyathaxonia* faunas may also reflect cold or cool water as suggested by the underlying diamictites, which were regarded as being glacial in origin (Jin 1994; Wopfner 1996; Wang & Sugiyama 2002). The occurrence of the *Cyathaxonia* faunas on the Cimmerian continent also reflects the glacial environment during the early Permian (Fig. 1). They are all consistent with widespread glacial deposits and high oxygen isotope values during this time (Fielding *et al.* 2008*a*, *b*; Chen *et al.* 2013; Fig. 3).

A distinct decrease in the abundance and diversity of rugose corals occurred at the end-Kungurian, characterized by the almost total disappearance of colonial and dissepimented solitary rugose corals in the CAU Realm (Fedorowski *et al.* 1999;

Fig. 3). However, this coral disappearance event has not been recognized in the Tethyan Realm, with no obvious change in the diversity of colonial and non-dissepimented solitary rugose genera (Wang *et al.* 2006*a*; Figs 1 & 3). The rugose corals are mainly composed of Waagenophyllinae and Wentzellophyllinae during the Guadalupian (Fig. 3). The end-Kungurian disappearance event may result from the northward movement of Pangaea into cooler water (Fedorowski *et al.* 1999; Fig. 2b). On the other hand, abundant colonial and large dissepimented solitary rugose corals, which are common in South China near the palaeoequator with a warm climate, occurred on the Cimmerian continent during the Guadalupian (Wang & Sugiyama 2002). Although the Cimmerian continent moved northward from the Cisuralian to the Guadalupian, they might be still located at a high latitude similar to that of the Cisuralian times (Fig. 2b). Hence, the occurrences of colonial and dissepimented solitary rugose genera on the Cimmeria were possibly triggered by the climate warming, coinciding with decreased glacial deposits and negative shifts in oxygen isotope values (Fielding *et al.* 2008*a*, *b*; Chen *et al.* 2013; Fig. 3).

During the Guadalupian–Lopingian transition, the abundance and diversity of the rugose corals distinctly declined, which was caused by the end-Guadalupian mass extinction event (Wang & Sugiyama 2000; Wang *et al.* 2006*a*; Fig. 3). This extinction event might have been triggered by the extrusion of Emeishan volcanism and the subsequent global regression, accompanied by negative and positive oxygen isotope excursions, respectively (Wang *et al.* 2006*a*; Chen *et al.* 2011, 2013; Fig. 3). During the Lopingian, rugose corals are dominated by the Cathaysian types (Waagenophyllinae) (Wang & Sugiyama 2002; Figs 1 & 3). All Permian rugose corals were eliminated at the end-Permian mass extinction, which was probably related to global climate warming in accordance with a prominent negative excursion in oxygen isotope, induced by the eruption of the Siberian Trap (Sobolev *et al.* 2011; Chen *et al.* 2013; Fig. 3).

Conclusions

Based on the composition and distribution of Permian rugose corals, two coral realms have been recognized: the Tethyan Realm and the CAU Realm, which are widely distributed in the eastern part of Pangaea including South China, Inner Mongolia–NE China, Japan, Peri-Gondwana (Cimmeria) and Timor, and in the western part of Pangaea including the Ural Mountains and adjacent areas, the Svalbard Archipelago, Greenland, North America, and Central and South America. The compositions of rugose corals between the Tethyan Realm and the CAU Realm are distinctly different during the Cisuralian, characterized by the coral family Kepingophyllidae and Waagenophyllidae in the Tethyan Realm, and the Durhaminidae and Kleopatrinidae in the CAU Realm. From the Guadalupian to Lopingian, the Tethyan Realm is represented by the family Waagenophyllidae in the Guadalupian and subfamily Waagenophyllinae in the Lopingian, whereas these coral families almost disappeared in the CAU Realm during this time, and only the genus *Waagenophyllum* occurred in North America. However, some genera are present in both the Tethyan and CAU realms, which may be caused by the migration of coral faunas across the Panthalassa Ocean using mid-ocean islands as dispersal points and/or being transported on moving ocean plates. Some sub-realms of rugose corals can be also identified, such as the Cathaysian Province and the Cimmerian Province in the Tethyan Realm, and the Diffingiina Province and the *Calophyllum* Province in the CAU Realm. The development of these coral provinces was controlled by the geographical barrier and isolated current circulation, resulting from the formation and dispersion of Pangaea.

The global rugose corals underwent several episodes of assemblage changes and extinctions during the Permian, which are closely related to geological events. During the Cisuralian, all of the colonial, dissepimented and non-dissepimented solitary rugose corals have high diversities and are globally distributed. An end-Sakmarian change event of rugose coral assemblages may have occurred globally, characterized by the change from the coral faunas of mixed Pennsylvanian (e.g. Bothrophyllidae and Petalaxidae) and Permian (e.g. Kepingophyllidae and Waagenophyllidae in the Tethyan Realm, Durhaminidae and Kleopatrinidae in the CAU Realm) taxa during the Asselian–Sakmarian to the faunas of typical Permian taxa as mentioned above in the Artinskian–Kungurian. This coral assemblage change at the end-Sakmarian may be related to a major global regression, consistent with widespread glacial deposits and high oxygen isotope values during this time.

A rugose coral extinction event occurred at the end of Kungurian, accompanied by a distinct decrease in the diversity of the colonial and dissepimented solitary rugose corals in the CAU Realm. This extinction event may have resulted from the northward movement of Pangaea into cooler water. However, this extinction event of rugose corals has not been recognized in the Tethyan Realm, which is dominated by Waagenophyllidae during the Guadalupian.

At the Guadalupian–Lopingian transition, the diversity of rugose corals obviously declined, with only a few rugose corals of the Waagenophyllinae developed in the Lopingian. This coral extinction event was caused by the end-Guadalupian mass extinction event, resulting from the extrusion of Emeishan volcanism and the subsequent global regression, accompanied by negative and positive oxygen isotope excursions, respectively. All Permian rugose corals were eliminated at the end-Permian mass extinction, induced by the global climate warming in accordance with a prominent negative excursion in oxygen isotope, probably resulting from the eruption of the Siberian Trap.

This work was financially supported by National Natural Science Foundation of China (grant nos 41290260, 41290263), the Strategic Priority Research Program (B) of the Chinese Academy of Sciences (XDB18000000) and the Ministry of Science and Technology Foundation Project. This manuscript was improved by the comments from Prof. Ian D. Somerville of Dublin University, one anonymous reviewer, and the editor Dr Spencer G. Lucas of New Mexico Museum of Natural History and Science.

References

ALI, J.R., THOMPSON, G.M., ZHOU, M.F. & SONG, X. 2005. Emeishan large igneous province, SW China. *Lithos*, **79**, 475–489.

ARCHBOLD, N.W. & SHI, G.R. 1996. Western Pacific Permian marine invertebrate palaeobiogeography. *Australian Journal of Earth Sciences*, **43**, 635–641.

ARETZ, M. 2010. Rugose corals from the upper Visean (Carboniferous) of the Jerada Massif (NE Morocco): taxonomy, biostratigraphy, facies and palaeobiogeography. *Palaeontologische Zeitschrift*, **84**, 323–344.

AUNG, A.K., FATT, N.T., NYEIN, K.K. & ZIN, M.H. 2013. New Late Permian rugose corals from Pahang, peninsular Malaysia. *Alcheringa*, **37**, 422–434.

BIRKENMAJER, K. & FEDOROWSKI, J. 1980. Corals of the Treskelodden Formation (Lower Permian) at Triasnuten, Hornsund, South Spitsbergen. *Studia Geologica Polonica*, **66**, 8–33.

CHEN, B., JOACHIMSKI, M.M., SUN, Y.D., SHEN, S.Z. & LAI, X.L. 2011. Carbon and conodont apatite oxygen isotope records of Guadalupian–Lopingian boundary sections: climatic or sea-level signal? *Palaeogeography, Palaeoclimatology, Palaeoecology*, **311**, 145–153.

CHEN, B., JOACHIMSKI, M.M. ET AL. 2013. Permian ice volume and palaeoclimate history: Oxygen isotope proxies revisited. *Gondwana Research*, **24**, 77–89.

CHOI, D.R. & FUJITA, T. 1970. On some Middle Permian fossils from the Shirahone limestone, Nagano prefecture, Japan. *Journal of the Faculty of Science, Hokkaido University, Series 4*, **14**, 365–381.

CHWIEDUK, E. 2007. Middle Permian rugose corals from the Kapp Starostin Formation, South Spitsbergen (Treskelen Peninsula). *Acta Geologica Polonica*, **57**, 281–304.

CHWIEDUK, E. 2009. Early Permian solitary rugose corals from Kruseryggen (Treskelodden Fm., Hornsund area, southern Spitsbergen). *Geologos*, **15**, 57–75.

CHWIEDUK, E. 2013. *Palaeogeographical and Palaeoecological Significance of the Uppermost Carboniferous and Permian Rugose Corals of Spitsbergen*. Wydawnictwo Naukowe UAM, Poznań.

DING, Y.J. 1983. On some new species of Early Permian tetracoralla from Xilin Gol Meng, Inner Mongolia. *Bulletin of the Tianjin Institute of Geology and Mineral Resources*, **8**, 105–122 [in Chinese with English abstract].

DING, Y.J. 1987. Some new species of Early Permian corals from Ulanqab Meng, Inner Mongolia. *Professional Papers of Stratigraphy and Palaeontology*, **21**, 277–292 [in Chinese with English abstract].

DING, Y.J., XIA, G.Y., DUAN, C.H., LI, W.G., LIU, X.L. & LIANG, Z.F. 1985. Study on the Early Permian stratigraphy and fauna in Zhesi district, Inner Mongolia. *Bulletin of the Tianjin Institute of Geology and Mineral Resources*, **10**, 1–244 [in Chinese with English abstract].

DOMEIER, M. & TORSVIK, T.H. 2014. Plate tectonics in the late Paleozoic. *Geoscience Frontiers*, **5**, 303–350.

DOUGLAS, L.A. 1936. A Permo-Carboniferous fauna from Southwest Persia (Iran). *Memoirs of the Geological Survey of India, Palaeontologica Indica, New Series*, **22**, 1–59.

ERWIN, D.H. 1995. Permian global bio-events. *In*: WALLISER, O.H. (ed.) *Global Events and Event Stratigraphy*. Springer, Berlin, 251–264.

EZAKI, Y. 1991. Permian corals from Abadeh and Julfa Iran, West Tethys. *Journal of the Faculty of Science, Hokkaido University*, **23**, 53–146.

EZAKI, Y. 1997. Cold-water Permian Rugosa and their extinction in Spitsbergen. *Boletin de la Real Sociedad Española de Historia Natural (Seccion Geologia)*, **92**, 381–388.

FAN, Y.N. 1988. *The Carboniferous System in Xizang (Tibet)*. Chingqing, Chongqing, China.

FEDOROWSKI, J. 1965. Lower Permian Tetracoralla of Hornsund, Vestspitsbergen. *Studia Geologica Polonica*, **17**, 1–173.

FEDOROWSKI, J. 1967. The Lower Permian Tetracoralla and Tabulata from Treskelodden, Vestspitsbergen. *Norsk Polarinstitutt Skrifter*, **142**, 5–44.

FEDOROWSKI, J. 1982*a*. Coral Thanatocoenoses and Depositional environments in the upper Treskelodden beds of the Hornsund area, Spitsbergen. *Palaeontologia Polonica*, **43**, 17–68.

FEDOROWSKI, J. 1982*b*. Some rugose corals from the Upper Permian of East Greenland. *Rapport Grønlands Geologiske Undersøgelse*, **108**, 71–91.

FEDOROWSKI, J. 1986. The rugose coral faunas of the Carboniferous/Permian boundary interval. *Acta Palaeontologica Polonica*, **31**, 47–70.

FEDOROWSKI, J. 1987. Upper Palaeozoic rugose corals from southwestern Texas and adjacent areas: Gaptank Formation and Wolfcampian corals. Part I. *Palaeontologia Polonica*, **48**, 1–271.

FEDOROWSKI, J. 1997. Diachronism in the development and extinction of Permian Rugosa. *Geologos*, **2**, 59–164.

FEDOROWSKI, J. & BAMBER, W. 2001. Guadalupian (Middle Permian) solitary rugose corals from the Degerbols and Trold Fiord formations, Ellesmere and Melville islands, Canadian Arctic Archipelago. *Acta Geologica Polonica*, **51**, 31–79.

FEDOROWSKI, J., BAMBER, W. & STEVENS, C.H. 1999. Permian corals of the Cordilleran–Arctic–Uralian Realm. *Acta Geologica Polonica*, **49**, 159–173.

FEDOROWSKI, J., BAMBER, E.W. & STEVENS, C.H. 2007. *Lower Permian Colonial Rugose Corals, Western and Northwestern Pangea: Taxonomy and Distribution.* National Research Council Research Press, Ottawa.

FIELDING, C.R., FRANK, T.D., BIRGENHEIER, L.P., RYGEL, M.C., JONES, A.T. & ROBERTS, J. 2008*a*. Stratigraphic imprint of the Late Palaeozoic Ice Age in eastern Australia: a record of alternating glacial and nonglacial climate regime. *Journal of the Geological Society, London*, **165**, 129–140, https://doi.org/10.1144/0016-76492007-036

FIELDING, C.R., FRANK, T.D. & ISBELL, J.L. 2008*b*. The late Paleozoic ice age – a review of current understanding and synthesis of global climate patterns. *Geological Society of America Special Papers*, **441**, 343–354.

FLÜGEL, H.W. 1973. Rugose Korallen aus dem oberen Perm Ost-Grönlands. *Verhandlungen der Geologischen Bundesanstalt*, **1**, 1–57.

FLÜGEL, H.W. 1989. Rugosa aus dem Perm des N-Karakorum und der Aghil-Kette. *Geologisch-Paläontologische Mitteilungen Innsbruck*, **17**, 101–117.

FLÜGEL, H.W. 1994. Rugosa aus dem 'Mittel'-Perm des Zentralen Elburz (Iran). *Abhandlungen der Geologische Bundesanstalt*, **50**, 97–113.

FONTAINE, H. & JUNGYUSUK, N. 1995. Permian corals from Chom Bung area west of Bangkok: their palaeogeographic significance. *CCOP Newsletter*, **20**, 23–26.

FONTAINE, H. & SUTEETHORN, V. 1988. Late Paleozoic and Mesozoic Fossils of West Thailand and their environments. *CCOP Technical Bulletin*, **20**, 1–135.

FONTAINE, H., SONGSIRIKUL, B. & TANSUWAN, V. 1979. A massive colony of Waagenophyllid from southern Peninsular Thailand. *CCOP Newsletter*, **6**, 14–18.

FONTAINE, H., SATTAYARAK, N. & SUTEETHORN, V. 1994. Permian corals of Thailand. *CCOP Technical Bulletin*, **24**, 1–170.

GROSSMAN, E.L. 2012. Oxygen isotope stratigraphy. *In*: GRADSTEIN, F.M., OGG, J.G. & SCHMITZ, M. (eds) *Geologic Time Scale 2012*. Cambridge University Press, London, 181–206.

GRUNT, T.A. & SHI, G.R. 1997. A hierarchical framework of Permian global marine biogeography. *In*: JIN, Y.G. & DINELEY, D. (eds) *Palaeontology and Historical Geology*. Proceedings of the 30th International Geological Congress, **12**, 2–17.

GUO, S.Z. 1980*a*. A general account of Early Permian Rugose coral fauna from the geosynclinal region of northeastern China. *Bulletin of the Shenyang Institute of Geological and Mineral Resources*, **1**, 103–114 [in Chinese with English abstract].

GUO, S.Z. 1980*b*. Rugosa. *In*: SHENGYANG INSTITUTE OF GEOLOLGICAL AND MINERAL RESOURCES (ed.) *Palaeontological Atlas of the Northeast China Paleozoic*. Geological Publishing House, Beijing, 106–153 [in Chinese].

GUO, S.Z. 1983. Middle and Upper Carboniferous rugose corals from southern Dahinganling (Great Khingan Mountains). *Acta Palaeontologica Sinica*, **22**, 220–229 [in Chinese with English abstract].

GUO, S.Z. 1991. Correlation of Palaeozoic coral fauna between Inner Mongolia – Northeast China and Japan. *In*: ISHII, K. (ed.) *PreJurassic Geology of Inner Mongolia, China*. Matsuya Insatsu, Osaka, 201–212.

HAQ, B.U. & SCHUTTER, S.R. 2008. A chronology of Paleozoic sea-level changes. *Science*, **322**, 64–68.

HE, X.Y. 1990. Permian Palaeontology of Ngari. *In*: YANG, Z.Y. & NIE, Z.T. (eds) *Palaeontology of Ngari, Tibet (Xizang)*. Chinese University of Geosciences Press, Wuhan, 1–380 [in Chinese with English abstract].

HE, X.Y. & WENG, F. 1982. EarlyPermian rugose corals from Ali (Ngari), northern Xizang (Tibet). *Earth Science Journal of Wuhan College of Geology*, **3**, 131–142 [in Chinese with English abstract].

HILL, D. 1948. The distribution and sequence of Carboniferous coral faunas. *Geological Magazine*, **85**, 121–148.

HUDSON, R.G.S. 1958. Permian corals from northern Iraq. *Palaeontology*, **1**, 174–192.

IGO, H. 1959. Note on some Permian corals from Fukuji, Hida Massif, central Japan. *Transactions and Proceedings of the Palaeontological Society of Japan*, **34**, 79–85.

IGO, H. 1988. A Permian rugose coral from Kamikagemori Chichibu City, Saitama prefecture, central Japan. *Transactions and Proceedings of the Palaeontological Society of Japan*, **152**, 656–658.

IGO, H. & ADACHI, S. 2000. Description of some Carboniferous corals from the Ichinotani Formation, Fukuji, Hida Massif, central Japan. *Scientific Report, Institute of Geoscience, University of Tsukuba, Series B*, **21**, 41–69.

IGO, H. & ADACHI, S. 2001. Rugose corals from the Kanoyama Limestone in the Kanto Mountains, Gunma Prefecture, Japan. *Memoirs of the National Science Museum*, **37**, 71–87.

IGO, H., KOIZUMI, H. & KANIWA, T. 2000. A Permian rugose coral, Yatsengia kuzuensis, from north of Kiryu in the Ashio mountains, Gunma prefecture, Japan. *Bulletin of the National Science Museum, Tokyo, Series C*, **26**, 79–86.

IGO, H., ADACHI, S. & IGO, H. 2001. Permian rugose corals from the Gozenyama Formation, Hinohara Vilage, Nishitama County, Tokyo. *Scientific Report, Institute of Geoscience, University of Tsukuba, Series B*, **22**, 125–133.

ILINA, T.G. 1997. Distribution, taxonomyand morphologyof Permian Rugosa of southeastern Pamir (Tadzhikistan). *Boletin de la Real Sociedad Española de Historia Natural (Seccion Geologia)*, **92**, 127–141.

ISBELL, J.L., HENRY, L.C. *ET AL*. 2012. Evaluations of glacial paradoxes during the late Paleozoic Ice Age using the concept of the equilibrium line altitude (ELA) as a control on glaciations. *Gondwana Research*, **22**, 1–19.

JIN, X.C. 1994. Sedimentary and palaeogeographic significance of Permo-Carboniferous sequences in Western Yunnan, China. *Geologisches Institit der Universitaet zu Koeln Sonderveroeffentlichungen*, **99**, 10–136.

KAMEI, T. 1957. On two Permian corals fom the Mizuyagadani Formation. *Journal of the Faculty of Liberal Arts and Sciences, Shinshu University*, **7**, 29–35.

KATO, M. 1972. Permian corals of Miharanoro. *Journal of the Faculty of Science, Hokkaido University, Series 4*, **15**, 501–512.

KORN, D., TITUS, A.L., EBBIGHAUSEN, V., MAPES, R.H. & SUDAR, M.N. 2012. Early Carboniferous (Mississippian) ammonoid biogeography. *Geobios*, **45**, 67–77.

KOSSOVAYA, O.L. 1996. Correlation of Uppermost Carboniferous and Lower Permian rugose coral zones from the Urals to Western North America. *Palaios*, **11**, 71–82.

KOSSOVAYA, O.L. 2007. Ecological aspects of upper Carboniferous–lower Permian '*Cyathaxonia* fauna' taxonomical diversity (the Urals). *In*: HUBMANN, B. & PILLER, W.E. (eds) *Fossil-corals and Sponges*. Proceedings of the 9th International Symposium on Fossil Cnidaria and Porifera, Graz, 2003. Austrian Academy of Sciences Press, Vienna, **17**, 383–406.

KOSSOVAYA, O.L. 2009. Artinskian–Wordian antitropical rugose coral associations: A palaeogeographical approach. *Palaeoworld*, **18**, 136–151.

KULLMAN, J. 1966. Goniatiten-Korallen-Vergesell-schaRungen im Karbon des Kantabrischen Gebirges (Nordspanien). *Neues Jahrbuch fur Geologie und Paliiontologie, Abhandlungen*, **125**, 443–446.

LEVEN, E.Ja. 1997. Permian Stratigraphyand Fusulinida of Afghanistan with their Paleogeographic and Paleotectonic Implications. *Geological Society of America, Specical Papers*, **316**, 1–134.

LIN, B.Y. 1983. *Lower Permian Stratigraphy and Coral Faunas from Both Flanks of Yarlung Zangbo River in Central-Southern Xizang (Tibet). Contribution to the Geologyof the Qinghai-Xizang (Tibet) Plateau 8*. Geological Publishing House, Beijing [in Chinese with English abstract].

LIN, B.Y., XU, S.Y. *ET AL*. 1995. *Rugosa and Heterocorallia*. Geological Publishing House, Beijing [in Chinese with English abstract].

MEI, S., HENDERSON, C.M. & WARDLAW, B.R. 2002. Evolution and distribution of the conodonts Sweetognathus and Iranognathus and related genera during the Permian, and their implications for climate change. *Palaeogeography, Palaeoclimatology, Palaeoecology*, **180**, 57–91.

METCALFE, I. 1991. Late palaeozoic and mesozoic palaeogeography of Southeast Asia. *Palaeogeography, Palaeoclimatology, Palaeoecology*, **87**, 211–221.

METCALFE, I. 1999. Gondwana dispersion and Asian accretion: an overview. *In*: METCALFE, I. (ed.) *Gondwana Dispersion and Asia Accretion*. A.A. Balkema, Rotterdam, 9–28.

MINATO, M. 1955. Japanese Carboniferous and Permian corals. *Journal of the Faculty of Science, Hokkaido University, Series 4*, **9**, 1–202.

MINATO, M. & KATO, M. 1965*a*. Durhaminidae. *Journal of the Faculty of Science, Hokkaido University, Series 4*, **13**, 11–86.

MINATO, M. & KATO, M. 1965*b*. Waagenophyllidae. *Journal of the Faculty of Science, Hokkaido University, Series 4*, **12**, 1–241.

MINATO, M. & KATO, M. 1980. Polythecalis und Lonsdaleiastraeaaus dem Perm des Kitakami–Gebirges, nordostlich Honshu, Japan. *Munstersche Forschungen zur Geologie und Palaeontologie*, **52**, 1–11 [in German].

MINATO, M. & ROWETT, C.L. 1968. Modes of reproduction in rugose corals. *Lethaia*, **1**, 175–183.

MINATO, M., HUNAHASI, M., WATANAVE, J. & KATO, M. 1979. *The Abean Orogeny*. Tokai University Press, Tokyo.

NELSON, S.J. & NELSON, E.R. 1985. Allochthonous Permian micro- and macrofauna, Kamloops area, British Columbia. *Canadian Journal of Earth Science*, **22**, 442–451.

NIE, Z.T., SONG, Z.M., JIANG, J.J. & LIANG, D.Y. 1993. Biota features of the Gondwana affinity facies and review of their stratigraphic ages in the western Yunnan. *Geoscience, Journal of Graduate School of China University of Geoscience*, **7**, 384–393 [in Chinese with English abstract].

QIAO, L. & SHEN, S.Z. 2014. Global paleobiogeography of brachiopods during the Mississippian – response to the global tectonic reconfiguration, ocean circulation, and climate changes. *Gondwana Research*, **26**, 1173–1185.

ROSS, C.A. & ROSS, J.R.P. 1987. Late Paleozoic sea levels and depositional sequences. *In*: ROSS, C.A. & HAMAN, D. (eds) *Timing and Depositional History of Eustatic Sequences: Constraints on Seismic Stratigraphy*. Cushman Foundation for Foraminiferal Research, Special Publications, **24**, 137–149.

RYGEL, M.C., FIELDING, C.R., FRANK, T.D. & BIRGENHEIER, L.P. 2008. The magnitude of Late Paleozoic glacioeustatic fluctuations: a synthesis. *Journal of Sedimentary Research*, **78**, 500–511.

SAKAGUCHI, S. & YAMAGIWA, N. 1958. The Late Paleozoic corals from the southern part of the Tanba district. *Memoirs of the Osaka University of Liberal Arts and Education*, **7**, 163–178.

SAKAGUCHI, S. & YAMAGIWA, N. 1973. Upper Permian coralline and foraminiferal fauna from Mt. Ibuki, southwestern Japan. *Bulletin of the National Science Museum*, **16**, 387–396.

SALTZMAN, M.R. & THOMAS, E. 2012. Carbon isotope stratigraphy. *In*: GRADSTEIN, F.M., OGG, J.G. & SCHMITZ, M. (eds) *Geologic Time Scale 2012*. Cambridge University Press, London, 207–232.

SENGOR, A.M.C. 1979. Mid-Mesozoic closure of Permian–Triassic Tethys and its implications. *Nature*, **279**, 590–593.

SENGOR, A.M.C. 1984. The Cimmeride orogenic system and the tectonics of Eurosia. *Geologica Society of America, Special Papers*, **195**, 1–82.

SCOTESE, C.R. & MCKERROW, W.S. 1990. Revised world maps and introduction. *In*: MCKERROW, W.S. & SCOTESE, C.R. (eds) *Paleozoic Paleogeography and Biogeography*. Geological Society, London, Memoirs, **12**, 1–21, https://doi.org/10.1144/GSL.MEM.1990.012.01.01

SEPKOSKI, J.J., JR. 1996. Patterns of Phanerozoic extinction: a perspective from global data bases. *In*: WALLISER, O.H. (ed.) *Global Events and Event Stratigraphy*. Springer-Verlag, Berlin, 35–51.

SHEN, S.Z., CAO, C.Q. *ET AL*. 2010. End-Permian mass extinction and palaeoenvironmental changes in Neotethys: evidence from an oceanic carbonate section in southwestern Tibet. *Global and Planetary Change*, **73**, 3–14.

SHEN, S.Z., JIN, J.S. & SHI, G.R. 2016. Ecosystem evolution in deep time: Evidence from the rich Paleozoic fossil records of China. *Palaeogeography, Palaeoclimatology, Palaeoecology*, **448**, 1–3.

SHI, G.R. & ARCHBOLD, N.W. 1998. Permian marine biogeography of SE Asia. *In*: HALL, R. & HOLLOWAY, J.D.

(eds) *Biogeography and Geological Evolution of SE Asia*. Backbuys Publishers, Leiden, 57–72.

Shi, G.R., Archbold, N.W. & Zhan, L.P. 1995. Distribution and characteristics of mixed (transitional) mid-Permian (Late Artinskian–Ufimian) marine faunas in Asia and their palaeogeographical implications. *Palaeogeography, Palaeoclimatology, Palaeoecology*, **114**, 241–271.

Shi, G.R., Fang, Z.J. & Archbold, N.W. 1996. An Early Permian brachiopod fauna of Gondwana affinity from the Baoshan Block, western Yunnan, China. *Alcheringa*, **20**, 81–101.

Shi, G.R., Weldon, E.A. & Pierson, R.R. 2010. *Permian stratigraphy, sedimentology and palaeontology of the southern Sydney Basin, south-east Australia – a field excursion guide*. Burwood Campus Printery, Deakin University, Melbourne, 1–72.

Smith, S. 1941. Some Permian corals from the Plateau Limestone of the southern Shan States, Burma. *Memoirs of the Geological Survey of India, Palaeontologica Indica, New Series*, **30**, 1–21.

Sobolev, S.V., Sobolev, A.V. *et al.* 2011. Linking mantle plumes, large igneous provinces and environmental catastrophes. *Nature*, **477**, 312–316.

Somerville, D.I. 1997. Biostratigraphy and biofacies of Upper Carboniferous–Lower Permian rugose coral assemblages from the Isfjorden area, central Spitsbergen. *Boletin de la Real Sociedad Española de Historia Natural (Seccion Geologia)*, **92**, 363–378.

Sorauf, J.E. 2004. Permian corals of Timor (Rugosa and Tabulate): history of collection and study. *Alcheringa*, **28**, 157–183.

Stanley, S.M. 2007. An analysis of the history of marine animal diversity. *Paleobiology*, **33**, 1–55.

Stevens, C.H. 2008*a*. Permian Colonial rugose corals from the Wrangellian Terraine in Alaska. *Journal of Paleontology*, **82**, 1043–4050.

Stevens, C.H. 2008*b*. Fasciculate rugose corals from Gzhelian and Lower Permian strata, Pequop Mountains, Northeast Nevada. *Journal of Paleontology*, **82**, 1190–1200.

Stevens, C.H. 2009. New occurrences of Permian corals from the Mccloud Belt in Western North America. *Palaeontologia Electronica*, **12**, 1–16.

Stevens, C.H. 2010. New Early Permian colonial rugose corals from the central Cordlleran miogeocline, U. S. A. *Journal of Paleontology*, **84**, 529–537.

Stevens, C.H. 2012. Distribution and Diversity of Carboniferous and Permian Colonial Rugose Coral Faunas in Western North America: clues for placement of Allochthonous Terranes. *Geosciences*, **2**, 42–63.

Stevens, C.H. & Rycerski, B. 1989. Early Permian colonial rugose corals from the Stikine River area, British Columbia, Canada. *Journal of Paleontology*, **63**, 158–181.

Stevens, C.H., Miller, M.M. & Nestell, M. 1987. A new Permian waagenophyllid coral from the Klamath Mountains, California. *Journal of Paleontology*, **61**, 690–699.

Svensen, H., Planke, S., Polozov, A.G., Schmidbauer, N., Corfu, F., Podladchikov, Y.Y. & Jamtveit, B. 2009. Siberian gas venting and the end-Permian environmental crisis. *Earth and Planetary Science Letters*, **277**, 490–500.

Ueno, K. 2003. The Permian fusulinoidean faunas of the Sibumasu and Baoshan blocks: their implications for the paleogeographic and paleoclimatologic reconstruction of the Cimmerian continent. *Palaeogeography, Palaeoclimatology, Palaeoecology*, **193**, 1–24.

Ueno, K., Sugiyama, T. & Nagai, K. 1996. Discovery of Permian foraminifers and corals from the Ratburi Limestone of the Phatthalung area, Southern Peninsular Thailand. *In*: Noda, H. & Sashida, K. (eds) *Professor Hisayoshi Igo Commemorative Volume on Geology and Paleontology of Japan and Southeast Asia*. Gakujutu Tosho Insatsu, Tokyo, 201–216.

Wang, X.D. & Sugiyama, T. 2000. Diversity and extinction patterns of Permian coral faunas of China. *Lethaia*, **33**, 285–294.

Wang, X.D. & Sugiyama, T. 2001. Middle Permian rugose corals from Laibin, Guangxi, South China. *Journal of Paleontology*, **75**, 758–782.

Wang, X.D. & Sugiyama, T. 2002. Permian coral faunas of eastern Cimmerian Continent and paleogeographical implications. *Journal of Asian Earth Scicences*, **20**, 589–597.

Wang, X.D. & Wang, X.J. 2007. Extinction patterns of Late Permian (Lopingian) corals in China. *Palaeoworld*, **16**, 31–38.

Wang, X.D., Sugiyama, T., Ueno, K. & Mizuno, Y. 1999. Peri-Gondwanan sequences of Carboniferous and Permian age in the Baoshan block, West Yunnan, Southwest China. *In*: Tathanasthien, B. & Rieb, S.L. (eds) *Proceedings of the International Symposium on Shallow Tethys 5*. Chiang Mai University, Chiang Mai, 88–100.

Wang, X.D., Ueno, K., Mizuno, Y. & Sugiyama, T. 2001. Late Paleozoic faunal, climatic, and geographic changes in the Baoshan block as a Gondwana-derived continental fragment in southwest China. *Palaeogeography, Palaeoclimatology, Palaeoecology*, **170**, 197–223.

Wang, X.D., Shen, S.Z., Sugiyama, T. & West, R.R. 2003. Late Paleozoic corals of Xizang (Tibet) and West Yunnan, southwestern China: successions and paleobiogeography. *Palaeobiogeography, Palaeoclimatology, Palaeoecology*, **191**, 385–397.

Wang, X.D., Wang, X.J., Zhang, F. & Zhang, F. 2006*a*. Diversity patterns of Carboniferous and Permian rugose corals in South China. *Geological Journal*, **41**, 329–343.

Wang, X.D., Sugiyama, T., Kido, E. & Wang, X.J. 2006*b*. Permian rugose coral faunas of Inner Mongolia–Northeast China and Japan: paleobiogeographical implications. *Journal of Asian Earth Sciences*, **26**, 369–379.

Wang, X.D., Zhang, Y.Q. & Lin, W. 2010. Carboniferous–Permian rugose coral Cyathaxonia faunas in China. *Science China Earth Sciences*, **53**, 1864–1872.

Wang, X.D., Lin, W., Shen, S.Z., Chaodumrong, P., Shi, G.R., Wang, X.J. & Wang, Q.L. 2013. Early Permian rugose coral Cyathaxonia faunas from the Sibumasu Terrane (Southeast Asia) and the southern Sydney Basin (Southeast Australia): paleontology and paleobiogeography. *Gondwana Research*, **24**, 185–191.

Wang, Z.J. & Liu, S.K. 1982. *Early Lower Permian Rugose Corals from the Saga, Zhongba and Namu Co Areas of Xizang. Contribution to the Geology of the Qinghai-Xizang (Tibet) Plateau 7*. Geological

Publishing House, Beijing [in Chinese with English abstract].

Waterhouse, J.B. 1982. An early Permian cool-water fauna from pebbly mudstones in south Thailand. *Geological Magazine*, **119**, 337–354.

Weidlich, O. & Flügel, H.W. 1995. Upper Permian (Murghabian) rugose corals from Oman (Ba'id area, Saih Hatat): Community structure and contribution to reefbuilding processes. *Facies*, **33**, 229–264.

Wilson, E.C. 1982. Wolfcampian rugose and tabulate corals (Coelenterata: Anthozoa) from the lower Permian Mccloud limestone of northern California. *Contributions in Science, Natural History Museum of Los Angeles County*, **337**, 1–90.

Wilson, E.C. 1985. Rugose corals (Coelenterata, Anthozoa) from the Lower Permian McCloud Limestone at Tombstone Mountain, northern California. *Contributions in Science, Natural History Museum of Los Angeles County*, **366**, 1–11.

Wilson, E.C. 1990. Permian corals of Bolivia. *Journal of Paleontology*, **64**, 60–78.

Wilson, E.C. 1991. Permian corals from the Spring Mountains, Nevada. *Journal of Paleontology*, **65**, 727–741.

Wilson, E.C. 1994. Early Permian corals from the Providence Mountains, San Bernardino County, California. *Journal of Paleontology*, **68**, 938–951.

Wilson, E.C. & Langenheim, R.L., Jr. 1993. Permian corals from Arrow Canyon, Clark County, Nevada. *Journal of Paleontology*, **67**, 935–945.

Wopfner, H. 1996. Gondwana origin of the Baoshanand Tengchong terranes of West Yunnan. *In*: Hall, R. & Blundell, D. (eds) *Tectonic Evolution of Southeast Asia*. Geological Society, London, Special Publications, **106**, 539–547, https://doi.org/10.1144/GSL.SP.1996.106.01.34

Wu, W.S. 1975. *The Coral Fossils from Qomolangma Feng Region*. A Report of Scientific Expedition in the Mount Jolmo Lungma Region (1966–1968) Palaeontology, Fasc. I, Science Press, Beijing [in Chinese].

Wu, W.S., Liao, W.H. & Zhao, J.M. 1982. *Palaeozoic rugose corals from Xizang (Tibet). Palaeontology of Xizang 4*. Science Press, Beijing [in Chinese with English abstract].

Wu, W.S., Stevens, C.H. & Bamber, E.W. 1985. New Carboniferous and Permian Tethyan and Boreal Corals from Northwestern British Columbia, Canada. *Journal of Paleontology*, **59**, 1489–1504.

Xia, G.Y. & Ding, Y.J. 1983. Early Permian fusulinids and corals in Deyanqimiao District, Sonid Youqi of Nei Mongol Region. *Bulletin of the Tianjin Institute of Geological and Mineral Resources*, **8**, 143–160 [in Chinese with English abstract].

Yabe, H. & Minato, M. 1945. A new species of *Wentzelella* from the Permian limestone near Iwaizaki, Kitakami district, northeast Japan. *Proceedings of the Imperial Academy of Japan*, **21**, 469–472.

Yamada, K. & Yamano, H. 1980. Find of Permian fossils from the Moribu formation, Hida mountains, central Japan. *Science Report of Kanazawa University*, **25**, 53–65.

Yamagiwa, N. 1961. The Permo-Carboniferous corals from the Atetsu Plateau and the coral faunas of the same age in the southwest Japan. Part 1. The Permo Carboniferous corals from the Atetsu Plateau. *Memoirs of the Osaka University of Liberal Arts and Education*, **10**, 77–114.

Yamagiwa, N. & Suzuki, Y. 1976. A new species of the genus *Iranophyllum* from the Permian Shirasaki Limestone at Yura-machi, Wakayama prefecture. *Bulletin of the National Science Museum, Series C*, **2**, 27–30.

Yamagiwa, N. & Tsuda, H. 1980. A new coral species from a pebble in the basal Limestone conglomerate of the Triassic Adoyama formation at Karasawa in the Kuzu area, Tochigi prefecture, Japan. *Bulletin of the National Science Museum, Series C*, **6**, 97–100.

Yamagiwa, N. & Yamano, A. 1990. A new species ofIpciphyllum from the Akasaka Limestone, central Japan. *Bulletin of the National Science Museum, Series C*, **16**, 119–125.

Yamagiwa, N., Asami, T. & Hosono, A. 1999. A redescription of Permian Rugosa *Waagenophyllum (Waagenophyllum) compactum* Minato and Kato, 1965. *Bulletin of the National Science Museum. Series C*, **25**, 105–110.

Yanagisawa, I. 1967. Geology and Paleontology of the Takakurayama–Yaguki Area, Youtsukura-cho, Fukushima prefecture. *Science Report of Tohoku University, Series 2*, **39**, 63–112.

Yang, S.P. & Fan, Y.N. 1982. *Carboniferous Strata and Fauna in Shenzha District, Northern Xizang (Tibet)*. Geological Publishing House, Beijing [in Chinese with English abstract].

Yang, W.P. 1999. Stratigraphic and phytogeographic palynologyof Late Paleozoic sediments in western Yunnan, China. *Science Report, Niigata University Series E (Geology)*, **14**, 15–99.

Yang, Z.Y. & Nie, Z.T. 1990. *Palaeontology of Ngari, Tibet (Xizang)*. The China Univ. Geosciences Press, Wuhan [in Chinese with English abstract].

Yokoyama, T. 1960. Permian corals from the Taishaku district, Hiroshima prefecture, Japan. *Transactions and Proceedings of the Palaeontological Society of Japan*, **38**, 238–248.

Yu, J.Z., Lin, Y.T. & Huang, Z.X. 1981. Early Permian corals from central Jilin. *Acta Palaeontologica Sinica*, **20**, 273–285 [in Chinese with English abstract].

Zhang, Y.C., Shi, G.R. & Shen, S.Z. 2013. A review of Permian stratigraphy, palaeobiogeography and palaeogeography of the Qinghai–Tibet Plateau. *Gondwana Research*, **24**, 55–76.

Zhao, J.M. 1991. Earlyand Middle Permian rugose corals from Gegyai County, NW Xizang (Tibet). *In*: Sun, D.L. & Xu, J.T. (eds) *Permian, Jurassic and Cretaceous Strata and Fossils from Ritu, Tibet*. Nanjing University Press, Nanjing, 101–126 [in Chinese].

Zhao, J.M. & Wu, W.S. 1986. Bulletin of Nanjing Institute of Geology and Palaeontology. *Academia Sinica*, **10**, 169–194 [in Chinese with English abstract].

Zhao, J.M. & Yang, D.R. 1985. Discovery of the Multimurinuscoral fauna from Xiuzhumuqinqi of Nei Mongol. *Acta Palaeontologica Sinica*, **24**, 441–448 [in Chinese with English abstract].

Permian ammonoid biostratigraphy

TATIANA B. LEONOVA

Borissiak Palaeontological Institute, Russian Academy of Sciences, Moscow, Russia
tleon@paleo.ru

Abstract: A brief historical review of ammonoid-based Permian biostratigraphy is performed. Changes in ammonoid associations are shown for each of nine Permian stages. The major correlation problems were discussed. A renewed ammonoid zonal scale is proposed.

The rapid evolution and wide distribution of ammonoids in ancient basins have led to their key role in the biostratigraphy of the Phanerozoic. They allow broad, sometimes global, correlations of Permian beds. The only disadvantage of their use is the relative scarcity of localities where this succession can be traced. Unfortunately, when dealing with fossils, we constantly face this problem.

Almost all stages of the Permian were originally established based on ammonoids. Owing to the uneven distribution of the sediments containing these fossils, and varying levels of knowledge, different parts of the ammonoid scale were substantiated in different regions: Early Permian (Cisuralian), in Cisuralia and in the Urals; Middle Permian (Guadalupian), in the western USA; and Late Permian (Lopingian), in Transcaucasia and South China. Traditionally, ammonoid 'zones' of the Lower and Middle Permian have most often been used as equivalents of stages (Glenister 1981), while for the Upper Permian these subdivisions are much more detailed.

Historical review of Permian ammonoid biostratigraphy

Almost 100 years ago Böse (1919) proposed a succession defined by the change in ammonoid associations, which was the first example of a scheme in the Permian. Two 'zones' of this scale have been used by stratigraphers for many decades: the *Perrinites* and Waagenoceras zones, corresponding to the Artinskian Stage *sensu lato* and the Wordian Stage. Miller (1938), based on material from North America, expanded the range and introduced two more zones. The resulting zonal scheme was as follows: *Properrinites* and *Perrinites*, for the Lower Permian; *Waagenoceras* and *Timorites*, for the Upper Permian.

Ruzhencev (1955) discussed the major stages of Permian ammonoids on a global scale and recognized seven successive assemblages: the Asselian, Sakmarian, Aktastynian, Baigendzhinian, Sicilian (or Wordian), Capitanian and Dzhulfian. On the basis of these stages, Glenister & Furnish (1961) recognized six stages: the Asselian, Sakmarian, Artinskian with Aktastynian and Baigendzhinian substages, Wordian, Capitanian and Dzhulfian. They correlated the main formations with ammonoids globally (Fig. 1).

Furnish (1966) introduced a new, Roadian, stage for the Uppermost of the Lower Permian. In addition, he subdivided the upper series into three stages: the Guadalupian, Chidruan and Dzhulfian. Later (Furnish 1973), he proposed a more refined scheme of the Permian system. In that scheme, each of the two series was subdivided into two subseries, and each of the subseries included three stages: Lower Permian – the Sakmarian (Asselian, Tastubian and Sterlitamakian stages) and the Artinskian (Aktastynian, Leonardian and Roadian stages); Upper Permian – the Guadalupian (Wordian, Capitanian and Amarassian stages) and the Dzhulfian (Arakian, Chidruan and Changhsingian stages) (Fig. 1).

Three years later, Ruzhencev (1976) analysed this scale, but, despite reasonable correlation with the evolution of ammonoids, its subdivisions are uneven. He proposed a less refined version of the global scale, including four stages each: Lower Permian, with the Asselian, Sakmarian, Artinskian and Roadian; and Upper Permian, with the Wordian, Capitanian, Dzhulfian and Changhsingian stages (Fig. 1).

All Lower Permian stages were established in the South Urals. The first Artinskian ammonoids were discovered in the mid-nineteenth century (Murchison *et al.* 1845). The Artinskian was the first stage recognized. Karpinsky (1874) proposed the name 'Artinskian Stage', by which he understood the 'transitional Permo-Carboniferous beds', and all Lower Permian ammonoids of the Urals were considered Artinskian. In his famous monograph on Artinskian ammonoids (Karpinsky 1890), he continued insisting on this interpretation of this subdivision. At the end of the nineteenth century, Krotov (1885) studied Artinskian ammonoids and

From: LUCAS, S. G. & SHEN, S. Z. (eds) 2018. *The Permian Timescale*. Geological Society, London, Special Publications, **450**, 185–203.
First published online December 8, 2016, https://doi.org/10.1144/SP450.7

Ruzhentsev, 1955		Glenister, Furnish, 1961			Furnish, 1973		Ruzhentsev, 1976	Leonova, Dmitriev, 1989	Jin *et al.*, 1997			Leonova, 2011a–Present paper
Stages		Stages, substages		Ammonoid Zones	Series, Stages		Stages	Stages	Epoch/Stage		Ammonoid Zones	Ammonoid Zones
Late Permian	Dzhufian	Dzhufian		*Cyclolobus*	**Dzhufian**	Changhsingian	Changhsingian	Dorashamian	**Lopingian**	Changhsingian	*Pseudotirolites* *Paratirolites-Shevyrevites* *Iranites-Phisonites*	*Pleuronodoceras multinodosum-* *Rotodiscoceras asiaticum* *Paratirolites kittli* *Phisonites triangularis*
						Chideru	Dzhufian	Dzhufian		Wuchiapingian	*Araxoceras-Konglingites* *Anderssonoceras* *Roadoceras-Doulingoceras*	*Araxoceras ventrosulcatum*
						Araksian						*Araxoceras latissimum*
	Capitanian	**Guadalupian**	Capitanian	*Timorites*	**Guadalupian**	Amarassian	Capitanian	Capitanian	**Guadalupian**	Capitanian	*Timorites*	*Eoaraxoceras ruzhencevi* – *Kingoceras kingi* *Timorites schucherti- Cibolites uddeni.*
						Capitanian						
	Sicilian (or Word)		Wordian	*Waagenoceras*		Wordian	Wordian	Wordian		Wordian	*Waagenoceras*	*Adrianites elegans* – *Waagenoceras dieneri*
								Roadian		Roadian	*Demarezites* *Stacheoceras discoidale*	*Daubichites goochi* – *Demarezites oyensi*
Early Permian	Baigendzhinian	**Artinskian**	Baigendzhinian	*Perrinites*	**Artinskian**	Roadian	Roadian					
						Leonardian	Artinskian	Kungurian	**Cisuralian**	Kungurian	*Pseudovidrioceras dunbari* *Propinacoceras busterense*	*Epijuresanites musalitini–Perrinites hilli*
								Artinskian		Artinskian	*Uraloceras fedorowi*	*Neocrimites fredericksi* – *Medlicottia orbignyana*
	Aktastinian		Aktastinian			Aktastinian					*Aktubinskia notabilis* – *Artinskia artiensis*	*Neoshumardites triceps*– *Metaperrinites vicinus*
	Sakmarian	Sakmarian		*Properrinites*	**Sakmarian**	Sterlitamakian	Sakmarian	Sakmarian		Sakmarian	*Sakmarites inflatus*	*Crimites subkrotowi*– *Properrinites cumminsi*
						Tastubian					*Svetlanoceras strigosum*	*Propopanoceras simense* – *Properrinites boesei*
	Asselian	Asselian				Asselian	Asselian	Asselian		Asselian	*Svetlanoceras serpentinum* *S. primore*	*Svetlanoceras strigosum* - *Emilites prosperus* *Sv. primore* – *Subperrinites bakeri*

Fig. 1. Permian ammonoid subdivisions from Ruzhencev (1955, 1976), Glenister & Furnish (1961), Furnish (1973), Leonova & Dmitriev (1989), Jin *et al.* (1997) and Leonova (2011*a*, this paper).

subdivided them into the Carboniferous (150 species) and Permian (53 species). Further studies of the Early Permian ammonoids of the Urals resulted in the development of a substantiated stage scale for this interval of geological history. Ruzhencev (1938, 1951, 1954) established three large successive ammonoid assemblages: the Asselian, Sakmarian and Artinskian, which served as the basis for the recognition of stages. The type section of the Asselian was established on the Assel River in the Orenburg Region of the South Urals (Ruzhencev 1950, 1954).

The ammonoid-based Carboniferous–Permian boundary was established based on the first appearance of four new families: Paragastrioceratidae (*Svetlanoceras*), Metalegoceratidae (*Juresanites*), Popanoceratidae (*Protopopanoceras*) and Perrinitidae (*Subperrinites*, *Properrinites*). In addition, seven new genera appeared in previously existing families: *Vanartinskia*, *Mescalites*, *Kargalites*, *Cardiella*, *Martoceras*, *Prostacheoceras* and *Tabantalites* (Bogoslovskaya 1984; Bogoslovskaya *et al.* 1995).

Cisuralian

The Carboniferous–Permian boundary is defined in the stratotype section on Aidaralash Creek in the South Urals (Kazakhstan). Ruzhencev (1950, 1952) was the first to study this section and its ammonoids, and was the author of the Orenburgian and Asselian stages. In this section, the species *Prouddenites terminalis*, *Artinskia irinae* and *Prothalassoceras bashkiricus* were identified by Ruzhencev as the latest Carboniferous (lower part of Bed 19), whereas *Neopronorites rotundus*, *Daixites antipovi*, *Artinskia kazakhstanica* and *Prothalassoceras serratum* (Bed 20) were identified as the earliest Asselian, and the boundary between the systems was drawn between the two beds.

In the 1980s–90s, the section was re-examined for ammonoids (Bogoslovskaya *et al.* 1995). The combined effort of the Russian–American working group resulted in a detailed substantiation of the position of the Carboniferous–Permian boundary in sections of the South Urals (Davidov *et al.* 1998). The Aidaralash section was adopted as the Carboniferous–Permian boundary stratotype by the International Commission on Stratigraphy (ICS), and the position of the boundary within Bed 19 was based on the first appearance of the conodont species *Streptognathodus isolatus*. The position of this boundary is somewhat lower than the previous boundaries (6.3 m below the fusulinid-based boundary and 26.8 m lower than the ammonoid-based level). Because no ammonoids have been found in this interval of 26.8 m, ammonoid workers did not object to the position of the boundary.

Asselian

Apart from the Urals, beds of the Asselian stage containing ammonoids are definitely found in other regions of the globe: for example, North America: in western Texas and New Mexico – the Neal Ranch Formation, the lower part of the Wolf Camp Formation; in north-central Texas – the Pueblo and Moran formations in the lower part of the Wichita Group; in northern Oklahoma – Red Eagle Limestone, the lower part of the Council Grove Group; and in central New Mexico – the Bursum Formation.

In these regions, the Asselian beds are overlain by Sakmarian beds containing fossil remains (Furnish 1973). The age of the Bursum Formation in New Mexico is debatable. Ammonoid workers consider the entire formation or at least its upper part to be Asselian (e.g. Furnish & Glenister 1971; Furnish 1973; Bogoslovskaya *et al.* 1999).

In many regions, the earliest Permian ammonoids are dated as Asselian–Sakmarian. In NE Russia (the Verkhoyansk Region, and the Kolyma–Omolon and Okhotsk massifs), ammonoids are found in the Khorokyt and Tuora-Sis formations, dated as late Asselian–Sakmarian (Andrianov 1985; Kutygin *et al.* 2002). In Arctic Canada, rare occurrences of ammonoids from the lower part of the Hare Fiord and Belcher Channel formations also resemble those of the Asselian–Sakmarian. Their age is controlled by conodonts and fusulinids (Nassichuk & Spinosa 1972; Nassichuk & Henderson 1986). In the Palaeotethyan Realm, the age of the series with ammonoids – the Tashkazyk Formation of Pamir and the Maping Formation in South China – is dated as Late Asselian–Sakmarian (Ruzhencev 1978; Zhou 1987). Apparently, this uncertainty largely results from insufficient knowledge of these regions.

The Asselian ammonoid assemblage comprises 33 genera, almost half of which continue from the Late Carboniferous (16 genera). Of these genera, three (*Pinnoceras*, *Glaphyrites* and *Neoglaphyrites*) occur only in the lower part of the Asselian Stage, seven (*Boesites*, *Daixites*, *Aristoceras*, *Prothalassoceras*, *Somoholites*, *Eoasianites* and *Emilites*) are known from the Asselian and Sakmarian stages, and six genera (*Metapronorites*, *Neopronorites*, *Artinskia*, *Neoaganides*, *Agathiceras* and *Almites*) occur in younger beds.

Only three genera are restricted to the Asselian. Of these, *Shikhanites* and *Protopopanoceras* are rare Uralian endemics, whereas *Mescalites* is somewhat more widespread, being found in America and, possibly, Australia (Furnish & Glenister 1971; Glenister *et al.* 1990*a*, *b*).

The remaining 14 genera are known from the Asselian and younger beds. Species of the genera *Svetlanoceras*, *Subperrinites*, *Properrinites*, *Juresanites* and *Tabantalites* are found in the Asselian and Sakmarian in some regions (the Urals, North America, Pamir and China). For the three genera *Vanartinskia*, *Martoceras* and *Bulunites*, an Asselian–Sakmarian age was indicated only because of the debatable interpretation of the age of the host rock: the Tashkazyk Formation of Pamir, and the Khorokyt and Tuora-Sis formations of the Verkhoyansk Region. The geographical distribution of these genera is very limited (Ruzhencev 1978; Andrianov 1985; Kutygin *et al.* 2002). Species of the genera *Sakmarites*, *Paragastrioceras*, *Kargalites*, *Cardiella*, *Prostacheoceras* and *Neoaganides* are relatively long ranging stratigraphically (the entire Early Permian and, for the last one, the whole of the Permian).

The interregional correlations are based on the closely related species of the widespread genera *Boesites*, *Svetlanoceras*, *Eoasianites*, *Juresanites*, *Tabantalites*, *Subperrinites*, *Properrinites* and *Emilites*, taking into account their level of morphological progress.

We propose the following taxa for the zonal subdivision of Asselian. For the lower part of the stage, we propose using *Svetlanoceras primore–Subperrinites bakeri* as zonal index taxa. Both species are the most primitive representatives of their genera. The former species is known from the Uralian and Palaeotethyan realms, and the latter from the North American and Arctic realms. The species *Svetlanoceras strigosum–Emilites prosperus* are proposed as zonal indexes for the upper part of the stage. The former represent a second stage in the phylogenetic lineage of the species of the genus *Svetlanoceras*, while the latter has a narrow stratigraphic distribution in the Tethyan Realm (Fig. 2).

Sakmarian

The Sakmarian Stage was separated from the Artinskian Stage by Ruzhencev (1938) based on the analysis of ammonoid data. The Sakmarian Stage is readily traced in all continents, and is correlated with the Somohole beds of Timor, the Wolf Camp, Hueco and Wichita formations of North America, and deposits of the Holmwood Shale, Nura Nura Member, Fossil Cliff Formation and Callytharra Formation of Western Australia (Ruzhencev 1951; Glenister *et al.* 1983; Archbold & Dickins 1991). At the same time, it should be noted that the ammonoid substantiation of the lower and upper limits is

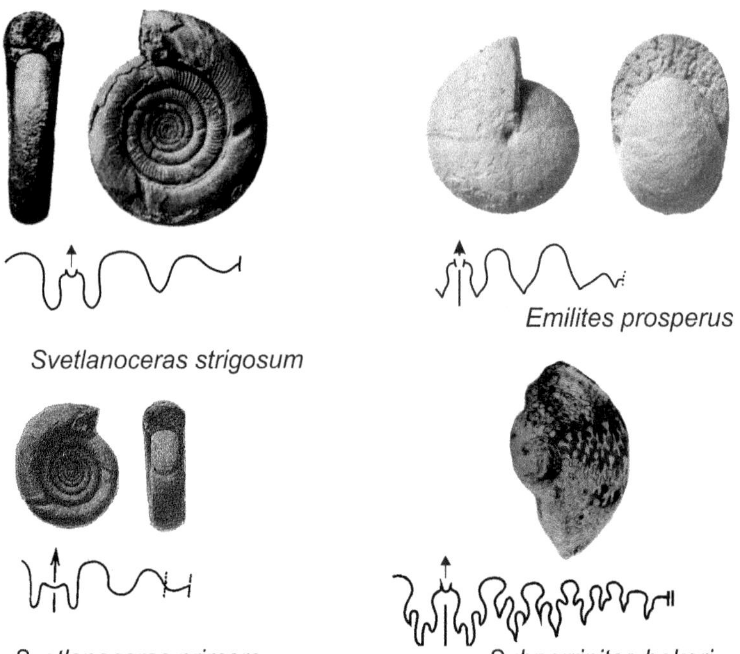

Fig. 2. Asselian zonal index species (sources of illustrations: Bogoslovskaya & Popov 1986; Plummer & Scott 1937; Ruzhencev 1952, 1978).

somewhat difficult. Only in the Urals, can the succession of the Asselian, Sakmarian and Early Artinskian assemblages be reliably traced. As noted above, ammonoid associations are described from several localities in the Arctic and Tethyan regions dated as the Late Asselian–Early Sakmarian.

The Sakmarian ammonoid assemblage is more diverse than the Asselian one, containing 42 genera, of which 24 continued from earlier beds. Fourteen genera complete their evolution in the Sakmarian. These are *Boesites*, *Daixites*, *Prothalassoceras*, *Somoholites*, *Eoasianites*, *Svetlanoceras*, *Juresanites*, *Subperrinites*, *Properrinites*, *Tabantalites*, *Emilites*, *Vanartinskia*, *Bulunites* and *Martoceras*, whereas 10 genera continued beyond the Sakmarian (*Metapronorites*, *Neopronorites*, *Sakmarites*, *Artinskia*, *Neoaganides*, *Agathiceras*, *Kargalites*, *Almites*, *Cardiella* and *Prostacheoceras*). Five genera are exclusively Sakmarian: *Synuraloceras* (Uralian endemic), *Leeites* (American endemic), *Andrianovia* and *Parametalegoceras* are geographically restricted, and *Propopanoceras* is cosmopolitan. Thirteen genera, the first appearance of which is fixed in the Sakmarian, are also known from younger deposits. These are *Parapronorites*, *Synartinskia*, *Medlicottia*, *Akmilleria*, *Propinacoceras*, *Miklukhoceras*, *Thalassoceras*, *Paragastrioceras*, *Uraloceras*, '*Stenolobulites*', *Metalegoceras*, *Nevadoceras* and *Crimites*.

The Sakmarian ammonoids are best represented in the South Urals, where they are represented by two assemblages. The lower (Tastubian) assemblage includes the following: *Sakmarites quadrilobatus*, *Metalegoceras distale* and *Propopanoceras simense*. The Tastubian assemblage also includes species continuing from the Late Asselian species: *Svetlanoceras strigosum*, *Prothalassoceras biforme* and *Tabantalites bifurcates*, among others. The upper (Sterlitamakian) assemblage is typified by *Sakmarites inflatus*, *Medlicottia semota*, *Juresanites kazakhorum*, *Metalegoceras gerassimovi* and *Crimites subcrotowi* (Ruzhencev 1951, 1952; Bogoslovskaya *et al.* 1995).

In other regions, it is impossible to recognize separate Early Sakmarian and Late Sakmarian assemblages. Broad correlations of the Sakmarian beds are based on species of the following genera: *Boesites*, *Daixites*, *Synartinskia*, *Artinskia*, *Medlicottia*, *Prothalassoceras*, *Juresanites*, *Metalegoceras*, *Paragastrioceras*, *Crimites* and *Propopanoceras*, as shown by the Khoridzh ammonoid assemblages (Leven *et al.* 1992). The species of these genera correspond to the Sakmarian species from the Urals in the level of the development of the suture and shell morphology. In addition, the comparison of the well-studied phylogenetic trees of the above genera has a high correlation potential.

For the lower substage, we recognize the *Propopanoceras simense–Properrinites boesei* Zone. The precise stratigraphic position of the former species is defined in the Urals, whereas the corresponding species are known from the American, Palaeotethyan and Australian areas, the latter species is known from America, and the corresponding species are known from the Arctic and Palaeotethyan realms. For the upper substage, we propose the *Crimites subcrotowi–Properrinites cumminsi* Zone. The former species is known from the Urals and in North America, while the corresponding species are known from the Palaeotethyan Realm; the latter species is established in North America, while the corresponding species are recognized in the Palaeotethyan Realm (in Timor) (Fig. 3).

Artinskian

Ruzhencev (1956) gave the most comprehensive characterization of the Artinskian Stage *sensu stricto*. Based on ammonoids, he subdivided it into two substages. It should be said that the lower, Aktastynian substage is well substantiated by ammonoids only in the Urals. In many other regions, it is not very easily delineated from the underlying Late Sakmarian Sterlitamakian substages (Nevada, Australia, Arctic Canada and the Verkhoyansk Region). The upper, Baigendzhinian, substage is well characterized by ammonoids in almost all continents. The Baigendzhinian ammonoid assemblage is extremely rich. It correlates with the assemblages from the upper part of the Yakhtashian Stage of Darvaz, from the Longyin Formation of South China, from the Jungle Creek Formation of Arctic Canada, from the upper part of the Echian Horizon of the Verkhoyansk Region, from the Skinner Ranch and Hess formations of Texas, and from the lower part of the Byro Group in Western Australia.

Altogether, 49 genera are known in the Artinskian Stage, of which 22 continued from earlier beds: 12 genera exclusively from the Sakmarian – *Parapronorites*, *Synartinskia*, *Akmilleria*, *Miklukhoceras*, *Propinacoceras*, *Medlicottia*, *Metalegoceras*, *Paragastrioceras*, *Uraloceras*, *Stenolobulites*, *Thalassoceras* and *Crimites*; and 10 genera from earlier beds – *Metapronorites*, *Neopronorites*, *Artinskia*, *Kargalites*, *Cardiella*, *Almites*, *Sakmarites*, *Prostacheoceras*, *Agathiceras* and *Neoaganides*. Almost all of these genera are of wide geographical distribution.

Six genera are restricted to the Artinskian, most of these are endemic: the Early Artinskian *Aktubinskia* and *Neoshumardites*; and the Late Artinskian *Artioceras*, *Paramedlicottia*, *Prosicanites* and *Darvasiceras*.

The evolution of a representative group of 21 genera, first appearing in the second half of

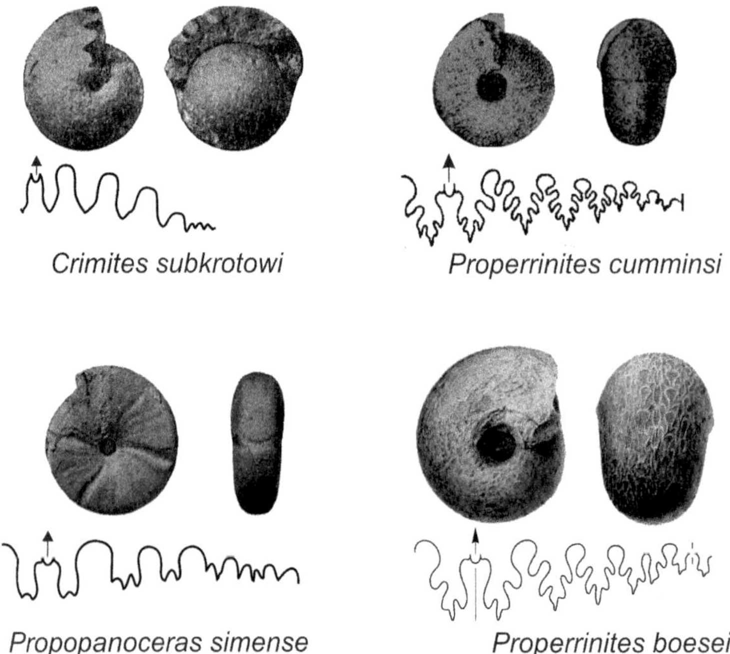

Fig. 3. Sakmarian zonal index species (sources of illustrations: Plummer & Scott 1937; Ruzhencev 1951).

the Artinskian, continued further. Eleven of these were restricted to the end-Artinskian–Kungurian: *Pseudohalorites*, *Metaperrinites*, *Shyndoceras*, *Waagenina*, *Artioceratoides*, *Veruzhites*, *Pamiropopanoceras*, *Eolegoceras*, *Atsabites*, *Pseudoshistoceras* and *Gobioceras*. Their geographical ranges are also relatively narrow. The remaining 10 genera are also known from younger beds: *Eothinites*, *Daraelites*, *Popanoceras*, *Neocrimites* and *Bamyaniceras* (genera of wide geographical distribution), and *Bransonoceras*, *Perrinites*, *Sosiocrimites*, *Parasicanites* and *Perrimetanites* (genera of the narrower geographical distribution).

The Artinskian Stage is subdivided into two substages: the Aktastynian and the Baigendzhinian. Trends in the evolution of the marine biota, in general, and ammonoids, in particular, are clearly different in these intervals of the geological history (Leven *et al.* 1996): hence, they are considered separately.

Aktastynian ammonoids are known in the South Urals in the Akstastynian Stage; in Arctic Canada from the 'Unnamed formation A'; in NE Russia from the Endybal–Echian Formation of the Verkhoyansk Region and from the Munudzhak Formation of the Kolyma–Omolon Massif; in Texas and New Mexico in the Clyde, Hueco and Wichita formations; and in Nevada in the Portuguese Springs Formation.

For the lower part of the Artinskian, we propose the *Neoshumardites triceps–Metaperrinites vicinus* Zone. The former species was established in the Urals, whereas closely related species occur in the Arctic Realm, and the latter is known from North America (Fig. 4).

Baigendzhinian ammonoids are known from the Urals, from the Baigendzhinian substage; from Arctic Canada, in the middle part of the Jungle Creek Formation; from the Verkhoyansk Region, in the upper part of the Echian Horizon of the Mys Formation; from Texas, in the Skinner Ranch and Hess (partly) formations, corresponding to the upper part of the Artinskian; from Pamir (Darvaz), in the upper part of the Yakhtashian Stage; from South China, in the Longyin Formation; from Timor, in theAtsabe Beds; and from Western Australia, in the lower part of the Byro Group.

Species of *Medlicottia*, *Bamyaniceras*, *Neocrimites*, *Eothinites*, *Popanoceras* and other genera with a wide geographical distribution and with a tendency towards increased morphological complexity are important for interregional correlation, which is largely based on the well-studied phylogenetic lineages of species of these genera, which are a good tool for geological dating because stages of morphological transformations relatively clearly define the successive time interval of the Sakmarian and Artinskian.

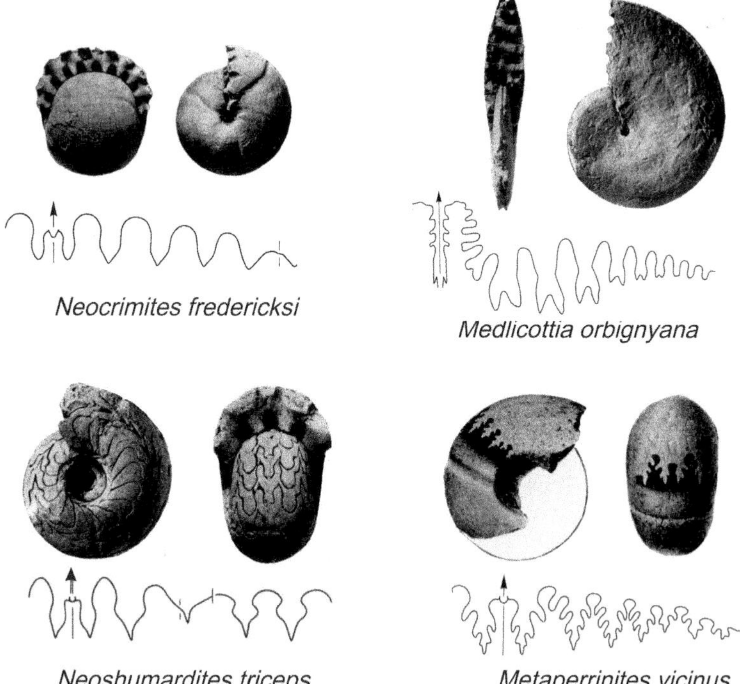

Fig. 4. Artinskian zonal index species (sources of illustrations: Miller & Furnish 1940; Ruzhencev 1956).

For the Upper Artinskian substage, I proposed the *Neocrimites fredericksi–Medlicottia orbignyana* Zone. Both of these species are found in the Urals, whereas the correlating species are known from the Palaeotethyan, American and Arctic realms, and from Australia (Fig. 4).

Kungurian

For a long time, the Kungurian Stage in the evolution of ammonoids was considered together with the Upper Artinskian (Baigenzhinian) (e.g. Ruzhencev 1955, 1960, 1976; Glenister & Furnish 1961, 1981). Only at the end of the last century was the Kungurian ammonoid assemblage considered as a separate unit. This resulted from the study of the very rich Bolorian ammonoid assemblage from Pamir, which is intermediate between the Baigendzhinian and Roadian assemblages (Leonova & Dmitriev 1989; Leven *et al.* 1992). The conclusion of the correspondence of the Bolorian Stage of Pamir to the Kungurian Stage was based on ammonoids from the Kochusu and Shindy formations of SE Pamir (Leonova & Dmitriev 1989). The subsequent study of the Late Yakhtashian ammonoids from Darvaz (Leven *et al.* 1992) supported this conclusion. A comparative analysis of the level of evolutionary advancement of the Uralian (Late Artinskian), Late Yakhtashian (also Late Artinskian) and Bolorian (Kungurian) taxa allowed the recognition of the age differences of these assemblages. The separation of the Kungurian Stage in the ammonoid evolution was based on an assemblage that included numerous genera and species. This period corresponds to a certain stage in the general history of the group and is relatively closely connected with the previous stage. At present, all ammonoid workers accept the existence of a separate Kungurian stage in ammonoid evolution (Furnish *et al.* 2009). Data on conodonts from the Bolorian Stage (the Kochusu Formation) of Pamir (*Mesogondolella intermedia*, *Neostreptognathodus sulcoplicatus*, *N. leonovae* and *Sweetognathus guizhouensis*) suggest a Late Irenian, or even Solikamskian, age (Kozur 1994; Chuvashov & Chernykh 2004).

Of 68 genera found in the Kungurian, 40 are also known from earlier deposits, and all of these genera were listed previously. Sixteen genera are restricted to the Kungurian (this number suggests an increased evolutionary momentum apparently related to an increased differentiation of Permian basins and an increased number of ecological niches).

Exclusively, Kungurian genera are known in each biogeographical realm. In the Arctic Realm, these are *Epijuresanites*, *Tumaroceras* and *Baraioceras*. In the Palaeotethyan Realm, they are particularly diverse. In Pamir, these are *Suakites*, *Istycoceras*, *Aksuites*, *Mapirites*, *Ripernites*, *Nepirrites*, *Pseudoemilites*, *Pamiritella*, *Pamirioceras* and *Pamirites*; in Timor, *Paraperrinites*; and *Zhonglupuceras* in South China. Fourteen genera, first appearing in the Kungurian, continued their evolution later in various regions of the Palaeotethyan and American realms. These are *Neouddenites*, *Yinoceras*, *Biarmiceras*, *Lianyuanoceras*, *Eohyattoceras*, *Pseudovidrioceras*, *Gaetanoceras*, *Aricoceras*, *Sicanites*, *Eumedlicottia*, *Metacrimites*, *Epadrianites*, *Hyattoceras* and *Stacheoceras*. It is now difficult to subdivide these genera into early and late Kungurian because ammonoid specialists only recently acknowledged the existence of a separate Kungurian stage in the evolution of the group.

In the stratotype region, Kungurian ammonoids are not very prominent, being represented by a few new species of paragastrioceratid genera: *Paragastrioceras kungurense*, *Uraloceras tchuvaschovi*, *U. sofronizkyi*, *U. alekense* and *Thalassoceras* sp. (Bogoslovskaya 1976; Chuvashov *et al.* 2002). The recently re-examined section in Mechetlino (Bashkortostan) contains ammonoids in the Saraninian Regional Stage alongside the Kungurian conodonts *Neostreptognathodus pnevi*. The occurrence of *Uraloceras tchuvaschovi* at the base of the Saraninian Regional Stage is particularly interesting (Chuvashov & Chernykh 2004; Boiko 2009, 2010).

Kungurian ammonoids have been reported from the northern regions of European Russia, in the Talata Formation (Pai Khoy) and Lek-Vorkuta Formation (Vaigach Island); and in the Verkhoyansk Region in the Tumara Formation. In Arctic Canada, the Kungurian Stage correlates with the Unnamed Formation, in the northern Richardson Mountains and, possibly, Sabine Bay; in Texas, it correlates with the Cathedral Mountain Formation and its equivalents; in Mexico, it correlates with the Las Sardinas and Puebla formations; and in Pamir, it correlates with the Kochusu, Shindy and Chelamcha formations. Kungurian ammonoids are found in Afghanistan, on Bamian Mountain; in Thailand, in Timor in the Bitauni Bed and in South China (Hunan Province), Chihsia Formation; in NW China, in the Esangan Formation; in the Far East, in the Kungurian of Primorye; in eastern Australia, in the Tiverton Formation; and in Western Australia, in the upper part of the Byro Group.

For the Kungurian Stage, I propose the *Epijuresanites musalitini–Perrinites hilli* Zone. The distribution of the former species is restricted to the Arctic Realm, and the latter to the American Realm, whereas similarly advanced species of the genera *Paraperrinites* and *Perrimetanites* are relatively common in the Palaeotethyan Realm (Fig. 5). Around 60% of previously existing genera became extinct at the Early–Middle Permian boundary.

Guadalupian

The stratotype of the Lower–Middle Permian boundary is fixed in the Guadalupe Mountains in the south-western USA. According to the decisions of the International Subcommission on Permian Stratigraphy, it is drawn by the first appearance of the conodont *Jinogondolella nankingensis* in the Roadian section (Jin *et al.* 1997). The boundary is not well substantiated by ammonoid data. The conodont-based boundary is drawn above the first appearance of ceratitids and above the entry of the Roadian ammonoid assemblage. The first representatives of the genera *Paraceltites*, *Glassoceras*, *Texoceras* and other exclusively Roadian taxa appear at

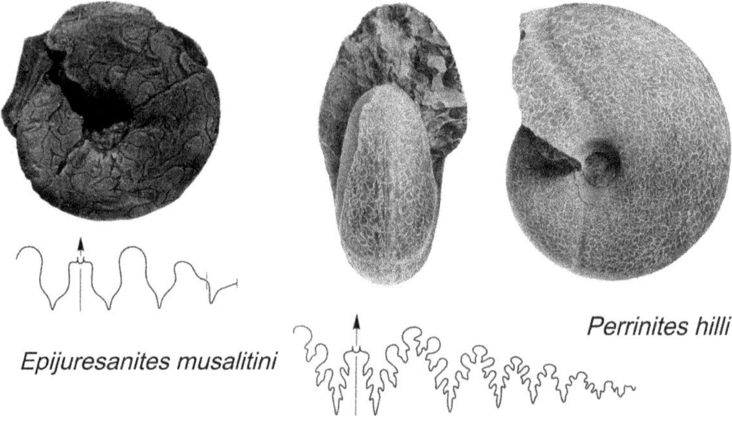

Fig. 5. Kungurian zonal index species (sources of illustrations: Furnish *et al.* 2009; Miller & Furnish 1940).

the base of the Road Canyon Formation, below the bed with the conodont *J. nankingensis*, the marker level for the boundary (Lambert *et al.* 2000). On the whole, the ammonoid assemblage of this formation is an entire association, which could not be subdivided into sub-assemblages, although that has been attempted (see Leven & Bogoslovskaya 2006). It seems incorrect to place the boundary between the series in the middle of the formation, subdividing a single assemblage strikingly differing from the Kungurian (the uppermost Early Permian) ammonoid assemblage. The fusulinids data also contradict the presently accepted boundary level. The exclusive use of a single fossil group (conodonts) by stratigraphers developing the standard scale reduced the objectivity of the boundaries. In my opinion, it is necessary to take into account results from other fossil groups found in this region (Leven 2001; Leven & Bogoslovskaya 2006).

Data from other regions also do not give a reason to underestimate ammonoid successions in the sections. At present, there are sufficient results allowing the recognition of the boundary between the lower and middle series of the Permian based on ammonoids almost globally.

In northern regions of the globe, ammonoids play a leading role in dating rocks and in global correlation of the Roadian beds. Conodonts used in the lower-latitude regions are found here extremely rarely, whereas fusulinids are completely absent. At present, the Roadian ammonoids are known from the following high-latitude regions of Arctic Canada (the Assistance Formation and, possibly, the Sabine Bay Formation) and NE Russia: Kharaulakh Anticlinorium (Kharaulakh), northern Verkhoyansk Region (Chin Formation), western Verkhoyansk Region (Delendzha Formation), Ayan–Yuryakh Anticlinorium and central Okhotsk Massif (Khuren Formation), central Omolon Massif (Omolon Formation), Novaya Zemlya (Kocherga and Gerka Formations), and the Volga–Uralian Region (Kazanian).

Substantial progress in defining this boundary in Cisuralia has been made in the past few years, after Roadian ammonoids were found in the Kazanian beds of this region (Leonova *et al.* 2002, 2005; Leonova 2006, 2007, 2011*a*; Barskov *et al.* 2014). Apart from ammonoids, the Kazanian conodonts *Kamagnathus khalimbadzhai* and *K. volgensis* were extracted and examined (Chernykh 2003; Chernykh & Silantiev 2004), which supported the correlation of the Kazanian Stage of the East European scale (EES) with the Roadian of the International scale.

In the Palaeotethyan and North American realms, the boundary between the lower and middle series of the Permian is clearly traced based on the appearance of the representative Roadian ammonoid assemblage, containing in particular the genus *Paraceltites*, the initial taxon of the new order Ceratitida; in the Arctic and Australian basins, the Roadian assemblage is relatively impoverished taxonomically, altogether five or six genera, but most of these genera (*Sverdrupites*, *Daubichites*, *Altudoceras* and *Anuites*) appeared at this level and it is used as a good marker of a high-ranked boundary.

Roadian

Originally, the Roadian Stage was introduced by Furnish (1966) for the uppermost beds of the Lower Permian. The stratotype was chosen in the first Limestone Member in the Word Formation–Road Canyon Formation (Glass Mountain), which has a clearly defined faunal characterization, primarily based on ammonoids. Furnish (1973) noted that beds of the Roadian Stage are widespread in the USA (Texas, New Mexico and Idaho) and in Mexico. Ruzhencev (1976) approved Furnish's scheme and proposed his own, slightly modified version of the global Permian scale. He suggested that the Roadian could be considered as the first stage of the Late Permian. Later, this interpretation received wide acceptance, and the Roadian Stage was firmly placed at the base of the Upper Permian. For many years, almost all workers correlated the Roadian with the Ufimian of the General stratigraphic scale and the Kubergandian of the Tethyan Permian scale. After the tripartite subdivision of the Permian was accepted, the Roadian was used as the lower stage of the Middle Permian, or Guadalupian of the global standard. The correlation with the stages of other scales remained unchanged until recently. The fundamental change of these views coincided with the discovery of ammonoids in the Kazanian stage of the Volga–Urals Region, with which it is currently correlated.

The Roadian assemblage is represented by 62 genera, half of which were inherited from the early Permian. Thirty-two genera first appeared at this time. In all regions, the generic composition of the ammonoid assemblage is renewed, and prominently in the Palaeotethyan and North American realms. The most essential event at the Lower and Upper Permian boundary was the appearance of the order Ceratitida (genus *Paraceltites*). The stratigraphic ranges of 19 genera (*Daubichites*, *Sverdrupites*, *Pseudosverdrupites*, *Anuites*, *Spirolegoceras*, *Peritrochia*, *Glassoceras*, *Eohyattoceras*, *Texoceras*, *Tongluceras*, *Erinoceras*, *Shangraoceras*, *Liuzhouceras*, *Aulacogastrioceras*, *Nodogastrioceras*, *Chekiangoceras Pseudometalegoceras*, *Pericycloceras* and *Guiyangoceras*) were restricted to the Roadian. Fourteen genera (*Sangzhites*, *Elephantoceras*, *Lanceoloboceras*, *Epithalassoceras*, *Altudoceras*, *Roadoceras*, *Shengoceras*, *Paratongluceras*,

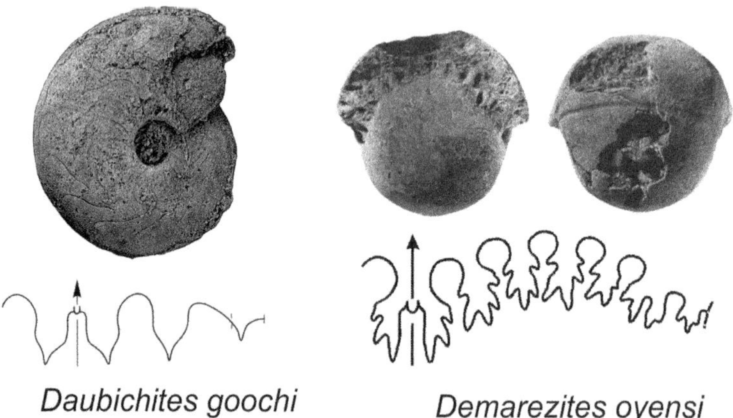

Fig. 6. Roadian zonal index species (sources of illustrations: Furnish *et al.* 2009; Glenister & Furnish 1961).

Demarezites, *Hyattoceras*, *Tauroceras*, *Palermites*, *Epadrianites* and *Paraceltites*) continued after the Roadian.

Roadian ammonoids are known from the following regions: from Arctic Canada (Assistance and van Hauen formations); from the Verkhoyansk Region (Delendzha Formation and its equivalents); from Novaya Zemlya (Kochergin Formation); from the Kirov region and Mari El Republic (East European Platform) (Nemda Formation); from Idaho and Wyoming of the USA (Phosphoria Formation); from Texas and New Mexico (Road Canyon Formation); from Mexico (Palo Quemado Formation); from South China (lower part of the Maokou-Kufeng Formation, Hutang Formation, Dongwuli Member); from North China (Inner Mongolia) (Shuang-Putang Formation); from Timor (Tae-Wei Beds); from Pamir and Afghanistan (Kubergandian) (Leonova 2011*b*); and from Western Australia, from deposits correlated to the Coolkilya Sandstone Formation.

Apart from the appearance of the order Ceratitida, the Roadian is also marked by the beginning of the most complex goniatitid family Cyclolobidae characterizing the Middle and Late Permian. The genera *Demarezites* and *Tongluceras* represent the first, most primitive stages in the evolution of the cyclolobids defining the Roadian Stage. The genera *Waagenoceras* and *Timorites* characterize the Wordian and Capitanian, respectively. *Cyclolobus* reached the maximum diversity in the Wuchiapingian, whereas *Changhsingoceras* completed the evolution of the family in the Changhsingian. Cyclolobids with their complex morphology and directed evolution are traditionally used for interregional correlations.

A common genus *Daubichites* (family Pseudogastrioceratidae) is similarly important for the correlation of the Roadian beds in almost all biogeographical regions.

For the Roadian Stage, I propose the *Daubichites goochi–Demarezites furnishi* Zone. The former species is known from the Arctic and Australian realms, and the comparable species are known from the Palaeotethyan and North American realms. All species of *Demarezites* are restricted to the Roadian, and are known from the North American and Palaeotethyan realms (Fig. 6).

Wordian

The Wordian Stage was named after the Word Formation in Texas, which corresponded to the classical *Waagenoceras* ammonoid zone (e.g. Böse 1919; Miller 1938; Miller & Furnish 1940; Furnish 1973; Glenister 1981). Almost all authors assigned the best-known ammonoid localities of SW regions of North America, Sicily (Sosio Beds) and Timor (Basleo Beds) to this stage. Rich localities of Wordian ammonoids are known in China. Until 2002, the Wordian Stage was correlated with the Kazanian Stage of the EES. At present, it is premature to draw conclusions about the correlation of the Wordian Stage with any stage of the EES, because no ammonoids younger than the Roadian are found in this region.

The Wordian Stage contains 54 ammonoid genera, 36 of which continued from the preceding assemblages. Eighteen genera appeared at that time: *Neogeoceras*, *Sosioceras*, *Anatsabites*, *Aristoceratoides*, *Epiglyphioceras*, *Adrianites*, *Doryceras*, *Sizilites*, *Clinolobus*, *Palermoceras*, *Neoaricoceras*, *Pseudagathiceras*, *Mongoloceras*, *Newellites*, *Mexicoceras*, *Jilingites*, *Cibolites* and *Waagenoceras* (index genus of this stage). The

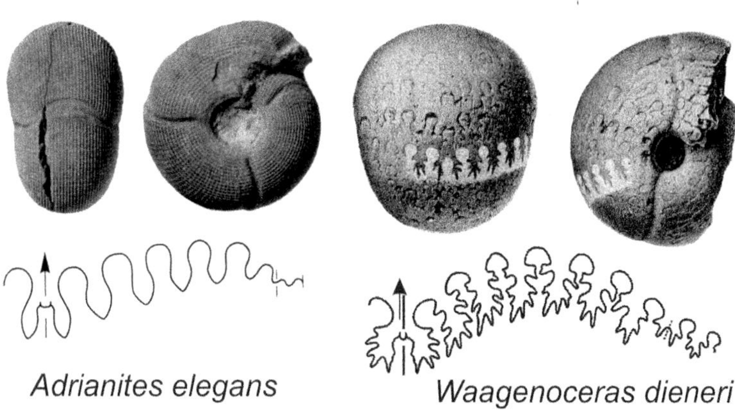

Fig. 7. Wordian zonal index species (sources of illustrations: Furnish *et al.* 2009; Miller & Furnish 1940).

stratigraphic ranges of most of these genera are restricted to the Wordian.

Wordian ammonoids are known from: Arctic Canada (Trold Fiord Formation) – *Neogeoceras*; Novaya Zemlya (Shadrov Formation) – *Neogeoceras*; Omolon Massif loose in the Chambin and Imtachan formations – *Paramexicoceras aldanicum* (no precise age is known) (Andrianov 1985; Kutygin *et al.* 2002); Texas (Word Formation and its equivalents); Mexico (Los Acros Formation); in western Canada (Cache Creek Formation); in Sicily (Sosio Beds); in Oman, Tunisia and Kurdistan (deposits correlated to the Sosio); in Timor (Basleo Beds); in South China (middle part of the Maokou Formation); in NE China, Jiling Province (Fanjiatun Formation); in Western China, in Tibet (Jiala Formation); and in the Crimea (Burnian Regional Stage).

The Wordian Stage was established as equivalent to the range of the *Waagenoceras* Genozone (Furnish 1973). Owing to the acceptance of a new international standard based on conodont phylogenies, the positions of the stage boundaries have been substantially changed, and their current position no longer agrees with the ammonoid data. The first appearance of the genus *Waagenoceras* is fixed considerably below the conodont-based lower boundary of the stage. While the Wordian Stage as understood by Furnish (1973) is readily recognizable in various regions of the world and has a comprehensive characterization, it is not well recognized in its new interpretation beyond Texas. Occurrences of the conodont species *Jinogondolella aserrata* outside the USA are only known from a few sections in South China: therefore, they cannot be used for correlations. If the Wordian Stage is considered as equal to the range of the *Jinogondolella aserrata* Zone, it is impossible to recognize an ammonoid association that would typify this particular interval. I accept the traditional interpretation of the Wordian Stage, which is characterized by a rich ammonoid assemblage (Furnish 1973). This interpretation allows the correlation of the Wordian deposits in different regions, independent of the presence of the conodont marker species *Jinogondolella aserrata*. In this paper, I propose the *Adrianites elegans*–*Waagenoceras dieneri* Zone. The former species is known from various regions of the Palaeotethys, and the second from the American Realm, whereas morphologically similar sutures have been indicated from several localities of the Palaeotethyan Realm (Fig. 7).

Capitanian

The Capitanian Stage was also recognized in southwestern regions of North America and corresponds to the *Timorites* Zone (e.g. Miller 1938; Miller & Furnish 1940; Glenister 1981). Classical localities of Capitanian ammonoids are in Texas and Mexico. New data on the Capitanian ammonoid occurrences in Japan, China and in the Far East were published in the second half of the twentieth century. Until now, there is no unequivocal answer to the question of the stage affinity of Amarassian. This unit is based on a rich ammonoid assemblage from the Upper Permian of Timor. Many authors (e.g. Furnish 1973; Bogoslovskaya 1990) recognized it as a separate stage. Furnish (1973, pp. 539–540) justifiably noted:

> [T]hose sediments have no definable stratigraphic relationships (boundaries with older and younger beds). Practically, the recognition of this stage in the type area, and in other portions of world, rests on a faunal basis; there are specific identities as well as evolutionary positions in a series of phylogenetic lineages. Many of these ammonoids can be regarded as diagnostic time indices at the specific level; for example, *Timorites* and *Cyclolobus* are intermediate between those from the

type Guadalupian in a continental western province (southwestern United States and northern Mexico) and those from Dzhulfian of the Tethys (Indian Ocean and Himalayas). Collectively, the ammonoid assemblage includes about a dozen lineages in which ancestral forms and/or descendants have been recognized; only one of these (Sundaitinae) is restricted to the Southeastern Province (Timor). Further, only a single important genus, the unique Araxoceratidae [genus *Eoaraxoceras*–TBL], is lacking from this faunal list on Timor.

The Amarassian assemblage in fact represents a separate stage in the ammonoid evolution readily recognizable by the evolutionary level and taxonomic diversity of forms composing it, but also because of the local distribution and unequivocal interpretation of the stratigraphic level of the host beds represented by isolated small exposure; its position in the general scheme remains unresolved. It seems more logical to assign it to the uppermost Capitanian, but, based on the appearance of the first representative of araxoceratids, it is considered by many authors (e.g. Spinosa *et al.* 1970; Spinosa & Glenister 2000) as the base of the Dzhulfian Stage (in modern interpretation, as the base of the Upper Permian). Spinosa & Glenister (2000) indicated the precise locality of *Eoaraxoceras* in the uppermost part of La Colorada Formation in Mexico. Taxa accompanying this species and those in the much lower horizons of the sections include *Timorites schucherti*, *Stacheoceras toumanskayae* and *Kingoceras kingi*. The species identified as '*Neocrimites sp.*' (apparently *Metacrimites*) and *Xenodiscus wanneri* appear in the middle of the upper part of this formation. The authors of that paper (Spinosa & Glenister 2000) placed the boundary between stages (and series of the Permian) tentatively within the Upper La Colorada Formation. In my opinion, this is not the best decision. Spinosa & Glenister (2000) compared the Mexican *Eoaraxoceras* with the Iranian specimens and showed that its first appearance in Abadeh is fixed at the Capitanian–Dzhulfian boundary. In the higher part of the Beds with '*Araxoceras*', around 10 m thick in this section, *Eoaraxoceras* occurs in association with other, but not primitive, araxoceratids (*Araxoceras*, *Vescotoceras*) and *Cyclolobus ruzhencevi*. Some other authors (Spinosa & Glenister 2000) assign *Cyclolobus ruzhencevi* to *Timorites*. The fact that the latter species belongs to the genus *Cyclolobus*, rather than to *Krafftoceras* or *Timorites*, is discussed by Leonova (2010), based on a special study of the holotype of this species. This suggests that the Beds with '*Araxoceras*' in Abadeh, as well as in Transcaucasia, contain a younger ammonoid assemblage than the Amarassian assemblage in Timor or Mexico. The taxonomic composition of the Amarassian and corresponding assemblages, their morphology and ecological structure are much closer to those of the Middle Permian than to the Late Permian evolutionary stage, in which the dominant role belonged to ceratitids. In any case, it is evident that the Amarassian ammonoid assemblage is borderline in age between the Middle and Upper Permian.

The Capitanian Stage contains 32 genera, of which 15 continued from the older beds. These are *Eumedlicottia*, *Neogeoceras*, *Neoaganides*, *Altudoceras*, *Roadoceras*, *Jilingites*, *Stacheoceras*, *Hyattoceras*, *Waagenoceras*, *Shengoceras*, *Mongoloceras*, *Metacrimites*, *Epadrianites*, *Paraceltites* and *Cibolites*.

Seventeen genera appeared at this level. Typically, Capitanian taxa include the following 11 genera: *Difuntites*, *Sundaites*, *Nodosageceras*, *Shouchangoceras*, *Strigogoniatites*, *Timorites*, *Angrenoceras*, *Neostacheoceras*, *Epitauroceras*, *Doulingoceras* and *Nielsenoceras*. *Eoraxoceras* is known only from the Capitanian and Wuchiapingian boundary beds. Five genera that appeared in the Capitanian, and more precisely in its upper part, continued to the Late Permian. These are *Syrdenites*, *Episageceras*, *Cyclolobus*, *Xenodiscus* and *Kingoceras*.

The Capitanian ammonoids are known from Texas (Capitan Formation and its equivalents) and from Mexico (La Difunta Formation). Synchronous assemblages are studied in South China from the upper part of the Maokou Formation, in the Far East (Chandalazian Regional Stage), in Japan (Ochiai Formation), in Timor (Amarassi Beds) and, probably, in Tibet (Langcuo Formation).

More refined zonal Tethyan schemes have been proposed for the Midian Stage, which approximately corresponds to the Capitanian, and for the Upper Permian stages (Zakharov & Pavlov 1986). The zonal scheme of the Midian Stage appears as follows (from bottom to top): *Metacrimites kropatchevae*, *Stacheoceras orientale*, *Xenodiscus subcarbonarius* and *Cyclolobus kiselevae* zones. The essential shortcoming of this scheme is the narrow geographical distribution of the suggested index species, as they reflect the zonal succession for the Far East region only.

For nearly 100 years, Capitanian deposits have been considered to correspond to the *Timorites* Zone. Certainly, representatives of this genus are most suitable as indexes of this stage. In this work, the major (larger) portion of this stage is the *Timorites schucherti*–*Cibolites uddeni* Zone. The former species is known from the Capitan Formation and the corresponding species are recorded from all regions where Capitanian deposits are found. The latter species is also known from the stratotype region, and also from South China. The presence of ceratitids in the upper part of the Middle Permian

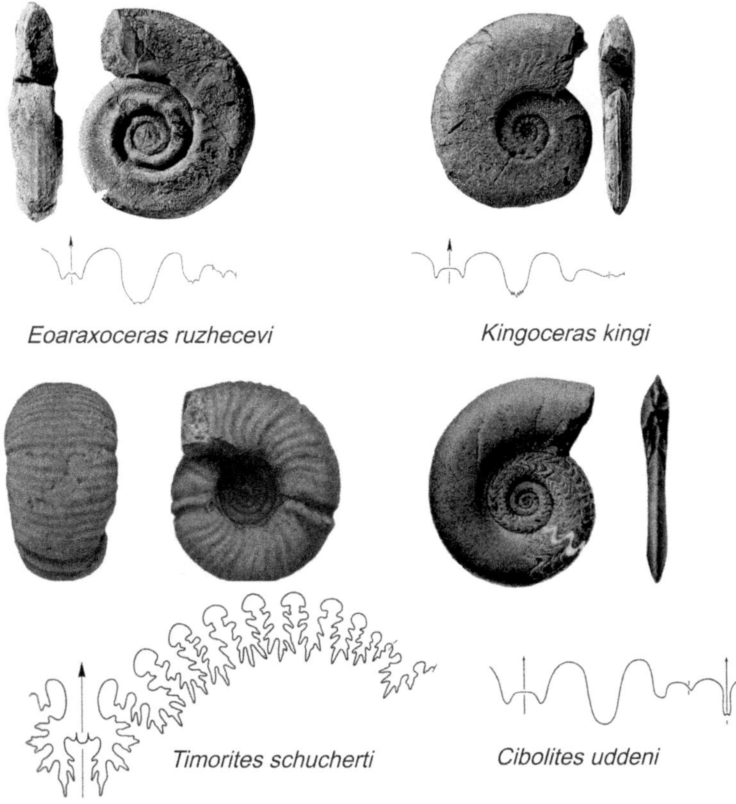

Fig. 8. Capitanian zonal index species (sources of illustrations: Furnish *et al.* 2009; Miller & Furnish 1940; Spinosa & Glenister 2000).

becomes very noticeable and this should be recorded in the accepted schemes. Apparently, another zone in the uppermost part of the stage should be added to this scheme *Eoaraxoceras ruzhencevi–Kingoceras kingi*. These species are known from the American and Palaeotethyan realms (Fig. 8).

Lopingian

In accordance with the international standard of the Permian System, the Upper Permian, or Lopingian, with stratotypes in South China, is composed of the Wuchiapingian and Changhsingian stages (Jin *et al.* 1997). Based on ammonoids, they are relatively precisely correlated to the Dzhulfian and Dorashamian stages established in Transcaucasia. Ruzhencev (1959, 1962, 1963, 1965) carefully studied the Dzhulfian and Dorashamian faunas, and concluded that these represent the terminal stage in the evolution of the Permian ammonoids. Recent studies in Iran (Korn *et al.* 2016) support this conclusion. The acceptance of the Chinese instead of the Transcaucasian stages as the international standard was mainly due to the difficulty in accessing the Transcaucasian stratotypes and localities because of the unstable political situation, rather than for scientific reasons.

Although ammonoids in the uppermost Permian South of China are relatively numerous, these units are recognized based on conodonts. The Meishan section, accepted as the stratotype of the Permian–Triassic, does not contain an uninterrupted succession of ammonoid assemblages. Other sections of this region that were also considered as candidates for a stratotype also do not contain detailed ammonoid data. As a specialist on this fossil group, it appears more reasonable to me to base correlations on the well-studied assemblages of Transcaucasia and Iran, and to date successions as the Dzhulfian and Dorashamian, which correlate with the Chinese Wuchiapingian and Changhsingian stages, respectively. As shown by international teams (Ghaderi *et al.* 2014; Korn *et al.* 2016), these deposits have a high potential for establishing a detailed and reliable succession of ammonoid assemblages of the Late Permian.

Upper Permian ammonoid localities are known in a few regions of the globe. They were first established in Kashmir, the Central Himalayans and in western Pakistan (the Salt Range). The most representative assemblages are known from Transcaucasia, Iran, South China and the Far East. There are also occasional occurrences of Late Permian ammonoids in Madagascar and in East Greenland, showing that the Palaeotethys was the main region for distribution of ammonoids in the Late Permian epoch.

Wuchiapingian

Altogether in the Wuchiapingian Stage there are 26 genera of ammonoids, of these 14 appeared at this level – pseudogastrioceratids: *Pseudogastrioceras*, *Metagastrioceras* and *Retiogastrioceras*; and ceratitids: *Araxoceras*, *Anderssonoceras*, *Anfuceras*, *Avushoceras*, *Dzhulfoceras*, *Julfotoceras*, *Lenticoceltites*, *Abadehceras*, *Pericarinoceras*, *Prototoceras* and *Pseudotoceras*. In addition, another 12 genera continue from the earlier beds – prolecanitids: *Eumedlicottia*, *Neogeoceras*, *Syrdenites* and *Episageceras*; gonititids: *Cyclolobus*, *Stacheoceras* and *Epadrianites*; ceratitids: *Cibolites*, *Eoaraxoceras*, *Kingoceras* and *Xenodiscus*; and a single tornoceratid: *Neoaganides*.

Wuchiapingian (Dzhulfian) ammonoids are known from Transcaucasia and Iran from the Dzhulfian Stage, in South China, from the Laoshan Shale, Loping, Wuchiaping and Fengtien; in the Far East, in the lower part Lyudyanzian Regional Stage; and in Japan, in the lower part of the Toyoma Formation.

In addition, a few ammonoids of Dzhulfian or Wuchiapingian age are known from several other localities. In Greenland, in Madagascar (Ambilobe Beds), in the Central Himalayas (Kuling Shale), Kashmir (Zewan Formation) and western Pakistan (Chideru Formation).

Ceratitids of the suborder Otoceratina became widespread in the Wuchiapingian. Araxoceratids were particularly abundant, and the first representatives appeared in America at the Middle–Late Permian boundary, and their main distribution is connected with the Tethyan Realm. Anderssonoceratids restricted to the South China Basin were also a characteristic group. Apart from ceratitids, there were several genera of goniatitids and prolecanitids, with retrogressive trends in evolution shown as a general impoverishment of their taxonomic composition and morphological simplification: for example, in the genus *Cyclolobus* (Leonova 2010).

In the Wuchiapingian (Dzhulfian) Stage of Palaeotethys, two zones are recognized: *Araxoceras latissimum* (for the lower part); and *Araxoceras ventrosulcatum* (for the upper part) (Kotlyar *et al.* 1983; Zakharov & Pavlov 1986; Shevyrev 1986) (Fig. 9).

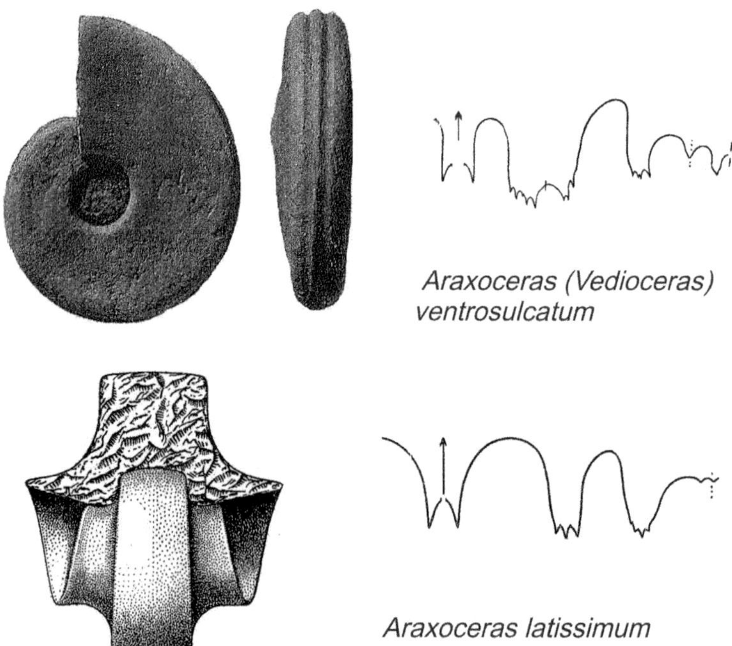

Fig. 9. Wuchiapingian zonal index species (sources of illustrations: Ruzhencev 1959, 1963).

Changhsingian

In general, the taxonomic diversity in the Changhsingian remained the same as in the preceding stage, 25 genera. Seven genera continued from the Wuchiapingian – prolecanitids: *Eumedlicottia* and *Neogeoceras*; tornoceratids: *Neoaganides*; goniatitids: *Pseudogastrioceras*, *Retiogastrioceras* and *Stacheoceras*; and ceratitids: *Xenodiscus* and *Julfotoceras*. Eighteen genera are restricted to the Changhsingian – gonititid: *Changhsingoceras*; and the ceratitids: *Abichites*, *Phisonites*, *Iranites*, *Shevyrevites*, *Dzhulfites*, *Paratirolites*, *Stoyanowites*, *Alibashites*, *Arasella*, *Pseudotirolites*, *Pentagonoceras*, *Pleuronodoceras*, *Tapashanites*, *Rotodiscoceras*, *Schizoloboceras*, *Meitianoceras* and *Penglaites*.

The area of distribution of Changhsingian ammonoids is even more strongly restricted; they are known only in the Palaeotethyan Realm.

Changhsingian (Dorashamian) ammonoids are known: in Transcaucasia, in the Dorashamian Stage and in Iran in the upper part of the Hambast Formation; in South China, in the Dalong and Changhsing, and the Meishan, assigned to the Changhsingian Stage; in the Far East, in the upper part of the Lyudzyanian Regional Stage of South Primorye; and in Japan, from the upper part of the Toyoma Formation.

The Changhsingian Stage is characterized by diverse assemblages of ceratitids of the families Xenodiscidae (*Xenodiscus*, *Phisonites*, *Iranites*, *Shevyrevites*, *Tapashanites* and *Penglaites*), Dzhulfitidae (*Dzhulfites*, *Paratirolites* and *Abichites*) and Pseudotirolitidae (*Pseudotirolites*, *Pleuronodoceras*, *Rotodiscoceras*, *Pentagonoceras* and *Schizoloboceras*), the development of which occurred almost exclusively in that age. Apart from the ceratitids at the very end of the Permian, there were the latest Medlicottiidae and Pseudogastrioceratidae (Goniatitina), and the last Cyclolobina *Changhsingoceras*, which appeared as a result of retrogressive development of the extremely advanced genus *Cyclolobus*. Several schemes of the zonal subdivision of the Changhsingian (Dorashamian) stage are proposed. For the Transcaucasia–Iran region, Zakharov (Kotlyar *et al.* 1983; Zakharov & Pavlov 1986) proposed six ammonoid zones: *Phisonites triangularis*, *Iranites transcaucasicus*, *Dzhulfites spinosus*, *Shevyrevites shevyrevi*, *Paratirolites kittli* and *Pleuronodoceras occidentale*. Based on a detailed bed-by-bed collection near Dzhulfa (NW Iran), Korn *et al.* (2016) subdivided the *Paratirolites* Beds of Shevyrev (1965), which are 4–5 m thick (Zakharov's *Paratirolites kittli* Zone), into eight zones: *Dzhulfites zalensis*, *Paratirolites trapezoidalis*, *Paratirolites kittli*, *Stoyanowites dieneri*, *Alibashites mojsisovicsi*, *Abichites abichi*, *Abichites stoyanowi* and *Arasella minuta*. The former assumption that the succession in this region lacks the uppermost Permian beds was shown to be incorrect. Shevyrev (1999), while describing the Transcaucasian section, noted that immediately above the beds assigned to the *Pleuronodoceras occidentale* Zone, which he considered to be definitely Permian, the *Claraia* Beds contained the conodont species *Hindeodus parvus*, which certainly indicates a Triassic age. A similar conclusion can be derived from the recent studies of international teams (Ghaderi *et al.* 2014; Korn *et al.* 2016). In South China, the following zonal ammonoid-based scheme was proposed: *Iranites–Phisonites*, *Paratirolites–Shevyrevites* and *Pseudostephanites–Tapashanites* in the lower substage; and *Pseudotirolites–Pleuronodoceras*, *Rotodiscoceras* and an unnamed zone in the upper (Zhao *et al.* 1978). The conodont species *Hindeodus parvus* marks the boundary between the systems in the South China stratotype. Unfortunately, the situation with ammonoids in this section is not entirely clear. A mixed ammonoid assemblage is recorded from two clay beds (bed numbers 25 and 26), 'intermediate' between the Permian and Triassic (Wang 1995). In the opinion of recognized experts on the Triassic (Dagys & Dagys 1988; Shevyrev 2000), they are representatives of Changhsingian biota. The Triassic taxa recorded from these beds are probably based on erroneous identifications owing to the poor preservation of the ammonites, the shells of which are deformed and mostly do not show visible sutures (Shevyrev 2000, p. 64). Yang (1985) considered the recognition of two ammonoid zones, *Tapashanites–Shevyrevites* and *Pseudotirolites–Rotodiscoceras* (Dalong Formation), to be reasonable.

I propose three zones for the Changhsingian: *Phisonites triangularis*, *Paratirolites kittli* and *Pleuronodoceras multinodosum–Rotodiscoceras asiaticum* (Fig. 10).

I accept the traditional view on the position of the Permian–Triassic boundary at the base of the *Otoceras* Zone, as I consider it to be the one best substantiated. However, there are also a number of difficulties with this. Shevyrev (1999, 2000) justifiably commented that there was no section known to show the complete succession of the marine beds at the Permian–Triassic boundary. In sections showing the uppermost Permian (these are Tethyan sections), the basal Triassic is absent. Conversely, sections with the lowermost Triassic exposed (Boreal sections) have a gap in the Permian, so in all cases part of the succession is missing.

Conclusion

As noted above, the officially accepted international Permian scale is based on conodont data.

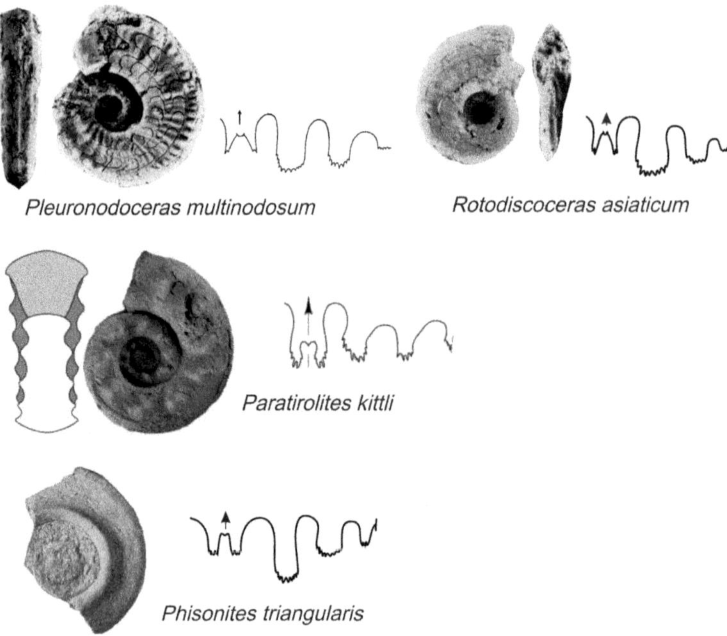

Fig. 10. Changhsingian zonal index species (sources of illustrations: Shevyrev 1965; Zhao *et al.* 1978).

Ammonoids are a significant, but auxiliary, taxon. In my opinion, this situation results primarily from the activity and persistence of conodont workers. To speak objectively, conodonts are no better or worse than ammonoids for correlations. The distribution of conodonts in different regions of the world is similarly uneven and they are also facies-dependent, and in addition their taxonomy and identifications are not resolved. It is evident that ammonoids have remained as one of the major tools for subdivision and correlation of Permian sedimentary successions.

What are the possibilities of ammonoids? On a global scale, all nine stages of the Permian are relatively completely characterized by this group, which are successfully used for biostratigraphy and interregional correlations. For each Permian stage, there is a characteristic ammonoid assemblage taking into account the phylogeny and position of taxa in stratigraphic successions. The main difficulty of making this a workable zonal scale is that the ammonoid localities in Permian sections are relatively rare, and sections with a well-represented change in species and genera, characterizing long intervals of geological time, are extremely rare. Assemblage zones are used in the Permian biostratigraphy. Ammonoids characterizing rock associations usually define their affinity to a stage or substage.

Unlike the Carboniferous stratigraphic scale, which includes a detailed succession of ammonoid genozones and, for some intervals, species zones, no substantiated or widely accepted zonation has been developed for the Permian System. The problem of the Permian ammonoid zonation has been discussed repeatedly in the literature. In recent decades, as a result of an increased volume of information, various authors have proposed several versions of the zonal subdivisions of the Permian (e.g. Jin *et al.* 1994, 1997; Bogoslovskaya *et al.* 1999). The first of the above scales proposed by an international team (Jin *et al.* 1994) was generally ill conceived (e.g. the Kubergandian Stage in that scheme preceded the Roadian, rather than being at the same stratigraphic level). After a short while, the same authors (Jin *et al.* 1997) proposed another version of the zonal scale, which was accepted as an international standard (Fig. 1). The main shortcoming of this version was the very narrow geographical range of taxa proposed as zonal indexes, which complicated their use for correlations. For example, for the Artinskian Stage, they used *Aktubinskia* (a single species of this genus is found in one locality of the Aktyubinsk Region of the South Urals). Another index species *Uraloceras fedorowi* is found on the western slope of the Urals; in addition, it was recognized from the Artinskian of Arctic Canada, but an increasing number of authors now doubt in the conspecificity of its representatives even within the Urals (Bogoslovskaya & Boiko 2002; Boiko 2009, 2010;

Borisenkov 2010). Similarly difficult is the use of the Kungurian species *Propinacoceras busterense*, which was described based on a poorly preserved specimen and unknown beyond Pamir (Fig. 1). The second scale (Bogoslovskaya *et al.* 1999) has also had a number of deficiencies that complicate correlations (Fig. 1). For example, the genera *Svetlanoceras* and *Juresanites* occur both in the Asselian and Sakmarian stages, which makes their use as a zonal index somewhat illogical. The geographical distribution of genera proposed as zonal indexes is extremely limited (only two specimens of *Protopopanoceras* are known from one locality for the Asselian; for the Artinskian Stage it is *Aktubinskia* – see the 'Artinskian' subsection earlier in this paper).

I proposed another version of the zonal subdivision of the Permian (Leonova 2011*a*), a somewhat modified version of which is accepted in this paper (Fig. 1). The main principles of the recognition of the species index zones include: a short interval of existence (not longer than one phase); a relatively wide (according to the existing data) geographical range; and a certain level of evolutionary development that allows a comparison of this species with other species of the genus (i.e. characters), and so allowing recognition of the relative isochronism of the host assemblages in different biogeographical regions. Usually for one interval of geological time, we propose using two index species, simultaneously existing in different biogeographical regions.

I thank Dr. Masayuki Ehiro (Tohoku University Museum, Japan) and Dr. Zhou Zuren (Nanjing Institute of Geology and Palaeontology, China) for critically reviewing the manuscript and providing helpful comments. I am also grateful to Dr. Svetlana Nikolaeva (Paleontological Institute RAS, Russia) for helping with translation of this paper. The study is supported by the Program 28 of the Presidium of the Russian Academy of Sciences 'Evolution of the Organic World and Planetary Processes'.

References

ANDRIANOV, V.N. 1985. *Permian and some Carboniferous Ammonoids of Northeastern Asia*. Nauka, Novosibirsk.

ARCHBOLD, N.W. & DICKINS, J.M. 1991. *Australian Phanerozoic Timescales. 6. A Standard for the Permian System in Australia*. Bureau of Mineral Resources Australia, Record, **1989**, (36).

BARSKOV, I.S., LEONOVA, T.B. & SHILOVSKY, O.P. 2014. Middle permian Cephalopods of the Volga–Ural Region. *Paleontological Zhurnal*, **48**, 1331–1414.

BOGOSLOVSKAYA, M.F. 1976. Kungurian ammonoids of Middle Cisuralia. *Paleontologicheskii Zhurnal*, **4**, 43–50.

BOGOSLOVSKAYA, M.F. 1984. Ammonoids. *In*: KOTLYAR, G.V. & STEPANOV, D.L. (eds) *Osnovnye cherty stratigrafii permskoi sistemy*. Nedra, Leningrad, 248–257.

BOGOSLOVSKAYA, M.F. 1990. Major trends of evolution and classification of Late Paleozoic ammonoids Marathonitaceae and Cyclolobaceae. *Iskopaemye tsefalopody, Trudy Paleontologicheskogo Instituta, Rossiiskoi Akademii nauk*, **243**, 70–86.

BOGOSLOVSKAYA, M.F. & BOIKO, M.S. 2002. Development and distribution of the Early Permian genus *Uraloceras* (Ammonoidea). *Paleontologicheskii Zhurnal*, **6**, 31–37.

BOGOSLOVSKAYA, M.F. & POPOV, A.V. 1986. New ammonoid species from the Carbon-Permian boundary deposits of the Southern Urals. *In*: PAPULOV, G.N. (ed.) *Pogranichnye otlozheniya karbona i permi Urala, Priuraliya i Srednei Azii*. Nauka, Moscow, 125–129 [in Russian].

BOGOSLOVSKAYA, M.F., LEONOVA, T.B. & SHKOLIN, A.A. 1995. The Carboniferous–Permian boundary and ammonoids from the Aidaralash section, Southern Urals. *Journal of Paleontology*, **69**, 288–301.

BOGOSLOVSKAYA, M.F., KUSINA, L.F. & LEONOVA, T.B. 1999. Classification and distribution of Late Paleozoic ammonoids. *In*: ROZANOV, A.YU. & SHEVYREV, A.A. (eds) *Fossil Cephalopods: Recent Advances in Their Study*. Paleontological Institute, Russian Academy of Sciences, Moscow, 89–124 [in Russian].

BOIKO, M.S. 2009. Ammonoids from the boundary deposits of the Artinskian and Kungurian stages in the Mechetlino section (Bashkortostan). *In*: LEONOVA, T.B., BARSKOV, I.S. & MITTA, V.V. (eds) *Sovremennye problemy izucheniya golovonogikh mollyuskov. Morfologiya, sistematika, evolyutsiya, ekologiya i biostratigrafiya*. PIN RAN, Moscow, 99–101.

BOIKO, M.S. 2010. The evolution of the Early Permian family Paragastrioceratidae (Ammonoidea) in the Urals. *Paleontologicheskii Zhurnal*, **3**, 31–37.

BORISENKOV, K.V. 2010. New records of Permian ammonoids on the Kobylka River (North Urals). *Paleontologicheskii Zhurnal*, **3**, 18–22.

BÖSE, E. 1919. The Permo-Carboniferous ammonoids of the Glass Mountains, west Texas, and their stratigraphical significance. *University of Texas Bulletin*, **1762**, 1–241.

CHERNYKH, V.V. 2003. Global correlation of the Artinskian and Kungurian Stages based on conodonts. *Litosfera*, **1**, 64–71.

CHERNYKH, V.V. & SILANTIEV, V.V. 2004. *Conodonts of the Kazanian stage of Middle Povolzhye. Doklady Vaserossiiskogo soveshcjaniya 'Struktura i status Vostochno-Evropeiskoi stratigraficheskoi shkaly Permskoi sistemy, usovershenstvovanie yarusnogo raschleneniya permskoi sistemy obshchei stratigraficheskoi shkaly*. Kazanskii Universitet, Kazan, 83–86.

CHUVASHOV, B.I. & CHERNYKH, V.V. 2004. Kungurian stage of the Cis-Uralian division: biostratigraphy and correlative potential. *Permophiles*, **44**, 12.

CHUVASHOV, B.I., CHERNYKH, V.V. & BOGOSLOVSKAYA, M.F. 2002. Biostratigraphic characterization of the stratotypes of the Lower Permian stages. *Stratigrafiya. Geologicheskaya Korrelyatsiya*, **10**, 3–19.

DAGYS, A.S. & DAGYS, A.A. 1988. Biostratigraphy of the lowermost Triassic and the boundary between Paleozoic and Mesozoic. *Memorie della Società Geologica Italiana*, **34**, 313–320.

Davidov, V.I., Glenister, B.F. *et al.* 1998. Proposal of Aidaralash as Global Stratotype Section and Point (GSSP) for base of the Permian System. *Episodes*, **21**, 13–18.

Furnish, W.M. 1966. Ammonoids of the Upper Permian Cyclolobus-zone. *Neues Jahrbuch für Geologie und Paläontologie Abhandlungen*, **125**, 265–296.

Furnish, W.M. 1973. Permian stage names. *In*: Logan, A. & Hills, L.V. (eds) *Permian and Triassic Systems and Their Mutual Boundary*. Canadian Society of Petroleum Geologists, Memoirs, **2**, 522–548.

Furnish, W.M. & Glenister, B.F. 1971. Permian Gonioloboceratidae (Ammonoidea). *Smithsonian Contribution to Paleobiology*, **3**, 301–312.

Furnish, W.M., Glenister, B.F., Kullmann, J. & Zhou, Z. 2009. *Treatise on Invertebrate Paleontology. Part L, Mollusca 4 (Revised), Volume 2: Carboniferous and Permian Ammonoidea (Goniatitida and Prolecanitida).* University of Kansas Press, Lawrence, KS.

Ghaderi, A., Leda, L., Schobben, M., Korn, D. & Ashouri, A.R. 2014. High-resolution stratigraphy of the Changhsingian (Late Permian) successions of NW Iran and the Transcaucasus based on lithological features, conodonts and ammonoids. *Fossil Record*, **17**, 41–57.

Glenister, B.F. 1981. Permian ammonoid 'zones'. *In*: House, M.R. & Senior, J.R. (eds) *The Ammonoidea*. Academic Press, New York, 386–396.

Glenister, B.F. & Furnish, W.M. 1961. The Permian ammonoids of Australia. *Journal of Paleontology*, **35**, 673–736.

Glenister, B.F. & Furnish, W.M. 1981. Permian ammonoids. *In*: House, M.R. & Senior, J.R. (eds) *The Ammonoidea*. Academic Press, New York, 49–64.

Glenister, B.F., Glenister, L.M. & Skwarco, S.K. 1983. Lower Permian Cephalopods from Western Irian Java, Indonesia. *Geological Research and Development Centre, Paleonotology Series*, **4**, 74–85.

Glenister, B.F., Backer, C., Furnish, W.M. & Dickins, J. 1990*a*. Late Permian ammonoid cephalopod, Cyclolobus from Western Australia. *Journal of Paleontology*, **64**, 399–402.

Glenister, B.F., Backer, C., Furnish, W.M. & Thomas, G.A. 1990*b*. Additional Early Permian ammonoid cephalopods from Western Australia. *Journal of Paleontology*, **64**, 392–399.

Jin, Y.G., Glenister, B.F., Kotlyar, G.V. & Sheng, J.Zh. 1994. An Operational scheme of Permian chronostratigraphy. *Paleoworld*, **4**, 1–13.

Jin, Y.G., Wardlaw, B.R., Glenister, B.F. & Kotlyar, G.V. 1997. Permian chronostratigraphic subdivisions. *Episodes*, **20**, 10–15.

Karpinsky, A.P. 1874. Geological studies in the Oregburg Region. *Zapiski Mieralogicheskogo obshchestva*, **2**, 212–310.

Karpinsky, A.P. 1890. *Ob ammoneyakh artinskogo yarusa i o nekotorykh skhodnykh s nimi kamennougolnykh formakh*. [On Ammonoidea of the Artinskian Stage and similar Carboniferous forms]. *Zapiski Mineralogicheskogo Obshchestva*, **2**, 1–192 [in Russian].

Korn, D., Ghaderi, A., Leda, L., Schobbena, M. & Ashouri, A.R. 2016. The ammonoids from the Late Permian *Paratirolites* Limestone of Julfa (East Azerbaijan, Iran). *Journal of Systematic Palaeontology*, **14**, 841–890.

Kotlyar, G.V., Zakharov, Yu.D. *et al.* 1983. *Pozdnepermskii etap evolyutsii organicheskogo mira, Dzhulfinskii i Dorashamskii yarusy SSSR [Evolution of the latest Permian biota, Dzhulfian and Dorashamian stages of the USSR]*. Nauke SSSR, Leningradskoe Otdelennie, Leningrad [in Russian].

Kozur, H. 1994. Permian pelagic and shallow water conodont zonation. *Permophiles*, **24**, 16–20.

Krotov, P.I. 1885. *Artinskian Stage. Geologica–Palaeontological Monograph of the Artinskian Sandstone*. Trudy obshchestva estestvoispytatelei pri Kazanskom universitete, **13**.

Kutygin, R.V., Budnikov, I.V., Byakov, A.S. & Klets, A.G. 2002. Beds with ammonoids of the Permian system of the Verkhoyansk Region. *Otechestvennaya Geologiya*, **4**, 66–71.

Lambert, L.L., Lehrmann, D.J. & Harris, M.T. 2000. Correlation of the Road Canyon and Cutoff Formations, West Texas, and its revelance to establishing an international Middle Permian (Guadalupian) series. *In*: Wardlow, B.R., Grant, R.E. & Rohr, D.M. (eds) *The Guadalupian Symposium*. Smithsonian Contributions to the Earth Sciences, **32**. Smithsonian Institution Press, Washington, DC, 153–183.

Leonova, T.B. 2006. Roadian ammonoids in northern regions of the globe. *In*: Rozhnov, S.V. (ed.) *Evolutsiya biosphery i bioraznoobraziya [Evolution of the Biosphere and Biodiversity: A Collection of Papers]*. KMK, Moscow, 540–551.

Leonova, T.B. 2007. Correlation of the Kazanian of the Volga–Urals with the Roadian of the global Permian scale. *Paleoworld*, **16**, 246–253.

Leonova, T.B. 2010. Revision of the Permian ammonoid family Cyclolobidae. *Paleontologicheskii Zhurnal*, **3**, 23–30.

Leonova, T.B. 2011*a*. Permian ammonoids: biostratigraphic, biogeographical and ecological analysis. *Paleontological Journal*, **45**, 1206–1312.

Leonova, T.B. 2011*b*. Kubergandian (Middle Permian) ammonoid assemblage of Southeastern Pamir (Tajikistan). Byulletene Moskovskogo Obshchestva Isopytatelei Prirody. *Otdel Geologichesky*, **86**, 21–31.

Leonova, T.B. & Dmitriev, V.Yu. 1989. *Early Permian Ammonoids of Southeastern Pamir*. Trudy Paleontologicheskogo Instituta Akademii Nauk SSSR, Moscow, **235**, 1–198 [in Russian].

Leonova, T.B., Esaulova, N.K. & Shilovsky, O.P. 2002. The first record of Kazanian ammonoids in the Volga–Urals Region. *Doklady Rossiiskoi Akademii Nauk*, **383**, 509–511.

Leonova, T.B., Kutygin, R.B. & Shilovsky, O.P. 2005. New data on the composition and evolution of the Permian Superfamily Popanocerataceae Hyatt, 1900 (Ammonoidea). *Paleontologicheskii Zhurnal*, **5**, 20–29.

Leven, E.Ya. 2001. On the possibility of using the global stage scale of the Permian system in the Tethys. *Stratigrafiya. Geologicheskaya Korrelyatsiya*, **9**, 15–29.

Leven, E.Ya. & Bogoslovskaya, M.F. 2006. Roadian Stage of the Permian and problems of its global correlation. *Stratigrafiya. Geologicheskaya Korrelyatsiya*, **14**, 67–78.

LEVEN, E.YA., LEONOVA, T.B. & DMITRIEV, V.YU. 1992. *Permian System of the Darvaz-Zaalai Zone of Pamir: Fusulinds, Ammonoids, Stratigraphy*. Nauka, Moscow [*Trudy Paleontologicheskogo Instituta, Rossiiskoi Akademii Nauk*, **253**].

LEVEN, E.YA., BOGOSLOVSKAYA, M.F., GANELIN, V.G., GRUNT, T.A., LEONOVA, T.B. & REIMERS, A.N. 1996. The restructuring of the marine biota in the middle of the Early Permian epoch. *Stratigrafiya. Geologicheskaya korrelyatsiya*, **4**, 61–70.

MILLER, A.K. 1938. Comparison of Permian ammonoid zones of Soviet Russia with those of North America. *Bulletin of the American Association of Petroleum Geologists*, **22**, 1014–1019.

MILLER, A.K. & FURNISH, W.M. 1940. *Permian Ammonoids of the Guadalupe Mountain Region and Adjacent Areas*. Geological Society of America, Special Papers, **26**.

MURCHISON, R., VERNEUIL, E. & KEYSERLING, A. 1845. *Geology de la Russie d'Europe et des montagnes de l'Oural. Vol. II. Paleontologie*. John Murray, Paris.

NASSICHUK, W.W. & HENDERSON, CH.M. 1986. Lower Permian (Asselian) ammonoids and conodonts from the Belcher Channel Formation, southwestern Ellesmere Island. *Current Research, Part B. Geological Survey of Canada*, **86–81B**, 411–416.

NASSICHUK, W.W. & SPINOSA, C. 1972. Early Permian (Asselian) ammonoids from the Hare Fiord Formation, northern Ellesmere Island. *Journal of Paleontology*, **46**, 536–544.

PLUMMER, F.B. & SCOTT, G. 1937. Upper Paleozoic ammonites in Texas. *Geology of Texas, Bulletin of Texas University*, **3701**, 1–516.

RUZHENCEV, V.E. 1938. Ammonoids of the Sakmarian stage and their stratigraphic significance. *Problemy Paleontologii*, **4**, 187–285.

RUZHENCEV, V.E. 1950. Upper Carboniferous ammonites of the Urals. *Trudy Paleontologicheskogo Instituta Akademii Nauk SSSR*, **29**, 220c.

RUZHENCEV, V.E. 1951. Lower Permian ammonites of the South Urals. 1. Ammonites of the Sakmarian Stage. *Trudy Paleontologicheskogo Instituta Akademii Nauk SSSR*, **33**, 1–188.

RUZHENCEV, V.E. 1952. Biostratigraphy of the Sakmarian Stage in the Aktyibinsk Region of the Kazakh SSR. *Trudy Paleontologicheskogo Instituta Akademii Nauk SSSR*, **42**, 1–85.

RUZHENCEV, V.E. 1954. Asselian stage of the Permian system. *Doklady Akademii Nauk SSSR*, **99**, 1079–1082.

RUZHENCEV, V.E. 1955. Main stratigraphic assemblages of ammonoids of the Permian systems. *Izvestiya Akademii Nauk SSSR. Seriya Biologicheskaya*, **4**, 120–132.

RUZHENCEV, V.E. 1956. Lower Permian ammonites of the Urals. П. Ammonites of the Artinskian Stage. *Trudy Paleontologicheskogo Instituta Akademii Nauk SSSR*, **60**, 1–271.

RUZHENCEV, V.E. 1959. Classification of the superfamily Otocerataceae. *Paleontologicheskii Zhurnal*, **2**, 56–67.

RUZHENCEV, V.E. 1960. Principles of the systematics, system and phylogeny of Paleozoic ammonoids. *Trudy Paleontologicheskogo Instituta Akademii Nauk SSSR*, **83**, 1–331.

RUZHENCEV, V.E. 1962. Classification of the family Araxoceratidae. *Paleontologicheskii Zhurnal*, **4**, 88–103.

RUZHENCEV, V.E. 1963. New data on the family Araxoceratidae. *Paleontologicheskii Zhurnal*, **3**, 56–64.

RUZHENCEV, V.E. 1965. Changes in the organic world at the Paleozoic–Mesozoic boundary. evolution and changes in marine organisms at the Paleozoic–Mesozoic boundary. *In*: RUZHENCEV, V.E. & SARYCHEVA, T.G. (eds) *Evolution and Change of Marine Organisms at the Paleozoic–Mesozoic Boundary*. Trudy Paleontologicheskogo Instituta Akademii Nauk SSSR, **108**, 117–134.

RUZHENCEV, V.E. 1976. Late Permian ammonoids in the Far East. *Paleontologicheskii Zhurnal*, **3**, 36–50.

RUZHENCEV, V.E. 1978. Asselian ammonoids in Pamir. *Paleontologicheskii Zhurnal*, **1**, 36–52.

SHEVYREV, A.A. 1965. Superorder Ammonoidea. *In*: RUZHENCEV, V.E. & SARYCHEVA, T.G. (eds) *Evolution and Change of Marine Organisms at the Paleozoic–Mesozoic Boundary*. Trudy Paleontologicheskogo Instituta Akademii Nauk SSSR, **108**, 166–182.

SHEVYREV, A.A. 1986. Triassic ammonoids. *Trudy Paleontologicheskogo Instituta Akademii Nauk SSSR*, **217**, 184c.

SHEVYREV, A.A. 1999. Lower Triassic boundary and it correlation in marine deposits. Paper 1. Boundary sections of the Tethys. *Stratigrafiya Geologicheskaya Korrelyatsiya*, **7**, 14–27.

SHEVYREV, A.A. 2000. Lower Triassic boundary and it correlation in marine deposits. Paper 2. Borela sections of the basal Triassic and their correlation with the boundary sections of the Tethys. *Stratigrafiya Geologicheskaya Korrelyatsiya*, **8**, 55–65.

SPINOSA, C. & GLENISTER, B.F. 2000. Ancestral Araxoceratinae (Upper Permian Ammonoidea) from Mexico and Iran in Guadalupian Symposium. *In*: WARDLOW, B.R., GRANT, R.E. & ROHR, D.M. (eds) *The Guadalupian Symposium*. Smithsonian Contributions to the Earth Sciences, **32**. Smithsonian Institution Press, Washington, DC, 397–406.

SPINOSA, C., FURNISH, W.M. & GLENISTER, B.F. 1970. Araxoceratidae, Upper Permian ammonoids from the Western Hemisphere. *Journal of Paleontology*, **44**, 730–736.

WANG, CH. 1995. Conodonts of Permian–Triassic boundary beds and biostratigraphic boundary. *Acta Palaeontologica Sinica*, **43**, 129–151.

YANG, F. 1985. Late Late Permian strata and their ammonoid zones in Southwest China. *Earth Science – Journal of Wuhan College of Geology*, **10**, 129–144.

ZAKHAROV, YU.D. & PAVLOV, A.M. 1986. *Permian cephalopods of Primorye and problem of the zonal subdivision of the Permian in the Tethyan Realm. In*: ZAKHAROV, YU.D. & ONOPRIENKO, YU.I. (eds) Korrelyatsiya permo-triasovykh otlozhenii vostoka SSSR. Izdatelstvo DVNTs SSSR, Vladivostok, 5–32 [in Russian].

ZHAO, J., LIANG, X. & ZHENG, ZH. 1978. *Late Permian Cephalopods of South China*. Science Press, Beijing.

ZHOU, Z. 1987. First discovery of Asselian ammonoid fauna in China. *Acta Palaeontologica Sinica*, **26**, 130–148.

Permian smaller foraminifers: taxonomy, biostratigraphy and biogeography

DANIEL VACHARD

1 rue des Tilleuls, 59152 Gruson, France
Daniel.Vachard@univ-lille1.fr

Abstract: This review has been undertaken in order to present some interpretations about the biostratigraphy of the smaller foraminifers belonging to four classes present during the Permian: Fusulinata, Miliolata, Nodosariata and Textulariata. Biostratigraphic markers of these classes are principally known in the orders and superfamilies of Lasiodiscoidea, Bradyinoidea and Globivalvulinoidea (Fusulinata), Cornuspirida (Miliolata), and in the entire class Nodosariata. The class Textulariata is too little known during the Permian to play a significant biostratigraphical role: nevertheless, the appearance of the order Verneulinida is probably an important bioevent. The main genera among the lasiodiscids are *Mesolasiodiscus*, *Lasiodiscus*, *Lasiotrochus*, *Asselodiscus*, *Pseudovidalina*, *Xingshandiscus* and *Altineria*; the bradyinoids *Bradyina* and *Postendothyra*; the globivalvulinoids *Globivalvulina*, *Septoglobivalvulina*, *Labioglobivalvulina*, *Paraglobivalvulina*, *Sengoerina*, *Dagmarita*, *Danielita*, *Louisettita*, *Paradagmarita*, *Paradagmaritopsis* and *Paremiratella*; the miliolates *Rectogordius*, *Okimuraites*, *Neodiscus*, *Multidiscus*, *Hemigordiopsis*, *Lysites*, *Shanita* and *Glomomidiellopsis*, and the tubiphytids and ellesmerellids, which might be specialized miliolate and cyanobacterium consortia, with reference to microstructures and phylogenies of these groups. The Nodosariata markers belong to *Nodosinelloides*, *Tezaquina*, *Polarisella*, *Geinitzina*, *Pachyphloia*, *Rectoglandulina*, first true *Nodosaria*, *Langella*, *Pseudolangella*, *Calvezina*, *Cryptoseptida*, *Cylindrocolaniella*, *Colaniella*, *Frondina* and *Ichthyofrondina*, but their lineages are too poorly understood to permit an accurate biostratigraphic use at the present time. The superfamily Geinitzinoidea is emended. Finally, palaeobiogeographical implications based on *Shanita*, *Colaniella* and *Altineria* are given.

During the nineteenth century and the first third of the twentieth century, there were few works devoted to the smaller foraminifers, except perhaps for the Nodosariata. From 1900 to 1930, some papers were published by Spandel (1901), Cherdyntsev (1914), Lange (1925) and Likharev (1939), and in the USA by Cushman, Galloway, Waters and Harlton. In the 1940s, a fundamental contribution was provided by Reichel (1945, 1946), especially in Greece and Cyprus. This author was present at the beginning of the Swiss school, with workers Brönnimann, Zaninetti, Martini and Jenny-Deshusses. D. Altıner, trained by this school, has written since the 1980s a fundamental work about the Permian lasiodiscoid, hemigordiid and globivalvulinoid foraminifers. Other micropalaeontologists working in Turkey also provided important contributions: Güvenç, Sellier de Civrieux, Dessauvagie, Lys and Okuyucu. In the 1960s, Russian studies predominated, with A.D. and K.V. Miklukho-Maklay, Reitlinger, Sosnina, Gerke, Kireeva, and Sosipatrova. In the 1980s, Japanese teams studied the palaeontology of their country, as well as in Iran, Pakistan and Thailand. During this century, the Russian school for the Permian smaller foraminifers has been reactivated by Filimonova (2008, 2010). G. Nestell, also known as Pronina, Vuks or Pronina-Nestell, wrote several very important studies (despite her questionable foraminiferal macroclassification, from families to classes). Representatives of the French and Italian schools were productive all around the world, including Lys, Vachard, Rettori, Angiolini and Gaillot (in chronological order). Chinese colleagues, Gu, Song and Zhang, published important recent contributions, especially about the Permian–Triassic boundary foraminifers, extending the classical studies of Ho, Lin and Wang.

Although considered as a subordinate biostratigraphic group, the smaller foraminifers may contribute to the biozonation of the carbonate and evaporite shallow environments with a wide distribution of the representatives of the Miliolata (e.g. Vachard *et al.* 2005; Gaillot & Vachard 2007; Koehrer *et al.* 2012 for the Khuff Formation in the Middle East; and Lucas *et al.* 2015; Vachard *et al.* 2015 for the Yeso Group of southern USA) or to deeper environments with the distribution of representatives of the Nodosariata (e.g. Gu *et al.* 2007; Zhang & Gu 2015) in South China; it is noteworthy that these shallowest and deepest biotopes were occupied only by foraminifers from the Permian (Vachard *et al.* 2010).

From: Lucas, S. G. & Shen, S. Z. (eds) 2018. *The Permian Timescale*. Geological Society, London, Special Publications, **450**, 205–252.
First published online December 8, 2016, https://doi.org/10.1144/SP450.1

Previous classifications of foraminifers

The current classifications of foraminifers are difficult to reconcile, both at the genus level and at the suprageneric level. However, the older classifications based on the wall microstructure have been challenged by the data and suggestions of molecular clocks (Rigaud *et al.* 2015*b* with references therein). In this biostratigraphic and practical paper, the wall microstructure is preserved as a preponderant criterion, even if the original microstructures are often difficult to identify because of the recrystallization of walls consisting of aragonite and high-Mg calcite; even low-Mg calcite transforms itself. The agglutinates are difficult to distinguish because of finely granular siliceous recrystallization. Initially, along with J. Gaillot, in trying to find a true difference between the classes Fusulinata and Textulariata, we separated the Fusulinata with calcareous agglutinates from the Textulariata with siliceous agglutinate (Gaillot & Vachard 2007; Vachard *et al.* 2010); however, it was observed that some representatives of the Fusulinata have a true siliceous agglutinated wall (e.g. palaeotextulariids, endothyranellins and endotebids). Currently, this subject is being examined by S. Rigaud (Rigaud *et al.* 2014, 2015*a*, *b*).

The foraminiferal molecular clock developed notably by J. Pawlowski (e.g. Pawlowski *et al.* 2013) is difficult to follow because it does not correspond to available observations in sampled materials. Even if the interest of this molecular classification is to be theoretical, it is manifestly incompatible with the fossil record in Palaeozoic rocks. Some consensual syntheses have been presented in Langer *et al.* (1993) and Rigaud *et al.* (2014, 2015*a*, *b*). Despite these divergent concerns, one palaeontological observation is, nevertheless, consistent with the molecular data: the transition from an agglutinated to a calcareous wall, and vice versa, occurred several times in the foraminiferal evolution, and in independent lineages (Bowser *et al.* 2006; Pawlowski *et al.* 2013): for example, when *Psammosphaera* (agglutinated) give rise to miliolates (porcelaneous), as well ammodiscids (agglutinated) (Pawlowski *et al.* 2013), and the miliolates (porcelaneous) in turn give *Miliammina* (agglutinated) (Fahrni *et al.* 1997).

Here, a proposed classification is developed according to the phylogenetic hypotheses of Vachard (1994), Gaillot & Vachard (2007), Vachard *et al.* (2010) and Hance *et al.* (2011). It differs significantly from previous systematics proposed by Mikhalevich (1980, 2013) and Loeblich & Tappan (1987, 1992) . The taxonomic rank of the foraminifera (phylum or subphylum) is still a matter of debate (see e.g. Adl *et al.* 2012) but, herein, the views of Cavalier-Smith (2002, 2003) are adopted.

Even if macroevolution remains very questionable, the lower units of the classification, especially at the generic level, seem to be objective and can be used confidently with their FAD (first appearance datum: i.e. the oldest of the successive local appearances of a taxon) and LAD (last appearance datum: i.e. the youngest of the successive disappearances of a taxon) for the biozonation. A complete taxonomical reference list would be huge and, therefore, the reader can refer to Loeblich & Tappan (1987), Vdovenko *et al.* (1993), Rauzer-Chernousova *et al.* (1996), Pronina-Nestell & Nestell (2001), M. K. Nestell *et al.* (2006), Gaillot & Vachard (2007), Karavaeva & Nestell (2007), G. P. Nestell *et al.* (2009) and Vachard *et al.* (2015).

Previous biostratigraphical data

The recent Permian scale was established thanks to the work of Jin *et al.* (1997). For a long period of time, two parallel scales for the foraminifers have existed: one for the Tethyan fusulinids (Leven 1967; Leven & Gorgij 2011), and another one for the Omolon Massif (NE Siberia) nodosariates (Pronina 1999*a*; Karavaeva & Nestell 2007). Both are, however, difficult to apply to other areas because of: (a) the regionalism of the two groups; (b) the development of the foraminiferal classifications; (c) the progress of the correlations with the conodonts scales; (d) the problems of definition of the Leven's stratotypes (Angiolini *et al.* 2015, 2016); (e) the introduction of concurrent regional stages (e.g. Hermagorian by Davydov *et al.* 2013); and (f) the precise, new correlations between the different biozonations (Henderson *et al.* 2012). The smaller foraminifers are actually little used, but four recent types of investigations could develop their knowledge: (1) the smaller foraminifers associated with the fusulinids described by Leven (1967, 1998, 2003) are currently revised by Filimonova (2010, 2013), even if some parts of these contributions may provoke questions (see Angiolini *et al.* 2015, 2016); (2) the microfaunas of smaller foraminifers associated with endemic North American fusulinids (e.g. *Polydiexodina*) have shown that more Tethyan smaller foraminifers than expected are present in Mexico, New Mexico and Texas (Vachard *et al.* 1992; Nestell & Nestell 2006; Nestell *et al.* 2006), and needs further investigation in other states of the USA and countries in the Americas; (3) the potential of the nodosariates was re-demontrated by Karavaeva & Nestell (2007); and (4) areas, unfavourable for fusulinids because of their deposits and types of environments, have been dated and biozoned recently thanks to smaller foraminifers (Lucas *et al.* 2015; Vachard *et al.* 2015).

Important papers on smaller foraminifers have recently been published: relating especially to understanding the Permian–Triassic Boundary (PTB) and its extinctions in central and eastern Tethys (i.e. South China and SE Asia: Fig. 1): Altıner (1999, 2001), Groves & Altıner (2005), Groves *et al.* (2005), Gaillot & Vachard (2007), Gaillot *et al.* (2009), Altıner & Özkan-Altıner (2010), Ueno *et al.* (2010), Wang *et al.* (2010), Kobayashi (2012*a*, *b*, 2013), Song *et al.* (2007, 2011, 2013) and Nestell *et al.* (2015).

The foraminiferal biostratigraphic scales are generally based on fusulinids (e.g. Leven 2003; Wilde 2006; Leven & Gorgij 2011; Henderson *et al.* 2012). After some attempts to establish the stratigraphical significance of smaller foraminifers, pioneering detailed biozonations have been proposed for the late Middle Permian and Late Permian (Capitanian–Lopingian) by Vachard *et al.* (1993*a*, *b*, 2002) and Pronina (1996). First biozonations by smaller foraminifers for the Cisuralian of North America have been published very recently (Lucas *et al.* 2015; Vachard *et al.* 2015).

Proposed classification of Permian smaller foraminifers

Phylum **Foraminifera** d'Orbigny, 1826
nom. translat. Cavalier-Smith,
2002 (subphylum) and 2003 (phylum)
Class **Fusulinata** Gaillot & Vachard, 2007

The members of the class Fusulinata have a homogeneous, microgranular, primary test wall of low-Mg calcite in which crystal units are optically unordered, more or less equidimensional and are only a few micrometres in size. The class Fusulinata contains eight orders: (1) Parathuramminida Bykova in Bykova & Polenova, 1955; (2) Tuberitinida Vachard *et al.*, 2015; (3) Earlandiida Loeblich & Tappan, 1982; (4) Archaediscida Poyarkov & Skvortsov, 1979 emend. Hance *et al.*, 2011; (5) Pseudopalmulida Mikhalevich, 1993; (6) Tournayellida Dain in Dain & Grozdilova, 1953; (7) Endothyrida Fursenko, 1958 (including Palaeotextulariina Hohenegger & Piller, 1975); and (8) Fusulinida Fursenko, 1958. It should be noted that this latter order is the only one designated less formally as fusulines, fusulinids, fusulinoideans or fusulinaceans, and is conventionally separated in the foraminiferal treatises from the smaller foraminifers (or non-fusulines) and often even described by other authors. The Permian smaller foraminifers also include representatives of the classes Miliolata, Nodosariata and Textulariata (see later). The Permian smaller foraminifers of the class Fusulinata belong to four orders: Tuberitinida, Earlandiida, Archaediscida and Endothyrida. Regarding Archaediscida, it is noteworthy that their Permian representatives only belong to the superfamily Lasiodiscoidea. In spite of this, many genera and even species of Archaediscoidea (e.g. *Permodiscus*, *Archaediscus*, *Kasakhstanodiscus*: see Davydov 1988) have been mentioned from the Permian in Russia, Tajikistan

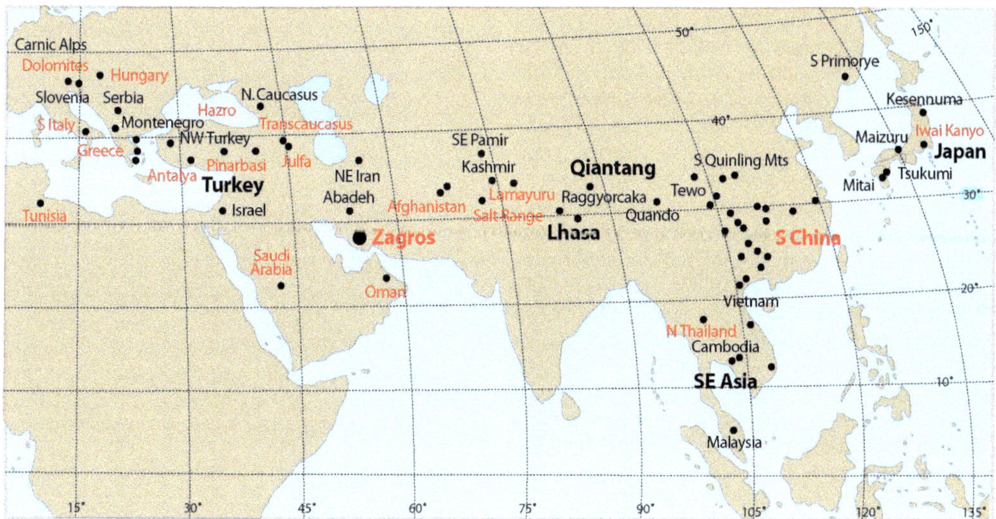

Fig. 1. Location and toponymy of the main Permian areas in the world (Transcaucasus, or Transcaucasia, were the former Soviet names for a geopolitical region encompassing all of Armenia and the majority of Georgia and Azerbaijan; the classical Permian outcrops of 'Transcaucasia' are at this time located in Armenia and Nakhchivan, an enclave of Azerbaijan; e.g. see Théry *et al.* 2007, fig. 3a).

and Uzbekistan. Although Pinard & Mamet (1998) have refuted this assignment and demonstrated that all these forms are, in reality, pseudovidalinids, the misinterpretations continue. For instance, an archaediscoid *Permodiscus sumsariensis*, which is a marker for the latest early Visean (biozone MFZ11B: Okuyucu *et al.* 2013), has recently been misinterpreted in the Cisuralian by Filimonova (2010). Definitively, the Archaediscoidea disappear in the lowermost early Moscovian (earliest Middle Pennsylvanian) and do not exist in the Permian.
The name Fusulinata Gaillot & Vachard, 2007 is valid for designating this class: even if Fusulinata was introduced by Maslakova (1990) and made explicit in Maslakova *et al.* (1995), it was, in both articles, for a subclass (because the unique class recognized in both articles is Foraminifera) and, moreover, without any description and/or discussion. In contrast, the name of the subclass introduced by Maslakova (1990) has priority over Fusulinana Vachard *et al.*, 2010 (see below), even if the spelling Fusulinana is more correct than the original Fusulinata (compared to the endings *ana* and *ata* used in the classifications of, e.g., V. Mikhalevich and G. Nestell).

Subclass **Afusulinana** Vachard *et al.*, 2010
Order **Tuberitinida** Vachard *et al.*, 2015
Family **Tuberitinidae** Miklukho-Maklay, 1958

Permian genera. Tuberitina Galloway & Harlton, 1928; *Eotuberitina* Miklukho-Maklay, 1958 (=*Diplosphaerina* Derville, 1952 part.); *Mendipsia* Conil & Longerstaey in Conil *et al.*, 1980.

Remarks. The Tuberitinidae are calcareous microfossils that have two stages of their cycle of life: diplospherin (free) and tuberitin (attached) (Conil *et al.* 1977). *Diplosphaerina sensu stricto* being the diplospherin stage of *Eotuberitina*, both genera are probably synonymous, and the former genus has priority over the second one. Despite their atypical palaeobiological characters, the tuberitinids probably belong to the foraminifers of the Afusulinana subclass due to their numerous, minute foramina and their wall microstructures, which are often dark-microgranular and hyaline-pseudofibrous, but this assignment may be discussed. The Afusulinana differ from the Parathuramminida by their biphased cycle of life, less conspicuous apertures and absence of homeomorphies with extant foraminifers (the Parathuramminida being homeomorphs of the extant and fossil genera, e.g. *Thurammina* Brady).

According to Vachard (1994) and Vachard *et al.* (2015), the Tuberitinidae constitute a homogeneous group. The genus *Polysphaerinella* Mamet differs from this group by its bilayered test and, thus, could belong to another family or subfamily. Moreover, other phylogenetic trends, depending on distinct types of walls, exist in the subfamilies Tubeporininae Zadorozhnyi & Yuferev and Altjusellinae Zadorozhnyi & Yuferev, and the genus *Tubesphaera* Vachard. All these phylogenetic trends could be considered as different families in the order Tuberitinida.

Biostratigraphy. The tuberitinids are not very useful for Permian biostratigraphy. An interesting thing to note is their survival after the PTB (Song *et al.* 2007, 2011; Okuyucu *et al.* 2014).

Biogeography. The tuberitinids have a cosmopolitan distribution from Late Devonian to latest Permian times, and the last representatives of the earliest Triassic are currently only known in South China (Song *et al.* 2007, 2009, 2011) and Turkey (Okuyucu *et al.* 2014).

Order **Earlandiida** Loeblich & Tappan, 1982
Superfamily **Earlandioidea** Loeblich & Tappan, 1982
Family **Earlandiidae** Cummings, 1955 emend. Vachard, 1994

Permian genera. Earlandia Plummer, 1930 (=*Aeolisaccus* Elliott, 1958 = *Decastronema* Golubic *et al.*, 2006 (part.) = *Hyperammina* Brady, 1878 (part.)); ?*Chitralina* Angiolini & Rettori, 1995; ?*Giraliarella* Crespin, 1958; ?*Rectoformata* Okuyucu, 2007.

Remarks. Some tubular, undivided microfossils from the Permian-Triassic times are discussed, but are often synonymized with *Earlandia*. The species *Aeolisaccus dunningtoni* Elliott, 1958 seems to be a representative of the group *Earlandia elegans*, in which the proloculus would be inconspicuous and/or broken (Gaillot & Vachard 2007 and references therein). *Decastronema* Golubic *et al.*, 2006, by its type species, is a junior synonym of the Cambrian–Triassic cyanobacterium *Proaulopora* Vologdin, 1932, but *Decastronema* also includes some *Earlandia* among its other species. However, many so-called Palaeozoic representatives of '*Hyperammina*' (*non Hyperammina* Brady, 1878, which is an extant genus) generally correspond to secondary silicified *Earlandia*, and are interesting elements for discussing the secreted, against agglutinated, types of walls (see Vachard *et al.* 2010, 2015; Nestell *et al.* 2015; and see 'Textulariata' later in this section).

Biostratigraphy. The family is known throughout the Palaeozoic and might survive up to the Cretaceous with the taxa *Earlandia*? and *Giraliarella*? of Arnaud-Vanneau (1980): for a discussion on this, see Vachard *et al.* (2010), Krainer & Vachard (2011), Hance *et al.* (2011) and Nestell *et al.* (2015). There was a widespread conviction that *Earlandia* was a

PTB survivor (e.g. Groves & Altıner 2005; Groves *et al.* 2005, 2007; Vachard *et al.* 2010; Krainer & Vachard 2011); this fact, nevertheless, has been recently disputed (Nestell *et al.* 2015). *Chitralina* and/or *Giraliarella* (two genera to revise and probably emendate) might be the ancestors of *Rectostipulina* Jenny-Deshusses, 1985 and/or *Tubulastella* Rigaud *et al.*, 2015*b* (see below), as well as *Earlandia* being the ancestor of the Nodosariata for many authors (Hohenegger & Piller 1975; Vachard 1994; Altıner & Savini 1997; Groves 1997; Groves *et al.* 2003; Vachard *et al.* 2010; Krainer & Vachard 2011; Rigaud *et al.* 2015*b*). *Rectoformata*, known in Turkey (Okuyucu 2007) and South China (Zhang *et al.* 2015), might be biostratigraphically important for dating the Tethyan assemblages.

Biogeography. Probably cosmopolitan during the Palaeozoic, the family seems then to be restricted to some Tethyan areas.

Superfamily **Caligelloidea** Reitlinger in Rauzer-Chernousova & Fursenko, 1959
nom. translat. Gaillot & Vachard, 2007
Family **Insolentithecidae** Loeblich & Tappan, 1986*a*
nom. translat. Gaillot & Vachard, 2007

Permian genera. Insolentitheca Vachard in Bensaid *et al.*, 1979; *Floritheca* Gaillot & Vachard, 2007.

Remarks. This atypical group of foraminifers, ontogenetically deformed probably because of a particular infaunal life, is principally known during the Carboniferous. Rare taxa subsist in the Permian; especially in the Cisuralian.

Biostratigraphy. As the Late Silurian–Late Permian caligelloid genera are poorly known, their precise biostratigraphic contribution remains small. *Insolentitheca*, which is essentially a late Mississippian–Pennsylvanian genus, is rare in the Early Cisuralian of Bolivia (Mamet 1996), and in the Kungurian of southern Turkey and Tajikistan (Vachard & Moix 2013; Angiolini *et al.* 2016). *Floritheca* is only known in the Lopingian of southern Iran (Zagros, Fars), Abu Dhabi (UAE) and the southern Chichibu Terrane (SW Japan) (Gaillot & Vachard 2007).

Biogeography. The insolentithecid genera are either cosmopolitan (as *Insolentitheca*) or more restricted (as *Floritheca*).

Order **Archaediscida** Poyarkov & Skvortsov, 1979 emend. Hance *et al.*, 2011
Suborder **Lasiodiscina** Gaillot & Vachard, 2007
Superfamily **Lasiodiscoidea** Reitlinger in Vdovenko *et al.*, 1993
Family **Lasiodiscidae** Reitlinger, 1956 emend. Gaillot & Vachard, 2007

Permian genera. Lasiodiscus Reichel, 1946; *Hemidiscus* Schellwien, 1898 emend. Vachard & Krainer, 2001*a*; *Lasiotrochus* Reichel, 1946; *Mesolasiodiscus* Rauzer-Chernousova & Chermnykh, 1990.

Remarks. This group has been largely neglected after the fundamental papers of Reichel (1946) and Reitlinger (1956). Recent studies of the ancestral lasiodiscids and howchiniids (Cózar *et al.* 2015) demonstrated the maximal complexity of the phylogenies of these groups. Consequently, the apparently well-known Permian lasiodiscids need further studies (1 & 2 in Fig. 2), peculiarly in North America where they are not yet known or are very questionable. Similarly, the Carboniferous representatives mentioned in North America (most often called *Monotaxinoides*) are misinterpreted miliolates (Krainer *et al.* 2015; Lucas *et al.* 2016*a*, *b*), except for '*Monotaxinoides priscus*' *sensu* Groves, 1992 (although this taxon most probably belongs to *Eolasiodiscus*), *Eolasiodiscus* sp. and *Mesolasiodiscus*? sp. (Lucas *et al.* 2016*b*).

Biostratigraphy. Hemidiscus sensu stricto seems to be restricted to the Cisuralian, but it is difficult to distinguish from the Pennsylvanian genus *Eolasiodiscus* or Mississippian genus *Hemidiscopsis* Cózar in Cózar *et al.*, 2015. *Mesolasiodiscus* was introduced as a Cisuralian transitional taxon between *Eolasiodiscus* and *Lasiodiscus*, but it seems to exist as early as the Middle (even Early?) Pennsylvanian: its FAD and LAD are poorly known, apparently late Asselian and Kungurian, respectively (Filimonova 2010). Similarly, the true FAD of *Lasiodiscus* is poorly known (many Cisuralian references to *Lasiodiscus tenuis* in the literature are, in reality, *Hemidiscus* due to the absence of umbilical fillings and sutural appendices), whereas its LAD is clearly latest Permian, perhaps even latest Changhsingian (Miklukho-Maklay 1954; Pronina-Nestell & Nestell 2001; Gu *et al.* 2007). *Lasiotrochus sensu stricto* seems to be endemic to the late Guadalupian–early Lopingian, from Greece to South China (many previous identifications are, in reality, misinterpreted *Mesolasiodiscus*: Vachard *et al.* 2002, pl. 1, figs 17 & 18; Lucas *et al.* 2016*b*; Vachard *et al.* unpublished data).

Biogeography. The Permian lasiodiscids, relatively common in the Palaeo- and Neotethys, are absent from North America.

Family **Pseudovidalinidae** Altıner, 1988

Permian genera. Pseudovidalina Sosnina, 1978 emend. Alipour & Vachard in Alipour *et al.*, 2013 (=*Raphconilia* Brenckle & Wahlman, 1996); *Altineria* Özdikmen, 2009 emend. Vachard *et al.*, in press; *Asselodiscus* Mamet & Pinard, 1992; *Xingshandiscus* Zheng, 1986.

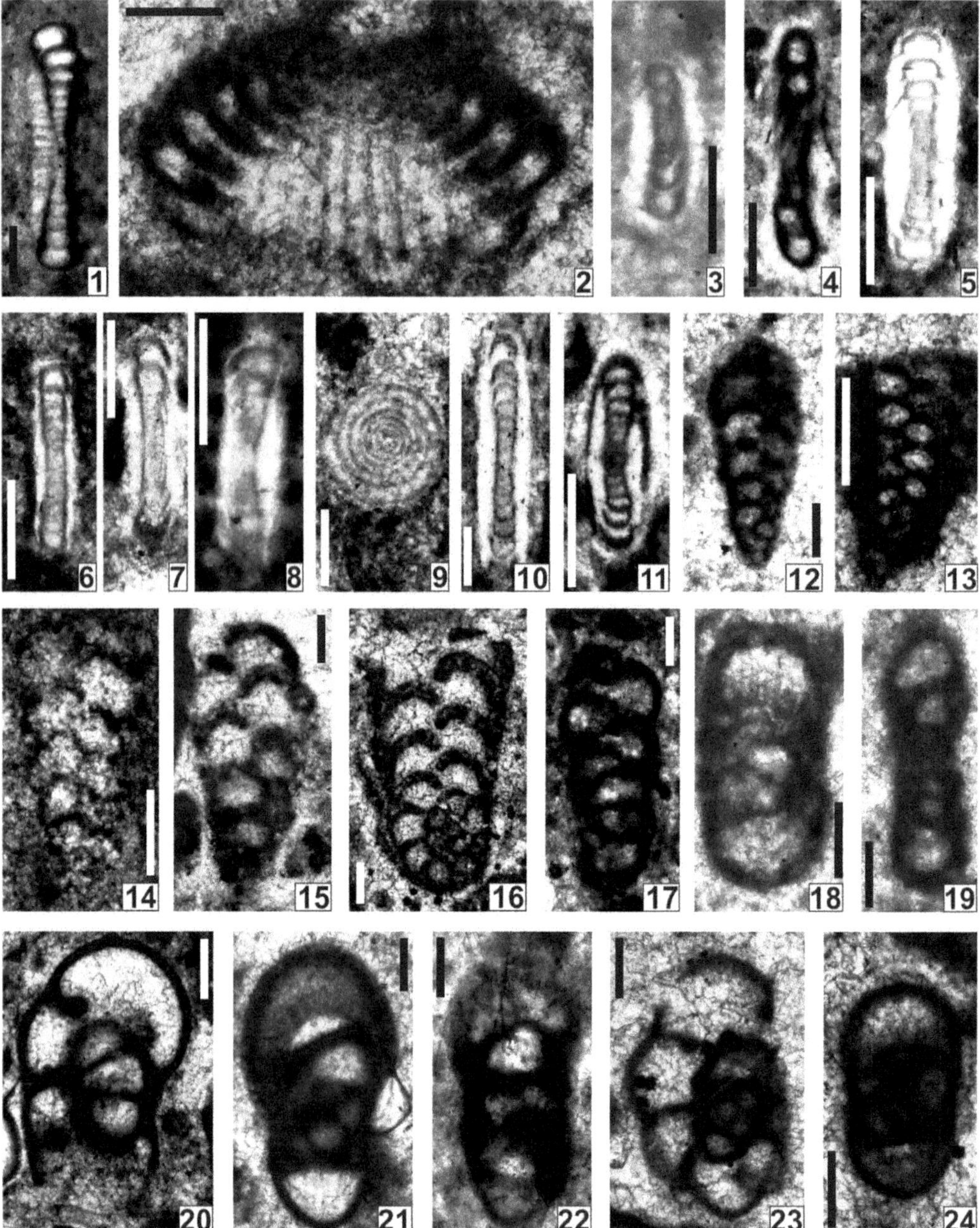

Fig. 2. Smaller foraminifers (Fusulinata) from Trogkofel (Late Cisuralian of the Carnic Alps, Austria). Scale bar = 0.100 mm, except for (2) where it is 0.250 mm. (**1**) *Lasiodiscus tenuis* Reichel, 1946. (**2**) *Mesolasiodiscus* sp. (**3**) & (**4**) Transition *Asselodiscus–Pseudovidalina*? sp. (**5**)–(**11**) *Pseudovidalina* spp. (**12**)–(**17**) *Spireitlina* spp. (**18**)–(**24**) *Endothyra* spp.

Remarks. This interesting group, still poorly known during the Permian, might provide, with further studies, more accurate biomarkers for the Tethyan Permian. The Pseudovidalinidae have a double Permian history (the second episode of which was illustrated by Altıner 1988). During the first episode, during the Late Pennsylvanian–Cisuralian, except for rare *Asselodiscus* (Davydov *et al.* 2001; Alipour *et al.* 2013), the genus *Pseudovidalina* was the unique representative of the family (3–11 in

Fig. 2). However, during this period, *Pseudovidalina* displays various evolutionary stages, eventually interpretable as subgenera or genera (*Raphconilia* is one of these stages), as is the case for Carboniferous homeomorphous archaediscoids (see the earlier remarks about the misinterpretations of pseudovidalinids as archaediscoids and the detailed discussion in Alipour *et al.* 2013). *Altineria* (pro *Angelina* Altıner, 1988 pre-occupied by a Cambrian trilobite) is sometimes considered as a junior synonym of *Xingshandiscus* (see Pinard & Mamet 1998). This synonymy is irrelevant because *Altineria* is unilayered pseudofibrous, whereas the type species of *Xingshandiscus* exhibits a bilayered wall with a dark-microgranular inner layer and a hyaline-pseudofibrous outer layer. Currently, we are revising the taxa of this family with R. Rettori, D. Altıner and V. Gennari.

Biostratigraphy. Pseudovidalina sensu lato is rare in the Gzhelian (Henderson *et al.* 1995; Davydov *et al.* 2001), relatively frequent within the Carboniferous–Permian boundary interval (=Orenburgian = Bursumian = Newwellian regional substages) and subsists up to the late Changhsingian. At the end of the Cisuralian, during the Kungurian, a first evolute pseudovidalinid genus called *Xingshandiscus* appears. During the Middle–Late Permian boundary interval, another evolute trend is represented by *Altineria*. The Capitanian–Wuchiapingian biostratigraphy and peri-Gondwanan palaeobiogeography of the *Altineria* lineage need further study. The possible FAD of *Altineria* might be early? Capitanian in Japan (Kobayashi 2006*a* reinterpreted), but this event most probably occurs between the LAD of *Yabeina* Deprat and the FAD of *Nanlingella* Rui & Sheng (Vachard *et al.* unpublished data). The type species, *Altineria alpinotaurica* (Altıner, 1988), is probably limited to the Capitanian–Wuchiapingian boundary interval, and, apparently, no late Wuchiapingian–Changhsingian representatives of the lineage are known.

Biogeography. Genera of this family are either cosmopolitan (e.g. *Asselodiscus* and *Pseudovidalina*) or limited to the peri-Gondwanan areas (*Altineria*).

Subclass **Fusulinana** Maslakova, 1990 nom. correct. Vachard *et al.*, 2010
Order **Endothyrida** Fursenko, 1958
Suborder **Endothyrina** Bogush, 1985
Superfamily ?**Endothyroidea** Glaessner, 1945
Family **Endothyridae** Rhumbler, 1895

Permian genera. Endothyra Phillips, 1846 *sensu* Brady, 1876 emend. China, 1965; *Endothyranella* Galloway & Harlton in Galloway & Ryniker, 1930; *Linendothyra* Mamet & Pinard, 1992; *Neoendothyra* Reitlinger, 1965; *Neoendothyranella* Nestell & Nestell, 2006; and *Planoendothyra* Reitlinger in Rauzer-Chernousova & Fursenko, 1959.

Remarks. The Endothyridae were very diversified during the Carboniferous. They remain present during the Permian, but are rare and without biostratigraphic importance, except perhaps for some species of *Neoendothyra*. The name Endothyroidea Glaessner, 1945 has priority over Endothyroidea Brady, 1884 nom. translat. Loeblich & Tappan, 1981 in designating this superfamily.

Biostratigraphy. Endothyra is primarily a late Tournaisian (MFZ6: Poty *et al.* 2006) to Permian genus. Its acme is Visean–Moscovian. Its LAD is Permian, but not yet precisely established (Early or Middle Permian?), owing to confusion in the literature with *Neoendothyra* (19–24 in Fig. 2). The majority of the so-called Triassic *Endothyra*, in reality, belong to Endotebidae (Vachard *et al.* 1994). *Planoendothyra sensu stricto* is mainly a Pennsylvanian genus, but its LAD might be Kungurian (Vachard *et al.* unpublished data). *Endothyranella*, which is essentially a Middle–Late Pennsylvanian genus, is represented by relatively atypical forms in the Cisuralian (Vachard 1980; Baryshnikov *et al.* 1982; Vachard & Krainer 2001*b*). The material of the Urals and Afghanistan displays a siliceous agglutinate wall (see 'Textulariata' later in this section). *Neoendothyra* exhibits its primitive, relatively atypical forms in the late Cisuralian (Vachard & Krainer 2001*b*, pl. 4, figs 6 & 15; Vachard *et al.* unpublished data) or the late Chihsian = Kungurian–Roadian) (Lin *et al.* 1990); its first typical forms seem to be early Roadian (Angiolini *et al.* 2015), and its acme is Capitanian–Wuchiapingian and up to the PTB (Song *et al.* 2007; Mohtat-Aghai *et al.* 2009). Triassic species have been reinterpreted as belonging to *Endoteba* (see later). *Linendothyra* seems to be more common in the late Capitanian–Lopingian times (Vachard unpublished data), and *Neoendothyranella* is Capitanian in age.

Biogeography. Endothyra is rare but cosmopolitan during the Cisuralian. Rare Permian *Planoendothyra* are known in Indonesia (Nguyen Duc Tien 1989*a*, pl. 11, fig. 1) and in the Carnic Alps (Vachard & Krainer 2001*a*; Vachard *et al.* unpublished data). Cisuralian *Endothyranella* are represented by relatively atypical forms in the Urals (Baryshnikov *et al.* 1982), Afghanistan (Vachard 1980) and Bolivia (Mamet 1996). *Neoendothyra* exhibits its primitive forms in South China, the Carnic Alps and the Canadian Artic (see Lin 1985; Pinard & Mamet 1998; Vachard & Krainer 2001*b*). The first typical forms are observed in the late Cisuralian (especially Kungurian) of the Carnic Alps (Vachard *et al.* unpublished data) and South China (Lin *et al.* 1990). The Capitanian–Wuchiapingian specimens are common on the Palaeotethyan and Panthalassan shelves: Armenia, Azerbaijan, Montenegro, Tunisia, Turkey, NW Caucasus, Iran (central Alborz), Oman, Afghanistan, Cambodia, Thailand, Malaya, Sumatra, South China, Texas, Mexico and Guatemala. Changhsingian last forms subsist in South China, Thailand, Malaysia, Primorye (in the far east of Russia) and Japan; and up to the PTB in South China and Iran. *Linendothyra* seems to be more common in eastern Tethys but exists in Oman (Vachard *et al.* 2002). *Neoendothyranella* is endemic to New Mexico because the specimen of Cambodia of Nguyen Duc Tien (1986*a*, pl. 4, fig. 2), assigned to this genus by Nestell & Nestell (2006), is most probably a *Linendothyra*.

Superfamily **Bradyinoidea** Reitlinger, 1950
nom. translat. Rauzer-Chernousova *et al.*, 1996
Family **Bradyinidae** Reitlinger, 1950 nom. translat. Reitlinger, 1958
Subfamily **Bradyininae** Reitlinger, 1950

Permian genera. Bradyina von Möller, 1878 (= *Bradyinelloides* Mamet & Pinard, 1992); *Postendothyra* Lin, 1984.

Remarks. There are two episodes in the history of Permian bradyinoids. At the beginning of this system, the last *Bradyina* (sometimes designated as *Bradyinelloides*) survive and are more or less cosmopolitan, whereas the terminal period is characterized by Tethyan *Postendothyra*, a form totally atypical because, in contrast to all other bradyinoids, it is small sized and with a simple, basal aperture.

Biostratigraphy. Bradyina is a late Mississippian–Early Cisuralian cosmopolitan genus. Its last representatives are late Cisuralian in age (Baryshnikov *et al.* 1982; Filimonova 2010). The FAD of *Postendothyra* is unknown; nevertheless, this genus becomes relatively common from the Capitanian to the Changhsingian in South China (Lin 1984; Song *et al.* 2009, 2011; Ueno *et al.* 2010; Zhang *et al.* 2015), Italy (Vachard & Miconnet 1990), Greece (Vachard *et al.* 1993*b*), Slovenia (Flügel *et al.* 1984), Crimea (Pronina & Nestell 1997), Cyprus (Nestell & Pronina 1997), Turkey (Zaninetti *et al.* 1981), Armenia and Azerbaijan (former 'Transcaucasia': Pronina 1988*a*; Kotlyar *et al.* 1989), Oman (Vachard *et al.* 2002), Afghanistan (Vachard unpublished data), Himalaya (Lys *et al.* 1980) and southern Tibet (Wang *et al.* 2010), Malaysia (Fontaine *et al.* 1994), NW Thailand (Fontaine *et al.* 1993), East Thailand (Fontaine *et al.* 1997), Cambodia (Nguyen Duc Tien 1986*b*), Primorye (Sosnina & Nikitina 1977), and Japan (Kobayashi 1986, 1997*c*, 2006*a*, 2012*a*, *b*).

Biogeography. Bradyina is cosmopolitan, at least up to the early Artinskian (Blazejowski 2009). *Postendothyra* is Palaeotethyan, Neotethyan and Panthalassan, and is unknown in North America (see Gaillot *et al.* 2009).

Superfamily **Palaeotextularioidea** Galloway, 1933
nom. translat. Habeeb, 1979
Family **Palaeotextulariidae** Galloway, 1933
nom. translat. Wedekind, 1937
Subfamily **Palaeotextulariinae** Galloway, 1933

Permian genera. Palaeotextularia Schubert, 1921 emend. Galloway & Ryniker, 1930; *Climacammina* Brady in Etheridge, 1873 emend. Cummings, 1956; *Cribrogenerina* Schubert, 1908; *Deckerella* Cushman & Waters, 1928*a*.

Remarks. Among the palaeotextulariids, the consistent continuity in morphology is remarkable (4–7 in Fig. 3), as it is in the tetrataxids (see below). The morphologies of *Climacammina* and *Deckerella* remain remarkably constant from the beginning to the end of their range (Vachard *et al.* 2010). *Cribrogenerina*, a unique genus that appeared in the Permian, differs only by lower chambers and more numerous openings in the cribrate aperture.

Biostratigraphy. Palaeotextularia is Middle Mississippian (late Visean: MFZ13 biozone) to latest Permian; nevertheless, after the Visean it is difficult to distinguish the adult forms of this genus from juvenile or neotenic *Climacammina*. The FAD of *Climacammina* is an important datum in the late Mississippian, but after that the morphology of *Climacammina* remains very stable during the Pennsylvanian and Permian times, up to the latest Changhsingian and the PTB (Gaillot & Vachard 2007; Vachard *et al.* 2010). *Deckerella* is Moscovian (Reitlinger 1950) to latest Changhsingian (Kotlyar *et al.* 1999; Wang *et al.* 2010; Ebrahim Nejad *et al.* 2015). *Cribrogenerina* is Middle–Late Permian.

Biogeography. Palaeotextularia and *Climacammina*, which are, first, Tethyan and Uralian genera during the late Mississippian, become cosmopolitan from the Early Pennsylvanian. *Deckerella* is rare but cosmopolitan throughout its range. *Cribrogenerina* is principally an eastern Tethyan and Panthalassan genus.

Superfamily **Endoteboidea** Vachard *et al.*, 2013
Family **Spireitlinidae** Vachard *et al.*, 2013

Permian genus. Spireitlina Vachard in Vachard & Beckary, 1991.

Remarks. The Spireitlinidae, due to their stratigraphic distribution, were suggested herein as the missing link between the Palaeotextulariidae and the

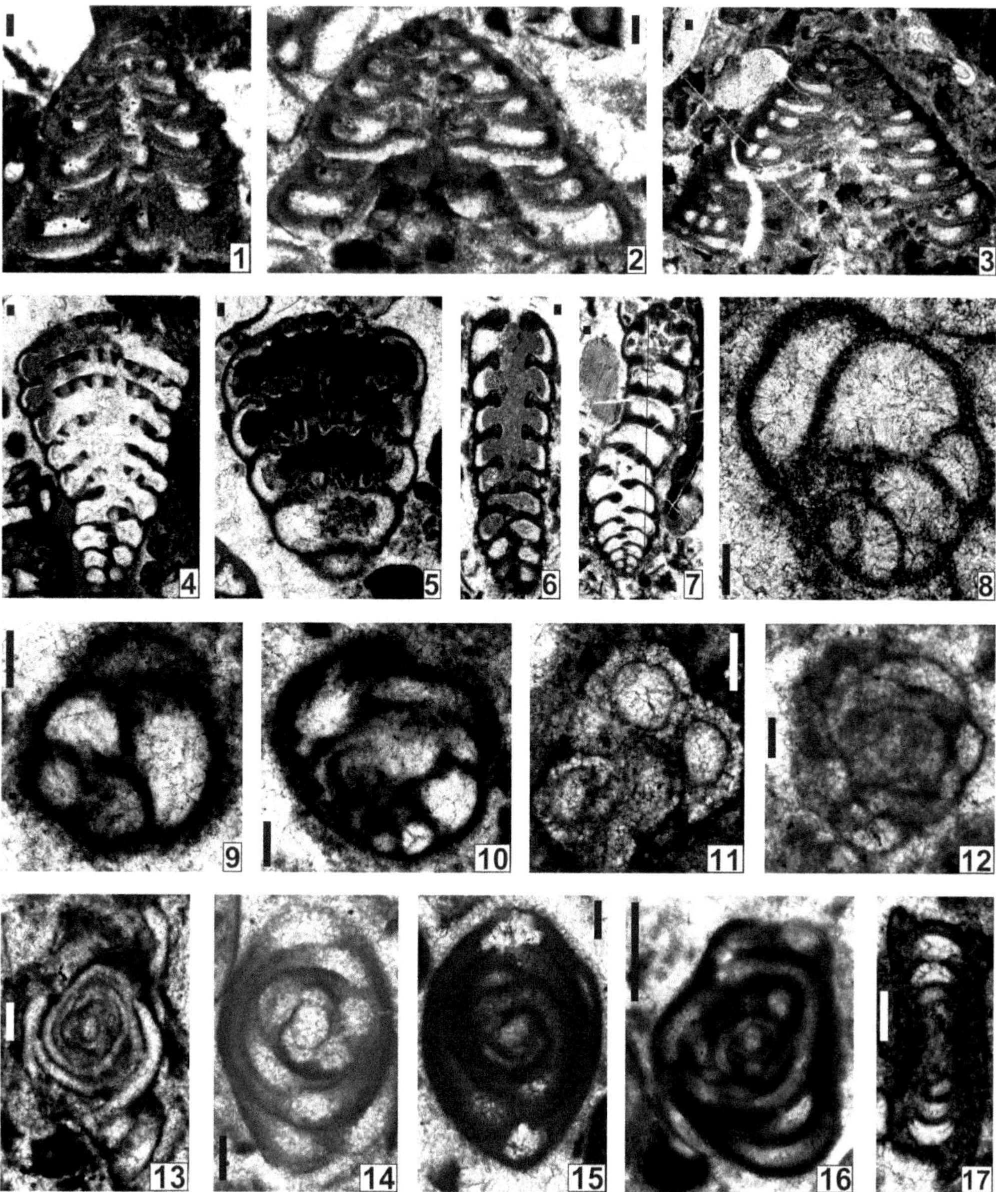

Fig. 3. Smaller foraminifers (Fusulinata and Miliolata) from Trogkofel (Late Cisuralian of the Carnic Alps, Austria). Scale bar = 0.100 mm. (**1**) & (**2**) *Tetrataxis* sp. (**3**) Transition *Tetrataxis–Abadehella* sp. (**4**) & (**5**) *Climacammina* spp. (**6**) & (**7**) ?*Deckerella* cf. *laheei* Cushman & Waters, 1928*a*. (**8**)–(**10**) *Globivalvulina* spp. (**11**) *Hedraites* sp. (**12**) *Hemigordiellina* sp. (**13**) *Hemigordius*? sp. (**14**)–(**15**) *Praeneodiscus* sp. (**16**) *Glomomidiella* sp. (**17**) *Hemigordius* sp.

Endotebidae (Vachard *et al.* 2012). *Spireitlina*, which appears during the Visean–Serpukhovian boundary interval, is relatively well represented and cosmopolitan from the Moscovian to the Artinskian, but shows a weak species diversification (12–18 in Fig. 2).

Biostratigraphy. *Spireitlina* is rare in the latest Visean–Serpukhovian of southern France (Pille 2008; Lucas *et al.* 2016*a*) and Tian Shan (Kulagina *et al.* 1992, reinterpreted by Vachard *et al.* 2012); it is relatively common and cosmopolitan from Early Pennsylvanian (Bashkirian) to Middle

Permian (Wordian) times (Vachard *et al.* 2012), and rare in Capitanian (Vachard & Miconnet 1990, pl. 1, fig. 18, dating reinterpreted here; Nestell & Nestell 2006; Nestell *et al.* 2006; Zhang *et al.* 2015) and Wuchiapingian (Ueno *et al.* 2010).

Biogeography. Spireitlina is cosmopolitan, and particularly common and constant in North America (e.g. Groves & Boardman 1999; Vachard & Krainer 2001*a*; Nestell & Nestell 2006; Nestell *et al.* 2006; Lucas *et al.* 2016*a*).

Family **Endotebidae** Vachard *et al.*, 1994

Permian genera. Endoteba Vachard & Razgallah, 1988; *Vachardella* Nestell & Nestell, 2006; ?Gen. and sp. indet. *A sensu* Kobayashi (2001); ?*Granuliferelloides*? *sensu* Kobayashi (2001); ?'*Haplophragmina*?' *sensu* Nestell & Nestell, 2006 (*non* Reitlinger, 1950).

Remarks. Being related with Palaeotextularioidea, via the Spireitlinidae, the Endotebidae might also belong to the Textulariida (Rigaud *et al.* 2014) (see below). As evidence of this assignment, it is noteworthy that *Endoteba* and *Vachardella* have occasionally a siliceous agglutinate (see Nestell *et al.* 2006, pl. 4, figs 10–13). *Endoteba* probably first occurs in the Artinskian or Kungurian (Vachard *et al.* 2002). It is abundant in the Capitanian. *Endoteba* is rare in the Late Permian and earliest Triassic, but reappears and diversifies in the Middle Triassic (Vachard *et al.* 1994; Rettori 1995). Except for the Nodosariata, *Endoteba* is one of the most advanced foraminifers (because it is multichambered) that crosses over the PTB, the most severe biotic turnover in the geological history of the Earth. Finally, Triassic biseriate stages reappear among the Endotebidae with *Malayspirina* Vachard in Fontaine *et al.*, 1988*a* (as a homeomorph of the late Tournaisian–early middle Visean genus *Eotextularia* Mamet).

Biostratigraphy. Endoteba is questionable in questionable Artinskian olistolites (Vachard *et al.* 2001*b*), and is known from Kungurian (Angiolini *et al.* 2016), Roadian (Kobayashi 2012*a*) and Capitanian (Vachard & Razgallah 1988; Vachard & Miconnet 1990). *Vachardella* is only known in the late Capitanian. The taxa in open nomenclature, *Haplophragmina*?, *Granuliferelloides*? and ?Gen. and sp. indet. *A*, are Kungurian–Changhsingian in age.

Biogeography. Endoteba, common in Tunisia (Vachard & Razgallah 1988; Ghazzay *et al.* 2015), is also known in Italy (Apennines: Vachard & Miconnet 1990; Sicily: Vachard *et al.* 2001*b*), Cyprus (Nestell & Pronina 1997), Turkey (Leven & Okay 1996; Moix *et al.* 2013; Şahin *et al.* 2014), Oman (Hauser *et al.* 2000; Vachard unpublished data), Tajikistan (South Pamirs: Angiolini *et al.* in press), Malaysia (Ishii *et al.* 1975, pl. 2, fig. 8, reinterpreted) and Japan (Kobayashi 2012*a*). *Vachardella* and '*Haplophragmina*?' are currently endemic to Texas and New Mexico (Nestell & Nestell 2006). *Granuliferelloides*? and ?Gen. and sp. indet. *A* were found in Tethys and western Panthalassa.

Superfamily **Tetrataxoidea** Haynes, 1981
Family **Tetrataxidae** Pokorny, 1958
Subfamily **Tetrataxinae** Galloway, 1933

Permian genera. Tetrataxis Ehrenberg, 1854 emend. Nestler, 1973; *Abadehella* Okimura & Ishii in Okimura *et al.*, 1975; *Polytaxis* Cushman & Waters, 1928*a*.

Remarks. Tetrataxis and *Polytaxis* are relatively common during the Permian, but their species repeat the Carboniferous morphologies (1 & 2 in Fig. 3) and, therefore, have no biostratigraphic value. *Abadehella*, with its endoskeleton, is a Permian genus that mimics the Mississippian genus *Valvulinella* Schubert, but there is no phylogenetical continuity between these two taxa.

Biostratigraphy. Tetrataxis is Early Mississippian (late Tournaisian, MFZ6: Poty *et al.* 2006) to latest Permian (late Changhsingian: Gaillot 2006; Song *et al.* 2009; Ueno & Tsutsumi 2009). *Polytaxis* is Early Pennsylvanian (Bashkirian)–Late Permian (Wuchiapingian) in age. The most interesting Permian tetrataxid is *Abadehella*, the FAD of which is poorly known. In the Carnic Alps, we observed (Vachard *et al.* unpublished data) secondarily septated tetrataxids: that is *Abadehella sensu lato* as early as the Kungurian (3 in Fig. 3). All these tetrataxids attain the PTB, but apparently do not cross through the PTB, as the *Tetrataxis* of the Triassic (e.g. described by Salaj *et al.* 1983) obviously differ both morphologically and microstructurally. The oldest occurrence of *Abadehella* dates back to the Artinskian in the Carnic Alps (Vachard *et al.* unpublished data), the Kungurian of New Mexico (Lucas *et al.* 2015, fig. 33.12 & 33.13) and, possibly, to the lower Cisuralian in South China (see Lin *et al.* 1990). The *Abadehella* reported from Israel, NW Caucasus, Armenia, Azerbaijan, central Iran and SE Pamir are probably all Capitanian–Lopingian in age (Okimura & Ishii 1981; Kotlyar *et al.* 1984; Kobayashi 1996, 1999; Pronina-Nestell & Nestell 2001; Orlov-Labkovsky 2004).

Biogeography. All tetrataxid genera seem to be cosmopolitan. Although for a long time considered as Tethyan, *Abadehella* is now also known to be cosmopolitan, based on its occurrence in Guerrero, Mexico (Vachard *et al.* 1992), and in Texas (Nestell *et al.* 2006) and New Mexico, in the USA (Nestell & Nestell 2006; Lucas *et al.* 2015).

Superfamily **Globivalvulinoidea** Hance *et al.*, 2011
Family **Globivalvulinidae** Reitlinger, 1950 emend. Gaillot & Vachard, 2007
Subfamily **Globivalvulininae** Reitlinger, 1950 (sic Globivalvulinae)

Permian genera. Globivalvulina Schubert, 1921; *Charliella* Altıner & Özkan-Altıner, 2001; *Labioglobivalvulina* Gaillot & Vachard, 2007; *Retroseptellina* Gaillot & Vachard, 2007; *Septoglobivalvulina* Lin, 1978 emend. Gaillot & Vachard, 2007.

Remarks. The evolution of the globivalvulinids was slow and weak during the Pennsylvanian–Cisuralian (8–10 in Fig. 3). Their differentiation began in the Kungurian, and accelerated during latest Capitanian–earliest Wuchiapingian time.

Biostratigraphy. Globivalvulina is Late Mississippian (earliest Serpukhovian)–latest Permian (Changhsingian), with a presence to confirm in the earliest Triassic (see *G. curiosa* Gaillot *et al.*, 2009; see also the double PTB event of Song *et al.* 2011, 2013). The Early Mississippian (latest Tournaisian) species, formerly called *Globivalvulina bristolensis* Reichel, 1946, is now excluded from *Globivalvulina*: it corresponds to another lineage and was renamed *Parabiseriella bristolensis* by Cózar & Somerville (2012). *Charliella* appears in the Wordian of South China (Zhang *et al.* 2015) and NW Iran (Ebrahim Nejad *et al.* 2015). It is mainly known in the Capitanian, but was also found in the late Wuchiapingian and early Changhsingian. *Labioglobivalvulina* is late Capitanian–Lopingian. *Septoglobivalvulina* is early?/late Capitanian–Changhsingian. *Retroseptellina* is questionable in the late Roadian of Thailand (Ueno & Sakagami 1993; nevertheless, in our opinion, these levels might be Capitanian in age) and is principally Wordian–Changhsingian.

Biogeography. Initially, a Palaeotethyan taxon, *Globivalvulina* became cosmopolitan after the late Bashkirian. *Charliella* is present in Turkey (NW Anatolia), Italy (Monte Facito), NW Iran (Ebrahim Nejad *et al.* 2015), Zagros, Fars and Abu Dhabi (Gaillot & Vachard 2007), Sumatra (as *Globivalvulina cyprica sensu* Nguyen Duc Tien 1986*b*, pl. 13, fig. 5 only), Cambodia (as *Globivalvulina cyprica* and *Globivalvulina* sp. B *sensu* Nguyen Duc Tien 1979, pl. 9, figs 10–12, and 1986*a*, pl. 4, figs 6 & 7), central Mexico (Vachard *et al.* 1992 as *Globivalvulina* ex gr. *cyprica*, pl. 6, fig. 7), western Texas, USA (Nestell & Nestell 2006 as *Crescentia migrantis*), and Late Permian of Thailand (Vachard unpublished data). *Charliella* is probably present in Japan (Okimura 1972; Kobayashi 2013, fig. 7.17? & 7.18), southern Tibet (Wang *et al.* 2010), Yunnan (Ueno *et al.* 2010) and Oman (Jebel Akhdar, 'late Dzhulfian') (Lys in Montenat *et al.* 1977; Vachard unpublished data). *Labioglobivalvulina* is found in NW Iran, Zagros and Fars (Iran), Hazro (Turkey), ?Armenia and Azerbaijan, Montenegro, Italy, Hungary, South China, central Japan, and northern Thailand (Gaillot & Vachard 2007) and Malaysia (Aw *et al.* 1977, pl. 43, fig. 19). *Septoglobivalvulina* only exists in South China, Oman, ?Armenia and Azerbaijan, Turkey (Lycian nappes, Hazro), Iran (Fars and Zagros, NW Iran), and Abu Dhabi (see Gaillot *et al.* 2009; Koehrer *et al.* 2012). *Retroseptellina* is widespread in the Palaeo- and Neotethys: Slovenia (Nestell *et al.* 2009 as *Paraglobivalvulina*? *globosa* (Wang)); Hungary (Théry *et al.* 2007); Greece (Baud *et al.* 1991; Grant *et al.* 1991; Altıner & Özkan-Altıner 1998); southern Turkey (Köylüoglu & Altıner 1989; Ünal *et al.* 2003; Gaillot & Vachard 2007; Şahin *et al.* 2014); northern Caucasus (Pronina-Nestell & Nestell 2001 as *Paraglobivalvulina globosa*); northern Italy and Iran (Mohtat-Aghai & Vachard 2005; Ebrahim Nejad *et al.* 2015); the Duhaysan Member of the Khuff Formation in Saudi Arabia (Vachard *et al.* 2005); Oman, Batain Plain (Vachard *et al.* 2002); Armenia and Azerbaijan (Kotlyar *et al.* 1989); Thailand and Malaysia (Yanagida *et al.* 1988; Fontaine *et al.* 1993, 1994); southern Tibet (Wang *et al.* 2010); South China (Lin *et al.* 1990; Song *et al.* 2007; Gaillot *et al.* 2009; Zhang *et al.* 2015 as *Globivalvulina*? sp., fig. 4QQ); New Zealand (Vachard & Ferrière 1991); and Japan (Kobayashi 2006*a*, 2012*a*, *b*, 2013).

Subfamily **Paraglobivalvulininae** Gaillot & Vachard, 2007

Permian genera. Paraglobivalvulina Reitlinger, 1965; *Paraglobivalvulinoides* Zaninetti & Jenny-Deshusses, 1985; *Urushtenella* Pronina-Nestell in Pronina-Nestell & Nestell, 2001.

Remarks. The globivalvulinin ancestors of the paraglobivalvulinins are probably *Septoglobivalvulina* and/or *Retroseptellina.* This subfamily evolves to the realization of complex, secondary, sutural apertures.

Biostratigraphy. Paraglobivalvulina is Capitanian–Changhsingian; *Paraglobivalvulinoides* is latest Changhsingian; *Urushtenella* is Changhsingian.

Biogeography. Paraglobivalvulina is present in Armenia and Azerbaijan, Israel, Turkey, Iran (Alborz, Zagros and Fars, NW Iran), NW Caucasus, Carnic Alps, Hungary, Greece, Cyprus, Oman, Salt Range, southern Tibet, South China, Thailand, Philippines, and Japan. *Paraglobivalvulinoides* was mentioned in Iran (Alborz and Zagros), ?Italy, Greece (Vachard *et al.* 1993*b*), NW Caucasus,

Himalaya, South China, Thailand, Malaysia, Japan (Gaillot *et al.* 2009 and references therein) and SE Pamirs (Kotlyar *et al.* 1999). *Urushtenella* is endemic to NW Caucasus (Pronina-Nestell & Nestell 2001): its presence in Zagros (Gaillot & Vachard 2007), southern Tibet (Wang *et al.* 2010) and, especially, in the Capitanian of Japan (Kobayashi 2012*a*) to be confirmed.

Subfamily **Dagmaritinae** Bozorgnia, 1973

Permian genera. Dagmarita Reitlinger, 1965; *Bidagmarita* Gaillot & Vachard in Gaillot *et al.*, 2009; *Crescentia* Ciarapica *et al.*, 1986; *Danielita* Altıner & Özkan-Altıner, 2010; *Louisettita* Altıner & Brönnimann, 1980 emend. Gaillot & Vachard, 2007; *Sengoerina* Altıner, 1999; ?*Labiodagmarita* Gaillot & Vachard, 2007.

Remarks. The FAD of *Dagmarita* is discussed. However, as indicated by Altıner & Özkan-Altıner (2010), the lineage *Globivalvulina cyprica–Sengoerina–Dagmarita–Danielita* in Turkey suggests that the evolutionary derivations of dagmaritin genera occurred very rapidly in the Capitanian.

The foraminiferal material of Zheng (1986), which is dated as Chihsian (i.e. Kungurian–early Roadian), displays some taxa that are more probably late Guadalupian or even early Lopingian in age, such as *Aulacophloia* Gaillot & Vachard (pl. 5, figs 22 & 23 of Zheng 1986); *Dagmarita* Reitlinger (pl. 5, fig. 8); *Geinitzina* cf. *taurica* Sellier de Civrieux & Dessauvagie (pl. 5, figs 9–13c); transition *Globivalvulina* Schubert–*Paraglobivalvulina* Reitlinger (pl. 5, fig. 41); *Crassiglomella* Gaillot & Vachard (pl. 6, figs 3 & 4); and *Multidiscus* Miklukho–Maklay (pl. 6, fig. 7a, b). In this condition, four hypotheses are possible for the interpretation of the proposed age of this microfauna: (a) the Chihsian age is misinterpreted; (b) the smaller foraminifers are reworked (and biostratigraphically mixed) in calciturbidites; (c) these smaller foraminifers are contained in calcareous olistolites of different ages mixed in a tectonic mélange; and (d) some taxa appear previously in South China and then migrate to western Tethys. At the moment, there are no geological arguments allowing one of these hypotheses in particular to be selected, but it is clear that the FAD of *Dagmarita* is younger than the Chihsian (see below).

Biostratigraphy. The FAD of unquestionable *Dagmarita* is late Roadian (=early Murgabian of Vachard 1980) or early Maokouan (Lin *et al.* 1990); the LAD is latest Changhsingian (Zhao *et al.* 1981; Lin *et al.* 1990; Wang *et al.* 2010). *Sengoerina* is late Capitanian (Altıner 1999; Nestell & Nestell 2006; Altıner & Özkan-Altıner 2010; Zhang *et al.* 2015) to latest Changhsingian (Song *et al.* 2007, 2009, 2011). The type species of *Crescentia* comes from a Triassic calciturbidite of the Monte Facito (Italy), where it is associated with abundant Capitanian foraminifers and algae (Ciarapica *et al.* 1986; Vachard & Miconnet 1990), as for the fusulinids *Neoschwagerina* and *Kahlerina*; nevertheless, they are mixed with rarer Changhsingian taxa, such as *Colaniella* (Jenny-Deshusses *et al.* 2000). Consequently, owing to its probable reworked character, the exact dating of the type species of *Crescentia* is debatable. Rare additional specimens of this genus exist in the Changhsingian of Himalaya (Lys *et al.* 1980) and Zagros (Gaillot & Vachard 2007), which reinforce the ambiguous dating. *Bidagmarita* might be limited to the late Changhsingian. *Danielita* is Capitanian or Lopingian, reworked in the Late Triassic. *Louisettita* probably has its FAD in the late Wuchiapingian in the Persian Gulf (see Gaillot & Vachard 2007; Mohtat-Aghai *et al.* 2009), but is generally limited to Changhsingian. *Labiodagmarita* is of Lopingian age (Gaillot & Vachard 2007).

Biogeography. Dagmarita is a common Palaeotethyan, Neotethyan and Panthalassan genus, with the following detailed distribution (from west to east): Tebaga, Tunisia (Vachard & Razgallah 1988; Ghazzay *et al.* 2015); Apennines, Italy (Panzanelli-Fratoni *et al.* 1987; Vachard & Miconnet 1990); Slovenia (Nestell *et al.* 2009); Montenegro (Pantic 1970); the Carnic Alps (Noé 1987); Hungary (Berczi-Makk 1992); Greece (Hydra: Wignall *et al.* 2012); Cyprus (Nestell & Pronina 1997); Turkey (western Turkey: Argyriadis *et al.* 1976; Lys & Marcoux 1978); eastern Taurus (Altıner 1981, 1984; Zaninetti *et al.* 1981; Köylüoglu & Altıner 1989); Hazro (Canuti *et al.* 1970 as *Palaeotextularia* sp.; updated in this study); Armenia and Azerbaijan (Reitlinger 1965; Kotlyar *et al.* 1984, 1989); Iran (central Alborz: Bozorgnia 1973; Jenny-Deshusses 1983; Mohtat-Aghai & Vachard 2003; central Iran, Abadeh: Okimura & Ishii 1981; Zagros and Fars: Gaillot & Vachard 2007; and NW Iran: Ebrahim Nejad *et al.* 2015); Oman (Koehrer *et al.* 2012); central Afghanistan (Vachard & Montenat 1981); Salt Range, Pakistan (Okimura 1988); Lamayuru, Ladakh, Himalaya (Lys *et al.* 1980); South China (Zhao *et al.* 1981; Lin *et al.* 1990; Gaillot *et al.* 2009); western Thailand (Fontaine *et al.* 1988*c*) and the Philippines (Fontaine *et al.* 1986); NWThailand (Caridroit *et al.* 1990); Malaysia (Fontaine *et al.* 1988*b*); Cambodia (Nguyen Duc Tien 1979, 1986*a*); Primorye (Sosnina in Sosnina & Nikitina 1977); and Japan (Kobayashi 1997*c*, 2006*b*, *c*, 2012*a*, *b*, 2013). *Dagmarita* was also mentioned in Texas, USA (Nestell *et al.* 2006). *Sengoerina* is known in Turkey, Greece (Hydra), South China and New Mexico, and doubtful in central Iran and northern

Afghanistan (Vachard unpublished data). *Crescentia* comes from Monte Facito (Italy); rare specimens also exist in Zagros (Gaillot & Vachard 2007), Himalaya (Lys *et al.* 1980) and Thailand (Fontaine *et al.* 1993); the so-called *Crescentia* of Texas, USA (Nestell & Nestell 2006) are misinterpreted (see above). *Bidagmarita* is known in the Salt Range, Pakistan, central Iran, NW Caucasus, the Lamayuru Block (Ladakh Himalaya), Malaysia and South China (Gaillot *et al.* 2009). *Danielita* is endemic to the Bursa area (Turkey). *Louisettita* is present in the Changhsingian of Turkey, Iran (Zagros, Julfa area), South China, and, questionably, of Afghanistan, Armenia and Azerbaijan; its FAD probably occurred in the Persian Gulf (see Gaillot & Vachard 2007; Mohtat-Aghai *et al.* 2009). *Labiodagmarita* only exists in Saudi Arabia (Vachard *et al.* 2005), Zagros, Fars, Abu Dhabi and Turkey (Gaillot & Vachard 2007).

Subfamily **Paradagmaritinae** Gaillot & Vachard, 2007

Permian genera. Paradagmarita Lys in Lys & Marcoux, 1978; *Paradagmacrusta* Gaillot & Vachard, 2007; *Paradagmaritella* Gaillot & Vachard, 2007; *Paradagmaritopsis* Gaillot *et al.*, 2009; *Paremiratella* Gaillot & Vachard, 2007.

Remarks. Described by Gaillot & Vachard (2007), this family has been critically revised by Altıner & Özkan-Altıner (2010): for example, Altıner (pers. comm. March 2016) suggests that the genera *Paradagmaritella*, *Paradagmaritopsis* and *Paremiratella* should be included in the subfamily Dagmaritinae. However, as these taxonomic problems do not modify the biostratigraphy and biogeography of the paradagmaritinins, the previous classification is followed herein.

Biostratigraphy. According to Kotlyar *et al.* (1989, table 1), the first representative of the subfamily appears in the latest Khachikian (interpreted by these authors as the latest Capitanian/Midian). However, this part of the Khachik Formation is, most probably, earliest Wuchiapingian in age (Leven 1998; and see below). Gaillot & Vachard (2007) indicated that the range of *Paradagmarita sensu stricto* is late Wuchiapingian-Changhsingian. Other genera of the subfamily have more restricted ranges: *Paradagmacrusta* is Changhsingian; *Paradagmaritella* is ?latest Capitanian–Changhsingian; *Paradagmaritopsis* is late Wuchiapingian–Changhsingian; and *Paremiratella* is Changhsingian.

Biogeography. The subfamily appeared probably in Armenia or in Azerbaijan (i.e. the former Transcaucasia), but, as indicated above, the corresponding datum of Kotlyar *et al.* (1989) is yet to be verified. The subfamily is Palaeotethyan and Neotethyan, and is principally known in Turkey (Taurus: Altıner 1984), Iran (Zagros: Gaillot & Vachard 2007; NW Iran: Ebrahim Nejad *et al.* 2015), Saudi Arabia (Vachard *et al.* 2005), Oman (Koehrer *et al.* 2012) and NW Caucasus (Pronina-Nestell & Nestell 2001), although it was mentioned from Italy to Japan. The so-called *Paradagmarita* from Afghanistan described by Vachard (1980) in reality belong to *Louisettita*; those from Thailand and Pakistan are very atypical; and those from Japan (Kobayashi 1997*c*, 2004) instead belong to *Paradagmaritopsis*, a genus also observed in southern China (Gaillot *et al.* 2009) and Oman (Koehrer *et al.* 2012). The other genera are more endemic: (1) *Paradagmacrusta* is only mentioned in Zagros and Fars (Iran); (2) *Paradagmaritella*, in Armenia or Azerbaijan, Fars, Zagros, Saudi Arabia, and, questionably, in NW Caucasus and Turkey; and (3) *Paradagmaritopsis* in Iran (Zagros and Fars), Japan and South China (Gaillot *et al.* 2009). *Paremiratella* is endemic to Zagros, Fars, Abu Dhabi and Oman (and questionable in Japan).

Class **Miliolata** Saidova, 1981
Order **Cornuspirida** Mikhalevich, 1980

All Late Palaeozoic miliolates exhibit a diversely coiled, undivided, tubular chamber (11–17 in Fig. 3 and Figs 4–6). Hence, they constitute a unique order: the Cornuspirida. Two suborders are distinguished here: the attached Nubeculariina and the free Cornuspirina. We agree with Gaillot & Vachard (2007) in speculating that all the Palaeozoic miliolates derive from a pseudolituotubellid Fusulinata at the end of the Visean, since the majority of the Pennsylvanian miliolates are calcivertellids (i.e. attached forms). However, contrary to the opinion of Gargouri & Vachard (1988), Vachard *et al.* (1993*a*) and Gaillot & Vachard (2007), who supposed that the hemigordiopsin cornuspirids gave rise to the involutinids, and in agreement with Groves & Altıner (2005), it is more likely that Triassic involutinids might have arisen from a very simple, planispiral evolute ancestor, which is either *Pseudoammodiscus* (consensual hypothesis), *Ammodiscus sensu* Nestell *et al.* (2015) or, most probably, *Postcladella* Krainer & Vachard, 2011.

The name Cornuspirida Mikhalevich, 1980 has priority over Cornuspirida Vdovenko *et al.*, 1993 and Hemigordiopsida Pronina, 1994, for designating this order of miliolates.

Palaeozoic miliolates are known from the latest Visean or earliest Serpukhovian, and are common during the Permian. Nubeculariina are abundant as early as the Serpukhovian–Bashkirian. Primitive typical cornuspirids (with the typical, well-preserved, amber-coloured test: see the discussion in Vachard *et al.* 2015) (see Figs 4 & 5, showing the initial aspect of the wall (Fig. 4) and its

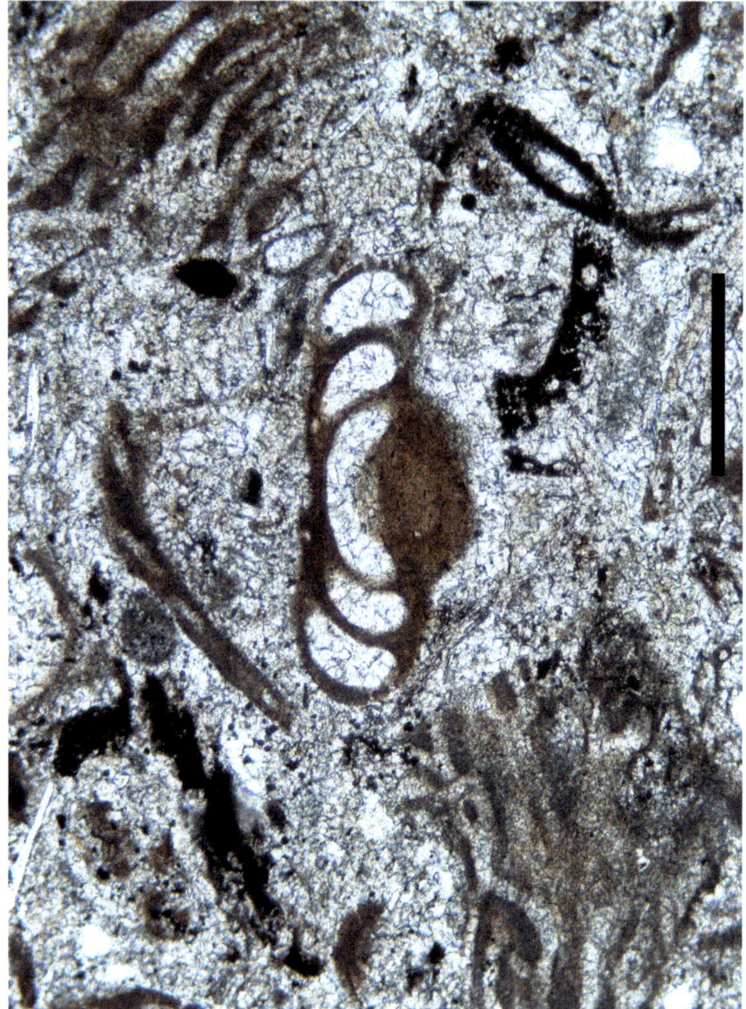

Fig. 4. Well-preserved *Uralogordius* (centre) and *Ellesmerella* (top left and bottom right) with amber-coloured wall characteristic of the extant Miliolata, to compare with the dark-microgranular wall of an earlandiid (top right). Scale bar = 0.250 mm. Early Artinskian of Garnitzenbach (Carnic Alps, Austria).

diagenetic transformations (Fig. 5)), *Cornuspira* and *Hemigordiellina*, are only known from the Bashkirian. More complex coilings appear with *Brunsiella* and then with *Hemigordius* in the Moscovian. The diversity is already important in the Cisuralian, but increases gradually during the Guadalupian where neodiscid and hemigordiopsid giant forms appear. The Capitanian–Lopingian is an episode of maximal diversity for the group.

Suborder **Nubeculariina** Jones in Griffith & Henfrey, 1875 nom. translat. Mikhalevich, 1988 nomen re-translat. herein

Family **Calcivertellidae** Reitlinger in Vdovenko *et al.*, 1993 emend. Gaillot & Vachard, 2007

Permian genera. Calcivertella Cushman & Waters, 1928*b*; *Ammovertella* Cushman, 1928; *Baryshnikovia* Reitlinger in Vdovenko *et al.*, 1993; *Calcitornella* Cushman & Waters, 1928*b*; *Hedraites* Henbest, 1963; *Hedraites*? *sensu* Lucas *et al.*, 2015; *Palaeonubecularia* Reitlinger, 1950; *Pseudoagathammina* Lin *et al.*, 1990; *Pseudospira* Reitlinger in Vdovenko *et al.*, 1993; *Pseudovermiporella* Elliott, 1958 emend. Henbest, 1963; *Tansillites* Nestell & Nestell, 2006; *Trepeilopsis* Cushman & Waters, 1928*b*; ?*Orthovertella*

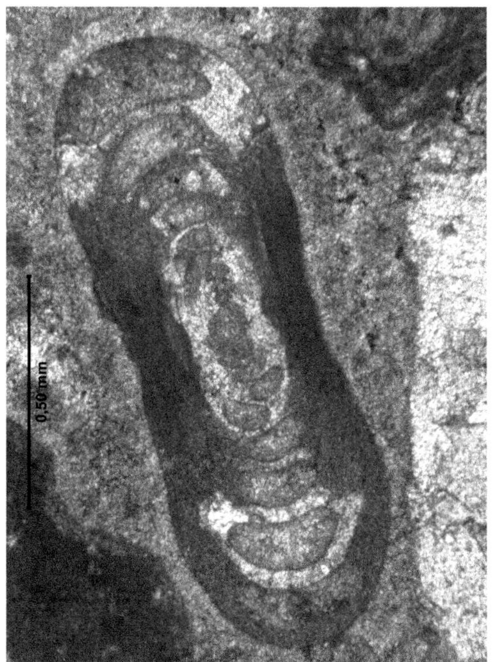

Fig. 5. *Uralogordius* sp., after diagenesis of the wall (cf. with Fig. 4); this wall became first dark, then whitish and neomicrosparitized. Scale bar = 0.500 mm. Kungurian of the stratotype of the regional Bolorian Stage (Pamirs; Tajikistan).

Cushman & Waters, 1928*b*; ?*Orthovertellopsis* Vachard *et al.*, 2015.

Remarks. Calcivertelloids seem to constitute the most primitive Miliolata, whereas the attachment is generally considered as a secondary trend among the foraminifers. *Orthovertella* and *Orthovertellopsis*, because of the growth of their tests, seem to have been attached during part of their life: therefore, they are questionably assigned to calcivertellids.

Biostratigraphy. All these genera are long ranged, *Calcivertella* is Late Mississippian (Serpukhovian)–Late Permian. *Calcitornella*, *Trepeilopsis*, *Ammovertella* and *Orthovertella* are Pennsylvanian–Permian. Other genera might be more restricted in time, but they are still poorly known. *Orthovertellopsis* is biostratigraphically poorly known (i.e. cited as Kungurian (late Leonardian), Early Permian, Permian, Early Kazanian (Middle Permian) or ?latest Capitanian, following the areas where it is present). *Tansillites* is late Capitanian. *Pseudoagathammina*, *Baryshnikovia* and *Pseudospira* also have a questionable Permian distribution. *Palaeonubecularia* is Serpukhovian–Changhsingian: perhaps Triassic in Europe and Taurus (Turkey). *Hedraites* (11 in Fig. 3) is Late Pennsylvanian–Cisuralian, *Hedraites*? *sensu* Lucas *et al.* (2015) is Kungurian, *Pseudovermiporella* is late Cisuralian (Artinskian or even latest Sakmarian)–latest Permian (late Changhsingian: Zhao *et al.* 1981; Flügel & Reinhardt 1989; Vachard *et al.* 2003; Mohtat-Aghai *et al.* 2009; Ebrahim Nejad *et al.* 2015), it is common in the Lopingian of southern Turkey (Hazro) and Zagros. Recently, we used the Artinskian FAD of *Pseudovermiporella* as a biozonal, regional marker in South China (Liu *et al.* in press).

Biogeography. Calcivertella and *Calcitornella* are cosmopolitan. *Trepeilopsis*, *Orthovertella* and *Ammovertella* are more rarely cited, but are probably also cosmopolitan. *Orthovertellopsis* exists in New Mexico, Australia (Carnarvon Basin, Canning Basin, Tasmania, and, perhaps, the Sydney Basin), South China, the Urals (Russia) and, perhaps, Texas (Vachard *et al.* 2015). *Pseudoagathammina* was described in South China, but was recently re-found in Tunisia (Ghazzay-Souli *et al.* 2015). *Tansillites* is endemic to New Mexico. *Baryshnikovia* and *Pseudospira* are sporadically mentioned; they are, perhaps, cosmopolitan, being currently known in Donbass, the Pre-Urals, Uzbekistan and New Mexico. *Palaeonubecularia* is cosmopolitan. *Hedraites* and *Hedraites*? *sensu* Lucas *et al.* (2015) are common in North America and rare in Tethys. *Pseudovermiporella* is cosmopolitan, and is especially common during the Lopingian in Turkey and Iran (Ebrahim Nejad *et al.* 2015).

Suborder **Cornuspirina** Jirovec, 1953 emend. Gaillot & Vachard, 2007
Superfamily **Cornuspiroidea** Mikhalevich, 1988
Family **Cornuspiridae** Schultze, 1854
Subfamily **Cornuspirinae** Schultze, 1854

Permian genera. Cornuspira Schultze, 1854 (=*Ammodiscus* Reuss, 1862 (part.); =*Pseudoammodiscus* Conil & Lys in Conil & Pirlet, 1970 (part.)); *Flectospira* Crespin & Belford, 1957; *Glomotrocholina* Nikitina in Sosnina & Nikitina, 1977; *Hemigordiellina* Marie in Deleau & Marie, 1961 emend. Vachard in Vachard & Beckary, 1991; *Hoyenella* Rettori, 1994 emend. Gaillot & Vachard, 2007 (=*Glomospirella* Plummer, 1945 (part.)); *Pilammina sensu* Vachard in Termier *et al.*, 1977 *non* Pantic, 1965; *Postcladella* Krainer & Vachard, 2011 (=*Ammodiscus* Reuss, 1862 (part.) = *Rectocornuspira* Warthin, 1930 (part.)); *Pseudohemigordius* Nestell & Nestell, 2006; *Streblospira* Crespin & Belford, 1957.

Remarks. Almost every genus of this family is discussed; especially *Hemigordiellina* (with *Glomospira diversa* Cushman & Waters, 1930 as type species) and *Glomospira* or *Pseudoglomospira*, as well as the quatuor *Cornuspira*, *Rectocornuspira*,

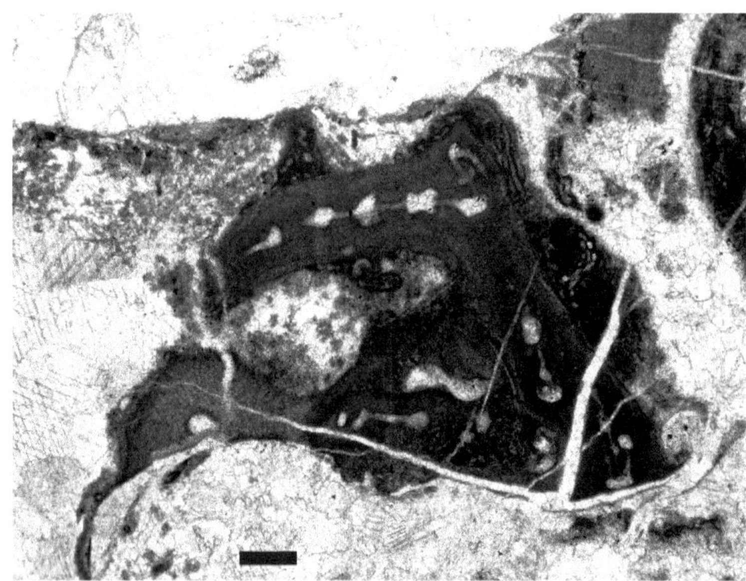

Fig. 6. A complex specimen of *Tubiphytes* ex gr. *obscurus* Maslov, 1956 showing a well-preserved series of sagittiform (arrow-shaped) chambers. Scale bar = 0.250 mm. Early Kungurian of Trogkofel (Carnic Alps, Austria).

Postcladella, *Pseudoammodiscus* or *Ammodiscus* (12 & 13 in Fig. 3) (e.g. Loeblich & Tappan 1987; Vachard *et al.* 2010; Krainer & Vachard 2011; Nestell *et al.* 2015). In the Permian, the genus *Cornuspira* was first confused with *Ammodiscus*, and then with *Pseudoammodiscus* (e.g. Groves & Boardman 1999; Pronina, 1999*a*, p. 184; Nestell & Nestell 2006) or again with *Ammodiscus* in the Triassic (Nestell *et al.* 2015). *Postcladella*, which has distinct morphological characters compared with *Ammodiscus*, *Cornuspira* and *Rectocornuspira* (see the discussion in Krainer & Vachard 2011; Vachard *et al.* 2015), was, however, recently re-synonymized with *Ammodiscus* (Nestell *et al.* 2015): this latter interpretation requires further intensive discussion. If it is admitted that the wall microstructure has a secondary importance, *Glomospira* can, in turn, replace *Pseudoglomospira* and *Hemigordiellina*, despite these genera, in my opinion, corresponding to three different classes: Textulariata, Fusulinata and Miliolata. *Rectocornupira* is a Pennsylvanian porcelaneous genus (Warthin 1930; Loeblich & Tappan 1987). This name was erroneously applied to : (1) dark-microgranular walled, uncoiled pseudoammodiscids of the Visean–Serpukhovian boundary interval (see Vdovenko *et al.* 1993); and (2) to Early Triassic *Postcladella kahlori* (Brönnimann *et al.*) (see Krainer & Vachard 2011). Recently, this latter species was reconsidered as possessing an agglutinated wall and reassigned to *Ammodiscus* (see Nestell *et al.* 2015). However, because of the terminal uncoiling of *Postcladella kahlori*, it is to be noted that the name *Rectoammodiscus* Reitlinger in Vdovenko *et al.*, 1993 would be formally more correct.

Other controversial genera are *Streblospira*, *Glomotrocholina*, *Pilammina*? and *Hoyenella*. The genus *Streblospira* is interpreted herein as different to the Triassic marker *Meandrospira*, as has been done by Vdovenko *et al.* (1993), but contrary to Loeblich & Tappan (1964, 1987) and Kalia *et al.* (2000). *Glomotrocholina* described in the Middle–Late Permian of the Primorye has never been re-found since its creation, although it occurs in reefal environments (G. Nestell pers. comm. March 2016), which are relatively common in the Middle Permian and where the foraminifers are abundant. *Glomotrocholina* is therefore difficult to identify and needs a revision because it probably differs significantly from its proposed diagnosis. Eventually, *Pseudohemigordius* and/or *Vissariotaxis*? *nativus* Nestell & Nestell in Nestell *et al.*, 2006 from the Capitanian of Texas might correspond to *Glomotrocholina* or to a homeomorph of this latter genus. *Pilammina*? sp., only mentioned by Vachard in Termier *et al.* 1977 and Gaillot & Vachard 2007, and strongly questioned by D. Altıner (pers. comm. March 2016), differs indeed from *Pilammina* Pantic, 1965, and is possibly only a minute species of *Hemigordiellina*. It is noteworthy that Palmieri (2004) proposed the new name *Flectospiroides* for a taxon previously called *Pilammina* by him, and of 'Late Kazanian–Midian (?)' (i.e. Middle Permian) age in Australia. The genus *Hoyenella sensu* Gaillot & Vachard, 2007 is also discussed because: (1) it could differ from *Hoyenella* Rettori, 1994, the distribution of which would only be Triassic (D. Altıner pers. comm. March 2016); and (2) it is homeomorph with the microgranular genus *Brunsia*, the porcelaneous genus *Brunsiella* and the agglutinated genus *Glomospirella*. Owing to the current trends, there is great temptation to return to *Glomospirella*, but, as indicated by Vachard *et al.* (2015), this genus has *Glomospira umbilicata* Cushman and Waters for the type species, which is a very large taxon, with a diameter of up to 1.00 mm. Its age is Middle–Late Pennsylvanian, whereas other species described in *Glomospirella* are small to very small (because they measure generally 0.25–0.40 mm), with a second part of coiling markedly planispiral and evolute, and are often described in the Permian and the Triassic. However, it is clear that these Permian glomospirellids have a porcelaneous wall, whereas the true microstructure of Triassic *Hoyenella* is not well established (porcelaneous, calcitic microgranular, aragonitic microgranular or an uncharacteristic and unknown wall?).

Biostratigraphy. Cornuspira is early?/late Serpukhovian (i.e. latest Mississippian) to Holocene. *Postcladella* is essentially an Early Triassic taxon (Krainer & Vachard 2011), but it is very rare in the late Capitanian and sporadic in the Late Permian of the Middle East (Shang *et al.* 2003). *Hemigordiellina* is Pennsylvanian and Permian; it appears as soon as the bed T in Tindouf (Morocco) (Cózar *et al.* 2014), which is probably late Serpukhovian in age, and earlier in Europe, if the poorly known genus *Warnantella* Conil & Lys in Conil *et al.*, 1977 is synonymous of *Hemigordiellina*. *Hoyenella* has a possible FAD in the early Sakmarian (see Vachard & Krainer 2001*b*, pl. 5, fig. 24): its acme is Early–Late Triassic. *Pilammina*? is rare in the late Guadalupian of Tebaga (Tunisia) (Termier *et al.* 1977), and the late Wuchiapingian of Zagros (Gaillot & Vachard 2007). *Glomotrocholina* is Middle or Late Permian. *Flectospira* is Artinskian, and *Streblospira* is ?Artinskian–Middle Permian in age.

Biogeography. Cornuspira and *Hemigordiellina* are probably cosmopolitan during the Permian. *Postcladella* is widespread in the Tethyan Early Triassic, but is very rare in the Middle East during the Permian. *Hoyenella* is cosmopolitan in the Permian, and is distributed, during its Triassic acme, from Western Europe to China and Japan. *Pilammina*? is endemic to Tunisia and Zagros. Other endemic genera are *Glomotrocholina* (Primorye) and *Flectospira* (Western Australia). The *Flectospira* of eastern Himalaya illustrated by Kalia *et al.* (2000) are misinterpreted and could correspond to every other cornuspirin. *Streblospira* is rare out of Australia, where it was described, but was mentioned in Bolivia (Mamet 1996), Armenia–Azerbaijan (Kotlyar *et al.* 1984), Oman (Vachard *et al.* 2002) and Turkey (Wignall *et al.* 2012; Şahin *et al.* 2014).

Subfamily **Agathammininae** Ciarapica, Cirilli & Zaninetti in Ciarapica *et al.*, 1987

Permian genera. Agathammina Neumayr, 1887; *Septagathammina* Lin, 1984; *Graecodiscus* Vachard in Vachard *et al.*, 1993*a*.

Remarks. The agathamminins seem to be the group that is transitional between the Palaeozoic, atypical miliolates and the typical Mesozoic–Recent miliolinins; hence, it is difficult to believe that, as in many molecular clocks, the miliolinins derive from the very primitive, monothalamous foraminifers.

Biostratigraphy. The FAD and LAD of *Agathammina* are poorly known (Pronina 1988*b*, 1999*a*), probably Artinskian–Kungurian and latest Changhsinghian, respectively. The genus is relatively common from the Capitanian to the Changhsingian throughout the Palaeotethys; many formerly admitted Triassic survivors are now assigned to another genus. The FAD might be Kungurian (see Vachard & Moix 2013). *Septagathammina* is late Chihsian–Wuchiapingian (Lin *et al.* 1990) or Changhsingian. *Graecodiscus* defined in the Lopingian might appear as early as the Roadian (Angiolini *et al.* 2015).

Biogeography. Agathammina is traditionally cited throughout the Palaeotethys, from Tunisia to Japan. Although mentioned for a relatively short time in North America (Nestell & Nestell 2006; Nestell *et al.* 2006), where it can be common (G. Nestell pers. comm. March 2016). *Septagathammina* is known in South China (Lin *et al.* 1990), Sumatra, Cambodia, Armenia, Azerbaijan, Crimea, NW Caucasus, ?Hungary, Zagros, Thailand and ?Primorye (Nestell & Nestell 2006; Gaillot & Vachard 2007). *Graecodiscus*, described in Greece, might be also present in Saudi Arabia (Vachard *et al.* 2005), New Mexico (Nestell & Nestell 2006), Tajikistan (Angiolini *et al.* 2015) and South China (Liu *et al.* in press).

Family **Hemigordiidae** Reitlinger in Vdovenko *et al.*, 1993

Subfamily **Hemigordiinae** Pronina, 1994

Permian genera. Hemigordius Schubert, 1908; *Arenovidalina* Ho, 1959 emend. Gaillot & Vachard, 2007 (=*Multidiscus* of the authors, part.); *Brunsiella* Reitlinger, 1950; *Neodiscopsis* Gaillot & Vachard, 2007; *Okimuraites* Reitlinger in Vdovenko *et al.*, 1993 emend. Ebrahim Nedjad *et al.* 2015 (=? *Brunsispirella* Gaillot & Vachard,

2007); *Rectogordius* Alipour & Vachard in Alipour *et al.*, 2013 (=*Neohemigordius* of the authors *non* Wang & Sun); ?*Triadodiscus* Piller, 1983.

Remarks. There exist several nomenclatural problems in this family. For example, as suggested by Nestell *et al.* (2009), *Brunsispirella* could be an unnecessary junior synonym of *Okimuraites*: however, both taxa differ from *Hemigordius* (17 in Fig. 3) by their smaller, more compressed tests with many planispiral, evolute whorls. *Triadodiscus* either belongs to this group (partly or entirely) or to the involutinids: this depends on whether the wall remains porcelaneous or is already aragonitic in composition. Similarly, according their true microstructure, the different species of *Arenovidalina* probably belong to several genera.

Biostratigraphy. The FAD of *Hemigordius* is Bashkirian. Its acme is Moscovian–Changhsingian. *Rectogordius* is Cisuralian (late Asselian–Artinskian) (Forke *et al.* 1998; Vachard & Krainer 2001*a*, *b*). *Brunsiella*, which appeared in the Early Pennsylvanian and is a probable ancestor of *Hemigordius*, was still mentioned in the Cisuralian by Groves & Boardman (1999). *Okimuraites* is late Capitanian–Lopingian. *Arenovidalina* is Late Pennsylvanian–Late Permian and Early–Late Triassic. *Neodiscopsis* is late Capitanian–Changhsingian.

Biogeography. Hemigordius is cosmopolitan from Moscovian to Changhsingian. *Rectogordius* is Early Permian (Late Asselian–Artinskian) of Iran, Arctic Canada, Donets, Afghanistan and the Carnic Alps (Forke *et al.* 1998; Vachard & Krainer 2001*a*, *b*). Consequently, its palaeobiogeographical range seems to be located in the northern Palaeotethys, in the shelf of the Uralian Ocean and in the northern part of North America. *Arenovidalina* is apparently cosmopolitan, and is known especially in Greece, Italy, Turkey, Armenia–Azerbaijan, Iran (Zagros, Fars), Saudi Arabia, Oman, Sumatra, Primorye, South China and Japan. *Neodiscopsis* was found in Turkey (Antalya nappes and Hazro) (Şahin *et al.* 2014), Oman (Koehrer *et al.* 2012), Iran (Zagros), South China and Japan (Gaillot & Vachard 2007).

Family **Neodiscidae** Lin, 1984 nom. translat. and emend. Gaillot & Vachard, 2007

Permian genera. Neodiscus Miklukho-Maklay, 1953 emend. Gaillot & Vachard, 2007; *Crassispirella* Gaillot & Vachard, 2007; *Crassiglomella* Gaillot & Vachard, 2007; *Glomomidiella* Vachard *et al.*, 2008; *Multidiscus* Miklukho-Maklay, 1953 emend. Gaillot & Vachard, 2007; *Neohemigordius* Wang & Sun, 1973 emend. Alipour *et al.*, 2013; *Olgaorlovella* Vachard *et al.*, 2015; *Praeneodiscus* Vachard *et al.*, 2015; *Uralogordius* Gaillot & Vachard, 2007 (=*Arenovidalina sensu* Baryshnikov *et al.*, 1982 non Ho, 1959); ?*Midiella* Pronina, 1988*b*.

Remarks. This family displays a trend to the gigantism (compared with the normal size of the Miliolata); another trend is the presence of buttresses, and a differentiated wall with an outer tectum (14–16 in Fig. 3 & Figs 4 & 5). *Midiella*, because of its sigmoidal coiling and its moderate size, could be transititional between the Hemigordiidae and the Neodiscidae.

Biostratigraphy. Neodiscus is Middle–Late Permian. *Glomomidiella* is relatively rare from the Kungurian to the Wordian, and has its acme from the Capitanian to the Changhsingian. *Olgaorlovella* and *Praeneodiscus* are currently limited to the Cisuralian (?Sakmarian–Kungurian). *Neohemigordius* and *Crassiglomella* are probably Capitanian–Changhsingian. *Multidiscus* might be late Cisuralian, but is most probably Wordian–Changhsingian. *Uralogordius* is principally known in the late Cisuralian (Baryshnikov *et al.* 1982; Vachard *et al.* unpublished data), but possibly survives into the Middle Permian. *Crassispirella* is Lopingian, and mainly Changhsingian. *Midiella* is present in all of the Permian; its acme is Capitanian–Changhsingian.

Biogeography. Neodiscus is present in Oman, Slovenia, Tunisia, NW Iran, central Iran, NW Caucasus, Greece, Turkey (Hazro, Himmetli), SE Pamirs, southern Tibet, South China and Japan. *Glomomidiella*, during its acme, was mentioned in Greece, Tunisia, Croatia, Serbia, Hungary, Italy, Armenia, Azerbaijan, Turkey, Iran, Oman, Himalaya, South China, Sumatra, Malaysia and Japan. *Olgaorlovella* and *Praeneodiscus*, described in New Mexico, might exist in Greece (as *Hemigordius* aff. *ovatus* Grozdilova *sensu* Vachard *et al.* 1993*b*, pl. 3, fig. 14), Japan and Bashkortostan (Russia). *Neohemigordius* is from South China and Japan (Kobayashi 2012*a*). *Crassiglomella* was encountered in South China, central Japan, Cambodia, Turkey (Himmetli), Iran (Julfa area, Zagros and Fars), Oman and Tunisia. *Multidiscus* is found in Sumatra, Italy, Slovenia, Greece, Israel, Turkey, Armenia–Azerbaijan, Saudi Arabia, Oman, Iran (Zagros, Fars), southern Tibet, South China, Japan and Texas. *Uralogordius* is present in the Urals, the Carnic Alps (Vachard *et al.* unpublished data), Armenia–Azerbaijan, northern Afghanistan, Japan and Mexico; *Crassispirella* in Persian Gulf (Iran), Turkey (Hazro), Saudi Arabia and South China. *Midiella* is distributed in the Palaeotethys and Neotethys: Turkey (Hazro), NW Iran, central Iran (Abadeh), Persian Gulf, Saudi Arabia, Oman, South China and Primorye.

Family **Baisalinidae** Loeblich & Tappan, 1986*a*

Permian genera. Baisalina Reitlinger, 1965; *Pseudobaisalina* Sosnina, 1983; *Nikitinella* Sosnina, 1983; *Pseudomidiella* Pronina-Nestell in Pronina-Nestell & Nestell, 2001; *Septigordius* Gaillot & Vachard, 2007.

Remarks. This family is characterized by a pseudosepation, remarkably irregularly sized and distributed compared with the other foraminifers. Some illustrations of Kobayashi (1988, pl. 2, figs 18 & 19, 2012*a*, pl. 3, 2013, fig. 8.38–8.42) are especially significant.

Biostratigraphy. Baisalina is late Capitanian–late Changhsingian; *Pseudobaisalina* and *Nikitinella* are Wordian and/or Capitanian; *Pseudomidiella* is Changhsingian (Pronina-Nestell & Nestell 2001); *Septigordius* is late Early Permian–latest Permian.

Biogeography. Baisalina is a mentioned in Armenia, Azerbaijan, Hungary, Greece, Turkey, NW and central Iran, Afghanistan, Myanmar, southern Tibet (Wang *et al.* 2010), South China (Zhang *et al.* 2015), Japan (Kobayashi 1988, 2012*a*, 2013), New Zealand (Vachard & Ferrière 1991) and Texas, USA (Nestell *et al.* 2006). *Nikitinella* and *Pseudobaisalina* are endemic to Primorye (Sosnina 1983). *Pseudomidiella* occurs in NW Caucasus (Pronina-Nestell & Nestell 2001), Slovenia (Nestell *et al.* 2009), Greece (Hydra: Vachard *et al.* 2008; Wignall *et al.* 2012). *Septigordius* seems to be Palaeotethyan, Neotethyan and Panthalassan: Primorye, Crimea, Cyprus, Turkey, Oman, Ladakh, South China, Japan and New Zealand (Gaillot & Vachard 2007 and references therein).

Family **Hemigordiopsidae** Nikitina, 1969 emend. Gaillot & Vachard, 2007

Permian genera. Hemigordiopsis Reichel, 1945 (=*Gansudiscus* Wang & Sun, 1973); *Glomomidiellopsis* Gaillot & Vachard, 2007; *Kamurana* Altıner & Zaninetti, 1977; *Lysites* Reitlinger in Vdovenko *et al.*, 1993; *Shanita* Brönnimann, Whittaker & Zaninetti, 1978.

Remarks. The Hemigordiopsidae is another giant family of the Miliolata during the Wordian?–Capitanian–Lopingian. It presents the first phenomenon of flosculinization (i.e. the thickenings of basal layers that reduce the height of chambers in miliolids and alveolinids) registered among the Miliolata (Gaillot & Vachard 2007; Vachard *et al.* 2010).

Biostratigraphy. Hemigordiopsis is rare in the Wordian (Zhang *et al.* 2015), common in the Capitanian and the Wuchiapingian (Vachard *et al.* 2003), and questionable in the Changhsingian (Altıner 1984). *Lysites* are probably restricted in the Capitanian–Wuchiapingian boudary interval. *Shanita* characterizes the Capitanian–Wuchiapingian boundary (Gaillot & Vachard 2007; Koehrer *et al.* 2012; Kolodka *et al.* 2012). Because of the association of *Shanita* with *Chusenella* in Oman, and *Neoschwagerina* in Iran, apparently preserved *in situ*, we consider *Shanita* as preferentially latest Capitanian rather than earliest Lopingian. *Glomomidiellopsis* is late Capitanian–Changhsingian. *Kamurana* is either late Capitanian or Wuchiapingian, but was mentioned in the latest Changhsingian of South China (Wignall & Hallam 1996, text fig. 4 p. 591) and southern Tibet (Wang *et al.* 2010).

Biogeography. Hemigordiopsis with synonym *Gansudiscus* were unquestionably mentioned in Cyprus, Montenegro, Greece, Tunisia, Turkey, Armenia, Oman, central Iran, peninsular Thailand, South China and Japan (see discussions in Gargouri & Vachard 1988; Nestell & Pronina 1997). *Lysites*, even if defined in Turkey, seems to be more common in eastern Tethys (South China and Thailand). Known from Turkey and Oman to Myanmar, Thailand and Yunnan (e.g. Brönnimann *et al.* 1978; Sheng & He 1983; Şengör *et al.* 1988; Dawson *et al.* 1993; Jin & Yang 2004), *Shanita* is considered here as characteristic of these peri-Gondwanan terranes which, after the opening of the Neotethys, migrated to the Cimmeria and Sibumasu microcontinents (Nestell & Pronina 1997). *Glomomidiellopsis* is known in Cambodia, Primorye, Hazro (Turkey), Zagros (Iran) and Oman, and is questionable in Myanmar. *Kamurana sensu stricto* seems endemic of Turkey, but, because often confused with *Neodiscus* and/or *Glomomidiellopsis*, was also mentioned in Italy (Panzanelli-Fratoni *et al.* 1987), Armenia–Azerbaijan (Pronina 1988*a*), South China (Wignall & Hallam 1996) and New Zealand (Vachard & Ferrière 1991).

Groups incertae sedis
possibly partly related to the Miliolata
Group 1: ellesmerellids

Permian genera. Ellesmerella Mamet & Roux in Mamet, Roux & Nassichuk, 1987 emend. Vachard & Krainer, 2001*b* (eventually junior synonym of *Osagia* and/or *Ottonosia*, which designate, in reality, biopisoliths composed of *Ellesmerella*; e.g. see Henbest 1963; Vachard *et al.* 2015).

Remarks. Girvanella permica Pia, 1937, the type species of *Ellesmerella*, was initially described as an alga or cyanobacterium. In contrast, it was proposed (Vachard & Krainer 2001*b*; Sanders & Krainer 2005; Vachard *et al.* 2015; Vachard *et al.* unpublished data) that it is most probably a representative of the family Nubeculariidae Jones, 1895 because of its amber-coloured well-preserved wall (Fig. 4), and because some nubeculariids can build biopisolites and bioconstructions (e.g. 'oolithes canabinnes', 'microreefs', *Osagia, Ottonosia*), which mimic oncoids or columnal stromatolites.

Biostratigraphy. Early Cisuralian (see Vachard & Krainer 2001*b*).

Biogeography. Probably cosmopolitan (especially in peri-Gondwanan and North American craton shelves).

Group 2: tubiphytids

Permian genera. Tubiphytes Maslov, 1956; *Epimonella* Vachard in Kolodka *et al.*, 2012; *Latitubiphytes* Vachard *et al.*, 2012; *Ramovsia* Kochansky-Devidé, 1973 (=*Dorudia* Jenny & Jenny-Deshusses, 1978); '*Tubiphytes*' *epimonellaeformis* Vachard *et al.*, 2015.

Remarks. No amber-coloured, well-preserved specimens were never encountered among the tubiphytids. However, relationships with the miliolates have been suggested for a long time (see Vachard *et al.* 2010 and references therein), especially for the Jurassic forms based to some similarities between the inner cavities of the tubiphytids and the shape of the chambers of some miliolates (Fig. 6). Another argument is that the first tubiphytids resemble the miliolate *Palaeonubecularia*. This latter genus was even proposed as a transitional form between the families Calcivertellidae and Tubiphytidae (Vachard & Krainer 2001*b*; Vachard *et al.* 2012). This transition possibly took place at the end of the Moscovian (Middle Pennsylvanian) (Vachard *et al.* 2012), and was probably induced by an increasingly close association (consortium) between miliolate foraminifers and cyanobacteria.

Biostratigraphy. Tubiphytes has its FAD in the latest Pennsylvanian (Chuvashov *et al.* 1993; Vachard *et al.* 2012) and its LAD in the Jurassic (Crescenti 1969; Senowbari-Daryan *et al.* 2008). *Latitubiphytes* is late Moscovian–Cisuralian, '*Tubiphytes*' *epimonellaeformis* is Kungurian; *Epimonella* is Kungurian–Capitanian; and *Ramovsia* is Cisuralian.

Biogeography. Tubiphytes and *Latitubiphytes* are cosmopolitan during the Permian. '*Tubiphytes*' *epimonellaeformis* is endemic of New Mexico. *Epimonella* has been found in Iran, southern Turkey and New Mexico. *Ramovsia* is only known in the Carnic Alps (Kochansky-Devidé 1973), Iran (Jenny & Jenny-Deshusses 1978) and southern Turkey (Vachard & Moix 2013).

Class **Nodosariata** Mikhalevich, 1993
Subclass **Nodosariana** Mikhalevich, 1993
Order **Nodosariida** Calkins, 1926

The nodosariates are distributed from late early Moscovian to Recent: their genera are either cosmopolitan or endemic. The Palaeozoic Nodosariida exhibit mostly a monolamellar wall of radiate calcite, with crystal *c*-axes perpendicular to the surface and without secondary lamination (Loeblich & Tappan 1987). The Palaeozoic Nodosariida are divided into two superfamilies, Nodosarioidea and Geinitzinoidea emend. herein, and 10 families: Syzraniidae Vachard in Vachard & Montenat, 1981; Protonodosariidae Mamet & Pinard, 1992; Geinitzinidae Bozorgnia, 1973; Robuloididae Reiss, 1963 nom. translat. Loeblich & Tappan, 1984; Partisaniidae Loeblich & Tappan, 1984; Frondinidae Gaillot & Vachard, 2007; Colaniellidae Fursenko in Rauzer-Chernousova & Fursenko, 1959; Nodosariidae Ehrenberg, 1838; Pachyphloiidae Loeblich & Tappan, 1984; and Ichthyolariidae Loeblich & Tappan, 1986*b*. The two superfamilies are distinguished by the presence of a primitive cylindrical aperture (Geinitzinoidea) or a typical stellate aperture (Nodosarioidea). The exact status of the ancient family Nodosellinidae and its possible priority on one of these names depends of the true status of its generotype *Nodosinella*; the revisions of *Nodosinella* by Cummings (1955), Foster *et al.* (1985) or Pinard & Mamet (1998) do not allow a decision to be made.

Superfamily **Geinitzinoidea** Loeblich & Tappan, 1984 emend. herein

Emended diagnosis. Test subcylindrical to subtriangular uniseriate, rarely coiled and lenticular, with a primitive round terminal aperture. The septa are complete in all the families, except for the Syzraniidae family (which exhibits a progressive lineage with undivided, then pseudo-septated and finally completely septated test). The round aperture is generally simple, areal and central. A wall of radial, hyaline, fibrous calcite, generally unilayered. Sometimes, the wall is bilayered with a dark-microgranular inner layer and a clear radial, hyaline-fibrous outer layer (Syzraniidae, Protonodosariidae) or, more rarely, the wall is unilayered and dark-radial (Frondinidae).

Remarks. The fundamental difference is the simple (not stellate) aperture. The superfamily was called Robuloidoidea Reiss, 1963 nom. translat. Loeblich & Tappan, 1984 emend. Gaillot & Vachard, 2007, but this name is misinterpreted because the name Geinitzinoidea Bozorgnia, 1973 translated in Loeblich & Tappan (1984, p. 20) preceeds the name Robuloidoidea translated in Loeblich & Tappan (1984, pp. 32–33). Numerous homeomorphs exist between the two superfamilies, Nodosarioidea and Geinitzinoidea, and even within the same genus (e.g. *Nodosaria*, *Dentalina* and *Lingulina*). The oldest hyaline-fibrous forms (Syzraniidae) appear in the late early Moscovian (Kashirian); the complete septation is Late Pennsylvanian in age; the acme of the group is Permian, but it also exists in the Triassic and even the Jurassic.

Family **Syzraniidae** Vachard in Vachard & Montenat, 1981

Permian genera. Syzrania Reitlinger, 1950; *Syzranella* Mamet & Pinard, 1992; *Amphoratheca* Mamet & Pinard, 1992; *Rectostipulina* Jenny-Deshusses, 1985 emend. Rigaud *et al.*, 2015*b*; *Tezaquina* Vachard in Vachard & Montenat, 1981 *non* Vachard, 1980 emend. Alipour & Vachard in Alipour *et al.*, 2013; *Vervilleina* Groves in Groves & Boardman, 1999.

Remarks. This family is currently unanimously accepted as the oldest family of the Nodosariata, which are consequently monophyletic (Vachard *et al.* 2010 and references therein). The Triassic family Tubulastellidae Rigaud *et al.*, 2015*b* is homeomorphous of the Syzraniidae, but with an aragonitic wall. As indicated above, the Syzraniidae display a progressive appearance of the septation among the nodosariates (1 & 2 in Fig. 7).

Biostratigraphy. Syzrania is Moscovian–Changhsingian. *Amphoratheca* and *Syzranella* are Late Pennsylvanian and Early Permian. *Rectostipulina* remains typically a Lopingian marker, although its divergence from *Syzrania* unquestionably occurs in the Capitanian (Gaillot & Vachard 2007; Song *et al.* 2009; Ebrahim Nejad *et al.* 2015; Zhang *et al.* 2015). It exists in the latest Changhsingian of southern Tibet (Wang *et al.* 2010) and in the lowermost Triassic beds in South China (Song *et al.* 2007), but it is absent in the coeval series of Italy (Groves *et al.* 2007). *Tezaquina* has its FAD in late or latest Gzhelian (Russia, the Canadian Arctic, the Carnic Alps, Darvas, northern Iran); its acme seems to be Asselian–Sakmarian; it is rare and/or doubtful in the other Permian stages of Russia, where these questionable representatives might belong to *Biparietata* (a homeomorphous genus with a multilayered wall microstructure, which is assigned here to the Protonodosariidae). *Vervilleina* is generally late Gzhelian–Cisuralian, but has been noted up to the latest Permian (Shang *et al.* 2003; Song *et al.* 2007).

Biogeography. Syzrania is cosmopolitan from the Moscovian to the Artinskian, it is then only Tethyan. *Amphoratheca* and *Syzranella* are probably cosmopolitan, but mainly noted in North America. *Rectostipulina* is known in Turkey, Slovenia, Cyprus, Greece, Afghanistan, Iran (Alborz, Zagros and Julfa areas), Armenia, Azerbaijan, Saudi Arabia, Ladakh, southern Tibet, South China, western Thailand and Cambodia. *Tezaquina* is first distributed in Russia, the Canadian Arctic, the Carnic Alps, Darvas and northern Iran. During its acme, *Tezaquina* is known in Donbass (Ukraine); the Perm area and Bashkortostan (Russia); northern Iran; Tezak (Afghanistan); Darvas (Tajikistan); Bolivia; and the Barents Sea, Canadian Arctic, Greenland and New Mexico. It is rare and/or doubtful in other Permian regions of Russia (the Urals, Pechora and Primorye) and, perhaps, confused with *Biparietata. Vervilleina* is probably cosmopolitan, but is especially known in North and South America (Groves & Boardman 1999; Wood *et al.* 2002), and central Palaeotethys (Groves & Boardman 1999; Filimonova 2010).

Family **Protonodosariidae** Mamet & Pinard, 1992 emend. Gaillot & Vachard, 2007

Subfamily **Protonodosariinae** Gaillot & Vachard, 2007

Permian genera. Protonodosaria Gerke, 1959 *non* 1952 emend. Sellier de Civrieux & Dessauvagie, 1965; *Biparietata* Zolotova in Zolotova & Baryshnikov, 1980; *Frondinodosaria* Sellier de Civrieux & Dessauvagie, 1965; *Nestellorella* Gaillot & Vachard, 2007; *Lingulonodosaria sensu* K. V. Miklukho-Maklay 1960*a* (*non* Silvestri, 1903, *nec sensu* Sellier de Civrieux & Dessauvagie, 1965); *Nodosinelloides* Mamet & Pinard, 1992; *Polarisella* Mamet & Pinard, 1992 emend. Gaillot & Vachard, 2007; '*Pseudoglandulina*' Cushman, 1929 (part.).

Remarks. Protonodosariidae, as proposed here, encompasses the uniseriate, tapering nodosariates with more or less hemispherical chambers and an aperture that is simple, and rarely with a neck (3–14 in Fig. 7). They differ from the Nodosariidae in the non-stellate aperture. Probably, this family is still poorly defined, with 'mixed morphologically different genera' (G. Nestell pers. comm. March 2016). The most questionable genus of this conventional family is *Pseudoglandulina*, which was used diversely among Permian taxa, either forgotten or provided with a variety of species (e.g. Filimonova 2010), and which was, in addition, interpreted as a junior synonym of *Pyramidulina* Fornasini (Loeblich & Tappan 1987).

Biostratigraphy. As for the family, the subfamily is latest Pennsylvanian–latest Triassic in age. *Protonodosaria* and *Nodosinelloides* have a long range: Late Pennsylvanian (Kasimovian)–latest Permian/earliest Triassic; the range of the genus *Polarisella*, as emendated by Gaillot & Vachard (2007), is even longer, from the Cisuralian to the Middle Triassic (Anisian) and possibly up to the Jurassic of Germany. In constrast, *Biparietata* would be limited to the Kungurian; and *Nestellorella* is questionably present in the Wordian of southern Oman (Angiolini *et al.* 2004), and most probably the Capitanian–late Changhsingian in

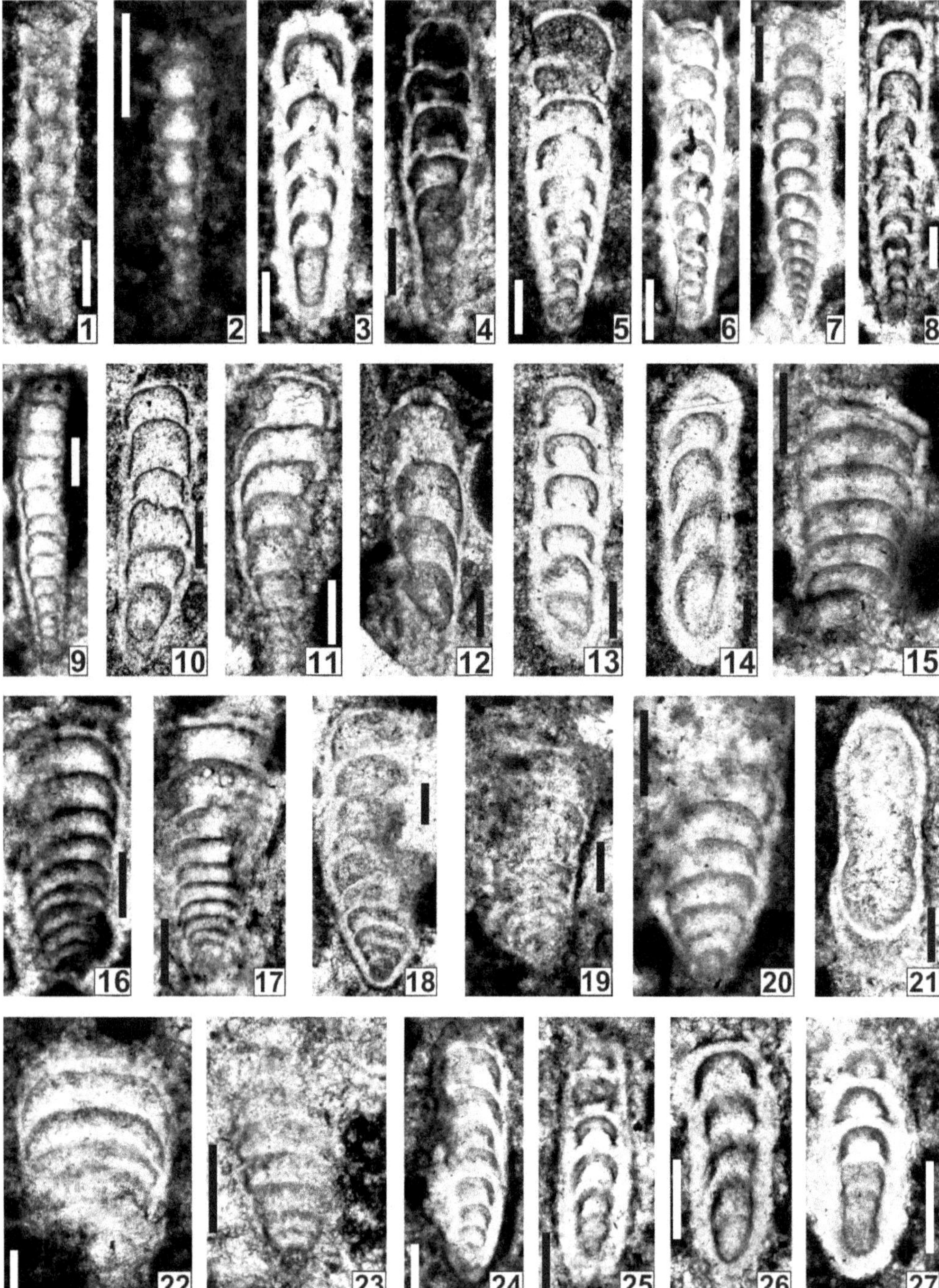

Fig. 7. Smaller foraminifers (Nodosariata) from Trogkofel (Late Cisuralian of the Carnic Alps, Austria). Scale bar = 0.100 mm. (**1**) *Tezaquina* sp. (**2**) *Vervilleina* sp. (**3**)–(**14**) *Nodosinelloides* spp. (**15**)–(**24**) *Geinitzina* spp. (**25**)–(**27**) *Pachyphloia* spp.

Armenia–Azerbaijan (Pronina 1988*a*; Kotlyar *et al.* 1989), the Capitanian in the central Pamirs (Dronov 2004), the late Capitanian in Hazro and the Lopingian of Zagros (Gaillot & Vachard 2007).

Biogeography. As for the family, the genera are partly cosmopolitan, partly Tethyan. *Protonodosaria*, *Nodosinelloides* and *Polarisella* are probably cosmopolitan. *Nestellorella* is Neotethyan in southern Oman (Angiolini *et al.* 2004), Armenia–Azerbaijan (Pronina 1988*a*; Kotlyar *et al.* 1989), in the Rushan–Pshart zone in the central Pamirs (Dronov 2004), Hazro (Turkey) and Zagros.

Subfamily **Langellinae** Gaillot & Vachard, 2007

Permian genera. *Langella* Sellier de Civrieux & Dessauvagie, 1965; *Pseudolangella* Sellier de Civrieux & Dessauvagie, 1965; ?*Cryptoseptida* Sellier de Civrieux & Dessauvagie, 1965; ?*Maichelina* Sosnina, 1977.

Remarks. Because of the reservations expressed above regarding the family, the individuality and composition of this subfamily can be discussed. *Eomarginulinella* and *Maichelina* are poorly known. Despite numerous misinterpretations of its type species *C. anatoliensis* Sellier de Civrieux & Dessauvagie, 1965, *Cryptoseptida* continues to be commonly used in the literature. It has been confused with *Pachyphloides* by Berczi-Makk (1996), *Polarisella* ex gr. *elabugae* by Kobayashi (1996), *Pseudolangella* by Kobayashi (2001, pl. 2, fig. 14) and *Aulacophloia* by Kobayashi (2002, fig. 9. 40, 41). We propose herein its assignment to Langellinae because of its increasing thickness of wall, the septa being thinner than the wall, the shape of the chambers and the type of aperture.

Biostratigraphy. *Langella*, rare in the Artinskian–Kungurian (=Longlingian of South China), has its acme in the late Chihsian and its LAD in the Changhsingian. *Pseudolangella* has a probably Sakmarian FAD and a Changhsingian LAD; it is questionable in the Early Permian, and has its acme in the Guadalupian (=Middle Permian). *Cryptoseptida* was described in the 'Calcaires à Bellerophon' of Turkey: that is, a composite unit probably late Guadalupian and Lopingian in age. It was never really re-found in the Lopingian except by Kotlyar *et al.* (1984) in the early Wuchiapingian.

Biogeography. *Langella* and *Pseudolangella* are relatively widespread in the Palaeo- and Neotethys (see Vachard 1990), from Italy to South China, and the Panthalassan in New Zealand (see the detailed distribution in Gaillot &Vachard 2007). *Maichelina* is endemic to Primorye. *Cryptoseptida* seems endemic to an area encompassing southern Turkey, Armenia and Azerbaijan (Sellier de Civrieux & Dessauvagie 1965; Kotlyar *et al.* 1984). Questionable *Cryptoseptida* were mentioned in South China (Wang & Ueno 2003), Japan (Kobayashi 1996, 2001, 2002) and Hungary (Berczi-Makk 1996).

?Family **Nodosinellidae** Rhumbler, 1895

Permian genus. *Nodosinella* Brady, 1876.

Remark. Despite many attempts, the old genus *Nodosinella* was neither correctly revised nor refound; it perhaps resembles *Protonodosaria* with a differentiated and/or recrystallized wall.

Biostratigraphy. Middle or Late Permian.

Biogeography. England; all the other localities have to be revised.

Family **Geinitzinidae** Bozorgnia, 1973

Permian genera. *Geinitzina* Spandel, 1901 (= ?*Neogeinitzina* K.V. Miklukho-Maklay, 1954); *Frondinodosaria* Sellier de Civrieux & Dessauvagie, 1965; *Geinitzinita* Sellier de Civrieux & Dessauvagie, 1965; *Gerkeina* Grozdilova & Lebedeva in Sosipatrova, 1969; *Howchinella* Sellier de Civrieux & Dessauvagie, 1965 emend. Palmieri in Foster *et al.*, 1985; *Lunucammina* Spandel, 1898 (=?*Frondicularia* Defrance in d'Orbigny, 1826 *sensu* Sellier de Civrieux & Dessauvagie, 1965); *Omoloniella* Karavaeva & Nestell, 2007; *Pachyphloides* Sellier de Civrieux & Dessauvagie, 1965; *Pseudotristix* Miklukho-Maklay, 1960*a* (=*Multifarina* Lin, 1984 =*Tristix sensu* Sellier de Civrieux & Dessauvagie *non* MacFadyen, 1941 (part.)); *Rectoglandulina sensu* Karavaeva & Nestell, 2007 *non* Loeblich & Tappan, 1955; *Reitlingeria* Pronina in Kotlyar *et al.*, 1989; *Spandelinoides* Cushman & Waters, 1928*c*; *Tauridia* Sellier de Civrieux & Dessauvagie, 1965; ?*Pseudonodosaria* Boomgaart 1949 (partially considered as synonymous of *Rectoglandulina*).

Remarks. Once again, the use of Recent genera for Permian taxa is questionable in this family and, except for the generotype *Geinitzina*, which is relatively well characterized (15–24 in Fig. 7), the other genera are discussed and/or poorly known. *Pseudonodosaria* is used by few authors for Permian taxa (e.g. Woszczynska 1987; Palmieri *et al.* 1994; Karavaeva & Nestell 2007; Filimonova 2010); it needs revision as does *Rectoglandulina*, which was abandoned by its originators (see Loeblich & Tappan 1987), but 'reinstated' recently by Karavaeva & Nestell (2007): further studies are required because the taxon described under this name by Gerke (1961) unquestionably exists and is noticeable in the Permian biostratigraphy. The genus *Lunucammina* remains equally mysterious, even after the long discussions of Loeblich & Tappan (1964, 1987), Sellier de Civrieux & Dessauvagie (1965) and K. V. Miklukho-Maklay (1960*b*). In my opinion, most species attributed to this genus in the literature belong to *Frondicularia sensu* Sellier de Civrieux & Dessauvagie, 1965. However, Palmieri *et al.* (1994) have indicated that many Russian species of *Frondicularia* correspond to Australian species of *Howchinella* and '*Ichthyolaria*' (see below). Owing to their concomitant occurrences in Zechstein deposits, *Nodosinella* and *Lunucammina* are possibly synonymous. As for *Nodosinella*, the genus *Tauridia* remains a very enigmatic taxon (Fig. 8); moreover, it was often mispelled as *Taurida* (e.g. Pinard & Mamet 1998; Groves 2000), and misinterpreted by Loeblich & Tappan (1987) as a synonymous of *Frondina*, the wall of which differs completely.

Biostratigraphy. The typical representatives of the family exist from the latest Pennsylvanian to the Triassic. *Geinitzina* is latest Pennsylvanian (Groves 2000, 2002) or earliest Permian (Groves & Wahlman 1997) to Changhsingian (Lin *et al.* 1990). The lowermost Triassic '*Geinitzina*' recently mentioned in Italy (Groves *et al.* 2007) are most probably representative of *Nestellorella* (fig. 10.5) and *Nodosaria* (fig. 10.6, 14, 15). The possible synonym *Neogeinitzina* is limited to the Changhsingian. *Geinitzinita*, which is generally a Mesozoic genus, is known by two or three species in the Changhsingian (Vuks 1984; Gu *et al.* 2007). *Pseudotristix* is generally cited in the Wuchiapingian (=Dzhulfian) from Turkey to southern China (including *Multifarina*, where it is indicated as Maokouan in age by Lin *et al.* 1990). Its FAD seems to be late Guadalupian in Sumatra, western Thailand, central Japan and Primorye, and its LAD is Changhsingian in age in Greece (Hydra), Israel, Armenia–Azerbaijan, Italy, Zagros, Cambodia, South China, North Thailand and Japan. *Pachyphloides* belongs to the Maokouan, Wuchiapingian? and Changhsingian. *Rectoglandulina* is probably homeomorphous of a taxon found in the Cretaceous and Paleocene–Holocene (?); during the Permian, it seems to be Kungurian–Changhsingian in age. The taxon '*Pseudonodosaria*', which is generally mentioned as late Cisuralian in age, is similarly homeomorphous of a Holocene taxon; it was indicated as extending 'beyond the P–T boundary elsewhere in South China' (Jin *et al.* 2000). The range of *Lunucammina* = *Frondicularia* is discussed; perhaps late Cisuralian–late Lopingian. *Howchinella* is Early Permian (early Cisuralian, Artinskian)–Late Permian (latest Changhsingian) (Gu *et al.* 2007; Karavaeva & Nestell 2007; Song *et al.* 2009; Zhang & Gu 2015). *Gerkeina* is Kungurian. *Omoloniella* is late Kungurian–Lopingian and perhaps Triassic. *Spandelinoides* is Late Pennsylvanian–Early Permian. The range of *Tauridia* is little known; perhaps from Cisuralian to Lopingian.

Biogeography. Geinitzina is cosmopolitan; its possible synonym *Neogeinitzina* is, however, only

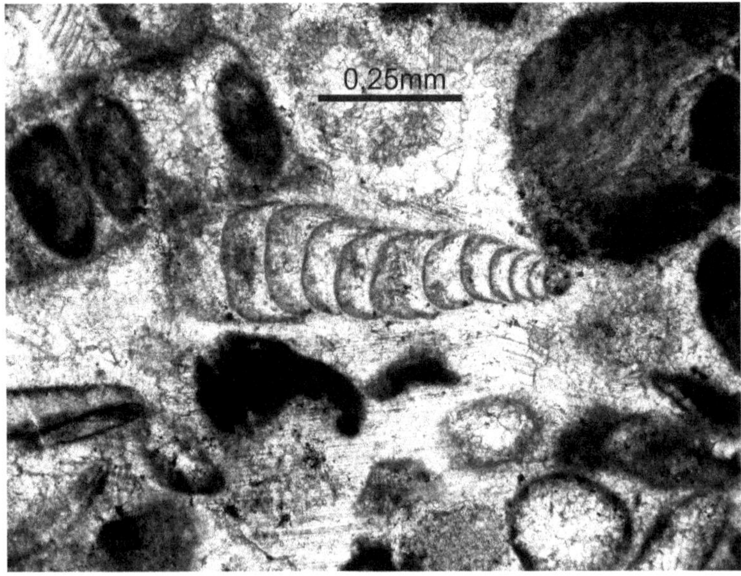

Fig. 8. A typical nodosariate, *Tauridia* n. sp. Scale bar = 0.250 mm. Late Artinskian of Zweikofel (Carnic Alps, Austria).

mentioned in NW Caucasus (K.V. Miklukho-Maklay 1954; Pronina-Nestell & Nestell 2001), South China (Wang 1974; Lin 1984; Lin *et al.* 1990; Zhang & Hong 2004) and NW Iran (Mohtat-Aghai *et al.* 2009). *Geinitzinita* is mentioned in Armenia–Azerbaijan (Vuks 1984) and South China (Gu *et al.* 2007). *Pseudotristix* is sporadically cited from Italy, Greece, Poland, Turkey, Israel, Armenia, Azerbaijan, Iran (Zagros, NW Iran), Sumatra, Cambodia, South China, North Thailand, Japan and Primorye. *Pachyphloides* is mentioned in NW Turkey, NW Caucasus, Zagros and South China. *Rectoglandulina* and *Pseudonodosaria* are probably cosmopolitan, but still poorly known. *Lunucammina* (=*Frondicularia* part.) was found in Germany, Italy, Hungary, Poland, Iran (Zagros and Fars), eastern Himalaya, Australia and ?Japan (where it was probably confused with *Frondina*). *Howchinella* is probably cosmopolitan, but relatively rare everywhere, except in the Far East of Russia and Australia. Similarly, *Tauridia*, although poorly known, is possibly cosmopolitan, as it is currently cited in Turkey, the Carnic Alps, the Urals and New Mexico. Endemic genera are *Spandelinoides* (Texas, USA), *Gerkeina* (the Urals) and *Omoloniella* (NE Siberia, Russia).

Family **Robuloididae** Reiss, 1963 nom. translat. Loeblich & Tappan, 1984

Permian genera. Robuloides Reichel, 1946; *Pararobuloides* Miklukho-Maklay, 1954; *Eocristellaria* Miklukho-Maklay, 1954; *Cryptomorphina* Sellier de Civrieux & Dessauvagie, 1965; *Gourisina* Reichel, 1946; *Hubeirobuloides* Lin, Li & Zheng in Lin *et al.*, 1990; ?*Calvezina* Sellier de Civrieux & Dessauvagie, 1965; ?*Eomarginulinella* Sosnina, 1969.

Remarks. This family encompasses all the planispirally coiled nodosariates of the Permian and, therefore, of the Palaeozoic. Hence, *Calvezina*, initially coiled, is related here to this family. *Gourisina*, very rarely re-found since its description, might correspond to a teratogenic form of *Pararobuloides*,especially the species *Gourisina rossica* K. V. Miklukho-Maklay, 1954.

Biostratigraphy. Robuloides, misinterpreted as 'Artinskian or younger' (Mamet 1996), is Capitanian–latest Changhsingian (and even earliest Triassic in South China: Song *et al.* 2007). *Pararobuloides* is Wuchiapingian–Changhsingian. *Hubeirobuloides* is late Capitanian–late Changhsingian. The genus *Eocristellaria* is relatively poorly known, with an acme probably located in the Late Permian, and a possible FAD in the Middle Permian (with some early Kazanian forms described under the name of '*Astacolus*'). *Calvezina*, the FAD of which is probably early Middle Permian, has its acme in the Capitanian; its LAD is late Changhsingian (Altıner 1984; Song *et al.* 2009). Its possible synonym *Eomarginulinella* is Middle Permian. *Cryptomorphina* is late Guadalupian–Changhsingian. *Gourisina* is only typical in the late Changhsinghian.

Biogeography. Robuloides was found in Greece (Hydra, Attica), Slovenia, Hungary, Cyprus, Turkey, Israel, NW Caucasus, Armenia, Azerbaijan, Saudi Arabia, central Iran, Afghanistan, the Salt Range, SE Pamirs, southern Tibet, South China, Primorye, Japan and New Zealand; this genus is absent from the Americas (contrary to Mamet 1996). *Pararobuloides* is known in Greece, NW Caucasus, Turkey, Oman, southern Tibet, Thailand and Japan. *Hubeirobuloides* is only illustrated in South China (Lin *et al.* 1990), NW Caucasus (Pronina-Nestell & Nestell 2001), Crimea (Pronina & Nestell 1997) and Japan (as *Pachyphloia*? sp.: Kobayashi 2001). *Eocristellaria* probably exists in Turkey, central Iran (Abadeh: Okimura & Ishii 1981), South China (Zhang & Gu 2015) and, perhaps, in Australia (Palmieri *et al.* 1994). *Calvezina* was encountered in Turkey, Greece, Alborz, Armenia, Azerbaijan, Himalaya, Primorye, ?Italy, ?Hungary, ?Malaysia, ?Japan and ?Texas. *Eomarginulinella* was mentioned (often as *Calvezina*) in Primoye (in the far east of Russia: Sosnina 1969), Armenia–Azerbaijan (Kotlyar *et al.* 1989; Pronina 1996), southern Turkey (Altıner 1981), New Mexico (Nestell & Nestell 2006) and Slovenia (Nestell *et al.* 2009). *Cryptomorphina* is known in Turkey (central Taurides and Hazro), Italy (Tesero section) and Iran (Zagros; NW Iran). *Gourisina* is known from Greece (Reichel 1946), NW Caucasus (K.V. Miklukho-Maklay 1954) and, questionably, from central Iran (Abadeh) (Baghbani 1993).

Family **Partisaniidae** Loeblich & Tappan, 1984

Permian genera. Partisania Sosnina, 1978 (=*Xintania* Lin, 1984); *Eoguttulina* Cushman & Ozawa, 1930 (part.); ?*Nodoinvolutaria* Wang in Lin, 1978.

Remarks. This poorly known family might be the ancestor of the Mesozoic polymorphinids (Rigaud *et al.* 2015*b*).

Biostratigraphy. The family has been cited in Wordian, Kazanian, Capitanian, Tatarian and Lopingian.

Biogeography. Palaeotethyan, Neotethyan and Panthalassan *Partisania* are known in Primorye, NW Thailand, South China, Philippines, Japan, Armenia, New Zealand, Oman, Cyprus, Crimea, Iran (Fars area) and New Mexico. *Eoguttulina* (or more probably a homeomorph of this genus) was very rarely cited in Russia (Pronina 1999*a*).

Nodoinvolutaria is Maokouan–late Changhsingian of South China (Lin *et al.* 1990; Wignall & Hallam 1996; Gu *et al.* 2007).

Family **Frondinidae** Gaillot & Vachard, 2007

Permian genera. Frondina Sellier de Civrieux & Dessauvagie, 1965 emend. Gaillot & Vachard 2007; *Ichthyofrondina* Vachard in Vachard & Ferrière, 1991 emend. Gaillot & Vachard, 2007.

Remarks. This family contains two small forms that have a dark-hyaline wall, an atypical character amongst the Nodosariata. They are interesting because they are probably important for the biozonation of the late Capitanian–Lopingian.

Biostratigraphy. Frondina and *Ichthyofrondina* are late Capitanian–Changhsinghian in age, and possibly exist up to the lowermost Triassic in Italy (Groves *et al.* 2007) and South China (Song *et al.* 2007). Furthermore, *Frondina* is questionably present up to the Cretaceous (Sellier de Civrieux & Dessauvagie 1965).

Biogeography. Frondina and *Ichthyofrondina* are Palaeotethyan and Neotethyan, and both occur in Turkey, Cyprus, Greece, Tunisia, Iran, NW Caucasus, Armenia, Azerbaijan, Saudi Arabia, Oman, southern Tibet, NW Thailand, South China and Japan. *Frondina* has been also mentioned in New Zealand (Vachard & Ferrière 1991), Saudi Arabia (Hughes 2005) and Slovenia (Nestell *et al.* 2009). In North America, both genera were never correctly identified; nevertheless, in El Antimonio (Sonora, Mexico), Brunner (1979) has illustrated two specimens that resemble *Frondina* (pl. 3, fig. 8) and *Ichthyofrondina* (pl. 3, fig. 3).

Family **Colaniellidae** Fursenko in Rauzer-Chernousova & Fursenko, 1959

Permian genera. Colaniella Likharev, 1939 (=*Pseudocolaniella* Wang, 1966; =*Paracolaniella* Wang, 1966); *Cylindrocolaniella* Loeblich & Tappan, 1985 (=*Wanganella* Sosnina in Kiparisova *et al.*, 1956 pre-occupied); *Pseudowanganella* Sosnina, 1983.

Remarks. This family is well characterized by the shape of its chambers; it encompasses few genera that rapidly evolved and totally disappear.

Biostratigraphy. Colaniella is latest Capitanian–late Changhsingian; *Cylindrocolaniella* and *Pseudowanganella* are Lopingian. *Pseudocolaniella* and *Paracolaniella*, which are very similar in shape to *Colaniella* and have the same stratigraphic distribution, are considered herein as junior synonyms of *Colaniella*.

Biogeography. Colaniella is essentially a Palaeotethyan and Panthalassan taxon (e.g. see Ishii *et al.* 1975; Vuks & Chediya 1986; Okimura 1988; Shang *et al.* 2003; Wang *et al.* 2010; Kobayashi 2013; Vachard 2014); it remains rare and only represented by primitive forms in the peri-Gondwanan margin of the Neotethys (Israel, Turkey, Iran, Saudi Arabia, Oman and Tunisia). *Cylindrocolaniella* and *Pseudowanganella* are endemic of Primorye because *Pseudowanganella*, illustrated by Nestell & Pronina (1997, pl. 1, figs 26 & 27) in Cyprus, corresponds, in our opinion, to a more primitive taxon; the Japanese forms assigned to *Wanganella* (Ishii *et al.* 1975; Okimura 1988) most probably being true *Colaniella* to compare, for example, with *Colaniella bozkiri* Çatal & Dager, 1974 from Taurus (southern Turkey).

Superfamily **Nodosarioidea** Norvang, 1957 nom. translat. Loeblich & Tappan, 1961
Family **Nodosariidae** Ehrenberg, 1838
Subfamily **Nodosariinae** Ehrenberg, 1838 nom. translat. Reuss, 1862

Permian genera. Nodosaria Lamarck, 1816 non 1812 (see synonyms in Loeblich & Tappan 1987; and those which are different in Karavaeva & Nestell 2007); *Dalongella* Gu *et al.*, 2007; '*Dentalina*' non? d'Orbigny, 1826; '*Lingulina*' d'Orbigny, 1826 emend. Sellier de Civrieux & Dessauvagie, 1965.

Remarks. Uniseriate, rectilinear, with hemispherical to elongate chambers and a stellate aperture. *Lingulina* is diversely interpreted, sometimes as a *Frondina*.

Biostratigraphy. Nodosaria is Permian–Recent. Karavaeva & Nestell (2007) indicated a Roadian FAD for the 'first true *Dentalina*-like forms', but the second, false and/or typical specimens of '*Dentalina*' are not well characterized and are probably homeomorphs of undeterminate genera. '*Lingulina*' is generally late Capitanian–Changhsingian in age (Pronina, 1999*b*), whereas *Dalongella* is Changhsingian. Apparently, the taxon '*Pseudolingulina*' spp. (sic) of Orlov-Labkovsky (2004, text-fig. 2), mentioned in the late Wuchiapingian–early Changhsingian of Israel, is a nomen nudum.

Biogeography. Nodosaria is cosmopolitan. *Dalongella* is endemic of South China. Permian '*Dentalina*' and '*Lingulina*' have been cited all around the world, especially in Russia, Western Europe and China.

Family **Pachyphloiidae** Loeblich & Tappan, 1984

Permian genera. Pachyphloia Lange, 1925; *Robustopachyphloia* Lin, 1980; *Aulacophloia* Gaillot & Vachard, 2007; *Sosninella* Sellier de Civrieux & Dessauvagie, 1965 nomen imperfectum.

Remarks. This typical Permian family is now relatively well known, except for the ancestral forms of *Pachyphloia*, which probably appear in the late Cisuralian (25–27 in Fig. 7).

Biostratigraphy. The complete range of *Pachyphloia* in the literature is Early Permian–Jurassic: however, *Pachyphloia* has its FAD in the Kungurian (Lin *et al.* 1990; Pinard & Mamet 1998; Groves 2000; Vachard & Krainer 2001*b*; Filimonova 2010). Cisuralian citations (e.g. Groves & Wahlman 1997) correspond most probably to oblique sections of *Geinitzina*. *Pachyphloia* has its acme in the Guadalupian–Lopingian (especially in Tunisia, Turkey and Iran); its LAD is most probably latest Permian (e.g. Lin *et al.* 1990; Pronina-Nestell & Nestell 2001; Shang *et al.* 2003; Wang & Ueno 2003; Groves *et al.* 2007). *Robustopachyphloia* is late Capitanian, Wuchiapingian and Changhsingian. *Aulacophloia* is rather early Changhsingian in age, but was mentioned in the late Changhsingian (Wang *et al.* 2010).

Biogeography. *Pachyphloia* is cosmopolitan. *Robustopachyphloia* is known from South China, central Japan, New Mexico, southern Turkey (Hazro), Zagros and Oman (Hauser *et al.* 2000; Gaillot & Vachard 2007), as well as the central Pamirs (Pronina 1996), Texas, USA (Nestell *et al.* 2006) and Tunisia (Ghazzay *et al.* 2015). *Aulacophloia* exists in Turkey, Saudi Arabia and southern Tibet. *Cryptoseptida* was described in the 'Calcaires à Bellerophon' of Turkey: that is, a composite unit probably Guadalupian and Lopingian in age, but *Cryptoseptida* was never really re-found – for example, its type species *C. anatoliensis* re-illustrated by Groves *et al.* (2005, fig. 23.26) is more likely to be *Langella ocarina* Sellier de Civrieux & Dessauvagie, 1965.

Family **Ichthyolariidae** Loeblich & Tappan, 1986*b*

Permian genus. *Ichthyolaria sensu* Sellier de Civrieux & Dessauvagie, 1965 (part.) *non* Wedekind, 1937.

Remarks. The presence of true *Ichthyolaria* (i.e. defined by the type species *Frondicularia bicostata* d'Orbigny, 1850), or simply of ichthyolariids, in the Permian seems to be very doubtful. Furthermore: (1) the Permian taxon is often misspelled *Ichtyolaria* or even *Icthyolaria*; and (2) the Permian *Ichthyolaria* are generally confused with the genus *Ichthyofrondina*, which differs in its wall microstructure.

Biostratigraphy. The genus *Ichthyolaria* and its probable homeomorphs are cited from the Permian to the Jurassic.

Biogeography. Permian homeomorphs of *Ichthyolaria* were mentioned in the Early Permian of Australia (Palmieri 1984; Palmieri *et al.* 1994); Middle–Late Permian of Turkey (Sellier de Civrieux & Dessauvagie 1965); Middle Permian of Cyprus (Nestell & Pronina 1997); Middle Permian of Russia, Volga–Kama region (Pronina 1998, 1999*a*), Taymir (Pronina 1999*a*) and NE Siberia (Karavaeva & Nestell 2007); late Guadalupian of New Mexico (Nestell & Nestell 2006) and South China (Gu *et al.* 2007; Zhang & Gu 2015).

Group incertae sedis misinterpreted as Nodosariata

It would take too long to discuss the history and the assignment of the genus *Sphairionia* Nguyen Duc Tien, 1989*b* non 1987 = *Pseudosphairionia* Pronina, 1996 (see e.g. Pronina 1996; Vachard *et al.* 2001*a*; Sone 2008). The genus is interesting because it is a biomarker of the early Midian (Pronina 1996). This regional substage is the equivalent of the early Capitanian (e.g. Henderson *et al.* 2012; Davydov & Arefifard 2013; Ghazzay *et al.* 2015). Moreover, *Sphairionia* is a good biogeographical marker, being probably only peri-Gondwanan and/or Cimmerian (see the Discussion later). Pronina (1996) attributed this genus to the nodosariates: however, its morphology totally differs (especially when the sections of the *Pseudosphairionia* type are integrated into the reconstruction of this microfossil, as in Vachard *et al.* 2001*a*). Furthermore, the microstructure of each layer of the bilayered wall differs from that of the nodosariates, whereas, as indicated by several authors (Vachard 1980; Mertmann 2000; Vachard *et al.* 2001*a*), this microstructure is more similar to that of the umbellacean microproblematica. The umbellacean, generally assigned to the Charophyta (e.g. Vachard 2000 and references therein), are known from the late Givetian to the Tournaisian (?early Visean), but similar forms have been mentioned up to the Miocene (Vachard *et al.* 1982). A Lazarus effect during the Permian is not excluded.

Class **Textulariata** Mikhalevich, 1980

The taxa originally considered to have an agglutinated test, like *Ammodiscus*, *Glomospira*, *Tolypammina*, *Hyperammina*, *Reophax* or *Ammobaculites* (e.g. Sosipatrova 1970; Ukharskaya 1970; Woszczynska 1987; Pronina 1998), should be assigned to Fusulinata and/or Miliolata genera (e.g. Vachard *et al.* 2010; Hance *et al.* 2011). Notice that when an agglutinated genus name existed in the literature, the earliest authors often gave the name of a Textulariata: for example, *Rectoseptournayella chappelensis* was initially called *Ammobaculites chappelensis* (see Vachard *et al.* 2010). Today, the trend is to give a Palaeozoic name (e.g. *Pseudoammodiscus*, *Pseudoglomospira*, *Earlandia* and so on) to these taxa: that is, to suggest that they possess a calcitic dark-migrogranular wall, not agglutinated. Recently, the pendulum has swung back, and the names *Ammodiscus* and *Hyperammina* have, again, been proposed (Nestell *et al.* 2015). This problem is therefore very complex.

As indicated by Rigaud *et al.* (2014), large benthic agglutinated and microgranular foraminifers show significant morphological and microstructural similarities. The two groups have yet to be regarded as being part of the distinct lineages on the basis of a gap exceeding 50 myr in the record of the first large alveolar agglutinated Mesozoic forms (Textulariida) and their last known Palaeozoic microgranular homologues (Fusulinana). In fact, both Fusulinana and the Textulariida may possess a microgranular test, potentially containing some optically discernible agglutinated grains and a keriothecal/alveolar texture. Three major hypotheses can explain these striking wall similarities: (1) the Textulariida and the Fusulinana have experienced a convergent evolution; (2) the Endotebidae belong to the Textulariida and have been erroneously included in the Fusulinana; and (3) all Fusulinana have agglutinated microparticles, the microgranular appearance of their wall being related to the nature and size of the grains that they incorporate. In the latter case, the Fusulinana would have survived the Permo-Triassic and the Triassic–Jurassic major extinction events and still have representatives among some lineages of Textulariata. In consequence, there would be no major extinct groups of foraminifers, and molecular phylogeny, combined with fossil data, should allow the foraminiferal evolution to be retraced to a very advanced level.

As a result, the Textulariata reappear, or seem to reappear, in the late Middle Permian, with the first Ataxophragmiidae (Gaillot 2006). Undoubtedly, the family is represented by well-identified loose specimens in the Kazanian (=Middle Permian) of Russia (Ukharskaya 1970).

Another correlation is possible: an inner organic lining exists among the Ataxophragmiidae. The first unquestionable inner organic linings of foraminifers, found in Palaeozoic sediments, are Late Permian in age (Stancliffe 1989; Groves *et al.* 2004). Owing to the monophyletism of the Rotaliata (e.g. Flakowski *et al.* 2005), the lineage admitted between Fusulinata and Rotaliata via the Nodosariata (Gaillot & Vachard 2007), and the connection between Fusulinata and Textulariata suggested herein, it seems difficult to integrate the Textuladiida in the Rotaliata as proposed by Mikhalevich (2003).

Order **Textulariida** Lankester, 1885

Trying to accurately differentiate the Fusulinata and the Textulariata, Vachard *et al.* (2010) have indicated that the siliceous agglutinate was exclusive to Textulariata, whereas the agglutinates of the Fusulinata, when they existed, were always calcareous. This statement is not entirely true because some Palaeotextularioidea, Endoteboidea and even Endothyroidea (e.g. *Endothyranella kamaica* Baryshnikov in Baryshnikov *et al.*, 1982) can exhibit a siliceous agglutinate wall (Fig. 9). Among the Palaeotextularioidea with siliceous agglutinates, the genus *Palaeobigenerina* Galloway, 1933 (type species: *Bigenerina geyeri* Schellwien, 1898) is probably the most typical. This genus is rarely used, probably because the original description of its apertures is erroneous. It was described with an agglutinating wall, but, in this case, the original description is probably exact. This agglutinated wall is obvious in the Carnic Alps (e.g. Vachard & Krainer 2001*b*, pl. 4, figs 10 & 14) and northern Afghanistan (Vachard unpublished data). The most developed siliceous agglutinate was probably illustrated in '*Deckerella* sp. 1' by Nestell *et al.* (2006, pl. 5, fig. 20). The name *Textularia* has been

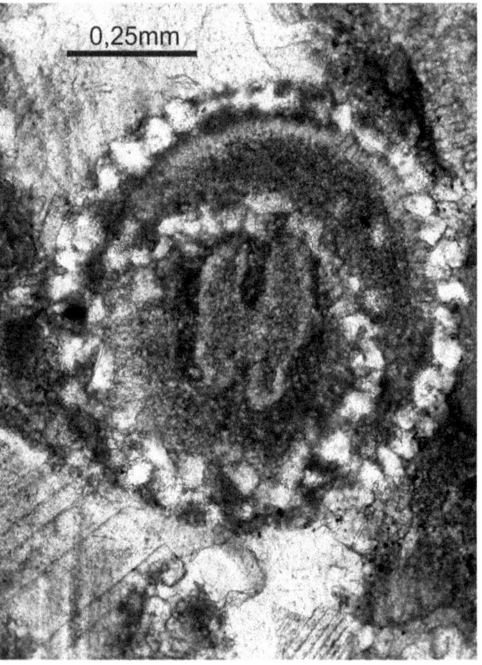

Fig. 9. Regularly arranged siliceous agglutinate in an oblique subtransverse section of a palaeotextulariid (the inner pseudo-fibrous hyaline layer is also visible). It should be noted that for Miklukho-Maklay (1956) only the pseudofibrous layer was interpreted as secreted, while the siliceous grains and the calcitic grains of the dark-microgranular layer were considered as agglutinated. Scale bar = 0.100 mm. Late Artinskian of Zweikofel (Carnic Alps, Austria).

recently used for naming Middle Permian (Kazanian) species (Ukharskaya 1970; Pronina 1998).

As proposed by Altıner & Özkan-Altıner (2010), it is possible that *Labiodagmarita* should be assigned to Palaeotextulariidae and not to Dagmaritidae. In contrast, the similar nomenclatural change, proposed by the same authors, for *Bidagmarita* is not acceptable because this latter genus possesses an unquestionable wall of dagmaritins.

Order **Verneuilinida** Mikhalevich & Kaminski in Mikhalevich, 2003

First Verneuilinida seem appear in the Kubergandy Formation (late Kungurian–early Roadian) of Tajikistan (Angiolini *et al.* 2015). Then, Verneuilinida are illustrated in the Capitanian of Oman (Vachard *et al.* 2002). Many species of *Verneuilinoides* and *Digitina* were described in the Middle Permian (Kazanian) of Russia (Ukharskaya 1970; Pronina 1998) and in the Permian of Australia (e.g. Crespin 1958). Sections of Verneuilinidae also correspond to '*Reophax* sp.' of Flügel *et al.* (1984, pl. 32, fig. 6) and *Climacammina* sp. (Flügel *et al.* 1991, pl. 43, fig. 8), both from the Middle Permian in Croatia and Sicily. See also Ataxophragmiidae indet. in Vachard *et al.* (2002, pl. 1, figs 1 & 2) from the Capitanian of Oman. *Monodorina* Lin *et al.*, 1990, described as coming from the Early Permian of South China, is most probably a Triassic species.

Order **Lituolida** Lankester, 1885

Vachardella, attributed here to the Endoteboidea, might alternatively be the first lituolid in the geological story.

Discussion: biostratigraphic and palaeobiogeographical implications

Biostratigraphy of the Permian smaller foraminifers

Smaller foraminifer biozones were introduced in the Early Permian formations of the Pamirs (Tajikistan) by Filimonova (2010), but it was impossible to identify these biozones either in our North American material (Fig. 10) or in our Tethyan material, probably because: (1) the evolution for all the groups has been very slow for Kasimovian–Artinskian times; and (2) many species are poorly defined.

During the Middle and Late Permian, the Miliolata and Nodosariata display the best biostratigraphic value of the smaller foraminifers owing to their quite rapid evolution and their strong diversification. Some small Fusulinata, such as Globivalvulinoidea, remain biostratigraphically interesting (see Altıner & Brönnimann 1980; Altıner 1999; Altıner & Özkan-Altıner 2001, 2010; Gaillot & Vachard 2007; Gaillot *et al.* 2009), as well as the Spireitlinoidea genus *Endoteba* (see Vachard *et al.* 2013) and Pseudovidalinidae. As recorded for the fusulinids, after recovery in the Wuchiapingian after the end-Guadalupian crisis, the Changhsingian was a period of increasing diversity of the genera of the smaller foraminifers.

A biostratigraphic scheme, proposed in this paper, is given in Figure 11.

Cisuralian. Many survivals from the Late Pennsylvanian are present and relatively numerous up to the Artinskian: *Endothyra*, *Bradyina*, *Tetrataxis*, calcivertellids and *Nodosinelloides*. The Asselian is characterized by oncolites built by *Ellesmerella* (formerly '*Girvanella*'), occasionally associated with *Claracrusta* Vachard, 1980 (incertae sedis algae) and cyanobacteria (rarely, true *Girvanella*). No genera of smaller foraminifers are characteristic for the Asselian; only species of *Bradyina*/*Bradyinelloides*, *Endothyra*, *Pseudovidalina*, *Hemidiscus* and *Tetrataxis* can be used in association when characterizing the Asselian. Amongst the Nodosariata, characteristic species belong to the genera *Nodosinelloides*, *Polarisella*, *Protonodosaria* and *Geinitzina* (the FAD of which occurs in the Carboniferous–Permian boundary interval).

Sakmarian. This period has been characterized by Blazejowski (2009) with species of *Calcitornella* and *Midiella*, and by Filimonova (2010) with species of *Geinitzina* and *Nodosinelloides*. Species of *Pseudovidalina* and *Bradyina* could be also useful.

Artinskian. The Artinskian sees the LAD of *Bradyina* and the possible replacement of *Protonodosaria* (or *Nodosinelloides*) by *Nodosaria*: that is, the appearance of the stellate aperture in this group. *Pseudovermiporella* appeared in the Sakmarian–Artinskian boundary interval and became rapidly common.

Kungurian. The main bioevents are the LAD of *Mesolasiodiscus* and *Xingshandiscus*. Various species of *Globivalvulina* are noticeable, such as *G. praegraeca* Vachard *et al.*, 2015, which heralds the relative gigantism of this genus in the Capitanian–Lopingian. First *Septoglobivalvulina* appear, or at least the transitional forms between *Globivalvulina* and *Septoglobivalvulina*. In association, giant *Geinitzina* and first atypical *Rectoglandulina* and *Pachyphloia* are found, and the FAD of *Nestellorella* occurs.

Roadian. Smaller foraminifers are poorly known during this period. New species of *Nodosaria*, first typical species of *Pachyphloia* and *Rectoglandulina* appear in many areas of western Palaeotethys, and probably all around the world. The gigantism of the Miliolata and Nodosariata could begin during

ZONATION	FORMATIONS	BIOMARKERS (and representative samples)	REGIONAL STAGES		INTERNATIONAL STAGES
Zone 15	SAN	*Olgaorlovella davydovi* *Tubiphytes epimonellaeformis* *Geinitzina indepressa*			late KUNGURIAN
Zone 14	ANDRES Fm	*Hemigordiellina* spp.			middle KUNGURIAN
Zone 13	GLORIETA Fm	barren of foraminifers		CATHEDRALIAN	
Zone 12		*Frondicularia* aff. *turae*	LEONARDIAN		
Zone 11	YESO Gr	*Ellesmerella rara* *Nestellorella*? sp.			early KUNGURIAN
Zone 10		*Globivalvulina novamexicana* *Orthovertellopsis protaeformis* *Glomomidiella infrapermica*			
Zone 9					late ARTINSKIAN
Zone 8	ROBLEDO Fm	*Pseudoreichelina* sp. *Geinitzina* sp. 2 [? *multicamerata*]			middle ARTINSKIAN
Zone 7		*Praeneodiscus sp.* *Globivalvulina novamexicana* *Globivalvulina praegraeca*		HESSIAN	early ARTINSKIAN
Zone 6		*Pseudovermiporella* spp.	WOLFCAMPIAN	LENOXIAN	latest SAKMARIAN
Zone 5		*Pachyphloia?* sp. *Geinitzina postcarbonica* *Nodosinelloides longissima*			late SAKMARIAN
Zone 4	Community Pit Fm	*Nodosinelloides pinardae* *Globivalvulina parapiciformis* *Geinitzina postcarbonica*		LENOXIAN	early SAKMARIAN
Zone 3		*Hedraites* sp.	NEALIAN		late ASSELIAN
Zone 2		*Geinitzina postcarbonica* *Nodosinelloides netjachewi*			middle ASSELIAN
Zone 1	Shalem Colony Fm	*Leptotriticites* sp. *Nodosinelloides longissima* *Tubiphytes* sp., *Geinitzina* sp. *Pseudovidalina* sp., *Pseudoschwagerina* sp.	BURSUMIAN		early ASSELIAN-latest GZHELIAN

Fig. 10. Cisuralian biozonation in New Mexico based on smaller foraminifers (according to Lucas *et al.* 2015; Vachard *et al.* 2015). *Abbreviations*: DAA, DAB and DAC designate samples of the sections A, B and C in the Doña Ana Mountains (DA) in New Mexico.

this period with the first true *Neodiscus*, *Graecodiscus*, *Agathammina* and *Langella*, amongst others.

Wordian. Wordian smaller foraminifers are also poorly known and often confused with the Capitanian taxa (partly because the two stages were previously linked under the name Murgabian or Murghabian). Giant Nodosariata diversify with *Langella*, *Calvezina* and large *Nodosaria*. The FAD of *Postendothyra* and *Hemigordiopsis*, if confirmed, might become characteristic of the Wordian.

Capitanian. Numerous FAD of smaller foraminifer genera occur during this period, and especially during the late Capitanian (El Capitan stratotype in the USA, Tebaga, Turkey, Thailand and South China). It is also the acme of *Hemigordiopsis* and *Endoteba*, especially in Tebaga (Tunisia). The first colaniellids and the first *Altineria* appear near the end of the late Capitanian.

Middle–Late Permian interval. A possible significant time to distinguish in the Permian history

CHRONOSTRATIGRAPHY	FUSULINATA			MILIOLATA	NODOSARIATA	OTHERS
	LASIODISCOIDEA	BRADYINOIDEA	GLOBIVALVULINOIDEA			
Earliest Triassic	-	-	*Globivalvulina?*	-	*Polarisella- Tezaquina?- Nodosinelloides?- Nodosaria*	
Late Changhsingian	LAD *Lasiodiscus*- LAD *Lasiotrochus*	-?	*Paraglobivalvulinoides Paradagmarita monodi*- LAD *Paremiratella- Louisettitta*	-	Evolved *Colaniella*	
Early Changhsingian	FAD *Lasiotrochus*	Second acme *Postendothyra*	*Labioglobivalvulina*- FAD *Paremiratella*- FAD *Paradagmaritopsis*- Prim. *Paradagmarita*	-	Acme *Frondina*- Acme *Ichthyofrondina*	*Floritheca*
Wuchiapingian	*Lasiodiscus* ex gr. 2	?	*Paraglobivalvulina*- FAD *Paradagmarita- Charliella*	*Glomomidiellopsis Okimuraites*	LAD *Frondina*- LAD *Ichthyofrondina*- primitive *Colaniella*- FAD *Rectostipulina*	
Capitanian/Wuchiapingian boundary interval	*Angelita austroalpina* n. comb.	?	FAD *Sengoerina*	*Shanita-Lysites*	Transition *Syzrania/ Rectostipulina*	
Late Capitanian	FAD of *Angelita- Lasiodiscus* ex gr. 1	?	*Dagmarita-Danielita*	*Hemigordiopsis*	*Cylindrocolaniella*	Acme *Endoteba*
Early Capitanian	?	First acme *Postendothyra*	Acme *Dagmarita*	*Miultidiscus*	*Tauridia- Langella- Pseudolangella*	*Sphaerionia*
Wordian	?	FAD *Postendothyra*	FAD? *Dagmarita*	FAD *Multidiscus?* FAD *Neodiscus?*	*Calvezina- Cryptoseptida*	
Roadian	?	?	?	-	FAD *Rectoglandulina*- FAD true *Pachyphloia*	
Kungurian	*Xingshandiscus*- LAD *Mesolasiodiscus*	?	FAD *Septoglobivalvulina*	-	FAD *Nodosaria- Rectoglandulina*- Acme *Geinitzina*	
Artinskian	*Pseudovidalina*- FAD ? *Lasiodiscus*	FAD of *Bradyina*	*Globivalvulina* spp.	FAD *Pseudovermiporella*	Transition *Nodosinelloides/Nodosaria*	FAD *Endoteba*
Sakmarian	Acme *Pseudovidalina- Hemidiscus*- FAD *Mesolasiodiscus*	*Bradyina* spp.	*Globivalvulina* spp.	*Ellesmerella*- tubiphytids- *Rectogordius*	FAD *Pachyphloia?*	
Asselian	*Pseudovidalina- Asselodiscus*	*Bradyina* spp.	*Globivalvulina* spp.	*Calcivertella- Ellesmerlla*-tubiphytids- *Cornuspira*	FAD *Tezaquina*- Acme *Nodosinelloides*- FAD *Geinitzina*- FAD *Polarisella*	

Fig. 11. Proposed biomarkers and bioevents among the smaller foraminifers during the Permian.

is the Capitanian–Wuchiapingian boundary interval, in areas such as El Capitan (USA), Turkey, Japan, Iran (Abadeh), Tunisia (Tebaga), Armenia and Azerbaijan (e.g. see, Gaillot & Vachard 2007; Ghazzay *et al.* 2015 and references therein). This period could be characterized by the following fusulinid events: the FAD of true *Codonofusiella*, the FAD of true *Reichelina* and the LAD of *Chusenella* – see, for example, the Khachik Formation in Armenia (Leven 1998), the *Sphaerulina* beds of Abadeh in Iran (Iranian–Japanese Research Group 1981; Kobayashi & Ishii 2003; Davydov & Arefifard 2013), the 'barren interval' of Isozaki *et al.* (2007) and the *Neoendothyra permica* Zone in Japan (Kobayashi 2012*b*). Some smaller foraminifers are also characteristic for this time interval: *Shanita*, *Lysites* and *Altineria alpinotaurica* (Altıner, 1988), and a possible transitional taxon between *Syzrania* and *Rectostipulina* (this latter genus having a Wuchiapingian FAD, as well as *Paradagmarita*: see Gaillot & Vachard 2007).

Wuchiapingian and early Changhsingian. Both of these periods in the Lopingian superstage share the same smaller foraminifers, which had already appeared in the late Guadalupian. *Langella*, *Pachyphloia*, *Pseudolangella*, primitive *Colaniella*, *Frondina* and *Ichthyofrondina* dominate among the Nodosariata; *Glomomidiellopsis*, *Pseudovermiporella* and *Multidiscus* among the Miliolata; and *Charliella* among the globivalvulinids.

Late Changhsingian. This substage is dominated by the FAD, and the short duration (because they disappear at the PTB), of *Paraglobivalvulinoides*, *Urushtenella*, advanced *Colaniella*, *Paradagmarita monodi* and *Louisettita* (Fig. 9, with the first and last occurrences in Hazro, SW Turkey).

Palaeogeography of the Middle–Late Permian smaller foraminifers

Many palaeobiogeographical reconstructions have been proposed for the Guadalupian and Wuchiapingian, especially based on the distributions of brachiopods (see Angiolini *et al.* 2013 and references therein), fusulinids (Kobayashi 1997*a*, *b*, 1999; Kobayashi & Ishii 2003; Ueno 2003; Colpaert *et al.* 2015) and smaller foraminifers (e.g. *Hemigordiopsis*, *Shanita*, *Abahedella*, *Paradagmarita* and *Colaniella*: see Şengör *et al.* 1988; Nestell & Pronina 1997; Altıner *et al.* 2000; Gaillot & Vachard 2007; Huang *et al.* 2007; Vachard 2014). The two borders of the Palaeotethys and Neotethys are difficult to discriminate palaeobiogeographically at this time because both are located in the subtropical realm, and because both Tethyan oceanic branches lead to a cul-de-sac and close westwards (Ghazzay *et al.* 2015). Consequently, all faunal and floral mixtures are possible. Palaeobiogeographical data on Panthalassa are very rare, except in Japan; however, this country itself is composed of numerous terranes that complicate the interpretations and reconstructions. In contrast, Cimmerian assemblages are generally characterized more accurately (Gaetani *et al.* 2009; Angiolini *et al.* 2015).

Since the remarkable pioneering work of Şengör *et al.* (1988), *Shanita* has always been confirmed as an important palaeobiogeographical marker (Sheng & He 1983; Nestell & Pronina 1997; Jin & Yang 2004; Gaillot & Vachard 2007; Huang *et al.* 2007; Ebrahim Nejad *et al.* 2015). It is sometimes assigned to Cimmerian terranes (Nestell & Pronina 1997; Ueno 2003), but we propose herein that *Shanita* is somewhat characteristic of these peri-Gondwanan terranes, initially peri-Gondwanan, which migrated to the Cimmeria and Sibumasu microcontinents after the opening of the Neotethys.

In addition, Vachard (2014) indicated how *Colaniella*, by its evolution and migration, in turn confirms this palaeogeography with: (1) the Salt Range (Pakistan) as the radiation centre; (2) a first migration along the peri-Gondwanan border with primitive species of *Colaniella*; and (3) a second migration to the peri-Laurentian northern boundary of the Palaeotethys, and the shelves bordering the eastern Tethys and western Panthalassa. During the late Changhsingian, owing to the rapid opening of the Neotethys between the peri-Gondwanan margin and the Cimmerian microcontinent, and the connection with the peri-Gondwanan areas and, consequently, with the Salt Range in Pakistan, the centre of speciation during the Wuchiapingian was interrupted (Fig. 12).

Currently, work is being carried out with a team of colleagues (Vachard, Rettori, Altıner and Gennari) on the palaeobiogeography of *Altineria*. At first, the palaeogeography of all the outcrops where *Altineria* has been recorded can appear complicated, as a result of the current data in the literature. The source of *Altineria* is the Taurus Mountains in southern Turkey; *Altineria* is also known, under the name *Angelina*, in Armenia (Pronina & Gubenko 1990; Pronina 1996) and Tunisia (Ghazzay *et al.* 2015; Ghazzay-Souli *et al.*

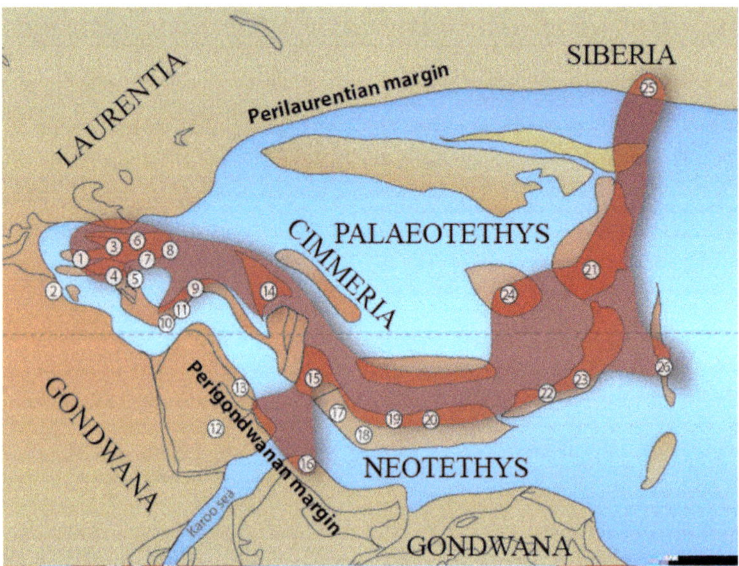

Fig. 12. Palaeobiogeographical distribution of the genus *Colaniella* during the Late Permian. The numbers indicated on the map, from 1 to 26, correspond to the localities containing species of *Colaniella*: (1) Sicily; (2) Tunisia; (3) Albania; (4) Montenegro; (5) Greece; (6) Crimea; (7) NW Turkey; (8) NW Caucasus; (9) central Turkey; (10) Cyprus; (11) Taurus (southern Turkey); (12) Saudi Arabia; (13) Zagros (southern Iran) and Oman; (14) Alborz and central Iran; (15) Central Mountains of Afghanistan; (16) Salt Range; (17) Kashmir; (18) Lamayuru (Ladakh, Himalaya); (19) western Tibet; (20) southern Tibet; (21) South China; (22) Thailand; (23) Malaysia; (24) Indochina (Vietnam); (25) Primorye; (26) Japan terranes (according to Vachard 2014; text-fig. 2). In the late Changhsingian, owing to the rapid opening of the Neotethys between the peri-Gondwanan margin (13 & 16) and the Cimmerian subcontinent (14–20), the connection with the peri-Gondwanan areas and, consequently, with the Salt Range in Pakistan (16), the centre of speciation during the Wuchiapingian, was interrupted.

2015). Our team has also found *Altineria* in limestones of regions that were alternatively assigned to peri-Gondwanan or Cimmerian areas, as for the NW part of Chios Island (Greece), NW Iran and the Abadeh area (central Iran). Since, at all of these localities, *Altineria* is confined to shallow carbonate platforms in the vicinity of bioconstructions with microbialites, *Tubiphytes*, inozoan calcisponges and richthofeniid brachiopods, it seems difficult that under these conditions *Altineria* may have migrated through the Neotethys up to the Cimmerian Block. Therefore, we suggest that Taurus, Armenia, NW Iran (see Ebrahim Nejad *et al.* 2015) and Chios (and the related Karaburun area), as well as Tunisia and the Abadeh area, could have been located on the same extended carbonate platform. Owing to the demonstrated location of Taurus and Tunisia on the peri-Gondwanan border, this carbonate platform therefore corresponds to the whole of the peri-Gondwanan margin (i.e. southern) of the Tethyan Ocean (eventually, in the yet to open Neotethys). This peri-Gondwanan margin supposedly also exists in Afghanistan (but more probably in the Kabul Block than in the Central Mountains, often abusively called the Helmend Block: see the discussion in Colpaert *et al.* 2015), in Karakoram, SE Pamirs, Tibet, Baoshan Block, western Myanmar and western Thailand. The presence of *Altineria* in these regions is possible, although not demonstrated. However, the presence of *Altineria* in Hubei (China) would be puzzling because this region is generally emplaced far from the Tethyan borders on the traditional palaeomaps. However, it this is easier to explain on the palaeomap of Gaillot & Vachard (2007), where the South China and Iran terranes are located closely together.

During the entire Permian, the southern part of the North American Craton (i.e. the modern states of Sonora, New Mexico and Texas) remained relatively isolated from the Laurentian and Gondwanan continents. The endemism of the smaller foraminifers was important during the late Cisuralian (=Wolfcampian) (Lucas *et al.* 2015; Vachard *et al.* 2015), where several Tethyan groups are apparently lacking and probably endemic taxa exist among the globivalvulinid and miliolate foraminifers and the gymnocodiacean algae (Vachard *et al.* 2015). In contrast, several studies (Vachard *et al.* 1992; Nestell & Nestell 2006; Nestell *et al.* 2006) have demonstrated that the endemism of the smaller foraminifers is lesser than for the large fusulinids, during the Middle Permian and, especially, the Capitanian. However, if some genera were found (e.g. *Partisania*, *Abadehella* and *Charliella*), other ones remain unknown (e.g. giant paraglobivalvulinins, baisalinids, hemigordiopsids and langellins). Finally, no Late Permian foraminifers are known in the Americas.

Conclusions

- Four classes of smaller foraminifers were present during the Permian: Fusulinata, Miliolata, Nodosariata and Textulariata
- Biostratigraphically, the most interesting groups are the Lasiodiscoidea, Bradyinoidea, Globivalvulinoidea among the Fusulinata; the Cornuspirida among the Miliolata; and the entire class of the Nodosariata. The class Textulariata is too poorly known during the Permian to permit the characterization of biostratigraphical markers; nevertheless, the appearance of the Verneuilida order is probably very important.
- The Lasiodiscoidea are poorly differentiated during the Cisuralian. *Pseudovidalina* gives rise to two evolute forms: *Xingshandiscus* during the Kungurian; and *Altineria* at the Capitanian–Wuchiapingian boundary interval. *Lasiodiscus* and *Lasiotrochus* display several species during the Late Guadalupian–Lopingian.
- The Bradyinoidea are separated in two lineages during the Permian: one of *Bradyina* and one of *Postendothyra*.
- The phylogeny of the globivalvulinids, from the Early Pennsylvanian to the late Cisuralian, was only that of the lineages of *Globivalvulina*. During the Guadalupian–Lopingian, they diversify with three new subfamilies: Paraglobivalvulininae, Dagmaritinae and Paradagmaritinae, the evolutionary trends of which are relatively well known in Turkey, Iran and peri-Gondwanan areas.
- The Miliolata remain poorly known, especially in the Cisuralian. It is difficult to understand if the genera recently described in the USA, *Praeneodiscus* and *Olgaorlovella*, have also a biostratigraphic importance in the Tethys. The evolution of tubiphytids and ellesmerellids is yet to be discussed, because these taxa are generally assigned to cyanobacteria and/or algae. The Middle-Late Permian is well characterized by giant genera: *Neodiscus*, *Neohemigordius*, *Baisalina*, *Glomomidiellopsis*, *Hemigordiopsis* and *Shanita*.
- The Permian families and genera of the Nodosariata, as for those of the Mesozoic and Cenozoic, are still poorly understood. The Cisuralian genera most interesting with respect to biostratigraphy are probably *Tezaquina*, *Nodosinelloides*, *Geinitzina* and *Polarisella*, and *Vervilleina*, although they present many problems of nomenclature. The Middle Permian is a period of generic diversification of *Pachyphloia*, *Calvezina*, *Cryptoseptida*, *Rectoglandulina*, true *Nodosaria*, *Langella* and *Pseudolangella*. The Late Permian is marked by evolved colaniellids and frondinids, whereas at the base of the

Triassic several very primitive genera, which appeared in the Cisuralian, *Tezaquina*, *Nodosinelloides* and *Polarisella*, reappear.

- Palaeogeographically, the most interesting markers have already been characterized: *Shanita* and *Colaniella*, but *Altineria* should be restudied (work in progress). All of these genera indicate the opening of the Neotethys and its consequence for foraminiferal evolution.
- *Shanita* is somewhat characteristic of these peri-Gondwanan terranes, which migrated to the Cimmeria and Sibumasu microcontinents after the opening of the Neotethys. Our hypothesis is that *Altineria*, more discrete and minute than *Shanita*, confirms such a palaeogeographical and geodynamic evolution. *Colaniella* has the Salt Range (Pakistan) as its radiation centre; a first migration along the peri-Gondwanan border with the primitive species of *Colaniella*; and a second migration to the peri-Laurentian northern boundary of the Palaeotethys and the shelves bordering the eastern Tethys and western Panthalassa.
- The southern part of the North American Craton (i.e. the modern states of Sonora, New Mexico and Texas) remained relatively isolated from the Laurentian and Gondwanan continents. The endemism of the smaller foraminifers, important during the late Cisuralian (=Wolfcampian), decreased during the late Guadalupian (Capitanian).

I wish to thank Galina and Merlynd Nestell and Demir Altıner, who improved the manuscript with their constructive criticisms and English corrections. I appreciated the friendly assistance of Elisabeth Locatelli, Sébastien Clausen (Villeneuve d'Ascq, France) and Karl Krainer (Innsbruck, Austria). Thanks to Drs Lucas and Shen for the invitation to write a paper and to Jessica Pollitt at GSL for the editing.

References

Adl, S., Simpson, A. *et al.* 2012. The revised classification of eukaryotes. *Journal of Eukaryotic Microbiology*, **59**, 429–493.

Alipour, Z., Hosseini-Nezhad, S.M., Vachard, D. & Rashidi, K. 2013. Biostratigraphy and description of smaller foraminifers of Dorud Group (uppermost Gzhelian-lower Sakmarian) in Tang-e Shamshirbor section Eastern Alborz, Iran. *Geological Journal*, **48**, 385–402.

Altiner, D. 1981. *Recherches stratigraphiques et micropaléontologiques dans le Taurus Oriental au NW de Pinarbasi (Turquie)*. PhD thesis, University of Geneva.

Altiner, D. 1984. Upper Permian foraminiferal biostratigraphy in some localities of the Taurus Belt. *In*: Tekeli, O. & Göncüoğlu, M.C. (eds) *Geology of the Taurus Belt: Proceedings of the International Symposium* 26–29 September 1983, Ankara, Turkey. MTA, Ankara, Turkey, 255–268.

Altiner, D. 1988. Pseudovidalinidae n. fam. and *Angelina* n. gen. from the Upper Permian of south and southeast Turkey. *Revue de Paléobiologie*, Special Volume, 2, 25–36.

Altiner, D. 1999. *Sengoerina argandi*, n. gen., n. sp., and its position in the evolution of late Permian biseriamminid foraminifers. *Micropaleontology*, **45**, 215–220.

Altiner, D. & Brönnimann, P. 1980. *Louisettita elegantissima*, n. gen. n. sp., un nouveau foraminifère du Permien supérieur du Taurus oriental (Turquie). *Notes du Laboratoire de Paléontologie de l'Université de Genève*, **6**, 39–43.

Altiner, D. & Özkan-Altiner, S. 1998. Baudiella stampflii n. gen., n. sp., and its position in the evolution of late Permian ozawainellid. *Revue de Paléobiologie*, **17**, 163–175.

Altiner, D. & Özkan-Altiner, S. 2001. *Charliella rossae* n. gen., n. sp., from the Tethyan realm: remarks on the evolution of late Permian biseriamminids. *Journal of Foraminiferal Research*, **31**, 309–314.

Altiner, D. & Özkan-Altiner, S. 2010. *Danielita gailloti* n. gen., n. sp., within the evolutionary framework of Middle–Late Permian dagmaritins. *Turkish Journal of Earth Sciences*, **19**, 497–512, https://doi.org/10.3906/yer-0905-4

Altiner, D. & Savini, R. 1997. New species of *Syzrania* from the Amazonas and Solimöes basins (North Brazil): remarks on the generic and suprageneric position of syzraniid foraminifers. *Revue de Paléobiologie*, **16**, 7–20.

Altiner, D. & Zaninetti, L. 1977. *Kamurana broennimanni*, n. gen., n. sp., un nouveau foraminifère porcelané perforé du Permien supérieur du Taurus oriental, Turquie. *Notes du Laboratoire de Paléontologie de l'Université de Genève*, **1**, 1–6.

Altiner, D., Özkan-Altiner, S. & Koçyigit, A. 2000. Late Permian foraminiferal biofacies belts in Turkey: palaeogeographic and tectonic implications. *In*: Bozkurt, E., Winchester, J.A. & Piper, J.D.A. (eds) *Tectonics and Magmatism in Turkey and the Surrounding Area*. Geological Society London, Special Publications, **173**, 83–96, https://doi.org/10.1144/GSL.SP.2000.173.01.04

Angiolini, L. & Rettori, R. 1995. *Chitralina undulata* gen. n. sp. n. (Foraminiferida) from the Late Permian of Karakorum (Pakistan). *Rivista Italiana di Paleontologia e Stratigrafia*, **100**, 477–491.

Angiolini, L., Crasquin-Soleau, S., Platel, J.P., Roger, J., Vachard, D., Vaslet, D. & Al-Husseini, M. 2004. Saiwan, Gharif and Khuff formations, Haushi-Huqf Uplift, Oman. *In*: Al-Husseini, M. (ed.) *Carboniferous, Permian and Early Triassic Arabian Stratigraphy*. GeoArabia, Special Publications, **3**, 149–183.

Angiolini, L., Crippa, G., Muttoni, G. & Pignatti, J. 2013. Guadalupian (Middle Permian) paleobiogeography of the Neotethys Ocean. *Gondwana Research*, **24**, 173–184.

Angiolini, L., Zanchi, A. *et al.* 2015. From rift to drift in South Pamir (Tajikistan): Permian evolution of a Cimmerian terrane. *Journal of Asian Earth Sciences*, **102**, 146–169.

ANGIOLINI, L., CAMPAGNA, M. *ET AL.* 2016. Brachiopods from the Cisuralian–Guadalupian of Darvaz, Tajikistan and implications for Permian stratigraphic correlations. *Palaeoworld*, **25**, 539–568, https://doi.org/10.1016/j.palwor.2016.05.006.

ARGYRIADIS, I., DE GRACIANSKY, P.C. & LYS, M. 1976. Datation de niveaux rouges dans le Permien marin péri-égéen. *Bulletin de la Société géologique de France, Serie 7*, **18**, 513–519.

ARNAUD-VANNEAU, A. 1980. Micropaléontologie, paléoécologie et sédimentologie d'une plate-forme carbonatée de la marge passive de la Téthys: l'Urgonien du Vercors septentrional et de la Chartreuse (Alpes occidentales). *Géologie Alpine, Mémoire*, **11**, 269–874.

AW, P.C., ISHII, K.-I. & OKIMURA, Y. 1977. On *Palaeofusulina* and *Colaniella* fauna from the upper Permian of Kelantan, Malaysia. *Transactions of the Proceedings of the Palaeontological Society of Japan*, **104**, 407–417.

BAGHBANI, D. 1993. The Permian sequence in the Abadeh region, Central Iran. *In*: NAIRN, A.E.M. & KOROTEEV, A.V. (eds) *Contributions to Eurasian Geology*. Occasional Publication Earth Sciences Research Institute,University of South Carolina, New Series, **9A–B**, (part II), 7–22.

BARYSHNIKOV, V.V., ZOLOTOVA, V.P. & KOSHEVA, V.F. 1982. *New Species of Artinskian Foraminifers of the Pre-Urals*. Akademiya Nauk SSSR, Uralskii Nauchnyi Tsentr, Institut Geologii i Geokhimii, preprint [in Russian].

BAUD, A., JENNY, C., PAPANIKOLAOU, D., SIDERIS, C. & STAMPFLI, G. 1991. New observations on Permian stratigraphy in Greece and geodynamic interpretation. *Bulletin of the Geological Society of Greece*, **25**, 187–206.

BENSAID, M., TERMIER, H., TERMIER, G. & VACHARD, D. 1979. Le Carbonifère (Viséen supérieur-Bachkirien) entre Bou Chber et Ich ou Mellal (Maroc Central). *Annales de la Société géologique du Nord*, **98**, 189–204.

BERCZI-MAKK, A. 1992. Midian (Upper Permian) Foraminifera from the large Mihalovits quarry at Nagyvisnyó (North-Hungary). *Acta Geologica Hungarica*, **35**, 27–38.

BERCZI-MAKK, A. 1996. Foraminifera of the Triassic formations of Alsó Hill (northern Hungary), part 3: Foraminifer assemblage of the basinal facies. *Acta Geologica Hungarica*, **39**, 413–459.

BLAZEJOWSKI, B. 2009. Foraminifers from the Treskelodden Formation (Carboniferous–Permian) of south Spitsbergen. *Polish Polar Research*, **30**, 193–230.

BOGUSH, O.I. 1985. Early Carboniferous foraminifers and stratigraphy from the western Siberian Plain. Akademiya Nauk SSSR, Sibirskoye Otdelenie, Trudy Instituta Geologii i Geofiziki, **619**, 49–68 [in Russian].

BOOMGAART, L. 1949. *Smaller Foraminifera from Bodjonegoro (Java)*. Smit & Dontje Publishers, Sappemeer, The Netherlands.

BOWSER, S.S., HABURA, A. & PAWLOWSKI, J. 2006. Molecular evolution of Foraminifera. *In*: KATZ, L. & BHATTACHARYA, D. (eds) *Genomics and Evolution of Microbial Eukaryotes*. Oxford University Press, Oxford, 78–93.

BOZORGNIA, F. 1973. *Paleozoic Foraminiferal Biostratigraphy of Central and East Alborz Mountains, Iran*. National Iranian Oil Company, Geological Laboratories Publications, **4**.

BRADY, H.B. 1876. A monograph of Carboniferous and Permian foraminifera (the genus *Fusulina* excepted). *Palaeontographical Society of London*, **30**, 1–166.

BRADY, H.B. 1878. On the Reticularian and Radiolarian Rhizopoda (Foraminifera and Polycystina) on the North Polar Expedition of 1875–76. *Annals and Magazine of Natural History*, Series 5, **1**, 425–440.

BRADY, H.B. 1884. Report of the foraminifera dredged by H.M.S. Challenger during the years 1873–76. *Reports on the Scientific Results of the Exploratory Voyage of the H.M.S. Challenger, Zoology*, **9**, 1–814.

BRENCKLE, P.L. & WAHLMAN, G.P. 1996. *Raphconilia*, a new name for the Upper Paleozoic pseudovidalinid foraminifer *Conilia* Brenckle and Wahlman, 1994. *Journal of Foraminiferal Research*, **26**, 184.

BRÖNNIMANN, P., WHITTAKER, J.E. & ZANINETTI, L. 1978. *Shanita* a new pillared miliolacean foraminifer from the Late Permian of Burma and Thailand. *Rivista Italiana di Paleontologia*, **84**, 63–92.

BRUNNER, P. 1979. Microfacies y microfosiles permotriásicos en el area El Antimonio, Sonora, Mexico. *Revista del Instituto Mexicano del Petróleo, México*, **11**, 6–41.

BYKOVA, E.V. & POLENOVA, E.N. 1955. *Devonian Foraminifers, Radiolarians and Ostracods of the Volga–Urals Region. Trudy VNIGRI*, **87** [in Russian; French translation, C.E.D.P. No. 1603. BRGM, Paris].

CALKINS, G.N. 1926. *The Biology of the Protozoa*. Lea & Febiger, Philadelphia, PA.

CANUTI, P., MARCUCCI, M. & PIRINI RADRIZZANI, C. 1970. Microfacies e microfaune nelle formazioni paleozoiche dell'anticlinale di Hazro (Anatolia sudorientale, Turchia). *Bolletino Societá Geologica Italiana*, **89**, 21–40.

CARIDROIT, M., FONTAINE, H., JONGKANJANASOONTORN, Y., SUTEETHORN, V. & VACHARD, D. 1990. First results of a paleontological study of Northwest Thailand. *In*: FONTAINE, H. (ed.) *Ten years of CCOP Research on the Pre-Tertiary of East Asia*. CCOP Technical Publications, **20**, 337–350.

ÇATAL, E. & DAGER, Z. 1974. Toros bölgesi permiyenine ait yeni bir *Colaniella* türünün tanımı (Description of a new *Colaniella* species from the Permian of Taurus region). *Türkije Jeoloji Kurumu Bülteni (Bulletin of the Geological Society of Turkey)*, **17**, 187–191.

CAVALIER-SMITH, T. 2002. The phagotrophic origin of eukaryotes and phylogenetic classification of Protozoa. *International Journal of Systematic and Evolutionary Microbiology*, **52**, 297–354.

CAVALIER-SMITH, T. 2003. Protists phylogeny and the high-level classification of Protozoa. *European Journal of Protistology*, **39**, 338–348.

CHERDYNTSEV, W. 1914. On a foraminiferal Permian fauna from western European Russia. *Trudy Obshchestva Estestvoispytateley pri Imperatorskomy Kazanskomy Universitety*, **46**, 3–88 [in Russian].

CHUVASHOV, B.I., SHUYSKY, V.P. & IVANOVA, R.M. 1993. Stratigraphical and facies complexes of the Paleozoic calcareous algae of the Urals. *In*: BARATTOLO, F., DE

Castro, P. & Parente, M. (eds) *Studies on Fossil Benthic Algae*. Bolletino della Societá Paleontologica Italiana, Modena, Special Volume, **1**, 93–119.

Ciarapica, G., Cirilli, S., Martini, R. & Zaninetti, L. 1986. Une microfaune à petits foraminifères d'âge permien remaniée dans le Trias moyen de l'Apennin méridional (Formation du Monte Facito, Lucanie occidentale); description de *Crescentia vertebralis* n. gen. n. sp. *Revue de Paléobiologie*, **5**, 207–215.

Ciarapica, G., Cirilli, S., Passeri, L., Trincati, E. & Zaninetti, L. 1987. 'Anidriti di Búrano' et formation du Monte Cetona (nouvelle formation), biostratigraphie des deux séries-types du Trias supérieur dans l'Apennin septentrional. *Revue de Paléobiologie*, **6**, 341–409.

Colpaert, C., Monnet, C. & Vachard, D. 2015. *Eopolydiexodina* (Middle Permian giant fusulinids of Afghanistan): biometry, morphometry, palaeobiogeography and end-Guadalupian events. *Journal of Asian Earth Sciences*, **102**, 127–145, https://doi.org/10.1016/j.jseaes.2014.10.028

Conil, R. & Pirlet, H. 1970. Le calcaire carbonifère du synclinorium de Dinant et le sommet du Famennien. *Congrès et Colloques de l'Université de Liège*, **55**, 47–63.

Conil, R., Groessens, E. & Lys, M. 1977. Etude micropaléontologique de la tranchée d'Yves-Gomezée (Tn 3c-V 1-V 2, Belgique). *Bulletin de la Société belge de Géologie*, **82**, 201–239.

Conil, R., Longerstaey, P.J. & Ramsbottom, W.H.C. 1980. Matériaux pour l'étude micropaléontologique du Dinantien de Grande-Bretagne. *Mémoires de l'Institut Géologique de l'Université de Louvain*, **30**, 1–115.

Cózar, P. & Somerville, I.D. 2012. Mississippian Biseriamminaceae and their evolutionary development. *Journal of Foraminiferal Research*, **42**, 216–233.

Cózar, P., Medina-Varea, P., Somerville, I.D., Vachard, D., Rodríguez, S. & Said, I. 2014. Foraminifers and conodonts from the late Viséan to early Bashkirian succession in the Saharan Tindouf Basin (southern Morocco): biostratigraphic refinements and implications for correlations in the western Palaeotethys. *Geological Journal*, **49**, 271–302.

Cózar, P., Sanz-López, J. & Blanco-Ferrera, S. 2015. Lasiodiscid foraminifers during the late Viséan to Serpukhovian in Vegas de Sotres section (Cantabrian Mountains, NW Spain). *Geobios*, **48**, 213–238.

Crescenti, U. 1969. Biostratigrafia delle facies mesozoiche dell'Appennino centrale. *Geologica Romana*, **8**, 15–40.

Crespin, I. 1958. *Permian Foraminifera of Australia*. Commonwealth of Australia, Department of National Development, Bureau of Mineral Resources, Geology and Geophysics, **48**.

Crespin, I. & Belford, D.J. 1957. New genera and species of foraminifera from the Lower Permian of Western Australia. *Contributions of the Cushman Foundation for Foraminiferal Research*, **8**, 73–76.

Cummings, R.H. 1955. *Nodosinella* Brady, 1876 and associated upper Paleozoic genera. *Micropaleontology*, **1**, 221–238.

Cummings, R.H. 1956. Revision of the upper Palaeozoic textulariid foraminifera. *Micropaleontology*, **2**, 201–242.

Cushman, J.A. 1928. Additional genera of the Foraminifera. *Contributions from the Cushman Laboratory for Foraminiferal Research*, **4**, 1–8.

Cushman, J.A. 1929. A Late Tertiary fauna of Venezuela and other related regions. *Contributions from the Cushman Laboratory for Foraminiferal Research*, **5**, 77–101.

Cushman, J.A. & Ozawa, Y. 1930. A monograph of the foraminiferal family Polymorphinidae, Reecnt and fossil. *Proceedings of the United States National Museum*, **77**, 1–195.

Cushman, J.A. & Waters, J.A. 1928*a*. The development of *Climacammina* and its allies in the Pennsylvanian of Texas. *Journal of Paleontology*, **2**, 119–130.

Cushman, J.A. & Waters, J.A. 1928*b*. Some Foraminifera from the Pennsylvanian and Permian of Texas. *Contributions from the Cushman Laboratory for Foraminiferal Research*, **4**, 31–55.

Cushman, J.A. & Waters, J.A. 1928*c*. Upper Paleozoic foraminifera from Sutton County,Texas. *Journal of Paleontology*, **2**, 358–371.

Cushman, J.A. & Waters, J.A. 1930. Foraminifera of the Cisco Group. *Bulletin University of Texas*, **3019**, 22–81.

Dain, L.G. & Grozdilova, L.P. 1953. Fossil foraminifers of the USSR: tournayellids and archaediscids. *Trudy VNIGRI*, **74**, 7–63 [in Russian; French translation, M. Jayet, BRGM, 1816, 1–124].

Davydov, V.I. 1988. Archaediscidae in the Upper Carboniferous and Lower Permian. *Revue Paléobiologie*, Special Volume, **2**, 39–46.

Davydov, V.I. & Arefifard, S. 2013. Middle Permian (Guadalupian) fusulinid taxonomy and biostratigraphy of the mid-latitude Dalan Basin, Zagros, Iran and their applications in paleoclimate dynamics and paleogeography. *GeoArabia*, **18**, 17–62.

Davydov, V.I., Nilson, I. & Stemmerik, L. 2001. Fusulinid zonation of the Upper Carboniferous Kap Jungersen and Foldedal Formations, southern Amdrup Land, eastern North Greenland. *Bulletin of the Geological Society of Denmark*, **48**, 31–77.

Davydov, V.I., Krainer, K. & Chernykh, V. 2013. Fusulinid biostratigraphy of the Lower Permian Zweikofel Formation (Rattendorf Group; Carnic Alps, Austria) and Lower Permian Tethyan chronostratigraphy. *Geological Journal*, **48**, 57–100.

Dawson, O., Racey, A. & Whittaker, J.E. 1993. The palaeoecological and palaeobiogeographical significance of Shanita (foraminifer) and associated foraminifera/algae from the Permian of peninsular Thailand. *In*: *International Symposium on Biostratigraphy on Mainland Southeast Asia: Facies & Paleontology*, 31 January–5 February 1993, Chiang Mai, Thailand, Volume **2**. Department of Geological Sciences, University of Chiang Mai, Chiang Mai, Thailand, 283–298.

Deleau, P. & Marie, P. 1961. Les Fusulinidés du Westphalien C du bassin d'Abadla et quelques autres foraminifères du Carbonifère algérien. *Publications Service Carte Géologique Algérie*, **25**, 45–160.

Derville, H. 1952. A propos de Calcisphères (rectification). *Comptes Rendus Sommaires de la Société Géologique de France*, **6**, 236–237.

D'ORBIGNY, A. 1826. Tableau méthodique de la classe des Céphalopodes. *Annales des Sciences Naturelles*, **7**, 245–314.

D'ORBIGNY, A. 1850. *Prodrome de paléontologie stratigraphique universelle des animaux mollusques et rayonnés, Volume 1*. Masson Publisher, Paris.

DRONOV, V.I. 2004. Carboniferous and Permian stratigraphy of the Rushan-Pshart Pamir (southern Rushan–Pshart Zone, southern Pshart Tectonic Block). *Doklady Akademii Nauk*, **395**, 791–793 [in Russian].

EBRAHIM NEJAD, E., VACHARD, D., SIABEGHODSY, A. & ABBASI, S. 2015. Middle–Late Permian (Murgabian–Djulfian) foraminifers of the northern Maku area (western Azerbaijan, Iran). *Palaeontologia Electronica*, **18.1.19A**, 1–63.

EHRENBERG, C.G. 1838. Über dem blossen Auge unsichtbare Kalkthierchen und Kieselthierchen als Hauptbestandtheile der Kreidegebirge. *Bericht über die zu Bekanntmachung geeigneten Verhandlungen der königlichen preussischen Akademie der Wissenschaften zu Berlin*, **3**, 192–200.

EHRENBERG, C.G. 1854. *Zur Mikrogeologie*. von Leopold Voss, Leipzig.

ELLIOTT, G.F. 1958. Fossil microproblematica from the Middle East. *Micropaleontology*, **4**, 419–428.

ETHERIDGE, R., JR. 1873. Notes on certain genera and species mentioned in the foregoing lists. Scotland Geological Survey, Memoir Explanation Sheet, **23**, appendix 2, 93–107.

FAHRNI, J.F., PAWLOWSKI, J., RICHARDSON, S., DEBENAY, J.P. & ZANINETTI, L. 1997. Actin suggests *Miliammina fusca* (Brady) is related to porcellaneous rather than to agglutinated foraminifera. *Micropaleontology*, **43**, 211–214.

FILIMONOVA, T.V. 2008. Smaller foraminifers from type sections of the Bolorian Stage, the Lower Permian of Darvaz. *Stratigraphy and Geological Correlation*, **16**, 22–41.

FILIMONOVA, T.V. 2010. Smaller foraminifers of the Lower Permian from Western Tethys. *Stratigraphy and Geological Correlation*, **18**, 687–811.

FILIMONOVA, T.V. 2013. Smaller foraminifers from the Permian of Central Iran. *Stratigraphy and Geological Correlation*, **21**, 18–35.

FLAKOWSKI, J., BOLIVAR, I., FAHRNI, J. & PAWLOWSKI, J. 2005. Actin phylogeny in foraminifera. *Journal of Foraminiferal Research*, **35**, 93–102.

FLÜGEL, E. & REINHARDT, J. 1989. Uppermost Permian reefs in Skyros (Greece) and Sichuan (China): implications for the Late Permian extinction event. *Palaios*, **4**, 502–518.

FLÜGEL, E., KOCHANSKY-DEVIDÉ, V. & RAMOVS, A. 1984. A Middle Permian calcisponge/algal/cement reef: Straza near Bled, Slovenia. *Facies*, **10**, 179–256.

FLÜGEL, E., DI STEFANO, P. & SENOWBARI-DARYAN, B. 1991. Microfacies and depositional structure of allochthonous carbonate base-of-slope deposits. The Late Permian Pietra di Salomone megablock, Sosio Valley (western Sicily). *Facies*, **25**, 147–186.

FONTAINE, H., NGUYEN DUC TIEN & VACHARD, D. 1986. Discovery of Permian limestone South of Tara Island in the Calamian Islands, Philippines. *United Nations ESCAP, CCOP Technical Bulletin*, **18**, 161–167.

FONTAINE, H., KHOO, H.P. & VACHARD, D. 1988*a*. Discovery of Triassic fossils at Bukit Chuping, in Gunung Sinyum area, and at Kota Jin, Peninsular Malaysia. *Journal of Southeast Asian Earth Sciences*, **2**, 145–162.

FONTAINE, H., LYS, M. & NGUYEN DUC TIEN 1988*b*. Some Permian corals from east Peninsular Malaysia, associated microfossils, paleogeographic significance. *Journal of Southeast Asian Earth Sciences*, **2**, 65–78.

FONTAINE, H., SUTEETHORN, V. *ET AL.* 1988*c*. Late Palaeozoic and Mesozoic fossils of West Thailand and their environments. *CCOP Technical Bulletin*, **20**, 1–107.

FONTAINE, H., SUTEETHORN, V. & VACHARD, D. 1993. Carboniferous and Permian limestones in Sop Pong area: unexpected lithology and fossils. *International Symposium on Biostratigraphy of Mainland Southeast Asia: Facies and Paleontology*, **2**, 319–336.

FONTAINE, H., BIN AMNAN, I., KHOO, H.P., TIEN, N.D. & VACHARD, D. 1994. *The Neoschwagerina and Yabeina–Lepidolina Zones in Peninsular Malaysia, and Dzhulfian and Dorashamian in Peninsular Malaysia, the Transition to the Triassic*. Geological Survey of Malaysia, Geological Papers, **4**.

FONTAINE, H., TANSUWAN, V. & VACHARD, D. 1997. Age of limestones associated with gypsum deposits in Northeast and Central Thailand. *CCOP Newsletter*, **21–22**, 6–7.

FORKE, H.C., KAHLER, F. & KRAINER, K. 1998. Sedimentology, microfacies and stratigraphic distribution of foraminifers of the Lower '*Pseudoschwagerina*' Limestone (Rattendorf Group, Late Carboniferous), Carnic Alps (Austria/Italy). *Senckenbergiana lethaea*, **78**, 1–39.

FOSTER, C.B., PALMIERI, V. & FLEMING, P.J.G. 1985. Plant microfossils, Foraminiferida, and Ostracoda, from the Fossil Cliff Formation (Early Permian, Sakmarian), Perth Basin, Western Australia. *Special Publications South Australia Department Mines and Energy*, **5**, 61–105.

FURSENKO, A.V. 1958. Main stages of development of foraminiferal fauna in the geological past. *Akademiya Nauk Beloruskoi SSR, Trudy Instituta Geologicheskikh Nauk*, **1**, 10–29 [in Russian].

GAETANI, M., ANGIOLINI, L. *ET AL.* 2009. Pennsylvanian–Early Triassic stratigraphy in the Alborz Mountains (Iran). *In*: BRUNET, M.-F., WILMSEN, M. & GRANATH, J.W. (eds) *South Caspian to Central Iran Basins*. Geological Society of London, Special Publications, **312**, 79–128, https://doi.org/10.1144/SP312.5

GAILLOT, J. 2006. *The Khuff Formation (Late Permian, Middle East): high-resolution biostratigraphy by means of foraminifers and algae, sequence stratigraphy, depositional environments, and new approach of the Permian/Triassic crisis*. PhD thesis, University of Lille.

GAILLOT, J. & VACHARD, D. 2007. The Khuff Formation (Middle East) and time-equivalents in Turkey and South China: biostratigraphy from Capitanian to Changhsingian times (Permian), new foraminiferal taxa and palaeogeographical implications. *Coloquios de Paleontología*, **57**, 37–223.

GAILLOT, J., VACHARD, D., GALFETTI, T. & MARTINI, R. 2009. New latest Permian foraminifers from Laren

(Guangxi Province, South China). Palaeobiogeographic implications. *Geobios*, **42**, 141–168.

GALLOWAY, J.J. 1933. *A Manual of Foraminifera*. James Furman Kemp Memorial Series, **1**. Principia Press, Bloomington, IN.

GALLOWAY, J.J. & HARLTON, B.H. 1928. Some Pennsylvanian Foraminifera of Oklahoma with special reference to the genus *Orobias*. *Journal of Paleontology*, **2**, 338–357.

GALLOWAY, J.J. & RYNIKER, C. 1930. Foraminifera from the Atoka Formation of Oklahoma. *Oklahoma Geological Survey Circular*, **21**, 1–36.

GARGOURI, S. & VACHARD, D. 1988. Sur *Hemigordiopsis* et d'autres Foraminifères porcelanés du Murghabien du Tebaga (Permien supérieur de Tunisie). *Revue de Paléobiologie*, Special Volume, **1**, 57–68.

GERKE, A.A. 1952. Permian microfauna of the Nordvik region and stratigraphic significance. *Trudy Nauchno-Issledovatel'skii Institut Geologii Arktiki (NIIGA)*, **28**, 1–209 [in Russian].

GERKE, A.A. 1959. On new Permian nodosarioid foraminifers and precisions on the diagnosis of *Nodosaria*. *Sbornik Statey po Paleontologii i Biostratigrafii, Nauchno-issledovatelskiy Institut Geologii Artiki (NIIGA)*, **17**, 41–59 [in Russian].

GERKE, A.A. 1961. Permian and Lower Mesozoic rectoglandulines from northcentral Siberia. *Sbornik statey po Paleontologii i Biostratigrafii*, **23**, 5–34 [in Russian].

GHAZZAY, W., VACHARD, D. & RAZGALLAH, S. 2015. Revised fusulinid biostratigraphy of the Middle–Late Permian of Jebel Tebaga (Tunisia). *Revue de Micropaléontologie*, **58**, 57–83.

GHAZZAY-SOULI, W., VACHARD, D. & RAZGALLAH, S. 2015. Carboniferous and Permian biostratigraphy by foraminifers and calcareous algae of Bir Mastoura (BMT-1) and related boreholes of southern Tunisia. *Revue de Micropaléontologie*, **58**, 239–265.

GLAESSNER, M.F. 1945. *Principles of Micropaleontology*. Melbourne University Press, Melbourne.

GOLUBIC, S., RADOICIC, R. & SEON-JOO, L. 2006. *Decastronema kotori* gen. nov., comb nov.; a mat-forming cyanobacterium on Cretaceous carbonate platforms and its modern counterparts. *Carnets de Géologie/Notebooks on Geology*, Article 2006/02.

GRANT, R.E., NESTELL, M.K., BAUD, A. & JENNY, C. 1991. Permian stratigraphy of Hydra Island, Greece. *Palaios*, **6**, 479–497.

GRIFFITH, J.W. & HENFREY, A. 1875. *The Micrographic Dictionary, Volume 1*. 3rd edn. Van Voorst, London.

GROVES, J.R. 1992. Stratigraphic distribution of non-fusulinacean foraminifers in the Marble Falls Limestone (Lower–Middle Pennsylvanian), western Llano region, central Texas. *In*: SUTHERLAND, P.K. & MANGER, W.L. (eds) *Recent Advances in Middle-Carboniferous Biostratigraphy*. A Symposium. Okhahoma Geological Survey, Circular, **94**, 145–161.

GROVES, J.R. 1997. Repetitive patterns of evolution in late Paleozoic foraminifers. *In*: ROSS, C.A., ROSS, J.R.P. & BRENCKLE, P.L. (eds) *Late Paleozoic Foraminifera, Their Biostratigraphy, Evolution, and Paleoecology, and the Mid-Carboniferous Boundary*. Cushman Foundation for Foraminiferal Research, Special Publications, **36**, 51–54.

GROVES, J.R. 2000. Suborder Lagenina and other smaller foraminifers from uppermost Pennsylvanian-lower Permian rocks of Kansas and Oklahoma. *Micropaleontology*, **46**, 285–326.

GROVES, J.R. 2002. Evolutionary origin of the genus *Geinitzina* (Foraminiferida) and its significance for international correlation near the Carboniferous–Permian boundary. *In*: HILLS, L.V., HENDERSON, C.M. & BAMBER, E.W. (eds) *Carboniferous and Permian of the World*. Canadian Society of Petroleum Geologists, Memoirs, **19**, 437–447.

GROVES, J.R. & ALTINER, D. 2005. Survival and recovery of calcareous foraminifera pursuant to the end-Permian mass extinction. *Comptes Rendus Paleovol*, **4**, 487–500.

GROVES, J.R. & BOARDMAN, H. 1999. Calcareous smaller foraminifers from the lower Permian Council Grove Group near Hooser, Kansas. *Journal of Foraminiferal Research*, **29**, 243–262.

GROVES, J.R. & WAHLMAN, G.P. 1997. Biostratigraphy and evolution of Late Carboniferous and Early Permian smaller foraminifers from the Barents Sea (offshore Arctic Norway). *Journal of Paleontology*, **71**, 758–779.

GROVES, J.R., ALTINER, D. & RETTORI, R. 2003. Origin and early evolutionary radiation of the order Lagenida (Foraminifera). *Journal of Paleontology*, **77**, 831–843.

GROVES, J.R., RETTORI, R. & ALTINER, D. 2004. Wall structures in selected Paleozoic Lagenide Foraminifera. *Journal of Paleontology*, **78**, 245–256.

GROVES, J.R., ALTINER, D. & RETTORI, R. 2005. *Extinction, Survival, and Recovery of Lagenide Foraminifers in the Permian–Triassic Boundary Interval, Central Taurides, Turkey*. Paleontological Society Memoir, **62**.

GROVES, J.R., RETTORI, R., PAYNE, J.L., BOYCE, M.D. & ALTINER, D. 2007. End-Permian mass extinction lagenide foraminifers in the Southern Alps (northern Italy). *Journal of Paleontology*, **81**, 415–434.

GU, S., FENG, Q. & HE, W. 2007. The last Permian deep-water fauna: latest Changhsingian small foraminifers from southwestern Guangxi, South China. *Micropaleontology*, **53**, 311–330.

HABEEB, K.H. 1979. SEM study of foraminiferal wall structure in the Lower Carboniferous Limestone. *In*: *Abstracts of Papers, Ninth International Congress of Carboniferous Stratigraphy and Geology*, 19–26 May 1979. University of Illinois, Urbana-Champaign, IL, 82–83.

HANCE, L., HOU, H.F. & VACHARD, D. 2011. *Upper Famennian to Visean Foraminifers and Some Carbonate Microproblematica from South China, Hunan, Guangxi and Guizhou*. Geological Science Press, Beijing.

HAUSER, M., VACHARD, D., MARTINI, R., MATTER, A., PETERS, T. & ZANINETTI, L. 2000. The Permian sequence reconstructed from reworked carbonate clasts in the Batain Plain (Northeastern Oman). *Comptes Rendus de l'Académie des Sciences*, **320**, 273–279.

HAYNES, J.R. 1981. *Foraminifera*. Wiley, New York.

HENBEST, L.G. 1963. Biology, mineralogy, and diagenesis of some typical late Paleozoic sedimentary foraminifera and algal-foraminiferal colonies. Cushman Foundation for Foraminiferal Research, Special Publications, **6**, 1–44.

HENDERSON, C.M., PINARD, S. & BEAUCHAMP, B. 1995. Biostratigraphic and sequence stratigraphic relationships of Upper Carboniferous conodont and foraminifer distribution, Canadian Arctic Archipelago. *Bulletin of Canadian Petroleum Geology*, **43**, 226–246.

HENDERSON, C.M., DAVYDOV, V.I. & WARDLAW, B.R. 2012. The Permian Period. *In*: GRADSTEIN, F.M., OGG, J.G., SCHMITZ, M.D. & OGG, G.M. (eds) *The Geologic Time Scale 2012*. Elsevier, Amsterdam, 653–679.

HO, J. 1959. Triassic Foraminifera from the Chialingchiang Limestone of South Szechuan. *Acta Paleontologica Sinica*, **7**, 387–418 [in Chinese].

HOHENEGGER, J. & PILLER, W. 1975. Wandstrukturen und Grossgliederung der Foraminiferen. *Sitzungsberichte der österreichischen Akademie der Wissenschaften, Mathematisch-naturwissenschaftliche Klasse, Abteilung I*, **184**, 67–96.

HUANG, H., YANG, X. & JIN, X. 2007. The *Shanita* fauna (Permian foraminifera) from Baoshan area, western Yunnan province, China. *Frontiers of Biology in China*, **2**, 114–124 [English translation from *Acta Palaeontologica Sinica*, 44, 544–555].

HUGHES, G.W. 2005. Saudi Arabian Permo-Triassic biostratigraphy, micropaleontology and palaeoenvironment. *In*: POWELL, A.J. & RIDING, J.B. (eds) *Recent Developments in Applied Biostratigraphy*. Micropalaeontological Society, Special Publications, **93**, 91–108.

IRANIAN–JAPANESE RESEARCH GROUP 1981. The Permian and the Lower Triassic systems in Abadeh region, Central Iran. *Memoirs Faculty Sciences, Kyoto University, Series Geology & Mineralogy*, **47**, 61–133.

ISHII, K.-I., OKIMURA, Y. & NAKAZAWA, K. 1975. On the genus *Colaniella* and its biostratigraphic significance. *Journal Geosciences, Osaka City University*, **19**, 107–129.

ISOZAKI, Y., KAWAHATA, H. & MINOSHIMA, K. 2007. The Capitanian (Permian) Kamura cooling event: the beginning of the Paleozoic-Mesozoic transition. *Palaeoworld*, **16**, 16–30.

JENNY, J. & JENNY-DESHUSSES, C. 1978. *Dorudia dorudensis* n. gen., n. sp., et les Tuberitininae du Permien de l'Elbourz oriental en Iran. *Note du Laboratoire de Paléontologie de l'Université de Genève*, **2**, 7–15.

JENNY-DESHUSSES, C. 1983. *Le Permien de l'Elbourz Central et Oriental (Iran): stratigraphie et micropaléontologie (foraminifères et algues)*. PhD thesis, University of Geneva.

JENNY-DESHUSSES, C. 1985. *Rectostipulina* n. gen. (=*Stipulina* Lys, 1978), a Upper Permian, incertae sedis organism of the central-eastern Tethys: morphological description and stratigraphic remarks. *Revue de Paléobiologie*, **4**, 153–158.

JENNY-DESHUSSES, C., MARTINI, R. & ZANINETTI, L. 2000. Discovery of the foraminifer Colaniella Likharev in the Upper Permian of the Sosio valley (Sicily). *Comptes Rendus de l'Académie des Sciences de Paris*, **330**, 799–804.

JIN, X. & YANG, X. 2004. Paleogeographic implications of the *Shanita-Hemigordius* fauna (Permian foraminifer) in the reconstruction of Permian Tethys. *Episodes*, **27**, 273–278.

JIN, Y., WARDLAW, B.R., GLENISTER, B.F. & KOTLYAR, G.V. 1997. Permian chronostratigraphic subdivisions. *Episodes*, **20**, 10–15.

JIN, Y.G., WANG, Y., WANG, W., SHANG, Q.H., CAO, C.Q. & ERWIN, D.H. 2000. Pattern of marine mass extinction near the Permian–Triassic boundary in South China. *Science*, **289**, 432–436.

JIROVEC, O. 1953. *Protozoologie*. Nakladatelstvi Ceskoslovenské Akademie Ved, Prague.

JONES, T.R. 1895. *A Monograph of the Foraminifera of the Crag, Part 2*. Monograph of the Palaeontographical Society, London, **49**, (230), 73–210.

KARAVAEVA, N.I. & NESTELL, G.P. 2007. Permian foraminifers of the Omolon Massif, northeastern Siberia, Russia. *Micropaleontology*, **53**, 161–211.

KALIA, P., KUMAR PANDE, P. & SINGH, T. 2000. Early Permian foraminifers from eastern Himalaya and their palaeobiogeographic significance. *Alcheringa*, **24**, 207–227.

KIPARISOVA, L.D., MARKOVSKY, B.P. & RADCHENKO, G.P. 1956. Materials for palaeontology, new families and genera. *Vsesoyuznogo Nauchno-issledovatelskogo Geologicheskii Institut (VSEGEI) novya seriya, Paleontologiya*, **12**, 1–354 [in Russian].

KOBAYASHI, F. 1986. Middle Permian foraminifers of the Gozenyama Formation, southern Kwanto Mountains, Japan. *Bulletin of the National Science Museum, Series C (Geology and Paleontology)*, **12**, 131–163.

KOBAYASHI, F. 1988. Middle Permian foraminifers of the Omi Limestone, Central Japan. *Bulletin of the National Science Museum, Series C (Geology and Paleontology)*, **14**, 1–35.

KOBAYASHI, F. 1996. Middle Triassic (Anisian) foraminifers from the Kaizawa Formation, southern Kanto Mountains, Japan. *Transactions and Proceedings of the Paleontological Society of Japan, New Series*, **183**, 528–539.

KOBAYASHI, F. 1997*a*. Middle Permian biogeography based on fusulinacean faunas. *In*: ROSS, C.A., ROSS, J.R.P. & BRENCKLE, P.L. (eds) *Late Paleozoic Foraminifera: Their Biogeography, Evolution, and Paleoecology; and Mid-Carboniferous Boundary*. Cushman Foundation for Foraminiferal Research, Special Publications, **36**, 73–76.

KOBAYASHI, F. 1997*b*. Middle Permian fusulinacean faunas and paleogeography of exotic terranes in the circum-Pacific. *In*: ROSS, C.A., ROSS, J.R.P. & BRENCKLE, P.L. (eds) *Late Paleozoic Foraminifera: Their Biogeography, Evolution, and Paleoecology; and Mid-Carboniferous Boundary*. Cushman Foundation for Foraminiferal Research, Special Publications, **36**, 77–80.

KOBAYASHI, F. 1997*c*. Upper Permian foraminifers from the Iwai–Kanyo area, West Tokyo, Japan. *Journal of Foraminiferal Research*, **27**, 186–195.

KOBAYASHI, F. 1999. Tethyan uppermost Permian (Dzhulfian and Dorashamian) foraminiferal faunas and their paleogeographic and tectonics implications. *Palaeogeography, Palaeoclimatology, Palaeoecology*, **150**, 279–307.

KOBAYASHI, F. 2001. Faunal analysis of Permian foraminifers of the Kuma Formation in the Kurosegawa Belt of west Kyushu, Southwest Japan. *News of Osaka Micropaleontologists*, Special Volume, **12**, 61–77.

KOBAYASHI, F. 2002. Lithology and foraminiferal fauna of the allochthonous limestone (Changhsingian) in the upper part of the Toyoma Formation in the South Kitakami Belt, northeast Japan. *Paleontological Research*, **6**, 331–342.

KOBAYASHI, F. 2004. Late Permian foraminifers from the limestone block in the Southern Chichibu Terrane of west Shikoku, SW Japan. *Journal of Paleontology*, **78**, 62–70.

KOBAYASHI, F. 2006*a*. Middle Permian foraminifers of Kaize, southern part of the Saku Basin, Nagano prefecture, central Japan. *Paleontological Research*, **10**, 179–194.

KOBAYASHI, F. 2006*b*. Latest Permian (Changhsingian) foraminifers in the Mikata area, Hyogo – Late Paleozoic and Early Mesozoic foraminifers of Hyogo, Japan, Part 3. *Nature and Human Activities*, **10**, 15–24.

KOBAYASHI, F. 2006*c*. Early Late Permian (Wuchiapingian) foraminifers in the Tatsuno area, Hyogo – Late Paleozoic and Early Mesozoic foraminifers of Hyogo, Japan, Part 4. *Nature and Human Activities*, **10**, 25–33.

KOBAYASHI, F. 2012*a*. Permian non-fusuline foraminifers of the Akasaka Limestone (Japan). *Revue de Paléobiologie, Genève*, **31**, 313–335.

KOBAYASHI, F. 2012*b*. Middle and Late Permian foraminifers from the Chichibu Belt, Takachiho area, Kyushu, Japan: implications for faunal events. *Journal of Paleontology*, **86**, 669–687.

KOBAYASHI, F. 2013. Late Permian (Lopingian) foraminifers from the Tsukumi Limestone, southern Chichibu Terrane of eastern Kyushu, Japan. *Journal of Foraminiferal Research*, **43**, 154–169.

KOBAYASHI, F. & ISHII, K. 2003. Permian fusulinaceans of the Surmaq Formation in the Abadeh region, central Iran. *Rivista Italiana di Paleontologia e Stratigrafia*, **109**, 307–337.

KOCHANSKY-DEVIDÉ, V. 1973. *Ramovsia limes* n. g., n. sp. (Problematica), ein Leitfossil der Grenzlandbänke (unteres Perm). *Neues Jahrbuch für Geologie, Paläontologie Monathefte*, **8**, 462–468.

KOEHRER, B., AIGNER, T., FORKE, H. & PÖPPELREITER, M. 2012. Middle to Upper Khuff (Sequences KS1 to KS4) outcrop-equivalents in the Oman Mountains: grainstone architecture on a some regional scale. *GeoArabia*, **17**, 59–104.

KOLODKA, C., VENNIN, E., VACHARD, D., TROCME, V. & GOODARZI, M.H. 2012. Timing and progression of the end-Guadalupian crisis in the Fars province (Dalan Formation; Kuh-e Gakhum, Iran) constrained by foraminifers and other carbonate microfossils. *Facies*, **58**, 131–153.

KOTLYAR, G.V., ZAKHAROV, YU.D. ET AL. 1984. *Late Permian Etaps of the Evolution of the Organic World, Dzhulfian and Dorashamian Stages in the USSR*. Akademiya Nauk SSSR, Dalnevostochnyi Nauchnyi Tsentr, Biologo-Pochvennyi Institut, Leningrad [in Russian].

KOTLYAR, G.V., ZAKHAROV, YU.D., KROPACHEVA, G.S., PRONINA, G.P., CHEDIYA, I.O. & BURAGO, V.I. 1989. Late Permian etaps of the evolution of the organic world, Midian Stage in the USSR. *Leningrad 'Nauka', Leningradskoe Otdelenie*, **1**, 1–177 [in Russian].

KOTLYAR, G.V., PRONINA, G.P., ZAKHAROV, YU.D. & NESTELL, M.K. 1999. Changhsingian of the northwestern Caucasus, southern Primorye and southeastern Pamirs. *Permophiles*, **35**, 18–22.

KÖYLÜOGLU, M. & ALTINER, D. 1989. Micropaléontologie (foraminifères) et biostratigraphie du Permien supérieur de la région d'Hakkari (SE Turquie). *Revue de Paléobiologie*, **8**, 467–503.

KRAINER, K. & VACHARD, D. 2011. The Lower Triassic Werfen Formation of the Karawanken Mountains (southern Austria) and its disaster survivor microfossils, with emphasis on *Postcladella* n. gen. (Foraminifera, Miliolata, Cornuspirida). *Revue de Micropaléontologie*, **54**, 59–85.

KRAINER, K., LUCAS, S.G., VACHARD, D. & BARRICK, J.E. 2015. The Pennsylvanian–Permian section at Robledo Mountain, Doña Ana County, New Mexico, USA. *In*: LUCAS, S.G. & DIMICHELE, W.A. (eds) *Carboniferous–Permian Transition in the Robledo Mountains, Southern New Mexico*. New Mexico Museum of Natural History and Science Bulletin, **65**, 9–41.

KULAGINA, E.I., RUMYANTSEVA, Z.C., PAZUKHIN, V.N. & KOCHETOVA, N.N. 1992. *Granitsa nizhnego-srednego Karbona na Yuzhnom Urale i Srednem Tyan-Shane*. Rossiiskaya Akademiya Nauk, Uralskoe Otdelenie, Bashkirskii Nauchnyi Tsentr, Institut Geologii, Moskva, Nauka, Moscow [in Russian].

LAMARCK, J.B. DE. 1812. *Extrait du cours de zoologie du Muséum d'Histoire Naturelle sur les animaux sans vertèbres*. D'Hautel, Paris.

LAMARCK, J.B. DE. 1816. *Tableau encyclopédique et méthodique des trois règnes de la nature. Partie 23 – Mollusques et Polypes divers*. Mme veuve Agasse, Paris.

LANGE, E. 1925. Eine mittelpermische Fauna von Guguk Bulat (Padanger Oberland, Sumatra). *Verhandelingen van het geologisch mijnbouwkunding Genootschap voor Nederland en Koloniën, geologisch serie*, **7**, 213–295.

LANGER, M.R., LIPPS, J.H. & PILLER, W.E. 1993. Molecular paleobiology of prtists: amplification and direct sequencing of foraminiferal DNA. *Micropaleontology*, **39**, 63–68.

LANKESTER, E.R. 1885. Protozoa. *In*: *Encyclopedia Britannica, Volume 19*. 9th edn. Encyclopædia Britannica, London.

LEVEN, E.YA. 1967. Stratigraphy and fusulinids of the Permian strata of Pamir. *Trudy Geologicheskogo Instituta Akademiya Nauk SSSR, Moscow*, **167**, 1–224 [in Russian].

LEVEN, E.YA. 1998. Permian fusulinids assemblages and stratigraphy of Transcaucasia. *Rivista Italiana di Paleontologia e Stratigrafia*, **104**, 299–328.

LEVEN, E.YA. 2003. Permian stratigraphy and fusulinids of the Tethys. *Rivista Italiana di Paleontologia i Stratiigrafia*, **109**, 267–280.

LEVEN, E.YA. & GORGIJ, M.N. 2011. Fusulinids and stratigraphy of the Carboniferous and Permian in Iran. *Stratigraphy and Geological Correlation*, **19**, 687–776.

LEVEN, E.YA. & OKAY, A.I. 1996. Foraminifera from the exotic Permo-Carboniferous limestone blocks in the Karakaya complex, northwestern Turkey. *Rivista Italiana di Paleontologia e Stratigrafia*, **102**, 139–174.

LIKHAREV, B.K. 1939. *Atlas of the Guide Species Among the Fossil Fauna of the USSR*. Tsentralnyi Nauchno-issledovatelskii Geologo-razvedochnyi Institut, Leningrad, **6** (Permian System), 1–268 [in Russian].

LIN, J.X. 1978. Carboniferous and Permian Foraminiferida. *In*: HUBEI INSTITUTE OF GEOLOGICAL SCIENCE (eds) *Paleontological Atlas of Central South China (Micropaleontological Volume)*. Geological Publishing House, Beijing, 10–43 [in Chinese].

LIN, J.X. 1980. On the age and stratigraphical significance of the genus *Gallowayinella*. *Bulletin of the Chinese Academy Geological Sciences*, Series 8, **1**, 37–46 [in Chinese].

LIN, J.X. 1984. Protozoa. *In*: YICHAN INSTITUTE OF GEOLOGY & MINERAL RESOURCES (eds) *Biostratigraphy of the Yangtze Gorge Area Chiefly, 3) Late Paleozoic Era*. Geological Publishing House, Beijing, 110–177 [in Chinese], 323–364 [in English].

LIN, J.X. 1985. Late Early Permian foraminifera and its paleoecology in Jiahe, Hunan. *Bulletin of the Yichang Institute of Geology and Mineral Resources, Chinese Academy of Geological Sciences*, **9**, 43–52 [in Chinese with English abstract].

LIN, J.X., LI, L.X. & SUN, Q.Y. 1990. *Late Paleozoic Foraminifers in South China, Volume 1*. Science Publication House, Beijing [in Chinese].

LIU, C., JAROCHOWSKA, E., DU, Y., VACHARD, D. & MUNNECKE, A. In press. Stratigraphical and $\delta^{13}C$ records of Permo-Carboniferous platform carbonates: responses to deep icehouse and icehouse–greenhouse transition. *Palaeogeography, Palaeoclimatology, Palaeogeography*.

LOEBLICH, A.R. & TAPPAN, H. 1955. *A Revision of Some Glanduline Nodosariidae (Foraminifera)*. Smithsonian Miscellaneous Collections, **126**, (3).

LOEBLICH, A.R. & TAPPAN, H. 1961. Suprageneric classification of the Rhizopodea. *Journal of Paleontology*, **35**, 245–330.

LOEBLICH, A.R. & TAPPAN, H. 1964. Sarcodina, chiefly 'Thecamoebians' and Foraminiferida. *In*: MOORE, R.C. (ed.) *Treatise of Invertebrate Paleontology, Part C, Protista 2, Volume 2*. Geological Society of America and the University of Kansas Press, Lawrence, KS, C1–C900.

LOEBLICH, A.R. & TAPPAN, H. 1981. Suprageneric revision of some calcareous Foraminiferida. *Journal of Foraminiferal Research*, **11**, 159–164.

LOEBLICH, A.R. & TAPPAN, H. 1982. Classification of the Foraminiferida. *In*: BROADHEAD, T.W. (ed.) *Foraminifera, Notes for a Short Course Organized by M.A. Buzas and B.K. Sen Gupta*. University of Tennessee, Department of Geological Sciences, Studies in Geology, **6**, 22–36.

LOEBLICH, A.R. & TAPPAN, H. 1984. Suprageneric classification of the Foraminiferida (Protozoa). *Micropaleontology*, **30**, 1–70.

LOEBLICH, A.R. & TAPPAN, H. 1985. *Cylindrocolaniella* a new name for the foraminiferal genus *Wanganella* Sosnina, 1956, non *Wanganella* Laserow, 1954. *Journal of Foraminiferal Research*, **15**, 218.

LOEBLICH, A.R. & TAPPAN, H. 1986*a*. Some new and redefined genera and families of Textulariina, Fusulinina, Involutinina and Miliolina (Foraminiferida). *Journal of Foraminiferal Research*, **16**, 334–346.

LOEBLICH, A.R. & TAPPAN, H. 1986*b*. Some new and revised genera and families of hyaline calcareous Foraminiferida (Protozoa). *Transactions of the American Microscopical Society*, **105**, 239–265.

LOEBLICH, A.R. & TAPPAN, H. 1987. *Foraminiferal Genera and Their Classification*. Van Nostrand Reinhold, New York.

LOEBLICH, A.R. & TAPPAN, H. 1992. Present status of foraminiferal classification. *In*: TAKAYANAGI, Y. & SAITO, T. (eds) *Studies in Benthic Foraminifera, Benthos' 90*. Tokai University Press, Tokyo, 93–102.

LUCAS, S.G., KRAINER, K. & VACHARD, D. 2015. The Lower Permian Hueco Group, Robledo Mountains, New Mexico (U.S.A.). *In*: LUCAS, S.G. & DIMICHELE, W.A. (eds) *Carboniferous–Permian Transition in the Robledo Mountains, Southern New Mexico*. New Mexico Museum of Natural History and Science Bulletin, **65**, 43–95.

LUCAS, S.G., KRAINER, K. & VACHARD, D. 2016*a*. The Pennsylvanian system in the Mud Springs Mountains, Sierra County, New Mexico. *New Mexico Museum of Natural History and Science Bulletin*, **69**, 1–58.

LUCAS, S.G., KRAINER, K. & VACHARD, D. 2016*b*. The Pennsylvanian section at Priest Canyon, southern Manzano Mountains, New Mexico. *In*: *New Mexico Geological Society, 67th Field Conference, Geology of the Belen Area*. New Mexico Geological Society, Socorro, NM, 69–95.

LYS, M. & MARCOUX, J. 1978. Les niveaux du Permien supérieur des Nappes d'Antalya (Taurides occidentales, Turquie). *Comptes Rendus Académie Sciences Paris*, **286**, série D, 1417–1420.

LYS, M., COLCHEN, M., BASSOULLET, J.P., MARCOUX, J. & MASCLE, G. 1980. La biozone à *Colaniella parva* du Permien supérieur et sa microfaune dans le bloc calcaire exotique de Lamayuru, Himalaya du Ladakh. *Revue de Micropaléontologie*, **23**, 76–108.

MACFADYEN, W.A. 1941. Foraminifera from the Green Ammonite Beds, lower Lias, of Dorset. *Philosophical Transactions of the Royal Society, London*, Series B, **231**, 1–73.

MASLAKOVA, N.I. 1990. Criteria for establishing higher taxa in foraminifers. *In*: MENNER, V.V. (ed.) *Systematics and Phylogeny of Invertebrates: The Criteria for Establishing Higher Taxa*. Izdateltsvo Nauka, Moscow, 22–27 [in Russian].

MASLAKOVA, N.I., GORBACHIK, T.N., ALEKSEEV, A.S., BARSKOV, I.S., GOLUBEV, S.N., NAZAROV, B.B. & PETRUCHEVSKAYA, M.G. 1995. *Micropalaeontology (Anonymous: Class Foraminifera, foraminifers, 13-111)*. Izdatelsvo Moskovskogo Universita, Moscow [in Russian].

MAMET, B. 1996. Late Paleozoic small foraminifers (endothyrids) from South America (Ecuador and Bolivia). *Canadian Journal of Earth Sciences*, **33**, 452–459.

MAMET, B. & PINARD, S. 1992. Note sur la taxonomie des petits foraminifères du Paléozoïque supérieur. *Bulletin de la Société belge de Géologie*, **99**, 373–398.

MAMET, B.L., ROUX, A. & NASSICHUK, W.W. 1987. Algues carbonifères et permiennes de l'Arctique canadien. *Geological Survey of Canada Bulletin*, **342**, 1–143.

MASLOV, V.P. 1956. Fossil calcareous algae from the USSR. *Trudy Instituta Geologichesnikh Nauk*,

Akademiya Nauk SSSR, **160**, 1–301 [in Russian; French translation BRGM No. 3517].

MERTMANN, D. 2000. Foraminiferal assemblages in Permian carbonates of the Zaluch Group (Salt Range and Trans Indus Ranges, Pakistan). *Neues Jahrbuch Paläontologie Monathefte*, **2000**, 129–146.

MIKHALEVICH, V.I. 1980. Systematics and evolution of the foraminifers in the light of some new data on their cytology and ultrastructure. *Trudy Zoologicheskogo Instituta, Akademiya Nauk SSSR*, **94**, 42–61 [in Russian].

MIKHALEVICH, V.I. 1988. System of the subclass Miliolata (Foraminifera). *Akademiya Nauk SSSR, Trudy Zoologicheskogo Instituta*, **184**, 77–110 [in Russian].

MIKHALEVICH, V.I. 1993. New higher taxa of the subclass Nodosariata (Foraminifera). *Zoosystematica Rossica*, **2**, 5–8.

MIKHALEVICH, V.I. 2003. System of the four foraminiferal subclasses with the agglutinated shell wall (Ammodiscana, Miliamminana, Hormosiniana, Textulariana) (Foraminifera). *Rossiiskaya Akademiya Nauk, Izvestiya Zoologicheskogo Instituta*, **7**, 5–50.

MIKHALEVICH, V.I. 2013. New insight into the systematics and evolution of the foraminifera. *Micropalaeontology*, **59**, 493–527.

MIKLUKHO-MAKLAY, A.D. 1953. On the systematics of the family Archaediscidae. *Ezhegodnik Vsesoyuznogo Paleontologischeskogo Obshchestva (1948–53)*, **14**, 127–131 [in Russian].

MIKLUKHO-MAKLAY, A.D. 1956. On the systematics of the Palaeozoic foraminifers. *Vestnik Leningradskogo Universiteta*, **6**, 58–66 [in Russian].

MIKLUKHO-MAKLAY, A.D. 1958. A new foraminiferal family, Tuberitinidae M.Maclay fam. nov. *Voprosy Mikropaleontologii*, **2**, 130–135 [in Russian].

MIKLUKHO-MAKLAY, K.V. 1954. Permian foraminifers from North Caucasus. *Trudy VSEGEI, Gosgeoltekhizdat*, **1**, 1–163 [in Russian; French translation BRGM No. 2683, 1–216].

MIKLUKHO-MAKLAY, K.V. 1960*a*. New Kazanian lagenids from the Russian Platform. *In*: MARKOVSKYI, B.P. (ed.) *New Species of Fossil Plants and Invertebrates from the USSR, pt. 1*. Trudy VSEGEI, Gosgeoltekhizdat, 153–161 [in Russian].

MIKLUKHO-MAKLAY, K.V. 1960*b*. On the genus *Lunucammina* Spandel, 1898 (foraminifers). *Ezhegodnik Vsesoyuznogo Paleontologicheskogo Obshchestva*, **32**, 5–12 [in Russian].

MOHTAT-AGHAI, P. & VACHARD, D. 2003. *Dagmarita shahrezahensis* n. sp. globivalvulinid foraminifer (Wuchiapingian, Late Permian, Central Iran). *Rivista Italiana di Paleontologia i Stratigrafia*, **109**, 37–44.

MOHTAT-AGHAI, P. & VACHARD, D. 2005. Late Permian foraminiferal assemblages from the Hambast region (Central Iran) and their extinctions. *Revista Española de Micropaleontología*, **37**, 205–227.

MOHTAT-AGHAI, P., VACHARD, D. & KRAINER, K. 2009. Transported foraminifera in Palaeozoic deep red nodular limestones exemplified by latest Permian *Neoendothyra* in the Zal section (Julfa area, NW Iran). *Revista Española de Micropaleontología*, **41**, 197–213.

MOIX, P., VACHARD, D., ALLIBON, J., MARTINI, R., WERNLI, R., KOZUR, H.W. & STAMPFLI, G.M. 2013. Palaeotethyan, Neotethyan and Hu Lu-Pindos series in the Lycian Nappes (SW Turkey): geodynamical implications. *In*: TANNER, L.H., SPIELMANN, J.A. & LUCAS, S.G. (eds) *The Triassic System*. New Mexico Museum of Natural History and Science, **61**, 401–444.

MONTENAT, C., DE LAPPARENT, A.F., LYS, M., TERMIER, H., TERMIER, G. & VACHARD, D. 1977. La transgression permienne et son substrat éocambrien dans le Jebel Akhdar, montagnes d'Oman, Péninsule Arabique. *Annales de la Société géologique du Nord*, **96**, 239–258.

VON MÖLLER, V. 1878. Die spiral-gewundenen Foraminiferen des russischen Kohlenkalks. *Mémoires de l'Académie Impériale des Sciences de St Pétersbourg*, 7th Series, **25**, 1–147.

NESTELL, G.P. & NESTELL, M.K. 2006. Middle Permian (Late Guadalupian) foraminifers from Dark Canyon, Guadalupe Mountains, New Mexico. *Micropaleontology*, **52**, 1–50.

NESTELL, G.P., MESTRE, A. & HEREDIA, S. 2009. First Ordovician Foraminifera from South America: a Darriwilian (Middle Ordovician) fauna from the San Juan Formation, Argentina. *Micropaleontology*, **55**, 329–344.

NESTELL, G.P., NESTELL, M.K. *ET AL.* 2015. High influx of carbon in walls of agglutinated foram inifers during the Permian–Triassic transition in global oceans. *International Geology Review*, **57**, 411–427, https://doi.org/10.1080/00206814.2015.1010610

NESTELL, M.K. & PRONINA, G.P. 1997. The distribution and age of the genus *Hemigordiopsis*. *In*: ROSS, C.A., ROSS, J.R.P. & BRENCKLE, P.L. (eds) *Late Paleozoic Foraminifera; Their Biostratigraphy, Evolution, and Paleoecology; and the Mid-Carboniferous Boundary*. Cushman Foundation for foraminiferal Research, Special Publications, **36**, 105–110.

NESTELL, M.K., NESTELL, G.P., WARDLAW, B.R. & SWEATT, M.J. 2006. Integrated biostratigraphy of foraminifers, radiolarians and conodonts in shallow and deep water Middle Permian (Capitanian) deposits of the 'Rader slide', Guadalupe Mountains, West Texas. *Stratigraphy*, **3**, 161–194.

NESTLER, H. 1973. The types of *Tetrataxis conica* Ehrenberg, 1854 and *Tetrataxis palaeotrochus* (Ehrenberg, 1854). *Micropaleontology*, **19**, 366–369.

NEUMAYR, M. 1887. Die natürlichen Verwandtschaftverhältnisse der schalentragenden Foraminiferen. *Sitzungsberichte der kaiserliche Akademie der Wissenschaften in Wien, Mathematisch-Naturwissenschaftliche Klasse*, **95**, 156–186.

NGUYEN DUC TIEN 1979. *Etude micropaléontologique (foraminifères) de matériaux du Permien du Cambodge*. PhD thesis, Université Paris Sud, Orsay.

NGUYEN DUC TIEN 1986*a*. Foraminifera and algae from the Permian of Kampuchea. *In*: FONTAINE, H., SUTEETHORN, V. *ET AL.* (eds) *The Permian of Southeast Asia*. CCOP Technical Bulletin, **18**, 116–137.

NGUYEN DUC TIEN 1986*b*. Foraminifera and algae from the Permian of Guguk Bulat and Silungkang, Sumatra. *In*: FONTAINE, H., SUTEETHORN, V. *ET AL.* (eds) *The Permian of Southeast Asia*. CCOP Technical Bulletin, **18**, 138–147.

NGUYEN DUC TIEN 1987. *Sphairionia sikuoides* gen. nov. sp. nov. a Permian incertae sedis organism. *11e*

Congrès International Stratigraphie et Géologie Carbonifère, 1–90, Beijing.

NGUYEN DUC TIEN 1989*a*. Lower Permian Foraminifera of Sumatra. *In*: FONTAINE, H. & GAFOER, S. (ed.) *The Pre-Tertiary Fossils of Sumatra and Their Environments*. CCOP Technical Publications, **19**, 71–93.

NGUYEN DUC TIEN 1989*b*. *Sphairionia sikuoides* nov. gen. nov. sp., a Permian incertae sedis organism. *11e Congrès International de Géologie du Carbonifère, Beijing 1987, Compte Rendu*, **3**, 73–78.

NIKITINA, A.P. 1969. Genus *Hemigordiopsis* (Foraminifera) in the Late Permian of Primorye. *Paleontologicheskii Zhurnal*, **3**, 63–69 [in Russian].

NOÉ, S.U. 1987. Facies and paleogeography of the marine Upper Permian and of the Permian-Triassic boundary in the southern Alps (Bellerophon Formation, Tesero Horizon). *Facies*, **16**, 89–142.

NORVANG, A. 1957. The foraminifera of the Lias series in Jutland, Denmark. *Meddeleser fra Dansk Geologisk Forening*, **13**, 1–135.

OKIMURA, Y. 1972. Permo-Carboniferous endothyraceans from Japan; Part 1: Biseriamminidae. *Transactions of the Proceedings of the Paleontological Society Japan*, **87**, 414–428.

OKIMURA, Y. 1988. Primitive colaniellid foraminiferal assemblage from the Upper Permian Wargal Formation of Salt Range, Pakistan. *Journal of Paleontology*, **62**, 715–723.

OKIMURA, Y. & ISHII, K.-i. 1981. Smaller foraminifera from the Abadeh Formation, Abadehian stratotype, central Iran. *Reports of the Geological Survey of Iran*, **49**, 7–27.

OKIMURA, Y., ISHII, K.-i. & NAKAZAWA, K. 1975. *Abadehella*, a new genus of tetrataxid Foraminifera from the Late Permian. *Memoirs of the Faculty of Science Kyoto University, Series Geology and Mineralogy*, **41**, 45–48.

OKUYUCU, C. 2007. New Middle Permian foraminifers (Chitralinidae) from the Karakaya Complex, in northwestern Turkey. *Comptes Rendus Palevol*, **6**, 311–319.

OKUYUCU, C., VACHARD, D. & GÖNCÜOĞLU, M.C. 2013. Refinements in biostratigraphy of the foraminiferal zone MFZ11 (middle Visean, Mississippian) in the Cebeciköy Limestone (Istanbul Terrane, NW Turkey), palaeogeographic implications. *Bulletin of Geosciences*, **88**, 621–645.

OKUYUCU, C., IVANOVA, D., BEDI, Y. & ERGEN, A. 2014. Discovery of an earliest Triassic, post-extinction foraminiferal assemblage above the Permian-Triassic boundary, Strandzha nappes, north-west Turkey. *Geological Quarterly*, **58**, 1–8, https://doi.org/10.7306/gq.1145

ORLOV-LABKOVSKY, O. 2004. Permian fusulinids (Foraminifera) of the subsurface of Israel: taxonomy and biostratigraphy. *Revista Española de Micropaleontología*, **36**, 389–406.

ÖZDIKMEN, H. 2009. Substitute names for some unicellular animal taxa (Protozoa). *Munis Entomology & Zoology*, **4**, 233–256.

PALMIERI, V. 1984. Permian foraminifera in the Bowen Basin, Queensland. *Queensland Geology*, **6**, 1–126.

PALMIERI, V. 2004. *Flectospiroides fusiformis* gen. et sp. nov. (Foraminiferida) from the Late Permian of the Bowen Basin, central Queensland. *Memoir of the Association of Australasian Palaeontologists*, **29**, 259–264.

PALMIERI, V., FOSTER, C.B. & BONDAREVA, E.V. 1994. First record of shared species of Late Permian small foraminiferids in Australia and Russia: time correlations and plate reconstructions. *AGSO Journal of Australian Geology and Geophysics*, **15**, 359–365.

PANTIC, S. 1965. *Pilammina densa* n. gen. n. sp. and other Ammodiscidae from the Middle Triassic in the Cr. Mnica (Montenegro). *Geoloski Vjesnik*, **18**, 189–192.

PANTIC, S. 1970. Lithostratigraphy and micropaleontology of the Middle and Upper Permian of Western Serbia. *Bulletin Institute Geology Geophysics Research (Geology), A*, **27**, 239–272.

PANZANELLI-FRATONI, R., LIMONGI, P., CIARAPICA, G., CIRILLI, S., MARTINI, R., SALVINI-BONNARD, G. & ZANINETTI, L. 1987. Les Foraminifères du Permien supérieur remaniés dans le 'Complexe Terrigène' de la formation triasique du Monte Facito, Apennin méridional. *Revue de Paléobiologie*, **6**, 293–319.

PAWLOWSKI, J., HOLZMANN, M. & TYSZKA, J. 2013. New supraordinal classification of Foraminifera: molecules meet morphology. *Marine Micropaleontology*, **100**, 1–10.

PHILLIPS, J. 1846. On the remains of microscopic animals in the rocks of Yorkshire. *Proceedings of the Geological and Polytechnic Society of the West Riding of Yorkshire (1844–45)*, **2**, 274–285.

PIA, J.von. 1937. Die wichtigsten Kalkalgen des Jungpaläozoikums und ihre geologische Bedeutung. *Compte Rendu du 2e Congrès Avancement Etudes de Stratigraphie du Carbonifère, Heerlen 1935*, **2**, 765–856.

PILLE, L. 2008. *Foraminifères et algues calcaires du Mississippien supérieur (Viséen supérieur-Serpukhovien): rôles biostratigraphique, paléoécologique et paléogéographique aux échelles locale, régionale et mondiale*. PhD thesis, Université de Lille.

PILLER, W.E. 1983. Remarks on the suborder Involutinina Hohenegger and Piller, 1977. *Journal of Foraminiferal Research*, **13**, 191–201.

PINARD, S. & MAMET, B. 1998. Taxonomie des petits foraminifères du Carbonifère supérieur-Permien inférieur du basin de Sverdrup, Arctique canadien. *Palaeontographica Canadiana*, **15**, 1–253.

PLUMMER, H.J. 1930. Calcareous foraminifera in the Brownwood shale near Bridgeport, Texas. *Bulletin of the University of Texas*, **3019**, 5–21.

PLUMMER, H.J. 1945. Smaller foraminifera in the Marble Falls, Smithwick, and lower Strawn strata around the Llano uplift in Texas. *Bulletin University of Texas*, **4401**, 209–271.

POKORNY, V. 1958. *Grundzüge der zoologischen Mikropaläontologie, Bd 1*. VEB Deutscher Verlag der Wissenschaften, Berlin.

POTY, E., DEVUYST, F.X. & HANCE, L. 2006. Upper Devonian and Mississippian foraminiferal and rugose coral zonations of Belgium and Northern France, a tool for Eurasian correlations. *Geological Magazine*, **143**, 829–857.

POYARKOV, B.V. & SKVORTSOV, V.P. 1979. *On methodological segregation of local epiboles and local biozones (exemplified by the Early Carboniferous of Tian Shan)*. Akademiya Nauk SSR, Dalvnevostochnyi

Nauchnyi Tsentr, Dalvnevostochnyi Geologicheskii Institut, Vladivostok, 5–27 [in Russian].

PRONINA, G.P. 1988*a*. The Late Permian smaller foraminifers of Transcaucasus. *Revue de Paléobiologie*, Special Volume, **1**, 89–96.

PRONINA, G.P. 1988*b*. Late Permian miliolates of Transcaucasia. *Trudy Zoologicheskogo Instituta, Akademiya Nauk SSSR*, **184**, 49–63 [in Russian].

PRONINA, G.P. 1994. Systematics and phylogeny of the order Hemigordiopsida (foraminifers). *Paleontologicheskii Zhurnal*, **1994**, 13–24 [in Russian].

PRONINA, G.P. 1996. Genus *Sphairionia* and its stratigraphic significance. *In*: *Reports of Shallow Tethys 4, International Symposium*, Albrechtsberg (Austria) 8–11 September 1994. *Supplement agli Annali dei Musei Civici di Rovereto, Sezione Archeologia, Storia e Scienze Naturali*, **11**, 105–118.

PRONINA, G.P. 1998. Foraminifers. *In*: ESAULOVA, N.K., LOZOVSKY, V.R. & ROZANOV, A.Yu. (eds) *Stratotypes and Reference Sections of the Upper Permian in the Regions of the Volga and Kama Rivers. International Symposium Upper Permian Stratotypes of the Volga Region*. GEOS, Moscow, 163–169 [in Russian].

PRONINA, G.P. 1999*a*. Correlations of Late Permian deposits of Boreal regions based on smaller foraminifers. *In*: *International Symposium Upper Permian Stratotypes of the Volga Region*. GEOS, Moscow, 182–188 [in Russian].

PRONINA, G.P. 1999*b*. New Late Permian species of smaller foraminifers from Transcaucasia. *Voprosy Paleontologii*, **11**, 4–14 [in Russian].

PRONINA, G.P. & GUBENKO, T.A. 1990. Upper Permian planoarchaediscids from Transcaucasia. *Paleontologicheskii Zhurnal*, **2**, 119–123 [in Russian; English translation: *Paleontological Journal*, 1990, 2, 119–123].

PRONINA, G.P. & NESTELL, M.K. 1997. Middle and Late Permian Foraminifera from exotic limestone blocks of the Alma river basin, Crimea. *In*: ROSS, C.A., ROSS, J.R.P. & BRENCKLE, P.L. (eds) *Late Paleozoic Foraminifera; Their Biostratigraphy, Evolution, and Paleoecology; and the Mid-Carboniferous Boundary*. Cushman Foundation for Foraminiferal Research, Special Publications, **36**, 111–114.

PRONINA-NESTELL, G.P. & NESTELL, M.K. 2001. Late Changhsingian foraminifers of the northwestern Caucasus. *Micropaleontology*, **47**, 205–234.

RAUZER-CHERNOUSOVA, D.M. & CHERMNYKH, B.A. 1990. *Mesolasiodiscus* gen. nov. – new data on the evolution of the Late Palaeozoic. *Paleontologicheskii Zhurnal*, **1990**, **1**, 121–125 [in Russian].

RAUZER-CHERNOUSOVA, D.M. & FURSENKO, A.V. 1959. *Fundamentals of Paleontology, General Part, Protozoans*. Akademiya Nauk SSSR, Moscow [in Russian].

RAUZER-CHERNOUSOVA, D.M., BENSH, F.R. *ET AL*. 1996. *Reference-Book on the Systematics of Paleozoic Foraminifers; Endothyrida and Fusulinoida*. Rossiiskaya Akademiya Nauk, Geologicheskii Institut, Moscow [in Russian].

REICHEL, M. 1945. Sur un Miliolidé nouveau du Permien de l'île de Chypre. *Verhandlungen der Naturforschenden Gesellschaft in Basel*, **56**, 521–530.

REICHEL, M. 1946. Sur quelques foraminifères nouveaux du Permien méditerranéen. *Eclogae Geologicae Helvetiae*, **38**, 524–560.

REISS, Z. 1963. Reclassification of perforate Foraminifera. *Bulletin of the Geological Survey of Israel*, **35**, 1–111.

REITLINGER, E.A. 1950. Middle Carboniferous foraminifers of the central part of Russian Platform (Fusulinidae excepted). *Trudy Instituta Geologicheskikh Nauk*, geologichevskaya seriya 6, **47**, 1–126 [in Russian; French translation BRGM No. 1456].

REITLINGER, E.A. 1956. New family Lasiodiscidae. *Voprosy Mikropaleontologii*, **1**, 69–78 [in Russian; French translation by Sigal, S. & Sigal, J., BRGM No. 1753, 1–16].

REITLINGER, E.A. 1958. On the questions of systematics and phylogeny of the superfamily Endothyroidea. *Voprosy Mikropaleontologii*, **2**, 53–73 [in Russian; French translation by Sigal, S. & Sigal, J. 1960, Editions Technip, Paris, 59–81].

REITLINGER, E.A. 1965. On the development of the foraminifera of the Late Permian and Early Triassic in Transcaucasia. *Voprosy Mikropaleontologii*, **9**, 45–70 [in Russian].

REUSS, A.E. 1862. Entwurf einer systematischen Zusammenstellung der Foraminiferen. *Sitzungsberichte der kaiserlichen Akademie der Wissenschaften in Wien, Mathematich-Naturwissenschaftlische Klasse (1862)*, **46**, 5–100.

RETTORI, R. 1994. Replacement name *Hoyenella*, gen. n. (Triassic Foraminiferida, Miliolina) for *Glomospira sinensis* Ho, 1959. *Bolletino della Società Paleontologica Italiana*, **33**, 341–343.

RETTORI, R. 1995. *Foraminiferi del Trias inferiore e medio della Tetide: revisione tassonomica, stratigrafia ed interpretazione filogenetica*. Université de Genève, Publications du Département de Géologie et Paléontologie, **18**.

RHUMBLER, L. 1895. Entwurf eines natürlichen Systems der Thalamophoren. *Nachrichten von der Gesellschaft der Wissenschaften zu Göttingen, Mathematisch–Physikalische Klasse*, **1895**, 51–98.

RIGAUD, S., VACHARD, D. & MARTINI, R. 2014. Unsuspected phylogenetic links between Paleozoic Fusulinana and Recent Textulariana. *In*: MARCHANT, M. & HROMIC, T. (eds) *International Symposium on Foraminifera Forams 2014*, 19–24 January 2014, Chile, *Abstract Volume*. Grzybowski Foundation Special Publication, **20**, 4.

RIGAUD, S., VACHARD, D., MARTINI, R. & RETTORI, R. 2015*a*. Agglutinated versus microgranular: end of a paradigm? *Journal of Systematic Palaeontology*, **13**, 75–95, https://doi.org/10.1080/14772019.2013.863232

RIGAUD, S., VACHARD, D., SCHLAGINTWEIT, F. & MARTINI, R. 2015*b*. New lineage of Triassic aragonitic Foraminifera and reassessment of the class Nodosariata. *Journal of Systematic Palaeontology*, **14**, 1–19, https://doi.org/10.1080/14772019.2015.1112846

ŞAHIN, N., ALTINER, D. & BÜLENT ERCENGIZ, M. 2014. Discovery of Middle Permian volcanism in the Antalya Nappes, southern Turkey: tectonic significance and global meaning. *Geodinamica Acta*, **25**, 286–304, https://doi.org/10.1080/09853111.2013.858949

SAIDOVA, KH.M. 1981. *On an Up-to-Date System of Supraspecific Taxonomy of Cenozoic Benthic*

Foraminifers. Institut Okeanologii P.P. Shirshova, Akademiya Nauk SSSR, Moscow [in Russian].

Salaj, J., Borza, K. & Samuel, O. 1983. *Triassic Foraminifers of the West Carpathians*. Geologicky Ustav Dionyza Stura, Bratislava, Slovakia.

Sanders, D. & Krainer, K. 2005. Taphonomy of Early Permian benthic assemblages (Carnic Alps, Austria): carbonate dissolution v. biogenic carbonate precipitation. *Facies*, **51**, 522–540, https://doi.org/10.1007/s10347-005-0065-6

Schellwien, E. 1898. Die Fauna des Karnischen Fusulinenkalks. Theil 2: Foraminifera. *Palaeontographica*, **44**, 237–282.

Schubert, R.J. 1908. Beiträge zu einer natürlichen Systematik der Foraminiferen. *Neues Jahrbuch für Mineralogie, Geologie und Paläontologie*, **25**, 232–260.

Schubert, R.J. 1921. Paläontologische Daten zur Stammesgeschichte der Protozoen. *Paläontologische Zeitschrift*, **3**, 129–188.

Schultze, M.S. 1854. *Über den Organismus der Polythalamien (Foraminiferen) nebst Bemerkungen über die Rhizopoden im allgemeinen*. Wielhem Engelmann, Leipzig.

Sellier de Civrieux, J.M. & Dessauvagie, T.F.J. 1965. Reclassification de quelques Nodosariidae, particulièrement du Permien au Lias. *Maden Tetkik ve Arama Enstitu sü Yayinlarindan (M.T.A.)*, **124**, 1–178.

Şengör, A.M.C., Altiner, D., Clin, A., Ustaösmer, T. & Hsü, K.J. 1988. Origin and assembly of the Tethyside orogenic collage at the expense of Gondwana Land. *In*: Audley-Charles, M.G. & Hallam, A. (eds) *Gondwana and Tethys*. Geological Society, London, Special Publications, **37**, 119–181, https://doi.org/10.1144/GSL.SP.1988.037.01.09

Senowbari-Daryan, B., Bucur, I., Schlagintweit, F., Sasaran, E. & Matyszkiewicz, J. 2008. Crescentiella, a new name for 'Tubiphytes' morronensis Crescenti, 1969: an enigmatic Jurassic–Cretaceous microfossil. *Geologica Croatica*, **61**, 185–214.

Shang, Q.H., Vachard, D. & Caridroit, M. 2003. Smaller foraminifera from the late Changhsingian (latest Permian) of southern Guangxi and discussion on the Permian-Triassic boundary. *Acta Micropalaeontologica Sinica*, **20**, 377–388.

Sheng, J.Z. & He, Y. 1983. Permian *Shanita-Hemigordius (Hemigordiopsis)* (Foraminifera) fauna in western Yunnan, China. *Acta Palaeontologica Sinica*, **22**, 55–59 [in Chinese with English abstract].

Silvestri, A. 1903. Linguloglanduline e Lingulonodosarie. *Atti della Pontifica Accademia Romana dei Nuovi Lincei, Roma (1902–03)*, **56**, 45–50.

Sone, M. 2008. Correction of the publication date of *Sphaerionia* Tiên, a possible foraminifer or *incertae sedis*. *Journal of Foraminiferal Research*, **38**, 271.

Song, H.J., Tong, J.N., Zhang, K.X., Wang, Q.X. & Chen, Z.Q. 2007. Foraminiferal survivors from the Permian–Triassic mass extinction in the Meishan section, South China. *Palaeoworld*, **16**, 105–119.

Song, H.J., Tong, J.N., Chen, Z.Q., Yang, H. & Wang, Y.B. 2009. End-Permian mass extinction of foraminifers in the Nanpanjiang Basin, South China. *Journal of Paleontology*, **83**, 718–738.

Song, H., Tong, J. & Chen, Z.Q. 2011. Evolutionary dynamics of the Permian-Triassic foraminifer size: evidence for Lilliput effect in the end-Permian mass extinction and its aftermath. *Palaeogeography, Palaeoclimatology, Palaeoecology*, **308**, 98–110.

Song, H.J., Wignall, P.B., Tong, J.N. & Yin, H.F. 2013. Two pulses of extinction during the Permian–Triassic crisis. *Nature Geoscience*, **6**, 52–56.

Sosipatrova, G.P. 1969. Foraminifers of the Starostin Suite of Spitsbergen. *Trudy NIIGA, Sbornik Statey po Paleontologii i Biostratigrafii*, **30**, 35–72 [in Russian].

Sosipatrova, G.P. 1970. Foraminifers. *In*: Menner, V.V., Sarycheva, T.G. & Chernyak, G.E. (eds) *Stratigraphy of the Carboniferous and Permian deposits of the Northern Verkhoyansk*. Trudy NIIGA, Vsesoyuznyi Aèrogeologicheskii Trest, **154**, 56–70 [in Russian].

Sosnina, M.I. 1969. New name *Eomarginulinella* for the genus *Marginulinella* Sosnina, 1967. *Paleontologicheskiy Zhurnal*, **1969**, 101 [in Russian].

Sosnina, M.I. 1977. Late Permian nodosariids of southern Primorye. *Ezhegodnik Vsesoyuznogo Paleontologicheskogo Obshchestva*, **20**, 10–31 [in Russian].

Sosnina, M.I. 1978. *On the foraminifers of the Late Permian Chandalaz horizon of Primorye*. Akademiya Nauk SSSR, Dalnevostochnyi Nauchniyi Tsentr, Institut Tektoniki i Geofiziki, Moscow, 24–42 [in Russian].

Sosnina, M.I. 1983. Some new Late Permian miliolids and nodosariids (foraminifers) from the southern Primorye. *Ezhegodnik Vsesoyuznogo Paleontologischeskogo Obshchestva*, **26**, 29–47 [in Russian].

Sosnina, M.I. & Nikitina, A.P. 1977. *Late Permian smaller foraminifers of Primoriye*. Akademiya Nauk SSSR, Dalnevostochnyi Nauchniyi Tsentr, Dalnevostochnyi Geologicheskii Institut, Moscow, 27–52 [in Russian].

Spandel, E. 1898. *Die Foraminiferen des deutschen Zechsteins (vorläufige Mitteilung) und ein zweifelhaftes mikrokopisches Fossil ebendaher*. Verlag des Verlags-Instituts 'General-Anzeiger', Nürnberg.

Spandel, E. 1901. *Die Foraminiferen des Permo-Karbons von Hooser, Kansas, Nord Amerika*. Festschrift der Naturhistorischen Gesellschaft, Nürnberg, 175–194.

Stancliffe, R.P.W. 1989. Microforaminiferal linings: their classification, biostratigraphy and paleoecology, with special reference to specimens from British Oxfordian sediments. *Micropaleontology*, **35**, 337–352.

Termier, G., Termier, H. & Vachard, D. 1977. Etude comparative de quelques Ischyrosponges. *Géologie Méditerranéenne*, **4**, 139–180.

Théry, J.M., Vachard, D. & Dransart, E. 2007. Late Permian limestones and Permian-Triassic Boundary: new biostratigraphic, palaeobiogeographic and geochemical data in Caucasus and Eastern Europe. *In*: Alvaro, J.J., Aretz, M., Boulvain, F., Munnecke, A., Vachard, D. & Vennin, E. (eds) *Palaeozoic Reeefs and Bioaccumulations: Climatic and Evolutionary Controls*. Geological Society, London, Special Publications, **275**, 255–274, https://doi.org/10.1144/GSL.SP.2007.275.01.16

Ukharskaya (= Ucharskaja), L.B. 1970. New Kazanian (Permian) arenaceous foraminifers of the Russian Platform. *Paleontologicheskii Zhurnal*, **4**, 468–476.

UENO, K. 2003. The Permian fusulinoidean faunas of the Sibumasu and Baoshan blocks: their implications for the paleogeographic and paleoclimatologic reconstruction of the Cimmerian Continent. *Palaeogeography, Palaeoclimatology, Palaeoecology*, **193**, 1–24.

UENO, K. & SAKAGAMI, S. 1993. Middle Permian foraminifers fauna from Ban Nam Suai Tha Sa-at, Changwat Loei, Northeast Thailand. *Transactions and Proceedings of the Paleontological Society of Japan*, **172**, 277–291.

UENO, K. & TSUTSUMI, S. 2009. Lopingian (Late Permian) foraminiferal faunal succession of a Paleo-Tethyan mid-oceanic carbonate buildup: Shifodong Formation in the Changning-Menglian Belt, West Ynnan, Southwest China. *Island Arc*, **18**, 69–93.

UENO, K., MIYAHIGASHI, A. & CHAOENTITIRAT, T. 2010. The Lopingian (Late Permian) of mid-oceanic carbonates in the Eastern Palaeotethys: stratigraphical outline and foraminiferal faunal [sic] succession. *Geological Journal*, **45**, 285–307.

ÜNAL, E., ALTINER, D., ÖMER YILMAZ, I. & ÖZKAN-ALTINER, S. 2003. Cyclic sedimentation across the Permian-Triassic boundary (Central Taurides, Turkey). *Rivista Italiana di Paleontologia i Stratigrafia*, **109**, 359–376.

VACHARD, D. 1980. Téthys et Gondwana au Paléozoïque supérieur; les données afghanes: biostratigraphie, micropaléontologie, paléogéographie. *Documents et Travaux IGAL, Institut Géologique Albert de Lapparent*, **2**, 1–463.

VACHARD, D. 1990. Fusulinoids, smaller foraminifera and pseudo-algae from southeastern Kelantan (Malaysia) and their biostratigraphic and paleogeographic value. *In*: FONTAINE, H. (ed.) *Ten years of CCOP Research on the Pre-Tertiary of East Asia*. CCOP Technical Publications, **20**, 43–167.

VACHARD, D. 1994. Foraminifères et moravamminides du Givétien et du Frasnien du domaine Ligérien (Massif Armoricain, France). *Palaeontographica, A*, **231**, 1–92.

VACHARD, D. 2000. On some umbellinids (carbonate microproblematica) from the Frasnian (Late Devonian) of Chah Riseh area (central Iran). *Annales de la Société Géologique du Nord,* 2nd series, **8**, 75–80.

VACHARD, D. 2014. *Colaniella*, wrongly named, well-distributed Late Permian nodosariate foraminifers. *Permophiles*, **60**, 16–24.

VACHARD, D. & BECKARY, S. 1991. Algues et foraminifères bachkiriens des coal balls de la Mine Rosario (Truebano, Léon, Espagne). *Revue de Paléobiologie*, **10**, 315–357.

VACHARD, D. & FERRIÈRE, J. 1991. Une association à *Yabeina* (foraminifère fusulinoïde) dans le Midien (Permien supérieur) de la région de Whangaroa (Baie d'Orua, Nouvelle-Zélande). *Revue de Micropaléontologie*, **34**, 201–230.

VACHARD, D. & KRAINER, K. 2001*a*. Smaller foraminifers of the Upper Carboniferous Auernig Group, Carnic Alps (Austria/Italy). *Rivista Italiana di Paleontologia e Stratigrafia*, **107**, 147–168.

VACHARD, D. & KRAINER, K. 2001*b*. Smaller foraminifers, characteristic algae and pseudo-algae of the latest Carboniferous/Early Permian Rattendorf Group, Carnic Alps (Austria/Italy). *Rivista Italiana de Paleontologia e Stratigrafia*, **107**, 169–195.

VACHARD, D. & MICONNET, P. 1990. Une association à Fusulinoïdes du Murghabien supérieur au Monte Facito (Appennin méridional, Italie). *Revue de Micropaléontologie*, **32**, 297–318.

VACHARD, D. & MOIX, P. 2013. Kubergandian (Roadian, Middle Permian) of the Lycian and Aladag Nappes (Southern Turkey). *Geobios*, **46**, 335–356, https://doi.org/10.1016/j.geobios.2013.02.002

VACHARD, D. & MONTENAT, C. 1981. Biostratigraphie, micropaléontologie et paléogéographie du Permien de la région de Tezak (Montagnes Centrales d'Afghanistan). *Palaeontographica, B*, **178**, 1–88.

VACHARD, D. & RAZGALLAH, S. 1988. Importance phylogénétique d'un nouveau foraminifère endothyroïde *Endoteba controversa* n. gen. n. sp. (Permien du Jebel Tebaga, Tunisie). *Geobios*, **21**, 805–811.

VACHARD, D., OTT D'ESTEVOU, P. & BRIEND, M. 1982. Un micro-organisme problématique du Miocène d'Espagne voisin des Umbellaceae paléozoïques: *Mitrigalia* n. gen. *Revue de Micropaléontologie*, **25**, 57–68.

VACHARD, D., OVIEDO, A., FLORES DE DIOS, A., MALPICA, R., BRUNNER, P., GUERRERO, M. & BUITRÓN, B.E. 1992. Barranca d'Olinalá: une coupe de référence pour le Permien du Mexique central; étude préliminaire. *Annales de la Société Géologique du Nord,* 2ème série, **2**, 153–160.

VACHARD, D., CLIFT, P. & DECROUEZ, D. 1993*a*. Une association à *Pseudodunbarula* (Fusulinoïde) du Permien supérieur (Djoulfien) remaniée dans le Jurassique d'Argolide (Grèce). *Revue de Paléobiologie*, **12**, 217–242.

VACHARD, D., MARTINI, R., ZANINETTI, L. & ZAMBETAKIS-LEKKAS, A. 1993*b*. Révision micropaléontologique (Foraminifères, Algues) du Permien inférieur (Sakmarien) et supérieur (Dorashamien) du Mont Beletsi (Attique, Grèce). *Bolletino della Societá Paleontologica Italiana*, **32**, 89–112.

VACHARD, D., MARTINI, R., RETTORI, R. & ZANINETTI, L. 1994. Nouvelle classification des Foraminifères endothyroïdes du Trias. *Geobios*, **27**, 543–557.

VACHARD, D., HAUSER, M., MARTINI, R., ZANINETTI, L., PETERS, T. & MATTER, A. 2001*a*. New algae and microproblematica with algal affinity from the Permian of the Aseelah Unit in the Batain Plain, East Oman. (Nouvelles algues et formes affines microproblématiques du Permien de l'Unité d'Aseelah) (Batain Plain, Est de l'Oman). *Geobios*, **34**, 375–404.

VACHARD, D., MARTINI, R. & ZANINETTI, L. 2001*b*. Earliest Artinskian (Early Permian) fusulinid reworked in the Triassic Lercara Formation (NW Sicily). *Journal of Foraminiferal Research*, **31**, 33–47.

VACHARD, D., HAUSER, M., MARTINI, R., ZANINETTI, L., MATTER, A. & PETERS, T. 2002. Middle Permian (Midian/Capitanian) fusulinid assemblages from the Aseelah Unit (Batain Group) in the Batain Plain, East-Oman: their significance to Neotethys paleogeography. *Journal of Foraminiferal Research*, **32**, 155–172.

VACHARD, D., ZAMBETAKIS-LEKKAS, A., SKOURTSOS, E., MARTINI, R. & ZANINETTI, L. 2003. Foraminifera, algae and carbonate microproblematica from the late Wuchiapingian/Dzhulfian (Late Permian) of Peloponnesus (Greece). *Rivista Italiana di Paleontologia i Stratigrafia*, **109**, 339–358.

Vachard, D., Gaillot, J., Vaslet, D. & Le Nindre, Y.M. 2005. Foraminifers and algae from the Khuff Formation (late Middle Permian–Early Triassic) of central Saudi Arabia. *GeoArabia*, **10**, 137–186.

Vachard, D., Rettori, R., Angiolini, L. & Checconi, A. 2008. *Glomomidiella* gen. n. (Foraminifera, Miliolata, Neodiscidae): a new genus from the late Guadalupian-Lopingian of Hydra Island (Greece). *Rivista Italiana di Paleontologia e Stratigrafia*, **114**, 349–361.

Vachard, D., Pille, L. & Gaillot, J. 2010. Palaeozoic Foraminifera: systematics, palaeoecology and responses to the global changes. *Revue de Micropaléontologie*, **53**, 209–254.

Vachard, D., Krainer, K. & Lucas, S. 2012. Pennsylvanian (Late Carboniferous) calcareous microfossils from Cedro Peak (New Mexico, USA); Part 1: Algae and Microproblematica. *Annales de Paléontologie*, **98**, 225–252.

Vachard, D., Krainer, K. & Lucas, S. 2013. Pennsylvanian (Late Carboniferous) calcareous microfossils from Cedro Peak (New Mexico, USA); Part 2: smaller foraminifers and fusulinids. *Annales de Paléontologie*, **99**, 1–42.

Vachard, D., Krainer, K. & Lucas, S. 2015. Late Early Permian (late Leonardian; Kungurian) algae, microproblematica, and smaller foraminifers from the Yeso Group and San Andres Formation (New Mexico; USA). *Palaeontologia Electronica*, **18.1.21A**, 1–77, palaeo-electronica.org/content/2015/1160-kungurian-of-new-mexico

Vachard, D., Rettori, R., Altiner, D. & Gennari, V. In press. The Permian foraminiferal family Pseudovidalinidae and the genus *Altineria* emend. herein. *Journal of Foraminiferal Research*.

Vdovenko, M.V., Rauzer-Chernousova, D.M., Reitlinger, E.A. & Sabirov, A.A. 1993. *Guide for the Systematics of Paleozoic Smaller Foraminifers*. Rossiiskaya Akademiya Nauk, Komissiya po Mikropaleontologii Nauka, Moscow [in Russian].

Vologdin, A.G. 1932. *Archaeocyaths of Siberia. Issue 2: Fauna of the Cambrian Limestones of Altaya*. State Scientific and Technical Geology Exploratory Publishing House, Moscow [in Russian].

Vuks, G.P. 1984. First discovery of *Geinitzinita* (foraminifer) in the Permian of Transcaucasia. *Paleontologicheskii Zhurnal*, **3**, 131–133 [in Russian].

Vuks, G.P. & Chediya, I.O. 1986. Foraminifers from the Lyudyansk Suite (southern Primorye). *In*: *Correlations of Permian–Triassic Deposits of Eastern USSR*. Academy of Sciences USSR, Far-Eastern Scientific Centre, Institute of Biology and Pedology, Vladivostok (Project IGCP No. 203, 82–88) [in Russian].

Wang, G.-L. 1966. On *Colaniella* and its two allied new genera. *Acta Palaeontologica Sinica*, **14**, 206–221 [in Chinese].

Wang, K.L. & Sun, X.F. 1973. Carboniferous and Permian Foraminifera of the Chinling Range and its geologic significance. *Acta Geologica Sinica*, **2**, 137–178 [in Chinese with English abstract].

Wang, Y. & Ueno, K. 2003. Latest Permian fusulinoideans and smaller foraminifers in the Meili section, Guangxi, S. China. *Permophiles*, **42**, 19–21.

Wang, Y., Ueno, K., Zhang, Y.C. & Cao, C.Q. 2010. The Changhsingian foraminiferal fauna of a Neotethyan seamount: the Gyanyima Limestone along the Yarlung–Zangbo Suture in southern Tibet, China. *Geological Journal*, **45**, 308–318.

Wang, Y.J. 1974. *A Handbook of the Stratigraphy and Paleontology in Southwest China*. Science Press, Beijing [in Chinese].

Warthin, A.S., Jr. 1930. Micropaleontology of the Wetumka, Wewoka, and Holdenville formations. *Bulletin of the Oklahoma Geological Survey*, **53**, 1–95.

Wedekind, P.R. 1937. *Einführung in die Grundlagen der historischen Geologie. Band II. Mikrobiostratigraphie die Korallen und Foraminiferenzeit*. Ferdinand Enke, Stuttgart.

Wignall, P.B. & Hallam, A. 1996. Facies change and the end-Permian mass extinction in S.E. Sichuan, China. *Palaios*, **11**, 587–596.

Wignall, P.B., Védrine, S. *et al.* 2012. Capitanian (Middle Permian) mass extinction and recovery in western Tethys: a fossil, facies, and $\delta^{13}C$ study from Hungary and Hydra Island (Greece). *Palaios*, **27**, 78–89.

Wilde, G.L. 2006. Pennsylvanian-Permian fusulinaceans of the Big Hatchet Mountains, New Mexico. *New Mexico Museum Natural History and Science Bulletin*, **38**, 1–311.

Wood, G.D., Groves, J.R., Wahlman, G.P., Brenckle, P.L. & Alemán, A.M. 2002. The paleogeographic and biostratigraphic significance of fusulinacean and smaller foraminifers, and palynomorphs from the Copacabana Formation (Pennsylvanian-Permian), Madre de Díos basin, Peru. *In*: Hills, L.V., Henderson, C.M. & Bamber, E.W. (eds) *Carboniferous and Permian of the World: XIV ICCP Proceedings*. Canadian Society of Petroleum Geologists, Memoirs, **19**, 630–664.

Woszczynska, S. 1987. Foaraminifera and ostracods from the carbonate sediments of the Polish Zechstein. *Acta Palaeontologica Polonica*, **32**, 115–205.

Yanagida, J., Sakagami, S. *et al.* 1988. Biostratigraphic study of Paleozoic and Mesozoic groups in central and northern Thailand, an interim report. *Geological Survey Division, Department of Mineral Resources, Bangkok*, 1–47.

Zaninetti, L. & Jenny-Deshusses, C. 1985. Les Paraglobivalvulines (foraminifères) dans le Permien supérieur téthysien; répartition géographique et description de *Paraglobivalvulinoides* n. gen. *Revue de Paléobiologie*, **4**, 343–346.

Zaninetti, L., Altiner, D. & Çatal, E. 1981. Foraminifères et biostratigraphie dans le Permien supérieur du Taurus oriental, Turquie. *Notes du Laboratoire de Paléontologie de l'Université de Genève*, **7**, 1–38.

Zhang, M. & Gu, S. 2015. Latest Permian deep-water foraminifers from Daxiakou, Hubei, South China. *Journal of Paleontology*, **89**, 448–464, https://doi.org/10.1017/jpa.2015.19

Zhang, Z., Wang, Y. & Zheng, Q.-F. 2015. Middle Permian smaller foraminifers from the Maokou

Formation at thcTieqiao section, Guangxi, South China. *Palaeoworld*, **24**, 263–276.

ZHANG, Z.H. & HONG, Z.Y. 2004. Smaller foraminiferal fauna from the Changhsing Formation of Datian, Fujian. *Acta Micropaleontologica Sinica*, **21**, 64–84 [in Chinese with English abstract].

ZHAO, J.-K., SHENG, J.-Z., YAO, Z.-Q., LIANG, X.-L., CHEN, C.-Z., LIN, R. & LIAO, Z.-T. 1981. The Changhsingian and Permian–Triassic boundary of South-China. *Bulletin Nanjing Institute Geology and Paleontology, Academia Sinica*, **2**, 58–69 [in Chinese with English abstract].

ZHENG, H. 1986. The smaller foraminifers faunas in Qixia stage (Early Permian) of Daxiakou, Xingshan County, Hubei Province. *Earth Science, Journal of Wuhan College of Geology*, **11**, 489–497 [in Chinese].

ZOLOTOVA, V.P. & BARYSHNIKOV, V.V. 1980. Foraminifers of the Kungurian Stage in the stratotype localities. *In*: RAUZER-CHERNOUSOVA, D.M. & CHUVASHOV, B.I. (eds) *Biostratigraphy of the Artinskian and Kungurian Stages of the Urals*. Akademiya Nauk SSSR, Uralskii Nauchnyi Tsentr, Sverdlovsk, 72–109 [in Russian].

Permian fusuline biostratigraphy

YI-CHUN ZHANG* & YUE WANG

State Key Laboratory of Palaeobiology and Stratigraphy, Nanjing Institute of Geology and Palaeontology, Chinese Academy of Sciences, 39 East Beijing Road, Nanjing 210008, China

**Correspondence: yczhang@nigpas.ac.cn*

Abstract: A review of Permian fusuline biostratigraphy is made in this paper in order to improve the correlation of Permian strata globally. Permian fusuline biostratigraphy in the Tethyan and Panthalassan regions can be correlated roughly because the fusulines had good faunal communications between these two regions. However, fusuline faunas from the North American Craton region were devoid of almost all neoschwagerinids and dominated exclusively by schwagerinids during the Guadalupian (Middle Permian) because of the blockage caused by the vast Pangaea supercontinent. This renders the correlation of Middle Permian biostratigraphy and chronostratigraphy between the Tethyan region and North American region challenging. Significant evolutionary key points in fusulines include the first occurrence of *Pseudoschwagerina* or *Sphaeroschwagerina* during the earliest Permian, first occurrence of *Pamirina* and *Misellina* during the Yakhtashian and Bolorian, and the extinction of all schwagerinids and neoschwagerinids by the end of the Midian.

Fusulines are large benthic foraminifers that flourished during the Carboniferous and Permian periods. Due to their rapid evolution, they have been recognized as index fossils in the subdivision and correlation of Carboniferous and Permian strata throughout the world (e.g. Sheng 1963; Ross & June 2003; Leven & Gorgij 2011*a*).

During the past decades, fusulines have been extensively studied in most parts of the world, such as South China (Sheng 1963; Xiao *et al.* 1986; Sheng & Jin 1994; Shi *et al.* 2012), the Akiyoshi Terrane in Japan (Ueno 1992; Ota 1997), the Mino Terrane in Japan (Zaw 1999; Kobayashi 2011), Darvaz in Uzbekistan (Leven & Scherbovich 1978; Leven *et al.* 1992), Pamirs in Tajikistan (Leven 1967), central Iran (Baghbani 1993; Kobayashi & Ishii 2003) and North America (Ross 1963; Wilde 1990, 2006; Yang & Yancey 2000; Stevens & Stone 2007). These studies have provided the basis for regional stratigraphical subdivisions and correlations. However, with the increasing knowledge of the biostratigraphy of fusulines and conodonts, some discrepancies and problems arise when trying to correlate the fusuline-based Tethyan timescale with the conodont-based international timescale, especially for the base-Asselian (Davydov 2013; Lucas 2013), base-Kungurian (Wang *et al.* 2011*a*; Davydov & Arefifard 2013) and base-Roadian (Henderson & Mei 2003; Leven & Gorgij 2011*a*; Davydov & Arefifard 2013; Angiolini *et al.* 2015).

This paper is designed to provide a review of the Permian biostratigraphy of fusulines across different regions, with particular emphasis on the correlations of Permian fusuline biostratigraphy between different regions. The Tethyan chronostratigraphic scale is preferred in this paper because it is the only chronostratigraphy designed that is based on the biostratigraphy of fusulines (Leven 2004).

Composition of Permian fusuline genera

Fusulines amount to more than 100 genera across the world. These genera exhibited different stratigraphic ranges and abundance during the Permian (Fig. 1). In general, the early Permian, middle Permian and late Permian were eras of pseudoschwagerinids, neoschwagerinids and small schubertellids, respectively, especially for fusulines in the Tethyan regions.

There are many internal evolutionary links among these genera. For example, fusulines during the Gzhelian of the latest Carboniferous were dominated by diverse *Triticites* assemblages (e.g. Davydov 1990*a*; Shi *et al.* 2012). During the early Permian, many pseudoschwagerinid fusulines, such as *Pseudoschwagerina*, *Sphaeroschwagerina* and *Robustoschwagerina*, flourished (Yang & Hao 1991). During the Yakhtashian, the appearance of *Levenella* and *Pamirina* was crucial because they are the ancestor genera of almost all neoschwagerinids and verbeekinids. By the end of the middle Permian, the fusulines suffered a great mass extinction event, resulting in the loss of all schwagerinids and neoschwagerinids (Jin *et al.* 1994; Ota & Isozaki 2006; Groves & Wang 2013). The remaining genera during the earliest Dzhulfian were all small ozawainellids, staffellids and schubertellids

From: LUCAS, S. G. & SHEN, S. Z. (eds) 2018. *The Permian Timescale*. Geological Society, London, Special Publications, **450**, 253–288.
First published online June 8, 2017, https://doi.org/10.1144/SP450.14

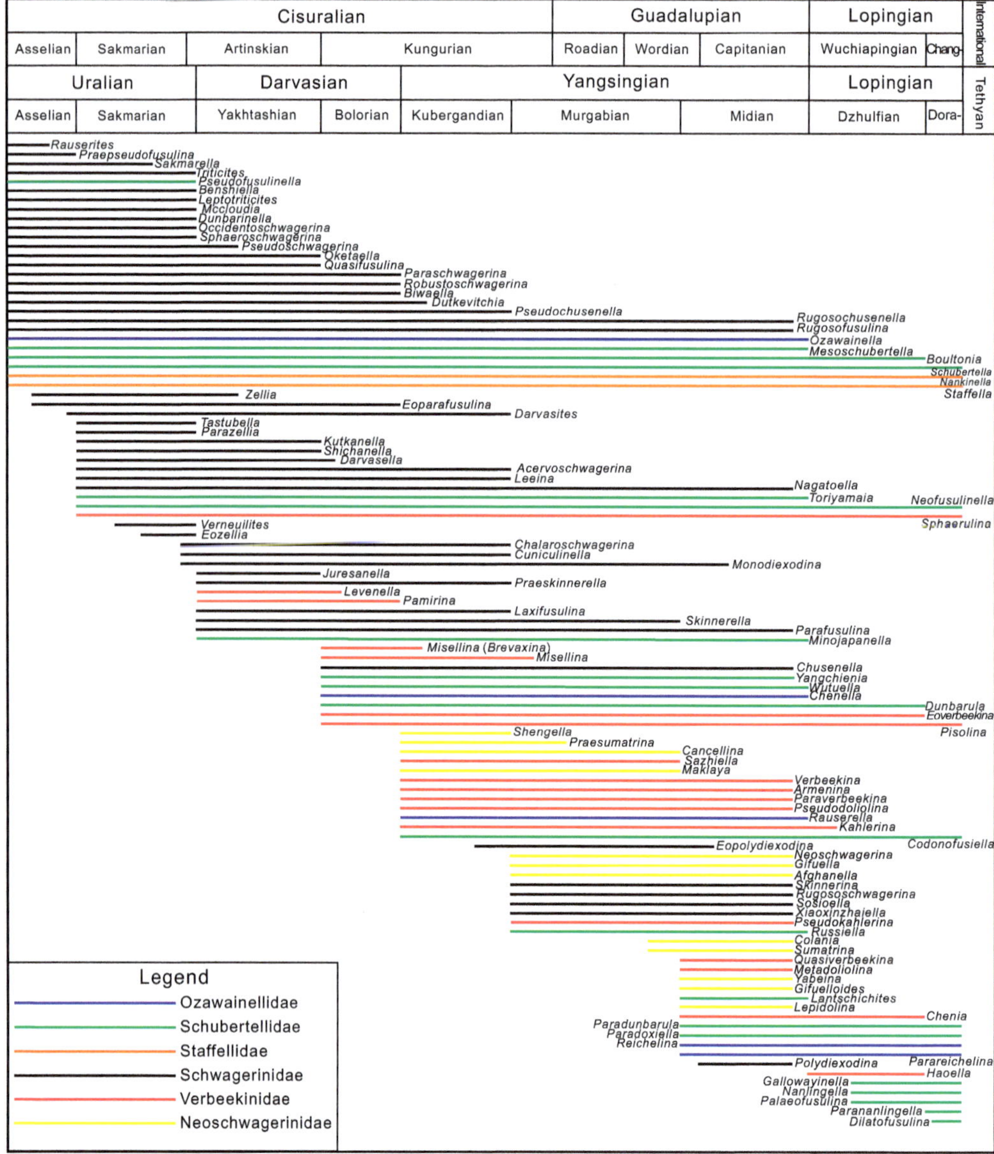

Fig. 1. Rough ranges of the main Permian fusuline genera. The taxonomy of the fusulines mainly follows Sheng *et al.* (1988).

(Ota & Isozaki 2006). The schubertellids experienced a new diversification during the late Dzhulfian that gave rise to *Gallowayinella*, *Nanlingella* and *Palaeofusulina*. Unfortunately, all fusulines died out at the end of Permian due to the severe mass extinction event (Jin *et al.* 2000; Shen *et al.* 2011). Overall, the history of fusulines during the Permian witnessed a process of origination, diversification and extinction. Most importantly, these evolution and extinction events have provided a solid basis for the biostratigraphic subdivisions.

Biostratigraphy of Permian fusulines

Fusulines are considered to have lived in warm and shallow-marine environments, which explained their wide distribution in tropical and subtropical regions during the Permian (Ross 1982; Vachard *et al.* 2010). However, because of geographical barriers and climatic gradients, the fusulines showed distinct palaeobiogeographical provinces (Ross 1967; Ozawa 1987; Kobayashi 1997*a*, 1999). Six broad palaeogeographical regions were recognized based on previous palaeobiogeographical studies (Ozawa 1987; Kobayashi 1997*a*) (Fig. 2). The most conspicuous palaeobiogeographical boundary was drawn between the Tethyan region and the North American Craton region because the vast Pangaea supercontinent hampered the faunal communications between these two regions. As a result, the fusuline fauna on each side of the supercontinent could not be correlated, especially in terms of middle and late Permian fusulines (Kobayashi 1997*a*, 1999). In order to facilitate the correlations of fusuline biostratigraphy between different regions, detailed descriptions of fusuline biostratigraphy from every block/region are provided here.

Eastern Tethyan region

The Eastern Tethyan region is centred on South China Block (Ozawa 1987). It consists of the South China Block, the Indochina Block, the North China Block, the Tarim Block, the North Qiangtang–Qamdo Block, seamounts within the Palaeotethys Ocean, Darvaz, north Afghanistan and southern Fergana, and the Urals and the Russian Platform (Fig. 3).

South China Block. The South China Block was located at equatorial regions during the Late Palaeozoic (Metcalfe 2013). This huge block has preserved a complete biostratigraphic zonation ranging from the Early Carboniferous to latest Permian (Sheng 1963; Xiao *et al.* 1986; L.X. Zhang *et al.* 1988; Sheng & Jin 1994).

The earliest Permian fusulines are marked by the appearance of pseudoschwagerinids in the Maping or Chuanshan Formation in South China, which differ from the underlying *Triticites*-bearing fusuline assemblages (e.g. Lin *et al.* 1979; Xiao *et al.* 1986; L.X. Zhang *et al.* 1988; Shi *et al.* 2009; Li *et al.* 2010). The Zisongian fusulines can be subdivided into two fusuline zones: the lower *Pseudoschwagerina uddeni* Zone (or *Pseudoschwagerina uddeni–P. texana* Zone) and the upper *Sphaeroschwagerina* Range Zone (Xiao *et al.* 1986; L.X. Zhang *et al.* 1988) (for the full species names see Appendix A). The lower *Pseudoschwagerina uddeni* Zone contains many species of *Pseudoschwagerina* and *Eoparafusulina.* The upper *Sphaeroschwagerina* Range Zone in the Houchang section in Guizhou Province is composed of three

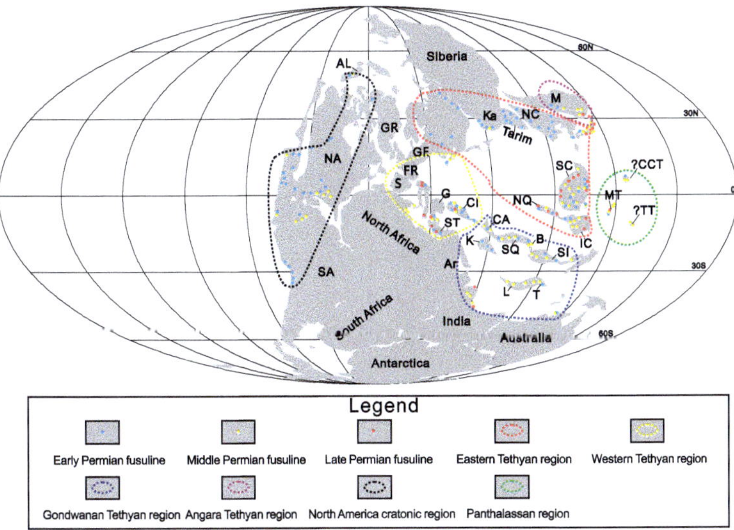

Fig. 2. Sketch map showing the distribution of fusulines during the Late Permian and the main palaeobiogeographical regions. The base map is modified from Scotese (2001). Abbreviations of tectonic blocks/regions are as follows, in alphabetical order: AL, Alaska; AR, Arabia Plate; B, Baoshan Block; CA, Central Afghanistan; CCT, Cache Creek Terrane in Canada; CI, Central Iran; FR, France; G, Greece; GE, Germany; GR, Greenland; IC, Indochina Block; K, Karakorum; Ka, Kazakhstan; L, Lhasa Block; M, Mongolia; MT, Mino Terrane in Japan; NA, North America; NC, North China; NQ, North Qiangtang Block; S, Spain; SA, South America; SC, South China; SI, Sibumasu Block; SQ, South Qiangtang Block; ST, Southern Turkey; T, Tengchong Block; TT, Torlesse Terrane in New Zealand.

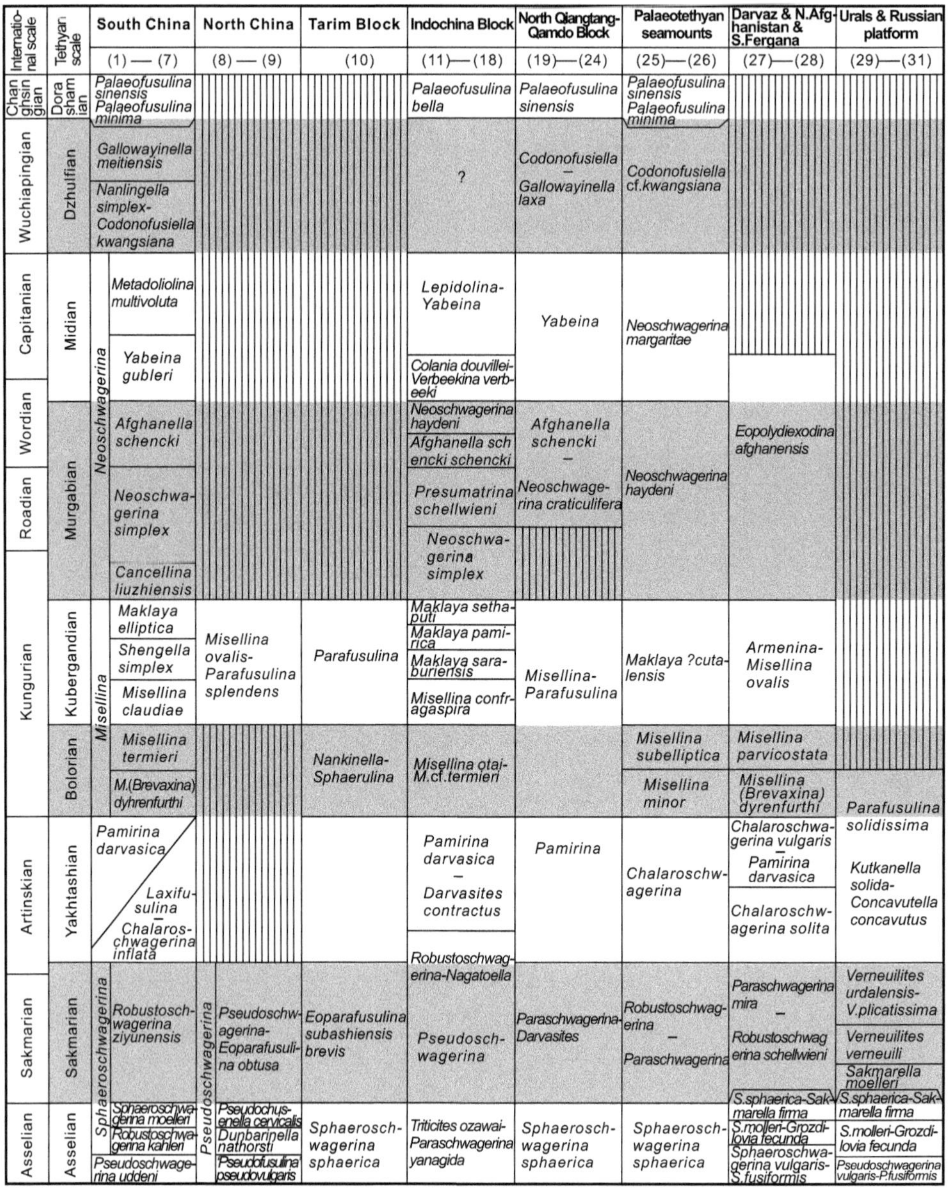

International scale	Tethyan scale	South China (1)—(7)	North China (8)—(9)	Tarim Block (10)	Indochina Block (11)—(18)	North Qiangtang-Qamdo Block (19)—(24)	Palaeotethyan seamounts (25)—(26)	Darvaz & N.Afghanistan & S.Fergana (27)—(28)	Urals & Russian platform (29)—(31)
Changhsingian	Dorashamian	Palaeofusulina sinensis; Palaeofusulina minima			Palaeofusulina bella	Palaeofusulina sinensis	Palaeofusulina sinensis; Palaeofusulina minima		
Wuchiapingian	Dzhulfian	Gallowayinella meitiensis; Nanlingella simplex-Codonofusiella kwangsiana			?	Codonofusiella – Gallowayinella laxa	Codonofusiella cf.kwangsiana		
Capitanian	Midian	*Neoschwagerina*: Metadoliolina multivoluta; Yabeina gubleri			Lepidolina-Yabeina; Colania douvillei-Verbeekina verbeeki	Yabeina	Neoschwagerina margaritae		
Wordian	Murgabian	*Neoschwagerina*: Afghanella schencki			Neoschwagerina haydeni; Afghanella schencki schencki	Afghanella schencki – Neoschwagerina craticulifera		Eopolydiexodina afghanensis	
Roadian	Murgabian	Neoschwagerina simplex; Cancellina liuzhiensis			Presumatrina schellwieni; Neoschwagerina simplex		Neoschwagerina haydeni		
Kungurian	Kubergandian	*Misellina*: Maklaya elliptica; Shengella simplex; Misellina claudiae	Misellina ovalis-Parafusulina splendens	Parafusulina	Maklaya sethaputi; Maklaya pamirica; Maklaya saraburiensis; Misellina confragaspira	Misellina-Parafusulina	Maklaya ?cutalensis	Armenina-Misellina ovalis	
Kungurian	Bolorian	*Misellina*: Misellina termieri; M.(Brevaxina) dyhrenfurthi		Nankinella-Sphaerulina	Misellina otai-M.cf.termieri		Misellina subelliptica; Misellina minor	Misellina parvicostata; Misellina (Brevaxina) dyrenfurthi	Parafusulina solidissima
Artinskian	Yakhtashian	Pamirina darvasica; Laxifusulina – Chalaroschwagerina inflata			Pamirina darvasica – Darvasites contractus; Robustoschwagerina-Nagatoella	Pamirina	Chalaroschwagerina	Chalaroschwagerina vulgaris – Pamirina darvasica; Chalaroschwagerina solita	Kutkanella solida-Concavutella concavutus
Sakmarian	Sakmarian	*Sphaeroschwagerina*: Robustoschwagerina ziyunensis	*Pseudoschwagerina*: Pseudoschwagerina-Eoparafusulina obtusa	Eoparafusulina subashiensis brevis	Pseudoschwagerina	Paraschwagerina-Darvasites	Robustoschwagerina – Paraschwagerina	Paraschwagerina mira – Robustoschwagerina schellwieni; S.sphaerica-Sakmarella firma	Verneuilites urdalensis-V.plicatissima; Verneuilites verneuili; Sakmarella moelleri; S.sphaerica-Sakmarella firma
Asselian	Asselian	Sphaeroschwagerina moelleri; Robustoschwagerina kahleri; Pseudoschwagerina uddeni	Pseudochusenella cervicalis; Dunbarinella nathorsti; Pseudofusulina pseudovulgaris	Sphaeroschwagerina sphaerica	Triticites ozawai-Paraschwagerina yanagida	Sphaeroschwagerina sphaerica	Sphaeroschwagerina sphaerica	S.moelleri-Grozdilovia fecunda; Sphaeroschwagerina vulgaris-S.fusiformis	S.moelleri-Grozdilovia fecunda; Pseudoschwagerina vulgaris-P.fusiformis

Fig. 3. Schematic framework of Permian fusuline biozones in the Eastern Tethyan region. The references are: (1) Xiao *et al.* (1986); (2) Z.H. Zhang *et al.* (1988); (3) Xu *et al.* (1986); (4) Sheng (1963); (5) Rui *et al.* (1984); (6) Sheng & Jin (1994); (7) Sheng & Zhang (1958); (8) Zhang (1983); (9) Xia (1981); (10) Zheng & Lin (1991); (11) Igo (1972); (12) Toriyama *et al.* (1965); (13) Dawson (1993); (14) Igo *et al.* (1993); (15) Toriyama *et al.* (1974); (16) Toriyama & Kanmera (1979); (17) Fontaine *et al.* (2002); (18) Ingavat *et al.* (1980); (19) Zhang *et al.* (2016); (20) Sichuan Regional Geological Survey & Nanjing Institute of Geology & Palaeontology (1982); (21) Liu (1993); (22) Niu *et al.* (2006); (23) Niu *et al.* (2010); (24) Zhao *et al.* (2006); (25) Ueno *et al.* (2003); (26) Ueno & Tsutsumi (2009); (27) Leven *et al.* (1992); (28) Leven (1997); (29) Rauzer-Chernousova (1965); (30) Forke (2002); (31) Chuvashov *et al.* (1990).

subzones: the *Robustoschwagerina kahleri* Subzone, the *Sphaeroschwagerina moelleri* Subzone and the *Robustoschwagerina ziyunensis* Subzone, in ascending order (Xiao *et al.* 1986). The *Robustoschwagerina kahleri* and *Sphaeroschwagerina moelleri* subzones show a stepwise increase of subglobular to globular pseudoschwagerinids, such as *Sphaeroschwagerina*, *Zellia* and *Robustoschwagerina*. In contrast, the *Robustoschwagerina ziyunensis* Subzone was dominated by large *Robustoschwagerina* species and abundant *Eoparafusulina* species (Xiao *et al.* 1986). The fusulines in the Zisongian Stage can be broadly correlated with those from the Maping Stage, in which the fusulines were divided into three zones, respectively: the *Pseudoschwagerina morsei–Robustoschwagerina xiaodushanica* Zone, the *Pseudoschwagerina parabeedei–Sphaeroschwagerina sphaerica* Zone and the *Pseudoschwagerina robusta–Zellia chengkungensis* Zone (Zhou *et al.* 1987).

The overlying fusuline assemblage is characterized by the appearance of *Pamirina* species synchronously in various areas in South China, such as the Zhengan area in Shanxi Province (Ding *et al.* 1989), the Longlin area in Guangxi Province (Huang & Zeng 1984) and the Ziyun area in Guizhou Province (Xiao *et al.* 1986). Xiao *et al.* (1986) established the *Pamirina darvasica* Zone, which was assigned to the Longlin Stage according to the Permian timescale of South China (Jin *et al.* 1999). The contemporaneous fusuline assemblages in outer shelf or slope environments, however, are often dominated by schwagerinids such as *Chalaroschwagerina* and *Laxifusulina*, which were recognized as the *Laxifusulina–Chalaroschwagerina inflata* Zone (Xu *et al.* 1986). Also, in the restricted platform environment, the fusulines are dominated by the *Pseudoendothyra* assemblage (Wu & Yang 1998).

The evolutionary trend, from *Pamirina* via *Misellina* (*Brevaxina*) to *Misellina*, has been reported in the Chihsia Formation and Houziguan limestone (Xu *et al.* 1986). The *Misellina* Range Zone established by Xiao *et al.* (1986) includes five subzones, respectively: the *Misellina* (*Brevaxina*) *dyhrenfurthi* Subzone, the *Misellina termieri* Subzone, the *M. claudiae* Subzone, the *Shengella simplex* Subzone and the *Maklaya elliptica* Subzone, in ascending order. These five subzones recorded a clear evolutionary progress of the following aspects: (1) the appearance of numerous parachomata in *Misellina* (*Brevaxina*); (2) the increase in size of *Misellina*; (3) the appearance of axial septula in *Shengella* and *Maklaya* subsequently. It should be noted, however, that in some places the underlying two subzones are often replaced by many schwagerinids, such as *Darvasites*, *Pseudochusenella*, *Praeskinnerella* and *Nagatoella*, which was named as the *Darvasites* Zone or *Pseudochusenella cushmani* Zone (originally the *Schwagerina cushmani* Zone) (Zhou 1982). The *Miselline claudiae* Subzone was widely distributed in the Chihsia Formation of South China (Wang 1978; Xiao *et al.* 1986; Ding *et al.* 1989). The *Shengella simplex* and *Maklaya elliptica* subzones also bear species of *Pseudodoliolina*, *Eoverbeekina* and *Neofusulinella*.

The overlying fusuline assemblage is dominated by neoschwagerinids, which were ascribed to the *Neoschwagerina* Zone (Xiao *et al.* 1986). This zone comprises the *Cancellina liuzhiensis* Subzone, the *Neoschwagerina simplex* Subzone, the *Afghanella schencki* Subzone, the *Yabeina gubleri* Subzone and the *Metadoliolina multivoluta* Subzone, in ascending order (Xiao *et al.* 1986). The *Cancellina liuzhiensis* Subzone is dominated by species of *Cancellina* and other genera, such as *Praesumatrina*, *Paraverbeekina*, *Pseudodoliolina*, *Verbeekina* and sporadic primitive *Neoschwagerina*. This subzone may correspond to the upper part of the *Cancellina* Subzone of Sheng (1963) or the *Praesumatrina neoschwagerinoides–Parafusulina multiseptata* Zone in the upper part of the Chihsia Formation in Tonglu in Zhejiang (Wang & Tang 1986). The *Neoschwagerina simplex* and *Afghanella schencki* subzones are dominated by diverse species of *Neoschwagerina*, *Afghanella*, *Verbeekina*, *Neofusulinella* and *Parafusulina*. The appearance of *Yabeina* marks the lower boundary of the *Yabeina gubleri* Subzone, which includes the species of *Yabeina*, *Neoschwagerina*, *Verbeekina*, *Chusenella*, *Sumatrina* and *Codonofusiella* (Xiao *et al.* 1986). The uppermost subzone in the *Neoschwagerina* Zone is the *Metadoliolina multivoluta* Zone, which is characterized by advanced species of *Metadoliolina*. However, with the demise of large-shelled neoschwagerinids in the uppermost part of this subzone, the fusuline assemblage is dominated by small schubertellids such as *Codonofusiella*, *Lantschichites* and *Dunbarula*. This subzone roughly corresponds to the upper part of the *Yabeina* Zone of Sheng (1963). The fusulines from this zone were also documented from the Yanqiao Formation in eastern South China (Sheng & Wang 1962).

The Late Permian fusuline assemblages were divided into four zones: the *Nanlingella simplex–Codonofusiella kwangsiana* Zone, the *Gallowayinella meitienensis* Zone, the *Palaeofusulina minima* Zone and the *Palaeofusulina sinensis* Zone, in ascending order (Sheng & Jin 1994). The *Nanlingella simplex–Codonofusiella kwangsiana* Zone is dominated by species of *Codonofusiella*, *Nankinella*, *Reichelina* and primitive *Nanlingella* (Rui *et al.* 1984). The decrease in *Codonofusiella* and the appearance of *Gallowayinella* is the feature of the *Gallowayinella meitienensis* Zone. The appearance of primitive *Palaeofusulina*, *Palaeofusulina minima*, marks the lower boundary of the *Palaeofusulina* Zone. Also, primitive *Palaeofusulina* occurred in both zones. The uppermost part of the fusuline zone is the *Palaeofusulina sinensis* Zone, which is widespread in the upper part of the Changxing Formation or Longdongchuan Formation in

South China (Sheng & Zhang 1958; Sheng 1963; Sun 1979). The former two zones are generally ascribed to Wuchiapingian/Dzhulfian, whereas the latter two are Changhsingian/Dorashamian in age (Wang *et al.* 1997; Jin *et al.* 2003; Leven 2004). However, the discovery of the *Palaeofusulina* assemblage at the base of the Heshan Formation in the Penglaitan section indicates an age of early Wuchiapingian (Wang & Jin 2006); further detailed taxonomic study is needed in the future.

North China Block. The North China Block has been a stable block since the beginning of the Palaeozoic (Shi 2006). This block underwent long-term erosion from the Late Ordovician, up until transgression took place during the Late Carboniferous (e.g. Wang *et al.* 1992). Fusulines were found in several limestone beds within the Taiyuan Formation in this block (Zhang 1983; Zhang *et al.* 1989; Jiang *et al.* 2001). They were often represented in the *Pseudoschwagerina* Zone (Zhang 1983; Zhang & Xia 1985; Zhang *et al.* 1989; Jiang *et al.* 2001) or *Sphaeroschwagerina* Zone (Song *et al.* 2014). The *Pseudoschwagerina* Zone in the Taiyuan Formation of the Xishan section near Taiyuan city was divided into four subzones, respectively: the '*Pseudofusulina*' *pseudovulgaris* Subzone, the *Dunbarinella nathorsti–Dunbarinella nathorsti laxa* Subzone, the *Pseudochusenella cervicalis* Subzone and the *Pseudoschwagerina texana–Eoparafusulina obtusa* Subzone, in ascending order (Zhang 1983), with the uppermost subzone being most probably of Sakmarian age (Zhang 1990).

Post-Sakmarian fusulines were not reported from the mainland of North China because of the retreat of seawater. However, in the northern margin of the North China Block, the Sanmianjing Formation in the Kangbao area of Hebei Province yields many fusulines referable to the *Misellina ovalis–Parafusulina splendens* assemblage (Xia 1981). Common fusulines in this assemblage comprise *Misellina ovalis*, *M. claudiae*, *Parafusulina splendens*, *P. yabei* and *Chusenella sinensis* (Sheng 1962; Xia 1981). This assemblage indicates a Kubergandian age.

Tarim Block. Shallow-water carbonates with diverse marine faunas prevailed in the Tarim Block during the Late Carboniferous and early Permian. The appearance of *Pseudoschwagerina* or *Sphaeroschwagerina* in the lower part of the Kangkeling Formation in the Keping area, in the lower part of the Xiaohaizi Formation in the Bachu area and in the middle part of the Tahaqi Formation in the SW Tarim Block marks the beginning of the Permian (Zhang 1963; Zheng & Lin 1991; Zhu 1997; Wang *et al.* 2011*b*). The Permian succession in the Subashi area near Keping City was subdivided into the following fusuline zones: the *Sphaeroschwagerina sphaerica* Zone, the *Eoparafusulina subashiensis brevis* Zone, the *Nankinella–Sphaerulina* Zone and the *Parafusulina splendens* Zone, in ascending order (Zheng & Lin 1991).

The lowermost *Sphaeroschwagerina sphaerica* Zone comprises species of *Sphaeroschwagerina*, *Pseudoschwagerina*, *Triticites*, *Paraschwagerina*, *Zellia*, *Quasifusulina*, *Schubertella* and *Eoparafusulina*, and is coeval with the *Pseudoschwagerina parasphaerica* Subzone in the Keping area (Zhang 1963), the *Praepseudofusulina–Sphaeroschwagerina* Zone in the Wuzunbulake area (Wang *et al.* 2011*b*), and the *Sphaeroschwagerina vulgaris* and *S. karnica* zones in the Paojianggou area (Zhu 1997). The age of the *Sphaeroschwagerina sphaerica* Zone is Asselian.

The *Eoparafusulina subashiensis brevis* Zone overlies the *Sphaeroschwagerina sphaerica* Zone. In this zone, many *Eoparafusulina* species have taken over the pseudoschwagerinids of the underlying zone (Zheng & Lin 1991). This *Eoparafusulina*-dominated assemblage was also documented in the Wuzunbulake, Bachu and Paojianggou areas (Zheng & Lin 1991; Zhu 1997; Wang *et al.* 2011*b*). The age of this assemblage is most likely to be Sakmarian.

The Balikelike Formation in the Subashi section records the youngest fusuline assemblage in the Tarim Block (Zheng & Lin 1991). This assemblage comprises species of *Chalaroschwagerina*, *Sphaerulina*, *Darvasites*, *Nankinella* and *Pisolina* in the lower part of the formation, and species of *Parafusulina* in the upper part. They were ascribed to the *Nankinella–Sphaerulina* and *Parafusulina* zones, respectively (Zheng & Lin 1991). The ages of these zones are most probably Bolorian and Kubergandian, respectively (Zheng & Lin 1991). Other areas in the Tarim Block deposited terrestrial sediments without any marine faunas during the middle and late Permian.

Indochina Block. The Indochina Block is bounded by the Song Ma Suture Zone in Vietnam and the Jinghong–Nan–Uttaradit–Sra Kaeo sutures in western Yunnan and central Thailand (Metcalfe 2013). This block was situated in the tropical zone of low latitude (Metcalfe 2013), which accounts for the flourishing Permian fusulines.

The Asselian fusulines are grouped into the *Triticites ozawai–Paraschwagerina yanagidai* assemblage in the Phetchabun area of central Thailand (Igo 1972) and the *Sphaeroschwagerina sphaerica–Rugosofusulina extensa* assemblage in the Terbat Formation of eastern Malaysia (Sakamoto & Ishibashi 2002). The *Triticites ozawai–Paraschwagerina yanagidai* assemblage was once considered to be of latest Gzhelian age (Igo *et al.* 1993). Since the typical Permian species *Pseudoschwagerina toriyamai* was also reported from the highest level of this zone (Igo 1972), the stratigraphic age of this zone should be at least partially Asselian.

A younger fusuline assemblage was reported from the limestones in the Noankowtok area of Thailand. It is represented by many species of *Pseudoschwagerina*, such as *P. sphaerica*, *P.* cf. *muongthensis rossica* and *P. turbida*, which indicates an age of Sakmarian according to Toriyama *et al.* (1965). A further younger fusuline fauna, composed of species of sporadic *Robustoschwagerina*, *Nagatoella* and *Toriyamaia* in the Saraburi area, was named as the *Robustoschwagerina–Nagatoella* assemblage of latest Sakmarian–earliest Yakhtashian age (Dawson 1993). The Yakhtashian fusulines were reported from the Changwat Loei area of NE Thailand (Igo *et al.* 1993). The fusulines of this age are mainly represented by *Pamirina darvasica*, *Darvasites contractus*, *Pseudoreichelina darvasica* and *Nankinella ?loiensis* (Igo *et al.* 1993).

The succeeding zones were established based on the evolution of neoschwagerinids, which are well exposed in the Khao Phrong Phrab and the Khao Khao areas of Sara Buri, central Thailand (Toriyama *et al.* 1974; Toriyama & Kanmera 1979; Ingavat *et al.* 1980). The lowermost zone was named the *Misellina otai–Misellina* cf. *termieri* Zone from the Khao Phrong Phrab area (Toriyama *et al.* 1974). Fusulines of this zone may also occur in the Wang Saphung area of northern Thailand, which comprise *Misellina termieri* and *Pravitoschwagerina thailandensis* (Toriyama 1982), indicating an age of Kubergandian according to Leven (2004). Kubergandian fusulines in the Phrong Phrab area were subdivided into the *Misellina confragaspira* Zone, the *Maklaya saraburiensis* Zone, the *M. pamirica* Zone and the *M. sethaputi* Zone, in ascending order (Toriyama *et al.* 1974). In these zones, species of *Maklaya*, *Pseudodoliolina*, *Neofusulinella*, *'Pseudofusulina'*, *Parafusulina*, *Chusenella* and *Neothailandina* were reported (Toriyama *et al.* 1974). Fusulines documented from the limestones in the NE of Salavan, southern Laos are equivalent to the *Misellina confragaspira* Zone (Fontaine *et al.* 1999). The succeeding fusuline zone is the *Neoschwagerina simplex* Zone, which is characterized by the appearance of many primitive species of *Neoschwagerina* such as *Neoschwagerina simplex*, *N. schuberti* and *N. tenuis* (Toriyama *et al.* 1974; Toriyama 1975), which indicate a Murgabian age. The upper part of this zone, however, is transformed to the *Afghanella megasphaerica–Neoschwagerina* cf. *kueichowensis* Zone in the Khao Khao area (Toriyama & Kanmera 1979). The overlying *Afghanella pesuliensis–Pseudodoliolina pseudolepida* Zone was considered to be contemporaneous with the *Praesumatrina schellwieni* Zone in the Khao Phrong Phrab area (Toriyama & Kanmera 1979). The succeeding fusulines in the Khao Khao area were subdivided into the *Afghanella schencki schencki* Zone and the *Neoschwagerina haydeni* Zone, which are dominated by species of *Neoschwagerina*, *Afghanella*, *Verbeekina*, *Chusenella* and *Pseudodoliolina*, but devoid of any *Yabeina* or *Lepidolina* species (Toriyama & Kanmera 1979). Thus, these two fusuline zones are most probably of late Murgabian age. In comparision, fusuline fossils of both zones are not preserved in the adjacent Khao Phrong Phrab area.

The next younger zone was named the *Colania douvillei–Verbeekina verbeeki* Zone by Ingavat *et al.* (1980). The fusulines in this zone are dominated by *Colania douvillei*, *Sumatrina annae*, *Verbeekina verbeeki*, *Schubertella simplex*, and a few species of *Reichelina* and *Rauserella* in the Khao Imot area (Ozawa 1970). Horizons 3 and 4 from the Khao Tham Yai limestone in the Khao Tham Yai area, northern Thailand consist of *Colania douvillei*, *Verbeekina verbeeki*, *Sumatrina* sp. and *Codonofusiella* sp., which are equivalent to the *Colania douvillei–Verbeekina verbeeki* Zone (Fontaine *et al.* 2002). The youngest middle Permian fusuline zone is the *Lepidolina multiseptata* Zone, which is widely distributed in the Indochina Block. The fusulines from this zone include species of *Lepidolina*, *Yabeina*, *Metadoliolina*, *Chusenella*, *Verbeekina*, *Reichelina*, *Russiella*, *Rauserella*, *Kahlerina* and *Dunbarula* (Ingavat *et al.* 1980). This zone has been widely documented in Thailand (Fontaine *et al.* 2002), Cambodia (Gubler 1935; Nguyen 1986) and East Malaysia (Igo 1966).

Dzhulfian fusulines have not yet been reported from the Indochina Block. Dorashamian fusulines, however, have been widely reported in this block. They were grouped into the *Palaeofusulina bella* Zone by Ingavat *et al.* (1980). The fusulines from this zone comprise *Palaeofusulina bella*, *P. sinensis*, *P. fusiformis*, *P. laxa*, *Codonofusiella kwangsiana*, *Gallowayinella guidingensis*, *Reichelina changhsingensis* and *R. cribroseptata*, and have been documented in northern Thailand, southern Cambodia and eastern Malaysia (e.g. Aw *et al.* 1977; Ueno & Sakagami 1991; Fontaine *et al.* 2013).

North Qiangtang–Qamdo Block. The North Qiangtang–Qamdo Block is sandwiched between the Jinshajiang Suture Zone in the north and the Longmu Co–Shuanghu Suture Zone in the south (Zhang *et al.* 2013*a*). This block was located in low-latitude areas during the Permian (Metcalfe 2013; Zhang *et al.* 2013*a*).

The earliest Permian fusulines in this block can be grouped into the *Sphaeroschwagerina sphaerica*

Zone, which was distributed in the Raggyorcaka Lake and Tanggula areas of northern Tibet, and in the Machala area in eastern Tibet (Sichuan Regional Geological Survey & Nanjing Institute of Geology and Palaeontology 1982; Liu 1993; Zhang *et al.* 2016). The fusulines in this zone comprise *Sphaeroschwagerina sphaerica*, *Sphaeroschwagerina parasphaerica*, *Pseudoschwagerina* sp. and *Montiparus tobensis*. The age of this zone is clearly Asselian (Zhang *et al.* 2016). The Sakmarian fusulines can only been recognized in eastern Tibet, where they consist of diverse species of *Paraschwagerina*, *Darvasites* and *Leeina* (Sichuan Regional Geological Survey & Nanjing Institute of Geology and Palaeontology 1982). These fossils are grouped as the *Paraschwagerina–Darvasites* assemblage here. The overlying fusuline assemblage in the North Qiangtang–Qamdo Block is represented by the Yakhtashian *Pamirina* Zone in the Tanggula area that includes *Pamirina pulchra* and *P. nobilis* (Liu 1993). Bolorian and Kubergandian fusulines are not easily differentiated in several areas in this block (Liu 1993; Niu *et al.* 2006). They are composed of *Sphaerulina hunanica*, *Nankinella orbicularia*, *Wutuella wutuensis*, *Misellina claudiae*, *Parafusulina yunanica* and *P. splendens*, which were ascribed to the *Misellina*–'*Schwagerina*' assemblage (Niu *et al.* 2006). Since species of '*Schwagerina*' are not well developed in this assemblage, it is renamed here as the *Misellina–Parafusulina* assemblage.

The equivalent of the *Neoschwagerina simplex* Zone in this block is dominated by clastic rocks lacking fusulines. The overlying strata were dominated by limestones named as the Jiushidaoban Formation in the Tanggula area of northern Tibet and the Jiaoga Formation in the Qamdo area in eastern Tibet (Liu 1993; Niu *et al.* 2010). Abundant fusuline fossils were found in these limestones, including *Neoschwagerina craticulifera*, *N. douvillei*, *Sumatrina annae*, *S. longissima* and *Afghanella schencki*. They were grouped into the *Afghanella schencki–Neoschwagerina craticulifera* assemblage (Niu *et al.* 2010), indicating an age of late Murgabian. A younger fusuline assemblage occurs in the upper part of the Jiushidaoban Formation and the Suojia Formation. It contains *Yabeina gubleri*, *Sumatrina annae* and *Chusenella schwagerinaeformis*, and was named as the *Yabeina* Zone (Liu 1993).

The late Permian strata, characterized by paralic facies in the lower part and limestones in the upper part, are distributed widely in the North Qiangtang–Qamdo Block (Sichuan Regional Geological Survey & Nanjing Institute of Geology and Palaeontology 1982). Two assemblages, the *Codonofusiella–Gallowayinella laxa* assemblage and the *Palaeofusulina sinensis* Zone, in ascending order, were established in the Labuchari Formation (Zhao *et al.* 2006), and refer to Dzhulfian and Dorashamian ages, respectively.

Seamounts within the Palaeotethys Ocean. Along the Changning–Menglian Suture Zone in western Yunnan, China and the Inthanon Zone in Thailand, a series of limestones developed that have been recognized to be seamounts located within the Palaeotethys Ocean (Ueno & Igo 1997; Ueno *et al.* 2003; Ueno & Tsutsumi 2009). The fusuline successions, ranging from Serpukhovian to latest Permian, were recorded in the Yutangzhai area of western Yunnan (Ueno *et al.* 2003; Ueno & Tsutsumi 2009). Early Permian fusulines are represented by *Sphaeroschwagerina sphaerica* and *Rugosofusulina latispiralis* in so-called Ytz 5, which indicates a late Asselian age (Ueno *et al.* 2003). The overlying fusuline assemblages Ytz 6 and Ytz 7 are dominated by *Robustoschwagerina–Paraschwagerina* and *Chalaroschwagerina* faunas, indicating Sakmarian and Yakhtashian ages, respectively (Ueno *et al.* 2003). Although the genera *Levenella* and *Pamirina* were not found, their descendent genera *Misellina–Maklaya–Neoschwagerina* lineage was well recorded in the limestones, indicating a Bolorian–Murgabian age (Ueno *et al.* 2003). The Midian fusuline assemblage consists only of *Neoschwagerina margaritae*, *N.* sp. and *Verbeekina verbeeki*, and lacks many of the other genera common in Tethys regions, such as *Yabeina*, *Lepidolina* and schwagerinid species (Ueno *et al.* 2003).

Late Permian fusulines were discovered in the Shifodong section. They were subdivided into three biozones: the *Codonofusiella* cf. *C. kwangsiana* Zone, the *Palaeofusulina minima* Zone and the *Palaeofusulina sinensis* Zone, in ascending order (Ueno & Tsutsumi 2009). The *Codonofusiella* cf. *C. kwangsiana* Zone is Dzhulfian in age as evidenced by the absence of either neoschwagerinids or *Palaeofusulina* species, whereas the latter two zones are Dorashamian in age because of the dominance of *Palaeofusulina* species (Ueno & Tsutsumi 2009).

Darvaz, north Afghanistan and southern Fergana. Darvaz, north Afghanistan and southern Fergana were located at the southern margin of Eurasia during Late Palaeozoic time (Angiolini *et al.* 2016). The fusulines reported from these blocks show a continuous succession from Late Carboniferous to the early Permian (e.g. Bensh 1972; Leven & Scherbovich 1978; Leven *et al.* 1992; Leven 1997). The middle Permian fusulines, however, were mainly reported from north Afghanistan (Leven 1997).

Asselian fusulines were grouped into three zones in Darvaz (Leven *et al.* 1992). They are the *Sphaeroschwagerina vulgaris–Sphaeroschwagerina fusiformis* Zone, the *Sphaeroschwagerina moelleri–Grozdilovia fecunda* Zone and the *Sphaeroschwagerina sphaerica–Sakmarella firma* Zones, in ascending order (Leven *et al.* 1992). These zones contain species of *Sphaeroschwagerina*, *Pseudoschwagerina*, *Rugosofusulina*, *Rugosochusenella*, *Dunbarinella*, *Quasifusulina*, *Grozdilovia*, *Sakmarella* and *Dutkevitchia* (Leven & Scherbovich 1978; Leven *et al.* 1992). The overlying zone was named as the *Paraschwagerina mira–Robustoschwagerina schellwieni* Zone of Sakmarian age (Leven *et al.* 1992). This zone is marked by the appearance of some new elements such as *Robustoschwagerina*, *Biwaella*, *Darvasites* and *Zellia*. During the Yakhtashian, the fusulines were grouped into the lower *Chalaroschwagerina solita* Zone and the upper *Chalaroschwagerina vulgaris–Pamirina darvasica* Zone (Leven *et al.* 1992). These two zones are typified by the occurrence of *Chalaroaschwagerina*, *Pamirina* and *Misellina* (*Brevaxina*). With the evolution of *Misellina* (*Brevaxina*) and *Misellina*, the *Misellina* (*Brevaxina*) *dyhrenfurthi* and *Misellina* (*Misellina*) *parvicostata* zones are established that indicate a Bolorian age, and the *Armenina–Misellina* (*Misellina*) *ovalis* Zone with additional Schwagerinidae genera such as *Skinnerella* and *Parafusulina* in the Kubergandian (Leven *et al.* 1992).

Murgabian and Midian fusulines are well documented in north Afghanistan (Ciry & Amiot 1965; Lys & Lapparent 1971; Vachard 1980; Leven 1997; Colpaert *et al.* 2015). They were assigned to the *Eopolydiexodina afghanensis* Zone because of its wide distribution of *Eopolydiexodina afghanensis* (Thompson 1946; Amiot *et al.* 1965). The associated genera include *Afghanella*, *Neoschwagerina*, *Verbeekina*, *Minojapanella*, *Sumatrina*, *Parafusulina* and *Codonofusiella* (Lys & Lapparent 1971).

The Urals and the Russian Platform. Late Carboniferous and Early Permian fusulines are widely distributed in the vast Russian Platform and Ural Mountains (e.g. Rauzer-Chernousova 1965; Mikhailova 1974; Alksne & Isakova 1980; Chuvashov *et al.* 1990; Davydov 1990*b*; Davydov *et al.* 1997).

The Asselian in the southern Urals was once considered to contain two zones, respectively, the lower *Sphaeroschwagerina moelleri–Grozdilovia fecunda* Zone and the upper *Sphaeroschwagerina sphaerica–Sakmarella firma* Zone (Rauzer-Chernousova 1965; Chuvashov *et al.* 1990). However, in the Pechora area of the northern Urals, the *Pseudoschwagerina vulgaris–Pseudoschwagerina fusiformis* Zone occurred below the first appearance of the *Sphaeroschwagerina* species (Mikhailova 1974; Papulov 1986; Chuvashov *et al.* 1990). Overall, Asselian fusulines in the Russian Platform and Urals are characterized by various species of *Pseudoschwagerina* and *Sphaeroschwagerina*, associated with *Grozdilovia*, *Sakmarella*, *Triticites*, *Daixina*, *Rugosofusulina* and *Jigulites*.

Sakmarian fusulines dominated by diverse pseudofusulinid species are mainly distributed in the southern Urals. The lower Sakmarian Tastubian horizon bears two zones, the *Sakmarella moelleri* Zone and the *Verneuilites verneuili* Zone, in ascending order (Rauzer-Chernousova 1965). These two zones are dominated exclusively by abundant pseudofusulinids and rare *Paraschwagerina* species. The upper Sakmarian Sterlitamaklan horizon is also dominated by pseudofusulinid species that have been grouped into the *Verneuilites urdalensis–Verneuilites plicatissima* Zone (Forke 2002).

The Yakhtashian fusulines are characterized by the *Kutkanella solida–Concavutella concavutus* Zone (Forke 2002). The pseudofusulinids in these faunas are hard to correlate with contemporaneous faunas in South China. The post-Yakhtashian fusulines in the Urals have diminished greatly, with only sporadic *Parafusulina solidissima* found in Bolorian strata in the Kungur area (Chuvashov *et al.* 1990).

Western Tethyan region

The Western Tethyan region was located at the western terminus of the Tethyan Ocean during the Permian. It covers Iran, Turkey, the Italian island of Sicily, Greece, Tunisia and the Carnic Alps (Fig. 4). According to Kobayashi (1997*a*), the Western Tethyan region differs from the Eastern Tethyan region in the absence of *Lepidolina* species during the middle Permian.

Iran. During the middle and late Permian, the main part of Iran (except for the Zagros Mountains) rifted away from the Gondwanan margin (Gaetani *et al.* 2009). These blocks included the Alborz Block in northern Iran, the Yazd and Tabas blocks in central Iran, and the Sanandaj-Sirjan Block in SW Iran (Leven & Gorgij 2011*a*).

Asselian fusulines were widely reported from the Emarat Formation in the Alborz Mountains (Gaetani *et al.* 2009), the Zaladou Formation in the Yazd Block and the Tabas Block (Leven & Taheri 2003; Leven & Gorgij 2006), and the Vazhnan Formation in the Sanandaj-Sirjan Block (Baghbani 1993; Leven & Gorgij 2011*b*). The fusuline genera in this stage include *Pseudoschwagerina*, *Sphaeroschwagerina*, *Praepseudofusulina*, *Ruzhenzevites* and *Anderssonites*. They were ascribed to the *Pseudoschwagerina–Praepseudofusulina* assemblage in the lower part and the *Sphaeroschwagerina–Eoparafusulina* assemblage in the upper part (Gaetani

International scale	Tethyan scale	Iran (1) — (3)	Turkey&Sicily &Tunisia (4) — (10)	Carnic Alps (11)—(13)
Changhsingian	Dorashamian	?		Palaeofusulina-Reichelina
Wuchiapingian	Dzhulfian	Reichelina media -Nanlingella	Paradunbarula (Shindella) sindensis -Palaeofusulina laxa	
		Codonofusiella		
Capitanian	Midian	Yabeina	Yabeina-Dunbarula	
		Chusenella abichi		
Wordian		Neoschwagerina occidentalis		
	Murgabian	Afghanella schencki		
		Eopolydiexodina persica		
Roadian		Neoschwagerina simplex	Praesumatrina ciryi	
Kungurian	Kubergandian	Cancellina	Cancellina-Armenina	
		Maklaya		
	Bolorian	Misellina		
Artinskian	Yakhtashian	Darvasites ordinatus	Chalaroschwagerina ?globosa	"Pseudofusulina" ex gr.fusiformis - Robustoschwagerina spatiosa
				Robustoschwagerina geyeri-Zellia heritschi-Paraschwagerina nitida-'Pseudofusulina' sp.
				Sphaeroschwagerina asiatica - Paraschwagerina paranitida - Zellia praeheritschi
Sakmarian	Sakmarian	?	Paraschwagerina pseudomira	Paraschwagerina mukhamedjarovica-Pseudodoschwagerina muongthensis
				Pseudoschwagerina aff.uddeni - Paraschwagerina pseudomira
Asselian	Asselian	Sphaeroschwagerina - Eoparafusulina	Dutkevitchia complicata	Sphaeroschwagerina carniolica - Pseudoschwagerina extensa
		Pseudoschwagerina - Praeschwagerina	Paraschwagerina sp.	

Fig. 4. Schematic framework of Permian fusuline biozones in the Western Tethyan region. References are: (1) Gaetani *et al.* (2009); (2) Kobayashi & Ishii (2003); (3) Baghbani (1993); (4) Kobayashi & Altiner (2008); (5) Vachard *et al.* (2001); (6) Vachard & Moix (2013); (7) Kobayashi & Altiner (2011); (8) Skinner & Wilde (1966); (9) Skinner & Wilde (1967); (10) Vachard *et al.* (2003); (11) Forke (2002); (12) Noe (1988); (13) Pasini (1985).

et al. 2009). These two assemblages are broadly comparable with the *Pseudoschwagerina*–'*Pseudofusulina*' Zone of Baghbani (1993).

Sakmarian fusulines are not reported from the Iranian blocks, except for the Kahmard Block which is included in the Gondwana Tethyan region in this paper. Yakhtashian fusulines were documented from an isolated limestone succession in the Abadeh area (Kobayashi & Ishii 2003). The fusulines were named as the *Darvasites ordinatus* Zone, which comprises *Darvasites ordinatus*, *Chalaroschwagerina* cf. *mengi*, *Leeina fusiformis* and *Pamirina* (*Levenella*) *leveni* (Kobayashi & Ishii 2003).

A younger fusuline fauna was reported by Leven & Gorgij (2008). It consists of *Skinnerella*, *Paraleeina* and *Misellina* in the lower part and *Kubergandella*, *Chusenella*, *Misellina* and *Armenina* in the upper part, suggesting a Bolorian–Kubergandian age (Leven & Gorgij 2008). These fusulines are equivalent to the *Misellina*, *Maklaya* and *Cancellina* zones of Baghbani (1993). Murgabian fusulines were grouped into the *Neoschwagerina simplex* Zone in the lower part and the *Eopolydiexodina persica* (or *Eopolydiexodina douglasi*) Zone in the upper part (Baghbani 1993; Kobayashi & Ishii 2003). The latter zone can be traced across several Peri-Gondwanan blocks and the southern Eurasia margin (Ueno 2003; Colpaert *et al.* 2015). The *Eopolydiexodina persica* Zone is overlain by the *Afghanella schencki* Zone of late Murgabian age (Kobayashi & Ishii 2003). This zone is dominated by many neoschwagerinids, such as *Afghanella*, *Sumatrina*, *Verbeekina* and *Pseudodoliolina* (Kobayashi & Ishii 2003). The overlying fusulines in the Abadeh area were grouped into the *Neoschwagerina occidentalis* Zone and *Chusenella abichi* Zone (Kobayashi & Ishii 2003). Common genera in these zones include *Neoschwagerina*, *Kahlerina*, *Wutuella*, *Sumatrina*, *Dunbarula* and *Chenella*, indicating a Midian age. Late Midian fusulines were reported from the upper part of the Abadeh Formation (Baghbani 1993). They are composed of *Yabeina*, *Metadoliolina*, *Reichelina*, *Kahlerina*, *Chusenella* and *Sichotenella*, and were assigned to the *Yabeina* Zone by Baghbani (1993).

Late Permian fusulines were divided into two zones: the lower *Codonofusiella* Zone and the upper *Reichelina media*–*Nanlingella* Zone (Baghbani 1993). Both zones were ascribed to a Dzhulfian age (Taraz *et al.* 1981; Baghbani 1993). The Dorashamian fusulines were not developed in the Abadeh area because the topmost part of the Hambast Formation was probably not suitable to accommodate fusulines but was favourable for ammonoids and conodonts (Taraz *et al.* 1981; Shen & Mei 2010).

Turkey, Sicily and Tunisia. The Tauride Block of southern Turkey, the Italian island of Sicily and southern Tunisia were located at the western terminus of the Palaeotethys Ocean (Şengör *et al.* 1984). The allochthonous Aladag unit in Turkey preserved a good carbonate sequence recording Early and Middle Permian fusulines (Kobayashi & Altiner 2008; Okuyucu 2008). The earliest Permian fusulines were represented by *Pseudoschwagerina robusta*, *Quasifusulina minina*, *Paraschwagerina* sp. and *Dutkevitchia complicata*, which were grouped into the lower *Paraschwagerina* sp. Zone and the upper *Dutkevitchia complicata* Zone (Kobayashi & Altiner 2008). The Sakmarian fusuline zone was named as the *Paraschwagerina pseudomira* Zone. It differs from the underlying zone in containing species of *Robustoschwagerina*, *Darvasites* and *Zellia* (Kobayashi & Altiner 2008; Okuyucu 2008). A slightly younger fusuline assemblage was reported from the Permian limestone boulders in the Lercara Formation in Sicily (Vachard *et al.* 2001). It is composed of *Chalaroschwagerina*? *globosa*, *Robustoschwagerina* cf. *schellwieni* and *Minojapanella*? sp., suggesting an early Yakhtashian age (Vachard *et al.* 2001).

The Bolorian fusulines are scarce in this region. The Kubergandian fusulines were found in the allochthonous limestones that constain *Cancellina*, *Armenina*, *Pseudodoliolina* and *Nankinella* (Vachard & Moix 2013). Murgabian fusulines contain *Praesumatrina ciryi*, *Verbeekina erki*, *Dunbarula protomathieui* and *Rauserella* ?sp. (Kobayashi & Altiner 2011). Midian fusulines were widely distributed in the island of Sicily and the Djebel Tebaga area of Tunisia (Skinner & Wilde 1966, 1967; Ghazzay *et al.* 2015). They were represented by *Yabeina*, *Dunbarula*, *Neoschwagerina*, *Yangchienia*, *Chusenella*, *Lantschichites*, *Rauserella* and *Kahlerina* (Skinner & Wilde 1966, 1967; Ghazzay *et al.* 2015). A similar assemblage was also found in the Ankara area of Turkey (Skinner 1969). This assemblage is referable to the *Yabeina–Dunbarula* assemblage here. By contrast, in eastern Turkey, the equivalent fusulines were named as the *Chusenella* Zone, which contains more species of *Chusenella* (Koyluoglu & Altiner 1989).

Late Permian fusulines were documented from southern Greece and southern Turkey (Lys & Marcoux 1978; Unal *et al.* 2003; Vachard *et al.* 2003). The fusulines were dominated by *Paradunbarula* (*Shindella*) *shindensis*, *Sphaerulina zisonzhengensis*, *Codonofusiella paradoxica*, *Palaeofusulina laxa*, *Reichelina cribroseptata* and *R.* sp. (Lys & Marcoux 1978; Unal *et al.* 2003; Vachard *et al.* 2003). They were named as the *Paradunbarula* (*Shindella*) *shindensis–Palaeofusulina laxa* Zone here.

The Carnic Alps. Permian fusulines occurred from the lower part of the Grenzland Formation, and contain the species of *Sphaeroschwagerina*, *Pseudoschwagerina*, *Paraschwagerina*, *Rugosofusulina* and *Triticites*. They were grouped in the *Sphaeroschwagerina carniolica–Pseudoschwagerina extensa* Zone and the *Pseudoschwagerina* aff. *uddeni–Paraschwagerina pseudomira* Zone (Forke 2002). According to Davydov *et al.* (2013), the age of the former zone is late Asselian, whereas that of the latter zone is Sakmarian. The overlying *Paraschwagerina mukhamedjarovica–Pseudoschwagerina muongthensis* Zone and *Sphaeroschwagerina asiatica–Paraschwagerina paranitida–Zellia praeheritschi* Zone in the upper part of the Grenzland Formation are different from the underlying two zones in the occurrence of *Zellia* and *Darvasites* (Forke 2002). The fusulines from the Upper '*Pseudoschwagerina*' limestone were ascribed to the *Robustoschwagerina geyeri–Zellia heritschi–Paraschwagerina nitida–*'*Pseudofusulina*' sp. Zone (Forke 2002). Similarly, the fusuline biostratigraphy from the equivalent Zweikofel Formation was divided into the *Sakmarella fluegeli–Zellia colanii* Zone, the *Sakmarella lubenbachensis–Robustoschwagerina nucleolata* Zone, the *Leeina pseudodivulgata–Chalaroschwagerina incomparabilis* Zone and the *Chalaroschwagerina solita floccose–Perigondwania forkii* Zone, in ascending order (Davydov *et al.* 2013). The age of these zones was considered to be Yakhtashian (Davydov *et al.* 2013). The fusulines in the overlying Trogkofel limestones bear *Biwaella* and *Robustoschwagerina*, which were grouped into the Yakhtashian *Leeina* ex gr. *fusiformis–Robustoschwagerina spatiosa* Zone (Forke 2002).

Middle Permian fusulines are unclear in the Carnic Alps. However, late Permian deposits developed well there. They are represented by the Bellerophon Formation and the lower Tesero Member of the Werfen Formation, representing the major second-rank transgressive–regressive cycles (Loriga *et al.* 1986). Fusulines were found in the upper part of the Bellerophon Formation and the lower part of the Werfen Formation (Loriga *et al.* 1986; Noe 1988). They consist of *Staffella* sp., *Nankinella quasihunanensis* and *Reichelina cribroseptata* of Dorashamian age (Pasini 1985; Noe 1988).

Gondwanan Tethyan region

The Gondwanan Tethyan region refers to both blocks derived from the northern Gondwanan margin during the early Permian and those areas in the Gondwana margin, such as southern Afghanistan, Central Pamir, South Pamir, Karakorum, the South Qiangtang and Lhasa blocks in northern Tibet, the Baoshan and Tengchong blocks in western Yunnan, the Sibumasu Block in SE Asia, the exotic limestone blocks within the Indus–YarlungTsangpo Suture Zone in southern Tibet, and the northern Gondwanan margin (Fig. 5). Because of the influence of Late Palaeozoic glaciation, there was no record of earliest Permian fusulines until the demise of glaciation from the latest Sakmarian.

South Afghanistan–Central and South Pamirs–Karakorum. The areas extended from the Kahmard Block of eastern Iran to Karakorum, through south Afghanistan, the Central Pamir Block and the

International scale	Tethyan scale	S.Afghanistan-Central and South Pamirs-Karakorum (1)—(4)	South Qiangtang (5)—(11)	Lhasa (12)—(16)	Baoshan (17)—(19)	Tengchong (20)—(22)	Sibumasu (23)—(26)	exotic limestone within Indus-Yarlung Tsangpo (14)(27)(28)	Northern Gondwanan margin (29)—(32)
Changhsingian	Dorashamian		?Palaeofusulina	Reichelina simplex					
Wuchiapingian	Dzhulfian	Codonofusiella-Reichelina	?Codonofusiella-Reichelina	Codonofusiella schubertelloides	?	?	Reichelina-Nanlingella	Reichelina pulchra-Dilatofusulina orthogonios	Nanlingella simplex
Capitanian	Midian	Neoschwagerina margaritae	Yabeina-Neoschwagerina	Lepidolina / Nankinella-Chusenella	Verbeekina / Sumatrina	Chusenella-Schwagerina / Nankinella-Chusenella	Rauserella-Kahlerina	Neoschwagerina fusiformis-Lantschichites minima	Neoschwagerina margaritae
Wordian	Murgabian	Neoschwagerina: Neoschwagerina schuberti	Eopolydiexodina		Eopolydiexodina	?	Afghanella-Sumatrina		Monodiexodina kattaensis / Neoschwagerina simplex
Roadian	Murgabian	Neoschwagerina: Neoschwagerina simplex	Neoschwagerina simplex		"Schwagerina" yunnanensis				
Kungurian	Kubergandian	Cancellina	Cancellina primigena / Parafusulina-Monodiexodina			?Cancellina	Misellina claudiae		Skinnerella
Kungurian	Bolorian	Leeina-Chlaroschwagerina					Monodiexodina shiptoni		Minojapanella-Parafusulina
Artinskian	Yakhtashian	Monodiexodina shiptoni	"Pseudofusulina"-Eoparafusulina		"Pseudofusulina"-Eoparafusulina	? / Eoparafusulina			Leeina kraffti
Sakmarian	Sakmarian	Pseudofusulina-Eoparafusulina							"Pseudofusulina"
Asselian	Asselian								

Fig. 5. Schematic framework of Permian fusuline biozones in the Gondwanan Tethyan region. References are: (1) Leven (1993); (2) Gaetani & Leven (2014); (3) Leven (1967); (4) Lys & Lapparent (1971); (5) Zhang *et al.* (2013*b*); (6) Nie & Song (1983*a*); (7) Nie & Song (1983*b*); (8) Zhang *et al.* (2012*a*); (9) Zhang (1991); (10) Cheng *et al.* (2005); (11) Wu & Lan (1990); (12) Zhang *et al.* (2010); (13) Zhu (1982*b*); (14) Wang *et al.* (1981); (15) Zhang Y.J. *et al.* (2014); (16) Chen *et al.* (1999); (17) Shi *et al.* (2011); (18) Huang *et al.* (2009); (19) Huang *et al.* (2015*b*); (20) Fan (1993); (21) Shi *et al.* (2017); (22) Shi *et al.* (2008); (23) Ingavat & Douglass (1981); (24) Ueno *et al.* (2015); (25) Ueno (2003); (26) Fontaine & Suteethorn (1988); (27) Zhang *et al.* (2009); (28) Wang *et al.* (2010); (29) Angiolini *et al.* (2006); (30) Hauser *et al.* (2000); (31) Vachard *et al.* (2002); (32) Pakistan–Japanese Working Group (1985).

South Pamir Block represent a huge continental slice rifting from the Gondwanan margin during the early Permian (Stampfli & Borel 2002; Davydov & Arefifard 2007; Leven *et al.* 2011; Angiolini *et al.* 2015).

The earliest fusuline record in these blocks is represented by the Kalaktash fusuline assemblage preserved in Central Pamir (Leven 1993), the eastern Hindu Kush (Gaetani & Leven 1993; Leven 2010), southern Afghanistan (Leven 1997) and the Kahmard Block of Iran (Davydov & Arefifard 2007; Leven *et al.* 2011; Leven & Gorgij 2013). This fauna is dominated by the '*Pseudofusulina*'–*Eoparafusulina* assemblage that consists of many special species such as '*Pseudofusulina*' *pamirensis*, '*Pseudofusulina*' *karapetovi karapetovi*, *Eoparafusulina pamirensis*, *E. regina* and *Neodutkevitchia insignis*. They were associated with some pseudoschwagerinids, such as *Robustoschwagerina psharti*, *Zellia nunosei* and *Sphaeroschwagerina zhongzanica*, in the Dangikalon Formation in Central Pamir (Leven 1993). The age of this assemblage in these blocks is considered to be late Sakmarian (Leven 1993).

A slightly younger fusuline zone was named as the *Monodiexodina shiptoni* Zone (Gaetani & Leven 2014). This zone was reported from the eastern Karakorum and eastern Hindu Kush (Leven 2010; Gaetani & Leven 2014). The associated fusulines in the eastern Hindu Kush include *Minojapanella*, *Darvasites* and *Chalaroschwagerina* (Leven 2010). The age of this assemblage is most probably Yakhtashian or early Bolorian (Leven 1967; Gaetani & Leven 2014). The *Monodiexodina shiptoni* Zone was succeeded by the *Leeina*–*Chalaroschwagerina* assemblage in the Shaksgam Valley of Karakorum, with its age being most likely Bolorian (Gaetani & Leven 2014). However, the equivalent assemblage in SE Pamir contains many species of *Misellina* and *Parafusulina* (Leven 1967). Kubergandian fusulines are widely distributed in South Pamir and Karakorum, where they were named as the *Cancellina* Zone (Leven 1967) or the *Cancellina*–*Pseudodoliolina*–*Parafusulina* assemblage (Gaetani & Leven 2014). The *Cancellina* Zone bears diverse genera such as *Cancellina*, *Parafusulina*, *Pseudodoliolina*, *Neofusulinella*, *Misellina*, *Armenina* and *Yangchienia* (Leven 1967). The succeeding fusuline zone developed in South Pamir is the *Neoschwagerina* Zone, which comprises *Neoschwagerina simplex*, *N. schuberti* and *N. margaritae* subzones, in ascending order (Leven 1967). These zones were well correlated with those from southern Afghanistan (Lys & Lapparent 1971). The ages of these zones range from Murgabian to Midian in light of the evolution of neoschwagerinids.

The late Permian fusuline fauna was documented in the Karabelesskaya Formation from SE Pamir (Leven 1967). It contains *Sphaerulina zisongzhengensis*, *Nankinella* sp., *Reichelina pulchra*, *Codonofusiella* sp. and *Paradunbarula pamirica*, which indicates a Dzhulfian–Dorashamian age.

South Qiangtang Block. The South Qiangtang Block was part of the Gondwanan continent before its departure from Gondwana during the Artinskian (Zhang *et al.* 2012*a*). Consequently, the fusulines did not appear until the demise of glaciation in this block: that is, since the Artinskian (Nie & Song 1983*a*; Zhang *et al.* 2013*b*). The earliest fusuline fauna in the South Qiangtang Block is represented by diverse fusulines in the Qudi Formation (Nie & Song 1983*b*; Zhang *et al.* 2013*b*). These fusulines are dominated by '*Pseudofusulina*' or *Eoparafusulina* elements of the typical Kalaktash fusuline fauna similar to those found in other Peri-Gondwanan blocks (Leven 1993). However, it should be noted that the Kalaktash fusuline elements such as '*Pseudofusulina*' *pamirensis*, *Neodutkevitchia insignis* and *N. tumidiscula* found in the central Qiangtang Block were associated with *Chalaroschwagerina vulgaris* and *C. globosa*, which indicate a Yakhtashian age (Zhang *et al.* 2013*b*).

The succeeding fusuline fauna in the South Qiangtang Block is dominated by *Parafusulina*–*Monodiexodina* fauna in the Tunglonggongba Formation in the Duoma area of the western Qiangtang Block (Nie & Song 1983*b*) or the lower part of the Lugu Formation (or Cainaha Formation) (Zhang 1991; Y.C. Zhang *et al.* 2014). Its equivalent fauna found in the central Qiangtang Basin is composed of *Cancellina primigena*, *Neofusulinella giraudi*, *Nankinella orbicularia* and *Chusenella schwagerinaeformis*, which can be roughly ascribed to the *Cancellina primigena* Zone of Kubergandian age (Zhang *et al.* 2012*b*). The succeeding fusuline fauna was represented by the *Neoschwagerina simplex* Zone of Murgabian age in the lower part of the Xianqian Formation (Zhang 1991) or the middle part of the Lugu Formation (unpublished data). The overlying fusuline assemblage in the Xianqian Formation is dominated by *Dunbarula*, *Yangchienia*, *Parafusulina* and *Chusenella* (Zhang 1991). The equivalent fusuline fauna in the Lugu Formation is dominated by *Eopolydiexodina* species (Cheng *et al.* 2005; Colpaert *et al.* 2015). The Longge Formation in the Duoma area yields a younger fusuline fauna comprising *Yabeina*, *Neoschwagerina*, *Verbeekina*, *Colania* and *Kahlerina*, and can be assigned to the *Yabeina*–*Neoschwagerina* assemblage of Midian (Liang *et al.* 1983; Nie & Song 1983*c*).

As far as the late Permian fusulines are concerned, Wu & Lan (1990) reported *Codonofusiella* and *Reichelina* from the Rehepan Formation and

Palaeofusulina from the Qinshuihe Formation in the west of the South Qiangtang Block. However, these fusulines have never been described nor illustrated. Thus, the validity of these assemblages needs to be confirmed in the future.

Lhasa Block. The Lhasa Block is bounded by the Yarlung Tsangpo Suture Zone to the south and the Bangong–Nujiang Suture Zone to the north. Because of the influence of glacio-marine deposits on the northern Gondwanan margin, the faunas in the Largar, Angjie formations and the lower part of the Xiala Formation are represented by cool-water biota, such as solitary corals, cool-water brachiopods and the conodont *Vjalovognathus*, but are devoid of any fusulines (Zhang *et al.* 2010, 2013*a*; Yuan *et al.* 2016). The fusulines occur in the middle part of the Xiala Formation or the Luobadui Formation (Zhu 1982*b*; Zhang *et al.* 2010). In the Xainza area, the fusulines in the Xiala Formation are dominated by diverse *Chusenella* and *Nankinella*, and, thus, were named as the *Nankinella–Chusenella* assemblage (Zhu 1982*a*; Zhang *et al.* 2010). Associated fusulines include *Staffella*, *Schwagerina*, *Armenina*, *Rugosochusenella*, *Neoschwagerina* and *Verbeekina* (Zhu 1982*a*; Zhang *et al.* 1985, 2010; Wang & Zhou 1986; Huang *et al.* 2007). The equivalent fusuline assemblage in the Luobadui Formation in the Lhunzhub area, however, is dominated by *Lepidolina* (originally described as *Yabeina*) (Wang *et al.* 1981; Zhu 1982*b*). Coexisting fusulines include *Dunbarula*, *Neoschwagerina*, *Verbeekina* and *Chusenella* (Wang *et al.* 1981; Zhu 1982*b*). The compositional difference between different regions in the Lhasa Block was considered to be the result of different palaeoenvironments (Zhang *et al.* 2010).

The distribution of late Permian fusulines in the Lhasa Block is restricted compared with that of middle Permian fusulines. According to limited reports in the Xainza area, the Dzhulfian fusulines are composed of *Codonofusiella schubertelloides* and *Reichelina* sp. (Y.J. Zhang *et al.* 2014). The advent of *Colaniella parva* in the top of the Xiala Formation indicates a possible Dorashamian age (unpublished data). The fusulines in this interval are dominated by *Reichelina changhsingensis* and *Codonofusiella* sp. The equivalent sequence in the Tsochen area in the Lhasa Block also bears *Reichelina simplex* according to Chen *et al.* (1999).

Baoshan Block. The Baoshan Block is bounded by the Lancangjiang Fault, the Kejie Fault and the Nandinghe Fault on the east, and by the Nujiang Fault on the west (Jin 2002). Similar to the South Qiangtang and Lhasa blocks, the lowermost Permian in this block is dominated by glacio-marine diamictites and pebbly mudstone, namely the Dingjiazhai Formation (Jin 2002; Jin *et al.* 2011). The earliest fusulines occurred in the uppermost part of the Dingjiazhai Formation, which are dominated by the *Eoparafusulina*–'*Pseudofusulina*' assemblage (Ueno 2003; Shi *et al.* 2011). This fauna indicates a Sakmarian–early Yakhtashian age and resembles contemporaneous assemblages in other peri-Gondwanan regions such as Central Pamir and Karakorum (Ueno *et al.* 2002; Shi *et al.* 2011).

The younger fusuline fauna was documented from the lower part the Shazipo Formation. These fusulines were well studied in the Xiaoxinzhai section and Bawei section in the southern Baoshan Block (Huang *et al.* 2009, 2015*a*, 2017). In the Xiaoxinzhai section, the fusulines constitute four zones, the '*Schwagerina*' *yunnanensis* Zone, the *Eopolydiexodina* Zone, the *Sumatrina* Zone and the *Verbeekina* Zone, in ascending order, with the former two zones belonging to the Murgabian and latter two zones belonging to the Midian (Huang *et al.* 2009, 2015*a*). The fusulines in the '*Schwagerina*' *yunnanensis* Zone are composed of schwagerinids, such as *Chusenella*, '*Schwagerina*' and *Parafusulina*, and some verbeekinids, such as *Pseudodoliolina* and *Verbeekina* (Huang *et al.* 2009). The *Eopolydiexodina* Zone can be distinguished from the underlying '*Schwagerina*' *yunnanensis* Zone by the dominance of *Eopolydiexodina* and other genera such as *Xiaoxinzhaiella* and *Neofusulinella* (Huang *et al.* 2009). According to Huang *et al.* (2015*b*), the neoschwagerinids and verbeekinids took over the schwagerinids in the overlying two zones. In the Bawei section, however, the fusulines are typified by the *Yangchienia–Nankinella* assemblage in the lower Shazipo Formation, and *Chusenella–Rugosofusulina* assemblage in the upper Shazipo Formation (Huang *et al.* 2015*b*, 2017). The compositional difference between the Bawei and Xiaoxinzhai sections is considered to be the result of different depositional environments (Huang *et al.* 2015*b*).

Tengchong Block. The Tengchong Block is situated to the west of the Baoshan Block (Jin 2002). The fusulines from this block have received limited study (Fan 1993; Shi *et al.* 2008, 2017). The earliest fusuline record in this block is characterized by the dominance of *Eoparafusulina* in the lower part of the Dadongchang Formation, which could be correlated with the Kalatkash fauna in the peri-Gondwanan regions (Shi *et al.* 2008). Fan (1993) reported the occurrence of *Cancellina* fauna in the upper part of the Guanyingshan Formation (equivalent to lower part of the Dadongchang Formation). However, this fauna was not found in later research (Shi *et al.* 2017). Middle Permian fusulines in the middle part of the Dadongchang Formation are typified by the dominance of the *Nankinella–Chusenella* assemblage, and those in the Yanpozi Formation in the southern Tengchong are dominated by *Chusenella–Schwagerina* according to Shi *et al.* (2017). These two assemblages are quite similar to those found in the Xainza area of the Lhasa Block.

Sibumasu Block. The Sibumasu Block proposed by Metcalfe (1984) includes the Baoshan and Tengchong blocks in western Yunnan, China, Shan State in eastern Myanmar, western Thailand, Peninsular Thailand, western Malaya and Sumatra. As the latest research has indicated that the Baoshan and Tengchong blocks may have not adjoined with the Shan State of eastern Myanmar during the Late Palaeozoic (Zhang *et al.* 2013*a*; Xie *et al.* 2016), the Sibumasu Block cited in this paper does not include the Baoshan and Tengchong blocks in western Yunnan.

The Sibumasu Block, like other Peri-Gondwanan blocks, was dominated by glacio-marine deposits during the earliest Permian (Metcalfe 2013). The earliest fusuline fauna is represented by the antitropical *Monodiexodina shiptoni* fauna in western Thailand (Ingavat & Douglass 1981). Its age was considered to be Bolorian or early Kubergandian (Ueno 2006). However, a nearly contemporaneous assemblage from the Tarn To Formation in the Yala area, southernmost Peninsular Thailand and the Kinta Valley in Malaysia yields abundant Tethyan fusulines ascribed to the *Misellina claudiae* Zone and *Leeina kraffti* Zone, in ascending order (Ishii 1966; Ueno *et al.* 2015). The appearance of these Tethyan elements was explained to be the result of a spike in warming in the Sibumasu Block (Ueno *et al.* 2015).

The middle Permian fusulines have a lower diversity. They are scattered in the Ratburi limestone in western Thailand and the Plateau limestone group in the Shan Plateau in eastern Myanmar. In some places, such as the Sakangyi area in eastern Myanmar and the Doi Pha Daeng area in eastern Thailand, the fusulines are composed of *Neoschwagerina*, *Afghanella* and *Sumatrina* (Fontaine & Suteethorn 1988; Zaw *et al.* 2011). Other genera in the Ratburi limestone include *Rauserella*, *Kahlerina*, *Minojapanella*, *Yangchienia*, *Eopolydiexodina*, *Rugososchwagerina*, *Chusenella* and *Nankinella* (Ueno 2003). *Eopolydiexodina* is a characteristic genus in the Sibumasu Block, which is closely linked to many other Peri-Gondwanan blocks (Ueno 2003).

The late Permian fusulines are more limited in the Sibumasu Block, and were reported from western Thailand. They contain a few genera such as *Reichelina*, *Nanlingella*, *Codonofusiella*, *Staffella* and *Nankinella* (Fontaine & Suteethorn 1988; Ueno 2003).

Exotic limestone blocks within the Indus–Yarlung Tsangpo Suture Zone. Some isolated blocks, scattered in the mélange of the Indus–Yarlung Tsangpo Suture Zone in southern Tibet, are known as the exotic limestone blocks (Shen *et al.* 2003). These limestones yield abundant fusulines ranging from Midian to latest Changhsingian in age (Lys *et al.* 1980; Wang *et al.* 1981, 2010; Wang & Ueno 2009; Zhang *et al.* 2009; Zhang 2010). Middle Permian fusulines were reported from the Xilanta Formation in the Gyanyima limestone block (Zhang *et al.* 2009; Zhang 2010) and the Lasaila limestone block in Zhongba County (Wang *et al.* 1981). All these fusulines bear similarities with South China, such as the presence of *Neoschwagerina*, *Verbeekina*, *Chusenella*, *Codonofusiella*, *Yangchienia* and *Lantschichites* (Wang *et al.* 1981; Zhang *et al.* 2009). However, some characteristic middle Permian fusulines have not yet been reported in these exotic limestone blocks, such as *Afghanella*, *Sumatrina*, *Yabeina* and *Lepidolina.* These fusulines are grouped into the *Neoschwagerina fusiformis–Lantschichites minima* assemblage.

Late Permian fusulines in these exotic limestone blocks were widely documented (Lys *et al.* 1980; Wang & Ueno 2009; Wang *et al.* 2010). The exact age of these fusulines is not easily determined due to a lack of *Palaeofusulina*, but they can be broadly ascribed to late Dzhulfian–Dorashamian. The fusulines in these blocks are dominated by *Reichelina*, *Parareichelina*, *Codonofusiella*, *Dilatofusulina* and *Nankinella* (Lys *et al.* 1980; Wang *et al.* 1981, 2010), and were grouped into the *Reichelina pulchra–Dilatofusulina orthogonios* assemblage by Wang *et al.* (2010).

Northern Gondwanan margin. The fusulines occurred in the Gondwanan margin such as Oman, Timor and Salt Range of Pakistan with gradual demise of the early Permian glaciation. In the margin of the Arabian plate, the earliest record of fusulines is represented by some '*Pseudofusulina*' species including '*P*'. *inobservabilis*, '*P*'. ex gr. *karapetovi karapetovi*, '*P*'. aff. *karapetovi tezakensis* and '*P*'. *licis* (Angiolini *et al.* 2006). Those fusulines indicate a late Sakmarian age, and represent the typical Kalaktash fusuline fauna reported from the Central Pamir (Leven 1993). The Yakhtashian and Bolorian fusulines were reported from the Aseelah units, which were originally deposited on the carbonate platform in the northern Arabian plate (Hauser *et al.* 2000). The Yakhtashian fusuline fauna consists of *Leeina kraffti*, *Schubertella* cf. *simplex* and *Minojapanella* cf. *elongata*, whereas Bolorian fusuline yields *Minojapanella* cf. *elongata*, *M.* cf. *parva* and *Skinnerella* ex gr. *japonica* (Hauser *et al.* 2000). The Yakhtashian and Bolorian fusulines were also documented from the Bua-bai limestone in Timor, where the fusulines consist of sporadic species of *Monodiexodina wanneri*, *Praeskinnerella* sp. and *Nankinella* sp. (Haig *et al.* 2017).

The succeeding fusuline assemblage in Oman is represented by *Skinnerella visseri* and *S. arabica* in the Qarari unit of the Batain group, indicating a Kubergandian age (Leven & Heward 2013). The Murgabian and Midian fusulines were reported from the Aseelah unit. They were represented by *Neoschwagerina simplex* in the lower and many advanced neoschwagerinids such as *Neoschwagerina margaritae*, *N. occidentalis* and *Colania douvillei* in the upper (Hauser *et al.* 2000; Vachard *et al.* 2002). The Murgabian fusulines also occurred in the Salt Range of Pakistan, where they were represented by only *Monodiexodina kattaensis* and *Codonofusiella laxa* (Douglass 1970). In Timor, the only Midian fusuline is represented by *Lantschichites weberi* from the Maubisse group (Schubert 1915; Thompson 1949).

The Lopingian fusulines were not reported from Oman, but sporadically occurred in the upper part of the Wargal Formation in the Salt Range, Pakistan. They comprise mainly *Nanlingella simplex*, *Reichelina* sp., and *Codonofusiella* aff. *schubertelloides* referable to the *Nanlingella simplex* Zone (Pakistan–Japanese Working Group 1985).

Angara Tethyan region

Angara Tethyan region refers to the blocks to the north of North China, which were located at relatively high latitude areas in the Northern Hemisphere (Ozawa 1987). It includes the Southern Kitakami and Hida terranes in northern Japan, Southern Primorye in eastern Russia and the Inner Mongolia area in northern China (Fig. 6). The blocks in this region are characterized by the occurrence of the antitropical genus *Monodiexodina* during the middle Permian.

Southern Kitakami Terrane. The Southern Kitakami Terrane, different from those pre-Cretaceous accretionary terranes in Japan, was always in a position proximal to the China continents during the Permian (Tazawa 1991; Kobayashi 1999). The earliest Permian sequence in the Southern Kitakami Terrane is represented by the Sakamotozawa Formation, which rests unconformably above the Upper Carboniferous strata (Kanmera & Mikami 1965). This formation was subdivided into four units, respectively the Sa and Sb members in the lower subformation and the Sc and Sd subformation in the upper subformation (Mikami 1965). Kanmera & Mikami (1965) established the *Zellia nunosei* and *Eoparafusulina langsonensis* zones in the Sb member of the Sakamotozawa Formation. These two zones consist of diverse fusuline species such as *Zellia nunosei*, *Eoparafusulina langsonensis*, *Nipponitella explicata* and *Quasifusulina tenuissima* (Kanmera & Mikami 1965). The fusulines contained in the Sa member include *Eoparafusulina* aff. *perplexa*, *Nipponitella explicata* and *Quasifusulina*? sp., which were considered to be similar to the fusuline elements in the overlying Sb member (Ueno *et al.* 2007). In the basal part of the Sakamotozawa in the Kamiyasse area of Kesennuma, however, the fusulines are composed of *Dutkevitchia*? *hindukushiensis*, *Shichanella* cf. *callosa*, *Pseudochusenella* ex gr. *cushmani*, *Nipponitella* sp. and *Eoparafusulina* sp. (Ueno *et al.* 2011). Thus, the fusulines in the Sa and Sb members in the Sakamotozawa Formation indicate a Sakmarian age (Watanabe 1991; Ueno *et al.* 2011). In the Sc and Sd members of the Sakamotozawa Formation, Kanmera & Mikami (1965) established the *Chalaroschwagerina vulgaris*, *Leeina fusiformis* and *Shichanella callosa* zones, respectively. Watanabe (1991) assigned a Yakhtashian age for these three zones. However, the upper part of the Sakamotozawa Formation contains elements of *Misellina* (Choi 1972; Ueno *et al.* 2009). So, the *Leeina fusiformis* and *Shichanella amgibua* zones in the Sd member of the Sakamotozawa Formation were considered to be Bolorian in age in terms of the presence of '*Pseudofusulina*' *dzamantalensis*, *Darvasites minatoi* and *Kubergandella*? sp. (Ueno *et al.* 2009).

Besides the Sakamotozawa Formation, the Midian fusulines were widely reported from the Southern Kitakami Terrane (Choi 1970, 1973; Kobayashi *et al.* 2009). They can be grouped into four zones, the *Sumatrina annae* Zone, the *Monodiexodina sutchanica* Zone, the *Parafusulina motoyoshi* Zone and the *Lepidolina shiraiwensis* Zone, in ascending order (Shiino *et al.* 2008; Kobayashi *et al.* 2009). The species *Sumatrina annae* was found in the Hoso-o Formation accompanying with *Pseudodoliolina pseudolepida* (Shiino *et al.* 2008). At the bottom of the overlying Kamiyasse Formation or equivalent beds in the Southern Kitakami Terrane, *Monodiexodina sutchanica* is highly concentrated (Ehiro & Misaki 2004; Kobayashi *et al.* 2009). The overlying fusuline elements consist of *Parafusulina motoyoshiensis*, *P. oyensis*, *Chusenella pseudocrassa*, *C. sinensis*, which were ascribed to the *Parafusulina motoyoshiensis* Zone (Kobayashi *et al.* 2009). The *Lepidolina shiraiwensis* fauna occurs in the upper part of the Kamiyasse Formation, which may correspond to the *Lepidolina multiseptata* Zone of Choi (1973). The fusulines in this zone include *Lepidolina shiraiwensis*, *L. multiseptata*, *Verbeekina verbeeki*, *Chusenella andersoni* and *Codonofusiella* sp. (Kobayashi *et al.* 2009).

The Late Permian fusulines are represented by scarce species such as *Nanlingella* cf. *meridionalis*,

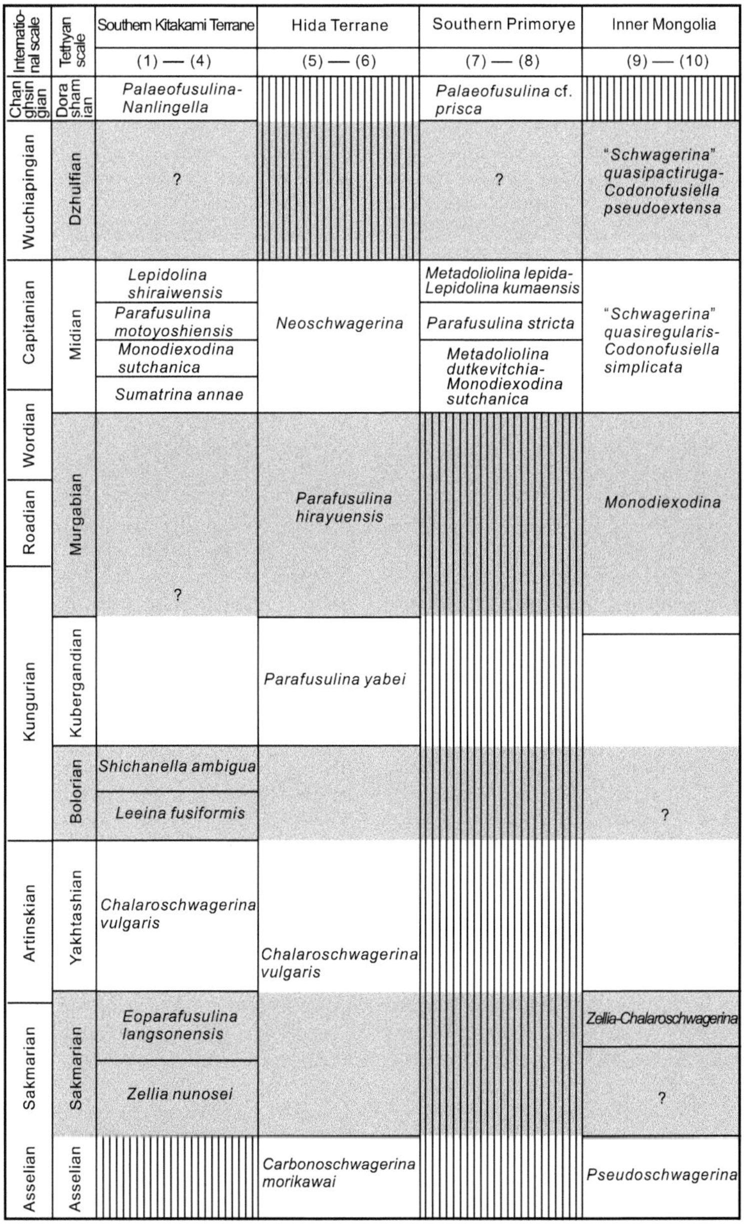

International scale	Tethyan scale	Southern Kitakami Terrane (1) — (4)	Hida Terrane (5) — (6)	Southern Primorye (7) — (8)	Inner Mongolia (9) — (10)
Changhsingian	Dorashamian	Palaeofusulina-Nanlingella		Palaeofusulina cf. prisca	
Wuchiapingian	Dzhulfian	?		?	"Schwagerina" quasipactiruga-Codonofusiella pseudoextensa
Capitanian	Midian	Lepidolina shiraiwensis Parafusulina motoyoshiensis Monodiexodina sutchanica Sumatrina annae	Neoschwagerina	Metadoliolina lepida-Lepidolina kumaensis Parafusulina stricta Metadoliolina dutkevitchia-Monodiexodina sutchanica	"Schwagerina" quasiregularis-Codonofusiella simplicata
Wordian Roadian	Murgabian	?	Parafusulina hirayuensis		Monodiexodina
Kungurian	Kubergandian		Parafusulina yabei		
	Bolorian	Shichanella ambigua Leeina fusiformis			?
Artinskian	Yakhtashian	Chalaroschwagerina vulgaris	Chalaroschwagerina vulgaris		
Sakmarian	Sakmarian	Eoparafusulina langsonensis Zellia nunosei			Zellia-Chalaroschwagerina ?
Asselian	Asselian		Carbonoschwagerina morikawai		Pseudoschwagerina

Fig. 6. Schematic framework of Permian fusuline biozones in the Angara Tethyan region. References are: (1) Kanmera & Mikami (1965); (2) Ueno *et al.* (2009); (3) Kobayashi *et al.* (2009); (4) Kobayashi (2002); (5) Igo (1957); (6) Igo (1959); (7) Kotlyar *et al.* (2006); (8) Ueno *et al.* (2005); (9) Sheng (1958); (10) Xia (1981).

Palaeofusulina sp., *Reichelina changhsingensis* and *Nankinella* sp. in the upper part of the Toyoma Formation (Kobayashi 2002). These fusulines overall indicate a Dorashamian age.

Hida Terrane. Like the South Kitakami Terrane, the Hida Terrane is located in the inner zone of Japan. The lowermost Permian fusulines are characterized by the *Carbonoschwagerina morikawai* Zone in the upper part of the Mizuyagadani Formation. There are species of *Carbonoschwagerina*, *Quasifusulina*, *Triticites* and *Rugosofusulina* (Igo 1957; Niikawa 1978). The overlying zone was documented in the lower part of the Hirayu Group, where the

Pseudoschwagerina Zone was covered by the *Chalaroschwagerina vulgaris* Zone (Igo 1959). This zone may range up to Bolorian because of the presence of *Misellina minor*. The succeeding zone is represented by the *Parafusulina yabei* Zone and the *Parafusulina hirayuensis* Zone, in ascending order (Igo 1959). The latter zone may range into Murgabian as evidenced by the dominance of typical *Parafusulina* and *Pseudodoliolina* (Igo 1959). The uppermost part of the Hirayu Group yields scarce *Neoschwagerina margaritae* and '*Schwagerina*' sp. that were grouped into the *Neoschwagerina* Zone (Igo 1959). This zone is probably Midian in age.

Southern Primorye. Southern Primorye area in Russia was composed of many terranes such as Voznesensk, Nakhimovsk and Okrainsk terranes (Kotlyar *et al.* 2006). The lower Permian strata in South Primorye are composed of terrestrial sediment with diverse plant fossils. The occurrence of fusulines starts as late as Midian (Kotlyar *et al.* 2006). Overall, the fusulines found in the Chandalaz Formation were grouped into three zones, respectively the *Metadoliolina dutkevitchi–Monodiexodina sutchanica* Zone, the *Parafusulina stricta* Zone and the *Metadoliolina lepida–Lepidolina kumaensis* Zone, in ascending order (Ueno *et al.* 2005; Kotlyar *et al.* 2006). The *Metadoliolina dutkevitchi–Monodiexodina sutchanica* Zone contains *Metadoliolina*, *Monodiexodina*, *Minojapanella*, *Lantschichites*, *Codonofusiella*, *Pseudofusulina*, *Parafusulina* and *Reichelina* (Sosnina 1981; Ueno *et al.* 2005; Kotlyar *et al.* 2006). The overlying *Parafusulina stricta* Zone is characterized by the appearance of *Lepidolina multiseptata* and rare *Yabeina* (Kotlyar *et al.* 2006). The *Metadoliolina lepida–Lepidolina kumaensis* Zone occurs in the upper part of the Chandalaz Formation, and is much more diverse than the underlying two zones. The common genera found in this zone include *Metadoliolina*, *Lepidolina*, *Sichotenella*, *Rauserella*, *Reichelina*, *Parareichelina*, *Minojapanella*, *Lantschichites*, *Codonofusiella*, *Chusenella*, *Parafusulina*, *Kahlerina* and *Pseudokahlerina* (Sosnina 1981; Kotlyar *et al.* 2006).

Late Permian fusulines were discovered in the Yastrebovka or equivalent formations in Southern Primorye (Sosnina 1981; Kotlyar *et al.* 2006). The fusulines consist of *Palaeofusulina* cf. *prisca*, *Shindella* sp., *Sphaerulina zisongzhengensis*, *Reichelina ulachensis* and *R. chorensis* (Sosnina 1981; Vuks & Chediya 1986; Kotlyar *et al.* 2006). Their age is Dorashamian in terms of abundant *Colaniella parva* in this assemblage (Kotlyar *et al.* 2006).

Inner Mongolia. Abundant fusulines were reported from the central part of Inner Mongolia of China, which tectonically belongs to the southern margin of the Altaid Belt (Shen *et al.* 2006). The Permian sequence is dominated by siliciclastic rocks and limestones beds (e.g. Han 1975; Xia 1981). The fusulines were found discretely in limestone beds. The lowermost zone is the *Pseudoschwagerina* Zone characterized by species of *Pseudoschwagerina*, *Triticites*, *Quasifusulina* and '*Schwagerina*'. It is of Asselian age and was reported in the Baiyunebo area (Sheng 1958; Han 1975). A younger fusuline fauna occurred in the Sonid right banner area yields *Eoparafusulina*, *Zellia*, *Chalaroschwagerina* and *Nipponitella*. This fauna probably suggests a late Sakmarian age. The overlying fusuline zone was named as the *Monodiexodina* Zone, which comprises *Monodiexodina*, *Parafusulina*, *Laxifusulina*, *Pseudodoliolina*, *Yangchienia* and *Armenina* (Xia 1981). Its age is broadly assigned to late Kubergandian to Murgabian. The upper two zones, respectively the '*Schwagerina*' *quasiregularis–Codonofusiella simplicata* Zone in the upper part of the Zhesi Formation and the '*Schwagerina*' *quasipactiruga–Codonofusiella pseudoextensa* Zone in the Yihewusu Formation, indicate a Murgabian to Midian age (Xia 1981). It is noteworthy that no neoschwagerinids have ever been reported in these two zones.

Panthalassan region

During the Permian, many seamounts developed in the Panthalassa Ocean. Those seamounts collapsed when they were amalgamated with continents surrounding the Panthalassan Ocean. So, the fusulines belonging to Panthalassan Province were dispersed in many allochthonous terranes in Japan, Canada, and New Zealand (Fig. 7).

Akiyoshi Terrane. The Akiyoshi Terrane in the Inner Zone of SW Japan represents the oldest accretionary complex in Japan (Ueno 1996). The Omi limestone in the south of Niigata Prefecture may represent the NE extremity of the Akiyoshi Terrane (Ueno & Nakazawa 1993). The Akiyoshi limestone in the Akiyoshidai area preserves a complete carbonate succession ranging from Early Carboniferous to middle Permian in the Panthalassa Ocean (Ueno 1996). The biostratigraphic schemes of Ota (1997) and Ueno (1996) are adopted in this paper.

The Permian starts from the appearance of *Pseudoschwagerina muongthensis*, which is associated with *Schubertella kingi*, '*Schwagerina*' *primigena* and '*S*'. *okafujii*. They were grouped in the *Pseudoschwagerina muongthensis* Zone (Ota 1997). This zone corresponds to the *Triticites simplex* Subzone of Toriyama (1954) because this subzone is also dominated by species of *Pseudoschwagerina* and *Triticites*. Its age is most likely to be Asselian. The overlying fusulines include

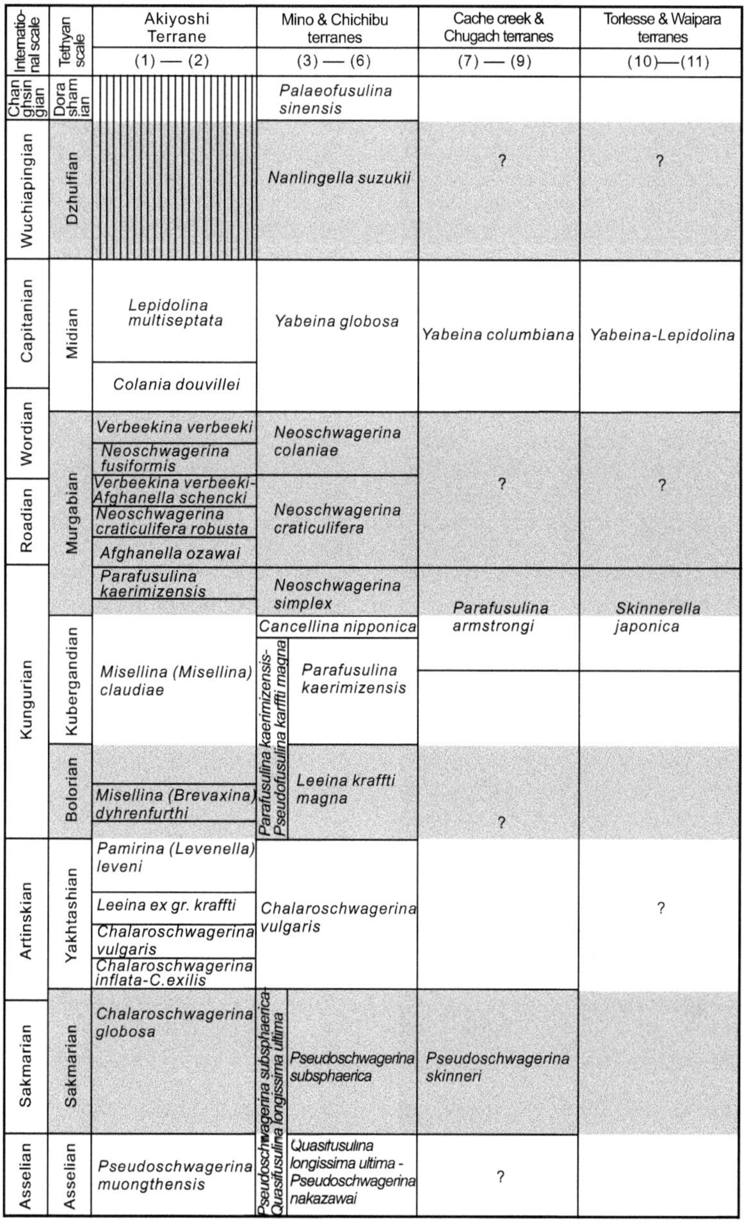

International scale	Tethyan scale	Akiyoshi Terrane (1) — (2)	Mino & Chichibu terranes (3) — (6)	Cache creek & Chugach terranes (7) — (9)	Torlesse & Waipara terranes (10)—(11)
Changhsingian	Dorashamian		Palaeofusulina sinensis		
Wuchiapingian	Dzhulfian		Nanlingella suzukii	?	?
Capitanian	Midian	Lepidolina multiseptata; Colania douvillei	Yabeina globosa	Yabeina columbiana	Yabeina-Lepidolina
Wordian	Murgabian	Verbeekina verbeeki; Neoschwagerina fusiformis	Neoschwagerina colaniae	?	?
Roadian	Murgabian	Verbeekina verbeeki-Afghanella schencki; Neoschwagerina craticulifera robusta; Afghanella ozawai	Neoschwagerina craticulifera		
Kungurian	Murgabian	Parafusulina kaerimizensis	Neoschwagerina simplex	Parafusulina armstrongi	Skinnerella japonica
Kungurian	Kubergandian	Misellina (Misellina) claudiae	Cancellina nipponica; Parafusulina kaerimizensis-Pseudofusulina kraffti magna: Parafusulina kaerimizensis		
Kungurian	Bolorian	Misellina (Brevaxina) dyhrenfurthi	Parafusulina kaerimizensis-Pseudofusulina kraffti magna: Leeina kraffti magna	?	
Artinskian	Yakhtashian	Pamirina (Levenella) leveni; Leeina ex gr. kraffti; Chalaroschwagerina vulgaris; Chalaroschwagerina inflata-C.exilis	Chalaroschwagerina vulgaris		?
Sakmarian	Sakmarian	Chalaroschwagerina globosa	Pseudoschwagerina subsphaerica-Quasifusulina longissima ultima: Pseudoschwagerina subsphaerica	Pseudoschwagerina skinneri	
Asselian	Asselian	Pseudoschwagerina muongthensis	Pseudoschwagerina subsphaerica-Quasifusulina longissima ultima: Quasifusulina longissima ultima - Pseudoschwagerina nakazawai	?	

Fig. 7. Schematic framework of Permian fusuline biozones in the Panthalassan region. References are: (1) Ota (1997); (2) Ueno (1996); (3) Nogami (1961); (4) Kobayashi (2011); (5) Ota & Isozaki (2006); (6) Kobayashi (1997*b*); (7) Sada & Danner (1992); (8) Thompson & Verville (1950); (9) Rui & Nassichuk (1996); (10) Leven & Grant-Mackie (1997); (11) Vachard & Ferriere (1991).

Chalaroschwagerina globosa and *Paraschwagerina* sp., which was ascribed to the *Chalaroschwagerina globosa* Zone.

The overlying fusulines were named as the *Chalaroschwagerina inflata–Chalaroschwagerina exilis* Zone with newly appeared species of *Biwaella* and *Mesoschubertella* (Ueno 1996). The succeeding fusulines were grouped into the *Chalaroschwagerina vulgaris* and *Leeina* ex gr. *kraffti* zones, which comprise mostly *Chalaroschwagerina*, pseudofusulinid species and some small taxa such as *Pseudoendothyra* and *Schubertella* (Ueno 1996). The above-mentioned three zones were assigned to the lower Yakhtashian (Ueno 1996). The overlying fusulines are dominated by small forms such as *Pamirina* (*Levenella*), *Minojapanella*, *Pseudoreichelina*, *Staffella* and *Pseudoendothyra*, which were named as the *Pamirina* (*Levenella*) *leveni* Zone. Because the species *Misellina* (*Brevaxina*) *nipponica* occurs in the upper part of this zone, the age of the *Pamirina* (*Levenella*) *leveni* Zone is clearly late Yakhtashian–lower Bolorian (Ueno 1996). The appearance of *Misellina* (*Brevaxina*) *dyhrenfurthi otai* marks a new zone in which primitive *Misellina* elements also occurred (Ueno 1996). The overlying zone was named as the *Misellina claudiae* Zone by Ueno (1996). It should be noted, however, that the *Misellina claudiae* Zone used here is different from those adopted widely in South China because many neoschwagerinids occurred in this zone in the Akiyoshi limestone such as *Maklaya pamirica* and *Neoschwagerina simplex* (Ueno 1996). The presence of the species *Neoschwagerina simplex* indicates a Murgabian age for the upper part of this zone.

The *Misellina claudiae* Zone is succeeded by the *Parafusulina kaerimizensis* Zone. In the latter zone, many schwagerinids occurred, including *Parafusulina kaerimizensis*, *P. lutugini* and *Chusenella alpina* (Ueno 1996). Above this zone, *Afghanella*, *Neoschwagerina*, *Verbeekina* and *Pseudodoliolina* appeared successively. They were assigned into the *Afghanella ozawai*, *Neoschwagerina craticulifera robusta*, *Verbeekina verbeeki–Afghanella schencki*, *Neoschwagerina fusiformis* and *Verbeekina verbeeki* zones, in ascending order (Ueno 1996).

The following fusuline zones include the *Colania douvillei* Zone and the *Lepidolina multiseptata* Zone, in ascending order (Ueno 1996). Common genera in these zones are *Verbeekina*, *Sumatrina*, *Kahlerina* and *Pseudokahlerina* (Ueno 1996). These two zones overall indicate a Midian age.

Mino and Chichibu terranes. Many seamounts developed in the Panthalassa Ocean during the Permian, and were later accreted to mainland Japan during the Jurassic–Early Cretaceous (Kobayashi 1999, 2011; Ueno *et al.* 2006). These seamounts were preserved as blocks of various sizes in the mélange, which were roughly included into the Mino Terrane or the Chichibu Terrane. The biostratigraphy of these limestone blocks has been thoroughly studied (e.g. Honjo 1959; Zaw 1999; Kobayashi 2011, 2012).

Among all the limestone blocks studied, the Akasaka limestone preserved a good record of fusuline zonations (e.g. Ozawa 1927; Zaw 1999). However, fusulines older than Kubergandian have not been discovered in the Akasaka limestone. These older fusulines have been reported from the Itsukaichi-Ome area of west Tokyo and the Atetsu Plateau (Nogami 1961; Kobayashi 2005). The early Permian fusulines were classified into the *Pseudoschwagerina subsphaerica–Quasifusulina longissima ultima*, *Chalaroschwagerina vulgaris* and *Parafusulina kaerimizensis–Leeina kraffti magna* zones, in ascending order (Nogami 1961). The former zone consists of the lower *Quasifusulina longissima ultima–Pseudoschwagerina nakazawai* Subzone and the upper *Pseudoschwagerina subsphaerica* Subzone, respectively, most probably indicating an Asselian–Sakmarian age. The overlying *Chalaroschwagerina vulgaris* Zone is composed of *C. vulgaris* and *C. globosa* of Yakhtashian age. The *Parafusulina kaerimizensis–Leeina kraffti magna* Zone also comprises the lower *Leeina kraffti magna* Subzone and the upper *Parafusulina kaeimizensis* Subzone (Nogami 1961). The latter subzone was considered to be equivalent to the *Parafusulina nakamigawai* Zone in the lower part of the Akasaka limestone, which indicates a Kubergandian age (Kobayashi 2011).

The overlying zone in the Akasaka limestone is the *Cancellina nipponica* Zone. It consists of *Cancellina*, *Codonofusiella*, *Neofusulinella*, *Toriyamaia* and *Pseudoreichelina* (Kobayashi 2011). The Murgabian fusulines in the Akasaka limestone were grouped into three zones according to the evolution of *Neoschwagerina*; they are the *Neoschwagerina simplex* Zone, the *N. craticulifera* Zone and the *N. colaniae* Zone from lower to upper. It is noticeable that *Gifuella* was found in both the *Neoschwagerina craticulifera* and *N. colaniae* zones, however, this genus was not found in the Akiyoshi Terrane. The appearance of *Yabeina* species above the *Neoschwagerina colaniae* Zone marks the beginning of the Midian age. It was ascribed to the *Yabeina globosa* Zone, which is identical to the *Yabeina igoi* Zone of Zaw (1999).

Many large fusulines went extinct across the Guadalupian–Lopingian boundary (Ota & Isozaki 2006). Thus, the remaining fusulines in the Dzhulfian were grouped into the *Nanlingella suzukii* Zone, which comprises small taxa such as *Nanlingella*, *Codonofusiella* and *Reichelina* (Ota & Isozaki 2006; Kobayashi 2011). The uppermost Permian fusulines have not been documented from the Akasaka limestone; but they were reported from the Iwai-Kanyo area in west Tokyo (Kobayashi 1997*b*). The fusulines found from the pebbles are referable to the *Palaeofusulina sinensis* Zone (Kobayashi 1997*b*).

Cache Creek and Chugach terranes. Along the western margin of the North American Craton some terranes developed extending from Alaska to central California in the USA. The fusulines are different from those of the North American Craton faunas in possessing abundant neoschwagerinids. These terranes were roughly ascribed to the Cache Creek Terrane and the Chugach Terrane (Ross 1995; Stevens *et al.* 1997; Johnston & Borel 2007; Kobayashi *et al.* 2007).

The fusulines from these two terranes can be subdivided into three assemblages. The lowest assemblage is characterized by the *Pseudoschwagerina skinneri* fauna, suggesting a Sakmarian age (Sada & Danner 1992). The slightly younger fusulines are dominated by *Parafusulina armstrongi* of late Kubergandian–Murgabian age (Thompson & Verville 1950). The youngest fauna is widely distributed in western Canada and the NW USA (Thompson & Wheeler 1942; Ross 1971; Rui & Nassichuk 1996; Kobayashi *et al.* 2007) or southern margin of Alaska (Stevens *et al.* 1997). These fusulines are named as the *Yabeina columbiana* Zone, which comprises *Yabeina*, *Lepidolina*, *Rauserella*, *Reichelina*, *Codonofusiella* and *Schwagerina*.

Torlesse and Waipara terranes in New Zealand. The Torlesse and Waipara terranes on the eastern margin of New Zealand are different from mainland New Zealand in yielding diverse warm-water Tethyan fusulines (Hornibrook 1951; Vachard & Ferriere 1991; Leven & Grant-Mackie 1997; Leven & Campbell 1998). These fusulines can be grouped into two assemblages. The older assemblage is characterized by the *Skinnerella japonica* assemblage of all *Skinnerella* species, with its age being late Kubergandian–Murgabian (Leven & Campbell 1998). The younger assemblage, however, consists of diverse fusulines such as *Yabeina*, *Reichelina*, *Kahlerina*, *Minojapanella*, *Sichotenella*, *Codonofusiella*, *Rauserella*, *Dunbarula*, *Chusenella*, *Neoschwagerina* and *Lepidolina* (Vachard & Ferrière 1991; Leven & Grant-Mackie 1997). This *Yabeina–Lepidolina* assemblage is Midian in age.

North American Craton region

North America and South America were located at the western margin of Pangaea during the Permian. As the Rheic Ocean closed during the Mississippian, there was little faunal communication between the Palaeotethys and North America (Qiao & Shen 2014). This results in a distinctive bioprovince in North America (Yancey 1975; Kobayashi 1997*a*).

The fusulines are widely distributed in both North and South America. In addition to the USA, some fusulines were also documented from Mexico (e.g. Vachard *et al.* 2000), Guatemala (e.g. Ross 1962), Venezuela (Thompson & Miller 1949), Peru (e.g. Wood *et al.* 2002) and Bolivia (e.g. Dunbar & Newell 1946). Overall, these faunas are broadly correlatable with each other. Thus, the biostratigraphy subdivision scheme of the USA serves as a criterion.

The Lower Permian chronostratigraphy was classified into the Wolfcampian Epoch and the Leonardian Epoch (Ross 1963). The Wolfcampian fusulines were grouped into three zones: the *Triticites–Thompsonites* Zone in the lower part, the *Triticites–Pseudoschwagerina* Zone in the middle and the *Pseudoschwagerina–Eoparafusulina* Zone in the upper part (Ross 1963). The lowest zone, which corresponds to the fusulines in the Bursumian Stage or the Newwellian Stage, contains *Thompsonites*, *Pseudofusulina* and *Leptotriticites*, and these were considered to be the earliest Permian fusulines (Wilde 1990, 2006; Lucas *et al.* 2015). However, because *Pseudoschwagerina* and *Sphaeroschwagerina* are widely considered to be indicators of the Permian, then it is better to place the uppermost part of the Newwellian Stage in the Permian because of the presence of *Pseudoschwagerina* in that interval (Wilde 2006). The succeeding zone is characterized by abundant *Pseudoschwagerina* species in the Middle Wolfcampanian (Nealian Stage) (Ross & Ross 1994). In this regard, the *Pseudoschwagerina uddeni–Occidentoschwagerina* Zone of Wilde (1984) is equivalent to the *Triticites–Pseudoschwagerina* Zone of Ross (1963). The overlying *Pseudoschwagerina–Eoparafusulina* Zone is characterized by the dominance of species of *Pseudoschwagerina*, *Paraschwagerina* and *Eoparafusulina* (Ross 1963).This zone is equivalent to the *Pseudoschwagerina convexa–Eoparafusulina linearis* assemblage of Ross & Ross (2003).

The fusulines of the Leonardian differ greatly from the underlying Wolfcampian fusulines in the appearance of the genus *Parafusulina* (Ross 1963). The lowermost part of the Leonardian faunas was named as the *Praeskinnerella crassitectoria–Praeskinnerella guembeli* Zone, whereas four other zones were subdivided based on the evolution of the *Parafusulina* species. They are the *Parafusulina allisonensis* Zone, the *P. deltoides* Zone, the *P. spissisepta* Zone and the *P. brooksensis–P. vidriensis* Zone, in ascending order (Ross 1960).

The biostratigraphic subdivision of the Guadalupian Series has been studied by many researchers (e.g. Dunbar & Skinner 1937; Wilde & Rudine 2000; Yang & Yancey 2000). The biostratigraphic subdivision scheme of Yang & Yancey (2000) is followed here.

The *Parafusulina bosei* Zone in the lower part of the Road Canyon Formation differs from the underlying Leonardian fusuline fauna in the appearance of genera such as *Rauserella* and *Yangchienia*. From this zone onwards, the biostratigraphic subdivision was based on the evolution of *Parafusulina* from the middle part of the Road Canyon Formation to the upper part of the Word Formation. It was classified into the *Parafusulina rothi*, *P. trumpyi*, *P. sellardsi* and *P. antimonioensis* zones, in ascending order (Yang & Yancey 2000). The former two zones belong to the Roadian and the latter two are of Wordian age.

The Capitanian Stage is typified by an abundance of the *Polydiexodina* species that was assigned into *Polydiexodina shumardi* Zone in the lower part of the Altuda Formation or Capitan limestone. Some genera, such as *Reichelina* and *Codonofusiella*, occurred from this zone (Yang & Yancey 2000). A spectacular feature during the Capitanian in Texas, USA was the occurrence of *Yabeina texana* in the Guadalupe Mountains (Skinner & Wilde 1955; Nestell & Nestell 2006). With the extinction of large fusulines such as *Yabeina* and *Polydiexodina*, the fusulines in the upper part of the Altuda Formation are dominated by smaller fusulines such as *Reichelina*, *Codonofusiella*, *Paradoxiella*, *Rauserella* and *Lantschichites* (Yang & Yancey 2000; Nestell & Nestell 2006). They were grouped into two zones: the *Reichelina lamarensis* Zone and the *Paraboultonia splendens* Zone, in ascending order (Yang & Yancey 2000). The *Reichelina lamarensis* Zone is equivalent to the *Paradoxiella* and *Reichelina lamarensis* zones of Wilde & Rudine (2000). Accordingly, the *Paraboultonia splendens* Zone of Yang & Yancey (2000) is equal to the *Paraboultonia–Lantschichites* Zone of Wilde & Rudine (2000).

Correlations of different Permian chronostratigraphic scales

Asselian

The Asselian Stage is characterized by the dominance of pseudoschwagerinids with loosely coiled outer volutions, such as *Pseudoschwagerina* and *Sphaeroschwagerina*. The transition of the *Triticites*-dominated assemblage in the Gzhelian of the Late Carboniferous to a *Pseudoschwagerina*- or *Sphaeroschwagerina*-dominated assemblage can be recognized in various regions, such as the Southern Urals (Rauzer-Chernousova 1965), Darvaz (Leven & Scherbovich 1978; Leven *et al.* 1992), South China (Xiao *et al.* 1986; Ding *et al.* 1989), North China (Wang *et al.* 1992), Tarim (Zheng & Lin 1991), Iran (Leven & Taheri 2003) and North America (Ross 1963; Ross & Ross 1994). Consequently, the base of the Asselian on the Tethyan scale corresponds to the base of the Zisongian Stage in South China and the base of the upper part of the Newwellian Stage (lower Wolfcampian) in North America (Fig. 8). In addition to *Pseudoschwagerina* and *Sphaeroschwagerina*, some Gzhelian (latest Carboniferous) genera, such as *Triticites*, *Quasifusulina* and *Rauserites*, can also range up into the Asselian.

Sakmarian

Sakmarian fusuline faunas share a lot of common genera with the underlying Asselian faunas, such as *Pseudoschwagerina*, *Rugosofusulina* and *Paraschwagerina*. That is possibly the reason why both the Asselian and the Sakmarian were grouped into the Zisongian Stage in South China (L.X. Zhang *et al.* 1988; Z.H. Zhang *et al.* 1988; Sheng & Jin 1994). Nevertheless, Sakmarian fusulines are characterized by the flourishing of species of *Eoparafusulina*, *Darvasites* and *Zellia* in the Tethyan region, such as the Carnic Alps (Forke 2002; Davydov *et al.* 2013), the Tarim Block (Zheng & Lin 1991), South China (L.X. Zhang *et al.* 1988; Shi *et al.* 2009), Darvaz (Leven *et al.* 1992) and north Afghanistan (Leven 1997). In North America, diverse species of *Eoparafusulina* and *Paraschwagerina* were documented in the upper part of the Wolfcampian strata such as the Leonox Hills Formation in Texas (Ross 1963), as well as the middle part of the Bird Spring Formation in southern California (Stevens & Stone 2007). Taking into account these similarities, the Sakmarian Stage in the Tethyan region is roughly correlated with the Lenoxian Stage of the upper Wolfcampian (Fig. 8). However, it should be mentioned that the upper part of the Lenoxian may fall within the Yakhtashian Stage of the Tethyan regions because of the presence of the genus *Chalaroschwagerina* (Wilde 1984). Additionally, it should be pointed out that the common Sakmarian fusuline genera, such as *Zellia* and *Darvasites*, in the Tethyan area were not reported in North America.

With regard to the correlations of the Sakmarian of the Tethyan scale with that of the international scale, the sections from the Carnic Alps help to resolve this correlation because of the presence of both fusulines and conodonts (Forke 2002). In those sections, the typical Artinskian conodont *Sweetognathus* aff. *whitei* and *Mesogondolella* cf. *bisseli* coexist with *Zellia*, *Robustoschwagerina* and *Pseudoschwagerina* of the Tethyan Sakmarian age (Forke 2002). Thus, the Tethyan Sakmarian Stage corresponds to the Sakmarian and lowermost Artinskian of the international scale (Fig. 8). The Sakmarian Stage of the Tethyan scale was replaced by the Hermagorian Stage, which is equal to the Sakmarian and the lower part of the Artinskian age of the international scale (Davydov *et al.* 2013)

Yakhtashian. The Yakhtashian fusulines can be differentiated from the underlying Sakmarian fusulines by the apperance of many characteristic genera such as *Chalaroschwagerina*, *Pamirina*

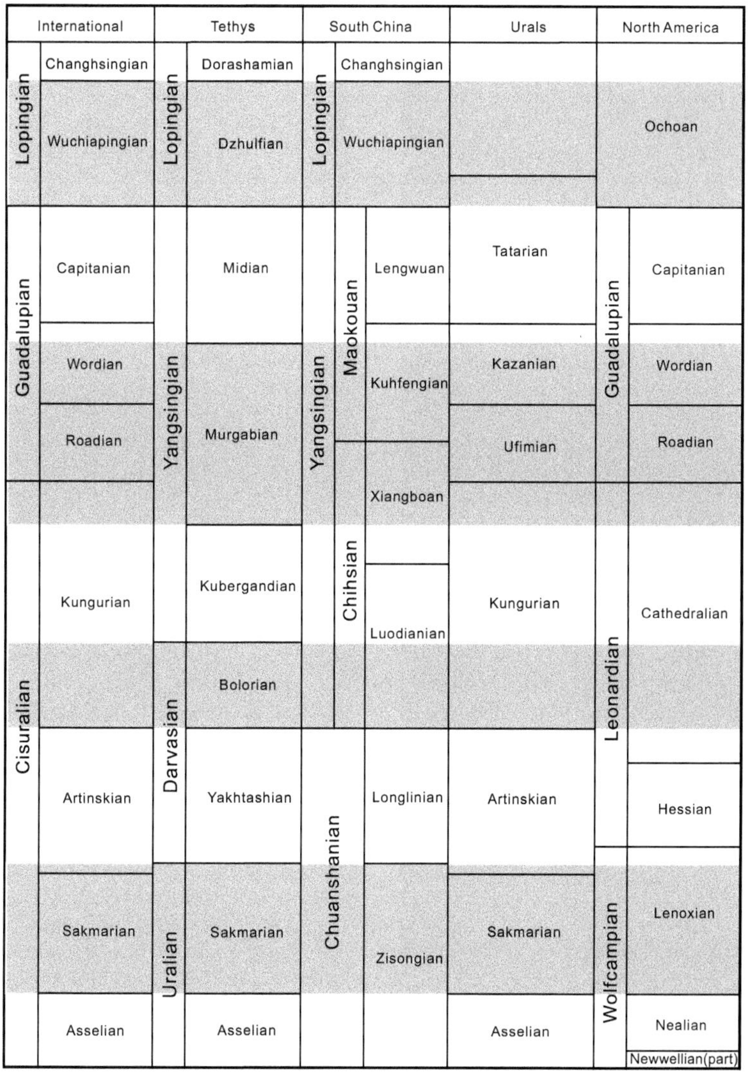

International		Tethys		South China			Urals	North America	
Lopingian	Changhsingian	Lopingian	Dorashamian	Lopingian		Changhsingian			Ochoan
	Wuchiapingian		Dzhulfian			Wuchiapingian			
Guadalupian	Capitanian	Yangsingian	Midian	Yangsingian	Maokouan	Lengwuan	Tatarian	Guadalupian	Capitanian
	Wordian		Murgabian			Kuhfengian	Kazanian		Wordian
	Roadian						Ufimian		Roadian
Cisuralian					Chihsian	Xiangboan			
	Kungurian		Kubergandian			Luodianian	Kungurian	Leonardian	Cathedralian
		Darvasian	Bolorian						
	Artinskian		Yakhtashian	Chuanshanian		Longlinian	Artinskian		Hessian
	Sakmarian	Uralian	Sakmarian			Zisongian	Sakmarian	Wolfcampian	Lenoxian
	Asselian		Asselian				Asselian		Nealian
									Newwellian(part)

Fig. 8. Correlations of different Permian chronostratigraphic subdivisions between different regions/scales.

(*Levenella*), *Pamirina* (*Pamirina*), *Praeskinnerella*, *Mesoschubertella* and *Toriyamaia*. The base of the Yakhtashian was defined by the occurrence of *Pamirina* (*Levenella*) or *Chalaroschwagerina* (Leven 2004). The *Chalaroschwagerina–Laxifusulina* fauna and the *Pamirina* fauna were considered to be contemporaneous but under a different depositional environment (Xia *et al.* 1986; Xu *et al.* 1986), although it is clear that the first occurrence of *Chalaroschwagerina* was earlier than that of *Pamirina* (*Levenella*) or *Pamirina* (Xu *et al.* 1986; Leven *et al.* 1992; Ueno 1996). The Longlinian Stage in South China is characterized by the *Pamirina–Darvasites* assemblage, which is contemporaneous with the Yakhtashian Stage of the Tethyan scale (Huang *et al.* 1982).

As mentioned earlier, the upper part of the Sakmarian of the Tethyan scale corresponds to the lowermost part of the Artinskian of the international scale. Consequently, the Yakhtashian Stage of the Tethyan scale is Artinskian in age.

Pamirina (*Levenella*) and *Pamirina* (*Pamirina*) have not been documented in America so far. This has hampered a direct correlation of chronostratigraphy between Tethys and North America. However, abundant *Chalaroschwagerina* occurs in the upper part of the Wolfcampian Series (Skinner & Wilde 1965; Magginetti *et al.* 1988; Ross & Ross

1994). In addition, the *Praeskinnerella* species, common in the Yakhtashian in the Tethyan region, are equal to the fusulines from the Hessian Stage of the Leonardian Series in the fusuline evolutionary stage (Ross & Ross 1997). Consequently, the Yakhtashian Stage broadly corresponds to the upper part of the Lenoxian Stage and the Hessian Stage (Leven 2004).

Bolorian

The evolution of *Pamirina* gave rise to *Misellina* (*Brevaxina*) and *Misellina*, which is the basis for the recognition of the base of the Bolorian Stage of the Tethyan scale (Leven 2004) or the Luodianian Stage of the South China scale (Sheng & Jin 1994). Common genera that prevailed in the Bolorian Stage include *Misellina* (*Brevaxina*), primitive *Misellina*, *Chalaroschwagerina*, *Darvasites*, *Leeina* and *Praeskinnerella* (Leven 2004). In correlation, the Luodianian of South China contains the *Misellina* (*Brevaxina*) *dyhrenfurthi*, *Misellina termieri*, *M. claudiae* and *Shengella simplex* zones, in ascending order (Xiao *et al.* 1986; Sheng & Jin 1994), which is longer than the Bolorian Stage. Consequently, the Bolorian Stage corresponds to the lower part of the Luodianian Stage in South China (Fig. 8).

The Bolorian of the Tethyan scale is hardly correlatable with the American scale owing to the lack of *Misellina* (*Brevaxina*) and *Misellina* in North America. In terms of the conodonts and ammonoids, the Bolorian Stage corresponds to part of the Cathedralian Stage of the American scale (Jin *et al.* 1997; Leven 2004).

Kubergandian

The Kubergandian Stage is typified by advanced forms of *Misellina*, such as *Misellina ovalis* and *M. claudiae*, and primitive neoschwagerinids, such as *Cancellina* successively (Leven 2004). Common genera found in the Kubergandian Stage include *Misellina*, *Armenina*, *Pseudodoliolina*, *Yangchienia*, *Maklaya*, *Cancellina*, *Nankinella*, *Neofusulinella*, *Pisolina*, *Sphaerulina* and *Parafusulina*. The Xiangboan Stage in South China contains two zones, the *Cancellina liuzhiensis* and the *Neoschwagerina simplex* zones, based on the sections from the Houchang area (Xiao *et al.* 1986). Consequently, the Kubergandian corresponds to the upper part of the Luodianian Stage and the lower part of the Xiangboan Stage.

The correlation of the Kubergandian Stage with the international scale has been a contentious issue (Leven 2004; Leven & Gorgij 2011*a*; Wang *et al.* 2011*a*; Angiolini *et al.* 2015). This stage was originally correlated with the Roadian Stage (Jin *et al.* 1997); however, the latest study of the strata in both South China and Japan confirmed that the conodont *Mesogondolella siciliensis*–*Sweetognathus guizhouensis* coexists with *Neoschwagerina simplex* of typical Murgabian species (Henderson & Mei 2003; Shen *et al.* 2013). These facts indicate that the Kubergandian Stage falls within the Kungurian Stage of the international scale. However, the Kubergandian Stage was correlated to the lower Roadian by Davydov *et al.* (2013) and to the upper Kungurian–early Roadian by Angiolini *et al.* (2015). More work is required in the future to verify their precise correlations.

The correlation of the Kubergandian Stage with the American scale is still not easily determined because the fusulines from North America are dominated by *Parafusulina* from the Cathedralian Stage to the Wordian Stage (Wilde 1990; Yang & Yancey 2000; Ross & June 2003). However, some common species such as *Parafusulina bakeri* documented from both South China and North America may indicate that the Kubergandian Stage can be correlated broadly with the upper part of the Cathedralian Stage (Z.H. Zhang *et al.* 1988; Ross & June 2003).

Murgabian. The Murgabian is characterized by the appearance of *Neoschwagerina*, which might be evolved from the *Maklaya* of the underlying Kubergandian Stage (Yang 1985; Leven 2004). In addition, the genus *Cancellina* evolved to *Presumatrina*, *Afghanella* and *Sumatrina* succeedingly during the early Murgabian (Yang 1985). Other genera include *Verbeekina*, *Armenina*, *Cancellina*, *Yangchienia*, *Chusenella* and *Parafusulina*, which are similar to the underlying Kubergandian Stage. However, the schwagerinid genus *Eopolydiexodina* developed well in bi-temperate regions in the Tethyan region during the Murgabian (Ueno 2003).

The Guadalupian Series of North America was adopted as the candidate Middle Permian chronostratigraphy in the international scale. However, a direct correlation between the Middle Permian Tethyan scale and the North American (international) scale is difficult because the Guadalupian Series in North America is dominated by abundant schwagerinids, such as *Parafusulina* and *Polydiexodina*, whereas in the Tethys it is dominated by abundant neoschwagerinids. However, conodonts and ammonites play important roles in this correlation. As mentioned previously, *Neoschwagerina simplex*, an index species of the base Murgabian, was found to coexist with Kungurian conodonts (Henderson & Mei 2003; Shen *et al.* 2013). Consequently, the lowermost Murgabian may be correlated with the upper part of the Kungurian of the international scale. With regard to the upper boundary of the Murgabian, both conodonts and ammonites proved that it falls within the Wordian Stage (e.g. Rui & Nassichuk 1996; Henderson & Mei

2003). However, this assignment was challenged by Angiolini *et al.* (2015), who claimed that the uppermost Murgabian fusulines coexist with Capitanian conodonts. Such a discrepancy requires more work in the future.

As previously stated, *Neoschwagerina simplex* occurred in the upper part of the Xiangboan Stage (Xiao *et al.* 1986). Thus, the lower part of the Murgabian is correlatable with the upper part of the Xiangboan Stage. Upwards, the Xiangboan Stage in China was succeeded by the Khufengian Stage (Sheng & Jin 1994). The Kuhfengian Stage was originated from the Kuhfeng Formation, which is characterized by ammonoids, conodonts, brachiopods and radiolarians (Sheng & Jin 1994). The equivalent strata in the Houchang section in the Guizhou Province bear two zones: the *Neoschwagerina craticulifera* Zone and the *N. margaritae* Zone (Xiao *et al.* 1986; L.X. Zhang *et al.* 1988). Because the *Neoschwagerina margaritae* Zone was regarded as the base Midian Stage (Leven 1996; Ueno *et al.* 2003), the upper boundary of the Murgabian Stage thus falls within the upper part of the Kuhfengian Stage. In conclusion, the Murgabian of the Tethyan scale corresponds to the upper Xiangboan Stage and the lower Kuhfeng Stage of South China (Fig. 8).

Midian. The base of the Midian Stage is characterized by the occurrence of the genus *Yabeina* (Leven 2004). This genus represents the most advanced forms of neoschwagerinids. In the Akasaka limestone of Japan, *Yabeina* occurs in the *Neoschwagerina margaritae* Zone (Zaw 1999). In South China, *Yabeina* first occurs in the upper part of the Houziguan limestone (Xiao *et al.* 1986). In addition, *Yabeina* or *Lepidolina* occur widely in Western Tethyan regions (e.g. Skinner & Wilde 1967), the Gondwanan Tethyan area (Zhu 1982*b*) and Panthalassan regions (e.g. Zaw 1999; Kobayashi *et al.* 2007; Kobayashi 2011). Common genera of the Midian Stage include *Yabeina*, *Lepidolina*, *Metadoliolina*, *Dunbarula*, *Kahlerina*, *Lantschichites*, *Reichelina*, *Codonofusiella* and *Paradoxiella*.

As mentioned above, the lower boundary of the Midian Stage corresponds to the upper Wordian of the international scale or the upper Kuhfeng Stage of the South China scale based on the evidence of coexisting conodonts and ammonoids (e.g. Rui & Nassichuk 1996; Henderson & Mei 2003). Thus, the middle and upper part of the Midian Stage is equal to the Lengwuan Stage of South China and the Capitanian Stage of the North American (international) scale because both the Lengwuan Stage and the Capitanian Stage possess *Yabeina*, *Reichelina*, *Kahlerina* and *Codonofusiella* typical of the Tethyan Midian Stage.

The upper boundary of the Midian Stage correlates well with the upper boundary of the Capitanian Stage or the Lengwuan Stage because this boundary marks the greatest mass extinction event for fusulines, especially for those schwagerinids and neoschwagerinids (Jin *et al.* 1994; Yang *et al.* 2004). A more detailed study has proved that the loss of schwagerinids and neoschwagerinds takes place prior to the Guadalupian–Lopingian boundary (Ota & Isozaki 2006; Groves & Wang 2013). This biostratigraphic information was well documented from both the Panthalassan region and the North American region, where many smaller fusulines, such as *Reichelina*, *Codonofusiella*, *Paradoxiella*, occurred in the latest Midian and latest Capitanian stages (Wilde & Rudine 2000; Yang & Yancey 2000; Nestell & Nestell 2006; Ota & Isozaki 2006).

Dzhulfian

Because of the extinction of the neoschwagerinids and schwagerinids during the latest Midian, the remaining fusulines in the Dzhulfian of the late Permian have reduced greatly in diversity. The most dominant genera during the Dzhulfian Stage are *Codonofusiella*, *Nankinella* and *Reichelina* (Sheng 1963; Rui 1979). In the meantime, the genera, such as *Gallowayinella*, *Tewoella*, *Paradunbarula* and *Palaeofusulina*, started to diversify (Leven 2004).

The stratotypes for the Wuchiapingian and Changhsingian stages are from South China (Jin *et al.* 2006*a*, *b*). The base of the Changhsingian Stage was defined by the first occurrence of *Clarkina wangi* (Jin *et al.* 2003). This conodont corresponds to the first occurrence of the fusuline *Palaeofusulina minima* (Jin *et al.* 2003). In addition, *Gallowayinella meitianensis*, underlying the *Palaeofusulina minima* Zone, coexists with *Clarkina orientalis* and *C. liangshanensis* of the upper Wuchiapingian (Wang *et al.* 1997). Consequently, the Dzhulfian Stage of the Tethyan scale is equal to the Wuchiapingian Stage of the South China scale (international scale).

Dorashamian

The Dorashamian Stage is the uppermost stage of the Permian. During this time, the genus *Palaeofusulina* has evolved to the advanced *Palaeofusulina sinensis* assemblage (Sheng & Rui 1984; Sheng & Jin 1994). The associated genera include *Nanlingella*, *Parananlingella*, *Nankinella*, *Sphaerulina*, *Reichelina*, *Staffella* and *Parareichelina* (Leven 2004). As addressed above, the Dorashamian Stage may correspond to the Changhsingian Stage of the South China scale (international scale) (Leven & Gorgij 2011*a*). All these fusulines became extinct by the end of the Dorashamian/Changhsingian because of the onset of the Permian/Triassic mass extinction event (Jin *et al.* 2000; Shen *et al.* 2011).

Summary

Fusulines are a kind of larger benthic foraminifer living in tropical and subtropical environments during the Permian. The rapid evolutionary rate and distinct biostratigraphic subdivision play an important role in global stratigraphic correlation.

Overall, there are several important evolutionary key points that permit the biostratigraphic boundaries to be defined. The advent of *Pseudoschwagerina* or *Sphaeroschwagerina* marks the base of the Asselian Stage. These pseudoschwagerinids gave rise to *Zellia* and *Robustoschwagerina* during the Sakmarian. During the Yakhtashian, *Levenella* and *Pamirina* originated from the Tethyan region and formed the basis of the neoschwagerinids. The following first occurrences of *Misellina (Brevaxina)*, advanced *Misellina*, *Neoschwagerina* and *Yabeina* define the base of the Bolorian, Kubergandian, Murgabian and Midian stages, respectively. Both neoschwagerinids and schwagerinids became extinct by the end of the Midian. The remaining fusulines during the Dzhulfian were small-sized *Codonofusiella* and *Reichelina*, and a few newly formed genera such as *Palaeofusulina* and *Gallowayinella*. *Palaeofusulina*, in particular, has evolved to *Palaeofusulina sinensis*, the index fossil for the Dorashamian/Changhsingian Stage. In summary, the evolutionary milestones of fusulines have formed the basis of fusuline biostratigraphy across the Tethyan region.

By contrast, the evolution of fusulines in the North American Craton region has a different story. The main difference between the Tethyan region and the North American Craton region lies in the faunal compositions from the Leonardian Stage to the Capitanian Stage. The fusulines of North America during this interval are characterized by the dominance of schwagerinids such as *Parafusulina* and *Polydiexodina*. However, *Parafusulina* plays a subordinate role from the Kubergandian to the Midian in the Tethyan region. Therefore, the direct and precise correlation of middle Permian fusuline biostratigraphy between the Tethyan region and the North American Craton region requires further detailed investigation.

We thank Prof. Daniel Vachard and Dr Yukun Shi for their critical reviews, which improved the manuscript greatly. We also thank Prof. Spencer Lucas and Prof. Shuzhong Shen for inviting our paper. This paper is jointly funded by the Strategic Priority Research Programme (B) of the Chinese Academy of Sciences (XDB18000000 and XDB03010102), the National Science Foundation of China (41420104003, 41290260, 41472029 and 41372024), the Ministry of Science and Technology of China (2015FY310100-9), and the Youth Innovation Promotion Association CAS for ZYC (2016282).

References

Alksne, A.E. & Isakova, T.N. 1980. O kompleksakh fuzulinid pogranichnyh otlozhenii Gzheliskogo i Asseliskogo yarusov nekotorykh razrezov yuzhnogo Urala i Russkoi Platformy. *Voprosy Mikropaleontologii*, **23**, 52–62 [in Russian].

Amiot, M., Ciry, R. *et al.* 1965. *Fossils of Karakorum and Chitral*. E.J. Brill, Leiden, The Netherlands.

Angiolini, L., Stephenson, M.H. & Leven, E.J. 2006. Correlation of the Lower Permian surface Saiwan Formation and subsurface Haushi limestone, Central Oman. *Geoarabia*, **11**, 17–38.

Angiolini, L., Zanchi, A. *et al.* 2015. From rift to drift in South Pamir (Tajikistan): Permian evolution of a Cimmerian terrane. *Journal of Asian Earth Sciences*, **102**, 146–169, https://doi.org/10.1016/j.jseaes.2014.08.001

Angiolini, L., Campagna, M. *et al.* 2016. Brachiopods from the Cisuralian–Guadalupian of Darvaz, Tajikistan and implications for Permian stratigraphic correlations. *Palaeoworld*, **25**, 539–568, https://doi.org/10.1016/j.palwor.2016.05.006

Aw, P.C., Ishii, K. & Okimura, Y. 1977. On Palaeofusulina–Colaniella fauna from the Upper Permian of Kelantan, Malaysia. *Transactions and Proceedings of the Palaeontological Society of Japan, New Series*, **1977**, (104), 407–417.

Baghbani, D. 1993. The Permian sequence in the Abadeh region, Central Iran. *In*: *Bibliography of South American Geology: Volume II, Bolivia, Brazil, Chile, Colombia*. Earth Sciences and Resources Institute (ERSI), University of South Carolina, Occasional Publications, 7–22.

Bensh, F.R. 1972. *Stratigrafiya i fuzulinidy verkhnego paleozoya yuzhnoi Fergany*. Akademiya nauk uzbekskoi SSR, institut geologii i geofiziki, Tashkent [in Russian].

Chen, Q.H., Wang, J.P., Wang, S.L. & Wu, K.Y. 1999. Discovery of the Upper Permian series in Cuoqin Basin, Xizang (Tibet) and its geological significance. *Chinese Science Bulletin*, **44**, 1520–1523.

Cheng, L.R., Li, C., Zhang, Y.C. & Wu, S.Z. 2005. The *Polydiexodina* (Fusulinids) fauna from central Qiangtang, Tibet, China. *Acta Micropalaeontologica Sinica*, **22**, 152–162.

Choi, D.R. 1970. Permian fusulinids from Imo, southern Kitakami Mountains, N. E. Japan. *Journal of the Faculty of Science, Hokkaido University, Series 4. Geology and Mineralogy*, **14**, 327–354.

Choi, D.R. 1972. Classification and phylogeny of genus *Misellina* with description of some *Misellina* from the Lower Permian in the Southern Kitakami Mountains, Japan. *Journal of the Faculty of Science, Hokkaido University, Series 4. Geology and Mineralogy*, **15**, 625–645.

Choi, D.R. 1973. Permian Fusulinids from the Setamai-Yahagi district, southern Kitakami Mountains, N.E. Japan. *Journal of the Faculty of Science, Hokkaido University, Series 4, Geology and Mineralogy*, **16**, 1–90.

Chuvashov, B.I., Dupina, G.V., Mizens, G.A. & Chernyh, V.V. 1990. *Opornye razrezy verhnego Karbona i nizhnei Permi zapadnogo sklona urala i priuraliya.*

Akademiya Nauk SSSR. Uraliskoe Otdelenie, Sverdlovsk, Russia [in Russian].

CIRY, R. & AMIOT, M. 1965. Sur Quelques Foraminifères Permiens D'Asie Centrale. *In*: AMIOT, M., CIRY, R. ET AL. (eds) *Fossils of Karakorum and Chitral*. E.J. Brill, Leiden, 127–133.

COLPAERT, C., MONNET, C. & VACHARD, D. 2015. Eopolydiexodina (Middle Permian giant fusulinids) from Afghanistan: biometry, morphometry, paleobiogeography, and end-Guadalupian events. *Journal of Asian Earth Sciences*, **102**, 127–145, https://doi.org/10.1016/j.jseaes.2014.10.028

DAVYDOV, V.I. 1990*a*. Clarification of the origin and phylogeny of *Triticites* and of the boundary of the middle and upper Carboniferous. *Paleontological Journal*, **2**, 39–51.

DAVYDOV, V.I. 1990*b*. Zanalinoe Delenie Gzheliskogo yarusa v Donbasse i Preddoneckom protibe po fuzulinidam. *In*: BOGDANOVA, T.N. & BUGROVA, E.M. (eds) *Problemy sovremennoi mikropaleontologii*. 'Nauka' Leningradskoe otdelenie, Leningrad, Russia, 52–63.

DAVYDOV, V.I. 2013. The GSSP at the Aidaralash section is solid and has no alternative. *Permophiles*, **58**, 13–15.

DAVYDOV, V.I. & AREFIFARD, S. 2007. Permian fusulinid fauna of Peri-Gondwanan affinity from the Kalmard region, east-central Iran and its significance for tectonics and paleogeography. *Palaeontologia Electronica*, **10**, 1–40.

DAVYDOV, V.I. & AREFIFARD, S. 2013. Middle Permian (Guadalupian) fusulinid taxonomy and biostratigraphy of the mid-latitude Dalan Basin, Zagros, Iran and their applications in paleoclimate dynamics and paleogeography. *GeoArabia*, **18**, 17–62.

DAVYDOV, V.I., SNYDER, W.S. & SPINOSA, C. 1997. Upper Paleozoic Fusulinacean Biostratigraphy of the Southern Urals. *Permophiles*, **30**, 11–14.

DAVYDOV, V., KRAINER, K. & CHERNYKH, V. 2013. Fusulinid biostratigraphy of the Lower Permian Zweikofel Formation (Rattendorf Group; Carnic Alps, Austria) and Lower Permian Tethyan chronostratigraphy. *Geological Journal*, **48**, 57–100, https://doi.org/10.1002/gj.2433

DAWSON, O. 1993. Fusuline foraminiferal biostratigraphy and carbonate facies of the Permian Ratburi Limestone, Saraburi, central Thailand. *Journal of Micropalaeontology*, **12**, 9–33, https://doi.org/10.1144/jm.12.1.9

DING, P.Z., JIN, T.A. & SUN, X.F. 1989. Carboniferous–Permian boundary of Xikou area of Zhen'an, south Shanxi, east Qinling Range. *Compte Rendu Congrès International de Stratigraphie et de Géologie du Carbonifère*, **2**, 199–206.

DOUGLASS, R.C. 1970. Morphologic study of fusulinids from the Lower Permian of West Pakistan. United States Geological Survey, Professional Papers, **643**, 1–11.

DUNBAR, C.O. & NEWELL, N.D. 1946. Marine Early Permian of the central Andes and its fusuline faunas. Part II. *American Journal of Science*, **244**, 457–491.

DUNBAR, C.O. & SKINNER, J.W. 1937. Permian Fusulinidae of Texas. *The University of Texas Bulletin*, **3701**, 517–825.

EHIRO, M. & MISAKI, A. 2004. Stratigraphic range of the genus *Monodiexodina* (Permian Fusulinoidea): additional data from the Southern Kitakami Massif, Northeast Japan. *Journal of Asian Earth Sciences*, **23**, 483–490.

FAN, J.C. 1993. Bryozoans of Late Carboniferous–early Early Permian in Tengchong area of western Yunnan. *Yunnan Geology*, **12**, 383–406.

FONTAINE, H. & SUTEETHORN, V. 1988. Late Palaeozoic and Mesozoic fossils of West Thailand and their environments. *CCOP Technical Bulletin*, **20**, 1–121.

FONTAINE, H., PRINYA, P. & VACHARD, D. 1999. A microfossil-rich Kubergandian (Lower Middle Permian) limestone of southern Laos. *Proceedings of the International Symposium on Shallow Tethys*, **5**, 170–178.

FONTAINE, H., SALYAPONGSE, S., TIEN, N.D. & VACHARD, D. 2002. The Permian of Khao Tham Yai area in Northeast Thailand. *In*: MANTAJIT, N. (ed.) *The Symposium on Geology of Thailand: 26–31 August 2002, Bangkok, Thailand*. Department of Mineral Resources, Bangkok, Thailand, 58–76.

FONTAINE, H., HOANG, T.T., KAVINATE, S., SUTEETHORN, V. & VACHARD, D. 2013. Upper Permian (Late Changhsingian) marine strata in Nan Province, northern Thailand. *Journal of Asian Earth Sciences*, **76**, 115–119, https://doi.org/10.1016/j.jseaes.2013.01.006

FORKE, H.C. 2002. Biostratigraphic subdivision and correlation of Uppermost Carboniferous/Lower Permian sediments in the Southern Alps: fusulinoidean and condont faunas from the Carnic Alps (Austria/Italy), Karavanke Mountains (Slovenia), and Southern Urals (Russia). *Facies*, **47**, 201–275.

GAETANI, M. & LEVEN, E. 1993. Permian stratigraphy and fusulinids from Rosh Gol (Chitral, E Hindu Kush). *Rivista Italiana di Paleontologia e Stratigrafia*, **99**, 307–326.

GAETANI, M. & LEVEN, E.J. 2014. The Permian succession of the Shaksgam valley, Sinkiang (China). *Italian Journal of Geosciences*, **133**, 45–62, https://doi.org/10.3301/IJG.2013.10

GAETANI, M., ANGIOLINI, L. ET AL. 2009. Pennsylvanian–Early Triassic stratigraphy in the Alborz Mountains (Iran). *In*: BRUNET, M.-F., WILMSEN, M. & GRANATH, J.W. (eds) *South Caspian to Central Iran Basins*. Geological Society, London, Special Publications, **312**, 79–128, https://doi.org/10.1144/SP312.5

GHAZZAY, W., VACHARD, D. & RAZGALLAH, S. 2015. Revised fusulinid biostratigraphy of the Middle–Late Permian of Jebel Tebaga (Tunisia). *Revue de Micropaleontologie*, **58**, 57–83, https://doi.org/10.1016/j.revmic.2015.04.001

GROVES, J.R. & WANG, Y. 2013. Timing and size selectivity of the Guadalupian (Middle Permian) fusulinoidean extinction. *Journal of Paleontology*, **87**, 183–196, https://doi.org/10.1666/12-076R.1

GUBLER, J. 1935. Les Fusulinidés Permiens de l'Indochine. *Mémoires de la Société Géologique de France*, **26**, 5–173.

HAIG, D.W., MORY, A.J. ET AL. 2017. Late Artinskian–Early Kungurian (Early Permian) warming and maximum marine flooding in the East Gondwana interior rift, Timor and Western Australia, and comparisons across East Gondwana. *Palaeogeography, Palaeoclimatology, Palaeoecology*, **468**, 88–121, https://doi.org/10.1016/j.palaeo.2016.11.051

HAN, J.X. 1975. Late Carboniferous fusulinids from the Amushan area, Wulanchabu district, Inner Mongolia. *Papers of Stratigraphy and Paleontology*, **1**, 132–203.

HAUSER, M., VACHARD, D., MARTINI, R., MATTER, A., PETERS, T. & ZANINETTI, L. 2000. The Permian sequence reconstructed from reworked carbonate clasts in the Batain Plain (northeastern Oman). *Comptes Rendus de l'Académie des Sciences, Series IIA – Earth and Planetary Science*, **330**, 273–279.

HENDERSON, C.M. & MEI, S.L. 2003. Stratigraphic v. environmental significance of Permian serrated conodonts around the Cisuralian–Guadalupian boundary: new evidence from Oman. *Palaeogeography, Palaeoclimatology, Palaeoecology*, **191**, 301–328, https://doi.org/10.1016/S0031-0182(02)00669-7

HONJO, S. 1959. Neoschwagerinids from the Akasaka Limestone (A Paleontological study of the Akasaka Limestone, 1st Report). *Journal of the Faculty of Science, Hokkaido University, Series 4, Geology and Mineralogy*, **10**, 111–162.

HORNIBROOK, N.B. 1951. Permian fusulinid foraminifera from the North Auckland Peninsula, New Zealand. *Transactions and Proceedings of the Royal Society of New Zealand*, **79**, 319–321.

HUANG, H., JIN, X.C. & YANG, X.N. 2007. Middle Permian fusulinids from the Xainza area of the Lhasa Block, Tibet. *Acta Palaeontologica Sinica*, **46**, 62–74 [in Chinese with English abstract].

HUANG, H., JIN, X.C., SHI, Y.K. & YANG, X.N. 2009. Middle Permian western Tethyan fusulinids from southern Baoshan block, western Yunnan, China. *Journal of Paleontology*, **83**, 880–896.

HUANG, H., JIN, X. & SHI, Y. 2015*a*. A *Verbeekina* assemblage (Permian fusulinid) from the Baoshan Block in western Yunnan, China. *Journal of Paleontology*, **89**, 269–280, https://doi.org/10.1017/jpa.2014.24

HUANG, H., SHI, Y. & JIN, X. 2015*b*. Permian fusulinid biostratigraphy of the Baoshan Block in western Yunnan, China with constraints on paleogeography and paleoclimate. *Journal of Asian Earth Sciences*, **104**, 127–144, https://doi.org/10.1016/j.jseaes.2014.10.010

HUANG, H., SHI, Y.K. & JIN, X.C. 2017. Permian (Guadalupian) fusulinids of Bawei Section in Baoshan Block, western Yunnan, China: biostratigraphy, facies distribution and paleogeographic discussion. *Palaeoworld*, **26**, 95–114, https://doi.org/10.1016/j.palwor.2016.03.004

HUANG, Z.X. & ZENG, X.L. 1984. The early Early Permian (Longlinian stage) fusulinid fauna from Longlin, Guangxi. *Earth Science – Journal of Wuhan College of Geology*, 11–24 [in Chinese with English abstract].

HUANG, Z.X., SHI, Y. & WEI, M.C. 1982. A new stratigraphic unit of Permian–Longlinian stage. *Journal of Chengdu College of Geology*, 63–73 [in Chinese with English abstract].

IGO, H. 1957. Fusulinids of Fukuji, southeastern part of the Hida Massif, central Japan. *Science Reports of the Tokyo Kyoiku Daigaku, Section C, Geology, Mineralogy and Geography*, **5**, 153–246.

IGO, H. 1959. Some Permian fusulinids from the Hirayu district, southeastern part of the Hida Massif, central Japan. *Science Reports of the Tokyo Kyoiku Daigaku, Section C, Geology, Mineralogy and Geography*, **6**, 231–254.

IGO, H. 1966. Some Permian fusulinids from Pahang, Malaya. *In*: KOBAYASHI, T. & TORIYAMA, R. (eds) *Geology and Palaeontology of Southeast Asia, Volume 13*. University of Tokyo Press, Tokyo, Japan, 30–38.

IGO, H. 1972. Fusulinacean fossils from Thailand. part 6. Fusulinacean fossils from north Thailand. *Geology and Palaeontology of Southeast Asia*, **10**, 63–116.

IGO, H., UENO, K. & SASHIDA, K. 1993. Lower Permian Fusulinaceans from Ban Phia, Changwat Loei, Northeastern Thailand. *Transactions and Proceedings of the Palaeontological Society of Japan, New Series*, **169**, 15–43.

INGAVAT, R. & DOUGLASS, R.C. 1981. Fusuline fossils from Thailand, part 14. The fusulinid genus *Monodiexodina* from Northwest Thailand. *Geology and Palaeontology of Southeast Asia*, **22**, 23–34.

INGAVAT, R., TORIYAMA, R. & PITAKPAIVAN, K. 1980. Fusuline zonation and faunal characteristics of the Ratburi limestone in Thailand and its equivalents in Malaysia. *Geology and Palaeontology of Southeast Asia*, **21**, 43–62.

ISHII, K.I. 1966. Preliminary notes of the Permian fusulinids of H. S. Lee Mine No. 8 limestone near Kampar, Perak, Malaya. *Journal of Geoscience, Osaka City University*, **9**, 145.

JIANG, H.C., WANG, M.Z., ZHANG, X.Q. & LI, Z.X. 2001. Study on the fusulinida fauna in the Taiyuan Formation from the east area of Jining coalfield, Shandong province. *Acta Palaeontologica Sinica*, **40**, 514–522 [in Chinese with English abstract].

JIN, X.C. 2002. Permo-Carboniferous sequences of Gondwana affinity in southwest China and their paleogeographic implications. *Journal of Asian Earth Sciences*, **20**, 633–646.

JIN, X.C., HUANG, H., SHI, Y.K. & ZHAN, L.P. 2011. Lithologic boundaries in Permian post-glacial sediments of the Gondwana-affinity regions of China: typical sections, age range and correlation. *Acta Geologica Sinica* (English edn), **85**, 373–386.

JIN, Y.G., ZHANG, J. & SHANG, Q.H. 1994. Two phases of the end-Permian Mass Extinction. *In*: EMBRY, A.F., BEAUCHAMP, B. & GLASS, D.J. (eds) *Pangea: Global Environments and Resources*. Canadian Society of Petroleum Geologists, Memoirs, **17**, 813–822.

JIN, Y.G., WARDLAW, B.R., GLENISTER, B.F. & KOTLYAR, G.V. 1997. Permian chronostratigraphic subdivisions. *Episodes*, **20**, 6–10.

JIN, Y.G., SHANG, Q.H., WANG, X.D., WANG, Y. & SHENG, J.Z. 1999. Chronostratigraphic subdivision and correlation of the Permian in China. *Acta Geologica Sinica*, **73**, 127–138.

JIN, Y.G., WANG, Y., WANG, W., SHANG, Q.H., CAO, C.Q. & ERWIN, D.H. 2000. Pattern of marine mass extinction near the Permian–Triassic boundary in South China. *Science*, **289**, 432–436.

JIN, Y.G., HENDERSON, C. ET AL. 2003. Proposal for the Global Stratotype Section and Point (GSSP) for the Wuchiapingian–Changhsingian Stage boundary

(Upper Permian Lopingian Series). *Permophiles*, **43**, 8–23.

JIN, Y.G., SHEN, S.Z. *ET AL.* 2006*a*. The Global Stratotype Section and Point (GSSP) for the boundary between the Capitanian and Wuchiapingian stage (Permian). *Episodes*, **29**, 253–262.

JIN, Y.G., WANG, Y., HENDERSON, C., WARDLAW, B.R., SHEN, S.Z. & CAO, C.Q. 2006*b*. The Global Boundary Stratotype Section and Point (GSSP) for the base of Changhsingian Stage (Upper Permian). *Episodes*, **29**, 175–182.

JOHNSTON, S.T. & BOREL, G.D. 2007. The odyssey of the Cache Creek terrane, Canadian Cordillera: implications for accretionary orogens, tectonic setting of Panthalassa, the Pacific superwell, and break-up of Pangea. *Earth and Planetary Science Letters*, **253**, 415–428.

KÖYLÜOGLU, M. & ALTINER, D. 1989. Micropaléontologie (Foraminifères) et biostratigraphie du Permien supérieur de la région d'Hakkari (SE Turquie). *Revue de Paléobiologie*, **8**, 467–503.

KANMERA, K. & MIKAMI, T. 1965. Fusuline zonation of the Lower Permian Sakamotozawa Series. *Memories of Faculty of Science, Kyushu University, Series D, Geology*, **16**, 275–320.

KOBAYASHI, F. 1997*a*. Middle Permian biogeography based on fusulinacean faunas. *In*: ROSS, C.A., ROSS, J.R.P. & BRENCKLE, P.L. (eds) *Late Paleozoic Foraminifera: Their Biogeography, Evolution, and Paleoecology; and the Mid-Carboniferous Boundary*. Cushman Foundation for Foraminifera Research, Special Publications, **36**, 73–76.

KOBAYASHI, F. 1997*b*. Upper Permian foraminifers from the Iwai-Kanyo area, west Tokyo, Japan. *Journal of Foraminiferal Research*, **27**, 186–195, https://doi.org/10.2113/gsjfr.27.3.186

KOBAYASHI, F. 1999. Tethyan uppermost Permian (Dzhulfian and Dorashamian) foraminiferal faunas and their paleogeographic and tectonic implications. *Palaeogeography, Palaeoclimatology, Palaeoecology*, **150**, 279–307.

KOBAYASHI, F. 2002. Lithology and foraminiferal fauna of allochthonous limestones (Changhsingian) in the upper part of the Toyoma Formation in the South Kitakami Belt, Northeast Japan. *Paleontological Research*, **6**, 331–342.

KOBAYASHI, F. 2005. Permian foraminifers from the Itsukaichi-Ome area, west Tokyo, Japan. *Journal of Paleontology*, **79**, 413–432, https://doi.org/10.1666/0022-3360(2005)0792.0.co;2

KOBAYASHI, F. 2011. Permian fusuline faunas and biostratigraphy of the Akasaka limestone (Japan). *Revue de Paléobiologie*, **30**, 431–574.

KOBAYASHI, F. 2012. Middle and Late Permian foraminifers from the Chichibu belt, Takachiho area, Kyushu, Japan: implications for faunal events. *Journal of Paleontology*, **86**, 669–687, https://doi.org/10.1666/11-049r.1

KOBAYASHI, F. & ALTINER, D. 2008. Late Carboniferous and Early Permian fusulinoideans in the central Taurides, Turkey: biostratigraphy, faunal composition and comparison. *Journal of Foraminiferal Research*, **38**, 59–73, https://doi.org/10.2113/gsjfr.38.1.59

KOBAYASHI, F. & ALTINER, D. 2011. Discovery of the Lower Murgabian (Middle Permian) based on Neoschwagerinids and Verbeekinids in the Taurides, southern Turkey. *Rivista Italiana di Paleontologia e Stratigrafia*, **117**, 39–50.

KOBAYASHI, F. & ISHII, K.I. 2003. Paleobiogeographic analysis of Yakhtashian to Midian fusulinacean faunas of the Surmaq formation in the Abadeh region, central Iran. *Journal of Foraminiferal Research*, **33**, 155–165, https://doi.org/10.2113/0330155

KOBAYASHI, F., ROSS, C.A. & ROSS, J.R.P. 2007. Age and generic assignment of *Yabeina columbiana* (Guadalupian Fusulinacea) in southern British Columbia. *Journal of Paleontology*, **81**, 238–253, https://doi.org/10.1666/0022-3360(2007)81[238:aagaoy]2.0.co;2

KOBAYASHI, F., SHIINO, Y. & SUZUKI, Y. 2009. Middle Permian (Midian) foraminifers of the Kamiyasse Formation in the Southern Kitakami Terrane, NE Japan. *Paleontological Research*, **13**, 79–99, https://doi.org/10.2517/1342-8144-13.1.079

KOTLYAR, G.V., BELYANSKY, G.C., BURAGO, V.I., NIKITINA, A.P., ZAKHAROV, Y.D. & ZHURAVLEV, A.V. 2006. South Primorye, Far East Russia – A key region for global Permian correlation. *Journal of Asian Earth Sciences*, **26**, 280–293.

LEVEN, E.J. 1967. Stratigrafiya i fuzulinidy perrmskikh otlozheniy Pamira. *Ademiya Nauk SSSR, Geologicheskiy Institut*, **167**, 1–224 [in Russian].

LEVEN, E.J. 1993. Early Permian fusulinids from the Central Pamir. *Rivista Italiana di Paleontologia e Stratigrafia*, **99**, 151–198.

LEVEN, E.J. 1996. The Midian Stage of the Permian and its boundaries. *Stratigraphy and Geological Correlation*, **4**, 540–551.

LEVEN, E.J. 1997. *Permian Stratigraphy and Fusulinida of Afghanistan with their Paleogeographic and Paleotectonic Implications*. Geological Society of America, Special Papers, **316**.

LEVEN, E.J. 2004. Fusulinids and Permian scale of the Tethys. *Stratigraphy and Geological Correlation*, **12**, 139–151.

LEVEN, E.J. 2010. Permian fusulinids of the east Hindu Kush and west Karakorum (Pakistan). *Stratigraphy and Geological Correlation*, **18**, 105–117.

LEVEN, E.J. & CAMPBELL, H.J. 1998. Middle Permian (Murgabian) fusuline faunas, Torlesse terrane, New Zealand. *New Zealand Journal of Geology and Geophysics*, **41**, 149–156.

LEVEN, E.J. & GORGIJ, M.N. 2006. Upper Carboniferous–Permian stratigraphy and fusulinids from the Anarak region, central Iran. *Russian Journal of Earth Sciences*, **8**, 1–25.

LEVEN, E.J. & GORGIJ, M.N. 2008. Bolorian and Kubergandian stages of the Permian in the Sanandaj-Sirjan zone of Iran. *Stratigraphy and Geological Correlation*, **16**, 455–466, https://doi.org/10.1134/s0869593808050018

LEVEN, E.J. & GORGIJ, M.N. 2011*a*. Fusulinids and stratigraphy of the Carboniferous and Permian in Iran. *Stratigraphy and Geological Correlation*, **19**, 687–776, https://doi.org/10.1134/s0869593811070021

LEVEN, E.J. & GORGIJ, M.N. 2011*b*. First record of Gzhelian and Asselian Fusulinids from the Vazhnan Formation (Sanandaj-Sirjan zone of Iran). *Stratigraphy and Geological Correlation*, **19**, 486–501.

LEVEN, E.J. & GORGIJ, M.N. 2013. Fusulinids from the lower Permian Chili Formation of the Rahdar section, Kalmard block, Central Iran. *Stratigraphy and Geological Correlation*, **21**, 408–420, https://doi.org/10.1134/s0869593813040060

LEVEN, E.J. & GRANT-MACKIE, J.A. 1997. Permian fusulinid foraminifera from Wherowhero Point, Orua Bay, Northland, New Zealand. *New Zealand Journal of Geology and Geophysics*, **40**, 473–486.

LEVEN, E.J. & HEWARD, A.P. 2013. Fusulinids from isolated Qarari limestone outcrops (Permian), occurring among Jurassic–Cretaceous Batain group (Batain Plain, Eastern Oman). *Rivista Italiana di Paleontologia e Stratigrafia*, **119**, 153–162.

LEVEN, E.J. & SCHERBOVICH, S.F. 1978. *Fuzulinidy i stratigrafiya assclyskogo yarusa Darvaza*. Izdatel Nauka, Moscow, Russia [in Russian].

LEVEN, E.J. & TAHERI, A. 2003. Carboniferous–Permian stratigraphy and fusulinids of East Iran. Gzhelian and Asselian deposits of the Ozbak-Kuh region. *Rivista Italiana di Paleontologia e Stratigrafia*, **109**, 399–415.

LEVEN, E.J., LEONOVA, T.B. & DMITRIEV, V.Y. 1992. Permi Darvaz-Zaalayskoy zony Pamira: fuzulinidy, ammonoidei, stratigrafiya. *Rossiyskaya Akademiya Nauk, Trudy Paleontologicheskogo Instituta*, **253**, 1–203 [in Russian].

LEVEN, E.J., NAUGOLNYKH, S.V. & GORGIJ, M.N. 2011. New findings of Permian marine and terrestrial fossils in Central Iran (The Kalmard Block) and their significance for correlation of the Tethyan, Uralian, and west European scales. *Rivista Italiana di Paleontologia e Stratigrafia*, **117**, 355–374.

LI, J.C., ZHANG, Z.H., LI, W.S. & HONG, Z.Y. 2010. Fusulinid fauna near the Carboniferous–Permian boundary in Shunchang, Fujian Province. *Geological Science and Technology Information*, **29**, 9–16.

LIANG, D.Y., NIE, Z.T., GUO, T.Y., XU, B.W., ZHANG, Y.Z. & WANG, W.P. 1983. Permo-Carboniferous Gondwana–Tethys facies in southern Karakoran Ali, Xizang (Tibet). *Earth Science*, **19**, 9–27 [in Chinese with English abstract].

LIN, J.X., PAN, S.S. & MENG, F.Y. 1979. Late Carboniferous and Early Lower Permian fusulinids from Jiahe, Hunan. *Acta Palaeontologica Sinica*, **18**, 561–571 [in Chinese with English abstract].

LIU, G.C. 1993. Age assignment of Kaixinling Group and Wuli group in the middle Tanggula Mountains. *Qinghai Geology*, **2**, 1–9 [in Chinese with English abstract].

LORIGA, C.B., NERI, C., PASINI, M. & POSENATO, R. 1986. Marine fossil assemblages from Upper Permian to Lowermost Triassic in the western Dolomites (Italy). *Memorie della Societa Geologica Italiana*, **34**, 5–44.

LUCAS, S.G. 2013. Reconsideration of the base of the Permian system. *New Mexico Museum of Natural History and Science Bulletin*, **60**, 230–232.

LUCAS, S.G., KRAINER, K. & VACHARD, D. 2015. The lower Permian Hueco Group, Robledo Mountains, New Mexico (U.S.A.). *New Mexico Museum of Natural History and Science Bulletin*, **65**, 43–95.

LYS, M. & LAPPARENT, A.F.D. 1971. Foraminifères et microfaciès du Permien de l'Afghanistan central. *Notes et Mémoires sur le Moyen-Orient*, **12**, 49–166.

LYS, M. & MARCOUX, J. 1978. Les niveaus du Permien supérieur des Nappes d'Antalya (Taurides occidentales, Turquie). *Comptes Rendus Hebdomadaires des Seances de l'Academie des Sciences*, **286**, 1417–1420.

LYS, M., COLCHEN, M., BASSOULLET, J.P., MARCOUX, J. & MASCLE, G. 1980. La biozone à *Colaniella parva* du Permien Supérieur et sa microfauna dans le bloc calcaire exotique de Lamayuru, Himalaya du Ladakh. *Revue de Micropaléontologie*, **23**, 76–108.

MAGGINETTI, R.T., STEVENS, C.H. & STONE, P. 1988. *Early Permian Fusulinids from the Owens Valley Group, East-Central California*. Geological Society of America, Special Papers, **217**, 1–61.

METCALFE, I. 1984. Stratigraphy, palaeontology and palaeogeography of the Carboniferous of Southeast Asia. *Mémoires de la Société Géologique de France*, **147**, 107–118.

METCALFE, I. 2013. Gondwana dispersion and Asian accretion: tectonic and palaeogeographic evolution of eastern Tethys. *Journal of Asian Earth Sciences*, **66**, 1–33, https://doi.org/10.1016/j.jseaes.2012.12.020

MIKAMI, T. 1965. The type Sakamotozawa Formation. *Journal of Geological Society of Japan*, **71**, 475–493 [in Japanese].

MIKHAILOVA, Z.P. 1974. *fuzulinidy verkhnego karbona pechorskogo priurad'ya*. Akademiya nauk SSSR komi filial institut geologii, Moscow, 1–116 [in Russian].

NESTELL, G.P. & NESTELL, M.K. 2006. Middle Permian (Late Guadalupian) foraminifers from Dark Canyon, Guadalupe Mountains, New Mexico. *Micropaleontology*, **52**, 1–50.

NGUYEN, D.T. 1986. Foraminifera and algae from the Permian of Kampuchea. *CCOP Technical Bulletin*, **18**, 116–137.

NIE, Z.T. & SONG, Z.M. 1983*a*. Fusulinids of Lower Permian Qudi Formation from Rutog of Xizang (Tibet), China. *Earth Science – Journal of Wuhan College of Geology*, **19**, 29–42 [in Chinese with English abstract].

NIE, Z.T. & SONG, Z.M. 1983*b*. Fusulinids of Lower Permian Tunlonggongba Formation from Rutog of Xizang. *Earth Science – Journal of Wuhan College of Geology*, **19**, 43–55 [in Chinese with English abstract].

NIE, Z.T. & SONG, Z.M. 1983*c*. Fusulinids of Lower Permian Maokouian Longge Formation from Rutog, Xizang (Tibet), China. *Earth Science – Journal of Wuhan College of Geology*, **19**, 57–68 [in Chinese with English abstract].

NIIKAWA, I. 1978. Carboniferous and Permian fusulinids from Fukuji, central Japan. *Journal of the Faculty of Science, Hokkaido University, Series 4. Geology and Mineralogy*, **18**, 533–610.

NIU, Z.J., DUAN, Q.F., WANG, J.X., BAI, Y.S., ZENG, B.F., TU, B. & BU, J.J. 2006. On the Gadikao Formation in Zhidoi and Zadoi areas, Southern Qinghai. *Journal of Stratigraphy*, **30**, 109–115 [in Chinese with English abstract].

NIU, Z.J., DUAN, Q.F., WANG, J.X., HE, L.Q. & BAI, Y.S. 2010. Early Permian (Cisuralian) lithostratigraphical succession in volcanic–sedimentary setting from southern Qinghai. *Earth Science – Journal of China University of Geosciences*, **35**, 11–21 [in Chinese with English abstract].

NOE, S. 1988. Foraminiferal ecology and biostratigraphy of the marine Upper Permian and of the

Permian–Triassic boundary in the southern Alps (Bellerophon Formation, Tesero Horizon). *Revue de Paléobiologie*, **S2**, 75–88.

NOGAMI, Y. 1961. Permische Fusuliniden aus dem Atetsu-Plateau Sudwestjapans. Teil 1.Fusulininae und Schwagerininae. *Memoirs of the College of Science, University of Kyoto, Series B*, **27**, 159–225.

OKUYUCU, C. 2008. Biostratigraphy and systematics of late Asselian–early Sakmarian (Early Permian) fusulinids (Foraminifera) from southern Turkey. *Geological Magazine*, **145**, 413–434, https://doi.org/10.1017/s0016756808004482

OTA, A. & ISOZAKI, Y. 2006. Fusuline biotic turnover across the Guadalupian–Lopingian (Middle–Upper Permian) boundary in mid-oceanic carbonate buildups: biostratigraphy of accreted limestone in Japan. *Journal of Asian Earth Sciences*, **26**, 353–368.

OTA, Y. 1997. Middle Carboniferous and Lower Permian fusulinacean biostratigraphy of the Akiyoshi limestone group, Southwest Japan. Part 1. *Bulletin of the Kitakyushu Museum of Natural History*, **16**, 1–97.

OZAWA, T. 1970. Variation and relative growth of *Colania douvillei* (Ozawa) from the Rat Buri limestone. *Geology and Palaeontology of Southeast Asia*, **8**, 19–40.

OZAWA, T. 1987. Permian fusulinacean biogeographical provinces in Asia and their tectonic implications. *In*: TAIRA, A. & TASHIRO, M. (eds) *Historical Biogeography and Plate Tectonic Evolution of Japan and Eastern Asia*. Terra Science Publishing, Tokyo, Japan, 45–63.

OZAWA, Y. 1927. Stratigraphical studies of the Fusulina Limestone of Akasaka, Province of Mino. *Journal of the Faculty of Science, Imperial University of Tokyo, Section 2, Geology, Mineralogy, Geography, Seismology*, **2**, 121–162.

PAKISTAN–JAPANESE WORKING GROUP 1985. Permian and Triassic systems in the Salt Range and Surghar Range, Pakistan. *In*: NAKAZAWA, K. & DICKINS, J.M. (eds) *The Tethys; Her Paleogeography and Paleobiogeography from Paleozoic to Mesozoic*. Tokai University Press, Tokyo, Japan, 221–312.

PAPULOV, G.N. 1986. *Pogranichnye otlozheniya karbona i Permi Urala, Priuraliya i Srednei Azii*. Akademiya nauk SSSR, Uralskii nauchnyi centr, Moscow, 3–151.

PASINI, M. 1985. Biostratigrafia con i foraminiferi del limite formazione di Werfen fra recoaro e la Val Badia (Alpi Meridionali). *Rivista Italiana di Paleontologia e Stratigrafia*, **90**, 481–510.

QIAO, L. & SHEN, S.Z. 2014. Global paleobiogeography of brachiopods during the Mississippian – Response to the global tectonic reconfiguration, ocean circulation, and climate changes. *Gondwana Research*, **26**, 1173–1185, https://doi.org/10.1016/j.gr.2013.09.013

RAUZER-CHERNOUSOVA, D.M. 1965. Foraminifery stratotipicheskogo Razreza sakmarskogo yarusa. *Akademiya nauk SSSR, Geologicheskii Institut*, **135**, 5–79 [in Russian].

ROSS, C.A. 1960. Fusulinids from the Hess Member of the Leonard Formation, Leonard Series (Permian), Glass Mountains, Texas. *Contributions from the Cushman Foundation for Foraminiferal Research*, **11**, 116–133.

ROSS, C.A. 1962. Permian foraminifera from British Honduras. *Palaeontology*, **5**, 297–306.

ROSS, C.A. 1963. *Standard Wolfcampian Series (Permian), Glass Mountains, Texas*. Geological Society of America, Memoirs, **88**.

ROSS, C.A. 1967. Development of Fusulinid (Foraminiferida) fauna realms. *Journal of Paleontology*, **41**, 1341–1354.

ROSS, C.A. 1971. New species of *Schwagerina* and *Yabeina* (Fusulinacea) of Wordian age (Permian) from northwestern British Columbia. *Geological Survey of Canada Bulletin*, **197**, 95–101.

ROSS, C.A. 1982. Paleobiology of fusulinaceans. *Third North American Paleontological Convention Proceedings*, **2**, 441–445.

ROSS, C.A. 1995. Permian fusulinaceans. *In*: SCHOLLE, P.A., PERYT, T.M. & ULMER-SCHOLLE, D.S. (eds) *The Permian of Northern Pangea. Volume 1: Paleogeography, Paleoclimates, Stratigraphy*. Springer, Berlin, 167–185.

ROSS, C.A. & JUNE, J.R.P. 2003. Fusulinid sequence evolution and sequence extinction in Wolfcampian and Leonardian Series (Lower Permian), Glass Mountains, West Texas. *Rivista Italiana di Paleontologia e Stratigrafia*, **109**, 281–306.

ROSS, C.A. & ROSS, J.R.P. 1994. Permian sequence stratigraphy and fossil zonation. *In*: EMBRY, A.F., BEAUCHAMP, B. & GLASS, D.J. (eds) *Pangea: Global Environments and Resources*. Canadian Society of Petroleum Geologists, Memoirs, **17**, 219–231.

ROSS, C.A. & ROSS, J.R.P. 1997. Hessian (Leonardian, Middle Lower Permian) depositional sequences and their Fusulinid Zones, west Texas. *In*: BRENCKLE, P. (ed.) *A Compendium of Upper Devonian–Carboniferous Type Foraminifers from the Former Soviet Union*. Cushman Foundation for Foraminifera Research, Special Publications, **38**, 119–124.

ROSS, C.A. & ROSS, J.R.P. 2003. Sequence evolution and sequence extinction: fusulinid biostratigraphy and species-level recognition of depositional sequences, Lower Permian, Glass Mountains, West Texas, USA. *In*: OLSON, H.C. & LECKIE, R.M. (eds) *Micropaleontologic Proxies for Sea-Level Change and Stratigraphic Discontinuities*. Society of Economic Paleontologists and Mineralogists, Special Publications, **75**, 317–359.

RUI, L. 1979. Upper Permian fusulinids from western Guizhou. *Acta Palaeontologica Sinica*, **18**, 271–297.

RUI, L. & NASSICHUK, W.W. 1996. Upper Permian (Wordian) fusulinaceans from the Cache Creek Terrane, northern British Columbia. *Canadian Journal of Earth Sciences*, **33**, 1022–1036.

RUI, L., ZHAO, J.M., MU, X.N., WANG, K.L. & WANG, Z.H. 1984. Restudy of Wujiaping limestone in Liangshan, Hanzhong, Shanxi Province. *Journal of Stratigraphy*, **8**, 179–193.

SADA, K. & DANNER, W.R. 1992. *Pseudoschwagerina skinneri*, N. sp. from near kamloops in British Columbia, Canada. *Transactions and Proceedings of the Palaeontological Society of Japan, New Series*, **167**, 1259–1263.

SAKAMOTO, T. & ISHIBASHI, T. 2002. Paleontological study of fusulinoidean fossils from the Terbat Formation, Sarawak, East Malaysia. *Memoirs of the Faculty of Science, Kyushu University, Series D, Earth and Planetary Sciences*, **31**, 29–57.

SCHUBERT, R.J. 1915. Die Foraminiferen des Jüngeren Paläozoikums von Timor. *Palaontologien von Timor*, **2**, 49–59.

SCOTESE, C.R. 2001. *Atlas of Earth History, Volume 1, Paleogeography*. PALEOMAP Project, Arlington, TX, USA.

ŞENGÖR, A.M.C., YILMAZ, Y. & SUNGURLU, O. 1984. Tectonics of the Mediterranean Cimmerides: nature and evolution of the western termination of Palaeo-Tethys. *In*: DIXON, J.E. & ROBERTSON, A.H.F. (eds) *The Geological Evolution of the Eastern Mediterranean*. Geological Society, London, Special Publications, **17**, 77–112, https://doi.org/10.1144/gsl.sp.1984.017.01.04

SHEN, S.Z. & MEI, S.L. 2010. Lopingian (Late Permian) high-resolution conodont biostratigraphy in Iran with comparison to South China zonation. *Geological Journal*, **45**, 135–161, https://doi.org/10.1002/gj.1231

SHEN, S.Z., CAO, C.Q., SHI, G.R., WANG, X.D. & MEI, S.L. 2003. Loping (Late Permian) stratigraphy, sedimentation and palaeobiogeography in southern Tibet. *Newsletters on Stratigraphy*, **39**, 157–179.

SHEN, S.Z., ZHANG, H., SHANG, Q.H. & LI, W.Z. 2006. Permian stratigraphy and correlation of Northeast China: a review. *Journal of Asian Earth Sciences*, **26**, 304–326, https://doi.org/10.1016/j.jseaes.2005.07.007

SHEN, S.Z., CROWLEY, J.L. *ET AL*. 2011. Calibrating the end-permian mass extinction. *Science*, **334**, 1367–1372, https://doi.org/10.1126/science.1213454

SHEN, S.Z., YUAN, D.X., HENDERSON, C.M., TAZAWA, J. & ZHANG, Y.C. 2013. Implications of Kungurian (Early Permian) conodonts from Hatahoko, Japan, for correlation between the Tethyan and international timescales. *Micropaleontology*, **58**, 505–522.

SHENG, J.Z. 1958. Some Upper Carboniferous fusulinids from the vicinity of Beiyin Obo, Inner Mongolia. *Acta Palaeontologica Sinica*, **6**, 35–50 [in Chinese with English abstract].

SHENG, J.Z. 1962. Chihsian fusulinids from Kangbao area, Hebei Province. *Acta Palaeontologica Sinica*, **10**, 426–432 [in Chinese with English abstract].

SHENG, J.Z. 1963. Permian fusulinids of Kwangsi, Kueichow and Szechuan. *Palaeontologia Sinica, Series B*, **10**, 1–247 [in Chinese with English abstract].

SHENG, J.Z. & JIN, Y.G. 1994. Correlation of Permian deposits in China. *In*: JIN, Y.G., UTTING, J. & WARDLAW, B.R. (eds) *Permian Stratigraphy, Environments and Resources, Palaeoworld 4*. Nanjing University Press, Nanjing, China, 14–113.

SHENG, J.Z. & RUI, L. 1984. Fusulinaceans from Upper Permian Changhsingian in Mingshan coal field of Leping, Jiangxi. *Acta Micropalaeontologica Sinica*, **1**, 30–46 [in Chinese with English abstract].

SHENG, J.Z. & WANG, Y.H. 1962. The fusulinids of the Maokou stage, southern Kiangsu. *Acta Palaeontologica Sinica*, **10**, 176–190 [in Chinese with English abstract].

SHENG, J.Z. & ZHANG, L.X. 1958. Fusulinids from the type-locality of the Changhsing limestone. *Acta Palaeontologica Sinica*, **6**, 205–214 [in Chinese with English abstract].

SHENG, J.Z., ZHANG, L.X. & WANG, J.H. 1988. *Fusulinids*. Science Press, Beijing, [in Chinese].

SHI, G.R. 2006. The marine Permian of East and Northeast Asia: an overview of biostratigraphy, palaeobiogeography and palaeogeographical implications. *Journal of Asian Earth Sciences*, **26**, 175–206, https://doi.org/10.1016/j.jseaes.2005.11.004

SHI, Y.K., JIN, X.C., HUANG, H. & YANG, X.N. 2008. Permian Fusulinids from the Tengchong Block, Western Yunnan, China. *Journal of Paleontology*, **82**, 118–127.

SHI, Y.K., LIU, J.R., YANG, X.N. & ZHU, L.M. 2009. Fusulinid faunas from the Datangian to Chihsian strata of the Zongdi section in Ziyun county, Guizhou Province. *Acta Micropalaeontologica Sinica*, **26**, 1–30 [in Chinese with English abstract].

SHI, Y.K., HUANG, H., JIN, X.C. & YANG, X.N. 2011. Early Permian fusulinids from the Baoshan Block, Western Yunnan, China and their paleobiogeographic significance. *Journal of Paleontology*, **85**, 489–501, https://doi.org/10.1666/10-039.1

SHI, Y.K., YANG, X.N. & LIU, J.R. 2012. *Early Carboniferous to Early Permian fusulinids from Zongdi Section in Southern Guizhou*. Science Press, Beijing [in Chinese with English abstract].

SHI, Y.K., HUANG, H. & JIN, X.C. 2017. Depauperate fusulinid faunas of the Tengchong block in western Yunnan, China and their paleogeographic and paleoenvironmental indications. *Journal of Paleontology*, **91**, 12–24, https://doi.org/10.1017/jpa.2016.122

SHIINO, Y., SUZUKI, Y. & KOBAYASHI, F. 2008. Middle Permian fusulines from the Hoso-o Formation in Kamiyasse area, southern Kitakami Mountains, Northeast Japan: their biostratigraphic implications. *Journal of Geological Society of Japan*, **114**, 200–205.

SKINNER, J.W. 1969. Permian Foraminifera from Turkey. *The University of Kansas Paleontological Contributions*, **36**, 1–14.

SKINNER, J.W. & WILDE, G.L. 1955. New Fusulinids from the Permian of West Texas. *Journal of Paleontology*, **29**, 927–940.

SKINNER, J.W. & WILDE, G.L. 1965. Permian biostratigraphy and fusulinid faunas of the Shasta lake area, northern California. *The University of Kansas Paleontological Contributions*, 1–98.

SKINNER, J.W. & WILDE, G.L. 1966. Permian fusulinids from Sicily. *The University of Kansas Paleontological Contributions*, **22**, 1–16.

SKINNER, J.W. & WILDE, G.L. 1967. Permian foraminifera from Tunisia. *The University of Kansas Paleontological Contributions*, **30**, 1–22.

SICHUAN REGIONAL GEOLOGICAL SURVEY & NANJING INSTITUTE OF GEOLOGY AND PALAEONTOLOGY 1982. *Stratigraphy and Palaeontology of Western Sichuan and Eastern Tibet, No.1*. Sichuan People's Publishing House, Chengdu, China.

SONG, J.J., SONG, H.B. & HU, B. 2014. Geologic age of the Taiyuan Formation in northwest Henan province: evidences from fusulinids (foraminifera). *Acta Micropalaeontologica Sinica*, **31**, 190–204.

SOSNINA, M.I. 1981. Nekotorye permskie fuzulinidy pal'nego vostoka. *In*: MOBZALEVSKAYA, E.A. & KOLOBOVA, I.M. (eds) *Ezhegodnik vsesouznogo paleontologicheskogo obschestva*. Leningradskoe otdelenie, Leningrad, Russia, 13–34 [in Russian].

STAMPFLI, G.M. & BOREL, G.D. 2002. A plate tectonic model for the Paleozoic and Mesozoic constrained by dynamic plate boundaries and restored synthetic oceanic isochrons. *Earth and Planetary Science Letters*, **196**, 17–33, https://doi.org/10.1016/s0012-821x(01)00588-x

STEVENS, C.H. & STONE, P. 2007. *The Pennsylvanian–Early Permian Bird Spring Carbonate Shelf, Southeastern California: Fusulinid Biostratigraphy, Paleogeographic Evolution, and Tectonic Implications*. Geological Society of America Special Papers, **429**, https://doi.org/10.1130/2007.2429

STEVENS, C.H., DAVYDOV, V.I. & BRADLEY, D. 1997. Permian Tethyan fusulinina from the Kenai peninsula, Alaska. *Journal of Paleontology*, **71**, 985–994.

SUN, X.F. 1979. Upper Permian fusulinids from Zhen'an of Shanxi and Tewo of Gansu, NW.China. *Acta Palaeontologica Sinica*, **18**, 163–168 [in Chinese with English abstract].

TARAZ, H., GOLSHANI, F. *ET AL*. 1981. The Permian and the Lower Triassic systems in Abadeh region, central Iran. *Memoirs of the Faculty of Science*, **47**, 61–133.

TAZAWA, J. 1991. Middle Permian brachiopod biogeography of Japan and adjacent regions in East Asia. *In*: ISHII, K.I., LIU, X.M., ICHIKAWA, K. & HUANT, B. (eds) *Pre-Jurassic Geology of Inner Mongolia, China*. Matsuya Insatsu, Osaka, Japan, 213–230.

THOMPSON, M.L. 1946. Permian fusulinids from Afghanistan. *Journal of Paleontology*, **20**, 140–157.

THOMPSON, M.L. 1949. The Permian fusulinids of Timor. *Journal of Paleontology*, **23**, 182–192.

THOMPSON, M.L. & MILLER, A.K. 1949. Permian fusulinids and cephalopods from the vicinity of the Maracaibo basin in northern South America. *Journal of Paleontology*, **23**, 1–24.

THOMPSON, M.L. & VERVILLE, G.J. 1950. Cache Creek fusulinids from Kamloops, British Columbia. *Contributions from the Cushman Foundation for Foraminiferal Research*, **1**, 67–70.

THOMPSON, M.L. & WHEELER, H.E. 1942. Permian Fusulinids from British Columbia, Washington and Oregon. *Journal of Paleontology*, **16**, 700–711.

TORIYAMA, R. 1954. Geology of Akiyoshi. *Memories of Faculty of Science, Kyushu University, Series D*, **4**, 39–97.

TORIYAMA, R. 1975. Fusuline fossils from Thailand, Part 9, Permian Fusulines from the Rat Buri limestone in the Khao Phlong Phrab area, Sara Buri, Central Thailand. *Memoirs of the Faculty of Science, Kyushu University, Series D, Geology*, **23**, 1–116.

TORIYAMA, R. 1982. Fusuline fossils from Thailand, part 15. Peculiar spirothecal structure of Schwagerinid from east of Wang Saphung, Changwat Loei Central North Thailand. *Geology and Palaeontology of Southeast Asia*, **23**, 1–7.

TORIYAMA, R. & KANMERA, K. 1979. Fusuline Fossils from Thailand, Part 12. Permian Fusulines from the Ratburi Limestone in the Khao Khao Area, Sara Buri, Central Thailand. *Geology and Palaeontology of Southeast Asia*, **20**, 23–94.

TORIYAMA, R., PITAKPAIV, K. & KANMERA, K. 1965. The Fusulinacean fossils of Thailand. *Memoirs of Faculty of Science, Kyushu University, Series D, Geology*, **17**, 1–69.

TORIYAMA, R., KANMERA, K., KAEWBAIDHOAM, S. & HONGNUSONTHI, A.-n. 1974. Biostratigraphic zonation of the Rat Buri limestone in the Khao Phlong Phrab area, Sara Buri, Central Thailand. *Geology and Palaeontology of Southeast Asia*, **14**, 25–81.

UENO, K. 1992. Verbeekinid and Neoschwagerinid Fusulinaceans from the Akiyoshi limestone group above the Parafusulina kaerimizensis zone, southwest Japan. *Transactions and Proceedings of the Palaeontological Society of Japan, New Series*, **165**, 1040–1069.

UENO, K. 1996. Late Early to Middle Permian fusulinacean biostratigraphy of the Akiyoshi limestone group, southwest Japan, with special reference to the verbeekind and neoschwagerinid fusulinicean biostratigraphy and evolution. *Museo Civico di Rovereto*, **S11**, 77–104.

UENO, K. 2003. The Permian fusulinoidean faunas of the Sibumasu and Baoshan blocks: their implications for the paleogeographic and paleoclimatologic reconstruction of the Cimmerian Continent. *Palaeogeography, Palaeoclimatology, Palaeoecology*, **193**, 1–24, https://doi.org/10.1016/S0031-0182(02)00708-3

UENO, K. 2006. The Permian antitropical fusulinoidean genus *Monodiexodina*: distribution, taxonomy, paleobiogeography and paleoecology. *Journal of Asian Earth Sciences*, **26**, 380–404, https://doi.org/10.1016/j.jseaes.2005.07.003

UENO, K. & IGO, H. 1997. Late Paleozoic foraminifers from the Chiang Dao area, northern Thailand. *Prace Państwowego Instytutu Geologicznego*, **157**, 339–358.

UENO, K. & NAKAZAWA, T. 1993. Carboniferous foraminifers from the lowermost part of the Omi limestone group, Niigata Prefecture, central Japan. *Science Reports of the Institute of Geoscience, University of Tsukuba, Section B: Geological Sciences*, **14**, 1–51.

UENO, K. & SAKAGAMI, S. 1991. Late Permian Fusulinacean fauna of Doi Pha Phlung, North Thailand. *Transactions and Proceedings of Palaeontological Society of Japan, New Series*, **164**, 928–943.

UENO, K. & TSUTSUMI, S. 2009. Lopingian (Late Permian) foraminiferal faunal succession of a Paleo-Tethyan mid-oceanic carbonate buildup: Shifodong Formation in the Changning-Menglian Belt, West Yunnan, Southwest China. *Island Arc*, **18**, 69–93, https://doi.org/10.1111/j.1440-1738.2008.00648.x

UENO, K., MIZUNO, Y., WANG, X.D. & MEI, S.L. 2002. Artinskian conodonts from the Dingjiazhai Formation of the Baoshan Block, West Yunnan, Southwest China. *Journal of Paleontology*, **76**, 741–750.

UENO, K., WANG, Y.J. & WANG, X.D. 2003. Fusulinoidean faunal succession of a Paleo-Tethyan oceanic seamount in the Changning–Menglian Belt, west Yunnan, southwest China: an overview. *The Island Arc*, **12**, 145–161.

UENO, K., SHI, G.R. & SHEN, S.Z. 2005. Fusulinoideans from the early Midian (late Middle Permian) *Metadoliolina dutkevitchi–Monodiexodina sutchanica* zone of the Senkina Shapka section, South Primorye, Far East Russia. *Alcheringa*, **29**, 257–273.

UENO, K., TAZAWA, J.I. & MIYAKE, Y. 2006. Middle Permian fusulinoideans from Hatahoko in the Nyukawa area, Gifu Prefecture, Mono Belt, central Japan. *Science Report of Niigata University (Geology)*, **21**, 47–72.

UENO, K., TAZAWA, J.I. & SHINTANI, T. 2007. Fusuline foraminifera from the basal part of the Sakamotozawa

Formation, south Kitakami Belt, Northeast Japan. *Science Report of Niigata University (Geology)*, **22**, 15–33.

UENO, K., SHINTANI, T. & TAZAWA, J.I. 2009. Fusuline foraminifera from the upper part of the Sakamotozawa Formation, South Kitakami Belt, Northeast Japan. *Science Report of Niigata University (Geology)*, **24**, 27–61.

UENO, K., SHINTANI, T. & TAZAWA, J.I. 2011. A fusuline fauna from the basal part of the Sakamotozawa Formation in the Kamiyasse area, South Kitakami Belt, Northeast Japan. *Science Reports of Niigata University (Geology)*, **26**, 23–41.

UENO, K., ARITA, M., MENO, S., SARDSUD, A. & SAESAENGSEERUNG, D. 2015. An Early Permian fusuline fauna from southernmost Peninsular Thailand: discovery of Early Permian warming spikes in the peri-Gondwanan Sibumasu Block. *Journal of Asian Earth Sciences*, **104**, 185–196, https://doi.org/10.1016/j.jseaes.2014.10.030

UNAL, E., ALTINER, D., YILMAZ, I.O. & OZKAN-ALTINER, S. 2003. Cyclic sedimentation across the Permian–Triassic boundary (Central Taurides, Turkey). *Rivista Italiana di Paleontologia e Stratigrafia*, **109**, 359–376.

VACHARD, D. 1980. Téthys et Gondwana au paléozoïque supérieur, les données Afghanes: biostratigraphie, micropaléontologie, paléographie. *Documents et travaux de l'Institut géologique Albert de Lapparent*, **2**, 1–463.

VACHARD, D. & FERRIERE, J. 1991. An assemblage with *Yabeina* (Fusulinid foraminifera) from the Midian (Upper Permian) of Whangaroa area (Orua Bay, New Zealand). *Revue de Micropaléontologie*, **34**, 201–230.

VACHARD, D. & MOIX, P. 2013. Kubergandian (Roadian, Middle Permian) of the Lycian and Aladağ Nappes (Southern Turkey). *Geobios*, **46**, 335–356, https://doi.org/10.1016/j.geobios.2013.02.002

VACHARD, D., FLORES DE DIOS, A., BUITRON, B.E. & GRAJALES, M. 2000. Biostratigraphie par fusulines des calcaires carboniferes et permiens de San Salvador Patlanoaya (Puebla, Mexique). *Geobios*, **33**, 5–33.

VACHARD, D., MARTINI, R. & ZANINETTI, L. 2001. Earliest Artinskian (Early Permian) fusulinids reworked in the Triassic Lercara Formation (NW Sicily). *Journal of Foraminiferal Research*, **31**, 33–47.

VACHARD, D., HAUSER, M., MARTINI, R., ZANINETTI, L., MATTER, A. & PETERS, T. 2002. Middle Permian (Midian) foraminiferal assemblages from the Batain Plain (Eastern Oman): their significance to Neotethyan Paleogeography. *Journal of Foraminiferal Research*, **32**, 155–172.

VACHARD, D., ZAMBETAKIS-LEKKAS, A., SKOURTSOS, E., MARTINI, R. & ZANINETTI, L. 2003. Foraminifera, Algae and Carbonate microproblematica from the Late Wuchiapingian/Dzhulfian (Late Permian) of Peloponnesus (Greece). *Rivista Italiana di Paleontologia e Stratigrafia*, **109**, 339–358.

VACHARD, D., PILLE, L. & GAILLOT, J. 2010. Palaeozoic Foraminifera: systematics, palaeoecology and responses to global changes. *Revue de Micropaleontologie*, **53**, 209–254, https://doi.org/10.1016/j.revmic.2010.10.001

VUKS, G.P. & CHEDIYA, I.O. 1986. Foraminifery ludyanzinskoi svity buhty neizvestiaya (Uzhnoe Primorye) [Lyudianza foraminifera of the Neizvestnaya Bay (South Primorye)]. *In*: ZAKHAROV, Y.D. & ONOPRIENKO, Y.I. (eds) *Korrelyatsiya permo-triasovykh otlozhenij vostoka SSSR [Correlation of Permo-Triassic Sediments of the East USSR]*. DVNC AN SSSR, Vladivostok, Russia, 82–88 [in Russian].

WANG, C.Y., QIN, Z.S., YUN, Y.K., ZHU, X.S., XU, D.Y. & CHEN, G.Y. 1997. Age of *Gallowayinella* and the lower limit of the Changhsingian stage based on conodonts. *Journal of Stratigraphy*, **21**, 100–108 [in Chinese with English abstract].

WANG, J.H. 1978. The boundary and biostratigraphy of the Chihsia Formation in Nanjing area. *Acta Stratigraphic Sinica*, **2**, 67–73 [in Chinese with English abstract].

WANG, J.H. & TANG, Y. 1986. Fusulinids from Chihsia Formation of Lengwu district in Tonglu County, Zhejiang. *Acta Micropalaeontologica Sinica*, **3**, 3–12 [in Chinese with English abstract].

WANG, Y. & JIN, Y.G. 2006. Radiation of the Fusulinoideans between the two phases of the end-Permian mass extinction, South China. *In*: RONG, J.Y., FANG, Z.J., ZHOU, Z.H., ZHAN, R.B. & YUAN, X.L. (eds) *Originations, Radiations and Biodiversity Changes – Evidences from the Chinese Fossil Record*. Science Press, Beijing, 503–516.

WANG, Y. & UENO, K. 2009. A new fusulinoidean genus Dilatofusulina from the Lopingian (Upper Permian) of southern Tibet, China. *Journal of Foraminiferal Research*, **39**, 56–65.

WANG, Y., UENO, K., ZHANG, Y.C. & CAO, C.Q. 2010. The Changhsingian foraminiferal fauna of a Neotethyan seamount: the Gyanyima limestone along the Yarlung–Zangbo Suture in southern Tibet, China. *Geological Journal*, **45**, 308–318.

WANG, Y., WANG, J., CHEN, J., WANG, W., SHEN, S. & HENDERSON, C.M. 2011*a*. Progress, problems and prospects on the stratigraphy and correlation of the Kungurian Stage, Early Permian (Cisuralian) Series. *Acta Geologica Sinica – English Edition*, **85**, 387–398, https://doi.org/10.1111/j.1755-6724.2011.00407.x

WANG, Y., WANG, W.J., ZHANG, Y.C., QI, Y.P., WANG, X.D. & LIAO, Z.T. 2011*b*. Late Carboniferous to Early Permian fusuline assemblages from the Wuzunbulake section, Keping area, Xinjiang. *Acta Palaeontologica Sinica*, **50**, 409–419 [in Chinese with English abstract].

WANG, Y.J. & ZHOU, J.P. 1986. New material of fusulinids from Xainza, Xizang. *Bulletin of Nanjing Institute of Geology and Palaeontology, Academia Sinica*, **10**, 141–156 [in Chinese with English abstract].

WANG, Y.J., SHENG, J.Z. & ZHANG, L.X. 1981. Fusulinids from Xizang of China. *In*: NANJING INSTITUTE OF GEOLOGY AND PALAEONTOLOGY (ed.) *Palaeontology of Xizang, Book 3*. Science Press, Beijing, China, 1–80 [in Chinese with English abstract].

WANG, Y.J., YUAN, X.Q. & GENG, G.C. 1992. A new advance of Carboniferous fusulinid study and preliminary exploration of the paleogeography in the Ordos Basin. *Acta Micropalaeontologica Sinica*, **9**, 127–150 [in Chinese with English abstract].

WATANABE, K. 1991. *Fusuline Biostratigraphy of the Upper Carboniferous and Lower Permian of Japan, with Special Reference to the Carbonifeorus–Permian Boundary*. Palaeontological Society of Japan, Special Papers, **32**.

Wilde, G.L. 1984. Systematics and the Carboniferous–Permian boundary. *Comptes Rendus du 9e Congrès International de Stratigraphie et de Géologie du Carbonifère, Washington et Champaign–Urbana 1979*, **2**, 543–558.

Wilde, G.L. 1990. Practical fusulinid zonation: the species concept; with Permian basin emphasis. *West Texas Geological Society, Bulletin*, **29**, 5–34.

Wilde, G.L. 2006. Pennsylvanian-Permian fusulinaceans of the Big Hatchet Mountains, New Mexico. *New Mexico of Natural History and Science Bulletin*, **38**, 1–331.

Wilde, G.L. & Rudine, S.F. 2000. Late Guadalupian biostratigraphy and Fusulinid faunas, Altuda Formation, Brewster County, Texas. *Smithsonian Contributions to the Earth Sciences*, **32**, 343–371.

Wood, G.D., Groves, J.R., Wahlman, G.P., Brenckle, P.L. & Aleman, A.M. 2002. The paleogeographic and biostratigraphic significance of fusulinacean and smaller foraminifers, and palynomorphs from the Copacabana Formation (Pennsylvanian–Permian), Madre de Dios Basin, Peru. *In*: Hills, L.V., Henderson, C.M. & Bamber, E.W. (eds) *Carboniferous and Permian of the World*. Canadian Society of Petroleum Geologists, Memoirs, **19**, 630–664.

Wu, R.Z. & Lan, B.L. 1990. New Late Permian strata from Northwest Tibet. *Journal of Stratigraphy*, **14**, 216–221 [in Chinese].

Wu, Z.P. & Yang, X.N. 1998. A study on different fusuinid fauna of the Changme stage, southwest China. *Acta Micropalaeontologica Sinica*, **15**, 37–47.

Xia, G.Y. 1981. Zonation of the Early Permian fusulinid-bearing strata in Inner Mongolia. *In*: *12th Annual Conference of the Palaeontological Society of China, Selected Papers*. Science Press, Beijing, China, 116–126.

Xia, G.Y., Li, J.X., Wang, Y.H. & Dong, W.L. 1986. The fusulinid zones and the boundary of Carboniferous–Permian in Longlin region, Guangxi. *Bulletin of Yichang Institute of Geology and Mineral Resources, CAGS*, **11**, 67–104.

Xiao, W.M., Wang, H.D., Zhang, L.X. & Dong, W.L. 1986. *Early Permian Stratigraphy and Faunas in Southern Guizhou*. The People's Publishing House of Guizhou, Guiyang, China [in Chinese with English abstract].

Xie, J.C., Zhu, D.C., Dong, G., Zhao, Z.D., Wang, Q. & Mo, X. 2016. Linking the Tengchong Terrane in SW Yunnan with the Lhasa Terrane in southern Tibet through magmatic correlation. *Gondwana Research*, **39**, 217–229, https://doi.org/10.1016/j.gr.2016.02.007

Xu, S.Y., Xia, G.Y., Yang, D.L., Li, J.X., Liang, Z.F., Li, L. & Zhang, Y.X. 1986. The Carboniferous–Permian boundary in Longlin region, Guangxi. *Bulletin of Yichang Institute of Geology and Mineral Resources, CAGS*, **11**, 1–32.

Yancey, T.E. 1975. Permian marine biotic provinces in North America. *Journal of Paleontology*, **49**, 758–766.

Yang, X.N. & Hao, Y.C. 1991. A study on ontogeny and evolution of Robustoschwagerinids (Permian Fusulinids). *Acta Palaeotologica Sinica*, **30**, 277–306 [in Chinese with English abstract].

Yang, X.N., Liu, J.R. & Shi, G.J. 2004. Extinction process and patterns of Middle Permian Fusulinaceans in southwest China. *Lethaia*, **37**, 139–147, https://doi.org/10.1080/00241160410005114

Yang, Z.D. 1985. Restudy of fusulinids from the 'Maokou limestone' (Permian) at Datieguan, Langdai Guizhou. *Acta Micropalaeontologica Sinica*, **2**, 307–335 [in Chinese with English abstract].

Yang, Z.D. & Yancey, T.E. 2000. Fusulinid biostratigraphy and paleontology of the Middle Permian (Guadalupian) strata of the Glass Mountains and Del Norte Mountains, West Texas. *Smithsonian Contributions to the Earth Sciences*, **32**, 185–259.

Yuan, D.X., Zhang, Y.C. *et al.* 2016. Early Permian conodonts from the Xainza area, central Lhasa Block, Tibet, and their palaeobiogeographical and palaeoclimatic implications. *Journal of Systematic Palaeontology*, **14**, 365–383, https://doi.org/10.1080/14772019.2015.1052027

Zaw, W. 1999. Fusuline biostratigraphy and paleontology of the Akasaka Limestone, Gifu prefecture, Japan. *Bulletin of the Kitakyushu Museum of Natural History*, **18**, 1–76.

Zaw, W., Aung, H.H. & Shwe, K.K. 2011. *Shanita thawtinti*, a new milioloid foraminifer from the Middle Permian of Myanmar. *Micropaleontology*, **57**, 125–138.

Zhang, L.X. 1963. Late Carboniferous fusulinids from Keping and adjacent regions in Xinjiang. *Acta Palaeontologica Sinica*, **11**, 200–227 [in Chinese with Russian abstract].

Zhang, L.X. 1991. Early–Middle Permian fusulinids from Ngari, Xizang (Tibet). *In*: Sun, D.L. & Xu, J.T. (eds) *Permian Jurassic and Cretaceous Strata and Palaeontology from Rutog Region, Xizang (Tibet)*. Nanjing University Press, Nanjing, China, 42–67 [in Chinese with English abstract].

Zhang, L.X., Rui, L., Wang, Z.H. & Li, H.Y. 1988. Fusulinids. *In*: Zhang, L.X., Rui, L. *et al.* (eds) *Permian Palaeontology from Southern Guizhou Province*. People's Publishing House of Guizhou, Guiyang, China, 1–123 [in Chinese with English abstract].

Zhang, L.X., Zhou, J.P., Niu, B.X. & Wang, H. 1989. Fusulinids from Late Carboniferous Taiyuan Formation in Zibo area, Shandong. *Acta Palaeontologica Sinica*, **28**, 803–818 [in Chinese with English abstract].

Zhang, Y.C. 2010. Late Middle Permian (Guadalupian) fusuline fauna from the Gyanyima area of Burang County, Tibet, China and its paleobiogeographic implications. *Acta Palaeontologica Sinica*, **49**, 231–250 [in Chinese with English abstract].

Zhang, Y.C., Wang, Y. & Shen, S.Z. 2009. Middle Permian (Guadalupian) Fusulines from the Xilanta Formation in the Gyanyima area of Burang County, southwestern Tibet, China. *Micropaleontology*, **55**, 463–486.

Zhang, Y.C., Cheng, L.R. & Shen, S.Z. 2010. Late Guadalupian (Middle Permian) fusuline fauna from the Xiala Formation in Xainza County, central Tibet: implication of the rifting time of the Lhasa Block. *Journal of Paleontology*, **84**, 955–973, https://doi.org/10.1666/10-005.1

Zhang, Y.C., Shen, S.Z., Shi, G.R., Wang, Y., Yuan, D.X. & Zhang, Y.J. 2012*a*. Tectonic evolution of the Qiangtang Block, northern Tibet during the Late Cisuralian (Late Early Permian): evidence from fusuline fossil records. *Palaeogeography, Palaeoclimatology, Palaeoecology*, **350–352**, 139–148, https://doi.org/10.1016/j.palaeo.2012.06.025

Zhang, Y.C., Wang, Y., Zhang, Y.J. & Yuan, D.X. 2012*b*. Kungurian (Late Cisuralian) fusuline fauna from the Cuozheqiangma area, northern Tibet and its palaeobiogeographical implications. *Palaeoworld*, **21**, 139–152, https://doi.org/10.1016/j.palwor.2012.09.001

Zhang, Y.C., Shi, G.R. & Shen, S.Z. 2013*a*. A review of Permian stratigraphy, palaeobiogeography and palaeogeography of the Qinghai–Tibet Plateau. *Gondwana Research*, **24**, 55–76, https://doi.org/10.1016/j.gr.2012.06.010

Zhang, Y.C., Wang, Y., Zhang, Y.J. & Yuan, D.X. 2013*b*. Artinskian (Early Permian) fusuline fauna from the Rongma area in northern Tibet: palaeoclimatic and palaeobiogeographic implications. *Alcheringa: An Australasian Journal of Palaeontology*, **37**, 529–546, https://doi.org/10.1080/03115518.2013.805484

Zhang, Y.C., Shi, G.R., Shen, S.Z. & Yuan, D.X. 2014. Permian Fusuline Fauna from the Lower Part of the Lugu Formation in the Central Qiangtang Block and its Geological Implications. *Acta Geologica Sinica – English Edition*, **88**, 365–379, https://doi.org/10.1111/1755-6724.12202

Zhang, Y.C., Shen, S.Z., Zhai, Q.G., Zhang, Y.J. & Yuan, D.X. 2016. Discovery of a Sphaeroschwagerina fusuline fauna from the Raggyorcaka Lake area, northern Tibet: implications for the origin of the Qiangtang Metamorphic Belt. *Geological Magazine*, **153**, 537–543.

Zhang, Y.J., Zhu, T.X., Yuan, D.X. & Zhang, Y.C. 2014. The discovery of Wuchiapingian fossils of Xiala Formation from Xainza area, Tibet and its significance. *Journal of Stratigraphy*, **38**, 411–418 [in Chinese with English abstract].

Zhang, Z.C. 1983. Biostratigraphy of fusulines in the Upper Carboniferous Taiyuan Formation in Xishan, Taiyuan. *Journal of Stratigraphy*, **7**, 272–278 [in Chinese with English abstract].

Zhang, Z.C. 1990. A restudy of Late Carboniferous fusulinaceans from the western hills of Taiyuan. *Acta Micropalaeontologica Sinica*, **7**, 95–122 [in Chinese with English abstract].

Zhang, Z.C. & Xia, G.Y. 1985. Fusulinid zonation of the Upper Carboniferous Shanxi Formation, southeastern Shanxi. *Regional Geology of China*, **12**, 53–61 [in Chinese with English abstract].

Zhang, Z.G., Chen, J.R. & Yu, H.J. 1985. Early Permian stratigraphy and character of fauna in Xainza district, northern Xizang (Tibet), China. *In*: CGQXP Editorial Committee, Ministry of Geology and Mineral Resources PRC (ed.) *Contribution to the Geology of the Qinhai–Xizang (Tibet) Plateau (16)*. Geological Publishing House, Beijing, China, 117–138 [in Chinese with English abstract].

Zhang, Z.H., Wang, Z.H. & Li, C.Q. 1988. *A Suggestion for Classification of Permian in South Guizhou*. Guizhou People's Publishing House, Guiyang, China [in Chinese with English abstract].

Zhao, B., Duan, L.L. & Chen, H.X. 2006. New data on palaeontology of Late Permian in the Ulanul Hu area, northern Qiangtang Basin, NW China. *Acta Micropalaeontologica Sinica*, **23**, 392–398 [in Chinese with English abstract].

Zheng, Y.T. & Lin, J.X. 1991. Foraminifera and fusulinids. *In*: Southern Xinjiang Petroleum Prospecting Corporation, Xinjiang Petroleum Administration Bureau & Exploration and Development Research Institute, Jianghan Petroleum Administration Bureau (eds) *Sinian to Permian Stratigraphy and Palaeontology of the Tarim Basin, Xinjiang (II), Kalpin–Bachu Region*. The Petroleum Industry Press, Beijing, China, 158–190 [in Chinese with English abstract].

Zhou, T.M., Sheng, J.Z. & Wang, Y.J. 1987. Carboniferous–Permian boundary beds and fusulinid zones at Xiaodushan, Guangnan, eastern Yunnan. *Acta Micropalaeontologica Sinica*, **4**, 123–160 [in Chinese with English abstract].

Zhou, Z.R. 1982. Earliest Permian *Schwagerina cushmani* fusulinid fauna from southeastern Hunan. *Acta Palaeontologica Sinica*, **21**, 225–248 [in Chinese with English abstract].

Zhu, X.F. 1982*a*. Lower Permian fusulinids from Xainza County, Xizang (Tibet). *In*: CGQXP Editorial Committee, Ministry of Geology and Mineral Resources PRC (ed.) *Contribution to the Geology of the Qinhai–Xizang (Tibet) (7)*. Geological Publishing House, Beijing, China, 110–133 [in Chinese with English abstract].

Zhu, X.F. 1982*b*. Lower Permian fusulinids from Lhunzhub County, Xizang (Tibet). *In*: CGQXP Editorial Committee, Ministry of Geology and Mineral Resources PRC (ed.) *Contribution to the Geology of the Qinhai–Xizang (Tibet) (7)*. Geological Publishing House, Beijing, China, 136–148 [in Chinese with English abstract].

Zhu, Z.L. 1997. Carboniferous and Permian fusulinid zonation in southwestern margin of the Tarim basin. *Acta Palaeontologica Sinica*, **36**, 104–115 [in Chinese with English abstract].

Global Permian brachiopod biostratigraphy: an overview

SHU-ZHONG SHEN

State Key Laboratory of Palaeobiology and Stratigraphy, Nanjing Institute of Geology and Palaeontology, Chinese Academy of Sciences, 39 East Beijing Road, Nanjing, Jiangsu, 210008, China

szshen@nigpas.ac.cn

Abstract: Establishing a Permian brachiopod biochronological scheme for global correlation is difficult because of strong provincialism during the Permian. In this paper, a brief overview of brachiopod successions in five major palaeobiogeographical realms/zones is provided. For Gondwanaland and peri-Gondwanan regions including Cimmerian blocks, *Bandoproductus* and *Punctocyrtella* (or *Cyrtella*) are characteristic of the lower Cisuralian, as is *Cimmeriella* for the middle Cisuralian. As the Cimmerian blocks continued drifting north during the late Kungurian, accompanied by climate amelioration, contemporaneous brachiopods inhabiting these blocks showed a distinct shift from cold-water to mixed or warm-water affinities. However, coeval brachiopods in the Northern Transitional Zone (NTZ) are characterized by warm-water faunas and are associated with fusulinids in the lower Cisuralian. The Guadalupian brachiopods of the NTZ were clearly mixed between the Boreal and palaeoequatorial affinities. The end-Guadalupian is marked by the disappearance of a few characteristic genera, such as *Vediproductus*, *Neoplicatifera* and *Urushtenoidea*, in the Palaeotethyan region. The onset of the end-Permian mass extinction in the latest Changhsingian is clearly exhibited by the occurrence of the dwarfed and thin-shelled brachiopods commonly containing *Paracrurithyris*.

Brachiopods are one of the most abundant and highly diverse benthic groups in the Palaeozoic oceans. The Permian was the last period that brachiopods maintained their high diversity and abundance during the Phanerozoic. Following the end-Permian mass extinction, the benthic marine ecosystem underwent a remarkable reorganization, changing from the brachiopod-dominated Palaeozoic Fauna to the mollusk-dominated Modern Fauna (Sepkoski 1984). Brachiopod body size was also significantly reduced in the aftermath of the end-Permian mass extinction (He *et al.* 2007; Twitchett 2007; Zhang & Payne 2012; Zhang *et al.* 2015; Shi *et al.* 2016). Four highly diverse brachiopod orders (Orthida, Orthotetida, Productida and Spiriferida) were totally wiped out by the extinction and the diversity of other orders were also significantly reduced (Shen & Shi 1996; Shi & Shen 2000; Harper & Rong 2001; Rong & Shen 2002; Chen *et al.* 2005; Shen *et al.* 2006; Curry & Brunton 2007; He *et al.* 2015).

Brachiopods have a short larval stage that swim for a few hours to a few weeks. Once they settle down, brachiopods mostly anchor on the seafloor and do not migrate unless they attach onto some moving objects. So, brachiopods have been widely used to evaluate their palaeobiogeographical distribution because they are sensitive to thermal gradients (Stehli 1957; Waterhouse & Carter 1975; Powell *et al.* 2015) and geographical barriers (e.g. Shi & Archbold 1995*a*; Shen *et al.* 2009, 2013*b*; Angiolini *et al.* 2013*a*; Ke *et al.* 2016), but relatively less has been presented for their biochronological correlation in a global or regional sense due to strong provinciality and relatively low evolutionary rates.

Generally, three palaeolatitude-related realms can be widely recognized during the Permian, namely the Boreal Realm in the northern hemisphere, the Gondwanan Realm in the southern hemisphere and the Palaeoequatorial Realm between them (Grunt & Shi 1997). Among them, the Palaeoequatorial Realm may be divided into the Palaeotethyan Subrealm in the east and the North American Subrealm in the west, separated by Pangea. Likewise, provinces have also been recognized within each of the realms in different stages (Bambach 1990; Shi *et al.* 1995; Shen *et al.* 2000*a*, 2009, 2013*b*; Shen & Shi 2000, 2004; Shi & Grunt 2000; Ke *et al.* 2016). Typically, every realm/subrealm/province contains many characteristic taxonomic elements useful for regional correlations, but it is relatively difficult to achieve robust correlations between realms. However, two transitional zones between these three realms were developed during the Permian (Shi *et al.* 1995), and these two transitional zones contain characteristic mixed brachiopod faunas between the two antitropical realms and the Palaeoequatorial Realm, which may provide gateways to realize the correlation between these three different realms, as demonstrated by Shi (2006).

Like many other benthic fossil groups, Permian brachiopod faunas were highly provincial and facies-dependent, and are thus not suited to serve

From: LUCAS, S. G. & SHEN, S. Z. (eds) 2018. *The Permian Timescale*. Geological Society, London, Special Publications, **450**, 289–320.
First published online December 9, 2018, https://doi.org/10.1144/SP450.11

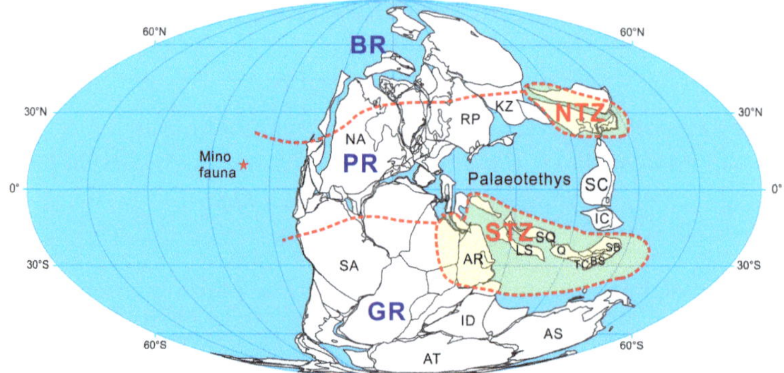

Fig. 1. Global palaeobiogeography based on brachiopods during the Permian. The base map is from Christophore Scotese (http://www.scotese.com/). BR, Boreal Realm; PR, Palaeoequatorial Realm; GR, Gondwanan Realm; NTZ, Northern Transitional Zone; STZ, Southern Transitional Zone; KZ, Kazankstan; RP, Russian Platform; NA, North America; SC, South China; IC, Indochina; SA, South America; AR, Arabian Plate; ID, Indian Plate; AT, Antarctic; AS, Australia; LS, Lhasa Block; SQ, South Qiangtang Block; Q, Qomdo Block; TC, Tengchong Block; BS, Baoshan Block; SB, Sibumasu Block.

as a global standard for biostratigraphic correlation. Some brachiopod successions reviewed by Waterhouse (1976) are still applicable for correlation. However, numerous new brachiopod faunas have been described after Waterhouse (1976) and the Permian timescale has been greatly revised based on conodonts and high-precision geochronology. Thus, it is timely to review the brachiopod biostratigaphical framework based on the new international Permian timescale (Henderson *et al.* 2012; Shen *et al.* 2013*a*; Shen & Henderson 2014). In this paper, an overview of the temporal distribution and global/interregional correlation based on the Permian brachiopods is presented (Fig. 1).

Compared to some other fossils groups, such as conodonts, fusulinids and ammonids, brachiopods are not particularly age-sensitive. Therefore, it is difficult to precisely determine the ages of brachiopod assemblages and realize their intercontinental correlation without other constrains. However, brachiopods are commonly associated with conodonts, fusulinids and ammonoids in the Palaeoequatorial Realm and the two transitional zones, so the ages and correlation of brachiopod assemblages discussed below to some extent rely on their associated age-sensitive fossils, including conodonts, ammonoids and fusulinids. As for the brachiopod assemblages between the two polar and two transitional zones, some bipolar/bitemperate forms (e.g. *Jakutoproductus*, *Kochiproductus* and *Spiriferella*) are useful for correlation (Shi & Grunt 2000; Shi 2006). In addition, some robust radioisotopic ages derived from ash beds intercalated with brachiopod faunas have become available in recent years, greatly aiding the age determination and correlation of some high-latitude Permian brachiopod biochronozes (Gulbranson *et al.* 2010; Césari *et al.* 2011; Mory *et al.* 2012; Metcalfe *et al.* 2015; Barbolini *et al.* 2016).

The Boreal Realm

Western Canada

The Permian brachiopod succession in western Canada has been well studied from the Jungle Creek Formation between the Ettrain Formation below and the Tahkandit Formation above in the Yukon Territory (Fig. 2). Four members were recognized by Waterhouse (2013). Member A is dated as Gzhelian and members B–D dated as Asselian. The brachiopod succession in this area was originally established by Bamber & Waterhouse (1971) and subsequently refined by Shi & Waterhouse (1996). Four Permian brachiopod zones from the Jungle Creek Formation were recognized. The lowest *Tomiopsis–Attenuatella* Zone was assigned to the Asselian. This is followed by the *Yakovlevia transversa* and *Ogilviecoelia inflata* zones: both were assigned to the Sakmarian because they are associated with the ammonoid *Uraloceras* sp. B resembling *U. bilobatus* (Shi & Waterhouse 1996). Shi (1995) provided a global review of all *Yakovlevia* species and considered that the *Yakovlevia transversa* Zone in the Yukon Territory is generally of early Sakmarian, but Waterhouse (2013) recently considered it to be of Asselian age. The overlying *Jakutoproductus verchoyanicus* Zone is associated with Artinskian ammonoids in the Urals and the fusulinid *Eoparafusulina yukonensis*, and overlies the horizon containing the ammonoids *Properrinites*, *Medlicottia* and *Somoholites* in the Peel River

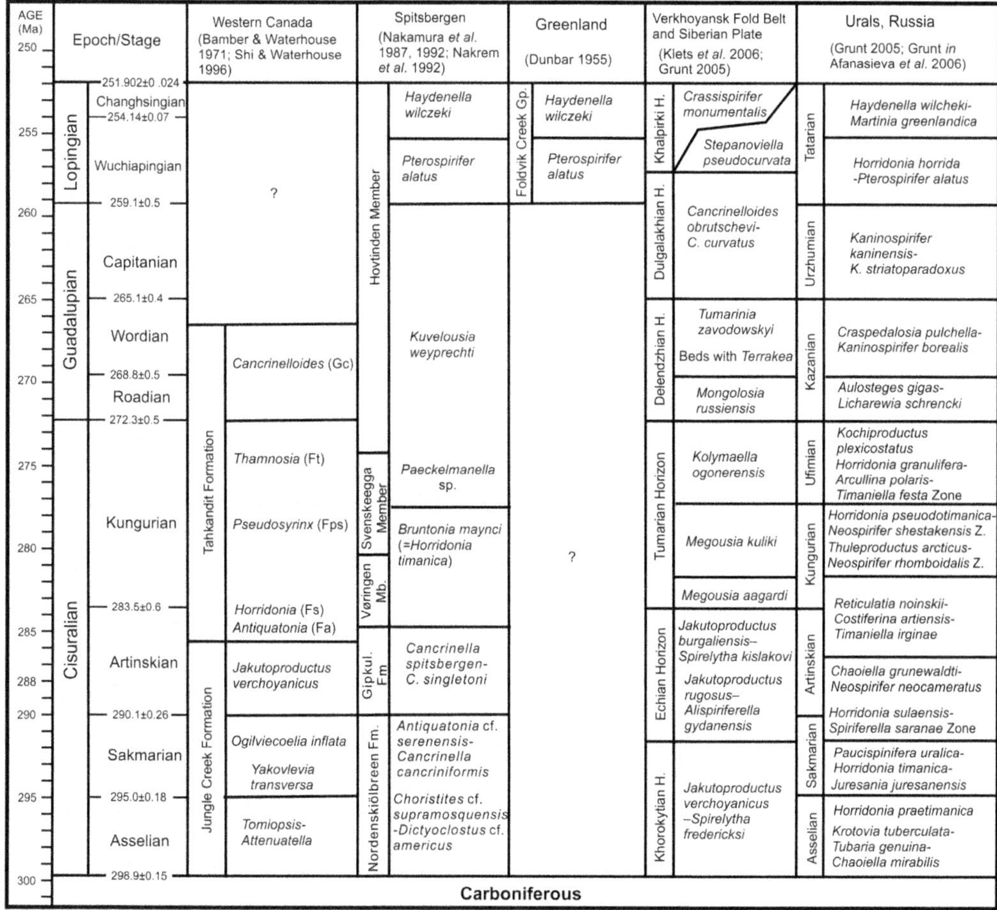

Fig. 2. Revised brachiopod successions in the Boreal Realm. The main references are shown in the columns. The Permian timescale is revised from Shen *et al.* (2013*a*).

in Yukon Territory, and therefore is assigned to the early Artinskian (Bamber & Waterhouse 1971; Shi & Waterhouse 1996). This zone is widely distributed in Arctic Russia and restricted to the Artinskian Echii Horizon in the Verkhoyansk region (Klets *et al.* 2006).

The overlying Tahkandit Formation contains four brachiopod zones in the Tatonduk River area (Bamber & Waterhouse 1971). The lower two zones, the *Antiquatonia* (Fa) and *Horridonia* (=*Sowerbina*, Fs) zones, are associated with the fusulinid *Schwagerina jenkinsi* Zone of late Artinskian age (Ross 1967). The third zone, the *Pseudosyrinx* (Fps) Zone, contains some common elements including *Neochonetes*, *Waagenoconcha*, *Terrakea* and *Yakovlevia mammata*, which are also common in the Kungurian Asistance Formation in the northern Richardson Mountains. The *Thamnosia* (Ft) Zone of the Tahkandit Formation contains *Anemonaria*, large *Yakovlevia* and *Horridonia*. Both upper zones are assigned to the Kungurian Stage.

The youngest brachiopod zone in the Yukon Territory was named the *Cancrinelloides* (Gc) Zone from the upper Tahkandit Formation, which was assigned to the Kazanian (Roadian–Wordian) age. This zone is characterized by the genera *Neochonetes* and *Cancrinelloides*. In the Jungle Creek area, it is found in the uppermost Tahkandit Formation, and contains *Neochonetes*, *Kuvelousia* and *Yakovlevia*. In the northern Richardsons, *Cancrinelloides* occurs with *Kuvelousia sphiva*, *Terrakea arctica* and *Yakovlevia duplex* (Bamber & Waterhouse 1971).

Spitsbergen and Greenland

Permian brachiopods with typical Boreal affinity are very abundant in Spitsbergen and Greenland

(Fig. 2). Uppermost Carboniferous and Lower Cisuralian faunal assemblages were reviewed by Nakrem *et al.* (1992). Brachiopods are common in the carbonates of the Tyrrellfjellet Member of the Nordenskiölbreen Formation in Svalbard. Asselian–?Sakmarian brachiopod assemblages are represented by *Choristites* cf. *supramosquensis*, *Dictyoclostus* cf. *americus* below the Brucebyen beds containing abundant *Schwagerina* spp., and *Sphaeroschwagerina*? sp. A, *Antiquatonia* cf. *serenensis*, *Cancrinella cancriniformis* and *Martinia semiglobosa* in the upper part (Limestone B). The former is more like an assemblage of latest Carboniferous, and the latter has a restricted range of the *Schwagerina* Limestone in the Urals. These brachiopod assemblages are associated with the conodonts *Streptognathodus elongatulus*, which also ranges from latest Carboniferous to early Cisuralian. A probably slightly younger brachiopod fauna was found in the Gipshuken Formation on the SW ridge of Cowantoppen. Gobbett (1963) reported the brachiopods *Cancrinella spitsbergeniana*, *C. singletoni*, *Linoproductus konincki* and *L. cora*, amongst others, from this formation. The conodont *Neostreptognathodus* cf. *pequopensis* is present in the upper part of the formation: thus, the brachiopods are most likely Artinskian in age. An international geological expedition was carried out and a large number of brachiopods were collected by the Japanese–Norwegian Research Group (Nakamura *et al.* 1992). However, the studies of these brachiopods have never been completed. Nakamura *et al.* (1987, 1992) produced some brief reports and proposed a brachiopod succession for the Kapp Starostin Formation of western Spitsbergen. Five brachiopod zones, the *Horridonia timanica* (it has now been reassigned to *Bruntonia maynci* by Angiolini & Long (2008)), *Paeckelmanella* sp., *Kuvelousia weyprechti*, *Pterospirifer alatus* and *Haydenella wilczeki* zones in ascending order, were established. However, their age assignments were questioned (Stemmerick 1988). The lowest *Bruntonia maynci* (=*Horridonia timanica*) Zone from the Vøringen Member (*Spirifer* Limestone) was characterized particularly by the abundance of productids, including *Bruntonia maynci*, *Archboldevia impressa*, *Yakovlevia mammata*, *Waagenoconcha irginae* and *Chaoiella neoinflata*, and was considered to be correlative with the Kungurian of the Urals (Nakamura *et al.* 1987; Nakrem *et al.* 1992). However, Mei & Henderson (2001) reassigned some of the figured conodont specimens of Szaniawski & Malkowski (1979) to *N.* ?*ruzhencevi* or *Sweetognathus clarki* and considered the conodont fauna to be late Artinskian in age rather than Kungurian or early Roadian. A recent study of the collection from the the Vøringen Member at Skansen, Billefjorden stored in the Natural History of Museum in London also suggests a Kungurian age (Angiolini & Long 2008). The overlying Svenskegga Member, together with the Hovtinden Member of the Kapp Starostin Formation, was previously named as 'Brachiopod Chert', which was assigned to the late Kungurian and early Guadalupian, and a long hiatus was inferred between the Kapp Starostin Formation and the overlying Lower Triassic Vardebuka Formation by Nakamura *et al.* (1987, 1992). However, Nakamura *et al.* (1987, 1992) distinguished two brachiopod zones in the Foldvik Creek Group, East Greenland, which are comparable to the *Kuvelousia weyprechti* Zone and the upper *Pterospirifer alatus* Zone of the Hovtinden Member in western Spitsbergen. The diagnostic species of these two zones in East Greenland are *Waagenoconcha payeri*, *Arctitreta kempei* and *Yakovlevia greenlandica* for the *Kuvelousia weyprechti* Zone, and *Pterospirifer alatus*, *Pleurohorridonia scoresbyensis*, *Choristites soderberghi* and *Odontospirifer mirabilis* for the *Pterospirifer alatus* Zone (Nakamura *et al.* 1987). Furthermore, *Liosotella spitzbergiana*, *Paeckelmannia toulai* and *Kochiproductus plexicostatus*, which are confined to the uppermost zones in the Kapp Starostin Formation, are also found in East Greenland. Thus, brachiopods from the *Haydenella wilczeki* and *Pterospirifer alatus* zones suggest a firm correlation with the upper part of the *Martinia* limestone containing the Lopingian ammonoid *Cyclolobus* in East Greenland. In addition, the Lopingian conodont *Mesogondolella rosenkrantzi* was reported from the *Martinia* Limestone (Mei & Henderson 2001). Therefore, the brachiopods of the Kapp Starostin Formation of the Hovtinden Member in western Spitsbergen are comparable with the Foldvik Creek Group of East Greenland, and at least the uppermost two zones are Lopingian in age (Waterhouse 1976; Stemmerick 1988; Shen *et al.* 2005; Angiolini & Long 2008). This is also consistent with the chemostratigraphical studies on end-Permian mass extinction in Spitsbergen and Greenland, which show that the Permian–Triassic boundary is continuous in these areas (Wignall *et al.* 1998; Twitchett *et al.* 2001; Shen *et al.* 2005).

Verkhoyansk Fold Belt and Siberian Plate

Permian deposits are widely distributed in the Verkhoyansk–Okhotsk region where they are divided into six different horizons (Fig. 2) (Klets *et al.* 2006). The lowest Khorokytian Horizon (Asselian–lower Sakmarian) has a thickness from 250 to 400 m. The brachiopod *Jakutoproductus verchoyanicus*–*Spirelytha fredericksi* Zone was recognized from this horizon. It overlies the horizon with the Upper Carboniferous *Verchojania expositus* Assemblage and is also present in the Lower

Permian (Cisuralian) of the Kolyma–Omolon region (Ganelin & Kotlyar 1984; Ganelin 1990; Ganelin & Biakov 2006) and Taimyr (Ustritsky & Tschernjak 1963; Klets 2005; Klets *et al.* 2006). A similar *Jakutoproductus*-dominated brachiopod assemblage has been reported in central and NE Mongolia (Manankov *et al.* 2006). The overlying Echian Horizon comprises two brachiopod assemblages: the *Jakutoproductus rugosus–Alispiriferella gydanensis* Zone in the lower and *Jakutoproductus burgaliensis–Spirelytha kislakovi* Zone in the upper. Associated with these two brachiopod zones are the ammonoid *Uraloceras subsimense*, *Eotumaroceras endybalense* and *E. subyakutorum* zones in descending order. The lower two ammonoid zones contain species-typical elements of Artinskian age in the Urals. Similar Artinskian brachiopods of these zones have been found in the coeval deposits of Taimyr, the Kolyma–Omolon regions and Transbaikalia (Ustritsky & Tschernjak 1963; Kotlyar & Popeko 1967; Ganelin 1990; Kotlyar *et al.* 2002; Klets 2005; Klets *et al.* 2006). The Russian Kungurian and Ufimian are represented by the Tumarian Horizon in this region. Three brachiopod assemblages, including the *Megousia aagardi*, *M. kuliki* and *Kolymaella ogonerensis* assemblages, in ascending order are recognized (Kotlyar *et al.* 2004*a*; Klets *et al.* 2006). They are associated with the Kungurian ammonoids *Tumaroceras yakutorum* and *Epijuresanites mesalitini*. The brachiopod assemblages have been found not only in the coeval deposits of the Taimyr and Kolyma–Omolon regions, but also in the Kungurian Stage of the Urals.

The Lower Delendzhian Subhorizon contains the brachiopod *Mongolosia russiensis* Assemblage, and beds with *Terrakea* in association with the ammonoids *Sverdrupites harkeri* and *S. baraiensis*, which indicate a Roadian age (Leonova *et al.* 2002). Within the Upper Delendzhian Subhorizon, the *Tumarinia zavodowskyi* Zone is recognized, which is highly likely to be of middle–upper Guadalupian age. The overlying Dulgalakhian Horizon is of middle Tatarian age (probably Capitanian–early Wuchiapingian). It includes the brachiopod *Cancrinelloides obrutschevi–C. curvatus* Zone. The Khalpirki Horizon is of upper Tatarian (Lopingian) age and includes the brachiopod *Crassispirifer monumentalis* Zone (Klets *et al.* 2006) or the *Stepanoviella pseudocurvata* Zone (Grunt *in* Afanasieva *et al.* 2006). Brachiopod species typical of the Khalpirki Horizon are *Marginalosia*? *magna* Abramov & Grigorieva, *Biplatyconcha lungersgauzeni* (Solomina), *Grantonia grandis* Solomina & *Crassispirifer monumentalis* Abramov & Grigorieva. The genera *Marginalosia* and *Biplatyconcha* are also common in the Lopingian of the peri-Gondwanan region (Waterhouse 1978; Shen *et al.* 2000*b*, 2003*b*). Both zones were assigned to the upper Guadalupian (Manankov *et al.* 2006), but a Lopingian age is also possible because the uppermost brachiopod *Stepanoviella pseudocurvata* Zone is correlative with the ammonoid *Cyclolobus kullingi* Zone in the Boreal Realm (Grunt *in* Afanasieva *et al.* 2006).

The Urals, Russia

Brachiopod successions are not well established and data are not available after the seawater withdrew from the southern part of the Urals seaway. However, a continuous post-Artinskian brachiopod succession was established in the Kanin peninsula in the northern Urals (Grunt 2005; Grunt *in* Afanasieva *et al.* 2006), and in the Kozhim and Tab-Yu sections of northern Cis-Urals (Kotlyar *et al.* 2004*a*). The succession in the Kanin peninsula is different to the succession in the Verkhoyansk Fold Belt (Fig. 2). The brachiopod assemblages are mostly composed of Boreal cold-water or bipolar elements, such as *Bruntonia* (or *Horridonia*), *Kochiproductus* and *Tuleproductus*, which are similar to those from Spitsbergen, Greenland and Novya Zemlya. Four regional brachiopod zones were proposed in the Asselian and Sakmarian stages (Grunt 2005). Three assemblages in the Artinskian, four assemblages in the Russian Kungurian and Ufimian, three assemblages of the Kazanian and Urzhumian, and two assemblages of the Tatarian were recognized by Grunt (2005) and Grunt *in* Afanasieva *et al.* (2006). Among them, the *Kaninospirifer kaninensis–K. striatoparadoxus* Assemblage is correlative with the *Cancrinelloides obrutshevi* Assemblage of the Guadalupian, and the *Horridonia horrida–Pterospirifer alatus* and the *Haydenella wilcheki–Martinia greenlandica* assemblages are correlative with the Lopingian *Cancrinelloides curvatus* and *Stepanoviella pseudocurvata* assemblages in the Verkhoyansk Fold Belt (Grunt 2005; Grunt *in* Afanasieva *et al.* 2006).

However, the Kungurian brachiopod succession in the Cis-Urals (Kotlyar *et al.* 2004*a*) is somewhat similar to those of the Verkhoyansk Fold Belt and provides some support for correlation with Spitsbergen. The Kungurian (=Russian Kungurian plus Ufimian) Stage contains four different brachiopod zones. They are the *Bruntonia maynci* (=*Horridonia timanica*)–*Megousia aagardi*, *Megousia kuliki–Bruntonia maynci*, *M. kuliki*–'*Horridonia granulifera*' and *Striapustula koninckiana* zones in ascending order (Kotlyar *et al.* 2004*a*). The lower two brachiopod zones probably suggest that the similar zones in Spitsbergen characterized by *Bruntonia maynci* may be also of Kungurian age in contrast to the age indicated by the conodonts of Mei & Henderson (2001).

The Northern Transitional Zone (NTZ)

Inner Mongolia, northern China

One of the most characteristic palaeobiogeographical phenomena during the Permian was the development of two transitional zones in which mixed faunas between the palaeoequatorial warm-water and antitropical cold-water faunas developed. In the north, this zone has been well recognized from the upper Artinskian to the upper Guadalupian in the Central Asian Orogenic Belt (CAOB) between the Sino-Korean Block and the Siberian Plate. It was named the Sino-Mongolian–Japanese transitional zone by Tazawa (1991, 1998) and has been subsequently extensively discussed (Shi *et al.* 1995; Shen & Shi 2000, 2004; Shi & Grunt 2000; Shi 2006; Shen *et al.* 2009, 2013*b*).

Brachiopods are extremely abundant in this zone, and they are well represented by the brachiopod faunas in the Zhesi area of Inner Mongolia, North China (Grabau 1931; Duan & Li 1985; Shi *et al.* 2002*b*; Wang & Zhang 2003) and the Solonker Zone in southern Mongolia (Manankov *et al.* 2006; Manankov 2012) and Japan (see below) (Fig. 3). A large number of the uppermost Carboniferous–lower Cisuralian brachiopods were recorded from the Amushan Formation containing the fusulinids *Triticites*, *Pseudoschwagerina* and *Eoparafusulina* (Li & Gu 1976). They are dominated by the brachiopods *Choristites*, *Dictyoclostus*, *Marginifera* and *Enteletes*, but no brachiopod assemblage has been established. The brachiopods from the Amushan Formation are overwhelmingly composed of Tethyan or cosmopolitan elements in association with fusulinids. Very few Boreal/bipolar brachiopods (e.g. *Jakutoproductus ordinatus*) have been recorded from the formation, which suggests that the NTZ had not yet developed during the early

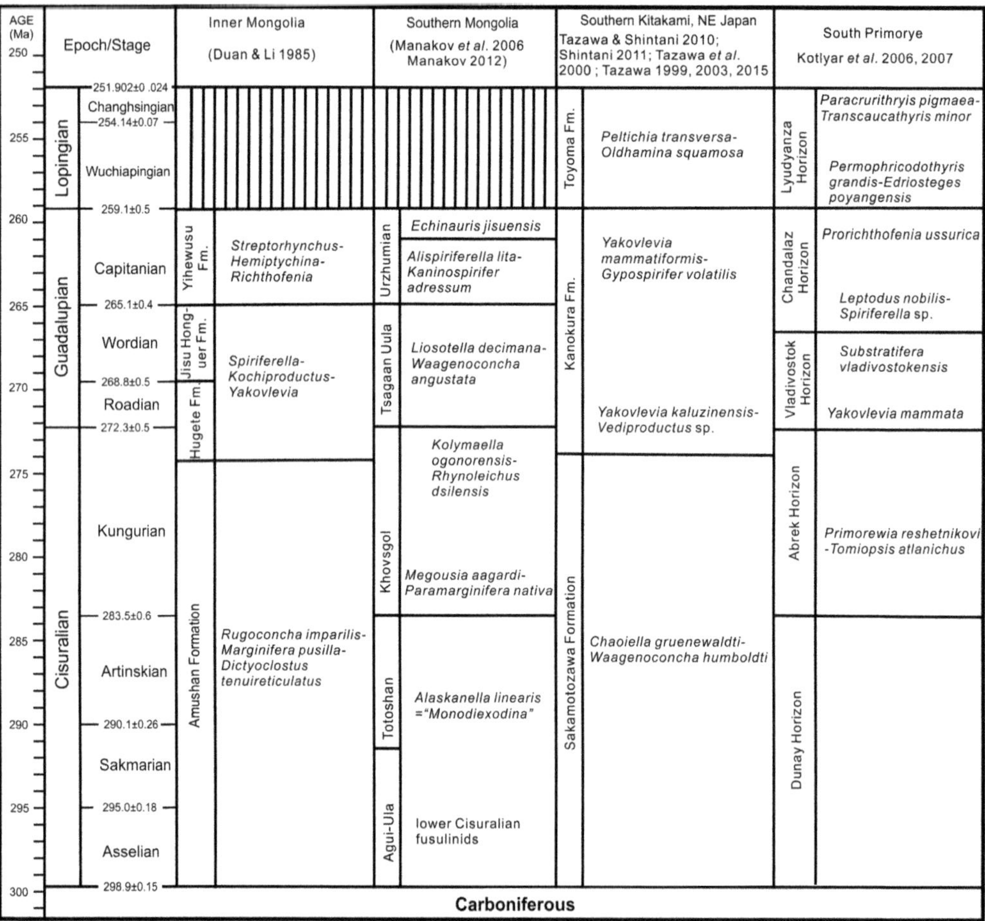

Fig. 3. Revised brachiopod successions in the Northern Transitional Zone. The main references are shown in the columns. The Permian timescale is revised from Shen *et al.* (2013*a*).

Cisuralian. This was most likely because the NTZ was still located in the palaeotropical area during the early Cisuralian. The Amushan Formation is generally characterized by the *Rugoconcha imparilis–Marginifera pusilla–Dictyoclostus tenuireticulatus* Assemblage. Many elements of this assemblage are also present in the lower Cisuralian of North China and South China.

As the NTZ moved further northwards, the Permian mixed brachiopod assemblages began to occur in the Hugete and Jisu Honguer formations. Two assemblages, the *Spiriferella–Kochiproductus–Yakovlevia* (SKY) Assemblage from the uppermost part of the Hugete and Jisu Honguer formations, and the *Streptorhynchus–Hemiptychina–Richthofenia* (SHR) Assemblage from the Yihewusu Formation, were established by Duan & Li (1985). The former was assigned to the uppermost Kungurian *Cancellina* Zone, but was considered to be of lower Murgabian (Roadian or lower Wordian) based on a review of the fusulinid *Monodiexodina* fauna in NE China (Ueno 2006) and conodonts (C.Y. Wang *et al.* 2004). The overlying SHR assemblage was assigned to the Capitanian age based on the conodonts from the Jisu Honguer Formation (C.Y. Wang *et al.* 2004). These two assemblages were subsequently over-split into five different assemblages (Wang & Zhang 2003). They are the *Yakovlevia mammata–Megousia aagardi*, *Alispiriferella neimongolensis–Spiriferella magna*, *Waagenoconcha* (*Yazengoconcha*) *neimengolica–Spiriferella salteri*, *Yakovlevia gigantica–Rhombospirifer zhesiensis* and *Richthofenia cornuformis–Enteletes andrewsi–Notothyris nucleolus* assemblages in ascending order. The first two assemblages were assigned to the upper Kungurian, the third and fourth assemblages to the Wordian, and the uppermost assemblage from the Yihewusu Formation to the Capitanian (Wang & Zhang 2003).

Permian brachiopods disappeared in the western part of the NTZ after the Guadalupian following the closure of the Central Asian Ocean (CAO) in the Tianshan–Junggar and Mongolian zone. However, possible Lopingian deposits containing marine fossils have been recently reported from the eastern Inner Mongolian area, but no brachiopods were found there (Zhang *et al.* 2014).

Southern and SE Mongolia

Permian brachiopods with a mixed Boreal–Cathaysian character have also been documented from the Solonker Zone in southern Mongolia (Fig. 3). The lower Cisuralian Agui–Ula Horizon (Asselian and Sakmarian) of southern Mongolia is characterized by its numerous fusulinids, including the *Daixina diafana gobiensis*, *Pseudoschwagerina uddeni* and *Sphaeroschwagerina sphaerica gigas* zones, which are comparable with those of the Amushan Formation in Inner Mongolia of North China. This is overlain by strata with the '*Monodiexodina linearis*' fauna in the Totoshan Horizon, which was assigned to the upper Sakmarian–Artinskian (Manankov *et al.* 2006; Manankov 2012). However, Ueno (2006) reassessed the true *Monodoixeodina* fauna in the NTZ and considered that they are referable to the lower Murgabian (Wordian). No brachiopod is known from the lower Cisuralian in southern Mongolia. The lowest brachiopod assemblage was established in the Khovsgol Horizon. This is the *Megousia aagardi–Paramarginifera nativa* Assemblage, which is followed by the *Kolymaella ogonrensis–Rhynocleichus dsilensis* Assemblage. Both were assigned to the Kungurian. The Guadalupian (Middle Permian) includes the *Liosotella decimana–Waagnoconcha angustata* (Roadian), *Alispiriferella lita–Kaninospirifer adressum* (Wordian) and *Echinauris jisuensis* (Capitanian) assemblages (Manankov *et al.* 2006; Manankov 2012). No Lopingian brachiopods have been found from southern Mongolia.

South Primorye, Russian Far East

A Permian brachiopod succession with mixed character has also been documented from South Primorye, in the Russian Far East (Kotlyar *et al.* 2006, 2007), and reviewed by Shi (2006). The *Primorewia reshetnikovi–Tomiopsis atlanichus* Zone from the middle part of the Abrek Formation was assigned to the Kungurian. The Guadalupian began with the *Yakovlevia mammata* Zone in the Vladivostok Horizon (Roadian–lower Wordian). This is followed by the upper Wordian *Substratifera vladivostokensis* Zone in the lowest part of the Chandalaz Horizon containing the conodont *Jinogondolella* cf. *aserrata*. Common brachiopods in this zone are *Derbyia grandis*, *Waagenoconcha kryshtofovochi*, *W. maliavkini*, *Leptodus* sp., *Transennatia* sp., *Spiriferella* sp., *Yakovlevia* sp. and *Alispiriferella lita*, which show an obvious mixed character between the Boreal and Palaeotethyan affinities. The middle and upper parts of the Chandalaz Horizon contain, respectively, the *Leptodus nobilis–Spiriferella rajah* Zone and the *Prorichthofenia ussurica* Zone. These two zones are totally comparable with those of the Yihewusu Formation in the Jisu Honguer area of Inner Mongolia, North China. They contain *Tyloplecta yangtzeensis*, *Haydenella kiangsiensis*, *Echinauris jisuensis*, *Leptodus nobilis* and *Spinomarginifera lopingensis*, in addition to the Boreal/bipolar genera *Yakovlevia*, *Kochiproductus* and *Waagenoconcha*. The overlying *Permophricodothyris grandis–Edriosteges poyangensis* Zone from the lower part of the Lyudyanza Horizon is of Wuchiapingian age because it contains the ammonoid *Cyclolobus kiselevae*, the fusulinid *Codonofusiella kwangsiana* and the conodont *Clarkina* ex gr. *orientalis* (Kotlyar *et al.* 2007). The upper part of the Lyudyanza Horizon contains typical Changhsingian brachiopods including *Paracrurithryis pigmaea* and *Transcaucathyris minor* in association with the ammonoids *Changhsingoceras*, *Tapashanites* and *Sinoceltites* (Kotlyar *et al.* 2006).

Japan

Brachiopods are also abundant in various terranes with different palaeogeographical affinities in Japan, of which the southern Kitakami Belt is the most extensively studied area in Japan for its Permian brachiopods. The Permian sequence in this area was traditionally divided into three formations (the Sakamotozawa, Kanokura and Toyomo formations in ascending order) (Tazawa 1976). More than 30 brachiopod species have been reported from the lower part of Sakamotozawa Formation (Asselian and Sakmarian) in the southern Kitakami Belt (Tazawa & Shintani 2010; Shintani 2011; Tazawa 2015), and they are overwhelmingly dominated by warm-water and cosmopolitan elements such as *Chaoiella gruenewaldti*, *Meekella uralica*, *Orthothetina curvata* and *Derbyia dorsisulcata*, and the only brachiopod species with a possibly cold-water affinity is *Waagenoconcha asiatica*. This brachiopod fauna is associated with the early Cisuralian fusulinids, which are comparable with the Cisuralian Amushan Formation in NE China (Tazawa 2015).

Similar to the brachiopod faunas in Inner Mongolia, NE China and the Solonker Zone in southern Mongolia, brachiopod faunas from the Guadalupian in the southern Kitakami area are characterized by a typical mixed fauna between the Boreal and Tethyan realms. Some typical Boreal-type or antitropical brachiopods such as *Yakovlevia kaluzinensis*, *Kochiproductus* sp., *Gypospirifer kobiyami*, *Kaninospirifer* sp. and *Spiriferella keilhavii* are associated with many warm-water species such as *Leptodus nobilis*, *Vediproductus* sp., *Derbyia grandis* and *Enteletes acutiplicatus* (Tazawa 1999, 2003; Tazawa *et al.* 2000; Tazawa & Ibaraki 2001; Tazawa & Araki 2013). This mixed brachiopod fauna is basically comparable with the Zhesi fauna (Tazawa *et al.* 2001; Wang & Zhang 2003; Shi 2006) and the upper Kungurian–Guadalupian brachiopod faunas in the Solonker Zone of southern Mongolia (Manankov *et al.* 2006), which suggests that the southern Kitakami Belt may belong to the eastern extension of the NTZ during the Guadalupian.

The CAO has been well documented to have been progressively closed from west to east during the Permian (Xiao *et al.* 2003). Therefore, there is no marine deposit with brachiopods recorded from NE China and southern Mongolia after Guadalupian. However, Lopingian brachiopods are still widely reported from the southern Kitakami area in NE Japan. Typical cold-water 'Wuchiapingian' brachiopods, including *Anemonaria pseudohorrida*, *Kochiproductus okutadamiensis*, *Yakovlevia mammatiformis* and *Spiriferella magna*, were reported from the Otori Formation at Okutadami, Joetsu Belt, central Japan (Tazawa 2011). However, this fauna seems comparable with the Guadalupian brachiopod faunas in NE China and southern Mongolia. Similar 'Late Permian' mixed brachiopod fauna, including *Yakovlevia mammatiformis*, *Oldhamina* cf. *lianyangensis*, *Gypospirifer volatilis* and *Alispiriferella lita*, were also reported from the Takakurayama Formation in the Abukuma Mountains, NE Japan (Tazawa 2008). Changhsingian brachiopods were widely reported from the Toyoma Formation in the southern Kitakami Belt, and are dominated by typical Palaeotethyan elements such as *Oldhamina squamosa*, *Peltichia transversa*, *Spinomarginifera lopingensis*, *Geyerella ofunatoensis* and *Fusichonetes* sp. in association with rare antitropical elements such as *Megousia auriculata* and *Attenuatella* sp. (Tazawa 1975, 2012; Tazawa & Miyake 2011).

A similar Permian mixed brachiopod succession has been reported from the Hida Gaien and Kurosegawa belts in central Japan (Tazawa 2001; Shi 2006) (Fig. 3).

North Pamirs

In addition to the above-mentioned areas, Permian brachiopods possibly belonging to the NTZ are also present in the north Pamirs, which tectonically belongs to the Kunlun Arc developed along the northern Palaeotethyan margin. Angiolini *et al.* (2016) described a complicated Permian succession in the north Pamirs in Tajikistan containing abundant fusulinids, some brachiopods and conodonts. The uppermost Yakhtashian–Bolorian Safetdara Formation contains the brachiopod *Larispirifer* sp., the conodont *Sweetognathus modulatus* and abundant late Artinskian–Kungurian fusulinids. The upper Bolorian–Kubergandian Gundara Formation, consisting of sandstones, shales and limestones, yields rich silicified brachiopods, including *Gundaria insolita*, *Hemileurus politus*, *Posicomta gundarensis*, *Orbicoelia* sp., *Spiriferellina* sp., *Paraspiriferina* sp., *Fredericksolasma lata* and *F. rhomboidalis*. However, a complete brachiopod succession has not yet been established in this area (Angiolini *et al.* 2016).

The Palaeoequatorial Realm

South China

South China contains the most diverse and complete Permian brachiopod succession in the world. Wang *et al.* (1981) reviewed the palaeobiogeography of the Palaeozoic brachiopod succession for the first time, and proposed the Southern, Central and Northern provinces, of which the Central Province is represented by South China. Ten assemblages were proposed from the Upper Carboniferous to the Permian. The present Asselian and Sakmarian stages were assigned to the Upper Carboniferous during the 1980s in China. Brachiopods in these two stages mostly comprise forms succeeded from the Upper Carboniferous, which are dominated by spiriferides and productides such as *Choristites*, *Brachythyrina*, *Avonia*, *Dictyoclostus*, *Echinoconchus* and *Juresania*. Most of those genera disappeared towards the end of the Sakmarian (upper Zisongian) and a few forms range into the Artinskian (=South Chinese Longlinian) Stage. Three assemblages were summarized. They are the Asselian *Protanidanthus–Choristites jigulensis* Assemblage, plus the Sakmarian *Dictyoclustus uralicus* and *Liraplecta richthofeni–Choristites pavlovi* assemblages (Wang *et al.* 1981). The Lower Permian (Cisuralian) brachiopod fauna in South China is well represented by the Longlin fauna described by Li *et al.* (1987), which consists of 155 species of 67 genera. It is dominated by productides, of which a majority belongs to the overtoniids, echinoconchids, buxtoniids, marginiferids and dictyoclostids. Three assemblages were established by Li *et al.* (1987). They are the *Proanidanthus enaagardi–Buxtonia–Linoproductus* Assemblage corresponding to the fusulinid *Pseudoschwagerina* Zone, the *Choristites–Eolyttonia* Assemblage and the *Orthotichia magnifica–Compressoproductus* Assemblage in ascending order. The former two assemblages belong to Asselian and Sakmarian, and the latter belongs to the Artinskian Stage.

Brachiopods from the Yangsingian (=Chihsian plus Maokouan) to Lopingian are extremely abundant in South China. A succession with five brachiopod assemblages from the Chihsian to the Changhsingian was described by Zhan & Li (1979), and this was revised to seven by Wang *et al.* (1981). The *Orthotichia chekiangensis* Assemblage can be readily recognized in the lower part of the Chihsia Formation in South China. The common genera of the Lower Cisuralian, such as *Avonia*, *Juresania*, *Buxtonia*, *Rugoconcha* (=*Plicatiferina*), *Echinoconchus*, *Choristites* and *Neospirifer*, disappeared in this assemblage. This assemblage is associated with the fusulinid *Misellina claudiae* and the conodont *Sweetognathus whitei* at the Tieqiao section in Laibin City, Guangxi (Wang 2002; Shen *et al.* 2007). Therefore, it belongs to the upper Artinskian or lower Kungurian stage. The lower and middle parts of the Chihsia Formation contain a characteristic and widely distributed brachiopod assemblage dominated by *Tyloplecta nankingensis* (Jin & Fang 1985; Zeng *et al.* 1995). This assemblage represents the greater part of the Chihsia Formation (Kungurian) and is widely distributed in South China. The upper part of the Chihsia Formation and the greater part of the Maokou Formation contain a few characteristic forms including, for example, '*Cryptospirifer*' *omeishanensis*, *Magniderbyia magnifica*, *Monticulifera sinensis*, *Neoplicatifera huangi* and *Vediproductus punctatiformis*. All of these species have been also reported from different blocks in the Palaeotethys. Among them, '*Cryptospirifer*' has been widely reported from Iran, Turkey, and the Qiangtang and Baoshan blocks (Nakamura & Golshani 1981; Shi & Shen 2001; Jin & Zhan 2008; Shen *et al.* 2016). *Vediproductus* has been reported from the coeval horizons in Transcaucasia, Malaysia (Campi *et al.* 2005). *Monticulifera* has been reported from the Qilian Mountains in North China. Thus, the '*Cryptospirifer*'*–Vediproductus–Neoplicatifera* Assemblage generally represents a widely correlative brachiopod assemblage from the upper Kungurian to the lower and middle Guadalupian in the Palaeotethys. In the upper part of the Maokou Formation (upper Guadalupian) in South China, some elements of this assemblage, including *Vediproductus punctatiformis*, *Neoplicatifera huangi* and *Monticulifera sinensis*, continue to be present. In addition, a characteristic upper Guadalupian genus, *Urushtenoidea*, began to occur, but it disappeared before the Guadalupian–Lopingian boundary (Shen & Shi 2009). It is noteworthy that some of the most common species in the Lopingian in South China such as *Spinomarginifera lopingensis*, *Transennatia gratiosa*, *Tyloplecta yangtzeensis* and *Haydenella kiangsiensis* began to occur in the uppermost part of the Maokouan, which suggests that the changeover of brachiopods during the pre-Lopingian crisis in South China was in the late Guadalupian rather than in the early Lopingian (Liang 1990; Shen & Shi 1996, 2009; Shen *et al.* 2006; Shen & Zhang 2008).

After the pre-Lopingian crisis, some characteristic Yangsingian brachiopods, including, for example, *Monticulifera*, *Neoplicatifera*, *Urushtenoidea* and '*Cryptospirifer*', disappeared, but their disappearance process occurred over a long time in the late Guadalupian (Jin *et al.* 1994; Shen & Shi 1996, 2002, 2009; Shen *et al.* 2006; Clapham *et al.* 2009). Brachiopods in the Lopingian can be readily distinguished from the Maokouan in South China. Based on the collections from the Wuchiaping, Lungtan and Changhsing formations in western Guizhou, South China, Liao (1980*b*) proposed two assemblages for the Wuchiapingian Stage and one for the Changhsingian Stage. They are the *Edriosteges poyangensis*, *Permophricodothyris grandis–Orthothetina ruber* and *Peltichia zigzag–Paryphella* *sulcatifera* assemblages in ascending order. *Transennatia gratiosa*, *Edriosteges poyangensis* and *Tyloplecta yangtzeensis* are also extremely abundant and widespread in the Wuchiapingian in the Palaeotethys. The general Changhsingian *Peltichia zigzag–Paryphella sulcatifera* Assemblage was further divided into two types of assemblages with respect to different lithofacies (Liao 1980*a*). The intra-platform basinal siliceous facies is characterized by small and thin-shelled brachiopods such as *Fusichonetes* (=*Tethyochonetes*) (Wu *et al.* 2016), *Paryphella* and *Spinomarginifera*, which were named the *Paryphella–Fusichonetes–Paracrurithyris* Assemblage or deep-water facies (Chen *et al.* 2006; He *et al.* 2014), while the shelf carbonate facies is characterized by large brachiopods such as *Peltichia*, *Meekella*, *Orthothetina*, *Oldhamina*, *Leptodus*, *Edriosteges*. This succession has been refined based on a series of studies on the Changhsingian brachiopods in South China. Three assemblages, the *Derbyia guangdongensis–Oldhamina squamosa–Orthothetina eusarkos*, *Peltichia transversa–Perigeyerella costellata* and *Fusichonetes pigmaea–Neochonetes* (*Huangichonetes*) *substrophomenoides–Notothyris crassa* assemblages in ascending order, were established for the Changhsingian Stage in South China (Shen & He 1994; Shen & Shi 2007).

The topmost part of the Changhsingian and basal Triassic is characterized by a brachiopod assemblage with small size and thin shell. Some brachiopods, in particular (e.g. *Fusichonetes pigmaea*, *Paracrurithyris pigmaea*, *Spinomarginifera chenyaoyanensis* and *Prelissorhynchia pseudoutah*), are abundant in the uppermost Permian, as well as in the end-Permian mass extinction interval (Shen & Shi 1996; Chen *et al.* 2005; He *et al.* 2014; Zhang *et al.* 2016). They are usually associated with numerous lingulid species (Peng *et al.* 2007). This dwarfed brachiopod assemblage can be seen in the end-Permian mass extinction interval all over the world (Fig. 4).

North Iran and Transcaucasia

Tectonically, both north Iran and Transcaucasia belong to the Cimmerian blocks that rifted from the northern peri-Gondwanan region (Angiolini *et al.* 2013*a*, *b*). However, available Permian brachiopods show more palaeobiogeographical affinity with those of the Palaeoequatorial Realm. No typical Gondwanan-type Permian brachiopod faunas have been reported. This is probably because the Iranian Block in western Palaeotethys rifted away from Gondwana earlier than other typical Cimmerian blocks. Permian brachiopods are very abundant in north Iran and Transcaucasia. The Cisuralian Asselian–lower Sakmarian brachiopods were recorded from the Dorud Group of the Alborz Mountains (Assereto 1963; Fantini Sestini 1965*a*; Assereto *et al.* 1973) and have recently been updated by Angiolini & Stephenson (2008). Surprisingly, the lower Cisuralian brachiopods from north Iran are characterized by *Reticulatia uralica*, *Calliprotonia* sp., *Juresania dorudensis*, *Linoproductus dorotheevi*, *Cancrinella cancriniformis* and *Larispirifer fantinisestinii*, and show strong affinities with the coeval faunas of the Urals and of the Russian Platform to the north, and to a lesser extent to the Trogkofel Limestone (Carnic Alps) in the west, but very little affinity with those of the Gondwanan and peri-Gondwanan regions (Angiolini *et al.* 2007; Angiolini & Stephenson 2008). However, the fauna also clearly shows some links with those of South China in terms of the presence of the genera *Reticulatia*, *Juresania*, *Linoproductus* and *Cancrinella*.

Guadalupian brachiopods in Iran were first reported by Fantini Sestini (1965*b*). Thirty-three brachiopod species were described recently from the Ruteh Limestone of north Iran (Crippa & Angiolini 2012). Three discrete biozones were established based on the Unitary Association method. They are the *Squamularia* sp. B–*Martinia bassa* at the base, the *Haydenella kiangsiensis–Neochonetes* (*Nongtaia*) *asseretoi* in the middle and the *Rostranteris exilis–R. gemmellaroi* Biozone at the top of the formation, which are recognized in most investigated sections (Crippa & Angiolini 2012). The brachiopod fauna from the Ruteh Limestone is generally comparable with those of the Guadalupian Gnishik Formation in Transcaucasia (Ruzhentsev & Sarytcheva 1965; Kotlyar & Zakharov 1989) in terms of the presence of *Vediproductus punctatiformis* (=*Vediproductus vediensis*) and *Orthothetina vediensis*. However, these biozones are rarely distinguishable in the eastern part of Palaeotethys (e.g. South China). Nevertheless, the presence of *Vediproductus punctatiformis* in the middle part of the Ruteh Formation strongly suggests that the Ruteh Limestone is correlative with the Maokouan of South China (Fig. 4).

Lopingian brachiopods are highly diverse and very abundant in the Albortz (north Iran), NW Iran and Transcaucasia (Ruzhentsev & Sarytcheva 1965; Angiolini & Carabelli 2010; Ghaderi *et al.* 2014). According to Ruzhentsev & Sarytcheva (1965), the *Araxilevis* Bed of Transcaucasia is overlain by the *Oldhamina* and *Haydenella* beds of the Dzhulfian in association with numerous *Araxathyris* and *Transcaucathyris*. The brachiopods from the Julfa beds of NW Iran were recently re-studied by Ghaderi *et al.* (2014). They established three biozones: that is, the *Araxilevis intermedius*, *Permophricodothyris ovata* and *Haydenella kiangsiensis* biozones in ascending order. Forty-eight species

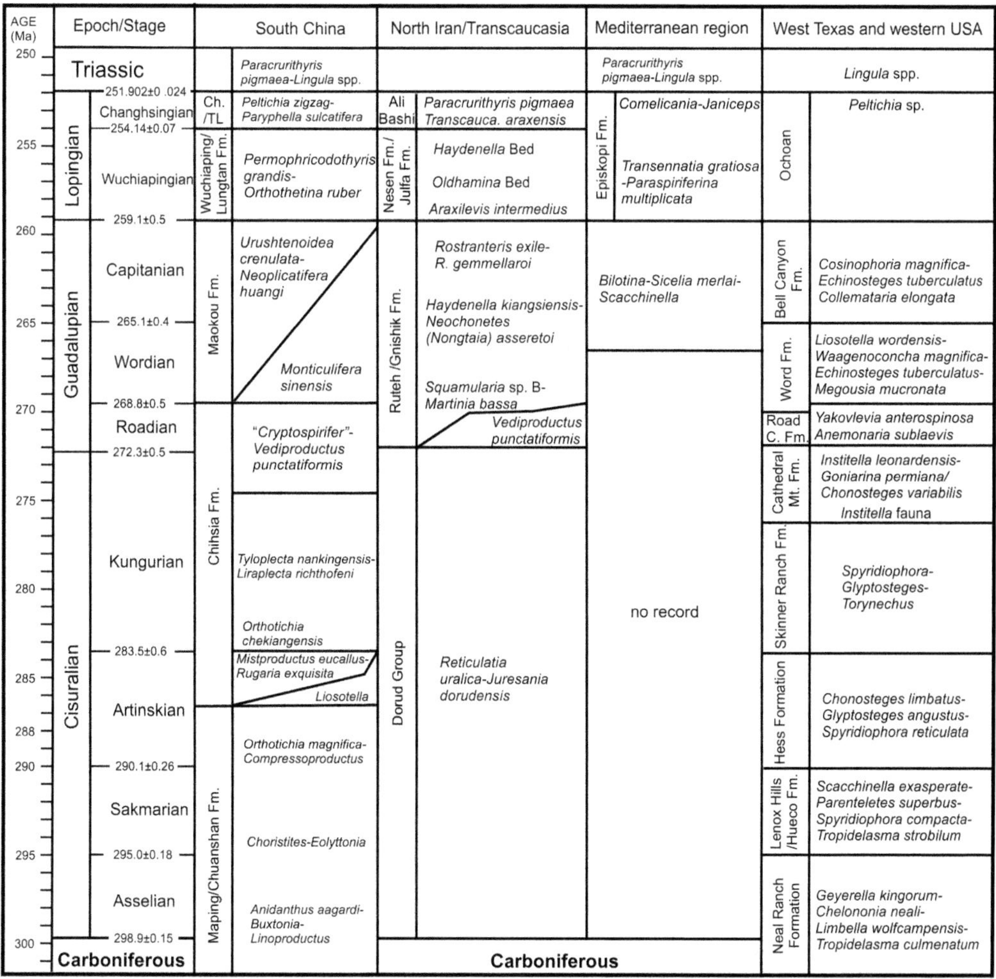

Fig. 4. Revised brachiopod successions in the Palaeoequatorial Realm. The Permian timescale is revised from Shen *et al.* (2013*a*). Ch/TL, Changhsing/Talung Fm.

of 31 genera from the lower member of the Nesen Formation of north Iran were also described recently (Angiolini & Carabelli 2010). Two biozones, the *Tyloplecta persica* and *Araxilevis intermedius* zones, are recognized in the Dzhulfian (=Wuchiapingian) and can be correlative with the *Araxilevis* Bed of Transcaucasia (Ruzhentsev & Sarytcheva 1965; Ghaderi *et al.* 2014). The lower member of the Nesen Formation shares many brachiopods of the Wuchiapingian of South China (e.g. *Tyloplecta yangtzeensis*, *Perigeyerella costellata*, *Tschernyschewia typical* and *Spinomarginifera* spp.). The Changhsingian in the Alborz Mountains of north Iran contains the *Permophricodothyris ovata* and *Enteletes lateroplicatus* zones. Both are very common elements in the entire Lopingian in the Palaeotethys and palaeoecologically controlled, and are therefore more difficult for precise interregional correlation. The Dorashamian (Changhsingian) in Transcaucasia contains much fewer brachiopods, and they are dominated by some small pedicle species such as *Transcaucathyris araxensis* and *Paracrurithyris pigmaea* (Garbelli *et al.* 2014), which may indicate a relatively deep-water setting. These species are common in the deep-water facies in the Changhsingian and the basal Triassic of South China. Ruzhentsev & Sarytcheva (1965) reported the *Gruntallina* Biozone for this interval, but it is rarely found in north Iran.

Upper Changhsingian brachiopods in association with the foraminifer *Colaniella parva* are very abundant in the NW Caucasus Mountains, Russia (Kotlyar *et al.* 2004*b*). The Nikitin Limestone is characterized by the *Tyloplecta yangtzeensis–Haydenella* *kiangsiensis–Spinomarginifera lopingensis* Assemblage. This assemblage is very similar to the Wuchiapingian assemblages of South China. The overlying Urushten reef limestone is characterized by the *Alphaneospirifer anshunensis–Spinomarginifera kueichowensis–Transennatia gratiosa* Assemblage, which is generally comparable with those of the Changhsingian in South China (Kotlyar *et al.* 2004*b*).

The distinct *Araxilevis* Bed in the lower part of the Dzhulfian in north Iran can be traced in the lower part of Unit 5 of the Hambast Formation in central Iran, but this bed has never been reported from the equivalents in South China.

Mediterranean region

A Guadalupian brachiopod fauna was described from units III–V of the shallow-water carbonate succession of South Tunisia (Verna *et al.* 2010). It is composed of 29 taxa, including, for example, *Bilotina*, *Scacchinella variabilis*, *Sicelia merlai* and *Permophricodothyris caoli*. The brachiopod fauna is Wordian or Capitanian in age because it is associated with the fusulinid *Chusenella rabatei* and the conodont *Sweetognathus hanzhongensis* (Verna *et al.* 2010).

Similar Guadalupian (Wordian) brachiopods were reported from the allochthonous Upper Unit of Chios (Greece) by Grant (1993, 1995) and Angiolini *et al.* (2005*b*). These brachiopod species are endemic (Grant 1993, 1995) or widely distributed elements with long ranges (Angiolini *et al.* 2005*b*), and are therefore of limited use for inter-regional correlation. Permian sequences are well developed in the Hydra Island, Greece (Grant *et al.* 1991), but no brachiopods have been described from the Cisuralian so far. Richard Grant had a large collection from the Upper Permian of Hydra Island (Greece), but, unfortunately, only a few orthothetid brachiopods had been described from this collection (Grant 1995) before he passed away. Twenty-three brachiopod species have been described from the Episkopi Formation in Hydra Island (Shen & Clapham 2009). Some very common Lopingian species such as *Transennatia gratiosa*, *Cathaysia chonetoides*, *Haydenella kiangsiensis* and *Paraspiriferina multiplicata* suggest that they have a good biogeographical link between western and eastern Palaeotethys (Shen & Shi 2000). The fauna is associated with the Wuchiapingian conodonts *Clarkina leveni* and *C. orientalis*.

Changhsingian brachiopods in the Mediterranean region have been reported from the Carnic Alps, the Dolomites and northern Hungary. They are characterized by an athyridid assemblage represented by *Comelicania* and *Janiceps* in the topmost part of the Bellerophon Formation, which is immediately below the *Hindeodus praeparvus* Zone (Posenato 1998, 2001, 2009, 2010). In the Permian–Triassic boundary interval, a small brachiopod assemblage including *Spinomarginifera* sp., *Orthothetina ladina*, *Ombonia tirolensis* and *Paracrurithyris pigmaea* (=*Orbicoelia tschernyschewi*) was reported from the basal beds of the Gerennavár Limestone in northern Hungary (Posenato *et al.* 2005). This assemblage is associated with some Triassic-type bivalves, similar to the uppermost Changhsingian mixed fauna 1 or 2 of South China described by Sheng *et al.* (1984). The *Comelicania* Assemblage has not been found from the equivalent horizon in South China.

West Texas, North America

A great number of silicified brachiopods were described from the Permian in the Glass Mountains of west Texas, North America in a series of monographs (Cooper & Grant 1972*b*, 1974, 1975, 1976*a*, *b*, 1977). Most of those brachiopods were silicified and their internal structures are clearly shown, facilitating the recognition of a large number of new taxa. Consequently, these faunas are very different to others at a global scale, which in part may reflect a monographic effect. Although numerous brachiopods species have been described from west Texas, brachiopod biostratigraphy has not been clearly established.

The lowest Permian (Asselian) brachiopods are mainly from the Wolfcampian Neal Ranch Formation, which contains the fusulinid *Pseudoschwagerina uddeni* in the base of the formation (Ross 1963). Most of the brachiopod species are endemic and, therefore, are of only limited value for global correlation. Common species and good indicators of the lower Wolfcampian in Texas include *Geyerella kingorum*, *Chelononia neali*, *Limbella wolfcampensis*, *Tropidelasma culmenatum* and *Parenteletes cooperi*. In addition, a typical cold-water genus, *Kochiproductus*, is also present. This is a widely distributed genus in the Boreal Realm. Sakmarian brachiopods are mainly from the Lenox Hills and Hueco formations. Most of the species range is from the underlying Neal Ranch Formation. *Scacchinella exasperate*, *Parenteletes superbus*, *Spyridiophora compacta* and *Tropidelasma strobilum* are good representatives of this stage. Artinskian brachiopods in the *Sweetognathus whitei* Zone are mainly from the Taylor Ranch Member of the Hess Formation. Some new inclusions, including *Acritosia peculiaris*, *Chonosteges limbatus*, *Glyptosteges angustus* and *Spyridiophora reticulate*, occurred. Many Sakmarian species persist through to the Hess Formation. The Skinner Ranch Formation was assigned to the lower Wolfcampian by Cooper & Grant (1972*a*), but the patch reef identified by Cooper & Grant (1964) actually comprised displaced limestone blocks deposited in a slope environment. Therefore, this formation has been reassigned to the Kungurian based on fusulinids and conodonts (Ross 1986; Rohr *et al.* 2000). The Skinner Ranch Formation contains the most abundant and highly diverse Kungurian brachiopods in the Glass Mountains of west Texas. Some peculiar and readily recognized productides such as *Spyridiophora*, *Glyptosteges*, *Torynechus* are characteristic of the Skinner Ranch Formation. There is a considerable faunal change at the boundary between the Skinner Ranch and the overlying Cathedral Mountains Formation which is marked by the appearance of the brachiopod *Institella*. The *Institella* fauna is characteristic of the Cathedral Mountains Formation in west Texas, which is late Kungurian in age. Some characteristic species such as *Institella leonardensis*, *Goniarina permiana*, *Chonosteges variabilis* and *Loxophragmus ellipticus* are also reported from the Mino Belt (Fig. 1) in central Japan, which suggests that the limestone blocks containing brachiopods with the North American affinity were from Panthalassa (Shen *et al.* 2011).

The Guadalupian Series in west Texas contains more abundant and highly diverse brachiopods than the Wolfcampian and Leonardian. The lower Guadalupian (Roadian) brachiopods are represented by those from the Road Canyon Formation. Characteristic elements include, amongst others, *Yakovlevia anterospinosa*, *Anemonaria sublaevis*, *Megousia mucronata*, *Chonosteges costellatus* and *Edriosteges multispinosus*. The brachiopod fauna in the Road Canyon Formation show some distinct mixed characters between the Boreal and Palaeoequatorial realms, which is generally comparable with those of the NTZ. The Word Formation still contains the mixed fauna and many are characteristic of the middle Guadalupian (e.g. *Liosotella wordensis*, *Cyclacantharia kingorum*, *Waagenoconcha magnifica*, *Xenosteges quadratus*, *Echinosteges tuberculatus*, *Cartorhium orbiculatum*, *Megousia mucronata* and *Yakovlevia intermedia*) in association with the ammonoid *Waagenoceras*. Although most species of the Wordian fauna are endemic, many genera are very common in the Maokouan of South China and the NTZ.

The youngest Permian brachiopod fauna in west Texas is from the Bell Canyon Formation in west Texas, which is of Capitanian age in terms of the presence of *Timorites* and the Capitanian conodonts. Characteristic elements include, for example, *Cosinophoria magnifica*, *Echinosteges tuberculatus* and *Collemataria elongata*. The former two were also reported from the exotic limestone blocks in the Jurassic mélange of Japan (Tazawa & Shen 1997; Shen *et al.* 2011). After the Capitanian, seawater withdrew from the Delaware Basin and the Guadalupian Series is overlain by the evaporate deposits of the Ochoan Series of North America (Fig. 4).

Guadalupian brachiopods are also abundant in the Phosphoria Formation of the western USA (Girty 1910; Cooper & Grant 1975), but more cold-water elements (e.g. *Yakovlevia multistriata*, *Megousia waagenianus* and *Bathymyonia nevadensis*) are present, which suggests that the Phosphoria Basin north of west Texas was more affected by coastal cold-water currents from the northern Boreal sea.

Lopingian marine deposits in the western USA may be present because the transgression began in the latest Changhsingian. Some specimens such as *Peltichia* were collected from the Hogup Mountains near Salt Lake City, but have not been described (Fig. 4).

The Southern Transitional Zone (STZ)

The Cimmerian Belt contains a number of microplates, most of which were detached from the northern peri-Gondwanan margin in the Cisuralian during the opening of the Neotethys Ocean (Sengör 1979). These microplates display a distinct faunal shift from typical cold-water Gondwanan affinity through a transitional character to a typical warm-water Cathaysian affinity when they drifted northwards during the Permian (Shi *et al.* 1995; Shi & Archbold 1995*b*, 1998). These microplates include the Baoshan, Tengchong, Karakorum, Sibumasu, the SE Pamirs, southern Qiangtang and Lhasa blocks (Figs 1, 5). Tectonically, north Iran was a Cimmerian block, but the brachiopod faunas throughout the Permian contain more affinity with those of the STZ are therefore not discussed in the Palaeoequatorial Realm. By contrast, Turkey and Oman do not belong to the Cimmerian blocks tectonically (Angiolini *et al.* 2013*b*). However, the brachiopod faunas there show some transitonal characters between the Gondwanan and Palaeoequatorial realms, so they are discussed herein.

Baoshan and Tengchong blocks

Lower Permian brachiopods from the Dingjiazhai Formation in the Baoshan Block in western Yunnan, SW China have been extensively studied (Fang 1994; Fang & Fan 1994; Shi *et al.* 1996; Shen *et al.* 2000*c*). Three assemblages were established and are mainly composed of cold-water elements. They are the *Bandoproductus qingshuigouensis–Marginifera semigratiosa*, *Punctocyrtella australis–Punctospirifer afghanus* and *Callytharrella dongshanpoensis* assemblages in ascending order.

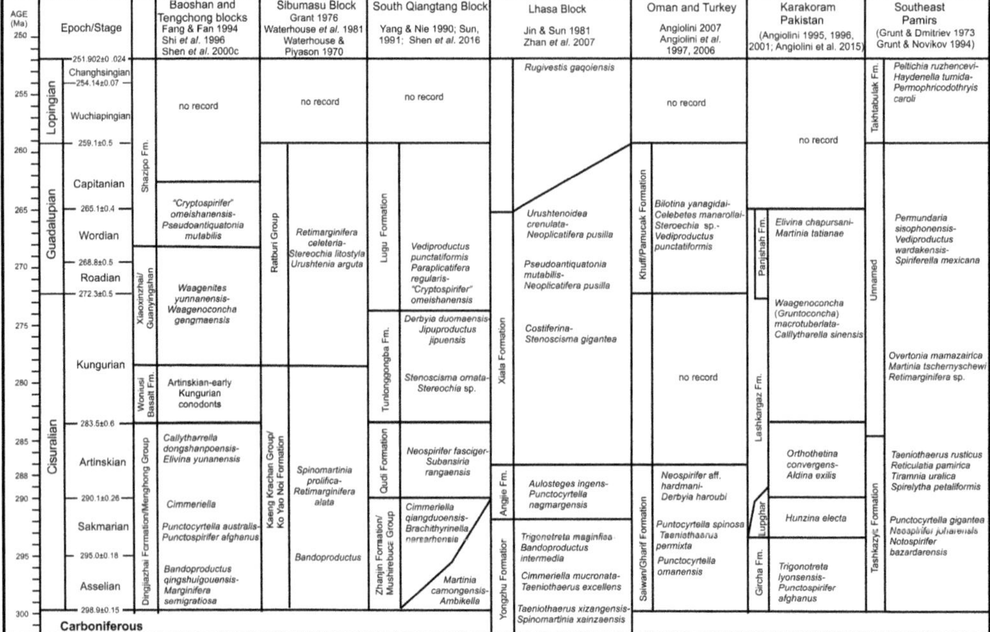

Fig. 5. Revised brachiopod successions in the Southern Transitional Zone. The main references are shown in the columns. The Permian timescale is revised from Shen *et al.* (2013*a*).

The *Bandoproductus* fauna has been widely reported from the Bowen and Gympie basins in eastern Australia (Waterhouse 1987, 2015*b*), southern Thailand (Waterhouse 1982; Shi *et al.* 2002*a*, *b*), Malaysia (Shi *et al.* 1997), northern India (Reed 1928; Archbold *et al.* 1996), the Himalaya Tethys Zone and the Lhasa Block of Tibet (Zhan *et al.* 2007). It is a distinctive assemblage of Asselian age in the peri-Gondwanan region. Another characteristic genus in the peri-Gondwanan region is *Cimmeriella*, which was previously identified as '*Globiella*' or '*Stepanoviella*' (Fang 1994). The *Cimmeriella* fauna may be slightly younger than the *Bandoproductus* fauna, but is occasionally associated with *Bandoproductus* (e.g. western Yunnan: Shen *et al.* 2000*c*). The presence of the *Bandoproductus–Punctocyrtella* (or *Cyrtella*) and the *Cimmeriella* faunas suggests that the Baoshan Block was still attached to the Gondwanaland during the early and middle Cisuralian because this assemblage has been widely reported from northern India, Kashmir and Western Australia (Shen *et al.* 2000*c*; W. Wang *et al.* 2004). The overlying *Callytharrella dongshanpoensis* Assemblage is characterized by the co-occurrences of *Nantanella elegantula* and *Elivina yunnanensis* in association with the conodont *Sweetognathus whitei* (Ueno *et al.* 2002) and the fusulinids *Pseudofusulina* and *Eoprarafusulina* (Ueno 2003), which suggest that the palaeoclimate was possibly ameliorated due to either northwards drifting of the blocks or climate warming in the peri-Gondwanan region. The *Callytharrella* fauna is characteristic of the Callytharra Formation of the Carnarvon Basin in Western Australia (Archbold *et al.* 1993), but the Callytharra Formation was assigned to the late Sakmarian (Sterlitamakian) based on its correlation with the Fossil Cliff Member of the Holmwood Shale of the Perth Basin, which contains the ammonoid *Metalegoceras* (Glenister *et al.* 1993). However, the base of precise correlation between the Callytharra Formation of the Carnarvon Basin and the Fossil Cliff Member of the Perth Basin is still unclear.

No brachiopods have been reported from the Woniusi Basalt Formation. However, conodonts from the limestone interlayers indicate that the Woniusi Basalt Formation is probably of upper Artinskian age (W. Wang *et al.* 2004). The Xiaoxinzhai Formation in the Baoshan Block contains abundant *Waagenites yunnanensis*, *Waagenoconcha gengmaensis*, *Leptodus nobilis*, *Stereochia litostyla* and *Permophricodothyris indica*, which show a distinct late Kungurian–early Guadalupian mixed fauna. This brachiopod fauna is associated with many solitary corals including *Thomasiphyllum* (Fang & Fan 1994). A slightly younger brachiopod fauna was reported from the *Eopolydiexodina* Zone of the Shazipo Formation (Shi & Shen

2001). The presence of '*Cryptospirifer*' *omeishanensis* suggests a distinct link with the brachiopod faunas of South China. *Pseudoantiquatonia mutabilis* was recorded from the *Chusenella* Zone in the Xiala Formation of central Lhasa Block. Therefore, the Shazipo fauna is clearly of middle Guadalupian. To date, no brachiopods have been reported from the Lopingian in the Baoshan Block.

There are very few brachiopods recorded from the equivalents of the Dingjiazhai Formation of the Baoshan Block in the Tengchong Block. *Spinomartinia* was recorded from the Menghong Group in the Tengchong Block, which suggests that the Menghong Group is probably of early Cisuralian age. An identical brachiopod fauna with that of the Xiaoxinzhai Formation of the Baoshan Block was recorded from the Guanyingshan Formation in the Tengchong Block. This fauna also consists of abundant *Waagenites yunnanensis*, *Stereochia litostyla* and *Spirigerella minuta*, amongst others, suggesting that the Tengchong Block was palaeogeographically very close to the Baoshan Block during the late Kungurian and Guadalupian (Shen *et al.* 2009).

Sibumasu Block

Permian brachiopods of the Sibumasu Block are well represented by faunas from southern Thailand, Malaysia (Waterhouse & Piyasin 1970; Grant 1976; Waterhouse *et al.* 1981; Shi *et al.* 1995, 1997; Sone *et al.* 2001; Campi *et al.* 2002, 2005; Sone & Leman 2005) and Sumatra (Crippa *et al.* 2014). The Permian sequence in southern Thailand was subdivided into two groups: the widely distributed Ratiburi Limestone Group in the upper group and the diamictic Kaeng Krachan Group or Phuket Group in the lower. A typical lower Cisuralian brachiopod fauna containing *Bandoproductus* was recorded from the diamictites of the Kaeng Krachan Group in southern Thailand (Waterhouse 1982). Waterhouse *et al.* (1981) described a large number of brachiopods from the Ko Yao Noi Formation in southern Thailand, which is equivalent to, or slightly younger than, the Cisuralian Kaeng Krachan Group. The *Spinomartinia prolifica* Zone was established based on its abundance from the Ko Yao Noi Formation, which was considered to be close in age to a part of the Chihsian of South China (Waterhouse *et al.* 1981). However, more recent studies indicate that the *Spinomartinia* fauna has been found in the Baoshan and Tengchong blocks in western Yunnan, the Lhasa Block Tibet, and Malaysia, most of which point to an early–middle Cisuralian age (Waterhouse 1982; Shen *et al.* 2000*c*). Lower Permian brachiopods were also reported from the Lower Permian Mengkarang Formation in Sumatra in Indonesia. The brachiopod assemblage consists of six taxa including *Neochonetes* (*Neochonetes*) *carboniferus*, *Marginifera* (*Arenaria*) sp., *Reticulatia* sp., *Stereochia* aff. *irianensis*, ?*Protoanidanthus* sp. and *Cancrinella* sp. They are associated with the fusulinids *Pseudoschwagerina* cf. *afghanensis*, *P. meranginensis*, *Eoparafusulina* ?*haydeni* and *Pseudofusulina rutschi*. Therefore, the brachiopod fauna is of Sakmarian.

Permian brachiopods are extremely abundant and highly diverse in the Ratburi Limestone in southern Thailand (Waterhouse & Piyasin 1970; Grant 1976). The basal part of the Ratburi Limestone contains the *Monodiexodina* fauna and is of Kungurian age (Waterhouse *et al.* 1981; Ueno 2003). The key part of the Ratburi Limestone is dated as Guadalupian (Wordian) based on fusulinids, and the uppermost part of the Ratburi Limestone contains possibly Wuchiapingian fusulinids. The brachiopod succession in the Ratburi Limestone is poorly defined because they were all collected from isolated limestone blocks. The Ratburi brachiopod fauna is composed of more than 100 species and is dominated by Tethys-type or temperate elements (e.g. *Perigeyerella*, *Leptodus*, *Richthofenia*, *Urushtenia*, *Meekella* and *Transennatia*), mixed with some antitropical elements such as *Retimarginifera*, *Spiriferella* and *Stereochia*. It is generally comparable with those from the Amb Formation of the Salt Range, Pakistan and the Shazipo Formation of the Baoshan Block.

Permian brachiopods have been widely reported from Peninsular Malaysia. The possibly Asselian–Sakmarian *Bandoproductus* was recorded from the Singa Formation in NW Malaysia (Shi *et al.* 1997). The Guadalupian brachiopods are highly diverse and show a close palaeobiogeographical link with those of South China. Characteristic genera include *Vediproductus punctatiformis*, *Permianella typica*, *Orthothetina ruber*, *Permundaria perplexa*, *Crenispirifer alpheus* and *Leptodus richthofeni* (Campi *et al.* 2000, 2002, 2005; Sone *et al.* 2003). To date, no Lopingian brachiopods have been recorded from Malaysia.

South Qiangtang Block

Cisuralian brachiopod succession in the western part of the South Qiangtang Block was first established by Liang *et al.* (1983) and subsequently updated by Yang & Nie (1990). The *Martinia camongensis–Ambikella* Assemblage from the Zhanjin Formation is probably Asselian or Sakmarian in age because it is associated with the bivalve *Eurydesma* fauna, which is widely recognized in Gondwanaland during this time. The *Neospirifer fasciger–Subansiria rangaensis* Assemblage from the Qudi Formation was previously assigned to the Sakmarian (Yang & Nie 1990). However, fusulinids from the Qudi Formation indicate an Artinskian age (Zhang *et al.* 2012). The *Stenoscisma ornata–Stereochia* sp. (= '*Costiferina*' *sinensis*) Assemblage from the Tunlonggongba Formation is assigned to the Kungurian because it is associated with the fusulinid *Monodiexodina* fauna. The *Derbyia duomaensis–Jipuproductus jipuensis* Assemblage from the upper part of the Tunlonggongba Formation is most probably of Kungurian age because *Monodiexodina* continued to exist in the brachiopod assemblage (Yang & Nie 1990).

In the eastern part of the Rutog area of western Tibet, four different brachiopod assemblages were documented from the Mushirebuca Group (Sun 1991). The *Comuquia mushirebuca–Acolosia lenticular* and *Cimmeriella qiangduoensis–Brachythyrinella narsarhensis* assemblages from the middle and upper parts of the Mushirebuca Group are most likely of early to middle Cisuralian age based on the presence of the *Cimmeriella* fauna. The *Orbicoelia fraterculus–Phricodothyris bullata* Assemblage from the upper part of the Mushirebuca Group is correlative with the assemblage of the Qudi Formation and the two assemblages from the Chainaha Formation are correlative with those of the Tunlonggongba Formation (Sun 1991).

Brachiopods from the upper Kungurian or lower Guadalupian in the central part of the South Qiangtang Block are characterized by the presence of numerous *Vediproductus punctatiformis*, *Paraplicatifera regularis* and rare '*Cryptospirifer*' *omeishanensis*. They show a strong affinity with those from the upper part of the Chihsia Formation or the lower part of the Maokou Formation of South China, which suggests that the palaeoclimate of the Qiangtang Block was significantly ameliorated during the early Guadalupian along with the northwards drift of the block (Shen *et al.* 2016). To date, upper Guadalupian and Lopingian brachiopods have not been reported from the South Qiangtang Block.

Lhasa Block

The brachiopod succession of the Lhasa Block is closely comparable with those of the other Cimmerian blocks rather than the Himalaya Tethys Zone after the early Cisuralian. The typical early Cisuralian *Bandoproductus* and *Cimmeriella* faunas were found from the Pangduoi Group (Jin & Sun 1981) and the Langmaria Formation in the Xianza area of the central Lhasa Block (Yang & Fan 1983). The latter also contains *Punctocyrtella nagmargensis*, a species commonly occurring in the *Bandoproductus* faunas that have been widely reported from the peri-Gondwanan region. More recently, four brachiopod assemblages, the *Taeniothaerus xizangensis–Spinomartinia xainzaensis*, *Cimmeriella mucronata–Taeniothaerus excellens*, *Trigonotreta maginfica–Bandoproductus intermedia* in the middle and upper parts of the Yongzhu Formation, and the *Aulosteges ingens–Punctocyrtella nagmargensis* Assemblage from the Angjie Formation in ascending order, were established in the Xianza area (Zhan *et al.* 2007). The first assemblage was assigned to the Upper Carboniferous Bashkirian. However, the genera *Spinomartinia*, *Taeniothaerus*, *Trigonotreta*, *Callytharrella* and *Sulciplica* all suggest an early Cisuralian age, if they are correctly identified. The second and third assemblages with *Bandoproductus* and *Cimmeriella* suggest an early to middle Cisuralian age. The *Aulosteges ingens–Punctocyrtella nagmargensis* Assemblage from the Angjie Formation may be of Sakmarian or early Artinskian age because *P. nagmargensis* has been widely reported from the equivalents in the peri-Gondwanan region (e.g. Kashmir, India, Oman and the SE Pamirs) and Cimmerian blocks (Reed 1932; Angiolini *et al.* 1997; Shen *et al.* 2000*c*). The Angjie Formation is overlain by the Xiala Formation in the central Lhasa Block. The lower part of the Xiala Formation (also called the Ria Formation) contains a cold-water brachiopod-dominated assemblage consisting of *Costiferina spiralis*, *Calliomarginatia orientalis* and *Spiriferella salteri* (Zhan & Wu 1982). This assemblage belongs to the upper Kungurian because a diverse conodont fauna including abundant *Mesogondolella idahoensis* and *Vjalovognathus nicolli* was found from the equivalent horizon (Yuan *et al.* 2016). By contrast, the upper part of the Xiala Formation contains a typical warm-water brachiopod fauna. Common elements include, for example, *Urushtenoidea crenulata*, *Neoplicatifera pusilla*, *Leptodus nobilis*, *Haydenella minuta* and *Permophricodothyris elegantula*. *Pseudoantiquatonia mutabilis* is also present (Zhan & Wu 1982). It shows a strong affinity with the late Guadalupian brachiopod faunas from the Shazipo Formation of the Baoshan Block and the Maokou Formation of South China (Shi & Shen 2001).

Lopingian (Changhsingian) brachiopods have been rarely recorded from the Lhasa Block. *Transennatia gratiosa*, ?*Peltichia* sp., *Spinomarginifera* sp., and *Crenispirifer* cf. *dzhulfensis* have been reported from the Lielonggou Formation at Qiusong, a locality close to Lhasa City. This fauna is probably of Changhsingian age because it is overlain by the Lower Triassic strata (Sun *et al.* 1981).

Oman and Turkey

Upper Asselian–Sakmarian brachiopod faunas have been reported from the Saiwan Formation in the Huqf area of Oman and subsurface equivalents (Angiolini *et al.* 1997, 2006; Angiolini 2007). They are suitably comparable with the cold-water assemblages from the middle and upper parts of the Yongzhu Formation in the Lhasa Block with respect of the presence of *Taeniothaerus permixta*, *Trigonotreta* sp., *Punctocyrtella* aff. *nagmargensis* and *Subansiria* sp. A similar brachiopod fauna containing *Taeniothaerus permixta* and *Punctocyrtella spinosa* in association with the fusulinid *Pseudofusulina* ex gr. *karapetovi karapetovi* was also documented from the Gharif Formation in the subsurface of the same region. At least three different biozones, *Punctocyrtella omanensis*, *P. spinosa–Taeniothaerus permixta* and *Neospirifer* aff. *hardmani–Derbyia haroubi* in ascending order, are recognized both in the subsurface and at the surface (Angiolini *et al.* 2006). Brachiopods from the Artinskian and Kungurian strata in this area are still under study (Angiolini pers. comm.). Guadalupian brachiopods were documented from the Khuff Formation in SE Oman (Yanagida & Pillevuit 1994; Angiolini & Bucher 1999). The brachiopod fauna is characterized by *Bilotina yanagidai*, *Celebetes manarollai*, *Calliprotonia* sp. and *Steroechia* sp. The fauna is generally comparable with the Guadalupian brachiopods from the Ratburi Limestone of southern Thailand (Waterhouse & Piyasin 1970; Grant 1976). To date, no Lopingian brachiopods have been found in Oman.

No Cisuralian brachiopods have been reported from Turkey, which was also a part of the STZ during the Permian. However, very abundant Guadalupian brachiopods were described from the lower–middle parts of the Pamucak Formation in western Taurus, Turkey by Verna *et al.* (2011). The brachiopod fauna from west Turkey shows a considerable similarity to the Wordian fauna of SE Oman, Chios, north Iran and southern Thailand, and also a strong affinity with the Cimmerian blocks (Angiolini *et al.* 2013*a*), Transcaucasia and, to a lesser degree, South China in terms of the presence of *Chonetella* sp., *Spinomarginifera helica*, *S. spinocostata*, *Bilotina yanagidai*, *Vediproductus punctatiformis*, *Permophricodothyris caroli* and *Perigeyerella miriae*. However, Verna *et al.* (2011) considered that the fauna is clearly peri-Gondwanan in terms of its palaeogeographical position of western Turkey.

Karakoram Range, Pakistan

Five brachiopod associations were established in the Karakoram Range, Pakistan based on extensive field collections (Angiolini 1995, 1996, 2001). The oldest association was named the *Trigonotreta lyonsensis–Punctospirifer afghanus* association characterized by low diversity and a dominance of spiriferids. The association is largely comparable with the assemblages from the lower part of the Dingjiazhai Formation in the Baoshan Block. It is assigned to an Asselian or early Sakmarian age. The overlying *Hunzina electa* association is from the Sakmarian Lupghar Formation in the Hunza Valley and Member 1 of the Lashkargaz Formation of Chitral (Angiolini 1995). The third association, named the *Orthothetina convergens–Aldina exilis* association, was found from the middle part of the Lashkargaz Formation in western Karakorum, and was assigned to the Kungurian based on the coexistence of the fusulinid *Pseudofusulina norikurensis krafftiformis* and *Darvasites* cf. *zulumartensis*. The fourth association from Member 4 of the Lashkargaz Formation was named the *Waagenoconcha (Gruntoconcha) macrotuberculata–Callytharrella sinensis* Assemblage. This assemblage was subdivided into five unitary associations (Angiolini 1996) that all belong to the upper Kungurian–Roadian (Angiolini *et al.* 2015, 2016). The youngest brachiopod association in the Karakoram Range was collected from Member 2 of the Panjshah Formation, and named the *Elivina chapursani–Martinia tatianae* Assemblage. The assemblage was considered to be of late Wordian age and correlative with the assemblage of the Gnishik Formation in Transcaucasia (Angiolini 2001) (Fig. 5).

SE Pamirs

The SE Pamirs possess a nearly complete Permian succession containing abundant fusulinids, brachiopods and some conodonts (Leven *et al.* 1997; Angiolini & Vachard 2015). Lower Permian brachiopods were reported from the Tashkazyk Formation (Grunt & Dmitriev 1973; Grunt & Novikov 1994) and discussed by Angiolini *et al.* (2013*b*). A brachiopod assemblage consisting of, for example, *Permochonetes pamiricus*, *Punctocyrtella gigantea*, *Neospirifer joharensis* and *Notospirifer bazardarensis* is present in the lower part of the formation, and most likely Sakmarian in age based on associated fusulinids (Leven *et al.* 1997). The upper part of the formation is characterized by *Taeniothaerus rusticus*, *Reticulatia pamirica*, *Tiramnia uralica* and *Spirelytha petaliformis*. This assemblage is of late Sakmarian or early Artinskian age (Grunt & Novikov 1994). The brachiopods include *Overtonia mamazairica*, *Martinia tschernyschewi* and *Retimarginifera* sp., and they were found in association with the fusulinids *Misellina*, *Cancellina* and *Neoschwagerina simplex*. The brachiopods are not distinctive in terms of age determination, but the associated fusulinids indicate that they are mostly of Kungurian age (Angiolini *et al.* 2015).

Late Kungurian or Guadalupian mixed-affinity brachiopod faunas have been described from central Afghanistan (Termier *et al.* 1974) where they were found locally associated with the ammonoid *Stacheoceras*. Characteristic brachiopod elements include, for example, *Waagenoconcha abichi*, *Permundaria sisophonensis*, *Dicystoconcha lapparenti*, *Vediproductus wardakensis*, *Spiriferella mexicana* and *Marginifera typica*, of which *Vediproductus* and *Permundaria* are common in the upper Kungurian and Guadalupian in the Palaeotethys.

Changhsingian brachiopods are very abundant in the upper Takhtabulak Formation in the SE Pamirs (Grunt & Dmitriev 1973; Kotlyar *et al.* 2004*b*), where they were found in association with the conodont *Clarkina subcarinata* and the foraminifer *Colaniella parva*. The most characteristic brachiopods include *Peltichia ruzhencevi*, *Haydenella tumida*, *Uncinunellina timorensis*, *Permophricodothryis caroli*, *Compressoproductus mongolicus* and *Strophalosiina multicostata*. The fauna bears significant similarities with the Wuchiapingian brachiopod faunas of South China, but the conodonts suggest an early Changhsingian age if they correctly identified (Grunt & Dmitriev 1973; Kotlyar *et al.* 2004*b*).

The Gondwanan Realm

Brachiopod successions in the Gondwanan Realm have been well established and widely used for correlation. However, the age determinations of these brachiopod faunas with the international Permian timescale are not well controlled owing to the absence or rare presence of conodonts and fusulinids. In some palaeogeographically closely connected regions (e.g. eastern Australia and New Zealand), several different brachiopod-based biostratigraphical zonation schemes have been proposed (Briggs 1998; Waterhouse 2008, 2015*a*, *b*; Waterhouse & Shi 2013). Although brachiopod and palynological zones have been proposed in Western Australia, eastern Australia and also in Argentina, the biostratigraphic ages of these zones are still problematic when they are caliberated with recently published high-precision U–Pb chemical abrasion-thermal ionization mass spectrometry (CA-TIMS) dates (Gulbranson *et al.* 2010; Césari *et al.* 2011; Mory *et al.* 2012; Metcalfe *et al.* 2015; Barbolini *et al.* 2016) (Fig. 6).

Eastern Australia

Permian brachiopods have been extensively documented from the Bowen, Sydney, Gympie and Tasmanian basins of eastern Australia. These faunas are characterized by typical cold-water Gondwanan elements, and, as a result, their correlations with those of the Palaeotethyan and Boreal regions have been difficult (Fig. 6). High-resolution brachiopod successions including more than 10 different zones in eastern Australia have been summarized by different authors (Runnegar & McClung 1975; Archbold & Dickins 1996; Briggs 1998; Archbold 1999; Waterhouse 2008, 2011, 2015*a*, *b*; Waterhouse & Shi 2013). The lowest Permian (Asselian) brachiopod zone is traditionally characterized by the species of *Lyonia* (e.g. *Lyonia lyoni* and *L. bourkei*), which is associated with the bivalve *Eurydesma* fauna (Briggs 1998), but the fauna has no known direct stratigraphic relationship with other faunas (Waterhouse 2008). The overlying brachiopod zones above the Asselian have been mostly established based on the species of a few genera of Strophalosioidea or Linoproductoidea (e.g. *Echinalosia*, *Strophalosia*, *Pseudostrophalosia* and *Terrakea*) and spiriferids (*Trigonotreta*, *Sulciplica* and *Ingelarella*). The Asselian Stage is represented by three zones, the *Strophalosia concentrica* Zone in the lower, the *S. subcircularis* Zone in the middle (Waterhouse 2011, 2015*a*, *b*; Cisterna & Shi 2014) and the *Bandoproductus walkomi* Zone (or the *Bandoproductus macrospinosa* Zone of Waterhouse 2011, 2015*a*, *b*) in the upper. The former two were mainly recognized in Tasmania. The *Bandoproductus walkomi* Zone was assigned to the lower Artinskian by Archbold (1999), and to the upper Sakmarian (Sterlitamakian), lower Sakmarian (Waterhouse 2011) or upper Asselian by other studies (Briggs 1998; Angiolini *et al.* 2005*a*; Waterhouse & Shi 2010, 2013; Waterhouse 2015*a*). Cisterna & Shi (2014) named the *Trigonotreta–Tomiopsis* Assemblage, which is equivalent to the *Bandoproductus macrospinosa* Zone. The overlying brachiopod biozones include the *Ingelarella strzeleckii*, *Echinalosia curtosa–E. warwicki* and *E. preovalis* zones, and were assigned to the Artinskian (Briggs 1998). The species *Ingelarella strzeleckii* was revised as *Monklandia gympiensis* by Waterhouse (1998) and the previous '*Ingelarella strzeleckii*' Zone was recently reassigned to the Asselian by Waterhouse (2015*b*). The uppermost *Echinalosia preovalis* Zone was assigned to the Kungurian by Archbold (1999). However, recent high-precision dates of ash beds in eastern Australia (Metcalfe *et al.* 2015) changed the ages of these brachiopod zones greatly. One ash bed from 11 m below the top of the Rowan Formation is 271.60 ± 0.08 Ma, which indicates that the uppermost part of the Rowan Formation is earliest Roadian in age, and thus pushes the overlying Branxton Formation containing the brachiopod *Echinalosia preovalis* Zone from the basal part to an early Guadalupian (probably Roadian) age if they are correctly correlated and identified. This is significantly older than the Kungurian suggested by Waterhouse & Shi (2013). However, Waterhouse (2001) considered that *Echinalosia preovalis* reported by Dickins (1981) should belong to another new species, *Echinalosia floodi*, and that its stratigraphic position is

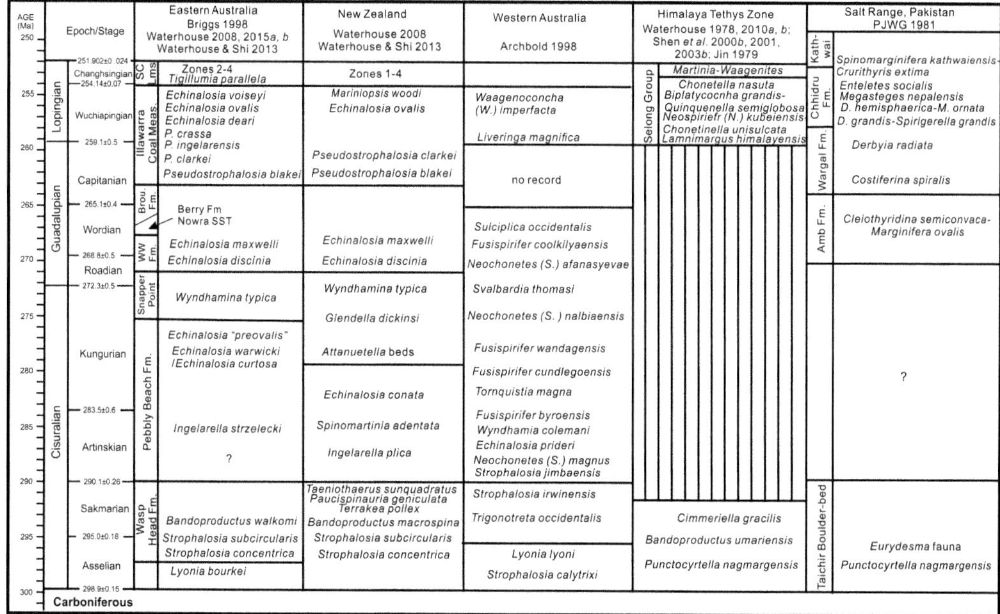

Fig. 6. Revised brachiopod successions in the Gondwanan Realm. The main references are shown in the columns. The Permian timescale is revised from Shen *et al.* (2013*a*). Zones 1–4 in New Zealand are the *Paucispinauria verecunda, Spinomartinia spinosa, Marginalosia planata, Wairakiella rostrata* zones in ascending order (Waterhouse & Shi 2013).

older than the zones of *Echinalosia maxwelli* and younger than the *Echinalosia preovalis–Terrakea dickinsi–Ingelarella plica* faunas. Briggs (1998, fig. 4) suggested that the Rowan Formation is equivalent to the upper part of Pebbly Peach Formation (Artinskian) in the Sydney Basin. If the brachiopod *Echinalosia preovalis* Zone is generally equivalent in different basins of eastern Australia, the entire overlying Snapper Point Formation containing the brachiopod *Echinalosia curtosa* Zone in the lower part of the formation is of Guadalupian age based on the high-precision U–Pb radioisotopic dates (Metcalfe *et al.* 2015). However, this correlation conflicts with the high-precision date (263.51 ± 0.05 Ma) from the top of the Broughton Formation, which is of Wordian age (Metcalfe *et al.* 2015). If the Snapper Point Formation is Roadian or Wordian, then there is very little time to accommodate the three other litholgical units (the Wandrawandian Formation, Nowra Sandstone and Berry Formation in ascending order) between the Snapper Point and the Broughton formations. Above the Snapper Point Formation, the Wandrawandian Formation contains the *Echinalosia maxwelli, E. davidi* and *E. discinia* zones in ascending order (Briggs 1998; Archbold 1999). These three brachiopod biozones belong to the latest Kungurian or early Guadalupian (Waterhouse & Shi 2013). The Broughton and Mulbring formations in the Sydney Basin, containing the brachiopod *Echinalosia robusta*/*hanloni*, *E. runegari*, *E. wassi* zones, and the lower part of the Vane Subgroup, containing the *Pseudostrophalosia blakei* and *P. clarkei* zones, are very probably of late Wordian or Capitanian age based on the high-precision radioistopic date (263.51 ± 0.05 Ma) from the Broughton Formation (Metcalfe *et al.* 2015) and the Guadalupian–Lopingian boundary in the upper part of the Vane Subgroup (Waterhouse & Shi 2010, 2013). All other brachiopod zones above the upper part of the Vane Subgroup are of Wuchiapingian age as U–Pb radioisotopic dates from the ash beds in the middle part of the Jerrys Plains Subgroup indicate that this subgroup is of lower–middle Wuchiapingian age (Metcalfe *et al.* 2015). The youngest Permian marine macro-faunas in eastern Australia were found in the upper South Curra Limestone and Tamaree Formation near Gympie in SE Queensland, which was previously represented by the *Plekonella multicostata* Zone dated as Kungurian by Runnegar & Ferguson (1969). Recently, Waterhouse (2015*a*) proposed four brachiopod zones for the Changhsingian Stage in the Gympie Basin of eastern Australia. They are the *Tigillumia parallela*, *Spinomartinia spinosa*, *Marginalosia planata* and *Wairakiella rostrata* zones in ascending order. There is no brachiopod reported from the Changhsingian Coal Measures in the Bowen and Sydney basins of eastern Australia. Clearly, there is a need to constrain the ages of the brachiopod assemblages in the Sydney and Bowen basins with further detailed geochronology of ash beds.

New Zealand

The Permian brachiopod succession in New Zealand is similar, in part, to that of eastern Australia, some of which can be found in both areas (Waterhouse 2008, 2015*a*, *b*). However, Waterhouse (2008) assigned those zones to different ages. The three Sakmarian brachiopod zones (Briggs 1998) – that is, the *Strophalosia concentrica*, *S. subcircularis* and *Bandoproductus macrospina* zones – are assigned to the Asselian by Waterhouse (2008) and Waterhouse & Shi (2010), of which the *Bandoproductus macrospina* Zone is unknown in New Zealand. The Sakmarian Stage is represented by the *Terrakea pollex*, *Paucispinauria geniculata* and *Taeniothaerus subquadratus* zones in ascending order according to Waterhouse (2008, 2015*a*, *b*), but these three zones may be of Artinskian age (e.g. Archbold 1999). Artinskian brachiopod representatives are different between eastern Australia and New Zealand. The *Ingelarella plica* Zone is equivalent to the *Notostrophia homeri* and *N. zealandicus* subzones of New Zealand. This zone is succeeded by the *Spinomartinia adentata* and *Echinalosia conata* zones (Waterhouse 2008). The latter was subsequently assigned to the Kungurian (Waterhouse & Shi 2010, 2013). The *Echinalosia conata* Zone is overlain by the *Attenuatella* beds in New Zealand, which are not present in eastern Australia. The overlying brachiopod zones from the Kungurian to the Guadalupian are generally comparable with each other between eastern Australia and New Zealand (Waterhouse 2008, table 4). The *Echinalosia discinia* Zone is possibly the lowest zone of the Guadalupian, and the *Echinalosia havilensis* (=*Tigillumia havilensis*) and *Pseudostrophalosia clarkei* zones are the youngest zones of the Guadalupian of New Zealand (Waterhouse 2008). The lower Changhsingian *Plekonella multicostata* Zone of the Gympie Basin in eastern Australia is also found in the Glendale Limestone with *Maitaia rechmanni* and *Paucispinauria verecunda* in New Zealand. This brachiopod zone is above the *Pseudostrophalosia clarkei* Zone in the Mangarewa Formation of Wairaki Downs, New Zealand, and is followed by the *Spinomartinia spinosa* and *Marginalosia planata* zones in the overlying Hilton Limestone. The topmost Wairaki Breccia contains the *Wairakiella rostrata* Zone, which represents the youngest brachiopod zone in the Permian of New Zealand (Waterhouse 2002). These zones are completely consistent with those of the Gympie Basin of eastern Australia (Waterhouse 2015*b*).

Western Australia

Permian brachiopod successions have been established in four basins (Carnarvon, Canning, Bonaparte and Perth) of Western Australia, of which the Carnarvon Basin contains the most detailed brachiopod zones and numerous key international correlation points based on the associated ammonoids and conodonts. Similar to those of eastern Australia, the lowest part of the Permian Lyons Group contains the *Lyonia lyoni* Zone in the Carnarvon Basin, which is of Asselian age. Below the *Lyonia lyoni* Zone is the *Strophalosia calytrixi* Zone created by Waterhouse (2011), which is considered to be of late Asselian age and thus pushes the overlying *Lyonia lyoni* Zone into the early Sakmarian, being correlative with the *Bandoproductus macrospinosa* Zone in eastern Australia (Waterhouse 2011, table 2). The Carrandibby Formation and a part of the Callytharra Formation of the Sakmarian Stage contain the *Trigonotreta occidentalis* (=*Neilotreta occidentalis*) and the *Strophalosia irwinensis* zones in the Carnarvon, Perth and Canning basins, which are considered to be correlative with the *Strophalosia concentrica* and *S. subcircularis* zones of the Sydney Basin in eastern Australia. The *Strophalosia irwinensis* Zone is associated with the ammonoids *Metalegoceras* sp. and *Uraloceras irwinense* (Archbold 1998), and the cold-water conodont *Vjalovognathus australis* (Nicoll & Metcalfe 1998). The lower Artinskian contains the *Strophalosia jimbaensis* Zone, associated with the conodont *Vjalovognathus australis* in the Carnarvon Basin. The upper Artinskian contains three brachiopod zones, all of which are associated with the conodont *Vjalovognathus shindyensis*. This conodont species is considered to be Kungurian in age because of its association with *Mesogondolella idahoensis* (Nicoll & Metcalfe 1998). It is also possible that the three upper Artinskian brachiopod zones from the Coyrie Formation in the Carnarvon Basin are much younger than the age previously assigned. The Kungurian Stage in the Carnarvon Basin contains a succession of six brachiopod zones. These brachiopod zones are associated with the ammonoid *Bamyaniceras australe* in the lower part and *Daubichites* in the upper part. Previously, there have been no reliable Guadalupian brachiopod faunas reported from Western Australia. However, a recent high-precision U–Pb zircon chemical abrasion- isotope dilution-thermal ionization mass spectrometry (CA-ID-TIMS) age range of the ash-fall tuffs in seams from the Lightjack Formation is given as 268.63–270.14 Ma, suggesting a Roadian–Wordian age for the brachiopod *Neochonetes* (*Sommeriella*) *afanasyevae* Zone (Mory *et al.* 2012). Two Lopingian brachiopod zones in association with the ammonoid *Cyclolobus persulcatus*, the *Liveringa magnifica* Zone in the lower and *Waagenoconcha imperfecta* Zone in the upper were recognized from the Hardman Formation in the Canning Basin and the Upper Marine beds of the Bonaparte Basin of Western Australia (Archbold 1998).

Himalaya Tethys Zone, the Salt Range of Pakistan and northern India

In the Himalaya Tethys Zone and the peri-Gondwanan region, typical lower Cisuralian (Sakmarian) brachiopods are characterized by the *Cimmeriella* (=previous '*Stepanoviella*') fauna from the Jilong Formation in the Mt Everest area and the Lower and Middle Manzhongrong formations in Mayang of the Rutog area in SW Tibet (Jin 1979; Yang & Nie 1990). *Spinomartinia mayangensis* is also common in the Mayang Formation in this area. This fauna is generally comparable with its counterparts reported in Western Australia, the Baoshan Block of western Yunnan, central Lhasa and SW Tibet. Another lower Cisuralian brachiopod fauna from the Agglomeratic Slate of Kashmir is characterized by containing some widely distributed elements in the peri-Gondwanan region such as *Punctocyrtella nagmargensis*, *Trigonotreta stokesi*, *T. australis*, *Taeniothaerus brenensis* and *Neospirifer fasciger* (Bion 1928; Reed 1932). Therefore, a Gondwanan cold-water brachiopod fauna characterized by the *Bandoproductus–Punctocyrtella* Assemblage was widely distributed in the peri-Gondwanan region and the Cimmerian blocks when they were attached to Gondwanaland during the early Cisuralian.

Although brachiopods from the Selong Group in southern Tibet were previously assigned to the Chihsian (Kungurian) and Maokouan (Guadalupian) (Zhang & Jin 1976; Xia & Zhang 1992), subsequent studies have suggested that the brachiopods of the Selong Group are basically comparable with those from the Senja Formation in NW Nepal, and were reassigned to the Lopingian (Waterhouse 1978, 2010*a*; Shen *et al.* 2000*b*, 2001, 2003*a*, *b*; Shi *et al.* 2010). Thus, no reliable Guadalupian brachiopod fauna is known from the Himalayan region, which probably suggests a long hiatus between the Cisuralian and the Lopingian in this region. Lopingian brachiopods have also been extensively documented from the Senja Formation in NW Nepal, the Kuling Shale of India, the Selong Group at Selong Xishan and Qubu sections in southern Tibet.

At the Selong Xishan section, three brachiopod assemblages were recognized. The *Marginalosia–Composita* and *Chonetella nasuta* assemblages are most probably of Wuchiapingian age. Many species of these two assemblages, such as *Marginalosia kalikotei*, *Spiriferella rajah*, *S. nepalensis*, *Neospirifer (N.) kubeiensis*, *Retimarginifera xizangensis* and *Lamnimargus himalayensis*, are also present in the Senja Formation in NW Nepal and the Kuling Shale of northern India. The topmost *Martinia*–'*Waagenites*' (now *Fusichonetes*) Assemblage is of Changhsingian age as indicated by the presence of the conodont *Mesogondolella sheni* (Mei 1996; Shen & Jin 1999). Two brachiopod assemblages from the Qubuerga Formation in the Mt Everest area, the *Neospirifer (Neospirifer) kubeiensis–Chonetinella unisulcata* Assemblage in the lower and the *Biplatyconcha grandis–Quinquenella semiglobosa* Assemblage in the upper, were established (Shen *et al.* 2003*b*). The brachiopods of these two assemblages are generally comparable with those of the Senja Formation in NW Nepal and the Selong Group in the Selong Xishan section of southern Tibet, which places them in the Wuchiapingian Stage.

Permian brachiopods are extremely abundant and highly diverse in the Salt Range of Pakistan as described by a series of monographs (Waagen 1882–85; Reed 1944). Several lower Cisuralian brachiopods associated with the typical Gondwanan *Eurydesma* fauna were reported from the Talchir Boulder-bed series of the Salt Range, Pakistan (Reed 1936). The Guadalupian brachiopods of the Salt Range are known from the Amb Formation and the lower part of the Wargal Formation. They exhibit some transitional characteristics between the Gondwanan and Tethyan affinities. The brachiopods from the Wordian Amb Formation in association with the fusulinid *Monodiexodina kattaensis* were named as the *Cleiothyridina semiconvaca* Zone. This zone contains *Marginifera ovalis*, *Neospirifer marcoui*, *Strophalosia tenuispina* and *Neochonetes (Sommeriella) strophomenoides* (Pakistan–Japanese Working Group 1985). The overlying Wargal Formation ranges from the Wordian to the Wuchiapingian in age based on fusulinids (Pakistan Japanese Working Group 1985) and conodonts (Wardlaw & Mei 1999). The basal unit (Unit 1) of the Wargal Formation is characterized by the *Costiferina spiralis* Zone. Other common species include *Spirigerella grandis*, *Dielasma truncata* and *Strophalosia tenuispira*. This zone belongs to the lower part of the fusulinid *Chusenella* sp. Zone, which is of Wordian age. Unit 4a in the upper part of the Wargal Formation contains the *Derbyia radiata* Zone. Unit 4b of the topmost part of the Wargal Formation, the Kalabagh Member and the lowest part of the Chhidru Formation belong to the *Derbyia grandis–Spirigerella grandis* Zone. Since Unit 4b contains the Wuchiapingian conodonts and the Kalabagh Member contains the Wuchiapingian ammonoid *Cyclolobus*, these two brachiopod zones were assigned to the Wuchiapingian (Pakistan–Japanese Working Group 1985). The middle and upper parts of the Chhidru Formation include the *Derbyia hemisphaerica–Marginifera ornata*, *Megasteges nepalensis* and *Enteletes socialis* zones in ascending order. These three brachiopod zones are most likely of Changhsingian age. A brachiopod assemblage consisting of some small elements, and represented by *Spinomarginifera kathwaiensis*, *Crurithyris extima* (probably = *Paracrurithyris pigmaea*) and *Martinia acutomarginiformis*, was found in the Kathwai Member of the Mianwali Formation, which is the youngest assemblage of the Permian in the Salt Range. The brachiopods of the Kathwai Member are generally comparable with those of the *Martinia*–'*Waagenites*' Assemblage at Selong Xishan, southern Tibet (Shen & Jin 1999).

Typical Gondwanan-type brachiopod faunas are also known from the Lower Mangzhongrong Formation of Mayang in SW Tibet. It is characterized by the *Cimmeriella* fauna of lower to middle Cisuralian age. The overlying Middle and Upper Mangzhongrong formations have yielded, amongst others, *Spiriferella rajah*, *Megasteges nepalensis*, *Taeniothaerus subquadratus*, *Fusispirifer plicatus*, *Lamnimargus himalayensis* and *Wyndhamia circularis*. The brachiopod fauna in Mayang, SW Tibet is the same as those found in the Selong Group, the Qubuerga Formation and the Senja Formation of the Himalaya Tethys Zone in southern Tibet and Nepal (Yang & Nie 1990).

The northern part of India also yields typical Gondwanan-type brachiopods, but only lower Cisuralian brachiopods have been reported. The brachiopod fauna contains *Bandoproductus umariensis*, *Trigonotreta orientalis*, *Punctocyrtella nagmargensis* and *Taeniothaerus kashimirica*, and can be reliably correlated with the lower Cisuralian brachiopod faunas in the peri-Gondwanan region (Singh & Archbold 1993; Archbold *et al.* 1996).

South America

South America contains continuous Carboniferous–Permian sequences with abundant plant fossils and marine fossils in the Cisuralian. The brachiopod succession is generally comparable with those from eastern and Western Australia. The widely distributed *Eurydesma–Lyonia* fauna in the basal part of the Permian in eastern and Western Australia was recorded from the Permian Itararé Group in the Paraná Basin of Brazil, suggesting an Asselian–early Sakmarian bio-correlation between Brazil and Australia (Taboada *et al.* 2016). ^{206}Pb–^{238}U dating of a tuffaceous horizon of the Hardap Shale Member in the Carboniferous–Permian Dwyka Group containing the *Eurydesma* fauna in southern Namibia is 297.1 ± 1.8 Ma (Stollhofen *et al.* 2008), suggesting that the *Eurydesma* fauna in South Africa is at least partly Asselian in age. The *Eurydesma* fauna was also recorded in eastern Argentina and is characterized by containing two brachiopod species, *Tivertonia pillahuincensis* and *Tomiopsis harringtoni*. Two biozones, the *Tivertonia jachalensis–Streptorhynchus inaequiornatus* Biozone in the lower and *Costatumulus amosi* (formerly *Cancrinella* cf. *farleyensis*) Biozone in the upper, from the Carboniferous–Permian of the Calingasta–Uspallata Sub-basin, Argentine Precordillera, were proposed. The lower *Tivertonia–Streptorhynchus* biozones were previously considered to be from the upper Asselian in the Precordillera; and the upper *Costatumulus* Biozone was estimated to be late Sakmarian–early Artinskian in age (Cisterna *et al.* 2002; Taboada 2010). However, high-precision radioisotopic dates provided by Gulbranson *et al.* (2010) from beds linked to the transgression bearing the *Tivertonia–Streptorhynchus* fauna from the Tupe and Río del Peñón formations in the Río Blanco Basin have yielded ^{206}Pb–^{238}U ages of 312.82 ± 0.11 and 310.63 ± 0.1 Ma, thus constraining the *Tivertonia–Streptorhynchus* Biozone to the Moscovian (also see Césari *et al.* 2011). This age has recently been supported by the faunal and biostratigraphic relationships between the *Tivertonia jachalensis–Streptorhynchus inaequiornatus* and *Costatumulus amosi* faunas (Taboada 2014). A lower Cisuralian *Cimmeriella–Jakutoproductus* Assemblage was also documented from the Tepuel-Genoa Basin in Patagonia, which is considered to be equivalent to the *Costatumulus amosi* Biozone (Taboada & Pagani 2010). This assemblage is arguably of latest Carboniferous or Early Permian age. No brachiopods have been reported from the Permian strata above the upper Artinskian in South America since seawater withdrew from this continent during the late Cisuralian.

Concluding remarks

More than 6100 species and 760 genera of Brachiopoda have been reported from the Permian sequences all over the world. Although global palaeobiogeography (e.g. Shen *et al.* 2009, 2013*b*; Angiolini *et al.* 2013*a*; Ke *et al.* 2016), diversity pattern (e.g. Shen & Shi 2002), palaeoclimate (Waterhouse & Shi 2010, 2013), and palaeoecology and environmental implications (Clapham & James 2008; He *et al.* 2015; Shi *et al.* 2016) based on Permian brachiopods have been extensively documented, it is difficult to review all the Permian brachiopod assemblages in such a short paper and to establish their global correlations with certainty (Figs 2–6). This is probably the reason why relatively little attention has been paid to the global biochronological correlation based on brachiopod assemblages. The main brachiopod successions briefly reviewed in this paper represent the updated knowledge of the brachiopod biostratigraphy in numerous different geographical areas since Waterhouse (1976), but it is still much less precise than the scales provided by the conodonts, fusulinids and ammonoids. Recently, high-precision radioisotopic dates (e.g. Gulbranson *et al.* 2010; Mory *et al.* 2012; Metcalfe *et al.* 2015; Barbolini *et al.* 2016) from the brachiopod-bearing strata have helped to constrain the ages of some of brachiopod assemblages (e.g. eastern Australia and South America).

To create a widely applicable brachiopod-based bio-chronostratigraphy, the first requirement is to provide a clear and generally consistent taxonomic approach for the brachiopods. However, such a basic requirement has not yet been met. Numerous previously established brachiopod taxa still require a review and an update. For example, some of these taxonomic revisions have been carried out by Waterhouse (2010*b*, 2013), which has resulted in the creation of a large number of new brachiopod taxa. Shen *et al.* (in press) have recently reviewed 208 Permian brachiopod genera based on type species from China, of which more than 30% (64 genera) are regarded as subjective or objective synonyms. Numerous new species and new genera were established based on the form species concept. One example is that of a very distinctive widespread brachiopod fauna in the lower and middle Cisuralian of the Gondwanan Realm represented by the *Bandoproductus–Punctocyrtella* and the *Cimmeriella* faunas. However, the species of *Cimmeriella* were previously assigned to a few different genera (e.g. *Stepanoviella*, *Globiella*, *Linoproductus*), and the species of *Punctocyrtella* were assigned, for example, to *Cyrtella*, *Syringothyris* and *Tuotalania*. Many species previously assigned to *Cancrinella*, *Costatumulus* and *Cancrinelloides*, amongst others, should also be re-studied.

As far as the strong endemism of Permian brachiopods is concerned, the brachiopod successions of the two transitional zones (Shi *et al.* 1995) are very useful in biochronological correlation because they contain brachiopods with both bipolar/bitemperate and palaeoequatorial affinities, and are commonly associated with conodonts, fusulinids and ammonoids. Some genera (e.g. *Kochiproductus*, *Waagenoconcha*, *Wyndahamia* and *Yakovlevia*) with bipolar/bitemperate distributions may also have a potential for correlation between the two antitropical realms (Shi & Grunt 2000; Shi 2006; Taboada & Pagani 2010).

A few brachiopod changeover events related to the palaeoclimate changes during the Permian may provide some value for wide biochronological correlation. The *Bandoproductus–Punctocyrtella* fauna in association with the bivalve *Eurydesma* fauna was widely distributed in the lower and middle Cisuralian (Asselian–Artinskian) in Gondwanaland, the Cimmerian blocks and South America, and has been regarded as the indicator of the acme stage of the Late Palaeozoic Ice Age (Shi *et al.* 1996; Shi & Archbold 1998; Shen *et al.* 2000*c*; Angiolini *et al.* 2005*a*; Taboada & Pagani 2010; Taboada *et al.* 2016). In addition to the typical early Cisuralian cold-water brachiopod fauna, brachiopods also show some changes in response to other several local or regional glaciations (Waterhouse & Shi 2010, 2013). However, the precise ages and their correlation of the brachiopod zones still need to be refined. The palaeoclimate amelioration after the Late Palaeozoic Ice Age has recently been refined to the late Kungurian (Chen *et al.* 2013; Yuan *et al.* 2016), a distinct brachiopod shift in composition from cold-water elements to warm-water or transitional elements occurred in the peri-Gondwanan region and the Cimmerian blocks (Zhan *et al.* 2007; Shen *et al.* 2016). In the NTZ, the occurrences of the *Yakovlevia* faunas may suggest that this zone entirely moved to a relatively higher palaeolatitude during the beginning of the Guadalupian. The Permian–Triassic extinction interval is characterized by an assemblage with small-sized and thin-shelled elements highlighted by *Paracrurithyris pigmaea* in the latest Changhsingian and basal Triassic, which may be assigned to a few different genera (e.g. *Crurithyris*, *Attenuatella* and *Orbicoelia*). The distinct body size reduction around the Permian–Triassic boundary indicates the onset of the end-Permian mass extinction (He *et al.* 2015; Shi *et al.* 2016; Zhang *et al.* 2016).

I would thank J.B. Waterhouse and A.C. Taboada for their kind reviews and numerous very useful comments. G.R. Shi, L. Angiolini and Roger Pierson also provided some very useful suggestions to improve the manuscript. This work was supported by NSFC (grant numbers 41290260, 41420104003) and the Strategic Priority Research Program (B) of the Chinese Academy of Sciences (XDB1 8030400).

References

Afanasieva, G.A., Bogoslovskaia, M.V. *et al.* 2006. *Late Permian of the Kanin Peninsula*. Nauka, Moscow [in Russian].

Angiolini, L. 1995. Permian brachiopods from Karakorum (Pakistan), Part 1. (With Appendix – New brachiopod taxa from the Bolorian–Murgabian/Midian of Karakorum). *Rivista Italiana di Paleontologia e Stratigrafia*, **101**, 165–214.

Angiolini, L. 1996. Permian brachiopods from Karakorum (Pakistan), Part 2. *Rivista Italiana di Paleontologia e Stratigrafia*, **102**, 3–26.

ANGIOLINI, L. 2001. Permian brachiopods from Karakorum (Pakistan), Part 3. *Rivista Italiana di Paleontologia e Stratigrafia*, **107**, 307–344.

ANGIOLINI, L. 2007. Quantitative palaeoecology in the *Pachycyrtella* Bed, Early Permian of Interior Oman. *Palaeoworld*, **16**, 233–245.

ANGIOLINI, L. & BUCHER, H. 1999. Taxonomy and quantitative biochronology of Guadalupian brachiopods from the Khuff Formation, Southeastern Oman. *Geobios*, **32**, 665–699.

ANGIOLINI, L. & CARABELLI, L. 2010. Upper Permian brachiopods from the Nesen formation, North Iran. *Special Papers in Palaeontology*, **84**, 41–90.

ANGIOLINI, L. & LONG, S.L. 2008. The ENS Collection: a systematic study of brachiopods from the Lower Permian Vøringen Member, Kapp Starostin Formation, Spitsbergen. *Proceedings of the Royal Society of Victoria*, **120**, 75–103.

ANGIOLINI, L. & STEPHENSON, M.H. 2008. Early Permian brachiopods and palynomorphs from the Dorud Group, Alborz Mountains, north Iran: evidence of their palaeobiogeographic affinities. *Fossils and Strata*, **54**, 117–132.

ANGIOLINI, L. & VACHARD, D. 2015. The Permian sedimentary successions of the Pamir mountains, Tajikistan. *Permophiles*, **62**, 15–18.

ANGIOLINI, L., BUCHER, H. ET AL. 1997. Early Permian (Sakmarian) brachiopods from southeastern Oman. *Geobios*, **30**, 379–405.

ANGIOLINI, L., BRUNTON, H. & GAETANI, M. 2005*a*. Early Permian (Asselian) brachiopods from Karakorum (Pakistan) and their palaeobiogeographical significance. *Palaeontology*, **48**, 69–86.

ANGIOLINI, L., CARABELLI, L. & GAETANI, M. 2005*b*. Middle Permian brachiopods from Greece and their palaeobiogeographical significance: new evidence for a Gondwanan affinity of the Chios Island Upper Unit. *Journal of Systematic Palaeontology*, **3**, 169–185.

ANGIOLINI, L., STEPHENSON, M.H. & LEVEN, E.J. 2006. Correlation of the Lower Permian surface Saiwan Formation and subsurface Haushi limestone, Central Oman. *Geoarabia*, **11**, 17–38.

ANGIOLINI, L., GAETANI, M., MUTTONI, G., STEPHENSON, M.H. & ZANCHI, A. 2007. Tethyan oceanic currents and climate gradients 300 myr ago. *Geology*, **35**, 1071–1074.

ANGIOLINI, L., CRIPPA, G., MUTTONI, G. & PIGNATTI, J. 2013*a*. Guadalupian (Middle Permian) paleobiogeography of the Neotethys Ocean. *Gondwana Research*, **24**, 173–184.

ANGIOLINI, L., ZANCHI, A., ZANCHETTA, S., NICORA, A. & VEZZOLI, G. 2013*b*. The Cimmerian geopuzzle: new data from South Pamir. *Terra Nova*, **25**, 352–360.

ANGIOLINI, L., ZANCHI, A. ET AL. 2015. From rift to drift in South Pamir (Tajikistan): Permian evolution of a Cimmerian terrane. *Journal of Asian Earth Sciences*, **102**, 146–169.

ANGIOLINI, L., CAMPAGNA, M. ET AL. 2016. Brachiopods from the Cisuralian–Guadalupian of Darvaz, Tajikistan and implications for Permian stratigraphic correlations. *Palaeoworld*, **25**, 539–568, https://doi.org/10.1016/j.palwor.2016.05.006

ARCHBOLD, N.W. 1998. Marine biostratigraphy and correlation of the West Australian Permian basins. *In*: PURCELL, P.G. & PURCELL, R.R. (eds) *The Sedimentary Basins of Western Australia 2: Proceedings of Petroleum Exploration Society of Austrialia Symposium*. Petroleum Exploration Society of Austrialia, Perth, 141–151.

ARCHBOLD, N.W. 1999. Permian Gondwanan correlations: the significance of the Western Australian marine Permian. *Journal of African Earth Sciences*, **29**, 63–75.

ARCHBOLD, N.W. & DICKINS, J.M. 1996. Permian (Chart 6). *In*: YOUNG, G.C. & LAURIE, J.R. (eds) *An Australian Phanerozoic Timescale*. Oxford University Press, London, 127–135.

ARCHBOLD, N.W., THOMAS, G.A. & SKWARKO, S.K. 1993. Brachiopods. *In*: SKWARKO, S.K. (ed.) *Palaeontology of the Permian of Western Australia*. Geological Survey of Western Australia, Bulletin, **136**, 45–51.

ARCHBOLD, N.W., SHAH, S.C. & DICKINS, J.M. 1996. Early Permian brachiopod faunas from Peninsular India: their Gondwanan relationships. *Historical Biology*, **11**, 125–135.

ASSERETO, R. 1963. Il Trias in Lombardia (Studi Geologici e Paleontologici); IV. Fossili dell'Anisico Superiore della Val Camonica. *Rivista Italiana di Paleontologia e Stratigrafia*, **69**, 3–123.

ASSERETO, R., BOSELLINI, A., FANTINI SESTINI, N. & SWEET, W.C. 1973. The Permian–Triassic boundary in the Southern Alps (Italy). *In*: LOGAN, A. & HILLS, L.V. (eds) *The Permian and Triassic Systems and Their Mutual Boundary*. Canadian Society of Petroleum Geologists, Memoirs, **2**, 176–199.

BAMBACH, R.K. 1990. Late Palaeozoic provinciality in the marine realm. *In*: MCKERROW, W.S. & SCOTESE, C.R. (eds) *Palaeozoic Palaeogeography and Biogeography*. Geological Society, London, Memoirs, **12**, 307–324, https://doi.org/10.1144/GSL.MEM.1990.012.01.30

BAMBER, E.W. & WATERHOUSE, J.B. 1971. Carboniferous and Permian stratigraphy and paleontology, northern Yukon Territory, Canada. *Bulletin of Canadian Petroleum Geology*, **19**, 29–250.

BARBOLINI, N., BAMFORD, M.K. & RUBIDGE, B. 2016. Radiometric dating demonstrates that Permian spore–pollen zones of Australia and South Africa are diachronous. *Gondwana Research*, **37**, 241–251, https://doi.org/10.1016/j.gr.2016.06.006

BION, H.S. 1928. *The Fauna of the Agglomeratic Slate Series of Kashmir, with an Introductory Chapter by C. S. Middlemiss*. Geological Survey of India Memoirs, Palaeontologia Indica, **12**.

BRIGGS, D.J.C. 1998. *Permian Productidina and Strophalosiidina from the Sydney-Bowen Basin and New England Orogen; Systematics and Biostratigraphic Significance*. Memoirs of the Association of Australasian Palaeontologists, **19**.

CAMPI, M.J., SHEN, S.Z., LEMAN, M.S. & SHI, G.R. 2000. First record of *Permianella* He & Zhu, 1979 (Permianellidae; Brachiopoda) from Peninsular Malaysia. *Alcheringa*, **24**, 37–43.

CAMPI, M.J., SHI, G.R. & LEMAN, M.S. 2002. The *Leptodus* shales of central Peninsular Malaysia: distribution, age and palaeobiogeographical affinities. *Journal of Asian Earth Sciences*, **20**, 703–717.

CAMPI, M.J., SHI, G.R. & LEMAN, A.S. 2005. Guadalupian (Middle Permian) brachiopods from Sungal Toh, a *Leptodus* shale locality in the Central Belt of Peninsular Malaysia – Part I: Lower horizons. *Palaeontographica, Abteilung A: Palaozoologie–Stratigraphie*, **273**, 97–160.

CÉSARI, S.N., LIMARINO, C.O. & GULBRANSON, E.L. 2011. An Upper Paleozoic bio-chronostratigraphic scheme for the western margin of Gondwana. *Earth Science Reviews*, **106**, 149–160.

CHEN, B., JOACHIMSKI, M.M. ET AL. 2013. Permian ice volume and palaeoclimate history: oxygen isotope proxies revisited. *Gondwana Research*, **24**, 77–89.

CHEN, Z.Q., CAMPI, M.J., SHI, G.R. & KAIHO, K. 2005. Post-extinction brachiopod faunas from the Late Permian Wuchiapingian coal series of South China. *Acta Palaeontologica Polonica*, **50**, 343–363.

CHEN, Z.Q., SHI, G.R., YANG, F.Q., GAO, Y.Q., TONG, J.N. & PENG, Y.Q. 2006. An ecologically mixed brachiopod fauna from Changhsingian deep-water basin of South China: consequence of end-Permian global warming. *Lethaia*, **39**, 79–90.

CISTERNA, G.A. & SHI, G.R. 2014. Lower Permian brachiopods from Wasp Head Formation, Sydney Basin, southeastern Australia. *Journal of Paleontology*, **88**, 531–544.

CISTERNA, G.A., SIMANAUSKAS, T. & ARCHBOLD, N.W. 2002. Permian brachiopods from the Tupe Formation, San Juan Province, Precordillera, Argentina. *Alcheringa*, **26**, 177–200.

CLAPHAM, M.E. & JAMES, N.R. 2008. Paleoecology of Early–Middle Permian marine communities in eastern Australia: response to global climate change in the aftermath of the Late Paleozoic ice age. *Palaios*, **23**, 738–750.

CLAPHAM, M.E., SHEN, S.Z. & BOTTJER, D.J. 2009. The double mass extinction revisited: reassessing the severity, selectivity, and causes of the end-Guadalupian biotic crisis (Late Permian). *Paleobiology*, **35**, 32–50.

COOPER, G.A. & GRANT, R.E. 1964. New Permian stratigraphic units in Glass Mountains, West Texas. *American Association of Petroleum Geologists Bulletin*, **48**, 1581–1588.

COOPER, G.A. & GRANT, R.E. 1972*a*. Dating and correlating the Permian of the Glass Mountains in Texas. *Bulletin of Canadian Petroleum Geology*, **20**, 727–741.

COOPER, G.A. & GRANT, R.E. 1972*b*. *Permian Brachiopods of West Texas, I.* Smithsonian Contributions to Paleobiology, **14**.

COOPER, G.A. & GRANT, R.E. 1974. *Permian Brachiopods of West Texas, II.* Smithsonian Contributions to Paleobiology, **15**.

COOPER, G.A. & GRANT, R.E. 1975. *Permian Brachiopods of West Texas, III.* Smithsonian Contributions to Paleobiology, **19**.

COOPER, G.A. & GRANT, R.E. 1976*a*. *Permian Brachiopods of West Texas, IV.* Smithsonian Contributions to Paleobiology, **21**.

COOPER, G.A. & GRANT, R.E. 1976*b*. *Permian Brachiopods of West Texas V.* Smithsonian Contributions to Paleobiology, **24**.

COOPER, G.A. & GRANT, R.E. 1977. *Permian Brachiopods of West Texas, VI.* Smithsonian Contributions to Paleobiology, **32**.

CRIPPA, G. & ANGIOLINI, L. 2012. Guadalupian (Permian) brachiopods from the Ruteh Limestone, North Iran. *GeoArabia*, **17**, 125–176.

CRIPPA, G., ANGIOLINI, L., VAN WAVEREN, I., CROW, M.J., HASIBUAN, F., STEPHENSON, M.H. & UENO, K. 2014. Brachiopods, fusulines and palynomorphs of the Mengkarang Formation (Early Permian, Sumatra) and their palaeobiogeographical significance. *Journal of Asian Earth Sciences*, **79**, 206–223.

CURRY, G.B. & BRUNTON, C.H.C. 2007. Stratigraphic distribution of brachiopods. *In*: SENDEN, P.A. (ed.) *Treatise on Invertebrate Paleontology, Part H, Brachiopoda, Volume 6, Supplement.* Geological Society of America, Boulder, CO & University of Kansas, Lawrence, KS, 2901–3081.

DICKINS, J.M. 1981. A Permian Invertebrate fauna from the Warwick area, Queensland. *Bureau of Mineral Resources (BMR) Bulletin of Australian Geology and Geophysics*, **209**, 24–43.

DUAN, C.H. & LI, W.G. 1985. Brachiopoda. *In*: DING, Y.J., XIA, G.Y., DUAN, C.H. & LI, W.G. (eds) *Study on the Early Permian Stratigraphy and Fauna in Zhesi District, Nei Mongol Zizhiqu (Inner Mongolia), Bulletin of the Tianjin Institute of Geology and Mineral Resources, the Chinese Academy of Geological Sciences*, **10**, 99–144 [in Chinese with English abstract].

DUNBAR, C.O. 1955. Permian brachiopod faunas of central East Greenland. *Meddelelser om Grönland*, **110**, 1–169.

FANG, R.S. 1994. The discovery of cold-water brachiopod *Stepanoviella* fauna in Baoshan region and its geological significance. *Yunnan Geology*, **13**, 264–277 [in Chinese with English abstract].

FANG, R.S. & FAN, J.C. 1994. *Middle to Upper Carboniferous–Early Permian Gondwana Facies and Palaeontology in Western Yunnan.* Yunnan Science and Technology Press, Kunming [in Chinese with English summary].

FANTINI SESTINI, N. 1965*a*. The geology of the Upper Djadjerud and Lar Valleys (North Iran), II, Palaeontology. *Brachiopods from Dorud Formation. Rivista Italiana di Paleontologia e Stratigrafia*, **71**, 773–788.

FANTINI SESTINI, N. 1965*b*. The geology of the Upper Djadjerud and Lar Valleys (North Iran), II, Palaeontology. Bryozoans, brachiopods and molluscs from Ruteh limestone (Permian). *Rivista Italiana di Paleontologia e Stratigrafia*, **71**, 13–108.

GANELIN, V.G. 1990. Biostratigraphiya permskikh otlozhenii Omolonskogo massiva [Biostratigraphy of Permian sediments in the Omolon Massif]. *In*: KASHIK, D.S. (ed.) *Opornyi Razrez Permi Omolonskogo Massiva [The Permian Reference Section in the Omolon Massif].* Nauka, Leningrad, 102–117 [in Russian].

GANELIN, V.G. & BIAKOV, A.S. 2006. The Permian biostratigraphy of the Kolyma–Omolon region, Northeast Asia. *Journal of Asian Earth Sciences*, **26**, 225–234.

GANELIN, V.G. & KOTLYAR, G.V. 1984. Brakhiopody. *In*: KOTLJAR, G.V. & STEPANOV, D.L. (eds) *Osnovnye Cherty Stratigrafii Permskoi Sistemy SSSR. [Main Features of Stratigraphy of Permian System in the USSR].* Vsesoiuznyi Nauchno-Issledovatel'skii Geologicheskii Institut (VSEGEI), Trudy Nauk, St Petersburg, 257–262 [in Russian].

Garbelli, C., Angiolini, L. *et al.* 2014. Additional brachiopod findings from the Lopingian succession of the Ali Bashi Mountains, NW Iran. *Rivista Italiana di Paleontologia e Stratigrafia*, **120**, 119–126.

Ghaderi, A., Garbelli, C., Angiolini, L., Ashouri, A.R., Korn, D., Rettori, R. & Gharaie, M.H.M. 2014. Faunal change near the end-Permian extinction: the brachiopods of the Ali Bashi Mountains, NW Iran. *Rivista Italiana di Paleontologia e Stratigrafia*, **120**, 27–59.

Girty, G.H. 1910. *The Fauna of the Phosphate Beds of the Park City Formation in Idaho, Wyoming, and Utah.* United States Geological Survey Bulletin, **436**.

Glenister, B.F., Rogers, F.S. & Skwarko, S.K. 1993. Ammonoids. *In*: Skwarko, S.K. (ed.) *Palaeontology of the Permian of Western Australia.* Geological Survey of Western Australia, Bulletin, **136**, 54–63.

Gobbett, D.J. 1963. Carboniferous and Permian brachiopods of Svalbard. *Norsk Polarinstitutt, Skrifter*, **127**, 1–201.

Grabau, A.W. 1931. *The Permian of Mongolia. Natural History of Central Asia. Volume IV. Central Asiatic Expeditions.* American Museum of Natural History, New York.

Grant, R.E. 1976. *Permian Brachiopods from Southern Thailand.* Paleontological Society Memoir, **9**. *Journal of Paleontology*, **50**, 1–269.

Grant, R.E. 1993. Permian brachiopods from Khios Island, Greece. *Journal of Paleontology*, **67**, 1–21.

Grant, R.E. 1995. Upper Permian brachiopods of the Superfamily Orthotetoidea from Hydra Island, Greece. *Journal of Paleontology*, **69**, 655–670.

Grant, R.E., Nestell, M.K., Baud, A. & Jenny, C. 1991. Permian stratigraphy of Hydra Island, Greece. *Palaios*, **6**, 479–497.

Grunt, T.A. 2005. Global and East European stage scales of the Permian System and their validity for extratropical sedimentation zones. *Stratigraphical and Geological Correlation*, **13**, 41–55.

Grunt, T.A. & Dmitriev, V.Y. 1973. Permskie brakhiopody Pamira [Permian Brachiopoda of the Pamir]. *Akademiia Nauk SSSR, Paleontologicheskii Institut, Trudy*, **136**, 1–212 [in Russian].

Grunt, T.A. & Novikov, V.P. 1994. Biostratigraphy and biogeography of the Early Permian in the Southeastern Pamirs. *Stratigraphy and Geological Correlation*, **2**, 331–339.

Grunt, T.A. & Shi, G.R. 1997. A hierarchical framework of Permian global marine biogeography. *In*: Jin, Y.G. & Dineley, D. (eds) *Proceedings of the 30th International Geological Congress, Volume 12. Palaeontology and Historical Geology.* VSP, Beijing, 2–17.

Gulbranson, E.L., Montanez, I.P., Schmitz, M.D., Limarino, C.O., Isbell, J.L., Marenssi, S.A. & Crowley, J.L. 2010. High-precision U–Pb calibration of Carboniferous glaciation and climate history, Paganzo Group, NW Argentina. *Geological Society of America Bulletin*, **122**, 1480–1498.

Harper, D.A.T. & Rong, J.Y. 2001. Palaeozoic brachiopod extinctions, survival and recovery: patterns within the rhynchonelliformeans. *Geological Journal*, **36**, 317–328.

He, W.H., Shi, G.R. *et al.* 2007. Brachiopod miniaturization and its possible causes during the Permian–Triassic crisis in deep water environments, South China. *Palaeogeography, Palaeoclimatology, Palaeoecology*, **252**, 145–163.

He, W.H., Shi, G.R. *et al.* 2014. Changhsingian (latest Permian) deep-water brachiopod fauna from South China. *Journal of Systematic Palaeontology*, **12**, 907–960.

He, W.H., Shi, G.R. *et al.* 2015. Late Permian marine ecosystem collapse began in deeper waters: evidence from brachiopod diversity and body size changes. *Geobiology*, **13**, 123–138.

Henderson, C.M., Davydov, V.I. & Wardlaw, B.R. 2012. The Permian Period. *In*: Gradstein, F.M., Ogg, J.G., Schmitz, M.D. & Ogg, G.M. (eds) *The Geological Time Scale 2012.* Elsevier, Amsterdam, 653–680.

Jin, X.C. & Zhan, L.P. 2008. Spatial and temporal distribution of the *Cryptospirifer* fauna (Middle Permian brachiopods) in the Tethyan realm and its paleogeographic implications. *Acta Geologica Sinica (English Edition)*, **82**, 1–16.

Jin, Y.G. 1979. Animal fossils from Jilong Formation (Permian) at the northern slope of the Mount Jolmolungma region. *In*: *A Report of Scientific Expedition in the Mount Jolmolungma Region (1966–1968).* Science Press, Beijing, 93–102 [in Chinese].

Jin, Y.G. & Fang, R.S. 1985. Early Permian brachiopods from the Kuangshan Formation in Luliang County, Yunnan with notes on paleogeography of South China during the Liangshanian Stage. *Acta Palaeontologica Sinica*, **24**, 216–228 [in Chinese with English abstract].

Jin, Y.G. & Sun, D.L. 1981. Palaeozoic brachiopods from Xizang. *In*: *Palaeontology of Xizang, Book III.* Scientific Expedition to the Qinghai–Xizang Plateau Series. Science Press, Beijing, 127–176 [in Chinese with English abstract].

Jin, Y.G., Zhang, J. & Shang, Q.H. 1994. Two phases of the end-Permian mass extinction, Pangea: global environments and resources. *In*: Embry, A.F., Beauchamp, B. & Glass, D.J. (eds) Canadian Society of Petroleum Geologists, Memoirs, **17**, 813–822.

Ke, Y., Shen, S.Z., Shi, G.R., Fan, J.X., Zhang, H., Qiao, L. & Zeng, Y. 2016. Global brachiopod palaeobiogeographical evolution from Changhsingian (Late Permian) to Rhaetian (Late Triassic). *Palaeogeography, Palaeoclimatology, Palaeoecology*, **448**, 4–25.

Klets, A.G. 2005. *Upper Palaeozoic of Marginal Seas of Angarida.* Academic Publishing House ‘Geo’, Novosibirsk [in Russian].

Klets, A.G., Budnikov, I.V., Kutygin, R.V., Biakov, A.S. & Grinenko, V.S. 2006. The Permian of the Verkhoyansk–Okhotsk region, NE Russia. *Journal of Asian Earth Sciences*, **26**, 258–268.

Kotlyar, G.V. & Popeko, L.I. 1967. Biostratigrafiia, mshanki i brakhiopody verkhnego paleozoiia Zabaikal'ia (Biostratigraphy, bryozoans and brachiopods of Upper Paleozoic of Transbaical). *Zapiski, Trudy Otdeleniia Geologii*, **28**, 1–257 [in Russian].

Kotlyar, G.V. & Zakharov, Y.D. (eds). 1989. *Pozdnepermskii etap Evoliutsii Organicheskogo Mira [Evolution of the Latest Permian Biota].* Nauka, Leningrad [in Russian].

KOTLYAR, G.V., KURILENKO, A.V., BIAKOV, A.S. & POPEKO, L.I. 2002. Permian System. *In*: OLEINIKOV, A.N. (ed.) *Atlas Faunyi I Flory Paleozoya–Mesozoya Zabaikaliya [Atlas of Faunas and Floras of the Paleozoic–Mesozoic of Zabaikal]*. Nauka, Novosibirsk, 217–314 [in Russian].

KOTLYAR, G.V., KOSSOVAYA, O.L., SHISHLOV, S.B., ZHURAVLEV, A.V. & PUKHONTO, S.K. 2004*a*. Boundary between Permian series in diverse sedimentary facies of north European Russia: constraints of event stratigraphy. *Stratigraphy and Geological Correlation*, **12**, 460–484.

KOTLYAR, G.V., ZAKHAROV, Y.D. & POLUBOTKO, I.V. 2004*b*. Late Changhsingian fauna of the northwestern Caucasus Mountains, Russia. *Journal of Paleontology*, **78**, 513–527.

KOTLYAR, G.V., BELYANSKY, G.C., BURAGO, V.I., NIKITINA, A.P., ZAKHAROV, Y.D. & ZHURAVLEV, A.V. 2006. South Primorye, Far East Russia – A key region for global Permian correlation. *Journal of Asian Earth Sciences*, **26**, 280–293.

KOTLYAR, G.V., SHEN, S.Z., KOSSOVAYA, O.L. & ZHURAVLEV, A.V. 2007. Middle Permian (Guadalupian) biostratigraphy in South Primorye, Russian Far East and correlation with Northeast China. *Palaeoworld*, **16**, 173–189.

LEONOVA, T.B., ESAULOVA, N.K. & SHILOVSKY, O.P. 2002. First discovery of Kazanian ammonoids in the Volga–Urals region. *Doklady Akademii Nauk SSSR*, **383**, 509–511 [in Russian].

LEVEN, E.J., STEVENS, C.H. & BAARS, D.L. 1997. *Permian Stratigraphy and Fusulinida of Afghanistan with their Paleogeographic and Paleotectonic Implications*. Geological Society of America, Special Papers, **316**.

LI, L. & GU, F. 1976. Brachiopoda (Carboniferous–Permian). *In*: *Paleontological Atlas of Northern China–Inner Mongolia Part*. Geological Publishing House, Beijing, 228–306 [in Chinese].

LI, L., YANG, D.L. & FENG, R.L. 1987. Late Carboniferous–Early Permian brachiopods in Longlin, Guangxi and its margin. *Bulletin of the Yichang Institute of Geology and Mineral Resources*, **11**, 199–266 [in Chinese with English abstract].

LIANG, D.Y., NIE, Z.T., GUO, T.Y., ZHANG, Y.Z., XU, B.W. & WANG, W.P. 1983. Permo-Carboniferous Gondwana–Tethys facies in southern Karakoram Ali, Xizang (Tibet). *Earth Science*, **19**, 9–28 [in Chinese].

LIANG, W.P. 1990. *Lengwu Formation of Permian and its Brachiopod Fauna in Zhejiang Province*. People's Republic of China Ministry of Geology and Mineral Resources, Geological Memoirs. Geological Publishing House, Beijing [in Chinese with English abstract].

LIAO, Z.T. 1980*a*. Brachiopod assemblages from the Upper Permian and Permian–Triassic boundary beds, South China. *Canadian Journal of Earth Sciences*, **17**, 289–295.

LIAO, Z.T. 1980*b*. Upper Permian brachiopods from western Guizhou. *In*: *Stratigraphy and Palaeontology of Upper Permian Coal-bearing Formations in Western Guizhou and Eastern Yunnan, China*. Science Press, Beijing, 241–277 [in Chinese].

MANANKOV, I.N. 2012. Brachiopods, biostratigraphy, and correlation of the Permian marine deposits of Mongolia. *Paleontological Journal*, **46**, 1325–1349.

MANANKOV, I.N., SHI, G.R. & SHEN, S.Z. 2006. An overview of Permian marine stratigraphy and biostratigraphy of Mongolia. *Journal of Asian Earth Sciences*, **26**, 294–303.

MEI, S.L. 1996. Restudy of conodonts from the Permian–Triassic boundary beds at Selong and Meishan and the natural Permian–Triassic boundary. *In*: WANG, H.Z. & WANG, X.L. (eds) *Centennial Memorial Volume of Professor Sun Yunzhu: Stratigraphy and Palaeontology*. China University of Geosciences Press, Wuhan, 141–148.

MEI, S.L. & HENDERSON, C.M. 2001. Evolution of Permian conodont provincialism and its significance in global correlation and paleoclimate implication. *Palaeogeography, Palaeoclimatology, Palaeoecology*, **170**, 237–260.

METCALFE, I., CROWLEY, J.L., NICOLL, R.S. & SCHMITZ, M. 2015. High-precision U–Pb CA-TIMS calibration of Middle Permian to Lower Triassic sequences, mass extinction and extreme climate-change in eastern Australian Gondwana. *Gondwana Research*, **28**, 61–81.

MORY, A., CROWLEY, J., NICOLL, R.S., METCALFE, I., MANTLE, D., MUNDIL, R. & BACKHOUSE, J. 2012. Wordian (Middle Permian) U–Pb CA-IDTIMS isotopic ages from the Lightjack Formation, Canning Basin, Western Australia. *In*: *Proceedings of the 34th International Geological Congress (IGC): Unearthing our Past and Future – Resourcing Tomorrow*, Brisbane, Australia, 5–10 August 2012. Australian Geosciences Council, Brisbane, 3386.

NAKREM, H.A., NILSSON, I. & MANGERUD, G. 1992. Permian biostratigraphy of Svalbard (Arctic Norway); a review. *International Geology Review*, **34**, 933–959.

NAKAMURA, K. & GOLSHANI, F. 1981. Notes on the Permian brachiopod genus *Cryptospirifer*. *Journal of the Faculty of Science, Hokkaido University, Series 4, Geology and Mineralogy*, **20**, 67–77.

NAKAMURA, K., KIMURA, G. & WINSNES, T.S. 1987. Brachiopod zonation and age of the Permian Kapp Starostin Formation (Central Spitsbergen). *Polar Research*, **5**, 207–220.

NAKAMURA, K., TAZAWA, J.I. & KUMON, F. 1992. Permian brachiopods of the Kapp Starostin Formation, west Spitsbergen. *In*: NAKAMURA, K. (ed.) *Investigations on the Upper Carboniferous–Upper Permian Succession of West Spitsbergen, 1989–91*. Hokkaido University, Sapporo, Japan, 77–95.

NICOLL, R.S. & METCALFE, I. 1998. Early and Middle Permian conodonts from the Canning and southern Carnarvon basins, Western Australia: their implications for regional biogeography and palaeoclimatology. *Proceedings of the Royal Society of Victoria*, **110**, 419–459.

PAKISTAN–JAPANESE WORKING GROUP 1985. Permian and Triassic systems in the Salt Range and Surghar Range, Pakistan. *In*: NAKAZAWA, K. & DICKINS, J.M. (eds) *'The Tethys'. Her Paleogeography and Paleobiogeography from Paleozoic to Mesozoic*. Tokai University Press, Tokyo, 221–312.

PENG, Y.Q., SHI, G.R., GAO, Y.Q., HE, W.H. & SHEN, S.Z. 2007. How and why did the Lingulidae (Brachiopoda) not only survive the end-Permian mass extinction but also thrive in its aftermath?

Palaeogeography, Palaeoclimatology, Palaeoecology, **252**, 118–131.

POSENATO, P., PELIKAN, P. & HIPS, K. 2005. Bivalves and brachiopods near the Permian–Triassic boundary from the Bukk Mountains (Balvany-North section, Northern Hungary). *Rivista Italiana di Paleontologia e Stratigrafia*, **111**, 215–232.

POSENATO, R. 1998. The genus *Comelicania* Frech, 1901 (Brachiopoda) from the Southern Alps: morphology and classification. *Rivista Italiana di Paleontologia e Stratigrafia*, **104**, 43–68.

POSENATO, R. 2001. The athyridoids of the transitional beds between Bellerophon and Werfen formations (uppermost Permian, Southern Alps, Italy). *Rivista Italiana di Paleontologia e Stratigrafia*, **107**, 197–226.

POSENATO, R. 2009. Survival patterns of macrobenthic marine assemblages during the end-Permian mass extinction in the western Tethys (Dolomites, Italy). *Palaeogeography, Palaeoclimatology, Palaeoecology*, **280**, 150–167.

POSENATO, R. 2010. Marine biotic events in the Lopingian succession and latest Permian extinction in the Southern Alps (Italy). *Geological Journal*, **45**, 195–215.

POWELL, M.G., MOORE, B.R. & SMITH, T.J. 2015. Origination, extinction, invasion, and extirpation components of the brachiopod latitudinal biodiversity gradient through the Phanerozoic Eon. *Paleobiology*, **41**, 330–341.

REED, F.R.C. 1928. A Permo-Carboniferous marine fauna from the Umaria coal-field. *Geological Survey of India Records*, **60**, 367–398.

REED, F.R.C. 1932. *New Fossils from Agglomeratic Slate of Kashmir*. Geological Survey of India Memoirs, Palaeontologia Indica, **20**.

REED, F.R.C. 1936. *Some Fossils from the Eurydesma and Conularia Beds (Punjabian) of the Salt Range*. Geological Survey of India Memoirs, Palaeontologia Indica, **23**.

REED, F.R.C. 1944. *Brachiopoda and Mollusca from the Productus Limestones of the Salt Range*. Geological Survey of India Memoirs, Palaeontologia Indica, **23**.

ROHR, D.M., WARDLAW, B.R., RUDINE, S.F., HANEEF, M., HALL, A.J. & GRANT, R.E. 2000. Guidebook to the Guadalupian symposium. *In*: WARDLAW, B.R., GRANT, R.E. & ROHR, D.M. (eds) *The Gaudalupian Symposium*. Smithsonian Contributions to Earth Sciences, **32**, 5–36.

RONG, J.Y. & SHEN, S.Z. 2002. Comparative analysis of the end-Permian and end-Ordovician brachiopod mass extinctions and survivals in South China. *Palaeogeography, Palaeoclimatology, Palaeoecology*, **188**, 25–38.

ROSS, C.A. 1963. *Standard Wolfcampian Series (Permian), Glass Mountains, Texas*. Geological Society of America, Memoirs, **88**.

ROSS, C.A. 1967. Late Paleozoic Fusulinacea from Northern Yukon Territory. *Journal of Paleontology*, **41**, 709–725.

ROSS, C.A. 1986. Paleozoic evolution of southern margin of Permian basin. *Geological Society of America Bulletin*, **97**, 536–554.

RUNNEGAR, B. & FERGUSON, J. 1969. Stratigraphy of the Permian and Lower Triassic marine sediments of the Gympie District, Queensland. *University of Queensland, Department of Geology, Papers*, **6**, 247–281.

RUNNEGAR, B. & MCCLUNG, G. 1975. A Permian time scale for Gondwanaland. *In*: CAMPBELL, K.S.W. (ed.) *Gondwana Geology, Papers Presented at the Third Gondwana Symposium, Canberra, Australia, 1973*. Australian National University Press, Canberra, 425–441.

RUZHENTSEV, V.E. & SARYTCHEVA, T.G. 1965. *The Development and Change of Marine Organisms at the Paleozoic and Mesozoic Boundary*. Akademiia Nauk SSSR, Paleontologicheskii Institutt Trudy, Moscow [in Russian].

SENGÖR, A.M.C. 1979. Mid-Mesozoic closure of Permo-Triassic Tethys and its implications. *Nature*, **279**, 590–593.

SEPKOSKI, J.J.J. 1984. A kinetic model of Phanerozoic taxonomic diversity III. Post-Paleozoic families and mass extinctions. *Paleobiology*, **10**, 246–267.

SHEN, S.Z. & CLAPHAM, M.E. 2009. Wuchiapingian (Lopingian, Late Permian) brachiopods from the Episkopi Formation of Hydra Island, Greece. *Palaeontology*, **52**, 713–743.

SHEN, S.Z. & HE, X.L. 1994. Brachiopod assemblages from the Changxingian to lowermost Triassic of southwest China and correlations over the Tethys. *Newsletters on Stratigraphy*, **31**, 151–165.

SHEN, S.Z. & HENDERSON, C.M. 2014. Progress of the Permian timescale. *In*: ROCHA, R., PAIS, J., KULLBERG, J.C. & FINNEY, S. (eds) *STRATI 2013*. Springer International, Cham, Switzerland, 447–451.

SHEN, S.Z. & JIN, Y.G. 1999. Brachiopods from the Permian–Triassic boundary beds at the Selong Xishan section, Xizang (Tibet), China. *Journal of Asian Earth Sciences*, **17**, 547–559.

SHEN, S.Z. & SHI, G.R. 1996. Diversity and extinction patterns of Permian Brachiopoda of South China. *Historical Biology*, **12**, 93–110.

SHEN, S.Z. & SHI, G.R. 2000. Wuchiapingian (early Lopingian, Permian) global brachiopod palaeobiogeography: a quantitative approach. *Palaeogeography, Palaeoclimatology, Palaeoecology*, **162**, 299–318.

SHEN, S.Z. & SHI, G.R. 2002. Paleobiogeographical extinction patterns of Permian brachiopods in the Asian–Western Pacific region. *Paleobiology*, **28**, 449–463.

SHEN, S.Z. & SHI, G.R. 2004. Capitanian (Late Guadalupian, Permian) global brachiopod palaeobiogeography and latitudinal diversity pattern. *Palaeogeography, Palaeoclimatology, Palaeoecology*, **208**, 235–262.

SHEN, S.Z. & SHI, G.R. 2007. *Lopingian (Late Permian) Brachiopods from South China, Part 1: Orthotetida, Orthida and Rhynchonellida*. Bulletin of the Tohoku University Museum, **6**, 1–102.

SHEN, S.Z. & SHI, G.R. 2009. Latest Guadalupian brachiopods from the Guadalupian/Lopingian boundary GSSP section at Penglaitan in Laibin, Guangxi, South China and implications for the timing of the pre-Lopingian crisis. *Palaeoworld*, **18**, 152–161.

SHEN, S.Z. & ZHANG, Y.C. 2008. Earliest Wuchiapingian (Lopingian, late Permian) brachiopods in southern Hunan, South China: implications for the pre-Lopingian crisis and onset of Lopingian recovery/radiation. *Journal of Paleontology*, **82**, 924–937.

SHEN, S.Z., ARCHBOLD, N.W. & SHI, G.R. 2000*a*. Changhsingian (Late Permian) brachiopod palaeobiogeography. *Historical Biology*, **15**, 121–134.

SHEN, S.Z., ARCHBOLD, N.W., SHI, G.R. & CHEN, Z.Q. 2000*b*. Permian brachiopods from the Selong Xishan section, Xizang (Tibet), China – Part 1: stratigraphy, Strophomenida, Productida and Rhynchonellida. *Geobios*, **33**, 725–752.

SHEN, S.Z., SHI, G.R. & ZHU, K.Y. 2000*c*. Early Permian brachiopods of Gondwana affinity from the Dingjiazhai Formation of the Baoshan Block, western Yunnan, China. *Rivista Italiana di Paleontologia e Stratigrafia*, **106**, 263–282.

SHEN, S.Z., ARCHBOLD, N.W., SHI, G.R. & CHEN, Z.Q. 2001. Permian brachiopods from the Selong Xishan section, Xizang (Tibet), China. Part 2: palaeobiogeographical and palaeoecological implications, Spiriferida, Athyridida and Terebratulida. *Geobios*, **34**, 157–182.

SHEN, S.Z., CAO, C.Q., SHI, G.R., WANG, X.D. & MEI, S.L. 2003*a*. Lopingian (Late Permian) stratigraphy, sedimentation and palaeobiogeography in southern Tibet. *Newsletters on Stratigraphy*, **39**, 157–179.

SHEN, S.Z., SHI, G.R. & ARCHBOLD, N.W. 2003*b*. Lopingian (Late Permian) brachiopods from the Qubuerga Formation at the Qubu section in the Mt. Qomolangma region, southern Tibet (Xizang), China. *Palaeontographica Abteilung A – Stuttgart*, **268**, 49–101.

SHEN, S.Z., TAZAWA, J. & SHI, G.R. 2005. Carboniferous and Permian Rugosochonetidae (Brachiopoda) from West Spitsbergen. *Alcheringa*, **29**, 241–256.

SHEN, S.Z., ZHANG, H., LI, W.Z., MU, L. & XIE, J.F. 2006. Brachiopod diversity patterns from Carboniferous to Triassic in South China. *Geological Journal*, **41**, 345–361.

SHEN, S.Z., WANG, Y., HENDERSON, C.M., CAO, C.Q. & WANG, W. 2007. Biostratigraphy and lithofacies of the Permian System in the Laibin–Heshan area of Guangxi, South China. *Palaeoworld*, **16**, 120–139.

SHEN, S.Z., XIE, J.F., ZHANG, H. & SHI, G.R. 2009. Roadian–Wordian (Guadalupian, Middle Permian) global palaeobiogeography of brachiopods. *Global and Planetary Change*, **65**, 166–181.

SHEN, S.Z., TAZAWA, J. & MIYAKE, Y. 2011. A Kungurian (Early Permian) Panthalassan brachiopod fauna from Hatahoko in the Mino Belt, central Japan. *Journal of Paleontology*, **85**, 553–566.

SHEN, S.Z., SCHNEIDER, J.W., ANGIOLINI, L. & HENDERSON, C.M. 2013*a*. The International Permian Timescale: March 2013 update. *In*: LUCAS, S.G., DIMICHELE, W.A., BARRICK, J.E., SCHNEIDER, J.W. & SPIELMANN, J.A. (eds) *The Carboniferous–Permian Transition. New Mexico Museum of Natural History and Science Bulletin*, **60**, 411–416.

SHEN, S.Z., ZHANG, H., SHI, G.R., LI, W.Z., XIE, J.F., MU, L. & FAN, J.X. 2013*b*. Early Permian (Cisuralian) global brachiopod palaeobiogeography. *Gondwana Research*, **24**, 104–124.

SHEN, S.Z., JIN, Y.G., ZHANG, Y. & WELDON, E.A. In press. Permian brachiopod genera with type species of China. *In*: RONG, J.Y., JIN, Y.G., SHEN, S.Z. & ZHAN, R.B. (eds) *Phanerozoic Brachiopod Genera from China*. Science Press, Beijing.

SHEN, S.Z., SUN, T.R., ZHANG, Y.C. & YUAN, D.X. 2016. An upper Kungurian/Lower Guadalupian (Permian) brachiopod fauna from the Southern Qiangtang Block in Tibet and its palaeobiogeographical implications. *Palaeoworld*, **25**, 519–538, https://doi.org/10.1016/j.palwor.2016.03.006

SHENG, J.Z., CHEN, C.Z. ET AL. 1984. Permian–Triassic boundary in middle and eastern Tethys. *Journal of the Faculty of Science, Hokkaido University, Series 4: Geology and Mineralogy*, **21**, 133–181.

SHI, G.R. 1995. The Late Palaeozoic brachiopod genus *Yakovlevia* Fredericks, 1925 and the *Yakovlevia transversa* Zone, Northern Yukon Territory, Canada. *Proceedings of the Royal Society of Victoria*, **107**, 51–71.

SHI, G.R. 2006. The marine Permian of East and Northeast Asia: an overview of biostratigraphy, palaeobiogeography and palaeogeographical implications. *Journal of Asian Earth Sciences*, **26**, 175–206.

SHI, G.R. & ARCHBOLD, N.W. 1995*a*. Palaeobiogeography of Kazanian–Midian (Late Permian) Western Pacific brachiopod faunas. *Journal of Southeast Asian Earth Sciences*, **12**, 129–141.

SHI, G.R. & ARCHBOLD, N.W. 1995*b*. Permian brachiopod faunal sequence of the Shan–Thai terrane: biostratigraphy, palaeobiogeographical affinities and plate tectonic/palaeoclimatic implications. *Journal of Southeast Asian Earth Sciences*, **11**, 177–187.

SHI, G.R. & ARCHBOLD, N.W. 1998. Permian marine biogeography of SE Asia. *In*: HALL, R. & HOLLOWAY, D.J. (eds) *Biogeography and Geological Evolution of SE Asia*. Backbuys Publishers, Leiden, 57–72.

SHI, G.R. & GRUNT, T.A. 2000. Permian Gondwana–Boreal antitropicality with special reference to brachiopod faunas. *Palaeogeography, Palaeoclimatology, Palaeoecology*, **155**, 239–263.

SHI, G.R. & SHEN, S.Z. 2000. Asian–western Pacific Permian Brachiopoda in space and time: biogeography and extinction patterns. *In*: YIN, H.F., DICKENS, M., SHI, G.R. & TONG, J.N. (eds) *Permo-Triassic Evolution of Tethys and Western Circum-Pacific*. Elsevier, London, 327–352.

SHI, G.R. & SHEN, S.Z. 2001. A biogeographically mixed, Middle Permian brachiopod fauna from the Baoshan Block, western Yunnan, China. *Palaeontology*, **44**, 237–258.

SHI, G.R. & WATERHOUSE, J.B. 1996. Lower Permian brachiopods and molluscs from the upper Jungle Creek Formaiton, Northern Yukon Territory, Canada. *Geological Survey of Canada, Bulletin*, **424**, 1–241.

SHI, G.R., ARCHBOLD, N.W. & ZHAN, L.P. 1995. Distribution and characteristics of mixed (transitional) mid-Permian (Late Artinskian–Ufimian) marine faunas in Asia and their palaeogeographical implications. *Palaeogeography, Palaeoclimatology, Palaeoecology*, **114**, 241–271.

SHI, G.R., FANG, Z.J. & ARCHBOLD, N.W. 1996. An Early Permian brachiopod fauna of Gondwanan affinity from the Baoshan block, western Yunnan, China. *Alcheringa*, **20**, 81–101.

SHI, G.R., LEMAN MOHD, S. & TAN, B.K. 1997. Early Permian brachiopods from the Singa Formation of Langkawi Island, northwestern peninsular Malaysia; biostratigraphical and biogeographical implications. *In*: DHEERADILOK, P., HINTHONG, C. ET AL. (eds)

Proceedings of the International Conference on Stratigraphy and Tectonic Evolution of Southeast Asia and the South Pacific. Department of Mineral Resources, Bangkok, 62–72.

SHI, G.R., RAKSAKULWONG, L. & CAMPBELL, H.J. 2002*a*. Early Permian brachiopods from central and northern Peninsular Thailand. *In*: HILLS, L.V., HENDERSON, C.M. & BAMBER, E.W. (eds) *Carboniferous and Permian of the World.* Canadian Society of Petroleum Geologists, Memoirs, **19**, 596–608.

SHI, G.R., SHEN, S.Z. & TAZAWA, J. 2002*b*. Middle Permian (Guadalupian) brachiopods from the Xiujimqinqi area, Inner Mongolia, northeast China, and their palaeobiogeographical and palaeogeographical significance. *Paleontological Research*, **6**, 285–297.

SHI, G.R., WATERHOUSE, J.B. & MCLOUGHLIN, S. 2010. The Lopingian of Australasia: a review of biostratigraphy, correlations, palaeogeography and palaeobiogeography. *Geological Journal*, **45**, 230–263.

SHI, G.R., ZHANG, Y.C., SHEN, S.Z. & HE, W.H. 2016. Nearshore–offshore basin species diversity and body size variation patterns in Late Permian (Changhsingian) brachiopods. *Palaeogeography, Palaeoclimatology, Palaeoecology*, **448**, 96–107.

SHINTANI, T. 2011. Orthotetoids from the Lower Permian (Sakmarian) of the Nagaiwa–Sakamotozawa area, South Kitakami Belt, NE Japan. *Science Report of Niigata University (Geology)*, **26**, 73–90.

SINGH, T. & ARCHBOLD, N.W. 1993. Brachiopoda from the Early Permian of the eastern Himalaya. *Alcheringa*, **17**, 55–75.

SONE, M. & LEMAN, M.S. 2005. Permian linoproductoid brachiopod *Permundaria* from Bera South, Peninsular Malaysia. *Journal of Paleontology*, **79**, 601–606.

SONE, M., LEMAN, M.S. & SHI, G.R. 2001. Middle Permian brachiopods from central Peninsular Malaysia – faunal affinities between Malaysia and west Cambodia. *Journal of Asian Earth Sciences*, **19**, 177–194.

SONE, M., METCALFE, I. & LEMAN, M.S. 2003. Palaeobiogeographic from implications of Middle Permian brachiopods Johore (Peninsular Malaysia). *Geological Magazine*, **140**, 523–538.

STEHLI, F.G. 1957. Possible Permian climatic zonation and its implications. *American Journal of Science*, **255**, 607–618.

STEMMERICK, L. 1988. Discussion of brachiopod zonation and age of the Permian Kapp Starostin Formation, central Spitsbergen, Arctic Ocean. *Polar Research*, **6**, 179–180.

STOLLHOFEN, H., WERNER, M., STANISTREET, I.G. & ARMSTRONG, R.A. 2008. Single-zircon U–Pb dating of Carboniferous–Permian tuffs, Namibia, and the intercontinental deglaciation cycle framework. *In*: FIELDING, C.R., FRANK, T.D. & ISBELL, J.L. (eds) *Resolving the Late Paleozoic Ice Age in Time and Space*. Geological Society of America, Special Papers, **441**, 83–96.

SUN, D.L. 1991. Permian (Sakmarian–Artinskian) brachiopod fauna from Gegyai County, northwestern Xizang (Tibet) and its biogeographic significance. *In*: SUN, D.L. & XU, J.T. (eds) *Permian Jurassic and Cretaceous Strata and Palaeontology from Rutog Region, Xizang (Tibet).* Nanjing University Press, Nanjing, 215–275 [in Chinese with English abstract].

SUN, D.L., HU, Z.X. & CHEN, T.E. 1981. Discovery of the Upper Permian strata in Lhasa, Xizang. *Journal of Stratigraphy*, **5**, 139–142 [in Chinese with English abstract].

SZANIAWSKI, H. & MALKOWSKI, K. 1979. Conodonts from the Kapp Starostin Formation (Permian) of Spitsbergen. *Acta Palaeontologica Polonica*, **24**, 231–257.

TABOADA, A.C. 2010. Mississippian–Early Permian brachiopods from western Argentina: tools for middle- to high-latitude correlation, paleobiogeographic and paleoclimatic reconstruction. *Palaeogeography, Palaeoclimatology, Palaeoecology*, **298**, 152–173.

TABOADA, A.C. 2014. New brachiopod records and considerations on the *Tivertonia–Streptorhynchus* (Moscovian) and *Costatulumus amosi* (Sakmarian–Artinskian) faunas from Western Argentina: the key sections at Quebrada Agua del Jagüel and Quebrada Santa Elena revisited. *Ameghiniana*, **51**, 226–242.

TABOADA, A.C. & PAGANI, M.A. 2010. The coupled occurrence of *Cimmeriella–Jakutoproductus* (Brachiopoda, Productidina) in Patagonia: implications for Early Permian high to middle paleolatitudinal correlations and paleoclimatic reconstruction. *Geologica Acta*, **8**, 513–534.

TABOADA, A.C., NEVES, J.P., WEINSCH TZ, L.C., PAGANI, M.A. & SIMÕES, M.G. 2016. *Eurydesma–Lyonia* fauna (Early Permian) from the Itararé group, Paraná Basin (Brazil): A paleobiogeographic W–E Trans-Gondwanan marine connection. *Palaeogeography, Palaeoclimatology, Palaeoecology*, **449**, 431–454.

TAZAWA, J. 1975. Uppermost Permian fossils from the southern Kitakami Mountains, Northeast Japan. *Geological Society of Japan Journal*, **81**, 629–640.

TAZAWA, J. 1976. The Permian of Kesennuma, Kitakami Mountains: a preliminary report. *Earth Science*, **30**, 175–185.

TAZAWA, J. 1991. Middle Permian brachiopod biogeography of Japan and adjacent regions in East Asia. *In*: ISHII, K.I., LIU, X.M., ICHIKAWA, K. & HUANT, B. (eds) *Pre-Jurassic Geology of Inner Mongolia, China, Volume 7.* Matsuya Insatsu, Osaka, 213–230.

TAZAWA, J. 1998. Pre-Neogene tectonic divisions and Middle Permian brachiopod faunal provinces of Japan. *Proceedings of the Royal Society of Victoria*, **110**, 281–288.

TAZAWA, J. 1999. Boreal-type brachiopod *Yakovlevia* from the Middle Permian of Japan. *Paleontological Research*, **3**, 88–94.

TAZAWA, J. 2001. Middle Permian brachiopods from the Moribu Area, Hida Gaien Belt, central Japan. *Paleontological Research*, **5**, 283–310.

TAZAWA, J. 2003. *Kochiproductus* and *Leptodus* (Brachiopoda) from the Middle Permian of the Obama area, South Kitakami Belt, northeast Japan. *Science Report of Niigata University (Geology)*, **18**, 25–39.

TAZAWA, J. 2008. Brachiopods from the Upper Permian Takakurayama Formation, Abukuma Mountains, northeast Japan. *Science Report of Niigata University (Geology)*, **23**, 13–53.

TAZAWA, J. 2011. Late Permian (Wuchiapingian) brachiopod fauna from Okutadami, central Japan: systematics, palaeobiogeography and tectonic implications. *Paleontological Research*, **15**, 168–180.

TAZAWA, J. 2012. Late Permian (Changhsingian) brachiopod fauna from Nabekoshiyama in the Kesennuma area, South Kitakami Belt, northeast Japan. *Science Report of Niigata University (Geology)*, **27**, 15–50.

TAZAWA, J. 2015. Early Permian (Sakmarian) brachiopods from the Nagaiwa–Sakamotozawa area, South Kitakami Belt, northeastern Japan, Part 3: productidina. *Science Report of Niigata University (Geology)*, **30**, 39–55.

TAZAWA, J. & ARAKI, H. 2013. Four brachiopod species newly described from the Middle Permian of Kesennuma, South Kitakami Belt, northeast Japan. *Science Report of Niigata University (Geology)*, **28**, 1–14.

TAZAWA, J. & IBARAKI, Y. 2001. Middle Permian brachiopods from Setamai, the type locality of the Kanokura Formation, southern Kitakami Mountains, northeast Japan. *Science Report of Niigata University (Geology)*, **16**, 1–33.

TAZAWA, J. & MIYAKE, Y. 2011. Late Permian (Changhsingian) brachiopod fauna from Maeda in the Ofunato area, South Kitakami Belt, NE Japan. *Science Report of Niigata University (Geology)*, **26**, 1–22.

TAZAWA, J. & SHEN, S.Z. 1997. Middle Permian brachiopods from Hiyomo, Mino Belt, central Japan: their provincial relationship with North America. *Science Report of Niigata University, Series E, Geology and Mineralogy*, **12**, 1–17.

TAZAWA, J. & SHINTANI, T. 2010. A Lower Permian mixed Boreal–Tethyan brachiopod fauna from the Nagaiwa–Sakamotozawa area, south Kitakami Belt, NE Japan. *Science Report of Niigata University (Geology)*, **25**, 51–62.

TAZAWA, J., TAKIZAWA, F. & KAMADA, K. 2000. A Middle Permian Boreal–Tethyan mixed brachiopod fauna from Yakejima, southern Kitakami Mountains, NE Japan. *Science Report of Niigata University (Geology)*, **15**, 1–21.

TAZAWA, J., SHEN, S.Z. & SHI, G.R. 2001. Middle Permian brachiopods from the Dongujimqinqi area, Inner Mongolia, China. *Science Report of Niigata University, Series E, Geology and Mineralogy*, **16**, 35–45.

TERMIER, G., TERMIER, H., DE LAPPARENT, A.F. & MARIN, P. 1974. Monographie du Permo-Carbonifere de Wardak (Afghanistan central). *Documents des Laboratoires de Geologie, Lyon, Hors Serie*, **2**, 1–167.

TWITCHETT, R.J. 2007. The Lilliput effect in the aftermath of the end-Permian extinction event. *Palaeogeography, Palaeoclimatology, Palaeoecology*, **252**, 132–144.

TWITCHETT, R.J., LOOY, C.V., MORANTE, R., VISSCHER, H. & WIGNALL, P.B. 2001. Rapid and synchronous collapse of marine and terrestrial ecosystems during the end-Permian biotic crisis. *Geology*, **29**, 351–354.

UENO, K. 2003. The Permian fusulinoidean faunas of the Sibumasu and Baoshan blocks: their implications for the paleogeographic and paleoclimatologic reconstruction of the Cimmerian continent. *Palaeogeography, Palaeoclimatology, Palaeoecology*, **193**, 1–24.

UENO, K. 2006. The Permian antitropical fusulinoidean genus *Monodiexodina:* Distribution, taxonomy, paleobiogeography and paleoecology. *Journal of Asian Earth Sciences*, **26**, 380–404.

UENO, K., MIZUNO, Y., WANG, X.D. & MEI, S.L. 2002. Artinskian conodonts from the Dingjiazhai Formation of the Baoshan Block, west Yunnan, southwest China. *Journal of Paleontology*, **76**, 741–750.

USTRITSKY, V.I. & TSCHERNJAK, G.E. 1963. Biostratigrafiia i brakhiopody verkhnego paleozoia Taimyra [Biostratigraphy and brachiopods of the Upper Paleozoic of Taimir]. *Nauchno-Issledovatel'skii Institut Geologii Arktiki (NIIGA), Trudy*, **134**, 1–139 [in Russian].

VERNA, V., ANGIOLINI, L. ET AL. 2010. Guadalupian brachiopods from Djebel Tebaga de Medenine, South Tunisia. *Rivista Italiana di Paleontologia e Stratigrafia*, **116**, 309–349.

VERNA, V., ANGIOLINI, L., BAUD, A., CRASQUIN, S. & NICORA, A. 2011. Guadalupian brachiopods from western Taurus, Turkey. *Rivista Italiana di Paleontologia e Stratigrafia*, **117**, 51–104.

WAAGEN, W. 1882–85. *Salt Range Fossils, Volume I, Productus Limestone Fossils*. Geological Survey of India Memoirs, Palaeontologia Indica, **13**.

WANG, C.W. & ZHANG, S.M. 2003. *The Zhesi Brachiopod Fauna*. Geological Publishing House, Beijing [in Chinese].

WANG, C.Y. 2002. Permian conodonts from Laibin and Heshan, Guangxi. *Bulletin of Nanjing Institute of Geology and Palaeontology, Academia Sinica*, **15**, 180–190 [in Chinese with English abstract].

WANG, C.Y., WANG, P. & GUO, L.W. 2004. Conodonts from the Permian Jisu Honguer (Zhesi) Formation of Inner Mongolia, China. *Geobios*, **37**, 471–480.

WANG, W., DONG, Z.Z. & WANG, C.Y. 2004. The conodont ages of the Dingjiazhai and Woniusi formations in the Baoshan area, western Yunnan. *Acta Micropalaeontologica Sinica*, **21**, 273–282 [in Chinese with English abstract].

WANG, Y., JIN, Y.G. ET AL. 1981. Stratigraphic distribution of Brachiopoda in China. *In*: TEICHERT, C., LIU, L. & CHEN, P.J. (eds) *Paleontology in China, 1979*. Geological Society of America, Special Papers, **187**, 97–105.

WARDLAW, B.R. & MEI, S.L. 1999. Refined conodont biostratigraphy of the Permian and lowest Triassic of the Salt and Khizor Ranges, Pakistan. *In*: YIN, H.F. & TONG, J.N. (eds) *Proceedings of the International Conference on Pangea and the Paleozoic–Mesozoic Transition*. China University of Geosciences Press, Wuhan, 154–156.

WATERHOUSE, J.B. 1976. *World Correlations for Permian Marine Faunas*. University of Queensland, Department of Geology, Papers, **7**, 1–232.

WATERHOUSE, J.B. 1978. *Permian Brachiopoda and Mollusca from North-west Nepal. Palaeontographica, Abteilung A*, **160**, 1–175.

WATERHOUSE, J.B. 1982. An early Permian cool-water fauna from pebbly mudstones in south Thailand. *Geological Magazine*, **119**, 337–354.

WATERHOUSE, J.B. 1987. Late Palaeozoic mollusca and correlations from the South-East Bowen basin, east Australia. *Palaeontographica, Abteilung A*, **198**, 129–233.

WATERHOUSE, J.B. 1998. Ingelarelloidea (Spiriferida: Brachiopoda) from Australia and New Zealand, and reclassification on Ingelarellidae and Notospiriferidae. *Earthwise*, **1**, 1–46.

WATERHOUSE, J.B. 2001. Late Paleozoic Brachipoda and Mollusca chiefly from Wairaki Downs, New Zealand; with notes on Scyphozoa and Triassic ammonoids and new classifications of Linoproductoidea (Brachiopoda) and Pectinida (Bivalvia). *Earthwise*, **3**, 1–195.

Waterhouse, J.B. 2002. The stratigraphic succession and structure of Wairaki Downs, New Zealand and its biostratigraphy of New Zealand and marine Permian of eastern Australia and Gondwana. *Earthwise*, **4**, 1–260.

Waterhouse, J.B. 2008. Golden spikes and black flags – macro-invertebrate faunal zones for the Permian of east Australia. *Proceedings of the Royal Society of Victoria*, **120**, 345–372.

Waterhouse, J.B. 2010*a*. Lopingian (Late Permian) stratigraphy of the Salt Range, Pakistan and Himalayan region. *Geological Journal*, **45**, 264–284.

Waterhouse, J.B. 2010*b*. New Late Paleozoic brachiopods and molluscs. *Earthwise*, **9**, 1–134.

Waterhouse, J.B. 2011. Origin and evolution of Permian brachiopods of Australia. *Memoirs of the Association of Australasian Palaeontologists*, **41**, 205–228.

Waterhouse, J.B. 2013. The evolution and classification of Productida (Brachiopoda). *Earthwise*, **10**, 1–575.

Waterhouse, J.B. 2015*a*. Early Permian Conulariida, Brachiopoda and Mollusca from Homevale, central Queensland. *Earthwise*, **11**, 1–391.

Waterhouse, J.B. 2015*b*. The Permian faunal sequence at Gympie, south-east Queensland, Australia. *Earthwise*, **12**, 1–201.

Waterhouse, J.B. & Carter, B.G.F. 1975. Global distribution and character of Permian biomes based on brachiopod assemblages. *Canadian Journal of Earth Sciences*, **12**, 1085–1146.

Waterhouse, J.B. & Piyasin, S. 1970. Mid-Permian brachiopods from Khao Phrik, Thailand. *Palaeontographica, Abteilung A*, **135**, 83–197.

Waterhouse, J.B. & Shi, G.R. 2010. Evolution in a cold climate. *Palaeogeography, Palaeoclimatology, Palaeoecology*, **298**, 17–30.

Waterhouse, J.B. & Shi, G.R. 2013. Climatic implications from the sequential changes in diversity and biogeographic affinities for brachiopods and bivalves in the Permian of eastern Australia and New Zealand. *Gondwana Research*, **24**, 139–147.

Waterhouse, J.B., Pitakpaivan, K. & Mantajit, N. 1981. Early Permian brachiopods from Ko Yao Noi and near Krabi, southern Thailand. *In*: Waterhouse, J.B., Pitakpaivan, K. & Mantajit, N. (eds) *The Permian Stratigraphy and Palaeontology of Southern Thailand*. Geological Survey of Thailand, Memoirs, **4**, 43–213.

Wignall, P.B., Morante, R. & Newton, R. 1998. The Permo-Triassic transition in Spitsbergen: $\delta^{13}C_{org}$ chemostratigraphy, Fe and S geochemistry, facies, fauna and trace fossils. *Geological Magazine*, **135**, 47–62.

Wu, H.T., He, W.H., Zhang, Y., Yang, T.L., Xiao, Y.F., Chen, B. & Weldon, E.A. 2016. Palaeobiogeographic distribution patterns and processes of Neochonetes and Fusichonetes (Brachiopoda) in the late Palaeozoic and earliest Mesozoic. *Palaeoworld*, **25**, 508–518, https://doi.org/10.1016/j.palwor.2016.08.002

Xia, F.S. & Zhang, B.G. 1992. Age of Selong group in Xishan, Selong, Xizang and the Permian–Triassic boundary. *Journal of Stratigraphy*, **16**, 256–263 [in Chinese with English abstract].

Xiao, W.J., Windley, B.F., Hao, J. & Zhai, M.G. 2003. Accretion leading to collision and the Permian Solonker suture, Inner Mongolia, China: termination of the central Asian orogenic belt. *Tectonics*, **22**, 1–21.

Yanagida, J. & Pillevuit, A. 1994. Permian brachiopods from Oman. *Kyushu University, Faculty of Science, Memoirs (Earth and Planetary Sciences), D*, **28**, 61–99.

Yang, S.P. & Fan, Y.N. 1983. Carboniferous brachiopods from Xizang (Tibet) and their faunal provinces. *Contributions to the Geology of the Qinghai–Xizang (Tibet) Plateau*, **11**, 265–285 [in Chinese with Englsih abstract].

Yang, Z.Y. & Nie, Z.T. 1990. *Paleontology of Ngari, Tibet (Xizang)*. The China University of Geosciences Press, Wuhan [in Chinese with English summary].

Yuan, D.X., Zhang, Y.C. et al. 2016. Early Permian conodonts from the Xainza area, central Lhasa Block, Tibet, and their palaeobiogeographical and palaeoclimatic implications. *Journal of Systematic Palaeontology*, **14**, 365–383.

Zeng, Y., He, X.L. & Zhu, M.L. 1995. *Brachiopod Communities and Their Succession and Replacement in the Permian of Huayingshan Area*. China University of Mining and Technology Press, Xuzhou [in Chinese with English summary].

Zhan, L.P. & Li, L. 1979. The distribution of Permian brachiopod faunas in China. *In*: Wang, D.F. (ed.) *Geological Paper Collections for International Exchange, Volume 2, Stratigraphy and Palaeontology*. Geological Publishing House, Beijing, 104–115 [in Chinese].

Zhan, L.P. & Wu, R.Y. 1982. Early Permian brachiopods from Xainza District, Xizang (Tibet). *Contributions to the Geology of the Qinghai–Xizang (Tibet) Plateau*, **7**, 86–109 [in Chinese with English abstract].

Zhan, L.P., Yao, J.X., Ji, Z.S. & Wu, G.C. 2007. Late Carboniferous–Early Permian brachiopod fauna of Gondwanic affinity in Xainza County, northern Tibet, China: revisited. *Geological Bulletin of China*, **26**, 54–72 [in Chinese with English abstract].

Zhang, S.X. & Jin, Y.G. 1976. Late Paleozoic brachiopods from the Mount Jolmo Lungma Region. *In*: *A Report of Scientific Expedition in the Mount Jolmo Lungma Region (1966–68)*. Science Press, Beijing, 159–271 [in Chinese].

Zhang, Y., Shi, G.R. et al. 2016. Significant pre-mass extinction animal body-size changes: evidences from the Permian–Triassic boundary brachiopod faunas of South China. *Palaeogeography, Palaeoclimatology, Palaeoecology*, **448**, 85–95.

Zhang, Y.C. & Payne, J.L. 2012. Size–frequency distributions along a latitudinal gradient in Middle Permian fusulinoideans. *PLoS One*, **7**, e38603.

Zhang, Y.C., Shen, S.Z., Shi, G.R., Wang, Y., Yuan, D.X. & Zhang, Y.J. 2012. Tectonic evolution of the Qiangtang Block, northern Tibet during the Late Cisuralian (Late Early Permian): evidence from fusuline fossil records. *Palaeogeography, Palaeoclimatology, Palaeoecology*, **350–352**, 139–148.

Zhang, Y.S., Tian, S.G. et al. 2014. Discovery of marine fossils in the upper part of the Permian Linxi Formation in Lopingian, Xingmeng area, China. *Chinese Science Bulletin*, **59**, 62–74.

Zhang, Z.X., Augustin, M. & Payne, J.L. 2015. Phanerozoic trends in brachiopod body size from synoptic data. *Paleobiology*, **41**, 491–501.

Permian palynostratigraphy: a global overview

MICHAEL H. STEPHENSON

British Geological Survey, Keyworth, Nottingham NG12 5GG, UK
mhste@bgs.ac.uk

Abstract: Permian palynostratigraphic schemes are used primarily to correlate coal- and hydrocarbon-bearing rocks within basins and between basins, sometimes at high levels of biostratigraphic resolution. Up to now, their main shortcoming has been the lack of correlation with schemes outside the basins, coalfields and hydrocarbon fields that they serve, and chiefly a lack of correlation with the international Permian scale. This is partly because of phytogeographical provinciality from the Guadalupian onwards, making correlation between regional palynostratigraphic schemes difficult. However, local high-resolution palynostratigraphic schemes for regions are now being linked either by assemblage-level quantitative taxonomic comparison or by the use of single well-characterized palynological taxa that occur across Permian phytogeographical provinces. Such taxa include: *Scutasporites* spp., *Vittatina* spp., *Weylandites* spp., *Lueckisporites virkkiae*, *Otynisporites eotriassicus* and *Converrucosisporites confluens*. These palynological correlations are being facilitated and supplemented with radiometric, magnetostratigraphic, independent faunal and strontium isotopic dating.

Palynostratigraphy is the use of palynomorphs (defined as organic-walled microfossils 5–500 µm in diameter) in correlating and assigning relative ages to rock strata. As such, it is a branch of biostratigraphy and follows the rules of biostratigraphic practice: for example, those set out by Rawson *et al.* (2002).

The Permian, falling between 252.2 and 298.9 Ma, was a period of intense change in which the giant continent of Pangea as a whole moved north, and in which, through the early part of the Period, a transition from icehouse to greenhouse conditions occurred (e.g. Fielding *et al.* 2008), alongside the decline in coal swamps and the establishment of widespread evaporite deposits (Henderson *et al.* 2012). The end of the Period saw a major extinction of fauna such as fusulinacean foraminifers, trilobites, rugose and tabulate corals, blastoids, acanthodians, placoderms, and pelycosaurs; a dramatic reduction in bryozoans, brachiopods, ammonoids, sharks, bony fish, crinoids, eurypterids, ostracodes and echinoderms (Henderson *et al.* 2012); and, although many conifers (e.g. glossopterids, cordaites) became extinct at the end of the Permian, there is no evidence of major extinction in the plants (Gradstein & Kerp 2012). Amongst the most important changes in land plants is the replacement, near the end of the Carboniferous, of arborescent lycophytes by arborescent tree ferns; arborescent lycophytes only persisted into the Guadalupian in China. The arborescent horsetails also declined by the end of the Carboniferous. In the Permian, a great variety of new seed plant groups appeared such as cycads, ginkgos, voltzialean conifers and glossopterids. The latter are important biostratigraphic markers for the Permian of Gondwana and include several hundred species. It is estimated that by the Lopingian about 60% of the world's flora consisted of seed plants (Gradstein & Kerp 2012).

These large-scale evolutionary changes in plants, filtered by local and regional effects, are responsible for the palynological succession that provides opportunities for subdivision on which palynostratigraphic schemes are built. However, the pronounced phytogeographical differentiation of the Permian has a powerful effect on palynostratigraphy, such that schemes differ considerably across Pangea and correlation between schemes is even now tentative or incomplete. In the Gondwana phytogeographical province, for example, it is difficult to correlate to the standard Permian stages; and the Carboniferous–Permian and Permian–Triassic boundaries are not precisely correlateable into Gondwana basins using palynology (Stephenson 2008*a*).

Until recently, progress in correlation was hampered by the lack of fundamental stratigraphic standards such as stage Global Stratigraphic Sections and Points (GSSPs); however, since 1997 (Jin *et al.* 1997; Henderson *et al.* 2012) a number of GSSPs have been established within the Pennsylvanian–Permian succession, the most important of which is the basal Permian GSSP at Aidaralash Creek in the southern Urals (Jin *et al.* 1997; Henderson *et al.* 2012), and the basal Triassic GSSP at Meishan section D, Changxing County, Zhejiang Province, South China (Yin *et al.* 2001). Since these developments, there have also been other

From: Lucas, S. G. & Shen, S. Z. (eds) 2018. *The Permian Timescale*. Geological Society, London, Special Publications, **450**, 321–347.
First published online December 8, 2016, https://doi.org/10.1144/SP450.2

advances contributing to the precision and utility of palynological biostratigraphy in this interval, including radiometric and faunal dating of palynological biozones, and limited high-resolution correlation between continents using a well-defined palynological species.

Palynological research in the Permian is extensive, being partly driven by exploration for coal (e.g. in India and Australia), and oil and gas (e.g. in the Middle East, South America, Australia and the Barents Sea), but has tended to be regional or local in focus (see Truswell 1980). A number of authors (Bharadwaj 1969; Kemp 1975; Bharadwaj & Srivastava 1977; Balme 1980; Truswell 1980; Utting & Piasecki 1995; Warrington 1996; Price 1997; Playford & Dino 2005; Azcuy *et al.* 2007; Stephenson 2008*a*) have attempted to summarize the research or to correlate the main biozones across regions, but correlation has been tentative. Among the difficulties acknowledged by these previous reviewers are disparate stratigraphic and taxonomic methods practised in different parts of the world, and different standards of documentation of palynological data.

The approach taken in this paper is to survey the palynostratigraphic schemes in the main phytogeographic provinces and then to attempt synthesis; and so the focus is on palynostratigraphy not taxonomy. Given the plethora of palynological literature on this interval, the review is necessarily selective. Most recent published palynostratigraphic schemes (e.g. since 2000) have emanated from South American and Middle Eastern basins. In the following account, age assignments related to these and other schemes reflect those of the original authors but may not necessarily use modern chronostratigraphic nomenclature, thus a variety of stratigraphic stage and other nomenclature is used in this paper. For the convenience of the reader, a chart showing correlations of the main chronostratigraphic subdivisions used internationally is shown in Figure 1.

Permian palynostratigraphic schemes use pollen and spores almost exclusively. While it is recognized that marine palynomorphs (acritarchs) may be present in Permian rocks, no study has sought to produce a palynostratigraphy based purely on

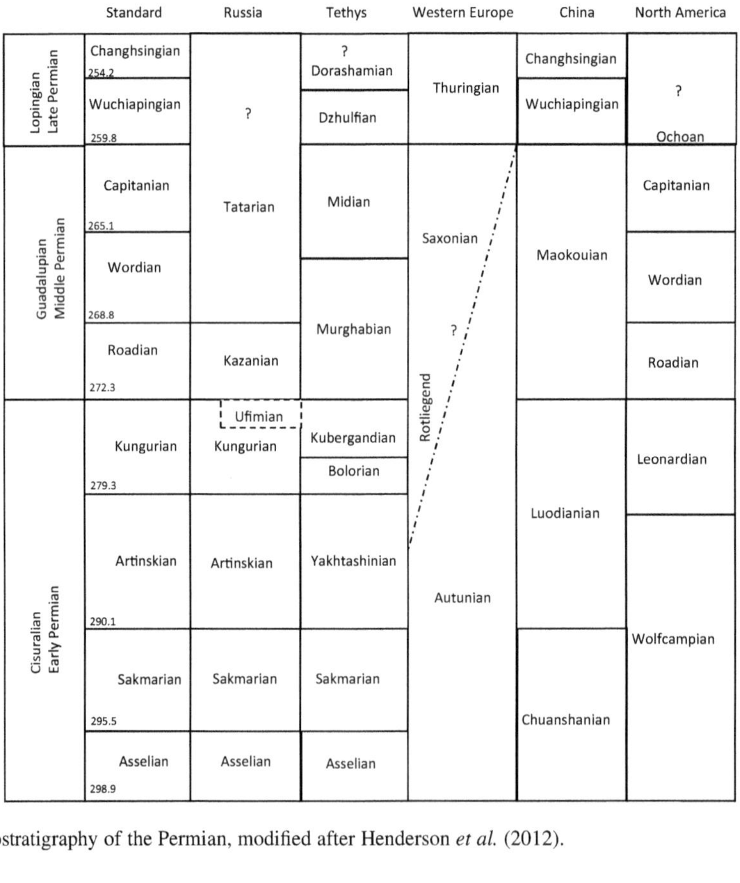

	Standard	Russia	Tethys	Western Europe	China	North America
Lopingian Late Permian	Changhsingian 254.2	?	? Dorashamian	Thuringian	Changhsingian	?
	Wuchiapingian 259.8		Dzhulfian		Wuchiapingian	Ochoan
Guadalupian Middle Permian	Capitanian 265.1	Tatarian	Midian	Saxonian	Maokouian	Capitanian
	Wordian 268.8		Murghabian	?		Wordian
	Roadian 272.3	Kazanian				Roadian
Cisuralian Early Permian	Kungurian 279.3	Ufimian Kungurian	Kubergandian Bolorian	Rotliegend		Leonardian
	Artinskian 290.1	Artinskian	Yakhtashinian	Autunian	Luodianian	Wolfcampian
	Sakmarian 295.5	Sakmarian	Sakmarian		Chuanshanian	
	Asselian 298.9	Asselian	Asselian			

Fig. 1. Chronostratigraphy of the Permian, modified after Henderson *et al.* (2012).

Permian acritarchs, although they may show future potential (e.g. Lei *et al.* 2013).

The range of morphology seen in palynomorphs in the Permian is illustrated simply in Figure 2. To improve readability, names of authors of taxa are excluded from the main text of the paper, but the main taxa and their authorship are listed in Appendix A.

Phytogeography of the Permian

Phytogeographical provinciality makes correlation difficult because it tends to reduce the number of taxa in common between assemblages in different provinces. In general, it seems reasonable to expect geographical parity between palaeobotanical provinces and palynological provinces since plants and palynomorphs are biologically linked (Hart 1970). However, palynomorphs are subject to much wider distribution than plant remains; similar or identical palynomorphs may be produced by unrelated plants (Meyen 1969); and taphonomic factors may affect the preservation of palynomorphs and plant macrofossils differently (Utting & Piasecki 1995). Balme (1970), Sullivan (1965), Turnau (1978) and Van der Zwan (1981) surveyed the hazards of the reconstruction of palaeophytogeographical provinces by palynology. The value of pollen and spore taxa as indices of low-rank plant taxa is limited because the plant affinities of most Palaeozoic spore and pollen genera and species are unknown, and because botanical and palynological taxonomy are independent of one another. Despite this, the broad palynological characteristics of a region at a certain time are thought to be representative of the high-rank palaeobotanical characteristics of that region (Utting & Piasecki 1995).

In broad terms, there was a gradual diversification of phytogeographical provinces from relatively uniform palaeophytogeography in the Devonian to maximum provinciality in the Lopingian (Cleal & Thomas 1991) when four main provinces existed: Gondwana, Euramerica, Angara and Cathaysia (Fig. 3) (Utting & Piasecki 1995). Palaeobotanically, the Gondwana phytogeographical province is distinct from other provinces in the Permian because of the abundance of the glossopterids

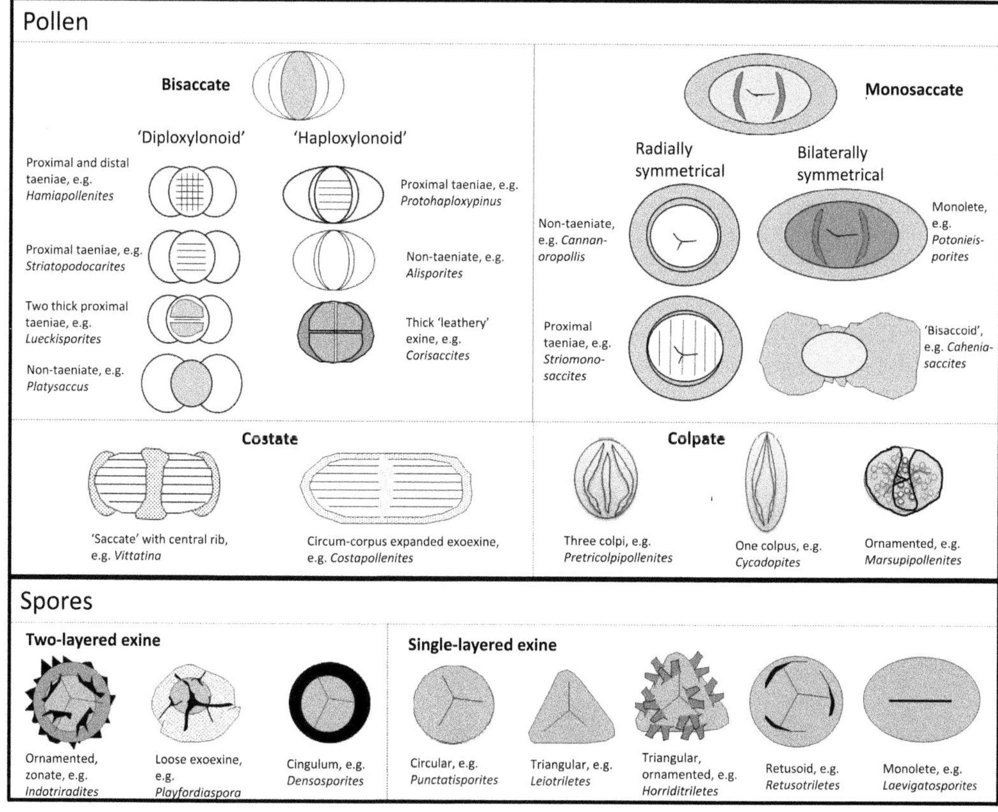

Fig. 2. Range of morphology in Permian palynomorphs.

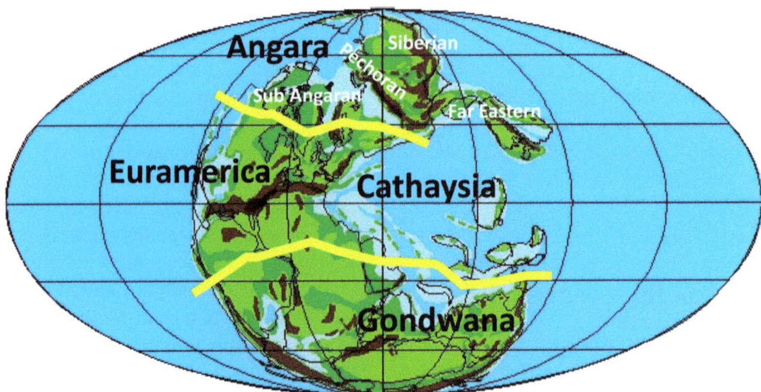

Fig. 3. Phytogeographical provinces of the Permian on a Wordian reconstruction (from the Scotese Palaeomap Project); palaeophytogeography after Utting & Piasecki (1995).

whose leaf impressions (*Glossopteris* and *Gangamopteris*) and more rare associated fruiting bodies (e.g. *Plumsteadia* and *Scutum*) are characteristic. Some conifers, sphenophytes and ferns were also present (Cleal & Thomas 1991). In the northern parts of Gondwana, marattialean ferns and lycophytes were present (Cleal & Thomas 1991). Palynologically, the Gondwana province is, perhaps, the most distinct because of its diversity of taeniate and monosaccate pollen, and the occurrence of restricted genera not yet recorded outside the province: for example, *Guttulapollenites*, *Microbaculispora*, *Dulhuntyispora* and *Corisaccites* (Truswell 1980). However, mixing of floras between Gondwana, Euramerica and Cathaysia are suggested by palaeobotanical studies (Archangelsky & Wagner 1983), and such mixing has been observed by palynologists (e.g. Kar & Jain 1975; Kaiser 1976; Eshet 1990; Nader *et al.* 1993*a*, *b*).

Palaeobotanically, the Euramerican province is distinguished by its abundant conifers, marattialean ferns and pteridosperms (Cleal & Thomas 1991). In the Lopingian, aridification caused the vegetation to become sparser and of lower diversity (Schweitzer 1986). Palynologically, a similar impoverishment of species occurs in the Lopingian (Pattison *et al.* 1973; Smith *et al.* 1974). British and west European assemblages are characterized by an abundance of conifer pollen of *Lueckisporites virkkiae*, and by the less abundant genera *Protohaploxypinus*, *Striatoabieites*, *Striatopodocarpites*, *Taeniaesporites*, *Vittatina*, *Falcisporites* and *Klausipollenites* (Smith *et al.* 1974).

According to Utting & Piasecki (1995), Angara was dominated by a cool-temperate flora containing diverse herbaceous sphenophytes and Cordaitales, with the most northerly part (Siberia) having a cold-temperate climate that spread to most of Angara by the Lopingian. According to Hart (1970), the palynology of Angara is characterized by the abundance of trilete spores rather than by bisaccate and taeniate bisaccate pollen, which are common in the other Permian phytogeographical provinces. Monosaccate pollen of *Cordaitina* is common in the lower parts of the Permian in Angara, whereas the monocolpate pollen *Cycadopites* is common in the upper parts.

Palaeobotanically, the Cathaysia province is distinguished by its gigantopterids, noeggerathialean-like progymnosperms and plants with cycad-like foliage; however, lycophytes, sphenophytes and pteridosperms were also present (Cleal & Thomas 1991). Both palaeobotanists and palynologists report a similarity between Cisuralian floras of Cathaysia and Pennsylvanian floras of Euramerica (Kaiser 1976; Cleal & Thomas 1991; Utting & Piasecki 1995). Kaiser (1976) interpreted the relict flora as being due to the continuing palaeotropical conditions of Cathaysia, which provided a refuge for the tropical vegetation of the Pennsylvanian swamps of Euramerica.

Apart from the Cisuralian abundance of Pennsylvanian Euramerican palynological taxa in Cathaysia, there are also endemic palynological taxa that characterize Cathaysia: for example, *Nixispora* and *Patellisporites* (Utting & Piasecki 1995). The Carboniferous 'relict flora' of Cathaysia appears to persist into the Lopingian in eastern Yunnan, China (Ouyang 1982), where diverse assemblages of genera such as *Torispora*, *Crassispora*, *Triquitrites* and *Laevigatosporites* occur.

Palynostratigraphy

Euramerica

The phytogeographical province of Euramerica is now represented by the areas west of the Ural

Mountains, Europe, parts of North Africa and North America, and contains the historical type area for the Permian and the present type area for the Early Permian (Cisuralian) in the southern Urals, including the base Permian GSSP at Aidaralash Creek, southern Urals.

Knowledge of the palynostratigraphy of the Permian of the southern Urals is hampered by the lack of work since the 1980s. Most work before that date was published in Russian, but was conveniently summarized by Hart (1970) and Warrington (1996). However, the work was regionally based and not gathered into a Urals-wide palynostratigraphic scheme, perhaps because other palaeontological groups (e.g. ammonoids and fusulinacean foraminifers) already provided ample resolution for stratigraphic subdivision.

Only a short, preliminary palynostratigraphic study has been carried out at Aidaralash Creek (Dunn 2001). Dunn (2001) described a section of approximately 50 m of strata from 24.2 m below, to 26 m above, the Carboniferous–Permian boundary. The assemblages contain abundant *Vittatina* (particularly *Vittatina costabilis*) and taeniate bisaccate pollen. The non-taeniate bisaccate pollen *Limitisporites monstruosus* is common throughout and spores are rare. No marked palynological change occurs at the Carboniferous–Permian boundary at the base of Bed 19.5 where the conodont *Streptognathodus isolatus* first appears (Davydov *et al.* 1998).

Faddeeva (1980) in a larger survey of southern Urals palynology considered that Gzhelian (latest Carboniferous) and Asselian assemblages are distinguished by changes in proportions of spores and pollen, in that the Gzhelian assemblages have common spores and few *Vittatina* specimens and pollen, while Asselian assemblages have common saccate pollen (e.g. *Cordaitina* and *Potonieisporites*) and *Vittatina*. A similar distinction is apparent to the SW in the Donetz Basin according to Inosova *et al.* (1975).

Later assemblages from the Sakmarian of the Urals are dominated by saccate pollen, while Artinskian assemblages contain common spores including *Tuberculatosporites* and *Granulatisporites*, as well as pollen (*Cordaitina* and *Cycadopites*) (Hart 1970). According to Hart (1970), the Kungurian of the Urals is characterized by taeniate bisaccate pollen, and the Kazanian by more diverse assemblages of monosaccate and taeniate bisaccate pollen and an increase in non-taeniate bisaccate pollen. Faddeeva (1980) also noted a change at the base of the Kazanian, including an increase in taeniate and non-taeniate taxa, as well as the appearance of *Lueckisporites*, *Taeniaesporites* and *Vesicaspora*.

Utting *et al.* (1997), in an important paper that constitutes one of the few modern surveys of Russian Permian palynostratigraphy, compared assemblages from the Ufimian and Kazanian type areas in the southern Urals around Perm and west towards Kazan, with the sub-Angaran palynological biozones of the Sverdrup Basin of the Canadian Arctic. The Ufimian, an original Russian stage recently abandoned by the All Russian Stratigraphic Commission (see Henderson *et al.* 2012), may form part of the upper Kungurian, and the Kazanian is considered equivalent to the Roadian (Fig. 1). Well-preserved palynological assemblages occur in the Ufimian and Kazanian, but Utting *et al.* (1997) recovered no palynomorphs from the Tatarian of the area.

Utting *et al.* (1997) found that many taxa range through the Ufimian and Kazanian: for example, *Florinites luberae*, *Cordaitina uralensis*, *Alisporites plicatus*, *Limitisporites monstruosus*, *Hamiapollenites tractiferinus* and *Weylandites striatus*. Several taxa appear first in the lower Kazanian (e.g. *Lueckisporites virkkiae*), while *Crucisaccites ornatus*, *Weylandites cincinnatus* and *Hamiapollenites bullaeformis* disappear in the lower Kazanian. Taxa of note that are common in the Ufimian and Kazanian sequences include *Protohaploxypinus perfectus*, *Weylandites striatus*, *Alisporites plicatus*, *Limitisporites monstruosus* and *Hamiapollenites tractiferinus* (see Utting *et al.* 1997, fig. 7). At the suprageneric level, it appears that Ufimian assemblages are more spore-rich than those of the Kazanian; the latter is characterized by abundant taeniate bisaccate pollen, and common non-taeniate and polyplicate pollen (see Utting *et al.* 1997, fig. 8).

Although Utting *et al.* (1997) did not recover palynomorphs from the Tatarian in the southern Urals, they referred to other studies (see Utting *et al.* 1997, p. 5) that characterize the Tatarian as being dominated by species of *Protohaploxypinus*, *Vittatina* and *Florinites luberae*. Gomankov (1992) refers to the presence in the lower Tatarian of *Scutasporites unicus*, which is similar to *Scutasporites nanuki* from the Wordian of the Sverdrup Basin, Canada (see later discussion); and Utting *et al.* (1997) noted some similarities between the Russian and Canadian assemblages, although they also noted quantitative differences (see Utting *et al.* 1997, fig. 8), which they attributed to palaeoclimatic differences.

Gomankov *et al.* (1998) described few palynological differences between the Tatarian and Kazanian, but identified four broad assemblage zones for the Tatarian. In ascending order, Zone I is characterized by *Weylandites* and *Lueckisporites virkkiae*, and species of *Protohaploxypinus*, *Taeniaesporites* and *Vittatina*, as well as forms related to *Vittatina* (e.g. *Ventralvittatina* and *Duplivittatina*). Zone II lacks *Ventralvittatina* and *Duplivittatina*, and contains the first appearance of spore specimens similar to the distinctive *Limatulasporites fossulatus*, as well as *Cladaitina* and *Cordaitina*. Zone III contains fewer specimens of *Cladaitina* and *Cordaitina*, but also contains rare *Scutasporites unicus*. Zone IV sees the first appearance of *Cedripites priscus*, as well as sculptured monolete spores of the genus *Punctatosporites*.

Much of the European sediments of the Cisuralian to the west of the southern Urals were deposited in restricted basins and are therefore difficult to correlate with beds of similar age in Russia (Utting & Piasecki 1995). Doubinger (1974) and Clayton *et al.* (1977) defined biozones based in the Carboniferous–Permian rocks of France and Germany. Balme (1980) related the base of the Disaccites Striatiti (DS) Zone of Clayton *et al.* (1977) to the base of Unit III of Western Australia (see later discussion) due to the common expansion of the taeniate bisaccate pollen at these horizons. Doubinger *et al.* (1987) and Jerzykiewicz (1987) described a similar abundance of taeniate bisaccate pollen in the Autunian (latest Pennsylvanian–Cisuralian: Fig. 1) of the Lodève Basin, France, and the Intrasudetic Basin of SW Poland, respectively.

Hochuli (1985) documented four biozones based in the Carboniferous–Permian rocks of NE Switzerland. Hochuli (1985) recognized the *Angulisporites splendidus–Latensina trileta* (ST) and the *Potonieisporites novicus/bharadwaji–Cheiledonites major* (NBM) biozones of Clayton *et al.* (1977), which range from late Stephanian A to Stephanian D; and two higher biozones (a lower *Vittatina costabilis* (VCI) Zone and an upper *Vittatina costabilis* (VCII) Zone) that Hochuli (1985) considered together to be equivalent to the *Vittatina costabilis* (VC) Zone of Clayton *et al.* (1977) (Fig. 4).

Visscher (1980) and Edwards *et al.* (1997) indicated that a long hiatus (the 'post-Variscan interval') exists between the Autunian and the Thuringian (Lopingian) of Europe (Fig. 1), although Schaarschmidt (1980) reported a Kungurian–Kazanian assemblage from Germany. Apart from this assemblage, palynological assemblages from the Guadalupian of Europe appear to be rare: however, palynological assemblages from the Lopingian rocks of Western Europe have been intensively studied (Visscher 1973), but have been shown to be, palynologically, rather homogeneous throughout the sequence (Grebe & Schweitzer 1962; Schaarschmidt 1963; Clarke 1965; Smith *et al.* 1974); the only notable change being a gradual overall impoverishment of the palynoflora occurring in the upper parts of the sequence (Pattison *et al.* 1973). Although several important taxonomic studies emanate from the European Lopingian (e.g. Klaus 1963; Schaarschmidt 1963), possibilities for palynostratigraphic subdivision lie with the variations within the *Lueckisporites virkkiae* palynodemes (lineages) of Visscher (1971) and with changes in the relative abundances of suprageneric groups in, for example,

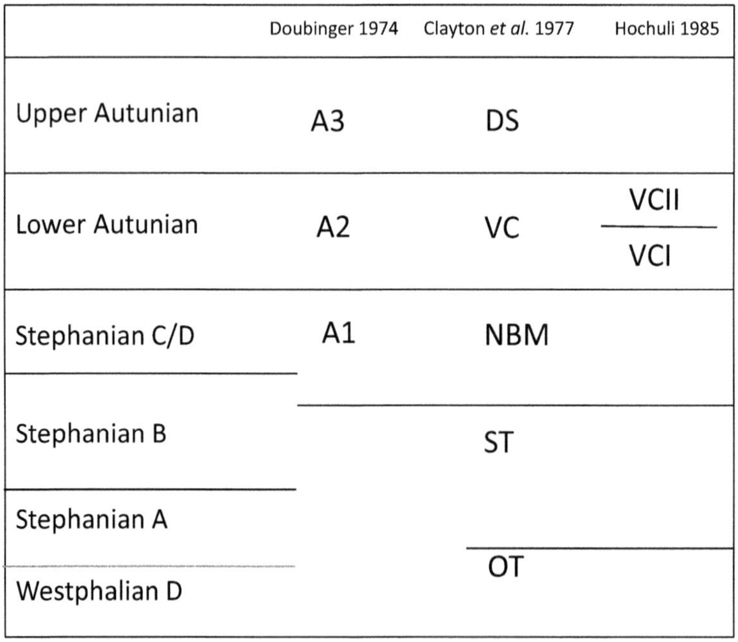

	Doubinger 1974	Clayton *et al.* 1977	Hochuli 1985
Upper Autunian	A3	DS	
Lower Autunian	A2	VC	VCII / VCI
Stephanian C/D	A1	NBM	
Stephanian B		ST	
Stephanian A			
Westphalian D		OT	

Fig. 4. Correlations of western European Carboniferous–Permian palynostratigraphic schemes, after Hochuli (1985).

the Zechstein Sea Basin (Pattison *et al.* 1973). Visscher (1971) erected six biozones within the Thuringian that are based on a palynodeme of *Lueckisporites virkkiae*.

Recent studies of European Thuringian sedimentary rocks have concentrated on the palaeoenvironment (e.g. Bercovici *et al.* 2009) or on the palynological character of the Permian–Triassic boundary in the Alps (Looy *et al.* 2001; Twitchett *et al.* 2001; Spina *et al.* 2015). In North America, only a small number publications, which concentrate on taxonomy, emanate from the Guadalupian and Lopingian, including those of Wilson (1962), Jizba (1962), Tschudy & Kosanke (1966) and Clapham (1970).

In North Africa, a palynostratigraphic scheme for a Carboniferous–Permian transition sequence was defined by Brugman *et al.* (1988) from cuttings samples from borehole A1-19 in NE Libya. On the basis of the ranges of taxa, Brugman *et al.* (1988) proposed two assemblage biozones: a lower Assemblage Zone A and an upper Assemblage Zone B. The top of the lowest biozone was defined by Brugman *et al.* (1988) as the stratigraphic level of the disappearance of *Lycospora pusilla*, a spore type particularly characteristic of the Euramerican coal belts of the Pennsylvanian. Brugman & Visscher (1988) and Brugman *et al.* (1988) believed this level to approximate to the Permian–Carboniferous boundary. The main difference across the boundary between Assemblage Zone A and an upper Assemblage Zone B is the loss in the latter of taxa of Euramerican affinity, including *Endosporites ornatus*, *Crassispora kosankei* and *Spelaeotriletes* spp.

Brugman & Visscher (1988) also identified Sakmarian–?Ufimian assemblages in A1-19. Assemblages containing *Lueckisporites virkkiae* was recorded from Tunisia by Kilani-Mazraoui *et al.* (1990).

Angara

Angara was divided into four by Utting & Piasecki (1995), including the Sub-Angara, Far Eastern, Pechoran and Siberia subprovinces (Fig. 3). Sub-Angara occupied a more southerly position in Permian palaeogeographical reconstructions, while Siberia was to the NE, at high Permian latitudes.

The palynostratigraphy of Sub-Angara is the best studied of these subprovinces as a result of work carried out in the 1980s and 1990s in the Canadian Arctic Sverdrup Basin by the Canadian Geological Survey, and work related to oil exploration in the Barents Sea and Svalbard.

Utting (1989) established five preliminary palynological biozones in the Sverdrup Basin from shallow basin margin facies; later, Utting (1994) refined the ages and detailed characteristics of two of the biozones: the Wordian *Ahrensisporites thorsteinssonii Scutasporites nanuki* and the Roadian *Alisporites plicatus–Jugasporites compactus* concurrent range zones (Fig. 5). The ages of the biozones are based mainly on ammonoids, conodonts, brachiopods and foraminifers from the associated fossiliferous beds of shallow and deep basinal facies.

Utting (1994) considered that the two Sverdrup Basin biozones were Sub-Angaran in character and correlateable with assemblages in western Canada, Alaska, Greenland, Svalbard, the Barents Sea, the Pechora Basin and the northern Russian Platform. However, the Wordian *Ahrensisporites thorsteinssonii–Scutasporites nanuki* Concurrent Range Zone differs from assemblages further south in the Urals in the type area of the Kazanian Stage (see the previous discussion). The Sverdrup Basin assemblages are dominated by trilete spores, non-taeniate and taeniate bisaccate pollen, and polyplicate taxa. Utting's detailed studies do not show large differences in the overall suprageneric character of assemblages across the Roadian and Wordian (see Utting 1994, fig. 7), but a number of taxa are confined to particular biozones including *Crinalites sabinensis*, *Cladaitina kolodae* and *Sverdrupollenites agluatus* to the *Alisporites plicatus–Jugasporites compactus* Concurrent Range Zone, and *Striatoabieites borealis* and the eponymous species to the *Ahrensisporites thorsteinssonii–Scutasporites nanuki* Current Range Zone. The latter biozone differs markedly from the succeeding Griesbachian *Tympanicysta stoschiana–Striatoabieites richteri* Assemblage Zone, which was accounted for by Utting (1994) as being due to a sedimentary hiatus and climatic differences.

Mangerud (1994) established two biozones for the Cisuralian–Guadalupian in the offshore Norway Finnmark Platform: the *Dyupetalum* sp.–*Hamiapollenites bullaeformis* Assemblage Zone of ?Kungurian–Ufimian age, and the Kazanian–?Tatarian *Scutasporites* cf. *unicus–Lunatisporites* spp. Assemblage Zone, as well as recognizing an older biozone previously established in Spitsbergen by Mangerud & Konieczny (1993) – the Cisuralian *Hamiapollenites tractiferinus* Assemblage. The *Hamiapollenites tractiferinus* Assemblage is dominated by species of *Vittatina* including *V. costabilis* and *V. saccata*, but also contains taeniate bisaccate pollen including *Protohaploxypinus* and *Striatopodocarpites*; and distally taeniate bisaccate pollen such as *Hamiapollenites tractiferinus* and *H. bullaeformis*. The *Dyupetalum* sp.–*Hamiapollenites bullaeformis* and *Scutasporites* cf. *unicus–Lunatisporites* spp. assemblage zones are similar except for the occurrence in the upper biozone of *Scutasporites* cf. *unicus* and the occurrence of *Hamiapollenites bullaeformis* in the lower biozone. According to Mangerud (1994) and Nilsson *et al.* (1996), the

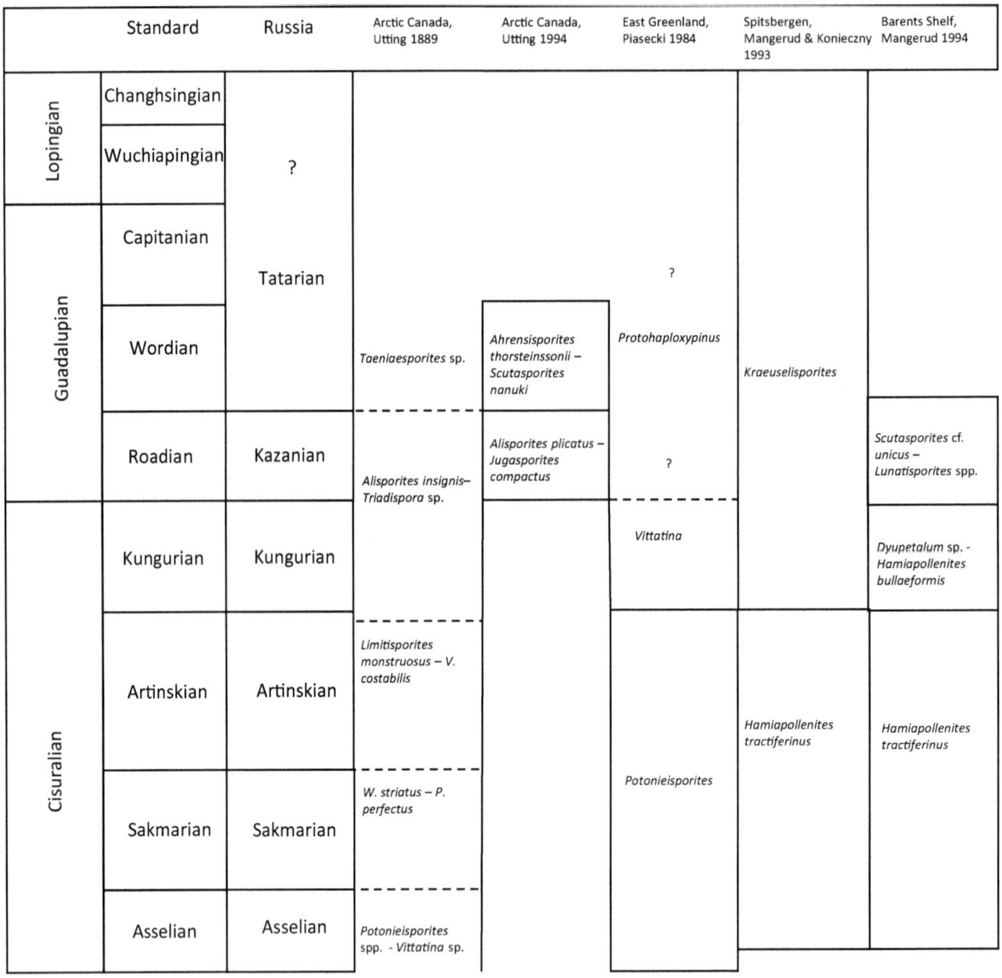

Series	Standard	Russia	Arctic Canada, Utting 1889	Arctic Canada, Utting 1994	East Greenland, Piasecki 1984	Spitsbergen, Mangerud & Konieczny 1993	Barents Shelf, Mangerud 1994
Lopingian	Changhsingian	?					
	Wuchiapingian						
Guadalupian	Capitanian	Tatarian			?		
	Wordian		*Taeniaesporites* sp.	*Ahrensisporites thorsteinssonii – Scutasporites nanuki*	*Protohaploxypinus*	*Kraeuselisporites*	
	Roadian	Kazanian	*Alisporites insignis– Triadispora* sp.	*Alisporites plicatus – Jugasporites compactus*	?		*Scutasporites* cf. *unicus – Lunatisporites* spp.
Cisuralian	Kungurian	Kungurian			*Vittatina*		*Dyupetalum* sp. - *Hamiapollenites bullaeformis*
	Artinskian	Artinskian	*Limitisporites monstruosus – V. costabilis*			*Hamiapollenites tractiferinus*	*Hamiapollenites tractiferinus*
	Sakmarian	Sakmarian	*W. striatus – P. perfectus*		*Potonieisporites*		
	Asselian	Asselian	*Potonieisporites* spp. - *Vittatina* sp.				

Fig. 5. Sub-Angara palynostratigraphic schemes, adapted from Utting (1989, 1994) and Mangerud (1994).

Finnmark Platform and other Barents Sea sequences can be correlated across the northern Atlantic, to Spitsbergen, Greenland and the Canadian Arctic (Fig. 5) (Mangerud & Konieczny 1993; Utting 1994).

There is very little modern palynological work on the basins to the north of Sub-Angara. Hart (1970) and Utting & Piasecki (1995) summarized palynological work from the Taimyr, Kuznets and Tungus areas in Siberia, concluding that the main difference between the lower and upper parts of the Permian is that monosaccate pollen are replaced by monosulcate pollen such as *Cycadopites*. To the east of Angara, in the Far Eastern subprovince, Utting & Piasecki (1995) summarized the generalized palynology in the Xinjiang, Tianshan, Kunlun and Junggar areas. More recent work by Zhu *et al.* (2005) compared the Junggar Basin with the Tarim Basin, which belonged to the Euramerican Province at least prior to the Cisuralian. According to Zhu *et al.* (2005), the Junggar Basin Permian contains mainly terrestrial sediments and the Lower (1), Middle (2) and Upper (3) Permian palynofloras are characterized, respectively, by: (1) the overwhelming dominance of taeniate bisaccate pollen; (2) the high content of *Cordaitina* and taeniate bisaccate pollen; and (3) by the appearance of many newly evolved forms with a 'Mesophytic aspect'. These latter include pollen and spores of 'advanced conifers and ferns' including distinctive taxa such as *Lueckisporites virkkiae*, *Scutasporites xinjiangensis*, *Klausipollenites schaubergeri*, *Falcisporites zapfei*, *Eucommiidites*, *Dictyophyllidites* and abundant small bisaccate pollen (e.g. *Vitreisporites*). Zhu *et al.* (2005, fig. 2) showed stratigraphic ranges of important taxa in the Permian sequences of the Junggar and Tarim basins.

Metcalfe *et al.* (2009) focused on the Permian–Triassic non-marine sequence at Dalongkou and Lucaogou in the Junggar Basin, and defined three assemblages, two of Lopingian age and an upper assemblage of probable Early Triassic age. The oldest is the *Tuberculatosporites homotubercularis–Potonieisporites* sp. Q assemblage. Apart from the eponymous species, this assemblage contains monosaccate pollen, including *Cordaitina uralensis*, and rare specimens of *Scutasporites* cf. *unicus*. The youngest Permian assemblage, the *Klausipollenites schaubergeri–Reduviasporonites chalastus–Syndesmorion stellatum* assemblage, is again dominated by the eponymous taxa, but also contains common non-taeniate bisaccate pollen, abundant algal remains (see Foster & Afonin 2006) and common *Scutasporites* cf. *unicus*. Other taxa include *Lueckisporites virkkiae*, *Lunatisporites transversundatus*, *Lunatisporites pellucidus*, *Platysaccus queenslandi*, *Alisporites splendens* and *Striatoabieites richteri*. Among spores making their appearance in this assemblage are *Leptolepidites jonkeri*, *Limatulasporites fossulatus* and *Naumovaspora striata*. The probable alga *Reduviasporonites chalastus* is also common.

The upper assemblage is the *Lundbladispora foveota–Pechorosporites disertus–Otynisporites eotriassicus* assemblage. Of note here is the presence of the distinctive megaspore *Otynisporites eotriassicus*. This taxon is considered a useful marker for the base of the Triassic because it occurs in both marine and non-marine sections in Greenland, Italy, Russia and Poland (Foster & Afonin 2005).

Cathaysia

Tha Cathaysia province is associated with the South China and Indochina blocks of the eastern Palaeotethys Ocean, including the present-day regions of Yunnan, Shanxi, Meishan and Hunan.

In eastern Yunnan, Ouyang (1982) described a Lopingian *Torispora gigantea–Patellisporites meishanensis* assemblage dominated by spores similar to many that characterize the Pennsylvanian of Euramerica (a so-called Carboniferous ‘relict flora’), with a rather small representation of gymnosperm pollen. The succeeding *Yunnanospora radiata–Gardenasporites* assemblage from the Lopingian Changhsing Formation contains more common taeniate bisaccate pollen and other gymnosperm pollen.

Liu *et al.* (2008) described palynological assemblages from the Cathaysian province of Shanxi, North China, establishing four assemblage zones, in ascending order, the *Torispora securis–Torispora laevigata* (SL), the *Torispora verrucosa–Pachetisporites kaipingensis* (VK), the *Thymospora thiessenii–Striatosporites heyleri* (TH) and the *Platysaccus minus–Gulisporites cochlearius* (MC) assemblage zones. Liu *et al.* (2008) considered that the SL Assemblage Zone correlates approximately with the western European SL Zone of Clayton *et al.* (1977: Westphalian C–D). The VK Assemblage Zone is believed to span the Carboniferous–Permian boundary. According to Liu *et al.* (2008), the upper TH and MC assemblage zones are of Cisuralian age. Liu *et al.* (2011) described Pennsylvanian–Lopingian megaspores from the same sections, establishing four megaspore assemblage zones extending from the Pennsylvanian to the Lopingian. This palynostratigraphic scheme is correlated with the assemblage zones of Liu *et al.* (2008). The lowest biozone is Carboniferous, but the succeeding *Bentzisporites margaritatus–Spencerisporites radiatus* (MS) assemblage zone is Kasimovian–Roadian in age based on fusulinids and conodonts. The *Biharisporites grosstriletus* (G) assemblage zone is of Wordian–Capitanian age, while the highest of the assemblage zones of Liu *et al.* (2011), the *Biharisporites* cf. *foskettensis* (F) assemblage zone, is believed to be of Wuchiapingian age. Liu *et al.* (2011) commented that several Carboniferous megaspores characteristic of Euramerica persist into the Guadalupian in Shanxi, North China, indicating that a warm and humid climate prevailed in this area during the Pennsylvanian–Roadian, whereas the climate of Euramerica had already become arid by the end of the Carboniferous. This warm and humid climate in northern China made it a refuge for some typically Euramerican Carboniferous plants.

The only study of the Permian–Triassic boundary sequence in the Cathaysian province and of the GSSP of the basal Triassic is that of Ouyang & Utting (1990). The Changhsing Formation yields a low-diversity assemblage dominated by acritarchs, but contains rare *Klausipollenites* sp., scolecodonts and *Reduviasporonites chalastus*. The lower part of the succeeding Griesbachian Chinglung Formation contains *Lueckisporites virkkiae*, *Klausipollenites schaubergeri*, *Alisporites* cf. *nuthallenis*, *Protohaploxypinus* spp., *Weylandites* sp., *Cedripites* spp. and *Reduviasporonites chalastus*, as well as ‘taxa of Triassic aspect’ (e.g. *Aratrisporites* cf. *yunnanensis*). Species of *Aratrisporites* have previously been used to correlate the base of the Triassic (e.g. see Metcalfe *et al.* 2009), although strong evidence for their reliability is lacking because species of *Aratrisporites* are known from the Carboniferous of the Middle East and North Africa (e.g. *Aratrisporites saharaensis*; Loboziak *et al.* 1986). Nevertheless, *Aratrisporites* cf. *yunnanensis* first appears 2.7 m above the base of the Triassic at Meishan according to Ouyang & Utting (1990), and thus this taxon might be taken as a local marker for the base of the Triassic.

Gondwana

Of the four phytogeographical provinces, Gondwana underwent the greatest changes through the Permian chiefly because the landmass of Gondwana was comprehensively ice-bound at the beginning of the period and then swiftly underwent deglaciation in the Cisuralian.

The Late Palaeozoic glaciation of Gondwana probably comprised three distinct episodes (Isbell *et al.* 2003); the third and last glaciation was geographically widespread and had continental ice sheets spanning the Carboniferous–Permian boundary through the Moscovian to the Artinskian (Fielding *et al.* 2008). The extent of palaeobotanical isolation and the poverty of flora and fauna during this glacial period mean that palynological assemblages are particularly difficult to relate to the international stages and to the Carboniferous–Permian boundary, with the result that even now Carboniferous and Permian palynological assemblages are difficult to distinguish in Gondwana (Stephenson 2008*a*).

The sedimentary rocks deposited at this time occur in cratonic basins that are now spread widely apart across Australia, India, Antarctica, southern and central Africa, and South America. The Permian rocks of these cratonic basins contain commercial deposits of coal (e.g. in India, Australia, southern Africa and South America), and oil and gas (in Australia, South America and the Middle East), and thus are amongst the most intensely studied Permian sequences in the world. The need in exploration to access stratigraphic sequences through borehole material (cuttings, core and sidewall core) has meant that palynology is also, by far, the most important biostratigraphic tool used in locally correlating sequences in the Gondwana Permian, mainly for resource exploitation rather than for academic study. Thus, palynostratigraphic schemes tend to be locally focused, although attempts have been made recently to correlate more widely across Gondwana, and from Gondwana to the international Permian scale. Because of this local focus, this part of the review concentrates on parts of Gondwana, and then attempts a synthesis at the end of the section.

South America. Palynostratigraphy has progressed most recently in four regions: the Tarija and Chacoparaná basins in northern Argentina, the Paganzo and other basins in central western Argentina, the Claromecó Basin in eastern Argentina, and the Paraná and Amazonas basins in Brazil. For reviews of earlier literature pertaining to South America see Azcuy (1980) and Archangelsky *et al.* (1980). Referring generally to South American basins, Azcuy (1980) and Archangelsky *et al.* (1980) considered Pennsylvanian palynological assemblages (e.g. Palynozone I of Azcuy 1980) (Fig. 6) to be dominated by monosaccate pollen (e.g. *Potonieisporites*) and zonate spores (e.g. *Ancistrospora* and *Lundbladispora*), with taeniate bisaccate pollen being rare. Permian assemblages (e.g. Palynozone III of Azcuy 1980) contain more taeniate bisaccate pollen, *Vittatina* and *Cristatisporites*. Superimposed on these general trends are a large number of other more subtle changes that indicate variation between South American basins and which have allowed a number of separate basin-specific palynostratigraphic schemes to evolve (e.g. see Vergel 1993; Archangelsky & Vergel 1996; Césari & Gutiérrez 2000; Playford & Dino 2000, 2002; di Pasquo 2003; Di Pasquo *et al.* 2003; Souza & Marques-Toigo 2003; Souza *et al.* 2003; Pérez Loinaze & Césari 2004; Souza 2006; Balarino 2014) (Fig. 6).

In general, the biostratigraphy of the basins is difficult to relate to the international stages of the Carboniferous and Permian (Archangelsky *et al.* 1980) because of the scarcity of marine faunas: however, indirect comparisons with sequences containing palynomorphs and cosmopolitan faunas and floras (e.g. Marques-Toigo 1974; Cisterna *et al.* 2011) allow some tentative dates to be assigned. Since 2007, the most marked progress has been made in integrating radiometric dates with palynological biozones, allowing limited – not always reconcilable – calibration of the latter with the international scale. Amongst the most important of these studies are those of Césari (2007), Guerra-Sommer *et al.* (2008), Césari *et al.* (2011), Mori *et al.* (2012) and di Pasquo *et al.* (2015).

In the first of the studies, Césari (2007) noted radiometric dates in the San Rafael Basin in central western Argentina and in the Paraná Basin in southern Brazil that suggested numerical ages for biozones established by Césari & Gutiérrez (2000) and Souza & Marques-Toigo (2003) in those basins, respectively. Thus, the *Lueckisporites–Weylandites* Assemblage Biozone of Césari & Gutiérrez (2000) in the San Rafael Basin contains a horizon dated at 266.3 $\pm$ 0.8 Ma (Wordian), while the *Lueckisporites virkkiae* Interval Biozone of Souza & Marques-Toigo (2003) in the Paraná Basin contains a dated horizon of 278.4 $\pm$ 2.2 Ma (Kungurian).

Guerra-Sommer *et al.* (2008) reported an age of 285.4 $\pm$ 8.6 Ma (Artinskian) within the Paraná Basin Faxinal coal seam, which is assigned to the *Hamiapollenites karooensis* Sub-biozone of the *Vittatina costabilis* Interval Biozone of Souza & Marques-Toigo (2003). Mori *et al.* (2012) noted a date of 281 $\pm$ 3.4 Ma (Artinskian) for another horizon within the *Lueckisporites virkkiae* Interval Biozone of the Paraná Basin in the Candiota coal mine.

Césari *et al.* (2011, fig. 6) produced a helpful synthesis of the palynostratigraphy and radiometric dating of the Carboniferous and Cisuralian sequence

System or part	South America Azcuy (1980)	Chaco-Paraná, Argentina Playford &Dino (2002) Vergel (1993)		NW Argentina Césari & Gutiérrez (2000)	N Argentina, Bolivia Tarija Basin, Di Pasquo (2003)	E Argentina, Claromecó Basin, Balarino (2014)	Brazil, Paraná Basin Souza & Marques-Toigo (2003)	Brazil, Amazonas Basin Playford & Dino (2000)
Upper Permian		?		?	?	T. toreutos –R. chalastus Assemblage Zone		T. toreutos –R. chalastus Assemblage Zone
Middle Permian		*Striatites* Zone		LW			L. virkkiae	
Lower Permian	Palynozone IV	*Cristatisporites* Zone	Upper		L. virkkiae	C.confluens – V. vittifera Assemblage Zone	V. costabilis	V. costabilis
			Middle		?			
?	Palynozone III		Lower	FS	V. costabilis			
Upper Carb.	Palynozone II	*Potonieisporites-Lundbladispora* Zone	Upper	DM	TB		C. monoletus	R. cephalata
					MR			S. heyleri
								I. unicus
			Lower		BC		A. cristatus	S. incrass.
	Palynozone I							S. triangulus

Fig. 6. Correlation of South American palynostratigraphic schemes after Stephenson (2008*a*), Balarino (2014) and di Pasquo *et al.* (2015). Biozone codes: Césari & Gutiérrez (2000) DM, *Raistrickia densa–Convolutispora muriornata*; FS, *Fusacolpites fusus–Vittatina subsaccata*; LW, *Lueckisporites–Weylandites*; di Pasquo (2003) BC, *Dictyotriletes bireticulatus–Cristatisporites chacoparanensis*; MR, *Granulatisporites micronodosus–Reticulatisporites reticulatus*; TB, *Marsupipollenites triradiatus–Lundbladispora braziliensis*.

across Argentina and Brazil correlating the San Rafael and Paraná basin biozones and using radiometric dates to relate South American palynological biozones to those of Namibia and Australia.

di Pasquo *et al.* (2015) gave radiometric dates from five volcanic ash beds within the Cisuralian Copacabana Formation in central Bolivia (Tarija Basin). The five dates (cited as preliminary and published only in the non-peer reviewed Permian ICS Newsletter *Permophiles*, **53**, Supplement 1) are 298, 295.4–295.1 and 293 Ma (for two ash layers approximately 25 m apart stratigraphically), and 292.1–291.3 Ma. According to di Pasquo *et al.* (2015), these dates suggest an Asselian age for the '*Vittatina costabilis* assemblage' and an Asselian–Sakmarian age for the '*Lueckisporites virkkiae* assemblage' of di Pasquo *et al.* (2015, fig. 4). This latter date is clearly inconsistent with other dates for the *Lueckisporites virkkiae* Interval Biozone of Souza & Marques-Toigo (2003) in the Paraná Basin (e.g. see Césari *et al.* 2011).

The accuracy of radiometric dates is important for discussion of the stratigraphic occurrence of *Lueckisporites virkkiae*, and inaccuracies inherent in radiometric dating may be the cause of apparent discrepancy in its first appearance, otherwise it is possible that *Lueckisporites virkkiae* has a diachronous first occurrence strongly reducing its value as a possible Euramerica–Gondwana 'bridging taxon'.

Another possibility is that *Lueckisporites virkkiae* is being misidentified or that the conception of the taxon being used by taxonomists is too wide. The original concept of *Lueckisporites virkkiae* Potonié & Klaus (1954) was of a diploxylonoid bisaccate pollen grain with a wide separation of sacci (e.g. see Potonié & Klaus 1954, text-fig. 5, pl. 10, fig. 3; Klaus 1963, p. 300). Clarke (1965) allowed more haploxylonoid specimens within *Lueckisporites virkkiae* in his emendation of the species, referring to these as 'variant B'. di Pasquo *et al.* (2015) did not illustrate the specimens that they attribute to *Lueckisporites virkkiae*, but the specimen illustrated by Mori *et al.* (2012, fig. 3j) is strongly haploxylonoid and lacks evidence of a prominent cappa or exoexinal taeniae. It appears closer to *Corisaccites alutas*. Haploxylonoid specimens of *Lueckisporites virkkiae* (using the conception of Clarke 1965) are difficult to separate from *C. alutas*, although Venkatachala & Kar (1966) regarded *C. alutas* as 'subsaccate', and subsequent authors have described *C. alutas* as having poorly inflated 'leathery' sacci, the exoexine of which is structurally indistinguishable from that of the corpus (see Stephenson 2008*b*).

To maintain the value of *Lueckisporites virkkiae* as a biostratigraphical marker may mean rejecting the emendation of Clarke (1965) and retaining the original concept of *Lueckisporites virkkiae* Potonié & Klaus, 1954 as a diploxylonoid bisaccate pollen grain with a wide separation of sacci (see Stephenson 2008*b*). It may also be valuable to start comparative studies between South American Gondwanan and Euramerican localities focusing on the genera *Lueckisporites* and *Corisaccites*.

Such uncertainties over correlation mean that, at present, it is not possible to accurately correlate South American biozones to the international scale (Fig. 6). However, Vergel (1993) and Playford & Dino (2002) correlated the Carboniferous–Permian boundary in the Chacoparaná Basin to the boundary between the *Potonieisporites–Lundbladispora* Zone and the *Cristatisporites* Zone.

Africa. Much palynological research in southern African Permian sequences has been aimed at the correlation of coal seams (Falcon *et al.* 1984; Aitken 1994; Millsteed 1999): however, more general schemes have been proposed (Hart 1969; Falcon 1975; Anderson 1977; MacRae 1988).

In the most detailed and extensive study covering more than 35 boreholes, four surface localities and more than 1000 samples, Anderson (1977) erected seven biozones for the Permian. The lowest, Zone 1 (Anderson 1977, fig. 8), contains common *Microbaculispora*, monosaccate pollen (grouped mainly under *Vestigisporites*) and non-taeniate bisaccate pollen (grouped mainly under *Pityosporites*). Anderson (1977) considered Zone 1 to be Sakmarian in age. The succeeding Zone 2 (Sakmarian–Artinskian), divided into four subzones, is chiefly distinguished by its higher numbers of zonate spores, although the highest subzone of Zone 2, and Zone 3 (Artinskian), are characterized by very abundant non-taeniate bisaccate pollen. Taeniate bisaccate pollen become common only in the highest subzone of Anderson's Zone 3.

Anderson's Zone 4 (Artinskian–Wordian) contains common taeniate bisaccate pollen with small numbers of proximal taeniae which were referred to *Lueckisporites* by Anderson (1977), but would be referred by modern taxonomists to *Corisaccites*, *Guttulapollenites* and, possibly, *Taeniaesporites*, as well as to *Lueckisporites*. Zone 5 (Capitanian) is marked by common *Vittatina*. Amongst other changes, a marked drop in *Vittatina* marks the base of Zone 6. Zones 6 and 7 (Lopingian) are similar, but are distinguished by the differing proportions of their constituent species (Anderson 1977).

The work of Anderson (1977) was not only based on a large palynological database, but also on extensive description, and particularly illustration, of important taxa. Backhouse (1991) used these data for a comparison and correlation with palynological biozones established in the Western Australian Collie Basin (Fig. 7). This study remains one of the most rigorous intercontinental palynological correlations yet completed for the Gondwana phytogeographical province, and, although correlation to the international stages is necessarily tentative, the

System/Stage		Karoo Basin biozones, Anderson (1977) Strat	Biozone	Collie Basin biozones, Backhouse (1991)
Permian	?Ufimian	Upper Ecca	4d	*Protohaploxypinus rugatus*
				Didecitriletes ericianus
	?Kungurian		4c	*Dulhuntyispora granulata*
		Middle Ecca	3d	*Microbaculispora villosa*
			3c	*Praecolpatites sinuosus*
			3b	
	Artinskian		3a	*Microbaculispora trisina*
		Lower Ecca	2d	*Striatopodocarpites fusus*
			2c	
			2b	*Pseudoreticulatispora pseudoreticulata*
	Sakmarian		2a	
		Dwyka	1	*Converrucosisporites confluens*
?	Asselian			Stage 2
Carboniferous				

Fig. 7. Correlation of Karoo and Collie Basin biozones, after Backhouse (1991).

correlation of lithological units between Western Australia and southern Africa is probably robust.

Other studies have also applied Australian palynostratigraphic schemes in southern Africa (Millsteed 1993, 1999; Stephenson & McLean 1999). In east and central Africa, palynostratigraphic schemes were developed by Hart (1965*a*) and Utting (1978). Extensive taxonomic studies in that region were carried out by Bose & Kar (1966, 1976) and Bose & Maheshwari (1968).

Unlike South America, only a few ash layers provided opportunities for direct dating of palynological biozones in Africa. However, Stephenson (2009) recovered palynological assemblages of the *Converrucosisporites confluens* Oppel Zone from the Ganigobis Shale Member of Namibia, close to Ash layer IIb, which was radiometrically dated as 302.0 ± 3.0 Ma (Pennsylvanian: Gzhelian or Kasimovian) (see Bangert 2000). Thus, the *Converrucosisporites confluens* Oppel Zone may range earlier than previously thought and may influence age considerations of non-calibrated palynological biozones, such as those of Backhouse (1991) discussed earlier.

Recent work on the palynology of South African coal seams (Götz & Ruckwied 2014, Ruckwied *et al.* 2014) concentrated on using perceived climatic changes expressed in palynological sequences to correlate across South African basins. Barbolini & Bamford (2014) established two biozones in the Mmamantswe coalfield (Lower Ecca) in Botswana.

Arabia, the Middle East, Pakistan and India. The broad character of Cisuralian assemblages in the southern parts of the Middle East, and Pakistan and India, are similar to those of Africa and South America in that monosaccate pollen and simple trilete spores dominate the lowest glacigene horizons, and these are succeeded by horizons with more diverse assemblages containing taeniate and non-taeniate bisaccate pollen, cycad pollen, and fern spores. Above this level in the late Cisuralian and Guadalupian, the assemblages of Arabia, the Middle East and Pakistan begin to diverge from those of the more southerly part of Gondwana probably because of more rapid climate warming due to the fact that these areas were already of lower latitude and were rapidly moving north (e.g. Stephenson *et al.* 2008*b*).

In Arabia, there are palynostratigraphic schemes for Oman (Besems & Schuurman 1987; Love 1994; Stephenson & Osterloff 2002; Penney *et al.* 2008; Stephenson *et al.* 2008*a*) and Saudi Arabia (Stump & van der Eem 1995); and these, as well as in-house oil company schemes, have been synthesized by Stephenson *et al.* (2003) and Stephenson (2006) to a standard palynozonation for the entire Arabian Peninsula, which also has application in the wider Middle East (e.g. see Nader *et al.* 1993*a*; Stolle 2007; Angiolini & Stephenson 2008; Jan *et al.* 2009; Stephenson *et al.* 2013; Stephenson & Powell 2013, 2014). The chief unifying features of the standard palynozonation that occur throughout the Arabian Peninsula are, towards the base, the presence of common monosaccate pollen, as well as taxa such as *Anapiculatisporites concinnus* (characterizing Oman and Saudi Arabia Palynological Zone 1 (OSPZ1): ?Pennsylvanian) (Fig. 8), and above this the appearance of *Microbaculispora* and *Horriditriletes* (characterizing OSPZ2: ?Pennsylvanian–Asselian).

The base of OSPZ3 is defined chiefly by the abrupt increase in the small non-taeniate bisaccate pollen *Pteruchipollenites indarraensis* from approximately 10 to 50 or 60% of assemblages (Stephenson 2015). Other taxa that occur first consistently in OSPZ3 are the taeniate bisaccate pollen *Striatopodocarpites cancellatus* and *S. fusus*, and the taeniate 'circumstriate' pollen taxa, such as *Circumstriatites talchirensis* and *Striasulcites tectus*. *Kingiacolpites subcircularis* is common throughout

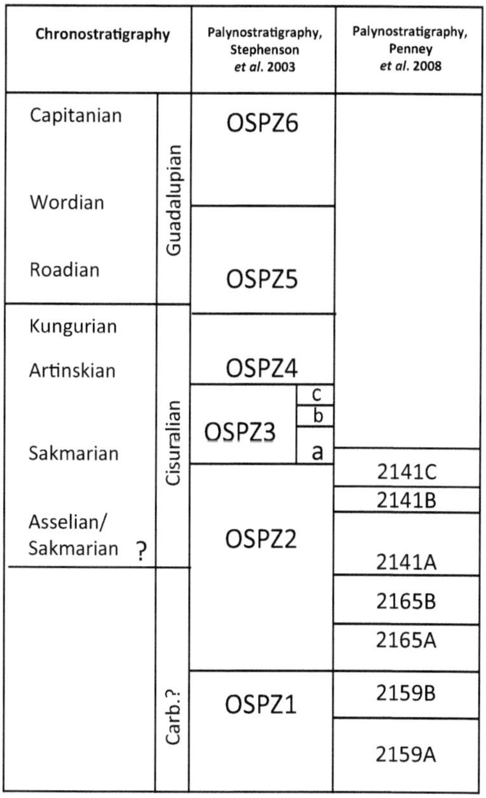

Chronostratigraphy		Palynostratigraphy, Stephenson *et al.* 2003		Palynostratigraphy, Penney *et al.* 2008
Capitanian	Guadalupian	OSPZ6		
Wordian	Guadalupian	OSPZ6 / OSPZ5		
Roadian	Guadalupian	OSPZ5		
Kungurian	Cisuralian	OSPZ5 / OSPZ4		
Artinskian	Cisuralian	OSPZ4		
Sakmarian	Cisuralian	OSPZ3	c	
			b	
			a	
		OSPZ2		2141C
				2141B
Asselian/ Sakmarian ?	Cisuralian			2141A
	Carb.?			2165B
				2165A
	Carb.?	OSPZ1		2159B
				2159A

Fig. 8. Palynostratigraphic schemes of the Arabian peninsula.

this biozone, occasionally reaching 50% of assemblages. OSPZ3 is believed to be Sakmarian in age (Stephenson 2015).

The chief distinguishing characteristic of OSPZ4 (?Artinskian–Kungurian: Stephenson *et al.* 2003) is the common occurrence of *Barakarites rotatus, K. subcircularis* and *P. limpidus*, although *C. alutas, P. amplus, Plicatipollenites* spp., *Striatopodocarpites* spp., *S. tectus, Strotersporites* spp., *Vesicaspora* spp., *Vittatina costabilis and Vittatina* spp. are also present. OSPZ5 (?Roadian–Wordian) is characterized by, amongst others, *Distriatites insolitus, Playfordiaspora cancellosa* and *Thymospora opaqua*; and OSPZ6 (?Wordian–Capitanian: Stephenson 2008*b*) is characterized by ?*Florinites balmei* and *Protohaploxypinus uttingii*. A high-resolution palynozonation mainly aimed at correlation in hydrocarbon basins of Oman can be correlated with the OSPZ scheme (Penney *et al.* 2008) (Fig. 8)

Palynological data on higher parts of the sequence to the Permian–Triassic transition is not available in the western part of the Middle East because lithologies are not suitable for palynomorph preservation. However, to the east of the region in the Salt Range of Pakistan, this part of the Permian sequence is represented by rocks that contain both palynomorphs and well-preserved fauna. Balme (1970), in a pioneering taxonomic work on the Salt Range, defined many of the taxa of value in palynostratigraphy of Permian–Triassic Tethyan and Gondwanan palynostratigraphy, although Balme (1970) did not define biozones himself. More recently, Hermann *et al.* (2012) defined two latest Permian assemblages in the Salt Range ('Chhidru 1' and 'Chhidru 2') present in the uppermost Chhidru Formation (Changhsingian). Chhidru 1 (the *Protohaploxypinus* spp.–*Weylandites* spp. Association) is dominated by gymnosperm pollen, including common *Klausipollenites* spp. (including *K. schaubergeri*) and *Protohaploxypinus* spp., (including *P. limpidus*), as well as *Weylandites lucifer* and *Alisporites* spp. Spores are very rare. Chhidru 2 (the *Kraeuselisporites wargalensis–Protohaploxypinus* spp. Association) is characterized by abundant cavate trilete spores such as *Kraeuselisporites wargalensis* and *Kraeuselisporites* spp. According to Hermann *et al.* (2012), the range of *K. wargalensis* is restricted to Chhidru 2. The top of Chhidru 2 is characterized by the last occurrences of *Klausipollenites schaubergeri* and *Lueckisporites* spp. (including *Lueckisporites virkkiae*).

In India, much of the taxonomic work in the 1960s was directed at correlation between coal seams, and not at correlation with other Gondwana sequences or international stages of the Permian. Major taxonomic studies were confined mainly to one or other of the 'lithostratigraphic stages' of the peninsular Indian Permian (in ascending order: Talchir, Karharbari, Barakar, Barren Measures and Raniganj) and did not span larger parts of the sequence (Truswell 1980). However, attempts have been made to synthesize data into palynostratigraphic schemes for the entire Indian Permian (e.g. Tiwari & Tripathi 1992) and some limited correlations of Indian palynostratigraphy with international stages have been made (Tiwari 1996; Vijaya 1996).

The biozones of Tiwari & Tripathi (1992) are well defined, but are not dated using the international scale. The earliest of these biozones, the *Potonieisporites neglectus* Biozone, from the lowest glacigene sediments of the Talchir Formation is dominated by monosaccate pollen, but lacks taeniate bisaccate pollen. The base of the succeeding *Plicatipollenites gondwanensis* Biozone is marked by diversification and the appearance of taeniate bisaccate pollen, and the biozone above (*Parasaccites korbaensis* Biozone) in the upper Talchir Formation by fern spores such as *Microbaculispora tentula.* The base of the succeeding *Crucisaccites monoletus* Biozone of the Karhabari Formation is marked by the first appearances of the eponymous taxon, as well as taxa such as *Marsupipollenites*. The ages of these biozones are considered to range from the Asselian to Sakmarian (Vijaya 1996). The succeeding biozones of Tiwari & Tripathi (1992) are, in order of deceasing age, the *Scheuringipollenites barakarensis, Faunipollenites varius, Densipollenites indicus, Gondisporites raniganjensis* and *Densipollenites magnicorpus* biozones. Tiwari (1996) attempted to date 'palynoevents' through the Indian Permian: for example, the decline of monosaccates (Palynoevent 3) was considered to be late Artinskian, and the first appearance of *Densipollenites* (Palynoevent 4) as Kungurian–Ufimian – although these were not related to the biozones of Tiwari & Tripathi (1992).

Despite the fact that there has been some consolidation of schemes in India, local palynostratigraphic schemes continue to be developed in the cratonic basins. In Andhra Pradesh, Aggarwal & Jha (2013) defined eight Permian palynological biozones for the Godavari Graben, linking these with climate changes, although not to the international Permian scale. Jha *et al.* (2014), also working in the Andhra Pradesh Godavari Graben, identified two assemblages (Palynoassemblage-I and -II) dated as Lopingian. In northern India in Kashmir, Tewari *et al.* (2015) identified Permian–Triassic assemblages broadly correlated to the *Densipollenites magnicorpus* and *Klausipollenites decipiens* biozones of peninsular India (see Tiwari & Tripathi 1992).

Australia. Australia has the best-documented Cisuralian sequences of Gondwana and, owing to the presence of rare marine intervals, some calibration of palynological biozones with international stages has been possible using marine macrofaunas (e.g. Leonova 1998; Archbold 1999). Later palynological biozones in the Guadalupian and Lopingian are more difficult to date due to the lack of marine fauna (in coal and red beds), although there are tie points (Foster & Archbold 2001): for example, the record of a single specimen of the ammonoid *Cyclolobus persulcatus*, from the Cherrabum Member of the Hardman Formation, in the Canning Basin, Western Australia (Foster & Archbold 2001), dated as 'post-Guadalupian' by Glenister *et al.* (1990) and 'Capitanian–Dzhulfian' by Leonova (1998). Such tie points allow spot dates to be applied to the otherwise well-developed palynostratigraphy.

A feature of palynostratigraphy in the Australian Permian is significant endemism, with the result that separate palynozonations developed in western and eastern Australian basins in the earliest period of palynological investigation (Fig. 9). In the west, eight assemblage biozones (units I–VIII), ranging in age from Pennsylvanian to Lopingian, were recognized; while, in the east, five assemblage biozones (stages 1–5) encompass the same interval (Evans 1969; Kemp *et al.* 1977; Foster 1979; Truswell 1980; Price 1983, 1997). The precise relationship between these schemes, and the ages of the biozones, remains speculative (see Jones & Truswell 1992) (Fig. 9), but there are broad similarities across the continent (see Balme 1980, text-fig. 4).

In the lower parts of the glacigene sequence (Stage 1), radially and bilaterally symmetrical monosaccate pollen is dominant, and taeniate and

System/Stage (Ages after Backhouse (1991))	Eastern Australia: Kemp *et al.* (1977)	Eastern Australia: Price (1983, 1997)	Western Australia: Balme (1980)	Western Australia: Backhouse (1991)
Permian	*Protohaploxypinus reticulatus*			
?Ufimian	Stage 5	APP5	Unit VII	*D. parvithola*
?Ufimian	Stage 5	APP4	Unit VII	*P. rugatus*
?Ufimian	Stage 5	APP4	Unit VII	*D. ericianus*
?Kungurian	Stage 5	APP4	Unit VII	*D. granulata*
?Kungurian	Stage 4	APP3	Unit VII	*M. villosa*
?Kungurian	Stage 4	APP3	Unit VI	
?Kungurian	Stage 4	APP3	Unit V	*P. sinuosus*
Artinskian	Stage 3	APP2	Unit V	*M. trisina*
Artinskian	Stage 3	APP2	Unit IV	*S. fusus*
Sakmarian	Stage 3	APP2	Unit III	*P. pseudo.*
Sakmarian	Stage 2	APP1	Unit II	*C. confl.*
? Asselian	Stage 2	APP1	Unit II	
Carboniferous ?Gzhelian	Stage 2	APP1	Unit II	
?Kazimovian	Stage 1			

Fig. 9. Correlation of eastern and western Australian palynostratigraphic schemes, after Kemp *et al.* (1977), Foster (1979), Balme (1980), Backhouse (1991) and Price (1983, 1997).

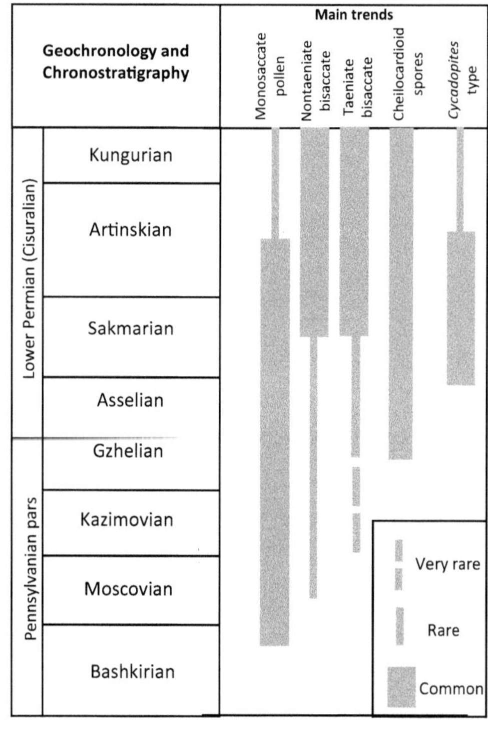

Fig. 10. Main trends in palynomorph types across the Gondwana phytogeographical province from the Pennsylvanian to the Cisuralian, after Stephenson (2008*a*). Correlation of events to the international scale is approximate.

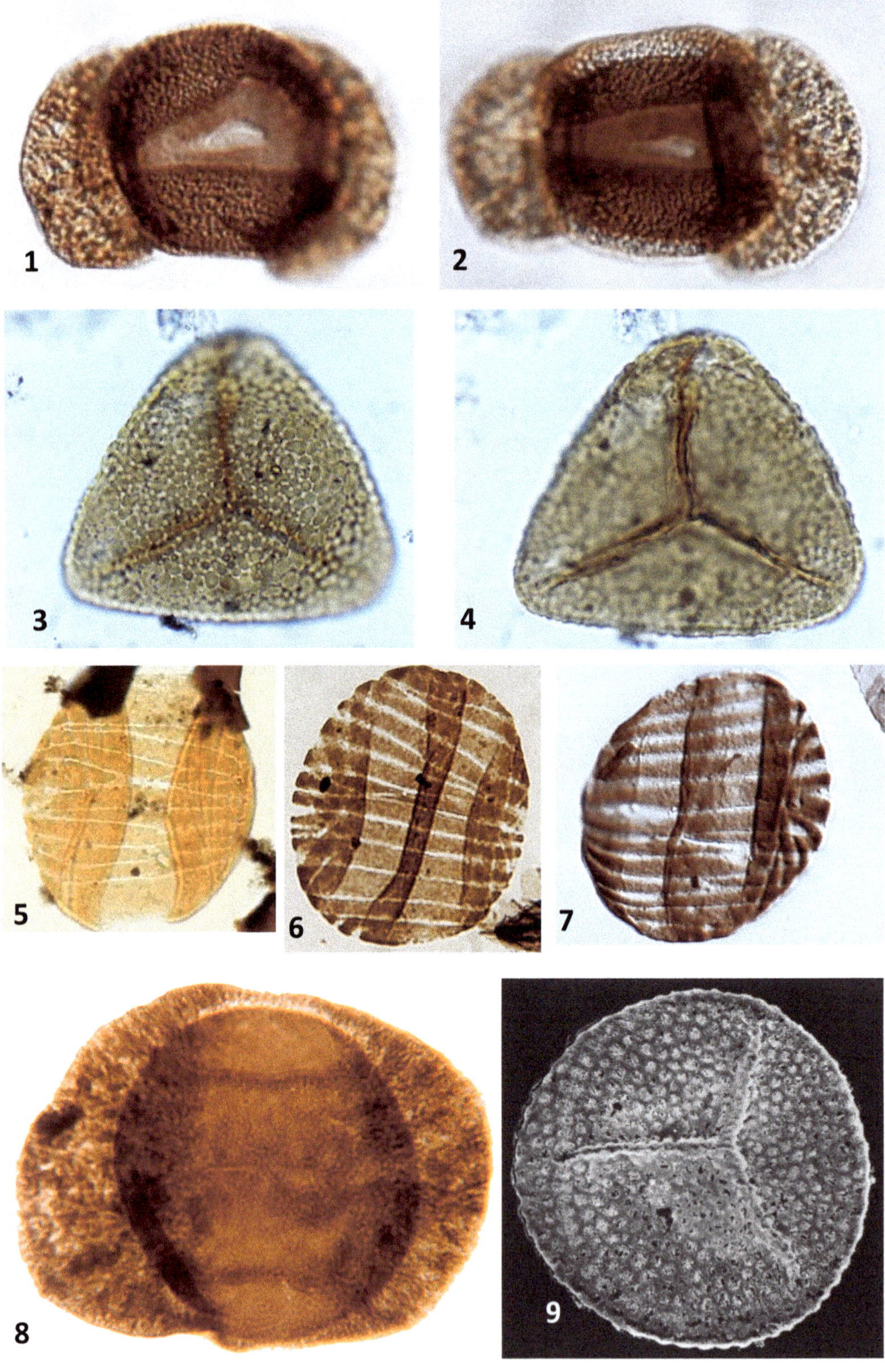
1
2
3
4
5
6
7
8
9

non-taeniate bisaccate pollen is subordinate. In Stage 2 above, small triangular fern spores (cheilocardioid spores) and *Cycadopites* become common; while, around the bases of Stage 3, taeniate and non-taeniate bisaccate pollen become common. In Stage 4, a diversity of pollen and spores develops, including distinctive colpate pollen such as *Praecolpatites sinuosus*; while Stage 5 is partly characterized by the very distinctive spore genus *Dulhuntyispora*, which is distinguished by expanded blister-like exoexine in each of the inter-radial areas.

Antarctica. Early work in Antarctica was carried out by Balme & Playford (1967) and Kemp (1973). Kyle (1977) erected two informal palynological biozones from study of a wide range of Antarctica localities. Her *Parasaccites* Zone is dominated by monosaccate pollen, along with rare bisaccate taeniate pollen, and, locally, high frequencies of *Microbaculispora tentula* and *Cycadopites cymbatus*. According to Truswell (1980), this biozone compares closely with the upper part of Stage 2 in Australia. The succeeding *Protohaploxypinus* Zone of Kyle (1977) is distinguished by abundant bisaccate pollen, including many taeniate taxa. In gross composition and species content, this biozone resembles Stage 4 assemblages from eastern Australia. Work following that of Kyle (1977) has tended to apply Australian palynostratigraphy to Antarctica, rather than develop more sophisticated local palynostratigraphic schemes (e.g. Farabee *et al.* 1990; Larrsson *et al.* 1990; Lindström 1995*a*, *b*; Lindström & McLoughlin 2007).

Synthesis of Gondwana palynostratigraphy. It is clear that there are quantitative and qualitative similarities in the late Pennsylvanian and Cisuralian that allow correlation across Gondwana with reasonable accuracy; after this period, however, palynological characteristics of the various parts of Gondwana begin to diverge. For example, Guadalupian and Lopingian palynological assemblages from Oman and Saudi Arabia are very different to those of Australia. Similarly, the Guadalupian palynology of South America differs considerably to that of Australia. These palynological differences, generated in the late Cisuralian following deglaciation of the Gondwana continent, are probably due to the northwards drift of Gondwana through the Permian, and the different order in which the various parts of Gondwana moved through latitudinal belts.

For the Cisuralian, the main palynological events that can be tracked with reasonable certainty through all basins of Gondwana appear to be the appearance and diversification patterns of: (1) monosaccate pollen; (2) cheilocardioid spores (such as *Microbaculispora*); (3) *Cycadopites* pollen; and (4) taeniate and non-taeniate bisaccate pollen. The precise ages of these events in relation to the international scale are not known, and recent data from radiometric dating are still being collected and assimilated (e.g. Stephenson 2009; Césari *et al.* 2011; di Pasquo *et al.* 2015) (Fig. 10).

Having said this, a few high-resolution correlations within Gondwana that involve younger Permian rocks have been possible based on close taxonomic comparison and accurate stratigraphic information: for example, the correlation of the Collie Basin, Western Australia by Backhouse (1991) with the South African biozones of Anderson (1977) (see Fig. 7). Dino & Playford (2002) and Stephenson (2008*a*) identified taxa that are common to South America and Australia in the Carboniferous–Permian. *Converrucosisporites confluens* occurs across most Gondwana sequences: Australia (Foster & Waterhouse 1988), India (Srivastava & Bhattacharyya 1996), Argentina (Archangelsky & Gamerro 1979), Uruguay (Beri & Goso 1996), Antarctica (Lindström 1995*b*), and Oman and Saudi Arabia (Stephenson *et al.* 2003). *Ahrensisporites cristatus* is considered to have a near-synchronous first appearance in the early Pennsylvanian, while

Fig. 11. Photomicrographs of taxa of possible biostratigraphical value between and within phytogeographical provinces of the Permian. (1) *Lueckisporites virkkiae*, F46, V-619, 2187; from the basal Khuff clastics, Saudi Arabia; width of specimen approximately 50 μm. (2) *Lueckisporites virkkiae*, E42, V-619, 2187; from the basal Khuff clastics, Saudi Arabia; width of specimen approximately 50 μm. (3) *Converrucosisporites confluens*, V42, MM-151, 3167, distal focus; from the Al Khlata Formation, Oman; width of specimen approximately 40 μm. (4) *Converrucosisporites confluens*, V42, MM-151, 3167, proximal focus; from the Al Khlata Formation, Oman; width of specimen approximately 40 μm. (5) *Weylandites* sp. slide: 1563, EF: V30.3; from the Lower Karoo sequence, Kalahari Karoo Basin, Botswana; courtesy Alain Le Hérissé; specimen featured in Modie & Le Hérissé (2009); width of specimen approximately 35 μm. (6) *Vittatina* sp., M27, RA-2, 4256; from the Al Khlata Formation, Oman; width of specimen approximately 35 μm. (7) *Vittatina* sp., M27, TL-16, 982, from the Al Khlata Formation, Oman; width of specimen approximately 35 μm; (8) *Scutasporites unicus* , collection number GBA 2010/013/0027 (holotype); single grain preparation, slide No. 404, England-finder K34r; Gröden Formation = Grödner Sandstein (Arenaria di Val Gardena); courtesy of Prof. Hans Egger; specimen featured in Draxler (2010); width of specimen approximately 50 μm. (9) *Otynisporites eotriassicus*; Otyn IG 1 borehole, depth 817.5 m; Balic Formation (Lower Buntsandstein), Induan; courtesy of Dr Teresa Marcinkiewicz; specimen featured in Marcinkiewicz *et al.* (2014); width of specimen approximately 400 μm.

the last appearances of *Psomospora detecta*, *Spelaeotriletes ybertii* and *Rattiganispora apiculata* were also considered to be possibly coeval in both continents (Dino & Playford 2002).

Discussion and conclusions

The purpose of palynostratigraphy, like biostratigraphy generally, is to provide correlation. Correlation helps palynologists, stratigraphers and geologists to relate sedimentary rocks deposited in one place to those in another, and to relate geological resources or events to each other and to other important geological or scientific phenomena.

In the Permian, the most important reasons for correlation related to resource extraction are to describe coal and hydrocarbon resources, to place them into a regional framework, and to help to understand how to find more coal and hydrocarbons. Probably, the bulk of work that has been done on Permian rocks in palynostratigraphy has been in this area. There have been several tens of palynostratigraphy papers published on coal measures stratigraphy in India, South America, Australia and southern Africa, with the result that coal measures sequences (e.g. Foster 1979), and even individual coal seams, can be correlated (e.g. Aitken 1994). Perhaps the work that has led to the highest-resolution palynostratigraphy (in terms of being able to recognize and distinguish the smallest divisions of stratigraphy and time) has been for the hydrocarbons industry, particularly in the Middle East where very large accumulations of oil and gas occur in the Permian. These schemes are capable of biostratigraphically characterizing rock units of only a few metres in thickness across oil and gas fields or even basins (e.g. Stephenson & Osterloff 2002; Penney *et al.* 2008; Stephenson *et al.* 2008*a*). In Australia also, the work of Price (1983, 1997) has been widely used for Permian onshore oil, gas, unconventional hydrocarbons and coal exploration.

Although these palynostratigraphic schemes related to resource extraction have been very successful, their main shortcoming has been a lack of correlation with schemes outside the basins, coalfields and hydrocarbon fields that they serve, and chiefly a lack of correlation with the international Permian scale. In the past, a lack of fundamental stratigraphic standards (e.g. GSSPs) for the Permian hampered this kind of correlation, but these now substantially exist (see Henderson *et al.* 2012) and local and regional palynostratigraphic schemes should be fitted into the wider Permian scale.

In general, also, standards of recording of palynological data and accessibility of data are higher, allowing easier comparison between palynostratigraphic schemes. Some problems still exist in the precision of taxonomic classification in different areas such that the utility of potentially useful widely correlateable palynological taxa are reduced (e.g. *Lueckisporites virkkiae*).

The benefits of a better integrated general palynostratigraphy are very great scientifically because there are numerous events of global scientific interest in the Permian: for example, the timing and order of deglaciation events across the continent of Gondwana that can, arguably, only be established through a Gondwana-wide palynostratigraphic scheme (see Stephenson *et al.* 2007), large-scale patterns of floral migration across continents, and the detailed characteristics and timing of mass extinction events within the Permian and at the Permian–Triassic boundary.

Perhaps the most important generalization that can be made about the palynostratigraphy of the Permian is that phytogeographical provinciality is strong throughout the Permian and particularly from the Guadalupian onwards, as predicted by palaeobotanical studies. This makes correlation between regional palynostratigraphic schemes difficult. These differences appear to be more than could be accounted for by differing taxonomic concepts, although these may contribute to the extent of the apparent differences between phytogeographical provinces. For these reasons, it is unlikely that a single comprehensive palynostratigraphic scheme for the Permian globally will ever be developed, or at least not one that contains inherent precision that will be valuable for resolving Earth events. This paper does not, therefore, suggest such a scheme, nor even suggest unified schemes, for the phytogeographical provinces of the Permian.

What are likely to be possible are high-resolution palynostratigraphic schemes for regions (which substantially already exist) that can be linked by tie points. These tie points will be provided by precise assemblage-level quantitative taxonomic comparison or by the use of single well-characterized palynological taxa ('bridging taxa') that occur across Permian phytogeographical provinces.

Perhaps the most important of these taxa for the Permian include (Fig. 11): (1) *Scutasporites* spp., *Vittatina* spp., *Weylandites* spp. and *Lueckisporites virkkiae* (for correlation between the Euramerican and Angaran provinces); and (2) *Vittatina* spp., *Weylandites* spp. and *Lueckisporites virkkiae* (for correlation between the Euramerican and Gondwana provinces). Correlation between the Cathaysian and other phytogeographical provinces remains a problem because few 'bridging taxa' have been identified. This may represent a lack of effort in palynological studies in linking Cathaysian and other phytogeographical provinces, and the isolation of much Chinese palynology in that it is published only in Chinese.

None of the Permian GSSPs involve palynological definitions, which may be problematic given the importance of palynology in correlation in the commercial and academic worlds. However, there appear to be taxa that occur at GSSPs or well-dated boundary sections that could be used to correlate those boundaries. For example, *Aratrisporites* spp. and *Otynisporites eotriassicus* may be useful to correlate the Permian–Triassic boundary into non-marine sections or sections without radiometric dates. *Converrucosisporites confluens* may be useful in correlating the Carboniferous–Permian boundary (Fig. 11).

The Director of the British Geological Survey (NERC) is thanked for permission to publish this paper. Dr Mercedes di Pasquo is acknowledged for providing access to valuable reference materials. Prof. Clinton Foster and an anonymous reviewer provided constructive comments. Dr Alain Le Hérissé (University of Brest), Prof. Hans Egger (Geological Survey of Austria) and Dr Teresa Marcinkiewicz are thanked for providing images of specimens in Figure 11.

Appendix A

Species names and authors

Ahrensisporites cristatus Playford & Powis, 1979
Alisporites cf. *nuthallenis* Clarke, 1965
Alisporites plicatus Jizba, 1962
Alisporites splendens (Jizba) Foster, 1979
Anapiculatisporites concinnus Playford, 1962
Aratrisporites cf. *yunnanensis* Ouyang & Li, 1980
Aratrisporites saharaensis Loboziak *et al.*, 1986
Barakarites rotatus (Balme & Hennelly) Bharadwaj & Tiwari, 1964
Brevitriletes cornutus (Balme & Hennelly) Backhouse, 1991
Cedripites priscus Balme, 1970
Circumstriatites talchirensis Lele & Makada, 1972
Cladaitina kolodae Utting, 1994
Converrucosisporites confluens (Archangelsky & Gamerro) Playford & Dino, 2002
Cordaitina uralensis (Luber) Samoilovich, 1953
Corisaccites alutas Venkatachala & Kar, 1966
Crassispora kosankei (Potonié & Kremp) Bharadwaj, 1957
Crinalites sabinensis Utting, 1994
Crucisaccites ornatus (Samoilovich) Dibner, 1971
Cycadopites cymbatus (Balme & Hennelly) Segroves, 1970
Distriatites insolitus Bharadwaj & Salujah, 1964
Endosporites ornatus Wilson & Coe, 1940
Falcisporites zapfei (Potonié & Klaus) Leschik, 1956
Florinites luberae Samoilovich, 1953
Florinites? balmei Stephenson & Filatoff, 2000
Hamiapollenites bullaeformis (Samoilovich) Jansonius, 1962
Hamiapollenites tractiferinus (Samoilovich) Jansonius, 1962
Horriditriletes ramosus (Balme & Hennelly) Bharadwaj & Salujah, 1964
Horriditriletes tereteangulatus (Balme & Hennelly) Backhouse, 1991
Kingiacolpites subcircularis Tiwari & Moiz, 1971
Klausipollenites schaubergeri (Potonié & Klaus) Jansonius, 1962
Kraeuselisporites wargalensis Balme, 1970
Leiotriletes virkkii Tiwari, 1965
Leptolepidites jonkeri (Jansonius) Yaroshenko & Golubeva, 1991
Limatulasporites fossulatus (Balme) Helby & Foster in Foster, 1979
Limitisporites monstruosus (Luber & Valtz) Hart, 1965*b*
Lueckisporites virkkiae (Potonié & Klaus) emend. Clarke, 1965
Lunatisporites pellucidus (Goubin) Helby ex de Jersey, 1972
Lunatisporites transversundatus (Jansonius) Fischer, 1979
Lycospora pusilla (Ibrahim) Somers, 1972
Microbaculispora tentula Tiwari, 1965
Naumovaspora striata Jansonius, 1962
Otynisporites eotriassicus Fuglewicz, 1977
Platysaccus queenslandi de Jersey, 1962
Playfordiaspora cancellosa (Playford & Dettmann) Maheshwari & Banerji, 1975
Praecolpatites sinuosus (Balme & Hennelly 1956) Bharadwaj & Srivastava, 1969
Protohaploxypinus amplus (Balme & Hennelly) Hart, 1964
Protohaploxypinus limpidus (Balme & Hennelly) Balme & Playford, 1967
Protohaploxypinus perfectus (Naumova ex Kara-Murza) Samoilovich, 1953
Protohaploxypinus uttingii Stephenson & Filatoff, 2000
Pseudoulatispora pseudoreticulata (Balme & Hennelly) Bharadwaj & Srivastava, 1969
Psomospora detecta Playford & Helby, 1968
Pteruchipollenites indarraensis (Segroves) Foster, 1979
Rattiganispora apiculata Playford & Helby emend. Playford, 1986
Reduviasporonites chalastus (Foster) Elsik, 1999
Scutasporites nanuki Utting, 1994
Scutasporites unicus Klaus, 1963
Scutasporites xinjiangensis (Hou & Wang) Ouyang *et al.* 2003
Spelaeotriletes triangulus Neves & Owens, 1966
Spelaeotriletes ybertii (Marques-Toigo) Playford & Powis, 1979
Striasulcites tectus Venkatachala & Kar, 1968
Striatoabieites multistriatus (Balme & Hennelly) Hart, 1964
Striatoabieites richteri (Klaus) Hart, 1964
Striatopodocarpites cancellatus (Balme & Hennelly) Hart, 1964

Striatopodocarpites fusus (Balme & Hennelly) Potonié, 1958
Sverdrupollenites agluatus Utting, 1994
Thymospora opaqua Singh, 1964
Vittatina costabilis Wilson, 1962
Vittatina saccata (Hart) Jansonius, (1962)
Weylandites cincinnatus (Luber ex. Varyukhina) Utting, 1994
Weylandites lucifer (Bharadwaj & Salujah) Foster, 1975
Weylandites striatus (Luber) Utting, 1994

References

AGGARWAL, N. & JHA, N. 2013. Permian palynostratigraphy and palaeoclimate of Lingala-Koyagudem Coalbelt, Godavari Graben, Andhra Pradesh, India. *Journal of Asian Earth Sciences*, **64**, 38–57.

AITKEN, G.A. 1994. Permian palynomorphs from the No. 5 seam, Ecca Group, Witbank/Highvelt Coalfields, South Africa. *Palaeontologia Africana*, **31**, 97–109.

ANDERSON, J.M. 1977. The biostratigraphy of the Permian and Triassic. Part 3. A review of Gondwana Permian palynology with particular reference to the northern Karoo Basin of South Africa. *Memoirs of the Botanical Survey of South Africa*, **41**, 1–133.

ANGIOLINI, L. & STEPHENSON, M.H. 2008. Early Permian brachiopods and palynomorphs from the Dorud Formation, Alborz Mountains, north Iran: evidence for their palaeobiogeographic affinities. *Fossils and Strata*, **54**, 117–132.

ARCHANGELSKY, S. & GAMERRO, J.C. 1979. Palinología del Paleozoico Superior en el subsuelo de la cuenca Chacoparanense, República Argentina 1. Estudio sistemático de los palinomorfos de tres perforaciones de la Provincia de Córdoba. *Revista Española de Micropaleontología*, **11**, 417–478.

ARCHANGELSKY, S. & VERGEL, M. 1996. Paleontología, bioestratigrafía y paleoecología. *In*: *El sistema Pérmico en la Republica Argentina y en la Republica Oriental del Uruguay*. Academia Nacional de Ciencias, 1996, Córdoba, Argentina, 40–44.

ARCHANGELSKY, S. & WAGNER, R.H. 1983. *Glosspteris anatolica sp.* nov. from uppermost Permian strata in south east Turkey. *Bulletin of the British Museum (Natural History), Geology*, **37**, 81–91.

ARCHANGELSKY, S., AZCUY, C.L., PINTO, I.D., GONZÁLEZ, C.R., MARQUES-TOIGO, M., RÖSLER, O. & WAGNER, R.H. 1980. The Carboniferous and Early Permian of the South American Gondwana area: a summary of the biostratigraphic information. *Actas del II Congreso Argentino de Paleontología y Bioestratigrafía y I Congreso Latinoamericano de Paleontología 1978*, **4**, 257–269.

ARCHBOLD, N.W. 1999. Permian Gondwana correlations: the significance of the western Australian marine Permian. *Journal of African Earth Sciences*, **29**, 63–75.

AZCUY, C.L. 1980. A review of the early Gondwana palynology of Argentina and South America. *In*: *Proceedings of the IV International Palynological Conference*, 1976, Lucknow, India, Volume 2. Birbal Sahni Institute, Lucknow, India, 175–185.

AZCUY, C., BERI, Á. *ET AL.* 2007. Bioestratigrafía del Paleozoico Superior de América del Sur: Primera Etapa de Trabajo Hacia una Nueva Propuesta Cronoestratigráfica. *In*: *Comité Organizador de la XI Reunião de Paleobotânicos Paleobotânicos y por el Comité Organizador del XIII Simposio Argentino de Paleobotánica y Palinología (Bahía Blanca)*, 2006, Argentina. Asociación Geológica Argentina Serie D: Special Publication, **11**.

BACKHOUSE, J. 1991. Permian palynostratigraphy of the Collie Basin, Western Australia. *Review of Palaeobotany and Palynology*, **67**, 237–314.

BALARINO, M.L. 2014. Permian palynostratigraphy of the Claromecó Basin, Argentina. *Alcheringa*, **38**, 1–21.

BALME, B.E. 1970. Palynology of Permian and Triassic strata in the Salt Range and Surghar Range, West Pakistan. *In*: KUMMEL, B. & TEICHERT, C. (eds) *Stratigraphic Boundary Problems: Permian and Triassic of West Pakistan*. University Press of Kansas, Department of Geology Special Publications, **4**, 306–453.

BALME, B.E. 1980. Palynology and the Carboniferous–Permian boundary in Australia and other Gondwana countries. *Palynology*, **4**, 43–56.

BALME, B.E. & PLAYFORD, G. 1967. Late Permian plant microfossils from the Prince Charles Mountains, Antarctica. *Revue de Micropaléontologie*, **10**, 179–192.

BANGERT, B. 2000. *Tephrostratigraphy, petrography, geochemistry, age and fossil record of the Ganigobis Shale Member and associated glaciomarine deposits of the Dwyka Group, Late Carboniferous, southern Africa*. Unpublished PhD thesis, Bayerischen Julius-Maximilians-Universität Würzburg, Germany.

BARBOLINI, N. & BAMFORD, M.K. 2014. Palynology of an Early Permian coal seam from the Karoo Supergroup of Botswana. *Journal of African Earth Sciences*, **100**, 136–144.

BERCOVICI, A., DIEZ, J.B. *ET AL.* 2009. A palaeoenvironmental analysis of Permian sediments in Minorca (Balearic Islands, Spain) with new palynological and megafloral data. *Review of Palaeobotany and Palynology*, **158**, 14–28.

BERI, A. & GOSO, C.A. 1996. Análisis palinológico y estratigráfico de la Fm. San Gregorio (Pérmico Inferior) en el área de los Cerros Guazunambi, Cerro Largo, Uruguay. *Revista Española de Micropaleontología*, **28**, 67–79.

BESEMS, R.E. & SCHUURMAN, W.M.L. 1987. Palynostratigraphy of Late Paleozoic glacial deposits of the Arabian Peninsula with special reference to Oman. *Palynology*, **11**, 37–53.

BHARADWAJ, D.C. 1957. The spore flora of the Velener Schichten (Lower Westphalian D) in the Ruhr Coal Measures. *Palaeontographica B*, **102**, 110–138.

BHARADWAJ, D.C. 1969. Lower Gondwana Formations. *In*: *6th Congrès International de Stratigraphie et de Géologie du Carbonifère*. 1967, Sheffield. Compte Rendu, **1**, 255–274.

BHARADWAJ, D.C. & SALUJAH, S.K. 1964. Sporological Study of Seam VIII in Raniganj Coalfield, Bihar, India. Part I: Description of the sporae dispersae. *Palaeobotanist*, **12**, 181–215.

BHARADWAJ, D.C. & SRIVASTAVA, S.C. 1969. Some new miospores from Barakar Stage, Lower Gondwana, India. *Palaeobotanist*, **17**, 220–229.

BHARADWAJ, D.C. & SRIVASTAVA, S.C. 1977. Gondwana palynology – a historical review. *In*: VENKATACHALA, B.S. & SASTRI, V.V. (eds) *IV Colloquim on Indian Micropalaeontology and Stratigraphy 1974–1975*. Proceedings of the Instititute of Petroleum Exploration, Oil and Natural Gas Commission, Dehradun. Institute of Petroleum Exploration, Oil & Natural Gas Commission, Dehradun, 150–156.

BHARADWAJ, D.C. & TIWARI, R.S. 1964. On two monosaccate genera from Barakar Stage of India. *Palaeobotanist*, **12**, 139–146.

BOSE, M.N. & KAR, R.K. 1966. Palaeozoic sporae dispersae from Zaire (Congo) I. Kindu\Kalima and Walikale regions. *Annales Musee Royal de L'Afrique Centrale, Série 8*, **53**, 1–168.

BOSE, M.N. & KAR, R.K. 1976. Mezozoic sporae dispersae from Zaire I. Haute-Lueki Series. *Annales Musee Royal de L'Afrique Centrale, Série 8*, **78**, 1–40.

BOSE, M.N. & MAHESHWARI, H.K. 1968. Paleozoic sporae dispersae from Congo – VII. Coal measures near Lake Tanganyika, south of Albertville. *Annales Musee Royal de L'Afrique Centrale, Série 8*, **60**, 1–116.

BRUGMAN, W.A. & VISSCHER, H. 1988. Permian and Triassic palynostratigraphy of northeast Libya. *In*: EL ARNAUTI, A., OWENS, B. & THUSO, B. (eds) *Subsurface Palynostratigraphy of Northeast Libya*. Garyounis University Publications, Benghazi, Libya, 157–169.

BRUGMAN, W.A., LOBOZIAK, S. & VISSCHER, H. 1988. The problem of the Carboniferous-Permian boundary in north east Libya from a palynological point of view. *In*: EL ARNAUTI, A., OWENS, B. & THUSO, B. (eds) *Subsurface Palynostratigraphy of Northeast Libya*. Garyounis University Publications, Benghazi, Libya, 151–155.

CÉSARI, S.N. 2007. Palynological biozones and radiometric data at the Carboniferous–Permian boundary in western Gondwana. *Gondwana Research*, **11**, 529–536.

CÉSARI, S.N. & GUTIÉRREZ, P.R. 2000. Palynostratigraphy of Upper Palaeozoic sequences in central-western Argentina. *Palynology*, **24**, 113–146.

CÉSARI, S.N., LIMARINO, C.O. & GULBRANSEN, E.L. 2011. An Upper Paleozoic biochronostratigraphic scheme for the western margin of Gondwana. *Earth-Science Reviews*, **106**, 149–160.

CISTERNA, G.S., STERREN, A.F. & GUTIÉRREZ, P. 2011. The Carboniferous-Permian boundary in the central western Argentinean basins: paleontological evidences. *Andean Geology*, **38**, 349–370.

CLAPHAM, W.B. 1970. Permian miospores from the Flowerpot Formation of Western Oklahoma. *Micropaleontology*, **16**, 15–36.

CLARKE, R.F.A. 1965. British Permian saccate and monosulcate miospores. *Palaeontology*, **8**, 322–354.

CLAYTON, G., COQUEL, R., DOUBINGER, J., GUEINN, K.J., LOBOZIAK, S., OWENS, B. & STREEL, M. 1977. Carboniferous miospores of Western Europe: illustration and zonation. *Mededilingen Rijks Geologische Dienst*, **29**, 1–71.

CLEAL, C.J. & THOMAS, B.A. 1991. Carboniferous and Permian palaeogeography. *In*: CLEAL, C.J. (ed.) *Plant Fossils in Geological Investigation: The Palaeozoic*. Ellis Horwood, Chichester, 154–181.

DAVYDOV, V.I., GLENISTER, B.F., SPINOSA, C., RITTER, S.C., CHERNYKH, V.V., WARDLAW, B.R. & SNYDER, W.S. 1998. Proposal of Aidaralash creek as Global Stratotype Section and Point (GSSP) for base of the Permian System. *Episodes*, **21**, 11–18.

DE JERSEY, N.J. 1962. Triassic spores and pollen from the Ipswich Coalfield. *Geological Survey of Queensland Publications*, **307**, 1–18.

DE JERSEY, N.J. 1972. Triassic miospores from the Esk beds. *Geological Survey of Queensland Publications*, **357**, 1–40.

DIBNER, A.F. 1971. Cordaitales pollen of Angaraland. *Uchenye Zapiski, Nauchno Issledovatelski Instituti Geologii Arctiki, Paleontologia i Bioestratigrafia*, **32**, 5–66 (In Russian with English summary).

DINO, R. & PLAYFORD, G. 2002. Miospores common to South American and Australian Carboniferous sequences: stratigraphic and phytogeographic implications. *In*: HILLS, L.V., HENDERSON, C.M. & BAMBER, E.W. (eds) *Carboniferous and Permian of the World*. Canadian Society of Petroleum Geologists, Memoirs, **19**, 336–359.

DI PASQUO, M. 2003. Avances sobre palinología, bioestratigrafía y correlación de los Grupos Macharetí y Mandiyutí Neopaleozoico de la Cuenca Tarija, Provincia de Salta, Argentina. *Ameghiniana*, **40**, 3–32.

DI PASQUO, M., AZCUY, C.L. & SOUZA, P.A. 2003. Palinología del Carbonífero Superior del Subgrupo Itararé en Itaporanga, Cuenca Paraná, Estado de São Paulo, Brasil. Parte 2: sistemática de polen y significado paleoambiental y estratigráfico. *Ameghiniana*, **40**, 277–296.

DI PASQUO, M., GRADER, G.W., ISAACSON, P., SOUZA, P.A., IANNUZZI, R. & DÍAZ- MARTÍNEZ, E. 2015. Global biostratigraphic comparison and correlation of an early Cisuralian palynoflora from Bolivia. *Historical Biology*, **27**, 868–897.

DOUBINGER, J. 1974. Études palynologiques dans L'Autunien. *Review of Palaeobotany and Palynology*, **17**, 21–38.

DOUBINGER, J., ODIN, B. & CONRAD, G. 1987. Les associations sporopolliniques du Permien continental du bassin de Lodève (Hérault, France): caracterisation de L'Autunien supérieur, du 'Saxonien' et du Thuringien. *Annales de la Société Géologique du Nord*, **106**, 103–109.

DRAXLER, I. 2010. Upper Permian spore holotypes of W. Klaus from the Southern Alps (Dolomites, Italy) in the collections of the Geological Survey of Austria. *Jahrbuch der Geologischen Bundesanstalt*, **150**, 85–99.

DUNN, M.T. 2001. Palynology of the Carboniferous – Permian boundary stratotype, Aidaralash Creek, Kazakhstan. *Review of Palaeobotany and Palynology*, **116**, 175–194.

EDWARDS, R.A., WARRINGTON, G., SCRIVENER, S.C., JONES, N.S., HASLAM, H.W. & AULT, L. 1997. The Exeter Group, south Devon, England: a contribution to the early post-Variscan stratigraphy of northwest Europe. *Geological Magazine*, **134**, 177–197.

ELSIK, W.C. 1999. *Reduviasporonites* Wilson 1962: synonymy of the fungal organism involved in the Late Permian crisis. *Palynology*, **23**, 37–41.

ESHET, Y. 1990. *Paleozoic–Mezozoic Palynology of Israel. I: Palynological Aspects of the Permo-Trias*

Succession in the Subsurface of Israel. Geological Survey of Israel Bulletin, **81**, 1–57.

EVANS, P.R. 1969. Upper Carboniferous and Permian palynological stages and their distribution in eastern Australia. *In*: *Gondwana Stratigraphy: IUGS Symposium*, 1976, Argentina. *UNESCO, Earth Sciences*, **2**, 41–54.

FADDEEVA, I.Z. 1980. Regularities in the changes of miospore complexes in stratotype sections of the East European platform and the Urals. *In*: *Proceedings of the Fourth International Palynological Conference*, 1976–1977, Lucknow, Volume 2. Birbal Sahni Institute of Palaeobotany, Lucknow, India, 844–848.

FALCON, R.M.S. 1975. Palynostratigraphy of the Lower Karroo sequence in Sebungwe District, Mid Zambezi Basin, Rhodesia. *Palaeontologia Africana*, **18**, 1–29.

FALCON, R.M.S., PINHEIRO, H. & SHEPERD, P. 1984. The palynobiostratigraphy of the major coal seams in the Witbank Basin with lithostratigraphic, chronostratigraphic and palaeoclimatic implications. *Comunicações dos Serviços Geológicos de Portugal*, **70**, 215–243.

FARABEE, M.J., TAYLOR, E.L. & TAYLOR, T.N. 1990. Correlation of Permian and Triassic palynomorph assemblages from the central Transantarctic Mountains, Antarctica. *Review of Palaeobotany and Palynology*, **65**, 257–265.

FIELDING, C.R., FRANK, T.D., BIRGENHEIER, L.P., RYGEL, M.C., JONES, A.T. & ROBERTS, J. 2008. Stratigraphic record and facies associations of the late Paleozoic ice age in eastern Australia (New South Wales and Queensland). *In*: FIELDING, C.R., FRANK, T.D. & ISBELL, J.L. (eds) *Resolving the Late Paleozoic Ice Age in Time and Space*. Geological Society of America, Special Papers, **441**, 41–57.

FISCHER, M.J. 1979. The Triassic palynofloral succession in the Canadian Arctic Archipelago. *American Association of Stratigraphic Palynologists Contributions Series*, **5B**, 83–100.

FOSTER, C.B. 1975. Permian plant microfossils from the Blair Atholl Coal Measures Central Queensland, Australia. *Palaeontographica B*, **154**, 121–171.

FOSTER, C.B. 1979. *Permian Plant Microfossils of the Blair Atholl Coal Measures, Baralaba Coal Measures and Basal Rewan Formation of Queensland*. Geological Survey of Queensland Publications, **372**.

FOSTER, C.B. & AFONIN, S.A. 2005. Abnormal pollen grains: an outcome of deteriorating atmospheric conditions around the Permian–Triassic boundary. *Journal of the Geological Society, London*, **162**, 653–659, https://doi.org/10.1144/0016-764904-047

FOSTER, C.B. & AFONIN, S.A. 2006. Syndesmorion gen. nov. – a coenobial alga of Chlorococcalean affinity from the continental Permian–Triassic deposits of Dalongkou section, Xinjiang Province, China. *Review of Palaeobotany and Palynology*, **6**, 1–8.

FOSTER, C.B. & ARCHBOLD, N.W. 2001. Chronologic anchor points for the Permian Early Triassic of the eastern Australian basins. *In*: WEISS, R.H. (ed.) *Contributions to Geology and Palaeontology of Gondwana in Honour of Helmut Wopfner*. Geological Institute, University of Cologne, Cologne, Germany, 175–199.

FOSTER, C.B. & WATERHOUSE, J.B. 1988. The Granulatisporites confluens Oppel Zone and Early Permian marine faunas from the Grant Formation on the Barbwire Terrace, Canning Basin, Australia. *Australian Journal of Earth Sciences*, **35**, 135–157.

FUGLEWICZ, R. 1977. New species of megaspores from the Trias of Poland. *Acta Palaeontologica Polonica*, **22**, 405–431.

GLENISTER, B.F., BAKER, C., FURNISH, W.M. & DICKENS, J.M. 1990. Late Permian ammonoid cephalopod from Western Australia. *Journal of Palaeontology*, **64**, 399–402.

GOMANKOV, A.V. 1992. The interregional correlation of the Tatarian and the problem of the Permian upper boundary. *International Geology Review*, **34**, 1015–1020.

GOMANKOV, A.V., BALME, B.E. & FOSTER, C.B. 1998. Tatarian palynology of the Russian Platform: a review. *Proceedings of the Royal Society of Victoria*, **110**, 115–135.

GÖTZ, A.E. & RUCKWIED, K. 2014. Palynological records of the Early Permian postglacial climate amelioration (Karoo Basin, South Africa). *Palaeobiodiversity and Palaeoenvironments*, **94**, 229–235.

GRADSTEIN, S.R. & KERP, H. 2012. A brief history of plants on Earth. *In*: GRADSTEIN, F.M., OGG, J.G., SCHMITZ, M. & OGG, G. (eds) *The Geologic Time Scale 2012*. Elsevier, Amsterdam, 233–237.

GREBE, H. & SCHWEITZER, H.-J. 1962. Die sporae dispersae des niederrheinischen Zechsteins. *Fortschritte der Geologie von Rheinland und Westfalen, Vorausdruck*, **10**, 1–24.

GUERRA-SOMMER, M., CAZZULO-KLEPZIG, M., MENEGAT, R., FORMOSO, M.L.L., BASEI, M.A.S., BARBOZA, E.G. & SIMAS, M.W. 2008. Geochronological data from the Faxinal coal succession, southern Parana Basin, Brazil: a preliminary approach combining radiometric U–Pb dating and palynostratigraphy. *Journal of South American Earth Sciences*, **25**, 246–256.

HART, G.F. 1964. A review of the classification and distribution of the Permian miospore: Disaccate Striatiti. *5th International Congress of Stratigraphy and Geology of the Carboniferous, Comptes Rendus, Paris 1963*, **3**, 1171–1199.

HART, G.F. 1965*a*. Microflora from the Ketewaka–Mchuchuma Coalfield, Tanganyika. *Bulletin of the Geological Survey of Tanganyika*, **36**, 1–27.

HART, G.F. 1965*b*. *The systematics and distribution of Permian miospores*. Witwatersrand University Press, Johannesburg.

HART, G.F. 1969. The stratigraphic subdivision and equivalents of the Karroo sequence as suggested by palynology. *In*: *IUGS Gondwana Symposium Proceedings*, 1967, Buenos Aires, **1**, 23–35.

HART, G.F. 1970. The biostratigraphy of Permian palynofloras. *Geoscience and Man*, **1**, 89–131.

HENDERSON, C.H., DAVYDOV, V. & WARDLAW, B. 2012. The Permian period. *In*: GRADSTEIN, F.M., OGG, J.G., SCHMITZ, M. & OGG, G. (eds) *The Geologic Time Scale 2012*. Elsevier, Amsterdam, 653–679.

HERMANN, E., HOCHULI, P.A., BUCHER, H. & ROOHI, G. 2012. Uppermost Permian to Middle Triassic palynology of the Salt Range and Surghar Range, Pakistan. *Review of Palaeobotany and Palynology*, **169**, 61–95.

HOCHULI, P.A. 1985. Palynostratigraphische gleiderung und korrelation des Permo-Karbon der Nordoschweiz. *Eclogae Geologicae Helvetiae*, **78**, 719–831.

INOSOVA, K.I., KRUSINA, A.K. & SHVARTSMAN, E.G. 1975. The scheme of distribution of characteristic genera and species of microspores and pollen in Carboniferous and Lower Permian deposits. *In*: *Field Excursion Guidebook for the Donets Basin, Ministry of Geology of the Ukrainian SSR*. Nauka, Moscow [in Russian & English].

ISBELL, J.L., MILLER, M.F., WOLFE, K.L. & LENAKER, P.A. 2003. Timing of late Paleozoic glaciation in Gondwana: was glaciation responsible for the development of northern hemisphere cyclothems? *In*: CHAN, M.A. & ARCHER, A.W. (eds) *Extreme Depositional Environments: Mega End Members in Geologic Time*. Geological Society of America, Special Papers, **370**, 5–24.

JAN, I.U., STEPHENSON, M.H. & KHAN, F.R. 2009. Palynostratigraphic correlation of the Sardhai Formation (Permian) of Pakistan. *Review of Palaeobotany and Palynology*, **156**, 402–442.

JANSONIUS, J. 1962. Palynology of Permian and Triassic sediments, Peace River Area, Western Canada. *Palaeontographica B*, **110**, 35–98.

JERZYKIEWICZ, J. 1987. Latest Carboniferous (Stephanian) and Early Permian (Autunian) palynological assemblages from the Intrasudetic Basin, southwestern Poland. *Palynology*, **11**, 117–131.

JHA, N., SABINA, K.P., AGGARWAL, N. & MAHESH, S. 2014. Late Permian Palynology and depositional environment of Chintalapudi sub basin, Pranhita–Godavari basin, Andhra Pradesh, India. *Journal of Asian Earth Sciences*, **79**, 382–399.

JIN, Y., WARDLAW, B.R., GLENISTER, B.F. & KOTLYAR, G.V. 1997. Permian chronostratigraphic subdivisions. *Episodes*, **20**, 10–15.

JIZBA, K.M.M. 1962. Late Palaeozoic bisaccate pollen from the United States midcontinent area. *Journal of Paleontology*, **36**, 871–887.

JONES, M.J. & TRUSWELL, E.M. 1992. Late Carboniferous and Early Permian palynostratigraphy of the Joe Joe Group, southern Galilee Basin, Queensland, and implications for Gondwana stratigraphy. *Bureau of Mines and Mineral Resources Journal of Australian Geology and Geophysics*, **13**, 143–185.

KAISER, H. 1976. Die permische Mikroflora der Cathaysia-schichten von nordwest-Schansi, China. *Palaeontographica B*, **159**, 83–157.

KAR, R.K. & JAIN, K.P. 1975. Libya, a probable part of Gondwanaland. *Geophytology*, **5**, 72–80.

KEMP, E.M. 1973. Permian flora from the Beaver Lake area, Prince Charles Mountains, Antarctica 1. Palynological examination of samples. *Bulletin of the Bureau of Mineral Resources, Geology and Geophysics, Australia*, **126**, 7–12.

KEMP, E.M. 1975. Palynology of Late Palaeozoic glacial deposits of Gondwanaland. *In*: CAMPBELL, K.W. (ed.) *Gondwana Geology*. Australian National University Press, Canberra, 125–134.

KEMP, E.M., BALME, B.E., HELBY, R.J., KYLE, R.A., PLAYFORD, G. & PRICE, P.L. 1977. Carboniferous and Permian palynostratigraphy in Australia and Antarctica: a review. *Bureau of Mines and Mineral Resources Journal of Australian Geology and Geophysics*, **2**, 177–208.

KILANI-MAZRAOUI, F., RAZGALLAH-GARGOURI, S. & MANNAI-TAYECH, B. 1990. The Permo-Triassic of Southern Tunisia – Biostratigraphy and palaeoenvironment. *Review of Palaeobotany and Palynology*, **66**, 273–291.

KLAUS, W. 1963. Sporen aus dem südalpinen Perm. *Jahrbuch der Geologischen Bundesanstalt, Wien*, **106**, 229–363.

KYLE, R.A. 1977. Palynostratigraphy of the Victoria Group of south Victoria Land, Antarctica. *New Zealand Journal of Geology and Geophysics*, **20**, 1081–1102.

LARRSSON, K., LINDSTRÖM, S. & GUY-OHLSON, D. 1990. An early Permian palynoflora from Milorgfjella, Dronning Maud Land, Antarctica. *Antarctic Science*, **2**, 331–344.

LEI, Y., SERVAIS, T. & FENG, Q. 2013. The diversity of the Permian phytoplankton. *Review of Palaeobotany and Palynology*, **198**, 145–161.

LELE, K.M. & MAKADA, R. 1972. Studies in the Talchir flora of India – 7. Palynology of the Talchir Formation in the Jayanti coalfield, Bihar. *Geophytology*, **2**, 41–73.

LEONOVA, T.B. 1998. Permian ammonoids of Russia and Australia. *Proceedings of the Royal Society of Victoria*, **110**, 157–162.

LESCHIK, G. 1956. Sporen aus dem Salzton des Zechsteins von Neuhof (bei Fulda). *Palaeontographica B*, **100**, 122–142.

LINDSTRÖM, S. 1995*a*. Early Late palynostratigraphy and palaeo-biogeography of Vestfjella, Dronning Maud Land, Antarctica. *Review of Palaeobotany and Palynology*, **86**, 157–173.

LINDSTRÖM, S. 1995*b*. Early Permian palynostratigraphy of the Northern Heimefrontfjella Mountain Range, Dronning Maud Land, Antarctica. *Review of Palaeobotany and Palynology*, **89**, 359–415.

LINDSTRÖM, S. & MCLOUGHLIN, S. 2007. Synchronous palynofloristic extinction and recovery after the end-Permian event in the Prince Charles Mountains, Antarctica: implications for palynofloristic turnover across Gondwana. *Review of Palaeobotany and Palynology*, **145**, 89–122.

LIU, F., ZHU, H. & OUYANG, S. 2008. Late Carboniferous–Early Permian palynology of Baode (Pao-te-chou) in Shanxi Province, North China. *Geological Journal*, **43**, 487–510.

LIU, F., ZHU, H. & OUYANG, S. 2011. Taxonomy and biostratigraphy of Pennsylvanian to Late Permian megaspores from Shanxi, North China. *Review of Palaeobotany and Palynology*, **165**, 135–153.

LOBOZIAK, S., CLAYTON, G. & OWENS, B. 1986. Aratrisporites saharaensis sp. nov.: a characteristic Lower Carboniferous miospore species of North Africa. *Geobios*, **19**, 497–503.

LOOY, C.V., TWITCHETT, R.J., DILCHER, D.L., VAN KONIJNENBURG-VAN CITTERT, J.H.A. & VISSCHER, H. 2001. Life in the end-Permian dead zone. *Proceedings of the National Academy of Sciences of the United States of America*, **98**, 7879–7883.

LOVE, C.F. 1994. The palynostratigraphy of the Haushi Group (Westphalian–Artinskian) in Oman. *In*: SIMMONS, M.D. (ed.) *Micropalaeontology and Hydrocarbon Exploration in the Middle East*. Chapman & Hall, London, 23–39.

MACRAE, C.S. 1988. *Palynostratigraphical Correlation Between the Lower Karoo Sequence of the Waterburg and Pafuri Coal Basins and the Hammanskraal Plant*

Macrofossil Locality, RSA. Memoirs of the Geological Survey of South Africa, **75**.

Maheshwari, H.K. & Banerji, J. 1975. Lower Triassic palynomorphs from the Maitur Formation, West Bengal, India. *Palaeontographica B*, **152**, 149–190.

Mangerud, G. 1994. Palynostratigraphy of the Permian and lowermost Triassic succession, Finnmark Platform, Barents Sea. *Review of Palaeobotany and Palynology*, **82**, 317–349.

Mangerud, G. & Konieczny, R.M. 1993. Palynology of the Permian succession of Spitsbergen, Svalbard. *Polar Research*, **12**, 65–93.

Marcinkiewicz, T., Fijalkowska-Mader, A. & Pienkowski, G. 2014. Megaspore zones of the epicontinental Triassic and Jurassic deposits in Poland – overview. *Biuletyn Pañstwowego Instytutu Geologicznego*, **457**, 15–42.

Marques-Toigo, M. 1974. Some new species of spores and pollens of Lower Permian age from the San Gregorio Formation in Uruguay. *Anais da Academia Brasiliera de Ciêncas*, **46**, 601–616.

Metcalfe, I., Foster, C.B., Afonin, S.A., Nicoll, R.S., Mundil, R., Xiaofeng, W. & Lucas, S.G. 2009. Stratigraphy, biostratigraphy and C-isotopes of the Permian–Triassic non-marine sequence at Dalongkou and Lucaogou, Xinjiang Province, China. *Journal of Asian Earth Sciences*, **36**, 503–520.

Meyen, S.V. 1969. New data on relationship between Angara and Gondwana Late Palaeozoic floras. *In*: *Gondwana Stratigraphy, IUGS Symposium*, 1967, Buenos Aires. UNESCO, Paris, 141–157.

Millsteed, B.D. 1993. Palynological evidence for the age of the Permian Karroo coal deposits near Vereeniging, Northern Orange Free State, South Africa. *South African Journal of Geology*, **97**, 15–20.

Millsteed, B.D. 1999. Palynology of the Early Permian coal-bearing deposits near Vereeniging, Free State, South Africa. *Bulletin of the Council for Geoscience*, **124**, 1–77.

Modie, B.N. & Le Hérissé, A. 2009. Late Palaeozoic palynomorph assemblages from the Karoo Supergroup and their potential for biostratigraphic correlation, Kalahari Karoo Basin, Botswana. *Bulletin of Geosciences*, **84**, 337–358.

Mori, A.L.O., Souza, P.A., Marques, J.C. & Lopes, R.C. 2012. A new U–Pb zircon age dating and palynological data from a Lower Permian section of the southernmost Paraná Basin, Brazil: biochronostratigraphical and geochronological implications for Gondwanan correlations. *Gondwana Research*, **21**, 654–669.

Nader, A.D., Khalaf, F.H. & Hadid, A.A. 1993*a*. Palynology of the Permo-Triassic boundary in Borehole Mityah-1, south west Mosul City, Iraq. *Mu'tah Journal Research and Studies*, **8**, 223–280.

Nader, A.D., Khalaf, F.H. & Yousif, R.A. 1993*b*. Palynology of the upper part of the Ga'ara Formation in the western Iraqi desert. *Mu'tah Journal Research and Studies*, **8**, 77–137.

Neves, R. & Owens, B. 1966. Some Namurian camerate miospores from the English Pennines. *Pollen Spores*, **8**, 337–360.

Nilsson, I., Mangerud, G. & Mørk, A. 1996. Permian stratigraphy of the Svalis Dome, south-western Barents Sea. *Norsk Geologisk Tidsskrift*, **76**, 127–146.

Ouyang, S. 1982. Upper Permian and Lower Triassic palynomorphs from eastern Yunnan, China. *Canadian Journal of Earth Sciences*, **19**, 68–80.

Ouyang, S. & Li, Z. 1980. *Upper Carboniferous Spores from Shuo Xian, Northern Shanxi*. Papers for the 5th International Palynological Conference, Nanjing Institute of Geology and Paleontology, Academia Sinica, Special Publication, 1–16.

Ouyang, S. & Utting, J. 1990. Palynology of Upper Permian and Lower Triassic rocks, Meishan, Changxing County, Zhejiang Province, China. *Review of Palaeobotany and Palynology*, **66**, 65–103.

Ouyang, S., Wang, Z., Zhan, J.Z. & Zhou, Y.X. 2003. *Palynology of Carboniferous–Permian Strata in Northern Xinjiang*. University of Science and Technology of China Press, Hefei, Anhui (in Chinese with English summary).

Pattison, J., Smith, D.B. & Warrington, G. 1973. A review of Late Permian and Early Triassic biostratigraphy in the British Isles. *In*: Logan, A. & Hills, L.V. (eds) *The Permian and Triassic Systems and their Mutual Boundary*. Canadian Society of Petroleum Geologists, Memoirs, **2**, 220–260.

Penney, R.A., Al Barram, I. & Stephenson, M.H. 2008. A high resolution palynozonation for the Al Khlata Formation (Pennsylvanian to Lower Permian), South Oman. *Palynology*, **32**, 213–231.

Pérez Loinaze, V.S. & Césari, S.N. 2004. Palynology of the Estratos de Mascasín, Upper Carboniferous. Paganzo Basin, Argentina: systematic descriptions and stratigraphic considerations. *Revista Española de Micropaleontología*, **36**, 407–438.

Piasecki, S. 1984. Preliminary palynostratigraphy of the Permian-Lower Triassic sediments in Jameson Land and Scoresby Land, East Greenland. *Bulletin of the Geological Society of Denmark*, **32**, 139–144.

Playford, G. 1962. Lower Carboniferous microfloras of Spitzbergen. *Palaeontology*, **5**, 550–618.

Playford, G. 1986. Morphological and preservational variation of *Rattiganispora* Playford and Helby, 1968, from the Australian Carboniferous. *Pollen Spores*, **28**, 83–96.

Playford, G. & Dino, R. 2000. Palynostratigraphy of the upper Palaeozoic strata (Tapajós Group) Amazonas Basin, Brazil: Part Two. *Palaeontographica B*, **255**, 87–145.

Playford, G. & Dino, R. 2002. Permian palynofloral assemblages of the Chaco-Paraná Basin, Argentina: systematics and stratigraphic significance. *Revista Española de Micropaleontología*, **34**, 235–288.

Playford, G. & Dino, R. 2005. Carboniferous and Permian Palynostratigraphy. *In*: Koutsoukos, E. (ed.) *Applied Stratigraphy*. Springer, Dordrecht, The Netherlands, 101–121.

Playford, G. & Helby, R. 1968. Spores from a Carboniferous section in the Hunter Valley, NSW. *Journal of the Geological Society of Australia*, **15**, 103–119.

Playford, G. & Powis, G.D. 1979. Taxonomy and distribution of some trilete spores in Carboniferous strata of the Canning Basin, Western Australia. *Pollen Spores*, **21**, 371–394.

Potonié, R. 1958. Synopsis der Gattungen der Sporae dispersae. II. Teil: Sporites (Nachträge), Saccites, Aletes,

Praecolpates, Polyplicates, Monocolpates. *Beihefte zum Geologischen Jahrbuch*, **31**, 1–114.

POTONIÉ, R. & KLAUS, W. 1954. Einige sporegattungen des alpinen Salzgebirges. *Geologisches Jahrbuch*, **68**, 517–546.

PRICE, P.L. 1983. A Permian palynostratigraphy for Queensland. *In*: FOSTER, C.B. (ed.) *Permian Geology of Queensland*. Geological Society of Australia, Brisbane, 155–211.

PRICE, P.L. 1997. Permian to Jurassic palynostratigraphic nomenclature of the Bowen and Surat basins. *In*: GREEN, P. (ed.) *The Surat and Bowen Basins, Southeast Queensland*. Queensland Department of Mines and Energy, Brisbane, 137–178.

RAWSON, P.F., ALLEN, P.M. *ET AL*. 2002. *Stratigraphical Procedure*. Geological Society, London, Professional Handbook.

RUCKWIED, K., GÖTZ, A.E. & JONES, P. 2014. Palynological records of the Permian Ecca Group (South Africa): utilizing climatic icehouse–greenhouse signals for cross basin correlations. *Palaeogeography, Palaeoclimatology, Palaeoecology*, **413**, 167–172.

SAMOILOVICH, S.R. 1953. Pollen and spores from the Permian deposits of the Cherdya and Aktyubinsk areas, Cis-Urals. *Oklahoma Geological Survey Circular*, **56**, 1–103.

SCHAARSCHMIDT, F. 1963. Spores und hystricosphaerideen aus dem Zechstein von Büdingen in der Wetterau. *Palaeontographica B*, **113**, 38–91.

SCHAARSCHMIDT, F. 1980. Pollen flora and vegetation at the end of the Lower Permian. *In*: *Proceedings of the IV International Palynological Conference*, 1976–77, Lucknow, India, Volume **2**. Birbal Sahni Institute, Lucknow, 750–752.

SCHWEITZER, H.-J. 1986. The land flora of the English and German Zechstein sequences. *In*: HARWOOD, G.M. & SMITH, D.B. (eds) *The English Zechstein and Related Topics*. Geological Society, London, Special Publications, **22**, 31–54, https://doi.org/10.1144/GSL.SP.1986.022.01.04

SEGROVES, K.L. 1970. Permian spores and pollen grains from the Perth Basin, Western Australia. *Grana*, **10**, 43–73.

SINGH, H.P. 1964. A miospore assemblage from the Permian of Iraq. *Palaeontology*, **7**, 240–265.

SMITH, D.B., BRUNSTROM, R.G.W., MANNING, P.I., SIMPSON, S. & SHOTTON, F.W. 1974. *A Correlation of the Permian Rocks of the British Isles*. Geological Society, London, Special Reports, **5**.

SOMERS, Y. 1972. Révision du genre *Lycospora* Schopf, Wilson & Bentall – Microfossiles Organiques du Paléozoïque. 5, Spores. *Centre National de la Recherche Scientifique*, 11–110.

SOUZA, P.A. 2006. Late Carboniferous palynostratigraphy of the Itararé Subgroup northeastern Paraná Basin, Brazil. *Review of Palaeobotany and Palynology*, **138**, 9–29.

SOUZA, P.A. & MARQUES-TOIGO, M. 2003. An overview on the palynostratigraphy of the Upper Paleozoic strata of the Brazilian Paraná Basin. *Revista del Museo Argentino Ciencias Naturales, Nueva Serie*, **5**, 205–214.

SOUZA, P.A., PETRI, S. & DINO, R. 2003. Late Carboniferous palynology from the Itararé Subgroup (Paraná Basin) at Araçoiba da Serra, São Paulo State, Brazil. *Palynology*, **27**, 39–74.

SPINA, A., CIRILLI, S., UTTING, J. & JANSONIUS, J. 2015. Palynology of the Permian and Triassic of the Tesero and Bulla sections (Western Dolomites, Italy) and consideration about the enigmatic species Reduviasporonites chalastus. *Review of Palaeobotany and Palynology*, **218**, 3–14.

SRIVASTAVA, S.C. & BHATTACHARYYA, A.P. 1996. Palynofloral assemblages from Permian sediments, West Siang District, Araunachal Pradesh, India. *In*: *Gondwana Nine: Ninth International Gondwana Symposium*, 10–14 January 1994, Hyderabad, India. A.A. Balkema, Rotterdam, The Netherlands, 261–268.

STEPHENSON, M.H. 2006. Stratigraphic Note: update of the standard Arabian Permian palynological biozonation; definition and description of OSPZ5 and 6. *GeoArabia*, **11**, 173–178.

STEPHENSON, M.H. 2008*a*. Review of the palynological biostratigraphy of Gondwanan Late Carboniferous to Early Permian glacigene successions. *In*: FIELDING, C.R., FRANK, T.D. & ISBELL, J.L. (eds) *Resolving the Late Paleozoic Ice Age in Time and Space*. Geological Society of America, Special Papers, **441**, 304–320.

STEPHENSON, M.H. 2008*b*. Spores and pollen from the Middle and Upper Gharif members (Permian) of Oman. *Palynology*, **32**, 157–183.

STEPHENSON, M.H. 2009. The age of the Carboniferous–Permian Converrucosisporites confluens Oppel Biozone: new data from the Ganigobis Shale Member, Dwyka Group, Namibia. *Palynology*, **33**, 55–63.

STEPHENSON, M.H. 2015. Bisaccate pollen from the Early Permian OSPZ3a Sub-Biozone of the Lower Gharif Member, Oman. *Review of Palaeobotany and Palynology*, **212**, 214–225.

STEPHENSON, M.H. & MCLEAN, D. 1999. International correlation of Early Permian palynofloras from the Karoo sediments of Morupule, Botswana. *South African Journal of Geology*, **102**, 3–14.

STEPHENSON, M.H. & OSTERLOFF, P.L. 2002. *Palynology of the Deglaciation Sequence Represented by the Lower Permian Rahab and Lower Gharif Members, Oman*. American Association of Stratigraphic Palynologists, Contribution Series, **40**.

STEPHENSON, M.H. & POWELL, J.H. 2013. Palynology and alluvial architecture in the Permian Umm Irna Formation, Dead Sea, Jordan. *GeoArabia*, **18**, 17–60.

STEPHENSON, M.H. & POWELL, J.H. 2014. Selected spores and pollen from the Permian Umm Irna Formation, Jordan, and their stratigraphic utility in the Middle East and North Africa. *Rivista Italiana di Paleontologia e Stratigrafia*, **120**, 145–156.

STEPHENSON, M.H., OSTERLOFF, P.L. & FILATOFF, J. 2003. Palynological biozonation of the Permian of Oman and Saudi Arabia: progress and challenges. *GeoArabia*, **8**, 467–496.

STEPHENSON, M.H., ANGIOLINI, L. & LENG, M.J. 2007. The Early Permian fossil record of Gondwana and its relationship to deglaciation: a review. *In*: WILLIAMS, M., HAYWOOD, A.M., GREGORY, F.J. & SCHMIDT, D.N. (eds) *Deep-Time Perspective on Climate Change:*

Marrying the Signal from Computer Models and Biological Proxies. The Micropalaeontological Society, Special Publications. Geological Society, London, 103–122.

Stephenson, M.H., Al Rawahi, A. & Casey, B. 2008*a*. Correlation of the Al Khlata Formation in the Mukhaizna Field, Oman, based on a new downhole, cuttings-based palynostratigraphic scheme. *GeoArabia*, **13**, 45–75.

Stephenson, M.H., Angiolini, L. *et al*. 2008*b*. Abrupt environmental and climatic change during the deposition of the Early Permian Haushi limestone, Oman. *Palaeogeography, Palaeoclimatology, Palaeoecology*, **270**, 1–18.

Stephenson, M.H., Jan, I.U. & Al-Mashaikie, S.Z.A.K. 2013. Palynology and correlation of Carboniferous–Permian glacigene rocks in Oman, Yemen and Pakistan. *Gondwana Research*, **24**, 203–211.

Stolle, E. 2007. Regional Permian Palynological correlations: southeast Turkey–northern Iraq. *Comunicações Geológicas*, **94**, 125–143.

Stump, T.E. & van der Eem, J.G. 1995. The stratigraphy, depositional environments and periods of deformation of the Wajid outcrop belt, southwestern Saudi Arabia. *Journal of African Earth Sciences*, **21**, 421–441.

Sullivan, H.J. 1965. Palynological evidence concerning the regional differentiation of Upper Mississippian floras. *Pollen Spores*, **7**, 539–563.

Tewari, R., Ram-Awatar, *et al*. 2015. The Permian–Triassic palynological transition in the Guryul Ravine section, Kashmir, India: implications for Tethyan–Gondwanan correlations. *Earth-Science Reviews*, **149**, 53–66.

Tiwari, R.S. 1965. Miospore assemblage in some coals of the Barakar Stage (Lower Gondwana) of India. *Palaeobotanist*, **13**, 168–214.

Tiwari, R.S. 1996. Palynoevent stratigraphy in Gondwana sequence of India. *In*: *Gondwana Nine: Ninth International Gondwana Symposium*, 10–14 January 1994, Hyderabad, India. A.A. Balkema, Rotterdam, The Netherlands, 3–19.

Tiwari, R.S. & Moiz, A.A. 1971. Palynological study of Lower Gondwana (Permian) coals from the Godavari Basin, India. *Palaeobotanist*, **19**, 95–104.

Tiwari, R.S. & Tripathi, A. 1992. Marker assemblage zones of spore and pollen species through the Gondwana Palaeozoic and Mezozoic sequence in India. *Palaeobotanist*, **40**, 194–236.

Truswell, E.M. 1980. Permo-Carboniferous palynology of Gondwanaland: progress and problems in the decade of 1980. *Bureau of Mines and Mineral Resources Journal of Australian Geology and Geophysics*, **5**, 95–111.

Tschudy, R.H. & Kosanke, R.M. 1966. Early Permian vesiculate pollen from Texas, USA. *Palaeobotanist*, **15**, 59–71.

Turnau, E. 1978. Spore zonation of uppermost Devonian and Lower Carboniferous deposits of western Pomerania. *Mededelingen rijks geologische dienst*, **30**, 1–35.

Twitchett, R.J., Looy, C.V., Morante, R., Visscher, H. & Wignall, P.B. 2001. Rapid and synchronous collapse of marine and terrestrial ecosystems during the end-Permian biotic crisis. *Geology*, **29**, 351– 354.

Utting, J. 1978. Lower Karroo pollen and spore assemblages from the coal measures and underlying sediments of the Siankondobo Coalfield, Mid Zambezi Valley, Zambia. *Palynology*, **2**, 53–68.

Utting, J. 1989. Preliminary palynological zonation of surface and subsurface sections of Carboniferous, Permian and lowest Triassic rocks, Sverdrup Basin, Canadian Arctic Archipelago. *Geological Survey of Canada Paper*, **89-1G**, 233–241.

Utting, J. 1994. *Palynostratigraphy of Permian and Lower Triassic Rocks, Sverdrup Basin, Canadian Arctic Archipelago*. Geological Survey of Canada Bulletin, **478**.

Utting, J. & Piasecki, S. 1995. Palynology of the Permian of northern Continents: a review. *In*: Scholle, P.A., Peryt, T.M. & Ulmer-Scholle, P.A. (eds) *The Permian of Northern Pangea, Volume 1. Palaeogeography, Palaeoclimates, Stratigraphy*. Springer, Berlin, 236–261.

Utting, J., Esaulova, N.K., Silantiev, V.V. &. Makarova, O.V. 1997. Late Permian palynomorph assemblages from Ufimian and Kazanian type sequences in Russia and comparison with Roadian and Wordian assemblages from the Canadian Arctic. *Canadian Journal of Earth Sciences*, **34**, 1–16.

Van der Zwan, C.J. 1981. Palynology, phytogeography and climate of the Lower Carboniferous. *Palaeogeography, Palaeoclimatology, Palaeoecology*, **33**, 279–310.

Venkatachala, B.S. & Kar, R.K. 1966. Corisaccites gen. nov., a new saccate pollen genus from the Permian of the Salt Range, West Pakistan. *Palaeobotanist*, **15**, 107–109.

Venkatachala, B.S. & Kar, R.K. 1968. Palynology of the Kathwai Shales, Salt Range, West Pakistan. 1. Shales 25ft above the Talchir Boulder Bed. *Palaeobotanist*, **16**, 156–166.

Vergel, M. del M. 1993. Palinoestratigrafía de la secuencia Neopalaeozoica de la Cuenca Chacoparanense, Argentina. *In*: *12th Congrès International de la Stratigraphie et Géologie du Carbonifère et Permien. Compte Rendus*, **1**, 201–212.

Vijaya 1996. Advent of Gondwana deposition on Indian Peninsula: a palynological reflection and relationship. *In*: *Gondwana Nine: Ninth International Gondwana Symposium*, 10–14 January 1994, Hyderabad, India. A.A. Balkema, Rotterdam, The Netherlands, 283–298.

Visscher, H. 1971. *The Permian and Triassic of the Kingscourt Outlier, Ireland*. Geological Survey of Ireland, Special Papers, **1**.

Visscher, H. 1973. The Upper Permian of Western Europe – a palynological approach to chronostratigraphy. *In*: Logan, A. & Hills, L.V. (eds) *The Permian and Triassic Systems and their Mutual Boundary*. Canadian Society of Petroleum Geologists, Memoirs, **2** , 200–219.

Visscher, H. 1980. Aspects of a palynological characterisation of Late Permian and Early Triassic 'standard' units of chronostratigraphical classification in Europe. *In*: *Proceedings of the IV International Palynological Conference*, 1976–77, Lucknow, India, Volume 2. Birbal Sahni Institute, Lucknow, 236–244.

Warrington, G. 1996. Palaeozoic spores and pollen (Chapter 18E) Permian. *In*: Jansonius, J. & McGregor, D.C. (eds) *Palynology: Principles and Applications, Volume 3*. American Association of

Stratigraphical Palynologists Foundation, Dallas, TX, 607–619.

WILSON, L.R. 1962. *Permian Plant Microfossils from the Flowerpot Formation, Greer County, Oklahoma.* Oklahoma Geological Survey, Circular, **49**.

WILSON, L.R. & COE, E.A. 1940. Descriptions of some unassigned plant microfossils from the Des Moines Series of Iowa. *American Midland Naturalist*, **23**, 182–186.

YAROSHENKO, O.P. & GOLUBEVA, L.P. 1991. *Miospores and stratigraphy of the lower Triassic of Pechora Syncline.* Trydy Geologicheskogo Instituta Akademii Nauk SSSR, Moscow (in Russian).

YIN, H., ZHANG, K., TONG, J., YANG, Z. & WU, S. 2001. The Global Stratotype Section and Point (GSSP) of the Permian–Triassic Boundary. *Episodes*, **24**, 102–114.

ZHU, H.-C., OUYANG, S., ZHAN, J.-Z. & WANG, Z. 2005. Comparison of Permian palynological assemblages from the Junggar and Tarim Basins and their phytoprovincial significance. *Review of Palaeobotany and Palynology*, **136**, 181–207.

A global review of Permian macrofloral biostratigraphical schemes

CHRISTOPHER J. CLEAL

Department of Natural Sciences, National Museum Wales, Cathays Park, Cardiff CF10 3NP, UK
chris.cleal@museumwales.ac.uk

Abstract: Separate biostratigraphical schemes have been developed for Permian macrofloras in the five main phytochoria (palaeokingdoms), reflecting the essential lack of overlap in taxonomic composition. In Europe two biozones are normally recognized, in North America three zones, in Cathaysia three or four zones, in Gondwana four zones and in Angara five zones. The stratigraphical resolution tends to be far less than that of palynology, and up to an order of magnitude coarser than the macrofloral biozones of the Pennsylvanian subsystem. This is probably due, at least in part, to the lack of rigor in the way that the Permian macrofloral zones have been defined. Nevertheless, the existing zones do provide evidence of the overarching trajectory of change in vegetation through the Permian Period, as it responded at all palaeolatitudes to a combination of climate change, large-scale volcanic eruptions and tectonically driven landscape changes.

The Permian Period was a time of marked changes in terrestrial vegetation. A combination of global climate change with the end of the Late Palaeozoic Ice Age (e.g. Chumakov & Zharkov 2003; Fielding *et al.* 2008; Montañez & Poulsen 2013; Frank *et al.* 2015) and major tectonically driven landscape disruption in the palaeotropical belt that caused large-scale biome migration (Hilton & Cleal 2007) and the replacement of the Palaeophytic by the Mesophytic 'evolutionary floras' (Cleal & Cascales-Miñana 2014). The period was eventually brought to an end by a major disruption to biotas – one of the few to affect both terrestrial and marine biotas (Cascales-Miñana *et al.* 2015). These vegetational dynamics could have the potential for the establishment of a detailed biostratigraphical model for Permian macrofloras, which given the extensive distribution of terrestrial deposits of this age would have a significant impact for local and global stratigraphical correlations. However, the different drivers of these vegetational dynamics (climatic, landscape and phylogenetic) have presented considerable challenges to establishing a robust biostratigraphical model (or models) for Permian macrofloras.

Nearly a quarter of a century ago, Cleal (1991) attempted a global review of Permian macrofloral biostratigraphy. The present contribution will re-examine this analysis in the light of subsequent new data and ideas. It would be impossible within the scope of a review paper such as this to give a detailed documentation of all of the palaeobotanical records relevant to the analysis; where possible, therefore, emphasis is mainly given to review papers or the most recent descriptive studies, where more complete bibliographies can be found. The review will be restricted to the fossil record of adpressions (compression–impression fossils).

Palaeobiogeographical context

There is a now a wide consensus that the Permian palaeobotanical record can be divided globally into between three and five broad geographical divisions (phytochoria) that Cleal (1991) referred to as palaeokingdoms – Euramerica, Angara, Gondwana, Cathaysia and North America (Fig. 1) (Chaloner & Meyen 1973; Vakhrameev *et al.* 1978; Meyen 1987; Allen & Dineley 1988; Cleal & Thomas 1991; Wnuk 1996; Rees *et al.* 2002). These phytochoria reflect a combination of latitudinal/climatic influence and, particularly in the lower palaeolatitudes, the transition between Palaeophytic and Mesophytic Evolutionary Floras *sensu* Cleal & Cascales-Miñana (2014).

Although there have been some records of Permian 'mixed-floras', including elements from more than one palaeokingdom (e.g. Berthelin *et al.* 2003, 2006; Broutin & Berthelin 2005; Sun 2006), these phytochoria mostly consist of fossil floras that are mutually exclusive, which makes any attempt at a global biostratigraphical scheme for the Permian at this time all but impossible. In this review, therefore, the same approach is taken as by Cleal (1991) and the stratigraphical distribution of fossils floras will be examined separately for each phytochorion.

Europe Palaeokingdom

As the name implies, this phytochorion covers most of Europe as far east as the Urals, together with NW Africa and eastern North America (Fig. 1). It broadly corresponds to the area that during Pennsylvanian times was covered by the palaeotropical wetlands known as coal swamps associated with the Variscan and Appalachian foreland areas and

From: Lucas, S. G. & Shen, S. Z. (eds) 2018. *The Permian Timescale*. Geological Society, London, Special Publications, **450**, 349–363.
First published online December 8, 2016, https://doi.org/10.1144/SP450.4

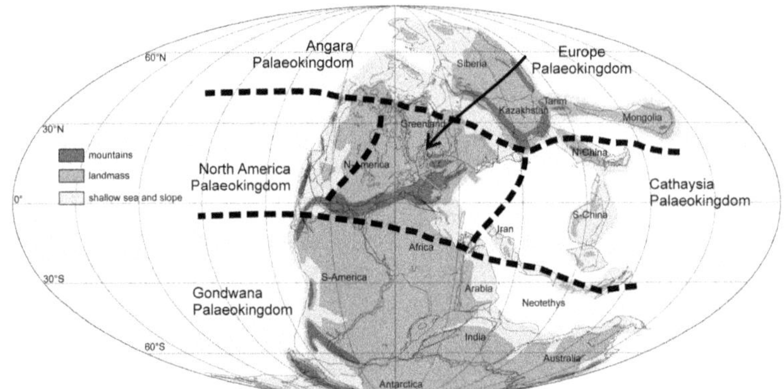

Fig. 1. Approximate palaeogeographical distribution of five main phytochoria (palaeokingdoms *sensu* Cleal 1991) recognized in the Permian Period. Plotted on a base map for the early Wordian Age, adapted from Lucas *et al.* (2006, fig. 2), which in turn had been based on the work of J. Golonka.

intra-montane basins. However, in latest Pennsylvanian–earliest Cisuralian times, the area became much drier as a result of a combination of landscape and climate changes. The result was not only what seems to have been a much sparser and lower diversity vegetation, but also low pH taphonomic conditions that did not favour the preservation of plant remains. Consequently, the Permian palaeobotanical record of this area is mostly very poor, so it is difficult to establish a coherent set of biozones.

Traditionally, the base of the Permian System in this area was taken to coincide with the base of the Autunian Stage, which in turn was usually taken to approximate to the appearance of floras characterized by various peltasperms (including *Autunia* and other 'callipteroids') and conifers (e.g. Jongmans & Gothan 1937). They are today referred to the *Autunia conferta* Zone (formerly the *Callipteris conferta* Zone) and represent the transition between Palaeophytic and Mesophytic vegetation as defined by Cleal & Cascales-Miñana (2014). However, it is now evident that this change in the floras was a result of environmental changes that were not isochronous across the area (e. g. Bouroz & Doubinger 1977; Wagner 1984; Broutin *et al.* 1990; Kerp 1996; Blake *et al.* 2002). The base of the Autunian Stage remains poorly defined, and has variously been assigned to the lower Cisuralian Subsystem (e.g. Roscher & Schneider 2005) and upper Pennsylvanian Subsystem (e. g. Wagner 1998; Heckel & Clayton 2006). Nevertheless, the *A. conferta* Biozone almost certainly extends up into at least the Asselian Stage and so can be taken as the lowest Permian macrofloral biozone for the Europe phytochorion. The zone is well documented in Germany (Barthel 1976; Kerp & Fichter 1985), France (Doubinger 1956; Galtier & Broutin 2008; Gand *et al.* 2013), Spain (Wagner & Álvarez-Vázquez 2010), Italy (Cassinis & Ronchi 2002; Ronchi *et al.* 2012), the Czech Republic (Opluštil *et al.* 2013), Romania (Popa 1999) and the Ukraine (Boyarina 2010).

Autunia conferta Zone floras are also known from eastern North America, most notably in parts of the Dunkard Group of the Central Appalachian Basin (Blake *et al.* 2002; Blake & Gillespie 2011; DiMichele *et al.* 2013); these deposits have also been dated as Asselian or possibly earliest Sakmarian in age (Schneider *et al.* 2013). As in Europe, there appears to be a transition from the more typical Late Pennsylvanian wetland floras (assignable to the *Danaeites* Zone of Read & Mamay 1964) to the *A. conferta* Zone floras representing better drained conditions, with the two zones interfingering (e.g. Fedorko & Skema 2013). The Early Permian macroflora reported by Ziegler *et al.* (2002) from Prince Edward Island (Canada) is almost certainly of similar age but, as it consists only of conifers, is not really compatible with the *A. conferta* Zone.

The stratigraphical age of the upper part of the *A. conferta* Zone is problematic, as using the floras themselves to establish correlation with the normally accepted chronostratigraphical scheme obviously has the potential for circular argument. However, there has been an attempt at correlation of these sequences based on a combination of insect and vertebrate biostratigraphy (Roscher & Schneider 2005, 2006), and, based on this, most of the strata that have yielded *A. conferta* Zone floras belong to the Asselian and Sakmarian stages. This means that the zone spans a considerable time interval of 10 Ma or more. However, attempts to subdivide it (e.g. Kozur 1978; Haubold 1980) have not proved widely applicable (see comments by Kerp 1988).

Other than some indeterminable roots and conifer shoots from red beds found in several places in

Europe (e.g. Thüringia in Germany), until recently the best-documented post-Sakmarian macrofloras in Europe are associated with the Zechstein and Kupferschiefer deposits of northern and central Europe from the British Isles to Poland and Hungary (Stoneley 1958; Schweitzer 1986; Bauer *et al.* 2013). These are dominated by conifers (*Ullmannia*, *Pseudovoltzia*), pteridosperms (*Germaropteris*), (?)ginkgophytes (*Sphenobaiera*) and (?)cycads (*Taeniopteris*), and have been dated as Lopingian (Roscher & Schneider 2005). These macrofloras have never previously been incorporated into a formal biostratigraphical scheme but, for the purposes of the present analysis, they will be referred to as the *Germaropteris martinsii* Zone (formerly *Lepidopteris martinsii* Zone: see Kustatscher *et al.* 2014).

Very similar *G. martinsii* Zone floras are known from the Gröden Sandstone (Val Gardena Formation) of northern Italy, which is also regarded as Lopingian in age (e.g. Visscher *et al.* 2001; Kustatscher *et al.* 2012, 2014). Also very similar, however, are the Tregiovo–Le Fraine and Collio macrofloras of north Italy (Visscher *et al.* 2001; Wachtler 2012), but this is of a rather older Artinskian–Kungurian age (Cassinis & Ronchi 2001), suggesting that these floras had their origins in Cisuralian times.

No *G. martinsii* Zone floras appear to have been discovered in the American part of the Europe Palaeokingdom.

North America Palaeokingdom

The recognition of a separate North America Palaeokingdom is based on the presence of a distinct set of fossil floras, probably representing vegetation of better-drained substrates. They are characterized by foliage very similar to *Gigantopteris* and allied forms, which are more normally associated with the Cathaysian Floras of China, although data on their cuticles or the reproductive structures of the parent plants remain very limited. They occur in a broad belt between Arizona and Oklahoma, in the SW USA (Fig. 1). Read & Mamay (1964) produced the most detailed biostratigraphical classification of these floras (see also Mamay & Read 1954), and this was essentially adopted by Cleal (1991), although with the zones given different names to reflect subsequent changes in taxonomic nomenclature. Three zones were recognized (Fig. 2):

- *Callipteris* spp. Zone – this is broadly similar to the *Autunia conferta* Zone of the Europe Palaeokingdom, dominated by peltasperms such as *Autunia*, and conifers. It occurs in the Chase Group of Kansas (e.g. Krings *et al.* 2005), and the Community Pit and Abo formations of New Mexico (e.g. DiMichele *et al.* 2004*a*, 2007; Falcon-Lang *et al.* 2015). These floras are have been correlated with the Asselian and, possibly, lower Sakmarian stages (Lucas *et al.* 2015). Similar floras have been referred to from the Archer City and Nocona formations of Texas (Tabor *et al.* 2013) but have not been described in detail (W.A. DiMichele pers. comm. 2016).
- *Gigantopteridium americanum* Zone – this is characterized by *G. americanum*, *Yuania taeniata*, *Glenopteris* spp. and *Supaia* spp. The zone occurs in the Petrolia, Waggoner Ranch and Neal Ranch formations of Texas (e.g. Mamay 1968, 1973, 1975, 1988; Glasspool

Stages	N. America	Europe	Cathaysia	Gondwana	Angara
Changhsingian			*Ullmannia*	"Zone V"	*Pecopteris oviformis*
Wuchiapingian		*Lepidopteris martinsii*	*Gigantonoclea hallei*		*Glottophyllum primaevum*
Capitanian				*Glossopteris*	
Wordian			*Cathaysiopteris whitei*		
Roadian					*Nephropsis lampadiformis*
Kungurian	*Cathaysiopteris yochelsonii*				
Artinskian	*Gigantopteridium americanum*		*Emplectopteridium alatum*	*Gangamopteris*	*Zamiopteris*
Sakmarian					
Asselian	*Callipteris* spp.	*Autunia conferta*	*Lepidodendron posthumii*	*Botrychiopsis*	*Evenkiella zamiopteroides*

Fig. 2. Approximate correlations between the macrofloral biozones identified in the five main phytochoria shown in Fig. 1 (palaeokingdoms *sensu* Cleal 1991) recognized in the Permian Period.

et al. 2013), the Cutler Formation of Utah (e.g. Mamay & Breed 1970), and the Hermit Shale Formation of Arizona and New Mexico (e.g. Mamay & Watt 1971). These floras are regarded as Sakmarian and possibly lower Artinskian in age (Lucas *et al.* 2015).

- *Cathaysiopteris yochelsonii* Zone – this is characterized by *Zeilleropteris wattii*, *Evolsonia texana* and *Delnortia abbottiae*; *Cathaysiopteris yochelsonii*, although present, is quite rare here (W.A. DiMichele pers. comm. 2016). This has only been reported from the Clear Fork and lower Pease River groups of Texas (e.g. Mamay 1966, 1986, 1989; Mamay *et al.* 1984, 1988, 2009; DiMichele *et al.* 2000, 2001, 2004*b*, 2011; Chaney & DiMichele 2007; Looy 2007). These intervals are dated as Kungurian (Lucas 2006) or possibly early Roadian (DiMichele *et al.* 2001).

Much of this classification had been based on work carried out in the early twentieth century, and Read & Mamay (1964, p. 13) clearly indicated that the situation might be more complicated than this classification might suggests, with the geographical and chronostratigraphical relationship between the zones not always being clear. More recent work has also suggested that there may be taxonomic problems with some of the taxa used to define the zones (e.g. DiMichele *et al.* 2005*a*; Krings *et al.* 2005). It is also now evident that the climate of the whole area was progressively becoming drier (Chaney & DiMichele 2007; Tabor *et al.* 2013), reflecting a combination of global climate changes and tectonically driven landscape changes, and that this was causing the change from vegetation favouring wetlands to that favouring better-drained substrates (DiMichele & Aronson 1992; DiMichele *et al.* 2004*a*, 2005*b*, 2006; DiMichele & Chaney 2005). This has resulted in the apparently 'early' appearance of plant fossil taxa in some of these American Permian floras, such as peltasprms and ullmanniacean-like conifers (e.g. DiMichele *et al.* 2000, 2001; Looy 2007; Falcon-Lang *et al.* 2015), which DiMichele & Aronson (1992) had earlier argued could be a terrestrial analogue of the onshore–offshore hypothesis recognized in marina faunas.

A meaningful biostratigraphical analysis of these American floras clearly needs far more data on the stratigraphical and geographical distribution of the taxa, and how they relate to the biome replacement taking place at the time. It would still be possible to develop a unified biostratigraphical scheme for the area as a whole, but this would be mainly reflecting the biome replacement – perhaps a useful tool for describing that biome replacement, but of relatively little use as a guide for chronostratigraphical correlation. It might alternatively be possible to develop separate biostratigraphical schemes for the different biomes, which would run in parallel, and possibly provide more robust evidence of correlations of the strata, but this will have to remain a task for the future.

Cathaysia Palaeokingdom

Until relatively recently our understanding of the stratigraphical distribution of Permian plant fossils in Cathaysia was rather limited. Other than the classic works of Halle (1927) and Li (1963), both based mainly on the floras of Shanxi in North China, most records of the stratigraphical distribution of taxa tended to be tabular depictions of presence/absence in lithostratigraphical intervals (e.g. 'Gu & Zhi' 1974; Shen 1995; Wang *et al.* 2005). Li (1963, 1964) established a series of biozones for the Shanxi floras and these have been refined by Wang (2010) using a more modern biostratigraphical approach. In the synthetic analysis by Cleal (1991), a simplified nomenclature for the zones was proposed, each based on a single species name, and this is followed here (Fig. 2):

- *Lepidodendron posthumii* Zone – this occurs in the middle and upper Taiyuan Formation, and lower Shanxi Formation in Shanxi (e.g. Wang & Pfefferkorn 2013) and Shaanxi (Wang 2010), the Zhutun Formation of Henan (Yang & Wang 2012), and is also probably represented by the Djambi flora of Indonesia (van Waveren *et al.* 2005, 2007). It may be regarded as a transitional flora containing taxa characteristic of both the Euramerica (e.g. medullosaleans and marattialean ferns) and Cathaysia (e.g. *Tingia*, *Cathaysiodendron*) realms. The lower part of the zone is generally regarded as Late Pennsylvanian in age. In the model of biome migration between Euramerica and Cathaysia (Hilton & Cleal 2007) during late Pennsylvanian and early Cisuralian times, the *L. posthumii* Zone would be equivalent to the *Autunia conferta* Zone of Europe, but with the former representing swamp vegetation and the latter more mesic vegetation.
- *Emplectopteridium alatum* Zone – this marks the start of the fully developed Cathaysian floras (the middle Sakmarian Shanxi Formation radiation of Stevens *et al.* 2011), with the appearance of taxa such as *Emplectopteridium*, *Emplectopteris* and *Plagiozamites*, and a marked increase in diversity of cycadophytes such as *Taeniopteris*. There are also rare occurrences of gigantopteroids such as *Gigantonoclea* and *Cathaysiopteris*. This zone is typical of the Shanxi Formation and the lower part of the Lower Shihhotse Formation of Shanxi (Wang 2010), parts of the the

Shenhou Formation in Henan (Yang 2006; Yang & Wang 2012) and the Yujiabeigou Formation of Nei Mongol (Huang 1993), and in Indonesia (van Waveren *et al.* 2007). Stevens *et al.* (2011) suggested that the increase in diversity seen in these floras reflects the last phase of the Late Palaeozoic Ice Age, when lower sea levels allowed colonization of previously shallow-marine environments.

- *Cathaysiopteris whitei* Zone – the base of this zone as originally recognized by Li (1963, 1964) probably also corresponds to the 'LSF' extinction event of Stevens *et al.* (2011), and is typically seen in the upper part of the Lower Shihhotse Formation of Shanxi. Wang *et al.* (2005) and Wang (2010) argued that its distinction from the underlying *E. alatum* Zone is weak and that they should be unified. However, Stevens *et al.* (2011) reported a loss of about 45% of the plant species at this level, so the zonal separation has been retained.
- *Gigantonoclea hallei* Zone (the *Otofolium* Zone of earlier authors) – this is re-named from the earlier classifications as *Otofolium* tends to be rare in North China (Shen 1995). Of the three fossil taxa used by Shen (1995) in his trinomial nomenclature, *G. hallei* was chosen as the epithet for this biozone as it is abundantly found in, although not restricted to, the zone. In Shanxi, it occurs in the Tianlongsi (formerly known as the Upper Shihhotse) Formation, the base of which, according to Stevens *et al.* (2011), is at about the base of the Capitanian Stage; it is also probably equivalent to the *Gigantonoclea cathaysiana–Monogigantopteris clathroreticulatus–Lobatannularia ensifolia–Fascipteris sinensis* Assemblage Zone reported from the Xiaofengkou Formation of Henan by Yang (2006). The start of the zone marks the peak development of the Cathaysian Flora (Wang 2010) with diverse seeds plants, especially among the cycadophytes and ginkgophytes, and the gigantopteroid complex. However, a number of taxonomic groups decline or totally disappear in the zonal interval, especially among the lycophytes, sphenophytes and ferns (e.g. Wang & Chen 2001). Based on this, Wang (1993) distinguished *Plagiozamites* and *Psygmophyllum* subzonal divisions, with a boundary corresponding to the 'mUSF' extinction event of Stevens *et al.* (2011), and probably reflecting the progressive change to more mesic/xeric vegetation. Another significant reduction in taxonomic diversity (the 'uUSF' event) was recorded in the Upper Tianlongsi Formation by Stevens *et al.* (2011), but this is mainly marked by the extirpation of taxa and so is not recognized biozonally.
- *Ullmannia* Zone – this occurs in the predominantly red-bed sequences of Shanxi and Shaanxi, of the Sunjiagou (formerly Shihchienfeng) Formation, which according to Stevens *et al.* (2011) approximately coincides with the Changhsingian Stage. Plant fossils tend to be relatively scarce, although Wang (1983) has reported a number of diverse assemblages. The flora is indicative of more xeric conditions, with greater numbers of conifer, cycadophyte and peltasperm remains, and a relative reduction in diversity and abundance of lycophytes, ferns and gigantopteroids. In the Yungaishan Formation of Henan, Yang (2006) divided the interval into two units (?subzones): *Monogigantonoclea colosasifolia–Pinnagigantonoclea mucronata–Lobatannularia heianensis* and *Pinnagigantopteris nicotianaefolia–Psygmophyllum multipartitum–Pseudohripidopsis brevicaulis* (Sub)Zones.

Although the relative sequence of these biozones has been well established over several parts of North China, their relationship to chronostratigraphy is less certain due to the scarcity of faunas to allow correlation with the internationally recognized series and stages. Based on comparisons with Permian fossil floras in Europe, Wang (2010) suggested that the base of the *E. alatum* Zone correlated with the base of the Artinskian Stage, the base of the *G. hallei* Zone with the base of the Capitanian Stage, and the base of the *Ullmannia* Zone with the base of the Changhsingian Stage: but, as the chronostratigraphical age of the European floras is far from certain, this correlation for the Chinese floras must be regarded as being very approximate. As pointed out by Stevens *et al.* (2011), however, a Capitanian age for what is referred to here as the *G. hallei* Zone appears to be borne out by palaeomagnetic evidence (the Illawara Reversal). Shen (1995) tentatively suggested correlations with some of the zone with the internationally recognized stages based on comparisons with the American Permian floras. However, as the plant distributions in the American Permian fossil floras are reflecting a large-scale biome replacement caused by local climatic changes, such a comparison as the basis for intercontinental correlation is far from robust. It should also be noted that this overall pattern of change in the fossil floras of North China is reflecting what was probably tectonically driven landscape and climate change (Wang 1996), so even within this area the biozones are likely to be at least partly diachronous.

A broadly similar pattern of Permian floras in NW China was also summarized by Shen (1995). However, the empirical data are not presented in detail, and any biostratigraphical analysis at this juncture is impossible.

The Permian floras of South China were also reviewed by Shen (1995), who recognized four assemblage zones. These zones differ in detail from those recognized in North China but, overall, represent a comparable transition of the parent vegetation. There is no equivalent of the *L. posthumii* Zone, but there are the following:

- *?Emplectopteris triangularis–Taeniopteris multinervis* and *Gigantonoclea fukiensis–Tingia carbonica* assemblages – these are the equivalents of the *E. alatum* and *C. whitei* zones. They are characterized by remains of plants of a Cathaysian aspect, with genera such as *Emplectopteris* and *Plagiozamites*, but lacking *Emplectopteridium* and *Neuropteris*. Floras of this zone are relatively rare and are most typically found in the Liangshan Formation in central South China.
- *Gigantopteris nicotiannaefolia–Lobatannularia multifolia* Assemblage – this is equivalent to the *Gigantonoclea hallei* Zone and marks when the gigantopteroids come to prominence in South China. The zone is particularly developed in the lower Lungtan Coal Formation of SE China, such as in Jiangsu Province (e.g. Yao & Taylor 1988; Yao & Liu 2004).
- *Gigantonoclea guizhouensis–Ullmannia* cf. *bronnii* Assemblage – this is the equivalent of the *Ullmannia* Zone. These floras are generally similar to those of the previous assemblage but of much lower diversity, showing in particular a marked decline, but not total disappearance, of the gigantopteroids. Representative floras occur in western Guizhou, eastern Yunnan and southern Sichuan (e.g. Zhao *et al.* 1980).

Gondwana Palaeokingdom

In the analysis by Cleal (1991), the Permian floras of Gondwana (Fig. 1) were divided into *Gangamopteris* and *Glossopteris* zones, broadly following the analysis by Schopf & Askin (1980). A very similar broad, two-fold division was also suggested by Archangelsky (1971), who referred to them as the Lubeckense and Bonetense floral assemblages, respectively (see also Archangelsky 1984, 1990; Archangelsky & Cúneo 1984). It was stated by Cleal (1991) that the boundary between the zones corresponded to the boundary between the lower and upper Permian, but it was acknowledged that independent dating of (and even correlation between) the terrestrial plant sequences containing the macrofloras was extremely weak.

The most detailed analysis of Permian plant distribution in Gondwana has been by Cúneo (1996), who recognized five zone-like units that were called 'stages'. However, the lower zone (I) now appear to be Carboniferous in age based on radiometric data (Rocha-Campos *et al.* 2006; Césari 2007; Guerra-Sommer *et al.* 2008; Césari *et al.* 2011; Simas *et al.* 2012). The result is, therefore, an essentially four-fold biozonal division for the Permian floras of Gondwana (Fig. 2). The following analysis is based on Cúneo (1996), but using formal biozonal names based on other studies:

- *Botrychiopsis* Zone (*sensu* Jasper *et al.* 2003, 2007) – this is roughly equivalent to Cúneo's (1996) Zone II; also to the *Glossopteris/Rhodeopteridium* Zone of Boardman *et al.* (2012*a*). Evidence of the exact age of this zone is unclear, although the Carboniferous–Permian boundary probably occurs somewhere near the base and it probably extends to near or just beyond the top of the Asselian Stage (Jasper *et al.* 2003). Although formed during the later phases of the Late Palaeozoic Ice Age, the zone seems to mark the onset of what was to become the post-glacial vegetation of Gondwana, with glossopterids, ginkgophytes, marattialeans and lycophytes. *Botrychiopsis* itself seems to be absent from India, southern Africa and Patagonia, possibly reflecting the somewhat higher palaeolatitiudes of these basins. It is best documented in the lower Rio Bonito Formation in the Paraná Basin of Brazil (Archangelsky & Cúneo 1984; Guerra-Sommer & Cazzulo-Klepzig 2000) and the Karharbari Formation of India (Singh & Chandra 1996; Singh 2000).
- *Gangamopteris* Zone (=Zone III of Cúneo 1996) – this represents the glossopterid/cordaitoid/marattialean/lycophyte-dominated, peat-forming forests that developed as humid, warm-temperate climates spread across much of Gondwana (Cúneo 1996). The zone is well represented in the lower La Golondrina Formation in Patagonia (Archangelsky 1959; the *Dizeugotheca waltonii* Zone *sensu* Archangelsky & Cúneo 1984), the middle–upper Rio Bonito Formation of the Paraná Basin in Brazil (Archangelsky & Cúneo 1984; Guerra-Sommer & Cazzulo-Klepzig 2000; Iannuzzi 2010; Boardman *et al.* 2012*b*; the *Glossopteris–Brasilodendron* Flora of Iannuzzi & Souza 2005), the Ecca Group in the Karroo Basin of South Africa (Anderson & Anderson 1985), the Barkar Formation in India (Singh 2000; Goswami & Singh 2013), the Sydney–Bowen Basin in Australia (Rigby 1983) and the Upper Weller Formation of Antarctica (Taylor *et al.* 1991; Retallack *et al.* 2006; Tewari *et al.* 2015). Retallack *et al.* (2005, 2006) divided this interval into the *Ottokaria* and *Plumstedia* zones, but the differentiating biostratigraphical criteria were not clearly stated, so this division has not been

followed here. It has been suggested that these floras are Sakmarian–Kungurian in age (e.g. Rocha-Campos *et al.* 2006; Cagliari *et al.* 2014).

- *Glossopteris* Zone (=Zone IV of Cúneo 1996) – this is marked by a significant reduction in taxonomic diversity among plant macrofossils (e.g. Retallack *et al.* 2006), possibly linked with a marked climatic drying (Cúneo 1996). The floras are generally of much lower diversity (e.g. Singh 2000); *Glossopteris* remains a widespread leaf fossil, but *Gangamopteris* is much rarer. They are best represented in the upper La Golondrina Formation in Patagonia (Archangelsky & Cúneo 1984); the Waterford Formation in South Africa may also belong here (Anderson & Anderson 1985). Cúneo (1996) placed the base of the zone in the Ufimian Stage (Kungurian Stage in the currently accepted chronostratigraphy). More recently, the zonal base has been interpreted as a vegetation change linked with the middle Capitanian Emeishan Large Igneous Province that also affected lower palaeolatitude vegetation (Bond *et al.* 2010; Stevens *et al.* 2011): in which case, the base of the zone would be middle–late Capitanian in age (Retallack *et al.* 2006; Bond *et al.* 2010; Stevens *et al.* 2011). However, this would appear to leave a substantial time gap between the *Gangamopteris* and *Glossopteris* zones, and in this analysis a late Kungurian age has been provisionally accepted.
- Zone V of Cúneo (1996) – this has never been assigned a name. It marks a return to forests with glossopterid-bearing leaves, also with sphenophytes and marattialeans. The best-documented floras of this zone are from the Sydney–Bowen Basin in Australia (McLoughlin 1992, 1994*a*, *b*), the Escourt Formation in South Africa (Anderson & Anderson 1985), in the Weller Coal Measures and Buckley Formation of Antarctica (Retallack *et al.* 2005, 2006), and the lower Kamthi Formation in India (Singh 2000; Goswami & Singh 2013).

There remain, however, significant problems with developing an overall model for Permian plant biostratigraphy in Gondwana owing to a lack of independent dating of the fossil beds that can result in circular reasoning when trying to establish interbasinal correlation. For a further discussion of these problems, see the analysis of the Paraná Basin floras of Argentina by Christiano-de-Souza & Ricardi-Branco (2015).

Angara Palaeokingdom

Cleal (1991) attempted to develop a biostratigraphical model for the Permian floras of Angara (Fig. 1) based on data from the Kuznesk Basin (Gorelova *et al.* 1973). Durante (1995) subsequently introduced a series of informal 'geofloras' that can be partially correlated with these zones. There have been no published alternative schemes, so this will be summarized below (Fig. 2):

- *Evenkiella zamiopteroides* Zone – this is marked by the appearance of the eponymous species, plus *Tschernovia kuznetskiana* Neuburg, *Nephropis rhomboidea* Neuburg, *Prynadaeopteris tunguscana* (Schmalhausen) Radchenko and *Annularia planifolia* Radchenko. It corresponds to the upper part of the 'pteridosperm–cordaite geoflora' of Durante (1995) and is correlated with the Asselian Stage, although on what basis is not clear.
- *Zamiopteris* Zone – this marks the onset of the fully developed 'cordaite geoflora' of Durante (1995), and is regarded as representing somewhat cooler-temperate conditions. It is marked by the appearance of the cordaitoid or dicranophyll eponymous fossil-genus, as well as the sphenophyte *Annularia rarifolia* Radchenko; also by the disappearance of the taxa whose appearance characterized the underlying zone. It appears to correlate with the Sakmarian–middle Kungurian stages.
- *Nephropsis lampadiformis* Zone – there is a floral turnover in the upper Kungurian Stage in the Kuznetsk Basin part of the 'cordaite geoflora' of Durante (1995), with the appearance of taxa such as *Cordaites minax* (Gorelova) Meyen, *Cordaites kuznetskianus* (Gorelova) Meyen and *Zamiopteris crassinervis* Gorelova, as well as the eponymous species. Although still representing an essentially cool-temperate vegetation, the zone sees the appearance of a number of mesophytic taxa such as *Rhipidopsis*, *Tomia* and *Yavorskiya*.
- *Glottophyllum primaevum* Zone – the base of this zone appears to coincide with the appearance of the 'fern–pteridosperm–cordaite geoflora' of Durante (1995) in the middle Guadalupian Series, and represents a period of climatic warming. It is marked by the appearance of sphenophytes such as *Koretrophyllites mollifolius* Radchenko, and a number of cordaitoids such as *Cordaites clercii* Zalessky and *Noeggerathiopsis incisa* Vakhrameev & Radchenko; the zone also sees a marked increase in abundance of peltasperm pteridosperms.
- *Pecopteris oviformis* Zone – this represents another flora turnover, probably in the lower–middle Lopingian Series, although exact correlations with the IUGS chronostratigraphy are not certain. The base of the zone is marked by the appearance of the eponymous fern species,

the sphenophytes *Koretrophyllites tomiensis* Radchenko and *Annularia lanceolata* Radchenko, and the cordaitopids *Cordaites insignis* (Radchenko) Meyen and *Crassinervia brevifolia* Radchenko. Based on the Durante (1995) model, this marks a further trend towards vegetation favoured by warmer climates and a greater emphasis of mesophytic elements.

Permian floras are also known over many other parts of Angara (Durante 1976; Meyen in Vakhrameev *et al.* 1978; Meyen 1982; Naugolnykh 2000), but there has been no published attempt at a biostratigraphical synthesis or zonation to incorporate them into the zonation developed in the Kuznetsk area. Although the general chronological trend from Palaeophytic to Mesophytic vegetation seems to be comparable in the different parts of Angara, independent dating of the various successions is weak so it is not possible to determine whether or not the changes are diachronous (contrast with the recent testing of floral zone boundaries in the Carboniferous of Euramerica by Opluštil *et al.* 2016).

Surrounding the main Angara area is an ecotonal belt often referred to as Subangara, stretching from the Pechora area in NW Siberia, through Kazakhstan, Mongolia and northern China (Meyen in Vakhrameev *et al.* 1978). The floristic relationships of the various areas within this belt have been subject to debate (Sun 1989; Cleal & Thomas 1991), and in the more eastern parts there is the added complication of the dynamic interaction with the Cathaysia floras (e.g. Wang 1989, 1996). This has made it difficult to develop any sort of coherent biostratigraphical model for the Subangaran floras, although Durante (1995) outlined a series of 'geofloras' similar to those she used in the Kuznets area. Best documented are the middle–late Permian floras from NW Siberia (e.g. Meyen 1983; Gomankov & Meyen 1986; Naugolnykh 2005, 2013). According to Meyen (1983), there are differences in the floras between the lower (Kungurian) Vorkutsk 'Series' and the overlying Pechorskya 'Series' (similar to the *Zamiopteris* Zone and *Glottophyllum primaevum* Zone of Kuznetsk, respectively).

The Permian floras of Kazakhstan are poorly documented but appear to be dominated by conifers, and endemic lycophytes and peltasperms (Salmenova 1979), which makes a comparison with the Kuznetsk biostratigraphy difficult.

Rather more attention has been paid to the Permian floras of northern China (Huang 1995), where three zonal intervals ('Plant assemblages') have been recognized:

- '*Crassinervia–Nephropsis–Zamiopteris* Assemblage' – this is characterized by relatively low-diversity floras with ferns and *Zamiopteris* species, such as found in the Tumenling and Daheshen formations in the Khingan Mountain region of the Inner Mongolia Administrative Area (Huang 1977). It is regarded as early Permian in age and may be a correlative of the *Zamiopteris* Zone floras of Kuznetsk.
- '*Callipteris–Supaia–Viatcheslva* Assemblage' – this occurs in the Sanjiaoshan and Hongshan formations of the Khingan Mountain region (Huang 1977), and the Tiemulike Formation of northern Xinjiang (Dou *et al.* 1983). Its base is marked by the appearance of various peltasperms such as *Supaia* and '*Callipteris*', and is likely to be of middle Permian age. It may correlate with the *Glottophyllum primaevum* Zone of Kuznetsk.
- '*Callipteris–Comia–Iniopteris* Assemblage' – this is best represented in the Changfangkou Group of northern Xinjiang (Dou *et al.* 1983) and is regarded by Huang (1995) as late Permian in age. The criteria for distinguishing this from the underlying zone is not clearly stated, but seems to be based on the appearance of Mesozoic-like taxa such as '*Nilssonia*' and '*Pterophyllum*'.

Conclusions

Since the Cleal (1991) review, some progress has been made in developing Permian macrofloral biostratigraphy in a modern sense, especially in parts of China (e.g. Yang 2006; Wang 2010; Stevens *et al.* 2011) and Gondwana (e.g. Jasper *et al.* 2003, 2007). However, there remains a general reluctance to define biozones formally using stratigraphical range charts; there is still a widespread tendency to compile macrofloral inventories for lithostratigraphical (or sometimes putative chronostratigraphical) units and then compare these between basins. This is contrary to accepted modern-day biostratigraphical procedures, which are based around the recognition of stratigraphical intervals defined exclusively on palaeontological criteria (Salvador 1994). Such units provide a far better empirical description of the fossil record for vegetation and palaeoclimate studies, and are far less likely to introduce circular argument into stratigraphical correlation. There are, undoubtedly, circumstances where range charts are not easily developed, and where there is perhaps just one or two plant-bearing horizons through a succession detailed range charts are not especially meaningful. However, there can be no doubt that if temporal studies of Permian plant distribution are to advance significantly, a more empirically robust biostratigraphical approach will be needed.

The approach that has been so widely adopted in Permian macrofloral biostratigraphy (i.e. comparing taxonomic inventories of pre-defined formations or stages) may explain the much coarser resolution compared with the macrofloral zones of the Pennsylvanian Subsystem: the Permian zones in China and Gondwana represent mean time intervals of 10–15 Ma (Fig. 2), in contrast to the 1–2 Ma of the Pennsylvanian macrofloral zones (e.g. Wagner 1984). This probably has more to do with the approach used than anything intrinsic in the macrofloral record, as Permian palynology provides a far better resolution compared with the macrofloras (see Stephenson, this volume, in press; E. Stolle, pers. comm., 2016) – again, it is worth comparing with the Pennsylvanian Subsystem, where the standard palynological zonation of Clayton *et al.* (1977) does not provide a significantly better resolution than the macrofloral zones of Wagner (1984).

We are, therefore, still a long way from being able to use the macrofloras for stratigraphical correlation in any detailed sense, especially intercontinental; the idea that you can, for instance, use the currently available Permian biostratigraphical data from North America as a guide to the relative ages of Chinese sequences (as has been suggested) is not really sustainable. However, this is not to say that the presently recognized Permian macrofloral biostratigraphy is totally valueless: it provides at least a general indication of vegetation change through time, as it responded to a combination of climate change, large-scale volcanic eruptions and tectonically driven landscape changes (Fig. 2). For instance, the marked increase in diversity in the middle Sakmarian Age may be reflecting a change to warmer, glacial-free climates. The changes in the Capitanian Age were perhaps linked with the eruption of the Emeishan Large Igneous Province (Stevens *et al.* 2011) and/or climatic warming; and the changes in the Chanhsingian Age, often represented by a reduction in taxonomic diversity, may be reflecting climatic stress and a foreshadowing of the pressure–temperature extinction event at the end of this age. Nevertheless, significant care has to be taken when trying correlate the various zonal boundaries between continents, as we cannot assume that vegetation changes in response to global climatic dynamics were coincident in high and low latitudes, or even in different parts of continental landmasses at the same latitudes. Before we can start making such assumptions, we need more robust independent dating of the changes observed in the plant fossil record.

I would like to thank Dr Spencer Lucas (New Mexico Museum of Natural History and Science) for inviting me to contribute this review and providing some background information, especially on the dating of some of the floras. I would also like to thank Dr Bill DiMichele (Smithsonian Institution) for his invaluable comments, especially on the North American floras, and to E. Stolle (EP Research, Ennigerloh-Westkirchen) for comments on the palynological background.

References

Allen, K.C. & Dineley, D.L. 1988. Mid-Devonian to mid-Permian floral and faunal regions and provinces. *In*: Harris, A.L. & Fettes, D.J. (eds) *The Caledonian–Appalachian Orogen*. Geological Society, London, Special Publications, **38**, 531–548, https://doi.org/10.1144/GSL.SP.1988.038.01.36

Anderson, J.M. & Anderson, H.M. 1985. *Prodromus of South African megafloras, Devonian to Lower Cretaceous*. Botanical Research Institute, Pretoria, South Africa.

Archangelsky, S. 1959. Estudio geológico y paleontológico del Bajo de La Leona (Santa Cruz). *Acta Geológica Lilloana*, **2**, 5–133.

Archangelsky, S. 1971. Las tafofloras del sistema Paganzo en la Republica Argentina. *Anais da Academia Brasileira de Ciências (Supplement)*, **43**, 67–88.

Archangelsky, S. 1984. Floras Neopaleozoicas del Gondwana y su zonación estratigráfica. Aspectos paleogeográficos conexos. *Comunicações dos Serviços Geológicos de Portugal*, **70**, 135–150.

Archangelsky, S. 1990. Plant distribution in Gondwana during the Late Palaeozoic. *In*: Taylor, T.N. & Taylor, E.L. (eds) *Antarctic Paleobiology. Its Role in the Reconstruction of Gondwana*. Springer, Berlin, 102–117.

Archangelsky, S. & Cúneo, R. 1984. Zonacion del Permico continetal de Argentina sobre la base de sus plantas fosiles. *In*: *Memoria III Congreso Latinoamericano Paleontologia*, 1984, Mexico. Universidad Nacional Autonoma de Mexico, Instituto de Geologia, Mexico City, 143–153.

Barthel, M. 1976. Die Rotliegendflora Sachsens. *Abhandlungen aus dem Staatliches Museum für Mineralogie und Geologie Dresden*, **24**, 1–190.

Bauer, K., Kustatscher, E. & Krings, M. 2013. The ginkgophytes from the German Kupferschiefer (Permian), with considerations on the taxonomic history and use of *Baiera* and *Sphenobaiera*. *Bulletin of Geosciences*, **88**, 539–556.

Berthelin, M., Broutin, J., Kerp, H., Crasquin-Soleau, S., Platel, J.P. & Roger, J. 2003. The Oman Gharif mixed paleoflora: a useful tool for testing Permian Pangea reconstructions. *Palaeogeography, Palaeoclimatology, Palaeoecology*, **196**, 85–98.

Berthelin, M., Stolle, E., Kerp, H. & Broutin, J. 2006. *Glossopteris anatolica* Archangelsky and Wagner 1983, in a mixed middle Permian flora from the Sultanate of Oman: comments on the geographical and stratigraphical distribution. *Review of Palaeobotany and Palynology*, **141**, 313–317.

Blake, B.M. & Gillespie, W.H. 2011. The enigmatic Dunkard macroflora. *In*: Harper, J.A. (ed.) *Geology of the Pennsylvanian–Permian in the Dunkard Basin. Guidebook for the 76th Annual Field*

Conference of Pennsylvania Geologists, Washington, PA. Field Conference of Pennsylvania Geologists, Inc., Middletown, PA, 103–143.

Blake, B.M., Cross, A.T., Eble, C.F., Gillespie, W.H. & Pfefferkorn, H.W. 2002. Selected plant megafossils from the Carboniferous of the Appalachian Region, eastern United States: geographic and stratigraphic distribution. *In*: *Carboniferous and Permian of the World. Proceedings of the XIV International Congress on Carboniferous and Permian stratigraphy*, August 17–21, 1999, Calgary, Alberta, Canada. Canadian Society of Petroleum Geologists, Memoirs, **19**, 259–335.

Boardman, D.R., Souza, P.A., Iannuzzi, R. & Mori, A.L. 2012*a*. Paleobotany and palynology of the Rio Bonito Formation (Lower Permian, Paraná Basin, Brazil) at the Quitéria Outcrop. *Ameghiniana*, **49**, 451–472.

Boardman, D.R., Iannuzzi, R., de Souza, P.A. & da Cunha Lopes, R. 2012*b*. Paleobotanical and palynological analysis of Faxinal Coalfield (Lower Permian, Rio Bonito Formation, Paraná Basin), Rio Grande Do Sul, Brazil. *International Journal of Coal Geology*, **102**, 12–25.

Bond, D.G.P., Hilton, J., Wignall, P.B., Ali, J.R., Stevens, L.G., Sun, Y. & Lai, X.L. 2010. The Middle Permian (Capitanian) mass extinction on land and in the oceans. *Earth-Science Reviews*, **102**, 100–106.

Bouroz, A. & Doubinger, J. 1977. Report on the Stephanian–Autunian boundary and on the contents of Upper Stephanian and Autunian in their stratotypes. *In*: Holub, V.M. & Wagner, R.H. (eds) *Symposium on Carboniferous Stratigraphy*. Geological Survey, Prague, 147–169.

Boyarina, N. 2010. Late Gzhelian pteridosperms with callipterid foliage of the Donets Basin, Ukraine. *Acta Palaeontologica Polonica*, **55**, 343–359.

Broutin, J. & Berthelin, M. 2005. Dynamic of settlement of mixed floras during the Permian in the Peri-Tethyan domain: paleogeographic and paleoclimatic significance. *In*: Lucas, S.G. & Zeigler, K.E. (eds) *The Nonmarine Permian*. New Mexico Museum of Natural History and Science Bulletin, **30**, 24–25.

Broutin, J., Doubinger, J., Farjanel, G., Freytet, P. & Kerp, H. 1990. Le renouvellement des flores au passage Carbonifère Permien: approches stratigraphiques, biologiques, sédimentologiques. *Comptes Rendus Académie des Sciences*, **321**, 1563–1569.

Cagliari, J., Lavina, E.L.C., Philip, R.P., Tognoli, F.M.W., Basei, M.A. & Faccini, U.F. 2014. New Sakmarian ages for the Rio Bonito formation (Paraná Basin, southern Brazil) based on LA-ICP-MS U–Pb radiometric dating of zircons crystals. *Journal of South American Earth Sciences*, **56**, 265–277.

Cascales-Miñana, B., Diez, J.B., Gerrienne, P. & Cleal, C.J. 2015. A palaeobotanical perspective on the great end-Permian biotic crisis. *Historical Biology*, **28**(8), 1–9, https://doi.org/10.1080/08912963.2015.1103237

Cassinis, G. & Ronchi, A. 2001. Permian chronostratigraphy of the Southern Alps (Italy) – an update. *In*: Weiss, R.H. (ed.) *Contributions to Geology and Palaeontology of Gondwana in Honour of Helmut Wopfner*. Geologisches Institut, Cologne, 77–88.

Cassinis, G. & Ronchi, A. 2002. The (late-) post-Variscan continental succession of Sardinia. *Rendiconti della Società Paleontologica Italiana*, **1**, 77–92.

Césari, S.N. 2007. Palynological biozones and radiometric data at the Carboniferous–Permian boundary in western Gondwana. *Gondwana Research*, **11**, 529–536.

Césari, S.N., Limarino, C.O. & Gulbranson, E.L. 2011. An Upper Paleozoic bio-chronostratigraphic scheme for the western margin of Gondwana. *Earth-Science Reviews*, **106**, 149–160.

Chaloner, W.G. & Meyen, S.V. 1973. Carboniferous and Permian floras of the northern continents. *In*: Hallam, A. (ed.) *Atlas of Palaeobiogeography*. Elsevier, Amsterdam, 169–186.

Chaney, D.S. & DiMichele, W.A. 2007. Paleobotany of the classic redbeds (Clear Fork Group Early Permian) of north central Texas. *In*: Wong, Th.E. (ed.) *Proceedings of the XVth International Congress on Carboniferous and Permian Stratigraphy*, 10–16 August 2003, Utrecht, The Netherlands. Royal Netherlands Academy of Arts and Sciences, Amsterdam, 357–366.

Christiano-de-Souza, I.C. & Ricardi-Branco, F.S. 2015. Study of the West Gondwana Floras during the Late Paleozoic: a paleogeographic approach in the Paraná Basin–Brazil. *Palaeogeography, Palaeoclimatology, Palaeoecology*, **426**, 159–169.

Chumakov, N.M. & Zharkov, M.A. 2003. Climate during the Permian–Triassic biosphere reorganizations. Article 2. Climate of the Late Permian and Early Triassic: general inferences. *Stratigraphy and Geological Correlation*, **11**, 361–375.

Clayton, G., Coquel, R., Doubinger, J., Gueinn, K.J., Loboziak, S., Owens, B. & Streel, M. 1977. Carboniferous miospores of western Europe: illustration and zonation. *Mededelingen Rijks Geologische Dienst*, **29**, 1–71.

Cleal, C.J. 1991. Carboniferous and Permian biostratigraphy. *In*: Cleal, C.J. (ed.) *Plant Fossils in Geological Investigation: The Palaeozoic*. Ellis Horwood, Chichester, 182–215.

Cleal, C.J. & Cascales-Miñana, B. 2014. Composition and dynamics of the great Phanerozoic Evolutionary Floras. *Lethaia*, **47**, 469–484.

Cleal, C.J. & Thomas, B.A. 1991. Carboniferous and Permian palaeogeography. *In*: Cleal, C.J. (ed.) *Plant Fossils in Geological Investigation: The Palaeozoic*. Ellis Horwood, Chichester, 154–181.

Cúneo, N.R. 1996. Permian phytogeography in Gondwana. *Palaeogeography, Palaeoclimatology, Palaeoecology*, **125**, 75–104.

DiMichele, W.A. & Aronson, R.B. 1992. The Pennsylvanian–Permian vegetational transition: a terrestrial analogue of the onshore–offshore hypothesis. *Evolution*, **46**, 807–824.

DiMichele, W.A. & Chaney, D.S. 2005. Pennsylvanian–Permian fossil floras from the Cutler Group, Cañon del Cobre and Arroyo del Agua areas, in northern New Mexico. *New Mexico Museum of Natural History and Science Bulletin*, **31**, 26–33.

DiMichele, W.A., Chaney, D.S., Dixon, W.H., Nelson, W.J. & Hook, R.W. 2000. An Early Permian coastal flora from the Central Basin Platform of Gaines County, West Texas. *Palaios*, **15**, 524–534.

DiMichele, W.A., Mamay, S.H., Chaney, D.S., Hook, R.W. & Nelson, W.J. 2001. An Early Permian flora with Late Permian and Mesozoic affinities from north-central Texas. *Journal of Paleontology*, **75**, 449–460.

DiMichele, W.A., Kerp, H. & Chaney, D.S. 2004*a*. Tropical floras of the Late Pennsylvanian–Early Permian transition: Carrizo Arroyo context. *In*: Lucas, S.G. & Zeigler, K.E. (eds) *Carboniferous–Permian Transition*. New Mexico Museum of Natural History and Science Bulletin, **25**, 105–110.

DiMichele, W.A., Hook, R.W., Nelson, W.J. & Chaney, D.S. 2004*b*. An unusual Middle Permian flora from the Blaine Formation (Pease River Group: Leonardian–Guadalupian Series) of King County, west Texas. *Journal of Paleontology*, **78**, 765–782.

DiMichele, W.A., Kerp, H., Krings, M. & Chaney, D.S. 2005*a*. The Permian peltasperm radiation: evidence from the southwestern United States. *In*: Lucas, S.G. & Ziegler, K.E. (eds) *The Nonmarine Permian*. New Mexico Museum of Natural History and Science Bulletin, **30**, 67–79.

DiMichele, W.A., Tabor, N.J. & Chaney, D.S. 2005*b*. Outcrop-scale environmental heterogeneity and vegetational complexity in the Permo-Carboniferous Markley Formation of north central Texas. *In*: Lucas, S.G. & Ziegler, K.E. (eds) *The Nonmarine Permian*. New Mexico Museum of Natural History and Science Bulletin, **30**, 60–66.

DiMichele, W.A., Tabor, N.J., Chaney, D.S. & Nelson, W.J. 2006. From wetland to wet spots: environmental tracking and the fate of Carboniferous elements in Early Permian tropical floras. *In*: Greb, S.F. & DiMichele, W.A. (eds) *Wetlands Through Time*. Geological Society of America, Special Papers, **399**, 223–248.

DiMichele, W.A., Chaney, D.S., Nelson, W.J., Lucas, S.G., Looy, C.V., Quick, K. & Wang, J. 2007. A low diversity, seasonal tropical landscape dominated by conifers and peltasperms: early Permian Abo Formation, New Mexico. *Review of Palaeobotany and Palynology*, **145**, 249–273.

DiMichele, W.A., Looy, C.V. & Chaney, D.S. 2011. A new genus of gigantopterid from the Middle Permian of the United States and China and its relevance to the gigantopterid concept. *International Journal of Plant Sciences*, **172**, 107–119.

DiMichele, W.A., Kerp, H., Sirmons, R., Fedorko, N., Skema, V., Blake, B.M. & Cecil, C.B. 2013. Callipterid peltasperms of the Dunkard Group, Central Appalachian Basin. *International Journal of Coal Geology*, **119**, 56–78.

Dou, Y., Sun, Z., Wu, S. & Gu, D. 1983. Palaeozoic plants. *In*: Xinjiang Team of Regional Geological Survey (eds) *Palaeontological Atlas of Northwest China, Xinjiang Uygur Automomous Region, Volume 2*. Geological Publishing House, Beijing, 561–614 [in Chinese].

Doubinger, J. 1956. Contribution à l'étude des flores Autuno-Stéphaniennes. *Mémoires de la Société Géologique de France, N. S.*, **75**, 1–180.

Durante, M.V. 1976. *Paleobotanicheskoe obosnovanie stratigrafii karbona i permi Mongolii*. Nauka, Moscow [Trudy Sovmestnaya Sovetsko-Mongol'skaya Geologicheskaya Ekspeditsiya, **19**].

Durante, M.V. 1995. Reconstruction of Late Paleozoic climatic changes in Angaraland according to phytogeographic data. *Stratigraphy and Geological Correlations*, **3**, 123–133.

Falcon-Lang, H.J., Lucas, S.G. *et al.* 2015. Early Permian (Asselian) vegetation from a seasonally dry coast in western equatorial Pangea: paleoecology and evolutionary significance. *Palaeogeography, Palaeoclimatology, Palaeoecology*, **433**, 158–173.

Fedorko, N. & Skema, V. 2013. A review of the stratigraphy and stratigraphic nomenclature of the Dunkard Group in West Virginia and Pennsylvania, USA. *International Journal of Coal Geology*, **119**, 2–20.

Fielding, C.R., Frank, T.D. & Isbell, J.L. 2008. The late Paleozoic ice age – a review of current understanding and synthesis of global climate patterns. *In*: Fielding, C.R., Frank, T.D. & Isbell, J.L. (eds) *Resolving the Late Paleozoic Ice Age in Time and Space*. Geological Society of America, Special Papers, **441**, 343–354.

Frank, T.D., Shultis, A.I. & Fielding, C.R. 2015. Acme and demise of the late Palaeozoic ice age: a view from the southeastern margin of Gondwana. *Palaeogeography, Palaeoclimatology, Palaeoecology*, **418**, 176–192.

Galtier, J. & Broutin, J. 2008. Floras from red beds of the Permian Basin of Lodève (Southern France). *Cuadernos de Geología Ibérica*, **34**, 57–72.

Gand, G., Galtier, J., Garric, J., Teboul, P.A. & Pellenard, P. 2013. Discovery of an Autunian macroflora and lithostratigraphic re-investigation on the western border of the Lodève Permian basin (Mont Sénégra, Hérault, France). Paleoenvironmental implications. *Comptes rendus de l'Académie des sciences*, **12**, 69–79.

Glasspool, I., Wittry, J., Quick, K., Kerp, H. & Hilton, J. 2013. A preliminary report on a Wolfcampian age floral assemblage from the type section for the Neal ranch Formation in the Glass Mountains, Texas. *In*: Lucas, S.G., DiMichele, W.A., Barrick, J.E., Schneider, J.W. & Spielmann, J.A. (eds) *The Carboniferous–Permian Transition*. New Mexico Museum of Natural History and Science Bulletin, **60**, 98–102.

Gomankov, A.V. & Meyen, S.V. 1986. *Tatarinovaya Flora (sostav i rasprostranenie v pozdnei Permi Evrazii)*. Ordena Trudovogo Krasnogo Znameni Geologichcskii Institut, Moscow [Trudy 401], Proceedings (in Russian).

Gorelova, S.G., Men'shikova, L.V. & Khalfin, L.L. 1973. *Fitostratigrafiya i opredelitel' rastenii Verkhnepaleozoskikh uglenosnikh otlozhenyi Kuznetskogo Basseina*. Sibirskogo Nauchno-Issledovatel'skogo Instituta Geologii, Geofiziki i Mineral'nogo Syr'ya, Kemorovo [Trudy 140].

Goswami, S. & Singh, K.J. 2013. Floral biodiversity and geology of the Talcher Basin, Orissa, India during the Permian–Triassic interval. *Geological Journal*, **48**, 39–56.

'Gu & Zhi'. 1974. *Palaeozoic Plants from China*. Scientific Press, Beijing [in Chinese].

Guerra-Sommer, M. & Cazzulo-Klepzig, M. 2000. Early Permian palaeofloras from southern Brazilian Gondwana: a palaeoclimatic approach. *Brazilian Journal of Geology*, **30**, 486–490.

GUERRA-SOMMER, M., CAZZULO-KLEPZIG, M., SANTOS, J.O.S., HARTMANN, L.A., KETZER, J.M. & FORMOSO, M.L.L. 2008. Radiometric age determination of tonsteins and stratigraphic constraints for the Lower Permian coal succession in southern Paraná Basin, Brazil. *International Journal of Coal Geology*, **74**, 13–27.

HALLE, T.G. 1927. Palæozoic plants from central Shanxi. *Palaeontologia Sinica, Series A*, **2**, 1–316.

HAUBOLD, H. 1980. Die biostratigraphische Gliederung des Rotliegenden (Permo-Siles) im mittleren Thüringer Wald. *Schriftenreihe für geologische Wissenschaften*, **16**, 331–356.

HECKEL, P.H. & CLAYTON, G. 2006. The Carboniferous System. Use of the new official names for the subsystems, series, and stages. *Geologica Acta*, **4**, 403–407.

HILTON, J. & CLEAL, C.J. 2007. The relationship between Euramerican and Cathaysian tropical floras in the Late Palaeozoic: palaeobiogeographical and palaeogeographical implications. *Earth-Science Reviews*, **85**, 85–116.

HUANG, B. 1977. *Permian Flora from the Southeastern Part of the Xiao Hinggan Ling (Lesser Khingan Mt.), NE China*. Geological Publishing House, Beijng [in Chinese].

HUANG, B. 1993. *Carboniferous and Permian Systems and Floras in the Da Hinggan Range*. Geological Publishing House, Beijing [in Chinese].

HUANG, B. 1995. Angara floras in Carboniferous and Permian of China and their relationship with Cathaysian Flora. *In*: XINGXUE, Li (ed.) *Fossil floras of China Through the Geological Ages (English Edition)*. Guangdong Science and Technology Press, Guangzhou, China, 224–243.

IANNUZZI, R. 2010. The flora of Early Permian coal measures from the Paraná Basin in Brazil: a review. *International Journal of Coal Geology*, **83**, 229–247.

IANNUZZI, R. & SOUZA, P.A. 2005. Floral succession in the Lower Permian deposits of the Brazilian Paraná Basin: an up-to-date overview. *In*: LUCAS, S.G. & ZIEGLER, K.E. (eds) *The Nonmarine Permian*. New Mexico Museum of Natural History and Science Bulletin, **30**, 144–149.

JASPER, A., GUERRA-SOMMER, M., CAZZULO-KLEPZIG, M. & MENEGAT, R. 2003. The *Botrychiopsis* genus and its biostratigraphic implications in southern Paraná Basin. *Anais da Academia Brasileira de Ciências*, **75**, 513–535.

JASPER, A., GUERRA-SOMMER, M., CAZZULO-KLEPZIG, M. & IANNUZZI, R. 2007. Biostratigraphic and paleoclimatic significance of *Botrychiopsis* fronds in the Gondwana realm. *In*: WONG, TH.E. (ed.) *Proceedings of the XVth International Congress on Carboniferous and Permian Stratigraphy*, 10–16 August 2003, Utrecht, The Netherlands. Royal Netherlands Academy of Arts and Sciences, Amsterdam, 379–388.

JONGMANS, W.J. & GOTHAN, W. 1937. Betrachtungen über die Egebnisse des zweiten Kongresses für Karbonstraigraphie. *Compte rendu 2e Congrès International de Stratigraphie et de Géologie du Carbonifère (Heerlen, 1935)*, **1**, 1–40.

KERP, J.H.F. 1988. Aspects of Permian palaeobotany and palynology. X. The west- and central-European species of the genus *Autunia* Krasser emend. Kerp (Peltaspermaceae) and the form-genus *Rhachiphyllum* Kerp (callipterid foliage). *Review of Palaeobotany and Palynology*, **54**, 249–360.

KERP, J.H.F. 1996. Post-Varsican late Palaeozoic Northern Hemisphere gymnosperms: the onset to the Mesozoic. *Review of Palaeobotany and Palynology*, **90**, 263–285.

KERP, J.H.F. & FICHTER, J. 1985. Die Makrofloren des saarpfälzischen Rotliegend (?Ober-Karbon–Unter-Perm). *Mainzer Geowissenschaftliche Mitteliungen*, **14**, 159–286.

KOZUR, H. 1978. Beiträge zur Stratigraphie des Perms. Teil III(1): Zur Korrelation der überwiegend kontinentalen Ablagerungen des obersten Karbons und Perms von Mittel- und Westeuropa. *Freiberger Forschungshefte, Reihe C*, **342**, 117–142.

KRINGS, M., KLAVINS, S.D., DIMICHELE, W.A., KERP, H. & TAYLOR, T.N. 2005. Epidermal anatomy of *Glenopteris splendens* Sellaerds nov. emend., an enigmatic seed plant from the Lower Permian of Kansas (U.S.A.). *Review of Palaeobotany and Palynology*, **136**, 159–180.

KUSTATSCHER, E., VAN KONIJNENBURG-VAN CITTERT, J.H., BAUER, K., BUTZMANN, R., MELLER, B. & FISCHER, T.C. 2012. A new flora from the Upper Permian of Bletterbach (Dolomites, N-Italy). *Review of Palaeobotany and Palynology*, **182**, 1–13.

KUSTATSCHER, E., BAUER, K., BUTZMANN, R., FISCHER, T.C., MELLER, B., VAN KONIJNENBURG-VAN CITTERT, J.H. & KERP, H. 2014. Sphenophytes, pteridosperms and possible cycads from the Wuchiapingian (Lopingian, Permian) of Bletterbach (Dolomites, Northern Italy). *Review of Palaeobotany and Palynology*, **208**, 65–82.

LI, X. 1963. Fossil plants of the Yuehmenkou Series, North China. *Palaeontologica Sinica*, **148**, 1–185.

LI, X. 1964. The succession of Upper Palaeozoic plant assemblages of North China. *Compte rendu 5e Congrès International de Stratigraphie et de Géologie du Carbonifère (Paris, 1963)*, **2**, 531–537.

LOOY, C.V. 2007. Extending the range of derived Late Paleozoic conifers: *Lebowskia* gen. nov. (Majonicaceae). *International Journal of Plant Sciences*, **168**, 957–972.

LUCAS, S.G. 2006. Global Permian tetrapod biostratigraphy and biochronology. *In*: LUCAS, S.G., CASSINIS, G. & SCHNEIDER, J.W. (eds) *Non-Marine Permian Biostratigraphy and Biochronology*. Geological Society, London, Special Publications, **265**, 65–93, https://doi.org/10.1144/GSL.SP.2006.265.01.04

LUCAS, S.G., SCHNEIDER, J. & CASSINIS, G. 2006. Nonmarine Permian biostratigraphy and biochronology: an introduction. *In*: LUCAS, S., CASSINIS, G. & SCHNEIDER, J.W. (eds) *Non-Marine Permian Biostratigraphy and Biochronology*. Geological Society, London, Special Publications, **265**, 1–14, https://doi.org/10.1144/GSL.SP.2006.265.01.01

LUCAS, S.G., KRAINER, K. *ET AL*. 2015. Progress report on correlation of nonmarine and marine Lower Permian strata, New Mexico, USA. *Permophiles*, **61**, 10–17.

MAMAY, S.H. 1966. *Tinsleya, A New Genus of Seed-bearing Callipterid Plants from the Permian of North-central Texas*. United States Geological Survey, Professional Papers, **523-E**.

MAMAY, S.H. 1968. *Russellites, New Genus, A Problematical Plant from the Lower Permian of Texas*. United States Geological Survey, Professional Papers, **593-I**.

MAMAY, S.H. 1973. *Archaeocycas* and *Phasmatocycas* – new genera of Permian cycads. *Journal of Research of the United States Geological Survey*, **1**, 687–689.

MAMAY, S.H. 1975. *Sandrewia*, n. gen., a problematical plant from the Lower Permian of Texas and Kansas. *Review of Palaeobotany and Palynology*, **20**, 75–83.

MAMAY, S.H. 1986. New species of Gigantopteridaceae from the Lower Permian of Texas. *Phytologia*, **61**, 311–315.

MAMAY, S.H. 1988. *Gigantoclea* in the lower Permian of Texas. *Phytologia*, **64**, 330–332.

MAMAY, S.H. 1989. *Evolsonia*, a new genus of Gigantopteridaceae from the Lower Permian Vale Formation, north-central Texas. *American Journal of Botany*, **76**, 1299–1311.

MAMAY, S.H. & BREED, W.J. 1970. *Early Permian plants from the Cutler Formation in Monument Valley, Utah*. United States Geological Survey, Professional Papers, **700-B**, 109–117.

MAMAY, S.H. & READ, C.B. 1954. Differentiation of Permian floras in the southwestern United States. *In*: *Compte Rendu, Eighth International Botanical Congress, Paris, Section 5*. French Committee for the Eighth International Botanical Congress, Paris-Nice, 157–158.

MAMAY, S.H. & WATT, A.D. 1971. An ovuliferous callipteroid plant from the Hermit Shale (Lower Permian) of the Grand Canyon, Arizona. *In*: *Geological Survey Research, Chapter C*. United States Geological Survey, Professional Papers, **750-C**, 48–51.

MAMAY, S.H., MILLER, J.M. & ROHR, D.M. 1984. Late Leonardian plants from west Texas: the youngest Paleozoic plant megafossils in North America. *Science*, **223**, 279–281.

MAMAY, S.H., MILLER, J.M., ROHR, D.M. & STEIN, W.E. 1988. Foliar morphology and anatomy of the gigantopterid plant *Delnortea abbottiae*, from the Lower Permian of Texas. *American Journal of Botany*, **75**, 1409–1433.

MAMAY, S.H., CHANEY, D.S. & DIMICHELE, W.A. 2009. *Comia*, a seed plant possibly of peltaspermous affinity: a brief review of the genus and description of two new species from the early Permian (Artinskian) of Texas, *C. greggii* sp. nov. and *C. craddockii* sp. nov. *International Journal of Plant Science*, **170**, 267–282.

MCLOUGHLIN, S. 1992. Late Permian plant megafossils from the Bowen Basin, Queensland, Australia: part 1. *Palaeontographica, Abteilung B*, **228**, 105–149.

MCLOUGHLIN, S. 1994*a*. Late Permian plant megafossils from the Bowen Basin, Queensland, Australia: part 2. *Palaeontographica Abteilung B*, **231**, 1–29.

MCLOUGHLIN, S. 1994*b*. Late Permian plant megafossils from the Bowen Basin, Queensland, Australia: part 2. *Palaeontographica Abteilung B*, **231**, 31–62.

MEYEN, S.V. 1982. The Carboniferous and Permian floras of Angaraland (a synthesis). *Biological Memoirs*, **7**, 1–110.

MEYEN, S.V. 1983. *Paleontologicheskii atlas Permskikh otlozhenii Pechorskogo Basseina*. Nauka, Leningrad.

MEYEN, S.V. 1987. *Fundamentals of Palaeobotany*. Chapman & Hall, London.

MONTAÑEZ, I.P. & POULSEN, C.J. 2013. The Late Paleozoic ice age: an evolving paradigm. *Annual Review of Earth and Planetary Sciences*, **41**, 629–656.

NAUGOLNYKH, S.V. 2000. Mixed Permian floras of Eurasia: a new concept with significance for paleophytogeographic reconstructions. *Paleontological Journal*, **34**, 599–5105.

NAUGOLNYKH, S.V. 2005. Upper Permian flora of Vjazniki (European part of Russia), its Zechstein appearance, and the nature of the Permian/Triassic extinction. *In*: LUCAS, S.G. & ZIEGLER, K.E. (eds) *The Nonmarine Permian*. Proceedings of the XVth International Congress on Carboniferous and Permian Stratigraphy, 10–16 August 2003, Utrecht, **30**, 226–242.

NAUGOLNYKH, S.V. 2013. Permian ferns of western Angaraland. *Paleontological Journal*, **47**, 1379–1462.

OPLUŠTIL, S., ŠIMŮNEK, Z., ZAJÍC, J. & MENCL, V. 2013. Climatic and biotic changes around the Carboniferous/Permian boundary recorded in the continental basins of the Czech Republic. *International Journal of Coal Geology*, **119**, 114–151.

OPLUŠTIL, S., SCHMITZ, M., CLEAL, C.J. & MARTÍNEK, K. 2016. A review of the Middle–Late Pennsylvanian west European regional substages and floral biozones, and their correlation to the Geological Time Scale based on new U–Pb ages. *Earth-Science Reviews*, **154**, 301–335.

POPA, M.E. 1999. The Early Permian megaflora from the Resita Basin, south Carpathians, Romania. *Acta Palaeobotanica*, **2**, (Suppl.), 47–57.

READ, C.B. & MAMAY, S.H. 1964. *Upper Paleozoic Floral Zones and Floral Provinces of the United States*. United States Geological Survey, Professional Papers, **454-K**, 1–35.

REES, P.M., ZIEGLER, A.M., GIBBS, M.T., KUTZBACH, J.E., BEHLING, P.J. & ROWLEY, D.B. 2002. Permian phytogeographic patterns and climate data/model comparisons. *Journal of Geology*, **110**, 1–31.

RETALLACK, G.J., JAHREN, A.H., SHELDON, N.D., CHAKRABARTI, R., METZGER, C.A. & SMITH, R.M.H. 2005. The Permian–Triassic boundary in Antarctica. *Antarctic Science*, **17**, 241–258.

RETALLACK, G.J., METZGER, C.A., GREAVER, T., JAHREN, A.H., SMITH, R.M. & SHELDON, N.D. 2006. Middle–Late Permian mass extinction on land. *Geological Society of America Bulletin*, **118**, 1398–1411.

RIGBY, J.F. 1983. The role of the *Glossopteris* Flora in biostratigraphy. A preliminary assessment in the Reid Dome Beds. *In*: MURRAY, C.G. (ed.) *Permian Geology of Queensland: Proceedings of the Symposium on the Permian Geology of Queensland*. Geological Society of Australia Queensland Division, Brisbane, 221–229.

ROCHA-CAMPOS, A.C., BASEI, M.A.S., NUTMAN, A.P. & SANTOS, P.D. 2006. SHRIMP U–Pb zircon geochronological calibration of the Late Paleozoic supersequence, Paraná Basin, Brazil. *In*: *V South American Symposium on Isotope Geology*, Punta del Este, Uruguay, 298–301.

RONCHI, A., KUSTATSCHER, E., PITTAU, P. & SANTI, G. 2012. Pennsylvanian floras from Italy: an overview of the main sites and historical collections. *Geologia Croatica*, **65**, 299–322.

ROSCHER, M. & SCHNEIDER, J.W. 2005. An annotated correlation chart for continental Late Pennsylvanian and

Permian basins and the marine scale. *In*: LUCAS, S.G. & ZIEGLER, K.E. (eds) *The Nonmarine Permian*. Proceedings of the XVth International Congress on Carboniferous and Permian Stratigraphy, 10–16 August 2003, Utrecht, **30**, 282–291.

ROSCHER, M. & SCHNEIDER, J.W. 2006. Permo-Carboniferous climate: Early Pennsylvanian to Late Permian climate development of central Europe in a regional and global context. *In*: LUCAS, S.G., CASSINIS, G. & SCHNEIDER, J.W. (eds) *Non-Marine Permian Biostratigraphy and Biochronology*. Geological Society, London, Special Publications, **265**, 95–136, https://doi.org/10.1144/GSL.SP.2006.265.01.05

SALMENOVA, K.Z. 1979. Osobennosti Permskoi floriu yzhnogo Kazakhstana i ee svyazi s sosednimi florami. *Paleontologiski Zhurnal*, **1979**, 119–127.

SALVADOR, A. (ed.) 1994. *International Stratigraphic Guide. A Guide to Stratigraphic Classification, Terminology and Procedure*. 2nd edn. International Union of Geological Sciences, Trondheim & Geological Society of America, Boulder, CO.

SCHNEIDER, J.W., LUCAS, S.G. & BARRICK, J.E. 2013. The Early Permian age of the Dunkard Group, Appalachian basin, USA, based on spiloblattinid insect biostratigraphy. *International Journal of Coal Geology*, **119**, 88–92.

SCHOPF, J.M. & ASKIN, R.A. 1980. Permian and Triassic floral biostrratigraphic zones of southern land masses. *In*: DILCHER, D.L. & TAYLOR, T.N. (eds) *Biostratigraphy of Fossil Plant. Successional and Paleoecological Analyses*. Dowden, Hutchinson & Ross, Stroudsburg, PA, 119–152.

SCHWEITZER, H.-J. 1986. The land flora of the English and German Zechstein sequences. *In*: HARWOOD, G.M. & SMITH, D.B. (eds) *The English Zechstein and Related Topics*. Geological Society, London, Special Publications, **22**, 31–54, https://doi.org/10.1144/GSL.SP.1986.022.01.04

SHEN, G. 1995. Permian floras. *In*: LI, Xingxue (ed.) *Fossil Floras of China Through the Geological Ages (English Edition)*. Guangdong Science and Technology Press, Guangzhou, China, 127–223.

SIMAS, M.W., GUERRA-SOMMER, M., CAZZULO-KLEPZIG, M., MENEGAT, R., SANTOS, J.O.S., FERREIRA, J.A.F. & DEGANI-SCHMIDT, I. 2012. Geochronological correlation of the main coal interval in Brazilian Lower Permian: radiometric dating of tonstein and calibration of biostratigraphic framework. *Journal of South American Earth Sciences*, **39**, 1–15.

SINGH, K.J. 2000. Plant biodiversity in the Mahanadi Basin, India, during the Gondwana period. *Journal of African Earth Sciences*, **31**, 145–155.

SINGH, K.J. & CHANDRA, S. 1996. Plant fossils from the exposure near Gopal Prasad Village, Talchir Coalfield, Orissa with remarks on the age of the bed. *Geophytology*, **26**, 69–75.

STEPHENSON, M.H. In press. Permian palynostratigraphy: a global overview. *In*: LUCAS, S.G. & SHEN, S.Z. (eds) *The Permian Timescale*. Geological Society, London, Special Publications, **450**, https://doi.org/10.1144/SP450.2

STEVENS, L.G., HILTON, J., BOND, D.P.G., GLASSPOOL, I.J. & JARDINE, P.E. 2011. Radiation and extinction patterns in Permian floras from North China as indicators of environmental and climate change. *Journal of the Geological Society, London*, **168**, 607–619, https://doi.org/10.1144/0016-76492010-042

STONELEY, H.M.M. 1958. The Upper Permian flora of England. *Bulletin of the British Museum (Natural History), Geology Series*, **3**, 293–337.

SUN, F.-S. 1989. On subdivisions of Angara Floral Province in light of cluster analysis. *Acta Palaeontologica Sinica*, **28**, 773–785 [in Chinese].

SUN, K.-Q. 2006. The Cathaysia Flora and the mixed Late Permian Cathaysian–Angaran Floras in East Asia. *Journal of Integrative Plant Biology*, **48**, 381–389.

TABOR, N.J., DIMICHELE, W.A., MONTAÑEZ, I.P. & CHANEY, D.S. 2013. Late Paleozoic continental warming of a cold tropical basin and floristic change in western Pangea. *International Journal of Coal Geology*, **119**, 177–186.

TAYLOR, E.L., CÚNEO, N.R. & TAYLOR, T.N. 1991. Permian and Triassic fossil forests from the central Transantarctic Mountains. *Antarctic Journal of the United States*, **26**, 23–24.

TEWARI, R., CHATTERJEE, S., AGNIHOTRI, D. & PANDITA, S.K. 2015. *Glossopteris* flora in the Permian Weller Formation of Allan Hills, South Victoria Land, Antarctica: implications for paleogeography, paleoclimatology, and biostratigraphic correlation. *Gondwana Research*, **28**, 905–932.

VAKHRAMEEV, V.A., DOBRUSKINA, I.A., MEYEN, S.V. & ZAKLINSSSKAJA, E.D. 1978. *Paläozoische und mesozoische Floren Eurasiens und die Phytogeographie dieser Zeit*. G. Fischer, Jena, Germany.

VAN WAVEREN, I.M., HASIBUAN, F. *ET AL*. 2005. Taphonomy, palaeobotany and sedimentology of the Mengkarang Formation (Early Permian, Jambi, Sumatra, Indonesia). *In*: LUCAS, S.G. & ZIEGLER, K.E. (eds) *The Nonmarine Permian*. Proceedings of the XVth International Congress on Carboniferous and Permian Stratigraphy, 10–16 August 2003, Utrecht, **30**, 333–341.

VAN WAVEREN, I.M., ISKANDAR, E.A.P., BOOI, M. & VAN KONIJNENBURG-VAN CITTERT, J.H.A. 2007. Composition and palaeogeographic position of the Early Permian Jambi flora from Sumatra. *Scripta Geologica*, **135**, 1–28.

VISSCHER, H., KERP, H., CLEMENT-WESTERHOF, J.A. & LOOY, C.V. 2001. Permian floras of the southern Alps. *Natura Bresciana*, **25**, 117–123.

WACHTLER, M. 2012. *The Latest Artinskian/Kungurian (Early Permian) Flora from Tregonio-Le Fraine in the Val di Non (Trentino, Northern Italy. Preliminary Researches*. Dolomythos Museum, Innichen, Italy.

WAGNER, R.H. 1984. Megafloral zones of the Carboniferous. *Compte rendu 9e Congrès International de Stratigraphie et de Géologie du Carbonifère (Washington, 1979)*, **2**, 109–134.

WAGNER, R.H. 1998. Nuevas tendencias en Paleobotánica. Implicaciones en el estudio de los cambios climáticos, Paleogeografía y Bioestratigrafía. *Monografias de la Academia de Ciencias Exactas Físicas Químicas y Naturales de Zaragoza*, **13**, 9–19.

WAGNER, R.H. & ÁLVAREZ-VÁZQUEZ, C. 2010. The Carboniferous floras of the Iberian Peninsula: a synthesis with geological connotations. *Review of Palaeobotany and Palynology*, **162**, 239–324.

WANG, J. 2010. Late Paleozoic macrofloral asemblages from Weibei Coalfield, with reference to vegetational change through the Late Paleozoic ice-age in the North China Block. *International Journal of Coal Geology*, **83**, 292–317.

WANG, J. & PFEFFERKORN, H.W. 2013. The Carboniferous–Permian transition on the North China microcontinent – Oceanic climate in the tropics. *International Journal of Coal Geology*, **119**, 106–113.

WANG, J., PFEFFERKORN, H.W., XU, A. & ZHANG, Yi. 2005. A note on Permian plant macrofossil biostratigraphy of China. *In*: LUCAS, S.G. & ZIEGLER, K.E. (eds) *The Nonmarine Permian.* Proceedings of the XVth International Congress on Carboniferous and Permian Stratigraphy, 10–16 August 2003, Utrecht, **30**, 352–355.

WANG, Z. 1983. New materials of fossil plants from Shiqianfeng Group in N. China. *Bulletin of the Geological Society, Tianjin*, **1**, 27–80 [in Chinese].

WANG, Z. 1989. Permian gigantic palaeobotanical events in North China. *Acta Palaeontologica Sinica*, **28**, 314–343.

WANG, Z. 1993. Evolutionary ecosystem of Permian–Triassic redbeds in North China: a historical record of global desertification. *In*: LUCAS, S.G. & MORALES, M. (eds) *The nonmarine Triassic*. Proceedings of the XVth International Congress on Carboniferous and Permian Stratigraphy, 10–16 August 2003, Utrecht, **3**, 471–476.

WANG, Z. 1996. Past global floristic changes: the Permian Great Eurasian Floral Interchange. *Palaeontology*, **39**, 189–217.

WANG, Z. & CHEN, A. 2001. Traces of arborescent lycopsids and dieback of the forest vegetation in relation to the terminal Permian mass extinction in North China. *Review of Palaeobotany and Palynology*, **117**, 217–243.

WNUK, C. 1996. The development of floristic provinciality during the Middle and Late Paleozoic. *Review of Palaeobotany and Palynology*, **90**, 5–40.

YANG, G. 2006. *The Permian Cathaysian Flora in Western Henan Province, China – Yuzhou Flora.* Geological Publishing House, Beijing [in Chinese and English].

YANG, G. & WANG, H. 2012. Yuzhou Flora – A hidden gem of the Middle and Late Cathaysian Flora. *Science China Earth Sciences*, **55**, 1601–1619.

YAO, Z.-Q. & LIU, L.-J. 2004. A new gigantopterid plant with cuticles from the Permian of South China. *Review of Palaeobotany and Palynology*, **131**, 29–48.

YAO, Z. & TAYLOR, T.N. 1988. On a new gleicheniaceous fern from the Permian of South China. *Review of Palaeobotany and Palynology*, **54**, 121–134.

ZHAO, X.-H., MO, Z.-G., ZHANG, S.-Z. & YAO, Z.-Q. 1980. Late Permian flora from western Guizhou and eastern Yunnan. *In*: NANJING INSTITUTE OF GEOLOGY AND PALAEONTOLOGY, ACADEMIA SINICA (eds) *Late Permian Coal Bearing Strata and Biota from Western Guizhou and Eastern Yunnan.* Science Press, Beijing, 70–122 [in Chinese].

ZIEGLER, A.M., REES, P.M. & NAUGOLNYKH, S.V. 2002. The Early Permian floras of Prince Edward Island, Canada: differentiating global from local effects of climate change. *Canadian Journal of Earth Sciences*, **39**, 223–238.

Late Pennsylvanian–Early Triassic conchostracan biostratigraphy: a preliminary approach

JOERG W. SCHNEIDER[1,2]* & FRANK SCHOLZE[1,2]

[1]*Technical University Bergakademie Freiberg, Cotta-Strasse 2, D-09596 Freiberg, Germany*

[2]*Kazan Federal University, Kremlyovskaya Street 18, 420008 Kazan, Russia*

**Correspondence: Joerg.Schneider@geo.tu-freiberg.de*

Abstract: Conchostracans are one of the most common fossil animal groups of continental deposits from late Palaeozoic to modern times. Their habitats have ranged from perennial lakes of the Carboniferous and Early Permian to seasonal playa lakes and temporary ponds from the late Early Permian into the Triassic, where they could form mass occurrences. This, together with relatively high speciation rates, makes them ideal guide fossils, especially in otherwise fossil-poor wet and dry red beds. Based on material and data collected since the 1980s from both surface outcrops and well cores in central Europe, a preliminary conchostracan zonation is proposed. We used a conservative approach, erecting assemblage zones comprising two or three species instead of species-range zones with only one or, sometimes, two forms. Assemblage zones are more robust and provide more reliability for each delineated time interval. Isotopically dated occurrences of conchostracan zone species, or co-occurrences of conchostracans, insect zone species and marine index fossils such as conodonts and fusulinids, allow us to correlate our assemblage zones with the marine Standard Global Chronostratigraphic Scale.

Conchostracans, or clam shrimps, are a paraphyletic grouping used traditionally in geosciences to include any fossil valves of branchiopod crustaceans belonging to the suborders Spinicaudata, Laevicaudata and Cyclestherida (e.g. Richter *et al.* 2007; Olesen 2009; Scholze *et al.* 2015, 2016), which are not readily distinguishable from their valve morphologies. Their valves, with a mean length of only 5 mm, are common and widely distributed in continental settings from the Devonian to the present. Conchostracans, as with branchiopods in general, have a very high distribution potential because of their minute, drought and/or frost resistant, wind-transportable dormant eggs (e.g. Kobayashi 1954; Graham & Wirth 2008; Vanschoenwinkel *et al.* 2008; for a detailed discussion see Scholze *et al.* 2016). They often form mass occurrences in lacustrine environments, including lakes, ponds and puddles, which they continue to inhabit today. Conchostracans are one of the most common fossil animal groups found in continental deposits from late Palaeozoic to modern times.

Based on our observations, a shift in habitat occurred from perennial lakes in the Carboniferous to seasonal playa lakes and temporary ponds and puddles during the late Early and the Middle Permian. The latter have remained their preferred habitats into modern times. This shift occurred because the palaeo-equatorial region changed in the Permian from having a tropical humid climate to a steppe-like, seasonally dry climate as a result of Permian–Triassic aridification (e.g. Roscher & Schneider 2006; Schneider *et al.* 2006; Tabor *et al.* 2008; Aref'ev *et al.* 2015). This produced a strong adaptive pressure on some groups, such as conchostracans, which reacted with elevated speciation rates. All of this together made conchostracans ideal guide fossils, especially for continental dry red beds deposited under the extreme climatic conditions of the Permian and Triassic. However, their use in biostratigraphy is hampered by two problems. The first one is the simple, functionally determined construction of their valves, poor in diagnostic features. This has led, and still leads, to an over-interpretation of the smallest variations in their valve anatomy. Thus, the smallest morphological differences, which could easily be caused simply by biological (e.g. intraspecific variation, ontogenetic stages), ecological and taphonomic factors, were and are used to define taxa, leading to taxonomic over-splitting as discussed by Scholze *et al.* (2015, 2016). The second problem derives from the distribution of species through time. Based on our experience, the majority of well-determined conchostracan species occur only during very short time intervals and without any recognizable forerunners. Consequently, evolutionary lineages with clearly defined first and last appearance data (FAD and LAD, respectively) of species, best suited for biostratigraphy, are generally unknown so far. Hence, only first and last occurrence data (FOD and LOD, respectively) are applicable.

From: LUCAS, S. G. & SHEN, S. Z. (eds) 2018. *The Permian Timescale*. Geological Society, London, Special Publications, **450**, 365–386.
First published online December 8, 2016, https://doi.org/10.1144/SP450.6

Because of the high reproduction rate of conchostracans as r-strategists, with one new generation each year, we conclude that speciation processes ran very fast. Thus, the FOD and LOD of a given species might be close to, if not nearly identical with, its FAD and LAD. However, the limited set of constructional elements of the valves could cause the appearance of homeomorphic forms at different times, which could then be erroneously interpreted as a long-lived species. Despite this, because of often missing alternatives in Late Carboniferous and Early–Middle Permian continental sediments, conchostracans might form a reliably applicable biostratigraphic tool for this time interval, as has already been shown in several publications for the latest Permian and parts of the Triassic (e.g. Tasch & Jones 1979; Ghosh *et al.* 1987; Ferreira-Oliveira & Rohn 2010; Kozur & Weems 2010, Scholze *et al.* 2015, 2016). In the following sections, a preliminary conchostracan biostratigraphy for the Late Carboniferous up to the Permian–Triassic transition of the Euramerican region is proposed to stimulate further research in this field.

Material and methods

Most data on conchostracan occurrences in the Pennsylvanian were acquired in the nineteenth and especially in the first half of the twentieth century because of the extensive prospecting and mining of coals, mainly in Europe and North America. Several attempts have been made to use the potential of conchostracans for Late Palaeozoic and Early Mesozoic biostratigraphy by, for example, Novozhilov (1946, 1958, 1970) for the territory of the former Soviet Union in eastern Europe and Asia, by Tasch (1958, 1960, 1987) for North America and Gondwana, by Ghosh *et al.* (1987) and Ghosh (2011) for India, by Jones & Chen (2000) for Australia, by Gallego (2001) for South America, and by Zhang *et al.* (1976) for China. In Germany, after the unpublished PhD thesis of Warth (1963), interest in conchostracan biostratigraphy was rekindled along with intensified hydrocarbon exploration in the late 1970s. In East Germany basic research, initiated and financed by the Central Geological Institute of the former German Democratic Republic (GDR), was focused on the taxonomy and biostratigraphy of insects, conchostracans, fishes and amphibians. Resulting biostratigraphic methods based on collections from surface outcrops were published (e.g. Schneider 1982; Kozur & Seidel 1983*a*, *b*; Martens 1983*a*; Werneburg 1989), but data from hydrocarbon exploration wells remained buried in confidential internal research reports with only one exception – Hoffmann *et al.* (1989).

The positive experiences with conchostracans for biostratigraphic dating and correlation of drilling cores were successfully applied again in the 1990s within the framework of a project of the Deutsche Wissenschaftliche Gesellschaft für Erdöl, Erdgas und Kohle e.V. (DGMK, German Society for Petroleum and Coal Science and Technology), which focused on the revision of the Late Carboniferous stratigraphy of about 45 exploration wells drilled in the north German part of the Variscan foredeep (Fig. 1) (Rößler 1996). As a result, a combined conchostracan–insect range chart for the Westphalian (Bashkirian–Moscovian) up to the Kungurian (late Early Permian) was proposed, based on archimylacrid and spiloblattinid blattoid (cockroach) insect lineage zones and conchostracan assemblage zones. Unfortunately, this DGMK research report (No. 459-3/3; Schneider & Rößler 1996) remained largely unknown because of its very limited distribution; only the Carboniferous part was later published (Schneider *et al.* 2005*a*). The conchostracan collection from drill cores at the Geological Institute of the Technical University Bergakademie Freiberg, which was the result of several projects since the start of the 1980s, was further expanded by subsequent projects, such as, for example, the re-evaluation of hydrocarbon source rock potentials at the southern border of the Variscan foredeep in East Germany (Schneider *et al.* 1998; Gaitzsch *et al.* 1999). This material (Fig. 1), as well as surface samples from ongoing fieldwork in western, central and eastern Europe, Asia, Arabia, North Africa, and North America, forms the database for the conchostracan zonation presented here. For the stratigraphical levels of these conchostracan occurrences see later in this paper.

The material is stored in the following collections, with their respective abbreviations used in the text:

- LANRW – Geologisches Landesamt Nordrhein-Westfalen, Krefeld, Paproth collection;
- SaM – Saarberg Museum, Schiffweiler, Warth collection;
- FG – TU Bergakademie Freiberg, Department of Palaeontology and Stratigraphy, Schneider and Scholze collection;
- UH – University Hamburg, Palaeontological collection, Reichert samples;
- NHMS – Museum of Natural History, Bertholdsburg Castle, Schleusingen;
- NMMNH – New Mexico Museum of Natural History, Albuquerque, Schneider collection.

In addition, the conchostracan collections of the following institutions have also been studied: the British Geological Survey, London, UK; the Jarzembowski collection, Bristol Museum, UK; the Fritsch collection, National Museum, Prague,

Czech Republic; Museum d' Histoire Naturelle, Autun, France; the Lapeiry collection, Museum Lodève, France; Museum of Natural Science, Humboldt University, Berlin, Germany; the Senckenberg Natural History collections, Dresden, Germany.

Methodologically, we use here assemblage zones instead of taxon-range zones based on single guide forms as, for example, applied by Kozur (1993) and Martens (1982, 2012). The problem with using single guide forms as zone species is the unknown full stratigraphic range of a respective species, even if we assume that the FOD and LOD of this species coincide nearly with its FAD and LAD (see earlier in this paper). Assemblage zones based on the co-occurrence of two or more species can provide much more reliable time limits. Nevertheless, it is recognized that only rarely are even two species found together on a single bedding plane, because one conchostracan species as a pioneer organism could occupy almost exclusively a local habitat, leaving no space for others. In a larger sedimentary unit of several metres to decametres in thickness, which represents in continental deposits a relatively short time (several tens of thousands of years), at most two or three species could be sampled (e.g. Tasch 1964; Scholze *et al.* 2015; Schneider *et al.* 2016), enabling the application of assemblage zones.

For our stratigraphy, we use the classic Heerlen West European subdivisions for the Westphalian that divides it into Westphalian A–D (abbreviated as WA–WD) because the conchostracans from coal mines and well cores, stored in older collections, were biostratigraphically designated to these levels by co-occurring plant fossils (e.g. Josten & Amerom 1999). Wagner (1984), Wagner & Álvarez-Vázquez (2010) and Knight & Wagner (2014) have redefined and renamed these units, as well as attempted to link them to the marine Standard Global Chronostratigraphic Scale (SGCS). Recently, Opluštil *et al.* (2016) improved the definition of the continental West European Stages and Substages of Wagner & Álvarez-Vázquez (2010) through detailed reinvestigation of macrofloral biozones in the Central and Western Bohemian basins of the Czech Republic, and correlated them with floral biozones of the Donets Basin (Opluštil *et al.* 2016, fig. 25). In addition, new high-precision U–Pb single-crystal zircon ages from the Bohemian basins are used to calibrate the macrofloral biozones and continental stages, respectively, with the mixed marine–continental section of the Donets Basin (Opluštil *et al.* 2016, fig. 26), which is well dated and correlated to the SGCS by marine fossils (conodonts, foraminifera) and isotopic ages (Davydov *et al.* 2010; Schmitz & Davydov 2012). As a result, a highly improved macrofloral zonation for the Westphalian and its substages, as well as for large parts of the Stephanian, is now established and, for the first time, very precisely linked to the SGCS (Opluštil *et al.* (2016, fig. 28). In the following, we try to adapt our stratigraphic data as much as possible to this new scheme. However, it must be borne in mind that macrofloral zones of the late Stephanian and the Permian are not as precisely defined as in the Westphalian and early Stephanian because increased climate cyclicity during this time interval caused rapid vertical and lateral changes in the habitats of the flora, and therefore in the distribution of plants in space and time.

Stephanian and Permian conchostracan occurrences and assemblage zones, respectively, are as far as possible calibrated by cross-correlation with other biostratigraphic zonations, such as the combined insect and amphibian zonation of Schneider & Werneburg (2006, 2012). Again, for the Stephanian, we use the classical West European subdivision into Stephanian A, B and C (abbreviated as StA–StC) because this subdivision is well supported by the biostratigraphic zonations of Schneider & Werneburg (2006, 2012). In some cases, calibration of conchostracan-bearing sections to the SGCS is possible by isotopic ages (see below). In the best cases, even though currently still rare, we could link the conchostracan assemblage zones with insect lineage zones and also with the marine conodont zones, as shown below, for example, for the Bursum Formation of New Mexico (Schneider *et al.* 2004; Lucas *et al.* 2013, 2016). For the post-Stephanian part, we do not use terms such as Autunian, Saxonian and Thuringian, which are now obsolete, as discussed by Kozur (1978), Broutin *et al.* (1999), Schneider (1996, 2001) and Lucas *et al.* (2006). Instead, we use the lithostratigraphic terms for the corresponding horizons (i.e. the formation names in the respective basins), and try to link these to the SGCS using the regrettably still very scattered isotopic ages (see below and Fig. 4).

A still largely unresolved problem is the classification of Late Palaeozoic conchostracans, especially those of the latest Pennsylvanian and the Early Permian. This is mainly because of insufficiently preserved older type specimens, as well as the differing description styles and a widely differing terminology of carapace morphology (for details see Scholze & Schneider 2015). Here, we primarily follow the classification of Martens and Kozur for several Permian genera and species, which have reached some agreement after lengthy discussion (e.g. Kozur *et al.* 1981; see references below for the respective taxa). In the case of taxa erected by Novozhilov, we follow the profound revision of the Novozhilov-type specimens based on methods developed by our Freiberg team (Stoyan *et al.* 1994) and undertaken by Goretzki (2003). Latest Permian–earliest Triassic conchostracans were

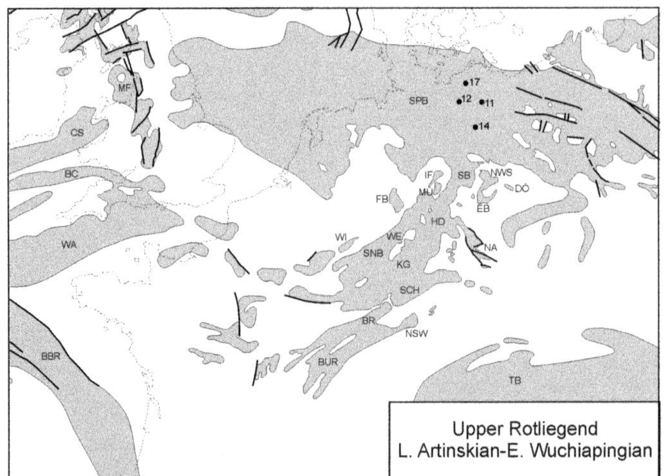
Upper Rotliegend
L. Artinskian-E. Wuchiapingian
MF
CS
BC
WA
BBR
SPB
17
12
11
14
IF
SB
NWS
DÖ
MU
FB
EB
HD
WI
WE
SNB
KG
NA
SCH
BR
NSW
BUR
TB

Stephanian/Lower Rotliegend
L. Gzhelian-M. Artinskian
MF
RU
VFD
NGVC
EL
19
21
WX
WA
CH
FL
WEI
DÖ
IF
SB
EB
NS
IS
KP
CA
TF
FR
WCB
BLG
BKG
ZO
SNB
KG
SCH
BR
NSW
AU
BU
BUR
BMO
SV
ST
TB
BBR
LO

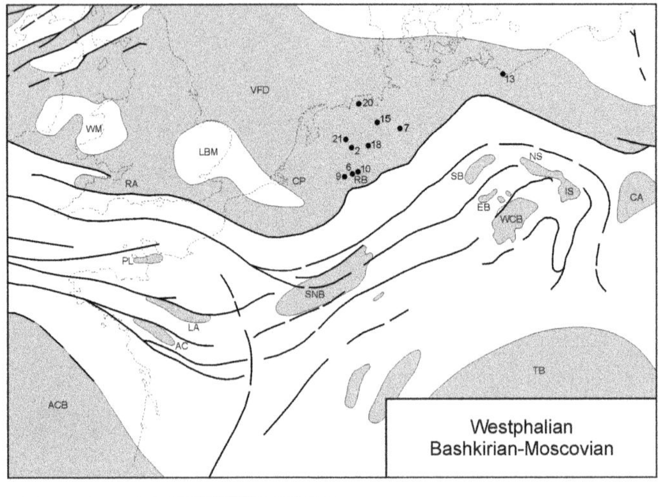
Westphalian
Bashkirian-Moscovian
VFD
13
20
15
7
WM
LBM
21
18
2
6
10
9
RB
RA
CP
SB
NS
IS
CA
EB
WCB
PL
SNB
LA
AC
ACB
TB
0
250
500
750
1000 km

recently revised by Scholze *et al.* (2015, 2016) based on an improved concept for the objective (semi-)quantitative and, therefore, well-reproducible description of fossil conchostracan valves (Scholze & Schneider 2015). To avoid the introduction of new taxon names before a comprehensive revision of Late Carboniferous and Early Permian conchostracans is carried out, we use here informal names derived from discovery localities and designated as 'forms' for potentially new species.

Conchostracan zonation from the Carboniferous (Late Pennsylvanian) to the Permian–Triassic transition

Besides our own observations, we use data from different sources as follows. Pruvost (1919, 1930) published the first detailed information on conchostracan occurrences in the French, Belgian and German parts of the paralic Variscan foredeep (also called the Variscan Foreland Basin) of Westphalian age, much of which remains valid today. Defrise-Gussenhoven & Pastiels (1957) added further information. Wehrli (1931, 1933, 1938) described some freshwater faunas, including conchostracans, from several coal mines in the Lower Saxony part of the Variscan foredeep; unfortunately, his collection was lost during the World War II. Böger & Fiebig (1962) summarized the then current knowledge on conchostracan distribution from the Late Namurian (Early Bashkirian) up to the late Westphalian C (Bolsovian, Middle Moscovian) of the Lower Saxony part of the Variscan foredeep, and compared this with adjacent occurrences in France and Belgium. For the Inner Variscan so-called limnic or intramontane basins, most of the information on the Stephanian (Kasimovian–Gzhelian) occurrences was delivered by Waterlot (1934), Guthörl (1934) and Warth (1963), in his unfortunately unpublished PhD thesis based on the Saar–Lorain Basin (Fig. 2). For the Early Permian basins of central Europe, Martens (1982, 1983*a*, *b*, 1984, 1994) provided comprehensive data on the stratigraphic distribution of conchostracans in surface outcrops. Holub & Kozur (1981) and Kozur & Sittig (1981) added some further data and dealt with synonyms. Lastly, Martens (2012) published a summary of his conchostracan studies. Unfortunately, his biostratigraphic conclusions (Martens 2012, table 1) are in large part inconsistent with other biostratigraphic data, and even with lithostratigraphic data, as well as with isotopic ages of conchostracan-bearing sequences. That is partly due to unresolved taxonomic problems and partly due to Martens' presumptions of the age of the collection horizons, to which the species designations were adapted by him. In the following we try to minimize and, in some cases, to eliminate errors in the lithostratigraphic, biostratigraphic and taxonomic designation of conchostracan taxa. Again, what we present here is a preliminary result, which should stimulate further, more exhaustive research on both the classification and the distribution of Pennsylvanian and Permian conchostracans through time and space.

(1) *Palaeolimnadiopsis pruvosti*–*Pseudestheria* cf. *striata* assemblage zone (1 & 2 in Fig. 2)

Definition and range. FOD of the zone species up to the FOD of the zone species of the following *Palaeolimnadiopsis* form Pudagla–*Pseudestheria* form Ibbenbüren assemblage zone. WA to WA–WB

Fig. 1. Palaeogeographical maps showing the European basins mentioned in the text. According to three instances of basin reorganization, the disappearance of older basins and the origin of new basins in the European Variscides, three maps are shown (after Ziegler 1990). The numbers indicate the position of the respective wells from which the conchostracan samples originate. Basins and massifs: AC, Ancenis Basin; ACB, Aquitaine–Cantabrian Basin; AU, Autun Basin; BBR, Bay of Biscay Rift; BC, Bristol Channel Basin; BKG, Boskovice Graben; BLG, Blanice Graben; BMO, Blancy-Montceau les Mines Basin; BR, Breisgau Basin; BU, Bourbon ÍArchambault Basin; BUR, Burgundy Trench; CA, Carpathian Basin; CH, Channel Basin; CP, Campine Basin; CS, Celtic Sea Basin; DÖ, Döhlen Basin; EB, Erzgebirge Basin; EL, Emsland Low; FB, Franconian Basin; FL, Flechting Block; FR, Frankenberg Bay; HD, Hessian Depression; IF, Ilfeld Basin; IS, Intrasudetic Basin; KG, Kraichgau Basin; KP, Krkonoše–Piedmont Basin; LA, Laval Basin; LBM, London–Brabant Massif; LO, Lodève Basin; MF, Manx–Furness Basin; MÜ, Mühlhausen Basin; NA, Naab Basin; NGVC, North German Volcanite Complex; NS, North Sudetic Basin; NSW, North Switzerland Basin; NWS, Northwest Saxony Basin; PL, Plessis Basin; RA, Radstock Basin; RB, Ruhr Basin; RÜ, Rügen Basin; SB, Saale Basin; SCH, Schramberg Basin; SNB Saar–Nahe/Saar–Lorraine Basin; SPB, Southern Permian/Central European Basin; ST, Saint Etienne Basin; SV, Salvan–Dorénaz Basin; TB, Tethys Basin; TF, Thuringian Forest Basin; VFD, Variscan Foredeep Basin; WA, Western Approaches Basin; WCB, Western and Central Bohemian Basin; WE, Wetterau Basin; WEI, Weissig Basin; WI, Wittlich Basin; WM, Welsh Massif; WX, Wessex Basin; ZÖ, Zöbingen Basin. Wells: **1**, Apeldorn Z2; **2**, De Lutte 6; **3**, Eberswalde 2; **4**, Frenswegen 5; **5**, Grüneberg 3/76; **6**, Hervest 1 and 2; **7**, Hoya Z1; **8**, Jessen 2z/61; **9**, Löhnen 1; **10**, Markenkamp 2; **11**, Mirow 1/74; **12**, Parchim 1/68; **13**, Pudagla 1/68; **14**, Rhinow 5/71; **15**, Sagermeer Nord Z1; **16**, Schadewalde 2/75; **17**, Schwaan 1/76; **18**, UB 414 (Kar 2491); **19**, Varnhorn Z7; **20**, Victorbur Z1; **21**, Wielen Z4; **22**, WISBAW 860/79.

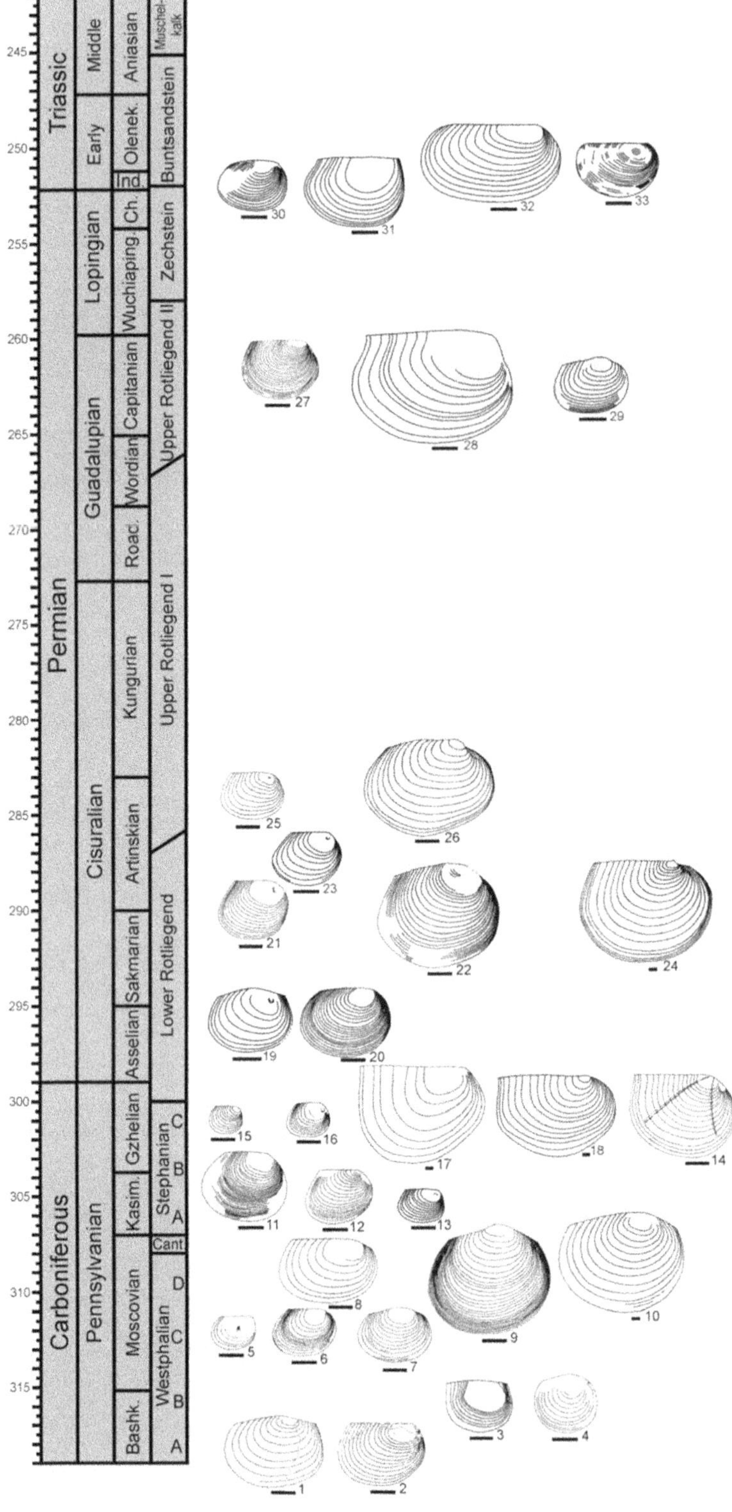
245
250
255
260
265
270
275
280
285
290
295
300
305
310
315
Triassic
Middle
Early
Anisian
Olenek.
Ind.
Muschel-kalk
Buntsandstein
Permian
Lopingian
Ch.
Wuchiaping.
Zechstein
Guadalupian
Capitanian
Wordian
Road.
Upper Rotliegend II
Upper Rotliegend I
Cisuralian
Kungurian
Artinskian
Sakmarian
Asselian
Lower Rotliegend
Carboniferous
Pennsylvanian
Gzhelian
Kasim.
Moscovian
Bashk.
Stephanian
C
B
A
Cant
Westphalian
D
C
B
A
1
2
3
4
5
6
7
8
9
10
11
12
13
14
15
16
17
18
19
20
21
22
23
24
25
26
27
28
29
30
31
32
33

transition (Duckmantian–Early Bolsovian; Late Bashkirian–Early Moscovian).

Accompanying form. *Leaia tricarinata minima* Pruvost 1919.

Comments. The specimens described as *Estheria dawsoni* Jones 1870 by Pruvost (1919) from Nord-Pas-de-Calais, France were later determined as *Palaeolimnadiopsis pruvosti* sp. nov. by Raymond (1946). In addition, Raymond (1946, p. 273) included in this new species forms that were described by Tchernychev (1928) from the Westphalian of the Donetsk Basin. *Estheria dawsoni* Jones 1870 was primarily based on specimens from Early Carboniferous limestones of Horton Bluff, Nova Scotia (but see Raymond 1946). Later, the name was erroneously used by several authors, such as Böger & Fiebig (1962), to determine conchostracans of Early Westphalian age without proof of species identity.

Occurrences. Both zone species are known from the Variscan foredeep of northern France, Pas-de-Calais Basin, Vicoigne and Anzin formations

Fig. 2. Zone index species of the conchostracan assemblage zones for the Late Pennsylvanian–Early Triassic: (**1**) *Palaeolimnadiopsis pruvosti* – LANRW Löhnen 1-3, well Löhnen 1, Ruhr area, Voort Horizon, Bochum Formation, WA, Bashkirian, Variscan foredeep; (**2**) *Pseudestheria* cf. *striata* – LANRW Löhnen 1/1, well Löhnen 1, Ruhr area, Voort Horizon, Bochum Formation, WA, Bashkirian, Variscan foredeep; (**3**) *Palaeolimnadiopsis* form Pudagla – FG 681/Pud 1/86, well Pudagla 1/68 by 5,456.00 m, ?Wieck Formation, WA–WB, Late Bashkirian–Early Moscovian, NE part of the Variscan foredeep; (**4**) *Pseudestheria* form Ibbenbüren – LANRW Kar 2491, well UB 414 (Kar 2491), Ibbenbühren, Domina Horizon, Horst Formation, WB, Moscovian, Lower Saxony part of the Variscan foredeep; (**5**) *Megasitum* form Markenkamp – LANRW Mar 2, well Markenkamp 2, 986.2–986.7 m, Dorsten Formation, WC, Moscovian, Lower Saxony part of the Variscan foredeep; (**6**) *Pseudestheria* form Hoya – LANRW ZFH 6, well Hoya Z1, core 4580.7–4593.2 m, Dorsten Formation, WC, Moscovian, Lower Saxony part of the Variscan foredeep; (**7**) *Pseudestheria* form Hervest – LANRW He 2/4, well Hervest 2, 872.65 m, Dorsten Formation, WC, Moscovian, Lower Saxony part of the Variscan foredeep; (**8**) *Anomalonema reumauxi* – LANRW Wielen Z4/1, well Wielen Z4, 2647.7–2664.7 m, Osnabrück Formation, WD, Moscovian, Lower Saxony part of the Variscan foredeep; (**9**) *Pseudestheria simoni* – FG 681/PdC 3, Pas-de-Calais Basin, Sallaumines, Faisceau de Du Soich and Faisceau d'Edouard, WD, Variscan foredeep, Pas-de-Calais Basin, northern France; (**10**) ?*Palaeolimnadiopsis freysteini* – FG 681 Zw 3/1, Oelsnitz town, Zwickau Formation, WD, Zwickau–Oelsnitz Basin, Moscovian, East Germany; (**11**) *Pseudestheria limbata* – SaM D/545/4, Merchweiler village, lower Ottweiler Subgroup, Göttelborn Formation, StA, Kasimovian, Saar–Nahe Basin; (**12**) *Pseudestheria rimosa* – SaM D/Mi/1, Michelsberg village, lower Ottweiler Subgroup, Göttelborn Formation, StA, Kasimovian, Saar–Nahe Basin; (**13**) *Lioestheria* form Köllerbach – SaM D/490/2, Köllner Mühle, lower Ottweiler Subgroup, Göttelborn Formation, StA, Kasimovian, Saar–Nahe Basin; (**14**) *Rostroleaia* form WISBAW – FG 681 W 860/79/35/42, well WISBAW 860/79, Mansfeld Subgroup, Stephanian, Gzhelian, Saale Basin; (**15**) *Pseudestheria* sp. M – FG Sw 2/75/1, well Schadewalde 2/75, Mansfeld Subgroup, late Stephanian, Gzhelian, Saale Basin; (**16**) *Lioestheria* form Frenswegen – LANRW Fw 5/29/1, well Frenswegen 5, at 2735.0 m, late Stephanian, Lower Saxony part of the Variscan foredeep; (**17**) *Palaeolimnadiopsis* form Jessen – FG 681 Js2z/61/3, well Jessen 2z/61, 637.4 m, Mansfeld Subgruppe, Siebigerode Formation, StC, Gzhelian, Saale Basin; (**18**) *Palaeolimnadiopsis wettinensis* – after Laspeyres (1870, pl. 16, fig. 2), Fischer shaft, Wettin town, Mansfeld Subgroup, Siebigerode Formation, Wettin Subformation, StC, Gzhelian, Saale Basin; (**19**) *Lioestheria paupera* – after Holub & Kozur (1981, pl.4, fig. 1), Cesky Brod, Skalka, StC; (**20**) *Pseudestheria palaeoniscorum* – after Martens (1983*a*, fig. 12), MNG-3625-6-1, Homigtal near Breitenbach village, Goldlauter Formation, Sakmarian, Thuringian Forest Basin; (**21**) *Lioestheria extuberata* – FG O/LO/5, Lochbrunnen near Oberhof village, Lower Oberhof Formation, Sakmarian–Artinskian, Thuringian Forest Basin; (**22**) *Pseudestheria* form Oberhof – after Martens (1983*a*, fig. 25), Lochbrunnen near Oberhof village, Lower Oberhof Fomation, Sakmarian–Artinskian, Thuringian Forest Basin; (**23**) *Palaeolimnadiopsis obenaueri* – after Martens (1984, fig. 3), Meisenheim Formation, Humberg bank (Lebach geods), Artinskian, Saar–Nahe Basin; (**24**) *Lioestheria oboraensis* – FG 681 Leina 2, Leinatal near Finsterbergen village, Upper Oberhof Formation, Artinskian, Thuringian Forest Basin; (**25**) *Lioestheria andreevi* – FG 681 Ta1, Bromacker near Tambach-Dietharz town, Tambach Formation, latest Artinskian, Thuringian Forest Basin; (**26**) *Pseudestheria* form Wilhelmsthal after Martens (1983*a*, fig. 30), MNG-3507-1, Wilhelmsthal near Eisenach town, Eisenach Formation, Kungurian–Rodian, Thuringian Forest Basin; (**27**) *Pseudestheria graciliformis* – FG 681 Ho/Kon1, Konberg quarry, Rothenschirmbach village, Upper Hornburg Formation, Early Capitanian, Saale Basin; (**28**) *Palaeolimnadiopsis* form Rhinow – FG681 118036/1b, well Rhinow 5/71, Hannover Formation, Early Wuchiapingian, Southern Permian Basin; (**29**) *Pseudestheria* form Lieth – UH, salt dome Lieth in Schleswig-Holstein, Hannover Formation, Early Wuchiapingian, Southern Permian Basin; (**30**) *Euestheria gutta* – after Kozur & Seidel (1983*a*, pl. 5, fig. 1), Caaschwitz quarry at the Läuseberg hill, Fulda Formation ('brittle shale'), Induan, Central European Basin; (**31**) *Palaeolimnadiopsis vilujensis* – after Scholze *et al.* (2015, fig. 14A), FG 618/7b, Caaschwitz quarry at the Galgenberg, basal Calvörde Formation, Induan, Central European Basin; (**32**) *Magniestheria mangaliensis* – after Scholze *et al.* (2016, fig. 7B), FG 618/4, Nelben clay pit, Calvörde Formation, Induan, Central European Basin; (**33**) *Cornia germari* – after Scholze *et al.* (2016, fig. 7B), NHMS-WT1436, Kraftsdorf, Bernburg Formation, Induan, Central European Basin.

(Pruvost 1919), Belgium (Raymond 1946) and Germany (Zeller 1987). *Palaeolimnadiopsis pruvosti* seems to be restricted to the WA (e.g. Zeller 1987), for example, in well Löhnen 1, Ruhr area, Germany, Voort Horizon, Bochum Formation, WA. Copeland 1957 (p. 40, pl. 4, fig. 7) reported fragmentary preserved specimens of this species from the Maritime Provinces, Canada, (?)Riversdale Group, WA. Occurrences of the species from the WB reported by Böger & Fiebig (1962, pl. 2) (2 in Fig. 2) and Fiebig (1966) are misidentifications. *Pseudestheria striata* (Münster 1840, p. 280, pl. 159, fig. 19) was primarily described from Dinantian deposits near Hof in Bavaria. The type material has never been reinvestigated in modern times. We use *striata* here in the sense of Pruvost (1919) and designate it as cf. *striata*, which has a range from WA up into the WB, and possibly also into the WC.

(2) *Palaeolimnadiopsis* form Pudagla–*Pseudestheria* form Ibbenbüren assemblage zone (3 & 4 in Fig. 2; Fig. 3a)

Definition and range. FOD of the zone species up to the FOD of the zone species of the following *Megasitum* form Markenkamp–*Pseudestheria* form Hoya assemblage zone. Latest WA and WB (latest Langsettian and Duckmantian; Late Bashkirian–Early Moscovian).

Accompanying form. Pseudestheria cf. *striata* (Münster 1840) (see 2 in Fig. 2).

Comments. Because of minor coal seams and therefore limited mining activity, the fossil content of the WB is not well known.

Occurrences. NE part of the Variscan foredeep, Usedom Island, well Pudagla 1/68 by 5456.00 m, ?Wieck Formation, WA–WB; Lower Saxony part of the Variscan foredeep, well UB 414 (Kar 2491), Ibbenbüren, Domina Horizon, Horst Formation, WB.

(3) *Megasitum* form Markenkamp–*Pseudestheria* form Hoya assemblage zone (5 & 6 in Fig. 2)

Definition and range. FOD of the zone species up to the FOD of the zone species of the following *Anomalonema reumauxi–Pseudestheria simoni* assemblage zone. WC without its upper part (Bolsovian, Middle Moscovian).

Accompanying forms. Pseudestheria form Hervest (7 in Fig. 2).

Occurrences. Lower Saxony part of the Variscan foredeep, well Markenkamp 2, 986.2–986.7 m, Dorsten Formation, WC; well Hervest 1, 848.5–848.7 m, Midgard seam, Upper Dorsten Formation, WC; well Hervest 2, 872.65 m, Dorsten Formation, WC; Mine Franz Haniel, Bottrop, Midgard seam, Upper Dorsten Formation, WC; well Hoya Z1, core 4580.7–4593.2 m, Dorsten Formation, WC; well Victorbur Z1, 5319.0 m and 5359.6–5360.6 m, WC; fragments of conchostracans that belong possibly to this assemblage in well Sagermeer Nord Z1, 4256 m, WC/WD.

(4) *Anomalonema reumauxi–Pseudestheria simoni* assemblage zone (8 & 9 in Fig. 2; Fig. 3b)

Definition and range. FOD of the zone species up to the FOD of the zone species of the following *Pseudestheria limbata–Pseudestheria rimosa–Lioestheria warthi* assemblage zone. Late WC and WD (Late Asturian and Cantabrian; Late Moscovian and Early Kasimovian).

Accompanying forms. ?*Palaeolimnadiopsis freysteini* (Geinitz 1855) (10 in Fig. 2 & Fig. 3c)

Comments. The widespread *A. reumauxi* (Pruvost 1911) is easily recognizable, even as fragments, by its typical ornamentation consisting of regularly spaced, tiny cones along its growth lines. Slightly deformed *A. reumauxi* from the shale above the Upper Kittanning coal, Desmoinesian (WC–WD), Clearfield County, Pennsylvania, were erroneously described by Tasch (1960, pp. 288–289, pl. 42, fig. 2a–c) as a new species, *A. williamsii* sp. nov. According to Böger & Fiebig (1962), the first occurrence of *Pseudestheria simoni* (Pruvost 1911) should be situated in the WA, but, as is so far verifiable, this is based on misidentifications, such as in Defrise-Gussenhoven & Pastiels (1957). Based on Pruvost (1911, 1919), this species occurs first in the Late WC. Large conchostracans similar to ?*Palaeolimnadiopsis freysteini* and up to 1.5 cm or more in length appear sporadically throughout the latest Westphalian up into the Permian, as for example, *P. wettinensis* (Laspeyres 1870); see below and 18 in Figure 2. Their classification from family down to species level needs more study in the future.

Occurrences. The zone species are known from the Variscan foredeep, Pas-de-Calais Basin, northern France, Assise de Bruay (= Bruay Formation), uppermost Faisceau d'Ernestine, latest WC, as well as from the Faisceau de Dusoich and Faisceau d'Edouard of the same formation, WD (Pruvost 1911, 1919); from the UK in the Radstock Basin, Writhlington Formation, WD (Jarzembowski 2004; Jarzembowski & Schneider 2007); from Lower Saxony, Lembeck Formation, Siegfried seams,

WC (Fiebig & Groscurth 1984) and well Wielen Z4, 2647.7–2664.7 m, Osnabrück Formation, WD; from North America, Pennsylvania, Clearfield County, Upper Kittanning coal, Allegheny Formation, Desmoinesian, Pennsylvanian, WC–WD (Tasch 1960). ?*Palaeolimnadiopsis freysteini* (Geinitz 1855) is known from the type locality Zwickau-Oberhohndorf, Zwickau Sub-basin of the Erzgebirge Basin, upper Zwickau Formation, Scherbenkohlen seam, WD (to Cantabrian?) and from the Karl-Liebknecht mine, Lugau, Oelsnitz Sub-basin of the Erzgebirge Basin, WD (Schneider *et al.* 2005*b*); Variscan foredeep, Lower Saxony, well De Lutte 6, 3163.5 m, WD.

(5) *Pseudestheria limbata–Pseudestheria rimosa–Lioestheria* form Köllerbach assemblage zone (11, 12 & 13 in Fig. 2)

Definition and range. FOD of the zone species up to the FOD of the zone species of the following *Pseudestheria minima–Lioestheria* form Frenswegen–*Palaeolimnadiopsis* form Jessen assemblage zone. StA and ?StB (Barruelian–Saberian; Middle Kasimovian–Early Gzhelian).

Accompanying form. Leaia baentschiana Beyrich 1864.

Comments. The type specimens of *P. limbata* (Goldenberg 1877) and *P. rimosa* (Goldenberg 1877) are most probably different ontogenetic stages of the same species (Waterlot 1934; Goretzki 2003).

The fossil content of the StB is not well investigated because of a lack of workable coal seams (see below).

Occurrences. P. limbata/rimosa after Guthörl (1934) and Martens (2012) Saar–Nahe Basin, lower Ottweiler Subgroup, StA (Barruelian, Kasimovian). *Lioestheria* form Köllerbach is known from the lower Ottweiler Subgroup, Göttelborn Formation, StA (Barruelian, Kasimovian) of the same basin. Kozur *et al.* (1992) reported conchostracans from the Tinajas Member of the Atrasado Formation in the famous Kinney Brick Quarry, Manzanita Mountains, New Mexico (Lucas & Allen 2011). They are not well preserved, but appear in the curvature of the posterior margin, of the arrangement and number of growth lines, the straight dorsal margin and the anterior position of the umbo very close to *P. rimosa*. *Lioestheria carinacurvata* Martens & Lucas 2005 from the stratotype of the Tinajas Member compares in most features to *Lioestheria* form Köllerbach (see 13 in Fig. 2), but differs from it in a 'radial rib' on the larval valve. A reinvestigation has to show whether this 'rib' is a taphonomic effect only and both forms are otherwise conspecific. *Leaia baentschiana* Beyrich 1864 is reported from the Saar–Nahe Basin, lower Ottweiler Subgroup, StA (Barruelian, Kasimovian) by Guthörl (1934) and Warth (1963), from the Cinera–Matallana coalfield (Spain), Cascajo Formation, StB (Wagner 1971), and from North America, lower Conemaugh Formation, late Desmoinesian, Cantabrian/?StA (Martens 1986 based on Raymond 1946). Leaiaid conchostracans are widespread and can form mass occurrences (*Leaia* horizons) in Late Pennsylvanian deposits, but their classification is complicated because of intraspecific variability (?ecospecies), and because even minor deformations of the valves very strongly modify the appearance of diagnostic features.

(6) *Pseudestheria* sp. M–*Lioestheria* form Frenswegen–*Palaeolimnadiopsis* form Jessen assemblage zone (15, 16 & 17 in Fig. 2)

Definition and range. FOD of the zone species up to the FOD of the zone species of the following *Lioestheria paupera–Pseudestheria palaeoniscorum* assemblage zone. StB and StC, Late Gzhelian–?earliest Asselian.

Accompanying forms. Palaeolimnadiopsis wettinensis (Laspeyres 1870) (18 in Fig. 2); *Pseudestheria* cf. *limbata* and *Rostroleaia* form WISBAW (14 in Fig. 2).

Comment. Because of the dominance of wet red-bed facies and minor, non-workable coal seams, the fossil content of the StB has not been well investigated. StB and StC therefore cannot be distinguished thus far based on conchostracans.

Occurrences. Pseudestheria sp. M Martens 1983*a* (15 in Fig. 2) – Thuringian Forest Basin, Möhrenbach Formation, StC; Krkonoše Basin, Ploužnice Horizon, Semily Formation, StB/C; Saar–Nahe Basin, Breitenbach Formation, StC; Saale Basin, Siebigerode Formation, Wettin Subformation, StC (Schneider *et al.* 2005*c*); Saale Basin, well WISBAW 860/79, at 169.4 m, Siebigerode Formation, late Stephanian; well Schadewalde 2/75, 2.146.5–2148.0 m, latest Stephanian–lowermost Rotliegend; well Jessen 2z/61, 634.0–653.0 m (together with leaiaid conchostracans), 949.3 m and 963.0–974.0 m, Mansfeld Subgroup, StB–StC; Variscan foreland basin, Lower Saxony, Ems depression, well De Lutte 6, at 2780.4–2781.3 m *Ps.* sp. M very common, (?)higher Stephanian. *Lioestheria* form Frenswegen (16 in Fig. 2) – Variscan foreland basin, Lower Saxony, Ems depression, well Frenswegen 5, 2735.0 m; well De Lutte 6, 2954.55 m (together with leaiaid conchostracans); well Apeldorn Z2, 4137.0 m and 4036.55–4440.0 m, higher

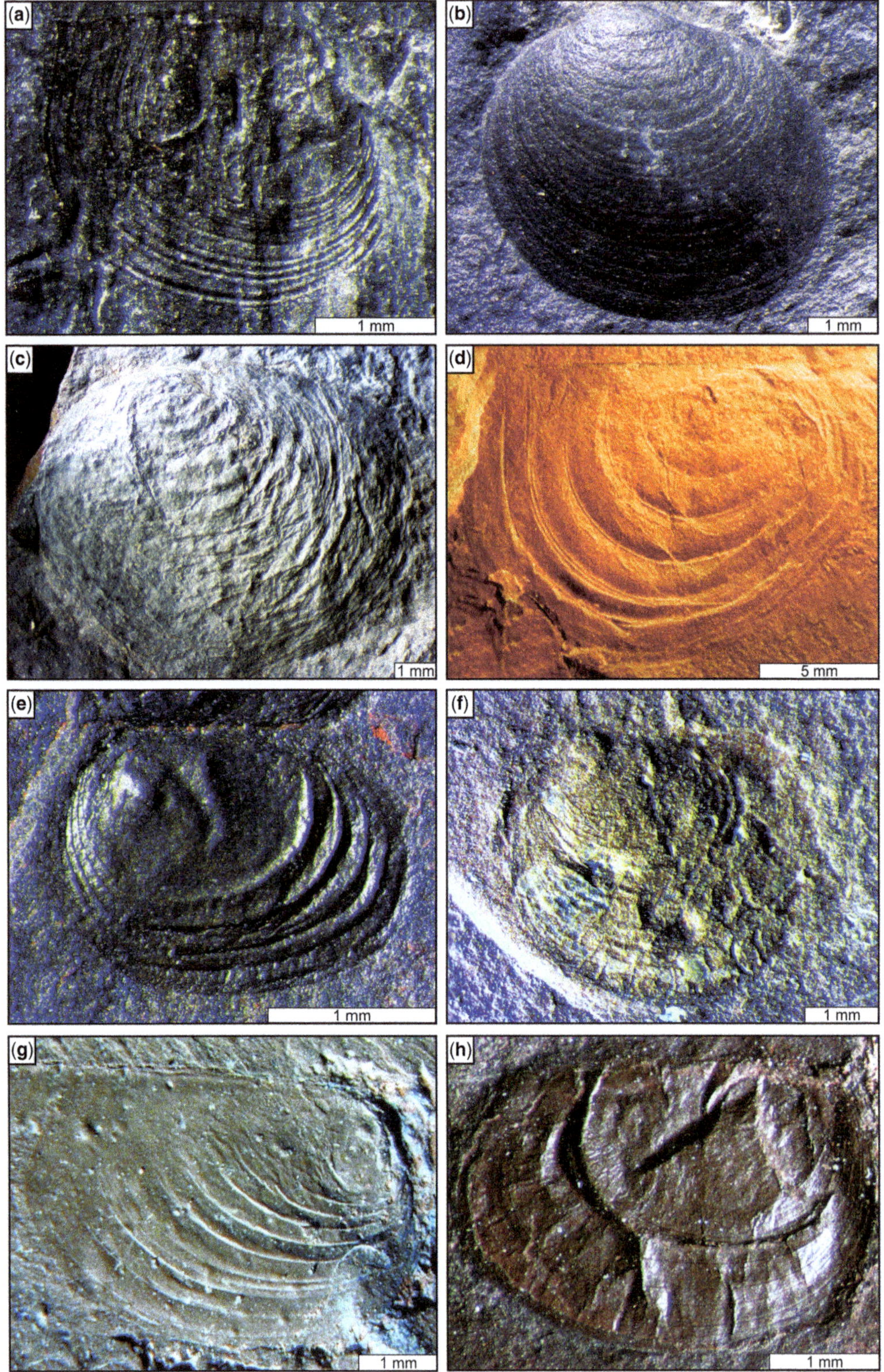
(a)
1 mm
(b)
1 mm
(c)
1 mm
(d)
5 mm
(e)
1 mm
(f)
1 mm
(g)
1 mm
(h)
1 mm

Stephanian; well Varnhorn Z7, 4088.7–4089.7 and 4146.0–4147.0 m, higher Stephanian. *Palaeolimnadiopsis* form Jessen (17 in Fig. 2) – well Jessen 2z/61, 637.4 m, StC (Rößler 1996, pl. 5, fig. 4); *Pseudestheria* cf. *limbata*, were found in the Saale Basin, well Jessen 2z/61, 949.3 m (StB/C) and 963.0–974.0 m and in the Donetsk Basin, Mironovskaja svita, p5 coal and black shale, Late Gzhelian. *Rostroleaia* form WISBAW (14 in Fig. 2) is known from the wells WISBAW 860/79 and Jessen Js2z/80, Stephanian C, Upper Mansfeld Subgroup, Siebigerode Formation of the Saale Basin.

P. wettinensis (18 in Fig. 2) from the Siebigerode Formation, Wettin Subformation, StC, of the Saale Basin was primarily described as *Leaia wettinensis* by Laspeyres (1870). But what he interpreted as 'radial folds' typical of *Leaia* is an artefact caused by slight deformation. *Montcestheria orri* Vannier *et al.* 2003 from latest Stephanian–earliest Permian, Assise de Montceau, Blanzy-Montceau Basin, France, is a junior synonym of *P. wettinensis* (Laspeyres 1870). Rare specimens of palaeolimnadiopsids of 1.3 cm length from the Lower Ottweiler Subgroup, StA, described by Warth (1963) as *P. obenaueri*, may belong to a forerunner species of *P. wettinensis*.

The Bursum Formation, earliest Asselian, of the Lucero basin, New Mexico, contains common small pseudestheriids close to *Pseudestheria* sp. M, additionally Schneider *et al.* (2004, p. 255, fig. 8D, E) reported a small *Lioestheria* that is identical with *Lioestheria* form Frenswegen, and Schneider *et al.* (2016) document exceptional large conchostracans identical to *Palaeolimnadiopsis* form Jessen (17 in Fig. 2; Fig. 3d).

(7) *Lioestheria paupera–Pseudestheria palaeoniscorum* assemblage zone (19 & 20 in Fig. 2)

Definition and range. FOD of the zone species up to the FOD of the zone species of the following *Lioestheria extuberata–Lioestheria oboraensis–Pseudestheria* form Oberhof assemblage zone. Lower Rotliegend, Asselian–Sakmarian.

Comments. *Estheria* (*Lioestheria*) *lallyensis* Depéret & Mazeran 1912 was synonymized by Holub & Kozur (1981) with *Lioestheria paupera* (Fritsch 1901). Martens (2012) synonymized *Lioestheria parvula* Martens 1983*a* with *Lioestheria lallyensis* Depéret & Mazeran 1912 emend. Kozur *et al.* 1981. Consequently, the *Lioestheria lallyensis* zone and the *Lioestheria parvula* zone of Martens (1983*b*) fall together and match with the *Lioestheria paupera* zone of Holub & Kozur (1981). *Pseudestheria convexa* Martens 1983*a* and *Pseudestheria breitenbachensis* Martens 1983*a* are junior synonyms of *Pseudestheria palaeoniscorum* (Fritsch 1901); both of Martens' species do not, neither lithostratigraphically nor biostratigraphically, come from the Stephanian (Late Pennsylvanian) Gehren Subgroup of the Thuringian Forest Basin as reported by Martens (1983*a*) up to Martens (2012), but from the Lower Goldlauter Formation, Early Sakmarian, as shown in several publications from Schneider *et al.* (1982) up to Lützner *et al.* (2012).

Occurrences. Thuringian Forest Basin, Ilmenau, Manebach and Goldlauter formations; Saar–Nahe Basin, Lower Glan Subgroup, Qirnbach–lowermost Meisenheim formations; Krkonoše basin, Vrchlaby Formation, Rudnik Horizon; Autun Basin, Grès de Lally; Saale Basin, well Schadewalde 2/75, 1825.0 m; Southern Permian Basin, well Grüneberg 3/76, well Eberswalde 2, Grüneberg Formation.

(8) *Lioestheria extuberata–Lioestheria oboraensis–Pseudestheria* form Oberhof assemblage zone (21, 22 & 23 in Fig. 2; Fig. 3e)

Definition and range. FOD of the zone species up to the FOD of the zone species of the following *Lioestheria andreevi–Pseudestheria* form Wilhelmsthal

Fig. 3. Some representative Late Pennsylvanian–Early Triassic conchostracans from Europe. (**a**) *Palaeolimnadiopsis* form Pudagla, compacted cast of a right valve – FG 681/Pud 1/86, well Pudagla 1/68 by 5456.00 m, WA–WB, Late Bashkirian–Early Moscovian, NE part of the Variscan foredeep; (**b**) *Pseudestheria simoni*, left valve – FG 681/PdC 3, Pas-de-Calais Basin, Sallaumines, Faisceau de Dusoich and Faisceau d'Edouard, WD, Moscovian, Variscan foredeep, Pas-de-Calais Basin, northern France; (**c**) ?*Palaeolimnadiopsis freysteini*, cast of a left valve – FG 681 Zw3/1, Oelsnitz town, Zwickau Formation, WD, Moscovian, Zwickau–Oelsnitz Sub-basin of the Erzgebirge Basin, East Germany; (**d**) *Palaeolimnadiopsis* form Jessen, imprint of a right valve – FG 681 Js2z/61/3, well Jessen 2z/61, 637.4 m, Mansfeld Subgroup, Siebigerode Formation, StC, Gzhelian, Saale Basin, Germany; (**e**) *Lioestheria extuberata*, cast of a left valve – FG O/LO/2, Lochbrunnen near Oberhof village, Lower Oberhof Formation, Sakmarian–Artinskian, Thuringian Forest Basin; (**f**) *Pseudestheria graciliformis*, cast of a right valve – FG 681H/Kon1, Konberg quarry, Rothenschirmbach village, Upper Hornburg Formation, Early Capitanian, Saale Basin; (**g**) *Palaeolimnadiopsis* form Rhinow, cast of a right valve – FG 681 118036/1b, well Rhinow 5/71, Hannover Formation, Early Wuchiapingian, Southern Permian Basin; (**h**) *Palaeolimnadiopsis vilujensis*, cast of a right valve – FG 618/7b, Caaschwitz quarry at the Galgenberg hill, Calvörde Formation, Induan, Central European Basin.

assemblage zone. Higher Lower Rotliegend (Late Sakmarian–Late Artinskian).

Accompanying form. Palaeolimnadiopsis obenaueri (Guthörl 1931) (24 in Fig. 2) is in the Saar–Nahe Basin in places common in the uppermost Meisenheim Formation, Humberg bank ('Lebach geodes').

Comments. Lioestheria pseudotenella Martens 1983*a*, type locality Lochbrunnen near Oberhof, Thuringian Forest Basin, is a junior synonym of *Lioestheria extuberata* (Jones & Woodward 1899) that has been reported from the same locality near Oberhof by Holub & Kozur (1981). Consequently, the *L. extuberata* zone of Holub & Kozur (1981) coincides with the *L. pseudotenella* zone of Martens (1983*b*). *Pseudestheria* n. sp. L in Martens (1983*b*) is called here *Pseudestheria* form Oberhof. *Lioestheria oboraensis* Holub & Kozur 1981 (24 in Fig. 2), wrongly described as *L. lallyensis* by Martens (1982), is regarded as transitional between *L. extuberata* (21 in Fig. 2; Fig. 3e) and *L. andreevi* by Holub & Kozur (1981) (25 in Fig. 2) and in the present paper.

Occurrences. Thuringian Forest Basin, ?uppermost Goldlauter Formation and Lower Oberhof Formation; Saar–Nahe Basin, upper Glan Subgroup, middle Meisenheim and Disibodenberg formations. *Lioestheria oboraensis* appears in the upper Oberhof Formation of the Thuringian Forest Basin and in the Schadewalde 2/75 well, 1462.0–1477.0 m, of the Saale Basin; the type locality is the famous Obora insect horizon in the Boskovice Graben, Letovice Formation, late Artinskian.

Martens & Lucas (2005) reported conchostracans from the NMMNH locality 3922, Placitas, of the Abo Formation in the Sandoval County of New Mexico. They were described as *Lioestheria* cf. *M. monticula* (*Lioestheria* cf. *L. monticula* in the figure caption of their fig. 8) and compared with *Lioestheria monticula* Martens 1983*a*. The latter is a junior synonym of *Lioestheria andreevi* (Zaspelova 1968) and was described under this name from the Tambach Formation, Artinskian, of the Thuringian Forest Basin by Holub & Kozur 1981. Beyond the specimens figured by Martens & Lucas (2005, their figs 7 & 8), only their figure 8F resembles *L. andreevi* (25 in Fig. 2) or, much better, *Lioestheria extuberata* (21 in Fig. 2). The forms figured in figures 7C, D and 8C, D of Martens & Lucas (2005) are similar in the number of growth lines, the straight dorsal margin, the strong curvature of the posterior margin, the small larval valve and the anterior position of the umbo with *Pseudestheria* form Oberhof, Sakmarian (22 in Fig. 2). Different, however, is the much smaller size of the largest, adult forms, as in Martens & Lucas (2005, their figs 7D & 8C, D). Possibly two species are represented in the material from Placitas. To answer these questions, a reinvestigation of the Placitas conchostracans, and of more and better preserved specimens of *Pseudestheria* form Oberhof, is desirable.

(9) Lioestheria andreevi–Pseudestheria form Wilhelmsthal assemblage zone (25 & 26 in Fig. 2)

Definition and range. FOD of the zone species up to the FOD of the zone species of the following *Pseudestheria graciliformis–Palaeolimnadiopsis* form Rhinow–*Pseudestheria* form Lieth assemblage zone. Higher Lower Rotliegend and Upper Rotliegend I (late Artinskian–Rodian/Wordian).

Comments. In the Tambach Formation, Thuringian Forest Basin, among typical *L. andreevi* appear single specimens still close to the transitional species *P. oboraensis*. The same is observed in the Wadern Formation, Sponheim Subformation, Sobernheim Horizon of the Saar–Nahe Basin. *Pseudestheria* form Wilhelmsthal (= *Pseudestheria* n.sp. W Martens 1983*a*) (26 in Fig. 2) of the Eisenach Formation, Thuringian Forest/Werra Basin, as well as *Pseudestheria* n.sp. A Martens 1983*a*, Michelbach Formation, Kraichgau Basin, *Pseudestheria fritschi* Kozur & Sittig 1981 and *Protolimnadia*? *sulzbachensis* Kozur & Sittig (1981), Kraichgau Basin, Upper Rotliegend Michelbach Formation, Sulzbach Subformation are very close together, in addition to *Pseudestheria brevis* Raymond 1946 of the Wellington Formation of Oklahoma. The latter are often preserved in great detail in 3D and show a prominent node on the tip of the larval valve. Unfortunately, this feature is not visible in the variably compacted valves of the aforementioned species.

Specimens described as *Protolimnadia kowalczyki* Kozur 1992 from the Upper Rotliegend lower Bleichenbach Formation of the Wetterau Basin are, in our opinion, only juvenile individuals of *Palaeolimnadiopsis* form Wilhelmsthal.

Occurrences. Kimjinskaja svita, NW Central Kazakhstan (after Holub & Kozur 1981); Thuringian Forest Basin, ?Upper Oberhof Formation, Rotterode, Tambach and Eisenach formations; Saar–Nahe Basin, Nahe Subgroup, Standenbühl Formation, Sobernheim Member; ?Kraichgau Basin, Michelbach Formation; ?Wetterau Basin, Bleichenbach Formation; ?Lodève Basin, Salagou Formation; Saale Basin, well Schadewalde 2/75, 1462.0–1482.0 m; Southern Permian Basin, Müritz Subgroup, wells Schwaan 1/76, Mirow 1/74 and Parchim 1/68 (Hoffmann *et al.* 1989).

(10) *Pseudestheria graciliformis–Palaeolimnadiopsis* form Rhinow–*Pseudestheria* form Lieth assemblage zone (17, 28 & 29 in Fig. 2; Fig. 3f, g)

Definition and range. FOD of the zone species up to the FOD of the zone species of the following *Euestheria gutta–Palaeolimnadiopsis vilujensis* fauna. Upper Rotliegend II (Capitanian–early Wuchiapingian).

Comments. In the late Cisuralian, the sedimentary record in all western European basins became increasingly patchy (see Fig. 4). The Guadalupian is mostly represented by hiatuses and, even if sediments occur, their real age is highly questionable. Vulcanites, and therefore isotopic ages, are completely missing in this interval; only the Illawarra reversal provides a time control, if it is detected. In large parts of central Europe, continental sedimentation is interrupted by the early Wuchiapingian marine Zechstein transgression (Legler *et al.* 2005; Legler & Schneider 2008). Therefore, conchostracans reappear and become common again only in the playa deposits of the uppermost Zechstein (Changhsingian). Consequently, it remains an open question as to whether the (10) assemblage zone follows without a gap after the (9) assemblage zone, and if (10) really is consecutively followed by the (11) assemblage zone.

Occurrences. P. graciliformis Martens 1983*a* (27 in Fig. 2), Saale Basin, Upper Hornburg Formation; *Palaeolimnadiopsis* form Rhinow (28 in Fig. 2), Southern Permian Basin, well Rhinow 5/71, Hannover Formation (Legler & Schneider 2008); *Pseudestheria* form Lieth, Southern Permian Basin, surface outcrop at the salt dome Lieth, Hannover Formation (Legler *et al.* 2005).

Late Permian assemblage zones of Kozur & Weems (2010)

Kozur & Weems (2010) defined two assemblage zones for the Late Permian, the *Falsisca*

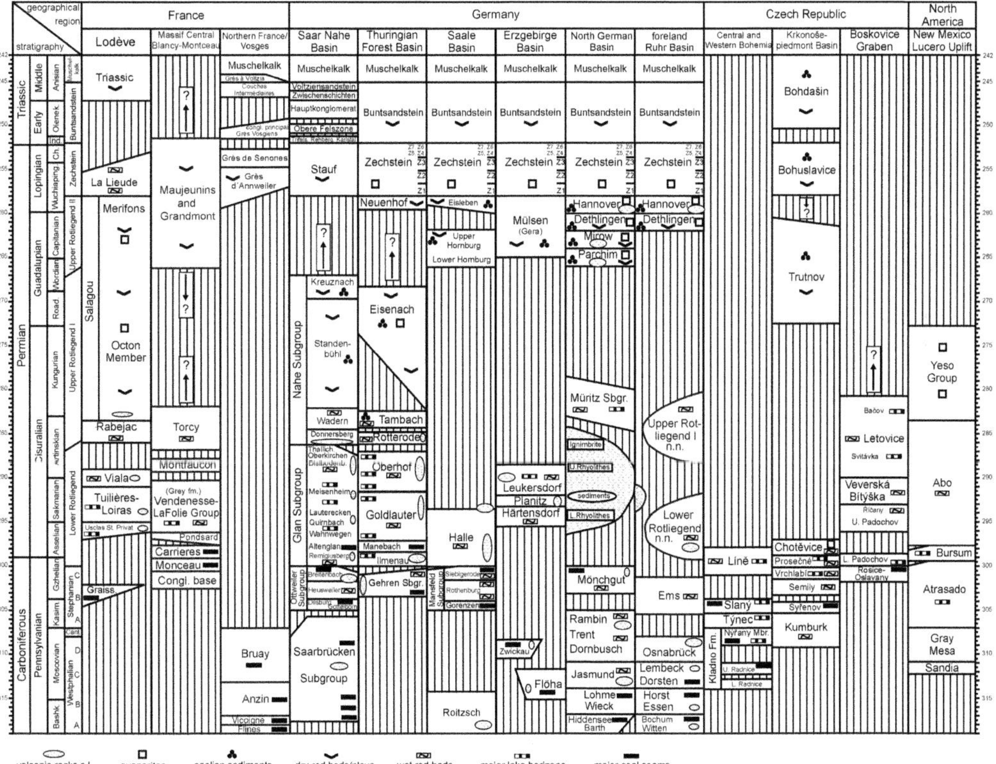

Fig. 4. Correlation chart of the basins with their respective formations in the European Variscides mentioned in the text. Correlations are based on data from Schneider *et al.* (2006), Roscher & Schneider (2006), Schneider & Werneburg (2006, 2012), Schneider & Romer (2010), Lucas *et al.* (2013), Michel *et al.* (2015) and Opluštil *et al.* (2016).

zavjalovi–Tripemphigus minutus zone (Changhsingian) and the *Falsisca turaica–Falsisca zavjalovi* zone (lower part of the Late Changhsingian). In the present study, these zones are not used because they were previously defined by index taxa that have taxonomic problems at the genus and species levels. As discussed by Scholze *et al.* (2016, p. 188, '*Falsisca* problem'), the genus *Falsisca* (= *Limnadia* (*Falsisca*) Novozhilov 1970) was correctly considered by Webb (1978) to be a younger synonym of *Palaeolimnadiopsis* Raymond 1946. The species '*Falsisca*' *zavjalovi* (= *Limnadia* (*Falsisca*) *zavjalovi* Novozhilov 1970) was interpreted by Goretzki (2003, p. 133) as adult individuals of *Palaeolimnadiopsis brevis* Novozhilov 1958, which was described from the Vetlugian Regional Stage of Early Triassic (Induan) age. Therefore, this species is not confined to a Late Permian (Changhsingian) age in the sense of Kozur & Weems (2010). The second species, *Tripemphigus minutus* (= *Trinodus minutus* Liu 1987), was considered by Kozur & Weems (2010, p. 362) to be a possible younger synonym of *Tripemphigus sibiricus* Novozhilov 1965. Based on their valve morphologies, however, it seems more likely that *Tripemphigus sibiricus* and *Tripemphigus minutus* instead belong to *Pseudestheria* and *Polygrapta* (or juvenile *Palaeolimnadiopsis*), respectively. In our opinion, their real phylogenetic relationships require further investigation. However, as *Trinodus minutus* was originally described by Liu (1987, p. 107) to be of Triassic age, the Late Permian (Changhsingian) age for the *Falsisca zavjalovi–Tripemphigus minutus* zone according to Kozur & Weems (2010) is highly questionable. The third species, *Falsisca turaica* (= *Limnadia turaica* Novozhilov 1965), was originally defined on extremely poorly preserved holotype and paratype material. According to Goretzki (2003, p. 144), this species is not reproducible and therefore should be considered as a nomen nudum. Owing to the taxonomic inconsistencies of the index taxa of the *Falsisca zavjalovi–Tripemphigus minutus* zone and the *Falsisca turaica–Falsisca zavjalovi* zone of Kozur & Weems (2010), both zones cannot be used as reliable indicators of a Late Permian age. Instead, an alternative definition of conchostracan assemblage zones is currently in development, based on newly collected material from Late Permian sections of the Zhukov Ravine locality near the city of Gorokhovets in central Russia and from the Dalongkou section in the Xinjiang Province of NW China (field campaigns by FS and JWS in 2014–2015).

(11) *Euestheria gutta* interval (30 in Fig. 2)

Definition and range. FOD of *Euestheria gutta* (Lutkevitch 1937) up to the FOD of the following *Cornia germari* (Beyrich 1857) interval. Early Triassic (Early Induan).

Accompanying forms. Palaeolimnadiopsis vilujensis Varentsov 1955 (31 in Fig. 2; Fig. 3h), *Magniestheria mangaliensis* (Jones 1862) (32 in Fig. 2), *Cornia germari* (33 in Fig. 2) and *Rossolimnadiopsis* sp. (Scholze *et al.* 2015).

Comments. '*Falsisca postera*' (= *Falsisca eotriassica postera* Kozur & Seidel 1983*a*, *b*) of the *Falsisca postera* zone (Kozur & Weems 2010) was regarded as synonymous with *Palaeolimnadiopsis vilujensis* Varentsov 1955 by Scholze *et al.* (2016, p. 188) and '*Falsisca eotriassica*' (= *Falsisca eotriassica eotriassica* Kozur & Seidel 1983*a*, *b*) of the *Falsisca eotriassica* zone (Kozur & Weems 2010) is synonymized with *Magniestheria mangaliensis* (Scholze *et al.* 2016).

Both the *Euestheria gutta* fauna and the following *Cornia germari* fauna are not currently defined as zones as the real stratigraphic ranges of the respective species are currently under study (Scholze *et al.* 2016).

Occurrences. Euestheria gutta and *Palaeolimnadiopsis vilujensis* occur in the Moscow Syncline, Vokhmian Regional Stage (Induan) within the *Tupilakosaurus wetlugensis* zone, Vokhma Formation (Scholze *et al.* 2015); *Euestheria gutta* occurs in South China, western Guizhou Province, Kayitou Formation (e.g. Chu *et al.* 2013) and together with *Palaeolimnadiopsis vilujensis* in the Central European Basin, upper Fulda Formation (uppermost Zechstein Group) to Calvörde Formation (Lower Buntsandstein Subgroup) (Scholze *et al.* 2015, 2016). *Rossolimnadiopsis* Novozhilov 1958 was recently reported from the Early Triassic Nimra Member, Ma'in Formation, Dead Sea region, Jordan (Abu Hamad *et al.* 2015); occurrences of *Magniestheria mangaliensis* (Jones 1862) were reported by Liu & He (2000) as '*Euestheria mangaliensis*', together with marine bivalves in the Sunjiagou Formation, Early Triassic (Induan) of China.

(12) *Cornia germari* interval (33 in Fig. 2)

Definition and range. FOD of *Cornia germari*. The upper stratigraphic range of this interval is currently in revision and therefore preliminarily confined here by the LOD of *Cornia germari*. Early Triassic (Induan).

Accompanying form. Magniestheria mangaliensis (Jones 1862) (32 in Fig. 2) – see (11) earlier.

Comments. Vertexia tauricornis Lutkevitch 1941 and *Molinestheria seideli* Kozur 1980, used by Kozur & Weems (2010) for the definition of the

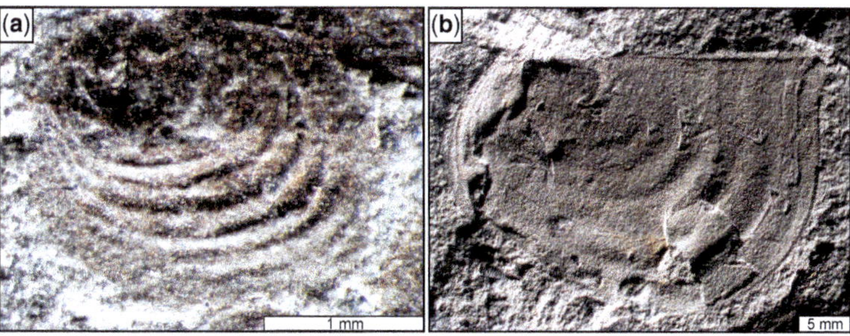

Fig. 5. *Lioestheria* form Frenswegen (**a**) and *Palaeolimnadiopsis* form Jessen (**b**) from depositional sequence 3 of the Bursum Formation, Red Tanks Member, of the Lucero Basin, New Mexico. Dated by conodonts and insects as transitional Gzhelian–Asselian (Lucas *et al.* 2013; Schneider *et al.* 2013). (a) Cast of a left valve, NMMNH 54083 (Schneider collection); and (b) combined cast of left and right valve, NMMNH 54039 (Schneider collection); both from Carrizo Arroyo, NMMNH locality number 3437.

Molinestheria seideli zone, have a spine on the external side of the umbo similar to *Cornia germari* (Beyrich 1857) and therefore were interpreted by Scholze *et al.* (2016) as a younger synonyms of the latter.

Occurrences. Moscow Syncline, Vokhmian Regional Stage (Induan), Vokhma Formation (Scholze *et al.* 2015); Central European Basin, uppermost Calvörde Formation and Bernburg Formation, together with radially ribbed conchostracans of the genus *Estheriella* Weiss 1875 (e.g. Kozur & Seidel 1983*a*, *b*; Kozur & Weems 2010; Scholze *et al.* 2016); occurrences of *Cornia* of Early Triassic age were previously reported from drill cores of the Arács Marl Formation, Werfen Group, Hungary (Kozur & Mock 1993), from the Goudikeng Formation, Junggar Basin, NW China (Liu 1993), from the Babii Kamen section, Mal'tsevo Formation, Kuznetsk basin, Siberia (Defretin-Lefranc 1965), and mentioned from Early Triassic deposits in Greenland (e.g. Defretin-Lefranc 1965; Kozur & Weems 2010).

Links of conchostracan and insect zones to the SGCS and a cross-check of data

One of the still largely unresolved problems is the connection of the continental regional stages, as well as biozones, to the marine stages of the Standard Global Chronostratigraphic Scale (SGCS). A solution could be gathered in two conceivable ways: by isotopic age correlations, and by direct biostratigraphic links by means of co-occurrences of marine and non-marine index fossils. Attempts to link the insect and amphibian zones via isotopic ages to the marine SGCS have been made since Lützner *et al.* (2007) and Schneider & Werneburg (2006). Unfortunately, isotopic ages (mainly Ar/Ar: e.g. Lippolt *et al.* 1984) within this time interval are too imprecise to be of real use. In the following, we try to connect the conchostracan zonation from recent isotopic ages and from co-occurrences of marine and nonmarine index fossils to the marine SGCS. For these connections we use only direct evidence derived from isotopically dated occurrences of conchostracan zone species or from co-occurrences of conchostracans, insect zone species and marine index fossils, such as conodonts and fusulinids. In addition, in these cases we make a cross-check of the conchostracan zonation with the insect zonation to test the reliability of the methods.

Pseudestheria sp. M–*Lioestheria* form Frenswegen–*Palaeolimnadiopsis* form Jessen assemblage zone, StB and StC, late Gzhelian–?earliest Asselian

This assemblage zone straddles two spiloblattinid zones: the *Sysciophlebia rubida–Syscioblatta lawrenceana* zone; and the following *Sysciophlebia euglyptica–Syscioblatta dohrni* zone. In the Ploužnice Horizon, Semily Formation, of the Krkonoše Basin, *Pseudestheria* sp. M co-occurs with *Sysciophlebia rubida* and *Syscioblatta lawrenceana*. The type horizon of *S. lawrenceana* is the Lawrence shale of the homonymous formation. The Lawrence Formation belongs to the Cass cyclothem at the base of the Virgilian and is assigned to the *Streptognathodus zethus* zone at the very base of the Virgilian, which corresponds to latest Kasimovian (Barrick *et al.* 2008, 2013*a*, *b*; Heckel *et al.* 2011). This position is confirmed by the foregoing spiloblattinid zone in which *Syscioblatta allegheniensis* forms the forerunner of *S. lawrenceana* and *S. grata* is the forerunner of *S. rubida* (Schneider & Werneburg 2012; Schneider *et al.* 2013). The superposition of this zone species is proven by co-occurrences with marine fossils in the Tinajas Member of the Atrasado Formation of central New Mexico (Lucas & Allen 2011). In the Kinney Brick quarry, *Syscioblatta allegheniensis* form K was identified (Schneider in Lucas & Allen 2011). The conodont fauna from unit 1, the marine limestone immediately below the insect beds, contains *Idiognathodus corrugatus* and *I. cherryvalensis*, which suggests an assignment to the *Idiognathodus confragus* Zone of the North America mid-continent region (Dennis cyclothem: Heckel 2013; Lucas & Allen 2011; Barrick *et al.* 2013*a*), that is roughly lower–middle Missourian or middle Kasimovian in age. The occurrence of *Sysciophlebia* cf. *S. grata* in the black shales of the Tinajas Member at Cerros de Amado, Socorro County, NMMNH locality 4667, supports this age designation (Lerner *et al.* 2009). This black shale is determined as most likely belonging to the *Streptognathodus gracilis* conodont zone of the middle Missourian (middle Kasimovian) (Barrick *et al.* 2013*b*). Conchostracans of the Tinajas Member at Cerros de Amado (Martens & Lucas 2005) and in the Kinney Brick Quarry (Kozur *et al.* 1992), which resemble strongly *P. rimosa* and *Lioestheria* form Köllerbach, support the correlation of the Stephanian A–?B *Pseudestheria limbata–Pseudestheria rimosa–Lioestheria* form Köllerbach assemblage zone to the Kasimovian of the marine scale.

Unfortunately, the most recently published isotopic ages of the type localities of two insect zone species in the Czech Republic (Opluštil *et al.* 2016) are in (minor) conflict with the logic sequences of insect zone species (Schneider & Werneburg 2012), which is well supported by marine index fossils. The age of the Mšec tuff below the type horizon of *S. grata* Schneider 1982, the Hredle Member in the middle part of the Slany Formation of the Western and Central Bohemian basins, has been determined as 302.47 ± 0.08 by chemical abrasion isotope-dilution thermal ionization mass spectrometry (CA-ID-TIMS), which is early Gzhelian age. The time difference between the *S. grata* of the type horizon in the early Gzhelian and *S.* cf. *grata* in the middle Kasimovian of New Mexico is about 2.5 myr. At the present state of non-marine–marine correlation, this is relatively minor but needs to be further investigated.

The second discrepancy concerns the age of the *Sysciophlebia rubida–Syscioblatta lawrenceana* zone. The type horizon of *S. rubida* Schneider 1982 is the Ploužnice lake horizon, Semily Formation, of the Krkonoše–Piedmont Basin. Opluštil *et al.* (2016) correlated them with the more than 50 km distant Klobuky lake horizon of the Line Formation in the Central and Western Bohemian basins. An ash bed within the latter horizon gives a CA-ID-TIMS age of 298.97 ± 0.09 Ma, which is just at the Gzhelian–Asselian boundary of 298.9 Ma. The type horizon of *Syscioblatta lawrenceana*, which co-occurs with *S. rubida* in the Ploužnice lake horizon (Schneider & Werneburg 2012), is the Lawrence Shale (see earlier) at the very base of the Virgilian, which corresponds to latest Kasimovian. The time difference between the latest Kasimovian and the assumed isotopic age of the Ploužnice lake horizon after Opluštil *et al.* (2016) is roughly 4 myr, which is too much even in continental sections and needs to be explained. Most possibly, the lithostratigraphic correlation of the Ploužnice horizon with the Klobuky horizon is wrong; a less likely cause is the misidentification of the zone species.

Pseudestheria sp. M and *Palaeolimnadiopsis* form Jessen are known from the Siebigerode Formation of the Saale Basin, which is the type horizon of both the zone species of the *Sysciophlebia euglyptica–Syscioblatta dohrni* zone (Schneider & Werneburg 2006, 2012). Based on vulcanite intrusions into the Wettin Subformation, the upper part of the Siebigerode Formation, the age of this zone is estimated as roughly 300 Ma (Schneider *et al.* 2013). Most recently this is supported by the 299 ± 3.2 Ma U–Pb SHRIMP age of a tuff bed in the upper Siebigerode Formation (K. Stanek pers. comm. 2015).

Forms identical to *Pseudestheria* sp. M, *Lioestheria* form Frenswegen and *Palaeolimnadiopsis* form Jessen occur in the Bursum Formation of the Lucero basin, New Mexico, within depositional sequence 3 (see above and Fig. 5). This sequence is designated to the *Sysciophlebia ilfeldensis–Spiloblattina weissigensis*–insect zone, which straddles the Gzhelian–Asselian boundary (Lucas *et al.* 2013; Schneider *et al.* 2013). Conodonts of the *Streptognathodus nevaensis* zone obtained from the top of sequence 3 yield an Early–Middle Asselian age (Barrick in Lucas *et al.* 2013), which confirms the dating by insect zonation and links the *Pseudestheria* sp. M–*Lioestheria* form Frenswegen–*Palaeolimnadiopsis* form Jessen assemblage zone to the SGCS.

Conclusions and summary

Plant macrofossils are well suited for the definition and correlation of the regional European continental substages of the Westphalian and Early Stephanian in the Euramerican palaeoequatorial belt, as recently demonstrated by Opluštil *et al.* (2016), but the biostratigraphic precision declines with increased climate cyclicity beginning in the late Westphalian and accelerates during the Stephanian (Roscher & Schneider 2006; cf. Opluštil *et al.* 2016). By the latest Stephanian and Early Permian, plant macrofossil zonations cease to be useful because the constitution of plant assemblages depends at this time mainly on local edaphic conditions and microclimate, respectively (e.g. Kerp 1996, 2000; DiMichele *et al.* 2008; DiMichele 2014). With the change from wet red beds to playa-like dry red beds during the late Early Permian and the Middle and Late Permian in the Euramerican region, plants became too rare to be of use for biostratigraphy. Alternatively, insects, especially the common blattoids (cockroaches), could be of use in biostratigraphy up to the late Early Permian. A preliminary archimylacrid insect zonation for the whole Westphalian was proposed by Schneider *et al.* (2005*a*) and Jarzembowski & Schneider (2007), but this remains of limited use owing to the much more common plant macrofossils of this time. In contrast, the spiloblattinid insect zonation (e.g. Schneider 1982; Schneider & Werneburg 2006, 2012) has emerged as the most suitable method for high-resolution biostratigraphy and correlations in the whole of Euramerica, and even to the marine SGCS. Similarly, high-resolution biostratigraphy is achieved with the temnospondyl amphibian (branchiosaurid) zonation of Werneburg (1989) and Werneburg & Schneider (2006, 2012). However, the amphibian zonation is only applicable up to the middle Early Permian because of the increasing aridification, and the disappearance of perennial lakes and ponds: that is, the disappearance of habitats to which branchiosaurid amphibians were adapted. Playa deposits could be locally rich in insects, such as, for example, the Middle–Late Permian Lodève playa (Bethoux *et al.* 2002; Schneider *et al.* 2006), but these insects remain biostratigraphically of little importance thus far. Therefore, the best applicable tool for wet and dry red beds is conchostracan biostratigraphy.

As shown earlier, we use a conservative approach to the assemblage zones that cover larger time intervals than the seemingly high-resolution conchostracan species-range zones of Holub & Kozur (1981), Martens (1982, 2012) and Kozur (1989). The mean time resolution of our 12 zones is roughly 5 Ma, which corresponds more or less to the mean duration of the Pennsylvanian and Permian stages. Unfortunately, because of an increasingly discontinuous fossil record during the late Early Permian and especially during the Middle Permian in central and western Europe, the time resolution deteriorates to 15 Ma in late Early–Middle Permian times. Middle–Late Permian conchostracan biostratigraphy, however, will be considerably improved by ongoing investigations based on the continuous Middle Permian–Early Triassic fossiliferous continental sections of the East European Platform (Golubev 2000; Golubev *et al.* 2012; Scholze *et al.* 2015).

We thank Birgit G. Gaitzsch, Ronny Rößler, Ute Gebhardt, Ralf Werneburg, Jürgen Wunderlich, Frank Körner, Jörg Goretzki, Frederik Spindler, Shuzhong Shen, Valeriy K. Golubev and Vladimir V. Silantiev for shared investigations of conchostracan-bearing sections in the framework of various projects. St. Schretzenmayr is thanked for a research project on the southern border of the Variscan foredeep. Vladlen Lozovsky supported the revision of the Novozhilov and other Russian collections, and gave valuable information on the stratigraphy of Russian conchostracan occurrences. We have to thank Andreas Brosig for his support in the compilation of the correlation chart, and for the graphics in Figures 1 and 4. Allan J. Lerner is thanked for improving the English in this paper. We thank S.G. Lucas and an anonymous reviewer for comments and suggestions that improved this paper. This project aims to contribute to IGCP 630 'Permian–Triassic climatic and environmental extremes and biotic response'. JWS thanks the German Research Foundation (grant numbers SCHN 408/2, SCHN 408/5, SCHN 408/7, SCHN 408/8, SCHN 408/10, SCHN 408/12 and SCHN 408/21) and FS is grateful for financial support from grant number SCHN 408/22-1. In addition, the research work was supported by a subsidy from the Russian Government to support the Programme of Competitive Growth of Kazan Federal University among the World's Leading Academic Centres. This publication is dedicated to Thomas Martens on the occasion of his 65th birthday in 2017: he has spent a large part of his life working on the investigation of Early Permian conchostracans.

References

ABU HAMAD, A., SCHOLZE, F., SCHNEIDER, J.W., GOLUBEV, V.K., VOIGT, S., UHL, D. & KERP, H. 2015. First occurrence of the Permian–Triassic enigmatic conchostracan *Rossolimnadiopsis* Novozhilov, 1958 from the Dead Sea, Jordan – preliminary report. *Permophiles*, **62**, 22–25.

AREF'EV, M.P., KULESHOV, V.N. & POKROVSKIIA, B.G. 2015. Carbon and oxygen isotope composition in upper Permian–lower Triassic terrestrial carbonates of the East European platform: a global ecological crisis against the background of an unstable climate. *Doklady Earth Sciences*, **460**, 11–15.

BARRICK, J.E., HECKEL, P.H. & BOARDMAN, D.R. 2008. Revision of the conodont *Idiognathodus simulator* (Ellison, 1941), the marker species for the base of the Late Pennsylvanian global Gzhelian Stage. *Micropaleontology*, **54**, 125–137.

BARRICK, J.E., LAMBERT, L.L., HECKEL, P.H., ROSCOE, S. & BOARDMAN, D.R. 2013*a*. Midcontinent Pennsylvanian conodont zonation. *Stratigraphy*, **10**, 55–72.

BARRICK, J.E., LUCAS, S.G. & KRAINER, K. 2013*b*. Conodonts of the Atrasado Formation (uppermost Middle to Upper Pennsylvanian), Cerros De Amado region, Central New Mexico, U.S.A. *In*: LUCAS, S.G., NELSON, W.J. ET AL. 2013. *The Carboniferous–Permian Transition in Central New Mexico*. New Mexico Museum of Natural History and Science Bulletin, **59**, 239–252.

BETHOUX, O., NEL, A., GAND, G., LAPEYRIE, J. & GALTIER, J. 2002. Discovery of the genus *Iasvia* Zalessky, 1934 in the Upper Permian of France (Lodève Basin)

(Orthoptera, Ensifera, Oedischiidae). *Géobios*, **35**, 293–302.

BEYRICH, E. 1864. Über Leaia Leidyi var. Baentschiana. *Zeitschrift der Deutschen Geologischen Gesellschaft*, **16**, 363.

BEYRICH, W. 1857. Verhandlungen der Gesellschaft: protokoll der Juni-Sitzung. *Zeitschrift der Deutschen Geologischen Gesellschaft*, **9**, 374–377.

BÖGER, H. & FIEBIG, H. 1962. Conchostracen im flözführenden Oberkarbon des niederrheinisch-westfälischen Steinkohlengebietes. *Paläontologische Zeitschrift, H.-Schmidt-Festband*, **36**, 8–24.

BROUTIN, J., CHATEAUNEUF, J.J., GALTIER, J. & RONCHI, A. 1999. L'Autunian d'Autun reste-t-il une référence pour les dépôts continentaux du Permien inférieur d'Europe? Apport desdonnées paléobotaniques. *Géologie de la France*, **2**, 17–31.

CHU, D., TONG, J., YU, J., SONG, H. & TIAN, L. 2013. The Conchostracan fauna from the Kayitou Formation of western Guizhou, China. *Acta Palaeontologica Sinica*, **52**, 265–280.

COPELAND, M.J. 1957. *The Arthropod Fauna of the Upper Carboniferous Rocks of the Maritime Provinces*. Geological Survey of Canada, Memoirs, **286**.

DAVYDOV, V.I., CROWLEY, J.L., SCHMITZ, M.D. & POLETAEV, V.I. 2010. High-precision U–Pb zircon age calibration of the global Carboniferous time scale and Milankovitch band cyclicity in the Donets Basin, eastern Ukraine. *Geochemistry, Geophysics, Geosystems*, **11**, 1–22.

DEFRETIN-LEFRANC, S. 1965. Etude et révision de Phyllopodes Conchostracés en provenance d'U.R.S.S. *Annales de la Société Géologique du Nord*, **85**, 15–48.

DEFRISE-GUSSENHOVEN, E. & PASTIELS, A. 1957. *Contribution à l'étude biométrique des Lioestheriidae du Westphalien supérier*. Publication de ĺAssociation pour ĺEtude de la Paléontologie et de la Stratigraphie Houillères, **31**.

DEPÉRET, C. & MAZERAN, P. 1912. Les Estheria du Permien d'Autun. *Bulletin de la Société d'Histoire Naturelle d'Autun*, **25**, 165–174.

DIMICHELE, W.A. 2014. Wetland-dryland vegetational dynamics in the Pennsylvanian ice age tropics. *International Journal of Plant Sciences*, **175**, 123–164.

DIMICHELE, W.A., KERP, H., TABOR, N.J. & LOOY, C.V. 2008. The so-called 'Paleophytic–Mesophytic' transition in equatorial Pangea – Multiple biomes and vegetational tracking of climate change through geological time. *Palaeogeography, Palaeoclimatology, Palaeoecology*, **268**, 152–163.

FERREIRA-OLIVEIRA, L.G. & ROHN, R. 2010. Leaiid conchostracans from the uppermost Permian strata of the Paraná Basin, Brazil: chronostratigraphic and paleobiogeographic implications. *Journal of South American Earth Sciences*, **29**, 371–380.

FIEBIG, H. 1966. Ausbildung und Faunenführung des marinen Ägir-Niveaus (Basis Westfal C1) in der Lippe-Mulde des Niederrheinisch–Westfälischen Steinkohlengebietes. *Fortschritte in der Geologie von Rheinland und Westfalen*, **13**, 203–242.

FIEBIG, H. & GROSCURTH, J. 1984. Das Westfal C im nördlichen Ruhrgebiet. *Fortschritte in der Geologie von Rheinland und Westfalen*, **32**, 257–267.

FRITSCH, A. 1901. *Fauna der Gaskohle und der Kalksteine der Permformation Böhmens, Bd 4.* Selbstverlag, Prague, 65–101.

GAITZSCH, B., RÖSSLER, R., SCHNEIDER, J.W. & SCHRETZENMAYR, St. 1999. Neue Ergebnisse zur Verbreitung potentieller Muttergesteine im Karbon von Nord- und Mitteldeutschland. *Geologisches Jahrbuch A*, **149**, 25–58.

GALLEGO, O.F. 2001. Conchostracofauna sudamericana del Paleozoico y Mesozoico: estado actual del conocimiento. Parte I: Argentina y Chile. *Acta Geologica Leopoldensia*, **24**, 311–328.

GEINITZ, H.B. 1855. *Die Versteinerungen der Steinkohlenformation in Sachsen*. Engelmann, Leipzig.

GHOSH, S.C. 2011. *Estheriids (fossil Conchostraca) of Indian Gondwana*. Palaeontologica Indica, New Series, **54**.

GHOSH, S.C., ASHIM, D., NANDI, A. & MUKHOPADHAYA, S. 1987. Estheriid zonation in the Gondwana. *The Palaeobotanist*, **36**, 143–153.

GOLDENBERG, F. 1877. Fauna Saraeponta Fossilis. *Die fossilen Thiere aus der Steinkohlenformation von Saarbrücken*, **2**, 1–54.

GOLUBEV, V.K. 2000. The faunal assemblages of Permian terrestrial vertebrates from Eastern Europe. *Paleontology Journal*, **34**, S211–S224.

GOLUBEV, V.K., SENNIKOV, A.G., MINIH, A.V., MINIH, M.G., KUHTINOV, D.A., BALABANOV, Y.P. & SILANTEV, V.V. 2012. The Permian–Triassic boundary in the south-east of Moscow Syneclise. *In*: IVANOV, A.V. (ed.) *Problems of Paleoecology and Historical Geoecology. Compilation of Scientific Materials of the All-Russian Scientific Conference Dedicated to the 80th Anniversary of Professor Vitaly Georgiyevich Ochev*. Saratov State Technical University Press, Saratov, 144–150 [in Russian].

GORETZKI, J. 2003. *Biostratigraphy of conchostracans: a key for the interregional correlations of the continental Palaeozoic and Mesozoic – computer-aided pattern analysis and shape statistics to classify groups being poor in characteristics*. PhD thesis, Technical University Bergakademie Freiberg.

GRAHAM, T.B. & WIRTH, D. 2008. Dispersal of large branchiopod cysts: potential movement by wind from potholes on the Colorado Plateau. *Hydrobiologia*, **600**, 17–27.

GUTHÖRL, P. 1931. *Estheria drummi* n. sp. und *Estheria obenaueri* n. sp. (Crustacea, Phyllopoda) aus den Lebacher Schichten des Saarländischen Rotliegend. *Jahresberichte und Mitteilungen des Oberrheinischen Geologischen Vereins, N.F.*, **20**, 80–83.

GUTHÖRL, P. 1934. Die Arthropoden aus dem Karbon und Perm des Saar-Nahe-Pfalz-Gebietes. *Abhandlungen der Preußischen Geologischen Landesanstalt, N.F.*, **164**, 1–219.

HECKEL, P.H. 2013. Pennsylvanian stratigraphy of Northern Midcontinent Shelf and biostratigraphic correlation of cyclothems. *Stratigraphy*, **10**, 3–39.

HECKEL, P.H., BARRICK, J.E. & ROSSCOE, S.J. 2011. Conodont-based correlation of marine units in lower Conemaugh Group (Late Pennsylvanian) in Northern Appalachian Basin. *Stratigraphy*, **8**, 253–269.

HOFFMANN, N., KAMPS, H.-J. & SCHNEIDER, J. 1989. Neuerkenntnisse zur Biostratigraphie und Paläodynamik des Perms in der Norddeutschen Senke – ein Diskussionsbeitrag. *Zeitschrift für Angewandte Geologie*, **35**, 198–207.

Holub, V. & Kozur, H. 1981. Revision einiger Conchostraken-Faunen des Rotliegenden und biostratigraphische Auswertung der Conchostraken des Rotliegenden. *Geologisch-Paläontologische Mitteilungen Innsbruck*, **11**, 39–94.

Jarzembowski, E.A. 2004. Atlas of animals from the Late Westphalian of Writhlington, United Kingdom. *Geologica Balcanica*, **34**, 47–50.

Jarzembowski, E.A. & Schneider, J.W. 2007. The stratigraphical potential of blattodean insects from the late Carboniferous of southern Britain. *Geological Magazine*, **144**, 1–8.

Jones, P.J. & Chen, P.-J. 2000. Carboniferous and Permian Leaioidea (Branchiopoda: Conchostraca) from Australia: taxonomic revision and biostratigraphic implications. *Records of the Australian Museum*, **52**, 223–244.

Jones, T.R. 1862. *A Monograph of the Fossil Estheriae.* Palaeontographical Society, London.

Jones, T.R. 1870. On some bivalved Entomostraca from the coal-measures of South Wales. *Geological Magazine, London*, **7**, 214–220.

Jones, T.R. & Woodward, H. 1899. Contributions to fossil Crustacea. *Geological Magazine, N. S.*, **6**, 388–395.

Josten, K.-H. & Amerom, H.W.J. 1999. *Die Pflanzenfossilien im Westfal D, Stefan und Rotliegend Norddeutschlands.* Fortschritte der Geologie in Rheinland und Westfalen, **39**.

Kerp, H. 1996. Post-Variscan late Palaeozoic Northern Hemisphere gymnosperms: the onset to the Mesozoic. *Review of Palaeobotany and Palynology*, **90**, 263–285.

Kerp, H. 2000. The modernization of landscapes during the Late Paleozoic–Early Mesozoic. *Paleontology Society Papers*, **6**, 80–113.

Knight, J.A. & Wagner, R.H. 2014. Proposal for the recognition of a Saberian Substage in the mid-Stephanian (West European chronostratigraphic scheme). *Freiberger Forschungshefte C*, **548**, 179–195.

Kobayashi, T. 1954. Fossil estherians and allied fossils. *Journal of the Faculty of Science, Tokyo University*, **9**, 1–192.

Kozur, H. 1978. Beiträge zur Stratigraphie des Perm. Teil III (1). Zur Korrelation der überwiegend kontinentalen Ablagerungen des obersten Karbons und Perms von Mittel- und Westeuropa. *Freiberger Forschungshefte C*, **342**, 117–142.

Kozur, H. 1980. Die Conchostraken-Fauna der mittleren Bernburg-Formation (Buntsandstein) und ihre stratigraphische Bedeutung. *Zeitschrift für Geologische Wissenschaften*, **8**, 885–903.

Kozur, H. 1989. Biostratigraphic zonations in the Rotliegendes and their correlations. *Acta Musei Reginaehradecensis A*, **22**, 15–30.

Kozur, H. 1992. *Protolimnadia kowalczyki* n. sp., Eine wichtige Conchostracen-Art aus dem Oberrotliegenden Mitteleuropas. *Geologisch-Paläontologische Mitteilungen Innsbruck*, **18**, 77–78.

Kozur, H. & Seidel, G. 1983*a*. Revision der Conchostracen-Faunen des unteren und mittleren Buntsandsteins. Teil I. *Zeitschrift für Geologische Wissenschaften*, **11**, 289–417.

Kozur, H. & Seidel, G. 1983*b*. Die Biostratigraphie des unteren und mittleren Buntsandstein unter besonderer Berücksichtigung der Conchostracen. *Zeitschrift für Geologische Wissenschaften*, **11**, 4229–4464.

Kozur, H. & Sittig, E. 1981. Das ‘Estheria’ tenella-Problem und zwei neue Conchostracen-Arten aus dem Rotliegenden von Sulzbach (Senke von Baden-Baden, Nordschwarzwald). *Geologisch Paläontologische Mitteilungen Innsbruck*, **11**, 1–36.

Kozur, H., Lucas, S.G. & Hunt, A.P. 1992. Preliminary report on Late Pennsylvanian Conchostraca from the Kinney Brick Quarry, Manzanita Mountains, New Mexico. *New Mexico Bureau of Mines and Mineral Resources Bulletin*, **138**, 123–126.

Kozur, H.W. 1993. Range charts of conchostracans in the Germanic Buntsandstein. *In*: Lucas, S.G. & Morales, M. (eds) *The Nonmarine Triassic. New Mexico Museum of Natural History and Science Bulletin*, **3**, 249–253.

Kozur, H.W. & Mock, R. 1993. The importance of conchostracans for the correlation of continental and marine beds. *In*: Lucas, S.G. & Morales, M. (eds) *The Nonmarine Triassic. New Mexico Museum of Natural History and Science Bulletin*, **3**, 261–266.

Kozur, H.W. & Weems, R.E. 2010. The biostratigraphic importance of conchostracans in the continental Triassic of the northern hemisphere. *In*: Lucas, S.G. (ed.) *The Triassic Timescale.* Geological Society, London, Special Publications, **334**, 315–417, https://doi.org/10.1144/SP334.13

Kozur, H.W., Martens, T. & Pacaud, G. 1981. Revision von ‘*Estheria*’ (*Lioestheria*) *lallyensis* Depéret and Mazeran, 1912 und ‘*Euestheria*’ *autunensis* Raymond, 1946. *Zeitschrift für Geologische Wissenschaften*, **9**, 1437–1447.

Laspeyres, H. 1870. Das fossile Phyllopoden-Genus, Leaia R. Jones. *Zeitschrift der Deutschen Geologischen Gesellschaft*, **22**, 733–746.

Legler, B. & Schneider, J.W. 2008. Marine ingressions in context to one million years cyclicity of Permian red-beds (Upper Rotliegend II, Southern Permian Basin, Northern Germany). *Palaeogeography, Palaeoclimatology, Palaeoecology*, **267**, 102–114.

Legler, B., Gebhardt, U. & Schneider, J.W. 2005. Late Permian non-marine–marine transitional profiles in the central southern Permian Basin, northern Germany. *International Journal of Earth Sciences*, **94**, 851–862.

Lerner, A.J., Lucas, S.G. *et al.* 2009. The biota and paleoecology of the Upper Pennsylvanian (Missourian) Tinajas locality, Socorro County, New Mexico. *In*: Lueth, V.W., Lucas, S.G. & Chamberlin, R.M. (eds) *Geology of the Chupadera Mesa.* New Mexico Geological Society, 60th Annual Field Conference Guidebook, Socorro, New Mexico, 7–10 October 2009, 267–280.

Lippolt, H.J., Hess, J.C. & Burger, K. 1984. Isotopische Alter von pyroklastischen Sanidinen aus Kaolin-Kohlentonsteinen als Korrelationsmarken für das mitteleuropäische Oberkarbon. *Fortschritte in der Geologie von Rheinland und Westfalen*, **32**, 119–150.

LIU, S. 1987. Some Permian–Triassic conchostracans and their significance of geological age from the middle area Tianshan Mountains. *Professional Papers of Stratigraphy and Palaeontology*, **18**, 92–116.

LIU, S. 1993. Some Permian–Triassic conchostracans from the Mid-Tianshan Mts. of China and the significance of their geological dating. *In*: LUCAS, S.G. & MORALES, M. (eds) *The Nonmarine Triassic. New Mexico Museum of Natural History and Science Bulletin*, **3**, 277–278.

LIU, S.W. & HE, Z.J. 2000. Marine conchostracans from the 'Sunjiagou Formation' of Qishan, Shaanxi. *Acta Palaeontologica Sinica*, **39**, 230–236.

LUCAS, S.G., CASSINIS, G. & SCHNEIDER, J.W. 2006. Nonmarine Permian biostratigraphy and biochronology: an introduction. *In*: LUCAS, S.G., CASSINIS, G. & SCHNEIDER, J.W. (eds) *Non-Marine Permian Biostratigraphy and Biochronology*. Geological Society, London, Special Publications, **265**, 1–14, https://doi.org/10.1144/GSL.SP.2006.265.01.01

LUCAS, S.G. & ALLEN, B.D. 2011. Precise age and biostratigraphic significance of the Kinney Brick Quarry Lagerstätte, Pennsylvanian of New Mexico, USA. *Stratigraphy*, **8**, 7–27.

LUCAS, S.G., BARRICK, J., KRAINER, K. & SCHNEIDER, J.W. 2013. The Carboniferous-Permian boundary at Carrizo Arroyo, Central New Mexico, USA. *Stratigraphy*, **10**, 153–170.

LUCAS, S.G., BARRICK, J.E., KRAINER, K. & SCHNEIDER, J.W. 2016. Pennsylvanian-Permian boundary at Carrizo Arroyo, Central New Mexico. *New Mexico Geological Society Guidebook, 67th Field Conference*, Geology of the Belen Area, 303–311.

LUTKEVITCH, E.M. 1937. O nekotorych Phyllopoda SSSR [On some Phyllopoda of the USSR]. *Ezhegodnik Vsesoyuznogo Paleontologicheskogo Obshchestva*, **11**, 60–70 [in Russian].

LUTKEVITCH, E.M. 1941. Phyllopoda permskich otlozhenij evropejskoj chasti SSSR [Phyllopoda from the Permian of the European part of the USSR]. *Paleontologija SSSR*, **5**, 1–44 [in Russian].

LÜTZNER, H., LITTMANN, S., MÄDLER, J., ROMER, R.L. & SCHNEIDER, J.W. 2007. Stratigraphic and radiometric age data for the continental Permocarboniferous reference-section Thüringer Wald, Germany. *In*: WONG, T.E. (ed.) *Proceedings of the XVth International Congress on Carboniferous and Permian Stratigraphy*, 2003, Utrecht. Royal Netherlands Academy of Arts and Sciences, Amsterdam, 161–174.

LÜTZNER, H., ANDREAS, D., SCHNEIDER, J.W., VOIGT, S. & WERNEBURG, R. 2012. Stefan und Rotliegend im Thüringer Wald und seiner Umgebung. *In*: Deutsche Stratigraphische Kommission: stratigraphie von Deutschland X. Rotliegend. Teil I: innervariscische Becken. *Schriftenreihe der Deutschen Gesellschaft für Geowissenschaften*, **61**, 418–487.

MARTENS, T. 1982. Zur Taxonomie und Biostratigraphie neuer Conchostraken-Funde (Phyllopoda, Crustacea) aus dem Permokarbon und der Trias von Mitteleuropa. *Freiberger Forschungshefte C*, **375**, 49–82.

MARTENS, T. 1983*a*. Zur Taxonomie und Biostratigraphie der Conchostraca (Phyllopoda, Crustacea) des Jungpaläozoikums der DDR, Teil I. *Freiberger Forschungshefte C*, **382**, 7–105.

MARTENS, T. 1983*b*. Zur Taxonomie und Biostratigraphie der Conchostraca (Phyllopoda, Crustacea) des Jungpaläozoikums der DDR, Teil II. *Freiberger Forschungshefte C*, **384**, 24–48.

MARTENS, T. 1984. Zur Taxonomie und Biostratigraphie der Conchostraca (Phyllopoda, Crustacea) des Rotliegenden (oberstes Karbon bis Perm) im Saar-Nahe-Gebiet (BRD). *Freiberger Forschungshefte C*, **391**, 35–57.

MARTENS, T. 1986. Zur Taxonomie und Biostratigraphie der Conchostraca (Phyllopoda, Crustacea) des Oberkarbon und Perm der USA – Teil II. *Abhandlungen und Berichte des Museums der Natur Gotha*, **13**, 55–60.

MARTENS, T. 1994. Die Conchostraken des Oberkarbon und Perm – Übersicht der Gattungen und Arten. *Abhandlungen und Berichte des Museum der Natur Gotha*, **18**, 53–62.

MARTENS, T. 2012. Biostratigraphie der Conchostraca (Branchiopoda, Crustacea) des Rotliegend. In: Deutsche Stratigraphische Kommission, Stratigraphie von Deutschland X. Rotliegend. Teil I: innervariscische Becken. *Schriftenreihe der Deutschen Gesellschaft für Geowissenschaften*, **61**, 98–109.

MARTENS, T. & LUCAS, S.G. 2005. Taxonomy and biostratigraphy of conchostraca (Branchiopoda, Crustacea) from two nonmarine Pennsylvanian and lower Permian localities in New Mexico. *In*: LUCAS, S.G. & ZEIGLER, K.E. (eds) *The Nonmarine Permian*. New Mexico Museum of Natural History and Science Bulletin, **30**, 208–213.

MICHEL, L.A., TABOR, N.J., MONTAÑEZ, I.P., SCHMITZ, M.D. & DAVYDOV, V.I. 2015. Chronostratigraphy and paleoclimatology of the Lodève Basin, France: evidence for a pan-tropical aridification event across the Carboniferous–Permian boundary. *Palaeogeography, Palaeoclimatology, Palaeoecology*, **430**, 118–131.

MÜNSTER, G. 1840. Muschelthiere der Vorwelt. *In*: MÜNSTER, G. & GOLDFUSS, A. (eds) *Petrefacta Germaniae, part 2*. List & Francke, Leipzig.

NOVOZHILOV, N.I. 1946. Novye Phyllopoda iz permskich i triasovych otlozhenij Nordvik– Chatangskogo rajona (New Phyllopoda from the Permian and Triassic deposits of the Nordwick–Khatanga region). *Nedra Arktiki*, **1**, 172–202 [in Russian].

NOVOZHILOV, N.I. 1958. Conchostraca de la super famille des Limnadiopseidea superfam. nov. *In*: NOVOZHILOV, N. (ed.) *Recueil d'articles sur les phyllopodes conchostracés. Annales du Service d'Information Geologique du Bureau de Recherches géologiques, géophysiques, et minières*, **26**, 95–127.

NOVOZHILOV, N.I. 1965. Novye dvustvorchatye listonogie korvunchanskoj serii nizhnej Tunguski. *In*: MOLIN, V.A. & NOVOZHILOV, N.I. (eds) *Dvustvorchatye listonogie permoi i triasa severa SSSR*. Akademija Nauk SSSR, Komi Filial, Institut Geologii, Syktyvkar, 45–56.

NOVOZHILOV, N.I. 1970. *Vymershie Limnadioidei. Conchostraca–Limnadioidea [Extinct Limnadioidei. Conchostraca–Limnadioidea]*. Nauka, Moscow [in Russian].

OLESEN, J. 2009. Phylogeny of Branchiopoda (Crustacea) – character evolution and contribution of uniquely preserved fossil. *Arthropod Systematics & Phylogeny*, **67**, 3–39.

OPLUŠTIL, S., SCHMITZ, M., CLEAL, C.J. & MARTÍNEK, K. 2016. A review of the Middle-Late Pennsylvanian west European regional substages and floral biozones, and their correlation to the Global Time Scale based on new U–Pb ages. *Earth-Science Reviews*, **154**, 301–335.

PRUVOST, P. 1911. Note sur les entomostracés bivalves du terrain houiller du Nord de la France. *Annales de la Société Géologique du Nord*, **40**, 60–80.

PRUVOST, P. 1919. *Introduction à l'étude du terrain du Nord et du Pas-de-Calais, la faune continentale du terrain houiller du Nord de la France*. Mémoires pour servir à léxplication de la carte géologique détaillée de la France. Imprimerie Nationale, Paris.

PRUVOST, P. 1930. La Faune continentale du terrain houiller de la Belgique. *Mémoires du Musée Royal dHistoiré Naturelle de Belgique*, **44**, 103–282.

RAYMOND, P.E. 1946. The genera of fossil Conchostraca: an order of bivalved Crustacea. *Bulletin of the Museum of Comparative Zoology at Harvard College*, **96**, 218–307.

RICHTER, S., OLESEN, J. & WHEELER, W.C. 2007. Phylogeny of Branchiopoda (Crustacea) based on a combined analysis of morphological data and six molecular loci. *Cladistics*, **23**, 301–336.

ROSCHER, M. & SCHNEIDER, J.W. 2006. Early Pennsylvanian to Late Permian climatic development of central Europe in a regional and global context. *In*: LUCAS, S.G., CASSINIS, G. & SCHNEIDER, J.W. (eds) *Nonmarine Permian Chronology and Correlation*. Geological Society, London, Special Publications, **265**, 95–136, https://doi.org/10.1144/GSL.SP.2006.265.01.05

RÖSSLER, R. 1996. *Stratigraphie des Oberkarbon: Litho- und Biofaziesmuster des kontinentalen Oberkarbon und Rotliegend in Norddeutschland*. DGMK-Berichte, Forschungsbericht 459-3/3.

SCHMITZ, M.D. & DAVYDOV, V.I. 2012. Quantitative radiometric and biostratigraphic calibration of the global Pennsylvanian–Early Permian time scale. *Geological Society of America Bulletin*, **124**, 549–577.

SCHNEIDER, J. 1982. Entwurf einer Zonengliederung für das euramerische Permokarbon mittels der Spiloblattinidae (Blattodea, Insecta). *Freiberger Forschungshefte C*, **375**, 27–47.

SCHNEIDER, J. 1996. Biostratigraphie des kontinentalen Oberkarbon und Perm im Thüringer Wald, SW-Saale-Senke – Stand und Probleme. *Beiträge zur Geologie von Thüringen, N.F.*, **3**, 121–151.

SCHNEIDER, J. & RÖSSLER, R. 1996. *Stratigraphie des Oberkarbons: Biostratigraphie der Rotfolgen*. DGMK-Berichte, Forschungsbericht 459-3/3.

SCHNEIDER, J., WALTER, H. & WUNDERLICH, J. 1982. Zur Biostratinomie, Biofazies und Stratigraphie des Unterrotliegenden der Breitenbacher Mulde (Thüringer Wald). *Freiberger Forschungshefte C*, **366**, 65–84.

SCHNEIDER, J.W. 2001. Rotliegend Stratigraphy – principles and problems. *Beiträge zur Geologie von Thüringen, N.F.*, **8**, 7–42 [in German].

SCHNEIDER, J.W. & ROMER, R. 2010. The Late Variscan Molasses (Late Carboniferous to Late Permian) of the Saxo-Thuringian Zone. *In*: LINNEMANN, U., KRONER, U. & ROMER, R.L. (eds) *Pre-Mesozoic Geology of Saxo-Thuringia – From the Cadomian Active Margin to the Variscan Orogen*. Schweizerbart, Science Publishers, Stuttgart, 323–346.

SCHNEIDER, J.W. & WERNEBURG, R. 2006. Insect biostratigraphy of the European Late Carboniferous and Early Permian. *In*: LUCAS, S.G., CASSINIS, G. & SCHNEIDER, J.W. (eds) *Non-Marine Permian Biostratigraphy and Biochronology*. Geological Society, London, Special Publications, **265**, 325–336, https://doi.org/10.1144/GSL.SP.2006.265.01.15

SCHNEIDER, J.W. & WERNEBURG, R. 2012. Biostratigraphie des Rotliegend mit Insekten und Amphibien. *In*: Deutsche Stratigraphische Kommission, Stratigraphie von Deutschland X. Rotliegend. Teil I: innervariscische Becken. *Schriftenreihe der Deutschen Gesellschaft für Geowissenschaften*, **61**, 110–142.

SCHNEIDER, J.W., SCHRETZENMAYR, St. & GAITZSCH, B.G. 1998. Rotliegend reservoirs at the margin of the Southern Permian Basin. Field trip F2; Field Trips to the EAGE Workshop Leipzig 1998. *Leipziger Geowissenschaften*, **7**, 15–44.

SCHNEIDER, J.W., LUCAS, S.G. & ROWLAND, J.M. 2004. The blattida (insecta) fauna of Carrizo Arroyo, New Mexico – biostratigraphic link between marine and non-marine Pennsylvanian/Permian boundary profiles. *New Mexico Museum of Natural History and Science Bulletin*, **25**, 247–261.

SCHNEIDER, J.W., GORETZKI, J. & RÖSSLER, R. 2005*a*. Biostratigraphisch relevante nicht-marine Tiergruppen im Karbon der variscischen Vorsenke und der Innensenken. *In*: WREDE, V. (ed.) *Stratigraphie von Deutschland, Oberkarbon. Courier Forschungsinstitut Senckenberg*, **254**, 103–118.

SCHNEIDER, J.W., HOTH, K., GAITZSCH, B.G., BERGER, H.J., STEINBORN, H., WALTER, H. & ZEIDLER, M. 2005*b*. Carboniferous stratigraphy and development of the Erzgebirge Basin, East Germany. *Zeitschrift der deutschen Gesellschaft für Geowissenschaften*, **156**, 431–466.

SCHNEIDER, J.W., RÖSSLER, R., GAITZSCH, B.G., GEBHARDT, U. & KAMPE, A. 2005*c*. Saale-Senke. *In*: WREDE, V. (ed.) *Stratigraphie von Deutschland, Oberkarbon. Courier Forschungsinstitut Senckenberg*, **254**, 419–440.

SCHNEIDER, J.W., KÖRNER, F., ROSCHER, M. & KRONER, U. 2006. Permian climate development in the northern peri-Tethys area the Lodève basin, French Massif Central, compared in a European and global context. *Palaeogeography, Palaeoclimatology, Palaeoecology*, **240**, 161–183.

SCHNEIDER, J.W., LUCAS, S.G. & BARRICK, J.E. 2013. The Early Permian age of the Dunkard Group, Appalachian basin, U.S.A., based on spiloblattinid insect biostratigraphy. *International Journal of Coal Geology*, **119**, 88–92.

SCHNEIDER, J.W., LUCAS, S.G., TRÜMPER, S., STANULLA, C. & KRAINER, K. 2016. Carrizo Arroyo, Central New Mexico – a new Late Palaeozoic taphotype of arthropod fossillagerstätte. *New Mexico Geological Society Guidebook*, **67**, 107–116.

SCHOLZE, F. & SCHNEIDER, J.W. 2015. Improved methodology of 'conchostracan' (Crustacea: Branchiopoda) classification for biostratigraphy. *Newsletter on Stratigraphy*, **48**, 287–298.

SCHOLZE, F., GOLUBEV, V.K., NIEDŹWIEDZKI, G., SENNIKOV, A.G., SCHNEIDER, J.W. & SILANTIEV, V.V. 2015. Early Triassic Conchostracans (Crustacea: Branchiopoda) from the terrestrial Permian–Triassic boundary sections in the Moscow syncline. *Palaeogeography, Palaeoclimatology, Palaeoecology*, **429**, 22–40.

SCHOLZE, F., SCHNEIDER, J.W. & WERNEBURG, R. 2016. Conchostracans in continental deposits of the Zechstein-Buntsandstein transition in central Germany: taxonomy and biostratigraphic implications for the position of the Permian–Triassic boundary within the Zechstein Group. *Palaeogeography, Palaeoclimatology, Palaeoecology*, **449**, 174–193.

STOYAN, D., FRENZ, M., GORETZKI, G. & SCHNEIDER, J. 1994. Tests zur formstatistischen Klassifikation von Conchostraken (Crustacea, Branchiopoda) mittels Prokrustesanalyse. *Freiberger Forschungshefte C*, **452**, 153–162.

TABOR, N.J., MONTAÑEZ, I.P., SCOTESE, C.R., POULSEN, C.J. & MACK, G.H. 2008. Paleosol archives of environmental and climatic history in paleotropical western Pangea during the latest Pennsylvanian through Early Permian. *In*: FIELDING, C.R., FRANK, T.D. & ISBELL, J.L. (eds) *Resolving the Late Paleozoic Ice Age in Time and Space*. Geological Society of America, Special Papers, **441**, 291–303.

TASCH, P. 1958. Novojilov's classification of fossil conchostracans – a critical evaluation. Part I. Family Leaiidae. *Journal of Paleontology*, **32**, 1094–1106.

TASCH, P. 1960. Conchostracan genus Anomalonema in the American Pennsylvanian. *Journal of Paleontology*, **34**, 285–289.

TASCH, P. 1964. Periodicity in the Wellington Formation of Kansas and Oklahoma. *Kansas Geological Survey Bulletin*, **169**, 481–495.

TASCH, P. 1987. *Fossil Conchostraca of the Southern Hemisphere and Continental Drift*. Geological Society of America, Memoirs, **165**.

TASCH, P. & JONES, P.J. 1979. *Carboniferous, Permian, and Triassic Conchostracans of Australia: Three New Studies*. Bureau of Mineral Resources, Geology and Geophysics, Canberra, Bulletin, **185**.

TCHERNYCHEV, B. 1928. Nouvelles données sur les Phyllopoda et les Xiphosura du bassin du Donetz. *Bulletins du Commité Géologiqe Leningrad*, **47**, 519–533.

VANNIER, J., THIÉRY, A. & RACHEBOEUF, P.R. 2003. Spinicaudatans and Ostracods (Crustacea) from the Montceau Lagerstätte (Late Carboniferous, France): morphology and palaeoenvironmental significance. *Palaeontology*, **46**, 999–1030.

VANSCHOENWINKEL, B., GIELEN, S., VANDEWAERDE, H., SEAMAN, M. & BRENDONCK, L. 2008. Relative importance of different dispersal vectors for small aquatic invertebrates in a rock pool metacommunity. *Ecography*, **31**, 567–577.

VARENTSOV, I.M. 1955. O sostave i rasprostranenii roda dvustvorchatyh listonogikh rakoobraznyh Palaeolimnadiopsis v paleozoe (On the composition and distribution of the bivalved phyllopod crustacean genus Palaeolimnadiopsis in the Palaeozoic). *Doklady Akademii Nauk S.S.S.R.*, **104**, 310–312 [in Russian].

WAGNER, R.H. 1971. The stratigraphy and structure of the Ciñera-Matallana coalfield (prov. León, N.W. Spain). *Trabajos de Geología*, **4**, 385–429.

WAGNER, R.H. 1984. Megafloral zones of the Carboniferous. *Compte rendu 9e Congrès International de Stratigraphie et de Géologie du Carbonifère, 1979, Washington*, **2**, 109–134.

WAGNER, R.H. & ÁLVAREZ-VÁZQUEZ, C. 2010. The Carboniferous floras of the Iberian Peninsula: a synthesis with geological connotations. *Review of Palaeobotany and Palynology*, **162**, 239–324.

WARTH, M. 1963. *Conchostraken (Crustacea, Phyllopoda) und Ostrakoden des saarländischen Stefan*. PhD thesis, E.-Karls-University, Tübingen.

WATERLOT, G. 1934. *Etudes des gites mineraux de la France. Bassin houiller de la Sarre et de la Lorraine. II. Faune fossile, Etude de la faune continentale du terrain houiller Sarro-lorrain*. Imprimerie L. Danel (par Ministere des travaux publics, France), Lille.

WEBB, J.A. 1978. A new Triassic *Palaeolimnadiopsis* (Crustacea: Conchostraca) from the Sydney Basin, New South Wales. *Alcheringa*, **2**, 261–267.

WEHRLI, H. 1931. Die Fauna der westfälischen Stufen A und B der Bochumer Mulde zwischen Dortmund und Kamen (Westfalen). *Palaeontographica*, **74**, 93–143.

WEHRLI, H. 1933. Die carbonische Süßwasserfauna der Zeche Baldur (Westfalen). *Neues Jahrbuch für Mineralogie, Geologie und Paläontologie – Abhandlungen*, **69B**, 172–188.

WEHRLI, H. 1938. Die Gliederfüßer (Arthropoden), mit Ausnahme der Insekten. *In*: KUKUK, P. (ed.) *Geologie des Niederrheinisch-Westfälischen Steinkohlengebietes*. Springer, Berlin, 128–132.

WEISS, W. 1875. 1. Protokoll der Juli-Sitzung. *Zeitschrift der Deutschen Geologischen Gesellschaft*, **27**, 709–712.

WERNEBURG, R. 1989. Labyrinthodontier (Amphibia) aus dem Oberkarbon und Unterperm Mitteleuropas – Systematik, Phylogenie und Biostratigraphie. *Freiberger Forschungshefte C*, **436**, 7–57.

WERNEBURG, R. & SCHNEIDER, J.W. 2006. Amphibian biostratigraphy of the European Permo-Carboniferous. *In*: LUCAS, S.G., CASSINIS, G. & SCHNEIDER, J.W. (eds) *Non-Marine Permian Biostratigraphy and Biochronology*. Geological Society, London, Special Publications, **265**, 201–215, https://doi.org/10.1144/GSL.SP.2006.265.01.09

ZASPELOVA, V.S. 1968. Novye pozdnepaleozojskie fillopody Centralnogo Kazachstan. In: *Novye vidy drevnych rastenij i bespozvonocnych SSSR*, [New Late Palaeozoic phyllopopods of Central Kazakhstan. In: *New species of fossil plants and invertebrates of the USSR*], **2**, 227–233 [in Russian].

ZELLER, M. 1987. Das produktive Karbon am Niederrhein. *Natur am Niederrhein*, **2**, 55–61.

ZHANG, W.-T., CHEN, P.-J. & SHEN, Y.-B. 1976. *Fossil Conchostraca of China*. Science Press, Beijing.

ZIEGLER, P.A. 1990. *Geological Atlas of Western and Central Europe 1990*. Shell Internationale Petroleum Maatschappij B.V.

Outline of a Permian tetrapod footprint ichnostratigraphy

SEBASTIAN VOIGT[1]* & SPENCER G. LUCAS[2]

[1]*Urweltmuseum GEOSKOP, Burg Lichtenberg (Pfalz), Burgstrasse 19, D-66871 Thallichtenberg, Germany*

[2]*New Mexico Museum of Natural History and Science, 1801 Mountain Road N.W., Albuquerque, NM 87104, USA*

**Correspondence: s.voigt@pfalzmuseum.bv-pfalz.de*

Abstract: Tetrapod footprints are among the most common fossil remains in continental Permian strata and thus are of biostratigraphic interest. Based on the vertical distribution of the 13 best-known Permian tetrapod ichnotaxa, three footprint biochrons are suggested for the period: (1) *Dromopus* – latest Carboniferous (approximately Gzhelian) to late Early Permian (approximately Artinskian), representing ichnoassemblages dominated by tracks of temnospondyls, reptiliomorphs, pelycosaurs and early diapsids; (2) *Erpetopus* – late Early Permian (approximately Kungurian) to late Middle Permian (approximately Capitanian), representing ichnoassemblages dominated by tracks of non-diapsid eureptiles; and (3) *Paradoxichnium* – Late Permian (Wuchiapingian and Changhsingian), representing ichnoassemblages dominated by tracks of medium- and large-sized parareptiles, non-diapsid eureptiles and early saurians. This is the most conservative ichnostratigraphic concept, and it may be possible to refine it to almost stage-level resolution by future comprehensive analysis, especially of Permian captorhinomorph and therapsid footprints. Other major tasks to improve Permian tetrapod footprint ichnostratigraphy include enhanced knowledge of Middle Permian tetrapod footprints, and clarification of the palaeoenvironmental factors that may control the distribution of tetrapod footprints in space and time.

Permian tetrapod footprints play a key role in the history of vertebrate ichnology and were the subject of the first published account of vertebrate ichnofossils almost 200 years ago (Anon. 1828; Pemberton *et al.* 1996; Lockley & Meyer 2000). Since then, tetrapod footprints have been found in Permian stratal successions of 21 countries on six continents (Fig. 1): (1) Africa – Morocco (Voigt *et al.* 2010, 2011*a*, *b*; Hminna *et al.* 2012), Niger (Smith *et al.* 2015), South Africa (De Beer 1986; Smith 1993; De Klerk 2002) and Tunisia (Newell *et al.* 1976); (2) Asia – Turkey (Gand *et al.* 2011); (3) Europe – Austria (Niedermayr & Scheriau-Niedermayr 1980; Voigt & Marchetti 2014), the Czech Republic (Geinitz 1861; Holub & Kozur 1981), France (Gand 1988; Gand *et al.* 2000; Gand & Durand 2006), Germany (Pabst 1908; Haubold 1984; Voigt 2005, 2012), the UK (Haubold & Sarjeant 1973, 1974), Hungary (Kaszap 1968; Barabás-Stuhl 1975; Haas 2001), Italy (Conti *et al.* 1977; Ceoloni *et al.* 1988; Marchetti 2014), Poland (Czyzewska 1955; Voigt *et al.* 2012), Russia (Schneider *et al.* 1992; Tverdokhlebov *et al.* 1997; Lucas *et al.* 1999; Gubin *et al.* 2003; Surkov *et al.* 2007), Serbia (Jovanovic 2012, 2013) and Spain (Gand *et al.* 1997; Voigt & Haubold 2015); (4) North America – Canada (Mossman & Place 1989; Calder *et al.* 2004; van Allen *et al.* 2005) and the USA (Cotton *et al.* 1995; Haubold *et al.* 1995; Lucas & Hunt 2005; Voigt & Lucas 2015*a*); (5) Oceania – Australia (Warren 1997); and (6) South America – Argentina (Melchor & Sarjeant 2004; Krapovickas *et al.* 2015) and Brazil (Costa da Silva *et al.* 2012).

The abundance of Permian tetrapod footprints disguises their uneven distribution in space and time, as the vast majority of tracks come from Early Permian (Cisuralian) beds that almost exclusively represent equatorial regions of Pangaea (Figs 1 & 2). Apart from the imbalanced record, Permian tetrapod footprints have a great potential to become a useful tool in stratigraphic analyses of continental deposits because the period experienced several major steps in the evolution of terrestrial tetrapods. In particular, these are the descent of anamniote reptiliomorphs (Lucas 2005, 2006; Kissel 2010; Clack & Milner 2015), the transition from 'pelycosaur'- to non-mammalian therapsid-dominated synapsids (Kemp 2006, 2012), and the diversification and increasing body size of parareptiles (Tsuji & Müller 2009), as well as the radiation of non-diapsid (='captorhinomorph') (O'Keefe *et al.* 2006) and diapsid eureptiles (Müller & Reisz 2006; Reisz *et al.* 2011) that are reflected in late Palaeozoic terrestrial tetrapod ichnoassemblages. A number of vertebrate ichnologists have tried to take advantage of this fact by proposing from simple

From: Lucas, S. G. & Shen, S. Z. (eds) 2018. *The Permian Timescale*. Geological Society, London, Special Publications, **450**, 387–404.
First published online December 9, 2016, https://doi.org/10.1144/SP450.10

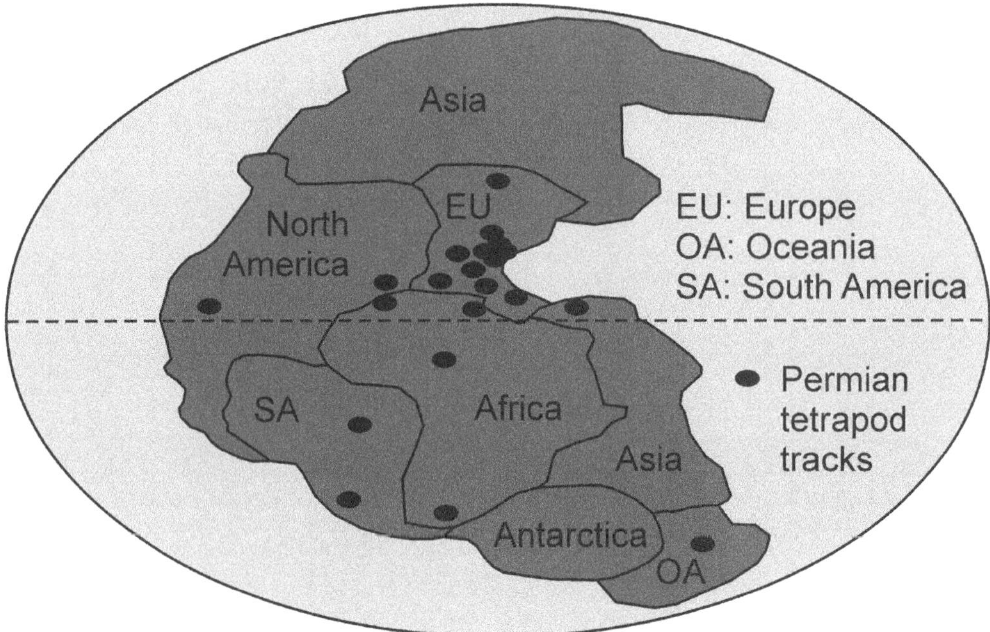

Fig. 1. Permian Pangaea and the distribution of Permian tetrapod tracksites.

to detailed schemes for subdividing Permian strata using fossil tetrapod footprints (Pabst 1908; Schmidt 1959; Haubold & Katzung 1972; Boy & Fichter 1988; Gand 1988; Voigt 2005; Gand & Durand 2006; Lucas 2007; Avanzini *et al.* 2011; Voigt & Lucas 2013). Most of the previous work suffers from ichnotaxonomical oversplitting and the overvaluation of local footprint records. By focusing on widely distributed and well-documented ichnotaxa, we intend to reboot the analysis of Permian tetrapod footprints for stratigraphic purposes.

Permian tetrapod footprint ichnotaxa and trackmaker relationships

Almost 400 different scientific names were introduced and used for Permian tetrapod footprints during the first two centuries of research (Haubold 2000). This count includes a huge number of synonyms that result from naming footprints based on extramorphological variations, the stratigraphic age of the track-bearing beds or even the geographical position of track localities (Haubold 1996, 2000; Voigt 2005). Instead, tetrapod footprint ichnotaxa should be named on the basis of anatomically controlled characters of the imprint morphology and trackway pattern only. Following this ichnotaxonomic concept, there are currently 13 valid ichnogenera of Permian tetrapod footprints. For the purpose of the present paper, we designate an ichnogenus to be valid if there is representative material known from at least three different localities. Here, we briefly characterize the 13 valid ichnogenera of Permian tetrapod footprints.

Permian	Africa	Asia	Europe	North America	Oceania	South America
Late	Morocco, Niger, South Africa, Tunisia		France, Germany, Great Britain, Hungary, Italy, Poland, Russia		Australia	Brazil
Middle			France			
Early	Morocco	Turkey	Austria, Czech Republic, France, Germany, Great Britain, Hungary, Italy, Poland, Russia, Serbia, Spain	Canada, USA		Argentina

Fig. 2. Generalized stratigraphic distribution of Permian tetrapod tracks.

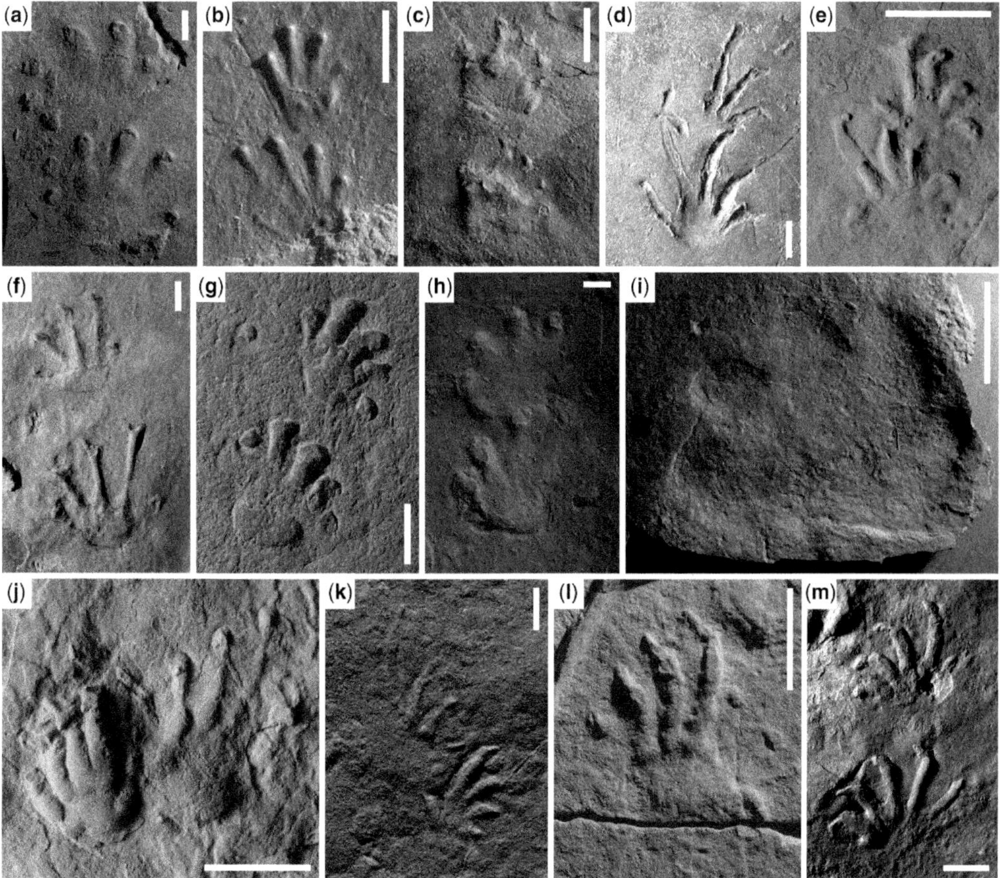

Fig. 3. Permian tetrapod footprint ichnotaxa: (**a**) *Amphisauropus* Haubold, 1970; (**b**) *Batrachichnus* Woodworth, 1900; (**c**) *Dimetropus* Romer & Price, 1940; (**d**) *Dromopus* Marsh, 1894; (**e**) *Erpetopus* Moodie, 1929; (**f**) *Hyloidichnus* Gilmore, 1927; (**g**) *Ichniotherium* Pohlig, 1892; (**h**) *Limnopus* Marsh, 1894; (**i**) *Pachypes* Leonardi *et al.*, 1975; (**j**) *Paradoxichnium* Müller, 1959; (**k**) *Rhynchosauroides* Maidwell, 1911; (**l**) *Tambachichnium* Müller, 1954; (**m**) *Varanopus* Moodie, 1929. Photographed tracks show manus-pes couples, except for (i) and (l), which illustrate imprints of an isolated manus and pes, respectively. Scale bars are 1 cm (in a, b, d–f, h & m), 5 cm (in c, g, j & l) and 10 cm (in i).

Amphisauropus *Haubold, 1970*

Pentadactyl manus and pes imprints up to 70 mm long (Fig. 3a). Manus one-quarter shorter than pes imprints, distinctly wider than long. Manual digits short, digit length increases from I to IV, digit V a little bit longer than I; palm more than half as long as imprint length. Pes imprint as long as wide, IV is the longest digit and V is slightly longer than II. Manus and pes imprints with distinct pads at the base of digit I. Trackways often with manus strongly rotated inwards and pes imprints rotated outwards; digit and tail/body traces common. Potential trackmakers are anamniote reptiliomorphs, especially Seymouriamorpha (Haubold 2000; Voigt 2005).

Batrachichnus *Woodworth, 1900*

Tetrapod track with pentadactyl pes and tetradactyl manus imprints less than 40 mm in length (Fig. 3b). Both imprints with indistinctly impressed proximal margin and distally rounded, straight digits. Manus imprints as long as wide, digits short with III the longest and IV shorter than II. Pes imprints longer than wide, digit length increases from I to IV, and V is shorter than II. Pedal digit IV about 1.5–2 times as long as the sole. Manus imprints often slightly turned inwards, pes imprints usually subparallel to the trackway midline. Tail/body traces may be present. Tracks of this ichnogenus are referred to small anamniote tetrapods, especially temnospondyls (Haubold 1996; Voigt 2005; Stimson *et al.* 2012).

Dimetropus *Romer & Price, 1940*

Tracks of quadrupedal tetrapods with pentadactyl manus and pes imprints that may exceed 200 mm in length (Fig. 3c). Imprints with relatively short digits, and proximally extended palm and sole. Digits increase in length from I to IV, V as long as III in the pes imprint, and V is as long as II in the manus imprint. Manus imprints as long as wide, pes imprints about one-quarter longer than wide. Imprints often show semi-spherical impressions of the metatarsal and metacarpal phalangeal pads. Digit tips more or less acute; trackways with imprints either aligned parallel to the trackway midline or slightly turned outwards. Tail/body traces common. *Dimetropus* seems to be the track of most 'pelycosaurs', except for varanopids and derived caseids (Haubold 1971*b*, 1973, 1984, 2000; Fichter 1979, 1983*b*; Voigt 2005).

Dromopus *Marsh, 1894*

Pentadactyl manus and pes imprints up to 80 mm in length (Fig. 3d). Palm/sole short, digits long, slender and distally tapered. Manus and pes imprints very similar in shape; both imprints are about one-third longer than wide, but manus about one-quarter shorter than the pes imprint. Digit length increases from I to IV, IV is highly elongated and V is about as long as II. Primary marginal overstepping of the manus imprint by the pes common. Digit dragging and tail/body traces only in tracks that were made on slippery ground. *Dromopus* is probably the track of various lizard-like Permian parareptiles and eureptiles, such as bolosaurids and araeoscelids (Haubold 1971*b*, 1973, 1984, 2000; Fichter 1979, 1983*a*; Voigt 2005).

Erpetopus *Moodie, 1929*

Pentadactyl manus and pes imprints up to 30 mm in length, average length 10–12 mm (Fig. 3e). Digit length increases from I to IV, and V is as long as I or II. Digits are thin, long and straight with acute terminations. Distal parts of digits I–IV are curved inwards, the distal part of digit V is curved outwards. Subrectangular palm/sole reaches about one-third of the total imprint length. Wide range of trackway pattern and orientation of imprints with respect to the trackway midline. Most likely trackmakers are small captorhinomorphs (Haubold 1971*b*; Haubold & Lucas 2001, 2003; Bernardi & Avanzini 2011; Voigt *et al.* 2013*a*).

Hyloidichnus *Gilmore, 1927*

Tracks of quadrupedal tetrapods with pentadactyl imprints up to 80 mm in length (Fig. 3f). Manus and pes imprints similar in size and digit proportions. Both imprints are about as long as wide with long, straight and slender digits. Digits increase in size from I to IV, V is about as long as I. Digits have tapering tips that are curved inwards in digits I–IV, but rotated outwards in digit V. Bifurcated digit tips (I–IV) are common. Short, indistinct palm/sole impression. Pes imprints slightly and manus imprints usually strongly rotated inwards with respect to the trackway midline. Tail or body traces not observed. Tracks of *Hyloidichnus* are referred to captorhinids (Haubold 1971*b*; Gand 1988; Gand & Durand 2006) or, even more specifically, to moradisaurine capthorinids (Voigt *et al.* 2009, 2010).

Ichniotherium *Pohlig, 1892*

Pentadactyl manus and pes imprint with enlarged, blunt digit tips (Fig. 3g). Pes imprints up to 200 mm in length, about as long as wide, and the sole characterized by large, oval to circular shaped pad. Manus imprints distinctly wider than long and about one-fifth shorter than the pes imprints. Digit length increases from I to IV, V shorter than II or as long as III depending on ichnospecies. Trackways with pes imprints rotated outwards to positioned forwards, manus imprints subparallel to the trackway midline or slightly to moderately rotated inwards. Primary marginal overstepping may occur. Tail or body traces very rare and always discontinuous. Based on the excellent vertebrate body and trace fossil record of the Early Permian Tambach Formation of central Germany, *Ichniotherium* is unambiguously referred to diadectomorph trackmakers (Berman *et al.* 1998, 2004; Fichter 1998; Voigt & Haubold 2000; Voigt 2005; Voigt *et al.* 2007).

Limnopus *Marsh, 1894*

Tracks of quadrupedal tetrapods with tetradactyl manus imprints (Fig. 3h). Tracks are identical to *Batrachichnus* except for the fourth digit of the manus imprint, which is about as long as digit II in *Limnopus*, but about as long as digit I in *Batrachichnus*. In addition, digits are relatively shorter in *Limnopus*. Total pes imprint length may reach up to 200 mm. Potential trackmakers are large-size temnospondyls, such as eryopids of Early Permian time (Haubold 1971*b*, 1973, 1984, 2000; Fichter 1979, 1983*a*; Tucker & Smith 2004; Voigt 2005; Voigt & Lucas 2015*b*).

Pachypes *Leonardi* et al., *1975*

Tracks of quadrupedal tetrapods with pentadactyl manus and pes imprints up to 400 mm in length

(Fig. 3i). Manus and pes imprints differ in size and shape. Both imprints with short and sturdy, distally rounded digits, as well as well-developed sole and palm. Pes imprints about as long as wide; digits increase in length from I to IV, and V is about as long as I. Manus imprints about one-quarter wider than long; digits I–IV almost sub-equal in length, V significantly shorter. Trackways with pes imprints directed forwards and manus imprints rotated inwards. No tail or body traces. *Pachypes* is usually referred to pareiasaurian trackmakers (Leonardi *et al.* 1975; Conti *et al.* 1977; Haubold 2000; Valentini *et al.* 2008, 2009; Voigt *et al.* 2010).

Paradoxichnium *Müller, 1959*

Lacertoid tracks with pentadactyl manus and pes imprints up to 110 mm in length (Fig. 3j). Manus imprints more than one-third shorter than pes imprints. Both imprints morphologically similar, except for a relatively shorter digit IV in the manus imprint. Digit length increases moderately from I to IV, V about as long as II. Low interdigital angles, I–V range from about 30° to 40°. Trackways with manus imprints rotated slightly inwards and pes imprints rotated outwards. Potential trackmakers are relatively large-sized neodiapsids, especially basal archosauromorphs, such as Protorosauridae (Haubold 1971*b*, 1973, 2000; Voigt 2012).

Rhynchosauroides *Maidwell, 1911*

Tracks of quadrupedal tetrapods with pentadactyl manus and pes imprints up to about 80 mm in length (Fig. 3k). Imprints are typical lacertoid and morphologically very similar to *Dromopus* except for three features: (1) the manus imprint is relatively smaller in *Rhynchosauroides*; (2) *Rhynchosauroides* has a digitigrade pes imprint and a plantigrade manus imprint, which is the exact opposite of *Dromopus*; and (3) *Rhynchosauroides* includes trackways that show a lateral overstep of the manus imprint by the pes, whereas the pes imprint has never been observed to be placed in front of the manus imprint in *Dromopus*. Potential trackmakers of Permian *Rhynchosauroides* are Pre-Mesozoic saurians: that is, basal lepidosauromorphs and archosauromorphs (Haubold 1966, 1971*a*, *b*; Avanzini & Renesto 2002; Klein *et al.* 2010).

Tambachichnium *Müller, 1954*

Quadrupedal trackway with pentadactyl manus and pes imprints up to 90 mm in length (Fig. 3l). Manus and pes imprints with similar proportions, both imprints longer than wide, showing slender digits with distinct claw impressions. The relative lengths of digits correspond to I < II ≃ V < III < IV. Digits I and V are often incompletely preserved. Trackway pattern highly variable, pes may be impressed behind the manus, partially overstepping the manus or even is placed in front of the manus imprint. Pes imprint is often rotated slightly outwardly; manus imprint set is in the direction of movement. No accessories such as digit dragging or tail/body traces are known. Digit and imprint proportions are most similar to the autopodes of varanopid synapsids (Voigt 2005; Voigt & Lucas 2015*b*).

Varanopus *Moodie, 1929*

Tracks of quadrupedal tetrapods with pentadactyl manus and pes imprints up to 45 mm in length (Fig. 3m). Manus imprint is slightly wider than long, pes imprint as long as wide and about one-fifth longer than the manus imprint. Digit length increases from I to IV, V is as long as II or III in the pes imprint, but shorter than II in the manus imprint. All digits have acute distal ends (distinct claw impressions). Palm/sole remarkably short, measures about one-fifth of the total imprint length. Pes imprints more or less parallel to the trackway midline, manus imprints are usually turned slightly inwards. Marginal primary overstepping may occur; tail/body traces have never been observed. Based on the digit proportions and the short sole/palm, early captorhinomorphs with a relatively long pedal digit V are the most likely *Varanopus* trackmakers (Fichter 1979; Voigt 2005; Voigt *et al.* 2009).

These fossil footprint ichnotaxa cover a wide range of potential trackmakers among Permian terrestrial tetrapods: temnospondyls, seymouriamorph and diadectomorph reptiliomorphs, 'pelycosaurian-grade' synapsids, and small-sized and large-sized parareptiles, as well as non-diapsid and diapsid eureptiles (Fig. 4). Nevertheless, there is a significant lack of knowledge with regard to the tetrapod footprint record of Permian lepospondyls, therapsids and procolophonids. As footprints of supposed Permian lepospondyls are rare and small (e.g. Voigt & Lucas 2015*a*, *b*), their ichnological characters have not yet been satisfyingly determined. Supposed therapsid tracks seem relatively common in Middle–Late Permian deposits (e.g. Newell *et al.* 1976; Conti *et al.* 1977; De Beer 1986; Ceoloni *et al.* 1988; Smith 1993; Fichter 1994; Gand *et al.* 2000; De Klerk 2002; Gand & Durand 2006), but the material currently lacks a comprehensive and comparative analysis. Tracks of Permian procolophonids are still obscure (Klein *et al.* 2015), although there is potential material among undescribed Lopinigian tracks from Germany, Morocco and Poland.

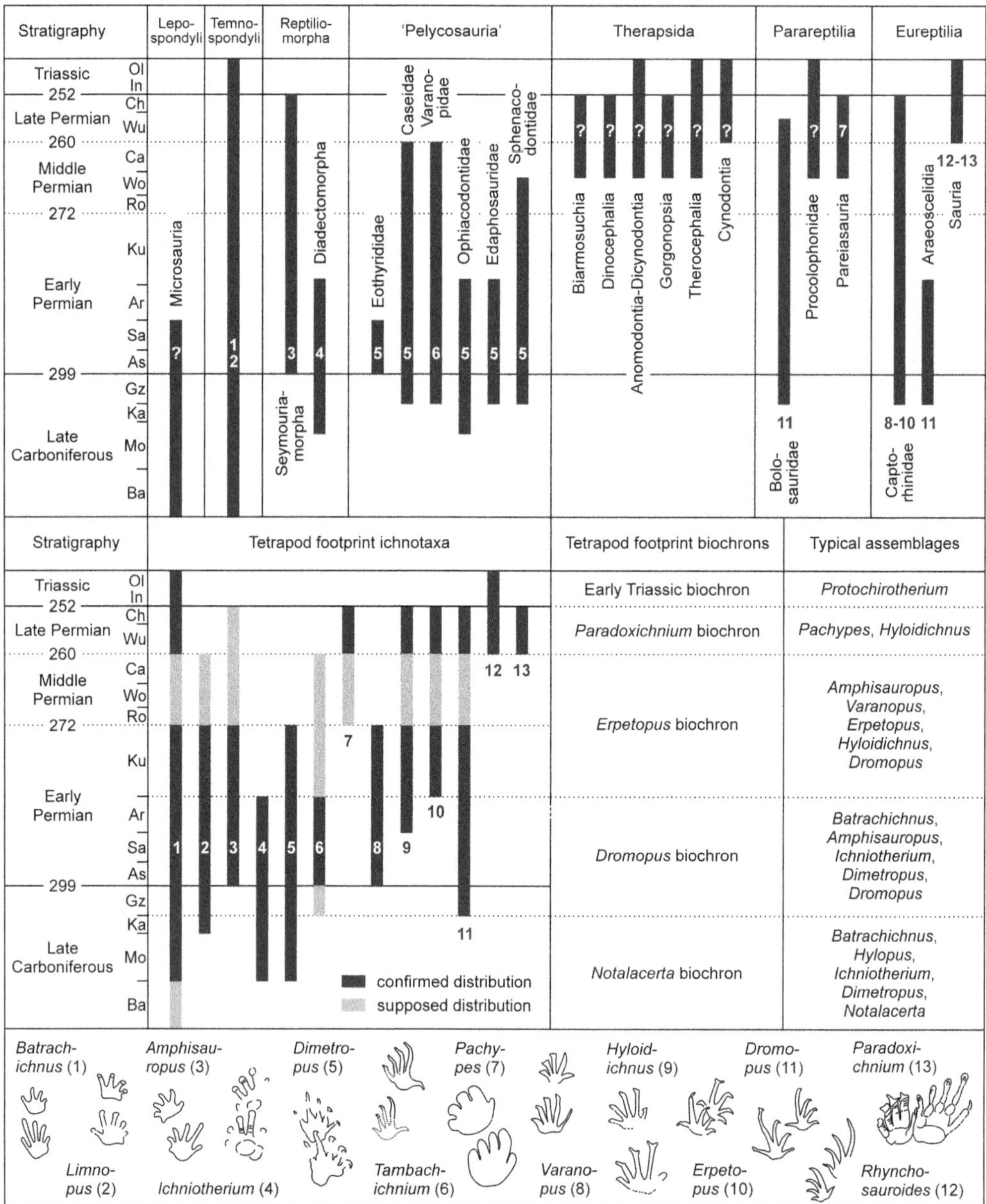

Fig. 4. Stratigraphic distribution of Permian terrestrial tetrapod taxa and related footprint ichnotaxa (compiled from Tsuji & Müller 2009; Reisz *et al.* 2011; Huttenlocker & Rega 2012; Kemp 2012; Ezcurra *et al.* 2014; Schoch 2014). The three suggested Permian tetrapod footprint biochrons – *Dromopus*, *Erpetopus* and *Paradoxichnium* – approximately correspond to the chronostratigraphic series early (Cisuralian), middle (Guadalupian) and late (Lopingian) Permian. For abbreviations of the chronostratigraphic stages see ICS (2015).

Stratigraphic distribution of Permian tetrapod footprints

The stratigraphic ranges of the valid Permian tetrapod footprint ichnotaxa are difficult to assess because the precise age of the footprint-bearing strata is often poorly constrained. Despite these problems, we will shortly discuss the first and last occurrences of the valid ichnotaxa and, wherever possible, compare it to the stratigraphic range of the body fossils of the inferred trackmakers (Fig. 4).

Batrachichnus tracks have been described from Carboniferous strata of North America and Europe (Butts 1891; Woodworth 1900; Matthew 1905; Haubold & Sarjeant 1973, 1974; Voigt 2005, 2012; Stimson *et al.* 2012). The same morphotype of tetrapod tracks is also known from Late Permian strata of Russia (Tverdokhlebov *et al.* 1997) and Morocco (Voigt *et al.* 2015), and there are even similar tracks known from Early Triassic beds (Klein & Lucas 2010, fig. 57). Thus, *Batrachichnus* is expected to range through the entire Permian Period. The apparent morphological stasis is not surprising, considering that this kind of tetrapod footprint refers to a plesiomorphic autopodial structure of a diverse and long-lived group, the temnospondyls (Schoch 2014).

Limnopus first occurs in the Late Carboniferous (Marsh 1894; Haubold & Sarjeant 1973, 1974; Voigt 2005, 2012). Its stratigraphically youngest record comprises unquestionable material from the late Early Permian of Italy (Marchetti *et al.* 2013, 2015; Marchetti 2014) and ambiguous material from the Middle Permian of France (Gand & Durand 2006). A Late Carboniferous–Middle Permian distribution of *Limnopus* is in accordance with the stratigraphic range of the most likely trackmakers, eryopids (Werneburg & Berman 2012; Schoch 2014).

Amphisauropus is almost exclusively known from Early Permian strata (Voigt 2015). The first occurrence might be coincident with the base of the Permian Period by track records from the Brule Formation of Nova Scotia, eastern Canada (van Allen *et al.* 2005) and the Ilmenau Formation of the Thuringian Forest, central Germany (Voigt 2005, 2012). *Amphisauropus* has been mentioned based on a single manus–pes imprint from the supposed latest Permian Ikakern Formation of Morocco (Hminna *et al.* 2012). If correct, this would be the only non-Early Permian record of the ichnotaxon. Terrestrial seymouriamorphs, the most likely trackmakers of *Amphisauropus*, are known from Early, as well as Late, Permian deposits (Berman *et al.* 2000; Klembara 2011).

The record of *Ichniotherium* extends down almost to the early Late Carboniferous (Voigt & Ganzelewski 2010), whereas the stratigraphically youngest occurrences come from late Early Permian deposits of New Mexico (Voigt & Lucas 2015*b*). This kind of track has also been reported from the Late Permian Val Gardena Sandstone Formation of Italy (Conti *et al.* 1977; Ceoloni *et al.* 1988; Avanzini *et al.* 2001), but, according to ongoing revisions, the related material most probably represents therapsid tracks that do not fit the characteristics of *Ichniotherium*. Diadectomorphs as probable trackmakers of *Ichniotherium* are only known from Late Carboniferous and Early Permian strata (Berman *et al.* 2004; Reisz 2007; Kennedy 2010; Kissel 2010). The recently proclaimed diadectid *Alveusdectes* from the Late Permian of China (Liu & Bever 2015) does not interfere with this view, as the advanced and differentiated dentition with canines of this taxon relates the Chinese specimen to therapsids rather than to diadectomorphs.

Dimetropus is well known from Late Carboniferous deposits in Africa (Lagnaoui *et al.* 2014) and Europe (Haubold & Sarjeant 1973, 1974; Voigt 2005, 2012; Voigt & Ganzelewski 2010). The stratigraphically youngest occurrence of the ichnotaxon is from the late Early Permian Wellington Formation of Oklahoma, USA (Sacchi *et al.* 2014). A Late Carboniferous–Early Permian range of *Dimetropus* largely coincides with the stratigraphic distribution of potential trackmakers (i.e. eothyridids, basal caseids, ophiacodontids, edaphosaurids and sphenacodontids: Kemp 2006, 2012; Huttenlocker & Rega 2012).

Tambachichnium is hitherto only known from Early Permian strata of France (undescribed material from the Rabejac Formation of the Lodève Basin), Germany (Müller 1954; Haubold 1971*b*; Haubold & Stapf 1998; Voigt 2005, 2012), Colorado (Voigt *et al.* 2005) and New Mexico (Lucas *et al.* 2013*b*; Voigt & Lucas 2015*b*). This ichnotaxon is also expected to be found in Middle Permian ichnoassemblages (see Gand *et al.* 1995 for an ambiguous record from SE France), as skeletal remains of the supposed varanopid trackmakers have been recorded in strata of Late Carboniferous–late Guadalupian age (Campione & Reisz 2010).

Pachypes has been reported from the Late Permian type locality in Italy (Leonardi *et al.* 1975; Conti *et al.* 1977; Valentini *et al.* 2009), and the Late Permian of Morocco (Voigt *et al.* 2010) and Niger (Smith *et al.* 2015). Moreover, Valentini *et al.* (2009) synonymized Late Permian tracks from Russia (Gubin *et al.* 2003; Surkov *et al.* 2007) with *Pachypes*. Thus, the ichnotaxon is relatively well known from uppermost Permian strata. As pareiasaurians, the most likely trackmakers of *Pachypes*, have a temporal range from the Middle to Late Permian (Tsuji 2011; Tsuji *et al.* 2013), a future discovery of these tracks in Guadalupian tetrapod ichnoassemblages is to be expected.

Varanopus is known from basal Permian deposits of Canada (van Allen *et al.* 2005) and the Czech Republic (unpublished specimen in the collection of the Natural History Museum Vienna, Austria), and is a common track in Early Permian deposits of Europe (Gand 1988; Haubold & Stapf 1998; Voigt 2005, 2012; Gand & Durand 2006; Marchetti 2014; Marchetti *et al.* 2015; Voigt & Haubold 2015) and North America (Gilmore 1927, *Hylopus hermitanus* = *Varanopus*; Lucas *et al.* 2001, 2013*a*, *b*; Voigt *et al.* 2005). Among the last occurrences of the ichnogenus seem to be tracks from the late Early Permian of France (Gand 1988; Gand & Durand 2006), Spain (Voigt & Haubold 2015) and Italy (Marchetti 2014; Marchetti *et al.* 2015). The supposed trackmakers, basal captorhinomorphs with a relatively long pedal digit V (Voigt *et al.* 2009), may have existed beyond the Early Permian, but at present there is no relevant skeletal record to demonstrate this.

Hyloidichnus is known from Early Permian (Gand 1988; Gand *et al.* 1997; Gand & Durand 2006; Marchetti *et al.* 2013, 2015; Marchetti 2014; Voigt & Haubold 2015; Voigt & Lucas 2015*a*), as well as Late Permian, strata (Ceoloni *et al.* 1988; Voigt *et al.* 2010). All mentioned Early Permian records are from beds of probable Artinskian–Kungurian age. There is, however, a single record from the lower part of the Abo Formation in the Fra Cristobal Mountains of southern New Mexico (Lucas *et al.* 2012) that could be older and thus shift the first appearance of the ichnotaxon to the early Early Permian. The first peak abundance of *Hyloidichnus* in the late Early Permian is in accordance with the radiation of moradisaurine capthorinids, which are the most likely trackmakers of this kind of tetrapod footprint (O'Keefe *et al.* 2006; Voigt *et al.* 2009).

Erpetopus is known from Early Permian strata of France (Haubold & Lucas 2001, 2003), Italy (Haubold & Lucas 2001, 2003; Santi 2007; Marchetti *et al.* 2013, 2015; Marchetti 2014) and the USA (Haubold & Lucas 2001, 2003; Lucas *et al.* 2013*a*), as well as Late Permian deposits of Morocco (Hminna *et al.* 2012; Voigt *et al.* 2015). Tracks of this ichnogenus first appear almost simultaneously in the late Early Permian (for a more detailed discussion see below: *Erpetopus* biochron). *Erpetopus* is very generally referred to small captorhinomorph trackmakers (Haubold 1971*b*; Haubold & Lucas 2001, 2003; Bernardi & Avanzini 2011; Voigt *et al.* 2013*a*), so that the body fossil record in this case does not help to constrain the temporal range of these tracks.

Dromopus is the most common and most widely distributed kind of Permian tetrapod track, with numerous reports from strata of Cisuralian, as well as Lopingian, age (Haubold 1971*b*; Gand 1988; Haubold *et al.* 1995; Haubold 1996, 2000; Voigt 2005, 2012; Marchetti *et al.* 2015). Although there is just one ambiguous record from possible Middle Permian deposits (Gand 1988; Gand & Durand 2006), the diversity and temporal range of potential trackmakers among lacertoid parareptiles and eureptiles raises the expectation that *Dromopus* will be found in Permian strata of any age (Reisz *et al.* 2011; Modesto *et al.* 2015).

Palaeozoic *Rhynchosauroides* is only known from the Late Permian Val Gardena Sandstone Formation of the Southern Alps, Italy (Conti *et al.* 1977; Haubold 2000; Avanzini *et al.* 2001; Valentini *et al.* 2007), although the ichnogenus has very extensive Triassic records. A Late Permian distribution of the ichnogenus is in agreement with the temporal range of potential trackmakers (i.e. basal lepidosauromorphs and archosauromorphs: Ezcurra *et al.* 2014).

Paradoxichnium has long been known just from the holotype, a trackway with well-preserved imprints from the terrestrial marginal facies of the Late Permian Zechstein of central Germany (Müller 1959; Haubold 1971*b*; Voigt 2012). Ceoloni *et al.* (1988) assigned some tracks from the Late Permian Val Gardena Sandstone Formation with a question mark to the ichnogenus *Paradoxichnium.* According to preliminary results of an ongoing revision of the Val Gardena tetrapod ichnoassemblage, these tracks are much more similar to *Hyloidichnus* than *Paradoxichnium.* The Val Gardena tetrapod ichnoassemblage includes, however, a few tracks that can unambiguously be assigned to the latter ichnogenus. The most recent record of *Paradoxichnium* is from marginal-marine red beds of the Zechstein in SW Germany (Voigt *et al.* 2015). Thus, *Paradoxichnium* seems to be restricted to Late Permian deposits. This is in accordance with the temporal range of the most likely trackmakers (i.e. protorosaurid archosaurmorphs: Gottmann-Quesada & Sander 2009; Voigt 2012).

Permian tetrapod footprint biochronology

Based on the vertical distribution of the 13 best-known Permian tetrapod ichnotaxa, we suggest a subdivision of the period into three footprint biochrons (Figs 4–7). From base to top these are the *Dromopus*, *Erpetopus* and *Paradoxichnium* biochrons. The bases of these biochrons are defined by the first appearances of the respective ichnogenus.

The *Dromopus* biochron lasts for about 20 myr and covers the stratigraphic interval of the latest Carboniferous (approximately Gzhelian) up to the late Early Permian (approximately Artinskian) (Lucas 2007). *Dromopus* was introduced for tracks from the Virgilian Howard Limestone Formation of the Wabaunsee Group of Kansas, USA (Marsh 1894; Haubold 1971*b*). This assemblage is considered to be the stratigraphically earliest record of the ichnogenus and thus defines the base of the *Dromopus* biochron. Tetrapod footprint assemblages of the *Dromopus* biochron are the most common and best documented of all Permian vertebrate trace fossils. They are dominated by temnospondyl (*Batrachichnus*, *Limnopus*), reptiliomorph anamniote

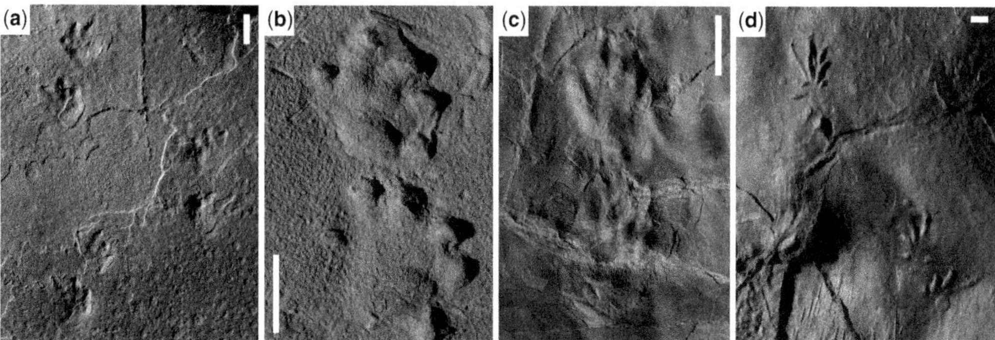

Fig. 5. Tetrapod footprints from the early Early Permian of the Intra-Sudetic Basin in southern Poland illustrating a typical ichnoassemblage of the *Dromopus* biochron (after Voigt *et al.* 2012): (**a**) *Amphisauropus*, trackway of three manus-pes sets; (**b**) *Ichniotherium*, left manus-pes set; (**c**) *Dimetropus*, left manus-pes set; and (**d**) *Dromopus*, two manus-pes sets. Scale bars are 1 cm (in a & d) and 5 cm (in b & c).

(*Amphisauropus*, *Ichniotherium*), 'pelycosaurian'-grade synapsid (*Dimetropus*, *Tambachichnium*) and lacertoid tracks of small parareptiles or early diapsids (*Dromopus*) (Figs 4 & 5). Tracks of early captorhinomorphs (*Varanopus*, *Hyloidichnus*) are a minor component. Among others, this biochron includes some famous late Carboniferous–Early Permian tetrapod footprint-bearing sections, such as those of the Boskovice, Krkonose-Piedmont and Intra-Sudetic basins of the Czech Republic and Poland (Geinitz 1861; Pabst 1908; Czyzewska 1955; Holub & Kozur 1981; Haubold 1984; Voigt *et al.* 2012), the French Lodève Basin (Tuilières-Loiras and Viala formations: Gand 1988; Gand & Durand 2006), the Thuringian Forest Basin of central Germany (Georgenthal–Tambach formations: Haubold 1985; Voigt 2005, 2012), the Cantabrian and Pyrenean basins of Spain (Peranera and Sagra formations: Gand *et al.* 1997; Voigt & Haubold 2015), the Souss, Khenifra and Tiddas basins in

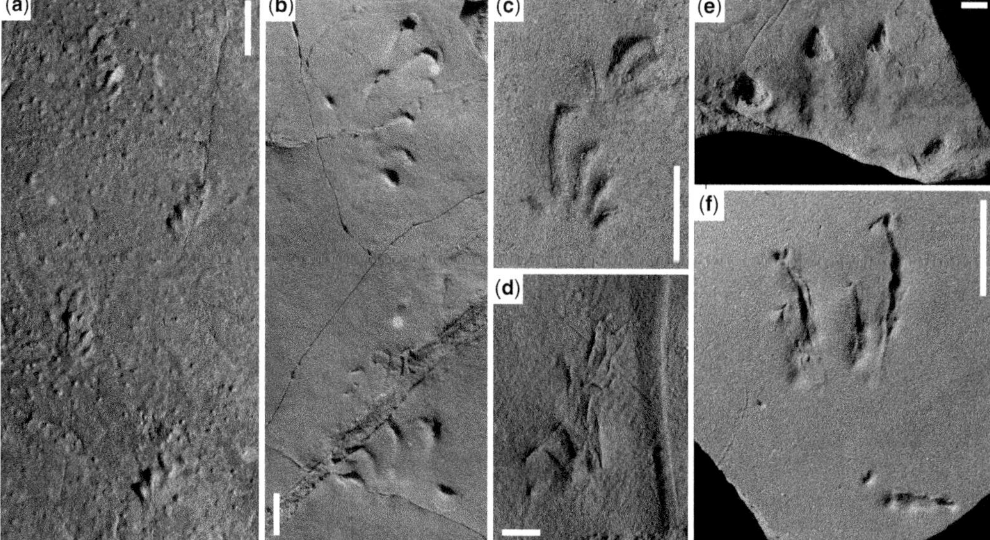

Fig. 6. Tetrapod footprints from the late Early Permian Arroyo de Alamillo Formation in central New Mexico, USA, illustrating a typical ichnoassemblage of the *Erpetopus* biochron (after Lucas *et al.* 2013*a*): (**a**) *Batrachichnus*, trackway of four manus-pes sets; (**b**) *Varanopus*, incomplete trackway of two manus-pes sets; (**c**) *Erpetopus*, left manus-pes set; (**d**) & (**e**) captorhinomorph tracks indet., left manus-pes set and incomplete imprint of right pes or manus; and (**f**) *Dromopus*, right manus-pes set. Scale bars are 1 cm.

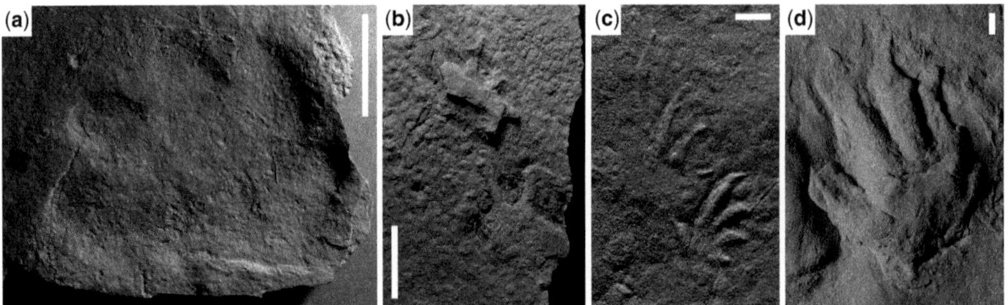

Fig. 7. Tetrapod footprints from the Late Permian Val Gardena Sandstone Formation of the Southern Alps, Italy, illustrating a typical ichnoassemblage of the *Paradoxichnium* biochron: (**a**) *Pachypes*, right manus; (**b**) *Hyloidichnus*, left manus-pes set; (**c**) *Rhynchosauroides*, left manus-pes set; and (**d**) *Paradoxichnium*, left manus. Scale bars are 5 cm (in a & b) and 1 cm (in c & d).

central Morocco (Hmich *et al.* 2006; Voigt *et al.* 2011*a*, *b*), and the late Carboniferous–Early Permian red beds of Nova Scotia and Prince Edward Island of maritime Canada (Mossman & Place 1989; Calder *et al.* 2004; van Allen *et al.* 2005; Brink *et al.* 2012), as well as numerous localities in the American Southwest (Hermit Formation, Arizona; Abo, Robledo Mountains and Sangre de Cristo formations, New Mexico; Maroon Formation, Colorado; Cotton *et al.* 1995; Haubold *et al.* 1995; Lucas & Hunt 2005; Voigt *et al.* 2005; Voigt & Lucas 2015*a*, *b*) (Fig. 8).

The *Erpetopus* biochron lasted for about 24 myr and covers the stratigraphic interval of the late Early Permian (approximately Kungurian) up to the late Middle Permian (approximately Capitanian). *Erpetopus* was introduced for tracks from the Choza Formation of the Clear Fork Group of Texas, USA (Moodie 1929; Haubold & Lucas 2001, 2003). The type locality and several other late Early Permian occurrences of *Erpetopus* in southern France (Rabejac Formation: Haubold & Lucas 2001, 2003), Italy (Collio and Pizzo del Diavolo formations: Marchetti *et al.* 2013, 2015; Marchetti 2014) and New Mexico (Arroyo de Alamillo Formation: Lucas *et al.* 2013*a*) compete for the first appearance of the ichnogenus (Figs 6 & 8). The only available direct radioisotopic age for these strata (283.1 ± 0.6–279.8 ± 1.1 Ma for the Collio Basin of the Italian Southern Alps: Schaltegger & Brack 2007) indicates a Kungurian age. A similar age is provided by recent radioisotopic

Stratigraphy		Czech Republic	France	Germany	Italy	Russia	Spain	Morocco Niger	South Africa	Tunisia	Canada USA	Footprint Biochron
Late Permian	Ch	Zech-stein	La Lieude	Zech-stein	Val Gardena	Russian Platform	Ribera d'Urgellet	Ikakern	Karoo Basin	Cheguimi		*Paradoxichnium*
	Wu							Moradi				
260												
Middle Permian	Ca		Salagou									*Erpetopus*
	Wo											
	Ro											
272												
Early Permian	Ku		Rabejac		Collio *s.l.*						Choza/ A.d.A.	
	Ar						Peranera Sagra	Tiddas			Hermit Abo Brule	*Dromopus*
	Sa	Sudetes Bohemia Moravia	Viala T.-Loiras	Thurin-gian Forest Basin		Northern Cau-casus		Khenifra Souss				
	As											
299												
Carboniferous	Gz											

Fig. 8. Approximate stratigraphic range of Permian tetrapod footprint-bearing strata that may help to refine the Permian tetrapod footprint ichnostratigraphic scheme. Explanation: Czech Republic/Poland – Sudetes, Bohemia, Moravia (Intra-Sudetic, Krkonoše-Piedmont, Boskovice basins); France – T.-Loiras (Tuilières-Loiras Formation); Germany – Zechstein (marginal terrestrial facies); Italy – Collio *s.l.* (Collio, Pizzo del Diavolo and Tregiovo formations); Morocco – Tiddas, Khenifra and Souss basins; Russia – Russian Platform (Kirov, Orenburg and Vologda regions); Canada/USA – A.d.A. (Arroyo de Alamillo Formation). For abbreviations of the chronostratigraphic stages, see ICS (2015).

dating of volcanic ash layers 150 m above the Rabejac Formation of the French Lodève Basin (284.4 ± 0.13 Ma: Michel *et al.* 2015).

In New Mexico, USA, footprint assemblages of the *Dromopus* and *Erpetopus* biochrons are stratigraphically superposed (e.g. Voigt & Lucas 2015*a*). These are footprint assemblages from the Robledo Mountains Formation and the bulk of the Abo Formation (*Dromopus* biochron) overlain by the uppermost Abo Formation and lower Yeso Group (Arroyo de Alamillo Formation) assemblages of the *Erpetopus* biochron. Marine strata that intertongue with and overlie these assemblages contain age-diagnostic foraminiferans which indicate that the boundary of the *Dromopus* and *Erpetopus* biochrons is very close to the base of the Kungurian (Lucas *et al.* 2015; Vachard *et al.* 2015).

Based on these data, we propose the base of the Kungurian as the approximate first appearance date of *Erpetopus*. The *Erpetopus* biochron is characterized by ichnoassemblages with abundant and diverse tracks of non-diapsid eureptiles (at least two different ichnospecies of *Varanopus*, *Hyloidichnus* and *Erpetopus*: Haubold & Lucas 2001, 2003; Lucas *et al.* 2013*a*; Marchetti 2014; Marchetti *et al.* 2015; Voigt & Haubold 2015) distinct from lacertoid tracks of small parareptiles or diapsids (*Dromopus*) (Figs 4 & 6). Seymouriamorph (*Amphisauropus*) and 'pelycosaurian'-grade synapsid tracks (*Dimetropus*) are less abundant, and temnospondyl tracks (*Batrachichnus*, *Limnopus*) extremely rare. Therapsid, huge parareptile and varanopid synapsid tracks (*Tambachichnium*) may be part of the supposedly Middle Permian tetrapod ichnoassemblages of the Lodève and Esterel basins of France (Gand 1988; Gand *et al.* 1995; Gand & Durand 2006), but this material is difficult to evaluate without careful revision and, especially, additional material for comparative analysis. Therefore, the knowledge of ichnoassemblages of the *Erpetopus* biochron is currently restricted to occurrences of probably exclusive latest Early Permian (Kungurian) age in France, Italy and the USA (Fig. 8).

The *Paradoxichnium* biochron lasts for about 8 myr and covers the Late Permian (Wuchiapingian and Changhsingian). *Paradoxichnium* has been introduced for tracks from marginal-marine deposits of the German Zechstein, more precisely in a horizon correlated with the second most basal evaporation cycle of the Southern Permian Basin (Staßfurt Formation, Z2: Müller 1959; Ullrich 1964). The Staßfurt Formation is considered to be late Wuchiapingian in age (Szurlies 2013). The recent find of *Paradoxichnium* from marginal-marine Zechstein red beds of the Annweiler Formation in SW Germany may be approximately the same age (Voigt *et al.* 2015). The third occurrence of *Paradoxichnium* is in the Val Gardena Sandstone Formation of the Italian Southern Alps, for which palynological analyses have suggested a late Capitanian–Changhsingian age (Pittau 2005) or just a Lopingian age (Posenato 2010; Kustatscher *et al.* 2012).

Owing to these age constraints, we propose the base of the Late Permian as the approximate first appearance date of *Paradoxichnium*. The *Paradoxichnium* biochron is dominated by pareiasaurian (*Pachypes*), captorhinomorph (*Hyloidichnus*, *Erpetopus*) and parareptilian–neodiapsid lacertoid (*Dromopus*), as well as early saurian tracks (*Rhynchosauroides*, *Paradoxichnium*) (Figs 4 & 7). Anamniote tracks are extremely rare (Tverdokhlebov *et al.* 1997; Voigt *et al.* 2015), whereas therapsid tracks play a key role, although most of the material is not yet adequately studied or even part of a valid ichnotaxonomy (Newell *et al.* 1976; Conti *et al.* 1977; Ceoloni *et al.* 1988; Smith 1993; Gand *et al.* 2000; De Klerk 2002; Voigt *et al.* 2015). Most typical ichnoassemblages of the *Paradoxichnium* biochron have been known from Italy (Val Gardena Sandstone Formation: Conti *et al.* 1977; Ceoloni *et al.* 1988), Morocco (Ikakern Formation: Voigt *et al.* 2010, 2015; Hminna *et al.* 2012) and Niger (Smith *et al.* 2015) (Fig. 8). Several other horizons and places hitherto provided just low-diversity ichnoassemblages (Müller 1959; Newell *et al.* 1976; Smith 1993; Tverdokhlebov *et al.* 1997; De Klerk 2002; Gubin *et al.* 2003; Surkov *et al.* 2007; Fortuny *et al.* 2011; Voigt *et al.* 2015) or diverse ichnoassemblages that need further study in order to allow conclusive comparisons (Gand *et al.* 2000).

Perspectives

The proposed Permian tetrapod footprint biochronozonation represents a conservative ichnostratigraphic concept that, following future analysis, provides scope for significant refinement. For example, the *Dromopus* biochron provides potential for subdivision by fixing the first appearance of *Amphisauropus*, *Hyloidichnus* and two distinct *Varanopus* ichnospecies (Lucas *et al.* 2013*a*; Marchetti 2014; Marchetti *et al.* 2015; Voigt & Haubold 2015). Totally unexplored is the stratigraphic value of Permian therapsid tracks. Other potential time markers are the first appearance of procolophonid and chirotheriid tracks in the Middle and Late Permian (Bernardi *et al.* 2015; Klein *et al.* 2015). Subject to sufficient data, a stage-level stratigraphic resolution of the Permian by tetrapod footprints would appear realistic. At present, biochronological organization of the Permian tetrapod-body-fossil record recognizes 10 biochrons (Lucas 2005, 2006), whereas there are nine marine stages recognized for the Permian. If our prediction is correct, Permian footprint biochronology thus could subdivide the Permian about as well as do tetrapod body fossils, although much work lies ahead to reach this ambitious goal.

In order to be able to develop a more effective Permian ichnostratigraphic scale, it is vital to both enhance the knowledge of Middle Permian tetrapod footprints and to apply an ichnotaxonomy that is solely based on anatomically controlled features of the imprint morphology and trackway pattern (Haubold 1996). The latter will be a key for understanding the abundant record of supposed therapsid footprints already known from France, Italy, Tunisia, South Africa and other countries (e.g. Newell *et al.* 1976; Conti *et al.* 1977; Ceoloni *et al.* 1988; Smith 1993; Gand *et al.* 2000; Voigt *et al.* 2015). Moreover, this will provide a reliable basis to revise the numerous occurrences with late Early Permian–Late Permian tetrapod footprints preserved in aeolian sandstones, such as the Coconino Sandstone of Arizona (Gilmore 1926, 1927, 1928), the Cornberg Sandstone of Germany (Schmidt 1959; Fichter 1994), the sandstones of the Yaciemiento Los Rayunos Formation in Argentina (Melchor 2001; Krapovickas *et al.* 2015), and the Corncockle and Lochabriggs sandstones of Scotland (McKeever & Haubold 1996). It is ironic that these first scientifically reported fossil tetrapod footprints on Earth (Anon. 1828; Pemberton *et al.* 1996; Lockley & Meyer 2000) might be the last we will understand in terms of their faunistic and stratigraphic meaning.

Finally, it is a major task to increase the knowledge of the influence of the palaeoenvironment (e.g. climate, facies, vegetation, topography and latitude) as a controlling factor for the distribution of tetrapod footprints in space and time. Basic studies with respect to Permian tetrapod footprints focus especially on the geographical distribution of seymouriamorph, diadectomorph and 'pelycosaurian'-grade synapsid tracks (Haubold & Katzung 1978; Hunt & Lucas 1998, 2006; Voigt & Lucas 2012; Voigt *et al.* 2013*b*). Future progress on Permian tetrapod ichnofacies will hopefully help to understand apparently isolated footprint records (e.g. *Robledopus macdonaldi* from the late Early Permian of New Mexico: Voigt *et al.* 2013*a*) in an otherwise remarkably uniform Permian tetrapod ichnofauna.

Conclusions

Based on the above, we offer the following conclusions:

- Tetrapod footprints are among the most common and widespread fossils in continental Permian strata, and have a great potential for Permian biostratigraphy and biochronology.
- Based on the vertical distribution of the 13 best-known Permian tetrapod ichnotaxa, three Permian footprint biochrons can be recognized: the *Dromopus* (oldest), *Erpetopus* and *Paradoxichnium* (youngest) biochrons.
- The *Dromopus* biochron is latest Carboniferous (approximately Gzhelian) to late Early Permian (approximately Artinskian), and it encompasses ichnoassemblages dominated by tracks of temnospondyls, reptiliomorphs, pelycosaurs and early diapsids.
- The *Erpetopus* biochron is late Early Permian (approximately Kungurian) to late Middle Permian (approximately Capitanian), and encompasses ichnoassemblages dominated by tracks of non-diapsid eureptiles.
- The *Paradoxichnium* biochron is Late Permian (Wuchiapingian and Changhsingian), and it includes ichnoassemblages dominated by tracks of medium- and large-sized parareptiles, non-diapsid eureptiles and early saurians.
- This conservative biochronology may be refined to almost stage-level resolution by future comprehensive analysis, especially of Permian captorhinomorph and therapsid footprints.
- Other major tasks to improve Permian tetrapod footprint ichnostratigraphy include augmenting our knowledge of Middle Permian tetrapod footprints and clarification of the palaeoenvironmental factors that control the distribution of tetrapod footprints in space and time.

This work is based on the study of approximately 20 000 Palaeozoic tetrapod footprints in more than 120 public and private collections worldwide. We are grateful to numerous colleagues in Argentina, Austria, Canada, the Czech Republic, France, Germany, Hungary, Italy, Morocco, Russia, South Africa, Spain, Tunisia, the UK and the USA who supported us by providing access to collections, the organization of joint fieldwork and many inspiring discussions. Special thanks go to two anonymous reviewers for their constructive comments.

References

Anon. 1828. Notice about Dumfriesshire footprints. *London and Paris Observer*, **141**, 93–94.

Avanzini, M. & Renesto, S. 2002. A review of *Rhynchosauroides tirolicus* Abel, 1926 ichnospecies (Middle Triassic: Anisian–Ladinian) and some inferences on *Rhynchosauroides* trackmaker. *Rivista Italiana di Paleontologia e Stratigrafia*, **108**, 51–66.

Avanzini, M., Ceoloni, P. et al. 2001. Permian and Triassic tetrapod ichnofaunal units of northern Italy: their potential contribution to continental biochronology. *Natura Bresciana*, **25**, 89–107.

Avanzini, M., Bernardi, M. & Nicosia, U. 2011. The Permo-Triassic tetrapod faunal diversity in the Italian Southern Alps. *In*: Dar, I.A. & Dar, M.A.

(eds) *Earth and Environmental Sciences*. InTech, Rijeka, 591–608.

Barabás-Stuhl, Á. 1975. Adatok a dunántúli újpaleozóos képződmények biosztratigráfiájához (Contribution to the biostratigraphy of the Upper Paleozoic in Transdanubia). *Földtani közlöny*, **105**, 320–334.

Berman, D.S., Sumida, S.S. & Martens, T. 1998. *Diadectes* (Diadectomorpha, Diadectidae) from the Early Permian of central Germany, with description of a new species. *Annals of Carnegie Museum*, **67**, 53–93.

Berman, D.S., Henrici, A.C., Sumida, S.S. & Martens, T. 2000. Redescription of *Seymouria sanjuanensis* (Seymouriamorpha) from the Lower Permian of Germany based on complete, mature specimens with a discussion of paleoecology of the Bromacker locality assemblage. *Journal of Vertebrate Paleontology*, **20**, 253–268.

Berman, D.S., Henrici, A.C., Kissel, R.A., Sumida, S.S. & Martens, T. 2004. A new diadectid (Diadectomorpha), *Orobates pabsti*, from the Early Permian of central Germany. *Bulletin of Carnegie Museum of Natural History*, **35**, 1–36.

Bernardi, M. & Avanzini, M. 2011. Locomotor behavior in early reptiles: insights from an unusual *Erpetopus* trackway. *Journal of Paleontology*, **85**, 925–929.

Bernardi, M., Klein, H., Petti, F.M. & Ezcurra, M.D. 2015. The origin and early radiation of archosauriforms: integrating the skeletal and footprint record. *PLoS One*, **10**, e0128449.

Boy, J. & Fichter, J. 1988. Zur Stratigraphie des höheren Rotliegend im Saar-Nahe-Becken (Unter-Perm; SW-Deutschland) und seiner Korrelation mit anderen Gebieten. *Neues Jahrbuch für Geologie und Paläontologie Abhandlungen*, **176**, 331–394.

Brink, K.S., Hawthorn, J.R. & Evans, D.C. 2012. New occurrences of *Ichniotherium* and *Striatichnium* from the Lower Permian Kildare Capes Formation, Prince Edward Island, Canada: palaeoenvironmental and biostratigraphic implications. *Palaeontology*, **55**, 1075–1090.

Butts, E. 1891. Recently discovered footprints of the amphibian age in the Upper Coal Measure Group of Kansas City, Missouri. *Kansas City Scientist*, **5**, 17–19.

Calder, J.H., Baird, D. & Urdang, E.B. 2004. On the discovery of tetrapod trackways from Permo-Carboniferous redbeds of Prince Edward Island and their biostratigraphic significance. *Atlantic Geology*, **40**, 217–226.

Campione, N.E. & Reisz, R.R. 2010. *Varanops brevirostris* (Eupelycosauria: Varanopidae) from the Lower Permian of Texas, with discussion of varanopid morphology and interrelationships. *Journal of Vertebrate Paleontology*, **30**, 727–746.

Ceoloni, P., Conti, M.A., Mariotti, N., Mietto, P. & Nicosia, U. 1988. New Late Permian tetrapod footprints from Southern Alps. *Memorie della Società Geologica Italiana*, **34**, 45–65.

Clack, J.A. & Milner, A.R. 2015. *Basal Tetrapoda: Handbook of Palaeoherpetology, Part 3A1*. Dr. Friedrich Pfeil, Munich.

Conti, M.A., Leonardi, G., Mariotti, N. & Nicosia, U. 1977. Tetrapod footprints of the Val Gardena Sandstone (North Italy). Their paleontological, stratigraphic and paleoenvironmental meaning. *Palaeontographia Italica*, **70**, 1–91.

Costa da Silva, R., Sedor, F.A. & Fernandes, A.C.S. 2012. Fossil footprints from the Late Permian of Brazil: an example of hidden biodiversity. *Journal of South American Earth Sciences*, **38**, 31–43.

Cotton, W., Hunt, A.P. & Cotton, J. 1995. Paleozoic vertebrate tracksites in eastern North America. *New Mexico Museum of Natural History and Science Bulletin*, **6**, 189–211.

Czyzewska, T. 1955. Reptile trackways from the Permian of Wambierzyce in Lower Silesia (Poland). *Acta Geologica Polonica*, **5**, 131–160 [in Polish: Tropy gadów permskich z Wambierzyc (Dolny Śląsk)].

De Beer, C.H. 1986. Surface markings, reptilian footprints and trace fossils on a palaeosurface in the Beaufort Group near Fraserburg, C. P. *Annale van die Geologiese Opname Republiek van Suid-Afrika*, **20**, 129–140.

De Klerk, W.J. 2002. A dicynodont trackway from the *Cistecephalus* assemblage zone in the Karoo, east of Graaff-Reinet, South Africa. *Palaeontologia Africana*, **38**, 73–91.

Ezcurra, M.D., Scheyer, T.M. & Butler, R.J. 2014. The origin and early evolution of Sauria: reassessing the Permian saurian fossil record and the timing of the crocodile–lizard divergence. *PLoS One*, **9**, e89165.

Fichter, J. 1979. *Aktuopaläontologische Studien zur Lokomotion rezenter Urodelen und Lacertilier sowie paläontologische Untersuchungen an Tetrapodenfährten des Rotliegenden (Unter-Perm) SW-Deutschlands*. Unpublished PhD thesis, Johannes-Gutenberg University, Mainz.

Fichter, J. 1983*a*. Tetrapodenfährten aus dem saarpfälzischen Rotliegenden (?Ober-Karbon – Unter-Perm; Südwest-Deutschland), Teil I: Fährten der Gattungen *Saurichnites, Limnopus, Ichniotherium, Amphisauroides, Protritonichnites, Gilmoreichnus, Hyloidichnus* und *Jacobiichnus*. *Mainzer Geowissenschaftliche Mitteilungen*, **12**, 9–121.

Fichter, J. 1983*b*. Tetrapodenfährten aus dem saarpfälzischen Rotliegenden (?Ober-Karbon – Unter-Perm; SW-Deutschland), Teil II: Fährten der Gattungen *Foliipes, Varanopus, Ichniotherium, Dimetropus, Palmichnus*, cf. *Chelichnus*, cf. *Laoporus* und *Anhomoiichnum*. *Mainzer Naturwissenschaftliches Archiv*, **21**, 125–186.

Fichter, J. 1994. Permische Saurierfährten. *Philippia*, **7**, 61–82.

Fichter, J. 1998. Bericht über die Bergung einer 20 t schweren Fährtenplatte aus dem Tambacher Sandstein (Unter Perm) des Thüringer Waldes und erste Ergebnisse ichnologischer Studien. *Philippia*, **8**, 147–208.

Fortuny, J., Bolet, A., Sellés, A.G., Cartanyà, J. & Galobart, À. 2011. New insights on the Permian and Triassic vertebrates from the Iberian Peninsula with emphasis on the Pyrenean and Catalonian basins. *Journal of Iberian Geology*, **37**, 65–86.

Gand, G. 1988. *Les traces de vertébrés tétrapodes du Permien français*. PhD thesis, Université de Bourgogne Edition Centre des Sciences de la Terre, Dijon.

Gand, G. & Durand, M. 2006. Tetrapod footprint ichno-associations from French Permian basins. Comparisons with other Euramerican ichnofaunas.

In: LUCAS, S.G., CASSINIS, G. & SCHNEIDER, J.W. (eds) *Non-Marine Permian Biostratigraphy and Biochronology*. Geological Society, London, Special Publications, **265**, 157–177, https://doi.org/10.1144/GSL.SP.2006.265.01.07

GAND, G., DEMATHIEU, G. & BALLESTRA, F. 1995. La palichnofaune de vertébrés tétrapodes du Permien supérieur de l'Esterel (Provence, France). *Palaeontographica*, **A235**, 97–139.

GAND, G., KERP, H., PARSONS, C. & MARTÍNEZ-GARCÍA, E. 1997. Palaeoenvironnemental and stratigraphic aspects of animal traces and plant remains in Spanish Permian red beds (Peña Sagra, Cantabrian Mountains, Spain). *Geobios*, **30**, 295–318.

GAND, G., GARRIC, J., DEMATHIEU, G. & ELLENBERGER, P. 2000. La palichnofaune de vertébrés tétrapodes du Permien supérieur du bassin de Lodève (Languedoc-France). *Palaeovertebrata*, **29**, 1–82.

GAND, G., TÜYSÜZ, O. *ET AL*. 2011. New Permian tetrapod footprints and macroflora from Turkey (Çakraz Formation, northwestern Anatolia): biostratigraphic and palaeonenvironmental implications. *Comptes Rendus Palevol*, **10**, 617–625.

GEINITZ, H.B. 1861. *Dyas oder die Zechsteinformation und das Rothliegende (Permische Formation zum Theil)*. W. Engelmann, Leipzig.

GILMORE, G.W. 1926. Fossil footprints from the Grand Canyon I. *Smithsonian Miscellaneous Collections*, **77**, 1–41.

GILMORE, G.W. 1927. Fossil footprints from the Grand Canyon II. *Smithsonian Miscellaneous Collections*, **80**, 1–78.

GILMORE, G.W. 1928. Fossil footprints from the Grand Canyon III. *Smithsonian Miscellaneous Collections*, **80**, 1–16.

GOTTMANN-QUESADA, A. & SANDER, P.M. 2009. A redescription of the early archosauromorph *Protorosaurus speneri* Meyer, 1832, and its phylogenetic relationships. *Palaeontographica*, **A287**, 123–220.

GUBIN, Y.M., GOLUBEV, V.K., BULANOV, V.V. & PETUCHOV, S.V. 2003. Pareiasaurian tracks from the Upper Permian of Eastern Europe. *Paleontological Journal*, **37**, 514–523.

HAAS, J. 2001. *Geology of Hungary*. Eötvös University Press, Budapest.

HAUBOLD, H. 1966. Therapsiden- und Rhynchocephalen-Fährten aus dem Buntsandstein Südthüringens. *Hercynia*, **3**, 147–183.

HAUBOLD, H. 1970. Versuch der Revision der Amphibien-Fährten des Karbon und Perm. *Freiberger Forschungshefte*, **C260**, 83–117.

HAUBOLD, H. 1971*a*. Die Tetrapodenfährten des Buntsandsteins. *Paläontologische Abhandlungen*, **AIV**, 395–548.

HAUBOLD, H. 1971*b*. Ichnia Amphibiorum et Reptiliorum fossilium. *Encyclopedia of Palaeoherpetology*, **18**, 1–124.

HAUBOLD, H. 1973. Die Tetrapodenfährten aus dem Perm Europas. *Freiberger Forschungshefte*, **C285**, 5–55.

HAUBOLD, H. 1984. *Saurierfährten. Die Neue Brehm-Bücherei*. Ziemsen, Wittenberg.

HAUBOLD, H. 1985. Stratigraphische Grundlagen des Stefan C und Rotliegenden im Thüringer Wald. *Schriftenreihe für Geologische Wissenschaften*, **23**, 1–110.

HAUBOLD, H. 1996. Ichnotaxonomie und Klassifikation von Tetrapodenfährten aus dem Perm. *Hallesches Jahrbuch für Geowissenschaften*, **B18**, 23–88.

HAUBOLD, H. 2000. Tetrapodenfährten aus dem Perm – Kenntnisstand und Progress 2000. *Hallesches Jahrbuch für Geowissenschaften*, **B22**, 1–16.

HAUBOLD, H. & KATZUNG, G. 1972. Die Abgrenzung des Saxon. *Geologie*, **21**, 884–910.

HAUBOLD, H. & KATZUNG, G. 1978. Paleoecology and palaeoenvironments of tetrapod footprints from the Rotliegend (Lower Permian) of Central Europe. *Paleogeography, Palaeoclimatology, Palaeoecology*, **23**, 307–323.

HAUBOLD, H. & LUCAS, S.G. 2001. Die Tetrapodenfährten der Choza Formation (Texas) und das Artinsk-Alter der Redbed-Ichnofaunen des Unteren Perm. *Hallesches Jahrbuch für Geowissenschaften*, **B23**, 79–108.

HAUBOLD, H. & LUCAS, S.G. 2003. Tetrapod footprints of the Lower Permian Choza Formation. *Paläontologische Zeitschrift*, **77**, 247–261.

HAUBOLD, H. & SARJEANT, W.A.S. 1973. Tetrapodenfährten aus den Keele und Enville Groups (Permokarbon: Stefan und Autun) von Shropshire und South Staffordshire, Großbritannien. *Zeitschrift für Geologische Wissenschaften*, **1**, 895–933.

HAUBOLD, H. & SARJEANT, W.A.S. 1974. Fossil vertebrate footprints and the stratigraphical correlation of the Keele and Enville Beds of the Birmingham Region. *Proceedings of the Birmingham Natural History Society*, **22**, 257–268.

HAUBOLD, H. & STAPF, H. 1998. The Early Permian tetrapod track assemblage of Nierstein, Standenbühl Beds, Rotliegend, Saar–Nahe Basin, SW-Germany. *Hallesches Jahrbuch für Geowissenschaften*, **B20**, 17–32.

HAUBOLD, H., HUNT, A.P., LUCAS, S.G. & LOCKLEY, M.G. 1995. Wolfcampian (Early Permian) vertebrate tracks from Arizona and New Mexico. *New Mexico Museum of Natural History and Science Bulletin*, **6**, 135–165.

HMICH, D., SCHNEIDER, J.W., SABER, H., VOIGT, S. & EL WARTITI, M. 2006. New continental Carboniferous and Permian faunas of Morocco – implications for biostratigraphy, palaeobiogeography and palaeoclimate. *In*: LUCAS, S.G., CASSINIS, G. & SCHNEIDER, J.W. (eds) *Non-Marine Permian Biostratigraphy and Biochronology*. Geological Society London, Special Publications, **265**, 297–324, https://doi.org/10.1144/GSL.SP.2006.265.01.14

HMINNA, A., VOIGT, S., SABER, H., SCHNEIDER, J.W. & HMICH, D. 2012. On a moderately diverse continental ichnofauna from the Permian Ikakern Formation (Argana Basin, Western High Atlas, Morocco). *Journal of African Earth Sciences*, **68**, 15–32.

HOLUB, V. & KOZUR, H. 1981. Revision einiger Tetrapodenfährten des Rotliegenden und biostratigraphische Auswertung der Tetrapodenfährten des obersten Karbon und Perm. *Geologisch-Paläontologische Mitteilungen Innsbruck*, **11**, 149–193.

HUNT, A.P. & LUCAS, S.G. 1998. Vertebrate ichnofaunas of New Mexico and their bearing on Early Permian tetrapod ichnofacies. *New Mexico Museum of Natural History and Science Bulletin*, **12**, 63–65.

HUNT, A.P. & LUCAS, S.G. 2006. Permian tetrapod ichnofacies. *In*: LUCAS, S.G., CASSINIS, G. & SCHNEIDER,

J.W. (eds) *Non-Marine Permian Biostratigraphy and Biochronology*. Geological Society London, Special Publications, **265**, 137–156, https://doi.org/10.1144/GSL.SP.2006.265.01.06

Huttenlocker, A.K. & Rega, E. 2012. The paleobiology and bone microstructure of pelycosaurian-grade synapsids. *In*: Chinsamy-Turan, A. (ed.) *Forerunners of Mammals: Radiation, Histology, Biology*. Indiana University Press, Bloomington, IN, 91–119.

ICS. 2015. *International Stratigraphic Chart*. International Commission on Stratigraphy, http://www.stratigraphy.org/index.php/ics-chart-timescale

Jovanovic, M. 2012. The first record of tetrapod tracks in Permian alevrolites of vicinity of Donji Milanovac (Eastern Serbia). *Permophiles*, **56**, 44–46.

Jovanovic, M. 2013. Amphibia and Reptilia of Permian red sandstones (red-beds) from Glavica hill near Donji Milanovac (Eastern Serbia). *Permophiles*, **57**, 20–22.

Kaszap, A. 1968. *Korynichnium sphaerodactylum* (Pabst) in the Permian of Balatonrendes. *Földtani Közlöny*, **98**, 429–433.

Kemp, T.S. 2006. The origin and early radiation of the therapsid mammal-like reptiles: a palaeobiological hypothesis. *Journal of Evolutionary Biology*, **19**, 1231–1247.

Kemp, T.S. 2012. The origin and radiation of therapsids. *In*: Chinsamy-Turan, A. (ed.) *Forerunners of Mammals: Radiation, Histology, Biology*. Indiana University Press, Bloomington, IN, 3–28.

Kennedy, N.K. 2010. Description of the postcranial skeleton of *Limnoscelis* (Diadectomorpha: Limnoscelidae) from the Upper Pennsylvanian of northern New Mexico and central Colorado. *New Mexico Museum of Natural History and Science Bulletin*, **49**, 211–220.

Kissel, R.A. 2010. *Morphology, phylogeny, and evolution of Diadectidae (Cotylosauria: Diadectomorpha)*. PhD thesis, University of Toronto.

Klein, H. & Lucas, S.G. 2010. Review of the tetrapod ichnofauna of the Moenkopi Formation/Group (Early-Middle Triassic) of the American Southwest. *New Mexico Museum of Natural History and Science Bulletin*, **50**, 1–67.

Klein, H., Voigt, S., Hminna, A., Saber, H., Schneider, J.W. & Hmich, D. 2010. Early Triassic Archosaur-dominated footprint assemblage from the Argana Basin (Western High Atlas, Morocco). *Ichnos*, **17**, 215–227.

Klein, H., Lucas, S.G. & Voigt, S. 2015. Revision of the? Permian-Triassic tetrapod ichnogenus *Procolophonichnium* Nopcsa, 1923 with description of the new ichnospecies *P. lockleyi*. *Ichnos*, **22**, 155–176.

Klembara, J. 2011. The cranial anatomy, ontogeny, and relationships of *Karpinskiosaurus secundus* (Amalitzky) (Seymouriamorpha, Karpinskiosauridae) from the Upper Permian of European Russia. *Zoological Journal of the Linnean Society*, **161**, 184–212.

Krapovickas, V., Marsicano, C.A., Mancuso, A.C., de la Fuente, M.S. & Ottone, E.G. 2015. Tetrapod and invertebrate trace fossils from aeolian deposits of the lower Permian of centralwestern Argentina. *Historical Biology*, **27**, 827–842.

Kustatscher, E., van Konijnenburg-van Cittert, J.H.A., Bauer, K., Butzmann, R., Meller, B. & Fischer, T.C. 2012. A new flora from the Upper Permian of Bletterbach (Dolomites, N-Italy). *Review of Palaeobotany and Palynology*, **182**, 1–13.

Lagnaoui, A., Voigt, S., Saber, H. & Schneider, J.W. 2014. First occurrence of tetrapod footprints from Westphalian strata of the Sidi Kassem Basin, central Morocco. *Ichnos*, **21**, 223–233.

Leonardi, P., Conti, M.A., Leonardi, G., Mariotti, N. & Nicosia, U. 1975. *Pachypes dolomiticus* n. gen. n. sp.; Pareiasaur footprint from the 'Val Gardena Sandstone' (Middle Permian) in the western Dolomites (N. Italy). *Atti della Academia Nazionale dei Lincei*, **57**, 221–232.

Liu, J. & Bever, G.S. 2015. The last diadectomorph sheds light on Late Palaeozoic tetrapod biogeography. *Biology Letters*, **11**, 20150100.

Lockley, M.G. & Meyer, C.A. 2000. *Dinosaur Tracks and Other Fossil Footprints of Europe*. Columbia University Press, New York.

Lucas, S.G. 2005. Permian tetrapod faunachrons. *New Mexico Museum of Natural History and Science Bulletin*, **30**, 197–201.

Lucas, S.G. 2006. Global Permian tetrapod biostratigraphy and biochronology. *In*: Lucas, S.G., Cassinis, G. & Schneider, J.W. (eds) *Non-Marine Permian Biostratigraphy and Biochronology*. Geological Society, London, Special Publications, **265**, 65–93, https://doi.org/10.1144/GSL.SP.2006.265.01.04

Lucas, S.G. 2007. Tetrapod footprint biostratigraphy and biochronology. *Ichnos*, **14**, 5–38.

Lucas, S.G. & Hunt, A.P. 2005. Permian tetrapod tracks from Texas. *New Mexico Museum of Natural History and Science Bulletin*, **30**, 202–206.

Lucas, S.G., Lozovsky, V.R. & Shishkin, M.A. 1999. Tetrapod footprints from Early Permian Redbeds of the Northern Caucasus, Russia. *Ichnos*, **6**, 277–281.

Lucas, S.G., Lerner, A.J. & Haubold, H. 2001. First record of *Amphisauropus* and *Varanopus* in the Lower Permian Abo Formation, central New Mexico. *Hallesches Jahrbuch für Geowissenschaften*, **B23**, 69–78.

Lucas, S.G., Krainer, K., Chaney, D.S., DiMichele, W.A., Voigt, S., Berman, D.S. & Henrici, A.C. 2012. The Lower Permian Abo Formation in the Fra Cristobal and Caballo mountains, Sierra County, New Mexico. *New Mexico Geological Society Guidebook*, **63**, 345–376.

Lucas, S.G., Krainer, K. & Voigt, S. 2013*a*. The Lower Permian Yeso Group in central New Mexico. *New Mexico Museum of Natural History and Science Bulletin*, **59**, 181–199.

Lucas, S.G., Nelson, W.J. et al. 2013*b*. Field guide to Carboniferous-Permian transition in the Cerros de Amado and vicinity, Socorro County, central New Mexico. *New Mexico Museum of Natural History and Science Bulletin*, **59**, 39–76.

Lucas, S.G., Krainer, K. & Vachard, D. 2015. The Lower Permian Hueco Group, Robledo Mountaisn, New Mexico (U.S.A.). *New Mexico Museum of Natural History and Science Bulletin*, **65**, 43–95.

Maidwell, F.T. 1911. Notes on footprints from the Keuper of Runcorn Hill. *Proceedings of the Liverpool Geological Society*, **11**, 140–152.

Marchetti, L. 2014. *Early Permian vertebrate ichnofauna from South Alpine Region (Northern Italy):*

ichnosystematics, paleoecology and stratigraphic meaning. PhD thesis, Università degli Studi di Padova Dipartimento di Geoscienze Padova.

Marchetti, L., Avanzini, M. & Conti, M.A. 2013. *Hyloidichnus bifurcatus* Gilmore, 1927 *Limnopus heterodactylus* (King, 1845) from the Early Permian of the Southern Alps (N Italy): a new equilibrium in the ichnofauna. *Ichnos*, **20**, 202–217.

Marchetti, L., Ronchi, A., Santi, G. & Voigt, S. 2015. The Gerola Valley site (Orobic Basin, Northern Italy): a key for understanding late Early Permian tetrapod ichnofaunas. *Palaeogeography, Palaeoclimatology, Palaeoecology*, **439**, 97–116.

Marsh, O.C. 1894. Footprints of vertebrates in the Coal Measures of Kansas. *American Journal of Science*, **48**, 81–84.

Matthew, G.F. 1905. New species and a new genus of batrachian footprints of the Carboniferous system in eastern Canada. *Proceedings and Transactions of the Royal Society of Canada*, **10**, 77–122.

McKeever, P.J. & Haubold, H. 1996. Reclassification of vertebrate trackways from the Permian of Scotland and related forms from Arizona and Germany. *Journal of Paleontology*, **70**, 1011–1022.

Melchor, R.N. 2001. Permian tetrapod footprints from Argentina. *Hallesches Jahrbuch für Geowissenschaften*, **B23**, 35–43.

Melchor, R.N. & Sarjeant, W.A.S. 2004. Small amphibian and reptile footprints from the Permian Carapacha Basin, Argentina. *Ichnos*, **11**, 1–22.

Michel, L.A., Tabor, N.J., Montañez, I.P., Schmitz, M.D. & Davydov, V.I. 2015. Chronostratigraphy and Paleoclimatology of the Lodève Basin, France: evidence for a pan-tropical aridification event across the Carboniferous–Permian boundary. *Palaeogeography, Palaeoclimatology, Palaeoecology*, **430**, 118–131.

Modesto, S.P., Scott, D.M., MacDougall, M.J., Sues, H.-D., Evans, D.C. & Reisz, R.R. 2015. The oldest parareptile and the early diversification of reptiles. *Proceedings of the Royal Society*, **B282**, 20141912.

Moodie, R.L. 1929. Vertebrate footprints from the red beds of Texas. *American Journal of Science*, **97**, 352–368.

Mossman, D.J. & Place, C.H. 1989. Early Permian fossil vertebrate footprints and their stratigraphic setting in megacyclic sequence II red beds, Prim Point, Prince Edward Island. *Canadian Journal of Earth Sciences*, **26**, 591–605.

Müller, A.H. 1954. Zur Ichnologie und Stratonomie des Oberrotliegenden von Tambach (Thüringen). *Paläontologische Zeitschrift*, **28**, 189–203.

Müller, A.H. 1959. Die erste Wirbeltierfährte (*Paradoxichnium problematicum* n.g. n.sp.) aus dem terrestrischen Zechstein von Thüringen. *Monatsberichte der Deutschen Akademie der Wissenschaften*, **1**, 613–623.

Müller, J. & Reisz, R.R. 2006. The phylogeny of early eureptiles: comparing parsimony and Bayesian approaches in the Investigation of a Basal Fossil Clade. *Systematic Biology*, **55**, 503–511.

Newell, N.D., Rigby, J.K., Driggs, A., Boyd, D.W. & Stehli, F.G. 1976. Permian reef complex, Tunisia. *Geology Studies*, **23**, 75–112.

Niedermayr, G. & Scheriau-Niedermayr, E. 1980. Eine Tetrapodenfährte aus dem Unter-Rotliegend von Kötschach in den westlichen Gailtaler Alpen, Kärnten – Österreich. *Annalen des Naturhistorischen Museums Wien*, **83**, 259–264.

O'Keefe, F.R., Sidor, C.A., Larsson, H.C.E., Maga, A. & Ide, O. 2006. Evolution and homology of the astragalus in early amniotes: new fossils, new perspectives. *Journal of Morphology*, **267**, 415–425.

Pabst, W. 1908. Die Tierfährten in dem Rotliegenden 'Deutschlands'. *Nova Acta Leopoldina*, **89**, 316–481.

Pemberton, S.G., Sarjeant, W.A.S. & Torrens, H.S. 1996. Footsteps before the flood: the first scientific report of vertebrate footprints. *Ichnos*, **4**, 321–323.

Pittau, P. 2005. The microflora. *In*: Pittau, P., Kerp, H. & Kustatscher, E. (eds) *The Bletterbach Canyon: 'Let us Meet Across the P/T Boundary' – Workshop on Permian and Triassic Palaeobotany and Palynology. Excursion Guide, Bozen,* 16–18 June 2005, Museo Scienze Naturali Alto Adige, Bozen, Italy, 9–19.

Pohlig, H. 1892. Altpermische Saurierfährten, Fische und Medusen der Gegend von Friedrichroda i. Thüringen. *In*: Anon. (ed.) *Festschrift 70. Geburtstag von Rudolf Leuckardt*. Engelmann, Leipzig, 59–64.

Posenato, R. 2010. Marine biotic events in the Lopingian succession and latest Permian extinction in the Southern Alps (Italy). *Geological Journal*, **45**, 195–215.

Reisz, R.R. 2007. The cranial anatomy of basal diadectomorphs and the origin of amniotes. *In*: Anderson, J.S. & Sues, H.-D. (eds) *Major Transitions in Vertebrate Evolution (Life of the Past)*. Indiana University Press, Bloomington, IN, 228–252.

Reisz, R.R., Modesto, J.P. & Scott, D.M. 2011. A new Early Permian reptile and its significance in early diapsid evolution. *Proceedings of the Royal Society*, **B278**, 3731–3737.

Romer, A.S. & Price, L.I. 1940. *Review of the Pelycosauria*. Geological Society of America, Special Papers, **28**.

Sacchi, E., Cifelli, R., Citton, P., Nicosia, U. & Romano, M. 2014. *Dimetropus osageorum* n. isp. from the Early Permian of Oklahoma (USA): a trace and its trackmaker. *Ichnos*, **21**, 1–18.

Santi, G. 2007. A short critique of the ichnotaxonomic dualism *Camunipes–Erpetopus*, Lower Permian ichnogenera from Europe and North America. *Ichnos*, **14**, 185–191.

Schaltegger, U. & Brack, P. 2007. Crustal-scale magmatic systems during intracontinental strike-slip tectonics: U, Pb and Hf isotopic constraints from Permian magmatic rocks of the Southern Alps. *International Journal of Earth Sciences*, **96**, 1131–1151.

Schmidt, H. 1959. Die Cornberger Fährten im Rahmen der Vierfüßler-Entwicklung. *Abhandlungen des Hessischen Landesamtes für Bodenforschung*, **28**, 1–137.

Schneider, J., Schamaev, M.I. & Walter, H. 1992. Paläobiogeographie und Stratigraphie von Tetrapoden- und Arthropoden-Fährten aus dem paralischen Oberkarbon und Perm (Gzhel/Assel) des Donez Bassins. *Freiberger Forschungshefte*, **C445**, 104–121.

Schoch, R.R. 2014. *Amphibian Evolution – The Life of Early Land Vertebrates*. Wiley-Blackwell, Hoboken, NJ.

Smith, R.M.H. 1993. Sedimentology and ichnology of floodplain paleosurfaces in the Beaufort Group (Late

Permian), Karoo sequence, South Africa. *Palaios*, **8**, 339–357.

SMITH, R.M.H., SIDOR, C.A., TABOR, N.J. & STEYER, J.S. 2015. Sedimentology and vertebrate taphonomy of the Moradi Formation of northern Niger: a Permian wet desert in the tropics of Pangaea. *Palaeogeography, Palaeoclimatology, Palaeoecology*, **440**, 128–141.

STIMSON, M., LUCAS, S.G. & MELANSON, G. 2012. The smallest known tetrapod footprints: *Batrachichnus salamandroides* from the Carboniferous of Joggins, Nova Scotia, Canada. *Ichnos*, **19**, 127–140.

SURKOV, M.V., BENTON, M.J., TWITCHETT, R.J., TVERDOKHLEBOV, V.P. & NEWELL, A.J. 2007. First occurrence of footprints of large therapsids from the Upper Permian of European Russia. *Palaeontology*, **50**, 641–652.

SZURLIES, M. 2013. Late Permian (Zechstein) magnetostratigraphy in Western and Central Europe. *In*: GĄSIEWICZ, A. & SŁOWAKIEWICZ, M. (eds) *Palaeozoic Climate Cycles: Their Evolutionary and Sedimentological Impact*. Geological Society, London, Special Publications, **376**, 73–85, https://doi.org/10.1144/SP376.7

TSUJI, L.A. 2011. *Evolution, morphology and paleobiology of the Pareiasauria and their relatives (Amniota: Parareptilia)*. PhD thesis, Humboldt-Universität Berlin, Mathematisch-Naturwissenschaftliche Fakultät.

TSUJI, L.A. & MÜLLER, J. 2009. Assembling the history of the Parareptilia: phylogeny, diversification, and a new definition of the clade. *Fossil Record*, **12**, 71–81.

TUCKER, L. & SMITH, M.P. 2004. A multivariate taxonomic analysis of the Late Carboniferous vertebrate Ichnofauna of Alveley, southern Shropshire, England. *Palaeontology*, **47**, 679–710.

TSUJI, L.A., SIDOR, C.A., STEYER, J.-S., SMITH, R.M., TABOR, N.J. & IDE, O. 2013. The vertebrate fauna of the Upper Permian of Niger-VII. Cranial anatomy and relationsships of *Bunostegos akokanensis* (Pareiasauria). *Journal of Vertebrate Paleontology*, **33**, 747–763.

TVERDOKHLEBOV, V.P., TVERDOKHLEBOVA, G.I., BENTON, M.J. & STORRS, G.W. 1997. First record of footprints of terrestrial vertebrates from the Upper Permian of the Cis-Urals, Russia. *Palaeontology*, **40**, 157–166.

ULLRICH, H. 1964. Zur Stratigraphie und Paläontologie der marin beeinflussten Randfazies des Zechsteinbeckens in Ostthüringen und Sachsen. *Freiberger Forschungshefte*, **C169**, 1–163.

VACHARD, D., KRAINER, K. & LUCAS, S.G. 2015. Kungurian (late Early Permian) algae, microproblematica, and smaller foraminifers from the Yeso Group and San Andres Formation (New Mexico; USA). *Palaeontologia Electronica*, **18.1.21A**, 1–77.

VALENTINI, M., CONTI, M.A. & MARIOTTI, N. 2007. Lacertoid footprints of the Upper Permian Arenaria di Val Gardena Formation (Northern Italy). *Ichnos*, **14**, 193–218.

VALENTINI, M., CONTI, M.A. & NICOSIA, U. 2008. Linking tetrapod tracks to the biodynamics, paleobiogeography, and paleobiology of their trackmakers: *Pachypes dolomiticus* Leonardi *et al.*, 1975, a case study. *Studi Trentini Scienze Naturali Acta Geologica*, **83**, 237–246.

VALENTINI, M., NICOSIA, U. & CONTI, M.A. 2009. A re-evaluation of *Pachypes*, a pareiasaurian track from the Late Permian. *Neues Jahrbuch für Geologie und Paläontologie Abhandlungen*, **251**, 71–94.

VAN ALLEN, H.E.K., CALDER, J.H. & HUNT, A.P. 2005. The trackway record of a tetrapod community in a walchian conifer forest from the Permo-Carboniferous of Nova Scotia. *New Mexico Museum of Natural History and Science Bulletin*, **30**, 322–332.

VOIGT, S. 2005. *Die Tetrapodenichnofauna des kontinentalen Oberkarbon und Perm im Thüringer Wald – Ichnotaxonomie, Paläoökologie und Biostratigraphie*. Cuvillier, Göttingen.

VOIGT, S. 2012. Tetrapodenfährten. *Schriftenreihe der Deutschen Gesellschaft für Geowissenschaften*, **61**, 92–106.

VOIGT, S. 2015. Der Holotypus von *Amphisauropus latus* Haubold, 1970 – ein besonderes Objekt permischer Tetrapodenfährten im Naturhistorischen Museum Schloss Bertholdsburg Schleusingen. *Semana*, **30**, 79–89.

VOIGT, S. & GANZELEWSKI, M. 2010. Toward the origin of amniotes: diadectomorph and synapsid footprints from the early Late Carboniferous of Germany. *Acta Palaeontologica Polonica*, **55**, 57–72.

VOIGT, S. & HAUBOLD, H. 2000. Analyse zur Variabilität der Tetrapodenfährte *Ichniotherium cottae* aus dem Tambacher Sandstein (Rotliegend, Unterperm, Thüringen). *Hallesches Jahrbuch für Geowissenschaften*, **B22**, 17–58.

VOIGT, S. & HAUBOLD, H. 2015. Permian tetrapod footprints from the Spanish Pyrenees. *Palaeogeography, Palaeoclimatology, Palaeoecology*, **417**, 112–120.

VOIGT, S. & LUCAS, S.G. 2012. Late Paleozoic Diadectidae (Cotylosauria: Diadectomorpha) of New Mexico and their potential preference for inland habitats. *Geological Society of America, Abstracts with Programs*, **44**, 90.

VOIGT, S. & LUCAS, S.G. 2013. Carboniferous-Permian tetrapod footprint biochronozonation. *New Mexico Museum of Natural History and Science Bulletin*, **60**, 444.

VOIGT, S. & LUCAS, S.G. 2015*a*. Permian tetrapod ichnodiversity of the Prehistoric Trackways National Monument (south-central New Mexico, U.S.A.). *New Mexico Museum of Natural History and Science Bulletin*, **65**, 153–167.

VOIGT, S. & LUCAS, S.G. 2015*b*. On a diverse tetrapod ichnofauna from Early Permian red beds in San Miguel County, north-central New Mexico. *New Mexico Geological Society Guidebook*, **2015**, 241–252.

VOIGT, S. & MARCHETTI, L. 2014. Über eine neue Fundstelle mit Tetrapodenfährten im Perm von Kötschach-Mauthen (Gailtaler Alpen, Kärnten). *Berichte der Geologischen Bundesanstalt*, **105**, 27–29.

VOIGT, S., SMALL, B. & SANDERS, F. 2005. A diverse terrestrial ichnofauna from the Maroon Formation (Pennsylvanian–Permian), Colorado: biostratigraphic and paleoecological significance. *New Mexico Museum of Natural History and Science Bulletin*, **30**, 342–351.

VOIGT, S., BERMAN, D.S. & HENRICI, A.C. 2007. First well-established track-trackmaker association of Paleozoic tetrapods based on *Ichniotherium* trackways and diadectid skeletons from the Lower Permian of

Germany. *Journal of Vertebrate Paleontology*, **27**, 553–570.

VOIGT, S., SABER, H., SCHNEIDER, J., HMINNA, A., HMICH, D. & KLEIN, H. 2009. Large imprints of Hyloidichnus Gilmore, 1927 from the Permian of Morocco in the light of captorhinid phylogeny and biogeography. *In*: *Abstract Volume, First International Congress on North African Vertebrate Palaeontology*, 25–27 May 2009, Marrakech, Université Cadi Ayyad, Marrakech, Morocco.

VOIGT, S., HMINNA, A., SABER, H., SCHNEIDER, J.W. & KLEIN, H. 2010. Tetrapod footprints from the uppermost level of the Permian Ikakern Formation (Argana Basin, Western High Atlas, Morocco). *Journal of African Earth Sciences*, **57**, 470–478.

VOIGT, S., LAGNAOUI, A., HMINNA, A., SABER, H. & SCHNEIDER, J.W. 2011*a*. Revisional notes on the Permian tetrapod ichnofauna from the Tiddas Basin, central Morocco. *Palaeogeography, Palaeoclimatology, Palaeoecology*, **302**, 474–483.

VOIGT, S., SABER, H., SCHNEIDER, J.W., HMICH, D. & HMINNA, A. 2011*b*. Late Carboniferous–Early Permian tetrapod ichnofauna from the Khenifra Basin, central Morocco. *Geobios*, **44**, 399–407.

VOIGT, S., NIEDŹWIEDZKI, G., RACZYŃSKI, P., MASTALERZ, K. & PTASZYŃSKI, T. 2012. Early Permian tetrapod ichnofauna from the Intra-Sudetic Basin, SW Poland. *Palaeogeography, Palaeoclimatology, Palaeoecology*, **313–314**, 173–180.

VOIGT, S., LUCAS, S.G., BUCHWITZ, M. & CELESKEY, M. 2013*a*. *Robledopus macdonaldi*, a new kind of basal eureptile footprint from the Early Permian of New Mexico. *New Mexico Museum of Natural History and Science Bulletin*, **60**, 445–459.

VOIGT, S., LUCAS, S.G. & KRAINER, K. 2013*b*. Coastal-plain origin of trace-fossil bearing red beds in the Early Permian of southern New Mexico, U.S.A. *Palaeogeography, Palaeoclimatology, Palaeoecology*, **369**, 323–334.

VOIGT, S., SABER, H., SCHNEIDER, J.W., HMINNA, A., LAGNAOUI, A. & KLEIN, H. 2015. Tetrapod footprint ichnodiversity of the Permian Ikakern Formation (Argana Basin, Western High Atlas, Morocco). *In*: *Abstract Volume, First International Congress on Continental Ichnology (ICCI)*, 21–25 April 2015, Université Chouaib Doukkali, El Jadida, Morocco.

WARREN, A. 1997. A tetrapod fauna from the Permian of the Sydney Basin. *Records of the Australian Museum*, **49**, 25–33.

WERNEBURG, R. & BERMAN, D.S. 2012. Revision of the aquatic eryopid temnospondyl *Glaukerpeton avinoffi* Romer, 1952, from the Upper Pennsylvanian of North America. *Annals of Carnegie Museum*, **81**, 33–60.

WOODWORTH, J.B. 1900. Vertebrate footprints on Carboniferous shales of Plainville, Massachusetts. *Geological Society of America Bulletin*, **11**, 449–454.

Permian tetrapod biochronology, correlation and evolutionary events

SPENCER G. LUCAS

New Mexico Museum of Natural History and Science, 1801 Mountain Road NW, Albuquerque, NM 87104-1375 USA

spencer.lucas@state.nm.us

Abstract: The most extensive Permian tetrapod (amphibian and reptile) fossil records from the western USA (New Mexico to Texas) and South Africa have been used to define 11 land vertebrate faunachrons (LVFs). These are, in ascending order, the Coyotean, Seymouran, Mitchellcreekian, Redtankian, Littlecrotonian, Kapteinskraalian, Gamkan, Hoedemakeran, Steilkransian, Platbergian and Lootsbergian. These faunachrons provide a biochronological framework with which to assign ages to, and correlate, Permian tetrapod fossil assemblages. Intercalated marine strata, radioisotopic ages and magnetostratigraphy were used to correlate the Permian LVFs to the standard global chronostratigraphic scale with varying degrees of precision. Such correlations identified the following significant events in Permian tetrapod evolution: a Coyotean chronofaunal event (end Coyotean); Redtankian events (Mitchellcreekian–Littlecrotonian); Olson's gap (late Littlecrotonian); a therapsid event (Kapteinskraalian); a dinocephalian extinction event (end Gamkan); and a latest Permian extinction event (Platbergian–Lootsbergian boundary). Problems of incompleteness, endemism and taxonomy, and the relative lack of non-biochronological age control continue to hinder the refinement and correlation of a Permian timescale based on tetrapod biochronology. Nevertheless, the global Permian timescale based on tetrapod biochronology is a robust tool for both global and regional age assignment and correlation. Advances in Permian tetrapod biochronology will come from new fossil discoveries, more detailed biostratigraphy and additional alpha taxonomic studies based on sound evolutionary taxonomic principles.

Permian tetrapod (amphibian and reptile) fossils are widely distributed (Fig. 1) and have long provided a basis for non-marine biostratigraphy and biochronology (see reviews by Lucas 1998*b*, 2002, 2004, 2005*d*, 2006). Here, I apply a formal tetrapod biochronology that recognizes the 11 time intervals or land vertebrate faunachrons (LVF) of the Permian proposed by Lucas (1998*a*, 2005*d*, 2006) to problems of correlation in the Permian tetrapod fossil record. This biochronology (Fig. 2) provides a tetrapod-based timescale that can be used to determine and discuss the temporal relationships of Permian tetrapod assemblages independent of other criteria. It also can be correlated with some precision to the standard global chronostratigraphic scale for the Permian, which is based on marine biostratigraphy. The global Permian tetrapod biochronology temporally organizes the Permian tetrapod fossil record to allow the identifcation of important events in Permian tetrapod evolution.

Some abbreviations, terminology and concepts

The standard global chronostratigraphic scale (SGCS) consists of the series and stages used as the framework for ordering geological time, sometimes referred to as the marine timescale. I make an important distinction between biostratigraphic datums and biochronological events. Biostratigraphic datums are the lowest occurrence (LO) and highest occurrence (HO) of a fossil in a stratigraphic section. Biochronological events are the first appearance datum (FAD) and last appearance datum of a taxon, i.e. its evolutionary origination and extinction, respectively. For biochronological definitions, it is hoped that the LO and the FAD of a taxon coincide, although, given the problems of sampling and facies, it is highly unlikely that this will often be the case.

LVFs are biochronological units and their beginnings are defined by biochronological events (Lucas 1998*a*, 2006, 2010). Each LVF begins with the FAD of a tetrapod taxon, here a genus. In so doing, the end of an LVF is defined by the beginning of the succeeding LVF, which is the FAD of another tetrapod taxon (Fig. 2). This is a precise way to define LVF boundaries, so LVFs are interval biochrons.

A distinctive assemblage of vertebrate fossils characterizes each LVF. This tetrapod assemblage is the primary basis for identifying the tetrapods that lived during the LVF. The index fossils of LVFs meet the criteria of true index fossils (temporally restricted, common, widespread and easily identified) and do not include endemic or rare taxa

From: LUCAS, S. G. & SHEN, S. Z. (eds) 2018. *The Permian Timescale*. Geological Society, London, Special Publications, **450**, 405–444.
First published online May 15, 2017, https://doi.org/10.1144/SP450.12

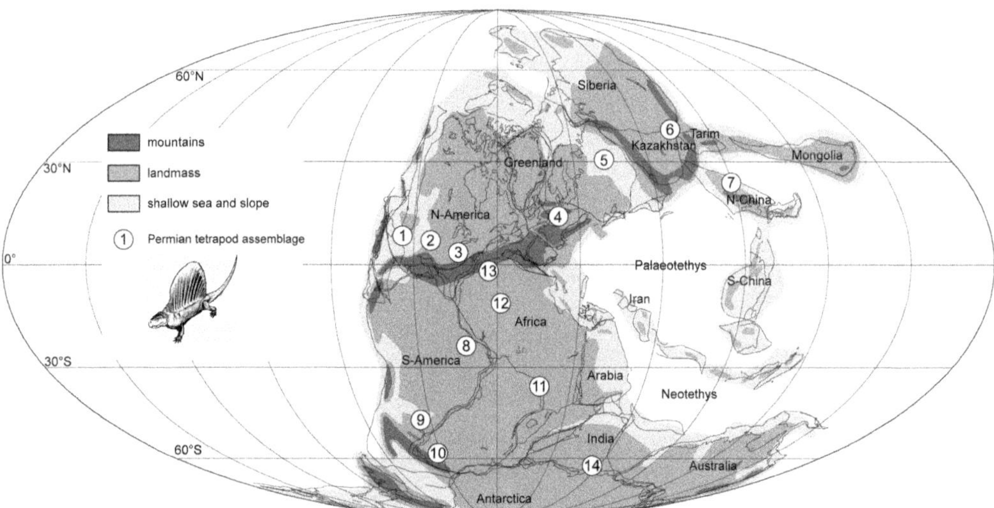

Fig. 1. Map of Permian Pangea at *c.* 270 Ma (after Lucas *et al.* 2006) showing principal tetrapod fossil localities. 1, western USA (New Mexico/Colorado); 2, western USA (Texas/Oklahoma); 3, eastern USA (Dunkard); 4, Western Europe (Rotliegend basins); 5, Russian Urals; 6, Junggur basin, China; 7, Ordos basin, China; 8, Parnaíba basin, Brazil; 9, Paraná basin, Brazil; 10, Karoo basin, South Africa; 11, Ruhuhu basin, Tanzania–Malawi; 12, Niger; 13, Morocco; 14, Pranhita-Godavari Valley, India.

that happen to be restricted to an LVF, usually as single records. The tetrapod biochronology of the Permian is a timescale that is independent of the SGCS (Fig. 2).

In this article, the genus is the operational taxonomic unit for biostratigraphy. This is because most species-level taxa of Permian tetrapods are meaningless for correlation because they are usually based on a single specimen or a local assemblage of well-preserved material and cannot be recognized at multiple localities (Williston 1915; Romer 1928). Some species of Permian tetrapod genera (such as *Seymouria* and *Bolosaurus*) are of use in correlation, however, and taxonomic revisions of some other genera (such as *Eryops* and *Dimetrodon*) should produce species-level taxa of value to biostratigraphy. Recent work by Brink & Reisz (2014), for example, has identified temporally successive tooth morphology in *Dimetrodon* species that may be of biostratigraphic value. Werneburg (1989) has argued that species lineages (chronoclines) provide a more precise biostratigraphy than genus-based correlations. I agree with Werneburg in principle, but am unable to construct meaningful species lineages for most of the Permian tetrapod genera that are of value to a global biochronology.

Value of vertebrate biochronology

Using a vertebrate biochronology that assigns ages and correlates based on the vertebrates themselves will free vertebrate biostratigraphers from attempting to correlate tetrapods directly to the SGCS. This is best exemplified by the North American land mammal 'ages' (LMAs), a set of biochronological units created by Wood *et al.* (1941) to organize Cenozoic, mammal-dominated assemblages, primarily from western North America. As an example, in that biochronological scheme a fossil mammal assemblage can be correlated to and within the Barstovian LMA based only on the known distribution of the mammal fossils. The question of correlating the Barstovian LMA to the SGCS is a separate problem not resolved by mammalian biochronology. Instead, magnetostratigraphy and radioisotopic ages indicate that the Barstovian spans *c.* 12.5–15.5 Ma, so it overlaps the Langhian and Serravalian marine stages of the Miocene on the SGCS (Tedford *et al.* 2004). No vertebrate biostratigrapher tries to make that correlation (which I term a cross-correlation) based on mammals and, in the literature of North American mammalian biochronology, Barstovian is almost always used with no reference to cross-correlation to the SGCS.

The important lesson is that a vertebrate biochronological scheme allows ages to be assigned and correlations to be determined by the vertebrates themselves. The correlation to the SGCS is a separate cross-correlation between the vertebrate biochronology and marine biochronology that relies on other data (e.g. palynostratigraphy,

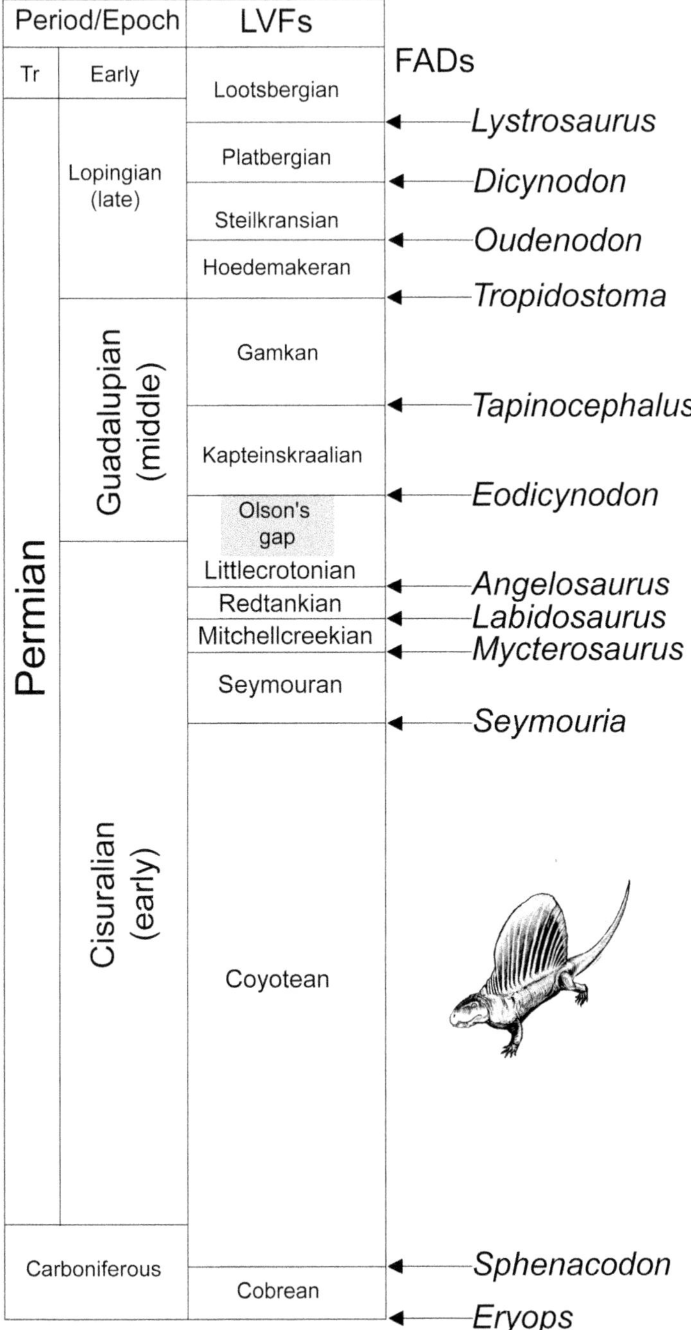

Fig. 2. Permian timescale based on tetrapod evolution showing the land vertebrate faunachrons (LVFs) and the first appearance datums (FADs) of the tetrapod genera that define the beginning of each LVF.

magnetostratigraphy or radioisotopic age) to be completed. Sometimes a terrestrial/freshwater fossil vertebrate is found displaced in marine deposits, which allows some direct cross-correlation of vertebrate taxa to the SGCS, but this cross-correlation typically relies wholly on non-vertebrate fossil data.

Permian land vertebrate faunachrons

The subdivision of Permian time used here is a biochronological scheme of 11 LVFs (Fig. 2). Lucas (2006) presented the Permian LVFs in detail and the scheme needs little modification based on a decade of new data and analyses. Therefore, I do not review the Permian LVFs in detail; many pertinent published references are listed in Lucas (2006). Instead, I review the global fossil assemblages of Permian tetrapods to correlate these records using the biochronological timescale of LVFs.

Record of early Permian tetrapods

The most extensive fossil record of early Permian tetrapods is from the western USA (Fig. 1), primarily Texas, as well as New Mexico, Colorado, Utah, Oklahoma, Kansas and Nebraska. This fossil record provides the majority of what vertebrate palaeontologists know about early Permian tetrapod evolution. Lucas (2005*d*, 2006) based the Early Permian tetrapod biochronology on the New Mexican and Texas records, which I briefly review here. I then discuss other important early Permian tetrapod records from North America, Western Europe and northeastern Brazil.

New Mexico

Because very few Late Carboniferous vertebrates are known from Texas, our understanding of the biostratigraphic succession of vertebrates across the Carboniferous–Permian boundary in North America long relied on comparing disparate Pennsylvanian sites (e.g. Howard in Colorado or Linton in Ohio) with the early Permian record in Texas or elsewhere. Cañon del Cobre in the Chama basin of northern New Mexico is the only location in North America where substantial assemblages of vertebrate fossils closely bracket the Pennsylvanian–Permian boundary in a single stratigraphic section without significant facies changes. Lucas (2006) reviewed the placement of the Carboniferous–Permian boundary in the Cañon del Cobre and at other New Mexican locations; Lucas *et al.* (2010, 2012*a*, *b*, 2013*b*, 2014, 2015*c*, 2016) and Berman *et al.* (2015) have since published new data on these tetrapods and their correlation.

The tetrapod fossils are from siliciclastic red beds of the Cutler Group and the partly equivalent Abo Formation, synorgenic deposits of the ancestral Rocky Mountain orogeny (Fig. 3). The Abo Formation can be traced from north to south across central New Mexico, where it overlies and interfingers southwards with marine strata (Hueco Group), thus allowing some direct correlation of the New Mexican Lower Permian tetrapods to the SGCS (Lucas *et al.* 2013*a*, 2015*a*, *b*, *c*; Vachard *et al.* 2015).

The New Mexican Carboniferous–Permian tetrapod record can be assigned to three LVFs.

(1) The oldest tetrapods, from the El Cobre Canyon Formation of the Cutler Group in the Cañon del Cobre in the Chama basin of northern New Mexico, include tetrapod taxa known elsewhere from Upper Pennsylvanian strata, such as *Desmatodon*, *Limnoscelis* and *Diasparactus*. They are associated with palynomorphs and megafossil plants indicative of a Late Pennsylvanian age (see references cited in Lucas *et al.* 2010). This is the tetrapod assemblage characteristic of the Cobrean LVF of Lucas (2006). Its beginning is defined by the FAD of the temnospondyl amphibian *Eryops*.

(2) The overlying tetrapod assemblage of the upper part of the El Cobre Canyon Formation in northern New Mexico, and tetrapod assemblages from the Red Tanks Member of the Bursum Formation and the Scholle Member of the Abo Formation to the south, are assigned to the Coyotean LVF of Lucas (2006). The FAD of the eupelycosaur *Sphenacodon* defines the beginning of the Coyotean and the characteristic assemblage is found in Cutler Group strata near Coyote and Arroyo del Agua in the Chama basin of northern New Mexico (Lucas 2006). Correlative tetrapod assemblages are from strata of the Bursum and Abo formations that extend as far south as Alamogordo and Truth or Consequences in southern New Mexico (Lucas *et al.* 2012*a*, *b*, 2014, 2016; Berman *et al.* 2015). These assemblages are mostly early Permian, but the Bursum Formation straddles the Carboniferous–Permian boundary (see following discussion). The New Mexican Coyotean tetrapod assemblages have long been correctly correlated to the tetrapod assemblages of the lower part of the Wichita Group in Texas and this provides a key overlap of the New Mexican and Texan tetrapod biochronology.

(3) In northern New Mexico, the LO of the seymouriamorph *Seymouria* is in the Arroyo del Agua Formation of the Cutler Group, stratigraphically above the Coyotean tetrapod assemblage of the El Cobre Canyon Formation. This provides another key point of correlation to the Texas section, where the LO of *Seymouria* is a proxy for its FAD, which defines the beginning of the Seymouran LVF.

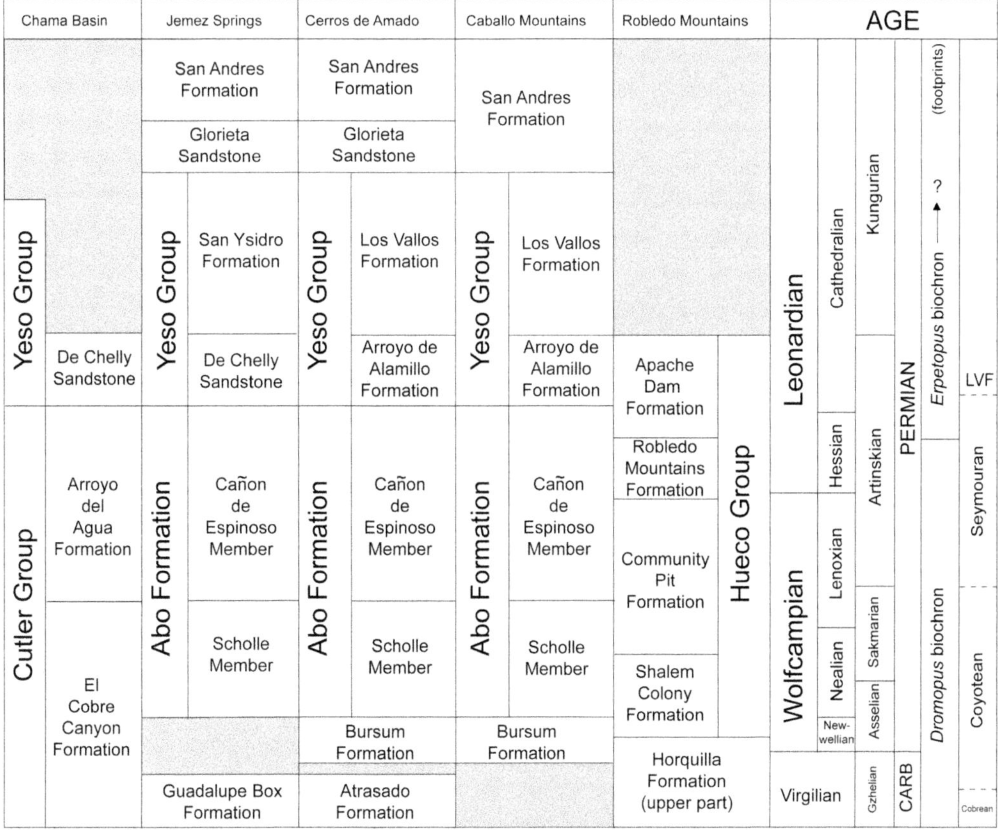

Fig. 3. Correlation of tetrapod-bearing latest Carboniferous–early Permian strata in New Mexico (after Lucas *et al.* 2015*b*). Columns proceed from north (left) to south (right) and thus from wholly non-marine lower Permian sections (Chama basin, Jemez Springs) to sections with intercalated marine strata. LVF, land vertebrate faunachron.

The New Mexican section thus contains three stratigraphically distinct and superposed tetrapod fossil assemblages that are the characteristic assemblages of the Cobrean and Coyotean LVFs, and an assemblage of the Seymouran LVF (Fig. 3). Stratigraphic data (discussed in more detail in the following sections) correlate these New Mexican tetrapods to marine strata of late Virgilian–early Leonardian age, which overlap the Gzhelian–Artinskian on the SGCS.

Texas

The most extensive lower Permian tetrapod record from the western USA is from north-central Texas (e.g. Romer 1928, 1935, 1958, 1974; Hook 1989; Sander *et al.* 2007). Fossil vertebrates have been collected from the non-marine Permian red beds in north-central Texas since the 1870s and have been extensively published (see historical review by Craddock & Hook 1989).

The lithostratigraphic nomenclature long applied to the Texas section was that of Plummer & Moore (1921), who named a series of rock formations of primarily marine origin, assigning them to the Cisco and Wichita groups of late Carboniferous–early Permian age (Fig. 4). Vertebrate palaeontologists, especially Romer (1928, 1935, 1958, 1974), readily placed vertebrate fossil localities in the Texas section into this lithostratigraphy. However, the localities bearing vertebrate fossils are primarily in non-marine red beds split by thin marine limestone–shale horizons that correlate to, but are lithologically distinct from, the formations named by Plummer & Moore (1921). Thus Hentz (1988) justifiably created a new lithostratigraphic nomenclature for the Texas Permian red beds and Lucas (2006) slightly modified that nomenclature to be consistent with accepted lithostratigraphic usage (Fig. 4). In the Texas section, Lucas (2006) recognized five superposed tetrapod assemblages of biochronological significance.

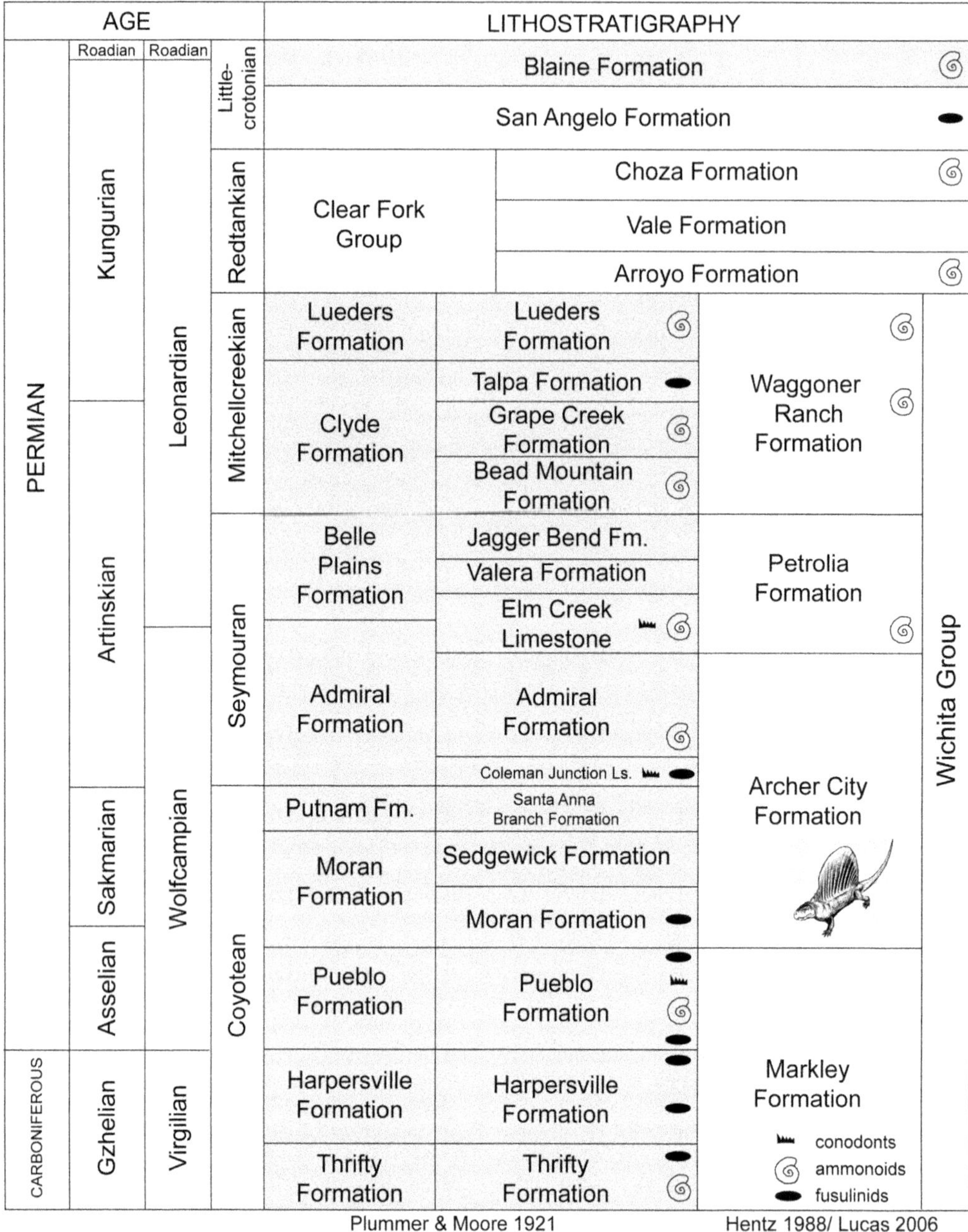

Fig. 4. Summary diagram of the Lower Permian section in north-central Texas showing cross-correlation of vertebrate biochronology and marine biostratigraphy based on fusulinids, ammonoids and conodonts. Formations on right of diagram are the tetrapod-bearing units. Shaded units are marine strata.

(1) The tetrapod assemblage from the Markley and lower part of the Archer City formations of the Wichita Group is of Coyotean age. This assemblage includes diverse temnospondyls (e.g. *Eryops*, *Edops*, *Neldasaurus*, *Zatrachys* and *Trimerorhachis*), a few microsaurs and nectrideans, anthracosaurs (*Archeria*), the diadectomorph *Diadectes*, the

eureptiles *Protorothyris* and *Romeria* and diverse eupelycosaurs (especially *Dimetrodon*, *Edaphosaurus* and *Stereophallodon*).

(2) The characteristic assemblage of the Seymouran LVF is from the upper Archer City ('Nocona') and Petrolia formations of the Wichita Group. This assemblage has temnospondyls similar to those of the Coyotean (but without *Edops* and *Neldasaurus*), the microsaurs *Carrolla* and *Pantylus*, few nectrideans, the anthracosaur *Archeria*, *Diadectes* and *Seymouria*, the eureptile *Protocaptorhinus*, diverse eupelycosaurs (including the LOs of *Secodontosaurus* and *Varanosaurus*), the diapsid *Araeoscelis* and the parareptile *Bolosaurus*.

(3) The characteristic Mitchellcreekian LVF tetrapod assemblage is from the Waggoner Ranch and Lueders formations. It has only a few temnospondyls (except for abundant armoured dissorophids), gymnarthrids, *Brachydectes*, an aïstopod, abundant nectrideans (especially *Diplocaulus*), *Archeria, Diadectes* and diverse eupelycosaurs.

(4) The characteristic Redtankian LVF assemblage is from the Clear Fork Group (Arroyo, Vale and Choza formations) or Clear Fork Formation (where the three constituent formations are not distinct mappable units; Nelson *et al.* 2013). It includes abundant *Brachydectes*, *Trimerorhachis* and *Diplocaulus*, as well as *Eryops*, *Trematops*, *Cacops*, *Trematopsis*, diverse dissorophids (including *Broiliellus*, *Aspidosaurus* and *Dissorophus*), diverse eureptiles (especially *Captorhinus, Captorhinikos, Captorhinoides* and *Labidosaurus*), *Seymouria*, *Diadectes*, diverse eupelycosaurs (including *Casea*, *Dimetrodon*, *Varanosaurus, Secodontosaurus* and *Edaphosaurus*) and *Araeoscelis*. Its characteristic assemblage is the classic Clear Fork Group chronofauna of Texas (Olson 1952).

(5) The youngest Permian tetrapod assemblage in Texas, from the San Angelo Formation (Olson & Beerbower 1953), is characteristic of the Littlecrotonian LVF. It includes the captorhinid *Rothianiscus,* the caseids *Caseoides, Cotylorhynchus* and *Angelosaurus*, the sphenacodontids *Steppesaurus* and *Tappensaurus* and the putative therapsid *Dimacrodon.* Olson (1962) later added these taxa to the San Angelo tetrapod assemblage: the temnospondyl *Slaugenhopia,* the captorhinid *Kahneria,* the sphenacodont *Dimetrodon,* the caseid *Caseopsis* and the 'therapsids' *Knoxosaurus, Gorgodon, Eosyodon, Driveria* and *Mastersonia.* Olson (1962) also reassigned *Tappenosaurus* and *Steppesaurus* to the Therapsida. Abundant and diverse caseids are characteristic of Littlecrotonian time. However, all the 'therapsid' taxa from this assemblage have been re-evaluated and deemed to be based on fragmentary eupelycosaur fossils (Parrish *et al.* 1986; Sidor & Hopson 1995; Lucas 2006).

The Texas section thus records a stratigraphically superposed succession of tetrapod assemblages that provides an excellent basis for early Permian tetrapod biostratigraphy. This biostratigraphy can be readily correlated to marine biostratigraphy (Fig. 4 and following text). This section lacks an extensive record of tetrapods across the Carboniferous–Permian boundary, but this is compensated for by the New Mexican section discussed earlier.

Recent publications on lower Permian tetrapods from the Texas section have not modified the stratigraphic ranges and age assessments (e.g. Modesto *et al.* 2007, 2014; Sander *et al.* 2007; Campione & Reisz 2010; Amson & Laurin 2011; Spindler 2015). Therefore I have not modified the tetrapod biochronology based on the Texas section presented by Lucas (2006).

Oklahoma, Kansas and Nebraska

Much less extensive early Permian tetrapod records in the western USA are known from Oklahoma (Olson 1967) and can be readily correlated to the Texas records (Fig. 5). In southern Oklahoma, the Oscar Group (especially the Waurika 1 locality) yields a tetrapod assemblage of Coyotean age that includes *Diplocaulus*, *Trimerorhachis*, *Eryops*, *Archeria*, *Pantylus*, *Ophiacodon*, *Dimetrodon* and *Edaphosaurus* (Olson 1967; Simpson 1979). In northern Oklahoma, the Wellington Formation (especially the Perry and Orlando localities) yields an extensive tetrapod assemblage of Seymouran age that includes *Trimerorhachis*, *Zatrachys*, *Seymouria*?, *Brachydectes*, *Eryops*, *Diplocaulus*, *Broiliellus*, *Diadectes, Archeria*, *Ophiacodon*, *Captorhinus*, *Edaphosaurus* and *Dimetrodon* (Olson 1967; Simpson 1979).

In southern Oklahoma, the tetrapod assemblage from the lower–middle Garber Formation (especially the South Grandfield and Northeast Frederick sites) is of Mitchellcreekian age and includes *Trimerorhachis*, *Tersomius*, *Brachydectes*, *Diplocaulus*, *Archeria*, *Diadectes*, *Captorhinus*, *Labidosaurikos*, *Ophiacodon*, *Dimetrodon* and *Araeoscelis*. The Richards Spur locality (a fissure-fill in Ordovician limestone) may also be of Mitchellcreekian age and includes *Phlegethontia*, *Doleserpeton*, *Cacops*, *Tersomius*, *Seymouria*, diverse gymnarthrids, *Captorhinus*, *Mycterosaurus*, *Bolosaurus* and a caseid (see discussion in following text).

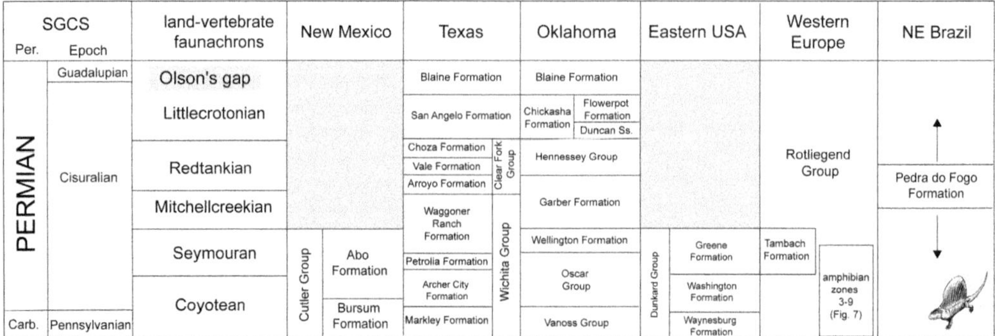

Fig. 5. Correlation of selected early Permian tetrapod assemblages to the land vertebrate faunachrons.

In Oklahoma, the LO of *Labidosaurus* is in the upper Garber Formation. The overlying Hennessey Group also yields a Redtankian assemblage that includes *Trematops*, *Tersomius*, *Trimerorhachis*, *Peroneodon*, *Brachydectes*, *Eryops*, *Captorhinus*, *Cotylorhynchus*, *Dimetrodon* and *Ophiacodon* (Olson 1967; Simpson 1979).

The Flowerpot Formation and the equivalent Chickasha Formation yield tetrapod assemblages of Littlecrotonian age. Olson & Barghusen (1962) described vertebrate fossils from two localities in the Flowerpot Formation in Kingfisher County, Oklahoma that yield *Cymatorhiza*, *Rothianiscus*, *Cotylorhynchus* and *Angelosaurus*.

Strata of the Chickasha Formation, which are laterally equivalent to the middle part of the Flowerpot Formation, yielded vertebrate fossils from about 20 localities, mostly in Blaine and Kingfisher counties, Oklahoma (Olson 1965). A single locality in McClain County, Oklahoma also yielded unidentified bone from the Duncan Sandstone (Olson 1965). The Chickasha assemblage includes *Cymatorhiza*, the amphibians *Diplocaulus, Nannospondylus* and *Fayella*, the captorhinomorph *Rothianiscus*, the caseids *Cotylorhynchus*, and *Angelosaurus* and the varanopid *Varanodon*. Olson (1974) described the supposed therapsid (actually a eupelycosaur) *Watongia* (cf. Reisz & Laurin 2004). The *Seymouria agilis* of Olson (1980) from the Chickasha Formation assemblage has been reassigned to the parareptile *Macroleter*, a genus previously known only from Russia (Reisz & Laurin 2001).

In Brown County, Kansas, the Robinson locality in the upper Virgilian Soldier Creek Shale Member of the Bern Limestone yields a lysorophid, *Diplocaulus*, *Cricotus*, a trimerorhachid and cf. *Platyhystrix* (Foreman & Martin 1988). This assemblage may be of Coyotean age. Also in Kansas, various localities in the upper Council Grove Group (especially those in the Speiser Shale) yield an assemblage that includes *Acroplous*, *Brachydectes*, *Diplocaulus*, *Trimerorhachis* and *Euryodus* (Foreman & Martin 1988; Englehorn *et al.* 2008). These tetrapods do not correlate precisely to the tetrapod biochronology, but marine biostratigraphy indicates that the tetrapod-bearing interval of the Council Grove Group is Wolfcampian (e.g. Boardman *et al.* 2009), so its tetrapods are Coyotean.

The Indian Cave Sandstone in Nemaha County, Nebraska has yielded *Ophiderpeton*, *Phlegethontia*, *Captorhinus*, *Denderpetron* and a eupelycosaur (Foreman & Martin 1988), an assemblage that may be of Coyotean age. In Richardson County, Nebraska, the Eskridge Formation yields *Acroplous*, *Brachydectes*, the dissorophoid *Plemmyradites*, a trimerorhachid, a microsaur, a diadectid and an edaphosaurid (Huttenlocker *et al.* 2005, 2007). Bracketing marine strata also indicate a Wolfcampian age, which is Coyotean.

Utah and Colorado

Permian tetrapod fossil assemblages from Utah and Colorado are much less extensive and diverse than those from New Mexico and Texas. In the Arizona–Utah borderland (principally Monument Valley), the Halgaito Formation (Cutler Group) yields a Coyotean tetrapod assemblage that includes *Diplocaulus*, *Phlegethonia*?, a trimerorhachid, *Eryops*, *Platyhystrix*, *Archeria*, a limnoscelid, *Limnoscelis*, *Ophiacodon*, *Edaphosaurus*, *Sphenacodon* and an araeoscelid? (e.g. Sumida *et al.* 1999*a*, *b*; Scott 2013).

Also in the Arizona–Utah borderland, the Organ Rock Shale (Cutler Group) yields a Seymouran tetrapod assemblage that includes *Seymouria*, *Eryops*, a trimerorhachid, a zatrachyid, the diadectomorphs *Tseajaia* and *Diadectes*, *Sphenacodon*, *Ophiacodon*, *Dimetrodon* and *Ctenospondylus* (Sumida *et al.* 1999*a*, *b*). The underlying Cedar Mesa Sandstone yields *Eryops* and *Sphenacodon* and could be either Coyotean or Seymouran in age.

In southwestern Colorado, the upper part of the Cutler Formation yields a Coyotean age assemblage that includes *Eryops*, *Platyhystrix*, a seymouriid, *Diadectes*, a captorhinid?, a haptodontid and '*Mycterosaurus*' (unreliable identification) (Lewis & Vaughn 1965; Wideman *et al.* 2005).

Pennsylvania–Ohio–West Virginia

The Dunkard Group is *c.* 343 m of mostly clastic rocks exposed in Pennsylvania, Ohio and West Virginia, USA. Lucas (2011, 2013*c*) reviewed the Permian tetrapod biostratigraphy and biochronological correlation of the Dunkard Group (Fig. 6). Fossil tetrapods from the Dunkard Group include lepospondyl and temnospondyl amphibians, diadectomorphs, primitive reptiles and eupelycosaurs.

These vertebrates represent two biostratigraphically distinct assemblages, one from the Waynesburg and Washington formations, and the other from the overlying Greene Formation (Fig. 6). The Dunkard vertebrate biostratigraphy supports correlation to the Coyotean and Seymouran LVFs. Tetrapod taxa from the Waynesburg and Washington formations include *Edops* and *Protorothyris*, Coyotean index taxa, as well as the characteristic Coyotean taxa *Trimerorhachis, Diadectes, Edaphosaurus* and *Dimetrodon*. Significantly, these Dunkard taxa are well known from the lower Archer City Formation in Texas, which is late Coyotean.

The Greene Formation contains the eupelycosaur *Ctenospondylus*, an index taxon of the Seymouran LVF. Vertebrate biochronology thus indicates that the Waynesburg and Washington formations are late Coyotean, whereas most of the Greene Formation is Seymouran (Figs 5 & 6).

Eastern Canada

Lower Permian red beds on Prince Edward Island in the eastern Canadian Maritimes yield *Eryops*, *Seymouria*, *Diadectes* and a eupelycosaur (Langston 1963; Spalding 1993), an assemblage of probable Seymouran age. The eupelycosaur, long ago named *Bathygnathus*, has been re-evaluated and deemed to be based on a specimen of *Dimetrodon*, which, if accepted, would give the little used *Bathygnathus* priority over the widely used *Dimetrodon* (Brink 2015; Brink *et al.* 2015).

Western Europe

The early Permian tetrapod record in Western Europe comes from non-marine deposits in a series of basins formed during the Hercynian orogeny, primarily in Germany, France and the Czech Republic, but also from related strata in Italy and Spain (Roscher & Schneider 2006) (Fig. 1). These are deposits of the classic Rotliegend Group in Germany and correlatives, and the tetrapod fossil assemblages are mostly from lacustrine facies. They are dominated by fossils of the newt-like amphibian families Branchiosauridae, Micromelerpetontidae and Discosauriscidae (e.g. Werneburg & Schneider 2006; Schoch & Milner 2008; Schneider & Werneburg 2012; Steyer *et al.* 2012). Some of the assemblages are also dominated by fish, which has led to the creation of some local biostratigraphy based on fish (e.g. Štamberg & Zajíc 2008; Zajíc 2014).

Some of the Rotliegend tetrapod assemblages (e.g. the lower *Protriton* and Gottlob horizons in the Thuringian forest and the Remigiusberg Formation in the Saar-Nahe basin of Germany; Voigt *et al.* 2014) are apparently of Coyotean age, but the lack of shared low-level tetrapod taxa between Europe and North America makes a direct tetrapod-based correlation impossible. The correlation must instead be based on other evidence, which indicates that most of the Rotliegend tetrapod assemblages are of late Virgilian–middle Wolfcampian age (e.g. Roscher & Schneider 2005), which means that they correlate to the Coyotean.

The latest version of a Rotliegend amphibian biostratigraphic zonation (Fig. 7) recognizes nine zones, ranging from Middle Pennsylvanian through early Permian, based on species chronoclines that provide correlations in the Czech Republic, Germany, France, Poland and Italy (Werneburg & Schneider 2006; Schneider & Werneburg 2012). This is a provincial biostratigraphy in the Rotliegend extensional basins of Europe, in which amphibian zones 3–9 mostly overlap Coyotean time (Fig. 5).

Steyer (2000) critiqued the Rotliegend amphibian biostratigraphy by arguing that taphonomic and palaeoecological factors control amphibian distributions more than the actual temporal ranges, and by critiquing the species-chronocline method of taxonomy. However, Steyer's assertions about palaeoecology and taphonomy remain undocumented and the species-chronocline method is the preferred method used in the micropalaeontological taxonomy of the fusulinids and conodonts, the two biostratigraphic workhorses of the Permian SGCS. In principle, an extensive record of European amphibians should be amenable to such methods.

European early Permian tetrapod assemblages not dominated by branchiosaurs are much less extensive and consist primarily of the Bromacker locality in the Tambach Formation of Germany and scattered records of amphibians and eupelycosaurs. Tetrapods from the Tambach Formation include the trematopid *Tambachia, Seymouria,* the

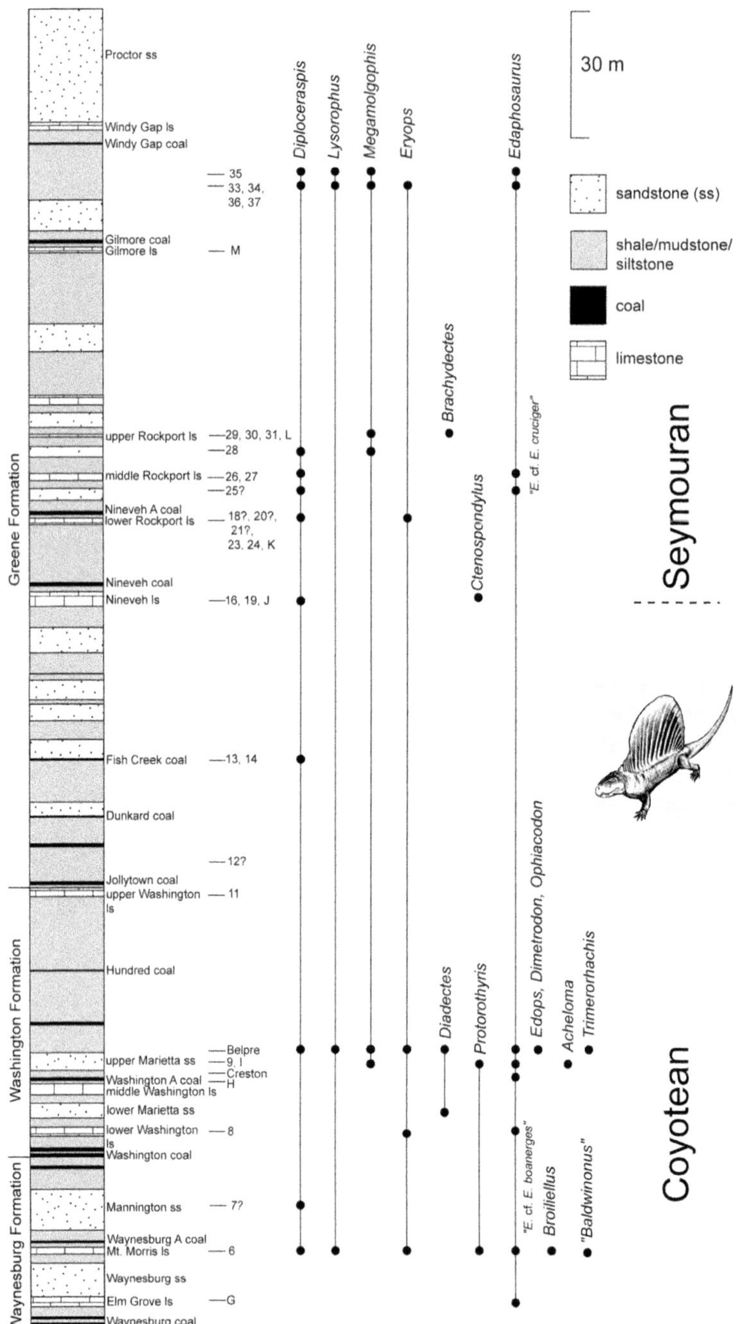

Fig. 6. Stratigraphy of the Dunkard Group and stratigraphic distribution of tetrapod genera. Uncertainty about the exact stratigraphic position of a locality is indicated with a question mark. Modified from Lucas (2013*c*).

eureptile *Thuringothyris*, the diadectomorphs *Diadectes* and *Orobates*, the bolosaurid *Eudibamus*, the varanopid *Tambacarnifex*, a caseid and *Dimetrodon* (Lucas 2006; Berman *et al.* 2013, and references cited therein). This assemblage is of Seymouran age.

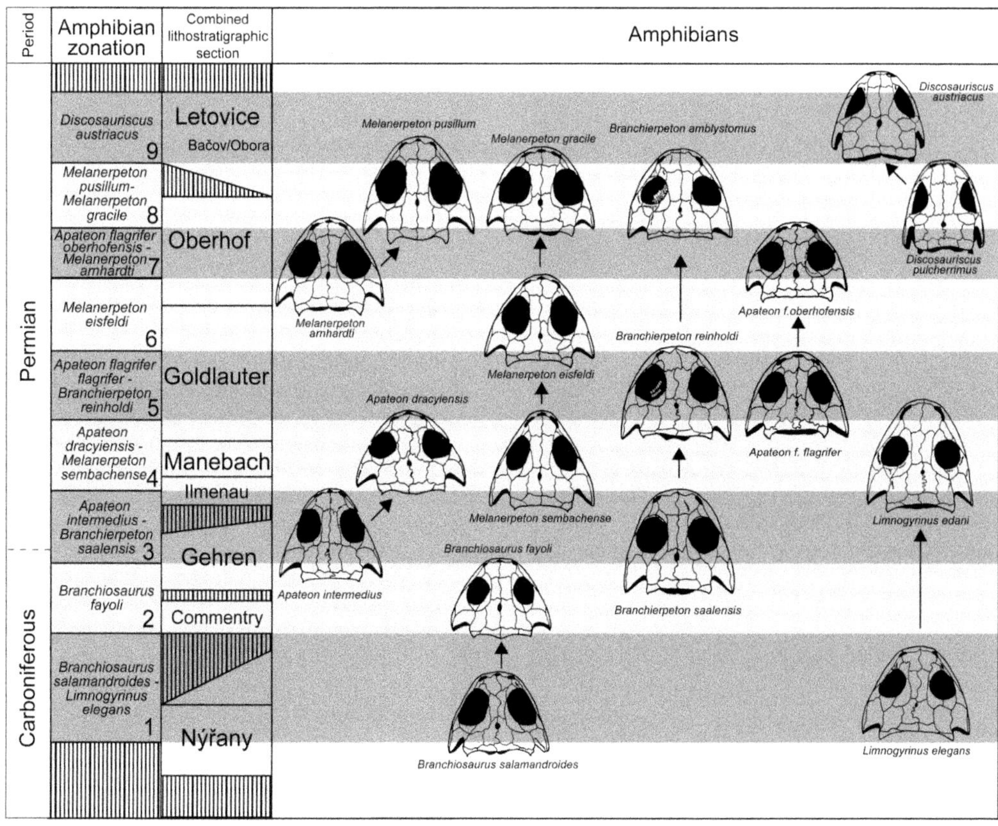

Fig. 7. Amphibian zonation for the European Permo-Carboniferous using species chronoclines of the aquatic families Branchiosauridae, Micromelerpetontidae and Discosauriscidae (after Schneider & Werneburg 2012).

The Bromacker locality is in siliciclastic red beds of the Tambach Formation and yields many genera of tetrapods found in the Texas section; therefore, it is readily correlated to the tetrapod biochronology (Fig. 5). Other non-branchiosaur lower Permian tetrapods from Western Europe, however, are mostly endemic eryopoids such as *Onchiodon* (e.g. Werneburg & Steyer 1999; Werneburg 2008; Schoch & Milner 2014) and endemic eupelycosaurs, such as the basal sphenacodontians *Pantelosaurus* and *Palaeohatteria* from Germany (Spindler 2013, 2015), or endemic or indeterminate caseids from Italy and France (e.g. Sigogneau-Russell & Russell 1974; Maddin *et al.* 2008; Reisz *et al.* 2011; Romano & Nicosia 2015; Spindler 2015) not directly correlateable to the tetrapod biochronology. Like the branchiosaurs, they are assigned ages on the SGCS based on other data, which indicate that the basal sphenacodontians are mostly of Coyotean age, whereas some of the caseids may be as young as Littlecrotonian (e.g. Ronchi *et al.* 2011).

Northeastern Brazil

The Permian Pedro do Fogo Formation in the Parnaíba basin of northeastern Brazil was long known as the source of one endemic temnospondyl amphibian, *Prionosuchus* (Price 1948). Cisneros *et al.* (2015*b*) described an assemblage of Permian tetrapods from the Pedro do Fogo Formation consisting of temnospondyls (two new, endemic genera), an indeterminate rhinesuchid and a captorhinomorph assigned to *Captorhinus*, but more likely to be *Captorhinikos* (S. Modesto, pers. comm. 2016). Cisneros *et al.* (2015*b*) also reviewed the diverse, rather contradictory and generally imprecise age data from the Pedra do Fogo Formation based on the correlation of palynomorphs, a fossil forest, selachians and the tetrapods, concluding that it is Kungurian.

Particularly important to that conclusion was equating the petrified forest in the Pedra do Fogo Formation to the famous Chemnitz forest in Germany, which is in volcaniclastic strata that have yielded a U–Pb age on detrital zircon of 278 ± 5 Ma (Rössler 2006). There is also a new date for the Chemnitz forest of 290.6 ± 1.8 Ma (Rössler *et al.* 2014), which is close to the estimated age of the Sakmarian–Artinskian boundary. However, correlation of the forests is not at all certain (see Rössler 2006) and other age indicators are equivocal or imprecise. Therefore the Pedra do Fogo tetrapods are certainly early Permian, but compelling evidence for a more precise age determination is not available (Fig. 5).

Olson's gap

At the early–middle Permian boundary, the basis for the LVFs shifts from North America to South Africa (Lucas 2006). Lucas (2004) recognized a global gap (Olson's gap) between the youngest North American Permian tetrapods (San Angelo Formation and equivalents) and what he considered to be the oldest, therapsid-dominated assemblages: those of Russian Zone I and the *Eodicynodon* Assemblage Zone of South Africa. This is a temporal gap between early Permian eupelycosaur-dominated tetrapod assemblages and middle–late Permian therapsid-dominated tetrapod assemblages.

Lucas (2004) explained in detail why the youngest North American Permian tetrapod assemblages (from the San Angelo, Flowerpot and Chickasha formations of Texas and Oklahoma) are late Leonardian (Kungurian) in age. In brief, this is because the marine strata intercalated in the San Angelo Formation yield Leonardian fusulinids, whereas the overlying strata at the base of the Blaine Formation (and the correlative San Andres Formation) yield ammonoids of Leonardian age (Fig. 8).

Lucas (2006) also accepted the argument (see later text) that the *Eodicynodon* Assemblage Zone of South Africa is probably the oldest therapsid-bearing assemblage because it has the most primitive therapsids (but see later discussion). Therefore he defined the beginning of the Kapteinskraalian by the FAD of *Eodicynodon* and considered the San Angelo assemblage to be older, characteristic of the Littlecrotonian LVF (Fig. 8). Thus the temporal gap in the tetrapod record is equivalent to the younger part of the Littlecrotonian LVF.

The only tetrapod assemblage that is clearly in Olson's gap is the Inta assemblage from the Pechora

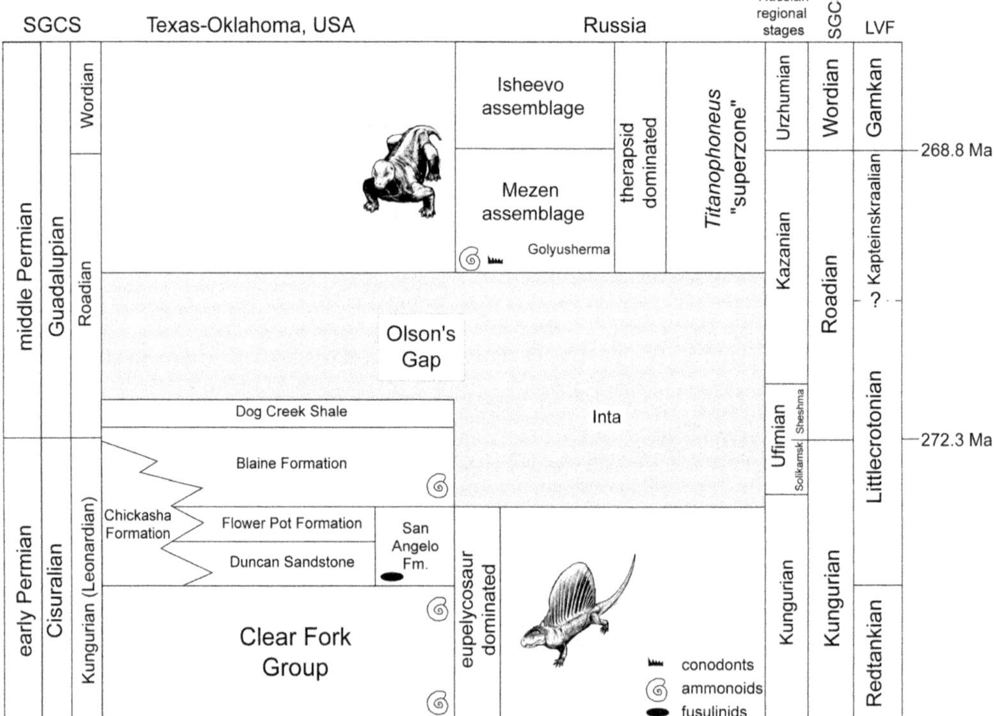

Fig. 8. Olson's gap is shown here as the hiatus between the youngest early Permian tetrapod assemblages of Texas–Oklahoma, USA and the oldest Russian Permian tetrapod assemblages. Modified from Lucas (2004). LVF, land vertebrate faunachron; SGCS, standard global chronostratigraphic scale.

basin in Russia. This assemblage is essentially an endemic amphibian fauna that resembles North American early Permian amphibians in its stage of evolution, but cannot be otherwise correlated based on tetrapod biochronology alone (Lucas 2004).

On the marine timescale, Olson's gap is equivalent to part of Kungurian–Roadian time (Fig. 8). The marine Blaine Formation overlies the youngest North American Permian tetrapod assemblage (those of Littlecrotonian age) and is late Leonardian (late Kungurian) at its base (Lucas 2004). The oldest age of the San Andres Formation, which is correlative to the Blaine Formation, is about middle Kungurian (*Neostreptognathodus prayi* conodont zone; Kerans *et al.* 1993), although the base of the Blaine Formation may be slightly younger in the Kungurian. Lucas (2004) reviewed conodont data that indicate a late Roadian–Wordian age for the marine Kazanian strata immediately beneath the stratigraphically lowest Russian therapsid locality at Golyusherma. Subsequent re-evaluation of the correlation of the Kazanian to the SGCS (Leven & Bogoslovskaya 2006; Leonova 2007) correlates it more closely to the Roadian, not to the Wordian. Nevertheless, in Russia, the Kazanian is younger than the underlying Ufimian Stage.

The Ufimian is divided into two horizons, with the Solikamsk overlain by the Sheshma (Fig. 8). Some workers, however, put the Solikamsk into the upper Kungurian (e.g. Kotlyar 1977; Klets *et al.* 2001) and others include the Sheshma in the Kazanian (e.g. Chuvashov *et al.* 2002; Kotlyar *et al.* 2004). As reviewed by Lozovsky *et al.* (2006), the Russian Interdepartmental Stratigraphic Committee abandoned the Ufimian, absorbing the Solikamsk horizon into the Kungurian and placing the base of the Middle Permian at the base of the Sheshma horizon, which was absorbed into the Kazanian. Lozovsky *et al.* (2006), nevertheless, argued for the retention of the Ufimian as a distinctive chronostratigraphic unit. They noted that the Committee's dismemberment of the Ufimian was largely a decision to allow the Russian regional stages to make a stage boundary (Kungurian–Kazanian) correspond to the Cisuralian–Guadalupian series boundary instead of having the series fall within the Ufimian. In other words, this was not a new correlation of the Ufimian, but instead simply a redefinition of the Kazanian downwards and the Kungurian upwards to eliminate the Ufimian. Nevertheless, there seems to be little disagreement on how to correlate the Ufimian horizons: the stage straddles the Cisuralian–Guadalupian series boundary, so that the Sheshma horizon is early Roadian (Fig. 8).

There are three ways by which vertebrate palaeontologists have attempted to bridge Olson's gap. The first is phylogenetically, by finding more primitive therapsids than are known after the gap. Nevertheless, identifying the most primitive therapsid(s) is confounded by a lack of agreement on early therapsid phylogeny (e.g. Kemp 2006, 2009; Benson 2012; Kammerer *et al.* 2013; Spindler 2014, 2015). For example, a supposed therapsid in Lower Permian strata, *Tetraceratops* from the Redtankian of Texas (see Laurin & Reisz 1996), is not considered to be a therapsid by most workers (e.g. Sidor & Hopson 1998; Conrad & Sidor 2001; Rubidge & Sidor 2001; Liu *et al.* 2009; although see Amson & Laurin 2011). Whether the most primitive therapsids are from South Africa, Russia or China is an interesting question for phylogeny and palaeobiogeography, but the judged 'primitiveness' of therapsids is not a reliable way to correlate them to the SGCS.

The second way to bridge the gap is to find tetrapod fossils from the gap – in other words, tetrapod fossil assemblages that correlate to the late Kungurian–early Roadian on the SGCS. Despite claims to the contrary, this has not happened. Thus, for example, Liu *et al.* (2009) claimed that the Dashankou locality in the Chinese Xidagou Formation fills the gap, but that assemblage is readily correlated to the Russian Isheevo assemblage and is thus of Gamkan age (Lucas 2006, and see later text). The correlation of Liu *et al.* (2009) was based on the idea that *Raranimus*, the new therapsid they named from Dashankou, is the most primitive therapsid, so their age assignment is based on phylogeny, or stage of evolution, and is thus questionable. Even if the Dashankou assemblage is older than Gamkan, there are no data to indicate that it is older than the South African *Eodicynodon* Assemblage Zone or the Russian Mezen assemblage (Abdala *et al.* 2008). There must be some actual data to correlate a tetrapod assemblage to Olson's gap, not just some idea about the primitiveness of its therapsids.

The third way to bridge the gap has been simply to not accept the evidence provided by marine biostratigraphy that the youngest Permian tetrapods from Texas–Oklahoma are beneath Kungurian strata and that the oldest Permian therapsid-dominated assemblages must be younger than that (e.g. Reisz & Laurin 2004; Lozovsky 2005; Benton 2012, answered by Lucas 2005*a*, 2013*d*). Thus Benton (2012, fig. 1) did not follow the marine biostratigraphic data, assigning a Roadian age to the San Angelo, Flowerpot and Chickasha formations of Texas–Oklahoma. He also moved the Ufimian down into the Kungurian (Cisuralian), citing the Russian Interdepartmental Stratigraphic Committee as authority, apparently not aware that they were not correlating the Ufimian to the Kungurian, but simply moving the Kungurian upwards to encompass the Ufimian (see earlier discussion). Benton (2012, fig. 1) thus mis-correlated the youngest North American Permian tetrapod assemblages upwards and the oldest Russian therapsid-dominated assemblages downwards, thereby 'closing' Olson's gap. The search for tetrapod assemblages that fill Olson's gap should continue, rather than not accepting the marine biostratigraphic data.

Record of middle–late Permian tetrapods

The Permian LVFs are what Lucas (2006) called a composite standard because the early Permian LVFs are based on tetrapod assemblages from the western USA, whereas those of the middle–late Permian are based on the tetrapod succession in the South African Karoo basin. Therefore the middle–late Permian tetrapod assemblages are correlated to the South African standard (Fig. 9).

South Africa

The middle–upper Permian tetrapod fossil record and its biostratigraphy in the Karoo basin of South Africa has long provided the classic succession of middle to late Permian tetrapod assemblages (Fig. 9). Karoo tetrapod fossils were discovered in 1838 and have been extensively studied and published since the 1850s.

Reviews by Keyser & Smith (1977–78), Rubidge (1995, 2005), Smith & Keyser (1995*a*, *b*, *c*, *d*), Kitching (1995), Rubidge (2005), Smith *et al.* (2012) and Viglietti *et al.* (2016) recognize six successive tetrapod assemblage zones in the Karoo basin based on tetrapods of middle–late Permian age (Fig. 9). Papers on the Karoo tetrapod record published during the last decade have had

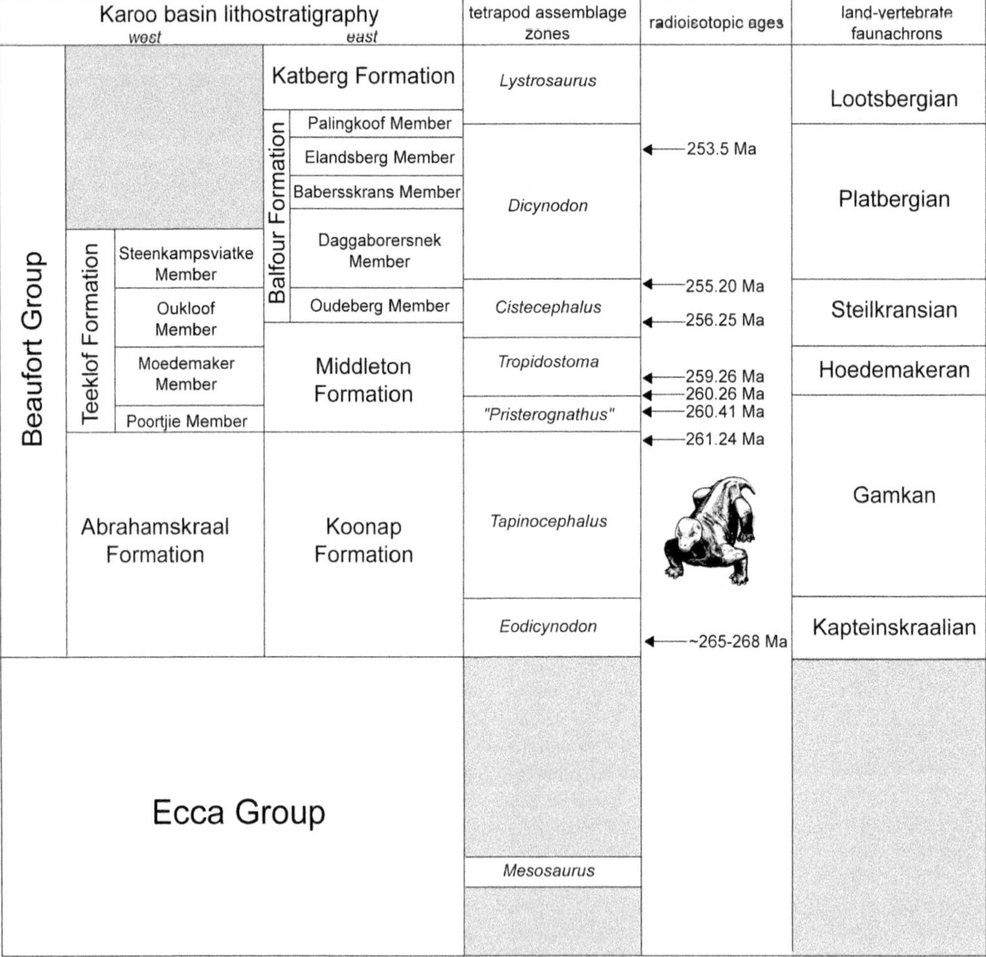

Fig. 9. Permian lithostratigraphy of the Karoo basin in South Africa (after Smith *et al.* 2012) correlated to the land vertebrate faunachrons, with radioisotopic ages of ash beds reported by Rubidge *et al.* (2013), Lanci *et al.* (2013), Day *et al.* (2015*a*) and Gastaldo *et al.* (2015).

little impact on the tetrapod biostratigraphy (e.g. Rubidge *et al.* 2006; Angielczyk & Rubidge 2013; Kammerer *et al.* 2015; Angielczyk *et al.* 2016). However, some important refinements and modifications have recently been reported (e.g. Day & Rubidge 2014; Jirah & Rubidge 2014; Day *et al.* 2015*a*, *b*; Viglietti *et al.* 2016).

Lucas (2006) recast five of the South African assemblage zones as biochronological units (LVFs) using the FAD of a widespread and characteristic tetrapod taxon to define the beginning of each faunachron. Earlier, Lucas (1998*b*) noted that the *Pristerognathus* Assemblage Zone lacks a useful LO to define its base; it was defined by Smith & Keyser (1995*b*) as little more than the upper part of the *Tapinocephalus* Assemblage Zone without dinocephalians. The recent discovery of dinocephalian fossils in the lower part of the *Pristerognathus* Assemblage Zone prompted a redefinition of the base of the zone upwards (Day *et al.* 2015*a*). I conclude that the *Pristerognathus* Assemblage Zone is a problematic biostratigraphic unit, recognized primarily by the absence of taxa, and should be abandoned.

The Karoo basin tetrapod assemblage zones are the basis of five LVFs equivalent to most of middle and late Permian time (Figs 2 & 9).

(1) The characteristic Kapteinskraalian LVF assemblage is from the lower Abrahamskraal Formation (Rubidge 1995). This is the *Eodicynodon* Assemblage Zone and includes temnospondyls, gorgonopsians, therocephalians, the dicynodont *Eodicynodon* and dinocephalians (Smith *et al.* 2012). Based on the stage of evolution, the *Eodicynodon* Assemblage Zone was long thought by many to be the oldest middle Permian tetrapod assemblage with therapsids; it is still thought to be the oldest such assemblage from Gondwana (e.g. Abdala *et al.* 2008; Smith *et al.* 2012). Liu *et al.* (2009) claimed that the Dashankou assemblage from China is the oldest therapsid assemblage, but this seems unlikely (see later discussion). Given that no early Permian tetrapod assemblage yields *bona fide* therapsids, it seems unlikely that the *Eodicynodon* Assemblage Zone is of early Permian age. Just how old it is in the middle Permian remains uncertain, however (see later discussion).

(2) The characteristic Gamkan LVF tetrapod assemblage is from the middle–upper Abrahamskraal and lower Teekloof formations (Fig. 9). It combines those of the *Tapinocephalus* Assemblage Zone of Smith & Keyser (1995*a*) and the '*Pristerognathus* Assemblage Zone' of Smith & Keyser (1995*b*) as redefined by Day *et al.* (2015*a*). It thus includes the temnospondyl *Rhinesuchus*, pareiasaurs (especially *Bradysaurus*), the varanopid *Heleosaurus*, diverse dinocephalians, anomodonts, dicynodonts (especially *Diictodon*), biarmosuchians, several gorgonopsians and therocephalians (see Smith *et al.* 2012).

(3) The characteristic Hoedemakeran LVF assemblage is from the middle Teekloof Formation (Fig. 9). It is much of the *Tropidostoma* Assemblage Zone (below the LO of *Cistecephalus*) of Smith & Keyser (1995*c*) and includes the temnospondyl *Rhinesuchus*, the pareiasaur *Pareiasaurus*, therocephalians (but no scylacosaurids), gorgonopsians and numerous dicynodonts, especially *Diictodon*, *Tropidostoma* and *Endothiodon* (Smith *et al.* 2012).

(4) The characteristic Steilkransian LVF tetrapod assemblage is from the upper Teekloof Formation (Fig. 9). It thus combines the uppermost *Tropidostoma* Assemblage Zone and the *Cistecephalus* Assemblage Zone (Smith & Keyser 1995*c*, *d*) and includes the temnospondyl *Rhinesuchus*, parareptiles, therocephalians, biarmosuchians and dicynodonts, especially *Diictodon*, *Cistecephalus*, *Emydops*, *Aulacephalodon* and *Oudenodon* (Smith *et al.* 2012).

(5) The characteristic Platbergian LVF tetrapod assemblage is from the uppermost Teekloof and the Balfour formations (Fig. 9). It combines the uppermost *Cistecephalus* Assemblage Zone and the *Dicynodon* Assemblage Zone (Kitching 1995), a tetrapod assemblage dominated by dicynodonts (especially *Dicynodon* and *Diictodon*) with a biarmosuchian, diverse gorgonopsians and therocephalians and cynodonts (Smith *et al.* 2012).

Dicynodon

At this point, a brief discussion of the taxonomy of *Dicynodon* is necessary to make clear how that name is used here. Kammerer *et al.* (2011) published a taxonomic revision of *Dicynodon* that split the genus into 11 genera, almost all of which are endemic to single basins. However, I do not use the taxonomy of Kammerer *et al.* (2011) because it is not based on a thorough evaluation of the morphological variation among fossils assigned to *Dicynodon*, so I use the name *Dicynodon* as it was used by Lucas (1997, 2006) and has been used by a variety of other workers, notably the taxonomic study of Cluver & Hotton (1981). I thus consider all the generic names created or resurrected by Kammerer *et al.* (2011) to be invalid. To evaluate completely the monographic revision published by

Kammerer *et al.* (2011) is beyond the scope of this article, so I just focus here on a few key points.

It is well accepted that to evaluate morphological variation for taxonomic purposes, those sources of variation that are due to taphonomic distortion/preservation, ontogeny, sexual dimorphism and epigenetics must be distinguished from features that can be used to distinguish taxa. However, the revision of Kammerer *et al.* (2011) of *Dicynodon* failed to do this. Thus Kammerer *et al.* (2011) focused on studying the type specimens of the named species that had been assigned to *Dicynodon*. From this, they extracted four 'morphotypes' (Fig. 10) and claimed that there are consistent differences between these morphotypes across 'a range of sizes and styles of taphonomic deformation, and thus [they are] unlikely to be the result of ontogenetic variation or distorton' (Kammerer *et al.* 2011, p. 4). Nevertheless, no data on ontogenetic variation is presented by Kammerer *et al.* (2011); indeed, they make no attempt to assess the variations in samples of *Dicynodon* from single

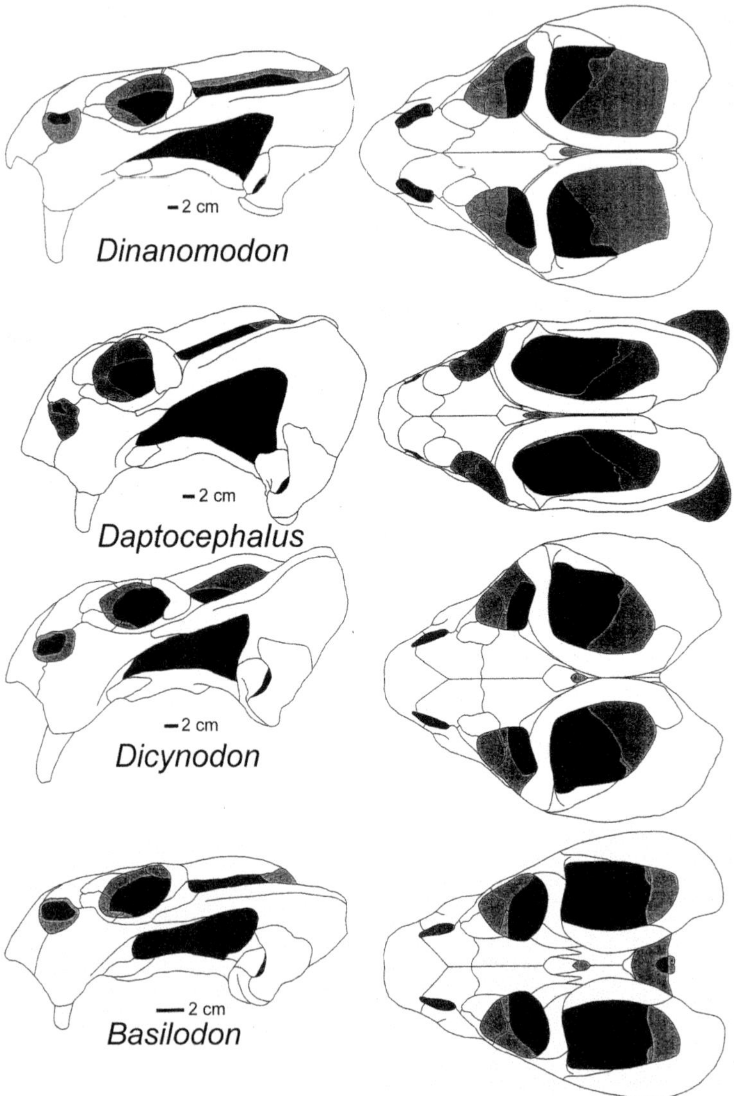

Fig. 10. The four 'morphotypes' of dicynodont skulls (left lateral views on left, dorsal views on right) recognized by Kammerer *et al.* (2011) in their taxonomic revision of *Dicynodon*. After Kammerer *et al.* (2011) reprinted by permission of the Society of Vertebrate Paleontology, www.vertpaleo.org.

localities, which would be the best way to examine ontogenetic and epigenetic variation, although it is clear from Kitching (1977) that such samples are available.

Despite the existence of a very large sample of *Dicynodon* – hundreds, perhaps thousands, of skulls – there are no metric data in Kammerer *et al.* (2011) beyond a morphometric analysis of the differences in snout shape between *Dicynodon lacerticeps* and '*Daptocephalus*' *leoniceps*. Given that Kammerer *et al.* (2011) attached great taxonomic significance to the relative length and width of the intertemporal bar, a feature readily measured, we would at least expect some metric data on the intertemporal bar. Yet, despite the absence of metric data and analysis of the morphological variation in samples, Kammerer *et al.* (2011, p. 135) claimed that 'we have documented and analysed inter- and intraspecific variation in "*Dicynodon*" using descriptive, phylogenetic and morphometric approaches'.

Kammerer *et al.* (2011) photographically illustrated the holotypes of the many species previously assigned to *Dicynodon*. That photographic atlas demonstrates that many of the type skulls are incomplete, poorly preserved and/or covered in concretion, and thus in need of preparation. Therefore it is not clear how Kammerer *et al.* (2011) could have evaluated many of the morphological characteristics of these fossils that they use to distinguish the taxa. As an example, the holotype of '*Basilodon*' *woodwardi*, which is one of their four 'morphotypes' (Fig. 10), is a concreted skull (mostly snout) fragment (Kammerer *et al.* 2011, fig. 151) lacking most of the characters that Kammerer *et al.* (2011, p. 136) list in the diagnosis of the genus. These questions arise: what is the 'morphotype' of *Basilodon* (Fig. 10) actually based on; and how were other specimens referred to that taxon? It looks more likely that *Basilodon woodwardi* is a *nomen dubium* based on an inadequate holotype specimen.

If we look at the four 'morphotypes' identified by Kammerer *et al.* (2011), they state that one of the characters that distinguishes them is differences in the intertemporal bar. However, Cluver & Hotton (1981, fig. 10) illustrated a specimen that they assigned to *Dicynodon* intermediate in intertemporal bar morphology between the *Dicynodon* and *Basilodon* morphotypes of Kammerer *et al.* (2011); they noted there is a significant variation in the intertemporal bar morphology. Given that the temporal muscles attach to the intertemporal bar, ontogenetic changes might influence the bar morphology and this could explain at least some of the differences between that morphology in the small morphotype *Basilodon* and the larger morphotypes *Daptocephalus* and *Dinanomodon* (Fig. 10). However, Kammerer *et al.* (2011) presented no analysis of ontogenetic changes in the cranial morphology of *Dicynodon*, so they did not evaluate this possibility.

The endemism of the many dicyonodont genera recognized by Kammerer *et al.* (2011) in what was formerly called *Dicynodon* is very significant. This endemism runs counter to the cosmopolitanism of other Permian dicynodonts – such as *Diictodon, Tropidostma, Oudenodon* and *Endothiodon* – and of the classic Early Triassic dicynodont *Lystrosaurus*. However, I am certain that a taxonomic revision of *Lystrosaurus* using the methods of Kammerer *et al.* (2011) would also split it into multiple endemic taxa.

I find it difficult to understand how Viglietti *et al.* (2016), using the taxonomy of Kammerer *et al.* (2011), reassigned to *Daptocephalus* the many specimens formerly assigned to *Dicynodon* from the South African *Dicynodon* Assemblage Zone. In other words, how were all the jaw fragments, skull fragments and otherwise incomplete or unprepared specimens formerly assigned to *Dicynodon* reassigned to *Daptocephalus*, a taxon for which an almost complete skull is needed to be identified by the taxonomic criteria of Kammerer *et al.* (2011)? Therefore I do not follow the renaming of the South African *Dicynodon* Assemblage Zone as the *Daptocephalus* Assemblage Zone by Viglietti *et al.* (2016). I also do not accept the conclusion of Viglietti *et al.* (2016, p. 153) that 'the biostratigraphic utility of *Da.[ptocephalus] leoniceps* and other South African dicynodontoids outside of the Karoo basin is limited due to basinal endemism at the species level and varying temporal ranges of dicyodontoids globally'.

The taxonomic revision of *Dicynodon* by Kammerer *et al.* (2011) is a good example of 'typological taxonomy' in that it primarily focused on type specimens to extract 'mophotypes' and did not adequately assess non-taxonomic sources of variation, such as ontogenetic or epigenetic variation in samples. The revision is, instead, what Lucas (2005*c*) called 'cladotaxonomic' – in effect, much of the morphological variation is looked at with the cladogram, assumed to represent a phylogenetic 'signal', and therefore deemed to be of taxonomic value. I think it is fair at this point to reiterate Lucas (2005*c*, p. 194), who stated that 'the amount of and significance of variation in the genus [*Dicynodon*] has never been fully documented and analyzed'. That was true in 2005, and it is true today. We need a taxonomic revision of *Dicynodon* based on sound principles of evolutionary taxonomy.

Other African records

The south-central and east African, as well as Malagasy, records of Permian tetrapods are in extensional basins referred to as the Karoo basins (e.g.

SGCS Per.	SGCS Epoch	land-vertebrate faunachrons	South Africa	Tanzania	Russia	Western Europe	China	Brazil	India
PERMIAN	Lopingian	Platbergian	Beaufort Group: Balfour Formation	Usili Formation	*Scutosaurus* superzone	Cutties Hillock/ Hopeman sandstones	lower Cangfanggou Group (Xinjiang)		
							Sunjiagou Formation (Shanxi)		Kundaram Formation
	Guadalupian	Steilkransian	Teekloof Formation	Ruhuhu Formation					
		Hoedemakeran							
		Gamkan	Abrahamskraal Formation		*Titanophoneus* superzone		Xidagou Formation (Gansu)	Rio do Rasto Formation (Morro de Pelado Member)	
		Kapteinskraalian		Mhukuru Formation					
	Cisuralian	Littlecrotonian	Olson's gap						

Fig. 11. Correlation of selected Middle–Late Permian tetrapod assemblages to the land vertebrate faunachrons. SGCS, standard global chronostratigraphic scale.

Catuneanu *et al.* 2005). Therefore it is not surprising that they contain similar tetrapod assemblages and stratigraphic successions, although none contain a middle–late Permian tetrapod fossil record as extensive as that found in South Africa. These Karoo basins are the Ruhuhu basin in Tanzania–Malawi, the Mid-Zambezi basin of Zimbabwe and Zambia, and the Majunga basin of Madagascar.

The Ruhuhu basin of Tanzania and Malawi has a long-studied, therapsid-dominated tetrapod record that has recently been re-evaluated to recognize three stratigraphic intervals in the Ruhuhu Formation of Gamkan–Steilkransian age, overlain by Steilkransian–Platbergian tetrapod records in the Usili Formation (Sidor *et al.* 2010; Simon *et al.* 2010; Angielczyk *et al.* 2014*a*) (Fig. 11). In the Malawi portion of the Ruhuhu basin, the Chiweta beds yield *Endothiodon*, *Oudenodon* and the biarmosuchian *Lende* and are of Steilkransian age (Jacobs *et al.* 2005; Kruger *et al.* 2015).

In the Mid-Zambezi basin of Zimbabwe and Zambia, the Madumabisa Mudstone Formation also yields tetrapod assemblages that range from Gamkan to Steilkransian in age (Boonstra 1946; Lee *et al.* 1997; Gay & Cruickshank 1999; Lepper *et al.* 2000; Smith *et al.* 2012; Sidor *et al.* 2014*a*, *b*; Angielczyk *et al.* 2014*b*).

Platbergian records of *Dicynodon* in the Karoo basins are in part of the Kawinga Formation in the Ruhuhu basin (Gay & Cruickshank 1999; Maisch & Gebauer 2005) and 'Horizon 5' in the Madumabisa Mudstone Formation in Zambia (King & Jenkins 1997; Smith *et al.* 2012). In the Majunga basin of Madagascar, the lower Sakamena Formation yields *Oudenodon*, *Rhinesuchus* and various endemic reptiles (Mazin & King 1991; Smith 2000) and is probably of Steilkransian age.

Permian tetrapod records in Saharan and northern Africa are much more problematic to correlate. In Morocco, the Tourbihine Member of the Ikakern Formation in the Argana basin has yielded temnopsondyl amphibians, a lepospondyl long termed *Diplocaulus minimus*, captorhinids and the derived pareiasaur *Arganaceras* (e.g. Dutuit 1988; Jalil & Dutuit 1996; Jalil & Janvier 2005; Steyer & Jalil 2009; Germain 2010). This assemblage is difficult to correlate because of its endemic and anachronistic taxa. Thus the '*Diplocaulus*' is closest to early Permian taxa (Germain 2010), whereas the captorhinids have been used to suggest a 'Kazanian' age (Jalil & Dutuit 1996). The pareiasaur, however, is judged to be closest to *Elginia*, so it may indicate a Platbergian age (Jalil & Janvier 2005).

A 10 m thick interval of the Moradi Formation in northern Niger has yielded a highly endemic tetrapod assemblage of temnospondyls (*Nigerpeton, Saharastega*), a captorhinid (*Moradisaurus*), a pareiasaur (*Bunostegos*) and a gorgonopsian, but no dicynodont (Sidor *et al.* 2005; Smith *et al.* 2015; Turner *et al.* 2015, and references cited therein). This assemblage lacks dinocephalians, so it is likely to be post-Gamkan in age, but otherwise is not precisely correlateable to the tetrapod biochronology.

Russia

In the Ural foreland basin, the Russian succession of middle–upper Permian tetrapod assemblages (Fig. 12) broadly correlates to the Karoo succession and has the advantage that not only can the lowermost (Kazanian) Russian tetrapods be tied to marine biostratigraphy, but the Illawara magnetostratigraphic event (see later discussion) has long been identified in the Russian Tatarian. This provides another way to correlate the Russian section to the SGCS. Virtually all of the genus-level taxa in the Russian section are endemic and thus of limited biostratigraphic value. This is why longstanding attempts to correlate the Russian tetrapod assemblages to coeval assemblages in Gondwana (especially in the South African Karoo) have largely been based on assessments of the stage of evolution, usually expressed as family-level correlations (e.g.

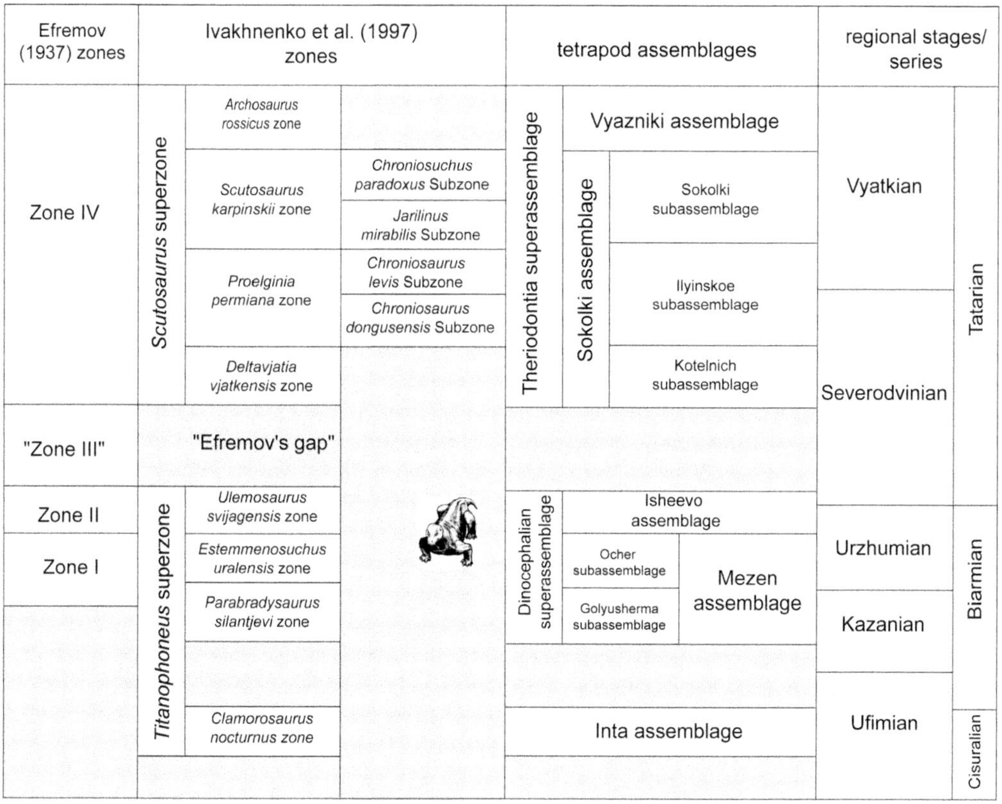

Efremov (1937) zones	Ivakhnenko et al. (1997) zones			tetrapod assemblages			regional stages/ series	
Zone IV	*Scutosaurus* superzone	*Archosaurus rossicus* zone		Theriodontia superassemblage	Vyazniki assemblage		Vyatkian	Tatarian
		Scutosaurus karpinskii zone	*Chroniosuchus paradoxus* Subzone		Sokolki assemblage	Sokolki subassemblage		
			Jarilinus mirabilis Subzone					
		Proelginia permiana zone	*Chroniosaurus levis* Subzone			Ilyinskoe subassemblage		
			Chroniosaurus dongusensis Subzone				Severodvinian	
		Deltavjatia vjatkensis zone				Kotelnich subassemblage		
"Zone III"	"Efremov's gap"							
Zone II	*Titanophoneus* superzone	*Ulemosaurus svijagensis* zone		Dinocephalian superassemblage	Isheevo assemblage		Urzhumian	Biarmian
Zone I		*Estemmenosuchus uralensis* zone			Mezen assemblage	Ocher subassemblage		
		Parabradysaurus silantjevi zone				Golyusherma subassemblage	Kazanian	
		Clamorosaurus nocturnus zone		Inta assemblage			Ufimian	Cisuralian

Fig. 12. Biostratigraphic zonation of Russian Permian tetrapods, their fossil assemblages, and correlation to the Russian regional stages and series (modified from Golubev 2000, 2005, 2015).

Ivakhnenko *et al.* 1997; Golubev 2005; Rubidge 2005; Smith *et al.* 2012), not on low-level (genus or species) taxonomic identity, so they are inherently imprecise. Therefore Lucas (2006) used the South African record as a more robust standard for middle–upper Permian tetrapod biostratigraphy than can be provided by the Russian record.

Efremov (1937) established the basic subdivision of the tetrapod assemblages of the Russian Permian section into four zones (I through IV, although note that III is a hiatus) (Fig. 12). Ivakhnenko *et al.* (1997; also see Golubev 1998, 2005, 2015) recast the Russian Permian tetrapod record as a biostratigraphic scheme that recognized two 'superzones', a *Titanophoneus* 'superzone' that includes Zones I and II of Efremov, and a *Scutosaurus* 'superzone' equivalent to Zone IV of Efremov (Fig. 12). These 'superzones' are divided into eight zones and two of the zones in the *Scutosaurus* Superzone are divided into subzones based on the succession of chroniosuchian temnospondyls (Golubev 1998, 2005, 2015). The summary of the Permian tetrapod record in the southeastern Urals (Tverdokhlebov *et al.* 2005) does not modify the existing scheme of tetrapod biostratigraphy in the Russian section.

The hiatus between Zones II and IV ('Efremov's gap') is marked by a dramatic change from the dinocephalian-dominated Isheevo assemblage to assemblages dominated by pareiasaurs and gorgonopsians (Kotelnich and younger assemblages). Sennikov (2014) has briefly reported an assemblage at Chuvashia Sundyr with dinocephalians that may fill part of this important gap in the Russian Permian tetrapod record.

The oldest Russian tetrapod assemblages of Kazanian age (Mezen assemblage and equivalents) yield basal anteosaurid dinocephalians and anomodonts (see Golubev 1998, 2005 for summaries). They have long been regarded as older than the South African *Tapinocephalus* zone (e.g. Boonstra 1971; Rubidge 2005; Smith *et al.* 2012), so they predate the beginning of the Gamkan LVF and are therefore of Kapteinskraalian age.

The Isheevo assemblage (Ivakhnenko *et al.* 1997) has long been correlated to the South African *Tapinocephalus* zone based on shared evolutionary counterparts in biarmosuchians, anteosaurid and tapinocephalid dinocephalians and anomodonts (e.g. Boonstra 1971; Chudinov 1975; Rubidge 2005; Smith *et al.* 2012). It is thus of Gamkan age (Fig. 11).

Above the hiatus, the *Proelginia permiana* Assemblage Zone contains *Tropidostoma* (interval of Kotelnich subassemblage) and *Oudenodon* (interval of Ilinskoye subassemblage) (Ivakhnenko *et al.* 1997), so these are likely to be of Hoedemakeran and Steilkransian age, respectively. Various localities of the upper Sokolki subassemblage and Vyazniki assemblage yield *Dicynodon* and are assigned a Platbergian age (Golubev 2000; Lucas 2005*c*).

Lucas (2006) made a mistake in stating that the LO of *Dicynodon* in the Russian section is just above the Urzhumian–Severodvinian boundary (it is much higher, in the Sokolki subassemblage and Vyatkian; Fig. 12). This created the incorrect appearance of a long Platbergian (Lucas 2006, fig. 11) that can now be seen to be much shorter (see later).

It was long thought that a substantial unconformity separated the Permian (Tatarian) strata in Russia from the overlying Lower Triassic strata (e.g. Lozovsky 1992; Wardlaw *et al.* 2004; Henderson *et al.* 2012). However, recent work on conchostracan biostratigraphy, vertebrate biostratigraphy and magnetostratigraphy indicates otherwise (e.g. Sennikov & Golubev 2006, 2012; Lozovsky & Kukhtinov 2007; Taylor *et al.* 2009; Kozur & Weems 2010; Lozovsky 2013*a*, *b*; Scholze *et al.* 2015). Thus there is no substantial hiatus between the Vyatkian (late Permian) and Vokhmian (Early Triassic) regional stages, with the Permo-Triassic boundary in a normal polarity magnetochron that Taylor *et al.* (2009) correlated to the Griesbachian N of Steiner (2006) (see later).

Western Europe

The Western European record of middle–late Permian tetrapods is of scattered occurrences, mostly of caseid eupelycosaurs and dicynodonts. Most taxa are endemic and there is no lengthy stratigraphic succession of tetrapod assemblages in any one area (see discussion by Mujal *et al.* 2016). One of the first reptile fossils described scientifically (in 1710) is *Protorosaurus* from the German Kupferschiefer (Gottman-Quesada & Sander 2009), which is at the base of the Zechstein Group and therefore considered to be Wuchiapingian in age on the SGCS (e.g. Roscher & Schneider 2005). This endemic taxon and others, of course, do not allow for correlation based on the tetrapods themselves.

A few records of *Dicynodon* from Western Europe are of Platbergian age (Fig. 11). These are from the Cutties Hillock Quarry, Elgin, Scotland and in the Cutties Hillock Sandstone Formation and the Hopeman Sandstone at Clashbach Quarry, Scotland (Benton & Walker 1985; King 1988; Clark 1999).

China

Chinese Permian tetrapods come from northern China, in the Junggar basin of Xinjiang and the Ordos basin, which is primarily in Ningxia, Nei Monggol, Gansu and Shanxi. They represent three temporally successive tetrapod assemblages (Lucas 2001, 2005*b*) (Fig. 11). The oldest is the *Biseridens* assemblage of Gamkan age from the Xidagou Formation in Gansu. It includes the temnospondyl *Anakamacops*, an *Intasuchus*-like temnospondyl, the anthracosaurs *Ingentidens* and *Phratochronis*, the bolosaur *Belebey* (also known from the Isheevo assemblage in Russia), a captorhinid, the dinocephalians *Sinophoneus* and *Stenocybus*, the therapsid *Raranimus* and the anomodont *Biseridens* (Liu *et al.* 2009, 2010; Liu 2013). As discussed earlier, I am not convinced that this assemblage is older than Gamkan, as argued by Liu *et al.* (2009) based on their perception of the stage of evolution of *Raranimus*.

The next youngest assemblage, the *Shihtienfenia* assemblage, is from the Shihezi Formation in Henan and the Sunjiagou Formation in Shanxi. It includes the chroniosuchian *Bystrowiana* and various pareiasaurs, especially *Shihtienfenia* (Lucas 2001, 2005*b*; Li & Liu 2013; Liu *et al.* 2014; Li *et al.* 2015; Benton 2016). The pareiasaurs are most similar to the characteristic Steilkransian pareiasaurs such as *Scutosaurus*, *Pareiasaurus* and *Pareiasuchus*, which suggests a tentative correlation (Lucas 2005*b*, 2006; Li *et al.* 2015) (Fig. 11).

The youngest Chinese Permian tetrapod assemblages are *Dicynodon*-dominated assemblages of Platbergian age. They are primarily from Xinjiang in northwestern China, from the lower Canfanggou Group in the Junggar basin and from the Turpan basin in the eastern part of the Tien Shan, especially at Taoshuyuan. The section on the southern flank of the Dalongkou anticline in the Junggur basin is particularly significant to regional vertebrate biostratigraphy across the Permo-Triassic boundary because it yields both the *Dicynodon* and *Lystrosaurus* assemblages in stratigraphic superposition and documents the stratigraphic overlap of *Dicynodon* and *Lystrosaurus* (Metcalfe *et al.* 2009, and references cited therein).

Dicynodon assemblage tetrapods from the Junggar basin are mostly dicynodonts assigned to *Dicynodon* or *nomina dubia* (Lucas 2001, 2005*b*, 2006). The only diagnostic Late Permian dicynodont specimen from the Junggar basin not assignable to *Dicynodon* is *Diictodon tienshanensis*. Therefore, in the upper Permian strata of the Junggar basin, only two dicynodont genera are known – the relatively abundant *Dicynodon* and the rare *Diictodon* (Angielczyk & Sullivan 2008) Other Chinese *Dicynodon* records are from the Sunan Formation, Gansu and the Naobaogou Formation, Nei Monggol, both in the Ordos basin (Lucas 1998*b*, 2001, 2005*b*; Li *et al.* 2000).

Southern Brazil

Permian tetrapod records in Brazil are mostly from the Paraná basin in the southern part of the country and they were most recently reviewed by Dias-da-Silva (2011) and Martinelli *et al.* (2016). The localities (assemblages) are from the sandstone-dominated, 250–300 m thick Morro de Pelado Member of the Rio do Rasto Formation. Brazilian vertebrate palaeontologists have not been able to present a lithostratigraphic organization of the localities, largely because they are in heavily vegetated country and some are far apart geographically (e.g. Dias-da-Silva 2011; Cisneros *et al.* 2012, 2015*a*; Martinelli *et al.* 2016; Pinheiro *et al.* 2016). Some workers have claimed that two of the localities, Posto Queimado and Aceguá, are stratigraphically at the same level and stratigraphically lower than a third locality, Serra de Cadeado (e.g. Cisneros *et al.* 2005). This supported the possibility of superposed Gamkan and Hoedemakeran age sites (e.g. Lucas 2006). However, I think a more cautious conclusion is to regard all of the Morro de Pelado Member tetrapod localities as Gamkan, given that they collectively yield dinocephalians, *Pareiasaurus* and *Endothiodon* from a relatively thin stratigraphic unit (Fig. 11). Nevertheless, actual stratigraphic data are needed to test this conclusion.

A sparse tetrapod assemblage from the Buena Vista Formation of Uruguay, formerly and tentatively considered to be Permian (Modesto *et al.* 2001; Piñeiro *et al.* 2003, 2004), has been re-evaluated and is considered more likely to be Early Triassic in age (Modesto & Botha-Brink 2010).

India and Laos

In the Pranhita-Godavari Valley of India, the Kundaram Formation yields a captorhinid and the dicynodonts *Endothiodon*, *Pristerodon*, *Emydops*, *Cistecephalus* and *Oudenodon*, an assemblage of Steilkransian age (Ray 1999, 2000, 2001) (Fig. 11).

North of the Mekong River in the Luang-Prabang basin of Laos, *Dicynodon* has been reported from the volcaniclastic 'purple claystone formation' (Battail *et al.* 1995; Battail 1997; Bercovici *et al.* 2015). This is a Platbergian age record.

Correlation to the SGCS

Intercalated non-marine and marine biostratigraphy

The most reliable way to cross-correlate Permian tetrapod biostratigraphy to the SGCS is by direct physical correlation between strata bearing tetrapod fossils and marine strata containing index fossils, namely fusulinids, ammonoids and/or conodonts. This is best achieved when the non-marine and marine strata are intercalated; the New Mexico and Texas sections are excellent examples of this (Figs 3 & 4). In some other locations, non-marine Permian tetrapod-bearing strata are above marine strata. The Russian section is a good example of this (Figs 8 & 12) and provides one of the few places where the middle–late Permian tetrapod biostratigraphy can be placed with confidence above a marine biostratigraphic datum.

The oldest early Permian cross-correlation is in New Mexico, where the mixed non-marine–marine Bursum Formation yields a tetrapod assemblage of Coyotean age (Harris *et al.* 2004). Conodont and fusulinid biostratigraphy indicates that the Bursum Formation straddles the Virgilian–Wolfcampian boundary (Lucas *et al.* 2013*a*) and, based on arguments by Lucas (2013*a*, *b*), I equate the Permian base to the Virgilian–Wolfcampian boundary. This means that the Coyotean straddles the Carboniferous–Permian boundary, although it encompasses only the very latest Virgilian; compare the stratigraphic distribution of Bursum tetrapods in Harris *et al.* (2004) with the distribution of Bursum conodonts in Lucas *et al.* (2013*a*).

Across New Mexico, the southwards intertonguing of the Abo Formation with the marine Hueco Group indicates that the Coyotean age tetrapod assemblages of the Abo Formation (almost totally from the lower Scholle Member) are middle to early late Wolfcampian (Nealian–early Lenoxian) in age (Fig. 3). The Seymouran-aged assemblage from the Arroyo del Agua Formation in northern New Mexico is not found in the Abo Formation, so it must be younger than early Lenoxian, close to the Wolfcampian–Leonardian boundary in age.

The Texas lower Permian red bed section represents fluvial deposition on a broad coastal plain between a Permian seaway to the west and a series of ancestral Rocky Mountain uplifts (Ouachita, Arbuckle and Wichita) to the east and NE (e.g. Brown 1973; Hentz 1988, 1989). The non-marine red beds intertongue with, and are laterally equivalent to, marine strata, allowing cross-correlation of the non-marine and marine biostratigraphy (see Lucas 2004, 2006 for a review of the published literature). This means it is possible to directly correlate a tetrapod biostratigraphy developed in the

Texas red beds with a marine biostratigraphy based largely on fusulinids and ammonoids, for which some conodont data are also available.

There are a number of key marine tiepoints in this correlation (Fig. 4).

(1) The Coleman Junction Limestone, immediately beneath the Seymouran characteristic assemblage, yields fusulinids of late Wolfcampian age. Wardlaw (2005) reported, but did not document, late Sakmarian conodonts from the Coleman Junction Limestone.
(2) The Elm Creek Limestone, intercalated with Seymouran tetrapod-bearing strata, yields ammonoids indicating an age very close to the Wolfcampian–Leonardian boundary. Wardlaw (2005) reported, but did not document, early Artinskian conodonts from the Elm Creek Limestone (see also Walsh & Barrick 2004).
(3) Ammonoid- and fusulinid-bearing strata are either intercalated or demonstrably laterally equivalent to the Waggoner Ranch Formation, which yields the Mitchellcreekian characteristic assemblage. These marine fossils cross-correlate the Mitchelcreekian to part of the early Leonardian.
(4) Leonardian ammonoids are known from the upper and lower part of the Clear Fork Group, which yields the characteristic tetrapod assemblage of the Redtankian LVF. This cross-correlates the Redtankian to part of the middle–upper Leonardian.
(5) Fusulinids from marine intercalations in the San Angelo Formation indicate that it is late Leonardian. It is overlain by the Blaine Formation, which has a late Leonardian ammonoid fauna at its base (see also earlier discussion). This means that the older part of the Littlecrotonian LVF, which has its characteristic tetrapod assemblage from the San Angelo Formation, is late Leonardian.

Thus the New Mexican and Texan sections precisely cross-correlate the early Permian LVFs to the North American regional marine chronostratigraphy. Correlation of the North American stages to the SGCS, however, has been inconsistent over the last two decades and this has been a source of confusion. Most of this is related to the definition and redefinition of the Artinskian and Kungurian stages using conodonts. Thus, in an earlier correlation, the Leonardian was equated to the Kungurian (Wardlaw *et al.* 2004). In more current correlations, the Leonardian = late Artinskian–Kungurian (Henderson *et al.* 2012) or the Leonardian = middle Artinskian–Kungurian (Davydov *et al.* 2013; Vachard *et al.* 2015); the latter correlation is used here. This, and the fact that the numerical calibrations of the Artinskian–Kungurian interval of the SGCS have varied, underlies some of the inconsistency in the ages assigned by the vertebrate palaeontologists who try to directly correlate Permian tetrapod assemblages to the SGCS (compare, for example, the contradictory numerical calibrations in Brink & Reisz 2014; LeBlanc *et al.* 2015).

The Russian Permian tetrapod-bearing section directly overlies late Kazanian marine strata (Figs 8 & 12). These strata, as discussed earlier, are of late Roadian age and indicate that the stratigraphically lowest Russian therapsid-dominated assemblages (Mezen assemblage) are no older than late Roadian (Fig. 8).

The Karoo section has no direct relationship with any marine strata. Nevertheless, South African workers (e.g. Rubidge 2005; Smith *et al.* 2012) have correlated the tetrapod assemblage zones to the SGCS as follows: *Eodicynodon* Assemblage Zone = late Wordian; *Tapinocephalus* Assemblage Zone = Capitanian; *Pristerognathus* Assemblage Zone = Capitanian; *Tropidostoma* Assemblage Zone = Wuchiapingian; *Cistecephalus* Assemblage Zone = Wuchiapingian; and *Dicyonodon* Assemblage Zone = Wuchiapingian–Changshingian. However, the basis for these correlations was never made clear and new data clarify them and raise some questions about the cross-correlation of the Karoo record to the SGCS (see later discussion).

Radioisotope ages

In the last decade, numerous radioisotopic ages directly tied to the Permian SGCS have become available for the earliest and latest Permian (e.g. Ramezani *et al.* 2007; Shen *et al.* 2010; Schmitz & Davydov 2012). However, in between, for much of the Permian, few reliable radioisotopic ages can be tied to the SGCS (e.g. Henderson *et al.* 2012). Thus a mathematical model is used to interpolate the stage boundary ages using all the ages, which include only a handful of Sakmarian, Artinskian, Wordian and Wuchiapingian ages (Henderson *et al.* 2012, fig. 24.10). A particularly important age in this calibration is a latest Wordian age of 265.3 ± 0.2 Ma from an ash bed in West Texas, which is very close in age to the Illawara palaeomagnetic event (see later).

In the non-marine section, there are numerous radioisotopic ages from igneous rocks intercalated with lower Permian Rotliegend Group strata in Europe. However, these are mostly ages of low precision, with error bars typically of ±2–5 Ma. Roscher & Schneider (2005) reviewed what they considered to be the more reliable Rotliegend radioisotopic ages to use them to calibrate the Rotliegend branchiosaur and blattoid (cockroach) biostratigraphy. The only numerical age that could directly calibrate the early Permian LVFs comes from rhyolite associated with the Tambach Formation, which contains the Seymouran Bromacker locality (see earlier). This is a U–Pb age on the Egersburg Rhyolite of 275 ± 4 Ma. However, on the current International Commission on Stratigraphy numerical timescale, this is a late Kungurian age, which is much younger than the age of the Seymouran based on intercalated marine index fossils (Fig. 4).

Roscher & Schneider (2005) did not use the age of the Egersburg Rhyolite to calibrate the age of the Tambach Formation, but, on biostratigraphic grounds, regarded it as older, close to 280 Ma, which is an early Kungurian age and closer to a reasonable estimate for part of Seymouran time. Furthermore, as Roscher & Schneider (2005, p. 288) well concluded, 'tectonic reactivations in the European Variscides and multiple Mesozoic thermal events have upset the geochronologic systems through large areas' of the European Rotliegend. Therefore I do not use any of the Rotliegend numerical ages to calibrate the early Permian LVFs.

Some recently published radioisotopic ages (Michel *et al.* 2015) from the Lodève basin in France appear to be more reliable. They indicate a numerical recalibration of the amphibian zonation of Werneburg & Schneider (2006; Schneider & Werneburg 2012) so that their zone nine (Fig. 7) is *c.* 290–293 Ma, older than the previous estimate of *c.* 283–286 Ma. This would make all of the Rotliegend amphibian zones of Permian age no younger than Sakmarian on the SGCS, but, again, the Lodève basin dates have no direct relationship with the Permian LVFs.

In Oklahoma, the Richards Spur locality yields tetrapod fossils from cave/karst-fill in an Ordovician limestone. A U–Pb age from a speleothem of 289 ± 0.68 Ma has been used as the numerical age of the tetrapod fossils in the cave/karst-fill (Woodhead *et al.* 2010). This is an early Artinskian age on the most recent Permian numerical timescale, although LeBlanc *et al.* (2015, fig. 8, based on Modesto *et al.* 2014) showed it as an older Sakmarian age.

Lucas (2006) regarded the Richards Spur tetrapod assemblage as Mitchellcreekian (note the presence of *Cacops* and *Mycterosaurus*), which should be younger than early Artinskian. However, there are two problems here: (1) given the nature of speleothem growth and karst/cave infilling (e.g. Self & Hill 2003), the numerical age of the speleothem must be older than the filling and its contained fossils, perhaps significantly older; and (2) there is no way to be certain that all the tetrapod fossils from the Richards Spur fill are of the same age – they could represent different episodes of infilling, even over millions of years, as do, for example, the Rhaeto-Liassic fissure fills of the UK, which yield tetrapods of a range of Late Triassic and Early Jurassic ages, some from the same fissure-fill (e.g. Lucas 2010).

Similar problems beset the Bally Mountain locality in Oklahoma, which is a fissure-fill with tetrapod fossils long considered correlative to the Richards Spur assemblage, although it has now produced the captorhinorph *Captorhinikos*, previously known from younger, Redtankian strata (LeBlanc *et al.* 2015). I therefore regard the speleothem numerical age from Richards Spur as, at best, an approximate age for the tetrapod assemblage, and I also consider such unstratified deposits as problematic for tetrapod biostratigraphy and biochronology (see discussion of similar problems by Lucas 2010 with regard to the British Triassic–Jurassic fissure-fill tetrapods).

Santos *et al.* (2006) reported a U–Pb age of 278.4 ± 2.2 Ma for the Irati Formation in Brazil, which is a Kungurian age on the current SGCS. The Irati Formation yields fossils of the marine reptile *Mesosaurus*, also known from the correlative Whitehill Formation of the Ecca Group in South Africa (e.g. Modesto 2006). The tetrapod assemblages of the Rio do Rasto Formation in Brazil and of the Beaufort Group in South Africa are stratigraphically well above these *Mesosaurus*-bearing units (Holz *et al.* 2010; Smith *et al.* 2012). Thus the Kungurian radioisotopic age significantly predates both tetrapod successions and provides only a broadly constrained maximum age for them.

In the Huab basin of northwestern Namibia, Warren *et al.* (2001) documented a temnospondyl amphibian fossil from the Gai-as Formation. The stratum that yielded this fossil overlies the *Mesosaurus*-bearing Huab Formation and, in turn, is overlain by two ash beds that yielded U–Pb ages of 272 ± 1.8 and 265 ± 2.5 Ma (dates reported by Warren *et al.* 2001, but with no presentation of analytical data). These are late Kungurian and Wordian ages on the current SGCS calibration and provide only an approximate upper age limit for a fossil of little biochronological significance.

Rubidge *et al.* (2013), Day *et al.* (2015*b*) and Gastaldo *et al.* (2015) have published a series of U–Pb ages on zircons in volcanic ashes that provide numerical calibration of various levels in the *Tapinocephalus, Tropidostoma, Cistecephalus* and *Dicynodon* Assemblage Zones of the South African Karoo. They indicate an age span of *c.* 261–252 Ma (Fig. 9). This indicates that the end of the Gamkan through Steilkransian is late Capitanian–Wuchiapingian in age, with the Capitanian–Wuchiapingian boundary close to the Gamkan–Hoedemakeran boundary.

As discussed earlier, the Russian Isheevo assemblage has long been biostratigraphically correlated to the South African *Tapinocephalus* Assemblage

Zone, so Isheevo is of Gamkan age. However, Isheevo also overlaps the Illawara magnetic event, so it is at least in part late Wordian, *c.* 265 Ma (see later). These data, and the South African U–Pb ages, suggest on face value that the Gamkan must be at least 4 myr long, *c.* 265–261 Ma. They also suggest that the Russian section does not record the younger part of Gamkan time, which is represented by the hiatus ('Efremov's gap') between the Isheevo and Kotelnich assemblages (Fig. 12). A third possibility is that the South African U–Pb ages of the late Gamkan are anomalously young. Here, I choose the option of a long Gamkan, in part influenced by the relatively great thickness (>1400 m) of the *Tapinocephalus* Assemblage Zone in the Karoo basin (Day & Rubidge 2014; Jirah & Rubidge 2014). However, there is the potential for erroneous U–Pb ages indicating a very long Gamkan, and this needs further evaluation.

Lanci *et al.* (2013) reported a series of U–Pb ages on zircons from volcanic ashes in the lower part of the Abrahamskraal Formation, strata that are part of the *Eodicynodon* Assemblage Zone. These dates have relatively wide error bars (they are the means of weighted data from large samples of zircon grains) of ±1.4–3.5 Ma and range from about 268 to 264 Ma (Fig. 9). They are associated with a magnetostratigraphy that identifies a normal polarity multichron at about 266 Ma in the lower Abrahamskraal Formation and three scattered, and much less well supported, normal polarity chrons stratigraphically lower in the Waterford Formation of the Ecca Group (also see Tohver *et al.* 2015). Lanci *et al.* (2013, fig. 7) identify the best supported of these three normals, their N3 in the Ecca Group, as the Illawara event, which, as they note, would make Illawara substantially older than currently thought, close in age to the Roadian–Wordan boundary. They equate the normal multichron low in the Abrahamskraal Formation with the Capitan N magnetochron of Steiner (2006), which would make that magnetochron *c.* 266 Ma, the generally inferred age of Illawara (Shen *et al.* 2013).

A simpler interpretation of the data of Lanci *et al.* (2013) that is consistent with the current numerical calibration of Illawara is to regard the normal multichron low in the Abrahamskraal Formation as the Illawara event and to treat their N3 normal multichron in the Ecca Group as a weakly supported geomagnetic excursion, or a previously unrecognized short, normal multichron that predates Illawara. If this is correct, then the lower Abrahamskraal Formation is close to the Wordian–Capitanian boundary and the *Eodicynodon* Assemblage Zone (and Kapteinskraalian LVF) is of Wordian age. However, if that correlation to the SGCS is accepted, then the *Eodicynodon* assemblage zone is younger than the oldest Russian therapsid-bearing assemblage (Golyusherma subassemblage), which is late Roadian. In that case, the oldest Russian therapsid-dominated assemblages would be Littlecrotonian, not Kapteinskraalian, in the tetrapod biochronology.

To further complicate this discussion, Fildani *et al.* (2009) reported U–Pb ages from the Ecca Group in the western Karoo basin that assign an age of 252.7 ± 2 Ma to an ash bed stratigraphically high in the Ecca Group (Laingsburg Formation). They thus identify the Permo-Triassic boundary within the Ecca Group, implying a great time transgression across the Karoo basin. Lanci *et al.* (2013) argued that such a great amount of time transgression is not reasonable and claim that the difference between the ages of Fildani *et al.* (2009) and those of Lanci *et al.* (2013) is methodological – Fildani *et al.* (2009) reported the ages of the 'youngest' zircon in a population and Lanci *et al.* (2013) reported mean-weighted data from a zircon population. The simpler conclusion is that the ages reported by Fildani *et al.* (2009) are anomalously young and evidently incorrect.

There are clearly contradictions among the numerical ages from the Karoo basin that should undermine our confidence in them. We would do well to remember that the Drakensburg Formation volcanics erupted and covered the Karoo section during the Jurassic and, for decades, no geologist believed that reliable palaeomagnetic data or numerical ages could be obtained from the Karoo succession because of the thermal event(s) to which it has been subjected. That may have been too sweeping a conclusion, but I do believe we need to view all the new Karoo numerical and magnetostratigraphic data with some caution until they are at least replicated, particularly by multiple laboratories.

Magnetostratigraphy

During the Pennsylvanian and most of the Permian, the magnetic field experienced few or no reversals, so it was of essentially reversed polarity for at least 50 myr – the Kiaman superchron. The field became active again in middle Permian time and the onset of this activity is called the Illawara event (e.g. Menning 2001; Steiner 2006; Isozaki 2009). Although long considered Capitanian, the Illawara event is now known to be of late Wordian age, slightly older than *c.* 265 Ma (Steiner 2006; Shen *et al.* 2013; Szurlies 2013). The Illawara event has been identified globally, including in Russia, Transcaucasia, Germany, Poland, the USA, Pakistan, China, Japan and Australia, and it provides a valuable tiepoint by which to cross-correlate the late Wordian to a level in the non-marine section.

Most significant to tetrapod biochronology is the identification of the Illawara event in the Russian

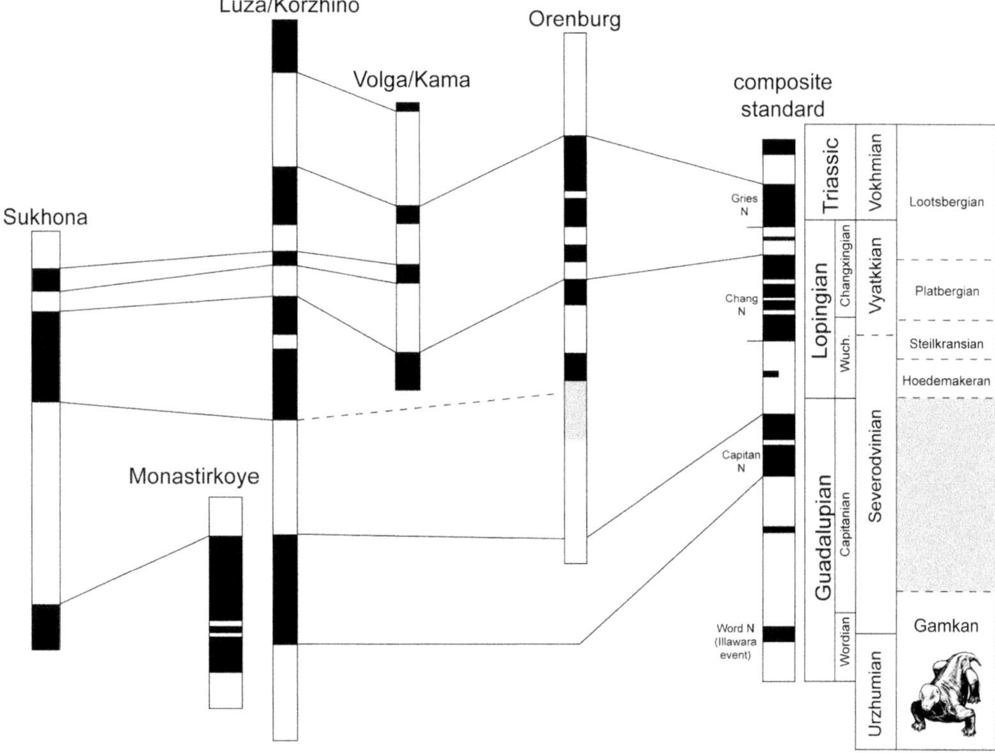

Fig. 13. Magnetostratigraphic sections of the Russian non-marine middle–upper Permian correlated to the Russian regional stages and the land vertebrate faunachrons (magnetostratigraphy after Taylor *et al.* 2009).

section, where it occurs above the Isheevo level near the Urzhumian–Severodvinian boundary (Fig. 13). This means Isheevo is at least in part late Wordian and the immediately overlying Kotelnich tetrapod level (above a hiatus) is very close in age to the Wordian–Capitanian boundary. The implications of these correlations have been discussed earlier in this paper.

Within resolution, the magnetostratigraphic correlation of the younger Russian tetrapod assemblages indicates that Kotelnich and Ilyinskoe are Wuchiapingian, whereas Sokolki and the Vyatkian are Changhsingian (Fig. 13). This is in agreement with the tetrapod biostratigraphic correlation of the Russian and Karoo sections, which also indicate that the Wuchiapingian–Changhsingian boundary should be between Ilinskoye and Sokolki (around the Steilkransian–Platbergian boundary).

Major events in Permian tetrapod evolution

Introduction

The temporal ordering of the global Permian record of tetrapod fossils presented here identifies a series of important junctures in Permian tetrapod evolution (Fig. 14). All of these events have previously been identified, but their timing and correlation have not always been well resolved. Therefore my goal is to draw attention to the age of these events and the remaining problems of timing and correlation. The search for causation is not my purpose here. However, once the ages of these events have been precisely established, their correlation to other causative events (e.g. climate or volcanism) can better be achieved.

Coyotean chronofaunal event

It has long been known that no substantial evolutionary turnover took place among tetrapods across the Carboniferous–Permian boundary. Thus the Coyotean tetrapod assemblages are a classic chronofauna of lepospondyl and temnospondyl amphibians, diadectomorphs, primitive amniotes and eupelycosaurs. Lucas (2006) drew attention to the long duration of the Coyotean, *c.* 15–20 Ma by the numerical calibration of the SGCS then available, and the current calibration still suggests a relatively long duration of *c.* 10 Ma (Fig. 14). This

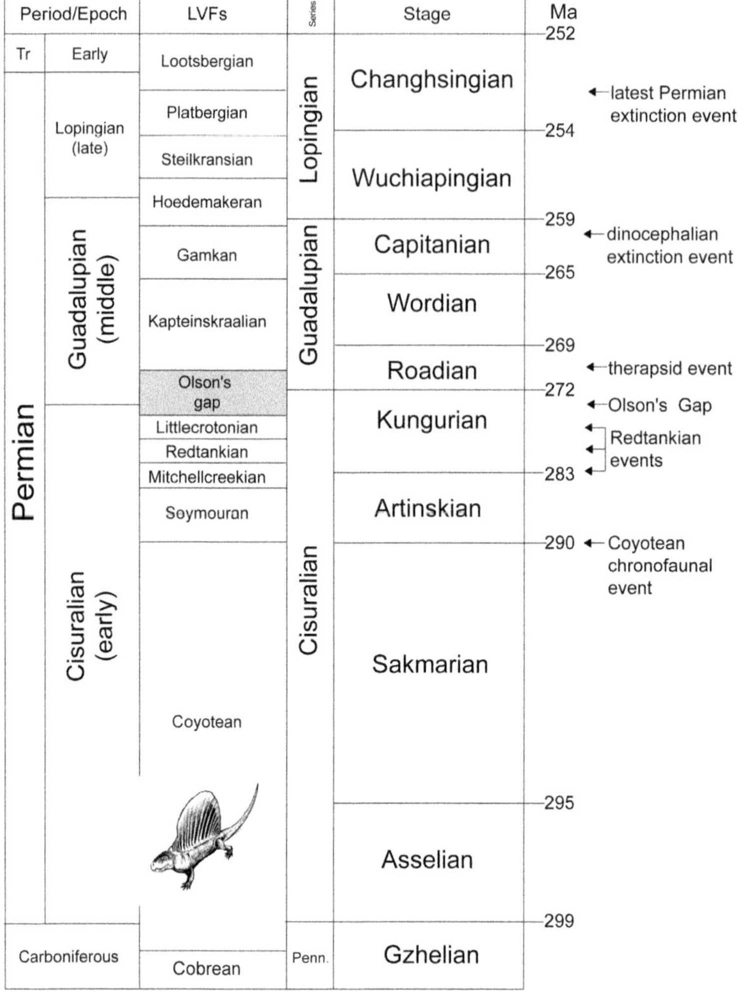

Fig. 14. Permian land vertebrate faunachrons (LVFs), their correlation to the standard global chronostratigraphic scale, numerical calibration and major events in Permian tetrapod evolution.

lengthy period of community/assemblage stasis merits investigation, as does the change to the much shorter succeeding early Permian LVFs that imply more rapid evolutionary turnover rates among tetrapods.

Redtankian events

Much evolutionary turnover took place late in the early Permian, mostly just before and during Redtankian time. Therefore I refer to this succession of events, which have often been viewed as a single event, as the Redtankian events (Lucas 2016). The oldest event is the extinction of lysorophians, 'microsaurs', anthracosaurs and ophiacodontid eupelycosaurs (by the end of the Mitchelcreekian) and this is followed by the extinction of trematopids, some captorhinomorphs, diadectomorphs, araeoscelids and edaphosaurids in the Texas–Oklahoma section in the early Redtankian (HOs are in the lower Clear Fork Group) (e.g. Lucas 2006; Kissel 2010; Bakker *et al.* 2013; Schoch & Milner 2014). However, the apparent recent discovery of a diadectomorph in the upper Permian of China (Liu & Bever 2015) indicates that the 'extinction' of diadectomorphs in Texas–Oklahoma is a local, not a global, event, although some workers believe that the Chinese specimen may actually be a therapsid (S. Modesto, pers. comm. 2016). Caseid synapsids diversified in Littlecrotonian time and there is an evident diversification with the LOs of moradisaurines and lanthanosuchoids in the Redtankian–Littlecrotonian interval (e.g. Maddin *et al.* 2008; Modesto *et al.* 2009, 2014, 2015; Tsuji *et al.* 2010, 2012; Ruta *et al.* 2011; MacDougall & Reisz 2014; LeBlanc *et al.* 2015; Romano & Nicosia 2015). However, cladograms of the parareptiles indicate long ghost lineages before these LOs, so their records may be extended back with further collecting.

Benton (1989) showed the Redtankian events as a mass extinction of amphibians and reptiles at the Sakmarian–Artinskian boundary. More recently, Sahney & Benton (2008) showed this extinction as a single major crash of tetrapod diversity at the Cisuralian–Guadalupian (Kungurian–Roadian) boundary, which they call Olson's extinction. To achieve this result, Sahney & Benton (2008) compressed all the extinctions of the Mitchellcreekian–Littlecrotonian and Olson's gap into one event. This is an example of what Lucas (1994) termed the compiled correlation effect, in which extinctions are artificially concentrated at stage boundaries as a result of poor temporal resolution, and is also predicated on the incorrect correlations later published by Benton (2012). Perhaps the term 'incorrect correlation effect' should be coined to refer to how incorrect correlations, such as those used by Sahney & Benton (2008) and Benton (2012), can create the appearance of a single extinction.

Bakker *et al.* (2013) refer to the Redtankian extinction of edaphosaurids and diadectomorphs as Olson's event. This is because Olson (1952; Olson & Vaughn 1970) first identified this evolutionary turnover in the Texas–Oklahoma section as a turnover in chronofaunas, attributing it to a change to a drier and more seasonally arid climate. Given the incomplete nature of the late early Permian tetrapod record outside Texas–Oklahoma, it is difficult to know whether the Redtankian events are geographically limited to the western USA or whether they are global in nature. The putative late Permian diadectomorph recently reported from China is certainly a cautionary note to those who would like to extrapolate all the extinctions and originations of the Redtankian events to the globe. However, there is some evidence from tetrapod footprint assemblages of a diversification of parareptiles during the late early Permian (e.g. Marchetti *et al.* 2015*a*, *b*), although this footprint-based diversification occurs in the New Mexico section near the top of the Abo Formation (Voigt & Lucas 2015), so it is close in age to the Seymouran–Mitchellcreekian boundary (Fig. 3).

Current stratigraphic resolution thus identifies a succession of events ranging from Mitchellcreekian through Littlecrotonian, which I call the Redtankian events (Fig. 14). Further study of the actual stratigraphic ranges of the taxa involved is needed to resolve further how many events, what events and the precise timing of the events that are combined here under one heading.

Olson's gap

Despite claims to the contrary, Olson's gap remains a hiatus in the global record of Permian tetrapods. The hiatus is equivalent to part of Kungurian–Roadian time, *c.* 2–3 Ma on the currently calibrated SGCS (Figs 8 & 14). The importance of the hiatus is the dramatic difference between tetrapod assemblages below and above Olson's gap. The origin and earliest diversification of therapsids clearly took place during the gap and the fossils documenting this diversification still remain to be found.

Therapsid event

What Bakker (1980) termed the Kazanian revolution is the sudden appearance of therapsid-dominated assemblages after Olson's gap, during what he regarded as Kazanian time. I prefer a more generic term for this major juncture in Permian tetrapod evolution: the therapsid event. Before Olson's gap, there were very few, or no, known therapsids, and the dominant synapsids were eupelycosaurs. After Olson's gap, therapsids were the dominant synapsids, with very rare eupelycosaurs.

Only a few of the major groups of early Permian amphibians – basal temnospondyls, stereospondyls and dissorophoids – extend across Olson's gap (Schoch 2014; Schoch & Milner 2014). However, most of the early Permian amphibians were gone by Littlecrotonian time (see earlier). By contrast, the parareptiles not only continue through, but also greatly diversify after, Olson's gap. The therapsid event thus not only encompasses the sudden appearance of therapsid-dominated assemblages, but also the significant diversification of parareptiles. This all begins by late Roadian time on the SGCS (Fig. 14).

Dinocephalian extinction event

A profound extinction of reptiles took place during the middle Permian, termed the dinocephalian extinction event by Lucas (2009*a*, *b*), and it was a more significant tetrapod extinction than that which took place close to the end of the Permian. This middle Permian (late Gamkan LVF) extinction is marked by the total disappearance of the dinocephalians, which were the first significant evolutionary diversification of the therapsids, a group of carnivores and herbivores that included some very large and specialized forms (Day *et al.* 2015*b*). Eupelycosaurs also disappeared at (or just before) this extinction (Modesto *et al.* 2011); large drops in the diversity of gorgonopsians and therocephalians took place and these latter groups did not recover diversity until substantially later in the Permian (during the Platbergian LVF). Younger (latest Gamkan–Hoedmakeran LVFs) post-extinction tetrapod assemblages lack dinocephalians and are initially characterized in the Karoo basin by a low diversity assemblage numerically dominated (*c.* 85% of all specimens) by the dicynodont *Diictodon*. The dinocephalian extinction event sets the stage for the takeover of much of the Permian tetrapod herbivore niche by dicynodonts.

In the Karoo basin of South Africa, where the extinction is best documented, it encompasses the total extinction of dinocephalians and the decimation of most therocephalians and gorgonospians. Day *et al.* (2015*b*) estimated a loss of 74–80% of generic richness at the top of the *Tapinocephalus* Assemblage Zone. Lucas (2009*a*, *b*), based on the age of the Isheevo assemblage (late Wordian, see earlier), considered the dinocephalian extinction event to be close in age to the Wordian–Capitanian boundary. However, the radioisotopic ages from the Karoo basin published by Rubidge *et al.* (2013) and Day *et al.* (2015*b*) indicate that the extinction level is very close to 260–261 Ma, so the extinction in the Karoo basin is a late Capitanian event on the SGCS.

Three different ideas about the timing of the Guadalupian marine extinctions are current: (1) a single mass extinction at the end of the Capitanian (Retallack *et al.* 2006); (2) a slightly earlier, within Capitanian, mass extinction (Bond *et al.* 2010); and (3) no mass extinction, but instead a steady decline in diversity from the Wordian through the Changhsingian (Clapham *et al.* 2009). Terrestrial plant extinctions have been viewed as either end-Capitanian (Retallack *et al.* 2006) or within Capitanian (Bond *et al.* 2010). The timing of Guadalupian extinctions on land and sea clearly needs further resolution.

The dramatic tetrapod extinction near the end of the Gamkan was evident at least as long ago as Boonstra (1968, 1969, 1971). He divided the South African *Tapinocephalus* zone (which is *c.* 1400 m thick; Jirah & Rubidge 2014) into three stratigraphically superposed assemblages, the youngest of which was later named the *Pristerognathus* zone. This subdivision showed most of the drop in diversity between the lower two intervals, with all dinocephalians gone by the top of the *Tapinocephalus* zone (Boonstra 1969, fig. 1). Existing stratigraphic data document a collapse of dinocephalian diversity through the middle–upper *Tapinocephalus* Assemblage Zone, not a single extinction (Day *et al.* 2015*a*). Biostratigraphic subdivision of the Abrahamskraal Formation and the equivalent Koonap Formation is now being undertaken (Day & Rubidge 2014; Jirah & Rubidge 2014) and should further refine our understanding of the duration and precise timing of the dinocephalian extinction event.

Latest Permian extinction event

The FAD of *Lystrosaurus* defines the beginning of the Lootsbergian LVF of Lucas (1998*a*, 2010). This biochronological datum was also long equated to the base of the Triassic or, more explicitly, the temporal succession from *Dicynodon* to *Lystrosaurus* was seen by vertebrate palaeontologists as the Permian–Triassic boundary. However, it became clear in the 1990s that there is a short stratigraphic overlap between the HO of *Dicynodon* and the LO of *Lystrosaurus*, both in the Karoo basin and in northwestern China (e.g. Rubidge *et al.* 1995; Hancox *et al.* 2002; Retallack *et al.* 2003; Lucas 2010). This led to some debate about whether to equate the Permo-Triassic boundary to the HO of *Dicynodon* or to the LO of *Lystrosaurus* (see Lucas 2009*a*, 2010 and references cited therein).

Lucas (2009*a*) challenged this correlation, arguing that magnetostratigraphic and chemostratigraphic data indicate that the LO of *Lystrosaurus* in the Karoo basin is slightly younger than the main marine extinction level and slightly older than the base of the Triassic defined at the GSSP at Meishan in southern China. Gastaldo *et al.* (2015) reported a new radioisotopic age and new magnetostratigraphic data from the Karoo basin to argue that the *Dicynodon–Lystrosaurus* turnover is even older in the late Permian than was suggested by Lucas (2009*a*). The reported radioisotopic age is a U–Pb age on an ash bed in the *Dicynodon* Assemblage Zone with a weighted mean average of 253.48 ± 0.15 Ma. Using this date as a datum, their new magnetostratigraphy places the LO of *Lystrosaurus* well down in the Changhsingian, correlative to the upper part of magnetochron Changhsing N of Steiner (2006). This decouples the Permo-Triassic boundary from the Platbergian–Lootsbergian boundary by at least 1 Ma.

Lucas (2009*b*) questioned the significance of the Permian–Triassic tetrapod extinctions, both in terms of the magnitude of the diversity collapse and the ecological severity of the extinction. Recent analyses have further undermined the concept of a single mass extinction of tetrapods at the Permo-Triassic boundary (Ruta *et al.* 2011; Bernardi *et al.* 2015; Roopnarine & Angielczyk 2016). The timing of the so-called end-Permian tetrapod extinction based on the data of Gastaldo *et al.* (2015) also indicates that this event is not coeval with the end-Permian marine extinctions, implying different causation for the events on land and sea.

Problems and prospectus

Incompleteness of the Permian tetrapod fossil record

Although the Permian record of tetrapods is both extensive and long studied, it is not without significant biases and imperfections, including the geographical restriction of early Permian tetrapods to the USA and Western Europe and the global gap in part of the middle Permian tetrapod fossil record. Recent discoveries and analyses of primitive therapsids are bridging the evolutionary break posed by Olson's gap (e.g. Kemp 2006; Abdala *et al.* 2008; Liu *et al.* 2009, 2010), but they are not demonstrably closing the temporal hiatus.

Early Permian tetrapod fossils are mainly restricted to the Pangean equatorial zone in North America and Western Europe (Milner 1993; Berman *et al.* 1997), although the recently discovered assemblage from northeastern Brazil (Cisneros *et al.* 2015*b*) is beginning to change this. The tetrapod record becomes more global in extent during the middle and late Permian, with substantial records in both Laurussia and Gondwana. Nevertheless, Permian tetrapod fossils (as is the case with most vertebrate fossils) are often found as isolated records or in bone beds (mass death assemblages) that lack stratigraphic range. The middle–upper Permian record of tetrapod fossils in the Karoo basin is a striking exception to this, but even in the Karoo most of the collected record lacks detailed stratigraphic provenance within a given assemblage zone, so that there is still a significant amount of stratigraphic imprecision. Both the geographical and stratigraphic inadequacies of the Permian tetrapod fossil record will always limit its utility in biostratigraphy and biochronology. New discoveries with precise stratigraphic data continue to be needed.

Endemism and index fossils

Most of the problems with developing a Permian tetrapod biostratigraphy and biochronology reduce to one problem: the rarity or lack of good Permian tetrapod index fossils. Good index fossils are easily identified, abundant and have a broad geographical (facies) range, but a short stratigraphic (temporal) range. Few, if any, Permian tetrapod genera or species meet these criteria. Most Permian tetrapod taxa (genera and species) are endemic to a single locality or to a geographically restricted region. In part, this is an artefact of taxonomy that requires much of a skull or skeleton to make many genus-level and all species-level identifications. As Williston (1915) and Romer (1928) noted, most Permian tetrapod species are specimen-specific: they are based on, and known from, one (or at most a few) exceptionally well-preserved fossils. This is why we are far from a species-level taxonomy of Permian tetrapods that is useful for biostratigraphy. The genus is the operational taxonomic unit of Permian tetrapod biostratigraphy.

A priori, most vertebrate palaeontologists have viewed Permian genera or higher level taxa as relatively cosmopolitan across Permian Pangea. A good demonstration of this was provided by the discovery of the Bromacker locality, a lower Permian bone bed in Germany. For more than a century, palaeontologists understood that North American red bed tetrapod assemblages differed strikingly from the age-equivalent upper Rotliegend assemblages in Europe, which are mostly branchiosaur-dominated and from carbonaceous, lacustrine facies. The Bromacker locality tetrapod assemblage is in fluvial red beds similar to the North American red bed facies and yields some of the same genera (e.g. *Seymouria*, *Diadectes* and *Dimetrodon*), indicative of the relative cosmopolitanism of some early Permian tetrapod genera.

However, there still appears to be some real endemism in the Permian tetrapod record. For example, the Permian tetrapod assemblage from the Moradi Formation of Niger is composed of endemic genera and is not readily correlated to other Permian tetrapod assemblages (see earlier discussion). On a larger scale, middle–late Permian tetrapod assemblages in the South African Karoo basin and the Russian Ural foreland overlap temporally, but they share few (if any) genera, so tetrapod-based correlations between South Africa and Russia rely on family-level taxa (e.g. Rubidge 2005). This classic endemism in two well-known tetrapod assemblages that are widely separated geographically may, in part, reflect provincial taxonomy.

Taxonomy and lineages

I regard the genus as the operational taxonomic unit for Permian tetrapod biostratigraphy and biochronology. This is because most species-level taxa of Permian tetrapods are meaningless for correlation because they are usually based on a single specimen or a local assemblage of well-preserved material and cannot be recognized at multiple localities. However, some species of early Permian tetrapod genera should be of value to biostratigraphy (Milner 1996). Werneburg (1989) has also argued that species lineages (chronoclines) provide a more precise biostratigraphy than genus-based correlations, but I am currently unable to construct meaningful species lineages for most of the Permian tetrapod genera that are of value to a global biochronology.

What Feduccia (1999) aptly called the 'jihad of cladism' has given rise to a type of alpha taxonomy based on cladistics (cladotaxonomy), which is undermining many widespread and long-recognized Permian tetrapod taxa useful to biostratigraphy. *Dicynodon* is the poster child of this problem (Angielczyk & Kurkin 2003; Kammerer *et al.* 2011; see earlier discussion). Thus what was long recognized as a single genus, *Dicynodon* (e.g. Cluver & Hotton 1981), has become multiple endemic genera with cladotaxonomic analysis, so that the one taxon can no longer be used for correlation. This is a large problem for vertebrate palaeontology, especially given that cladistics is a method of phylogenic reconstruction that is inconsistent with what we know of the evolutionary process and is therefore a questionable method (Vermeij 1999).

With regard to alpha taxonomy, taxonomic identity should be demonstrated by morphological similarity analysed within the context of population variation. Such an analysis will produce species-level taxa of potential biological significance that can be organized into genera. This is preferable to the typology inherent to cladotaxonomy, which, in the case of *Dicynodon*, recognizes 11 genera in what was formerly *Dicynodon*, based largely on little more than their perceived cladistic relationships. However, having said this, there still needs to be an extensive overhaul of the taxonomy of the genus *Dicynodon* to better assess its utility and the utility of its species in Permian biostratigraphy.

Non-biostratigraphic chronology

Integrated timescales, in which biostratigraphy, magnetostratigraphy, radioisotopic ages and chemostratigraphy provide calibration, are the most precise timescales because they use independent methods to evaluate correlations and datum points. However, there has long been a general lack of radioisotopic ages, magnetostratigraphy, isotopic data and other non-biostratigraphic means of correlating non-marine Permian strata. This review of radioisotopic ages and magnetostratigraphy, especially from the South African Karoo basin, indicates how many new, non-biostratigraphic data are available to calibrate the Permian tetrapod biochronology. However, it also reveals evident problems and contradictions inherent to these new data, which need further testing and replication to establish their reliability.

Prospectus

The global Permian timescale based on tetrapod evolution published by Lucas (2006) has been used by few vertebrate palaeontologists, who instead prefer to directly correlate Permian tetrapod fossil assemblages to the SGCS. However, I believe we need a non-marine Permian tetrapod biochronology with which to correctly sequence the history of tetrapod evolution on land. This would also free vertebrate palaeontologists from being confused by differing SGCS definitions and the inherent imprecision and uncertainty of many correlations of the Permian tetrapod fossil record to marine biostratigraphy. Advances in the scheme published in 2006 will come from new fossil discoveries, more detailed biostratigraphy and additional alpha taxonomic studies based on sound evolutionary taxonomic principles. As the work reviewed here demonstrates, the global Permian timescale based on tetrapod biochronology remains a robust tool for both global and regional age assignment and correlation.

Collaboration in the field and the museum with D. Berman, D. Chaney, S. Harris, A. Henrici, K. Krainer, D. Vachard and S. Voigt influenced the content of this paper. M. Celeskey drew the eupelycosaur and dinocephalian clip art used in many of the figures. M. Day provided literature and informative discussion. Comments by Sean Modesto and two anonymous reviewers corrected many shortcomings in the manuscript and are gratefully acknowledged.

References

Abdala, F., Rubidge, B.S. & van den Heever, J. 2008. The oldest therocephalians (Therapsida, Eutheriodonta) and the early diversification of Therapsida. *Palaeontology*, **51**, 1011–1024.

Amson, E. & Laurin, M. 2011. On the affinities of *Tetraceratops insignis*, an early Permian therapsid. *Acta Palaeontologica Polonica*, **56**, 301–312.

Angielczyk, K.D. & Kurkin, A.A. 2003. Has the utility of *Dicynodon* for upper Permian terrestrial biostratigraphy been overstated? *Geology*, **31**, 363–366.

Angielczyk, K.D. & Rubidge, B.S. 2013. Skeletal morphology, phylogenetic relationships and stratigraphic range of *Eosimops newtoni* Broom, 1921, a pylaecephalid dicynodont (Therapsida, Anomodontia) from the middle Permian of South Africa. *Journal of Systematic Palaeontology*, **11**, 191–231.

Angielczyk, K.D. & Sullivan, C. 2008. *Diictodon feliceps* (Owen, 1876), a dicynodont (Therapsida, Anomodontia) species with a Pangaean distribution. *Journal of Vertebrate Paleontology*, **23**, 788–802.

Angielczyk, K.D., Huertas, S. *et al.* 2014*a*. New dicynodonts (Therapsida, Anomodontia) and updated tetrapod stratigraphy of the Permian Ruhuhu Formation (Songea Group, Tuhuhu basin) of southern Tanzania. *Journal of Vertebrate Paleontology*, **34**, 1408–1426.

Angielczyk, K.D., Steyer, S.-J., Sidor, C.A., Smith, R.M.H., Whatley, R.L. & Tolan, S. 2014*b*. Permian and Triassic dicynodont (Therapsida; Anomodontia) faunas of the Luangwa basin, Zambia: taxonomic update and implications for dicynodont biogeography and biostratigraphy. *In*: Kammerer, C.F., Angielczyk, K.D. & Fröbisch, J. (eds) *Early Evolutionary*

History of the Synapsida. Springer, Dordrecht, 93–138.

ANGIELCZYK, K.D., RUBIDGE, B.S., DAY, M.O. & LIN, F. 2016. A reevaluation of *Brachyprosopus broomi* and *Chelydontops altidenalis*, dicynodonts (Therapsida, Anomodontia) from the middle Permian *Tapinocephalus* assemblage of the Karoo basin, South Africa. *Journal of Vertebrate Paleontology*, https://doi.org/10.1080/02724634.2016.1078342

BAKKER, R.T. 1980. Dinosaur heresy – dinosaur renaissance: why we need endothermic archosaurs for a comprehensive theory of bioenergetic evolution. *In*: THOMAS, R.D.K. & OLSON, E.C. (eds) *A Cold Look at the Warm-blooded Dinosaurs*. Westview Press, Boulder, 351–462.

BAKKER, R.T., TEMPLE, D.P. & ZOEHFELD, K. 2013. Baron Cuvier in north Texas: earliest classic extinction of large land herbivores, in the early Permian Clear Fork Group. *Geological Society of America, Abstracts with Programs*, **45**, 395.

BATTAIL, B. 1997. Les genres *Dicynodon* et *Lystrosaurus* (Therapsida, Dicynodontia) en Eurasie: une mise au point. *Geobios Memoire Speciale*, **20**, 39–48.

BATTAIL, B., DEJAX, J., RICHIR, P., TAQUET, P. & VE'RAN, M. 1995. New data on the continental Upper Permian in the area of Luang-Prabang, Laos. *Geological Survey of Vietnam Journal of Geology B*, **5-6**, 11–15.

BENSON, R.B.J. 2012. Interrelationships of basal synapsids: cranial and postcranial morphological partitions suggest different topologies. *Journal of Systematic Palaeontology*, **10**, 601–624.

BENTON, M.J. 1989. Mass extinctions among tetrapods and the quality of the fossil record. *Philosophical Transactions of the Royal Society of London B*, **325**, 369–386.

BENTON, M.J. 2012. No gap in the Middle Permian record of terrestrial vertebrates. *Geology*, **40**, 339–342.

BENTON, M.J. 2016. The Chinese pareiasaurs. *Zoological Journal of the Linnean Society*, **177**, 813–853, https://doi.org/10.1111/zoj.12389

BENTON, M.J. & WALKER, A.D. 1985. Palaeoecology, taphonomy and dating of Permo-Triassic reptiles from Elgin, north-east Scotland. *Palaeontology*, **28**, 207–234.

BERCOVICI, A., BOURQUIN, S. *ET AL.* 2015. Permian continental paleoenvironments in southeastern Asia: new insights from the Luang Prabang basin (Laos). *Journal of Asian Earth Sciences*, **50**, 197–211.

BERMAN, D.S., SUMIDA, S.S. & LOMBARD, R.E. 1997. Biogeography of primitive amniotes. *In*: SUMIDA, S.S. & MARTIN, K.L.M. (eds) *Amniote Origins*. Academic Press, San Diego, 85–139.

BERMAN, D.S., HENRICI, A.C., SUMIDA, S.S., MARTENS, T. & PELLETIER, V. 2013. First European record of a varanodontine (Synapsida: Varanopidae): member of a unique, early Permian upland paleoecosystem, Tambach Basin, central Germany. *In*: KAMMERER, C.F., ANGIELCZYK, K.D. & FRÖBISCH, J. (eds) *Early Evolutionary History of the Synapsida*. Springer, Dordrecht, 69–88.

BERMAN, D.S., HENRICI, A.C. & LUCAS, S.G. 2015. Pennsylvanian-Permian red bed vertebrate localities of New Mexico and their assemblages. *New Mexico Museum of Natural History and Science Bulletin*, **68**, 65–76.

BERNARDI, M., KLEIN, H., PETTI, F.M. & EZCURRA, M.D. 2015. The origin and early radiation of Archosauriformes: integrating the skeletal and footprint record. *Plos One*, https://doi.org/10.1371/journal.pone.0128449

BOARDMAN, D.R., WARDLAW, B.R. & NESTELL, M.K. 2009. Stratigraphy and conodont biostratigraphy of the uppermost Carboniferous and lower Permian from the North American Midcontinent. *Kansas Geological Survey Bulletin*, **255**, 1–42.

BOND, D.P.G., HILTON, J., WIGNALL, P.B., ALI, J.R., STEVENS, L.G., SUN, Y. & LAI, X. 2010. The middle Permian (Capitanian) mass extinction on land and in the oceans. *Earth-Science Reviews*, **102**, 100–116.

BOONSTRA, L.D. 1946. Report on some fossil reptiles from Gunyanka's Kraal, Busi Valley. *Proceedings and Transactions of the Rhodesian Scientific Association*, **41**, 46–49.

BOONSTRA, L.D. 1968. The terrestrial reptile fauna of *Tapinocephalus*-zone-age and Gondwanaland. *South African Journal of Science*, **64**, 199–204.

BOONSTRA, L.D. 1969. The fauna of the *Tapinocephalus* Zone (Beaufort Beds of the Karoo). *Annals of the South African Museum*, **56**, 1–73.

BOONSTRA, L.D. 1971. The early therapsids. *Annals of the South African Museum*, **59**, 17–46.

BRINK, K.S. 2015. Case 3695 *Dimetrodon* Cope, 1878 (Synapsida, Sphenacodontidae): proposed conservation by reversal of precedence with *Bathygnathus* Leidy, 1853. *Bulletin of Zoological Nomenclature*, **72**, 297–299.

BRINK, K.S. & REISZ, R.R. 2014. Hidden dental diversity in the oldest terrestrial apex predator *Dimetrodon*. *Nature Communications*, https://doi.org/10.1038/ncomms4269

BRINK, K.S., MADDIN, H.C., EVANS, D.C. & REISZ, R.R. 2015. Re-evaluation of the historic Canadian fossil *Bathygnathus borealis* from the early Permian of Prince Edward Island. *Canadian Journal of Earth Science*, **52**, 1109–1120.

BROWN, L.F., JR. 1973. Cisco depositional systems in north-central Texas. *The University of Texas Bureau of Economic Geology Guidebook*, **14**, 57–73.

CAMPIONE, N.E. & REISZ, R.R. 2010. *Varanops brevirostris* (Eupelycosauria: Varanopidae) from the lower Permian of Texas, with discussion of varanopid morphology and interrelationships. *Journal of Vertebrate Paleontology*, **30**, 724–726.

CATUNEANU, O., WOPFNER, H., ERIKSSON, P.G., CAIRNCROSS, B., RUBIDGE, B.S. & HANCOX, P.J. 2005. The Karoo basin of south-central Africa. *Journal of African Earth Sciences*, **43**, 211–253.

CHUDINOV, P.K. 1975. New facts about the fauna of the Upper Permian of the U.S.S.R. *Journal of Geology*, **73**, 117–130.

CHUVASHOV, B.I., CHERNYKH, V.V. *ET AL.* 2002. Progress report on the base of the Artinskian and base of the Kungurian by the Cisuralian Working Group. *Permophiles*, **41**, 13–16.

CISNEROS, J.C., ABDALA, F. & MALABARBA, M.C. 2005. Pareiasaurids from the Rio do Rasto Formation, southern Brazil: biostratigraphic implications for Permian faunas of the Paraná basin. *Revista Brasiliera de Paleontologia*, **8**, 13–24.

CISNEROS, J.C., ABDALA, F., ATAYMAN-GÜVEN, S., RUBIDGE, B.S., ŞENGÖR, A.M.C. & SCHULTZ, C.L. 2012. Carnivorous dinocephalian from the middle Permian of Brazil and tetrapod dispersal in Pangaea. *Proceedings of the National Academy of Sciences USA*, **109**, 115975109.

CISNEROS, J.C., ABDALA, F., JASHASHVILI, T., BUENO, A.de O. & DENTZIEN-DIAS, P. 2015*a*. *Tiarajudens eccentricus* and *Anomocephalus africanus*, two bizarre anomodonts (Synapsida, Therapsida) with dental occlusion from the Permian of Gondwana. *Royal Society Open Science*, **2**, 150090.

CISNEROS, J.C., MARSICANO, C. ET AL. 2015*b*. New Permian fauna from tropical Gondwana. *Nature Communications*, https://doi.org/10.1038/ncomms9676

CLAPHAM, M.E., SHEN, S. & BOTTJER, D.J. 2009. The double mass extinction revisited: reassessing the severity, selectivity, and causes of the end-Guadalupian biotic crisis (Late Permian). *Paleobiology*, **35**, 32–50.

CLARK, N.D.L. 1999. The Elgin marvel. *OUGS Journal*, **20**, 16–18.

CLUVER, M.A. & HOTTON, N. 1981. The genera *Dicynodon* and *Diictodon* and their bearing on the classification of the Dicynodontia. *Annals of the South African Museum*, **83**, 99–146.

CONRAD, J. & SIDOR, C.A. 2001. Re-evaluation of *Tetraceratops insignis* (Synapsida, Sphenacodontidae). *Journal of Vertebrate Paleontology*, **21**, 42A.

CRADDOCK, K.W. & HOOK, R.W. 1989. An overview of vertebrate collecting in the Permian System of north-central Texas. *In*: HOOK, R.W. (ed.) *Permo-Carboniferous Vertebrate Paleontology, Lithostratigraphy and Depositional Environments of North-central Texas*. Society of Vertebrate Paleontology, Austin, 40–46.

DAVYDOV, V., KRAINER, K. & CHERNYKH, V. 2013. Fusulinid biostratigraphy of the lower Permian Zweikofel Formation (Rattendorf Group; Carnic Alps, Austria) and Lower Permian Tethyan chronostratigraphy. *Geological Journal*, **48**, 57–100.

DAY, M.O. & RUBIDGE, B.S. 2014. A brief lithostratigraphic review of the Abrahamskraal and Koonap formations of the Beaufort Group, South Africa: towards a basin-wide stratigraphic scheme for the middle Permian Karoo. *Journal of African Earth Sciences*, **100**, 227–242.

DAY, M.O., GÜVEN, S., ABDALA, F., JIRAH, S., RUBIDGE, B. & ALMOND, J. 2015*a*. Youngest dinocephalian fossils extend the *Tapinocephalus* zone, Karoo basin, South Africa. *South African Journal of Science*, **111**, 1–5.

DAY, M.O., RAMEZANI, J., BOWRING, S.A., SADLER, P.M., ERWIN, D.H., ABDALA, F. & RUBIDGE, B.S. 2015*b*. When and how did the terrestrial mid-Permian mass extinction occur? Evidence from the tetrapod record of the Karoo basin, South Africa. *Proceedings of the Royal Society B*, **282**, 20150834.

DIAS-DA-SILVA, S. 2011. Middle-late Permian tetrapods from the Rio do Rasto Formation, southern Brazil: a biostratigraphic reassessment. *Lethaia*, https://doi.org/10.1111/j.1502-3931.2011.00263.x

DUTUIT, J.-M. 1988. *Diplocaulus minimus* n. sp. (Amphibia: Nectridea), lepospondyle de la formation d'Argana, dans l'Atlas occidental marocain. *Compte Rendus Academie de Science Paris*, **307**, 851–854.

EFREMOV, I.A. 1937. On the stratigraphic subdivision of the continental Permian and Triassic of the USSR on the basis of the fauna of early Tetrapoda. *Doklady Akademiya Nauk SSSR*, **16**, 121–126.

ENGLEHORN, J., SMALL, B.J. & HUTTENLOCKER, A. 2008. A redescription of *Acroplous vorax* (Temnospondyli; Dvinosauria) based on new specimens from the early Permian of Nebraska and Kansas, U.S.A. *Journal of Vertebrate Paleontology*, **28**, 291–305.

FEDUCCIA, A. 1999. 1,2,3 = 2,3,4: accommodating the cladogram. *Proceedings of the National Academy of Sciences USA*, **96**, 4740–4742.

FILDANI, A., WEISLOGEL, A. ET AL. 2009. U–Pb zircon ages from the southwestern Karoo basin, South Africa – implications for the Permian–Triassic boundary. *Geology*, **37**, 719–722.

FOREMAN, B.C. & MARTIN, L.D. 1988. A review of Paleozoic tetrapod localities of Kansas and Nebraska. *Kansas Geological Survey Guidebook*, **6**, 133–145.

GASTALDO, R.A., KAMO, S.L., NEVELING, J., GEISSMAN, J.N., BAMFORD, M. & LOOY, C.V. 2015. Is the vertebrate defined Permian–Triassic boundary in the Karoo basin, South Africa, the terrestrial expression of the end-Permian marine event? *Geology*, **43**, 939–942.

GAY, S.A. & CRUICKSHANK, A.R.I. 1999. Biostratigraphy of the Permian tetrapod faunas from the Ruhuhu Valley, Tanzania. *Journal of African Earth Sciences*, **29**, 195–210.

GERMAIN, D. 2010. The Moroccan diplocaulid: the last lepospondyl, the single one on Gondwana. *Historical Biology*, **22**, 4–39.

GOLUBEV, V.K. 1998. Revision of the late Permian chronosuchians (Amphibia, Anthracosauromorpha) from eastern Europe. *Paleontological Journal*, **32**, 390–401.

GOLUBEV, V.K. 2000. The faunal assemblages of Permian terrestrial vertebrates from eastern Europe. *Paleontological Journal*, **34**, 5211–5224.

GOLUBEV, V.K. 2005. Permian tetrapod stratigraphy. *New Mexico Museum of Natural History and Science Bulletin*, **30**, 95–99.

GOLUBEV, V.K. 2015. Dinocephalian stage in the history of the Permian tetrapod fauna of eastern Europe. *Paleontological Journal*, **49**, 1346–1352.

GOTTMAN-QUESADA, A. & SANDER, P.M. 2009. A redescription of the early archosauromorph *Protorosaurus speneri* Meyer, 1832, and its phylogenetic relationships. *Palaeontographica Abteilung A*, **287**, 123–220.

HANCOX, P.J., BRANDT, D., REIMOLD, W.U., KOERBEL, C. & NEVELING, J. 2002. *Permian-Triassic Boundary in Northwest Karoo Basin: Current Stratigraphic Placement, Implications for Basin Developmental Models, and the Search for Evidence of an Impact*. Geological Society of America, Special Papers, **356**, 429–444.

HARRIS, S.K., LUCAS, S.G., BERMAN, D.S. & HENRICI, A.C. 2004. Vertebrate fossil assemblage from the Upper Pennsylvanian Red Tanks Member of the Bursum Formation, Lucero uplift, central New Mexico. *New Mexico Museum of Natural History and Science Bulletin*, **25**, 267–283.

HENDERSON, C.M., DAVYDOV, V.I. & WARDLAW, B.R. 2012. The Permian Period. *In*: GRADSTEIN, F.M., OGG, J.G., SCHMITZ, M.D. & OGG, G.M. (eds) *The*

Geologic Time Scale 2012. Vol. 2. Elsevier, Amsterdam, 653–679.

HENTZ, T.F. 1988. *Lithostratigraphy and Paleoenvironments of Upper Paleozoic Continental Red Beds, North-central Texas: Bowie (New) and Wichita (Revised) Groups*. The University of Texas at Austin, Bureau of Economic Geology, Report of Investigations, **170**, 1–55.

HENTZ, T.F. 1989. Permo-Carboniferous lithostratigraphy of the vertebrate-bearing Bowie and Wichita Groups, north-central Texas. *In*: HOOK, R.W. (ed.) *Permo-Carboniferous Vertebrate Paleontology, Lithostratigraphy and Depositional Environments of North-Central Texas*. Society of Vertebrate Paleontology, Austin, 1–21.

HOLZ, M., FRANÇA, A.B., SOUZA, P.A., IANNUZZI, R. & ROHN, R. 2010. A stratigraphic chart of the Late Carboniferous/Permian succession of the eastern border of the Paraná basin, Brazil, South America. *Journal of South American Earth Sciences*, **29**, 381–399.

HOOK, R.W. 1989. Stratigraphic distribution of tetrapods in the Bowie and Wichita Groups, Permo-Carboniferous of north-central Texas. *In*: HOOK, R.W. (ed.) *Permo-Carboniferous Vertebrate Paleontology, Lithostratigraphy and Depositional Environments of North-central Texas*. Society of Vertebrate Paleontology, Austin, 47–53.

HUTTENLOCKER, A.K., PARDO, J.D. & SMALL, B.J. 2005. An earliest Permian nonmarine vertebrate assemblage from the Eskridge Formation, Nebraska. *New Mexico Museum of Natural History and Science Bulletin*, **30**, 133–143.

HUTTENLOCKER, A.K., PARDO, J.D. & SMALL, B.J. 2007. *Plemmyradytes shintoni*, gen. et. sp. nov., an early Permian amphibamid (Temnospondyli: Dissorophoidea) from the Eskridge Formation, Nebraska. *Journal of Vertebrate Paleontology*, **27**, 316–328.

ISOZAKI, Y. 2009. Illawara reversal: the fingerprint of a superflume that triggered Pangean breakup and the end-Guadalupian (Permian) mass extinction. *Gondwana Research*, **15**, 421–432.

IVAKHNENKO, M.F., GOLUBYEV, V.K., GUBIN, YU.M., KALANDADZE, N.N., NOVIKOV, I.V., SENNIKOV, A.G. & RAUTIAN, A.S. 1997. *Permskiye I Triasovyye tetrapody vostochnoi Evropy* [Permian and Triassic Tetrapods of Eastern Europe]. GEOS, Moscow [in Russian].

JACOBS, L.L., WINKLER, D.A., NEWMAN, J.D., GOMANI, E.M. & DEINO, A. 2005. Therapsids from the Permian Chiweta Beds and the age of the Karoo Supergroup in Malawi. *Palaeontologia Electronica*, Article Number 8.1.28A.

JALIL, N. & DUTUIT, J-M. 1996. Permian captorhinid reptiles from the Argana Formation, Morocco. *Palaeontology*, **39**, 907–918.

JALIL, N. & JANVIER, P. 2005. Les pareiasaurs (Amniota, Parareptila) du Permien superieur du basin d'Argana, Maroc. *Geodiversitas*, **27**, 35–132.

JIRAH, S. & RUBIDGE, B.S. 2014. Refined stratigraphy of the middle Permian Abrahamskraal Formation (Beaufort Group) in the southern Karoo basin. *Journal of African Earth Sciences*, **100**, 121–135.

KAMMERER, C., JANSEN, M. & FRÖBISCH, J. 2013. Therapsid phylogeny revisited. *Society of Vertebrate Paleontology 73rd Annual Meeting Abstracts*, 30 October–2 November, Los Angeles, California, 150.

KAMMERER, C.F., ANGIELCZYK, K.D. & FRÖBISCH, J. 2011. A comprehensive taxonomic revision of *Dicynodon* (Therapsida, Anomodontia) and its implications for dicynodont phylogeny, biogeography, and biostratigraphy. *Society of Vertebrate Paleontology Memoir*, **11**, 1–158.

KAMMERER, C.F., ANGIELCZYK, K.D. & FRÖBISCH, J. 2015. Redescription of *Digalodon rubidgei*, an emydopoid dicynodont (Therapsida, Anomodontia) from the late Permian of South Africa. *Fossil Record*, **18**, 43–55.

KEMP, T.S. 2006. The origin and early radiation of the therapsid mammal-like reptiles: a palaeobiological hypothesis. *Journal of Evolutionary Biology*, **19**, 1231–1247.

KEMP, T.S. 2009. Phylogenetic interrelationships and pattern of evolution of the therapsids. *Palaeontologica Africana*, **44**, 1–12.

KERANS, C., FITCHEN, W.M., GARDNER, M.H. & WARDLAW, B.R. 1993. A contribution to the evolving stratigraphic framework of middle Permian strata of the Delaware Basin, Texas and New Mexico. *New Mexico Geological Society Guidebook*, **44**, 175–184.

KEYSER, A.W. & SMITH, R.M.H. 1977–78. Vertebrate biozonation of the Beaufort Group with special reference to the western Karoo basin. *Annals of the Geological Survey of South Africa*, **12**, 1–35.

KING, G.M. 1988. Anomodontia. *Handbook of Paleoherpetology*, **17C**, 1–174.

KING, G.M. & JENKINS, I. 1997. The dicynodont *Lystrosaurus* from the upper Permian of Zambia: evolutionary and stratigraphical implications. *Palaeontology*, **40**, 149–156.

KISSEL, R. 2010. *Morphology, phylogeny, and evolution of Diadectidae (Cotylosauria: Diadectomorpha)*. PhD thesis, University of Toronto.

KITCHING, J.W. 1977. On the distribution of the Karoo vertebrate fauna. *Memoirs of the Bernard Price Institute for Palaeontological Research*, **1**, 1–131.

KITCHING, J.W. 1995. Biostratigraphy of the *Dicynodon* Assemblage Zone. *South African Committee for Stratigraphy Biostratigraphic Series*, **1**, 29–34.

KLETS, A.G., BUDNIKOV, I.B., KUTYGIN, R.V. & GRINENKO, V.S. 2001. The reference section of the lower-Upper Permian boundary beds in the Verkhoyansk region and its correlation. *Stratigraphy and Geological Correlation*, **9**, 247–262.

KOTLYAR, G.V. 1977. *Stratigraphic Dictionary of the USSR, Carboniferous, Permian*. Nedra, Leningrad [in Russian].

KOTLYAR, G.V., KOSSOVSKAYA, O.L., SHISHLOV, S.B., ZHRAVLEV, A.R. & PUKHONTO, S.K. 2004. Boundary between Permian Series in diverse sedimentary facies of north European Russia: constraints of event stratigraphy. *Stratigraphy and Geological Correlation*, **12**, 460–484.

KOZUR, H.W. & WEEMS, R.E. 2010. The biostratigraphic importance of conchostracans in the continental Triassic of the northern hemisphere. *In*: LUCAS, S.G. (ed.) *The Triassic Timescale*. Geological Society, London, Special Publications, **334**, 315–417, https://doi.org/10.1144/SP334.13

Kruger, A., Rubidge, B.S., Abdala, F., Chindebvu, E.G. & Jacobs, L.L. 2015. *Lende chiweta*, a new therapsid from Malawai, and its influence on burnetiamorph phylogeny and biogeography. *Journal of Vertebrate Paleontology*, https://doi.org/10.1080/02724634.2015.1008698

Lanci, L., Tohver, E., Wilson, A. & Flint, S. 2013. Upper Permian magnetic stratigraphy of the lower Beaufort Group, Karoo basin. *Earth and Planetary Science Letters*, **375**, 123–134.

Langston, W., Jr. 1963. Fossil vertebrates and the late Paleozoic red beds of Prince Edward Island. *National Museum of Canada Bulletin*, **187**, 1–36.

Laurin, M. & Reisz, R.R. 1996. The osteology and relationships of *Tetraceratops insignis*, the oldest known therapsid. *Journal of Vertebrate Paleontology*, **16**, 95–102.

Lee, M.S.Y., Gow, C.E. & Kitching, J. 1997. Anatomy and relationships of the pareiasaur *Pareiasuchus nasicornis* from the Upper Permian of Zambia. *Palaeontology*, **40**, 307–335.

LeBlanc, A.R.H., Brar, A.K., May, W. & Reisz, R.R. 2015. Multiple tooth-rowed captorhinids from the early Permian fissure fills of the Bally Mountain locality of Oklahoma. *Vertebrate Anatomy Morphology Palaeontology*, **1**, 35–49.

Leonova, T.B. 2007. Correlation of the Kazanian of the Volga-Urals with the Roadian of the global Permian scale. *Palaeoworld*, **16**, 246–253.

Lepper, J., Raath, M.A. & Rubidge, B.S. 2000. A diverse dinocephalian fauna from Zimbabwe. *South African Journal of Science*, **96**, 403–405.

Leven, E.Y. & Bogoslovskaya, M.F. 2006. The Roadian Stage of the Permian and problems of its global correlation. *Stratigraphy and Geological Correlation*, **14**, 164–173.

Lewis, G.F. & Vaughn, P.P. 1965. *Early Permian Vertebrates from the Cutler Formation of the Placerville Area Colorado*. US Geological Survey Professional Papers, **503-C**, 1–50.

Li, P., Cheng, Z. & Li, J. 2000. A new species of *Dicynodon* from upper Permian of Sunan, Gansu, with remarks on related strata. *Vertebrata PalAsiatica*, **38**, 147–157.

Li, X. & Liu, J. 2013. New specimens of pareiasaurs from the upper Permian Sunjiagou Formation of Liulin, Shanxi and their implications for the taxonomy of Chinese pareiasaurs. *Vertebrata PalAsiatica*, **51**, 199–204.

Li, X., Li, X., Jia, S. & Liu, J. 2015. The Jiyuan tetrapod fauna of the upper Permian of China: new pareiasaur material and the reestablishment of *Honania complicidentata*. *Acta Paleontologica Polonica*, **60**, 689–700.

Liu, J. 2013. Osteology, ontogeny, and phylogenetic position of *Sinophoneus yumenensis* (Therapsida, Dinocephalia) from the middle Permian Dashankou fauna of China. *Journal of Vertebrate Paleontology*, **33**, 1394–1407.

Liu, J. & Bever, G.S. 2015. The last diadectomorph sheds light on late Paleozoic tetrapod biogeography. *Biology Letters*, **11**, 20150100.

Liu, J., Rubidge, B. & Li, J. 2009. New basal synapsid supports Laurasian origin for therapsids. *Acta Palaeontologica Polonica*, **54**, 393–400.

Liu, J., Rubidge, B. & Li, J. 2010. A new specimen of *Biseridens qilanicus* indicates its phylogenetic position as the most basal anomodont. *Proceedings of the Royal Society B*, **277**, 285–292.

Liu, J., Xu, L., Jia, S., Pu, H. & Liu, X. 2014. The Jiyuan tetrapod fauna of the upper Permian of China – 2. Stratigraphy, taxonomical review, and correlation. *Vertebrata PalAsiatica*, **52**, 328–339.

Lozovsky, V.R. 1992. The Permian-Triassic boundary in continental series of Laurasia and its correlation with the marine scale. *International Geology Review*, **34**, 1008–1014.

Lozovsky, V.R. 2005. Olson's gap or Olson's bridge, that is the question. *New Mexico Museum of Natural History and Science Bulletin*, **30**, 179–184.

Lozovsky, V.R. 2013*a*. Permo-Triasoviy krizis I evo vozmozhnaya prichina [Permo-Triassic crisis and its principal cause]. *Byulletin Moskovskaya Obshchestaya Ispitateley Prirody Otdel Geologii*, **88**, 49–58.

Lozovsky, V.R. 2013*b*. The Permian Period ended with the impact of a 'Siberia' comet on Earth. *New Mexico Museum of Natural History and Science Bulletin*, **60**, 224–229.

Lozovsky, V.R. & Kukhtinov, D.A. 2007. Vyazinskovskiy Yarus – samoye molodoye podrazdeleniye verkhney Permi Evropeyskoy Rossii [Vaznikian Stage – youngest subdivision of the Upper Permian in European Russia]. *Byulletin Moskovskaya Obshchestaya Ispitateley Prirody Otdel Geologii*, **82**, 17–26.

Lozovsky, V.R., Minikh, M.G., Grunt, T.A., Kukhtinov, D.A., Ponomarenko, A.G. & Sukacheva, I.D. 2006. The Ufimian Stage of the East European scale: status, validity, and correlation potential. *Stratigraphy and Geological Correlation*, **17**, 602–614.

Lucas, S.G. 1994. *Triassic Tetrapod Extinctions and the Compiled Correlation Effect*. Canadian Society of Petroleum Geologists, Memoirs, **17**, 869–875.

Lucas, S.G. 1997. *Dicynodon* and late Permian Pangea. *In*: Wang, N. & Remane, J. (eds) *Stratigraphy: Proceedings of the 30th International Geological Congress*. Vol. 11. VSP International Science Publishers, Utrecht, 133–141.

Lucas, S.G. 1998*a*. Global Triassic tetrapod biostratigraphy and biochronology. *Palaeogeography, Palaeoclimatology, Palaeoecology*, **143**, 347–384.

Lucas, S.G. 1998*b*. Toward a tetrapod biochronology of the Permian. *New Mexico Museum of Natural History and Science Bulletin*, **12**, 71–91.

Lucas, S.G. 2001. *Chinese Fossil Vertebrates*. Columbia University Press, New York.

Lucas, S.G. 2002. Tetrapods and the subdivision of Permian time. *In*: Hills, L.V., Henderson, C.M. & Bamber, E.W. (eds) *Carboniferous and Permian of the World*. Canadian Society of Petroleum Geologists, Memoirs, **19**, 479–491.

Lucas, S.G. 2004. A global hiatus in the middle Permian tetrapod fossil record. *Stratigraphy*, **1**, 47–64.

Lucas, S.G. 2005*a*. Olson's gap or Olson's bridge: an answer. *New Mexico Museum of Natural History and Science Bulletin*, **30**, 185–186.

Lucas, S.G. 2005*b*. Age and correlation of Permian tetrapod assemblages from China. *New Mexico Museum of Natural History and Science Bulletin*, **30**, 187–191.

LUCAS, S.G. 2005c. *Dicynodon* from the Upper Permian of Russia: biochronological significance. *New Mexico Museum of Natural History and Science Bulletin*, **30**, 192–196.

LUCAS, S.G. 2005d. Permian tetrapod faunachrons. *New Mexico Museum of Natural History and Science Bulletin*, **30**, 197–201.

LUCAS, S.G. 2006. Global Permian tetrapod biostratigraphy and biochronology. *In*: LUCAS, S.G., CASSINIS, G. & SCHNEIDER, J.W. (eds) *Non-marine Permian Biostratigraphy and Biochronology*. Geological Society, London, Special Publications, **265**, 65–93.

LUCAS, S.G. 2009a. Timing and magnitude of tetrapod extinctions across the Permo-Triassic boundary. *Journal of Asian Earth Sciences*, **36**, 491–502.

LUCAS, S.G. 2009b. Global middle Permian reptile mass extinction: the dinocephalian extinction event. *Geological Society of America, Abstracts with Programs*, **41**, 360.

LUCAS, S.G. 2010. The Triassic timescale based on nonmarine tetrapod biostratigraphy and biochronology. *In*: LUCAS, S.G. (ed.) *The Triassic Timescale*. Geological Society, London, Special Publications, **334**, 447–500, https://doi.org/10.1144/SP334.15

LUCAS, S.G. 2011. Tetrapod fossils and the age of the upper Paleozoic Dunkard Group, Pennsylvania-West Virginia-Ohio. *In*: HARPER, J.A. (ed.) *Geology of the Pennsylvanian-Permian in the Dunkard Basin. Guidebook, 76th Annual Field Conference of Pennsylvania Geologists*, Pennsylvania Geological Survey and Pittsburgh Geological Society, 29 September–1 October, Washington, Pennsylvania, 174–180.

LUCAS, S.G. 2013a. Reconsideration of the base of the Permian System. *New Mexico Museum of Natural History and Science Bulletin*, **60**, 230–232.

LUCAS, S.G. 2013b. We need a new GSSP for the base of the Permian. *Permophiles*, **58**, 8–12.

LUCAS, S.G. 2013c. Vertebrate biostratigraphy and biochronology of the upper Paleozoic Dunkard Group, Pennsylvania-West Virginia-Ohio, USA. *International Journal of Coal Geology*, **119**, 79–87.

LUCAS, S.G. 2013d. Forum comment: no gap in the Middle Permian record of terrestrial vertebrates. *Geology*, https://doi.org/10.1130/G33734C.1

LUCAS, S.G. 2016. Early Permian tetrapod extinction events. *Permophiles*, **63**, 27–32.

LUCAS, S.G., SCHNEIDER, J.W. & CASSINIS, G. 2006. Non-marine Permian biostratigraphy and biochronology: an introduction. *In*: LUCAS, S.G., CASSINIS, G. & SCHNEIDER, J.W. (eds) *Non-marine Permian Biostratigraphy and Biochronology*. Geological Society, London, Special Publications, **265**, 1–14.

LUCAS, S.G., SCHNEIDER, J.W. & SPIELMANN, J.A. (eds) 2010. Carboniferous–Permian transition in Cañon del Cobre, northern New Mexico. *New Mexico Museum of Natural History and Science Bulletin*, **49**, 1–229.

LUCAS, S.G., KRAINER, K., CHANEY, D.S., DIMICHELE, W.A., VOIGT, S., BERMAN, D.S. & HENRICI, A.C. 2012a. The lower Permian Abo Formation in the Fra Cristobal and Caballo Mountains, Sierra County, New Mexico. *New Mexico Geological Society Guidebook*, **63**, 345–376.

LUCAS, S.G., HARRIS, S.K. ET AL. 2012b. Lithostratigraphy, paleontology, biostratigraphy and age of the upper Paleozoic Abo Formation near Jemez Springs, northern New Mexico, USA. *Annals of the Carnegie Museum*, **80**, 323–350.

LUCAS, S.G., BARRICK, J.E., KRAINER, K. & SCHNEIDER, J.W. 2013a. The Carboniferous–Permian boundary at Carrizo Arroyo, central New Mexico, USA. *Stratigraphy*, **10**, 153–170.

LUCAS, S.G., KRAINER, K., CHANEY, D.S., DIMICHELE, W.A., VOIGT, S., BERMAN, D.S. & HENRICI, A.C. 2013b. The lower Permian Abo Formation in central New Mexico. *New Mexico Museum of Natural History and Science Bulletin*, **59**, 161–179.

LUCAS, S.G., KRAINER, K., VOIGT, S., BERMAN, D.S. & HENRICI, A. 2014. The Lower Permian Abo Formation in the northern Sacramento Mountains, southern New Mexico. *New Mexico Geological Society Guidebook*, **65**, 287–302.

LUCAS, S.G., KRAINER, K. & VACHARD, D. 2015a. The lower Permian Hueco Group, Robledo Mountains, New Mexico. *New Mexico Museum of Natural History and Science Bulletin*, **65**, 43–96.

LUCAS, S.G., KRAINER, K. ET AL. 2015b. Progress report on correlation of nonmarine and marine Lower Permian strata, New Mexico, USA. *Permophiles*, **61**, 10–17.

LUCAS, S.G., KRAINER, K. ET AL. 2015c. Lithostratigraphy, biostratigraphy and sedimentology of upper Paleozoic Sangre de Cristo Formation, southwestern San Miguel County, New Mexico. *New Mexico Geological Society Guidebook*, **66**, 211–228.

LUCAS, S.G., KRAINER, K., OVIATT, C.G., VACHARD, D., BERMAN, D.S. & HENRICI, A.C. 2016. The Permian System at Abo Pass, central New Mexico (USA). *New Mexico Geological Society Guidebook*, **67**, 313–350.

MACDOUGALL, M.J. & REISZ, R.R. 2014. The first record of a nyctiphruretid parareptile from the early Permian of North America, with a discussion of parareptilian temporal fenestra. *Zoological Journal of the Linnean Society*, **172**, 616–630.

MADDIN, H.C., SIDOR, C.A. & REISZ, R.R. 2008. Cranial anatomy of *Ennatosaurus tecton* (Synapsida: Caseidae) from the middle Permian of Russia and the evolutionary relationships of Caseidae. *Journal of Vertebrate Paleontology*, **28**, 160–180.

MAISCH, M.W. & GEBAUER, E.V.I. 2005. Reappraisal of *Geikia locusticeps* (Therapsida: Dicynodontia) from the upper Permian of Tanzania. *Palaeontology*, **48**, 309–324.

MARCHETTI, L., RONCHI, A., SANTI, G. & VOIGT, S. 2015a. Revision of a classic site for Permian tetrapod ichnology (Collio Formation, Trompia). *Palaeogeography, Palaeoclimatology, Palaeoecology*, **433**, 140–155.

MARCHETTI, L., RONCHI, A., SANTI, G. & VOIGT, S. 2015b. The Gerola Valley site (Orobic Basin, Northern Italy): a key for understanding late early Permian tetrapod ichnofaunas. *Palaeogeography, Palaeoclimatology, Palaeoecology*, **439**, 97–116.

MARTINELLI, A.G., FRANCISCHINI, H., DENTZIEN-DIAS, P.C., SOARSE, M.B. & SCHULTZ, C.L. 2016. The oldest archosauromorph from South America: postcranial remains from the Guadalupian (mid-Permian) Rio do Rasto Formation (Paraná basin), southern Brazil. *Historical Biology*, 2016, https://doi.org/10.1080/08912963.2015.1125897

MAZIN, J.M. & KING, G.M. 1991. The first dicynodont from the late Permian of Malagasy. *Palaeontology*, **34**, 837–842.

MENNING, M. 2001. A Permian time scale 2000 and correlation of marine and continental sequences using the Illawarra reversal (265 Ma). *Natura Bresciana*, **25**, 355–362.

METCALFE, I., FOSTER, C.B., AFONIN, S.A., NICOLL, R.S., MUNDIL, R., WANG, X. & LUCAS, S.G. 2009. Stratigraphy, biostratigraphy and C-isotopes of the Permian–Triassic non-marine sequence at Dalongkou and Lucaogou, Xinjiang Province, China. *Journal of Asian Earth Sciences*, **36**, 503–520.

MICHEL, L.A., TABOR, N.J., MONTAÑEZ, I.P., SCHMITZ, M.D. & DAVYDOV, V.I. 2015. Chronostratigraphy and paleoclimatology of the Lodève basin, France: evidence for a pan-tropical aridification event across the Carboniferous–Permian boundary. *Palaeogeography, Palaeoclimatology, Palaeoecology*, **430**, 118–131.

MILNER, A.R. 1993. Biogeography of Palaeozoic tetrapods. *In*: LONG, J.A. (ed.) *Palaeozoic Vertebrate Biostratigraphy and Biogeography*. Belhaven Press, London, 324–353.

MILNER, A.R. 1996. Systematics of the genus *Eryops* (Amphibia: Temnospondyli) and its possible biostratigraphical value. *Journal of Vertebrate Paleontology*, **16**, 53A.

MODESTO, S.P. 2006. The cranial skeleton of the early Permian aquatic reptile *Mesosaurus tenuidens*: implications for relationships and palaeobiology. *Zoological Journal of the Linnean Society*, **146**, 345–368.

MODESTO, S. & BOTHA-BRINK, J. 2010. Problems of correlation of South African and South American tetrapod faunas across the Permian–Triassic boundary. *Journal of African Earth Sciences*, **57**, 242–248.

MODESTO, S., SIDOR, C.A., RUBIDGE, B.S. & WELMAN, J. 2001. A second varanopseid skull from the upper Permian of South Africa: implications for late Permian 'pelycosaur' evolution. *Lethaia*, **34**, 249–259.

MODESTO, S.P., SCOTT, D.M., BERMAN, D.S., MULLER, J. & REISZ, R.R. 2007. The skull and paleoecological significance of *Labidosaurus hamatus*, a captorhinid reptile from the lower Permian of Texas. *Zoological Journal of the Linnean Society*, **149**, 237–262.

MODESTO, S.P., SCOTT, D.M. & REISZ, R.R. 2009. A new parareptile with temporal fenestration from the middle Permian of South Africa. *Canadian Journal of Earth Science*, **46**, 9–20.

MODESTO, S.P., SMITH, R.M.H., CAMPIONE, N.E. & REISZ, R.R. 2011. The last 'pelycosaur': a varanopid synapsid from the *Pristerognathus* Assemblage Zone, middle Permian of South Africa. *Naturwissenschaften*, **98**, 1027–1034.

MODESTO, S.P., LAMB, A.J. & REISZ, R.R. 2014. The captorhinid reptile *Captorhinikos valensis* from the lower Permian Vale Formation of Texas, and the evolution of herbivory in eureptiles. *Journal of Vertebrate Paleontology*, **34**, 291–302.

MODESTO, S.P., SCOTT, D.M., MACDOUGALL, M.J., SUES, H.-D., EVANS, D.C. & REISZ, R.R. 2015. The oldest parareptile and the diversification of parareptiles. *Proceedings of the Royal Society B*, **282**, 20121912.

MUJAL, E., GRETTER, N. *ET AL.* 2016. Constraining the Permian/Triassic transition in continental environments: stratigraphic and paleontological record from the Catalan Pyrenees (NE Iberian Peninsula). *Palaeogeography, Palaeoclimatology, Palaeoecology*, **445**, 18–37.

NELSON, W.J., HOOK, R.W. & CHANEY, D.S. 2013. Lithostratigraphy of the lower Permian (Leonardian) Clear Fork Formation of north-central Texas. *New Mexico Museum of Natural History and Science Bulletin*, **60**, 286–311.

OLSON, E.C. 1952. The evolution of a Permian vertebrate chronofauna. *Evolution*, **6**, 181–196.

OLSON, E.C. 1962. Late Permian terrestrial vertebrates, U.S.A. and U.S.S.R. *Transactions of the American Philosophical Society, New Series*, **52**, 1–223.

OLSON, E.C. 1965. Chickasha vertebrates. *Oklahoma Geological Survey Circular*, **70**, 1–70.

OLSON, E.C. 1967. Early Permian vertebrates of Oklahoma. *Oklahoma Geological Survey Circular*, **74**, 1–111.

OLSON, E.C. 1974. On the source of the therapsids. *Annals of the South African Museum*, **64**, 27–46.

OLSON, E.C. 1980. The North American Seymouriidae. *In*: JACOBS, L.L. (ed.) *Aspects of Vertebrate History*. Museum of Northern Arizona Press, Flagstaff, 137–152.

OLSON, E.H. & BARGHUSEN, H. 1962. Vertebrates from the Flowerpot Formation, Permian of Oklahoma. *Oklahoma Geological Survey Circular*, **59**, 5–48.

OLSON, E.C. & BEERBOWER, J.R. 1953. The San Angelo Formation, Permian of Texas, and its vertebrates. *Journal of Geology*, **61**, 389–423.

OLSON, E.C. & VAUGHN, P.P. 1970. The changes of terrestrial vertebrates and climate during the Permian of North America. *Forma et Functio*, **3**, 113–138.

PARRISH, J.M., PARRISH, J.T. & ZIEGLER, A.M. 1986. Permian–Triassic paleogeography and paleoclimatology and implications for therapsid distribution. *In*: HOTTON, N., III, MACLEAN, P.D., ROTH, J.J. & ROTH, E.C. (eds) *The Ecology and Biology of Mammal-Like Reptiles*. Smithsonian Institution Press, Washington, 109–131.

PIÑEIRO, G., VERDE, M., UBILLA, M. & FERIGOLO, J. 2003. First basal synapsids ('pelycosaurs') from the upper Permian–lower Triassic of Uruguay, South America. *Journal of Paleontology*, **77**, 389–392.

PIÑEIRO, G., ROJAS, A. & UBILLA, M. 2004. A new procolophonoid (Reptilia, Parareptilia) from the upper Permian of Uruguay. *Journal of Vertebrate Paleontology*, **24**, 814–821.

PINHEIRO, F.L., FRANCA, M.A.G., LACERDA, M.B., BUTLER, R.J. & SCHULTZ, C.L. 2016. An exceptional fossil skull from South America and the origins of the archosauriform radiation. *Nature Scientific Reports*, **6**, 228127.

PLUMMER, F.B. & MOORE, R.C. 1921. Stratigraphy of the Pennsylvanian formations of north-central Texas. *University of Texas Bulletin*, **2132**, 1–237.

PRICE, L.I. 1948. Um anfíbio labirintodonte da Formacão Pedra de Fogo, Esado do Maranhão. *Boletim do DNPM/DGM*, **124**, 7–33.

RAMEZANI, J., SCHMITZ, M.D., DAVYDOV, V.I., BOWRING, S.A., SNYDER, W.S. & NORTHRUP, C.J. 2007.

High-precision U-Pb zircon age constraints on the Carboniferous–Permian boundary in the southern Urals stratotype. *Earth and Planetary Science Letters*, **256**, 244–257.

Ray, S. 1999. Permian reptile fauna from the Kundaram Formation, Pranhita-Godavari Valley, India. *Journal of African Earth Sciences*, **29**, 211–218.

Ray, S. 2000. Endothiodont dicynodonts from India. *Palaeontological Research*, **5**, 177–191.

Ray, S. 2001. Small Permian dicynodonts from India. *Paleontological Research*, **5**, 177–191.

Reisz, R.R. & Laurin, M. 2001. The reptile *Macroleter*, the first vertebrate evidence for correlation of upper Permian continental strata of North America and Russia. *Geological Society of America Bulletin*, **113**, 1229–1233.

Reisz, R.R. & Laurin, M. 2004. A reevaluation of the enigmatic Permian synapsid *Watongia* and its stratigraphic significance. *Canadian Journal of Earth Sciences*, **41**, 377–386.

Reisz, R.R., Maddin, H.C., Frobisch, J. & Falconet, J. 2011. A new large caseid (Synapsida, Caseasauria) from the Permian of Rodez (France), including reappraisal of '*Casea*' *rutena* Sigogneau-Russell and Russell 1974. *Geodiversitas*, **33**, 227–246.

Retallack, G.J., Smith, R.M.H. & Ward, P.D. 2003. Vertebrate extinction across Permian–Triassic boundary in Karoo basin, South Africa. *Geological Society of America Bulletin*, **115**, 1133–1152.

Retallack, G.J., Metzger, C.A., Greaver, T., Jahren, A.H., Smith, R.M.H. & Sheldon, N.D. 2006. Middle–late Permian mass extinction on land. *Geological Society of America Bulletin*, **118**, 1398–1411.

Romano, M. & Nicosia, U. 2015. Cladistic analysis of Caseidae (Casesauria, Synapsida): using the gap-weighting method to include taxa based on incomplete specimens. *Palaeontology*, **2015**, 1–22.

Romer, A.S. 1928. Vertebrate faunal horizons in the Texas Permo-Carboniferous red beds. *University of Texas Bulletin*, **2801**, 67–108.

Romer, A.S. 1935. Early history of Texas redbeds vertebrates. *Geological Society of America Bulletin*, **46**, 1597–1658.

Romer, A.S. 1958. The Texas Permian redbeds and their vertebrate fauna. *In*: Westoll, T.S. (ed.) *Studies on Fossil Vertebrates, Essays Presented to D.M.S. Watson*. Athlone Press, London, 157–179.

Romer, A.S. 1974. The stratigraphy of the Permian Wichita redbeds of Texas. *Breviora*, **427**, 1–29.

Ronchi, A., Sacchi, E., Romano, M. & Nicosia, U. 2011. A huge caseid pelycosaur from north-western Sardinia and its bearing on European Permian stratigraphy and palaeobiogeography. *Acta Palaeontologica Polonica*, **56**, 723–738.

Roopnarine, P.D. & Angielczyk, K.D. 2016. Community stability and selective extinction during the Permian–Triassic mass extinction. *Science*, **350**, 90–93.

Roscher, M. & Schneider, J.W. 2005. An annotated correlation chart for continental Late Pennsylvanian and Permian basins and the marine scale. *New Mexico Museum of Natural History and Science Bulletin*, **30**, 282–291.

Roscher, M. & Schneider, J.W. 2006. Permo-Carboniferous climate: early Pennsylvanian to Late Permian climate development of central Europe in a regional and global context. *In*: Lucas, S.G., Cassinis, G. & Schneider, J.W. (eds) *Non-Marine Permian Biostratigraphy and Biochronology*. Geological Society, London, Special Publications, **265**, 95–136, https://doi.org/10.1144/GSL.SP.2006.265.01.05

Rössler, R. 2006. Two remarkable petrified forests: correlation, comparison and significance. *In*: Lucas, S.G., Cassinis, G. & Schneider, J.W. (eds) *Non-Marine Permian Biostratigraphy and Biochronology*. Geological Society, London, Special Publications, **265**, 39–63, https://doi.org/10.1144/GSL.SP.2006.265.01.03

Rössler, R., Merbitz, M., Annacker, V., Luthardt, L., Noll, R., Neregato, R. & Rohn, R. 2014. The root systems of Permian arborescent sphenopsids: evidence from the Northern and Southern hemispheres. *Palaeontographica B*, **290**, 65–107.

Rubidge, B.S. 1995. Biostratigraphy of the *Eodicynodon* Assemblage Zone. *South African Committee for Stratigraphy, Biostratigraphic Series*, **1**, 3–7.

Rubidge, B.S. 2005. Re-uniting lost continents – fossil reptiles from the ancient Karoo and their wanderlust. *South African Journal of Geology*, **108**, 135–172.

Rubidge, B.S. & Sidor, C.A. 2001. Evolutionary patterns among Permo-Triassic therapsids. *Annual Review of Ecology and Systematics*, **32**, 449–480.

Rubidge, B.S., Johnson, M.R., Kitching, J.W., Smith, R.M.H., Keyser, A.W. & Groenewald, G.H. 1995. An introduction to the biozonation of the Beaufort Group. *South African Committee for Stratigraphy, Biostratigraphic Series*, **1**, 1–2.

Rubidge, B.S., Sidor, C.A. & Modesto, S.P. 2006. A new burnetiamorph (Therapsida: Biarmosuchia) from the middle Permian of South Africa. *Journal of Paleontology*, **80**, 740–749.

Rubidge, B.S., Erwin, D.H., Ramezani, J., Bowring, S.A. & De Klerk, W.J. 2013. High-precision temporal calibration of late Permian vertebrate biostratigraphy: U-Pb zircon constraints from the Karoo Supergroup, South Africa. *Geology*, **41**, 363–366.

Ruta, M., Cisneros, J.C., Liebrecht, T., Tsuji, L.A. & Muller, J. 2011. Amniotes through major biological crises: faunal turnover among parareptiles and the end-Permian mass extinction. *Palaeontology*, **2011**, 1–21.

Sahney, S. & Benton, M.J. 2008. Recovery from the most profound mass extinction of all time. *Proceedings of the Royal Society B*, **275**, 759–765.

Sander, M., Craddock, K., Johnson, G. & Loftin, J. 2007. *The Lower Permian of North Texas. Society of Vertebrate Paleontology, 67th Annual Meeting, 17–20 October, Field Trip Guidebook*, Austin, Texas.

Santos, R.V., Souza, P.A., de Alvarenga, C.J., Dantas, E.L., Pimentel, M.M., de Oliveira, C.G. & de Araújo, L.M. 2006. Shrimp U-Pb zircon dating and palynology of bentonitic layers from the Permian Irati Formation, Paraná basin, Brazil. *Gondwana Research*, **9**, 456–463.

Schmitz, M.D. & Davydov, V.I. 2012. Quantitative radiometric and bio-stratigraphic calibration of the Pennsylvanian–early Permian (Cisuralian) time scale and pan-Euramerican chronostratigraphic correlation. *Geological Society of America Bulletin*, **124**, 549–577.

SCHNEIDER, J.W. & WERNEBURG, R. 2012. Biostratigraphie des Rotliegend mit Insekten und Amphibien. *Schriftenreihe der Deutschen Gesellschaft fur Geowissenschaften*, **61**, 110–142.

SCHOCH, R.R. 2014. *Amphibian Evolution: The Life of Early Land Vertebrates*. Wiley-Blackwell, Chichester.

SCHOCH, R.R. & MILNER, A.R. 2008. The interrelationships and evolutionary history of the temnospondyl family Branchiosauridae. *Journal of Systematic Palaeontology*, **6**, 409–431.

SCHOCH, R.R. & MILNER, A.R. 2014. Temnospondyli I. *Handbook of Paleoherpetology*, **3A2**, 1–150.

SCHOLZE, F., GOLUBEV, V.K., NIEDZWIEDZKI, G., SCHNEIDER, J.W., SENNIKOV, A.G. & SILANTIEV, V.V. 2015. Early Triassic conchostracans (Crustacea: Branchiopoda) of the terrestrial Permian–Triassic boundary sections in the Moscow syncline. *Palaeogeography, Palaeoclimatology, Palaeoecology*, **429**, 22–40.

SCOTT, K.M. 2013. Carboniferous–Permian boundary in the Halgaito Formation, Cutler Group, Valley of the Gods and surrounding area, southeastern Utah. *New Mexico Museum of Natural History and Science Bulletin*, **60**, 389–409.

SELF, C.A. & HILL, C.A. 2003. How speleothems grow: an introduction to the ontogeny of cave minerals. *Journal of Cave and Karst Studies*, **65**, 130–151.

SENNIKOV, A.G. 2014. Stena dominuryushchikh tetrapod na rubezhe Paleozoya i Mesozoya//Razvitie zhizni v protsesse abioticheskykh izmeniy na Zemlya [Replacement of the dominant groups of tetrapods at the turn of the Paleozoic and Mesozoic//Development of life during the processes of abiotic changes on Earth]. *In*: RUSINEK, O.T. (ed.) *Proceedings of the Third All Russian Scientific Practical Congress*. Geographical Institute Press, Irkutsk, 204–208.

SENNIKOV, A.G. & GOLUBEV, V.K. 2006. Vyazniki biotic assemblage of the terminal Permian. *Paleontological Journal*, **40**, 475–481.

SENNIKOV, A.G. & GOLUBEV, V.K. 2012. On the faunal verification of the Permo-Triassic boundary in continental deposits of eastern Europe: 1. Ghorokovets-Zhukov Ravine. *Paleontological Journal*, **46**, 313–323.

SHEN, S., SCHNEIDER, J.W., ANGLIONI, L. & HENDERSON, C.M. 2013. The international Permian timescale: March 2013 update. *New Mexico Museum of Natural History and Science Bulletin*, **60**, 286–311.

SHEN, S.Z., HENDERSON, C.M. *ET AL.* 2010. High-resolution Lopingian (late Permian) timescale of South China. *Geological Journal*, **45**, 122–134.

SIDOR, C.A. & HOPSON, J.A. 1995. The taxonomic status of the upper Permian eotheriodont therapsids of the San Angelo Formation (Guadalupian), Texas. *Journal of Vertebrate Paleontology*, **15**, 53A.

SIDOR, C.A. & HOPSON, J.A. 1998. Ghost lineages and 'mammalness': assessing the temporal pattern of character acquisition in the Synapsida. *Paleobiology*, **24**, 254–273.

SIDOR, C.A., O'KEEFE, F.R. *ET AL.* 2005. Permian tetrapods from the Sahara show climate-controlled endemism. *Nature*, **434**, 886–889.

SIDOR, C.A., ANGIELCZYK, K.D., WEIDE, D.M., SMITH, R.M.H., NESBITT, S.J. & TSUJI, L.A. 2010. Tetrapod fauna of the lowermost Usili Formation (Songea Group, Ruhuhu basin) of southern Tanzania, with a new burnetiid record. *Journal of Vertebrate Paleontology*, **30**, 696–703.

SIDOR, C.A., ANGIELCZYK, K.D., SMITH, R.M.H., GOULDING, A.K., NESBITT, S.J., PEECOOK, B.R. & STEYER, J-S. 2014*a*. Tapinocephalids (Therapsida, Dinocephalia) from the Permian Madumabisa Mudstone Formation (lower Karoo, mid-Zambezi basin) of southern Zambia. *Journal of Vertebrate Paleontology*, **34**, 980–986.

SIDOR, C.A., ANGIELCZYK, K.D. *ET AL.* 2014*b*. Filling Olson's gap: a new Permian tetrapod assemblage from the Mid-Zambezi basin of southern Zambia. *Society of Vertebrate Paleontology 74th Annual Meeting Abstracts*, 5–8 November, Berlin, Germany, 230.

SIGOGNEAU-RUSSELL, D. & RUSSELL, D.E. 1974. Etude du premier caséidé (Reptilia, Pelycosauria) d'Europe occidentale. *Bulletin du Muséum National d'Histoire Naturelle*, **3**, 145–215.

SIMON, R.V., SIDOR, C.A., ANGIELCZK, K.D. & SMITH, R.M.H. 2010. First record of a tapinocephalid (Therapsida: Dinocephalia) from the Ruhuhu Formation (Songea Group) of southern Tanzania. *Journal of Vertebrate Paleontology*, **30**, 1289–1293.

SIMPSON, L.C. 1979. Upper Gearyan and lower Leonardian terrestrial vertebrate faunas of Oklahoma. *Oklahoma Geology Notes*, **39**, 3–21.

SMITH, R.M.H. 2000. Sedimentology and taphonomy of late Permian vertebrate fossil localities in southwestern Madagascar. *Palaeontologia Africana*, **36**, 25–41.

SMITH, R.M.H. & KEYSER, A.W. 1995*a*. Biostratigraphy of the *Tapinocephalus* Assemblage Zone. *South African Committee for Stratigraphy, Biostratigraphic Series*, **1**, 8–12.

SMITH, R.M.H. & KEYSER, A.W. 1995*b*. Biostratigraphy of the *Pristerognathus* Assemblage Zone. *South African Committee for Stratigraphy, Biostratigraphic Series*, **1**, 13–17.

SMITH, R.M.H. & KEYSER, A.W. 1995*c*. Biostratigraphy of the *Tropidostoma* Assemblage Zone. *South African Committee for Stratigraphy, Biostratigraphic Series*, **1**, 18–22.

SMITH, R.M.H. & KEYSER, A.W. 1995*d*. Biostratigraphy of the *Cistecephalus* Assemblage Zone. *South African Committee for Stratigraphy, Biostratigraphic Series*, **1**, 23–28.

SMITH, R.M.H., RUBIDGE, B. & VAN DER WALT, M. 2012. Therapsid biodiversity patterns and paleoenvironments of the Karoo basin, South Africa. *In*: CHINSAMY-TURAN, A. (ed.) *Forerunners of Mammals. Radiation. Histology. Biology*. Indiana University Press, Bloomington, 31–62.

SMITH, R.M.H., SIDOR, C.A., TABOR, N.J. & STEYER, J.S. 2015. Sedimentology and vertebrate taphonomy of the Moradi Formation of northern Niger: a Permian wet desert in the tropics of Pangaea. *Palaeogeography, Palaeoclimatology, Palaeoecology*, **440**, 128–141.

SPALDING, D.A.E. 1993. *Bathygnathus*, Canada's first 'dinosaur'. *Modern Geology*, **18**, 247–255.

SPINDLER, F. 2013. The Niederhäslich tetrapod assemblage (early Permian, Döhlen basin) from Germany – new insights into ecology, reptiliomorph diversity, and the biology of *Paleohatteria longicaudata* (basal Sphenacodontia). *Society of Vertebrate Paleontology 73rd Annual Meeting, Abstracts*, 30 October–2 November, Los Angeles, California, 218.

SPINDLER, F. 2014. Reviewing the question of the oldest therapsid. *Freiberger Forschungshefte*, **C548**, 1–7.

SPINDLER, F. 2015. *The basal Sphenacodontia – systematic revision and evolutionary implications*. PhD thesis, Technische Universität Bergakademie Freiberg, Germany.

ŠTAMBERG, S. & ZAJÍC, J. 2008. *Carboniferous and Permian Faunas and Their Occurrences in the Limnic Basins of the Czech Republic*. Museum of Eastern Bohemia at Hradec Králové, Hradec Králové.

STEINER, M.B. 2006. The magnetic polarity time scale across the Permian–Triassic boundary. *In*: LUCAS, S.G., CASSINIS, G. & SCHNEIDER, J.W. (eds) *Non-Marine Permian Biostratigraphy and Biochronology*. Geological Society, London, Special Publications, **265**, 15–38, https://doi.org/10.1144/GSL.SP.2006.265.01.02

STEYER, S. 2000. Are European Paleozoic amphibians good stratigraphic markers? *Bulletin Societe Géologique de France*, **171**, 127–135.

STEYER, S. & JALIL, N. 2009. First evidence of a temnospondyl in the late Permian of the Argana basin, Morocco. *Special Papers in Palaeontology*, **81**, 155–160.

STEYER, J.-S., SANCHEZ, S. *ET AL.* 2012. A new vertebrate Lagerstätte from the lower Permian of France (Franchesne, Massif Central): palaeoenvironmental implications for the Bourbon-l'Archambault basin. *Bulletin de la Societe Géologique de France*, **183**, 509–515.

SUMIDA, S.S., WALLISER, J.B.D. & LOMBARD, R.E. 1999*a*. Late Paleozoic amphibian-grade tetrapods of Utah. *Utah Geological Survey Miscellaneous Publications*, **99-1**, 21–30.

SUMIDA, S.S., LOMBARD, R.E., BERMAN, D.S. & HENRICI, A.C. 1999*b*. Late Paleozoic amniotes and their near relatives from Utah and northeastern Arizona, with comments on the Permian-Pennsylvanian boundary in Utah and Northern Arizona. *Utah Geological Survey Miscellaneous Publications*, **99-1**, 31–43.

SZURLIES, M. 2013. Late Permian (Zechstein) magnetostratigraphy in Western and Central Europe. *In*: GĄSIEWICZ, A. & ŁOWAKIEWICZ, M. (eds) *Palaeozoic Climate Cycles: Their Evolutionary and Sedimentological Impact*. Geological Society, London, Special Publications, **376**, https://doi.org/10.1144/SP376.7

TAYLOR, G.K., TUCKER, C. *ET AL.* 2009. Magnetostratigraphy of Permian/Triassic boundary sequences in the Cis Urals, Russia: no evidence for a major temporal hiatus. *Earth & Planetary Science Letters*, **281**, 36–47.

TEDFORD, R.H., ALBRIGHT, L.B., III *ET AL.* 2004. Mammalian biochronology of the Arikareean through Hemphillian interval (late Oligocene through early Pliocene epochs). *In*: WOODBURNE, M.O. (ed.) *Late Cretaceous and Cenozoic Mammals of North America: Biostratigraphy and Geochronology*. Columbia University Press, New York, 169–231.

TOHVER, E., LANCI, L., WILSON, A., HENSMA, J. & FLINT, S. 2015. Magnetostratigraphic constraints on the age of the lower Beaufort Group, western Karoo basin, South Africa, and a critical analysis of existing U-Pb geochronological data. *Geochemistry, Geophysics, Geosystems*, **16**, 3649–3663.

TSUJI, L.A., MÜLLER, J. & REISZ, R.R. 2010. *Microleter mckinzieorum gen.* et sp,. nov. from the lower Permian of Oklahoma: the basalmost parareptile from Laurasia. *Journal of Systematic Palaeontology*, **8**, 245–255.

TSUJI, L.A., MÜLLER, J. & REISZ, R.R. 2012. Anaomy of *Emeroleter levis* and the phylogeny of the nycterloter parareptiles. *Journal of Vertebrate Paleontology*, **32**, 45–67.

TURNER, M.L., TSUJI, L.A., IDE, O. & SIDOR, C.A. 2015. The vertebrate fauna of the upper Permian of Niger – IX. The appendicular skeleton of *Bunostegos akokanensis* (Parareptilia: Pareiasauria). *Journal of Vertebrate Paleontology*, https://doi.org/10.1080/02724 634.2014.994746

TVERDOKHLEBOV, V.P., TVERDOKHLEBOVA, G.I., MINIKH, A.V., SURKOV, M.V. & BENTON, M.J. 2005. Upper Permian vertebrates and their sedimentological context in the south Urals, Russia. *Earth Science Reviews*, **69**, 27–77.

VACHARD, D., KRAINER, K. & LUCAS, S.G. 2015. Late early Permian (late Leonardian; Kungurian) algae, microproblematica, and smaller foraminifers from the Yeso Group and San Andres Formation (New Mexico; USA). *Palaeontologia Electronica*, Article Number 18.1.21A.

VERMEIJ, G.J. 1999. A serious matter with character-taxon matrices. *Paleobiology*, **25**, 431–433.

VIGLIETTI, P.A., SMITH, R.M.H., ANGIELCZYK, K.D., KAMMERER, C.F., FRÖBISCH, J. & RUBIDGE, B.S. 2016. The *Daptocephalus* Assemblage Zone (Lopingian), South Africa: a proposed biostratigraphy based on a new compilation of stratigraphic ranges. *Journal of African Earth Sciences*, **113**, 153–164.

VOIGT, S. & LUCAS, S.G. 2015. Permian tetrapod ichnodiversity of the Prehistoric Trackways National Monument (south-central New Mexico, U.S.A.). *New Mexico Museum of Natural History and Science Bulletin*, **65**, 153–167.

VOIGT, S., FISCHER, J., SCHINDLER, J., WURRKE, M., SPINDLER, F. & RINEHART, L. 2014. On a potential fossil hotspot for Pennsylvanian-Permian non-aquatic vertebrates in central Europe. *Freiberger Forschungshefte*, **C548**, 39–44.

WALSH, T.R. & BARRICK, J.E. 2004. Conodonts from the Elm Creek Formation (Artinskian) of north-central Texas U.S.A. Fauna from a Permian intermittently restricted shallow shelf. *Geological Society of America, Abstracts with Programs*, **34**, A-26.

WARDLAW, B.R. 2005. Age assignment of the Pennsylvanian-early Permian succession of north central Texas. *Permophiles*, **46**, 21–22.

WARDLAW, B.R., DAVYDOV, V. & GRADSTEIN, F.M. 2004. The Permian Period. *In*: GRADSTEIN, F.M., OGG, J.G. & SMITH, A.G. (eds) *A Geologic Time Scale 2004*. Cambridge University Press, Cambridge, 249–270.

WARREN, A.A., RUBIDGE, B.S. *ET AL.* 2001. Oldest known stereospondylous amphibian from the early Permian of Namibia. *Journal of Vertebrate Paleontology*, **21**, 34–39.

WERNEBURG, R. 1989. Labyrinthodontier (Amphibia) aus dem Oberkarbon und Unterperm Mitteleuropas-Systematik, Phylogenie und Biostratigraphie. *Freiberger Forschungschrifte H*, **C436**, 7–57.

WERNEBURG, R. 2008. Der 'Manebacher Saurier' – ein neuer grosser Eryopide (*Onchiodon*) aus dem Rotliegend (Unter-Perm) des Thüringer Waldes. *Veröffentlichungen des Naturhistorischen Museums Schleusingen*, **22**, 3–40.

WERNEBURG, R. & SCHNEIDER, J.W. 2006. Amphibian biostratigraphy of the European Permocarboniferous. *In*: LUCAS, S.G., CASSINIS, G. & SCHNEIDER, J.W. (eds) *Non-Marine Permian Biostratigraphy and Biochronology*. Geological Society, London, Special Publications, **265**, 201–215, https://doi.org/10.1144/GSL.SP.2006.265.01.09

WERNEBURG, R. & STEYER, J-S. 1999. Redescription of the holotype of *Actinodon frossardi* (Amphibia, Temnospondyli) from the lower Permian of the Autun basin (France). *Geobios*, **32**, 599–607.

WIDEMAN, N.K., SUMIDA, S.S. & O'NEIL, M. 2005. A reassessment of the taxonomic status of the materials assigned to the early Permian tetrapod genera *Limnosceloides* and *Limnoscelops*. *New Mexico Museum of Natural History and Science Bulletin*, **30**, 358–362.

WILLISTON, S.W. 1915. New genera of Permian reptiles. *American Journal of Science*, **39**, 575–579.

WOOD, H.E., II, CHANEY, R.W., CLARK, J., COLBERT, E.H., JEPSEN, G.L., REESIDE, J.B., JR. & STOCK, C. 1941. Nomenclature and correlation of the North American continental Tertiary. *Geological Society of America Bulletin*, **52**, 1–48.

WOODHEAD, J., REISZ, R. ET AL. 2010. Speleothem climate records from deep time? Exploring the potential with an example from the Pemian. *Geology*, **38**, 455–458.

ZAJÍC, J. 2014. Permian fauna of the Krkonoše piedmont basin (Bohemian massif, central Europe). *Acta Musei Nationalis Pragae Serie B Historia Naturalis*, **70**, 131–142.

Index

Page numbers in *italics* refer to Figures. Page numbers in **bold** refer to Tables.